T0180717

Lecture Notes in Computer Science 12682

More information about this subseries at http://www.springer.com/series/7409

Christian S. Jensen · Ee-Peng Lim ·
De-Nian Yang · Wang-Chien Lee ·
Vincent S. Tseng · Vana Kalogeraki ·
Jen-Wei Huang · Chih-Ya Shen (Eds.)

Database Systems for Advanced Applications

26th International Conference, DASFAA 2021
Taipei, Taiwan, April 11–14, 2021
Proceedings, Part II

Editors
Christian S. Jensen
Aalborg University
Aalborg, Denmark

Ee-Peng Lim
Singapore Management University
Singapore, Singapore

De-Nian Yang
Academia Sinica
Taipei, Taiwan

Wang-Chien Lee
The Pennsylvania State University
University Park, PA, USA

Vincent S. Tseng
National Chiao Tung University
Hsinchu, Taiwan

Vana Kalogeraki
Athens University of Economics and Business
Athens, Greece

Jen-Wei Huang
National Cheng Kung University
Tainan City, Taiwan

Chih-Ya Shen
National Tsing Hua University
Hsinchu, Taiwan

ISSN 0302-9743 ISSN 1611-3349 (electronic)
Lecture Notes in Computer Science
ISBN 978-3-030-73196-0 ISBN 978-3-030-73197-7 (eBook)
https://doi.org/10.1007/978-3-030-73197-7

LNCS Sublibrary: SL3 – Information Systems and Applications, incl. Internet/Web, and HCI

This Springer imprint is published by the registered company Springer Nature Switzerland AG
The registered company address is: Gewerbestrasse 11, 6330 Cham, Switzerland

Preface

Welcome to DASFAA 2021, the 26th International Conference on Database Systems for Advanced Applications, held from April 11 to April 14, 2021! The conference was originally planned to be held in Taipei, Taiwan. Due to the outbreak of the COVID-19 pandemic and the consequent health concerns and restrictions on international travel all over the world, this prestigious event eventually happens on-line as a virtual conference, thanks to the tremendous effort made by the authors, participants, technical program committee, organization committee, and steering committee. While the traditional face-to-face research exchanges and social interactions in the DASFAA community are temporarily paused this year, the long and successful history of the events, which established DASFAA as a premier research conference in the database area, continues!

On behalf of the program committee, it is our great pleasure to present the proceedings of DASFAA 2021, which includes 131 papers in the research track, 8 papers in the industrial track, 8 demo papers, and 4 tutorials. In addition, the conference program included three keynote presentations by Prof. Beng Chin Ooi from National University of Singapore, Singapore, Prof. Jiawei Han from the University of Illinois at Urbana-Champaign, USA, and Dr. Eunice Chiu, Vice President of NVIDIA, Taiwan.

The highly selective papers in the DASFAA 2021 proceedings report the latest and most exciting research results from academia and industry in the general area of database systems for advanced applications. The quality of the accepted research papers at DASFAA 2021 is extremely high, owing to a robust and rigorous double-blind review process (supported by the Microsoft CMT system). This year, we received 490 excellent submissions, of which 98 full papers (acceptance ratio of 20%) and 33 short papers (acceptance ratio of 26.7%) were accepted. The selection process was competitive and thorough. Each paper received at least three reviews, with some papers receiving as many as four to five reviews, followed by a discussion, and then further evaluated by a senior program committee (SPC) member. We, the technical program committee (TPC) co-chairs, considered the recommendations from the SPC members and looked into each submission as well as the reviews and discussions to make the final decisions, which took into account multiple factors such as depth and novelty of technical content and relevance to the conference. The most popular topic areas for the selected papers include information retrieval and search, search and recommendation techniques; RDF, knowledge graphs, semantic web, and knowledge management; and spatial, temporal, sequence, and streaming data management, while the dominant keywords are network, recommendation, graph, learning, and model. These topic areas and keywords shed light on the direction in which the research in DASFAA is moving.

Five workshops are held in conjunction with DASFAA 2021: the 1st International Workshop on Machine Learning and Deep Learning for Data Security Applications (MLDLDSA 2021), the 6th International Workshop on Mobile Data Management,

Mining, and Computing on Social Networks (Mobisocial 2021), the 6th International Workshop on Big Data Quality Management (BDQM 2021), the 3rd International Workshop on Mobile Ubiquitous Systems and Technologies (MUST 2021), and the 5th International Workshop on Graph Data Management and Analysis (GDMA 2021). The workshop papers are included in a separate volume of the proceedings, also published by Springer in its Lecture Notes in Computer Science series.

We would like to express our sincere gratitude to all of the 43 senior program committee (SPC) members, the 278 program committee (PC) members, and the numerous external reviewers for their hard work in providing us with comprehensive and insightful reviews and recommendations. Many thanks to all the authors for submitting their papers, which contributed significantly to the technical program and the success of the conference. We are grateful to the general chairs, Christian S. Jensen, Ee-Peng Lim, and De-Nian Yang for their help. We wish to thank everyone who contributed to the proceedings, including Jianliang Xu, Chia-Hui Chang and Wen-Chih Peng (workshop chairs), Xing Xie and Shou-De Lin (industrial program chairs), Wenjie Zhang, Wook-Shin Han and Hung-Yu Kao (demonstration chairs), and Ying Zhang and Mi-Yen Yeh (tutorial chairs), as well as the organizers of the workshops, their respective PC members and reviewers.

We are also grateful to all the members of the Organizing Committee and the numerous volunteers for their tireless work before and during the conference. Also, we would like to express our sincere thanks to Chih-Ya Shen and Jen-Wei Huang (proceedings chairs) for working with the Springer team to produce the proceedings. Special thanks go to Xiaofang Zhou (DASFAA steering committee liaison) for his guidance. Lastly, we acknowledge the generous financial support from various industrial companies and academic institutes.

We hope that you will enjoy the DASFAA 2021 conference, its technical program and the proceedings!

February 2021

Wang-Chien Lee
Vincent S. Tseng
Vana Kalogeraki

Organization

Organizing Committee

Honorary Chairs

Philip S. Yu	University of Illinois at Chicago, USA
Ming-Syan Chen	National Taiwan University, Taiwan
Masaru Kitsuregawa	University of Tokyo, Japan

General Chairs

Christian S. Jensen	Aalborg University, Denmark
Ee-Peng Lim	Singapore Management University, Singapore
De-Nian Yang	Academia Sinica, Taiwan

Program Committee Chairs

Wang-Chien Lee	Pennsylvania State University, USA
Vincent S. Tseng	National Chiao Tung University, Taiwan
Vana Kalogeraki	Athens University of Economics and Business, Greece

Steering Committee

BongHee Hong	Pusan National University, Korea
Xiaofang Zhou	University of Queensland, Australia
Yasushi Sakurai	Osaka University, Japan
Lei Chen	Hong Kong University of Science and Technology, Hong Kong
Xiaoyong Du	Renmin University of China, China
Hong Gao	Harbin Institute of Technology, China
Kyuseok Shim	Seoul National University, Korea
Krishna Reddy	IIIT, India
Yunmook Nah	DKU, Korea
Wenjia Zhang	University of New South Wales, Australia
Guoliang Li	Tsinghua University, China
Sourav S. Bhowmick	Nanyang Technological University, Singapore
Atsuyuki Morishima	University of Tsukaba, Japan
Sang-Won Lee	SKKU, Korea

Industrial Program Chairs

Xing Xie	Microsoft Research Asia, China
Shou-De Lin	Appier, Taiwan

Demo Chairs

Wenjie Zhang	University of New South Wales, Australia
Wook-Shin Han	Pohang University of Science and Technology, Korea
Hung-Yu Kao	National Cheng Kung University, Taiwan

Tutorial Chairs

Ying Zhang	University of Technology Sydney, Australia
Mi-Yen Yeh	Academia Sinica, Taiwan

Workshop Chairs

Chia-Hui Chang	National Central University, Taiwan
Jianliang Xu	Hong Kong Baptist University, Hong Kong
Wen-Chih Peng	National Chiao Tung University, Taiwan

Panel Chairs

Zi Huang	The University of Queensland, Australia
Takahiro Hara	Osaka University, Japan
Shan-Hung Wu	National Tsing Hua University, Taiwan

Ph.D Consortium

Lydia Chen	Delft University of Technology, Netherlands
Kun-Ta Chuang	National Cheng Kung University, Taiwan

Publicity Chairs

Wen Hua	The University of Queensland, Australia
Yongxin Tong	Beihang University, China
Jiun-Long Huang	National Chiao Tung University, Taiwan

Proceedings Chairs

Jen-Wei Huang	National Cheng Kung University, Taiwan
Chih-Ya Shen	National Tsing Hua University, Taiwan

Registration Chairs

Chuan-Ju Wang — Academia Sinica, Taiwan
Hong-Han Shuai — National Chiao Tung University, Taiwan

Sponsor Chair

Chih-Hua Tai — National Taipei University, Taiwan

Web Chairs

Ya-Wen Teng — Academia Sinica, Taiwan
Yi-Cheng Chen — National Central University, Taiwan

Finance Chair

Yi-Ling Chen — National Taiwan University of Science and Technology, Taiwan

Local Arrangement Chairs

Chien-Chin Chen — National Taiwan University, Taiwan
Chih-Chieh Hung — National Chung Hsing University, Taiwan

DASFAA Steering Committee Liaison

Xiaofang Zhou — The Hong Kong University of Science and Technology, Hong Kong

Program Committee

Senior Program Committee Members

Zhifeng Bao — RMIT University, Vietnam
Sourav S. Bhowmick — Nanyang Technological University, Singapore
Nikos Bikakis — ATHENA Research Center, Greece
Kevin Chang — University of Illinois at Urbana-Champaign, USA
Lei Chen — Hong Kong University of Science and Technology, China
Bin Cui — Peking University, China
Xiaoyong Du — Renmin University of China, China
Hakan Ferhatosmanoglu — University of Warwick, UK
Avigdor Gal — Israel Institute of Technology, Israel
Hong Gao — Harbin Institute of Technology, China
Dimitrios Gunopulos — University of Athens, Greece
Bingsheng He — National University of Singapore, Singapore
Yoshiharu Ishikawa — Nagoya University, Japan

Nick Koudas	University of Toronto, Canada
Wei-Shinn Ku	Auburn University, USA
Dik-Lun Lee	Hong Kong University of Science and Technology, China
Dongwon Lee	Pennsylvania State University, USA
Guoliang Li	Tsinghua University, China
Ling Liu	Georgia Institute of Technology, USA
Chang-Tien Lu	Virginia Polytechnic Institute and State University, USA
Mohamed Mokbel	University of Minnesota Twin Cities, USA
Mario Nascimento	University of Alberta, Canada
Krishna Reddy P.	International Institute of Information Technology, India
Dimitris Papadias	The Hong Kong University of Science and Technology, China
Wen-Chih Peng	National Chiao Tung University, Taiwan
Evaggelia Pitoura	University of Ioannina, Greece
Cyrus Shahabi	University of Southern California, USA
Kyuseok Shim	Seoul National University, Korea
Kian-Lee Tan	National University of Singapore, Singapore
Yufei Tao	The Chinese University of Hong Kong, China
Vassilis Tsotras	University of California, Riverside, USA
Jianyong Wang	Tsinghua University, China
Matthias Weidlich	Humboldt-Universität zu Berlin, Germany
Xiaokui Xiao	National University of Singapore, Singapore
Jianliang Xu	Hong Kong Baptist University, China
Bin Yang	Aalborg University, Denmark
Jeffrey Xu Yu	The Chinese University of Hong Kong, China
Wenjie Zhang	University of New South Wales, Australia
Baihua Zheng	Singapore Management University, Singapore
Aoying Zhou	East China Normal University, China
Xiaofang Zhou	The University of Queensland, Australia
Roger Zimmermann	National University of Singapore, Singapore

Program Committee Members

Alberto Abelló	Universitat Politècnica de Catalunya, Spain
Marco Aldinucci	University of Torino, Italy
Toshiyuki Amagasa	University of Tsukuba, Japan
Ting Bai	Beijing University of Posts and Telecommunications, China
Spiridon Bakiras	Hamad Bin Khalifa University, Qatar
Wolf-Tilo Balke	Technische Universität Braunschweig, Germany
Ladjel Bellatreche	ISAE-ENSMA, France
Boualem Benatallah	University of New South Wales, Australia
Athman Bouguettaya	University of Sydney, Australia
Panagiotis Bouros	Johannes Gutenberg University Mainz, Germany

Stéphane Bressan	National University of Singapore, Singapore
Andrea Cali	Birkbeck University of London, UK
K. Selçuk Candan	Arizona State University, USA
Lei Cao	Massachusetts Institute of Technology, USA
Xin Cao	University of New South Wales, Australia
Yang Cao	Kyoto University, Japan
Sharma Chakravarthy	University of Texas at Arlington, USA
Tsz Nam Chan	Hong Kong Baptist University, China
Varun Chandola	University at Buffalo, USA
Lijun Chang	University of Sydney, Australia
Cindy Chen	University of Massachusetts Lowell, USA
Feng Chen	University of Texas at Dallas, USA
Huiyuan Chen	Case Western Reserve University, USA
Qun Chen	Northwestern Polytechnical University, China
Rui Chen	Samsung Research America, USA
Shimin Chen	Chinese Academy of Sciences, China
Yang Chen	Fudan University, China
Brian Chen	Columbia University, USA
Tzu-Ling Cheng	National Taiwan University, Taiwan
Meng-Fen Chiang	Auckland University, New Zealand
Theodoros Chondrogiannis	University of Konstanz, Germany
Chi-Yin Chow	City University of Hong Kong, China
Panos Chrysanthis	University of Pittsburgh, USA
Lingyang Chu	Huawei Technologies Canada, Canada
Kun-Ta Chuang	National Cheng Kung University, Taiwan
Jonghoon Chun	Myongji University, Korea
Antonio Corral	University of Almeria, Spain
Alfredo Cuzzocrea	Universitá della Calabria, Italy
Jian Dai	Alibaba Group, China
Maria Luisa Damiani	University of Milan, Italy
Lars Dannecker	SAP SE, Germany
Alex Delis	National and Kapodistrian University of Athens, Greece
Ting Deng	Beihang University, China
Bolin Ding	Alibaba Group, China
Carlotta Domeniconi	George Mason University, USA
Christos Doulkeridis	University of Piraeus, Greece
Eduard Dragut	Temple University, USA
Amr Ebaid	Purdue University, USA
Ahmed Eldawy	University of California, Riverside, USA
Sameh Elnikety	Microsoft Research, USA
Damiani Ernesto	University of Milan, Italy
Ju Fan	Renmin University of China, China
Yixiang Fang	University of New South Wales, Australia
Yuan Fang	Singapore Management University, Singapore
Tao-yang Fu	Penn State University, USA

Yi-Fu Fu	National Taiwan University, Taiwan
Jinyang Gao	Alibaba Group, China
Shi Gao	Google, USA
Wei Gao	Singapore Management University, Singapore
Xiaofeng Gao	Shanghai Jiaotong University, China
Xin Gao	King Abdullah University of Science and Technology, Saudi Arabia
Yunjun Gao	Zhejiang University, China
Jingyue Gao	Peking University, China
Neil Zhenqiang Gong	Iowa State University, USA
Vikram Goyal	Indraprastha Institute of Information Technology, Delhi, India
Chenjuan Guo	Aalborg University, Denmark
Rajeev Gupta	Microsoft India, India
Ralf Hartmut Güting	Fernuniversität in Hagen, Germany
Maria Halkidi	University of Pireaus, Greece
Takahiro Hara	Osaka University, Japan
Zhenying He	Fudan University, China
Yuan Hong	Illinois Institute of Technology, USA
Hsun-Ping Hsieh	National Cheng Kung University, Taiwan
Bay-Yuan Hsu	National Taipei University, Taiwan
Haibo Hu	Hong Kong Polytechnic University, China
Juhua Hu	University of Washington, USA
Wen Hua	The University of Queensland, Australia
Jiun-Long Huang	National Chiao Tung University, Taiwan
Xin Huang	Hong Kong Baptist University, China
Eenjun Hwang	Korea University, Korea
San-Yih Hwang	National Sun Yat-sen University, Taiwan
Saiful Islam	Griffith University, Australia
Mizuho Iwaihara	Waseda University, Japan
Jiawei Jiang	ETH Zurich, Switzerland
Bo Jin	Dalian University of Technology, China
Cheqing Jin	East China Normal University, China
Sungwon Jung	Sogang University, Korea
Panos Kalnis	King Abdullah University of Science and Technology, Saudi Arabia
Verena Kantere	National Technical University of Athens, Greece
Hung-Yu Kao	National Cheng Kung University, Taiwan
Katayama Kaoru	Tokyo Metropolitan University, Japan
Bojan Karlas	ETH Zurich, Switzerland
Ioannis Katakis	University of Nicosia, Cyprus
Norio Katayama	National Institute of Informatics, Japan
Chulyun Kim	Sookmyung Women's University, Korea
Donghyun Kim	Georgia State University, USA
Jinho Kim	Kangwon National University, Korea
Kyoung-Sook Kim	Artificial Intelligence Research Center, Japan

Seon Ho Kim	University of Southern California, USA
Younghoon Kim	HanYang University, Korea
Jia-Ling Koh	National Taiwan Normal University, Taiwan
Ioannis Konstantinou	National Technical University of Athens, Greece
Dimitrios Kotzinos	University of Cergy-Pontoise, France
Manolis Koubarakis	University of Athens, Greece
Peer Kröger	Ludwig-Maximilians-Universität München, Germany
Jae-Gil Lee	Korea Advanced Institute of Science and Technology, Korea
Mong Li Lee	National University of Singapore, Singapore
Wookey Lee	Inha University, Korea
Wang-Chien Lee	Pennsylvania State University, USA
Young-Koo Lee	Kyung Hee University, Korea
Cheng-Te Li	National Cheng Kung University, Taiwan
Cuiping Li	Renmin University of China, China
Hui Li	Xidian University, China
Jianxin Li	Deakin University, Australia
Ruiyuan Li	Xidian University, China
Xue Li	The University of Queensland, Australia
Yingshu Li	Georgia State University, USA
Zhixu Li	Soochow University, Taiwan
Xiang Lian	Kent State University, USA
Keng-Te Liao	National Taiwan University, Taiwan
Yusan Lin	Visa Research, USA
Sebastian Link	University of Auckland, New Zealand
Iouliana Litou	Athens University of Economics and Business, Greece
An Liu	Soochow University, Taiwan
Jinfei Liu	Emory University, USA
Qi Liu	University of Science and Technology of China, China
Danyang Liu	University of Science and Technology of China, China
Rafael Berlanga Llavori	Universitat Jaume I, Spain
Hung-Yi Lo	National Taiwan University, Taiwan
Woong-Kee Loh	Gachon University, Korea
Cheng Long	Nanyang Technological University, Singapore
Hsueh-Chan Lu	National Cheng Kung University, Taiwan
Hua Lu	Roskilde University, Denmark
Jiaheng Lu	University of Helsinki, Finland
Ping Lu	Beihang University, China
Qiong Luo	Hong Kong University of Science and Technology, China
Zhaojing Luo	National University of Singapore, Singapore
Sanjay Madria	Missouri University of Science & Technology, USA
Silviu Maniu	Universite Paris-Sud, France
Yannis Manolopoulos	Open University of Cyprus, Cyprus
Marco Mesiti	University of Milan, Italy
Jun-Ki Min	Korea University of Technology and Education, Korea

Jun Miyazaki	Tokyo Institute of Technology, Japan
Yang-Sae Moon	Kangwon National University, Korea
Yasuhiko Morimoto	Hiroshima University, Japan
Mirella Moro	Universidade Federal de Minas Gerais, Brazil
Parth Nagarkar	New Mexico State University, USA
Miyuki Nakano	Tsuda University, Japan
Raymond Ng	The University of British Columbia, Canada
Wilfred Ng	The Hong Kong University of Science and Technology, China
Quoc Viet Hung Nguyen	Griffith University, Australia
Kjetil Nørvåg	Norwegian University of Science and Technology, Norway
Nikos Ntarmos	University of Glasgow, UK
Werner Nutt	Free University of Bozen-Bolzano, Italy
Makoto Onizuka	Osaka University, Japan
Xiao Pan	Shijiazhuang Tiedao University, China
Panagiotis Papapetrou	Stockholm University, Sweden
Noseong Park	George Mason University, USA
Sanghyun Park	Yonsei University, Korea
Chanyoung Park	University of Illinois at Urbana-Champaign, USA
Dhaval Patel	IBM TJ Watson Research Center, USA
Yun Peng	Hong Kong Baptist University, China
Zhiyong Peng	Wuhan University, China
Ruggero Pensa	University of Torino, Italy
Dieter Pfoser	George Mason University, USA
Jianzhong Qi	The University of Melbourne, Australia
Zhengping Qian	Alibaba Group, China
Xiao Qin	IBM Research, USA
Karthik Ramachandra	Microsoft Research India, India
Weixiong Rao	Tongji University, China
Kui Ren	Zhejiang University, China
Chiara Renso	Institute of Information Science and Technologies, Italy
Oscar Romero	Universitat Politècnica de Catalunya, Spain
Olivier Ruas	Inria, France
Babak Salimi	University of California, Riverside, USA
Maria Luisa Sapino	University of Torino, Italy
Claudio Schifanella	University of Turin, Italy
Markus Schneider	University of Florida, USA
Xuequn Shang	Northwestern Polytechnical University, China
Zechao Shang	Univesity of Chicago, USA
Yingxia Shao	Beijing University of Posts and Telecommunications, China
Chih-Ya Shen	National Tsing Hua University, Taiwan
Yanyan Shen	Shanghai Jiao Tong University, China
Yan Shi	Shanghai Jiao Tong University, China
Junho Shim	Sookmyung Women's University, Korea

Hiroaki Shiokawa	University of Tsukuba, Japan
Hong-Han Shuai	National Chiao Tung University, Taiwan
Shaoxu Song	Tsinghua University, China
Anna Squicciarini	Pennsylvania State University, USA
Kostas Stefanidis	Tampere University, Finland
Kento Sugiura	Nagoya University, Japan
Aixin Sun	Nanyang Technological University, Singapore
Weiwei Sun	Fudan University, China
Nobutaka Suzuki	University of Tsukuba, Japan
Yu Suzuki	Nara Institute of Science and Technology, Japan
Atsuhiro Takasu	National Institute of Informatics, Japan
Jing Tang	National University of Singapore, Singapore
Lv-An Tang	NEC Labs America, USA
Tony Tang	National Taiwan University, Taiwan
Yong Tang	South China Normal University, China
Chao Tian	Alibaba Group, China
Yongxin Tong	Beihang University, China
Kristian Torp	Aalborg University, Denmark
Yun-Da Tsai	National Taiwan University, Taiwan
Goce Trajcevski	Iowa State University, USA
Efthymia Tsamoura	Samsung AI Research, Korea
Leong Hou U.	University of Macau, China
Athena Vakal	Aristotle University, Greece
Michalis Vazirgiannis	École Polytechnique, France
Sabrina De Capitani di Vimercati	Università degli Studi di Milano, Italy
Akrivi Vlachou	University of the Aegean, Greece
Bin Wang	Northeastern University, China
Changdong Wang	Sun Yat-sen University, China
Chaokun Wang	Tsinghua University, China
Chaoyue Wang	University of Sydney, Australia
Guoren Wang	Beijing Institute of Technology, China
Hongzhi Wang	Harbin Institute of Technology, China
Jie Wang	Indiana University, USA
Jin Wang	Megagon Labs, Japan
Li Wang	Taiyuan University of Technology, China
Peng Wang	Fudan University, China
Pinghui Wang	Xi'an Jiaotong University, China
Sen Wang	The University of Queensland, Australia
Sibo Wang	The Chinese University of Hong Kong, China
Wei Wang	University of New South Wales, Australia
Wei Wang	National University of Singapore, Singapore
Xiaoyang Wang	Zhejiang Gongshang University, China
Xin Wang	Tianjin University, China
Zeke Wang	Zhejiang University, China
Yiqi Wang	Michigan State University, USA

Raymond Chi-Wing Wong	Hong Kong University of Science and Technology, China
Kesheng Wu	Lawrence Berkeley National Laboratory, USA
Weili Wu	University of Texas at Dallas, USA
Chuhan Wu	Tsinghua University, China
Wush Wu	National Taiwan University, Taiwan
Chuan Xiao	Osaka University, Japan
Keli Xiao	Stony Brook University, USA
Yanghua Xiao	Fudan University, China
Dong Xie	Pennsylvania State University, USA
Xike Xie	University of Science and Technology of China, China
Jianqiu Xu	Nanjing University of Aeronautics and Astronautics, China
Fengli Xu	Tsinghua University, China
Tong Xu	University of Science and Technology of China, China
De-Nian Yang	Academia Sinica, Taiwan
Shiyu Yang	East China Normal University, China
Xiaochun Yang	Northeastern University, China
Yu Yang	City University of Hong Kong, China
Zhi Yang	Peking University, China
Chun-Pai Yang	National Taiwan University, Taiwan
Junhan Yang	University of Science and Technology of China, China
Bin Yao	Shanghai Jiaotong University, China
Junjie Yao	East China Normal University, China
Demetrios Zeinalipour Yazti	University of Cyprus, Turkey
Qingqing Ye	The Hong Kong Polytechnic University, China
Mi-Yen Yeh	Academia Sinica, Taiwan
Hongzhi Yin	The University of Queensland, Australia
Peifeng Yin	Pinterest, USA
Qiang Yin	Alibaba Group, China
Man Lung Yiu	Hong Kong Polytechnic University, China
Haruo Yokota	Tokyo Institute of Technology, Japan
Masatoshi Yoshikawa	Kyoto University, Japan
Baosheng Yu	University of Sydney, Australia
Ge Yu	Northeast University, China
Yi Yu	National Information Infrastructure Enterprise Promotion Association, Taiwan
Long Yuan	Nanjing University of Science and Technology, China
Kai Zeng	Alibaba Group, China
Fan Zhang	Guangzhou University, China
Jilian Zhang	Jinan University, China
Meihui Zhang	Beijing Institute of Technology, China
Xiaofei Zhang	University of Memphis, USA
Xiaowang Zhang	Tianjin University, China
Yan Zhang	Peking University, China
Zhongnan Zhang	Software School of Xiamen University, China

Pengpeng Zhao	Soochow University, Taiwan
Xiang Zhao	National University of Defence Technology, China
Bolong Zheng	Huazhong University of Science and Technology, China
Yudian Zheng	Twitter, USA
Jiaofei Zhong	California State University, East, USA
Rui Zhou	Swinburne University of Technology, Australia
Wenchao Zhou	Georgetown University, USA
Xiangmin Zhou	RMIT University, Vietnam
Yuanchun Zhou	Computer Network Information Center, Chinese Academy of Sciences, China
Lei Zhu	Shandong Normal Unversity, China
Qiang Zhu	University of Michigan-Dearborn, USA
Yuanyuan Zhu	Wuhan University, China
Yuqing Zhu	California State University, Los Angeles, USA
Andreas Züfle	George Mason University, USA

External Reviewers

Amani Abusafia
Ahmed Al-Baghdadi
Balsam Alkouz
Haris B. C.
Mohammed Bahutair
Elena Battaglia
Kovan Bavi
Aparna Bhat
Umme Billah
Livio Bioglio
Panagiotis Bozanis
Hangjia Ceng
Dipankar Chaki
Harry Kai-Ho Chan
Yanchuan Chang
Xiaocong Chen
Tianwen Chen
Zhi Chen
Lu Chen
Yuxing Chen
Xi Chen
Chen Chen
Guo Chen
Meng-Fen Chiang
Soteris Constantinou
Jian Dai
Sujatha Das Gollapalli
Panos Drakatos
Venkatesh Emani
Abir Farouzi
Chuanwen Feng
Jorge Galicia Auyon
Qiao Gao
Francisco Garcia-Garcia
Tingjian Ge
Harris Georgiou
Jinhua Guo
Surabhi Gupta
Yaowei Han
Yongjing Hao
Xiaotian Hao
Huajun He
Hanbin Hong
Xinting Huang
Maximilian Hünemörder
Omid Jafari
Zijing Ji
Yuli Jiang
Sunhwa Jo
Seungwon Jung
Seungmin Jung
Evangelos Karatzas

Enamul Karim
Humayun Kayesh
Jaeboum Kim
Min-Kyu Kim
Ranganath Kondapally
Deyu Kong
Andreas Konstantinidis
Gourav Kumar
Abdallah Lakhdari
Dihia Lanasri
Hieu Hanh Le
Suan Lee
Xiaofan Li
Xiao Li
Huan Li
Pengfei Li
Yan Li
Sizhuo Li
Yin-Hsiang Liao
Dandan Lin
Guanli Liu
Ruixuan Liu
Tiantian Liu
Kaijun Liu
Baozhu Liu
Xin Liu
Bingyu Liu
Andreas Lohrer
Yunkai Lou
Jin Lu
Rosni Lumbantoruan
Priya Mani
Shohei Matsugu
Yukai Miao
Paschalis Mpeis
Kiran Mukunda
Siwan No
Alex Ntoulas
Sungwoo Park
Daraksha Parveen
Raj Patel
Gang Qian
Jiangbo Qian
Gyeongjin Ra
Niranjan Rai
Weilong Ren
Matt Revelle
Qianxiong Ruan
Georgios Santipantakis
Abhishek Santra
Nadine Schüler
Bipasha Sen
Babar Shahzaad
Yuxin Shen
Gengyuan Shi
Toshiyuki Shimizu
Lorina Sinanaj
Longxu Sun
Panagiotis Tampakis
Eleftherios Tiakas
Valter Uotila
Michael Vassilakopoulos
Yaoshu Wang
Pei Wang
Kaixin Wang
Han Wang
Lan Wang
Lei Wang
Han Wang
Yuting Xie
Shangyu Xie
Zhewei Xu
Richeng Xuan
Kailun Yan
Shuyi Yang
Kai Yao
Fuqiang Yu
Feng (George) Yu
Changlong Yu
Zhuoxu Zhang
Liang Zhang
Shuxun Zhang
Liming Zhang
Jie Zhang
Shuyuan Zheng
Fan Zhou
Shaowen Zhou
Kai Zou

Contents – Part II

Text and Unstructured Data

Multi-label Classification of Long Text Based on Key-Sentences Extraction 3
Jiayin Chen, Xiaolong Gong, Ye Qiu, Xi Chen, and Zhiyi Ma

Automated Context-Aware Phrase Mining from Text Corpora. 20
Xue Zhang, Qinghua Li, Cuiping Li, and Hong Chen

Keyword-Aware Encoder for Abstractive Text Summarization 37
Tianxiang Hu, Jingxi Liang, Wei Ye, and Shikun Zhang

Neural Adversarial Review Summarization with Hierarchical Personalized Attention 53
Hongyan Xu, Hongtao Liu, Wenjun Wang, and Pengfei Jiao

Generating Contextually Coherent Responses by Learning Structured Vectorized Semantics. 70
Yan Wang, Yanan Zheng, Shimin Jiang, Yucheng Dong, Jessica Chen, and Shaohua Wang

Latent Graph Recurrent Network for Document Ranking 88
Qian Dong and Shuzi Niu

Discriminative Feature Adaptation via Conditional Mean Discrepancy for Cross-Domain Text Classification 104
Bo Zhang, Xiaoming Zhang, Yun Liu, and Lei Chen

Discovering Protagonist of Sentiment with Aspect Reconstructed Capsule Network 120
Chi Xu, Hao Feng, Guoxin Yu, Min Yang, Xiting Wang, Yan Song, and Xiang Ao

Discriminant Mutual Information for Text Feature Selection. 136
Jiaqi Wang and Li Zhang

CAT-BERT: A Context-Aware Transferable BERT Model for Multi-turn Machine Reading Comprehension 152
Cen Chen, Xinjing Huang, Feng Ji, Chengyu Wang, Minghui Qiu, Jun Huang, and Yin Zhang

Unpaired Multimodal Neural Machine Translation via Reinforcement Learning 168
Yijun Wang, Tianxin Wei, Qi Liu, and Enhong Chen

Multimodal Named Entity Recognition with Image Attributes and Image Knowledge 186
Dawei Chen, Zhixu Li, Binbin Gu, and Zhigang Chen

Multi-task Neural Shared Structure Search: A Study Based on Text Mining 202
Jiyi Li and Fumiyo Fukumoto

A Semi-structured Data Classification Model with Integrating Tag Sequence and Ngram 219
Lijun Zhang, Ning Li, Wei Pan, and Zhanhuai Li

Inferring Deterministic Regular Expression with Unorder and Counting 235
Xiaofan Wang

MACROBERT: Maximizing Certified Region of BERT to Adversarial Word Substitutions 253
Fali Wang, Zheng Lin, Zhengxiao Liu, Mingyu Zheng, Lei Wang, and Daren Zha

A Diversity-Enhanced and Constraints-Relaxed Augmentation for Low-Resource Classification 262
Guang Liu, Hailong Huang, Yuzhao Mao, Weiguo Gao, Xuan Li, and Jianping Shen

Neural Demographic Prediction in Social Media with Deep Multi-view Multi-task Learning 271
Yantong Lai, Yijun Su, Cong Xue, and Daren Zha

An Interactive NL2SQL Approach with Reuse Strategy 280
Xiaxia Wang, Sai Wu, Lidan Shou, and Ke Chen

Data Mining

Consistency- and Inconsistency-Aware Multi-view Subspace Clustering 291
Guang-Yu Zhang, Xiao-Wei Chen, Yu-Ren Zhou, Chang-Dong Wang, and Dong Huang

Discovering Collective Converging Groups of Large Scale Moving Objects in Road Networks 307
Jinping Jia, Ying Hu, Bin Zhao, Genlin Ji, and Richen Liu

Efficient Mining of Outlying Sequential Behavior Patterns 325
Yifan Xu, Lei Duan, Guicai Xie, Min Fu, Longhai Li, and Jyrki Nummenmaa

Clustering Mixed-Type Data with Correlation-Preserving Embedding 342
Luan Tran, Liyue Fan, and Cyrus Shahabi

Beyond Matching: Modeling Two-Sided Multi-Behavioral Sequences for Dynamic Person-Job Fit 359
Bin Fu, Hongzhi Liu, Yao Zhu, Yang Song, Tao Zhang, and Zhonghai Wu

A Local Similarity-Preserving Framework for Nonlinear Dimensionality Reduction with Neural Networks. 376
Xiang Wang, Xiaoyong Li, Junxing Zhu, Zichen Xu, Kaijun Ren, Weiming Zhang, Xinwang Liu, and Kui Yu

AE-UPCP: Seeking Potential Membership Users by Audience Expansion Combining User Preference with Consumption Pattern. 392
Xiaokang Xu, Zhaohui Peng, Senzhang Wang, Shanshan Huang, Philip S. Yu, Zhenyun Hao, Jian Wang, and Xue Wang

Self Separation and Misseparation Impact Minimization for Open-Set Domain Adaptation 400
Yuntao Du, Yikang Cao, Yumeng Zhou, Yinghao Chen, Ruiting Zhang, and Chongjun Wang

Machine Learning

Partial Modal Conditioned GANs for Multi-modal Multi-label Learning with Arbitrary Modal-Missing. 413
Yi Zhang, Jundong Shen, Zhecheng Zhang, and Chongjun Wang

Cross-Domain Error Minimization for Unsupervised Domain Adaptation 429
Yuntao Du, Yinghao Chen, Fengli Cui, Xiaowen Zhang, and Chongjun Wang

Unsupervised Domain Adaptation with Unified Joint Distribution Alignment 449
Yuntao Du, Zhiwen Tan, Xiaowen Zhang, Yirong Yao, Hualei Yu, and Chongjun Wang

Relation-Aware Alignment Attention Network for Multi-view Multi-label Learning 465
Yi Zhang, Jundong Shen, Cheng Yu, and Chongjun Wang

BIRL: Bidirectional-Interaction Reinforcement Learning Framework for Joint Relation and Entity Extraction 483
Yashen Wang and Huanhuan Zhang

DFILAN: Domain-Based Feature Interactions Learning via Attention Networks for CTR Prediction 500
Yongliang Han, Yingyuan Xiao, Hongya Wang, Wenguang Zheng, and Ke Zhu

Double Ensemble Soft Transfer Network for Unsupervised Domain Adaptation 516
Manliang Cao, Xiangdong Zhou, Lan Lin, and Bo Yao

Attention-Based Multimodal Entity Linking with High-Quality Images 533
Li Zhang, Zhixu Li, and Qiang Yang

Learning to Label with Active Learning and Reinforcement Learning 549
Xiu Tang, Sai Wu, Gang Chen, Ke Chen, and Lidan Shou

Entity Resolution with Hybrid Attention-Based Networks. 558
Chenchen Sun and Derong Shen

Information Retrieval and Search

MLSH: Mixed Hash Function Family for Approximate Nearest Neighbor Search in Multiple Fractional Metrics 569
Kejing Lu and Mineichi Kudo

Quantum-Inspired Keyword Search on Multi-model Databases 585
Gongsheng Yuan, Jiaheng Lu, and Peifeng Su

ZH-NER: Chinese Named Entity Recognition with Adversarial Multi-task Learning and Self-Attentions 603
Peng Zhu, Dawei Cheng, Fangzhou Yang, Yifeng Luo, Weining Qian, and Aoying Zhou

Drug-Drug Interaction Extraction via Attentive Capsule Network with an Improved Sliding-Margin Loss 612
Dongsheng Wang, Hongjie Fan, and Junfei Liu

Span-Based Nested Named Entity Recognition with Pretrained Language Model 620
Chenxu Liu, Hongjie Fan, and Junfei Liu

Poetic Expression Through Scenery: Sentimental Chinese Classical Poetry Generation from Images. 629
Haotian Li, Jiatao Zhu, Sichen Cao, Xiangyu Li, Jiajun Zeng, and Peng Wang

Social Network

SCHC: Incorporating Social Contagion and Hashtag Consistency for Topic-Oriented Social Summarization 641
Ruifang He, Huanyu Liu, and Liangliang Zhao

Image-Enhanced Multi-Modal Representation for Local Topic Detection from Social Media 658
Junsha Chen, Neng Gao, Yifei Zhang, and Chenyang Tu

A Semi-supervised Framework with Efficient Feature Extraction and Network Alignment for User Identity Linkage 675
Zehua Hu, Jiahai Wang, Siyuan Chen, and Xin Du

Personality Traits Prediction Based on Sparse Digital Footprints via Discriminative Matrix Factorization 692
Shipeng Wang, Daokun Zhang, Lizhen Cui, Xudong Lu, Lei Liu, and Qingzhong Li

A Reinforcement Learning Model for Influence Maximization in Social Networks 701
Chao Wang, Yiming Liu, Xiaofeng Gao, and Guihai Chen

A Multilevel Inference Mechanism for User Attributes over Social Networks 710
Hang Zhang, Yajun Yang, Xin Wang, Hong Gao, Qinghua Hu, and Dan Yin

Query Processing

Accurate Cardinality Estimation of Co-occurring Words Using Suffix Trees 721
Jens Willkomm, Martin Schäler, and Klemens Böhm

Shadow: Answering Why-Not Questions on Top-K Spatial Keyword Queries over Moving Objects 738
Wang Zhang, Yanhong Li, Lihchyun Shu, Changyin Luo, and Jianjun Li

DBL: Efficient Reachability Queries on Dynamic Graphs 761
Qiuyi Lyu, Yuchen Li, Bingsheng He, and Bin Gong

Towards Expectation-Maximization by SQL in RDBMS 778
Kangfei Zhao, Jeffrey Xu Yu, Yu Rong, Ming Liao, and Junzhou Huang

Correction to: Database Systems for Advanced Applications C1
Christian S. Jensen, Ee-Peng Lim, De-Nian Yang, Wang-Chien Lee, Vincent S. Tseng, Vana Kalogeraki, Jen-Wei Huang, and Chih-Ya Shen

Author Index 795

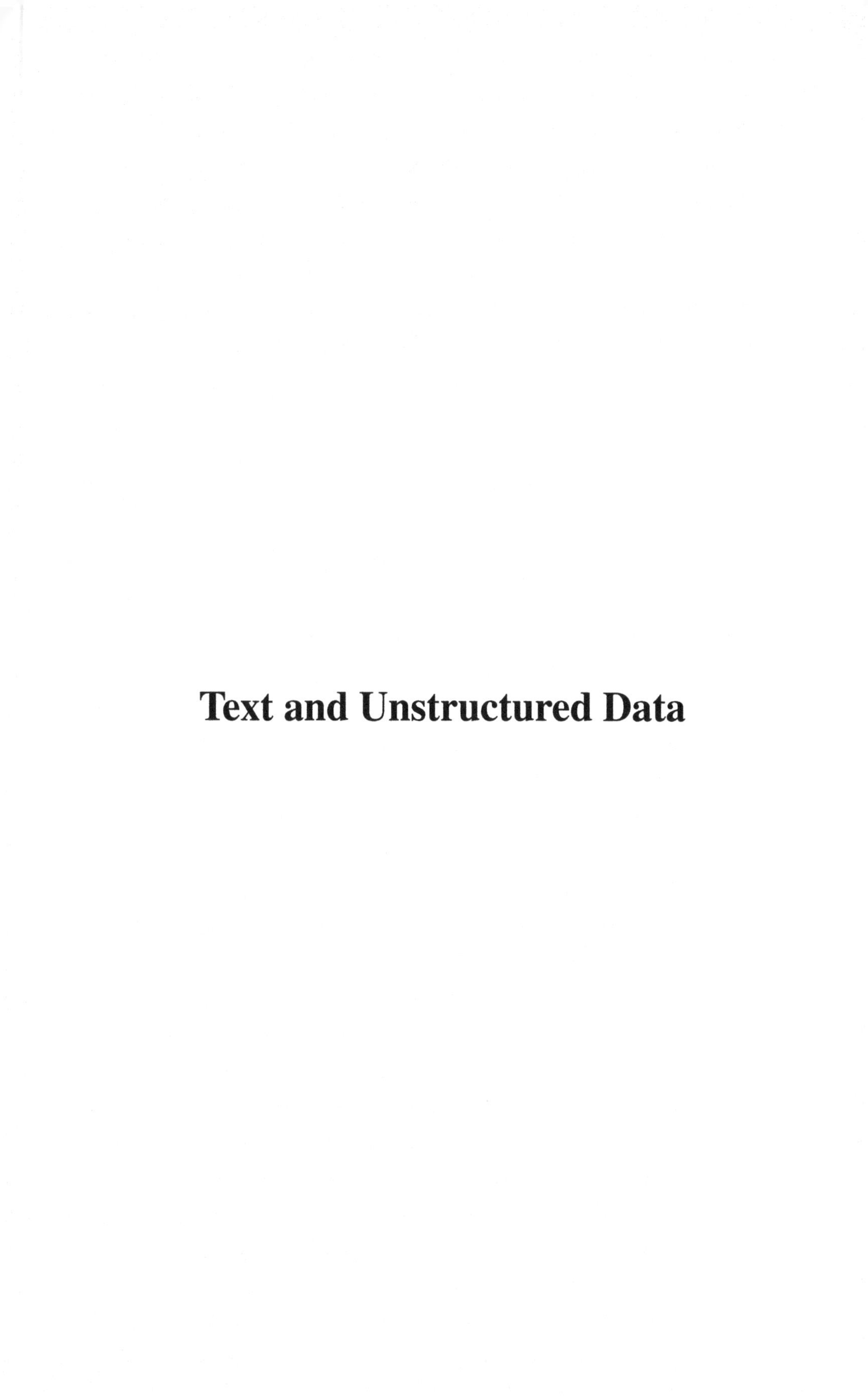

Text and Unstructured Data

Multi-label Classification of Long Text Based on Key-Sentences Extraction

Jiayin Chen[1], Xiaolong Gong[2], Ye Qiu[2], Xi Chen[1], and Zhiyi Ma[1,2](✉)

[1] Advanced Institute of Information Technology, Peking University, Hangzhou, China
{jychen,xchen}@aiit.org.cn

[2] School of Electronics Engineering and Computer Science, Peking University, Beijing, China
{xiaolgong,qiuye2014,mazhiyi}@pku.edu.cn

Abstract. Most existing works on multi-label classification of long text task will perform text truncation preprocessing, which leads to the loss of label-related global feature information. Some approaches that split an entire text into multiple segments for feature extracting, which generates noise features of irrelevant segments. To address these issues, we introduce key-sentences extraction task with semi-supervised learning to quickly distinguish relevant segments, which added to multi-label classification task to form a multi-task learning framework. The key-sentences extraction task can capture global information and filter irrelevant information to improve multi-label prediction. In addition, we apply sentence distribution and multi-label attention mechanism to improve the efficiency of our model. Experimental results on real-world datasets demonstrate that our proposed model achieves significant and consistent improvements compared with other state-of-the-art baselines.

Keywords: Multi-label classification · Key-sentences extraction · Sentence distribution

1 Introduction

Multi-label text classification (MTC) is an important task in the field of natural language processing (NLP) and text mining, which aims to train a model to classify a text into one or multiple labels. In recent years, more neural network classifiers [14,17,29,32] have been applied to multi-label classification (MC) tasks than traditional classifiers [6,23,30]. The main reason is that neural network approaches generalize adequately and perform robustly with the increase of available data.

In real-world scenarios, many classified texts are document-level, such as company introduction, legal case description, book abstract, etc. However, the

J. Chen and X. Gong—These authors contributed equally to this work and should be regared as co-first authors.

C. S. Jensen et al. (Eds.): DASFAA 2021, LNCS 12682, pp. 3–19, 2021.
https://doi.org/10.1007/978-3-030-73197-7_1

previous methods [1,2,17,32] have two major shortcomings on multi-label text classification of long text: **1)** Most methods adopt the trick of text truncation when the text is too long, which leads to the loss of label-related global feature information. Text truncation is the most common and most straightforward way for preprocessing long text in machine learning. Whether truncated at both ends or randomly, a large amount of global context information related to labels will inevitably be lost. Different with multi-class classification, the MTC task needs more label-related textual information. So how to take advantage of more useful information in the document-level text is something to consider. **2)** To solve the problems caused by text truncation, some methods [26,28] have typically applied segment approaches that split an entire text into multiple sentences and use encoding models (e.g., CNN, RNN, Transformer) to extract local features and finally concatenate those features to predict multi-label. Although these methods take into account global textual information, there are still some defects. First, these methods will generate a large number of noise features of irrelevant segments. Second, the contextual semantic information between sentences are discarded.

To our knowledge, few methods employ global context learning to learn representations for multi-label long text classification. The previous method mainly focused on extreme multi-label text classification (XMTC) [5,17,29,32] and hierarchical multi-label text classification (HMTC) [1,2,14,21,25], which solve the problems of the number of multi-label and the hierarchy between labels respectively. In fact, document-level text learning has been neglected in the above tasks, which is necessary for multi-label long text classification. The reason is that some sparse label-related features exist randomly in the text. In document-level text learning methods [4,22,26,27], most ideas divide one document into many sentences, then learn sentence-level representation, and finally concatenate those global context information for downstream specific tasks. These methods take advantage of each sentence representation by considering all sentences as equally important, while actually, many sentences are not associated with the labels.

To overcome these issues, we propose a novel neural multi-task learning framework named **KEMTC**, for modeling Multi-label Classification of Long Text Based on Key-sentences Extraction. We introduce key-sentences extraction task based on semi-supervised learning and learn global feature information of key-sentences to perform MTC task. In addition, we apply sentence distribution and multi-label attention mechanism [24,29] to further improve our model. Specifically, we first conduct a low-cost and little sentence-level annotation for each text and annotate the value (including *yes*, *no*) of each sentence, meaning whether it is related to the labels. Next, we apply the light pre-training model ALBERT [16] to encode sentences and form two sentence distributions to extract key-sentences. Finally, we concatenate key-sentences features with a filter gate, and then adopt AttentionXML [29] method for multi-label prediction. In our **KEMTC**, we perform key-sentences extraction and multi-label prediction tasks and form two losses as the overall optimization objective, which helps filter out irrelevant sentences and improve the prediction of MTC.

We conduct experiments on several real-world datasets to investigate the advantage of our model. Experimental results show that our method significantly and consistently outperforms other state-of-the-art models on all datasets and evaluation metrics. To summarize, we make the following contributions in this paper:

1. We propose a multi-task architecture which jointly trains a model to key-sentences extraction with distance square loss and multi-label classification task with cross-entropy loss, which can successfully mitigate the negative effect of having too many irrelevant sentences.
2. We apply the light pre-training model ALBERT to get two representation vectors of each sentence, one for sentence distribution and one for feature learning, and adopt the sentence distribution method with semi-supervised learning to simplify our model.
3. We use the multi-label attention mechanism to make the learned global features related to the labels more distinctive. What is more, we conduct efficient experiments on several real-world datasets, and our model significantly outperforms other baselines for MTC task.

2 Related Work

2.1 Multi-Label Learning

Multi-label learning studies the problem where each example is represented by a single instance while associated with *a set of* labels simultaneously [31]. During the past decade, significant amount of algorithms have been proposed to learning from multi-label data. Traditional methods can be divided into two categories. 1) *Problem transformation methods*: Representative algorithms include Binary Relevance [3], Calibrated Label Ranking [9], Random k-labelsets [23], and so on. 2) *Algorithm adaptation methods*: Representative algorithms include ML-kNN [30], ML-DT [10], and etc. In addition, deep learning methods have been widely applied to multi-label tasks, which include XML-CNN [17], AttentionXML [29], BERT-XML [32] for extreme multi-label text classification, HFT-CNN [21], Attention-based Recurrent Network [14], capsule networks [1,33], and hierarchical transfer learning [2] for hierarchical multi-label classification of text. The former methods rarely consider the relationship between labels, while the latter learn the hierarchical structure between labels. The above methods always perform text truncation when input text too long in MTC task, which leads to lose much helpful feature information. To avoid this problem, we adopt a multi-task framework that employs a shared network and two task-specific networks to derive a shared feature space. In key-sentences extraction, we employ the semi-supervised iterative sentence distribution method that simply and quickly extracts key-sentences. In multi-label prediction, we adopt the AttentionXML method that learns the relations between global features and labels to improve prediction.

2.2 Multi-Task Learning

In general, multi-task learning is learning about multiple related tasks simultaneously, where the training process shares some feature spaces in order to promote learning and improve the effect of generalization. Multi-task learning frameworks [8,18,19,28] have been employed successfully in various NLP tasks. He and W.S.[11] introduces a multi-task architecture to learn relation identification task and relation classification task, and mitigate the negative effect of having too many negative instances. Schindler and Knees [20] proposes an interactive multi-task learning network to address aspect-based sentiment analysis task. Hu [13] introduces multi-task learning framework to predict the attributes and charges of each simultaneously and combine these attribute-aware representations with an attribute-free fact representation to predict the final charges.

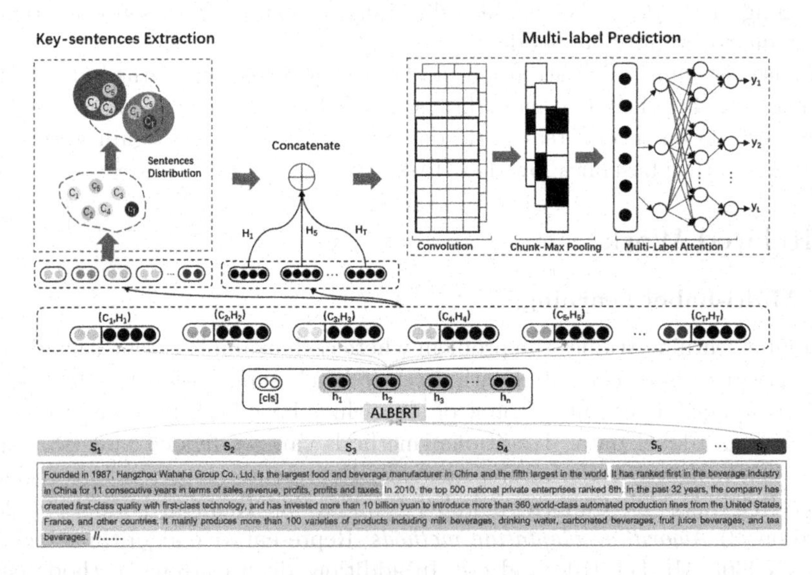

Fig. 1. An illustration of multi-label classification of long text based on key-sentences extraction.

In addition, some researchers have applied multi-task learning to solve multi-label classification problems. [4] proposes an RNN-based classifier that firstly perform sentence classification and summarization extraction, and finally make multi-label prediction. [20] presents an approach to multi-task representation learning using multi-label embedding. Unlike existing methods, our proposed model use sentence distribution methods for key-sentences extraction that filter out too many irrelevant sentences and employ an multi-label attention mechanism to capture the most relevant parts of texts to each label.

3 Model

In this section, we propose a novel model that jointly models key-sentence extraction task and multi-label classification task in a unified framework, as shown in Fig. 1. In the following parts, we first give definitions of key-sentence extraction and multi-label prediction. Then we describe the neural encoder of sentences based on ALBERT, key-sentences extraction with semi-supervised learning, multi-label prediction based multi-label attention. At last, we show the loss function of our model.

3.1 Task Definition

Key-Sentences Extraction. Given a long text sequence $X = \{x_1, x_2, .., x_n\}$, where n represents the sequence length, $x_i \in V$, and V is a fixed vocabulary. Cut text into sentences set $S = \{s_1, s_2, ..., s_T\}$ by rules, where T represents the number of sentences set. Form the key-Sentences distribution G_K and non key-sentences distribution G_N from the sentence set S and C_K, C_N respectively represent the center coordinates of the two distributions. Given an unknown sentence s_i, the extraction task is to determine which distribution s_i belongs to.

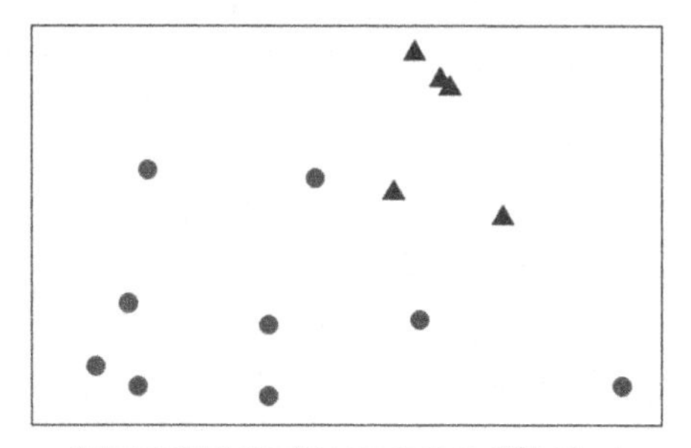

Sentence distribution for a sample text in CCD dataset

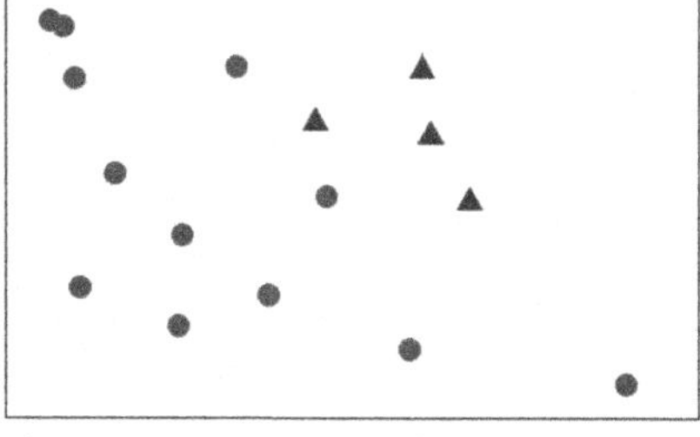

Sentence distribution for a sample text in LCD dataset

Fig. 2. The visualization of sentences distribution of two samples. The red triangles represent key-sentences and the blue dots represent non key-sentences. The details of two datasets will be presented in Sect. 4.

Multi-label Prediction. Set the label space with L labels $L = \{l_1, l_2, .., l_L\}$. Given the sequence X, the multi-label prediction is to assign a subset y containing m labels in the label space L to X. In our task, we consider each labels to be parallel, so the prediction task can be modeled as finding an optimal label y^* that maximizes the conditional probability $p(y|x)$, which is calculated as follows: $p(y|x) = \prod_{i=1}^{m} p(y_i|x)$.

3.2 Sentence Encoder

ALBERT Pre-training. In this work, we apply ALBERT to represent input text. The backbone of the ALBERT architecture is similar to BERT [7], but ALBERT has undergone technical improvements to enhance the overall effect of the model, mainly including factorized embedding parameterization, cross-layer parameter sharing, inter-sentence coherence loss. ALBERT is a lite BERT that is more suitable for downstream tasks.

Unlike many practitioners who use BERT models that have already been pre-trained on a wide corpus, we trained ALBERT models from scratch on our datasets to address the following issues. Firstly, each dataset contains a specific vocabulary that is not common within a general pretraining corpus, which leads to many out of vocabulary (OOV) words. Secondly, We use sentence representation to form sentence distributions that need to consider the context of the sentence. We retrain the inter-sentence task of ALBERT to help form more accurate distributions.

Sentence Representation. Given sentences set $S = \{s_1, s_2, ..., s_T\}$ from the long text X, we encode each sentence by ALBERT pre-training model. The standard architecture for sentence representation is to embed a $[CLS]$ token along with all additional inputs, yielding contextualized representations from the encoder. Assume $H_S = \{C, H\}$ is the last hidden layer corresponding to the $[CLS]$ token and input tokens, where $H = \{h_0, h_1, .., h_t\}$, t is the length of one sentence.

Given a sentence s_i, there are two forms of sentence representation, including C_i and $H_i = \{h_0, h_1, .., h_t\}$. The C_i is the whole representation of the sentence, and the H_i is the sequential representation of the sentence. We use C_i to form sentence distributions and H_i perform feature extraction through downstream fine-tuning. This simplifies the key-sentences extraction task, and learn more features to enhance the multi-label prediction.

3.3 Key-Sentences Extraction with Semi-supervised Learning

Key-Sentences Distribution. For multi-label, related sentences are similar, and similar sentences have the same distribution, as shown in Fig. 2. Based on this assumption, We encode the annotated sentences by ALBERT and get the sentence vector C to form two distributions: key-sentences distribution (G_K) and non key-sentences (G_C). Then applying the K-Means method, we calculate the centers (C_K, C_N) of the two distributions separately. $h_{cls}, C_K, C_N \in \mathbb{R}^{1 \times d_w}$, where d_w is the dimension of word vectors.

Algorithm 1. Distribution Iteration

1: **Update** C_K,C_N ;
2: initialize $C_K, C_N \longleftarrow G_K, G_N$ with labeled sentences
3: **for** s_i in batch_iter() **do**
4: compute dist(C_i, C_K), dist(C_i, C_N)
5: compute $r = p(y|C_i)$ follow Equation 1
6: **if** $r = 1$ **then**
7: add C_i to G_K
8: **else**
9: add C_i to G_N
10: $C_K, C_N \longleftarrow G_K, G_N$
11: **return** C_K, C_N;

For an unknown sentence s_i and its vector C_i, we calculate its distance from two centers, and then we can determine which distribution it belongs to.

$$
\begin{aligned}
dist(C_i, C_K) &= \sqrt{\sum_{j=1}^{d_w} (e_{s_j} - e_{c_j})^2} \\
p(y|C_i) &= \begin{cases} 1, & dist(C_i, C_K) < dist(C_i, C_N) \\ 0, & else \end{cases}
\end{aligned} \tag{1}
$$

where e_{s_j}, e_{c_j} are the coordinate values of the sentence and the center point, respectively. dist(C_i, C_N) is calculated in the same way. So key-sentences extraction task is to calculate unknown sentences which distribution center point close to.

Distribution Iteration. Due to the high cost of labeling key-sentences, we propose a distribution Iteration method to improve the learning space of a small number of labeled samples. Specifically, we add predicted sentences to their respective distributions and recalculate the center point in each batch-training process. This method can improve the prediction ability of key-sentences, and also avoid labeling pressure. The iteration process is shown in Algorithm 1.

3.4 Multi-label Prediction Based Multi-label Attention

Concatenate Layer. After key-sentences extraction, we concatenate key-sentences with a filter gate. For a set S from an input text, if num sentences are predicted as key-sentences, the concatenated vector $\boldsymbol{M}$ is as follows. Here, filter gate $u = \{u_1, u_2, ..., u_{num}\}$ for all sentences.

$$
\begin{aligned}
\boldsymbol{M} &= \sum_{i=0}^{T} u_i H_i \\
u_i &= \begin{cases} 1, & if\ s_i\ is\ key - sentence \\ 0, & else \end{cases}
\end{aligned} \tag{2}
$$

where $\sum$ denotes element-wise multiplication, and $\boldsymbol{M} \in \mathbb{R}^{(num \times t) \times d_w}$.

Global Feature Extraction. To extract global N-gram features, we apply a convolution filter $F \in \mathbb{R}^{g \times d_w}$ to a text region of g words $M_{i:i+g-1}$ to produce a new feature.

$$c_i = F \odot M_{i:i+g-1} + b \tag{3}$$

where $\odot$ denotes the component-wise multiplication, and $b \in \mathbb{R}$ is a bias term. Suppose f filters are used, and the f resulting feature maps are $\mathbf{c} = \{c^1, c^2, .., c^f\}$. Next, we apply the *Pooling* operation to each of these f feature maps to generate key-sentence feature vectors.

We adopt a Chunk-Max Pooling method in our model, which is inspired by XML [?]. Instead of generating only one feature per filter by max-pooling method, this method can capture richer information. In particular, we divide feature maps c into p chunks and take the maximum value within each chunk.

$$\boldsymbol{v} = [max(c_{1:\frac{pl}{p}}), .., max(c_{pl-\frac{pl}{p}+1:pl})] \tag{4}$$

which pl is the length of a feature map. The vector $\boldsymbol{v}=\{v_1, v_2, .., v_k\}$ contains both important features and position information about these local important features, where k is the number of N-gram features, and $k = f * p$.

Multi-Label Attention. Inspired by AttentionXML [29], we experiment the multi-label attention to predict and find it improves performance on our task. AttentionXML computes the (linear) combination of feature vectors v_i for each label through a multi-label attention mechanism to capture various intensive parts of a text. Compared with AttentionXML, our the label vector l_j get from ALBERT, and no parameter training is required. That is, the output of multi-label attention layer z_j of the j-th label can be obtained as follows:

$$\begin{aligned} a_{ij} &= \frac{exp(\langle v_i, l_j \rangle)}{\sum_{i=0}^{k} exp(\langle v_i, l_j \rangle)} \\ \mathbf{z}_j &= \sum_{i=0}^{k} a_{ij} v_i \end{aligned} \tag{5}$$

where a_{ij} is the normalized coefficient of v_i and l_j and also is the so-called attention parameters.

Fully Connected and Output Layer. In our model, there is one fully connected layer and one output layer. The same parameter values are used for all labels at the fully connected layer, to emphasize differences of attention among all labels. For label l_j, it will be calculated shown below.

$$\begin{aligned} \mathbf{o}_j &= tanh(\mathbf{W}^o \mathbf{z}_j + \mathbf{b}^o) \\ \mathbf{y}_j &= \sigma(\mathbf{W}^y \mathbf{o}_j + \mathbf{b}^y) \end{aligned} \tag{6}$$

Here, $\mathbf{W}^o$, $\mathbf{b}^o$ are shared parameters values in the fully connected layer. $\mathbf{W}^y$, $\mathbf{b}^y$ are weight matrix and bias parameters in the output layer.

3.5 Optimization

The training objective of our proposed model consists of two parts. The first one is to minimize the cross-entropy between predicted label distribution y and the ground-truth distribution $\hat{y}$. The other one is to minimize the square Loss about the distance from each sentence to two centers.

In multi-label prediction, we treat it as a binary classification task for each label. So we adapt the traditional extended cross-entropy loss. Specifically, consider the extended cross-entropy loss function as

$$\mathcal{L}_{label} = -\frac{1}{N}\sum_{i=1}^{N}\sum_{j=1}^{L} y_{ij}log(\hat{y}_{ij}) \tag{7}$$

where y_ij is the ground-truth label, and $\hat{y}_{ij}$ is prediction probability. N is the number of samples, and L is the number of labels.

In key-sentences extraction, We hope that some ambiguous sentences will also be added to the key-sentences distribution, so as to capture as much information as possible. To achieve this, we adjust the distance weights of the two distributions. Specifically, we formulate the extraction loss as:

$$\mathcal{L}_{extr} = \frac{1}{T}(\gamma\sum_{i=1}^{N} dist(C_i, C_K)^2 + (1-\gamma)\sum_{i=1}^{N} dist(C_i, C_N)^2) \tag{8}$$

where γ is a weight parameter to adjust sentence distributions bias, and T is the number of sentences. Here we set $\gamma = 0.55$.

Considering the two objectives, our final loss function $\mathcal{L}$ is obtained by adding $\mathcal{L}_{label}$ and $\mathcal{L}_{extr}$ as follows:

$$\mathcal{L} = \mathcal{L}_{label} + \alpha\mathcal{L}_{extr} \tag{9}$$

where α is a hyper-parameter to balance the weight of the two parts in the loss function. Here we set $\alpha = 0.5$.

4 Experiments

In order to verify the effectiveness of our model, we conduct experiments on real-world datasets and compare our model with state-of-the-art baselines.

4.1 Data

We use two real-world Chinese datasets to verify the effectiveness and practicability of our model:

- **CompanyCategoryDataset (CCD):** We construct a dataset of company entities from online open-source data[1], where each company entity may have

[1] https://github.com/cjymz886/ACNet.

Table 1. Quantitative characteristics of both datasets.

Datasets	CCD	LCD
Number of texts	12492	388784
Number of labels	17	202
Average length of texts	604	763
Max length of texts	4294	56694
Percentage of text longer than 1500	5.1%	2.6%

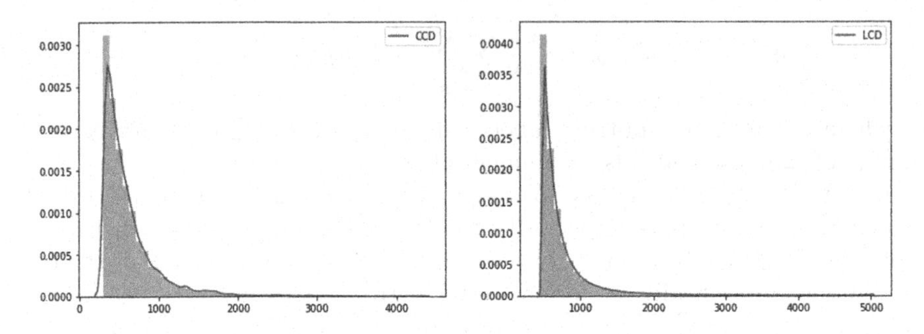

Fig. 3. The text length distribution of both datasets.

multiple industry category labels. The dataset includes 12492 company entities and 17 labels. Each company entity has *company_name*, *industry_category*, *abstract* common text attributes. We use *company_name* and *abstract* to form a document-level text for multiple *industry_category* prediction, where the average length of the texts is 604, and the maximum length is 4294. The label collection includes [*food and beverage, household life, medicine and health care,...*].

- **LegalCaseDataset (LCD):** We collect criminal cases published by the Chinese government from China Judgments Online[2]. Each case in the dataset is well structured and have *penalty_result*, *fact* attributes. As a case may have multi-penalty, we select the *fact* part of each case as our input. Specifically, the label collection includes [*fraud, theft, arson, intentional injury,...*].

The input text of these two datasets is relatively long, and the longest can reach 56694. The number of labels in the two datasets is 17, 202, respectively. Those datasets are randomly split by patients into 70/10/20 train, dev, test sets. The detailed statistics are shown in Table 1. In addition, we count the text length distribution of the two datasets in Fig. 3, which shows both distributions are similar to the long-tailed distribution, and samples are mainly concentrated in the 2000 range. The percentages of text longer than 1500 are 5.1%, 2.6%, respectively. According to these characteristics, we will truncate a small number of extremely long samples in the experiment to avoid a single sample taking too long to learn.

[2] http://wenshu.court.gov.cn.

Table 2. The statistics of annotations.

Datasets	CCD	LCD
Number of annotations per label	50	10
Total number of annotations	850	2020
Number of key-sentences	1231	5689
Number of non key-sentences	7789	21891

4.2 Baseline Models and Evaluation Metrics

Baseline Models. We employ several typical multi-label text classification models as baselines:

- **CNN-Kim** [15]: The CNN text classification model applies convolution-pooling operation to extract n-gram features.
- **LSTM** [12]: We implement a two-layer LSTM with a max-pooling layer as the local text encoder.
- **XML-CNN** [17]: The multi-label text classification uses dynamic max pooling and hidden bottleneck layer for taking multi-label co-occurrence patterns into account.
- **AttentionXML** [29]: The multi-label text classification proposes a multi-label attention mechanism to capture the most relevant part of the text to each label.
- **Bert-XML** [32]: Based on AttentionXML, it adapts the BERT architecture for encoding input texts and labels.

In those baselines, we adopt text truncation and set the maximum length of the text to 200. Besides, we also use 500-length text in CNN and LSTM, denoted as CNN-500 and LSTM-500.

Evaluation Metrics. We chose $P@k$ (Precision at k) [17] as our evaluation metrics for performance comparison since $P@k$ is widely used for evaluating the methods for MLC.

$$P@k = \frac{1}{k}\sum_{l=1}^{k} y_{rank(l)} \tag{10}$$

where $y \in \{0,1\}^L$ is the true binary vector, and *rank(l)* is the index of the l-th highest predicted label. In addition, we employ macro-F1 as another evaluation metric.

4.3 Experimental Settings

ALBERT Pre-training. We pretrain two different ALBERT architectures on two training set. The main purpose is to resolve some semantic gaps between open source pre-training models and training scenarios and to improve sentence

Table 3. Multi-label classification results of two datasets

Datasets	CCD				LCD			
Metrics	P@1	P@2	P@3	F1	P@1	P@2	P@3	F1
CNN	0.627	0.412	0.307	0.507	0.813	0.458	0.318	0.770
CNN-500	0.672	0.428	0.318	0.555	0.876	0.485	0.333	0.833
LSTM	0.613	0.390	0.289	0.493	0.793	0.425	0.299	0.755
LSTM-500	0.654	0.382	0.267	0.483	0.775	0.382	0.277	0.715
XML-CNN	0.705	0.443	0.338	0.570	0.884	0.493	0.343	0.836
AttentionXML	0.699	0.454	0.349	0.573	0.880	0.500	0.352	0.839
Bert-XML	0.707	0.523	0.424	0.608	**0.918**	0.513	0.407	**0.847**
KEMTC	**0.718**	**0.585**	**0.467**	**0.615**	0.910	**0.556**	**0.436**	0.843

representation. Note that we set max_seq_length = 50 in our pre-training models while *max_seq_length* supported by ALBERT is 512. The reason is that short sequence can obtain more accurate sentence representation, which is conducive to the generation and iteration of sentence distribution. In addition, the *sentence-order prediction* task in ALBERT focuses on modeling inter-sentence coherence that helps to judge similarity between sentences. Other detailed parameters include *hidden-size* = *768*, *layers* = *12*, *attention-heads* = *8*.

Key-Sentences Annotation. As mentioned in the previous, we propose to introduce key-sentences extraction to enhance multi-label prediction. Specifically, we annotate *yes/no* for each sentence to represent key-sentence and non key-sentence. For extracting key-sentences, we conduct a low-cost annotation over all datasets. Here, the low-cost annotation means we only need to annotate a few texts for each label in each dataset, specific to 50 in **CCD** and 10 in **LCD**.

Before annotation, we use rules to split the text into sentences, where the length of each sentence does not exceed 50. In practice, we adopt a looser annotating scheme: if a sentence has a fuzzy relationship with the label, it is also marked with *yes*. With these labeled sentences, we form two sentence distributions for key-sentences extraction. The detailed statistics are shown in Table 2.

Hyper-parameters. In all models, we set the size of word embedding size to 128, and the hidden state size of LSTM to 256. For the CNN based models, we set the filter kernel to [2,3,4] with 64 filters. We use Adam as the optimizer and set the learning rate to 0.001, the dropout rate to 0.5, and the batch size to 64. Considering that short sentence representation vectors are easier to form reasonable distributions, we set the maximum length of the segmented sentence to not exceed 50.

4.4 Results and Analysis

As shown in Table 3, we can observe that our **KEMTC** significantly and consistently outperforms all the baselines in **CCD** dataset, and achieves best per-

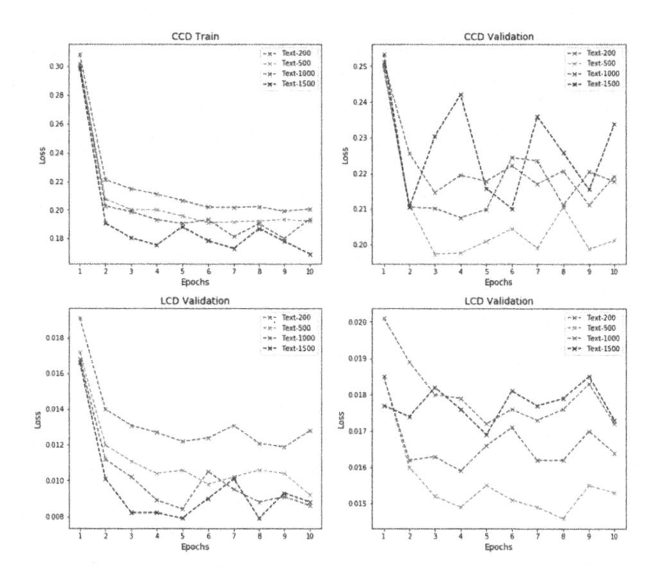

Fig. 4. Training loss and validation loss on both datasets with different text lengths.

Table 4. Experimental results of ablation test

Datasets	CCD				LCD			
Metrics	P@1	P@2	P@3	F1	P@1	P@2	P@3	F1
KEMTC	0.718	**0.585**	**0.467**	**0.615**	0.910	**0.556**	**0.436**	0.843
w/o key-sentences extraction	**0.722**	0.522	0.408	0.581	**0.917**	0.447	0.402	**0.849**
w/o multi-label attention	**0.722**	0.531	0.411	0.593	0.909	0.452	0.419	0.841

formance on the p@1 and p@2 indicators on both datasets, which indicates **KEMTC** improves the effect of multi-label classification and could captures more label-related information. With the increase of k value, the p@K indicator becomes worse in all existing methods, which demonstrates to classifying more labels, the model need to learn more label-related features, preferably distinguishing features. Conversely, **KEMTC** achieves promising improvements on the p@1 and p@2 (6.2%, 4.3%, and 4.3%, 2.9% absolutely on two datasets respectively), which shows key-sentences extraction task improving the efficiency of our model.

Bert-XML performs the best results among all baselines except our model, which demonstrates that the powerful pre-training model can achieve rich representation information. In addition, AttentionXML achieves higher performance than others except using the pre-training model. All these indicate that the pre-training method and the label attention mechanism have positive effects on our task, which also confirms the rationality of our model construction.

The experiment shows CNN-500 is better than CNN. Then there was the question of whether the longer the text, the better the model performs, so there is no need to segment the sentence. To test this hypothesis, We chose 4 different text lengths (including 200, 500, 1000, 1500) with the CNN model. As shown in Fig. 4, to a certain extent, increasing the length of the text can indeed improve the training effect and generalization ability of the model. However, when the input text is very long, the training effect is improved, but the verification loss is increased, which shows the model is easy to overfit and learns more irrelevant noise features. Therefore, when dealing with document-level text, it is necessary to segment sentences to capture global information.

4.5 Ablation Test

Our method is characterized by the incorporation of key-sentences extraction and multi-label attention mechanism. Thus, we design the ablation test respectively to investigate the effectiveness of these modules. When taken off key-sentences extraction, our method degrades into sentence segmentation and concatenating method based on CNN for multi-label classification. When taken off the attention mechanism, we replace attention mechanism with a fully connected layer with *sigmoid* activation function for all labels.

As shown in Table 4, we can observe that the performance drops obviously after removing the key-sentences extraction task or attention layer. The P@2, P@3 decreases by an average of 8% and 4%, respectively on both datasets. Therefore, it's obvious that both key-sentences extraction and multi-label attention mechanism play irreplaceable roles in our model.

4.6 Case Study

In this part, we select a representative sample to provide an intuitive illustration of how the key-sentences extraction help to promote the performance of multi-label classification of long text. In this case, many sentences are not related to

Example Sample-Company Industry Category **labels=['food and beverage', 'health care']**
S_1: 杭州娃哈哈集团有限公司创建于1987年，为中国最大，全球第五的食品饮料生产企业。
S_2: 目前在全国29省市建有58个基地近150家分公司，拥有总资产300亿元，员工近30000人。
S_3: 公司以一流的技术、一流的设备，一流的服务，打造出一流的品质......
S_7: 主要生产含乳饮料、饮用水、碳酸饮料、果汁饮料、茶饮料、保健食品、罐头食品、休闲食品等8大类100多个 品种的产品。
S_1: Founded in 1987, Hangzhou Wahaha Group Co., Ltd. is the largest food and beverage manufacturer in China and the fifth in the world.
S_2: At present, there are 58 bases and nearly 150 branches in 29 provinces and cities across the country, with total assets of 30 billion yuan and nearly 30,000 employees.
S_3: The company creates first-class quality with first-class technology, first-class equipment, first-class service...
S_7: It mainly produces more than 100 varieties of products in 8 categories, including milk beverages, drinking water, carbonated beverages, fruit juice beverages, tea beverages, health foods, canned foods, and snack foods.

Fig. 5. Visualization of key-sentences extraction

the company industry category, and the key-sentences related to the label are randomly distributed in the text. Moreover, it is common some words in the key-sentences that really affect the labels.

So we believe filtering out irrelevant sentences and learning feature words in key-sentences are essential in multi-label classification of this case. As shown in Fig. 5, our model correctly extracts key-sentences (S_1,S_7), and consequently recognizes the N-gram feature words (red font) associated with the labels. If the model does not learn the S_7 sentence, then *health foods* related to the *health care* label cannot be captured, so the prediction effect of this label is reduced, or even cannot be predicted. From this figure, we observe that the key-sentences extraction can improve the ability of the model to learn global useful features.

5 Conclusion

In this work, we focus on the task of multi-label classification of long text in two real application scenarios. To avoid the loss of global information and excessive noise features, we introduce key-sentences extraction into consideration and propose a novel multi-task learning model for multi-label long text classification. Specifically, our model uses the semi-supervised sentence distribution method to extract key-sentences and learns the global distinguishing features with the multi-label attention mechanism.

In the future, we will explore the following directions: **1)** There are more complicated data scenarios, such as categories of movies, categories of books, where sentences cannot be easily distinguished whether they are related to labels. Thus, it is challenging to handle this multi-label long text classification. **2)** In this work, we get sentence representation vectors from the pre-training model without considering the weight of words, while a word related to labels should be given more weight. How to obtain more accurate representations of key-sentences is expected to improve the interpretability of multi-label long text classification models.

Acknowledgments. We thank all the anonymous reviewers for their insightful comments. This work is supported by the National Natural Science Foundation of China (No. 61672046).

References

1. Aly, R., Remus, S., Biemann, C.: Hierarchical multi-label classification of text with capsule networks. In: Proceedings of the 57th Annual Meeting of the Association for Computational Linguistics: Student Research Workshop, pp. 323–330 (2019)
2. Banerjee, S., Akkaya, C., Perez-Sorrosal, F., Tsioutsiouliklis, K.: Hierarchical transfer learning for multi-label text classification. In: Proceedings of the 57th Annual Meeting of the Association for Computational Linguistics, pp. 6295–6300 (2019)
3. Boutell, M.R., Luo, J., Shen, X., Brown, C.M.: Learning multi-label scene classification. Pattern Recogn. **37**(9), 1757–1771 (2004)

4. Brandt, J.: Imbalanced multi-label classification using multi-task learning with extractive summarization. arXiv preprint arXiv:1903.06963 (2019)
5. Chalkidis, I., Fergadiotis, M., Malakasiotis, P., Androutsopoulos, I.: Large-scale multi-label text classification on eu legislation. arXiv preprint arXiv:1906.02192 (2019)
6. Chiang, T.H., Lo, H.Y., Lin, S.D.: A ranking-based knn approach for multi-label classification. In: Asian Conference on Machine Learning, pp. 81–96 (2012)
7. Devlin, J., Chang, M.W., Lee, K., Toutanova, K.: Bert: Pre-training of deep bidirectional transformers for language understanding. arXiv preprint arXiv:1810.04805 (2018)
8. Dong, H., Wang, W., Huang, K., Coenen, F.: Joint multi-label attention networks for social text annotation. In: Proceedings of the 2019 Conference of the North American Chapter of the Association for Computational Linguistics: Human Language Technologies, vol. 1 (Long and Short Papers), pp. 1348–1354 (2019)
9. Fürnkranz, J., Hüllermeier, E., Mencía, E.L., Brinker, K.: Multilabel classification via calibrated label ranking. Mach. Learn. **73**(2), 133–153 (2008)
10. Ghamrawi, N., McCallum, A.: Collective multi-label classification. In: Proceedings of the 14th ACM International Conference on Information and Knowledge Management, pp. 195–200 (2005)
11. He, R., Lee, W.S., Ng, H.T., Dahlmeier, D.: An interactive multi-task learning network for end-to-end aspect-based sentiment analysis. arXiv preprint arXiv:1906.06906 (2019)
12. Hochreiter, S., Schmidhuber, J.: Long short-term memory. Neural Comput. **9**(8), 1735–1780 (1997)
13. Hu, Z., Li, X., Tu, C., Liu, Z., Sun, M.: Few-shot charge prediction with discriminative legal attributes. In: Proceedings of the 27th International Conference on Computational Linguistics, pp. 487–498 (2018)
14. Huang, W., et al.: Hierarchical multi-label text classification: an attention-based recurrent network approach. In: Proceedings of the 28th ACM International Conference on Information and Knowledge Management, pp. 1051–1060 (2019)
15. Kim, Y.: Convolutional neural networks for sentence classification. arXiv preprint arXiv:1408.5882 (2014)
16. Lan, Z., Chen, M., Goodman, S., Gimpel, K., Sharma, P., Soricut, R.: Albert: A lite bert for self-supervised learning of language representations. arXiv preprint arXiv:1909.11942 (2019)
17. Liu, J., Chang, W.C., Wu, Y., Yang, Y.: Deep learning for extreme multi-label text classification. In: Proceedings of the 40th International ACM SIGIR Conference on Research and Development in Information Retrieval, pp. 115–124 (2017)
18. Liu, W.: Copula multi-label learning. In: Advances in Neural Information Processing Systems, pp. 6337–6346 (2019)
19. Maddela, M., Xu, W., Preoţiuc-Pietro, D.: Multi-task pairwise neural ranking for hashtag segmentation. arXiv preprint arXiv:1906.00790 (2019)
20. Schindler, A., Knees, P.: Multi-task music representation learning from multi-label embeddings. In: 2019 International Conference on Content-Based Multimedia Indexing (CBMI), pp. 1–6. IEEE (2019)
21. Shimura, K., Li, J., Fukumoto, F.: HFT-CNN: learning hierarchical category structure for multi-label short text categorization. In: Proceedings of the 2018 Conference on Empirical Methods in Natural Language Processing, pp. 811–816 (2018)
22. Tian, B., Zhang, Y., Wang, J., Xing, C.: Hierarchical inter-attention network for document classification with multi-task learning. In: IJCAI, pp. 3569–3575 (2019)

23. Tsoumakas, G., Vlahavas, I.: Random k-labelsets: an ensemble method for multilabel classification. In: Kok, J.N., Koronacki, J., Mantaras, R.L., Matwin, S., Mladenič, D., Skowron, A. (eds.) ECML 2007. LNCS (LNAI), vol. 4701, pp. 406–417. Springer, Heidelberg (2007). https://doi.org/10.1007/978-3-540-74958-5_38
24. Wang, H., Liu, W., Zhao, Y., Zhang, C., Hu, T., Chen, G.: Discriminative and correlative partial multi-label learning. In: IJCAI, pp. 3691–3697 (2019)
25. Xie, M.K., Huang, S.J.: Partial multi-label learning. In: Proceedings of the AAAI Conference on Artificial Intelligence, vol. 32 (2018)
26. Yang, P., Sun, X., Li, W., Ma, S., Wu, W., Wang, H.: SGM: sequence generation model for multi-label classification. arXiv preprint arXiv:1806.04822 (2018)
27. Yang, Z., Yang, D., Dyer, C., He, X., Smola, A., Hovy, E.: Hierarchical attention networks for document classification. In: Proceedings of the 2016 conference of the North American Chapter of the Association for Computational Linguistics: Human Language Technologies, pp. 1480–1489 (2016)
28. Ye, W., Li, B., Xie, R., Sheng, Z., Chen, L., Zhang, S.: Exploiting entity bio tag embeddings and multi-task learning for relation extraction with imbalanced data. arXiv preprint arXiv:1906.08931 (2019)
29. You, R., Zhang, Z., Wang, Z., Dai, S., Mamitsuka, H., Zhu, S.: Attentionxml: label tree-based attention-aware deep model for high-performance extreme multi-label text classification. In: Advances in Neural Information Processing Systems, pp. 5820–5830 (2019)
30. Zhang, M.L., Zhou, Z.H.: ML-KNN: a lazy learning approach to multi-label learning. Pattern Recogn. **40**(7), 2038–2048 (2007)
31. Zhang, M.L., Zhou, Z.H.: A review on multi-label learning algorithms. IEEE Trans. Knowl. Data Eng. **26**(8), 1819–1837 (2013)
32. Zhang, Z., Liu, J., Razavian, N.: Bert-xml: Large scale automated ICD coding using bert pretraining. arXiv preprint arXiv:2006.03685 (2020)
33. Zhu, Y., Kwok, J.T., Zhou, Z.H.: Multi-label learning with global and local label correlation. IEEE Trans. Knowl. Data Eng. **30**(6), 1081–1094 (2017)

Automated Context-Aware Phrase Mining from Text Corpora

Xue Zhang[1,2], Qinghua Li[1,2], Cuiping Li[1,2(✉)], and Hong Chen[1,2]

[1] Key Laboratory of Data Engineering and Knowledge Engineering, Renmin University of China, Beijing, China
{xue.zhang,qinghuali,licuiping,chong}@ruc.edu.cn
[2] School of Information, Renmin University of China, Beijing, China

Abstract. Phrase mining aims to automatically extract high-quality phrases from a given corpus, which serves as the essential step in transforming unstructured text into structured information. Existing statistic-based methods have achieved the state-of-the-art performance of this task. However, such methods often heavily rely on statistical signals to extract quality phrases, ignoring the effect of **contextual information**.

In this paper, we propose a novel context-aware method for automated phrase mining, ConPhrase, which formulates phrase mining as a sequence labeling problem with consideration of contextual information. Meanwhile, to tackle the global information scarcity issue and the noisy data filtration issue, our ConPhrase method designs two modules, respectively: 1) a topic-aware phrase recognition network that incorporates domain-related topic information into word representation learning for identifying quality phrases effectively. 2) an instance selection network that focuses on choosing correct sentences with reinforcement learning for further improving the prediction performance of phrase recognition network. Experimental results demonstrate that our ConPhrase outperforms the state-of-the-art approach.

Keywords: Phrase mining · Quality phrase recognition · Information extraction

1 Introduction

The explosive growth of unstructured text is becoming prohibitive. According to [1], the total volume of data will increase from 33 zettabytes in 2018 to 175 zettabytes in 2025, and 80% of them will be unstructured text data. The large amount of text data makes it difficult for people to access information efficiently. Therefore, advanced technologies for better extracting valuable information from the text are in great demand [2].

Phrase mining is an effective technique to obtain valuable information. It refers to extract quality phrases from large text corpora and transforms documents from unstructured text to structured information. Moreover, compared

C. S. Jensen et al. (Eds.): DASFAA 2021, LNCS 12682, pp. 20–36, 2021.
https://doi.org/10.1007/978-3-030-73197-7_2

with non-informative words (e.g. 'paper'), mining quality phrases would extract semantically meaningful word span as a whole semantic unit (e.g. 'support vector machine'). This semantic unit can further improve the understanding of texts and support various downstream applications, such as named entity recognition [3–5], keyword generation [6], topic modeling [7–9], and taxonomy construction [10].

Recent works mainly focus on general statistic-based methods, typically treat phrase mining as a binary classification problem. The representative approaches are SegPhrase [11] and AutoPhrase [12], which employ several raw frequency and rectified frequency signals to classify phrase candidates into quality phrases or non-quality ones. Such methods rely on statistical signals have achieved the state-of-the-art performance. However, statistical signals of phrases tend to less reliable in the medium (or small) corpus, fail to guide distinguishing high-quality phrases from inferior phrases, as the following example demonstrates.

Table 1. A hypothetical example of phrases' statistical signals in **database domain**

Text (with extracted phrase)	Freq.	Relative Freq. (PMI/OF/IDF)	Phrase? (assess by method)	Phrase? (assess by expert)
A [**relational database**] is the major type of database. A [**relational database**] stores and organizes data points that are related to one another.	800	0.06/0.0008/0.006	Yes	Yes
An implementation based on [**open source**] code for storage is released.	600	0.08/0.0006/0.005	Yes	No
A [**geographical database**] is the important type of database. A [**geographical database**] stores and organizes data points that are related to objects in space.	100	0.04/0.0001/0.003	No	Yes

* PMI: Pointwise Mutual Information; OF: Occurence Frequency; IDF: Inverse Document Frequency.

Example (Statistic-Based Phrase Mining). Consider a database corpus consisting of sentences with several frequencies shown in Table 1. The numbers are hypothetical but manifest the following key observations: (i) both quality and inferior phrases possess similar frequency signals (e.g. 'relational database' and 'open source'); (ii) quality phrases assessed by expert may have the lower statistical signal scores (e.g. 'geographical database'); (iii) quality phrases can occur under similar contexts (e.g. 'relational database' and 'geographical database').

In the above example, when statistical signals become less reliable, we observe that it will be hard for statistic-based methods to distinguish the quality phrase from the inferior one (i.e. 'geographical database', 'open source'). However, by observing the bold phrases in row 1 and row 3 of Table 1, we find that *similar contextual information* could help to capture high-quality phrases. For example, the contexts of 'geographic database' are very similar to other database types' context (e.g. 'relational database'), indicating it may be a quality phrase. Existing approaches for phrase mining usually utilize statistical signals to extract phrases, yet ignoring the contextual information of phrases. To consider the effect of context information, we propose a novel **context-aware method for automated phrase mining**, ConPhrase, which formulates phrase mining as a sequence labeling problem.

In proposed context-aware method ConPhrase, we aim to study the problem of automatically mining quality phrases from single sentences. Compared to the statistic-based method that extracts phrases from a corpus containing hundreds of documents, our method of identifying phrases from a single short sentence is generally more difficult. We need to solve the following two challenges:

Challenge 1: Lack of Global Information. For the problem of extracting phrases from a single sentence, the contextual information carried by the sentence itself is limited. For example, Table 1 shows the inferior phrase 'open source' and the quality phrase 'relational database' with their contextual information. We notice that without additional implicit information (e.g. global information that both phrases appear in the 'database' domain), it is difficult to distinguish which phrase is the high-quality one. Actually, high-quality phrases can reflect the key information of multiple or all documents in the corpus to a certain extent. The lack of key global information makes the phrases extracted from a single sentence not of high quality. For example, the mined phrase 'open source', in the database domain, is only a common word, not the quality phrase. Therefore, how to effectively mining quality phrases by incorporating domain-related global information is a challenging issue.

Challenge 2: Filtering Noisy Data. To effectively mining quality phrases from sentences, large amounts of annotated data is usually needed as a prerequisite. In the real-world text corpora, there have no ready-made labeled data for training models. Manual labeling is a common way to obtain golden training data, but it is often time-consuming and costly. Another alternative solution is to use the open knowledge bases (or dictionaries) as distant supervision (DS) to generate auto-labeled data [12]. However, unlike the expert-labeled data, the DS auto-labeled data usually contains noisy annotations, including missing labels, incorrect boundaries and types. Such noisy labeled data will cause the model to learn incorrect patterns, further affect the prediction performance of phrase recognition models. Therefore, how to select clean sentences from noisy data to improve the performance of phrase mining is another challenging issue.

To overcome these issues, our ConPhrase method designs two modules, respectively: 1) the topic-aware phrase recognition network that incorporates domain-related topic information into word representation learning for identifying quality phrases effectively. 2) the instance selection network that focuses on choosing correct sentences by reinforcement learning for further improving the prediction performance of phrase recognition network.

Specifically, to tackle the global information scarcity issue, we take domain-related topics into consideration. For a phrase in the single sentence, local context refers to the semantic information of all the words surrounding the phrase, which partially reflect the semantics of words. However, the topic information of entire corpus can cover the key expressions of multiple or all documents, which construct the prior knowledge of the domain and corresponding sub-domains. That is consistent with the goal of mining quality phrases: it is expected that the extracted quality phrases can reflect the key information of multiple texts. Motived by this, we design a topic-aware phrase recognition network (TPRNet). TPRNet utilizes the domain-related topic information to build a global context, and then integrates it into the local context (i.e. word representation) through the attention mechanism. Finally, TPRNet learns a topic-enriched word representation for identifying quality phrases effectively.

Further, to solve the noisy data filtration issue, we introduce an instance selection network (ISNet) with reinforcement learning, which learns a selection policy to choose correct sentences from the noisy labeled data. But in selecting clean sentences process, there has no absolutely reliable signal to evaluate whether the sequence labels annotated by DS are correct. So we need a trial-and-error process to explore a reliable selection policy. Besides, to make the exploration process converge as much as possible, ISNet uses high-scoring labeled sentences as the silver seed set into the selection policy to guide the selection process. To the end, ISNet learns a better selection policy to clean the noisy labeled data, and provide cleaned data to train TPRNet for further improving the prediction performance of quality phrase recognition.

The contributions of this work are as follows:

- We consider the effect of contextual information of phrases, propose a novel context-aware method for automated phrase mining, ConPhrase, which formulates phrase mining as a sequence labeling problem.
- To tackle the global information scarcity issue, we integrate topic information into word representation for effectively recognizing quality phrases.
- To solve the noisy data filtration issue, we introduce an RL-based selection policy that learns to select correct sentences from noisy labeled data.

2 Methodology

2.1 Problem Definition

We formally define the phrase mining task as follows: given a sequence of words $X = \{x_1, x_2, ..., x_L\}$, it aims to infer a sequence of labels $Y = \{y_1, y_2, ..., y_L\}$, where L is the length of the word sequence, $y_i \in Y$ is the label of the word x_i, each label indicates the boundary information by using IOB schema. B represents the word is **B**egin of a quality phrase, O and I represent **O**utside, **I**nside of the phrase. Here, 'fusidic acid' in Fig. 1 is a high-quality phrase. A text corpus containing many documents without sequence labels is available for mining quality phrases in our paper.

2.2 Overview

The overall process is shown in Fig. 1. Our ConPhrase consists of two components: Topic-aware Phrase Recognition Network (TPRNet) and Instance Selection Network (ISNet). To obtain training data, the ConPhrase first automatically labels domain data and computes the labeled score for each sentence. Then, TPRNet obtains word representation by incorporating domain-related topic information for recognizing quality phrases effectively. ISNet adopts a stochastic policy and samples an action at each state, which focuses on selecting correct instances from the auto-labeled data by reward computation offered by TPRNet. Obviously, the two components are intertwined together. TPRNet will be fine-tuned with the cleaned data provided by ISNet, and ISNet obtains rewards from TPRNet's prediction to guide the learning of a selection policy.

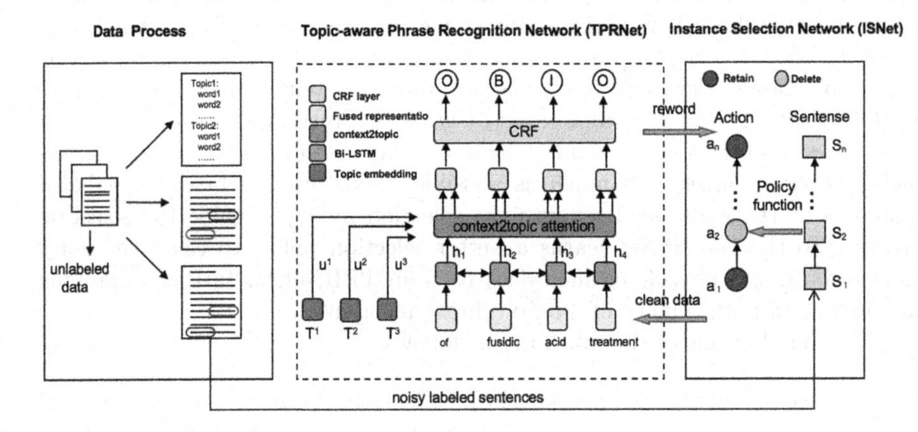

Fig. 1. Overview of our ConPhrase for mining quality phrases.

2.3 Data Process

Auto-Labeled Data Generation. One benefit of our proposed ConPhrase is that it can automatically assign weak labels to the unlabeled domain data, which helps enlarge the size of the training set with little cost. To obtain such labeled training data, we employ Distant Supervision (DS) [2,4] to automatically label the domain data by several approaches. There are two major sources to consider:

- Linguistic Analysis: We extract single words or word sequences as candidates if they matched the POS patterns, or are detected by pre-trained noun phrase chunking models [2].
- Knowledge Base: We extract single words or word sequences if they are detected by the DBPedia Spotlight linking model or domain dictionary [2,4].

The ConPhrase first assigns a labeled weight for each source, and then each source can vote score to phrase candidates. ConPhrase automatically selects high-scoring phrase candidates above threshold r for labeling one sentence through the majority voting. As a result, we can construct a noisy labeled training set D and obtain a high-scoring labeled sentence set D_l.

2.4 Topic-Aware Phrase Recognition Network (TPRNet)

To enrich local context for accurately recognizing high-quality phrases, we design a topic-aware phrase recognition network (TPRNet) by incorporating global topic information into consideration. Specifically, TPRNet first learns the domain-related topic information to build a global context. Then it employs the Context2Topic attention mechanism to supplement local context (i.e. word contextual representation) with global context information(i.e. topic representations), which is expected to bring prior knowledge for learning a good representation. Finally, TPRNet obtains a topic-enriched word representation for recognizing quality phrases effectively. Next, we detail how to utilize such topic information to learn a topic-enriched word representation.

Representation Layer. For each sentence X, we represent it as a list of vectors $W = \{w_1, w_2, ..., w_L\}$ by a look-up table, the length of sentence is L. Each representation vector consists of two parts: one is the word embedding which obtained from word2vec; the other is the character embedding which based on Convolutional Neural Networks (CNN). TPRNet then adopts Bi-LSTM to obtain word contextual representation and topic representations as follow:

Word Embeddings. To capture both past and future contextual information for each word, we employ Bi-LSTM to stack a forward LSTM and a backward LSTM, and to produce corresponding hidden states of a word in context (denoted as $\overrightarrow{h_t} = f_{LSTM}(w_t, h_{t-1})$ and $\overleftarrow{h_t} = f_{LSTM}(w_t, h_{t+1})$. Thus, the contextual representation for a word is $h_t = \left[\overrightarrow{h_t}, \overleftarrow{h_t}\right] \in \mathbb{R}^{2d}$, which forms a local context.

$$H = (h_1, h_2, ..., h_L) \in \mathbb{R}^{2d \times L} \tag{1}$$

Topic Embeddings. While the contextual representation above only considers the local context of each word, it is suggested that quality phrases should be relevant to the global information of the documents. To enrich the local content, we extract domain-related topics from the entire corpus as global information via the topic model, such as Latent Dirichlet Allocation(LDA) [13]. LDA is used to extract I topics and top N non-universal words t_n^i of correlated topic, which formally denoted each topic as $T^i = \left(t_1^i, t_2^i, ..., t_N^i\right)$, $i \in [1, I]$. Then, we use the same Bi-LSTM to represent topic sequences T^i by concatenating the last hidden states of two directions, i.e. $u^i = \left[\overrightarrow{h_N^i};\ \overleftarrow{h_1^i}\right] \in \mathbb{R}^{2d}$. Finally, all I corpus-level topic representations U forms the global context information.

$$U = \left(u^1, u^2, ..., u^I\right) \in \mathbb{R}^{2d \times I} \tag{2}$$

Context2topic Attention. Inspired by the attention mechanism in [14,15], we design a context2topic attention in TPRNet for linking and fusing local contextual information and global topic information.

The inputs to the layer are vector representations of words H and domain topics U. The outputs of the layer incorporate topic information into representation learning, further generate the topic-enriched word representation.

We first compute a similarity matrix $S \in \mathbb{R}^{L \times I}$ to score how well a context word and a topic match:

$$S_{li} = \varphi\left(H_{:l}, U_{:i}\right) \in \mathbb{R}^{L \times I} \tag{3}$$

where S_{li} indicates the semantic relatedness between l-th context word and i-th topic information. φ is a trainable scalar function that encodes the similarity between two input vectors, $H_{:l}$ is l-th column vector of H, and $U_{:i}$ is i-th column vector of U. We select $\varphi(h, u) = W^{\top}[h; h \circ u]$, where $W \in \mathbb{R}^{4d}$ is a trainable weight vector, $\circ$ is elementwise multiplication, $[;]$is vector concatenation.

Context2topic attention implies which topics are most relevant to each context word, and puts an attention on U to capture important topic information. For a hidden word h_l, the attention weights on topics T^i is defined as:

$$\alpha_l = softmax\left(S_{l:}\right) \in \mathbb{R}^{I} \tag{4}$$

Then, the attentive representation r_l is obtained by multiplying the weighted attention of every topic representation (i.e. $\alpha_l^i U_{:i}$) with the topic distribution of word sequences $\left(\beta^1, \beta^2, ..., \beta^I\right)$ inferred from LDA model:

$$r_l = \sum\nolimits_{i=1}^{I} \beta^i \alpha_l^i U_{:i} \tag{5}$$

Further, conditioned on $\{r_1, r_2, ..., r_L\}$, we generate the topic-enriched word representations H' over corresponding word hidden states H according to:

$$H' = \left(\left[h_1; r_1\right], \left[h_2; r_2\right], ..., \left[h_L; r_L\right]\right) \tag{6}$$

Phrase Recognition Layer. In phrase recognition phase, TPRNet uses classic sequence labeling model, including softmax layer and CRF layer, and predict each word's label Y':

$$P_{x_i, y_i} = softmax\left(W H_i' + b\right) \tag{7}$$

$$s(X, Y) = \sum\nolimits_{i=1}^{|X|} P_{x_i, y_i} + \sum\nolimits_{i=0}^{|X|} A_{y_i, y_{i+1}} \tag{8}$$

$$P(Y|X) = e^{s(X,Y)} / \sum\nolimits_{\widetilde{y} \in Y_X} e^{s(X, \widetilde{y})} \tag{9}$$

We estimate the score P_{x_i, y_i} for the word x_i being the label y_i, where label in set $\{B, I, O\}$. $A_{y_i, y_{i+1}}$ is the transition score from label y_i to y_{i+1} that is learned in CRF layer. During the training process, we minimize the negative log-likelihood function Eq. (9) of ground truth over all possible labeled sequences Y_X. For inference, we apply the Viterbi algorithm to maximize the score of Eq. (8) to extract quality phrases for each input sequence.

2.5 Instance Selection Network (ISNet)

We introduce an instance selection network (ISNet) with reinforcement learning, aims to selecting correct instances D' from the noisy auto-labeled dataset D for training a better phrase recognition model TPRNet.

Following [16], we cast instance selection as a reinforcement learning problem. The ISNet is trained as an agent, which interacts with the environment and makes a decision at the sentence-level. Based on the environment consisting of auto-labeled data and the TPRNet model, ISNet agent adopts a stochastic policy to samples an action (i.e. retain or delete) at each state. The action could decide whether the current sentence should be retained or deleted from the training data D. After all the selection are finished, the TPRNet provides a delayed reward to update agent.

Obviously, the above selection process is inefficient, because it needs to scan the entire training data before only updating the policy function once. To obtain more reward feedback and speed up the training process, we divide all sentence instances in dataset D into M small bags $\mathcal{B}$, i.e. $D = \left\{\mathcal{B}^1, \mathcal{B}^2, ..., \mathcal{B}^M\right\}$. For each instance in k-th bag $\mathcal{B}^k$, the ISNet agent takes action to select whether retains or deletes the sentence, according to the policy function. After completing all actions on a bag, TPRNet feedbacks a delayed reward. Then, the agent merges the correct sentence instances retained in each bag to build a cleaned dataset D', fine-tuning the TPRNet at the sentence-level.

Next, we suppose that one bag $\mathcal{B}$ consists of T sentences, and detail the procedures (i.e. state, action, policy, and reward) used in the ISNet as follow.

State. The state s_t encodes the current t-th instance and previous chosen instances of the bag. Formally,

$$s_t = [H_t^{'}; \overline{H'}_{1 \rightarrow t-1}],\ t\ \in\ [1, T] \tag{10}$$

where s_t represents the following information: 1) The vector representation of the current instance $H_t^{'}$, which is obtained from the topic-enriched representation of TPRNet by Eq. (6); 2) The representation of the previous instances $\overline{H'}_{1 \rightarrow t-1}$, which are the average of the vector representations of all previous sentences (before t step) chosen by the policy.

Action and Policy. Our selection policy is expected to select more instances annotated with correct labels for training TPRNet at the sentence-level. To implement this process, we adopt binary actions $a_t\ \in\ \{0, 1\}$ to indicate an instance can be chosen or not in a bag. The action space is $\{Retain, Delete\}$, where *Retain* indicates that the instance annotated with correct labels is retained in a bag, and *Delete* means that the instance is deleted and it has some incorrect word labels. The value of action a_t is sampled from a stochastic policy function π_Θ:

$$\pi_\Theta\left(a_t|s_t\right) = \sigma\left(W_\pi s_t + b_\pi\right) \tag{11}$$

Algorithm 1: Reinforcement Learning Algorithm of ConPhrase

Input: 1) Episode number L.
2) the noisy training dataset D, which is represented by a set of bags $D = \{\mathcal{B}^1, \mathcal{B}^2, ..., \mathcal{B}^M\}$. Each bag $\mathcal{B}^k$ contains T instances.
3) the TPRNet's parameter Φ; the ISNet's parameter Θ.

1 **for** *all episode* $e \leftarrow 1$ *to* L **do**
2 **for** *all bag* $\mathcal{B} \in D$ **do**
3 **for** *all time step* $t \leftarrow 1$ *to* T **do**
4 Compute the state vector s_t by the TPRNet ;
5 Sample an action $a_t \sim \pi_\Theta(a_t|s_t)$ from the selcetion policy π_Θ;
6 Compute the dealyed reward r_T of the selected $\mathcal{B}'$ by the TPRNet;
7 Update the parameter Θ of ISNet:
$\Theta \leftarrow \Theta + \alpha \sum_{t=1}^{T} (\gamma^{T-t} r_T) \nabla_\Theta log \pi_\Theta(a_t|s_t)$
8 Update the weights Θ of ISNet;
9 Update the parameter Φ of TPRNet using the cleaned data D';

where s_t denotes the state vector, $\pi_\Theta(a_t|s_t)$ is the probability of choosing the action a_t, and $\sigma(\cdot)$ is logistic function. We utilize a single-layer network to parametrize the selection policy with $\Theta = \{W_\pi, b_\pi\}$ to be learned.

Reward. The reward function is a feedback of all chosen sentences' quality to guide the selection policy learning. Once all the actions of current bag are sampled by the policy, a delayed reward at the terminal state T will be calculated from the TPRNet. Before the selection is finished, the rewards at other states are set to zero. Hence, the reward is described as follows:

$$r_t = \begin{cases} 0, & t < T \\ \frac{1}{|\mathcal{B}'|} \sum_{\langle instance, labels \rangle} log(P(Y|X)), & t = T \end{cases} \tag{12}$$

where $\mathcal{B}'$ is the set of selected sentences, and $P(Y|X)$ denotes the predicted phrase recognition probability of an instance $\langle instance, labels \rangle$, which is calculated by Eq. (9).

After all the actions sampling by the policy, TPRNet gives a delayed reward r_T to measure the quality of instances in $\mathcal{B}'$. However, such reward is not an absolutely reliable signal due to the noisy training data of TPRNet. Inspired by [17], we introduce a high-scoring labeled sentence set D_l (from chapter §2.3) as a silver seed set into the reward function, guiding the selection process more accurately. Therefore, for each bag $\mathcal{B}$, the silver seed set is added as chosen instances and used together with the instances in $\mathcal{B}'$ to calculate the terminal reward. All terminal rewards can supervise the ISNet to maximize the likelihood of all the instances in training dataset D.

Optimization. We optimize the parameters of ISNet by adopting policy gradient method [18], aiming to maximize the expected total reward of selections. The objective function is defined as:

$$J(\Theta) = \sum_{t=1}^{T} R_t \pi_\Theta(a_t|s_t) \tag{13}$$

For the bag $\mathcal{B}$, we sample a trajectory $\tau = (s_1, a_1, ..., s_T, a_T)$ determined by the current policy π_Θ and obtain a corresponding terminal reward r_T. Since ISNet only has a non-zero terminal reward r_T and $r_t = 0(t \in [1, T-1])$, the value R_t gradually decreases for states from s_T to s_1, namely $R_t = \sum_{t=1}^{T} \gamma^{T-t} r_t = \gamma^{T-t} r_T$. We can update the gradients of Θ by using the likelihood trick:

$$\nabla_\Theta J(\Theta) \approx \sum_{t=1}^{T} (\gamma^{T-t} r_T) \nabla_\Theta log \pi_\Theta(a_t|s_t) \tag{14}$$

where γ is discount factor ($\gamma \leq 1$).

2.6 Training Details

Since the TPRNet module and the ISNet module are intertwined together, they need to be trained jointly. We first pre-train the TPRNet on the auto-labeled data and use the gradient descent method to minimize its objective function (i.e., average negative log-likelihood of Eq. (9)). Then, after freezing the parameters of TPRNet, we pre-train the ISNet by using the delay rewards provided from TPRNet. At last, we jointly train all the two networks (detailed in Algorithm 1). In each episode, ISNet could select more correct sentences with the reward mechanism of the TPRNet. Meanwhile, TPRNet fine-tunes its phrase recognition performance using the cleaned data provided by ISNet.

3 Experiments

We conduct experiments on the BC5CDR dataset to evaluate and compare our proposed ConPhrase with other methods. We further investigate the effectiveness of incorporating topic information in TPRNet phase and the impact of RL-based selection policy.

3.1 Experimental Setup

Datasets. We evaluate our ConPhrase on the biomedical dataset BC5CDR [19]. Dataset BC5CDR annotates Chemical and Disease quality phrases and is proper for the phrase mining task. It consists of 1,500 PubMed articles, which has been separated into training set (500), development set (500), and test set (500). The dataset has 20,217 raw sentences and 28,787 quality phrases, including 15,935 Chemical and 12,852 Disease phrases, meet our requirements of phrase mining.

Only raw texts without labels are provided as the input of our ConPhrase and statistic-based approach, while the gold training set is not used. We use development set and test set for the model's early stopping and performance comparison, respectively.

Parameters and Model Training. The classical LDA with Gibbs Sampling technique is used to provide top-$N(N = 100)$ words of each topic for obtaining the topic information. In this paper, we set the number of topics to $I = 10$, and then remove universal words like "rats" and "who" in the corpus. Table 2 shows some of the topics for the BC5CDR corpus learned by the LDA model.

Table 2. Examples of topic words for the BC5CDR corpus.

No.	Topic words
T^1	Lithium, kidney, nephropathy, sodium, chronic, creatinine
T^2	Syndrome, liver, anemia, heparin, reaction, tacrolimus
T^3	Heart, cardiac, myocardial, ventricular, coronary , infarction
T^4	Pressure, hypotension, hypertension, plasma, arterial, infusion
T^5	Pain, neuropathy, muscle, complication, hyperalgesia, surgery

For TPRNet, LSTM hidden states are 200 dimensional, the optimizer is Adam with 10 mini-batch size and 5e–4 learning rate. We use dropout with 0.5 ratio after the input layer and the representation layer to relieve overfitting. For a better stability, we use gradient clipping of 5.0. Furthermore, we employ the early stopping in the development set. For ISNet, we set the max number of sentences in one bag to $T = 200$. The learning rate is 2e–5, and the discount factor is $\gamma = 0.9$.

Pre-trained Word Embeddings. For the biomedical dataset BC5CDR, we use the pre-trained 200-dimension word vectors from [20], which are trained on the whole PubMed abstracts, all the full-text articles from PubMed Central (PMC) and English Wikipedia.

Evaluation Metrics. We use the micro-averaged F1 score as the evaluation metric. Meanwhile, precision and recall are presented.

Baseline Approaches. We mainly compare with the state-of-the-art statistic-baed method AutoPhrase. For ablation test, we also compare ConPhrase with its two variants (i.e. TocPhrase and TocPhrase−) as described below:

- AutoPhrase is the state-of-the-art phrase mining technique, which combines the candidate phrase generation and quality estimation to extract salient phrases from text documents with little human labeling. We follow the settings recommended by the original paper and the released code.
- DS Match is our DS auto-labeled data generation method. Specifically, we apply it to the testing set directly to obtain quality phrases with exactly the same surface name. By comparing with it, we can check the improvements of neural models over DS noisy data.

- TocPhrase does not include the ISNet module in ConPhrase, directly training TPRNet on the auto-labeled data.
- TocPhrase− removes the ISNet module and topic representations in ConPhrase, and only utilizes word context to build the model compare with TocPhrase.

3.2 Experimental Results

Table 3. Phrase mining performance comparison. The **bold-faced scores** represent the best results among all methods

Method	BC5CDR		
	Precision	Recall	F1
AutoPhrase	70.26	60.27	64.89
Dictionary Match	94.52	57.31	71.35
TocPhrase-	90.29	60.59	72.52
TocPhrase	87.67	65.40	74.91
ConPhrase	86.02	67.75	**75.80**

We present F1, precision, and recall scores of different approaches on BC5CDR datasets in Table 3. From the table, one can observe that our ConPhrase achieves the best performance when there has only DS auto-labeled data, in terms of recall and F1 measurement. Even though AutoPhrase is based on several well-chosen frequency features, ConPhrase outperforms it in almost all metrics.

Unlike TocPhrase− which only considers the local context of a single sentence, TocPhrase incorporates global topic information of the domain corpus. Since the addition of informative topic representations, we can observe that a significant performance improvement. For instance, TocPhrase's recall rate and F1 increased by 4.5% and 2.4%, respectively, compared with TocPhrase−. Such observation illustrates that considering domain-related topic information during learning word representations can bring useful prior knowledge to assist quality phrase recognition and achieve performance improvements.

To select correct instances in noisy training data, the proposed ConPhrase introduces an instance selection component based on reinforcement learning techniques. The improvement from TocPhrase to ConPhrase demonstrates that applying reinforcement learning to guide clean data selection can boot the performance to best results of two metrics (67.75% Recall, 75.80% F1).

3.3 Impact of Topic Information

We have shown latent topic information useful for quality phrase recognition in Sect. 3.2. Here we further analyze the impact of topic information.

Table 4 shows the golden quality phrases in one sentence whose labels are wrongly predicted by TocPhase− but accurately predicted by TocPhase. Without implicit global information, TocPhrase− wrongly recognizes 'patient' and 'effect' as quality phrases. Similarly, new researchers who are not familiar with the biomedical domain will make the same mistakes as TocPhrase−. However, when researchers know that the phrases 'patient' and 'effect' have balanced attentive vectors for all I topics, they will realize that the phrases 'patient' and 'effect' are just common words in the biomedical field, not high-quality phrases. For the quality phrases 'pain' and 'Musculoskeletal', their attentive vectors pay more attention to the topic T^5 (see Table 2). Incorporating such topic information as the global context, the phrases 'pain' and 'Musculoskeletal' obtain the domain-related prior knowledge, can be easily recognized. This example verifies the critical impact of incorporating topic information in mining quality phrases.

Table 4. Examples: Only TocPhrase predicts the quality phrase accurately, while TocPhrase− fail to predict.

Example of quality phrase prediction											
Raw sent.:	Musculoskeletal	pain	may	be	an	important	side	effect	in	these	patients
TocPhrase−:	O	B	O	O	O	O	O	B	O	O	B
TocPhrase:	B	I	O	O	O	O	O	O	O	O	O

3.4 Effectiveness of Selection Policy

We then evaluate the effectiveness of the RL-based selection policy from two aspects. First, we evaluate whether the cleaned dataset provided by the instance selection module is useful for quality phrase recognition. Second, a case study is used to illustrate the results of the RL-based selection policy.

Quality Phrase Recognition on Cleaned Data. To evaluate the ability of our RL-based selection policy to select correct sentences, we conduct quality phrase recognition experiments on cleaned data. We first use an RL-based selection policy to select a high-quality sentence set D' from the noisy auto-labeled data D, and then train the TocPhrase on the cleaned data (denoted as TocPhrase+). In contrast with the original model (F1 score 74.91%) trained on a noisy data D, the performance of TocPhrase+ has some improvement, achieved 75.32% of F1 score. The result indicates that the model performs better on the selected data than the original data set. It also reveals that the RL-based selection policy possesses the ability to select the correct sentences, improving the quality of automatic labeling data and ensuring better phrase recognition performance.

Table 5. Examples of noisy labeled sentences deleted by selection policy. The words marked in blue are recoginized correctly in the gold labels. The words marked in red are recoginized wrongly by auto-labeled generation method.

Example of noisy labeled sentences										
Raw Sent.:	Known	causes	of	movement	disorders	were	eliminated	after	evaluation	.
Auto-label:	O	B	O	B	O	O	O	O	O	O
Gold-label:	O	O	O	B	I	O	O	O	O	O
Raw Sent.:	Mechanisms	of	FK	506-induced	hypertension	in	the	rat	.	
Auto-label:	O	O	O	O	B	O	O	B	O	
Gold-label:	O	O	B	I	I	O	O	O	O	

Case Study. The two sentences in Table 5 are filtered from the auto-labeled dataset through the RL-based selection policy. For instance, in the first example, this sentence wrongly labeled two words with incorrect types, such as the word 'cause' label with wrong type 'B', and word 'disorders' with 'O' type. The case indicates that the RL-based selection policy can effectively delete the incorrectly labeled instances.

4 Related Work

The study of phrase mining originates from the NLP community, with the closest works being noun phrase chunking and named entity recognition. A number of phrase mining algorithms have been proposed, and the process of extracting phrases usually divides into two steps: candidate phrases generation and phrase quality ranking.

The first step is to use heuristics to generate a list of candidate phrases. When these candidates are ready for further filtering, a large number of candidates will be generated in this step to increase the probability of retaining most of the correct phrases. The main methods of extracting candidate phrases include retaining word sequences that match certain part-of-speech tag patterns (for example, nouns, adjectives) [21], and extracting important n-gram or noun phrases [11,12,23,24].

The second step is to score the likelihood of each candidate phrase becoming a high-quality phrase in a given corpus. The sorted candidate list is returned as output, and downstream applications can use quality phrases that exceed a certain threshold in the sorted list. Here, the statistic-based methods are widely used, which typically treat phrase mining as a binary classification problem, and various types of learning methods and features have been explored [9,11,12,22–27].

Recent works mainly focus on utilizing statistical features of candidates acquired from a massive corpus to further estimate phrases' quality. Pitler [23] evaluates candidate phrases based on comparison of a phrase's frequency with its sub-(or /super-)phrases from web-scale corpus. Ahmed [8] uses several indicators, including frequency and statistical significance score (i.e. t-statistic), to assess candidate phrases. Li [9,25,26] proposes statistical signals based on χ^2-test to measure lexical collocation in a complete phrase. The representative approaches are SegPhrase [11] and AutoPhrase [12], which employ several raw frequency and rectified frequency features to classify phrase candidates into quality phrases or non-quality phrases. Such statistic-based methods have achieved the state-of-the-art performance of this task. However, statistical signals of phrases tend to less reliable in the medium (or small) corpus, and fail to guide distinguishing high-quality phrases from inferior ones.

Our proposed ConPhrase resolves the phrase mining task in entirely different ways, which joint processing of the candidate phrases generation and phrase quality ranking steps. Meanwhile, ConPhrase focuses on the effect of contextual information, and formulates phrase mining as a sequence labeling problem. Moreover, in the distantly supervised labeled data process, ConPhrase confronts with the noisy data filtration problem. We introduce reinforcement learning (RL) to our phrase mining task inspired by some works applying RL into distantly supervised tasks, such as stance detection [15,28] and relation extraction [16]. Thus, ConPhrase designs an ISNet module, which first selects correct sentences with reinforcement learning and then fine-tunes the TPRNet module with the cleaned data.

5 Conclusion

In this paper, we explore the problem that statistic-based phrase mining methods rely on statistical signals to extract phrases, ignoring the effect of contextual information. Therefore, we propose a novel context-aware method for automated phrase mining, ConPhrase, which formulates phrase mining as a sequence labeling problem. We also focus on two challenging issues of extracting phrases from a single short sentence: global information scarcity issue and noisy data filtration issue. ConPhrase overcomes the first issue by integrating topic information into word representation learning, and introduces an RL-based selection policy to choose clean data for resolving the second issue. Experimental results show that our ConPhrase is superior to the existing approach.

Acknowledgments. This work is supported by National Key R & D Program of China (No.2018YFB1004401) and NSFC under the grant No. 61772537, 61772536, 61702522, 61532021, 62072460.

References

1. Reinsel, D., Gantz, J., Rydning, J.: The digitization of the world from edge to core. IDC, Framingham, MA (2018)
2. Li, K., Zha, H., Su, Y., Yan, X.: Concept mining via embedding. In: 2018 IEEE International Conference on Data Mining (ICDM), pp. 267–276 (2018)
3. Liu, L., et al.: Empower sequence labeling with task-aware neural language model. In: Proceedings of the 32nd AAAI Conference on Artificial Intelligence, pp. 5253–5260 (2018)
4. Shang, J., Liu, L., Gu, X., Ren, X., Ren, T., Han, J.W.: Learning named entity tagger using domain-specific dictionary. In: Proceedings of the Conference on Empirical Methods in Natural Language Processing (EMNLP), pp. 2054–2064 (2018)

5. Safranchik, E., et al.: Weakly supervised sequence tagging from noisy rules. In: Proceedings of the AAAI Conference on Artificial Intelligence, pp. 5570–5578 (2020)
6. Chen, J., Zhang, X., Wu, Y., Yan, Z., Li, Z.: Keyphrase generation with correlation constraints. In: Proceedings of the Conference on Empirical Methods in Natural Language Processing (EMNLP), pp. 4057–4066 (2018)
7. Wang, C., et al.: A phrase mining framework for recursive construction of a topical hierarchy. In: Proceedings of the 19th ACM SIGKDD, pp. 437–445 (2013)
8. Ahmed, E.-K., Song, Y.L., Wang, C., Clare, R.V., Han, J.W.: Scalable topical phrase mining from text corpora. Proc. VLDB Endow. **8**(3), 305–316 (2014)
9. Li, B., Wang, B., Zhou, R., Yang, X.C., Liu, C.F.: A cluster-based iterative topical phrase mining framework. In: International Conference on Database Systems for Advanced Applications (DASFAA), pp. 197–213 (2016)
10. Shen, J.M., et al.: Hiexpan: task-guided taxonomy construction by hierarchical tree expansion. In: Proceedings of the 24th ACM SIGKDD International Conference on Knowledge Discovery & Data Mining, pp. 2180–2189 (2018)
11. Liu, J.L., Shang, J.B., Wang, C., Ren, X., Han, J.W.: Mining quality phrases from massive text corpora. In: Proceedings of the 2015 ACM SIGMOD International Conference on Management of Data, pp. 1729–1744 (2015)
12. Shang, J.B., Liu, J.L., Jiang, M., Ren, X., Voss, R.V., Han, J.W.: Automated phrase mining from massive text corpora. IEEE Trans. Knowl. Data Eng. **30**(10), 1825–1837 (2018)
13. Blei, D.M., Ng, A.Y., Jordan, M.I.: Latent dirichlet allocation. J. Mach. Learn. Res. **3**(1), 993–1022 (2003)
14. Seo, M., Kembhavi, A., Farhadi, A., Hajishirzi, H: Bidirectional attention flow for machine comprehension. In: Proceedings of the International Conference on Learning Representations (ICLR) (2017)
15. Wei, P., Mao, W., Chen, G.: A topic-aware reinforced model for weakly supervised stance detection. In: Proceedings of the 33rd AAAI Conference on Artificial Intelligence, pp. 7249–7256 (2019)
16. Feng, J., Huang, M., Zhao, L., Yang, Y., Zhu, X.: Reinforcement learning for relation classification from noisy data. In: Proceedings of the 32nd AAAI Conference on Artificial Intelligence, pp. 5779–5786 (2018)
17. Yang, Y., Chen, W., Li, Z., He, Z., Zhang, M.: Distantly supervised NER with partial annotation learning and reinforcement learning. In: Proceedings of the 27th International Conference on Computational Linguistics, pp. 2159–2169 (2018)
18. Sutton, R.S., McAllester, D.A., Singh, S.P., Mansour, Y.: Policy gradient methods for reinforcement learning with function approximation. In: Proceedings of the Conference on Neural Information Processing Systems, pp. 1057–1063 (1999)
19. Li, J., et al.: Biocreative V CDR task corpus: a resource for chemical disease relation extraction. Database (2016)
20. Pyysalo, S., Ginter, F., Moen, H., Salakoski, T., Ananiadou S.: Distributional semantics resources for biomedical text processing. In: Proceedings of the 5th International Symposium on Languages in Biology and Medicine, pp. 39–43 (2013)
21. Clahsen, H., Felser, C.: Grammatical processing in language learners. Appl. Psycholinguist. **27**(1), 3–42 (2006)
22. Deane, P.: A nonparametric method for extraction of candidate phrasal terms. In: Proceedings of the 43rd Annual Meeting of the Association for Computational Linguistics, pp. 605–613 (2005)
23. Pitler, E., Bergsma, S., Lin, D., Church, K.W.: Using web-scale n-grams to improve base NP parsing performance. In: Proceedings of the 23rd International Conference on Computational Linguistics (COLING), pp. 886–894 (2010)

24. Parameswaran, A.G., Garcia-Molina, H., Rajaraman, A.: Towards the web of concepts: extracting concepts from large datasets. PVLDB. **3**(1), 566–577 (2010)
25. Li, B., Yang, X., Wang, B., Cui, W.: Efficiently mining high quality phrases from texts. In: Proceedings of the 31st AAAI Conference on Artificial Intelligence, pp. 3474–3481 (2017)
26. Li, B., Yang, X., Zhou, R., Wang, B., Liu, C., Zhang, Y.: An efficient method for high quality and cohesive topical phrase mining. IEEE Trans. Knowl. Data Eng. **31**(1), 120–137 (2018)
27. Wang, L., et al.: Mining infrequent high-quality phrases from domain-specific corpora. In: Proceedings of the 29th ACM International Conference on Information & Knowledge Management, pp. 1535–1544 (2020)
28. Tian, S., Mo, S., Wang, L., Peng, Z.: Deep reinforcement learning-Based approach to tackle topic-aware influence maximization. Data Sci. Eng. **5**(1), 1–11 (2020)

Keyword-Aware Encoder for Abstractive Text Summarization

Tianxiang Hu, Jingxi Liang, Wei Ye(✉), and Shikun Zhang

The National Engineering Research Center for Software Engineering, Peking University, Beijing, China
wye@pku.edu.cn

Abstract. Text summarization aims to produce a brief statement covering main points. Human beings would intentionally look for key entities and key concepts when summarizing a text. Fewer efforts are needed to write a high-quality summary if keywords in the original text are provided. Inspired by this observation, we propose a keyword-aware encoder (KAE) for abstractive text summarization, which extracts and exploits keywords explicitly. It enriches word representations by incorporating keyword information and thus leverages keywords to distill salient information. We construct an attention-based neural summarizer equipped with KAE and evaluate our model extensively on benchmark datasets of various languages and text lengths. Experiment results show that our model generates competitive results comparing to state-of-the-art methods.

Keywords: Deep learning · Natural language processing · Text summarization

1 Introduction

Text summarization is a challenging task which aims to generate informative and non-redundant summary. Related techniques can mainly be classified into extractive methods and abstractive methods. Extractive summarization methods identify and concatenate relevant words from the original text, while abstractive methods try to express the main content in a condensed way, possibly using words that are not in the original text. Early studies explored various approaches including manually designed rules [35], syntactic tree pruning [19] and statistical machine translation techniques [3]. In this paper, we focus on abstractive text summarization.

Recently, neural network models have achieved an impressive performance in abstractive summarization task. Many of the works [27,36] benefit from the attention-based encoder-decoder framework [2], which is originally designed to tackle the machine translation problem. Firstly, an encoder converts an input

T. Hu and J. Liang—The first two authors contribute equally to this work.

C. S. Jensen et al. (Eds.): DASFAA 2021, LNCS 12682, pp. 37–52, 2021.
https://doi.org/10.1007/978-3-030-73197-7_3

sequence to a list of distributed representations, and then a decoder generates each output word by calculating a soft alignment over all input states before using their weighted combination as the current context state.

In machine translation, the attention mechanism calculates a soft alignment over all source words, which is preferable because the meaning of every word in the source text needs to be translated one way or another. In text summarization, however, many words in the source text do not provide useful information for the final summary.

In most cases, only a few words in the original text can capture most of the information, which we call primary information. Therefore, the rest of the words provide little additional information, which we call secondary information. Intuitively, it would be of great help if we can somehow identify these highlighted words in advance and tell a summarizer to focus on them. On the contrary, too much attention on secondary information would hinder a summarizer from focusing on the core concepts and could even be misleading, resulting in sub-optimal or even low-quality summaries. This is often the case with a standard attention-based decoder, whose attention mechanism distributes weights over the entire input sequence and may assign an undesirable amount of weights to words that only contain secondary information. Table 1 gives an example of the attention weight distribution of a attention-based decoder at each decoding step.

Table 1. Sum of the Top K soft alignment weights over an input text of 35 words when a decoder generates each output word in the leftmost column.

Output words	Top 2	Top 5	Top 8
Toyota	0.2981	0.5124	0.6486
are	0.3496	0.6389	0.7802
Banned	0.4157	0.6160	0.7301
for	0.3468	0.5963	0.7132
a	0.3484	0.6121	0.7681
Year	0.4868	0.6832	0.8127
<eos>	0.3726	0.5960	0.7209

In this example, there are 6 output words, which means the number of input words that contain primary information is close to 6. Yet, as shown in Table 1, the sum of top 8 out of 35 attention weights at any decoding step is only around 0.8, which indicates that about 20% attention weights are assigned to input words that only contain secondary information. This problem is more noticeable when the input text is longer, as the attention weight distribution is more scattered.

Although there are methods proposed to alleviate this problem such as local attention [23], selective gate [36] and stacked self-attention encoder [32], we argue that neural summarization models can benefit more from exploiting keywords explicitly. Humans would intentionally look for key entities and key concepts when summarizing a text, and these key entities and key concepts are often in the form of keywords. Keywords provide a natural way to narrow down primary information and filter out secondary information, so that the attention-based decoder can be more concentrated on primary information.

Due to aforementioned motivations, we propose a keyword-aware encoder (KAE) for abstract text summarization task in this paper. KAE consists of four encoders: a text encoder, a keyword encoder, a merge encoder and a refinement encoder. The text encoder generates word embeddings of the input text, while the keyword encoder generates keyword embeddings after the keywords are identified. Then for each input word, the merge encoder accomplish three tasks: (1) using its word embedding to calculate a weighted sum of keyword embeddings to form its keyword context embedding; (2) computing a merge gate that combines its word embedding and keyword context embedding into a keyword-aware embedding; (3) computing a selection gate that controls the information flow of the keyword-aware embedding. Finally, the refinement encoder fuses the output of the merge encoder at each time step to form state embeddings, which are used by an attention-based decoder to generate summary.

We use Term Frequency-Inverse Document Frequency (TF-IDF) to determine keywords in a text and then sort the keywords according to their post-order traversal order in the dependency tree. Despite of its simplicity, TF-IDF is a strong baseline method in many information retrieval tasks. And the dependency tree helps us with the inner connection between keywords. Our experiments show that TF-IDF works very well with our KAE model, though it can be replaced by any other keyword-extraction method.

To evaluate the effectiveness of our KAE model, we conducted extensive experiments on three benchmark datasets of various languages and text lengths, namely Gigaword in English, and LCSTS, TTNews in Chinese. Experiment results show that our approach outperforms the state-of-the-art methods and generates better-quality summaries.

2 Related Work

Thanks to the advancement of deep learning, many neural network-based approaches, especially attention-based sequence-to-sequence models, have been proposed to tackle the abstractive text summarization problem. [27] introduced an attention-based summarization (ABS) model which consists of an attention-based encoder and a neural language model (NNLM) decoder. The encoder models the input sentence by a convolutional neural network and the decoder is a standard feed-forward neural network.

Much attention has also been directed at introducing new structures to enhance the neural summarization model. [14] noticed that in many tasks, words in the output sequence could be found directly in the input sequence. Hence, they proposed CopyNet which combines word generation with a copying mechanism. [36] proposed Selective Encoding for Abstractive Sentence Summarization (SEASS) which adds a selective gate on the sentence representation to control

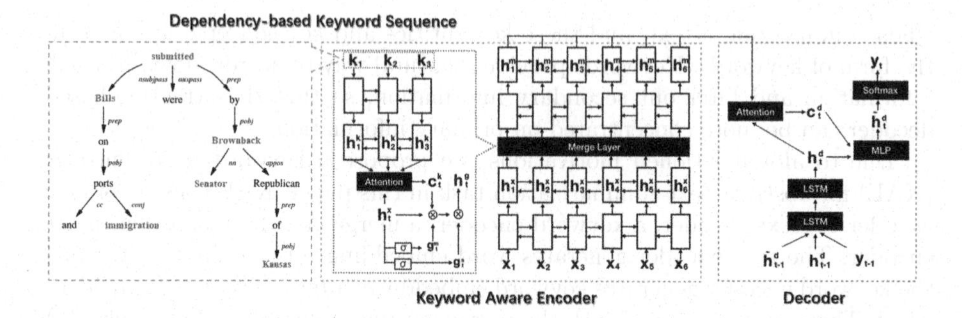

Fig. 1. Architecture of our KAE model. Keywords are in red. (Color figure online)

the information flow from the encoder to the decoder. [6] and [28] both introduced a coverage mechanism to prevent the model from generating repeated content. [21] proposed deep recurrent generative decoder (DRGD) which leverages the latent structure information of summaries to improve the summary quality. DRGD uses Variational Auto-Encoders (VAE) to learn the latent structure. [25] used rewards from policy gradient in reinforcement learning as a new objective function. They also proposed an intra-attention mechanism in the encoder to record previous attention weights and a sequential intra-attention in the decoder that takes into account all words that have been generated. [32] used a similar intra-attention mechanism as well. They proposed Transformer, which dumps the recurrent structure and relies entirely on self-attention mechanism. In the encoder, it uses positional embeddings to encode position information in the input sequence and uses a multi-head self-attention mechanism to utilize different positions of the input sequence to compute a representation of the input sequence. A similar structure is used in the decoder. Although Transformer was targeted at machine translation, it is a promising model to be applied to short-text summarization task. [4,33] and [11] first retrieves similar sample data and uses its summary as soft template to guide the process of summarization. [8] uses BERT as teacher model to distill bi-directional context information into summarization model.[31] proposed a graph-based attention mechanism in a hierarchical encoder-decoder framework. It combines traditional attention mechanisms and graph-based extractive summarization techniques to address the saliency issue in summarization on sentence level. Its motivation is quite similar to ours; however, our KAE model addresses the saliency issue mainly on word level.

3 Model Description

We propose to leverage keywords to enrich word embeddings and filter out secondary information. As shown in Fig. 1, the model consists of a keyword-aware encoder and a decoder, both of which use Long Short-Term Memory (LSTM) [15]. In KAE, there are four encoders: a keyword encoder, a text encoder, a merge encoder. Firstly, the keyword encoder and text encoder use multi-layered bidirectional LSTMs to process keyword sequence $\boldsymbol{k} = (k_1, ..., k_m)$ and input sequence $\boldsymbol{x} = (x_1, ..., x_l)$ respectively to build corresponding keyword embeddings $(\boldsymbol{h}_1^k, ..., \boldsymbol{h}_m^k)$ and word embeddings $(\boldsymbol{h}_1^x, ..., \boldsymbol{h}_l^x)$. The merge encoder uses $\boldsymbol{h}_i^x$ to compute the merge gate g_i^m and the selection gate g_i^s. Then, it computes keyword context embedding $\boldsymbol{c}_t^k$ by a weighted combination of keyword embeddings $(\boldsymbol{h}_1^k, ..., \boldsymbol{h}_m^k)$. After merging keyword information into word embeddings by the merge gate g_i^m and distilling primary information by the selection gate g_i^s, another LSTM-based reads the embedding sequence and produces state embeddings $(\boldsymbol{h}_1^m, ..., \boldsymbol{h}_l^m)$. The decoder is an attention-based decoder with input feeding [23], which computes a soft alignment over state embeddings at each decoding step to generate the summary $\boldsymbol{y}$ one word at a time.

3.1 Keywords Extraction

Keywords play an important role in writing a good summary. Although many methods to extract keywords exist, we use the simplest but an effective one, TF-IDF, to determine keywords, though any other method can be applied in our framework. We compute the TF-IDF value of each word using the whole training dataset (including the target summaries), and pick top K words with highest TF-IDF values as keywords. After examining the exact word overlap ratio with the target summary, we can see that the keyword sequence has more condensed information compared to the original input text, as shown in Table 2. We also pre-trained an keyword-extraction model using parallel corpus constructed from summarization dataset(mark the keyword showed in both original text and summary as keyword), but the results shows that TF-IDF is an simple but efficient method.

Table 2. Comparison of word overlap with target summary between input text and extracted keywords (top 50%) on Gigaword training set. The average overlap ratio is calculated as the average number of overlapped words divided by the average length of the sequence.

Statistics	Input text	Keywords
Average length	31.35	15.42
Average # word overlap	5.22	3.18
Average overlap ratio	16.65%	20.62%

3.2 Dependency-Based Keyword Sequence

One natural way to organize the keywords is sorting them according to their first appearance in the input text and remove keywords that do not belong to the source vocabulary $\mathcal{V}_s$. Every input text $\boldsymbol{x}$ has a corresponding keyword sequence $\boldsymbol{k} = (k_1, ..., k_m)$ where $m < l$.

Note that the keyword sequence generated in the above way is not a natural sentence anymore, and the inner connection among words could be missing. Since we want the order of the keyword sequence better characterize the original structure of source text, here we introduced the grammatical dependency tree (DT) to generate a more integral and reasonable keyword sequence. We first generate a dependency tree of the original text and then sort the keywords according to their post-order traversal order in the dependency tree. This operation is performed independently in each sentence.

3.3 Keyword-Aware Encoder

To get a semantic embedding, we employ a multi-layered bidirectional LSTM (BiLSTM) to encode the word sequence. At each layer, the BiLSTM consists of a forward LSTM and a backward LSTM. The forward LSTM reads the input sequence from left to right to get hidden states $(\overrightarrow{\boldsymbol{h}}_1, ..., \overrightarrow{\boldsymbol{h}}_e)$ and the backward LSTM reads the input sequence from right to left to generate hidden states $(\overleftarrow{\boldsymbol{h}}_1, ..., \overleftarrow{\boldsymbol{h}}_e)$.

$$\overrightarrow{\boldsymbol{h}}_i = \text{LSTM}(\text{input}_i, \overrightarrow{\boldsymbol{h}}_{i-1}), \overleftarrow{\boldsymbol{h}}_i = \text{LSTM}(\text{input}_i, \overleftarrow{\boldsymbol{h}}_{i+1}) \tag{1}$$

Concatenate $\overrightarrow{\boldsymbol{h}}_i$ and $\overleftarrow{\boldsymbol{h}}_i$, and we get $\boldsymbol{h}_i = [\overrightarrow{\boldsymbol{h}}_i; \overleftarrow{\boldsymbol{h}}_i]$. The text encoder and keyword encoder stack multiple BiLSTMs and read input text and extracted keywords respectively to generate word embeddings $(\boldsymbol{h}_1^x, ..., \boldsymbol{h}_l^x)$ and keyword embeddings $(\boldsymbol{h}_1^k, ..., \boldsymbol{h}_m^k)$.

For each $\boldsymbol{h}_t^x$, the merge encoder determines its relevance with keyword embedding $(\boldsymbol{h}_1^k, ..., \boldsymbol{h}_m^k)$ and generate a keyword context embedding $\boldsymbol{c}_t^k$ as a weighted combination of keyword embeddings.

$$\text{score}_{t,i}^k = \boldsymbol{h}_t^{xT} W_{xk} \boldsymbol{h}_i^k \tag{2}$$

$$\text{relevance}_{t,i} = \frac{\exp(\text{score}_{t,i}^k)}{\sum_{j=1}^m \exp(\text{score}_{t,j}^k)} \tag{3}$$

$$\boldsymbol{c}_t^k = \sum_{j=1}^m \text{relevance}_{t,j}\, \boldsymbol{h}_j^k \tag{4}$$

where W_{xk} is a weight matrix.

Next, the merge encoder uses $\boldsymbol{h}_t^x$ to compute the merge gate $\boldsymbol{g}_t^m$ and the selection gate $\boldsymbol{g}_t^s$.

$$\boldsymbol{g}_t^m = \sigma(W_m \boldsymbol{h}_t^x + \boldsymbol{b}_m) \tag{5}$$

$$\boldsymbol{g}_t^s = \sigma(W_s \boldsymbol{h}_t^x + \boldsymbol{b}_s) \tag{6}$$

where W_m and W_b are weight matrices, $\boldsymbol{b}_m$ and $\boldsymbol{b}_s$ are bias vectors, and σ denotes sigmoid activation function. The merge gate $\boldsymbol{g}_t^m$ controls how to merge $\boldsymbol{c}_t^k$

and $\boldsymbol{h}_t^x$, while the selection gate $\boldsymbol{g}_t^s$ distills primary information. After applying the two gates, the merge encoder outputs a new embedding sequence $(\boldsymbol{h}_1^g, ..., \boldsymbol{h}_l^g)$.

$$\boldsymbol{h}_t^{g'} = \boldsymbol{g}_t^m \odot \boldsymbol{h}_t^x + (1 - \boldsymbol{g}_t^m) \odot \boldsymbol{c}_t^k \tag{7}$$

$$\boldsymbol{h}_t^g = \boldsymbol{g}_t^s \odot \boldsymbol{h}_t^{g'} \tag{8}$$

where $\odot$ denotes element-wise multiplication.

A refinement encoder further uses a stacked BiLSTMs to fuse $(\boldsymbol{h}_1^g, ..., \boldsymbol{h}_l^g)$ into state embeddings $(\boldsymbol{h}_1^m, ..., \boldsymbol{h}_l^m)$ to be used by decoder. Its main purpose is to smooth out the embedding sequence outputted by the merge encoder.

Apart from using Eq. 5, we have also found it feasible to exploit the cosine similarity between keyword and word embeddings to compute the selection gate $\boldsymbol{g}_t^s$. It reduces the number of parameters to learn while achieving a better performance if a proper vocabulary size is used. For an input text $\boldsymbol{x} = (x_1, ..., x_l)$ and the corresponding keyword sequence $\boldsymbol{k} = (k_1, ..., k_m)$, we can construct a matrix M^d. The $\boldsymbol{g}_t^s$ now is a scalar and is obtained by picking the largest value in the corresponding row.

$$M_{i,j}^d = \frac{\text{emb}(x_i) \cdot \text{emb}(k_j)}{||\text{emb}(x_i)|| \ ||\text{emb}(k_j)||} \tag{9}$$

$$g_t^s = \max_{j \in [1,m]} M_{t,j}^d \tag{10}$$

where emb is a function to get word/keyword embedding.

3.4 Summary Decoder

Many variants of summary decoder have been proposed to improve the summary's quality. Following [23,30], we use an attention-based stacked LSTM decoder with input feeding so that we can keep our focus on KAE.

Due to space limitations, we do not describe the details of summary decoder here. The summary decoder generate a probability distribution $p(y_t|\boldsymbol{x}, \boldsymbol{y}_{<t})$.

3.5 Objective Function

Like most sequence-to-sequence models, our goal is to maximize the output summary probability given the input text and keywords. Hence, we use the negative log-likelihood to define our loss function

$$J(\theta) = -\frac{1}{|\mathcal{D}|} \sum_{(\boldsymbol{x},\boldsymbol{k},\boldsymbol{y}) \in \mathcal{D}} \log p(\boldsymbol{y}|\boldsymbol{x}, \boldsymbol{k}) \tag{11}$$

where $\mathcal{D}$ represents the training dataset, θ denotes all model parameters. We use the Adaptive Gradient [10] with mini-batch to learn the parameters θ.

4 Experiments

We have evaluated our KAE model on several benchmark datasets of various languages and text lengths. In this section, we will describe the datasets, baseline models, evaluation metrics, implementation details and experiment results.

4.1 Datasets

Gigaword is an English sentence summarization dataset constructed from the Annotated English Gigaword corpus by extracting the first sentence from each article with the headline to form a sentence-summary pair. We directly download the dataset used by [27]. This dataset contains about 3.8M sentence-summary pairs for training and 2,000 pairs for testing. However, among the 2,000 test pairs, there are empty titles and meaningless input sentences. We remove those pairs and end up with 1,943 pairs for testing.

LCSTS is a large-scale Chinese short-text summarization dataset collected from Sina Weibo, which contains 2.4M training text-summary pairs and 1,106 test pairs [17]. Test set and part of the training set are manually scored to indicate the relevance of the short text and the corresponding summary. Following [17], we remove those test pairs with score below 3 and end up with 725 pairs for testing. In our experiments we take the word-based approach and use LTP [5] to segment Chinese words. After word segmentation, the average length of the articles is 61 Chinese words.

TTNews is a Chinese long-text summarization dataset from NLPCC17 shared task 3 [18]. The training set contains 50K pairs of news articles and corresponding human-written summaries. This dataset also provides 2,000 pairs for testing. After word segmentation, the average length of the articles is 587 Chinese words.

4.2 Baselines

We compare our KAE model with the following baselines and state-of-the-art methods:

- **CopyNet** [14] introduces a copy mechanism. CopyNet dynamically chooses to generate words from the vocabulary or copy words from input text.
- **SEASS** [36] introduces selection gate to distill salient information for generating a summary.
- **DRGD** [21] uses VAE to model the latent structure of summaries.
- **TopicNHG** [34] applies Latent Dirichlet Allocation to assign topic labels to documents and build sequence-to-sequence models for each topic respectively.
- **s2s+att** We use stacked BiLSTMs with attention in the encoder-decoder framework as our baseline model.
- **Transformer** [32] is a neural machine translation model. It uses positional encoding, multi-head attention and self-attention to encode and decode texts.
- **Re^3Sum** [4] is a very competitive model that first retrieve a similar data sample from the training dataset, then use it as a soft-template to guide the generation of summary.

4.3 Evaluation Metric

Following previous works, we use the full-length F1 Rouge [22] to evaluate our model. The ROUGE measures the quality of a summary by its overlap with references in terms of unigram, bigram, longest common subsequence (LCS) etc. Full-length F1 ROUGE removes the length limit and penalizes longer summaries. We report evaluations of full-length F1 ROUGE-1 (unigram), ROUGE-2 (bigram) and ROUGE-L (LCS) on all datasets.

4.4 Implementation Details

We used PyTorch to implement our model. And we use Stanford Dependencies [1] to generate the dependency tree of the original text.

Model Parameters. For Gigaword, since this dataset has already substituted infrequent words with the UNKNOWN tag, we leave it untouched and get a source vocabulary and a target vocabulary of 120K and 69K distinct words respectively. Top 50% words in the sentence with highest TF-IDF values are used as keywords. We set the word embedding size to 300, the dimension of hidden states to 500, and the number of layers in LSTM to 2. Dropout [29] is applied with probability of 0.3. We also use the pretrained GloVe word embeddings [26].

For LCSTS, we remove infrequent words and obtain a 127K source vocabulary and a 53K target vocabulary. Top 25 words in the short text with highest TF-IDF values are used as keywords. All other settings on the LCSTS dataset are the same as those on Gigaword.

For TTNews, after removing infrequent words we obtain a 157K source vocabulary and a 22K target vocabulary. Top 25 words in the document with highest TF-IDF values are used as keywords. We set the number of layers in LSTM to 1. All other settings on the TTNews dataset are the same as those on LCSTS.

Training. All model parameters are initialized by a Gaussian distribution with Xavier scheme [13]. The batch size is set to 64, 64 and 4 for the Gigaword, LCSTS and TTNews dataset respectively. The learning rate is set to 0.15 with an initial accumulator value of 0.15. We halve the learning rate if the loss on the development set increases for 10 consecutive runs of 1000 batches. The gradient clipping with range [-5,5] is applied during training.

Decoding. During decoding, we use beam search with the beam size set to 6. Following [36], we average the ranking score along the beam path by dividing it by the number of generated words. To handle a generated UNKNOWN tag, we simply replace it with the source word that has the highest attention weight. We report experiment results of using beam search, and beam search with the replacing-the-unknown trick.

4.5 Evaluation

Table 3 summarizes the evaluation of full-length F1 Rouge-1, Rouge-2, and Rouge-L on the Gigaword test set. We have three observations here: 1) our KAE model outperforms all other models by a large margin, e.g., with improvements of 1.02 Rouge-L scores over Re^3Sum; 2) the full KAE model performs better than its variant KAE w/o DT which ordering keywords via a plain order, verifying the effectiveness of DT-based keywords organizing; and 3) those models aiming to filter out secondary information, including SEASS, Transformer, and our KAE model, generally achieve better results compared with the others.

Table 3. Full-length F1 ROUGE evaluation results on the English Gigaword test set.

Models	RG-1	RG-2	RG-L
s2s+att	34.68	17.10	32.10
Transformer	35.79	17.72	33.24
SEASS‡	36.15	17.54	33.63
DRGD‡	36.27	17.57	33.62
Re^3Sum	37.04	19.03	34.46
KAE (w/o DT)	**37.82**	**18.85**	**35.17**
KAE	**38.24**	**19.17**	**35.48**

The Rouge evaluation results on the LCSTS test set are shown in Table 4. Our KAE model also achieves the highest score. It is worth mentioning that TopicNHG actually trains five models, each for a topic, which makes it an ensemble method in effect. Without considering TopicNHG, models that alleviate the impact of secondary information all outperform the other models by a large margin.

On the long-text TTNews dataset, our model outperforms the baseline model s2s+att. The results are summarized in Table 5. In order to train Transformer on this dataset on our device, we have to reduce the number of layers and the dimension of hidden layers to make it work, which leads to a compromised result. Thus we choose not to report the result.

Table 4. Full-length F1 ROUGE evaluation results on the Chinese LCSTS test set.

Models	RG-1	RG-2	RG-L
CopyNet‡	35.00	22.30	32.00
s2s+att	36.23	24.12	33.63
DRGD‡	36.99	24.15	34.21
TpoicNHG‡	38.40	26.60	36.10
Transformer	38.96	26.00	35.64
KAE (w/o DT)	**41.31**	**27.94**	**38.06**
KAE	**41.70**	**28.18**	**38.36**

Table 5. Full-length F1 ROUGE evaluation results on the Chinese TTNews test set.

Models	RG-1	RG-2	RG-L
s2s+att	45.18	30.51	40.33
KAE (w/o DT)	51.26	37.31	46.07
KAE	51.71	37.69	46.30

5 Discussion

5.1 Visualization of Gates and Attention Weights

To gain a better understanding of our KAE model, we will first use an example to visualize the attention weights on keyword embeddings of each keyword context embedding and the behavior of merge gate and selection gate.

In Fig. 2, we can see that words in a sentence tend to attend to those keywords that have similar meanings to generate corresponding keyword context vectors. For example, the words "german," "chemical," "giant" and "hoechst" all pay attention to the keyword "hoechst." Furthermore, most function words like "to," "in," "with" produce keyword context embeddings by equally attending to each keyword.

Figure 3 shows that the selection gate (top) gives a free pass to those keywords and depresses secondary information flow. In comparison, the values of the merge gate (bottom) are less polarized. For some words that are not selected as keywords, it injects a fair amount of keyword context embedding into the word embedding.

5.2 Influence of Keyword Extraction Ratio

In order to analyze the influence of keywords on the quality of the summary, we conducted several experiments on Gigaword with different keyword extraction

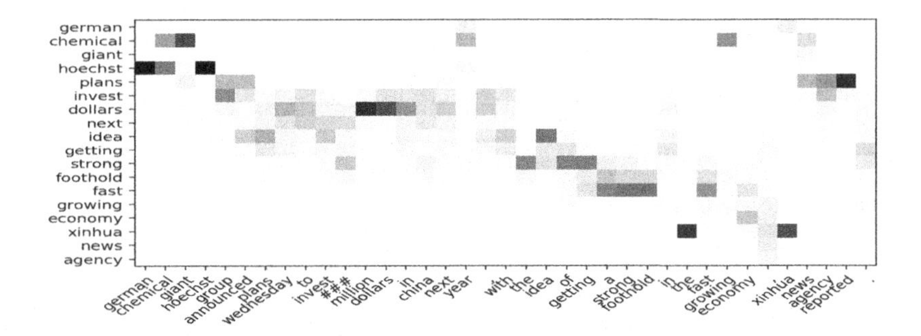

Fig. 2. A sample from the Gigaword test set illustrating the composition of keyword context embeddings. The sentence lies on the x-axis, while the extracted keywords are listed on the y-axis. The darker the square, the higher the attention weight.

Fig. 3. A sample from the Gigaword test set illustrating the information flow controlled by the selection gate (top) and merge gate (bottom). Since the gate at each time step is a high-dimensional vector, we calculate the mean value of all dimensions. The darker the square, the higher the value.

ratios and have observed the following trend: as the ratio increases from 0.1 to 0.5, the performance improves in terms of evaluation metrics; but when the ratio further increases from 0.5 to 0.9, the performance drops. This indicates that there is a delicate balance between too many and too few keywords extracted.

To further analyze the influences of different keyword extraction ratios, we define "Keyword Extraction Precision" as the ratio of the number of extracted keywords that occur in the generated summary to the length of the summary, and "Keyword Extraction Recall" as the ratio of the number extracted keywords that appear in the generated summary to the number of all extracted keywords. The calculated Keyword Extraction Precision and Recall of the generated summaries of KAE model as well as the golden summary with different keyword extraction ratios are shown in Fig. 4.

Firstly, we can see that the generated summary of KAE has higher "Keyword Extraction Precision" and "Keyword Extraction Recall" than golden summary at all ratios, which is conceivable since KAE focuses on information from keywords while humans may choose to rewrite some of the keywords according to personal habits.

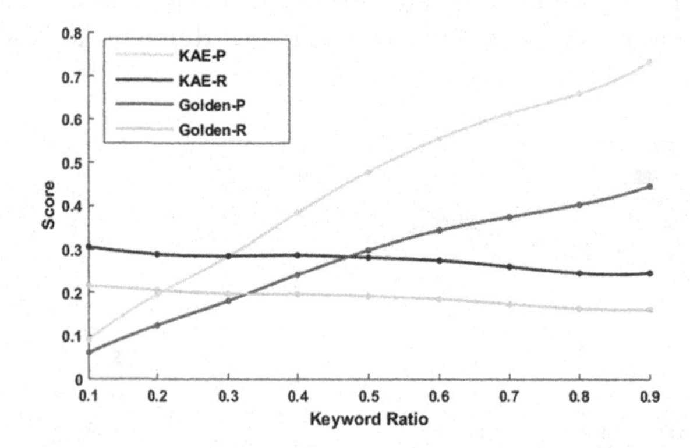

Fig. 4. Comparison of the generated summary of KAE and the golden summary in terms of Keyword Extraction Precision and Recall as the keyword extraction ratio varies from 0.1 to 0.9.

Secondly, we can see that the "Keyword Extraction Precision" increases as the keyword extraction ratio increase, while "Keyword Extraction Recall" decreases for both KAE and golden summary. This also indicates that, the amount of keyword to be considered matters, and it is necessary to use a balanced keyword extraction ratio in our KAE model.

5.3 Analysis of Content Selection Methods

Many recent works can be categorized as content selection methods which select key information first and then use it help generating summary. Content selection could be applied in different levels: word or entity level [12,20,24] which our model KAE also uses, sentence level [7,16,31] and section level [9]. How to select and utilize content are two key concerns.

In theory, any keyword extraction algorithm can replace TF-IDF in KAE. We did try using TextRank to extract keywords, and it yielded a result (37.87/18.95/35.01 in RG-1/2/L on Gigaword test set) similar to that of TF-IDF (38.24/19.17/35.48).

[31] proposed to extract linked entities and transform a list of entities into a topic embedding, whereas we propose to extract keywords, which may contain key entities, but we don't model the relations between entities explicitly. Their approach heavily depends on the quality of off-the-shelf entity linking system, while our approach does not rely on external knowledge.

[12] proposed to train a content selector first to predict which words in the input document may appear in the summary in a supervised fashion and then restrict the copy mechanism to attend to only these words. We propose to use unsupervised methods to extract keywords to guide the information flow during abstractive summarization, and the ratio of extracted keywords can be fine-tuned, which is a simper and more flexible way for content selection.

5.4 Case Study

We compare some summaries generated by our KAE model (with keyword ratio of 0.1 and 0.5). The source texts, golden summaries and generated summaries are listed in Table 6.

We observe that since KAE takes the keyword information into account explicitly, its generated summary would highly attend to the keyword information. But the ratio of keywords is also important. KAE (0.1) misses the important information "smoking causes cancer" in the first example and misses 'offseason" in the second, while KAE (0.5) successfully captures this information.

Table 6. Generated summaries from Gigaword test set.

S(1): vice president al gore thursday welcomed liggett group 's admission that smoking causes cancer and its decision to help state officials sue fellow cigarette makers .
Golden: gore welcomes liggett admission that smoking causes cancer UNK refiling
KAE(0.1): gore welcomes liggett admission
KAE(0.5): gore welcomes liggett group 's admission of smoking causes cancer
S(2): jason terry , hero of dallas ' game one national basketball association finals win over miami , may need off-season surgery on an injured thumb that he has been nursing for four months .
Golden: mavs hero terry may need surgery to fix injured thumb
KAE(0.1): dallas 's terry may need offseason surgery
KAE(0.5): terry may need surgery on injured thumb

6 Conclusion

In this work, we propose keyword-aware encoder (KAE), which extends RNN-based encoder and is used in the sequence-to-sequence framework. It merges keyword information into word representations and uses keywords to distill salient information. Experiments on benchmark datasets of various languages and lengths show that our KAE model significantly improves performance and is comparable to the latest SOTA that uses additional data for training. We will further discover the use of keyword in the text summarization task.

Acknowledgments. This research was supported by the National Key Research And Development Program of China (No.2019YFB1405802).

References

1. The stanford nlp group: Stanford dependencies. https://nlp.stanford.edu/software/stanford-dependencies.shtml
2. Bahdanau, D., Cho, K., Bengio, Y.: Neural machine translation by jointly learning to align and translate. arXiv preprint arXiv:1409.0473 (2014)
3. Banko, M., Mittal, V.O., Witbrock, M.J.: Headline generation based on statistical translation. In: Proceedings of the 38th Annual Meeting on Association for Computational Linguistics, pp. 318–325. Association for Computational Linguistics (2000)
4. Cao, Z., Li, W., Li, S., Wei, F.: Retrieve, rerank and rewrite: soft template based neural summarization. In: Proceedings of the 56th Annual Meeting of the Association for Computational Linguistics, ACL 2018, Melbourne, Australia, 15–20 July 2018, vol. 1: Long Papers, pp. 152–161 (2018)

5. Che, W., Li, Z., Liu, T.: Ltp: a Chinese language technology platform. In: Proceedings of the 23rd International Conference on Computational Linguistics: Demonstrations, pp. 13–16. Association for Computational Linguistics (2010)
6. Chen, Q., Zhu, X., Ling, Z., Wei, S., Jiang, H.: Distraction-based neural networks for document summarization. In: Proceedings of the Twenty-Fifth International Joint Conference on Artificial Intelligence, pp. 2754–2760. AAAI Press (2016)
7. Chen, Y.C., Bansal, M.: Fast abstractive summarization with reinforce-selected sentence rewriting. arXiv preprint arXiv:1805.11080 (2018)
8. Chen, Y., Gan, Z., Cheng, Y., Liu, J., Liu, J.: Distilling knowledge learned in BERT for text generation. In: Proceedings of the 58th Annual Meeting of the Association for Computational Linguistics, pp. 7893–7905. Association for Computational Linguistics (2020)
9. Cohan, A., et al.: A discourse-aware attention model for abstractive summarization of long documents. In: Proceedings of the 2018 Conference of the North American Chapter of the Association for Computational Linguistics: Human Language Technologies, vol. 2 (Short Papers), pp. 615–621 (2018)
10. Duchi, J., Hazan, E., Singer, Y.: Adaptive subgradient methods for online learning and stochastic optimization. J. Mach. Learn. Res. **12**(Jul), 2121–2159 (2011)
11. Gao, S., Chen, X., Li, P., Chan, Z., Zhao, D., Yan, R.: How to write summaries with patterns? learning towards abstractive summarization through prototype editing. In: Proceedings of the 2019 Conference on Empirical Methods in Natural Language Processing and the 9th International Joint Conference on Natural Language Processing, pp. 3739–3749 (2019)
12. Gehrmann, S., Deng, Y., Rush, A.M.: Bottom-up abstractive summarization. In: Proceedings of the 2018 Conference on Empirical Methods in Natural Language Processing, Brussels, Belgium, 31 October–4 November 2018, pp. 4098–4109 (2018)
13. Glorot, X., Bengio, Y.: Understanding the difficulty of training deep feedforward neural networks. In: Proceedings of the Thirteenth International Conference on Artificial Intelligence and Statistics, pp. 249–256 (2010)
14. Gu, J., Lu, Z., Li, H., Li, V.O.: Incorporating copying mechanism in sequence-to-sequence learning. In: Proceedings of the 54th Annual Meeting of the Association for Computational Linguistics, vol. 1: Long Papers), pp. 1631–1640 (2016)
15. Hochreiter, S., Schmidhuber, J.: Long short-term memory. Neural Comput. **9**(8), 1735–1780 (1997)
16. Hsu, W.T., Lin, C.K., Lee, M.Y., Min, K., Tang, J., Sun, M.: A unified model for extractive and abstractive summarization using inconsistency loss. arXiv preprint arXiv:1805.06266 (2018)
17. Hu, B., Chen, Q., Zhu, F.: LCSTS: a large scale Chinese short text summarization dataset. arXiv preprint arXiv:1506.05865 (2015)
18. Hua, L., Wan, X., Li, L.: Overview of the NLPCC 2017 shared task: single document summarization. In: Huang, X., Jiang, J., Zhao, D., Feng, Y., Hong, Yu. (eds.) NLPCC 2017. LNCS (LNAI), vol. 10619, pp. 942–947. Springer, Cham (2018). https://doi.org/10.1007/978-3-319-73618-1_84
19. Knight, K., Marcu, D.: Summarization beyond sentence extraction: a probabilistic approach to sentence compression. Artif Intell. **139**(1), 91–107 (2002)
20. Li, C., Xu, W., Li, S., Gao, S.: Guiding generation for abstractive text summarization based on key information guide network. In: Proceedings of the 2018 Conference of the North American Chapter of the Association for Computational Linguistics: Human Language Technologies, vol. 2 (Short Papers), pp. 55–60 (2018)

21. Li, P., Lam, W., Bing, L., Wang, Z.: Deep recurrent generative decoder for abstractive text summarization. In: Proceedings of the 2017 Conference on Empirical Methods in Natural Language Processing, pp. 2091–2100 (2017)
22. Lin, C.Y.: Rouge: A package for automatic evaluation of summaries. Text Summarization Branches Out (2004)
23. Luong, T., Pham, H., Manning, C.D.: Effective approaches to attention-based neural machine translation. In: Proceedings of the 2015 Conference on Empirical Methods in Natural Language Processing, pp. 1412–1421 (2015)
24. Pasunuru, R., Bansal, M.: Multireward reinforced summarization with saliency and entailment. In: Proceedings of the 2018 Conference of the North American Chapter of the Association for Computational Linguistics: Human Language Technologies, vol. 2 (Short Papers), pp. 646–653 (2018)
25. Paulus, R., Xiong, C., Socher, R.: A deep reinforced model for abstractive summarization. arXiv preprint arXiv:1705.04304 (2017)
26. Pennington, J., Socher, R., Manning, C.: Glove: global vectors for word representation. In: Proceedings of the 2014 Conference on Empirical Methods in Natural Language Processing (EMNLP), pp. 1532–1543 (2014)
27. Rush, A.M., Chopra, S., Weston, J.: A neural attention model for abstractive sentence summarization. In: Proceedings of the 2015 Conference on Empirical Methods in Natural Language Processing, pp. 379–389 (2015)
28. See, A., Liu, P.J., Manning, C.D.: Get to the point: summarization with pointer-generator networks. In: Proceedings of the 55th Annual Meeting of the Association for Computational Linguistics, vol. 1: Long Papers, pp. 1073–1083 (2017)
29. Srivastava, N., Hinton, G., Krizhevsky, A., Sutskever, I., Salakhutdinov, R.: Dropout: a simple way to prevent neural networks from overfitting. J. Mach. Learn. Res. **15**(1), 1929–1958 (2014)
30. Sutskever, I., Vinyals, O., Le, Q.V.: Sequence to sequence learning with neural networks. In: Advances in Neural Information Processing Systems, pp. 3104–3112 (2014)
31. Tan, J., Wan, X., Xiao, J.: Abstractive document summarization with a graph-based attentional neural model. In: Proceedings of the 55th Annual Meeting of the Association for Computational Linguistics, pp. 1171–1181 (2017)
32. Vaswani, A., et al.: Attention is all you need. In: Advances in Neural Information Processing Systems, pp. 6000–6010 (2017)
33. Wang, K., Quan, X., Wang, R.: Biset: bi-directional selective encoding with template for abstractive summarization. In: Proceedings of the 57th Conference of the Association for Computational Linguistics, ACL 2019, Florence, Italy, 28 July–2 August 2019, vol. 1: Long Papers, pp. 2153–2162 (2019)
34. Xu, L., Wang, Z., Liu, Z., Sun, M., et al.: Topic sensitive neural headline generation. arXiv preprint arXiv:1608.05777 (2016)
35. Zajic, D., Dorr, B.J., Lin, J., Schwartz, R.: Multi-candidate reduction: sentence compression as a tool for document summarization tasks. Inf. Process. Manag. **43**(6), 1549–1570 (2007)
36. Zhou, Q., Yang, N., Wei, F., Zhou, M.: Selective encoding for abstractive sentence summarization. In: Proceedings of the 55th Annual Meeting of the Association for Computational Linguistics, vol. 1: Long Papers), pp. 1095–1104 (2017)

Neural Adversarial Review Summarization with Hierarchical Personalized Attention

Hongyan Xu[1], Hongtao Liu[1], Wenjun Wang[1](✉), and Pengfei Jiao[2]

[1] College of Intelligence and Computing, Tianjin University, Tianjin, China
{hongyanxu,htliu,wjwang}@tju.edu.cn

[2] Center for Biosafety Research and Strategy, Law School, Tianjin University, Tianjin, China
pjiao@tju.edu.cn

Abstract. Review summarization aims to generate condensed text for online product reviews. Existing methods always focus on word-level representation of reviews and ignore different informativeness of different sentences in a review towards summary generation. In addition, the personalized information along with reviews (e.g., user/product and ratings) is also highly related to the quality of generated summaries. Hence, we propose a review summarization method with hierarchical personalized attention including a review encoder and a summary decoder. The encoder contains a sentence encoder to learn sentence representations with word-level attention, and a review encoder to learn review representations with sentence-level attention. Both the two attentions are of personalized paradigm whose attention vectors are derived from personalized information of input reviews instead of randomly initialized. Thus, our encoder could focus on important words and sentences in the input review. Then a summary decoder is employed to generate target summaries with hierarchical attention likewise, where the decoding scores are not only related to word information, but re-weighted by another sentence-level attention. We further design an adversarial discriminator which takes the generated summary and personalized information as inputs to force the generator adapting the generation policy accordingly. Extensive experimental results show the effectiveness of our method.

1 Introduction

In recommender systems, review summarization could generate brief summaries for product reviews. Recently, with the development of E-commerce platforms, this task has attracted more and more attention because it is able to not only help sellers understand the feedback of products quickly, but also help consumers learn the reviews of other users towards the target product and make more precise purchasing decisions.

Although text summarization [21–23] have been widely studied in natural language processing, review summarization in recommender system is quite different and challenging. In addition to the text in reviews, there are always other

C. S. Jensen et al. (Eds.): DASFAA 2021, LNCS 12682, pp. 53–69, 2021.
https://doi.org/10.1007/978-3-030-73197-7_4

personalized information along with reviews (e.g., user, product and rating) that could be utilized for summary generation. Recently, many works have been proposed in review summarization; these methods always are based on sequence-to-sequence framework, and focus on how to integrate the review with the personalized information for a better summary generation. Ma et al. [18] design a jointly end-to-end framework for improving text summarization and sentiment classification simultaneously by using ratings as sentiment label. Liu et al. [17] map the user and product ID to the fix-sized embedding and conduct a memory network for history reviews of the corresponding user and product to further capture the user and product information. Considering that different users care about different contents in the same review, Li et al. [10] design a user-aware sequence network for personalized summarization, which generates summaries by computing a score for each word in reviews to select more important words.

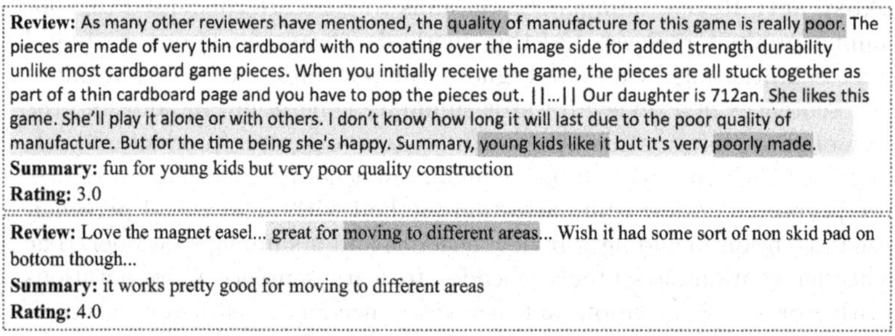

Review: As many other reviewers have mentioned, the quality of manufacture for this game is really poor. The pieces are made of very thin cardboard with no coating over the image side for added strength durability unlike most cardboard game pieces. When you initially receive the game, the pieces are all stuck together as part of a thin cardboard page and you have to pop the pieces out. ||...|| Our daughter is 712an. She likes this game. She'll play it alone or with others. I don't know how long it will last due to the poor quality of manufacture. But for the time being she's happy. Summary, young kids like it but it's very poorly made.
Summary: fun for young kids but very poor quality construction
Rating: 3.0

Review: Love the magnet easel... great for moving to different areas... Wish it had some sort of non skid pad on bottom though...
Summary: it works pretty good for moving to different areas
Rating: 4.0

Fig. 1. Two examples of online reviews and the corresponding summary and rating.

Although these methods have achieved great summary generation performance, there remain some crucial challenges. First, not only different words but the different sentences in a review are of different informativeness towards the summary generation. As shown in Fig. 1, some sentences of the review have strong semantic relevance with the user preferences and product characteristics and contribute more to the review representation. For example, in the first case, the sentences marked in green describe the main product characteristic, and the target user think the quality is poor but the young kids like it. In the second case, the second sentence shows the product characteristic that the target user cares about, i.e., *"great for moving to different areas"*. Besides, we observe that some words in the important sentence are more likely to be useful in the summary generation process. For example, the sentiment words *"poor"* and *"like"* in the first case are more salient than other words, because they indicate the user preference and sentiment towards the current product. Hence it is necessary to distinguish informative words and sentences from reviews both in the encoder and decoder phases. Besides, the personalized information (i.e., ratings, users and product information) have significant influence on the salient information identification. For example, different users have different preference and different products have different characteristic; hence, even if the same reviews may have different summaries under different users or products. As a result, we should fully exploit these personalized information for summary generation.

In this paper, we propose a Neural Adversarial Review Summarization method (NARS) with hierarchical attention. Our method is based on the encoder-decoder framework, which contains a review encoder to learn review representation from words and sentences, and a summary decoder to generate target summaries.

In the encoder, we propose to design a sentence encoder to learn sentence features from words, and a review encoder to learn review features from sentences. In the above two encoders, we apply personalized attention to select important words and sentences hierarchically. Different from general attention, the query vectors in our proposed personalized attention are derived from the personalized information along with the review. Hence, in our encode phase, we could learn more personalized representation for reviews via our hierarchical attention.

Our summary decoder is implemented based on Pointer Generator Network [22] which not only generates words from the fixed-size vocabulary but also copy words from the input review by probability. In addition to identify the keywords in the review, it is important to identify the important sentences, in which words should be more salient. Hence, we combine both personalized features of words and sentences in reviews to compute the alignment weights in the decode step (i.e., re-weighting the weights of words using the sentence weights). To further improve the quality of the generated summaries, we develop an adversarial training framework, where a discriminator is designed to evaluate the generated summary and force the summary decoder to generate high-quality summaries. During training, we update the generator and discriminator alternately to optimize the parameters in our model better.

The main contributions of our model are:

(1) We propose a neural adversarial review summarization model with hierarchical attention based on the encoder-decoder framework to generate brief summaries for online reviews.
(2) We propose to apply word- and sentence-level attention which integrates personalized information to focus on the important and informativeness words and sentences in reviews.
(3) Extensive experiments including performance evaluation and case studies validate the effectiveness of our proposed method, which could generate high-quality summaries.

2 Related Work

Review summarization aims to generate brief summaries for product reviews in many E-commerce platforms, and it is an important task in recommendation system. Previous researches [2,5] have demonstrated that abstractive methods perform better than the extractive methods [7] for review summarization. Thus in our paper we focus on abstractive generation works.

Table 1. Characteristics of different models. Especially, "Sentence-level" and "Word-level" denote the hierarchical structure of the source review. And, "Adversarial" denotes the adversarial training with discriminator to evaluate the generated summaries.

	S2S+Attn[1]	PGN[22]	HSSC[18]	memAttr[17]	USN[10]	Dual-View[3]	NARS
User ID	×	×	×	√	√	×	√
Product ID	×	×	×	√	×	×	√
Rating	×	×	√	×	×	√	√
Word-level	√	√	√	√	√	√	√
Sentence-level	×	×	×	×	×	×	√
Adversarial	×	×	×	×	×	×	√

Different from the text summarization methods [22], review summarization should take various personalized information into consideration, such as the user, product and the rating along with reviews. The user and product information have been proved to be helpful in many recommender system task, such as sentiment classification [4,26], text generation [6,15]. Chen et al. [4] conduct sentiment classification by incorporating the user and product information into a well-designed hierarchical neural network to capture crucial semantic components. Dong et al. [6] generate product reviews by conditioning on the given attribute information, e.g., ratings, users and products. Therefore, some methods propose to utilize the personalized information (e.g., ratings, user and item information) [10,11,14,18,25] to improve the quality of the generated summaries. Ma et al. [18] and Chan et al. [3] propose to use a jointly end-to-end neural network model for improving text summarization and sentiment classification by using ratings as sentiment label. Li et al. [14] conduct tips generation by considering persona information which is learned via applying adversarial variational auto-encoder on the history reviews and summaries of users and products. Li et al. [11] take the authors attributes (*e.g.*, gender, age and occupation) into account to generate personalized summaries for users towards the same reviews. However, these methods treat the source review as a long sequence while ignoring that different components contribute differently to summarization at both sentence-level and word-level. Therefore, in this paper, we try to capture the different usefulness of different components of source reviews by applying a hierarchical personalized attention. We list characteristics of several advanced methods and our model in Table 1.

Generative adversarial network ([8,12,20]) shows great advantage in text generation by leveraging discriminator to adjust the generation policy of the generator. For example, Li et al. [13] combine the maximum likelihood estimator with a summary quality estimator to make the generated summaries indistinguishable from the human-written ones. Hence, motivated by [20], we design an adversarial training framework to enhance the summary generation.

3 Proposed Method

In this section, we will first present the problem formulation, and then introduce our proposed model for review summarization. There are two main modules in our approach, i.e., a review encoder to learn review features from words and sentences, and a point-network based summary decoder to generate target summaries for source reviews. The overall architecture of our approach is shown in Fig. 2.

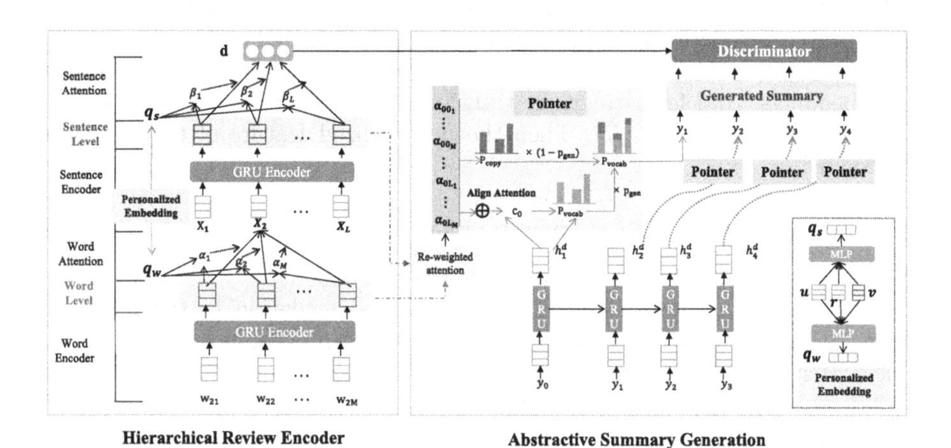

Fig. 2. The framework of our model. The left part is the review encoder which contains personalized sentence- and word-level attention. The right part is the summary decoder with re-weighted attention and the top is the discriminator for adversarial training.

3.1 Problem Formulation

Given an input review containing L sentences $S = \{X_1, X_2, \cdots, X_L\}$, along with the corresponding user and product information (u, v, r) where u is the user ID, v is the product ID and r is the rating given by user u to product v, our model aims to produce a summary $\hat{Y} = \{\hat{y}_1, \hat{y}_2, \hat{y}_3, \cdots, \hat{y}_{N'}\}$, where $N^{'}$ denote the number of words in a summary. Each sentence X_i is represented as $X_i = \{w_{i1}, w_{i2}, \cdots, w_{iM}\}$, where M denote the number of words in a sentence. In addition, we denote the gold summary sequence as $Y = \{y_1, y_2, y_3, \cdots, y_N\}$, where N is the length of the summary.

3.2 Review Encoder

In this section, we design a review encoder with word- and sentence-attention via integrating the personalized information to learn review representations. We will first introduce the personalized information (i.e., user, product and rating), which would be used to derive the attention vectors. Then the two hierarchical encoders will be described in details.

Personalized Information Embedding. In recommendation system, the user and product interactions indicate the user preferences and product characteristics, and the ratings given by users to products show the sentiment tendency of the corresponding reviews. In this paper, we treat the ratings, user and product features as **personalized information** of reviews, which can be regarded as the useful auxiliary information to learn more precise representations of source reviews. In this way, even if the contents in reviews are similar, the representations would be various while considering the personalized information.

Hence, given the personalized information (u, v, r) of the review, we first embed the corresponding user and product as real-valued ID embedding $\mathbf{u} \in \mathcal{R}^d$ and $\mathbf{v} \in \mathcal{R}^d$ respectively, where d is the vector dimension for the user and product ID embeddings. And for ratings, we transform the rating, ranging from 1 to 5, into a one-hot vector $\mathbf{r} \in \mathcal{R}^5$. Then the personalized feature of the review is denoted as:

$$\mathbf{p} = [\mathbf{u} : \mathbf{v} : \mathbf{r}] \,, \tag{1}$$

where : is the concatenation operator.

We propose to adopt hierarchical personalized attention in the following review encoder to indicate the importance of words and sentences of the source review. Hence, we derive the attention vectors from the personalized feature $\mathbf{p}$ via two multilayer perceptrons (MLP):

$$\mathbf{q_w} = \mathtt{MLP}(\mathbf{p}), \mathbf{q_s} = \mathtt{MLP}(\mathbf{p}), \tag{2}$$

where $\mathbf{q_w}$ is used as the word-level attention vector, and $\mathbf{q_s}$ is the sentence-level attention vector in the following hierarchical review encoder.

Personalized Word-Level Attention. Based on the learned personalized query vectors, we use a sentence encoder to learn the sentence representation by applying attention mechanism on the words in a sentence.

In fact, not all words contribute equally to the sentence representation for different users and products, and these words should get different attentions. Thus, we incorporate personalized information into attention mechanism to select these key words.

Given a sentence X_i in the input review, the words are firstly mapped to real-valued low-dimensional vectors via word embedding technology, and then fed into a bidirectional Gated Recurrent Unit (GRU) one by one:

$$\overrightarrow{\mathbf{h_i^w}} = \overrightarrow{\mathtt{GRU}}(\mathbf{w_i}), \overleftarrow{\mathbf{h_i^w}} = \overleftarrow{\mathtt{GRU}}(\mathbf{w_i}), \tag{3}$$

where $\mathbf{w_i}$ is the word embedding of the i-th word in the current sentence. We concatenate them to obtain the word representation, i.e., $\mathbf{h}_i^w = [\overrightarrow{\mathbf{h}}_i^w : \overleftarrow{\mathbf{h}}_i^w]$.

Then, we use the word-level attention vector $\mathbf{q_w}$ as query to calculate attention weight α_i for each word $\mathbf{h}_i^w$ and aggregate the vectors of these words to form the sentence representations $\mathbf{X_i}$:

$$\alpha_i = \mathtt{softmax}(\mathbf{w}_w^T \mathtt{tanh}(\mathbf{W}_{wq}\mathbf{q_w} + \mathbf{W}_{wh}\mathbf{h}_i^w)), \\ \mathbf{X_i} = \sum_{i=0}^{M} \alpha_i \mathbf{h}_i^w, \tag{4}$$

where $\mathbf{w}_w$, $\mathbf{W}_{wq}$, $\mathbf{W}_{wh}$ are matrix parameters of the model. And $\mathbf{X_i}$ is the final representation of the sentence X_i in the source review, which involves the personalized information and word-level informativeness.

Personalized Sentence-Level Attention. Based on the learned sentence representation, this part learns the review representation by applying the attention on the sentences in a review, which is then utilized to generate a summary in the decoder module and judge the quality of generated summaries in the discriminator module.

After obtaining the representations of the sentences in a review, in this section, we will utilize the sentence-level attention to learn the representations of the review. Likewise, different sentences are of different importance for review representation learning. Thus, we again incorporate the personalized information into attention mechanism to identify more salient sentences.

To get the review representation, we first define a bidirectional GRU to learn the relatedness among sentences in the review.

$$\overrightarrow{\mathbf{h_i^s}} = \overrightarrow{\mathtt{GRU}}(\mathbf{X_i}), \overleftarrow{\mathbf{h_i^s}} = \overleftarrow{\mathtt{GRU}}(\mathbf{X_i}), \tag{5}$$

We can obtain the high-level feature vector for each sentence: $\mathbf{h}_i^s = [\overrightarrow{\mathbf{h}}_i^s : \overleftarrow{\mathbf{h}}_i^s]$.

Then, we apply another personalized attention, and adopt $\mathbf{q_s}$ as query vector to calculate the salient score α_i for each sentence:

$$\alpha_i = \mathtt{softmax}(\mathbf{w}_s^T \mathtt{tanh}(\mathbf{W}_{sq}\mathbf{q_s} + \mathbf{W}_{sh}\mathbf{h}_i^s)) \\ \mathbf{d} = \sum_{i=0}^{L} \alpha_i \mathbf{h}_i^s \tag{6}$$

where $\mathbf{w}_s$, $\mathbf{W}_{sq}$, $\mathbf{W}_{sh}$ are model parameters, and $\mathbf{d}$ is the final personalized representation of the input source review.

3.3 Abstractive Summary Generation

Decoder with Hierarchical Attention. After obtaining the review representation $\mathbf{d}$ from the hierarchical review encoder, this module generates summary. We build the decoder upon the pointer generator network [22] with 2-layer GRU. First, we utilize the input review embedding $\mathbf{d}$ to initialize the hidden states $\mathbf{h}_0^d$ of the decoder:

$$\mathbf{h}_0^d = \mathbf{tanh}(\mathbf{W}_0\mathbf{d}), \tag{7}$$

where $\mathbf{W_0}$ is the model parameter. At each step t, the GRU receives the word embedding of the previous step $\mathbf{y_{t-1}}$ and output the hidden state $\mathbf{h_t^d}$.

Most previous review summarization works only consider the word-level importance in the attention mechanism between the encoder and the decoder. However, some product reviews are long and only a few sentences contribute significantly to the summarization. Thus, we design a hierarchical attention which can identify the salient information at both sentence-level and word-level. In detail, we first calculate the sentence-level score β_{ti} for the ith sentence in step t, based on the sentence representations $\mathbf{h}_i^s$.

$$\beta_{ti} = \mathbf{softmax}(\mathbf{W}_\beta^T \mathtt{tanh}(\mathbf{W}_{hs}\mathbf{h}_t^d + \mathbf{W}_{ts}\mathbf{h}_i^s)), \tag{8}$$

where $\mathbf{W}_\beta$, $\mathbf{W}_{hs}$ and $\mathbf{W}_{ts}$ are model parameters. For the jth word in the ith sentence, we re-weight the word-level attention weight α_{ti_j} with the corresponding sentence score β_{ti} to highlight the words in the important sentences.

$$\alpha_{ti_j} = \mathtt{softmax}(\beta_{ti}\mathbf{W}_\alpha^T \mathtt{tanh}(\mathbf{W}_{hw}\mathbf{h}_t^d + \mathbf{W}_{tw}\mathbf{h}_{ij}^w)), \tag{9}$$

where $\mathbf{W}_\alpha$, $\mathbf{W}_{hw}$ and $\mathbf{W}_{tw}$ are model parameters. Finally, the context vector $\mathbf{c}_t$ is computed as the weighted sum of the representations of words in the input review.

$$\mathbf{c}_t = \sum_{i=1}^{L}\sum_{j=1}^{M} \alpha_{ti_j}\mathbf{h}_{ij}^w. \tag{10}$$

Then, the vocabulary distribution P_{vocab} is computed by applying softmax function over the concatenation of decoder sate $\mathbf{h}_t^d$, the context vector $\mathbf{c}_t$ and the review representation $\mathbf{d}$:

$$P_{vocab} = \mathtt{softmax}(\mathbf{W}_v^T \mathtt{tanh}(\mathbf{W}_c[\mathbf{h}_t^d : \mathbf{c}_t : \mathbf{d}])), \tag{11}$$

where $\mathbf{W}_v$, $\mathbf{W}_c$ are model parameters. P_{vocab} is the probability distribution over all words in the fixed size vocabulary and our model predicts word y_t from it. Like See et al. [22], we also use the copy mechanism to copy out-of-vocabulary words from the input review. The generation probability P_{gen} at time t is calculated from the decoder hidden state $\mathbf{h}_t^d$, at embedding at previous step $\mathbf{y}_{t-1}$, context vector $\mathbf{c}_t$ and the review representation $\mathbf{d}$.

$$p_{gen} = \sigma(\mathbf{W}_g[\mathbf{h}_t^d : \mathbf{y}_{t-1} : \mathbf{c}_t : \mathbf{d}] + b_{gen}), \tag{12}$$

where $\mathbf{W_g}$ and b_{gen} are model parameters, and [:] denotes the concatenating operator. Then, p_{gen} is used to decide that the word $\hat{y}_t$ should be generated from the vocabulary with probability P_{gen} or be copied from the source with probability $1 - P_{gen}$.

$$P(\hat{y}_t) = p_{gen}P_{vocab} + (1 - p_{gen}) \sum_{i,j:w_{ij}=y_t} \alpha^{s_{ti_j}} \tag{13}$$

During training, we use the negative log-likelihood as the loss function(NLLLoss) in summary generation module:

$$\mathcal{L}_{\phi}(\hat{Y}|X) = \sum_{t=0}^{T} -logP(\hat{y}_t), \tag{14}$$

where T is the length of the generated summary.

Discriminator. After previous modules, we get the generated summary for the input review. Motivated by [13], in this module, we design an adversarial training framework to further enhance the summary generation. The gold summary Y is the positive sample and the generated summary $\hat{Y}$ is the negative sample. Especially, we feed the summary, the review and the corresponding personalized feature(i.e., review, rating, user and product embeddings) into the discriminator. As a result, the generated summary matches with the review and is consistent with the personalized features simultaneously.

In the adversarial training framework, the generator is the decoder with hierarchical attention which generates the summary based on the output of the review encoder. In discriminator, we employ a bidirectional GRU to learn the representation for input summaries.

$$\mathbf{h^Y} = \frac{\mathbf{1}}{\mathbf{T}} \sum_{\mathbf{t=1}}^{\mathbf{T}} [\overrightarrow{\mathbf{h}}_\mathbf{i}^\mathbf{Y} : \overleftarrow{\mathbf{h}}_\mathbf{i}^\mathbf{Y}]. \tag{15}$$

And for other feature, we employ the vector $\mathbf{d}$ from the review encoder. Then we feed the concatenation of the summary representations $\mathbf{h^Y}$ with the vector $\mathbf{d}$ into a multilayer perceptron and obtain a vector $\mathbf{s} \in \mathcal{R}^2$.

$$\mathbf{s} = f(\mathbf{h^Y}, \mathbf{d}), \tag{16}$$

where, f is a MLP network followed by a softmax layer and $\mathbf{s}$ is a binary label variable. Like [13], we use cross entropy as loss function $\mathcal{J}(\theta)$ to train the discriminator, where θ are parameters in the discriminator.

And we update the generator parameters ϕ via policy gradient [24] and treat the first dimension of $\mathbf{s}$ as the quality score, i.e., the reward in policy gradient $r = \mathbf{s}_{[0]}$.

$$\nabla \mathcal{J}_{\phi} = \sum_{t=1}^{T} \nabla logP(\hat{y}_t) \cdot r. \tag{17}$$

The policy parameters ϕ are updated as following:

$$\phi = \phi - \alpha_2 \nabla \mathcal{J}_{\phi}. \tag{18}$$

4 Experimental Setup

To validate the effectiveness of our method, we conduct extensive experiments, we will first introduce the experiment setups, including dataset description, baseline methods and experimental settings.

4.1 Datasets

To evaluate our method, we conduct experiments on four real-world datasets from Amazon[1]: **Toys & Games, Sports & Outdoors, Cloting, Shoes and Jewelry, Movies & TV**, the dataset statistics are shown in Table 2. Each data sample is consist of the user ID, product ID, rating, review text, and summary text. We build vocabulary for each dataset by selecting the high-frequency words from the review and summary text. In this paper, we only reserve the reviews given by active users to popular products, where each user and each product has at least 10 reviews. Following previous work [18], we randomly select 1000 samples for the validation set and test set separately, and the rest of the dataset are training set.

Table 2. Dataset statistics.

Dataset	Toys & Games	Sports & Outdoors	Clothing & Shoes	Movies & TV
Users	19,412	35,598	39,387	123,960
Products	11,924	18,357	23,033	50,052
Reviews	167,504	296,214	278,653	1,697,471

4.2 Baseline Methods

Here, we compare with many competitive summarization methods:

- **TextRank** [19]: is a famous extractive approach that ranks the sentences with the graph-based algorithm.
- **S2S+Attn** [1] is a sequence-to-sequence model with attention mechanism.
- **HSSC** [18]: a joint framework for abstractive summarization and sentiment classification.
- **PGN** [22]: a popular abstractive summarization method with copy mechanism to copy words from the input text.
- **memAttr** [17]: a neural review summarization method that leverages the user and product history reviews to enhance the model performance.
- **USN** [10]: a personalized review summarization model that generates summaries by designing a user-aware encoder and a user-specific vocabulary.
- **Dual-view** [3]:a very-recent dual-view model which introduces an inconsistency loss to make the generated summary have the same sentiment tendency with the input review.

[1] http://jmcauley.ucsd.edu/data/amazon/.

4.3 Experimental Settings

We build vocabulary for each dataset separately by reserving the high frequency words, meanwhile the hierarchical encoder and decoder module share the same vocabulary. For the hyper-parameters in our model, we tune them from the validation dataset. The hidden states of GRU is set to 512 (tuning in [64,256,512,1024]). The size of user and item ID embedding is set to 300 (tuning in [200, 300, 400]) and we use dropout with probability 0.3 for all datasets (tuning in [0.1, 0.3, 0.5, 0.7]). The encoder and decoder are both 2-layer GRU. We use the Adam [9] optimizer to train our model. The batch size is set to 48. For adversarial training, we pre-train the generator with the NLLLoss according to Eq. 14, like previous work [27].

Following the previous works, we utilize ROUGE [16] as our evaluation metric. ROUGE counts the number of overlapping units(i.e. n-gram) between the generated summaries and the references. In the experimental results, we report F-measures of ROUGE-1, ROUGE-2 and ROUGE-L.

5 Result and Discussion

5.1 Performance Evaluation

Table 3. ROUGE performance on the five datasets.

Dataset	Metric	TextRank	S2S+Attn	PGN	HSSC	memAttr	USN	Dual-view	NARS
Toys	Rouge-1	3.98	14.71	16.08	14.77	15.81	15.54	15.80	**19.21**
	Rouge-2	0.78	2.84	4.08	3.98	3.85	3.10	4.85	**5.12**
	Rouge-L	3.51	14.35	15.69	14.49	15.46	15.23	15.45	**18.56**
Sports	Rouge-1	3.91	15.12	16.35	15.44	17.50	14.93	16.63	**18.69**
	Rouge-2	0.72	3.86	4.99	4.08	5.54	5.08	5.12	**6.55**
	Rouge-L	3.42	14.98	16.25	15.25	17.37	14.81	16.30	**18.39**
Clothing	Rouge-1	2.53	14.13	16.10	15.86	17.15	16.24	16.03	**18.54**
	Rouge-2	0.45	3.25	4.71	4.89	4.96	4.95	5.02	**5.06**
	Rouge-L	2.27	14.09	15.97	15.81	16.90	16.05	15.90	**17.93**
Movie	Rouge-1	3.16	11.55	12.59	12.32	13.71	13.59	13.06	**15.63**
	Rouge-2	0.52	2.90	3.82	3.54	**4.27**	4.11	3.78	4.20
	Rouge-L	2.78	11.29	12.21	12.05	13.31	13.22	12.73	**14.22**

The comparison results of different methods are reported in Table 3. We have the following observations. First, abstractive methods outperform the extractive method (i.e., *TextRank*) with a large margin, which only extracts some original sentences from the source reviews. The reasons are that the extracted sentences in reviews are always noisy and the gold summaries usually cover content across multiple sentences.

Second, for abstractive methods, compared with the basic seq2seq model (e.g., *S2S+Attn*), the models with copy mechanism (e.g., *PGN*, *NARS*, etc.) perform better. This is because these methods solve the out-of-vocabulary problem by copying words from the input text. It is obvious that our method *NARS* outperform *HSSC* which only utilize the rating information to control the sentiment tendency of the generated summaries. The reason is that user and product attributes are crucial to the review modeling in the summarization. We can also see that our method *NARS* performs better than *memAttr*. It is because that our method considers the hierarchical structure of text in both encoder and decoder and utilizes adversarial training to generate high quality summaries.

Third, our approach *NARS* can consistently outperform all the baseline methods compared here. This is because our method considers the hierarchical structure of review rather than treats the review as a long sequence. In detail, we integrate the personalized information into the encoder to learn more comprehensive feature of the review (e.g., at word- and sentence- level), and conduct re-weighting the word importance by the sentence importance in the decoder to better identify the salient information. The experimental results demonstrate the effectiveness of our proposed method *NARS*.

5.2 Ablation Study

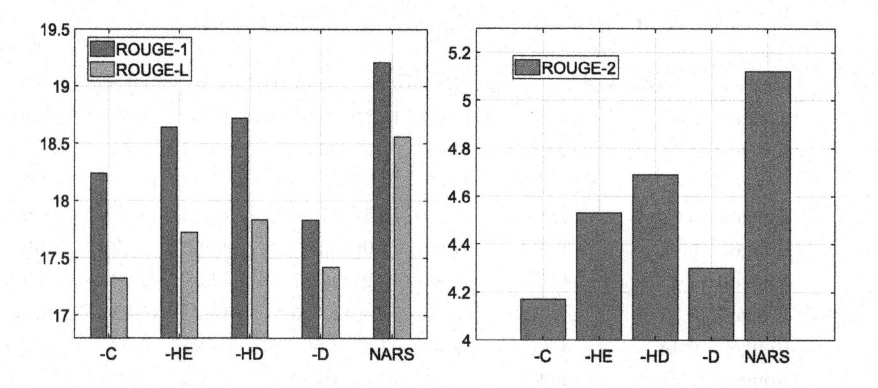

Fig. 3. Ablation study on the Toys & Games dataset.

There are several key components in our model and they play different roles in our model, i.e., the personalized attention vectors, word- and sentence-level attention in encoder/decoder, and the adversarial training part. In order to evaluate the effect of the different components of our method, we perform ablation experiments on the "Toys & Games" dataset and report the result in Fig. 3. We design four variants in our experiments:

(1) **"-C"** denotes that we remove the personalized attention vectors and adopt the same attention vectors for all users and products in the review encoder.

(2) **"-HE"** denotes that we remove the hierarchical attention in *review encoder* and learn review representation from words directly without considering the sentence-level information.
(3) **"-HD"** denotes that we remove the hierarchical attention in *summary decoder* and directly calculate the weights for words in the source review.
(4) **"-D"** denotes that we remove the discriminator in the adversarial training framework and only train the model with the negative log-likelihood loss.

As shown in Fig. 3, we can observe that removing any key component in our method would lead to a performance decline in terms of ROUGE-1, ROUGE-2 and ROUGE-L, i.e., the personalized information (**"-C"**), hierarchical attention in both the encoder and decoder (**"-HE"** and **"-HD"**), and the adversarial training (**"-D"**). The results validate that the effectiveness of our method and meet our motivation denoted above.

5.3 Case Study

In this section, we conduct two case studies to further demonstrate the method intuitively.

Recall that we incorporate the ratings as part of condition into the generator and discriminator to control the sentiment tendency of the generated summaries. Thus, we select some real cases under different ratings and list the result in Table 4. When the rating is larger than or equal to 3, gold summaries and generated summaries both contain the positive words (e.g., "great", "fun", "beautiful", etc.), while they both contain the negative words (e.g., "worst", "terrible" etc.) when the rating is lower than 3. However, our model performs

Table 4. Generated summaries for some reviews. In each row, the first line is the gold summary and the second line is the generated summary.

Rating	Summaries
5	**Beautiful high-quality puzzle**
	Beautiful puzzle
5	**Great family fun- no pictures though**
	Great family game
4	**Quality seems fine not too sure about the sizing**
	Good gloves but not for me
3	**Nice puzzle to learn numbers**
	Fun puzzle but not as the letter
2	**Cheesy**
	Not worth it
1	**Terrible terrible movie**
	One of the worst movies ever

relatively poorly on the samples whose ratings are lower than 3, because the proportion of these samples is much smaller in datasets.

Table 5. The generated summaries of our model and comparative methods

Review: This is basically a ker-plunk game played with bees and a bee hive... but still it's very fun and colorful. My wife uses it at school with her kids and they love it. It's easy to set up and

Use... and the kids the mechanical aspect of setting it up and letting it all fall down

S2S+Attn: fun for the whole family

HSSC: great game for the whole family

PGN: great game for kids

memAttr: fun and fun for kids

NARS: fun and colorful game for kids

Gold Summary: very fun, colorful easy to use toy

Review: These gloves feel solid and offer decent wrist support. Which is a good thing, because even

Though they're marked 34;large,34; there's no way I could get them on over wraps. They are

Bright pink, so maybe it's a women's size? I don't know. But I do recommend trying them on in a

store unless you want to risk having to send them back. That said, do try 'em – they're comfortable

To wear and seem built to last

S2S+Attn: good gloves

HSSC: good gloves

PGN: solid gloves

memAttr: solid gloves

NARS: good gloves but not for me

Gold Summary: quality seems fine; not sure about the sizing

In Table 5, we show a sample of generated summaries in our model and baseline methods on Toys & Games dataset. The result shows that the generated summaries of baselines are similar and always contains the high frequency word *great*. However, our model can select the salient words, e.g., word *colorful* in the first case. In the second case, we can see that the user bought the product just because that the product is cute, but the product is not suitable for the 6 month grandson. And our model could generate more precise summary All baselines just generate summaries ignoring the semantic information that covers across all the sentences. Our model performs better than baselines through the hierarchical attention network to compute the saliency weight score for each sentence of the source review.

5.4 Visualization of Attention

To validate that user and item information is able to select the important sentences and words in the input reviews, we visualize the attention from sentence-level and word-level of two samples from Toys and Sports datasets in Fig. 4. In each *Review*, every line is a sentence (sometimes a sentence covers two lines due to sentence length). In the first case, the first sentence and the second sentence

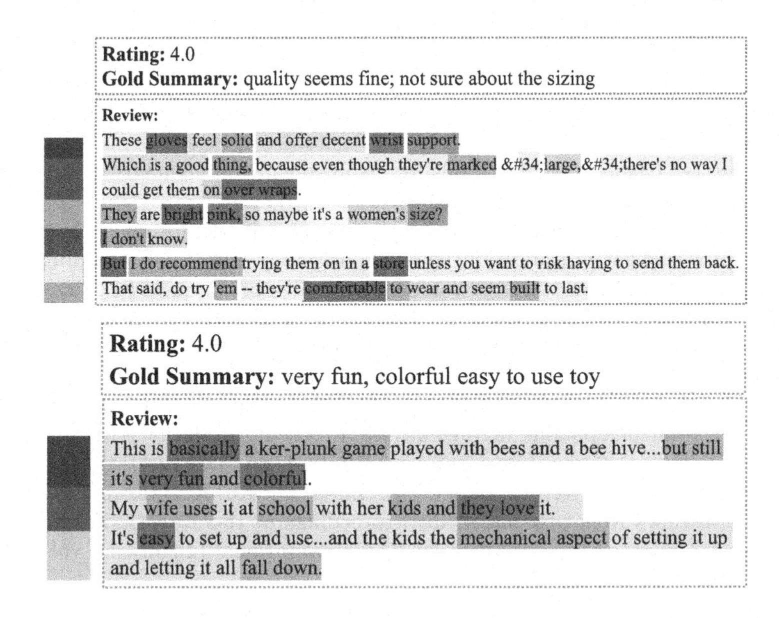

Fig. 4. Attention visualization. The blue denotes the sentence-level attention and the pink denotes the word-level attention in each sentence. (Color figure online)

get more attention at sentence-level, which represent 'quality seems fine" and the "size" problem, respectively. In the second case, compared with the gold summary, our model select the key words "fun" and "colorful" in the first sentence which is also selected as the important sentence. Especially, the result in Table 5 show that only the generated summary of our model contains the personalized word "colorful".

6 Conclusion

In this paper, we propose a neural adversarial review summarization method. Our model captures the different informativeness of different components of the review by applying a hierarchical attention at word-level and sentence-level in the review encoder and summary decoder. Especially, we utilize the user, the product and the rating, along with the review, as personalized attention query vector to identify the salient parts of the review. Besides, we design an adversarial training framework which makes the generated summaries controlled by the personalized information. The experimental results show that our model achieve the best performance than baselines.

Acknowledgement. This work was supported by the National Key R&D Program of China (2018YFC 0832100, 2020YFC0833303) and the National Natural Science Foundation of China (61902278).

References

1. Bahdanau, D., Cho, K., Bengio, Y.: Neural machine translation by jointly learning to align and translate. In ICLR (2015)
2. Carenini, G., Cheung, J.C.K., Pauls, A.: Multi-document summarization of evaluative text. Comput. Intell. **29**(4), 545–576 (2013)
3. Chan, H.P., Chen, W., King, I.: A unified dual-view model for review summarization and sentiment classification with inconsistency loss. In: Proceedings of the 43rd International ACM SIGIR Conference, SIGIR 2020, pp. 1191–1200 (2020)
4. Chen, H., Sun, M., Tu, C., Lin, Y., Liu, Z.: Neural sentiment classification with user and product attention. In: Proceedings of the 2016 Conference on Empirical Methods in Natural Language Processing, pp. 1650–1659 (2016)
5. Di Fabbrizio, G., Stent, A., Gaizauskas, R.: A hybrid approach to multi-document summarization of opinions in reviews. In: Proceedings of the 8th International Natural Language Generation Conference (INLG), pp. 54–63 (2014)
6. Dong, L., Huang, S., Wei, F., Lapata, M., Zhou, M., Xu, K.: Learning to generate product reviews from attributes. In: Proceedings of the 15th Conference of the European Chapter of the Association for Computational Linguistics, vol. 1, Long Papers, pp. 623–632 (2017)
7. Ganesan, K., Zhai, C.X., Han, J.: Opinosis: a graph-based approach to abstractive summarization of highly redundant opinions. In: 23rd International Conference on Computational Linguistics, Coling 2010 (2010)
8. Goodfellow, I., et al.: Generative adversarial nets. In: Advances in Neural Information Processing Systems, pp. 2672–2680 (2014)
9. Kingma, D.P., Ba, J.: Adam: a method for stochastic optimization. arXiv preprint arXiv:1412.6980 (2014)
10. Li, J., Li, H., Zong, C.: Towards personalized review summarization via user-aware sequence network. In: Proceedings of the AAAI Conference on Artificial Intelligence, vol. 33, pp. 6690–6697 (2019)
11. Li, J., Wang, X., Yin, D., Zong, C.: Attribute-aware sequence network for review summarization. In: Proceedings of the 2019 Conference on Empirical Methods in Natural Language Processing and the 9th International Joint Conference on Natural Language Processing (EMNLP-IJCNLP), pp. 2991–3001 (2019)
12. Li, P., Tuzhilin, A.: Towards controllable and personalized review generation. In: Proceedings of the 2019 Conference on Empirical Methods in Natural Language Processing and the 9th International Joint Conference on Natural Language Processing (EMNLP-IJCNLP), pp. 3228–3236 (2019)
13. Li, P., Bing, L., Lam, W.: Actor-critic based training framework for abstractive summarization. arXiv preprint arXiv:1803.11070 (2018)
14. Li, P., Wang, Z., Bing, L., Lam, W.: Persona-aware tips generation. In: The World Wide Web Conference, pp. 1006–1016. ACM (2019)
15. Li, P., Wang, Z., Ren, Z., Bing, L., Lam, W.: Neural rating regression with abstractive tips generation for recommendation. In: SIGIR, pp. 345–354. ACM (2017)
16. Lin, C.Y.: ROUGE: A package for automatic evaluation of summaries. In: Text Summarization Branches Out, pp. 74–81. Association for Computational Linguistics, Barcelona (2004)
17. Liu, H., Wan, X.: Neural review summarization leveraging user and product information. In: Proceedings of the 28th ACM International Conference on Information and Knowledge Management, pp. 2389–2392 (2019)

18. Ma, S., Sun, X., Lin, J., Ren, X.: A hierarchical end-to-end model for jointly improving text summarization and sentiment classification. In: Proceedings of the Twenty-Seventh International Joint Conference on Artificial Intelligence, pp. 4251–4257 (2018)
19. Mihalcea, R., Tarau, P.: Textrank: bringing order into text. In: Proceedings of the 2004 Conference on Empirical Methods in Natural Language Processing, pp. 404–411 (2004)
20. Mirza, M., Osindero, S.: Conditional generative adversarial nets. arXiv preprint arXiv:1411.1784 (2014)
21. Rush, A.M., Chopra, S., Weston, J.: A neural attention model for abstractive sentence summarization. In: Proceedings of the 2015 Conference on Empirical Methods in Natural Language Processing, pp. 379–389 (2015)
22. See, A., Liu, P.J., Manning, C.D.: Get to the point: Summarization with pointer-generator networks. In: Proceedings of the 55th Annual Meeting of the Association for Computational Linguistics, vol. 1: Long Papers, pp. 1073–1083 (2017)
23. Sutskever, I., Vinyals, O., Le, Q.V.: Sequence to sequence learning with neural networks. In: Advances in Neural Information Processing Systems, pp. 3104–3112 (2014)
24. Sutton, R.S., McAllester, D.A., Singh, S.P., Mansour, Y.: Policy gradient methods for reinforcement learning with function approximation. In: Advances in Neural Information Processing Systems, pp. 1057–1063 (2000)
25. Yang, M., Qu, Q., Shen, Y., Liu, Q., Zhao, W., Zhu, J.: Aspect and sentiment aware abstractive review summarization. In Proceedings of the 27th International Conference on Computational Linguistics, COLING 2018, pp. 1110–1120 (2018)
26. Yang, Z., Yang, D., Dyer, C., He, X., Smola, A., Hovy, E.: Hierarchical attention networks for document classification. In: Proceedings of the 2016 conference of the North American Chapter of the Association for Computational Linguistics: Human Language Technologies, pp. 1480–1489 (2016)
27. Yu, L., Zhang, W., Wang, J., Yu, Y.: Seqgan: sequence generative adversarial nets with policy gradient. In: Thirty-first AAAI Conference on Artificial Intelligence (2017)

Generating Contextually Coherent Responses by Learning Structured Vectorized Semantics

Yan Wang[1], Yanan Zheng[2(✉)], Shimin Jiang[1], Yucheng Dong[1], Jessica Chen[3], and Shaohua Wang[4]

[1] School of Information, Central University of Finance and Economics, Beijing, China
[2] School of Software, Tsinghua University, Beijing, China
[3] School of Computer Science, University of Windsor, Windsor, Canada
xjchen@uwindsor.ca
[4] Department of Informatics, New Jersey Institute of Technology, Newark, USA
davidsw@njit.edu

Abstract. Generating contextually coherent responses has been one of the most critical challenges in building intelligent dialogue systems. Key issues are how to appropriately encode contexts and how to make good use of them during the generation. Past works either directly use (hierarchical) RNN to encode contexts or use attention-based variants to further weight different words and utterances. They tend to learn dispersed focuses over all contextual information, which contradicts the facts that humans tend to respond to certain concentrated semantics of contexts. This leads to the results that generated responses are only show semantically related to, but not precisely coherent with the given contexts. To this end, this paper proposes a contextually coherent dialogue generation (ConDial) method by first encoding contexts into structured semantic vectors using self-attention, and then adaptively choosing key semantic vectors to guide the response generation. Based on the structured semantics, it also develops a calibration mechanism with a dynamic vocabulary during decoding, which enhances exact coherent expressions by adjusting word distribution. According to the experiments, ConDial shows better generative performance than state-of-the-arts and is capable of generating responses that not only continue the topics but also keep coherent contextual expressions.

Keywords: Dialogue generation · Contextual coherence · Structured vectorized semantics · Calibration mechanism

1 Introduction

Developing intelligent dialogue agents that can contextually coherently converse with humans is attracting more attention from academia and industry [16,21].

Y. Wang and Y. Zheng—Both are first authors with equal contributions.

C. S. Jensen et al. (Eds.): DASFAA 2021, LNCS 12682, pp. 70–87, 2021.
https://doi.org/10.1007/978-3-030-73197-7_5

Contextually coherent dialogues help mitigate user confusion, maintain topic continuity and keep long-term user engagement. Researchers have proposed various types of methods, including the retrieval-based [3], rule-based [3] and generative methods [11,12,18,22], based on which contextual coherence is researched. Among them, generative methods are proven the most promising, since they can automatically generate flexible responses without heavy manual handcrafts.

Generative dialogue methods train under the encoder-decoder framework using large-scale dialogue datasets. To be more specific, the encoder compresses necessary information of contexts into a fixed-size vector, conditioned on which the decoder generates responses word by word. Under the framework, one challenge for contextually coherent dialogues is how to precisely encode key information from contexts to guide coherent responses. (1) Some previous works directly use (hierarchical) RNN to encode contexts. For example, [14] proposed a recurrent language model based method, which is conditioned on past dialogue utterances that provide contextual information. [10] adopted a hierarchical recurrent encoder-decoder (HRED) method, where contexts are encoded using two RNNs. However, results show that they fail mainly because they "equally" encode words or utterances in contexts and key information can not be emphasized. (2) Other previous works use attention-based [1] variants to further assign different weights on contextual words or utterances. For example, [20] also treated context encoding as a hierarchical modeling process, particularly, it joined two-level attention mechanism, which considered the importance of tokens and utterances. However, such attention mechanism is originally designed for machine translation with alignment relationships between sources and targets, and is not suitable for dialogues, where sources and targets follow a centralized correspondence. To sum up, they are still far from *contextually coherent* that requires a tighter correspondence between a generated response and the counterpart of its context.

Table 1. Examples of contextually coherent dialogs with three semantic aspects. underline: "discussion of picnic" aspect; wave underline: "weather preference" aspect; dashed underline: "memories of pet" aspect; bold: contextually coherent local words. "(neg.)" shows a negative response that is only semantic related but not coherent with contexts.

Context	A:	I really like the weather today, and I miss old days when we were out for a picnic
	B:	Yes, I remember picnics are your favourite and you always take your dog Chico
	A:	What about going out for a **picnic** today?
Response	B:	Nice! Point Pelee National Park is best for **picnic**
(neg.) Response	B:	We plan to go hiking to enjoy the good day

On the contrary, as argued by authors in [15], responses from humans are determined more by certain semantic aspects indicated by word subsets, rather than disperse attention over every word. For example, as shown in Table 1, there could be more than one semantic aspects in given contexts (especially in multi-turn dialogue with long contexts), where there are three semantic aspects in context, i.e., *"preference of weather"*, *"discussion of picnic issue"*, *"memories of pet*" and each aspect is described by a set of key words. The response is generated by attending to the "*discussion of picnic issue*" aspect. Besides, the response also keeps the same keyword "*picnic*" with contexts. This example suggests us that simply (weighted) aggregating all words or disperse attention over all words does not appropriately encode contextual information. A more reasonable way of encoding contexts is to discover the multiple semantic aspects, and to attend to the most promising ones to respond. Moreover, keywords in contexts should be assigned extra importance to ensure local coherence.

This work aims to generate contextually coherent responses (as "response" in Table 1), instead of semantic related responses (as "neg. response" in Table 1). To this end, we initially propose the Contextually Coherent Dialogue Generation method (ConDial for short) as shown in Fig. 1. It architecturally employs the state-of-the-art conditional variational encoder-decoder (CVED) [13] as the backbone. Firstly, to enable the agent understand the multi-aspect semantics within long utterances, we develop a self-attention-based hierarchical encoder module to encode each utterance into multiple vectorized semantic aspects. Since each utterance is structurally separated into multiple parts, each part being encoded as a vector representing partial semantics, we named it as structured vectorized semantics (SVS for short). Secondly, to focus on the promising SVS aspects and mitigate the influence of irrelevant ones, the agent then employs an aligned attention mechanism to adaptively attend to each semantic aspect. Here we evaluate whether they are promising or not by calculating semantic distances between each semantic aspect and previously-given dialogue history. The key idea is that more promising SVS aspect keeps closer semantic distance to dialogue history. Thirdly, to strengthen contextually coherent expressions as much as possible, conditioned on SVS, the agent employs a calibration mechanism during decoding, where the probabilities of words within contexts are properly increased, such that they are more likely to be generated.

To sum up, main contributions of this work are as follows:

- We propose to learn structured multi-aspect vectorized semantics within long utterances and focus on the most promising ones to respond, such that dialogue agent can avoid simple aggregation or disperse attention over contexts and perform better in keeping contextual coherence.
- We integrate calibration mechanism to adjust word distribution when decoding, which strengthens coherent expressions in a more accurate way.
- We conduct extensive experiments to evaluate our method in various evaluation metrics and show superior performance over state-of-the-art methods.

2 Related Work

With the great advances in semantic learning of natural languages, more related work address ***how to encode contexts into real-valued vectors, such that response semantics could be well-guided accordingly.*** For example, The Seq2Seq dialogue model [17] initially organized contexts as a list of sequentially concatenated utterances, and directly encoded them through a recurrent encoder. Then, [10] proposed to organize contexts using a hierarchical architecture with two-level recurrent encoders. Moreover, [11] further introduced a stochastic latent variable at context level to improve the diversity of contextual information. Later, [20] introduced attention mechanism [1] and presented a hierarchical recurrent attention architecture to model contexts, taking both word-level attention and utterance-level attention into account. Built upon them, [21] further introduced the static and dynamic attention in context encoding, which weights the importance of each utterance using two attention mechanisms and obtains better context vectors. Existing work generally encode contexts by (weighted) aggregating representations of words or utterances or disperse attention, which contradicts the fact that humans-being usually receive contextual information from a centralized semantic perspective. Our work will take a step forward to build contextually coherent dialogue systems using centralized representation of contexts, instead of scatterly combining representations of words/utterances.

Another significant difference between our work and the above-mentioned work is that, our work also addresses ***how contextual information directly affects the decoding process, such that coherent words can be precisely maintained.*** There are mainly two types of works that address generating specific words. The first type generates specific types of words by using a fixed-size pre-defined vocabulary. For example, [23] proposed to generate responses that are not only grammatical but also emotionally consistent using a pre-defined external emotional vocabulary. [19] proposed to generate appropriate questions, where all words in each question are classified into one of three types, and each type is associated with a pre-defined external vocabulary. In such cases, the to-be-generated words are usually limited to a small fixed vocabulary and does not agree with the situation of generating context-related words that hold a constantly changing vocabulary. The second type relies on the copy mechanism [4]. They integrate regular decoder with the new copying mechanism which first chooses sub-sequences in the contexts and then puts them at proper places in the output responses. However, the use of whole context vectors to guide the prediction are proved to be not discriminative and sufficient enough. Our work overcomes the challenges of both types of methods, which uses a *dynamic vocabulary* to properly adjust probabilities of to-be-generated words, and uses *more precise partial contextual vector* to guide the prediction.

3 Method

3.1 Model Overview

The problem is formally defined as follows. At the T-th turn, given the dialogue history (also referred as dialogue contexts) $x = \{u_1, ...u_t..., u_T\}$, the model aims to generate a next response $y = u_{T+1}$ that is semantically coherent with the given contexts. Here each utterance is a sequence of discrete words with varying length $u_t = \{w_{t,1}, w_{t,2}, ..., w_{t,|u_t|}\}$. Specifically, **semantic coherence** is defined in detail, which means both *globally inheriting exact SVS aspects from contexts and locally keeping coherent expressions*, as exemplified in Table 1.

Figure 1 presents the overview architecture of our Contextually Coherent Dialogue method (ConDial for short). It uses the state-of-the-art conditional variational encoder-decoder (CVED) as the backbone, and specifically consists of three major components, i.e., the *Hierarchical Centralized Encoder* (HCE), the *Inference Network* (IN), and the *Context-calibrated Decoder* (CCD).

The model generally works as follows. (1) Given the contexts x, HCE first attempts to understand and extract its inherent semantics via a structured self-attention mechanism [7]. In this way, each utterance of contexts u_t is encoded into multiple vectors, where the vector m_t^i represents the i-th partial semantic aspect (denoted as SVS aspects). Then, it employs another standard attention mechanism [1] over the multiple SVS vectors, and adaptively attends to aspects with different weights α_t^i, which ultimately generates the utterance representation c_t. Finally, all utterance representations $c_t(1 \leq t \leq T)$ are properly aggregated to generate a dialogue context representation k_T. (2) The agent employs context representation k_T to perform inference over the latent variable, which models the distribution of high-level characteristics of dialogues from a stochastic perspective. The latent variable representation z is obtained by sampling from the learned stochastic distribution. (3) Last, conditioned on the last utterance representation c_T and the context-level latent variable representation z, the agent decodes new responses $\hat{y} = \{w_{T+1,1}, ..., w_{T+1,|u_{T+1}|}\}$ word by word. Particularly, during decoding, the learned SVS aspects are also used to adjust the output word distribution through a calibration mechanism, which guarantees coherent local expressions with contexts.

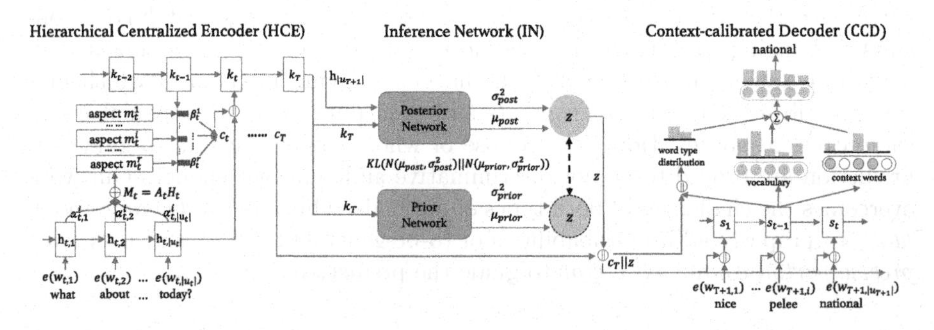

Fig. 1. The overview architecture of ConDial, consisting of three components: the hierarchical centralized encoder, the inference network and the context-calibrated decoder.

3.2 Hierarchical Centralized Encoder

The HCE takes contexts $x = \{u_1, ..., u_t, ..., u_T\}$ as inputs, and outputs last utterance representation c_T and the context representation k_T. The HCE structure is depicted as Fig. 2. HCE generally has three modules, including self-attended utterance encoder, connected attention, and context encoder. Details of each module are presented as follows.

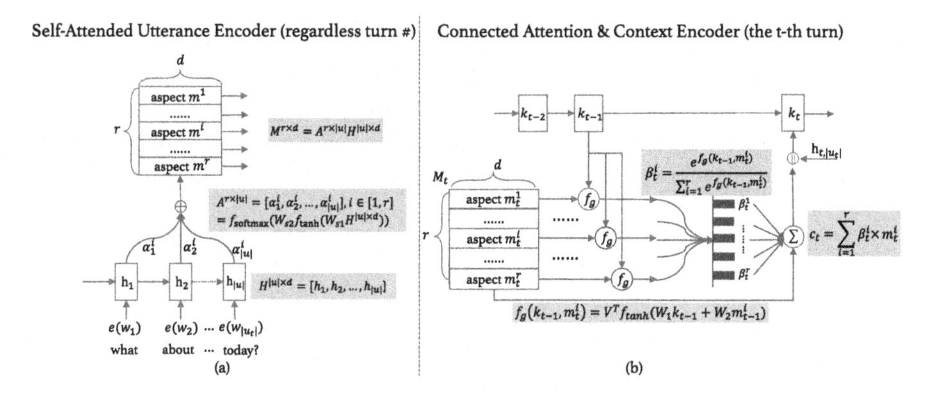

Fig. 2. HCE detailed structure, consisting of three modules, namely self-attended utterance encoder (subfigure a), connected attention and context encoder (subfigure b).

Self-attended Utterance Encoder Module. This module is intended for encoding utterances into utterance embeddings.

Firstly, given an utterance $u = \{w_{t,1}, w_{t,2}, ..., w_{t,|u|}\}$ (the subscript t indicating the number of utterance in contexts is omitted for simplicity), it is first encoded into a list of hidden states, denoted as $H = \{h_1, h_2, ..., h_{|u|}\}$, through a recurrent net with GRU unit. The h_i denotes the i-th utterance encoder hidden states, d means the hidden dimension, and $e(w_i)$ is the embedding of word w_i.

$$h_i = f_{\text{GRU}}(h_{i-1}, e(w_i)), H^{|u|\times d} = \{h_1, h_2, ..., h_{|u|}\}$$

Secondly, the hidden states are imposed with the self-attention mechanism to generate multiple semantic vectors. Given hidden states $H = \{h_1, h_2, ..., h_{|u|}\}$, a weight matrix A is first calculated by transforming H through multi-layer perceptrons, as follows.

$$A^{r\times|u|} = f_{\text{softmax}}(W_{s2} f_{\tanh}(W_{s1} H^T))$$

where W_{s1} and W_{s2} are weight matrices respectively with the shape of d_a-by-d and r-by-d_a, d_a is an intermediate dimension and r represents the number of possible semantic aspects. The weight matrix A has the shape of r-by-$|u|$. Then, by multiplying weight matrix A and hidden states H, we shall obtain the final utterance embedding M, with the shape of r-by-d.

$$M^{r\times d} = A^{r\times|u|} H^{|u|\times d}$$

Unlike previous work [10,11] where each utterance is encoded into one single vector, the final utterance embedding M has r different vectors. Each vector is of d-dimension and focuses on different partial aspects of the input utterance.

Connected Attention Module. This module aims to assign appropriate weights on multiple semantic vectors and generate the semantic-aspect-based utterance-level context. At the t-th turn, the connected attention module takes two inputs, including the t-th M_t from HCE and the (t-1)-th hidden states of context encoder module (referred to k_{t-1}). This module first computes a probability distribution over the r vectors within M_t and generates r weights $\beta = \{\beta_t^1, \beta_t^2, ..., \beta_t^r\}$. Then all the r semantic vectors within M_t are multiplied by corresponding weights and are combined together, forming semantic-aspect-based utterance-level context c_t. To be more specific, it is formalized as follows.

$$\beta_t^i = \frac{e^{f_g(k_{t-1}, m_t^i)}}{\sum_{i=1}^{r} e^{f_g(k_{t-1}, m_t^i)}}, i \in [1, r], \quad c_t = \sum_{i=1}^{r} \beta_t^i * m_t^i$$

where m_t^i represents the i-th semantic vector within M_t, $f_g(*)$ computes the similarity between both inputs. Here it uses a bilinear function $f_g(k_{t-1}, m_t^i) = V^T f_{\tanh}(W_1 k_{t-1} + W_2 m_t^i)$, where V^T,W_1 and W_2 are parameter matrices. c_t is computed by weighted sum combining all promising semantic aspects within the utterance, which agrees with our initial motivation of encoding utterance from the semantic perspective.

Context Encoder Module. The context encoder module further encodes inputs at the context level. It uses another GRU and we use k_t to represent its hidden states at the t-th step. At the t-th step, it takes three inputs, including the previous hidden states k_{t-1}, semantic-aspect-based utterance-level context C_t from connected attention module, and the traditional utterance embedding $h_{|u_t|}$, as follows.

$$k_t = f_{\text{GRU}}(k_{t-1}, h_{t,|u_t|} || c_t)$$

The last hidden states k_T of context encoder is considered as the summarization of all contextual information, i.e., a dialogue-level context.

3.3 Inference Network

We follow past studies [11,12] to use the same inference network structure shown in Fig. 1. Latent variables are assumed to take the form of Gaussian distribution, thus two parameters (mean and variance) remain to be inferred. For the prior distribution, it inputs context vector k_T, and outputs the mean $\mu_{\text{prior}}$and the variance σ^2_{prior} values via multi-layer perceptrons. For the posterior distribution, it takes both the context vector k_T and the encoded vector of the target response $h_{|u_{T+1}|}$ as inputs, and outputs the mean μ_{post} and the variance σ^2_{post} via another multi-layer perceptrons. The whole process is as follows.

$$\left[\mu_{\text{prior}}, \log \sigma^2_{\text{prior}}\right] = f_{\text{prior}}\ (k_T), z_{\text{prior}} \sim N(\mu_{\text{prior}}, \sigma^2_{\text{prior}})$$
$$\left[\mu_{\text{post}}, \log \sigma^2_{\text{post}}\right] = f_{\text{post}}(k_T, h_{|u_{T+1}|}), z_{\text{post}} \sim N(\mu_{\text{post}}, \sigma^2_{\text{post}})$$

During training, the latent variable is sampled from the posterior distribution. During inference phase, the latent variable is sampled from the prior distribution.

3.4 Decoder with Calibration Mechanism

For a general decoder, at each decoding step, each word is directly sampled from the vocabulary word distribution computed by decoder states. That is to say, the word distribution is only determined by the decoder states. However, we argue that not only the current decoder state but also the context-related information would have influences on the word distribution, especially for the appropriate SVS aspects of dialogue history.

This is illustrated by Table 1. When conditioning on "Nice! Point Pelee National Park is best for" and predicting the next word, there could be several words that are of high probability, such as "playing", "running" and "picnic" etc. However, since "picnic" has appeared in contexts and is the key information of crucial SVS aspects, its probability should be even higher. Therefore, to increase probabilities of key words that have been mentioned in contexts, we rewrite the decoding probability at the n-th step as follows.

$$\begin{aligned} P_\theta(y_n|y_{<n}, z, x) &= \sum_{t \in \{t_1, t_2\}} P_\theta(y_n, t|y_{<n}, z, x) \\ &= \sum_{t \in \{t_1, t_2\}} P_\theta(t|y_{<n}, z, x) P_\theta(y_n|t, y_{<n}, z, x) \end{aligned} \tag{1}$$

where $y_{<n}$ represents the first $n-1$ words of the response, and t is the word type distribution of current to-be-predicted word. Values t_1 and t_2 respectively are probabilities of being context-related and context-free (thus $t_1 + t_2 = 1$).

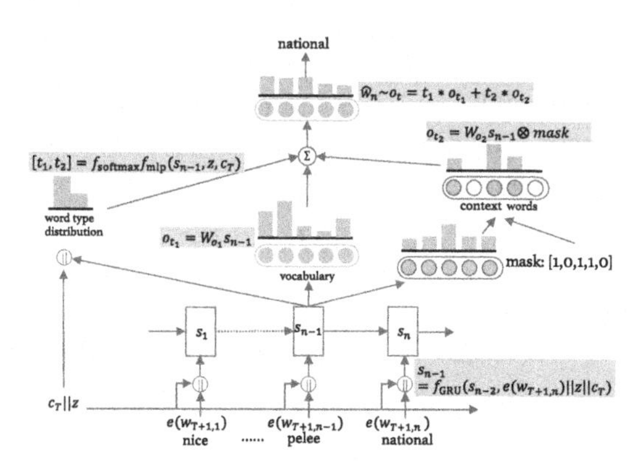

Fig. 3. Illustration of calibration mechanism.

The calibrated mechanism is depicted in Fig. 3. The decoder also uses a recurrent network, and we denote its hidden state at the n-th step as s_n. At the n-th decoding step, given previous decoder hidden state s_{n-1} and context-related information (including the latent variable sample z and the combined semantic vector of the last utterance c_T), the recurrent process is as follows.

$$
\begin{aligned}
s_n &= f_{\mathrm{GRU}}(s_{n-1}, e(w_n)||z||c_T) \\
[t_1, t_2] &= f_{\mathrm{softmax}}(f_{\mathrm{mlp}}(s_n, z, c_T)) \\
o_{t_1} &= W_{o_1} s_n, o_{t_2} = W_{o_2} s_n \otimes \text{mask} \\
\hat{w}_n &\sim o_t = t_1 * o_{t_1} + t_2 * o_{t_2}
\end{aligned}
$$

where W_{o_1}, W_{o_2} are linear weight matrices and mask is a vocabulary-size boolean vector, where context words are labeled 1 and others are 0. The mask can be directly obtained from original contexts.

3.5 Loss Function

The loss function of ConDial consists of multiple terms. In the next, we respectively introduce each optimization loss term in details.

Since ConDial is based on the CVED framework, it has the following evidence lower bound (ELBO) to be maximized.

$$
\mathcal{L}_{\mathrm{ELBO}} = -\mathbb{E}_{Q_\phi(z|x,y)}[\log P_\theta(y|z,x)] + \mathrm{KL}[Q_\phi(z|y,x)||P_\theta(z|x)] \tag{2}
$$

The ELBO has two sub-terms. The first part is the reconstruction error, which is calculated as the cross-entropy between the predicted word distribution and the expected distribution in the training set. The second part is the KL divergence, which is calculated by minimizing the differences between approximated posterior distribution and the prior distribution of latent variables.

Considering that the HCE is built upon the structured self-attention, as argued by [7], we introduce an additional penalty term, encouraging the multiple semantic vectors to learn different information. Therefore, the following loss term $\mathcal{L}_P$ is minimized, where A represents the weight matrix in self-attention module, $||*||_F$ means Frobenius norm and I is an identity matrix.

$$
\mathcal{L}_P = ||AA^\intercal - I||_F^2 \tag{3}
$$

Besides, we apply supervision on the word types. This loss term as follows is minimized, where t_y represents the true word type label, which is binary.

$$
\mathcal{L}_T = -\mathbb{E}[\log p(t = t_y|y_{<n}, x, z)] \tag{4}
$$

To sum up, ConDial is optimized by minimizing the following objective.

$$
\mathcal{L} = -\mathcal{L}_{\mathrm{ELBO}} + \mathcal{L}_P + \mathcal{L}_T \tag{5}
$$

4 Experiments

4.1 Experimental Settings

Datasets. Arguably collecting appropriate dialogue datasets is one of the challenges, and this is especially true when we focus on complicated dialogues that contain multiple semantic aspects in contexts [5]. Fortunately, previous works have contributed good-quality multi-turn datasets, such as DailyDialog[1] and Switchboard[2]. We formulate required data by treating the last utterance as response and concatenating the remaining ones as contexts. Dataset statistics are listed in Table 2.

1. **DailyDialog** covers 10 real-life topics (e.g., politics and finance). It totally consists of 13118 dialogue sessions. It is split into train, valid and test sets, with 10118, 1500 and 1500 dialogues. It has an average of 7.9 turns per dialogue, 14.6 words per utterance and a vocabulary with 17016 words.
2. **Switchboard** covers 70 open-domain topics. We randomly collect a subset (367380 dialogues) for experiments. It is split into train/valid/test sets with 357380, 5000 and 5000 dialogues. The average turns per dialogue is 6.0 and the average words per utterance is 15.3. The vocabulary is built from 20000 the most frequent words, covering 99.8% of the entire set of words.

Table 2. Statistics of datasets.

Datasets	#train	#valid	#test	# turns/dialog	#words/utterance	#vocab(%)
DailyDialog	10118	1500	1500	7.9	14.6	17016(100%)
Switchboard	357380	5000	5000	6.0	15.3	20000(99.8%)

Baseline Methods. Several baseline methods are selected for comparison. (1) **Seq2Seq** [17] adopts encoder-decoder framework and minimizes the cross-entropy in an end-to-end manner. The contexts are encoded using one single-layer RNN. (2) **HRED** [10] adopts a hierarchical recurrent encoder-decoder network. The contexts are encoded using two RNNs, where the first (utterance RNN) encodes each utterance into real-valued vector and the second (context RNN) takes utterance vectors as inputs and output context vectors. (3) **VHRED** [11] is built upon HRED and additionally incorporates latent variables to model high-level variation of contextual information. (4) **WSI** [16] shares the same architecture with HRED. Unlike HRED that takes the last hidden states of context RNN as context vectors, it takes weighted summarization of all hidden states as context vectors. (5) **HRAN** [20] is based on HRED and additionally joins two-level attention mechanism, which jointly models the importance of tokens and utterances respectively. (6) **Static**/Dynamic [21] are based on HRED and respectively add static and dynamic attention. Here we take Static, which performs better between the two, as a baseline.

[1] http://yanran.li/dailydialog.html.
[2] https://github.com/cgpotts/swda.

Training Setups. For fair comparison, we implement our ConDial and baselines using Tensorflow[3] and train them on the same machine with three 1080Ti GPUs. We use a validation set to tune parameters and finally measure metrics on the test set. We used the Adam optimizer with the learning rate initialized to 0.0005 and decayed under default settings. The batch size was set to 32/64 for DailyDialog/Switchboard. Both utterance encoder and decoder adopts single-layer unidirectional GRUs with 256 units. The context encoder uses a single-layer unidirectional GRU with 512 units. The latent variable dimension is 256 and the embedding size is 200. During training, we apply truncated back-propagation and gradient clipping with maximum gradient norm to be 5. During inferring, we use beam-search decoding with beam width 5. As reported in previous works [2], models with latent variables (VHRED and ConDial) would suffer from KL vanishing problem, where the latent samples are ignored and the whole models degenerate into auto-regressive models. We follow previous works to solve it with KL annealing technique [2] and bag-of-word loss [22]. All experiments are performed 10 times and the average results are reported to avoid randomness.

4.2 Automatic Metric-Based Evaluation

Metric Settings. We choose two types of metrics. First, from the word-level perspective, we follow previous work [10,11,22] to use PPL [8] and BLEU scores [9]. PPL is defined as the exponentiation of the word entropy, while BLEU score is based on the idea of modified n-gram precision (here we use normalized BLEU-1 to 2). Both measures to which degree the generated responses match golden standards verbatim. Smaller PPL and larger BLEU scores indicate better performance. Second, from the sentence-level perspective, we follow existing works to use embedding-based metrics: Average, Extrema and Greedy [8,11]. They describe the semantic fit between generated and golden responses, and larger values indicate better semantic similarities.

Results and Analysis. Table 3 shows the metric-based results, including both word-level and sentence-level results. Key observations are as follows.

Word-level Generative Performance. Firstly, we focus on validating detailed design of methods. (1) Comparing Seq2Seq and HRED, we can see HRED performs significantly better, with (10.82, 0.269, 0.0086) and (2.72, 0.0032,0.004) advantages for the three metrics (PPL, BLEU-1, BLEU-2) respectively on DailyDialog and Switchboard. It suggests that hierarchical modeling of contexts indeed improves the quality of generated responses, and validates the motivation of using a hierarchical structure as the backbone of HCE component. (2) Since WSI, HRAN, and Static are all based on the HRED model and directly incorporate attention mechanisms in different ways, by respectively comparing each of them with HRED, we find that none of them show stable advantages for all metrics. This validates our initial motivation that standard attention

[3] https://github.com/tensorflow/tensorflow.

mechanisms does not necessarily promote the quality of context encoding. Since they learn a dispersed attention over all words, it does not effectively highlight key contextual semantics while weakens unimportant information, which can also hurt response generation. (3) Additionally, we also find that VHRED stably outperforms HRED with (2.49,0.0008,0.0236) advantages on DailyDialog and (2.09,0.077,0.0506) advantages on Switchboard, proving the effectiveness of using latent variables to model the distribution of high-level semantics.

Secondly, the proposed ConDial generally achieves competitive word-level results than other methods, as is marked in bold. It proves that responses generated by ConDial best fits golden standards word by word. Besides, ConDial combines all designs mentioned above, and it also proves that the three techniques are compatible and can be combined together to achieve improved performance.

Table 3. Performance of response generation from word/sentence-level perspective.

	Models	#Params	PPL (KL)	BLEU-1	BLEU-2	Extrema	Average	Greedy
DailyDialog	Seq2Seq	13×10^6	44.49 (–)	0.2728	0.1790	0.2183	0.4173	0.4339
	HRED	19×10^6	33.67 (–)	0.2997	0.1876	0.2346	0.4158	0.4344
	VHRED	23×10^6	31.18 (3.78)	0.3005	0.2112	0.2635	**0.4549**	0.4479
	WSI	19×10^6	33.24 (–)	0.2804	**0.2404**	0.2040	0.3952	0.4190
	HRAN	21×10^6	45.48 (–)	0.2089	0.1132	0.2097	0.2681	0.3935
	Static	20×10^6	32.20 (–)	0.2440	0.2001	0.2182	0.2760	0.3952
	ConDial	23×10^6	**28.10 (3.92)**	**0.3103**	**0.2410**	**0.3066**	**0.4640**	**0.4682**
Switchboard	Seq2Seq	18×10^6	61.02 (–)	0.0628	0.0458	0.2126	0.2468	0.3199
	HRED	26×10^6	58.30 (–)	0.0660	0.0498	0.2207	0.2524	0.3446
	VHRED	31×10^6	56.21 (2.20)	**0.1430**	0.1004	0.1825	0.2303	0.3055
	WSI	26×10^6	63.30 (–)	0.0698	0.0402	0.1452	0.1680	0.2238
	HRAN	29×10^6	72.30 (–)	0.0609	0.0411	0.2015	0.2110	0.3080
	Static	26×10^6	62.45 (–)	0.0830	0.0460	0.1326	0.1662	0.2304
	ConDial	32×10^6	**42.20 (4.19)**	**0.1493**	**0.1180**	**0.2731**	**0.2613**	**0.3682**

Sentence-level Generative Performance. We observe that our ConDial method outperforms other baselines in terms of embedding-based metrics with comparable model size. It proves that responses from ConDial show better topic similarity with ground-truths, thus are more likely to semantically correspond to contexts. Specifically, Seq2Seq, WSI, HRAN and Static perform less well, as is predicted in previous subsection. For Seq2Seq, long contexts are encoded using a simple RNN, where long-distance information is easy to attenuate and there's no filtering of redundant information, thus resulting in poor performance. For WSI, HRAN and Static, such attention mechanisms lead to dispersed attention distribution over input contexts. ConDial and VHRED share the same variational auto-encoder architecture, but use different encoder and decoder, and we find that ConDial presents stably better results, which proves the effectiveness of HCE and Cali mechanisms.

4.3 Manual Evaluation

Metrics and Settings. For subjective evaluation, we compare our ConDial with baselines (including Seq2Seq, HRED, VHRED, WSI, HRAN, and Static), thus forming 6 comparison pairs, as Table 4 shows. We randomly sampled 100 contexts from the test set. We feed each context data into each comparison pair to generate responses (from two models), and generate the to-be-annotated sample in the form of (contexts, responsel, response2). Therefore, we totally have $100 \times 6 = 600$ to-be-annotated samples. Note that within each to-be-annotated sample, both responses are messed up in order, and only experiment designer knows the true order. We then let 3 volunteers who are not related to this work to annotate these samples by the following rules. (1) Each volunteer is asked to independently choose among win and loss (win: responsel is better; loss: response2 is better) in terms of certain factor. (2) Each volunteer is asked to consider two factors, including being fluent/grammatical (denoted as appropriate) and being contextually coherent. (3) We adopt the strategy of majority voting. Thus, for each comparison pair in terms of each factor, there are totally 100 to-be-annotated samples, each receiving 3 votes, corresponding result being reported in the form of percentage comparison, $(\frac{\#votes_1}{300} \times 100\%)$: $(\frac{\#votes_2}{300} \times 100\%)$ $(\#votes_1 + \#votes_2 = 300)$.

Table 4. Results of manual evaluation.

Comparison	Appropriate (100%)	Coherent (100%)
ConDial vs Seq2Seq	51.3:48.7	94.7:5.3
ConDial vs HRED	49.0:51.0	63.3:36.7
ConDial vs VHRED	49.3:50.7	70.0:30.0
ConDial vs WSI	76.7:23.3	85.3:14.7
ConDial vs HRAN	66.7:33.3	83.0:17.0
ConDial vs Static	63.0:37.0	86.7:13.3

Results and Analysis. Table 4 summarizes the results. As can be seen, ConDial outperforms other baselines significantly in the consistency metric, which agrees with the results of automatic evaluation. Putting aside the consistency factor and only considering appropriateness, results are as follows. Comparing with Seq2Seq, HRED, and VHRED, ConDial achieves comparable results, where the vote gap is less than 2%. Comparing with WSI, HRAN, and Static, ConDial shows competitive advantages, where the dominant vote gap is more than 20%.

Not considering much on consistency, we observe that the other models tend to generate responses that are shorter and simpler, and some even just generate general and safe responses, like "I don't know" and "that is right". Researchers have also ever reported similar observations [6,15]. They have less chances to make mistakes in grammar and syntactics. In order to keep contextual coherent, responses from ConDial are statistically longer and more information-rich (see Table 6), thus they are more likely to make mistakes in appropriateness.

4.4 Further Analysis of Our Method

Ablation Study. Firstly, to evaluate the effectiveness of HCE component, we experimented by replacing HCE with other different strategies of encoding contexts, including the standard RNN encoder (denoted as ConDial rpl. RNN-enc), the HRED encoder (denoted as ConDial rpl. HRED-enc), the WSI encoder (ConDial rpl. WSI-enc), the Static encoder (denoted as ConDial rpl. Static-enc). Secondly, to evaluate the effectiveness of Calibration mechanism, we experiment by removing it from ConDial (referred as ConDial rm. Cali) and using a general decoder. Considering the fact that automatic metrics cannot discriminate slight differences of generated responses, we use manual evaluation for ablation study, in order to provide subjective and discriminative results. We randomly select 120 contexts from DailyDialog. For each context, we let each model generate corresponding response, and then ask 3 volunteers to vote for the best one considering the above two factors. Thus there are totally $120 \times 3 = 360$ votes allocated among 6 models. Table 5 shows the results of ablation studies.

Table 5. Results of ablation study.

Model	Vote ratio ($\frac{\#\text{votes}}{\#\text{total votes}} \times 100\%$)
ConDial	21.1%
ConDial rm. Cali	19.4%
ConDial rpl. RNN-enc	8.3%
ConDial rpl. HRED-enc	16.1%
ConDial rpl. WSI-enc	18.7%
ConDial rpl. Static-enc	16.4%

We can observe ConDial get the highest votes. By removing calibration mechanism, the number of votes reduces by 1.7%, indicating that calibration mechanism have positive effects on response generation. However, the gap is small. It is worth noting results between ConDial and ConDial rm. Cali are largely affected by the dataset, where the gap would be larger if datasets contain more cases that require keeping coherent expressions, and vice versa. By replacing HCE with other encoding strategies, the number of votes drops to varying degree, proving the effectiveness of HCE. Among them, ConDial rpl. HRE+aggr shows the minimal drop (2.2% drop), ConDial rpl. SRE shows the maximal drop (12.8% drop), ConDial rpl. HRE+last & ConDial rpl. HRE+attn show similar drop ratio (around 5% drop). We also observe, replacing HCE (>2.2% drops) influences more than removing calibration mechanism (<2% drops), which indicates the HCE component plays the major role in contributing to the improvements.

Effects on Response Diversity. Intuitively, we know that making responses more coherent with given contexts would correspondingly reduce diversity since the contextual information restricts the semantic space of responses. Therefore, **assessing to which degree response diversity has been damaged remains to be an important question.** We use the widely-accepted Distinct-1/2 metrics [6] to measure the diversity of generated responses. They are calculated as the number of unique uni-grams or bi-grams in the predicted responses scaled by total number of tokens. Larger values means better diversity. Figure 4 shows results. Note that diversity analysis on both datasets achieve similar results, and here we take the results on DailyDialog for illustration.

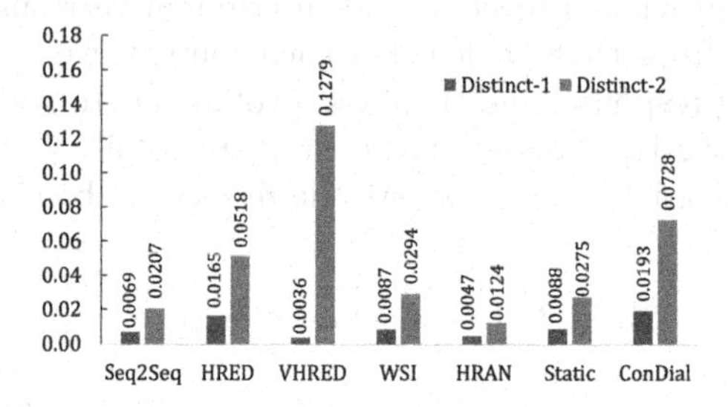

Fig. 4. Comparison of response diversity of the models.

Both VHRED and ConDial achieve more competitive results in diversity due to latent variable, Comparing VHRED with ConDial, responses from ConDial are less diverse than those from VHRED. As is expected, both diversity and contextual coherence hold a trade-off relationship. ConDial further improves coherence, which would naturally hurt diversity in a way. Fortunately, results show it is just a low degree reduce, and ConDial still significantly outperforms all the others (VHRED excluded) in diversity.

4.5 Case Study

We present several generated examples in Table 6. As we can see, Seq2Seq generates responses that are grammarly and syntactically good, but hardly carry semantic information. After modeling contexts with hierarchical architecture, HRED can generate responses that are longer with richer semantics, however, the responses are hardly semantically related to contexts. Built upon HRED, the WSI, HRAN, and Static incorporate weighted attention mechanisms in different ways. Correspondingly, we observe that their generated responses can have terms related to contextual words, however at a low frequency. Besides, the responses are not that informative as HRED. This could because, the ways they incorporate attention mechanism did not help select useful information, instead, reducing information indeed hurt the generative process. We observe that generated responses from ConDial are the most contextually coherent among all cases, where they show continuous topics and keep the same key words as contexts.

Table 6. Cases of generated responses from different models. ⇒ means change of turns. The words marked in bold are re-mentioned expressions. GT means the ground truth.

contexts	If you 'd like further **information** on this , there 's a guidebook in several languages on sale at the information desk .⇒do you have a **chinese edition** of this ?
Seq2Seq	Yes , i do
HRED	Yeah but there are various ways to find **chinese** money in the world
VHRED	Yes, it's very interesting. you'll have to pay for the final decision
WSI	Yes, but i have no idea of it
HRAN	I 'm sorry , but i do n't
Static	I 've never been a mango
ConDial	Yes, i do. i find it informative. the only thing is the excellent <unk> in this **information**
GT	I'm afraid we do n't . wo n't an english **edition** do ?
contexts	What can i do for you , miss ?⇒i am looking for a pair of **shoes**
Seq2Seq	What kind of color do you like ?
HRED	What size do you want to have ?
VHRED	How about this one ?
WSI	What do you think of your present one?
HRAN	I'm sorry about that
Static	What kind of juice do you prefer ?
ConDial	What kind of **shoes** are you looking for ?
GT	Do you like high heels or flat **shoes** ?
contexts	What do you think are the qualities a nurse should have ?⇒they are responsibility , skill , **patience** , quietness , and so on
Seq2Seq	I think you'll have to do it
HRED	I do n't know . what do you think ?
VHRED	I think you'll have to make a decision in the kitchen
WSI	I think that 's good
HRAN	What's the matter ?
Static	I think that they got into a fight and then decided to go to the school
ConDial	But i do n't know what you mean . what do you think about the **patience** ?
GT	What kind of medicines does the doctor usually prescribe to children who get highfever ?

5 Conclusion

In this paper, we proposed a contextually coherent dialogue model named ConDial. It was constructed with CVED as backbone model. Particularly, instead of encoding dialogue contexts from the word-level and utterance-level perspectives, it encoded dialogue contexts into structured semantic blocks and then used an attention mechanism to adaptively attend to multiple semantic aspects, using the hierarchical centralized encoder component. To further keep coherent words as contexts, a calibration mechanism is incorporated into decoding. We evaluated it on two datasets from multiple aspects and experiments show effectiveness of the proposed method compared to other state-of-the-arts. By encoding contexts from the semantic-level perspective, and using contexts to adjust decoding word distribution, ConDial is capable of generating contextually coherent responses.

Acknowledgements. The work was supported by the National Key Research and Development Program of China (No.2016YFB1001101)

References

1. Bahdanau, D., Cho, K., Bengio, Y.: Neural machine translation by jointly learning to align and translate. In: ICLR (2015)
2. Bowman, S.R., Vilnis, L., Vinyals, O., Dai, A.M., Józefowicz, R., Bengio, S.: Generating sentences from a continuous space. In: CoNLL, pp. 10–21. ACL (2016)
3. Chen, H., Liu, X., Yin, D., Tang, J.: A survey on dialogue systems: recent advances and new frontiers. SIGKDD Explor. **19**(2), 25–35 (2017)
4. Gu, J., Lu, Z., Li, H., Li, V.O.K.: Incorporating copying mechanism in sequence-to-sequence learning. In: ACL, no. 1. The Association for Computer Linguistics (2016)
5. Huang, M., Zhu, X., Gao, J.: Challenges in building intelligent open-domain dialog systems. ACM Trans. Inf. Syst. **38**(3), 21:1–21:32 (2020)
6. Li, J., Galley, M., Brockett, C., Gao, J., Dolan, B.: A diversity-promoting objective function for neural conversation models. In: HLT-NAACL, pp. 110–119 (2016)
7. Lin, Z., et al.: A structured self-attentive sentence embedding. In: ICLR. OpenReview.net (2017)
8. Liu, C., Lowe, R., Serban, I., Noseworthy, M., Charlin, L., Pineau, J.: How NOT to evaluate your dialogue system: an empirical study of unsupervised evaluation metrics for dialogue response generation. In: EMNLP, pp. 2122–2132 (2016)
9. Papineni, K., Roukos, S., Ward, T., Zhu, W.: Bleu: a method for automatic evaluation of machine translation. In: ACL, pp. 311–318. ACL (2002)
10. Serban, I.V., Sordoni, A., et al.: Building end-to-end dialogue systems using generative hierarchical neural network models. In: AAAI, pp. 3776–3784 (2016)
11. Serban, I.V., Sordoni, A., et al.: A hierarchical latent variable encoder-decoder model for generating dialogues. In: AAAI, pp. 3295–3301. AAAI Press (2017)
12. Shen, X., Su, H., Niu, S., Demberg, V.: Improving variational encoder-decoders in dialogue generation. In: AAAI, pp. 5456–5463. AAAI Press (2018)
13. Sohn, K., Lee, H., Yan, X.: Learning structured output representation using deep conditional generative models. In: NIPS, pp. 3483–3491 (2015)
14. Sordoni, A., et al.: A neural network approach to context-sensitive generation of conversational responses. In: HLT-NAACL, pp. 196–205 (2015)
15. Tao, C., et al.: Get the point of my utterance! learning towards effective responses with multi-head attention mechanism. In: IJCAI, pp. 4418–4424. ijcai.org (2018)
16. Tian, Z., Yan, R., Mou, L., Song, Y., Feng, Y., Zhao, D.: How to make context more useful? an empirical study on context-aware neural conversational models. In: ACL, no. 2, pp. 231–236. Association for Computational Linguistics (2017)
17. Vinyals, O., Le, Q.V.: A neural conversational model. In: ICML (2015)
18. Wang, W., Huang, M., Xu, X., Shen, F., Nie, L.: Chat more: deepening and widening the chatting topic via a deep model. In: SIGIR, pp. 255–264. ACM (2018)
19. Wang, Y., Liu, C., Huang, M., Nie, L.: Learning to ask questions in open-domain conversational systems with typed decoders. In: ACL, no. 1, pp. 2193–2203 (2018)

20. Xing, C., Wu, Y., Wu, W., Huang, Y., Zhou, M.: Hierarchical recurrent attention network for response generation. In: AAAI, pp. 5610–5617. AAAI Press (2018)
21. Zhang, W., Cui, Y., Wang, Y., Zhu, Q., Li, L., et al.: Context-sensitive generation of open-domain conversational responses. In: COLING, pp. 2437–2447 (2018)
22. Zhao, T., Zhao, R., et al.: Learning discourse-level diversity for neural dialog models using conditional variational autoencoders. In: ACL, pp. 654–664 (2017)
23. Zhou, H., Huang, M., Zhang, T., Zhu, X., Liu, B.: Emotional chatting machine: emotional conversation generation with internal and external memory. In: AAAI, pp. 730–739. AAAI Press (2018)

Latent Graph Recurrent Network for Document Ranking

Qian Dong[1,2(✉)] and Shuzi Niu[1(✉)]

[1] Institute of Software, Chinese Academy of Sciences, Beijing, China
dongqian19@mails.ucas.ac.cn, shuzi@iscas.ac.cn
[2] University of Chinese Academy of Sciences, Beijing, China

Abstract. BERT based ranking models are emerging for its superior natural language understanding ability. The attention matrix learned through BERT captures all the word relations in the input text. However, neural ranking models focus only on the text matching between query and document. To solve this problem, we propose a graph recurrent neural network based model to refine word representations from BERT for document ranking, referred to as **L**atent **G**raph **Re**current Network (LGRe for short). For each query and document pair, word representations are learned through transformer layer. Based on these word representations, we propose masking strategies to construct a bipartite-core word graph to model the matching between the query and document. Word representations will be further refined by graph recurrent neural network to enhance word relations in this graph. The final relevance score is computed from refined word representations through fully connected layers. Moreover, we propose a triangle distance loss function for embedding layers as an auxiliary task to obtain discriminative representations. It is optimized jointly with pairwise ranking loss for ad hoc document ranking task. Experimental results on public benchmark TREC Robust04 and WebTrack2009-12 test collections show that LGRe (The implementation is available at https://github.com/DQ0408/LGRe) outperforms state-of-the-art baselines more than 2%.

Keywords: Ad hoc retrieval · Graph neural network · Transformer

1 Introduction

Neural ranking models focus on learning query-document matching patterns, i.e., knowledge about search tasks. Recently, pretrained neural language models learn such knowledge from an extensive text collection and provide new opportunities for document ranking.

Word embedding [17] applied to document ranking is pretrained with a large corpus based on word co-occurrences within a window of the input text. Unlike such word embedding only encoding the local context, word representations learned from BERT are a function of the entire input text. Taking the concatenation of query and document as input, BERT is naturally fit for the search

C. S. Jensen et al. (Eds.): DASFAA 2021, LNCS 12682, pp. 88–103, 2021.
https://doi.org/10.1007/978-3-030-73197-7_6

task. The reason lies in that the attention matrix contains query-document interaction on the word level. In this sense, BERT based ranking model belongs to interaction based neural ranking models [7].

However, interaction based models only care for matching patterns between query and document. The attention matrix learned from BERT brings about additional word relations in the query-document matching process, such as query-query and document-document word relations. Whether these additional word relations are useful to derive the query-document relevance pattern remains unknown. One interesting observation is that long natural language queries perform better than short keyword queries for BERT based ranking models in document ranking tasks [2]. The problem that additional relations are dominant in the attention matrix is more serious for short queries.

To solve this problem, we propose a graph recurrent neural network based method to refine word embedding learned from BERT in the document ranking task. For each query and document pair, BERT first takes their concatenation as input and obtain word representations. Then, a masking method is adopted to construct a bipartite-core word graph as the matching between query and document, which is a masked attention matrix derived from learned word representations. Distinguished from explicitly defined graphs, the latent graph is learned from BERT for the following word representation refinement layer. To enhance word relations in this graph, word representations are updated through a gated recurrent unit. The final query and document pair representation is summarized from refined word representations and used for prediction. Pairwise ranking loss is a function of relevance scores. Moreover, a triangle distance loss is proposed as function of query, document and query-document pair representations to learn discriminative representations. Both loss functions are optimized jointly in an end-to-end manner. Experiments on public benchmark datasets Robust04 and WebTrack2009-12 are conducted to show the effectiveness of LGRe. Detailed implementations are further analyzed in experiments, such as the effect of additional word relations on query-document relations.

To sum up, our major contributions lie in the following aspects:

- We explore masking strategies applicable to building a bipartite-core word graph as the matching between query and document.
- We refine word representations on this graph through graph recurrent neural network to alleviate useless word relation's effect.
- We propose a triangle distance loss function for the embedding layer, which helps learn discriminative representations for the downstream ranking task.

2 Related Work

Here we briefly review some related studies in terms of neural ranking models without BERT, BERT based ranking models and Graph neural network.

2.1 Interaction Based Neural Ranking Models

Interaction based neural ranking models assume that relevance is in essence about the relation between input texts and it is more effective to learn from interactions rather than individual representations. They focus on designing the interaction function to produce the relevance score. Existing interaction functions are divided into two kinds: non-parametric and parametric interaction functions [7].

Traditional non-parametric interaction functions includes binary indicator, cosine similarity, dot product and radial basis function so on. DRMM [6] converts a local interaction matrix for the query-document word pair to a fixed-length matching histogram for relevance matching. MatchPyramid [16] produces a query-document relevance score by convolution operations over a query-document similarity matrix. Parametric interaction functions are to learn the similarity/distance function from data. For example, Conv-KNRM [4] uses convolutional neural network to represent n-grams of various lengths, matches them in a unified embedding space for the kernel pooling and learning-to-rank layers to generate the final ranking score. Arc-II [8] performs convolution and pooling on the word interaction between two sentences. In this sense, BERT based ranking models can also be treated as parametric interaction based neural ranking models.

2.2 Pretrained Neural Language Models for IR

Pretrained Neural Language Models (PNLM), e.g., BERT [5] and ELECTRA [1], have achieved state-of-the-art results in many NLP tasks. As mentioned above, it works for the ad hoc ranking because the attention matrix in BERT can be regarded as an interaction function. BERT based ranking models have been shown to be superior to neural ranking models without BERT.

BERT-MaxP [2] splits a document into overlapping passages. The neural ranker predicts the relevance score of each passage independently. Document score is the score of the best passage. CEDR [14] incorporates BERT's classification vector into existing neural models, such as DRMM [6] and Conv-KNRM [4]. PARADE [10] leverages passage-level representations to predict a document relevance score without passage independence assumption. Rather than fine tuning BERT-base on a Bing search log, PARADE improves performance by fine tuning on the MSMARCO passage ranking dataset. Other researches focus on how to improve the efficiency of PNLM in retrieval tasks. PreTTR [13] precomputes part of the document term representations at indexing time, and merge them with the query representation at query time to compute the final ranking score. DeepCT [3] maps the contextualized term representations from BERT into context-aware term weights for efficient passage retrieval.

2.3 Graph Neural Network

Graph neural network (GNN) has recently been widely studied in many fields because of its high-order relation capture ability. The information propagation step is key to obtain the hidden states of nodes (or edges) for GNN. According to different information propagation methods, GNN can be divided into convolution based, attention based and recursive based models so on [20]. Convolution based GNN, extending convolution operation to the graph domain, includes spectral approaches and spatial approaches. Through the attention mechanism, attention based GNN focuses on important nodes in the graph and important information of these nodes for the sake of improving the signal-to-noise ratio of the original data [18]. Recursive based GNN attempts to use the gate mechanism like GRU [11] in the propagation step to improve the long-term propagation of information across the graph structure.

3 Method

We first formalize the ad hoc document retrieval task. To overcome the inherent weakness of BERT for ranking, the network architecture of our proposed LGRe is described. Additionally, we propose a triangle distance loss function for better representations to aid the downstream ranking task. Both the triangle distance and pairwise ranking loss functions are optimized jointly.

3.1 Formalization

Ad hoc document retrieval task is to produce the ranking of documents in a corpus given a short query. There are Q queries $\{q_i\}_{i=1}^{Q}$ for training. Each query q is represented as a word sequence $s^q = \mathrm{w}_1^q, \mathrm{w}_2^q, \ldots, \mathrm{w}_m^q$ and also associated with a document set $D_q = \{(d_j, y_j)\}_{j=1}^{n_q}$. $y_j \in \{0, 1\}$ is the ground truth relevance label of document d_j. Non-relevant documents from D_q are denoted as D_q^- ($|D_q^-| = n_q^-$) and relevant documents denoted as D_q^+ ($|D_q^+| = n_q^+$). Document $d \in D_q$ is denoted as a word sequence $s^d = \mathrm{w}_1^d, \mathrm{w}_2^d, \ldots, \mathrm{w}_n^d$. How to model the text matching between query and document is key to neural ranking models.

3.2 Architecture

As mentioned before, BERT has such a natural advantage to become a ranker that its learned attention matrices model the query and document interaction. However, its disadvantage is also evident that these learned attention matrices describe all possible word relations without emphasis on the query-document word relations. To solve this problem, we mask these unnecessary word relations and introduce a refinement process over this masked graph. Through this refinement process, word representations will be more suitable to derive the relevance score between query and document. The whole architecture is depicted in Fig. 1.

Transformer Layer. For each query-document pair (q, d), two word sequences are concatenated, i.e. $s^{(q,d)} = [[\mathrm{CLS}], s^q, [\mathrm{SEP}], s^d, [\mathrm{SEP}]]$. Its input embedding $\mathbf{I}^{(q,d)}$ is derived from the sum of the word embedding and its corresponding position embedding. Then $\mathbf{I}^{(q,d)}$ is fed into BERT stacked with L identical layers. For

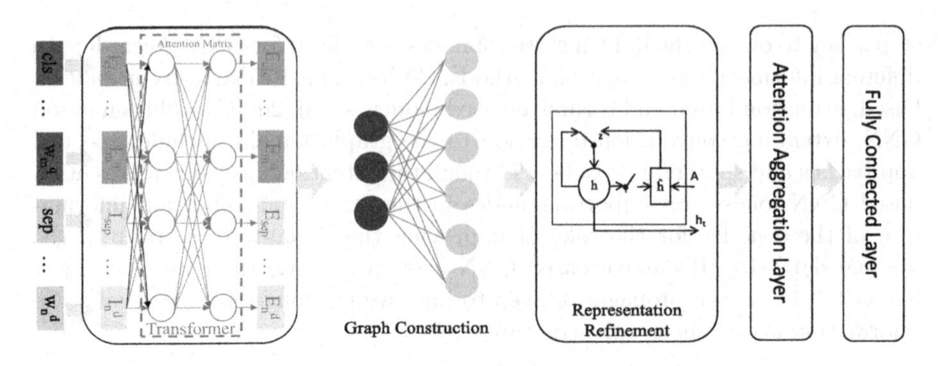

Fig. 1. Latent graph recurrent network architecture

example, $L = 12$ in BERT-base. For each word i at each layer $l = 1, \ldots, L$, its word representation $\mathbf{E}_l^{(q,d)}(i) \in \mathbb{R}^{d_k}$ is obtained by weighted summing representations of the other words in Eq. (2), d_k is the dimension of word representations.

$$\mathcal{A}_{l-1}^{(q,d)} = \text{softmax}(\frac{(\mathbf{W}_B\mathbf{E}_{l-1}^{(q,d)})(\mathbf{W}_B\mathbf{E}_{l-1}^{(q,d)})'}{\sqrt{d_k}}) \tag{1}$$

$$\mathbf{E}_l^{(q,d)}(i) = \mathbf{E}_{l-1}^{(q,d)}(i) + \sum_j \mathcal{A}_{l-1}^{(q,d)}(i,j)\mathbf{E}_{l-1}^{(q,d)}(j) \tag{2}$$

where $\mathcal{A}_{l-1}^{(q,d)}$ is the attention matrix learned in the $l-1$-th layer and $\mathbf{E}_0^{(q,d)} = \mathbf{I}^{(q,d)}$. Through this layer, we obtain L attention matrices $\{\mathcal{A}_l^{(q,d)}\}_{l=1}^L$ for the query-document pair (q, d). Each attention matrix $\mathcal{A}_l^{(q,d)}$ naturally models the query-document word interaction.

Bipartite-Core Word Graph Construction. These attention matrices also contain query-query and document-document word interaction, which are not obviously useful for query-document matching in the current ranking task. Particularly, some studies [2] show that these additional interactions may harm the retrieval performance. Here we propose to mask some relations to build a bipartite-core word graph to model the text matching between query and document. Intuitively, there are three different implementations. One extreme case is to keep all the word relations in the attention matrix without masking, which can be treated as the summation of query-document bipartite word adjacent matrix, full document word adjacent matrix and full query word adjacent matrix, referred to as *Full Word Graph.* The other extreme case is to keep only query-document bipartite word relations in the attention matrix and mask these other relations, namely *Query-Document Bipartite Word Graph.* In the middle, we keep some document neighbor word relations and query-document bipartite word relations in the attention matrix and obtain the summation of query-document bipartite word adjacent matrix and document neighbor word adjacent matrix, namely *Query-Document Bipartite and Neighbor Word Graph.* All three kinds of graphs are bipartite-core graphs. For each transformer layer l, we define a masking matrix $\mathcal{M}_l^{(q,d)}$ for each implementation, and the masked word adjacent matrix of the bipartite core graph is derived as Eq. (3), where ϵ is small enough.

$$\hat{\mathcal{A}}_l^{(q,d)} = \text{softmax}(\frac{(\mathbf{W}_A\mathbf{E}_l^{(q,d)})(\mathbf{W}_A\mathbf{E}_l^{(q,d)})'}{\sqrt{d_k}} + \epsilon(1 - \mathcal{M}_l^{(q,d)})) \tag{3}$$

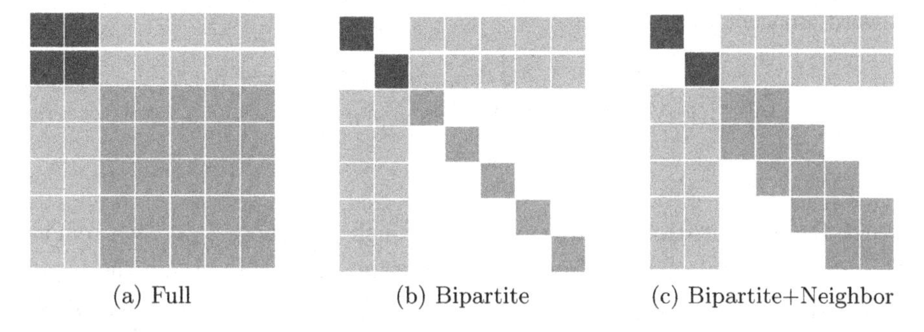

Fig. 2. Bipartite-core word graphs constructed from three strategies. Blue, green and grey color represent the word attention score between query and query, document and document, query and document separately. White means no word relation. (Color figure online)

- *Full Word Graph.* We keep all the word relations in $s^{(q,d)}$. No relations will be masked, so $\mathcal{M}_l^{(q,d)} = \mathbf{1}_{(m+n+3)\times(m+n+3)}$, where $\mathbf{1}_{(m+n+3)\times(m+n+3)}$ is a matrix with all elements 1. The masked attention matrix is shown in Fig. 2(a).
- *Query-Document Bipartite Word Graph.* In terms of query-document text matching, there are two types of words. We only keep all the relations between two types of words. As shown in Fig. 2(b), word relations within a query and a document are removed, and white means there are no edges between two corresponding nodes. The up-triangle masking matrix is obtained from Eq. (4) for $i \leq j$, and the down-triangle masking matrix is filled according the symmetry $\mathcal{M}_l^{(q,d)}(j,i) = \mathcal{M}_l^{(q,d)}(i,j)$. Thus the whole masking matrix $\mathcal{M}_l^{(q,d)}$ is applied in Eq. (3) to obtain the query-document bipartite word adjacent matrix.

$$\mathcal{M}_l^{(q,d)}(i,j) = \begin{cases} 1 & 1 \leq i \leq m, m+2 \leq j \leq m+n+2 \\ 1 & i = j \\ 0 & \text{otherwise} \end{cases} \tag{4}$$

- *Query-Document Bipartite and Neighbor Word Graph.* Word order information plays an important role in the search task, especially for the long document. Thus we take some local document word relations into consideration and keep its left and right r neighbors word relations. As shown in Fig. 2(c), the sliding window size is $2r+1$ over the document and nearly $2r+1$ diagonals are colored here. In bipartite attention matrix in Fig. 2(b), there is only one diagonal. The up-triangle masking matrix is set as Eq. (5) for $i \leq j$, and the down-triangle masking matrix is filled with $\mathcal{M}_l^{(q,d)}(j,i) = \mathcal{M}_l^{(q,d)}(i,j)$ according to its symmetric property. Thus the whole masking matrix $\mathcal{M}_l^{(q,d)}$ is applied in Eq. (3) to obtain the query-document bipartite and neighbor word adjacent matrix.

$$\mathcal{M}_l^{(q,d)}(i,j) = \begin{cases} 1 & 1 \leq i \leq m,\ m+2 \leq j \leq m+n+2 \\ 1 & m+2 \leq i \leq m+n+2, j = i, \ldots, i+r \\ 0 & \text{otherwise} \end{cases} \tag{5}$$

Word Representation Refinement. Through the above layer, we remove the unnecessary information from the word graph $\mathcal{A}_l^{(q,d)}$ and obtain the bipartite core graph $\hat{\mathcal{A}}_l^{(q,d)}$. Similarly, word representations $\mathbf{E}_l^{(q,d)}$ learned from BERT also need to be refined to separate the relevant information from these noisy relations. We use Gated Graph Neural Networks (GGNN) [11] to update word representations over the bipartite-core graph $\hat{\mathcal{A}}_l^{(q,d)}$. At each propagation step, GGNN aggregates neighbor word representations for each word in the graph $\hat{\mathcal{A}}_l^{(q,d)}$ and concatenates word representations from the last iteration and from neighborhood aggregation this iteration as the input embedding of Gated Recurrent Unit (GRU) in Eq. (7). This will help utilize high-order word relations to obtain fine-grained representations. Word attention matrix is computed according to Eq. (8). The query-document pair representation is aggregated as Eq. (9).

$$\mathbf{h}_0 = \mathbf{E}_l^{(q,d)} \tag{6}$$

$$\mathbf{h}_t = \text{GRU}([\mathbf{h}_{t-1}, \hat{\mathcal{A}}_l^{(q,d)}\mathbf{h}_{t-1}]) \tag{7}$$

$$\mathbf{h}_T^{att} = (\mathbf{W}_a\mathbf{h}_T) \cdot (\mathbf{W}_h\mathbf{h}_T)' \tag{8}$$

$$\mathbf{h}_l^{G_{q,d}} = [\text{sum}(\mathbf{h}_T^{att}) + \max(\mathbf{h}_T^{att}), \mathbf{E}_l^{(q,d)}(0)] \tag{9}$$

After T propagation steps, a final graph level representation for each query-document pair is learned denoted as $\mathbf{h}_l^{G_{q,d}}$ for each transformer layer l. Then it is fed into the last fully connected layer with weight matrix W_s to predict the relevance score $s_l(q,d)$ in Eq. (10). The final relevance score $f(q,d)$ is determined by the linear combination of all the relevance scores $\{s_l(q,d)\}_{l=1}^L$ in Eq. (11).

$$s_l(q,d) = \mathbf{W}_s\mathbf{h}_l^{G_{q,d}} + \mathbf{b}_s \tag{10}$$

$$f(q,d) = \mathbf{w}_f(s_l(q,d))_{1\times L} + b_f \tag{11}$$

3.3 Loss Function

To derive a robust ranking function, the pairwise ranking loss is usually used for optimization. Additionally, we introduce a metric learning task as an auxiliary task to learn discriminative representations.

From the embedding perspective, we propose a triangle distance loss to place constraints on query, document and query-document representations. Cosine distance [9] was first introduced to make examples with different labels separated from each other in the classification problem. Similarly treating query-document pair as an instance, we define the distance between query-document representations with different labels as this cosine distance, referred to as **pairwise cosine distance**. The pairwise cosine distance is computed for transformer and refinement layer respectively, whose query-document representations are $\mathbf{E}_L^{(q,d)}(0) = \mathbf{e}_L^{(q,d)}$ and $\mathbf{h}_L^{G_{q,d}}$ correspondingly. The distance summation of both layers is shown in Eq. (12). It only puts constraints on query-document representations in Fig. 3(b). We further split this unified query-document representation $\mathbf{E}_L^{(q,d)}$ into query representation $\mathbf{E}_L^q$ and document representation $\mathbf{E}_L^d$. Moreover, we define the **pointwise cosine distance** between a query and document representations in this ranking scenario as Eq. (13). This pointwise distance only puts constraints between a query and document representations without documents of different labels as Fig. 3(a). Neither pairwise nor pointwise distance will produce compact representations for query, document and query-document representations. So we propose a **triangle distance** to combine both pairwise and pointwise cosine distance as Eq. (14). As shown in Fig. 3(c), this triangle distance place constraints not only on the distance between a query and document representations but also on the distance between different documents.

$$\mathcal{C}_{\text{pair}}(q, D_q) = \frac{1}{2n_q^+ n_q^-} \sum_{\substack{d_+ \in D_q^+ \\ d_- \in D_q^-}} 2 + \cos(\mathbf{e}_L^{(q,d_+)}, \mathbf{e}_L^{(q,d_-)}) + \cos(\mathbf{h}_L^{G_{q,d_+}}, \mathbf{h}_L^{G_{q,d_-}}) \quad (12)$$

$$\mathcal{C}_{\text{point}}(q, D_q) = \frac{1}{n_q} \sum_{j=1}^{n_q} 1 + (1 - 2y_j) \cos(\mathbf{E}_L^q, \mathbf{E}_L^{d_j}) \quad (13)$$

$$\mathcal{L}_{\text{triangle}}(q, D_q) = \mathcal{C}_{\text{point}}(q, D_q) + \mathcal{C}_{\text{pair}}(q, D_q) \quad (14)$$

From the ranking perspective, we introduce a margin based pairwise ranking loss as follows

$$\mathcal{L}_{\text{rank}}(q, D_q) = \frac{1}{|D_q^+||D_q^-|} \sum_{d_+ \in D_q^+} \sum_{d_- \in D_q^-} \max(0, 1 - f(q, d_+) + f(q, d_-)). \quad (15)$$

We train both tasks in a multi-task learning framework with the optimization of $\lambda \mathcal{L}_{\text{triangle}}(q, D_q) + \mathcal{L}_{\text{rank}}(q, D_q)$.

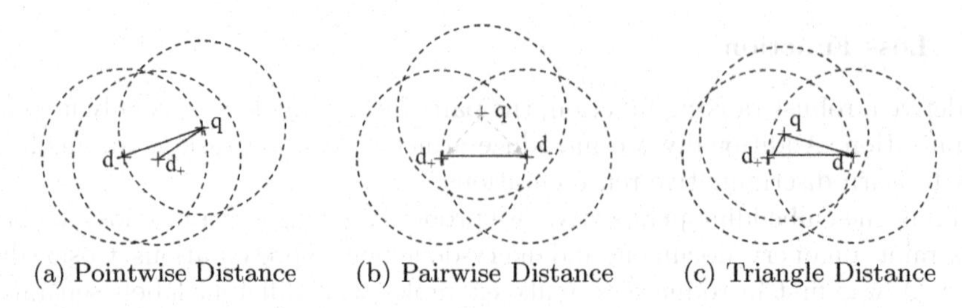

Fig. 3. Illustration of different constraints' effect on learned query/document representations

4 Experiments

We compare our proposed model LGRe with state-of-the-art baselines to investigate its effectiveness on two public benchmark datasets. Moreover, ablation studies for each component of LGRe are also explored.

4.1 Experimental Setting

Datasets. We use two TREC collections, Robust04 and WebTrack2009-12. Robust04 uses TREC discs 4 and 5[1], and WebTrack 2009-12 uses ClueWeb09b[2] as document collections. Note that the statistics are obtained only from the documents returned by BM25. Both data sets are white-space tokenized, lowercased, and stemmed using the Krovetz stemmer. Consistent with the baselines of the corresponding dataset, Robust04 uses Indri[3] for indexing, and WebTrack2009-12 uses Anserini [19] for indexing. Table 1 provides detailed information on these two data sets.

Table 1. Statistics of datasets.

	#Docs	Avg. Doc. Len	#Queries	Avg. Query Len	#Docs/Query
Robust04	37,500	428.2	250	3.62	150
WebTrack2009-12	19,590	1,393.0	200	2.64	100

Baselines. Three kinds of baselines are compared over these two datasets. (1) BM25: Candidate documents for each query are usually generated by BM25 in the first stage ranking. (2) Interaction based Neural Ranking Models (without BERT): DRMM [6] and ConvKNRM [4]. (3) BERT based Neural Ranking Models: Vanilla BERT, BERT-MaxP [2], CEDR-KNRM [14], and PARADE [10].

[1] 520k documents, https://trec.nist.gov/data_disks.html.
[2] 50M web pages, https://lemurproject.org/clueweb09/.
[3] http://www.lemurproject.org/indri.php.

Training Setting. For all BERT based baselines in our experiments, we make domain adaptation on MSMARCO.[4] Simple domain adaptation of BERT leads to a pre-trained model with both types of knowledge that can improve related search tasks where labelled data are limited [2]. Some performance results on Robust04 come from the paper aggregation site "Papers With Code".[5] Since WebTrack2009-12 does not have a unified data preprocessing pipeline similar to Robust04, we compare all baselines based on our data preprocessing pipeline.

Evaluation Setting. With the same division on both datasets, we use five fold cross validation with three folds for training, one fold for validation and one fold for test. The number of training epochs is 20 with batch size 32. The learning rate of BERT fine-tuning and LGRe is $1e-5$ and $5e-5$ respectively. λ is $1e-2$. All these hyperparameters are chosen according to performances in terms of the P@20 and nDCG@20 on the validation set, which are computed using script *trec_eval*.[6]

4.2 Effectiveness Analysis

The ranking performance of LGRe (bipartite and neighbor masking strategy + triangle distance) on both document ranking datasets is shown in Table 2. All the performances are averaged on five test sets for each dataset. Imp.% column in the table corresponds to the relative performance improvement of LGRe compared with each baseline. From Table 2, we observe the following phenomena.

Table 2. Ranking performance comparison among different models on Robust04 and WebTrack2009-12. Best results are in bold. The relative performance improvement is statistically significant with $p < 0.01$ in two-tailed paired t-test.

	Robust04				WebTrack2009-12			
Model	P@20	Imp.%	nDCG@20	Imp.%	P@20	Imp.%	nDCG@20	Imp.%
BM25	0.3123	53.38	0.4140	31.96	0.2805	27.95	0.1772	53.78
DRMM	0.2892	65.63	0.3040	79.70	0.3077	16.64	0.2015	35.24
Conv-KNRM	0.3408	40.55	0.3871	41.13	0.3155	13.76	0.213	27.93
Vanilla BERT	0.4042	18.51	0.4541	20.30	0.3253	10.32	0.254	7.28
BERT-MaxP	0.4277	11.99	0.4931	10.79	0.3373	6.40	0.2613	4.28
CEDR-KNRM	0.4667	2.64	0.5381	1.52	0.3481	3.10	0.2653	2.71
PARADE	0.4604	4.04	0.5399	1.19	–	–	–	–
LGRe	**0.479**	–	**0.5463**	–	**0.3589**	–	**0.2725**	–

[4] https://microsoft.github.io/TREC-2019-Deep-Learning.
[5] https://paperswithcode.com/sota/ad-hoc-information-retrieval-on-trec-robust04.
[6] https://trec.nist.gov/trec_eval.

(1) Compared with the best state-of-the-art baseline on each dataset, LGRe's relative performance gain is not less than 2% in terms of Precision@20. This improvement is statistically significant in the ranking task.
(2) Among all three kinds of baselines, BERT based ranking models achieve the best performance. One reason is that these interaction based ranking models without BERT usually derive the interaction matrix based on shallow pre-trained word embedding, such as word2vec [15]. These shallow word embedding only capture the local context, such as synonym, but cannot obtain complex or global patterns among words. This problem is solved by BERT with global word interactions, which makes it possible. The other reason is that interaction based ranking models like DRMM [6] predefine the query-document interaction matrix as input ignoring the query and document representation learning. All the interaction matrix, query and document representations are dynamically learned from data for BERT based ranking models. These learnable parameters make ranking models more flexible and suitable for different datasets.
(3) Compared with Vanilla BERT, LGRe's performance improvement agrees with our motivation that vanilla BERT has an inherent weakness though it naturally considers with the document ranking task. LGRe is mainly composed of BERT and word representation refinement process based on BERT. To a certain degree, LGRe's performance improvement also indicates the necessity of the following word refinement process in its architecture as Fig. 1.
(4) For all methods in Table 2 except DRMM [6], the ranking performance is higher on Robust04 than it on WebTrack2009-12. Dataset statistics show that the averaged query length is shorter and the averaged document number of each query is fewer on WebTrack2009-12. Fewer training instances may be one reason. So we will make a further study to verify the effect of query length on the ranking performance.

4.3 Ablation Study for Masking Strategy

Three candidate strategies for the bipartite-core word graph construction are compared: (1) Full Word Graph: denoted as LGRe(Full). (2) Query-Document Bipartite Word Graph: denoted as LGRe(Bipartite). (3) Query-Document Bipartite Core and Neighbor Word Graph: denoted as LGRe(Bipartite+Neighbor). Note that all the methods in Table 3 have the same setting except the masking strategy, such as adopting the pairwise ranking loss plus the triangle cosine distance loss as the loss function. Imp.% column means the relative performance improvement of each other method compared with LGRe(Full), i.e. no masking.

The primary comparison result in Table 3 is that masking some word relations in the attention matrix will bring about the performance gain. The relative performance gain is statistically significant, at least 0.5%. It indicates that some word relations, such as query-query and document-document, learned from BERT are noise for the query-document text matching problem. The masking

Table 3. Ranking performance comparisons with different masking strategies on Robust04

Model	P@20	Imp.%	nDCG@20	Imp.%
LGRe (Full)	0.471	–	0.5359	–
LGRe (Bipartite)	0.4764	1.15	**0.5447**	1.64
LGRe (Bipartite+Neighbor)	**0.4771**	1.3	0.5403	0.82

strategy for graph construction is essential for LGRe. Additionally, keeping document neighbor word relations does not always promote the ranking performance. The relative NDCG@20 decreases by 0.82% due to document neighbor word relations, although the relative P@20 increases by 0.15%. The introduction of document neighbor word relations makes adjacent word representations learned from the word graph much closer. This leads to a smaller distinction between relevant documents' representation, which are originally near to each other. That is why the addition of document neighbor word relation will increase the hit rate of relevant documents, and hurt the ranking of relevant documents.

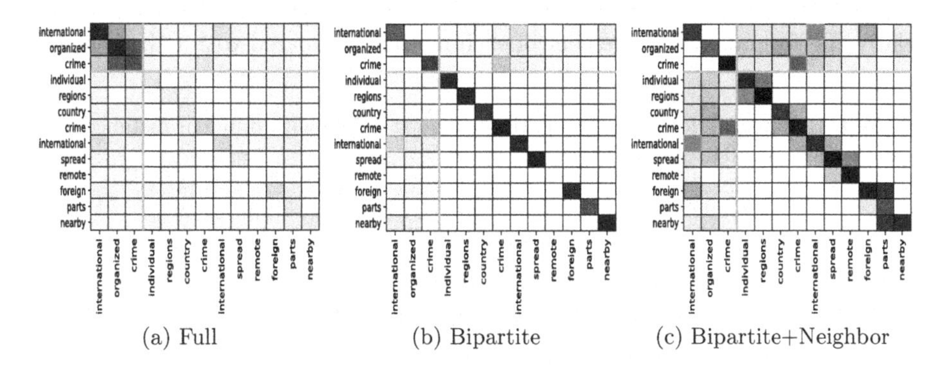

(a) Full (b) Bipartite (c) Bipartite+Neighbor

Fig. 4. Attention matrices learned from LGRe with different masking strategies. The green box represents exact term matching. The blue box represents synonym matching. The yellow line is the dividing line between query and document. (Color figure online)

For an intuitive understanding, we choose a specific query and document from Robust04. Query: "international, organized, crime". Document (stop words removed): "individual, regions, country, crime, international, spread, remote, foreign, parts, nearby". Attention matrices learned from different masking strategies are shown in Fig. 4. As we know, the meaning of short queries are vague, and forms of short queries are incomplete. For the full word graph in Fig. 4(a), the exact matching signals on "international" and "crime" is overwhelmed by many relations in documents. For the bipartite word graph in Fig. 4(b), the exact matching signals on 'international" and "crime" are obviously enhanced by masking query and document word relations. For Fig. 4(c), the addition of document neighbor word relations will promote the exact matching signals and strengthen the relation of semantically similar words, such as "international" and "foreign", "organized" and "country". Meanwhile, relating some unrelated words may become possible noise for the final ranking. This case study gives a better explanation of the performance gain of both LGRe (Bipartite) and LGRe (Bipartite+Neighbor) in Table 3.

4.4 Ablation Study for Distance Learning Task

We introduce the cosine distance learning task as the auxiliary task for document ranking in LGRe. Whether this task is an essential part will be studied here. If it is necessary, which distance definition among three kinds in the Loss function section is the best choice. We compare LGRe with different loss functions on Robust04: (1) LGRe+none: only the pairwise ranking loss without any distance loss. (2) LGRe+point: the linear combination of pairwise ranking loss and pointwise cosine distance loss. (3) LGRe+pair: the linear combination of pairwise ranking loss and pairwise cosine distance loss. (4) LGRe+triangle: the linear combination of pairwise ranking loss and triangle cosine distance loss. Experimental results are shown in Table 4. Imp.% column corresponds to the relative performance improvement of each method compared with LGRe+none.

Table 4. Ranking performance comparisons among LGRe with different distance definitions on Robust04

Model	P@20	Imp.%	nDCG@20	Imp.%
LGRe+none	0.4771	–	0.5403	–
LGRe+point	0.4769	−0.04	0.5419	0.30
LGRe+pair	0.4778	0.15	0.5427	0.44
LGRe+triangle	0.479	0.39	**0.5463**	1.11

In most cases, the auxiliary task, i.e. cosine distance learning task, plays a positive role in the document ranking problem in Table 4. The only exception is LGRe+point under the P@20 evaluation. Obviously, the relative performance gain for both LGRe+point and LGRe+pair is limited. However, the performance improvement from the combination of pointwise and pairwise cosine distance loss, i.e. triangle distance loss, is much higher than the summation of performance gains from pointwise and pairwise distance loss separately. This $1 + 1 > 2$ effect on ranking performances shows the advantage of triangle cosine distance loss. Whether the cosine distance loss will help learn discriminative and compact representations remains unknown. Thus, we analyze a specific query, and plot query and document representations through dimension reduction with t-sne[12] shown in Fig. 5.

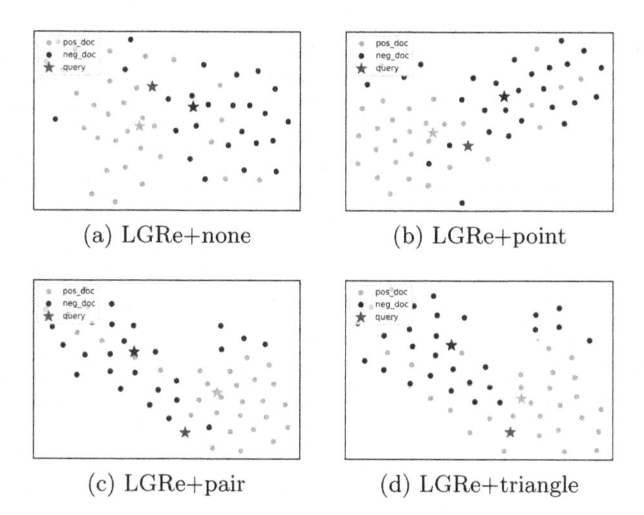

(a) LGRe+none (b) LGRe+point

(c) LGRe+pair (d) LGRe+triangle

Fig. 5. Query and document representations from LGRe with different losses. The pentagram means the mass center of each group.

Several results are obtained from Fig. 5. (1) (a) v.s. (b) and (c) and (d). The cosine distance learning task makes query, relevant and non-relevant document representations apart from each other. The reason lies that the embedding loss constrains representations directly, while the pairwise ranking loss takes indirectly effect on learned representations. (2) (b) v.s. (d). LGRe+point only defines a query and document point distance, and requires non-relevant document point far from and relevant document point near by the query point. This may lead to the problem in Fig. 5(b) that some relevant and non-relevant document points are mixed together. (3) (c) v.s. (d). LGRe+pair only defines a relevant and non-relevant document point distance, and requires non-relevant document points are far from relevant document points. This may lead to the problem in Fig. 5(c) that two kinds of distances from query to relevant and non-relevant document points respectively are not distinguishable. Generally, it is better to choose the triangle distance learning task as the auxiliary task to learn a discriminative representation for all the query, relevant and non-relevant documents.

4.5 Query Length Analysis

As mentioned before, one possible reason for the lower performance on WebTrack 2009-12 is shorter queries. To further explore the effect of query length on the ranking performance of BERT based ranking models, we conduct a group study on different query lengths. Robust04's queries are divided into two groups: one group with query length ≤ 3, the other group with query length > 3. The number of queries in two groups is 144 and 106 respectively. We randomly select 100 queries from each group, and randomly divide them into training, validation, and test set with a ratio of $8:1:1$. Performance comparisons on the test set with vanilla BERT and BM25 are shown in Table 5. Imp.% column represents the relative performance improvement of each other method compared with BM25.

Table 5. Ranking performance comparisons on two subsets of Robust04 with different query lengths.

	QLEN≤3				QLEN>3			
Model	P@20	Imp.%	nDCG@20	Imp.%	P@20	Imp.%	nDCG@20	Imp.%
BM25	0.3857	–	0.4689	–	0.425	–	0.4851	–
Vanilla BERT	0.3935	2.02	0.4729	0.85	0.4291	0.96	0.4876	0.52
LGRe	**0.4357**	12.96	**0.5058**	7.89	**0.44**	3.53	**0.493**	1.63

For all the methods, absolute performances on the shorter query subset are usually lower than these on the longer query subset. This suggests that document ranking for shorter queries is more difficult. Due to the concatenation of query and document pair as input, BERT models the global word interaction over the query-document text. This helps query words find their related words, which will alleviate the difficult short query problem to some degree. In this sense, both BERT based ranking models obtain higher performances gain on shorter queries than these on longer queries in Table 5. Due to the addition of the word representation refinement process, LGRe's relative performance improvement is much higher than vanilla BERT's. Compared with longer queries, the global word interaction learned from BERT is easier to generate a query-document representation submerging the query information. The refinement process of LGRe makes the query part emerge in the query-document representation.

5 Conclusion

To overcome the inherent weakness of BERT in the ranking task, we propose LGRe to refine word representations. We propose to mask the attention matrix from BERT to construct a bipartite-core word graph as the text matching between query and document. Then, word representations are updated through recurrent propagation steps to remove the useless information from original word embedding. Additional, triangle distance learning task is proposed to serve as the auxiliary task for document ranking.

Acknowledgement. This research work was funded by the National Natural Science Foundation of China under Grant No. 62072447.

References

1. Clark, K., Luong, M.T., Le, Q.V., Manning, C.D.: ELECTRA: pre-training text encoders as discriminators rather than generators. arXiv preprint arXiv:2003.10555 (2020)

2. Dai, Z., Callan, J.: Deeper text understanding for IR with contextual neural language modeling. In: Proceedings of the 42nd International ACM SIGIR, pp. 985–988 (2019)
3. Dai, Z., Callan, J.: Context-aware term weighting for first stage passage retrieval. In: Proceedings of the 43rd International ACM SIGIR, pp. 1533–1536 (2020)
4. Dai, Z., Xiong, C., Callan, J., Liu, Z.: Convolutional neural networks for soft-matching N-grams in ad-hoc search. In: Proceedings of the Eleventh ACM International Conference on Web Search and Data Mining, pp. 126–134 (2018)
5. Devlin, J., Chang, M.W., Lee, K., Toutanova, K.: BERT: pre-training of deep bidirectional transformers for language understanding. arXiv preprint arXiv:1810.04805 (2018)
6. Guo, J., Fan, Y., Ai, Q., Croft, W.B.: A deep relevance matching model for ad-hoc retrieval. In: Proceedings of the 25th ACM International on Conference on Information and Knowledge Management, pp. 55–64 (2016)
7. Guo, J., et al.: A deep look into neural ranking models for information retrieval. Inf. Process. Manage. **57**, 102067 (2019)
8. Hu, B., Lu, Z., Li, H., Chen, Q.: Convolutional neural network architectures for matching natural language sentences. In: Advances in Neural Information Processing Systems, pp. 2042–2050 (2014)
9. Li, Baoli, Han, Liping: Distance weighted cosine similarity measure for text classification. In: Yin, H., et al. (eds.) IDEAL 2013. LNCS, vol. 8206, pp. 611–618. Springer, Heidelberg (2013). https://doi.org/10.1007/978-3-642-41278-3_74
10. Li, C., Yates, A., MacAvaney, S., He, B., Sun, Y.: PARADE: passage representation aggregation for document reranking. arXiv preprint arXiv:2008.09093 (2020)
11. Li, Y., Tarlow, D., Brockschmidt, M., Zemel, R.: Gated graph sequence neural networks. arXiv preprint arXiv:1511.05493 (2015)
12. van der Maaten, L., Hinton, G.: Visualizing data using t-SNE. J. Mach. Learn. Res. **9**(Nov), 2579–2605 (2008)
13. MacAvaney, S., Nardini, F.M., Perego, R., Tonellotto, N., Goharian, N., Frieder, O.: Efficient document re-ranking for transformers by precomputing term representations. arXiv preprint arXiv:2004.14255 (2020)
14. MacAvaney, S., Yates, A., Cohan, A., Goharian, N.: CEDR: contextualized embeddings for document ranking. In: Proceedings of the 42nd International ACM SIGIR, pp. 1101–1104 (2019)
15. Mikolov, T., Chen, K., Corrado, G., Dean, J.: Efficient estimation of word representations in vector space. arXiv preprint arXiv:1301.3781 (2013)
16. Pang, L., Lan, Y., Guo, J., Xu, J., Wan, S., Cheng, X.: Text matching as image recognition. arXiv preprint arXiv:1602.06359 (2016)
17. Pennington, J., Socher, R., Manning, C.D.: GloVe: global vectors for word representation. In: Proceedings of the 2014 Conference on Empirical Methods in Natural Language Processing (EMNLP), pp. 1532–1543 (2014)
18. Veličković, P., Cucurull, G., Casanova, A., Romero, A., Lio, P., Bengio, Y.: Graph attention networks. arXiv preprint arXiv:1710.10903 (2017)
19. Yang, P., Fang, H., Lin, J.: Anserini: enabling the use of Lucene for information retrieval research. In: Proceedings of the 40th International ACM SIGIR, pp. 1253–1256 (2017)
20. Zhou, J., et al.: Graph neural networks: a review of methods and applications. arXiv preprint arXiv:1812.08434 (2018)

Discriminative Feature Adaptation via Conditional Mean Discrepancy for Cross-Domain Text Classification

Bo Zhang[1], Xiaoming Zhang[1(✉)], Yun Liu[2], and Lei Chen[3]

[1] School of Cyber Science and Technology, Beihang University, Beijing, China
{zhangnet,yolixs}@buaa.edu.cn
[2] State Key Laboratory of Software Development Environment, Beihang University, Beijing, China
[3] Shenzhen Research Institute of Big Data, Shenzhen, China

Abstract. This paper concerns the problem of Unsupervised Domain Adaptation (UDA) in text classification, aiming to transfer the knowledge from a source domain to a different but related target domain. Previous methods learn the discriminative feature of target domain in terms of noisy pseudo labels, which inevitably produces negative effects on training a robust model. In this paper, we propose a novel criterion Conditional Mean Discrepancy (CMD) to learn the discriminative features by matching the conditional distributions across domains. CMD embeds both the conditional distributions of source and target domains into tensor-product Hilbert space and computes Hilbert-Schmidt norm instead. We shed a new light on discriminative feature adaptation: the collective knowledge of discriminative features of different domains is naturally discovered by minimizing CMD. We propose Aligned Adaptation Networks (AAN) to learn the domain-invariant and discriminative features simultaneously based on Maximum Mean Discrepancy (MMD) and CMD. Meanwhile, to trade off between the marginal and conditional distributions, we further maximize both MMD and CMD criterions using adversarial strategy to make the features of AAN more discrepancy-invariant. To the best of our knowledge, this is the first work to definitely evaluate the shifts in the conditional distributions across domains. Experiments on cross-domain text classification demonstrate that AAN achieves better classification accuracy but less convergence time compared to the state-of-the-art deep methods.

Keywords: Unsupervised Domain Adaptation · Discriminative feature · Kernel method

1 Introduction

In practice, annotating sufficient training data is usually an expensive and time-consuming work for diverse application domains. *Unsupervised Domain Adaptation* (UDA) aims at solving this learning problem in the unlabeled *target*

C. S. Jensen et al. (Eds.): DASFAA 2021, LNCS 12682, pp. 104–119, 2021.
https://doi.org/10.1007/978-3-030-73197-7_7

domain by utilizing the abundant labeled data in an existing domain called *source* domain, even when these domains may have different distributions [20]. This technique has motivated research on cross-domain text classification where the knowledge in the source domain is transferred to the target domain.

One of the popular approaches of UDA discovers a domain-invariant representation by minimizing a distance metric of domain discrepancy, such as Maximum Mean Discrepancy (MMD) [24], or introducing adversarial learning [4,8]. However, this strategy fails to gain expected performance in the pipeline of Pre-trained Language Models (PLM) (e.g., BERT [6]) for cross-domain text classification. This is because the representations of BERT already display very domain-invariance characteristics compared to TextCNN-based model as shown in Fig. 1(a) and 1(b). PLMs have enjoyed tremendous success in Natural Language Processing (NLP) where transferable factors underlying different populations can be extracted efficiently. We observe that there is no explicit decision boundary to distinguish the two classes of the target data as shown in Fig. 1(b). To alleviate this problem, it is critical to explore class information to learn discriminative features of the target domain. Self-training [22,28,30] uses the classifier trained with source labeled data to generate pseudo labels for target domain. Thus, the high-confidence predictions are retained as the labels of unlabeled target data, over which the classifier is trained to capture the discriminative features. However, self-training learns discriminative features in terms of noisy hard labels generated by cluster algorithms [12] or predicted by classifiers [22,28], which inevitably produces negative effects on training a robust model.

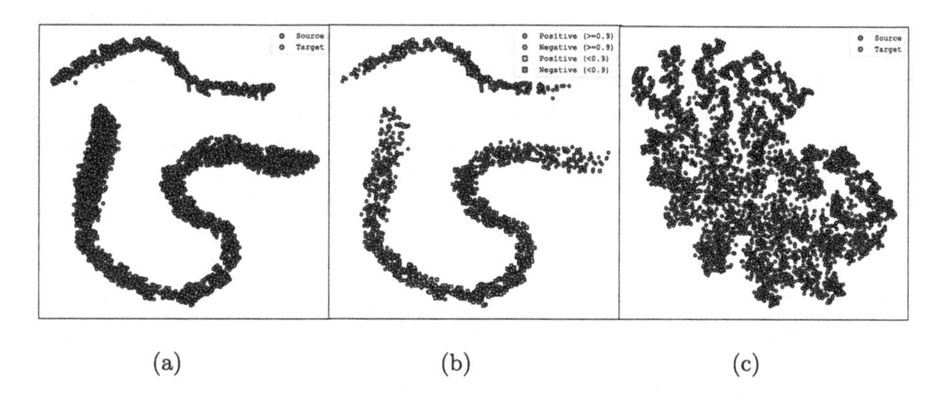

Fig. 1. Visualization of representations that are extracted from BERT-based and TextCNN-based [13] classifiers trained over the source data (using t-SNE [18]). Refer to Sect. 4 for experiment details. (a) BERT-based representations of source and target data; (b) BERT-based representations of target data w.r.t. 'Positive' and 'Negative' classes. The points are labeled with ○ and □ markers in terms of the predictions from the classifier. (c) TextCNN-based representations of source and target data.

In this paper, we adapt the discriminative feature to the target domain by matching conditional distributions across domains. The conditional distributions of different domains cover abundant class-related information of domains. The discriminative features are shared across different domains by minimizing the shifts in the conditional distributions. Recent works attempted to approximately reason about the discrepancy between conditional distributions in an implicit manner including constructing multiple conditional discriminators [21,29] and making an explicit hypothesis between conditional distributions [7]. However, the actual shifts in the conditional distributions still cannot be observed and evaluated intuitively. It is necessary to direct our attention on the way to measure how close the conditional distributions between source and target domains are, analogous to MMD and $\mathcal{A}$-distance for marginal distributions matching. To tackle the challenges, we define a conditional mean discrepancy (CMD) in tensor-product Hilbert space to measure the shifts in the conditional distributions explicitly. Compared to self-training, we shed a new light on discriminative feature adaptation: we discover the collective knowledge of discriminative features across different domains by minimizing CMD instead of leveraging noisy pseudo labels. Essentially, CMD dynamically decides whether the target data share the same labels with the source data according to a conditioning variable computed using all the data in the source and target domains.

In this paper, we present Aligned Adaptation Networks (AAN) to match both the marginal and conditional distributions across domains for UDA. Specifically, AAN is learnt by joint optimizing both MMD and CMD criterions across domains. CMD measures the Hilbert-Schmidt norm between Hilbert space embeddings of the conditional distributions of source and target data. Similar to MMD, a theorem is given to answer the question in what condition that CMD is equal to 0. We can draw a min-batch of samples to estimate the CMD criterion, and implement it efficiently via backpropagation. In addition, it is important to automatically balance the influence between marginal and conditional distributions [25]. We further maximize both two criterions using adversarial strategy to make the extracted features more *discrepancy-invariant*, i.e., AAN is insensitive to the weights between two criterions. We experiment under two Amazon review datasets for text classification with three backbones: fully connected networks, TextCNN and BERT models. We demonstrate that our model has obvious advantages of high accuracy, and stable and fast convergence in all settings.

2 Preliminary

2.1 Kernels and Hilbert Space Embedding

A popular approach to measure the similarity between structured objects is to use kernel methods. Let $\mathcal{X}$ be a non-empty set. A function $k : \mathcal{X} \times \mathcal{X} \mapsto \mathbb{R}$ is called a kernel if there exists a reproducing kernel Hilbert space (RKHS) $\mathscr{H}$ and a feature map $\phi : \mathcal{X} \mapsto \mathscr{H}$ such that $\forall \mathbf{x}, \mathbf{y} \in \mathcal{X}$,

$$k(\mathbf{x}, \mathbf{y}) \triangleq \langle \phi(\mathbf{x}), \phi(\mathbf{y}) \rangle_{\mathscr{H}}$$

where $\phi(\mathbf{x}) = k(\cdot, \mathbf{x}) \in \mathscr{H}$ is the explicit feature map of $\mathbf{x}$. In a RKHS $\mathscr{H}$, the kernel function $k(\cdot, \cdot)$ satisfies the reproducing property, i.e., $\langle f, \phi(\mathbf{x}) \rangle_{\mathscr{H}} = \langle f, k(\cdot, \mathbf{x}) \rangle_{\mathscr{H}} = f(\mathbf{x})$ for $\forall f \in \mathscr{H}$ and $\forall \mathbf{x} \in \mathcal{X}$.

The kernel embedding represents a probability distribution $P(\mathbf{x})$ by an element in RKHS endowed by a kernel k,

$$\mu_P \triangleq \mathbb{E}_P[\phi(\mathbf{x})] = \int_{\mathbf{x}\in\mathcal{X}} k(\cdot,\mathbf{x})\mathrm{d}P(\mathbf{x})$$

where the distribution is mapped to the expected feature maps. If $\mathbb{E}_P\sqrt{k(\mathbf{x},\mathbf{x}')} < \infty$, then μ_P exists and is an element in $\mathscr{H}$ [2]. Given a dataset $\mathcal{D}_{\mathbf{X}} = \{\mathbf{x}_i\}_{i=1}^n$ drawn from $P(\mathbf{x})$, the embedding μ_P can be estimated empirically using finite samples, i.e., $\hat{\mu}_P = \frac{1}{n}\sum_i \phi(\mathbf{x}_i)$. Maximum Mean Discrepancy (MMD) is a frequentist estimator for answering the question that two distributions are identical. Given distributions $P(\mathbf{x})$ and $Q(\mathbf{x})$, formally, MMD defines the following discrepancy, $M(P,Q) \triangleq \sup_{f\in\mathcal{F}} |E_P(f(\mathbf{x})) - E_Q(f(\mathbf{x}))| = \|\mu_P - \mu_Q\|_{\mathscr{H}}$ where $\mathcal{F}$ is a unit ball in $\mathscr{H}$. If $\mathscr{H}$ is a universal RKHS, then $M(P,Q) = 0$ if and only if $P = Q$ [2].

2.2 Hilbert Space Embedding of Conditional Distributions

Suppose $k : \mathcal{X}\times\mathcal{X} \mapsto \mathbb{R}$ and $l : \mathcal{Y}\times\mathcal{Y} \mapsto \mathbb{R}$ are the positive definite kernels with feature maps $\phi(\cdot)$ and $\psi(\cdot)$ for domains of X and Y, respectively that corresponds to RKHS $\mathscr{H}$ and $\mathscr{G}$. Let $\mathcal{U}_{Y|X} : \mathscr{H} \mapsto \mathscr{G}$ and $\mathcal{U}_{Y|\mathbf{x}} \in \mathscr{G}$ be the *conditional embeddings* of the conditional distributions $P(\mathbf{Y}|\mathbf{X})$ and $P(\mathbf{Y}|\mathbf{X} = \mathbf{x})$ respectively, and they satisfy two properties:

$$\mathcal{U}_{Y|\mathbf{x}} = \mathbb{E}_{Y|\mathbf{x}}[\psi(\mathbf{y})|X = \mathbf{x}] = \mathcal{U}_{Y|X}k(\mathbf{x},\cdot),$$

$$\mathbb{E}_{Y|\mathbf{x}}[g(\mathbf{y})|X = \mathbf{x}] = \langle g, \mathcal{U}_{Y|\mathbf{x}}\rangle_{\mathscr{G}}, \forall g \in \mathscr{G}.$$

The work [23] defines $\mathcal{U}_{Y|X}$ as $\mathcal{C}_{YX}\mathcal{C}_{XX}^{-1}$ that satisfies the above two properties simultaneously where $\mathcal{C}_{YX} : \mathscr{H} \mapsto \mathscr{G}$ is an uncentered cross covariance operator: $\mathcal{C}_{YX} = \mathbb{E}_{YX}[\psi(\mathbf{y})\otimes\phi(\mathbf{x})]$. Given a dataset $\mathcal{D} = \{(\mathbf{x}_i,\mathbf{y}_i)\}_{i=1}^n$ of size n drawn i.i.d. from $P(\mathbf{X},\mathbf{Y})$, the conditional embedding $\mathcal{U}_{Y|X}$ can be empirically estimated as

$$\hat{\mathcal{U}}_{Y|X} = \mathbf{\Psi}(\mathbf{\Phi}^{\mathrm{T}}\mathbf{\Phi} + \epsilon nI)^{-1}\mathbf{\Phi}^{\mathrm{T}}$$

where I is an identity matrix, $\mathbf{\Psi} = (\psi(\mathbf{y}_1),\cdots,\psi(\mathbf{y}_n))$, $\mathbf{\Phi} = (\phi(\mathbf{x}_1),\cdots,\phi(\mathbf{x}_n))$ and ϵ serves as regularization to hold the existence of $\mathcal{U}_{Y|X}$ in a continuous domain [19,23].

3 Proposed Model

In UDA, we are given a *source* domain $\mathcal{D}_s = \{(\mathbf{x}_i^s,\mathbf{y}_i^s)\}_{i=0}^m \subset \mathcal{X}\times\mathcal{Y}$ with m labeled samples, and a *target* domain $\mathcal{D}_{\mathbf{X}^t} = \{\mathbf{x}_j^t\}_{j=0}^n \subset \mathcal{X}$ with n unlabeled samples. Assume that the target domain shares the same label space with the source domain. The marginal distributions and conditional distributions of two domains are both different, i.e., $P(\mathbf{X}^s) \neq Q(\mathbf{X}^t)$ and $P(\mathbf{Y}^s|\mathbf{X}^s) \neq Q(\mathbf{Y}^t|\mathbf{X}^t)$.

The goal of this paper is to design a deep neural network $\mathbf{y} = f(\mathbf{x})$ to enable minimizing the target expected risk $R_t(f) = \mathbb{E}_{(x,y)\sim Q(\mathbf{X}^t,\mathbf{Y}^t)}[f(\mathbf{x}) \neq \mathbf{y}]$ by formally reducing the shifts in both the marginal and conditional distributions across domains simultaneously.

Formally, suppose $T : \mathcal{X} \mapsto \mathcal{Z}$ is a nonlinear operator (e.g., deep neural network) that transforms domain samples into a unified latent space $\mathcal{Z}$. Maximum Mean Discrepancy (MMD) as a kernel two-sample test statistic, has been widely applied to measure the discrepancy in marginal distributions $P(\mathbf{Z}^s)$ and $Q(\mathbf{Z}^t)$ over latent space $\mathcal{Z}$ [24]. Let $\phi : \mathcal{Z} \mapsto \mathscr{H}$ be feature map for space $\mathcal{Z}$ with corresponding RKHS $\mathscr{H}$. Let $\mathcal{D}_{\mathbf{X}^s} = \{\mathbf{x}_1^s, \cdots, \mathbf{x}_m^s\}$ and $\mathcal{D}_{\mathbf{X}^t} = \{\mathbf{x}_1^t, \cdots, \mathbf{x}_n^t\}$ be the sets of samples from distributions $P(\mathbf{X}^s)$ and $Q(\mathbf{X}^t)$. MMD and its empirical estimation over space $\mathcal{Z}$ are defined as:

$$
\begin{aligned}
M(P,Q) &\triangleq \|\mathbb{E}_P[\phi(T\mathbf{x})] - \mathbb{E}_Q[\phi(T\mathbf{x})]\|_{\mathscr{H}} \\
\hat{M}(P,Q) &\triangleq \left\| \frac{1}{m}\mathbf{\Phi}_s - \frac{1}{n}\mathbf{\Phi}_t \right\|_F
\end{aligned} \tag{1}
$$

where $\mathbf{\Phi}_s = (\phi(T\mathbf{x}_1^s), \cdots, \phi(T\mathbf{x}_m^s))$ and $\mathbf{\Phi}_t = (\phi(T\mathbf{x}_1^t), \cdots, \phi(T\mathbf{x}_n^t))$ are implicitly formed feature matrices, and $\|\cdot\|_F$ is the Frobenius norm. Using kernel tricks, we can estimate MMD only in term of kernel gram matrices,

$$
\hat{M}^2(P,Q) = \mathrm{tr}(\mathbf{KH}) \tag{2}
$$

$$
\mathbf{K} = \begin{bmatrix} \mathbf{K}_s & \mathbf{K}_{st} \\ \mathbf{K}_{ts} & \mathbf{K}_t \end{bmatrix}, \mathbf{H} = \begin{bmatrix} \frac{1}{m^2}\mathbf{1}_{m\times m} & \frac{-1}{mn}\mathbf{1}_{m\times n} \\ \frac{-1}{mn}\mathbf{1}_{n\times m} & \frac{1}{n^2}\mathbf{1}_{n\times n} \end{bmatrix}
$$

where $\mathbf{K}_s = \mathbf{\Phi}_s^T\mathbf{\Phi}_s$, $\mathbf{K}_t = \mathbf{\Phi}_t^T\mathbf{\Phi}_t$ and $\mathbf{K}_{ts} = \mathbf{\Phi}_t^T\mathbf{\Phi}_s$ are gram matrices, and $\mathbf{1}_{m\times n}$ is an $m \times n$ matrix with all elements equal to 1.

In this paper, our goal is to intuitively learn the discriminative features by reducing the shifts between two conditional distributions $P(\mathbf{Y}^s|\mathbf{Z}^s)$ and $Q(\mathbf{Y}^t|\mathbf{Z}^t)$ over latent space $\mathcal{Z}$. Long et al. [17] derive a criterion called Joint MMD (JMMD) to evaluate the discrepancy in distributions $P(\mathbf{Z}^{s1}, \cdots, \mathbf{Z}^{sN})$ and $Q(\mathbf{Z}^{t1}, \cdots, \mathbf{Z}^{tN})$ where $\mathbf{Z}^{s1}, \cdots, \mathbf{Z}^{sN}$ are activations of N task-specific layers in deep networks. They use JMMD to approximate the shifts in the joint distributions $P(\mathbf{X}^s, \mathbf{Y}^s)$ and $Q(\mathbf{X}^t, \mathbf{Y}^t)$. But, this method is heavily dependent on the specific architecture of deep networks and thus cannot apply to BERT-based networks directly. In addition, some shallow methods [16,25] use $P(\mathbf{Z}^t|\mathbf{Y}^t)$ to approximately reason about the conditional distribution $P(\mathbf{Y}^t|\mathbf{Z}^t)$. To date there is no explicit metric to evaluate the discrepancy between two conditional distributions.

Figure 2 shows the overview of our Aligned Adaptation Networks (AAN). First a new metric called conditional mean discrepancy (CMD) is proposed to learn the discriminative features over target domain by matching the conditional distributions, and then adversarial training is introduced to balance the influence between the marginal and conditional distributions, which is an important problem in domain adaptation [25].

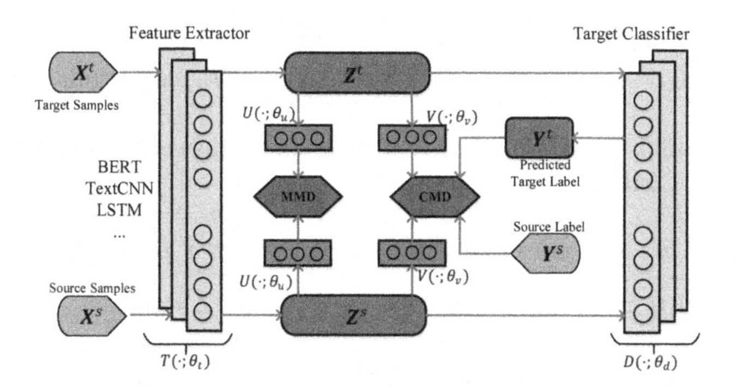

Fig. 2. The overview of Aligned Adaptation Networks (AAN). The green components U and V are introduced in the adversarial version of AAN (AAN-A).

3.1 Conditional Mean Discrepancy

Following the virtue of MMD, we use the Hilbert space embedding of conditional distributions to measure the discrepancy of two conditional distributions $P(\mathbf{Y}^s|\mathbf{Z}^s)$ and $Q(\mathbf{Y}^t|\mathbf{Z}^t)$. We call this metric Conditional Mean Discrepancy (CMD) defined as follows.

Definition 1. *Define a nonlinear operator* $T : \mathcal{X} \mapsto \mathcal{Z}$. *Suppose* $\phi(\cdot) : \mathcal{Z} \mapsto \mathscr{H}$ *is a feature map for domain of* $\mathcal{Z}$ *with corresponding RKHS's* $\mathscr{H}$. *We use parallel notation* $\psi(\cdot)$ *and* $\mathscr{G}$ *for domain of* $\mathcal{Y}$. *Let* $\mathcal{U}^s_{Y|Z} : \mathscr{H} \mapsto \mathscr{G}$ *and* $\mathcal{U}^t_{Y|Z} : \mathscr{H} \mapsto \mathscr{G}$ *be conditional mean embeddings of* $P(\mathbf{Y}^s|\mathbf{Z}^s)$ *and* $Q(\mathbf{Y}^t|\mathbf{Z}^t)$ *respectively. Let* $\mathcal{D}_s = \{(\mathbf{x}^s_i, \mathbf{y}^s_i)\}^m_{i=0} \subset \mathcal{X} \times \mathcal{Y}$ *and* $\mathcal{D}_t = \{(\mathbf{x}^t_j, \mathbf{y}^t_j)\}^n_{j=0} \subset \mathcal{X} \times \mathcal{Y}$ *be samples composed of i.i.d. observations obtained from* $P(\mathbf{X}^s, \mathbf{Y}^s)$ *and* $Q(\mathbf{X}^t, \mathbf{Y}^t)$ *respectively. We define Conditional Mean Discrepancy (CMD) over latent space* $\mathcal{Z}$ *and its average empirical estimator as*

$$C(P,Q) \triangleq \left\|\mathcal{U}^s_{Y|Z} - \mathcal{U}^t_{Y|Z}\right\|_{HS}$$

$$\hat{C}(P,Q) \triangleq \left\|\frac{1}{m}\mathbf{\Psi}_s(\mathbf{\Phi}_s^{\mathrm{T}}\mathbf{\Phi}_s + m\epsilon I_m)^{-1}\mathbf{\Phi}_s^{\mathrm{T}} - \frac{1}{n}\mathbf{\Psi}_t(\mathbf{\Phi}_t^{\mathrm{T}}\mathbf{\Phi}_t + n\epsilon I_n)^{-1}\mathbf{\Phi}_t^{\mathrm{T}}\right\|_F \quad (3)$$

where $\mathbf{\Phi}_s = (\phi(T\mathbf{x}^s_1), \cdots, \phi(T\mathbf{x}^s_m))$, $\mathbf{\Phi}_t = (\phi(T\mathbf{x}^t_1), \cdots, \phi(T\mathbf{x}^t_n))$, $\mathbf{\Psi}_s = (\psi(\mathbf{y}^s_1), \cdots, \psi(\mathbf{y}^s_m))$ *and* $\mathbf{\Psi}_t = (\psi(\mathbf{y}^t_1), \cdots, \psi(\mathbf{y}^t_n))$ *are implicitly formed feature gram matrices, and* $\|\cdot\|_{HS}$ *is the norm of Hilbert-Schmidt operator mapping from* $\mathscr{H}$ *to* $\mathscr{G}$.

Using kernel trick, the empirical estimation of $C(P,Q)$ is given by

$$\hat{C}^2(P,Q) = \mathrm{tr}(\mathbf{L}\tilde{\mathbf{K}}^{-1}\bar{\mathbf{K}}\tilde{\mathbf{K}}^{-1}) \quad (4)$$

$$\mathbf{L} = \begin{bmatrix}\mathbf{L}_s & \mathbf{L}_{st}\\ \mathbf{L}_{ts} & \mathbf{L}_t\end{bmatrix}, \tilde{\mathbf{K}} = \begin{bmatrix}m\tilde{\mathbf{K}}_s & \mathbf{0}\\ \mathbf{0} & n\tilde{\mathbf{K}}_t\end{bmatrix}, \bar{\mathbf{K}} = \begin{bmatrix}\mathbf{K}_s & -\mathbf{K}_{st}\\ -\mathbf{K}_{ts} & \mathbf{K}_t\end{bmatrix}$$

where $\mathbf{L}_s = \mathbf{\Psi}_s^T\mathbf{\Psi}_s$, $\mathbf{L}_t = \mathbf{\Psi}_t^T\mathbf{\Psi}_t$ and $\mathbf{L}_{st} = \mathbf{\Psi}_s^T\mathbf{\Psi}_t$ are the gram matrices for $\mathcal{Y}$ domain, and $\tilde{\mathbf{K}}_s = (\mathbf{K}_s + \epsilon m I_m)$ and $\tilde{\mathbf{K}}_t = (\mathbf{K}_t + \epsilon n I_n)$ are regularization gram matrices.

Further, similar to MMD, we give a theorem to answer the question in what condition that CMD is equal to 0.

Lemma 1. *If $\mathscr{G}$ is a universal RKHS, given a fixed point $\mathbf{z}_0 \in \mathcal{Z}$, $\left\|\mathcal{U}^s_{Y|\mathbf{z}_0} - \mathcal{U}^t_{Y|\mathbf{z}_0}\right\|_{\mathscr{G}} = 0$ if and only if $P(\mathbf{Y}^s|\mathbf{Z} = \mathbf{z}_0) = Q(\mathbf{Y}^t|\mathbf{Z} = \mathbf{z}_0)$.*

Theorem 1. *If $\mathscr{G}$ is a universal RKHS and T is bijective, $C(P, Q) = 0$ if and only if $P(\mathbf{Y}^s|\mathbf{X} = \mathbf{x}_0) = Q(\mathbf{Y}^t|\mathbf{X} = \mathbf{x}_0)$ for every fixed $\mathbf{x}_0$.*

Proof. **Necessity.** Since $C(P, Q) = 0$, for $\forall \mathbf{z}_0 \in \mathcal{Z}$, we have

$$\begin{aligned}\left\|\mathcal{U}^s_{Y|\mathbf{z}_0} - \mathcal{U}^t_{Y|\mathbf{z}_0}\right\|_{\mathscr{G}} &= \left\|\mathcal{U}^s_{Y|Z}k(\cdot, \mathbf{z}_0) - \mathcal{U}^t_{Y|Z}k(\cdot, \mathbf{z}_0)\right\|_{\mathscr{G}} \\ &\leq \left\|\mathcal{U}^s_{Y|Z} - \mathcal{U}^t_{Y|Z}\right\|_{\mathrm{HS}} \|k(\cdot, \mathbf{z}_0)\|_{\mathscr{G}} = C(P, Q)\, \|k(\cdot, \mathbf{z}_0)\|_{\mathscr{G}} = 0\end{aligned}$$

Since T is bijective, there exists a unique $\mathbf{x}_0 \in \mathcal{X}$ for $\forall \mathbf{z}_0 \in \mathcal{Z}$. According to Lemma 1, $P(\mathbf{Y}^s|\mathbf{X} = \mathbf{x}_0) = Q(\mathbf{Y}^t|\mathbf{X} = \mathbf{x}_0)$ holds for every fixed $\mathbf{x}_0$.
Sufficiency. Denote $\mathcal{U}_{Y|Z} = \mathcal{U}^s_{Y|Z} - \mathcal{U}^t_{Y|Z}$. It is obvious that $\mathcal{U}_{Y|Z} : \mathscr{H} \mapsto \mathscr{G}$ is a bounded linear operator and thus its Hilbert-Schmidt norm is defined as,

$$\|\mathcal{U}_{Y|Z}\|_{\mathrm{HS}} = \sup_{\|\mathbf{f}\|_{\mathscr{H}} < \infty} \frac{\|\mathcal{U}_{Y|Z}\mathbf{f}\|_{\mathscr{G}}}{\|\mathbf{f}\|_{\mathscr{H}}}, \forall \mathbf{f} \in \mathscr{H}$$

Let $\mathscr{H}_0 \triangleq span[\{k(\cdot, \mathbf{z})\}_{\mathbf{z}\in\mathcal{Z}}] \subseteq \mathscr{H}$ denote the pre-RKHS that is dense in $\mathscr{H}$ [1]. Each element $\mathbf{g}$ in $\mathscr{H}_0$ can be written as $\mathbf{g} = \sum_{\mathbf{z}\in\mathcal{Z}} \alpha_{\mathbf{z}} k(\cdot, \mathbf{z})$ where $(\alpha_{\mathbf{z}})_{\mathbf{z}\in\mathcal{Z}} \in \mathbb{R}^{|\mathcal{Z}|}$ is the coordinate of $\mathbf{g}$ in $\mathscr{H}_0$. Lemma 1 guarantees $\left\|\mathcal{U}^s_{Y|\mathbf{z}} - \mathcal{U}^t_{Y|\mathbf{z}}\right\| = 0$ for each $\mathbf{z} = T\mathbf{x} \in \mathcal{Z}$ under the condition $P(\mathbf{Y}^s|\mathbf{X} = \mathbf{x}) = Q(\mathbf{Y}^t|\mathbf{X} = \mathbf{x})$ for every fixed $\mathbf{x} \in \mathcal{X}$. Therefore, for each $\mathbf{g} \in \mathscr{H}_0$, the norm $\|\mathcal{U}_{Y|Z}\mathbf{g}\|_{\mathscr{G}}$ is equal to 0 according to

$$\left\|\mathcal{U}^s_{Y|Z}\mathbf{g} - \mathcal{U}^t_{Y|Z}\mathbf{g}\right\| = \left\|\sum_{\mathbf{z}\in\mathcal{Z}} \alpha_{\mathbf{z}}(\mathcal{U}^s_{Y|\mathbf{z}} - \mathcal{U}^t_{Y|\mathbf{z}})\right\| \leq \sum_{\mathbf{z}\in\mathcal{Z}} |\alpha_{\mathbf{z}}| \left\|\mathcal{U}^s_{Y|\mathbf{z}} - \mathcal{U}^t_{Y|\mathbf{z}}\right\| = 0$$

where the property $\mathcal{U}^s_{Y|Z}k(\cdot, \mathbf{z}) = \mathcal{U}^s_{Y|\mathbf{z}}$ is leveraged. Since $\mathscr{H}_0$ is dense in $\mathscr{H}$, for $\forall \mathbf{f} \in \mathscr{H}$, there exists a Cauchy sequence $\{\mathbf{g}_n\}$ in $\mathscr{H}_0$ converging to $\mathbf{f}$, i.e., $\lim_{n\to\infty} \mathbf{g}_n = \mathbf{f}$. We prove $\|\mathcal{U}_{Y|Z}\mathbf{f}\| = 0$ for any $\mathbf{f} \in \mathscr{H}$, because

$$\|\mathcal{U}_{Y|Z}\mathbf{f}\| = \left\|\lim_{n\to\infty} \mathcal{U}_{Y|Z}\mathbf{g}_n\right\| = \lim_{n\to\infty} \|\mathcal{U}_{Y|Z}\mathbf{g}_n\| = 0.$$

Therefore, $C(P, Q) = \|\mathcal{U}_{Y|Z}\|_{\mathrm{HS}} = 0$ holds.

Remark: Taking a close look on the objectives of MMD (Eq. (2)) and CMD (Eq. (4)), we can find some interesting connections. Intuitively, $\hat{M}^2(P,Q) = \sum_{i,j} \alpha_{ij} k(\mathbf{z}_i, \mathbf{z}_j)$ where α_{ij} is a real-valued weight including 3 fixed cases, i.e., $1/m^2$, $1/n^2$ and $-/mn$. Instead of applying uniform weights, CMD applies *non-uniform* weights β_{ij},

$$\hat{C}^2(P,Q) = \sum_{i,j} \beta_{ij} l(\mathbf{y}_i, \mathbf{y}_j) \tag{5}$$

where $\beta_{ij} = (\tilde{\mathbf{K}}^{-1}\bar{\mathbf{K}}\tilde{\mathbf{K}}^{-1})_{i,j}$ is, in turn, determined by the conditioning variables. These non-uniform weights reflect the effects of conditioning on Hilbert space embeddings. When the sample i comes from the target domain, CMD computes a dynamic weight β_{ij} to decide whether the sample i shares the same label with each sample j in the source domain. All previous deep transfer learning models have not addressed the issue of measurement of shifts in the conditional distributions $P(\mathbf{Y}^s|\mathbf{Z}^s)$ and $Q(\mathbf{Y}^t|\mathbf{Z}^t)$.

3.2 Aligned Adaptation Networks with Adversarial Learning

We propose an end-to-end Aligned Adaptation Network (AAN) with min-batch training to align both the marginal and conditional distributions across domains simultaneously. Specifically, the nonlinear operator $T(\cdot;\theta_T) : \mathcal{X} \mapsto \mathcal{Z}$ is modeled as a neural network to extract features. To match the marginal distributions over $\mathcal{Z}$, it is required to minimize MMD criterion with the parameters θ_T: $L_M(P,Q;\theta_T) = \hat{M}(P,Q)$. Meanwhile, the target classifier $D(\cdot;\theta_D) : \mathcal{Z} \mapsto \mathcal{Y}$ is constructed to predict labels for target samples. To align conditional distributions across domains, we minimize CMD criterion in terms of the datasets $\mathcal{D}_s = \{(\mathbf{x}_i^s, \mathbf{y}_i^s)\}$ and $\mathcal{D}_t = \{(\mathbf{x}_j^t, \hat{\mathbf{y}}_j^t = D(T(\mathbf{x}_j^t)))\}$ with the parameters θ_T, θ_D: $L_C(P,Q;\theta_T,\theta_D) = \hat{C}(P,Q)$. Overall, AAN aims at minimizing the source risk and the distribution discrepancies (MMD and CMD) simultaneously,

$$J(\mu) = \frac{1}{m}\sum_{i=1}^{m} L(D(T(\mathbf{x}_i^s)), \mathbf{y}_i^s; \theta_T, \theta_D) + \mu L_M(P,Q;\theta_T) + L_C(P,Q;\theta_T,\theta_D) \tag{6}$$

where $\sum_i L(\cdot,\cdot)$ is the source risk of source domain (e.g., cross-entropy loss) and μ is a hyperparameter to control the influence of two discrepancies. We can draw a min-batch of samples to estimate MMD and CMD criterions, and implement it efficiently via backpropagation. Note here that CMD supervises the predicted target label $\hat{\mathbf{y}}_j^t$ to move towards or away from source label $\mathbf{y}_i^s$ determined by non-uniform weights $\beta_{i,j}$ in Eq. (5) when the samples i and j comes from source and target domains respectively. Different from self-training methods [22,28,30], all the target data is used to learn the discriminative feature of target domain by substituting the artificial rules as dynamic weights. We explore the correlation between the features and target labels that can guide D to learn the unique patterns of the target domain in a unsupervised manner.

It is important to balance the influence between the marginal and conditional distributions adaptively [25]. The works [26,29] trade off the importance by evaluating an appropriate hyperparameter μ based on $\mathcal{A}$-distance. However, they may suffer from heavily evaluation offset of μ because the process of discrepancy balancing is separated from the transfer learning. To circumvent this issue, instead of directly searching for μ explicitly, we integrate the discrepancy balancing into transfer learning using adversarial training. Specifically, we multiple fully connected layers U parameterized by θ_U to MMD, i.e., $\mathbf{\Phi}_s = (\phi(U(T\mathbf{x}_1^s)), \cdots, \phi(U(T\mathbf{x}_m^s)))$ and $\mathbf{\Phi}_t = (\phi(U(T\mathbf{x}_1^t)), \cdots, \phi(U(T\mathbf{x}_n^t)))$ in Eq. (2) while another fully connected layers $V(\cdot; \theta_V)$ are multiplied to CMD in Eq. (4): $\mathbf{\Phi}_s = (\phi(V(T\mathbf{x}_1^s)), \cdots, \phi(V(T\mathbf{x}_m^s)))$ and $\mathbf{\Phi}_t = (\phi(V(T\mathbf{x}_1^t)), \cdots, \phi(V(T\mathbf{x}_n^t)))$. We maximize MMD and CMD with respect to these new parameters θ_U and θ_V to misalign two distribution discrepancies. Then, the objective $J(\mu)$ with parameters θ_T is minimized to obtain *discrepancy-invariant* features by confronting this misalignment such that AAN is insensitive to the weights between two criterions. Therefore, this leads to a new adversarial aligned adaptation network (AAN-A) as,

$$\min_{\theta_T, \theta_D} \max_{\theta_U, \theta_V} J(\mu = 1) \tag{7}$$

The adversarial version AAN-A is inspired by the work in [8,17], but differs in that we use MMD and CMD as discrepancy adversary to balance the importance of discrepancies adaptively, while [8,17] leverage adversarial training to distinguish different domains.

4 Experiments

We evaluate the Aligned Adaptation Networks with state of the art transfer learning and deep learning methods in the cross-domain text classification. Code is available at https://github.com/gregbuaa/aan_model.

4.1 Setup

Amazon-Review[1] is a benchmark dataset for domain adaptation in text classification task. Two versions of Amazon Review datasets are used to evaluate models. The work [3] provides a simplified Amazon-Review dataset (**Amazon-Feature**) comprising 3,996 samples with 400d feature vectors and 2 categories (positive and negative) collected from four distinct domains: *Books* (**B**), *DVD* (**D**), *Electronics* (**E**) and *Kitchen* (**K**). A larger dataset called **Amazon-Text** is also constructed from Amazon-Review with the same domains in Amazon-Feature to test the model performance for large-scale transfer learning. The review texts are divided into two categories according to user rating, i.e., positive (5 stars) and negative (1 star). There are 10,000 original review texts in each category and 20,000 texts in each domain. The notation $\mathbf{S} \rightarrow \mathbf{T}$ represents the transfer learning from the source domain $\mathbf{S}$ to target domain $\mathbf{T}$.

Baselines. For the bulk of experiments the following baselines are evaluated. The **Source-Only** model is trained only over source domain and tested

[1] http://jmcauley.ucsd.edu/data/amazon/.

over target-domain data while **Train-on-Target** model is trained and tested over target-domain data directly. We compare with conventional and state-of-the-art transfer learning and deep learning methods: Transfer Component Analysis (**TCA**) [20], Balanced Distribution Adaptation (**BDA**) [25], Geodesic Flow Kernel (**GFK**) [10], Deep Domain Confusion (**DDC**) [24], Domain Adversarial Neural Networks (**RevGrad**) [8] and Dynamic Adversarial Adaptation Network (**DAAN**) [29].

For Amazon-Feature dataset, the extractor $T(\cdot;\theta_T)$ is simply modeled as a typical 2-layer fully connected network (**MLP**) to transform 400 dimensional inputs into 50 dimensional latent feature vectors. Two types of networks are leveraged for Amazon-Text dataset to extract the latent features from original texts, i.e., **TextCNN** and **BertGRU**. TextCNN is a text conventional network proposed in [13] that consists of 150 conventional filters with 3 different window sizes. We also evaluate the performance of transfer learning on a pre-training language model, i.e., BERT [6]. We freeze BERT model and construct a 2-layer bi-directional GRU [5] to learn from the representations produced by BERT. A 2-layer fully connected network is leveraged to model the target classifier $D(\cdot,\theta_d)$ for all the datasets. For AAN-A, U and V are modeled as weight matrices. For MMD-based methods (e.g., TCA, BDA, GFK and DDC) and AAN, we adopt Gaussian kernel with bandwidth set to median pairwise squared distances on the training data [11]. DDC model that we construct in the experiment is same with our AAN model except for $\mu = 0.0$ in Eq. (6).

We implement all deep methods based on Pytorch framework, and BERT model is implemented and pre-trained by *pytorch-transformers*[2] [27]. We fix $\mu = 0.1$ for AAN. AAN is trained end-to-end using Adam optimizer [14] with batch size of 128 and learning rate of 0.001. Since easy to optimize, AANs are not required to adjust the learning rate dynamically or pre-train over source domain as the works [8,24,29] do. Classification accuracy is used as the evaluation metric.

4.2 Results

The classification accuracy results on the *Amazon-Feature* dataset for domain adaptation based on MLP are shown in Table 1, in which AAN and AAN-A outperform other comparison methods on most transfer tasks. Some of the observations and analysis are listed as follows. **(1)** The performance of traditional shallow transfer learning models (e.g., TCA, GFK and BDA) is poor and even worse than Source-Only model, i.e., negative transfer learning occurs in all transfer tasks for traditional models. The traditional models directly define kernel over original input vectors which are highly sparse in Amazon-Feature that the kernel function cannot capture sufficient features to measure the similarity. **(2)** Deep transfer learning models (e.g., DDC, RevGrad, DAAN and proposed AANs) substantially outperform both Source-Only model and traditional transfer learning models. It verifies the positive effect of embedding domain adaptation module into deep networks to reduce domain discrepancy. **(3)** AAN, DDC and RevGrad

[2] https://github.com/huggingface/transformers.

Table 1. Classification accuracy (%) on *Amazon-Feature* dataset using MLP Extractor.

Model	D→B	E→B	K→B	B→D	E→D	D→E	B→K	E→K	Avg
Source only	73.5	71.1	68.5	79.9	69.3	75.3	75.8	81.0	74.3
Train on target	81.7	81.7	81.7	82.3	82.3	85.5	85.8	85.8	83.4
TCA [20]	62.2	59.5	64.0	62.4	62.7	66.3	65.1	73.8	64.5
BDA [25]	62.7	58.7	62.5	64.3	62.1	67.0	63.4	74.5	64.4
GFK [10]	66.5	63.0	65.5	66.3	63.4	64.0	69.2	73.3	66.4
DDC [24]	77.7	74.8	73.1	79.6	77.8	**80.3**	78.5	83.5	78.2
RevGrad [8]	76.9	74.7	74.7	80.2	76.1	79.4	79.3	84.1	78.2
DAAN [29]	78.4	70.9	68.5	77.0	75.5	77.3	78.7	84.0	76.3
AAN (our)	77.3	74.9	73.7	**80.7**	75.5	78.6	**81.8**	**85.8**	78.5
AAN-A (our)	**78.7**	**75.9**	**74.8**	80.1	**78.4**	79.1	81.2	85.2	**79.2**

get the similar performance over Amazon-Feature while the accuracy of AAN-A is slightly 1.0% higher than DDC and RevGrad overall.

Table 2 shows the classification performance of deep transfer learning models based on TextCNN and BertGRU over a larger dataset *Amazon-Text*. For TextCNN extractor, we have the following analysis. **(1)** AAN and AAN-A achieve superior performance over previous methods by larger margins compared to small dataset Amazon-Feature. In addition to obtain domain-invariant features by reducing marginal discrepancy as DDC and RevGrad do, AAN also adapts the discriminative feature to the target domain via CMD criterion. We fully correct the shifts in the marginal and conditional distributions across domain. DDC, RevGrad and DAAN only focus on learning a shared representation and generalize a classifier over source domain to target domain. But, our AAN sheds a new light on optimizing the target classifier $D(\cdot;\theta_D)$ by supervising target labels explicitly using CMD criterion. This direct supervision also leads to obvious advantages of stable and fast convergence particularly in large dataset (refers to the next subsection). DAAN is required to construct multiple binary discriminators to distinguish class features where target samples are fed into all discriminators with the weights computed by the target classifier. It relies on the accuracy of target classifier where the model may be driven to instability especially during early training process. **(2)** Exception occurs when AAN is performed in task $\mathbf{D} \rightarrow \mathbf{E}$. The cause of exception is mainly because that the $\mu = 0.1$ is a bad hyperparameter to balance influence between two discrepancies. However, AAN-A substantially has a nice performance for all transfer tasks, which shows that adversarial strategy enables AAN-A to trade off between MMD and CMD adaptively.

By going from TextCNN to extremely deep BertGRU, we attain a more in-depth understanding of feature transferability. **(1)** BertGRU-based methods outperform TextCNN-based methods by large margins. This suggests that the

Table 2. Classification accuracy (%) on *Amazon-Text* dataset using TextCNN and BertGRU Extractors.

Model	E→B	K→B	B→D	E→D	K→D	B→E	D→E	D→K	Avg
Using *TextCNN* as extractor									
Source only	68.7	69.7	81.2	75.8	70.3	68.7	62.8	64.9	70.3
Train on target	83.7	83.7	89.1	89.1	89.1	85.4	85.4	85.5	86.4
DDC [24]	69.6	69.9	82.0	76.8	76.5	72.5	70.2	63.4	72.6
RevGrad [8]	71.7	72.0	81.9	78.5	68.8	70.2	69.2	69.4	72.7
DAAN [29]	73.3	65.1	83.0	76.1	73.1	73.5	67.9	68.1	72.5
AAN (our)	**74.5**	**73.3**	84.7	79.8	78.8	74.7	65.0	74.8	75.7
AAN-A (our)	73.1	70.5	**84.8**	**80.0**	**79.8**	**75.6**	**74.6**	**75.4**	**76.7**
Using *BertGRU* as extractor									
Source only	86.4	87.2	91.6	89.2	90.2	87.8	84.2	85.2	87.7
Train on target	93.2	93.2	94.9	94.9	94.9	92.6	92.6	94.4	93.8
DDC [24]	87.8	86.6	92.2	91.2	90.9	87.3	87.0	87.4	88.8
RevGrad [8]	87.5	83.7	92.7	90.5	88.2	85.0	87.2	86.6	87.7
DAAN [29]	88.7	85.7	92.0	89.8	90.4	85.5	86.6	88.8	88.4
AAN (our)	89.1	88.4	92.9	90.5	91.8	88.0	88.6	88.6	89.7
AAN-A (our)	**90.5**	**90.7**	**93.5**	**91.9**	**92.4**	**89.1**	**88.3**	**89.9**	**90.8**

pre-trained language (PLM) models, e.g., BERT, not only learn better representations for general natural language task but also learn more transferable representations for domain adaptation. PLMs not only adapt in fine-tuning task with small labeled dataset but only adapts in domain adaptation task without target labels. **(2)** The accuracy of DDC, RevGrad and DAAN are slightly higher than Bert-based Source-Only method. PLM models can be easily transferred to other domain by fine tuning over small labeled datasets and achieve a nice performance. The features directly generated from BERT already have very domain-invariance characteristics. It is useless to only reduce the marginal distribution discrepancy for DDC and RevGrad. **(3)** The gap between our AAN-A and BertGRU-based Train-on-Target model narrows to 3%, indicating that our AAN can learn the discriminative features across domains via CMD efficiently.

4.3 Analysis

Some other important analysis results are shown in Fig. 3.

Distribution Discrepancy. Figure 3(a) shows MMD and CMD values on task **E→D** and **B→K** with features of AAN-A, DDC and RevGrad extracted from Amazon-Feature dataset. We observe that the sum of MMD and CMD using AAN-A is much smaller than that using DDC and RevGrad, which validates that AAN successfully reduces the shifts in marginal and conditional distributions to learn more transferable representations. Furthermore, MMD roughly matches

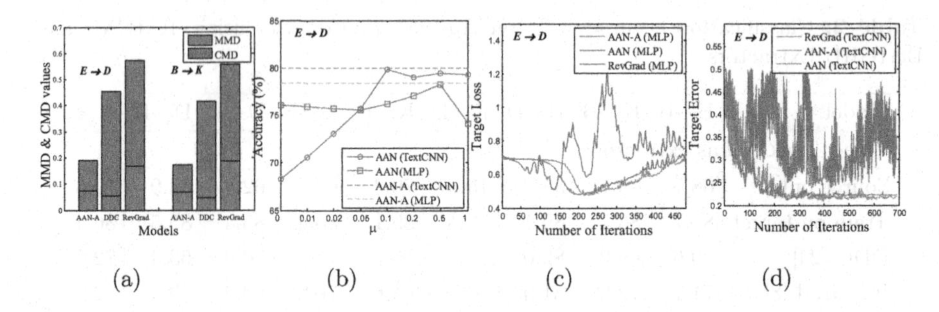

Fig. 3. Analysis. (a) MMD & CMD values on *Amazon-Feature*; (b) Parameter sensitivity of μ for AAN; (c) Target Loss w.r.t. the number of Iterations on *Amazon-Feature*; (d) Target Error w.r.t. the number of Iterations on *Amazon-Text* with TextCNN.

CMD for AAN-A while other methods have a great gap between MMD and CMD. It reveals the necessity of integrating adversarial training into AAN to balance the discrepancies adaptively.

Discrepancy Balancing. We check the sensitivity of AAN hyperparameter μ, i.e., an important factor to trade off the marginal and conditional distributions. Figure 3(b) shows the classification accuracy of AAN based on MLP and TextCNN respectively by varying $\mu \in \{0.0, 0.01, 0.02, 0.05, 0.1, 0.2, 0.5, 1.0\}$ for transfer task $\mathbf{E} \rightarrow \mathbf{D}$. The accuracy of AAN first increases and then decrease as μ varies. The result reveals two interesting observations. (1) AAN is very sensitive to μ for some transfer tasks where the accuracy may drop more than 10.0% when setting a worse μ. For different transfer tasks, AAN is required to select different μ to balance the influence between discrepancies. (2) AAN-A has the similar performance with AAN equipped with the best μ. It implies that AAN-A can trade off the marginal and conditional distribution discrepancies adaptively by introducing adversarial learning. This result further certifies that the adversarial learning enables AAN to obtain discrepancy-invariant features and thus enhance transferability without searching for an appropriate μ.

Convergence Performance. We testify convergence performance of AAN-A compared with RevGrad as they involve two different types of adversarial training procedures. Figure 3(c) and 3(d) show the target loss (i.e., cross-entropy loss on target domain) and target error on task $\mathbf{E} \rightarrow \mathbf{D}$ where the extractor T is set to MLP and TextCNN on Amazon-Feature and Amazon-Text respectively. The adversarial methods (AAN-A and RevGrad) converge slower than AAN. However, compared to RevGrad, AAN-A shows obvious advantages of stable and fast convergence and high accuracy. AAN-A and AAN have the similar convergence speed for small dataset (Amazon-Feature) and a large dataset (Amazon-Text) while RevGrad is required to train more iterations for larger datasets which is unfriendly to large-scale transfer learning task.

5 Related Work

Unsupervised Domain Adaptation (UDA) aims at learning a model which can generalize across different domains following different probability distributions [20]. Early works [10,20,25,26] adopt the shallow features to adapt the domain adaptation. They usually have three independent subtasks: 1) projecting original complex data (e.g., images and text) into general representations, 2) learning kernel parameters by minimizing defined domain discrepancy and 3) training a classifier (e.g., SVM) to predict the target labels. These methods cannot learn the domain-specific representations from complex data with the shallow extractors.

Domain adaptation methods equipped with powerful deep networks can remarkably boost transfer performance. Existing works mainly focus on how to learn domain-invariant features and discriminative features that are shared across different domains. Some methods correct the shifts in marginal distributions to obtain domain-invariant features. For examples, some works propose to learn the transferable features with deep networks by minimizing a distance metric of domain discrepancy [24], such as Maximum Mean Discrepancy (MMD), or maximizing the domain discriminator to make the features domain-invariant [8]. Unfortunately, only aligning marginal distributions may fail to gain expected performance particularly in the pipeline of PLM because the features of PLM already have very domain-invariance characteristics.

To learn discriminative features for UDA, self-training methods [9,22,28,30] are widely explored. To enhance the confidence level of pseudo labels, these works committed to improve the methods of pseudo-label selections including introducing mutual learning [9] and dual information maximization [28]. However, plenty of target data with low-confidence predictions are not utilized effectively, which also contribute to discriminative features equally. The other line of learning discriminative features is to match the conditional distributions across domains, which is known as a difficult problem. The work [7] introduces perturbation ϵ to the conditional distributions, i.e., $Q(\mathbf{Y}^t|\mathbf{X}^t) = P(\mathbf{Y}^s|\mathbf{X}^s) + \epsilon$ to increase flexibility. The works [15,17] learn a transfer network by aligning the joint distributions of multiple domain-specific layers across domains. DAAN [29] learns multiple binary classifiers to preserve class-related features to align conditional distributions. Our work mainly differs from these works by evaluating the discrepancy of conditional distributions explicitly in Hilbert embedding space and trading off the importance of marginal and conditional distributions without any hyperparameters.

6 Conclusion

This paper proposed a novel method of deep transfer learning, which enables end-to-end learning of transferable representations. We defined a new metric called Conditional Mean Discrepancy (CMD) to learn the discriminative features of the target domain. To the best of our knowledge, this is the first work to definitely evaluate the shifts in the conditional distributions. Based on MMD and

CMD criterions, Aligned Adaptation Network (AAN) was proposed to match both marginal and conditional distributions between different domains simultaneously. We also introduced the adversarial version of AAN to trade off these two criterions adaptively. Experiments testified the efficacy of the proposed approach in cross-domain text classification.

Acknowledgements. This work was supported in part by Fund of the State Key Laboratory of Software Development Environment and in part by the Open Research Fund from Shenzhen Research Institute of Big Data (No. 2019ORF01012).

References

1. Berlinet, A., Thomas-Agnan, C.: Reproducing Kernel Hilbert Spaces in Probability and Statistics. Springer, Dordrecht (2011). https://doi.org/10.1007/978-1-4419-9096-9
2. Borgwardt, K.M., Gretton, A., Rasch, M.J., Kriegel, H., Scholkopf, B., Smola, A.J.: Integrating structured biological data by kernel maximum mean discrepancy. Bioinformatics **22**(14), 49–57 (2006)
3. Chen, M., Xu, Z., Weinberger, K.Q., Sha, F.: Marginalized denoising autoencoders for domain adaptation. In: Proceedings of the 29th International Conference on International Conference on Machine Learning, pp. 1627–1634 (2012)
4. Chen, X., Sun, Y., Athiwaratkun, B., Cardie, C., Weinberger, K.: Adversarial deep averaging networks for cross-lingual sentiment classification. Trans. Assoc. Comput. Linguist. **6**, 557–570 (2018)
5. Cho, K., et al.: Learning phrase representations using rnn encoder-decoder for statistical machine translation. In: Proceedings of the 2014 Conference on Empirical Methods in Natural Language Processing (EMNLP), pp. 1724–1734 (2014)
6. Devlin, J., Chang, M., Lee, K., Toutanova, K.: BERT: pre-training of deep bidirectional transformers for language understanding. In: NAACL-HLT 2019, pp. 4171–4186. Association for Computational Linguistics (2019)
7. Fang, X., Bai, H., Guo, Z., Shen, B., Hoi, S., Xu, Z.: DART: domain-adversarial residual-transfer networks for unsupervised cross-domain image classification. Neural Netw. **127**, 182–192 (2020)
8. Ganin, Y., Lempitsky, V.: Unsupervised domain adaptation by backpropagation. In: International Conference on Machine Learning, pp. 1180–1189 (2015)
9. Ge, Y., Chen, D., Li, H.: Mutual mean-teaching: pseudo label refinery for unsupervised domain adaptation on person re-identification. In: 8th International Conference on Learning Representations, ICLR 2020, Addis Ababa, Ethiopia, 26–30 April 2020 (2020)
10. Gong, B., Shi, Y., Sha, F., Grauman, K.: Geodesic flow kernel for unsupervised domain adaptation. In: 2012 IEEE Conference on Computer Vision and Pattern Recognition, pp. 2066–2073. IEEE (2012)
11. Gretton, A., Borgwardt, K.M., Rasch, M.J., Schölkopf, B., Smola, A.: A kernel two-sample test. J. Mach. Learn. Res. **13**, 723–773 (2012)
12. Hermans, A., Beyer, L., Leibe, B.: In defense of the triplet loss for person re-identification. arXiv preprint arXiv:1703.07737 (2017)
13. Kim, Y.: Convolutional neural networks for sentence classification. In: EMNLP 2014, pp. 1746–1751. ACL (2014)

14. Kingma, D.P., Ba, J.: Adam: a method for stochastic optimization. arXiv e-prints arXiv:1412.6980 (December 2014)
15. Long, M., Cao, Z., Wang, J., Jordan, M.I.: Conditional adversarial domain adaptation. In: Advances in Neural Information Processing Systems, pp. 1640–1650 (2018)
16. Long, M., Wang, J., Ding, G., Sun, J., Yu, P.S.: Transfer feature learning with joint distribution adaptation. In: Proceedings of the IEEE International Conference on Computer Vision, pp. 2200–2207 (2013)
17. Long, M., Zhu, H., Wang, J., Jordan, M.I.: Deep transfer learning with joint adaptation networks. In: International Conference on Machine Learning, pp. 2208–2217 (2017)
18. Maaten, L., Hinton, G.: Visualizing data using t-SNE. J. Mach. Learn. Res. **9**, 2579–2605 (2008)
19. Muandet, K., Fukumizu, K., Sriperumbudur, B., Schölkopf, B., et al.: Kernel mean embedding of distributions: a review and beyond. Found. Trends Mach. Learn. **10**(1–2), 1–141 (2017)
20. Pan, S.J., Tsang, I.W., Kwok, J.T., Yang, Q.: Domain adaptation via transfer component analysis. IEEE Trans. Neural Netw. **22**(2), 199–210 (2010)
21. Pei, Z., Cao, Z., Long, M., Wang, J.: Multi-adversarial domain adaptation. In: 32nd AAAI Conference on Artificial Intelligence (2018)
22. Saito, K., Ushiku, Y., Harada, T.: Asymmetric tri-training for unsupervised domain adaptation. In: International Conference on Machine Learning, pp. 2988–2997 (2017)
23. Song, L., Huang, J., Smola, A., Fukumizu, K.: Hilbert space embeddings of conditional distributions with applications to dynamical systems. In: Proceedings of the 26th Annual International Conference on Machine Learning, pp. 961–968 (2009)
24. Tzeng, E., Hoffman, J., Zhang, N., Saenko, K., Darrell, T.: Deep domain confusion: maximizing for domain invariance. arXiv preprint arXiv:1412.3474 (2014)
25. Wang, J., Chen, Y., Hao, S., Feng, W., Shen, Z.: Balanced distribution adaptation for transfer learning. In: 2017 IEEE International Conference on Data Mining (ICDM), pp. 1129–1134. IEEE (2017)
26. Wang, J., Feng, W., Chen, Y., Yu, H., Huang, M., Yu, P.S.: Visual domain adaptation with manifold embedded distribution alignment. In: Proceedings of the 26th ACM International Conference on Multimedia, pp. 402–410 (2018)
27. Wolf, T., et al.: HuggingFace's transformers: state-of-the-art natural language processing. arXiv arXiv:abs/1910.03771 (2019)
28. Ye, H., Tan, Q., He, R., Li, J., Ng, H.T., Bing, L.: Feature adaptation of pre-trained language models across languages and domains for text classification. In: Empirical Methods in Natural Language Processing (2020)
29. Yu, C., Wang, J., Chen, Y., Huang, M.: Transfer learning with dynamic adversarial adaptation network. In: International Conference on Data Mining, pp. 778–786. IEEE (2019)
30. Zou, Y., Yu, Z., Liu, X., Kumar, B., Wang, J.: Confidence regularized self-training. In: Proceedings of the IEEE International Conference on Computer Vision, pp. 5982–5991 (2019)

Discovering Protagonist of Sentiment with Aspect Reconstructed Capsule Network

Chi Xu[1,3], Hao Feng[4], Guoxin Yu[1,2], Min Yang[5], Xiting Wang[6], Yan Song[7], and Xiang Ao[1,2(✉)]

[1] Key Lab of Intelligent Information Processing of Chinese Academy of Sciences (CAS), Institute of Computing Technology, CAS, Beijing 100190, China
[2] University of Chinese Academy of Sciences, Beijing 100049, China
[3] Beijing Institute of Computer Technology and Application, Beijing, China
{yuguoxin20g,aoxiang}@ict.ac.cn
[4] University of Electronic Science and Technology of China, Chengdu, China
[5] Shenzhen Key Laboratory for High Performance Data Mining, Shenzhen Institutes of Advanced Technology, Chinese Academy of Sciences, Beijing, China
[6] Microsoft Research Asia, Beijing, China
xitwan@microsoft.com
[7] The Chinese University of Hong Kong (Shenzhen),Shenzhen, China

Abstract. Most existing aspect-term level sentiment analysis (ATSA) approaches combined neural networks with attention mechanisms built upon given aspect to generate refined sentence representation for better predictions. In these methods, aspect terms are always provided in both training and testing process which may degrade aspect-level analysis into sentence-level prediction. However, the annotated aspect term might be unavailable in real-world scenarios which may challenge the applicability of the existing methods. In this paper, we aim to improve ATSA by discovering the potential aspect terms of the predicted sentiment polarity when the aspect terms of a test sentence are unknown. We access this goal by proposing a capsule network based model named CAPSAR. In CAPSAR, sentiment categories are denoted by capsules and aspect term information is injected into sentiment capsules through a sentiment-aspect reconstruction procedure during the training. As a result, coherent patterns between aspects and sentimental expressions are encapsulated by these sentiment capsules. Experiments on three widely used benchmarks demonstrate these patterns have potential in exploring aspect terms from test sentence when only feeding the sentence to the model. Meanwhile, the proposed CAPSAR can clearly outperform SOTA methods in standard ATSA tasks.

Keywords: Aspect-term level sentiment analysis · Capsule network · Sentiment-aspect reconstruction

---

C. Xu and H. Feng—Equally contributed.

C. S. Jensen et al. (Eds.): DASFAA 2021, LNCS 12682, pp. 120–135, 2021.
https://doi.org/10.1007/978-3-030-73197-7_8

1 Introduction

Aspect-level sentiment analysis is an essential building block of sentiment analysis [24]. It aims at extracting and summarizing the sentiment polarities of given aspects of entities, i.e. *targets*, from customers' comments. Two subtasks are explored in this field, namely Aspect-Term level Sentiment Analysis (ATSA) and Aspect-Category level Sentiment Analysis (ACSA). The purpose of ATSA is to predict the sentiment polarity with respect to given targets appearing in the text. For example, consider the sentence "*The camera of iPhone XI is delicate, but it is extremely expensive.*", ATSA may ask the sentiment polarity towards the given target "*camera*". Meanwhile, ACSA attempts to predict the sentiment tendency regarding a given target chosen from predefined categories, which may not explicitly appear in the comments. Take the same sentence as an example, ACSA asks the sentiment towards the aspect "*Price*" and derives a negative answer. In this paper, we aim at addressing the ATSA task.

Conventional approaches [5,15,19] incorporated linguistic knowledge, such as sentiment-lexicon, syntactic parser, and negation words, etc., and tedious feature engineering into the models to facilitate the prediction accuracy. Recently, supervised deep neural networks, e.g. Recurrent Neural Network (RNN) [36,49], Convolution Neural Network (CNN) [21,46] and attention mechanism [1,8,16,28–30,37,39,43] have shown remarkable successes without cumbersome feature designing. These models are able to effectively screen unrelated text spans and detect the sentiment context about the given target.

Despite these efforts, there is still a major deficiency in previous deep neural network based studies. Specifically, fully labeled aspect terms and their locations in sentence are explicitly required in both training and test process for recent methods, which may derive them degrade to sentence-level prediction and would fail for the test data without such annotations. To acquire aspect terms on predicted sentences, automatic aspect term detection may lead to error accumulation [45], and manually identifying is inefficient even infeasible. To support more authentic applications, it calls for an approach that is able to predict potential aspect-related sentiments based on the sentence and what it has learned from the training set. To this end, we propose a capsule network-based approach to remedy the above problem. Compared with previous studies, our method is able to explore featured sentiments so as to answer the question: "*What are the protagonists of the predicted sentiment polarity?*" We access this goal by leveraging the capsule network[1] [34], which has achieved promising results in computer vision [12,34], natural language processing [7,17,50–52] and recommendation tasks [23].

The core idea of capsule network is the unit named *capsule*, which consists of a group of neurons in which its activity vector can represent the instantiation parameters of a specific type of entity. The length of activity vector denotes the

[1] Here we refer to the capsule network proposed by [34]. Though the models in [44] and [45] also called capsule network in their papers, they are basically built upon RNN and attention mechanisms with distinct concepts and implementations.

probability that the entity exists and its orientation can encode the properties of the entity. Inspired by that, we propose CAPSAR (**CAP**sule network with **S**entiment-**A**spect **R**econstruction) framework by leveraging capsules to denote sentiment categories and enforce the potential aspect information as the corresponding properties. Specifically, during the training process, the sentence is first encoded with given aspects through a location proximity distillation. Then the encoded sentence representations are fed to hierarchical capsule layers and the final capsule layer represents all the concerned sentiment categories. To capture coherent patterns among aspects and sentimental expressions, the sentiment capsules are encouraged to encode the information about the aspect terms. We implement such procedure by reconstructing the aspect with the sentiment capsules. The reconstruction loss is taken as an additional regularization during the training. During the test phase, if the annotated aspect term is unseen by the model, CAPSAR can also make prediction and the potential aspect terms in the sentence could be detected by de-capsulizing the sentiment capsules. We evaluate the proposed methods on three widely used benchmarks. The results show the model has potential in unearthing aspect terms for new sentences, and it can also surpass SOTA baselines in standard aspect-term level sentiment analysis tasks.

2 Related Work

The related researches in literatures can be categorized as follows, including sentiment analysis based on neural network, aspect level sentiment classification and jointing learning methods for aspect level tasks.

Sentiment Analysis Based on Neural Network. Neural network approaches have achieved promising results on both document level [27,41,47] and sentence level [35] sentiment classification tasks without expensive feature engineering. Some works [10,11] even exploited available interactions between document level and sentence level sentiment classification. [44] firstly adopted capsules into document-level sentiment analysis, but their capsule is still based on RNN and attentions, which is different from the capsule designs in [34].

Aspect Level Sentiment Classification. Aspect level sentiment classification is an emerging essential research topic in the field of sentiment analysis. The purpose is to infer the polarity with respect to aspect phrase or predefined aspect categories within the text. [36,49] used multiple RNN layers to jointly model the relations between target terms and their left and right context. Attention-based methods were brought to this field by many researches [1,3,8,9,16,20,25,29,37,38,43] to exploit contextual and positional proximity of aspect terms for prediction and have achieved promising results. In addition, graph convolution networks (GCN) [48] were also utilized in this task. However, the final representation may still fail to capture the accurate sentiment due to target-sensitive problem [42] or because of the noise in data. Models based on convolution neural networks [21,46] are alternatives achieving competitive results, but some key information for modeling local meaning and overall sentiment may be blurred during the pooling operations. Recently, [45] proposed to use capsules to perform aspect-category level sentiment analysis. However as their previous work [44], the basic capsule module is based on attention mechanisms, which is entirely different with ours. And other works leveraging capsule network [7,17] for aspect-level sentiment analysis require explicit aspect annotation during prediction while our method does not require.

Joint Learning for Aspect Level Tasks. There are some recent joint learning methods striving to combine different aspect level tasks into a unified learning process. For example, some studies proposed to extract aspect terms and predict corresponding sentiment polarities in a pipeline or an integrated model. The pipeline models [14,33] are extract-then-classify processes and were proposed to solve the two tasks successively. For integrated models, [22,40] extracted aspect terms with polarities by collapsed tagging that is a unified tagging scheme to link two tasks. [26] considered the relationship between the two tasks and attempted to investigate useful information from one task to another. In addition, some emerging methods [2,11,31] proposed to extract opinion words in sentences as auxiliary information to further improve the performance of aspect level sentiment classification.

3 The CAPSAR Model

3.1 Model Overview

The overall architecture of CAPSAR is shown in Fig. 1. It starts from an embedding representation of words. In particular, we represent the i-th sentence in a dataset $\mathbb{D}$ with m sentences as $\{w_1^{(i)}, w_2^{(i)}, \ldots, w_{n_i}^{(i)}\}$, where $i \in [1, \ldots, m]$, n_i is the sentence length, and $w_{\cdot}^{(i)} \in W$ denotes a word where W is the set of vocabulary. The embedding layer encodes each word $w_t^{(i)}$ into a real-value word vector $x_t^{(i)} \in \mathbb{R}^{D_x}$ from a matrix $M \in \mathbb{R}^{|W| \times D_x}$, where $|W|$ is the vocabulary size and D_x is the dimension of word vectors. The sentence is encoded by a sequence encoder to construct a sentence representation. Next, the output of the sequence encoder is fed to 3-layer capsules. The up-most capsule layer contains C *sentiment capsules*, where C is the number of sentiment categories. The capsules layers are communicated with a simple yet effective *sharing-weight* routing algorithm.

During training, one objective of our model is to maximize the length of sentiment capsules corresponding to the ground truth since it indicates the likelihood of potential sentiments. Meanwhile, these active vectors of the sentiment capsules are used to model the connections between the considered aspect terms and its corresponding sentiment via an aspect reconstruction. The distance between the reconstructed aspect representation produced by sentiment capsules and the

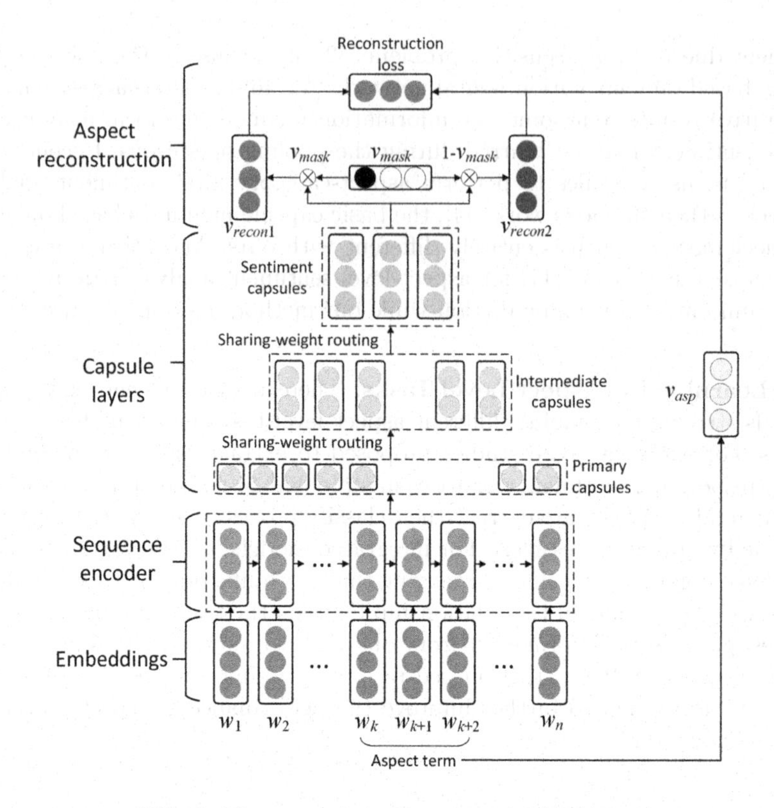

Fig. 1. The network architecture of CAPSAR.

given aspect embedding[2] is regarded as an additional regularization. In this manner, we encourage the sentiment capsules to learn the aspect information as their active vectors' orientations. In the test process, a sentiment capsule will be "active" if its length is above a user-specific threshold, e.g. 0.5. All others will then be "inactive". The sentiment prediction of a test sentence will be determined by the sentiment category associated with the active sentiment capsules.

3.2 Sequence Encoder

In our model, we adopt Bi-GRU as the sequence encoder for simplicity. For i-th sentence at step t, the corresponding hidden state $h_t^{(i)}$ are updated as follows.

$$h_t^{(i)} = \begin{bmatrix} \overrightarrow{h}_t^{(i)} \\ \overleftarrow{h}_t^{(i)} \end{bmatrix} = \begin{bmatrix} \overrightarrow{\text{GRU}}(x_t^{(i)}) \\ \overleftarrow{\text{GRU}}(x_t^{(i)}) \end{bmatrix}, \quad t = 1, \ldots, n_i, \tag{1}$$

where $h_t^{(i)}$ concatenates hidden states of the t-th word in the i-th sentence from both directions. Note that more advanced encoders such as LSTM [13]

[2] The aspect embedding is calculated by the average of the word embeddings that form the aspect term.

or BERT [4], can also be utilized as the sequence encoder. We will introduce how to combine CAPSAR with BERT in the following part.

3.3 Location Proximity with Given Aspect

In order to highlight potential opinion words that are closer to given aspect terms, we adopt a location proximity strategy, which is observed effective in [1, 21]. Specifically, we calculate relevance $l_t^{(i)}$ between the t-word and the aspect[3].

$$l_t^{(i)} = \begin{cases} 1 + \max(0, \alpha + n_i/\beta - |\gamma * (k - t)|) & t \leq n_i \\ 0 & t > n_i \end{cases} \quad (2)$$

where k is the index of the first aspect word, n_i is the sentence length, α, β and γ are pre-specified constants.

We use l to help the sequence encoder locate possible key words w.r.t the given aspect.

$$\hat{h}_t^{(i)} = h_t^{(i)} * l_t^{(i)}, t \in [1, n_i], i \in [1, m] \quad (3)$$

Based on Eq. 2 and 3, the salience of words that are distant to the aspect terms will be declined. Note that such location proximity could be optional in the test process when the annotated aspect terms are unavailable.

3.4 Capsule Layers with Sharing-Weight Routing

The capsule layers of CAPSAR consist of a primary capsule layer, an intermediate capsule layer and a sentiment capsule layer. The primary capsule layer contains a group of neurons which are constructed by the hidden vectors of the sequence encoder. Specifically, we simply perform convolutional operation over $h_{n_i}^{(i)}$ for i-th sentence and take its output to formulate the primary capsules. As a result, the primary capsules may contain the sentence coupled with aspect representations.

Next, the primary capsules are transformed into the intermediate layer and the subsequent sentiment capsule layer via a *sharing-weight* routing mechanism. Unlike the conventional dynamic routing algorithm in [34], our routing algorithm simultaneously keeps local-proximity information and significantly reduces training parameters.

The sharing-weight routing algorithm shares the weights between different children and the same parent. Specifically, we denote the output vector of a capsule i at level L and the total input vector of a capsule j at level $L+1$ as $p_i \in \mathbb{R}^{D_L}$ and $\tilde{q}_j \in \mathbb{R}^{D_{L+1}}$, respectively. We use a unified transformation weight matrix $W_j \in \mathbb{R}^{D_{L+1} \times D_L}$ for the capsule j at level $L+1$ to compute the prediction vectors, i.e. $\hat{p}_{j|i} \in \mathbb{R}^{D_{L+1}}$, for every possible child capsule i at level L. As a result, the total input $\tilde{q}_j$ of capsule j is updated as

$$\tilde{q}_j = \sum_i c_{ij} \hat{p}_{j|i}, \quad \hat{p}_{j|i} = W_j p_i \quad (4)$$

[3] t is possibly larger than n_i because of sentence padding.

where c_{ij} denotes coupling coefficients between capsule i and j and is initialized with equal probability. During the iterative dynamic routing process, c_{ij} is updated to $q_j \cdot \hat{p}_{j|i}$, where q_j is the output vector of capsule j computed by the squash function.

$$q_j = \frac{||\tilde{q}_j||^2}{1 + ||\tilde{q}_j||^2} \frac{\tilde{q}_j}{||\tilde{q}_j||} \tag{5}$$

Compared with the conventional dynamic routing algorithm, the sharing-weight routing algorithm clearly reduces the number of parameters and saves computational cost. For example, if two consecutive capsule layers have M and N capsules and the dimensions are D_L and D_{L+1} respectively, then the number of parameters to be learned in this layer will be reduced by $(M-1) \times N \times D_L \times D_{L+1}$ compared with the original routing algorithm in [34].

3.5 Model Training with Aspect Reconstruction

The training objective of this model is two-fold. On one hand, we aim to maximize the length of the correct sentiment capsules since it indicates the probability that the corresponding sentiment exists. To this end, we use a margin loss for every given sentence i

$$L_1^{(i)} = v_{mask}^{(i)} \max(0, m^+ - ||v_{prob}^{(i)}||)^2 + (1 - v_{mask}^{(i)}) \max(0, ||v_{prob}^{(i)}|| - m^-)^2 \tag{6}$$

Here $v_{prob}^{(i)} = (||q_1^{(i)}||, \cdots, ||q_C^{(i)}||)$ where $q_j^{(i)}$ denotes the output vector of sentiment capsule j for the sentence i. Each element in such $v_{prob}^{(i)}$ indicates the existence probability of the corresponding sentiment in sentence i; $v_{mask}^{(i)}$ is the mask for sentence i; m^+ and m^- are hyper-parameters.

On the other hand, we attempt to encourage the sentiment capsules to capture interactive patterns between the aspect and their corresponding sentiments. To this end, we utilize the output vectors of all the sentiment capsules $q_j (j \in [1, C])$ to participate in reconstructing the representation of aspect terms. Specifically, suppose v_{mask} is a one-hot mask[4] whose element representing the ground truth sentiment is 1, and the rest values are 0. Then we derive two vectors, namely v_{recon1} and v_{recon2}, through this mask and the sentiment capsules. First, we mask out all but the output vector of the correct sentiment capsule by v_{mask}. Then v_{recon1} is derived by transforming the masked output vector through a fully-connect layer. v_{recon2} can be derived in a similar manner where $1 - v_{mask}$ is used as the mask. We force both v_{recon1} and v_{recon2} have the same dimension of word embedding, i.e., $v_{recon1}, v_{recon2} \in \mathbb{R}^{D_x}$, and they are contributed to the aspect reconstruction during the training.

Suppose a given aspect embedding[5] of the sentence i is denoted as $v_{asp}^{(i)}$, another training objective in our CAPSAR is to minimize the distance between

[4] The dimension of v_{mask} is C.

[5] If there are more than one aspect in a same sentence, every aspect will be separately trained.

v_{asp} and v_{recon1}, and to maximize that between v_{asp} and v_{recon2}.

$$L_2^{(i)} = -v_{asp}^{(i)} \frac{v_{recon1}^{(i)}}{||v_{recon1}^{(i)}||} + v_{asp}^{(i)} \frac{v_{recon2}^{(i)}}{||v_{recon2}^{(i)}||} \tag{7}$$

Finally, the overall loss is the combination of $L_1^{(i)}$ and $L_2^{(i)}$, and a hyper-parameter λ is used to adjust the weight of L_2.

$$Loss = \sum_i (L_1^{(i)} + \lambda L_2^{(i)}) \tag{8}$$

For prediction, a sentence and an optional aspect in the sentence are fed to the network and the polarity attached to the sentiment capsule with the largest length will be assigned.

3.6 Combining CAPSAR with BERT

Our CAPSAR is meanwhile easily extended that utilizes the features learnt from large-scale pre-trained encoders, e.g. BERT [4]. An upgraded model, namely CAPSAR-BERT, is achieved by replacing the sequence encoder with BERT in CAPSAR. The other structures are kept the same. In this manner, the strength of BERT and the proposed structures could be combined.

4 Experiments

In this section, we verify the effectiveness of CAPSAR. Firstly, we verify the ability of CAPSAR on perceiving the potential aspect terms when they are unknown. Secondly, we investigate the performance of CAPSAR on standard ATSA tasks, where the aspect terms are always known for test sentences. Finally, we demonstrate the detailed differences of compared methods by case studies.

4.1 Datasets

Three widely used benchmark datasets are adopted in the experiment whose statistics are shown in Table 1. **Restaurant** and **Laptop** are from SemEval2014 Task 4[6], which contain reviews from Restaurant and Laptop domains, respectively. We delete a tiny amount of data with conflict labels which follows previous works [37,43]. **Twitter** is collected by [6] containing twitter posts. Though these three benchmarks are not large-scale datasets, they are the most popular and fair test beds for recent methods.

[6] http://alt.qcri.org/semeval2014/task4/index.php?id=data-and-tools.

Table 1. Statistics of datasets.

Dataset	Restaurant		Laptop		Twitter	
	Train	Test	Train	Test	Train	Test
Neg.	807	196	870	128	1562	173
Neu.	637	196	464	169	3124	346
Pos.	2164	728	994	341	1562	173

4.2 Compared Methods

We compare our method with several SOTA approaches.

ATAE-LSTM [43] appends aspect embedding with each input word embeddings. **TD-LSTM** [36] employs two LSTMs to model contexts of the targets and performs predictions based on the concatenated context representations. **IAN** [29] interactively learns the context and target representation. **MemNet** [37] feds target word embedding to multi-hop memory to learn better representations. **RAM** [1] uses recurrent attention to capture key information on a customized memory. **MGAN** [8] equips a multi-grained attention to address the aspect having multiple words or larger context. **ANTM** [30] adopts attentive neural turing machines to learn dependable correlation of aspects to context. **CAPSAR** and **CAPSAR-BERT** are models proposed in this paper. **BERT** [4] is also compared to show the improvement of CAPSAR-BERT.

4.3 Experimental Settings

In our experiments, we implement our method by Keras 2.2.4. The word embedding is initialized by Glove 42B [32] with dimension of 300. The max length for each sentence is set to 75. The batch size of training is 64 for 80 epochs, and Adam [18] with default setting is taken as the optimizer. For sequence encoder, we adopt the Bi-GRU with the dropout rate of 0.5 for CAPSAR. The pre-trained BERT of the dimension of 768 is used in CAPSAR-BERT. For hyper-parameters in Eq. 2, α, β, and γ are set to be 3, 10 and 1. For the capsule layers, there are 450 primary capsules with dimensions of 50, 30 intermediate capsules with dimensions of 150 and 3 sentiment capsules with dimensions of 300. The default routing number is 3. For hyper-parameters in the loss function (cf. Eq. 8), we set m^+, m^-, λ to be 1.0, 0.1 and 0.003 respectively. Evaluation metrics we adopt are Accuracy and Macro-F1, and the latter is widely used for recent ATSA tasks since it is more appropriate for datasets with unbalanced classes.

4.4 Results on Standard ATSA

Main Results. Table 2 demonstrates the performances of compared methods over the three datasets on the standard ASTA tasks. On such setting, every aspect term is known to all the models, and each model predicts the corresponding polarity for a given aspect term. Here we only consider the longest sentiment

Table 2. The average accuracy and macro F1-score on standard ATSA tasks. The results with '*' are retrieved from the papers of RAM, and other results of baselines are retrieved from corresponding papers.

	Models	Restaurant		Laptop		Twitter	
		Accuracy	F1	Accuracy	F1	Accuracy	F1
Baselines	ATAE-LSTM	0.7720	NA	0.6870	NA	NA	NA
	TD-LSTM	0.7560	NA	0.6810	NA	0.6662*	0.6401*
	IAN	0.7860	NA	0.7210	NA	NA	NA
	MemNet(3)	0.8032	NA	0.7237	NA	0.6850*	0.6691*
	RAM(3)	0.8023	0.7080	0.7449	0.7135	0.6936	0.6730
	MGAN	0.8125	0.7194	0.7539	**0.7247**	0.7254	0.7081
	ANTM	0.8143	0.7120	0.7491	0.7142	0.7011	0.6814
Ablation test	CAPSAR w/o R	0.8185	0.7216	0.7484	0.7039	0.7255	0.7067
	CAPSAR w/o H	0.8188	0.7226	0.7461	0.7054	0.7298	0.7080
	CAPSAR	**0.8286**	**0.7432**	**0.7593**	0.7221	**0.7368**	**0.7231**
Combine BERT	BERT	0.8476	0.7713	0.7787	0.7371	0.7537	0.7383
	CAPSAR-BERT	**0.8594**	**0.7867**	**0.7874**	**0.7479**	**0.7630**	**0.7511**

capsule is active. All the reported values of our methods are the average of 5 runs to eliminate the fluctuates with different random initialization, and the performance of baselines are retrieved from their papers for fair comparisons. The best performances are demonstrated in bold face. From the table, we observe CAPSAR has clear advantages over baselines on all datasets. Our model can outperform all the baselines by a large-margin on both evaluate measures except the F1 on Laptop dataset. Meanwhile, we also observe that CAPSAR-BERT further improves the performance of BERT. It demonstrates the advantages by combining CAPSAR with advanced pre-trained model.

Ablation Test. Next we perform ablation test to show the effectiveness of component of CAPSAR. We remove the aspect reconstruction and intermediate capsule layer, respectively, and derive two degrade models, namely **CAPSAR w/o R** and **CAPSAR w/o H**. Their performance on the three datasets are exhibited in Table 2, and we can observe the degraded model gives a clear weaker performance than CAPSAR, without the sentiment-aspect reconstruction. We conjecture such a regularizer might be able to learn the interactive patterns among the aspects with the complex sentimental expressions. Meanwhile, we also observe the advantage of the intermediate capsule layer, which illustrates the hierarchical capsule layers may be stronger in learning aspect-level sentiment features.

Case Studies. Then we illustrate several case studies on the results of ATSA task in Table 3. The predictions of CAPSAR, ANTM, MGAN and RAM are exhibited. We directly run their available models on test set to get the results. The input aspect terms are placed in the brackets with their true polarity labels

Table 3. Prediction examples of some of the compared methods. The abbreviations Pos, Neu and Neg in the table represent positive, neutral and negative. ✗ indicates incorrect prediction.

Sentence	CAPSAR	ANTM	MGAN	RAM
1. The [**chocolate raspberry cake**]$_{\text{Pos}}$ is heavenly - not too sweet , but full of [**flavor**]$_{\text{Pos}}$.	(Pos,Pos)	(Pos,Neg✗)	(Neg✗,Pos)	(Pos,Neg✗)
2. Not only was the sushi fresh , they also served other [**entrees**]$_{\text{Neu}}$ allowed each guest something to choose from and we all left happy (try the [**duck**]$_{\text{Pos}}$!	(Neu,Pos)	(Pos✗,Pos)	(Pos✗,Pos)	(Pos✗,Pos)
3. The [**baterry**]$_{\text{Pos}}$ is very longer .	Pos	Neg✗	Pos	Neg✗
4. [**Startup times**]$_{\text{Neg}}$ are incredibly long : over two minutes.	Neg	Neg	Pos✗	Pos✗
5. However I chose two day [**shipping**]$_{\text{Neg}}$ and it took over a week to arrive.	Neu✗	Neu✗	Pos✗	Pos✗

as subscripts. "Pos", "Neu" and "Neg" in the table represent positive, neutral and negative, respectively. Different targets are demonstrated by different colors, such as blue and red, etc. The context which may support the sentiment of targets is manually annotated and dyed with the corresponding color.

For instance, in the first sentence, one target is "*chocolate raspberry cake*" of which the sentiment is positive and the context "*is heavenly*" supports this sentiment. In this case, the results of CAPSAR, ANTM, and RAM are correct. For the second aspect "*flavor*", our method and MGAN predict correctly. We conjecture the reason is the sentence contains turns in its expression, which may confuse the existing neural network models. Similar observation is achieved on the second sentence in which only our method can predict different sentiments for distinct aspects.

Next we discuss a target-sensitive case which is shown by the third and the fourth sentences in the same table. The word "long" in the exhibited two sentences indicates entirely opposite sentiment polarities, since the expressed sentiment also depends on the considered aspects. These cases are challenge for algorithms to identify the sentiment of each sentence correctly. Among the demonstrated methods, only our approach can predict them all successfully. ANTM, as a strong competitor, can predict one of them correctly. The others fail to give any correct prediction. We do not claim our CAPSAR can perfectly address the target-sensitive cases, however, the results give an initial encouraging potential. We argue that it is the sentiment-aspect reconstruction in our model which makes aspect and its corresponding sentiment become more coupled, and it might be one of the essential reasons why our model achieves a

Table 4. The average Precision@1, Recall@5 and mAP on aspect term detection. The column "Avg. Aspect" and "Avg.SenLen" indicate average number of words on aspect terms in each sentence and average length of each sentence on the test set, respectively.

Datasets	Avg. Aspect	Avg. SenLen	Pre.@1	Rec.@5	mAP
Restaurant	2.76	16.25	0.8233	0.7884	0.7139
Laptop	2.54	15.79	0.6408	0.7557	0.6173

salient improvement on ATSA task. How to specifically explore capsule networks for target-sensitive sentiment analysis is out of the scope of this paper.

For error analysis, we find that all the listed models cannot predict correctly on the last sentence of Table 3. By looking closer to this sentence, we recognize that the sentiment polarity of this sentence comes from implicit semantics instead of its explicit opinion words. It indicates implicit semantics inference behind sentences is still a major challenge of neural network models, even exploiting capsule networks.

4.5 Results on Aspect Term Detection

Next, we investigate whether CAPSAR can detect potential aspect terms when they are unknown during the test. We use the trained model to predict every test sentence on Restaurant and Laptop datasets but we intentionally conceal the information about the aspect terms in the input. In another word, the model is only fed the test sentence without any other additional input. Then we decapsulize the sentiment capsule whose length is longer than 0.5 and compute normalized dot-product between its reconstructed vector and every word embedding in the test sentence. These dot-products can be regarded as the probabilities representing the possibility of the word to be part of an aspect term.

A sentence may simultaneously contain multiple sentiments (c.f. the 2nd sentence in Table 3), which derives more than one sentiment capsule whose length surpasses 0.5. As a result, we detect the potential aspect terms for every active sentiment capsule, respectively, in our evaluation. There could be more than one aspect terms for each sentiment category in the same sentence (c.f. the 1st sentence in Table 3). Hence we compute Precision@k, Recall@k and mean Average Precision (mAP) to comprehensively verify the effectiveness of CAPSAR on aspect term detection.

We retrain our CAPSAR five times and use the trained model to detect aspect terms on the test set. Table 4 shows the average results. From the table we observe that CAPSAR shows an encouraging ability to extract aspect terms even though they are unknown in new sentences. Meanwhile, the model achieves better performance on Restaurant dataset. We conjecture the reason is the Laptop dataset has more complicated aspect terms such as "Windows 7".

Finally, we visualize two test sentences in this evaluation shown as Fig. 2 and 3. Figure 2 demonstrates a case that only one sentiment capsule is active.

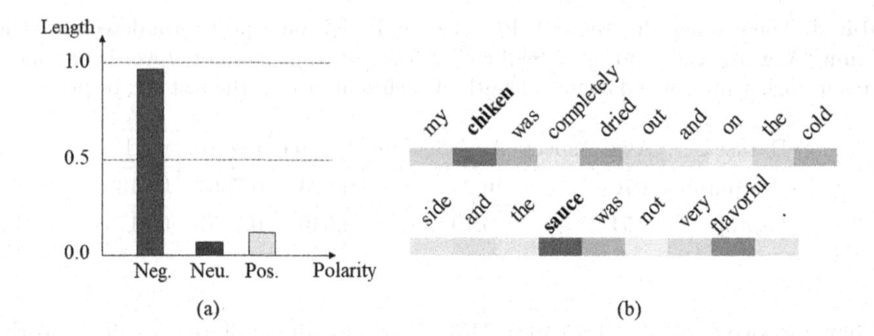

Fig. 2. The visualization of aspect term detection when single sentiment capsule is active. The real aspect terms are marked in bold face in the sub-figure (b).

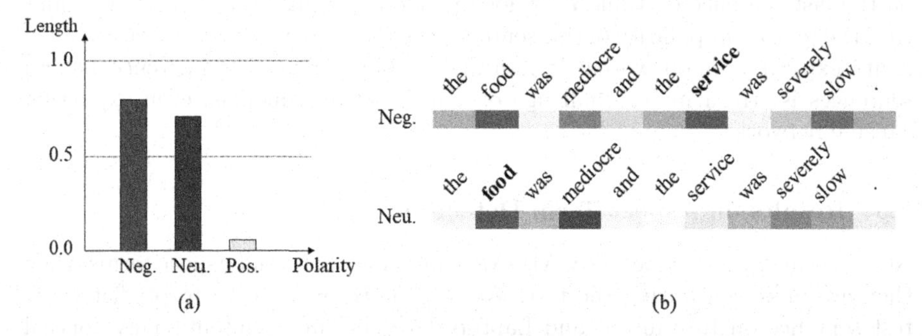

Fig. 3. The visualization of aspect term detection when multiple sentiment capsules are active. The real aspect terms are marked in bold face in the sub-figure (b), respectively.

The sentiment capsule length is exhibited in Fig. 2(a), and the corresponding dot-products are demonstrated as Fig. 2(b). The darker the color in Fig. 2(b), the higher value the dot-product is, which means the corresponding word is more likely to be a part of an aspect term. From this figure, we observe that the two real aspect terms hold much higher weights compared with the other words, which derives a correct detection. We can also obtain similar observations from the case shown as Fig. 3 in which two sentiment capsules are active. In this case, there are two aspect terms, namely "food" and "service". From the sentence, we observe that this review has a strong negative sentiment towards on "service" while has a neutral attitude to "food". After we de-capsulize the active sentiment capsules, we find both "service" and "food" are perceived by the negative capsule because of the overall negative sentiment of the whole sentence. While in the neutral capsule, only the corresponding aspect term "food" is highlighted.

5 Conclusion

In this paper, we proposed, CAPSAR, a capsule network based model for improving aspect-level sentiment analysis. The network is piled up hierarchical capsule layers equipped with a shared-weight routing algorithm to capture key features for predicting sentiment polarities. Meanwhile, the instantiation parameters of sentiment capsules are used to reconstruct the aspect representation, and the reconstruction loss is taken as a part of the training objective. As a consequence, CAPSAR could further capture the coherent patterns between sentiment and aspect information and is able to detect potential aspect terms by parsing the sentiment capsules when these aspect terms are unseen. Experimental results on three real-world benchmarks demonstrate the superiority of the proposed model.

Acknowledgments. This work is supported by the National Natural Science Foundation of China under Grant No. 61976204, 92046003 and U1811461, the Project of Youth Innovation Promotion Association CAS and Beijing Nova Program Z201100006820062. This work was also supported by the Natural Science Foundation of Chongqing under Grant No. cstc2019jcyj-msxmX0149. Min Yang is partially supported by Shenzhen Basic Research Foundation (No. JCYJ20200109113441941).

References

1. Chen, P., Sun, Z., Bing, L., Yang, W.: Recurrent attention network on memory for aspect sentiment analysis. In: EMNLP (2017)
2. Chen, Z., Qian, T.: Relation-aware collaborative learning for unified aspect-based sentiment analysis. In: ACL, pp. 3685–3694 (2020)
3. Cheng, J., Zhao, S., Zhang, J., King, I., Zhang, X., Wang, H.: Aspect-level sentiment classification with heat (hierarchical attention) network. In: CIKM, pp. 97–106 (2017)
4. Devlin, J., Chang, M.W., Lee, K., Toutanova, K.: BERT: pre-training of deep bidirectional transformers for language understanding. In: NAACL (2019)
5. Ding, X., Liu, B., Yu, P.S.: A holistic lexicon-based approach to opinion mining. In: WSDM (2008)
6. Dong, L., Wei, F., Tan, C., Tang, D., Zhou, M., Xu, K.: Adaptive recursive neural network for target-dependent Twitter sentiment classification. In: ACL (2014)
7. Du, C., et al.: Capsule network with interactive attention for aspect-level sentiment classification. In: EMNLP-IJCNLP (2019)
8. Fan, F., Feng, Y., Zhao, D.: Multi-grained attention network for aspect-level sentiment classification. In: EMNLP (2018)
9. He, R., Lee, W.S., Ng, H.T., Dahlmeier, D.: Effective attention modeling for aspect-level sentiment classification. In: COLING, pp. 1121–1131 (2018)
10. He, R., Lee, W.S., Ng, H.T., Dahlmeier, D.: Exploiting document knowledge for aspect-level sentiment classification. In: Proceedings of the 56th Annual Meeting of the Association for Computational Linguistics (Volume 2: Short Papers), Melbourne, Australia, pp. 579–585. ACL (July 2018)
11. He, R., Lee, W.S., Ng, H.T., Dahlmeier, D.: An interactive multi-task learning network for end-to-end aspect-based sentiment analysis. In: ACL, pp. 504–515 (2019)
12. Hinton, G.E., Sabour, S., Frosst, N.: Matrix capsules with em routing. In: ICLR (2018)
13. Hochreiter, S., Schmidhuber, J.: Long short-term memory. Neural Comput. **9**, 1735–1780 (1997)

14. Hu, M., Peng, Y., Huang, Z., Li, D., Lv, Y.: Open-domain targeted sentiment analysis via span-based extraction and classification. In: ACL, pp. 537–546 (2019)
15. Hu, M., Liu, B.: Mining and summarizing customer reviews. In: KDD (2004)
16. Huang, B., Ou, Y., Carley, K.M.: Aspect level sentiment classification with attention-over-attention neural networks. In: SBP-BRiMS (2018)
17. Jiang, Q., Chen, L., Xu, R., Ao, X., Yang, M.: A challenge dataset and effective models for aspect-based sentiment analysis. In: EMNLP (2019)
18. Kingma, D.P., Ba, J.: Adam: a method for stochastic optimization. In: ICLR (2015)
19. Kiritchenko, S., Zhu, X., Cherry, C., Mohammad, S.: NRC-Canada-2014: detecting aspects and sentiment in customer reviews. In: Proceedings of the 8th International Workshop on Semantic Evaluation, SemEval 2014 (2014)
20. Lei, Z., Yang, Y., Yang, M., Zhao, W., Guo, J., Liu, Y.: A human-like semantic cognition network for aspect-level sentiment classification. AAAI **33**, 6650–6657 (2019)
21. Li, X., Bing, L., Lam, W., Shi, B.: Transformation networks for target-oriented sentiment classification. In: ACL (2018)
22. Li, X., Bing, L., Li, P., Lam, W.: A unified model for opinion target extraction and target sentiment prediction. AAAI **33**, 6714–6721 (2019)
23. Li, C., Quan, C., Li, P., Qi, Y., Deng, Y., Wu, L.: A capsule network for recommendation and explaining what you like and dislike. In: SIGIR (2019)
24. Liu, B.: Sentiment analysis and opinion mining. In: Synthesis Lectures on Human Language Technologies (2012)
25. Liu, J., Zhang, Y.: Attention modeling for targeted sentiment. In: ACL, pp. 572–577 (2017)
26. Luo, H., Li, T., Liu, B., Zhang, J.: DOER: dual cross-shared RNN for aspect term-polarity co-extraction. In: ACL, pp. 591–601 (2019)
27. Luo, L., et al.: Beyond polarity: interpretable financial sentiment analysis with hierarchical query-driven attention. In: IJCAI (2018)
28. Luo, L., et al.: Unsupervised neural aspect extraction with sememes. In: IJCAI (2019)
29. Ma, D., Li, S., Zhang, X., Wang, H.: Interactive attention networks for aspect-level sentiment classification. In: IJCAI (2017)
30. Mao, Q., et al.: Aspect-based sentiment classification with attentive neural turing machines. In: IJCAI (2019)
31. Peng, H., Xu, L., Bing, L., Huang, F., Lu, W., Si, L.: Knowing what, how and why: a near complete solution for aspect-based sentiment analysis. In: AAAI, pp. 8600–8607 (2020)
32. Pennington, J., Socher, R., Manning, C.: GloVe: global vectors for word representation. In: EMNLP (2014)
33. Phan, M.H., Ogunbona, P.O.: Modelling context and syntactical features for aspect-based sentiment analysis. In: ACL, pp. 3211–3220 (2020)
34. Sabour, S., Frosst, N., Hinton, G.E.: Dynamic routing between capsules. In: NIPS (2017)
35. Socher, R., Pennington, J., Huang, E.H., Ng, A.Y., Manning, C.D.: Semi-supervised recursive autoencoders for predicting sentiment distributions. In: EMNLP (2011)
36. Tang, D., Qin, B., Feng, X., Liu, T.: Effective LSTMs for target-dependent sentiment classification. In: COLING (2016)
37. Tang, D., Qin, B., Liu, T.: Aspect level sentiment classification with deep memory network. In: EMNLP (2016)

38. Tang, J., et al.: Progressive self-supervised attention learning for aspect-level sentiment analysis. In: ACL, pp. 557–566 (2019)
39. Wang, B., Lu, W.: Learning latent opinions for aspect-level sentiment classification. In: AAAI (2018)
40. Wang, F., Lan, M., Wang, W.: Towards a one-stop solution to both aspect extraction and sentiment analysis tasks with neural multi-task learning. In: IJCNN, pp. 1–8 (2018)
41. Wang, J., Wang, Z., Zhang, D., Yan, J.: Combining knowledge with deep convolutional neural networks for short text classification. In: IJCAI (2017)
42. Wang, S., Mazumder, S., Liu, B., Zhou, M., Chang, Y.: Target-sensitive memory networks for aspect sentiment classification. In: ACL (2018)
43. Wang, Y., Huang, M., Zhao, L., et al.: Attention-based LSTM for aspect-level sentiment classification. In: EMNLP (2016)
44. Wang, Y., Sun, A., Han, J., Liu, Y., Zhu, X.: Sentiment analysis by capsules. In: The Web Conference (2018)
45. Wang, Y., Sun, A., Huang, M., Zhu, X.: Aspect-level sentiment analysis using AS-capsules. In: The Web Conference (2019)
46. Xue, W., Li, T.: Aspect based sentiment analysis with gated convolutional networks. In: ACL (2018)
47. Yang, Z., Yang, D., Dyer, C., He, X., Smola, A.J., Hovy, E.H.: Hierarchical attention networks for document classification. In: HLT-NAACL (2016)
48. Zhang, C., Li, Q., Song, D.: Aspect-based sentiment classification with aspect-specific graph convolutional networks. In: EMNLP, pp. 4560–4570 (2019)
49. Zhang, M., Zhang, Y., Vo, D.T.: Gated neural networks for targeted sentiment analysis. In: AAAI (2016)
50. Zhang, N., Deng, S., Sun, Z., Chen, X., Zhang, W., Chen, H.: Attention-based capsule networks with dynamic routing for relation extraction. In: EMNLP (2018)
51. Zhang, X., Li, P., Jia, W., Zhao, H.: Multi-labeled relation extraction with attentive capsule network. In: AAAI (2019)
52. Zhao, W., Ye, J., Yang, M., Lei, Z., Zhang, S., Zhao, Z.: Investigating capsule networks with dynamic routing for text classification. In: EMNLP (2018)

Discriminant Mutual Information for Text Feature Selection

Jiaqi Wang[1] and Li Zhang[1,2(✉)]

[1] School of Computer Science and Technology, Soochow University, Suzhou, China
zhangliml@suda.edu.cn

[2] Provincial Key Laboratory for Computer Information Processing Technology, Soochow University, Suzhou, China

Abstract. In text classification tasks, the high dimensionality of data would result in a high computational complexity and decrease the classification accuracy because of high correlation between features; so, it is necessary to execute feature selection. In this paper, we propose a Discriminant Mutual Information (DMI) criterion to select features for text classification tasks. DMI measures the discriminant ability of features from two aspects. One is the mutual information between features and the label information. The other is the discriminant correlation degree between a feature and a target feature subset based on the label information, which could be used for judging whether a feature is redundant in the target feature subset. Thus, DMI is a de-redundancy text feature selection method considering discriminant information. In order to prove the superiority of DMI, we compare it with the state-of-the-art filter methods for text feature selection and conduct experiments on two datasets: Reuters-21578 and WebKB. K-Nearest Neighbor (KNN) and Support Vector Machine (SVM) are taken as the subsequent classifiers. Experimental results shows that the proposed DMI has significantly improved the classification accuracy and F1-score of both Reuters-21578 and WebKB.

Keywords: Text classification · Feature selection · Mutual information · Discriminant information · Redundant features

1 Introduction

Today, text is a kind of carriers for information and can carry data through various files, such as news reports, product reviews, and blogs. To analyze text, it is

This work was supported in part by the Natural Science Foundation of the Jiangsu Higher Education Institutions of China under Grant No. 19KJA550002, by the Six Talent Peak Project of Jiangsu Province of China under Grant No. XYDXX-054, by the Priority Academic Program Development of Jiangsu Higher Education Institutions, and by the Collaborative Innovation Center of Novel Software Technology and Industrialization.

C. S. Jensen et al. (Eds.): DASFAA 2021, LNCS 12682, pp. 136–151, 2021.
https://doi.org/10.1007/978-3-030-73197-7_9

necessary to perform text processing. As a kind of text processing methods, text classification plays an important role in sentiment analysis, spam e-mail filtering, topic detection, and various real-word applications. Usually, machine learning-based text classification methods must change texts into a feature matrix using a bag-of-word model [17], so that the dimensionality of texts is relatively high. In order to avoid the curse of dimensionality and obtain a good performance, it is necessary to perform feature selection in text classification tasks.

The main goal of feature selection is to select the optimal feature subset from the original feature set that performs well in subsequent classification tasks. Generally, methods for feature selection can be divided into three categories [23]: wrapper, filter and embedded methods. Wrapper feature selection methods, initially proposed by Kohavi et al. [9], combine with classifiers and use the performance of classifiers as the selection criterion. Filter methods are the most commonly used ones to select the optimal feature subset by evaluating features with specific evaluation criteria that are independent of classifiers. The embedding methods combine the characteristics of the wrapper methods and the filter methods, which train machine learning algorithms to assign the weight coefficients to each feature and select the optimal feature subset from the original feature set. It is unavoidable to have redundancy features in high-dimensional data.

Text data has the characteristics of both high dimensionality and high sparsity. text feature selection is often based on document frequency or term count because text data is usually in form of feature matrix with elements representing the number of times a feature appears in the document [17]. Usually, filter methods are developed for text feature selection. Uysal and Gunal [19] designed criteria for measuring the importance of a term in corpus, and proposed a Distinguishing Feature Selector (DFS) based on those criteria. Rehman et al. [16] proposed a Relative Discrimination Criterion (RDC) that considers the frequency of each feature in the positive and negative samples and relies on the label information of a given dataset. Labani et al. [10] proposed a Multivariate Relative Discrimination Criterion (MRDC), which combines RDC with Pearson correlation coefficients for feature selection.

As a commonly used feature selection method, Mutual Information (MI) plays a great role in text feature selection and has achieved good results [3,14]. There are some different definitions on MI. For example, a common way to calculate mutual information is given in [3]. Xu et al. thought this way is in conflict with the definition of MI derived from information theory and introduced a novel way [21]. However, it is unnecessary to consider the absence of the term. In this paper, we adopt a way proposed in [18] and compare it with other ways in experiments.

In a general way, to consider the relationship between features, Peng et al. [13] provided the theoretical analysis of the relationships of dependency, relevance and redundancy. A heuristic minimal-Redundancy-Maximal-Relevance (mRMR) framework was proposed to minimize redundancy of features based on MI. Hoque et al. [8] also combined the feature-class mutual information with the average feature-feature mutual information to perform feature selection using Non-dominated Sorting Genetic Algorithm-II (NSGA-II). Lin et al. [11] took into account the feature dependency and the feature redundancy in the multi-label learning and proposed an evaluation measure that combines the max-dependency and min-redundancy (MDMR) with mutual information. However, MDMR needs to calculate the relevance between terms and categories under a given feature subset in each step of the feature selection process, which leads to a high time complexity. At the same time, MDMR calculates both the relevance of terms and the relevance of terms related to categories and deals with them by subtraction. In this way, the redundancy of terms may be weakened. In a word, as far as we know, existing MI-based methods that can eliminate the redundancy of features are not designed for text feature selection at present.

To effectively remove redundant features and select relevant features, this paper presents a novel text feature selection based on MI, called Discriminant Mutual Information (DMI) that can describe the discriminant ability of features. DMI is a de-redundancy text feature selection method considering discriminant information. First, DMI adopts MI to calculate the discriminant ability of features in terms of the relationship between features and the label information. Second, DMI judges the redundancy of features by calculating a discriminant correlation degree between a term and a target feature subset based on the label information, which takes into account the discriminant information. We conduct experiments on the Rueters-21578 and WebKB corpus to verify the proposed method.

The rest of this paper is organized as follows. Section 2 introduces the related work of text feature selection. Section 3 describes the structure of the proposed DMI. Section 4 gives experimental setting and results in details. Section 5 concludes this paper and prospects future research directions.

2 Related Work

2.1 Text Representation

Text is a set of ideographic symbols that has been used by humans to record and express information for a long time, which cannot be directly processed as the way like numerical data. Text representation is to quantify text data and express it into a format that can be recognized and processed by a computer. Common text representation methods include boolean model [1], Vector Space Model (VSM) [4], probabilistic model [6], and word embedding [12]. Boolean model based on boolean algebra represents only whether the feature word appears in a document. VSM, proposed by Salton et al. in the1970s, maps features and documents into a two-dimensional array and records the weights of each feature in each document. Probabilistic model is to represent text based on the probabilistic queuing theory and uses the conceptual correlation of term-term and term-document. The word embedding model generates a distributed representation of feature words by training a neural network, which well reflects the relationship between feature words and has good semantic characteristics.

In this paper, the bag-of-word model [17], one of the most commonly used vector space models, is adopted for text representation The bag-of-word model takes all feature words in all texts as features and treats each text as a sample. Assume that there is a data set of texts $X = \{(\boldsymbol{x}_1, y_1), (\boldsymbol{x}_2, y_2), \ldots, (\boldsymbol{x}_n, y_n)\}$, where n is the number of texts (samples), $\boldsymbol{x}_i = [x_{i1}, x_{i2}, \ldots, x_{if}]^T$ is the sample distribution of the i-th text, f is the number of features, $y_i \in \{1, \ldots, m\}$ is the label of $\boldsymbol{x}_i$, and m is the number of categories. Let $C = \{c_1, c_2, \cdots, c_m\}$ be the category set and $F = \{t_1, t_2, \cdots, t_f\}$ be the set of features. For term t_j, its feature representation is denoted as $\boldsymbol{f}_j = \{x_{1j}, x_{2j}, \ldots, x_{nj}\}$.

2.2 Mutual Information

In [3], MI can be used to select the optimal feature subset that is highly relevant to the category based on the mutual information of labels and document frequency of terms. Given term t_i, its mutual information in class c_k can be expressed as [3]:

$$MI(t_i, c_k) = \log_2 \frac{p(t_j, c_i)}{p(t_j)p(c_i)} \tag{1}$$

where $p(t_i)$ means the prior probability of term t_i, $p(c_k)$ means the prior probability of category c_k, and $p(t_i, c_k)$ is the joint probability of term t_i and category c_k. For multi-class classification tasks, there are generally two ways to calculate the mutual information of term t_i on C:

$$MI_{max}(t_i, C) = \max_{k=1}^{m} MI(t_i, c_k) \tag{2}$$

and

$$MI_{ave}(t_i, C) = \sum_{k=1}^{m} p(c_k) MI(t_i, c_k) \tag{3}$$

Xu et al. [21] considered the sum of mutual information of a term existing or not as the mutual information between the term and a category. In [21], the mutual information of term t_i on C is defined as follows:

$$MI(t_i, C) = \sum_{k=1}^{m} p(t_i, c_k) \log_2 \frac{p(t_i, c_k)}{p(t_i)p(c_k)} + \sum_{k=1}^{m} p(\tilde{t}_i, c_k) \log_2 \frac{p(\tilde{t}_i, c_k)}{p(\tilde{t}_i)p(c_k)} \tag{4}$$

where $\tilde{t}_i$ means that the term t_i is absence, and $p(\tilde{t}_i)$ is the prior probability of term $\tilde{t}_i$, $p(\tilde{t}_i, c_k)$ is the joint probability of term $\tilde{t}_i$ and category c_k.

We thought that it is superfluous to consider the absence of a term when calculating mutual information, so we adopt the MI algorithm proposed in [18], where the mutual information of term t_i on C can be calculated as:

$$MI(t_i, C) = \sum_{k=1}^{m} p(t_i, c_k) \log_2 \frac{p(t_i, c_k)}{p(t_i)p(c_k)} \tag{5}$$

2.3 mRMR

Mostly, MI provides the mutual information between features and labels. Considering the feature redundancy, Peng et al. [13] proposed the minimal-Redundancy-Maximal-Relevance (mRMR) framework to minimize redundancy and give an implementation based on mutual information. Although mRMR is not designed for text feature selection, it still can be used for text feature selection if text has been represented by applying VSM.

Let $S \subseteq F$ be a feature subset and C be the category set. The relevance of S and C is the average mutual information between S and C, which can be described as:

$$D(S,C) = \frac{1}{|S|} \sum_{t_i \in S} \sum_{c_k \in C} MI(t_i, c_k) \tag{6}$$

At the same time, the redundancy of S is determined by the mutual information between features in S, which can be calculated by

$$R(S) = \frac{1}{|S|^2} \sum_{t_i, t_j \in S} MI(t_i, t_j) \tag{7}$$

where

$$MI(t_i, t_j) = \log_2 \frac{p(t_i, t_j)}{p(t_i)p(t_j)} \tag{8}$$

which is modified from (1).

mRMR combines the above two constraints to form its evaluation criterion:

$$\max \phi(S,C) = D(S,C) - R(S) \tag{9}$$

which indicates that mRMR could get a feature subset with minimal redundancy and maximal relevance.

3 Discriminant Mutual Information

In this section, we propose a novel text feature selection method named discriminant mutual information. We first describe the preprocessing of texts, then present DMI, and finally give the algorithm description of DMI.

3.1 Text Preprocessing

Text needs to be changed to its numerical value representation that can be directly handled by feature selection algorithms. In our method, we express text as vector space using bas-of-words. In the text data, not all words are helpful for text classification tasks; so it is necessary to perform text preprocessing to reduce the amount of words.

The preprocessing of the text data mainly includes the removal of stop-words, stemming, and the removal of high frequency and low frequency words, etc. Removing stop-words is mainly to remove some common words that are not helpful for text classification, such as "a","the", "that", etc. These words often appear in the expression of text, but do not carry any practical meaning. Stemming aims to remove affixes from various forms of words and convert them into their root forms uniformly. For example, "stemming", "stemmer", and "stemmed" should be transformed into the form of the root "stem". Here, we use Porter's Stemmer [15], which is uses a set of rules for stemming. At the same time, high-frequency and low-frequency words should be removed [7]. High-frequency words often appear in each category and low-frequency words may appear only once or twice. These words do not carry information that is beneficial to the classification task. In the preprocessing stage, we reserve words with the frequency of appearing documents more than 3 and less than 25% of the number of total documents.

After this preprocessing stage, it is necessary to perform word statistic on the processed corpus to generate the feature space. Then, documents are sequentially mapped to the feature space to form a feature matrix, where each entry records whether a term appears in a document or not.

3.2 Discriminant Mutual Information

The traditional mutual information for text feature selection considers only the relationship between the term and the category. However, terms with similar meanings are usually similar in sample distributions. It is easy for MI to concentrate on selecting terms that are very close to each other in categorical distribution; so the redundancy between the selected features cannot be ignored.

Theoretically, eliminating redundant terms in the selection process is beneficial to subsequent classification tasks. In this paper, we propose a criterion based on the discriminant information of data for measuring the redundancy of terms. The criterion is to calculate the discriminant correlation degree of a given term and other term based on the label information. We have the following definition.

Definition 1. *Let $C = \{c_1, \cdots, c_m\}$ be the category set and $F = \{t_1, \cdots, t_f\}$ be the feature set. The discriminant correlation degree of term $t_i \in F$ and $t_j \in F$ on C is defined as:*

$$DC(t_i, t_j|C) = \sum_{k=1}^{m} \frac{p(t_i, t_j, c_k)}{p(c_k)} \tag{10}$$

where $p(t_i, t_j, c_k)$ is the jointly probability of term t_i, t_j, and category c_k, and $p(c_k)$ is the prior probability of category c_k.

$DC(t_i, t_j)$ means the relevance of t_i and t_j under all categories. The greater the value of $DC(t_i, t_j)$ is, the more possible t_i is redundant with t_j. The purpose of de-redundancy is to reduce the redundancy of a candidate term and the target feature subset.

Definition 2. *Let* $C = \{c_1, \cdots, c_m\}$ *be the category set and* $F = \{t_1, \cdots, t_f\}$ *be the feature set. If* $S \subseteq F$ *is the target feature subset, then the discriminant correlation degree of term* $t_i \in (F - S)$ *and* S *on* C *is defined as:*

$$DC(t_i, S|C) = \sum_{t_j \in S} DC(t_i, t_j|C) = \sum_{t_j \in S} \sum_{k=1}^{m} \frac{p(t_i, t_j, c_k)}{p(c_k)} \tag{11}$$

On one hand, one of principles for feature selection is to avoid the redundancy of the candidate term with the target feature subset as much as possible. Obviously, if the candidate term t_i makes $DC(t_i, S)$ great, then t_i is possibly redundant with terms in S. In this case, t_i could not be merged into the target feature subset. Therefore, we need the candidate term that is to minimize the redundancy of it and C on S. Namely,

$$\min_{t_i \in (F-S)} DC(t_i, S|C) \tag{12}$$

On the other hand, it requires maximizing the relevance of the candidate term and categories in feature selection tasks, which can be measured by mutual information. Namely,

$$\max_{t_i \in (F-S)} MI(t_i, C) \tag{13}$$

DMI provides a novel feature selection criterion by combining mutual information and discriminant correlation degree. Given the target feature subset S and category set C, we select the candidate term t_i with the maximal mutual information and the minimal discriminant correlation degree at the same time. Namely,

$$\max_{t_i \in (F-S)} DMI(t_i|S, C) = \lambda MI(t_i, C) - (1 - \lambda) DC(t_i, S|C) \tag{14}$$

where $0 < \lambda \leq 1$ is a parameter defined by users, which can balance the importance of $MI(t_i, C)$ and $DC(t_i, S)$. When $\lambda = 1$, DMI is reduced to MI.

The DMI algorithm is to iteratively find t^* that has the highest $DMI(t^*|S, C)$ with $DMI(t^*|S, C) \geq DMI(t_i|S, C)$, $t^*, t_i \in (F - S)$ and merge it into the target feature subset S. The algorithm description is shown in Algorithm 1. When the target feature subset S is an empty set, it is not necessary to judge the redundancy of candidate terms; thus, only the term with the highest mutual information is taken as the optimal term, as shown in Steps 4–8. The termination condition of DMI is that the target feature subset consists of *nSlct* terms, where *nSlct* is defined by users. In fact, the target feature subset can reflect the feature ranking of the first *nSlct* terms.

4 Experiments and Analysis

4.1 Datasets

In this section, we perform experiments on Reuters-21578 and WebKB corpora and compare our method with MI and the state-of-the-art filter methods for text feature selection.

Algorithm 1. DMI

Input: Text set X, category set C, the number of remained terms $nSlct$, and λ.
Output: Target feature subset S.

1: Initialize $S = \emptyset$ and $F = \{t_1, \ldots, t_f\}$;
2: **while** $|S| < nSlct$ **do**
3: **if** $S = \emptyset$ **do**
4: **for** $\forall t_i \in (F - S)$ **do**;
5: Calculate $MI(t_i, C)$ by Eq. (5);
6: **end for**
7: Merge t^* into S, or $S = S \cup \{t^*\}$, where $t^* = \arg\max_{t_i \in (F-S)} MI(t_i, C)$;
8: Remove t^* from F, or $F = F - \{t^*\}$;
9: **else**
10: **for** $\forall t_i \in (F - S)$ **do**
11: Calculate $DC(t_i, S|C)$ by Eq. (11);
12: Calculate $DMI(t_i|S, C) = \lambda MI(t_i, C) + (1 - \lambda) DC(t_i, S|C)$;
13: **end for**
14: Update $S = S \cup \{t^*\}$, where $t^* = \arg\max_{t_i \in (F-S)} DMI(t_i|S, C)$;
15: Update $F = F - \{t^*\}$;
16: **end if**
17: **end while**

The Reuters-21578 corpus is an economic news collected by Reuters news agency and is the most widely used dataset for text classification currently. The original Reuters-21578 dataset has a total of 135 categories. The category distribution of documents is relatively unbalanced and some documents contain multi-labels. Similar to the experimental setting in [2], we also select the subset of Reuters-21578 in our experiments. This subset consists of only single-label data and has 8 frequently occurring categories, called the R8 dataset. The information of this dataset is given in Table 1.

WebKB is a collection of webpages of some university computer science departments collected by the World Wide Knowledge Base project in 1997. These webpages are divided into four categories: course, project, faculty and student. Table 2 shows the data information of the WebKB dataset.

According to Sect. 3.1, we carry out the preprocessing of text data. In the processed texts, all stop-words are deleted, the words with a character length of less than 3 are deleted, and the words with too high and too low frequency are also deleted. At the same time, the stemming operation is performed. In experiments, we choose to retain the terms whose document frequency is greater than 3 and less than $n/4$. After texts are processed, the R8 dataset retains 4954 terms, and the WebKB dataset 6124 ones. Next, we represent these text data in the form of a feature matrix using the bag-of-words model.

4.2 Classifiers and Evaluation Measure

In experiments, we use Support Vector Machine (SVM) [20] and K-nearest Neighbor (KNN) [5] to classify the data after text feature selection, both of which are implemented by using the sklearn library in Python3.6. The accuracy and F1-score are used to evaluate classification results.

Table 1. Information of the R8 dataset

No.	#Category	#Training	#Test
1	Acq	1596	696
2	Crude	253	121
3	Earn	2840	1083
4	Grain	41	10
5	Interest	190	81
6	Money-Fx	206	87
7	Ship	108	36
8	Trade	251	75

Table 2. Information of the WebKB dataset

No.	#Category	#Training	#Test
1	Project	336	168
2	Course	620	310
3	Faculty	750	374
4	Student	1097	544

For binary classification tasks, the calculation way of accuracy is as follows:

$$Acc = \frac{TP + TN}{n'} \tag{15}$$

and that of F1-score as follows:

$$F1 = \frac{2TP}{2TP + FN + FP} \tag{16}$$

where TP is the number of positive samples that are classifies correctly (true positive), TN is the number of negative samples that are classifies correctly (true negative), FN is the number of negative samples that are classifies falsely (false positive), and n' is the total number of test samples.

For multi-category classification tasks, the accuracy and F1-score of category c_i is recorded as Acc_i and $F1_i$, respectively. The final the accuracy and F1-score are the average on all categories. Namely,

$$Acc = \frac{1}{m}\sum_{i=1}^{m} Acc_i \tag{17}$$

and

$$F1 = \frac{1}{m}\sum_{i=1}^{m} F1_i \tag{18}$$

where m is the total number of categories.

4.3 Experimental Results

Comparison of MI Algorithms: To compare three MI algorithms, we observe the classification accuracy obtained by KNN on the R8 dataset. The number of selected features takes values in the set $\{10, 20, 50, 100, 200, 500, 1000, 1500\}$.

Table 3. Accuracy obtained by KNN on the R8 dataset

Method	Number of selected features							
	10	20	50	100	200	500	1000	1500
MI (Eq. (5))	68.66	78.67	**89.95**	**89.40**	**89.36**	84.47	**82.78**	**80.45**
MI (Eq. (4))	**72.13**	**80.90**	89.04	89.17	88.58	**84.51**	82.73	80.4
MI (Eq. (1))	49.47	34.76	55.87	60.21	70.81	76.7	76.24	75.56

Table 3 show the comparison of the classification accuracy obtained by KNN on the R8 dataset, where the best results are in bold type and the second best ones are underlined. It is obvious that MI used in our method is more effective than others, which shows that it is correct for us to use Eq. (5) as the mutual information of text features.

Hyperparameter Analysis of DMI: DMI has a hyperparameter λ, which takes value in the interval $(0, 1]$. To observe the effect of λ on the performance of DMI, we set $\lambda \in \{0.1, 0.2, \cdots, 0.9\}$.

Figure 1 shows the curves of classification accuracy vs. λ obtained by SVM under different datasets, where the number of selected features varies from 50 to 1000. From Fig. 1, we can see that λ has a an great effect on the classification performance of DMI. When λ is very small, the relevance of terms and categories would be ignored so that we miss features with a high discriminant ability. If λ is very close to 1, the performance of DMI approaches to that of MI. Thus, an appropriate value for λ is required. For the R8 dataset, $\lambda > 0.3$ can provide a better performance. On the WebKB dataset, we should set $\lambda > 0.5$. To balance the two terms in Eq. (14), we set $\lambda = 0.5$ in the following experiments.

Comparison of DMI and MI: As mentioned before, MI is in the framework of DMI. If we set $\lambda = 1$ in Eq. (14), DMI is actually MI. Here, we observe the effect of the second term in Eq. (14) on the performance of algorithm by comparing DMI and MI.

The number of remained features varies from 10 to 3000 with a span of 10. The comparison curves of performance indexes vs. number of features are plotted in Figs. 2 and 3.

Figure 2 gives the experimental results on the R8 dataset, Figs. 2(a) and 2(c) are the results obtained by KNN, and Figs. 2(b) and 2(d) obtained by SVM. It

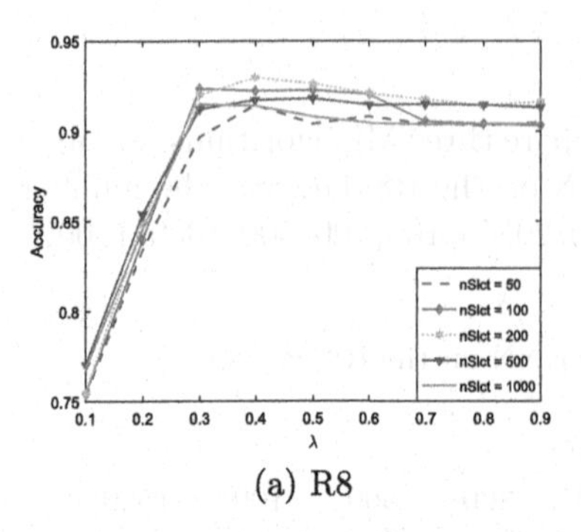

(a) R8

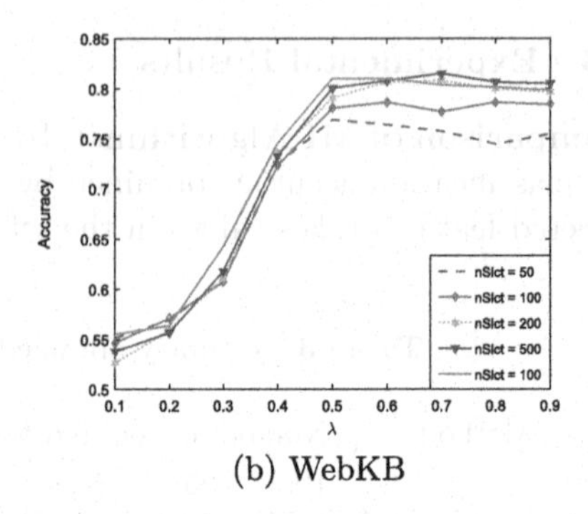

(b) WebKB

Fig. 1. Classification accuracy vs. λ obtained by SVM on the R8 dataset (a) and the WebKB dataset (b).

can be seen from Fig. 2 that it is necessary for text classification to select features. No matter what classifier is used and what performance index is adopted, the best classification performance is achieved when the number of features is less than 300 on the R8 dataset. Specifically, KNN with DMI reaches the best accuracy of 92.33% when the number of selected terms is 110, which is 2.83% higher than KNN with MI under the same dimension. SVM with DMI needs 140 terms to achieve its best accuracy of 93.01%, which is higher than SVM with MI by 1.83% under the same dimension. From the figures about F1-score, we have a similar conclusion. DMI is much better than MI when both classifier are applied to, which indicates that the discriminant correlation degree indeed plays an important role in feature selection.

Figure 3 gives the experimental results on the WebKB dataset, Figs. 3(a) and 3(c) are the results obtained by KNN, and Figs. 3(b) and 3(d) obtained by SVM. The necessity of feature selection can be obviously observed from both Figs. 3(a) and 3(c) when KNN is the subsequent classifier. DMI is rather superior to MI in both accuracy and F1-score. Although Figs. 3(b) and 3(d) show that the performance indexes seem to be increased as increasing the number of selected features, the best performance is achieved at 700 terms that is less than 3000; thus, feature selection is an essential stage.

The experimental results lead us to conclude that the effectiveness of DMI is due to eliminating the redundancy of terms using the discriminant correlation degree. The classical MI cannot remove the redundancy of terms so that MI is worse than DMI in the comparison of classification performance.

Comparison of More Methods. To further validate the effectiveness of the proposed method, we compare it with more related methods, including mRMR [13], MDMR [11], MI [18], RDC [16], MRDC [10], DFS [19] and Information Gain (IG) [22]. IG is a feature evaluation criterion in the ID3 decision tree, which can be applied to text feature selection. Other compared methods are mentioned in Sect. 1.

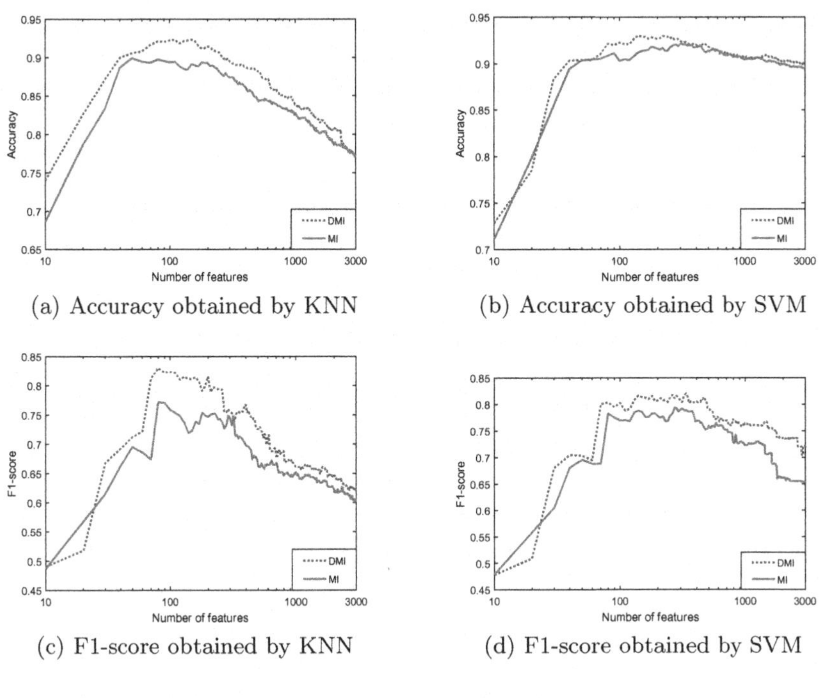

(a) Accuracy obtained by KNN

(b) Accuracy obtained by SVM

(c) F1-score obtained by KNN

(d) F1-score obtained by SVM

Fig. 2. Classification results on the R8 dataset

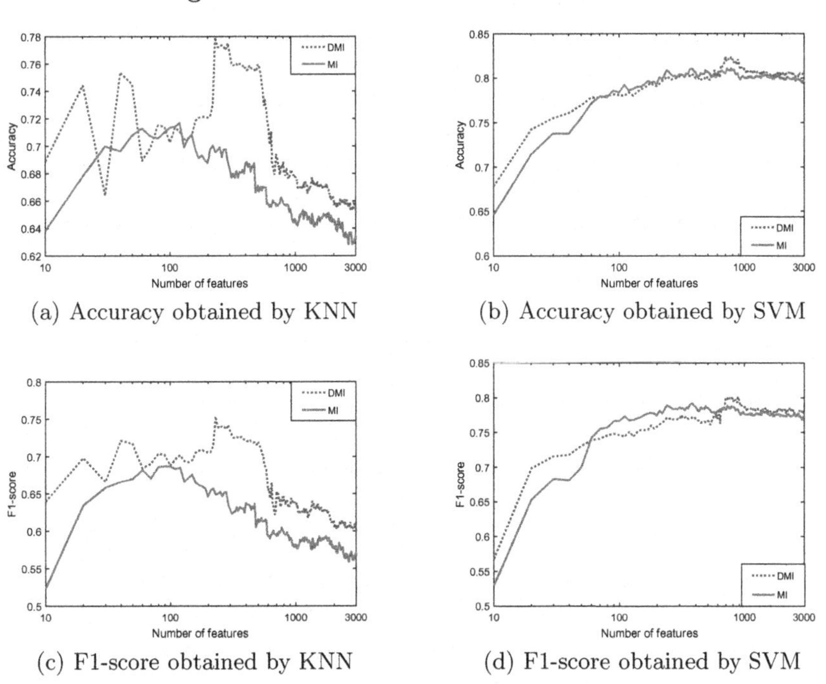

(a) Accuracy obtained by KNN

(b) Accuracy obtained by SVM

(c) F1-score obtained by KNN

(d) F1-score obtained by SVM

Fig. 3. Classification results on the WebKB dataset

Above experiment results tell us that the maximum number of selected feature is less than 1000 on both datasets. In the following experiments, the number of selected features takes values only in the set $\{10, 20, 50, 100, 200, 500, 1000, 1500\}$.

Table 4. Accuracy obtained by KNN on the R8 dataset

Method	Number of selected features							
	10	20	50	100	200	500	1000	1500
DMI	**74.00**	**82.60**	**90.59**	**92.23**	**91.46**	**88.21**	**83.83**	**81.96**
MI	68.66	78.67	89.95	89.40	89.36	84.47	82.76	80.45
mRMR	69.76	72.59	78.03	87.67	87.62	88.76	82.37	79.53
MRDC	52.99	75.38	70.35	81.96	82.91	86.25	82.64	80.81
RDC	64.92	71.36	84.74	86.34	86.71	84.79	83.10	80.90
DFS	49.47	49.66	50.43	53.77	61.44	62.40	63.13	65.65
IG	49.47	49.47	49.61	50.80	51.16	53.54	53.95	54.77

Tables 4 and 5 show the comparison of the classification accuracy obtained by KNN and SVM on the R8 dataset, respectively, where the best results are in bold type and the second best ones are underlined. On the R8 dataset, Tables 4 and 5 show that the best results can be achieved under most cases. When using KNN as a classifier, DMI can achieve the best accuracy for all sizes of feature subset among compared methods. The highest accuracy Moreover, KNN with DMI reaches the highest accuracy of 92.23% that is 2.17% higher than the second highest KNN with MI when the number of selected feature is 100. When using SVM as a classifier, DMI is superior to others for sizes of 50, 100, 200 and 500. SVM with DMI achieves the highest accuracy of 92.65% among all feature numbers.

Table 5. Accuracy obtained by SVM on the R8 dataset

Method	Number of selected features							
	10	20	50	100	200	500	1000	1500
DMI	72.86	78.48	**90.41**	**92.28**	**92.65**	**91.82**	90.82	90.86
MI	71.12	**79.85**	90.36	90.32	91.64	91.41	90.45	90.22
mRMR	65.97	73.18	80.68	87.80	89.26	91.64	**92.01**	**91.18**
MRDC	**73.41**	75.06	77.94	81.82	84.33	89.31	88.76	88.99
RDC	65.10	73.32	83.14	88.76	91.59	91.69	91.27	90.73
DFS	49.47	49.38	50.11	54.87	61.63	62.13	62.31	65.05
IG	49.47	49.25	53.13	53.95	55.37	57.74	59.21	60.44

Table 6. Accuracy obtained by KNN on the WebKB dataset

Method	Number of selected features							
	10	20	50	100	200	500	1000	1500
DMI	**68.91**	**74.43**	**74.50**	70.27	**72.06**	**75.93**	**67.77**	**67.12**
MI	63.83	67.84	70.77	**71.35**	68.77	67.12	64.40	64.54
mRMR	46.79	57.95	61.25	61.39	65.47	67.26	64.40	63.90
MRDC	59.24	59.03	60.74	62.46	61.46	62.54	63.25	64.68
RDC	66.83	69.48	69.91	67.41	68.05	66.12	65.04	64.18
DFS	34.31	46.99	52.36	51.36	57.31	64.76	65.26	65.47
IG	38.97	38.97	38.97	38.97	39.90	41.76	31.09	41.83

Table 7. Accuracy obtained by SVM on the WebKB dataset

Method	Number of selected features							
	10	20	50	100	200	500	1000	1500
DMI	67.84	**74.21**	76.93	**78.08**	79.08	80.09	**81.02**	**80.59**
MI	64.61	71.42	75.50	78.37	79.30	**80.73**	80.01	80.23
mRMR	57.02	58.09	60.24	65.83	69.34	75.07	79.15	80.16
MRDC	58.95	61.89	63.32	64.68	65.11	63.47	64.76	68.19
RDC	**71.13**	72.99	**77.01**	77.36	**80.01**	80.09	80.52	79.87
DFS	41.98	47.21	55.37	57.31	66.26	76.50	79.37	80.37
IG	38.97	38.97	38.97	38.97	38.25	40.69	41.62	39.97

Tables 6 and 7 show the comparison of the classification accuracy of obtained by KNN and SVM on the WebKB dataset, respectively, where the best results are in bold type and the second best ones are underlined. When using the KNN classifier, DMI has the highest accuracy when the number of features is 500, the value is 75.93%. Among other compared methods, the highest accuracy in Table 6 is 71.35% obtained by MI, when the feature number is 100. When using the SVM classifier, DMI achieves the best classification with 1000 terms, which is 0.29% higher than the second best method MI.

In summary, experimental results shown in Tables 4–7 show that DMI on the R8 dataset and WebKB dataset has been improved to a certain extent.

5 Conclusion

In this paper, we aim at the issue of redundancy in the feature subset selected by the method based on mutual information. First of all, we choose MI to measure the mutual information of text features. After that, we design the discriminant correlation degree for judging redundancy of a term on a given feature subset and combine it with MI to propose DMI. Experiments are conducted on Reuters-21578 and WebKB datasets. Two classifiers KNN and SVM are used to classify the selected feature subsets. First, we compare different MI ways and show that the one used in DMI can achieve the best performance in most instances. Second, we compare DMI with MI on two datasets. Experiment results indicate that DMI is much better than MI, which is due to the term of discriminant correlation degree. Finally, we compare DMI with more related methods, such as RDC, DFS, and mRMR. When using KNN, DMI is obviously superior to others under most of feature sizes. Although SVM with DMI is not always better than other methods under all levels, it still can achieve the best performance among all methods and all sizes.

Although DMI is satisfactory in our experiments, it ignores the information of term counts when evaluating terms. In future research, we will introduce term counts into DMI by assigning weights to terms that appear different times in documents, and then construct a more powerful text feature selection algorithm.

References

1. Allahyari, M., et al.: A brief survey of text mining: classification, clustering and extraction techniques, CoRR abs/1707.02919 (2017)
2. Cardoso-Cachopo, A.: Improving methods for single-label text categorization. Ph.D. Thesis, Instituto Superior Tecnico, Universidade Tecnica de Lisboa (2007)
3. Church, K.W., Hanks, P.: Word association norms, mutual information and lexicography. In: Hirschberg, J. (ed.) Proceedings of the 27th Annual Meeting of the Association for Computational Linguistics, University of British Columbia, Vancouver, BC, Canada, 26–29 June 1989, pp. 76–83. ACL (1989). https://doi.org/10.3115/981623.981633
4. Clark, S.: Vector space models of lexical meaning. In: The Handbook of Contemporary Semantic Theory, pp. 493–522 (2015)
5. Cunningham, P., Delany, S.J.: k-nearest neighbour classifiers, vol. 34, pp. 1–7 (2007)
6. Feng, G., Li, S., Sun, T., Zhang, B.: A probabilistic model derived term weighting scheme for text classification. Patt Recogn. Lett. **110**, 23–29 (2018)
7. Forman, G.: A pitfall and solution in multi-class feature selection for text classification. In: Brodley, C.E. (ed.) Proceedings of the 21st International Conference on Machine Learning, ICML 2004, Banff, Alberta, Canada, 4–8 July 2004. ACM International Conference Proceeding Series, vol. 69. ACM (2004). https://doi.org/10.1145/1015330.1015356
8. Hoque, N., Bhattacharyya, D.K., Kalita, J.K.: MIFS-ND: a mutual information-based feature selection method. Exp. Syst. Appl. **41**(14), 6371–6385 (2014). https://doi.org/10.1016/j.eswa.2014.04.019
9. Kohavi, R., John, G.H., et al.: Wrappers for feature subset selection. Artif. Intell. **97**(1–2), 273–324 (1997)
10. Labani, M., Moradi, P., Ahmadizar, F., Jalili, M.: A novel multivariate filter method for feature selection in text classification problems. Eng. Appl. Artif. Intell. **70**, 25–37 (2018). https://doi.org/10.1016/j.engappai.2017.12.014
11. Lin, Y., Hu, Q., Liu, J., Duan, J.: Multi-label feature selection based on max-dependency and min-redundancy. Neurocomputing **168**, 92–103 (2015). https://doi.org/10.1016/j.neucom.2015.06.010

12. Mikolov, T., Chen, K., Corrado, G., Dean, J.: Efficient estimation of word representations in vector space. arXiv preprint arXiv:1301.3781 (2013)
13. Peng, H., Long, F., Ding, C.: Feature selection based on mutual information criteria of max-dependency, max-relevance, and min-redundancy. IEEE Trans. Pattern Anal. Mach. Intell. **27**(8), 1226–1238 (2005)
14. Peng, H., Fan, Y.: Feature selection by optimizing a lower bound of conditional mutual information. Inf. Sci. **418**, 652–667 (2017). https://doi.org/10.1016/j.ins.2017.08.036
15. Porter, M.F.: An algorithm for suffix stripping. Program **14**(3), 130–137 (1980). https://doi.org/10.1108/eb046814
16. Rehman, A., Javed, K., Babri, H.A., Saeed, M.: Relative discrimination criterion-a novel feature ranking method for text data. Exp. Syst. Appl. **42**(7), 3670–3681 (2015)
17. Sebastiani, F.: Machine learning in automated text categorization. ACM Comput. Surv. (CSUR) **34**(1), 1–47 (2002)
18. Tang, L., Duan, J., Xu, H., Liang, L.: Mutual information maximization based feature selection algorithm in text classification. Comput. Eng. Appl. **44**(13), 130–133 (2008). (in Chinese)
19. Uysal, A.K., Gunal, S.: A novel probabilistic feature selection method for text classification. Knowl. Based Syst. **36**, 226–235 (2012)
20. Vapnik, V.N.: Statistical learning theory. In: Encyclopedia of the Sciences of Learning, vol. 41, no. 4, p. 3185 (1998)
21. Xu, Y., Jones, G., Li, J., Wang, B., Sun, C.: A study on mutual information-based feature selection for text categorization. J. Comput. Inf. Syst. **3**(3), 1007–1012 (2007)
22. Yang, Y., Pedersen, J.O.: A comparative study on feature selection in text categorization. In: Proceedings of the 14th International Conference on Machine Learning, ICML 1997, pp. 412–420. Morgan Kaufmann Publishers Inc., San Francisco (1997)
23. Zhang, X., Wu, G., Dong, Z., Crawford, C.: Embedded feature-selection support vector machine for driving pattern recognition. J. Franklin Inst. **352**(2), 669–685 (2015)

CAT-BERT: A Context-Aware Transferable BERT Model for Multi-turn Machine Reading Comprehension

Cen Chen[1], Xinjing Huang[2], Feng Ji[3], Chengyu Wang[3], Minghui Qiu[3], Jun Huang[3], and Yin Zhang[2(✉)]

[1] Ant Group, Hangzhou, China
chencen.cc@antfin.com
[2] Zhejiang University, Hangzhou, China
{huangxinjing,zhangyin98}@zju.edu.cn
[3] Alibaba Group, Hangzhou, China
{zhongxiu.jf,chengyu.wcy,minghui.qmh,huangjun.hj}@alibaba-inc.com

Abstract. Machine Reading Comprehension (MRC) is an important NLP task with the goal of extracting answers to user questions from background passages. For conversational applications, modeling the contexts under the multi-turn setting is highly necessary for MRC, which has drawn great attention recently. Past studies on multi-turn MRC usually focus on a single domain, ignoring the fact that knowledge in different MRC tasks are transferable. To address this issue, we present a unified framework to model both single-turn and multi-turn MRC tasks which allows knowledge sharing from different source MRC tasks to help solve the target MRC task. Specifically, the Context-Aware Transferable Bidirectional Encoder Representations from Transformers (CAT-BERT) model is proposed, which jointly learns to solve both single-turn and multi-turn MRC tasks in a single pre-trained language model. In this model, both history questions and answers are encoded into the contexts for the multi-turn setting. To capture the task-level importance of different layer outputs, a task-specific attention layer is further added to the CAT-BERT outputs, reflecting the positions that the model should pay attention to for a specific MRC task. Extensive experimental results and ablation studies show that CAT-BERT achieves competitive results in multi-turn MRC tasks, outperforming strong baselines.

Keywords: Machine reading comprehension · Question answering · Transfer learning · Pre-trained language model

1 Introduction

Conversational search [22,31], a way of seeking information through conversations, has become a heated topic in the filed of Information Retrieval (IR). The

C. Chen and X. Huang—Equal contribution.

C. S. Jensen et al. (Eds.): DASFAA 2021, LNCS 12682, pp. 152–167, 2021.
https://doi.org/10.1007/978-3-030-73197-7_10

core task of conversational search is to answer user questions in a multi-turn scenario. In the literature, such task can modeled as multi-turn Machine Reading Comprehension (MRC) [27], whose goal is to answer user questions based on a given passage, by means of multi-turn interactions between machines and users.

According to previous research, there are mainly two challenges faced by multi-turn MRC [27]. i) Some questions that users raise in the dialogue are unanswerable as the questions may belong to a wrong topic, or the information from which the answers can the extracted is missing in the passage. Thus, the answers to this type of questions can be categorized as "CANNOT ANSWER". ii) As a question raised by users usually depends on previous answers sent by machines (i.e., chatbots), modeling the dialogue history is important and challenging for answering the question in the multi-turn setting. Therefore, many phenomena may occur, such as co-references and omissions. For example, to answer the question "What happened to him?", where "him" is covered in the previous answer "Mr. David found the dog was lost and became very sad at that moment", one has to know that "he" refers to "Mr. David".

To address the above-mentioned challenges, recent studies consider incorporating contextual information into MRC models. Typical methods include prepending previous questions and answers [32], adding history answer markers to the passage [16], or using attention mechanisms to select the dialogue history [17]. There are also studies applying context-aware neural networks such as recurrent neural networks and graph neural networks to convey the information in past turns [2,6,11,28]. However, these methods often ignore the fact that knowledge in many kinds of MRC tasks are transferable. To be more specific. both multi-turn and single-turn MRC tasks share some commonalities, such as unanswerable question recognition and knowledge reasoning. The knowledge learned from one MRC task may benefit the learning of other MRC tasks, especially when the tasks are closely related. Hence, it is crucial to leverage *transfer learning* to capture the shared knowledge from different multi-turn and single-turn MRC tasks for mutual reinforcement of the model performance.

To better leverage the cross-domain, cross-task knowledge, we present a unified framework to solve both single-turn and multi-turn MRC tasks, named Context-Aware Transferable Bidirectional Enoceder Representations from Transformers (CAT-BERT). The overview CAT-BERT framework is shown in Fig. 1. Inspired by the recent success of pre-trained language models, we extend Bidirectional Encoder Representations from Transformers (BERT) [4] to consider both history questions and answers to the model the contextual information. Thus, the learned text representations are more robust across different MRC tasks. Observing the fact that different MRC tasks may possess some unique task-dependent attributes [9], we further augment our model with a task-specific attention layer to capture the task-level importance of different layer outputs.

To the best of our knowledge, our study is the first to present a unified framework for both multi-turn and single-turn MRC tasks. Our framework can also be easily combined with other tasks by multi-task learning.

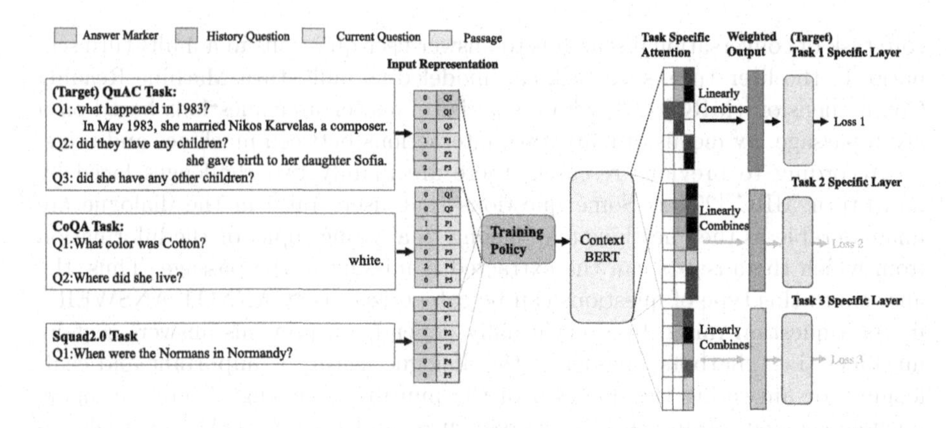

Fig. 1. An overview of the proposed Context-Aware Transferable BERT framework, which unifies three tasks, i.e., two multi-turn MRC tasks (QuAC [3] and CoQA [19]), and one single-turn MRC task (SQuAD 2.0 [18]). In the middle part, the context-aware BERT backbone is employed as the shared encoder, where the index in the input representation is the history answer index. The training policy selects the training data to feed into the context-aware BERT backbone, and then pass the data to task-specific attention and output layers to generate task-specific outputs.

We need to further claim that although multi-task learning has been recently studied for MRC (e.g., MultiQA [21], MT-DNN [10], MT-SAN [25]), CAT-BERT differs from these approaches in the following two perspectives. i) We focus on multi-turn MRC and propose a unified framework that can bridge the gaps between multi-turn and single-turn MRC tasks. ii) We seek to boost the performance of the MRC task in the target domain and better capture the transferable knowledge from other domains by considering task-specific attention.

To summarize, the contributions of this work are three-fold:

- We are the first to propose a unified framework named CAT-BERT for jointly learning multi-turn and single-turn MRC tasks. This sheds the light on how to leverage knowledge from large-scale single-turn MRC datasets to boost the performance of models for multi-turn MRC tasks.
- We propose a task-specific attention mechanism to model the task dependencies on each layer of CAT-BERT. Qualitative experiments show the attention weights learned are insightful and intuitive.
- Our method achieves competitive results in the QuAC leaderboard - a large-scale multi-turn MRC benchmark dataset. Extensive experiments demonstrate our method is effective. The model ablation studies show the importance of different integral parts of our model.

The reminder of this paper is summarized as follows. Section 2 briefly introduces the related work. The techniques of the CAT-BERT model is elaborated in Sect. 3. Experimental results are reported in Sect. 4. Finally, we draw the conclusion and discuss the future work in Sect. 5.

2 Related Work

In this section, we present a brief summarization on the related work of CAT-BERT, including the MRC task and transfer learning.

2.1 Machine Reading Comprehension

Our work is closely related to the MRC task. Unlike the typical question answering task [1,23,24], MRC [29] is a task to understand a given passage and use the passage to answer user questions. Different from single-turn MRC, we specifically focus on the multi-turn setting, where the user and the system interacts multiple times. The main challenging for multi-turn MRC is modeling the rich context of the multi-turns of human-machine interaction. In the literature, SDNet [32] takes the contexts into consideration by appending the history questions and answers to the inputs. HAE [16] adopts the marker to indicate the positions of history answers in the passage. HAM [17] further employs attention mechanisms to select the related history questions. However, these methods may fail when the context dependencies are more complicated.

There are also studies trying to model the contextual information using neural networks such as Recurrent Neural Networks (RNNs) and Graph Neural Networks (GNNs). For example, GraphFlow [2] views the relations between context words in each turn as a graph, and applies GNN to capture the information flow. FlowQA [6] employs RNNs to convey word representations of past turns and incorporates them with the current turn's representations. FlowDelta [28] further extends the FlowQA model to explicitly model the information gains by delta operations. MC^2 [30] adopts convolution neural networks to better capture the flow information in a more fine-grained manner with three perspectives. We notice that these studies only focus on one single domain for the MRC task, while we unify single-turn and multi-turn MRC tasks in different domains.

2.2 Transfer Learning

Moreover, our work is closely related to transfer learning, as we consider the joint learning of multiple MRC tasks in various domains. There are some studies to adopt multi-task and transfer learning to address MRC. The study in [20] transfers models trained on large span-level QA datasets to sentence-level QA datasets. MT-SAN [25] is a multi-task learning framework for MRC. The results of MT-SAN shows that performance on the target task can be improved by knowledge transfer. MT-DNN [10] further extends this idea to natural language understanding by multi-task training of a series of different tasks such as sentiment analysis, text matching and MRC. Li et al. [8] extend a similar method for the task of story ending prediction. Apart from these methods, MultiQA [21] is an empirical investigation of transfer learning in ten single-turn MRC tasks. The paper shows that training on multiple MRC datasets can make the underlying model more general and robust. However, these works do not consider multi-turn MRC tasks yet.

With the rapid development of deep neural networks, knowledge transferred from unsupervised tasks can be used for learning task-specific models. For instance, pre-trained word embeddings such as Word2vec [13] and Glove [14] are the key components for NLP tasks. With deeper models and more data, large-scale pre-trained language models such as ELMO [15], BERT [5], ALBERT [7], RoBERTa [12] and XLNet [26] show their effectiveness on many downstream NLP tasks. Different from the existing studies that address general NLP tasks, our study proposes a unified framework for both single-turn and multi-turn MRC tasks. We further design the CAT-BERT model to leverage information from source MRC tasks to help the learning of the target MRC task.

3 The CAT-BERT Model

In this section, we start with the task description. After that, we introduce the CAT-BERT model and its transfer learning procedure.

3.1 Task Description and Overall Framework

The CAT-BERT model is designed to address the following problem. Let $P = [w_{p_1}, w_{p_2}, \ldots, w_{p_i}]$ be the input passage, where w_{p_i} stands for i-th word in the passage. The history question answer pairs are represented as:

$$history = [(Q_1, A_1), (Q_2, A_2), \ldots, (Q_{n-1}, A_{n-1})] \quad (1)$$

where Q_i and A_i denote the question and the answer in the i-th turn. Given the passage P, the history question answer pairs *history* and the question Q_n in n-th turn, our goal is to predict the correct answer span $\hat{A}_n$ in the passage. Note that *history* is specifically employed to model the multi-turn MRC task. If there is no *history*, the problem setting will become the normal single-turn MRC task.

Figure 1 shows the high-level overview of the framework. It can be referred to as Context-Aware Transferable BERT, CAT-BERT for short. In this model, we design a unified input representations for both single-turn and multi-turn MRC tasks. The context-aware BERT model backbone is employed as the shared encoder for each task, with model modifications to handle both history questions and answers. After that, a token-wise task-specific attention is introduced to model the task dependencies on each layer. Finally, a dynamic training policy is adopted to train the model for multi-task learning of these tasks.

3.2 Context-Aware BERT Encoding

As shown in Fig. 2, our model augments the original BERT model with context modeling and a task-specific attention based transfer learning framework. Details of the model are introduced in the subsequent sections.

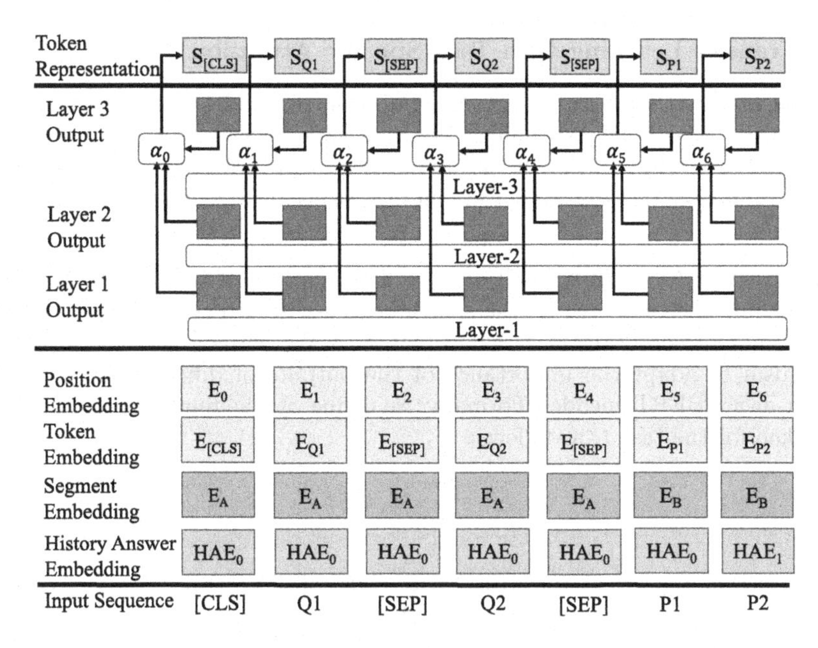

Fig. 2. The details of the context-aware BERT. We showcase an example of a BERT model with 3 layers. α_i is denoted as the token-wise layer attention for i-th token.

Modeling History Answers. Following [16], we introduce the History Answer Embedding (HAE) technique to the BERT model, in order to model history answers. Here, every token in the passage has an embedding index. If the embedding index of a token is non-zero, it means that this token is a part of the answer. For example, if the token "it" belongs to the answer of last third question, its embedding index is set to 3. Then, all embedding vectors of each token, including token embeddings, position embeddings, segment embeddings and history answer embeddings, will be summed together. Then the summed vector sequences serve as the input of the context-aware BERT encoder. For all questions and passage words that have not been used as an answer, the embedding index is 0.

Modeling History Questions. Besides history answers, it is also important to incorporate history questions. We consider a simple strategy to append the latest k history questions to the current n-th question. The history questions are separated by the special symbol `[SEP]`. For example, when k is 2, we append the previous two questions, in the format of followings:

$$\texttt{[CLS]}\, Q_n \,\texttt{[SEP]}\, Q_{n-1} \,\texttt{[SEP]}\, Q_{n-2} \,\texttt{[SEP]}\, P \,\texttt{[SEP]} \tag{2}$$

where Q_n and P refer to the tokens of the current n-th question and the passage.

3.3 Transfer Learning with Task-Specific Attention

We then present the transfer learning component for multi-task learning of different MRC tasks. Briefly speaking, the framework learns multi-turn and single-turn MRC tasks simultaneously, where all tasks share the context-aware BERT but with different task-specific layers and task-specific attention weights.

Task-Specific Attention. To learn the dependencies of tasks on specific layers, we equip the model with task-specific token-wise attention. We denote the i-th token representation in j-th layer as H_{ij}. We employ the soft attention mechanism to adapt the importance of the outputs of different levels in the context-aware BERT encoder. Formally, we define S_i^t, the final representation of i-th token for the task t, as follows:

$$S_i^t = \sum_j \alpha_{ij}^t H_{ij}, \tag{3}$$

where $t \in \{T_1, T_2, \ldots, T_k\}$ (i.e., the MRC task collection). α_{ij}^t is the attention weight corresponding to i-th token at j-th layer for the task t.

The attention weights are then defined as follows:

$$\alpha_{ij}^t = \frac{e^{H_{ij}^t * W_t + b_j^t}}{\sum_j e^{H_{ij}^t * W_t + b_j^t}}, \tag{4}$$

$$\sum_j \alpha_{ij}^t = 1. \tag{5}$$

Note that b_j^t in the above formula can be viewed as the layer bias. It is designed for helping the attention module to know the layer depth in the neural network, which plays a similar role to the position embeddings in the original token representations. Meanwhile, the attention weights are task-specific, which are essential for the model to capture the unique characteristics for different tasks.

We further denote the output from the shared context-aware BERT encoder as the matrix $S^t \in \mathbb{R}^{d \times m}$, where d is the dimension of each token's output vector and m is the length of input sequence. We add two output layers on S^t to predict the start position and end position of the answer spans. Formally, we have:

$$P_t^s = Softmax(W_s^t S^t + b_t^s), \tag{6}$$

$$P_t^e = Softmax(W_e^t S^t + b_t^e), \tag{7}$$

where t is the task index. W_s^t, $W_e^t \in R^{1 \times d}$, b_s^t and $b_e^t \in R^{1 \times 1}$ are the corresponding projection matrices and bias terms. s and e stand for the start and end positions of the answer spans, respectively. After we obtain the probabilities P_s^t and P_e^t for each word as the start and end positions of the answer span, during the inference phase, top c words with the highest probabilities are selected to form valid answer candidates.

Algorithm 1. CAT-BERT Training Procedure

Require: Batched context enhanced training examples $B = \{B^1, B^2, \ldots, B^K\}$ from the task set $\{T_1, T_2, \ldots, T_k\}$, where $B^t = \{B_1^t, B_2^t, \ldots, B_p^t\}$
Ensure: The CAT-BERT model M
1: Freeze parameters in task-specific attention and output layers. Set other parameters (w_t) to be trainable.
2: **while** *steps* $< N_1$ **do**
3: Sample a task t from a pre-defined task distribution.
4: Read a batch B_p^t from B^t.
5: Run through the CAT-BERT model to obtain the task-specific loss L_t.
6: Calculate the gradients $\nabla_{w_t} L_t$.
7: Update the parameters $w_t = w_t$ - $\lambda \nabla_{w_t} L_t$ where λ is the learning rate.
8: **end while**
9: Freeze the parameters of the context-aware BERT encoder. Set parameters in task-specific attention and output layers (w_t') to be trainable.
10: **while** *steps* $< N_2$ **do**
11: Sample a task t from a pre-defined task distribution.
12: Read a batch B_p^t from B^t.
13: Run through the CAT-BERT model to obtain the task-specific loss L_t.
14: Calculate the gradients $\nabla_{w_t'} L_t$.
15: Update the parameters $w_t' = w_t' - \lambda \nabla_{w_t'} L_t$.
16: **end while**

Learning Objectives. For a given MRC task, we adopt the negative log likelihood as the loss function. Formally, the sample-wise loss function for the start position is:

$$Loss_s^t = -\log P_{s_i}^t \tag{8}$$

The sample-wise loss for the end position $Loss_e^t$ can be obtained in a similar way. Hence, the total loss of the task t is the sum of two prediction losses, i.e.

$$Loss^t = \frac{Loss_s^t + Loss_e^t}{2}. \tag{9}$$

For simplicity, we omit all the regularization terms in the loss functions.

3.4 Dynamic Training Policy

The training policy is defined as a probability distribution for each MRC task, which can also be viewed as the coefficient weights of different tasks. By utilizing the dynamic training policy, our framework can be more flexible to handle different tasks. For ease of implementation, we adopt a simple strategy in this work, where we sample data from each task with equal probability. We leave the design and analysis of complicated training policies as future work.

Here we explain how to transfer knowledge from source MRC tasks to the target MRC task. The procedure is also shown in Algorithm 1. The whole process has two stages: (1) multi-task training and (2) task-specific fine-tuning:

Multi-task Training. We select a task t according to the training policy, and read the batch from the task t to do a forward pass. Then we make a backward pass and update all the parameters except task-specific attention parameters. This is achieved by simply set the attention weights α as fixed, where we set α_{ij}^{t} as 1 if j is the last layer's index, and 0 otherwise. This helps to train a shared context-aware BERT encoder.

Task-Specific Fine-Tuning. For this stage, we fix the parameters in the context-aware BERT encoder and only update the token-wise task-specific attention and task-specific output layers. This stage seeks to tune task-specific parameters to capture task-specific characters so as to boost the end-task performance.

4 Experiments

In this section, we conduct extensive experiments to examine our model performance. Firstly, we show that the CAT-BERT model is highly effective for multi-turn MRC. Next, we conduct experiments to examine the benefits brought by transfer learning and our context modeling method. Finally, we qualitatively evaluate the learned task-specific attention weights, and discuss the insightfulness of task-specific attention.

4.1 Datasets

In this work, all the experiments are conducted on three public MRC datasets: QuAC [3], SQuAD 2.0 [18] and CoQA [19]. The statistics of these datasets are shown in Table 1. We take SQuAD 2.0 [18] and CoQA [19] as source domain datasets and QuAC [3] as the target domain dataset. Both QuAC and CoQA are famous datasets for multi-turn conversational MRC tasks, thus multi-turn interaction knowledge learned from CoQA can be potentially transferred to QuAC. Below, we briefly introduce the three datasets:

- QuAC: The QuAC dataset aims to simulate the information-seeking scenario in real life. It contains 14k dialogues and 100k question-answers pairs in total. The passages are collected through crowdsourcing from one single domain in Wikipedia.
- CoQA: The CoQA dataset also belongs to conversational question answering, which contains 127k question-answer pairs and 8k conversations. Text passages are selected from seven different domains in Wikipedia. The abstractive answers and the supporting evidence are also provided.
- SQuAD 2.0: The SQuAD 2.0 dataset focuses on single-turn MRC. It augments the version 1.0 of the SQuAD dataset with additional 50k negative question answers.

Table 1. The statistics of QuAC, CoQA and SQuAD 2.0 datasets. Both QuAC and CoQA are multi-turn MRC datasets, while SQuAD is a singe-turn MRC dataset.

	QuAC			CoQA			SQuAD 2.0		
	Train	Dev	Test	Train	Dev	Test	Train	Dev	Test
Questions	83,568	7,354	7,353	108,647	7983	-	130,31	11,873	-
Dialogues	11,567	1,000	1,002	7199	500	-	19,035	1,204	-
Questions/dialogue	7.2	7.4	7.3	15.1	16.0	-	6.8	9.9	-
Tokens/question	6.5	6.5	6.5	5.5	5.5	-	9.9	10.1	-
Tokens/answer	15.1	12.3	12.3	9.3	9.2	-	3.2	3.2	-
Avg. tokens/passage	397	440	446	276	266	-	117	127	-
% Unanswerable	20.2	20.2	20.2	19.0	13.2	-	33.4	50.1	-

4.2 Experimental Setup

We follow the evaluation settings used in QuAC[1] to examine our method and all the baselines. We adopt three metrics to evaluate our model: the word-level F1 score measures the overlap between the prediction and gold answers, HEQQ refers to the percentage of questions in which the model exceeds human, and HEDD measures the percentage of dialogues where the model exceeds human.

In the experiments, we set the learning rate as 3e−5, the batch size as 12, and the max sequence length as 512. The training step is 24k for single task, and we double the training steps if we add another task. For the BERT-WWM model[2] on the three-task setting, the learning rate is set to 2e−5 and the training step is 48k. We sample batches from tasks with equal probability (which is the training policy). The max answer length is set to 50. For QuAC, CoQA and SQuAD 2.0 tasks, we append the token "CANNOT ANSWER" and "UNKNOWN" to the end of the passage. All the models are implemented with TensorFlow and trained with NVIDIA Tesla V100 GPU.

4.3 Overall Results

Table 2 shows the CAT-BERT performance on the QuAC test set[3]. Overall, our model achieves competitive results on the leaderboard, outperforming some strong baselines include BERT-FlowDelta, ConvBERT, BertMT, etc. For the results of history answer embeddings, we suggest readers to refer to the paper [16], as it makes a full comparison with the effects caused by different turns in history answer embeddings.

Note that there are two methods using data augmentation strategies achieve better results on the leaderboard. We will also consider such data augmentation strategies in near future as well. We also note that there is an concurrent study

[1] https://s3.amazonaws.com/my89public/quac/scorer.py.
[2] It refers to the BERT model with whole word masking.
[3] For fair comparison, we omit the ensemble methods and those methods with data augmentation in the QuAC leaderboard.

Table 2. A comparison of the proposed model and methods from the QuAC leaderboard. † means the score is copied from leaderboard. Note that, our final model was originally named as TransBERT in the leaderboard. To avoid confusion with other models, we name it as CAT-BERT in this work.

Methods	F1	HEQQ	HEQD	Total
BiDAF++ †	50.2	43.3	2.2	95.7
BiDAF++ w/2-Context †	60.1	54.8	4.0	118.9
BERT+HAE †	62.4	57.8	5.1	125.3
FlowQA†	64.1	59.6	5.8	129.5
GraphFlow†	64.9	60.3	5.1	130.3
BERT w/2-context†	64.9	60.2	6.1	131.2
HAM†	65.4	61.8	6.7	133.9
zhiboBERT†	67.0	63.5	8.6	139.1
ConvBERT†	68.0	63.5	9.1	140.6
BertMT†	68.9	65.2	8.9	143.0
Context-Aware-BERT †	69.6	65.7	8.1	143.4
BERT-FlowDelta†	67.8	63.6	**12.1**	143.5
CAT-BERT (Our model)	**71.4**	**68.1**	10.0	**149.5**

History-Att-TransBERT that achieves slightly better results, which shows the helpfulness of transfer learning. However, due to the lack of the published paper and the source code of the model, it is difficult to assess their method and compare our method with theirs.

4.4 Comparison of Transfer Policies

In this section, we compare the impacts of different transfer learning policies and the task-specific output layer. In Table 3, we conduct experiments using three types of transfer learning policies with different choices of source-domain tasks. We denote the sequential task learning setting as Seq and the mixed task learning as Mix. Our approach can be viewed as a mixed task training policy augmented with task-specific output layers, denoted as Co. From the results, we can see that our method achieves the best scores among all policies under the same tasks. Comparing with mixed task learning, our method also attains a better performance, due to the design of the task-specific output layer.

From the results, we can also find that sequential learning can obtain a little higher results than the mix training setting in F1 and HEQQ. However, in sequential learning, the training order does matter. The performance drops especially when a different type of task is inserted between two same tasks. Readers can observe the results of Seq (CoQA-SQuAD 2.0-QuAC) v.s. Seq (SQuAD 2.0-CoQA-QuAC). Additionally, a good training order requires prior knowledge. Thus, it might be easier yet beneficial to incorporate task-specific output lay-

Table 3. The experimental results of different transfer policies are presented. The task order in Seq stands for the task learning order. -L and -W refer to results obtained from BERT-Large and BERT-WWM models, respectively.

Policy	Tasks	F1	HEQQ	HEQD
None	QuAC	65.8	61.8	7.2
Seq	CoQA, QuAC	67.6	63.9	8.2
Seq	SQuAD 2.0, QuAC	67.2	63.2	8.6
Seq	CoQA & SQuAD 2.0 & QuAC	68.1	64.3	7.7
Seq	SQuAD 2.0 & CoQA & QuAC	**68.3**	**64.7**	**9.3**
Mix	CoQA & QuAC	67.6	63.4	8.4
Mix	SQuAD 2.0 & QuAC	66.8	62.9	**8.8**
Mix	QuAC & SQuAD 2.0 & CoQA	**68.4**	**64.5**	8.3
Co	CoQA & QuAC	67.9	64.3	8.9
Co	SQuAD 2.0 & QuAC	67.8	64.0	**9.6**
Co	QuAC & SQuAD 2.0 & CoQA	**68.7**	**65.0**	9.4
Co-L	QuAC & SQuAD 2.0 & CoQA	70.2	66.5	9.9
Co-W	QuAC & SQuAD 2.0 & CoQA	**73.1**	**69.9**	**13.3**

Table 4. The effects of task-specific attention mechanism. (w/o attn) means without the using of attention mechanisms.

Model	F1	HEQQ	HEQD	Total
CAT-BERT-12	68.6	64.8	9.2	142.6
CAT-BERT-6	68.7	65.0	9.4	143.1
CAT-BERT-3	**68.7**	**65.1**	**9.6**	**143.4**
CAT-BERT (w/o attn)	68.4	64.6	8.3	141.3
CAT-BERT-WWM-24	73.2	70.0	13.1	149.9
CAT-BERT-WWM-12	73.3	70.1	12.9	156.3
CAT-BERT-WWM-6	**73.3**	**70.1**	**13.4**	**156.8**
CAT-BERT-WWM-3	73.3	70.1	13.0	156.4
CAT-BERT-WWM (w/o attn)	73.1	69.9	13.3	156.3

ers in the mix policy to capture the task differences, so the potential negative transfer brought by the other tasks can be reduced. We also show the improvements made by employing better pre-trained language models. The results show that increasing the model size (see Co-L) and adopting the whole word masking technique for BERT (see Co-W) can improve all metrics greatly.

We notice that (SQuAD 2.0 & QuAC) always achieves better HEQD than (CoQA & QuAC). This means that the single-turn dataset SQuAD helps more than the multi-turn dataset CoQA for the QuAC task, although QuAC belongs to the category of the multi-turn MRC task. Benefiting from our unified framework, we can easily train a model to learn the shared knowledge between single-turn and multi-turn MRC tasks.

4.5 The Benefit from the Attention Mechanism

Table 4 shows the results with respect to the different numbers of layers that employ the attention mechanism. The upper part shows the performance of attention applied to the last 12, 6, 3 layers. The model size of the CAT-BERT backbone is the same as the BERT-Base model. The bottom part shows the performance of attention mechanism, with the backbone changed to the BERT-Large model with whole word masking. We can observe that, with the incorporation of the token-wise attention technique, the performance scores using both BERT-Base (denoted as CAT-BERT) and BERT-Large-WWM (denoted as CAT-BERT-WWM) as backbones are improved. For example, for CAT-BERT-WWM, F1 improves from 73.1 to 73.3, HEQQ from 69.9 to 70.1. This shows it is beneficial to incorporate the attention mechanism to capture the token-wise task-specific information to further improve the model performance.

Furthermore, we visualize the attention scores of the last three layers from our final model CAT-BERT-WWM on a randomly chosen example for both QuAC and SQuAD 2.0 tasks, as shown in Fig. 3. On the QuAC task (left), most of the attention weights are close to 1 on the last two layers; while on the SQuAD 2.0 task (right), the larger attention weights appear only in the last layer. The figure further demonstrates the necessity to introduce the task-specific token level attention mechanism to our framework to deal with the task differences among various MRC tasks.

4.6 Error Analysis

To analyze the shortcomings of our model, we randomly sample 50 wrong answers from predictions. The main errors can be categorized into two types:

- **Logical Error.** A typical error of the model is that the internal semantic changes in the passage are sometimes ignored. For example, the question is "Did Davies recover?", with two descriptions provided: "He subsequently collapsed after a drug overdose and was taken to hospital", and then "Ray recovered from his illness as well as his depression". The model only regards the first description as the answer and ignores the second description. This type of errors contributes mostly to the poor performance.
- **Indirect Description.** Although some answers are contained in the passages, they may be indirectly described, where complicated reasoning may be required for answering those implicit questions. For example, the question is "How profitable was the last album?" and the gold answer should be "the biggest-selling German music act in history". But the model gives a wrong prediction "CANNOT ANSWER". In this case, it is necessary to enhance the reasoning ability of the model, which is a non-trivial task.

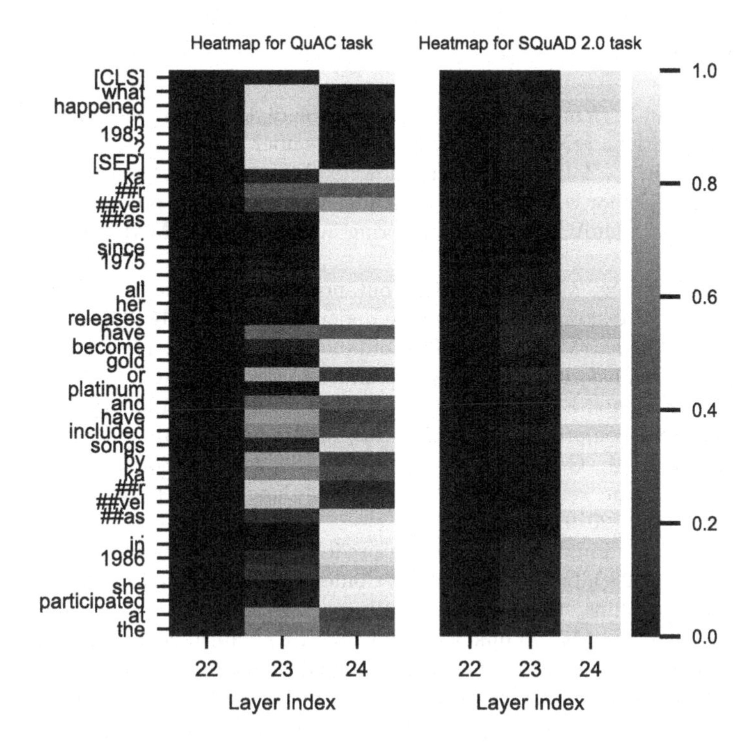

Fig. 3. The visualization of task-specific attention scores in the last three layers from the CAT-BERT-WWM model. The left is from the QuAC task, with the right from the SQuAD 2.0 task.

5 Conclusion and Future Work

In this work, we propose a deep BERT-based transfer learning model named CAT-BERT to unify the learning of multi-turn and single-turn MRC tasks. In this model, a task-specific token-wise attention mechanism is proposed to capture the dependencies on different layers for each task. Extensive evaluation results show thee proposed method is effective and achieves competitive results. Qualitative results also demonstrate that the attention weights learned by the model are insightful.

Acknowledgments. We would like to thank anonymous reviewers for their valuable comments. This work was partially sponsored by the NSFC projects (No. 61402403, No. 62072399, No. U19B2042), Chinese Knowledge Center for Engineering Sciences and Technology, MoE Engineering Research Center of Digital Library, Alibaba-Zhejiang University Joint Institute of Frontier Technologies, and the Fundamental Research Funds for the Central Universities. Any opinions, findings, and conclusions or recommendations expressed in this material are those of the authors and do not necessarily reflect those of the sponsors.

References

1. Chen, H., Liu, X., Yin, D., Tang, J.: A survey on dialogue systems: recent advances and new frontiers. arXiv:1711.01731 [cs] (November 2017)
2. Chen, Y., Wu, L., Zaki, M.J.: GraphFlow: exploiting conversation flow with graph neural networks for conversational machine comprehension, pp. 1230–1236 (2020)
3. Choi, E., et al.: QuAC: question answering in context. In: EMNLP, pp. 2174–2184 (2018)
4. Devlin, J., Chang, M.W., Lee, K., Toutanova, K.: BERT: pre-training of deep bidirectional transformers for language understanding. CoRR (2018)
5. Devlin, J., Chang, M.W., Lee, K., Toutanova, K.: BERT: pre-training of deep bidirectional transformers for language understanding. NAACL (2018)
6. Huang, H.Y., Choi, E., Yih, W.t.: FlowQA: grasping flow in history for conversational machine comprehension, CoRR abs/1810.06683 (2018)
7. Lan, Z., Chen, M., Goodman, S., Gimpel, K., Sharma, P., Soricut, R.: ALBERT: a lite BERT for self-supervised learning of language representations, CoRR abs/1909.11942 (2019)
8. Li, Z., Ding, X., Liu, T.: Story ending prediction by transferable BERT. In: Proceedings of the 28th International Joint Conference on Artificial Intelligence, IJCAI 2019, Macao, China, 10–16 August 2019, pp. 1800–1806 (2019)
9. Liu, N.F., Gardner, M., Belinkov, Y., Peters, M.E., Smith, N.A.: Linguistic knowledge and transferability of contextual representations. In: Proceedings of the 2019 Conference of the North American Chapter of the Association for Computational Linguistics: Human Language Technologies, NAACL-HLT 2019, pp. 1073–1094 (2019)
10. Liu, X., He, P., Chen, W., Gao, J.: Multi-task deep neural networks for natural language understanding. In: Proceedings of the 57th Conference of the Association for Computational Linguistics, ACL 2019, Florence, Italy, 28 July–2 August 2019, Volume 1: Long Papers, pp. 4487–4496 (2019)
11. Liu, X., Shen, Y., Duh, K., Gao, J.: Stochastic answer networks for machine reading comprehension. In: Proceedings of the 56th Annual Meeting of the Association for Computational Linguistics (Volume 1: Long Papers), , Melbourne, Australia, pp. 1694–1704. Association for Computational Linguistics (July 2018)
12. Liu, Y.: RoBERTa: a robustly optimized BERT pretraining approach. CoRR (2019)
13. Mikolov, T., Sutskever, I., Chen, K., Corrado, G.S., Dean, J.: Distributed representations of words and phrases and their compositionality. Adv. Neural. Inf. Process. Syst. **26**, 3111–3119 (2013)
14. Pennington, J., Socher, R., Manning, C.D.: GloVe: global vectors for word representation. In: Proceedings of the 2014 Conference on Empirical Methods in Natural Language Processing, EMNLP 2014, A meeting of SIGDAT, a Special Interest Group of the ACL, Doha, Qatar, 25–29 October 2014, pp. 1532–1543 (2014)
15. Peters, M.E., et al.: Deep contextualized word representations. In: Walker, M.A., Ji, H., Stent, A. (eds.) Proceedings of the 2018 Conference of the North American Chapter of the Association for Computational Linguistics: Human Language Technologies, NAACL-HLT 2018, pp. 2227–2237. Association for Computational Linguistics (2018)
16. Qu, C., Yang, L., Qiu, M., Croft, W.B., Zhang, Y., Iyyer, M.: BERT with history answer embedding for conversational question answering. In: SIGIR, pp. 1133–1136 (2019)

17. Qu, C., et al.: Attentive history selection for conversational question answering. In: CIKM, pp. 1391–1400 (2019)
18. Rajpurkar, P., Jia, R., Liang, P.: Know what you don't know: unanswerable questions for SQuAD. In: Gurevych, I., Miyao, Y. (eds.) Proceedings of the 56th Annual Meeting of the Association for Computational Linguistics, ACL 2018, Volume 2: Short Papers, Melbourne, Australia, 15–20 July 2018, pp. 784–789. Association for Computational Linguistics (2018)
19. Reddy, S., Chen, D., Manning, C.D.: CoQA: a conversational question answering challenge, CoRR abs/1808.07042 (2018)
20. Sun, Y., Cheng, G., Qu, Y.: Reading comprehension with graph-based temporal-casual reasoning. In: Proceedings of the 27th International Conference on Computational Linguistics, Santa Fe, New Mexico, USA, pp. 806–817. Association for Computational Linguistics (August 2018)
21. Talmor, A., Berant, J.: MultiQA: an empirical investigation of generalization and transfer in reading comprehension, CoRR abs/1905.13453 (2019)
22. Trippas, J.R., Spina, D., Cavedon, L., Joho, H., Sanderson, M.: Informing the design of spoken conversational search: perspective paper. In: CHIIR (2018)
23. Wang, R., Wang, M., Liu, J., Chen, W., Cochez, M., Decker, S.: Leveraging knowledge graph embeddings for natural language question answering. In: Proceedings of the 24th International Conference on Database Systems for Advanced Applications, DASFAA 2019, pp. 659–675 (January 2019)
24. Wu, H., Tian, Z., Wu, W., Chen, E.: An unsupervised approach for low-quality answer detection in community question-answering. In: Candan, S., Chen, L., Pedersen, T.B., Chang, L., Hua, W. (eds.) DASFAA 2017. LNCS, vol. 10178, pp. 85–101. Springer, Cham (2017). https://doi.org/10.1007/978-3-319-55699-4_6
25. Xu, Y., Liu, X., Shen, Y., Liu, J., Gao, J.: Multi-task learning with sample re-weighting for machine reading comprehension. In: Proceedings of the 2019 Conference of the North American Chapter of the Association for Computational Linguistics: Human Language Technologies, NAACL-HLT 2019, pp. 2644–2655 (2019)
26. Yang, Z., Dai, Z., Yang, Y., Carbonell, J.G., Salakhutdinov, R., Le, Q.V.: XLNet: generalized autoregressive pretraining for language understanding, CoRR abs/1906.08237 (2019)
27. Yatskar, M.: A qualitative comparison of CoQA, SQuAD 2.0 and QuAC. In: NAACL-HLT, pp. 2318–2323 (2019)
28. Yeh, Y.T., Chen, Y.N.: FlowDelta: modeling flow information gain in reasoning for conversational machine comprehension, CoRR abs/1908.05117 (2019)
29. Zhang, X., Yang, A., Li, S., Wang, Y.: Machine reading comprehension: a literature review, CoRR abs/1907.01686 (2019)
30. Zhang, X.: MC^2: Multi-perspective convolutional cube for conversational machine reading comprehension. In: ACL, pp. 6185–6190 (2019)
31. Zhang, Y., Chen, X., Ai, Q., Yang, L., Croft, W.B.: Towards conversational search and recommendation: system ask, user respond. In: CIKM (2018)
32. Zhu, C., Zeng, M., Huang, X.: SDNet: contextualized attention-based deep network for conversational question answering, CoRR abs/1812.03593 (2018)

Unpaired Multimodal Neural Machine Translation via Reinforcement Learning

Yijun Wang[1], Tianxin Wei[2], Qi Liu[1], and Enhong Chen[1](✉)

[1] Anhui Province Key Laboratory of Big Data Analysis and Application, University of Science and Technology of China, Hefei, China
wyjun@mail.ustc.edu.cn, {qiliuql,cheneh}@ustc.edu.cn

[2] Department of Computer Science and Technology, University of Science and Technology of China, Hefei, China
rouseau@mail.ustc.edu.cn

Abstract. End-to-end neural machine translation (NMT) heavily relies on parallel corpora for training. However, high-quality parallel corpora are usually costly to collect. To tackle this problem, multimodal content, especially image, has been introduced to help build an NMT system without parallel corpora. In this paper, we propose a reinforcement learning (RL) method to build an NMT system by introducing a sequence-level supervision signal as a reward. Based on the fact that visual information can be a universal representation to ground different languages, we design two different rewards to guide the learning process, i.e., (1) the likelihood of generated sentence given source image and (2) the distance of attention weights given by image caption models. Experimental results on the Multi30K, IAPR-TC12, and IKEA datasets show that the proposed learning mechanism achieves better performance than existing methods.

Keywords: Neural machine translation · Multimodal · Reinforcement learning

1 Introduction

End-to-end neural machine translation (NMT) has shown its superiority on several resource-rich language pairs [2,10,28], which is mainly attributed to the quality and scale of available parallel corpora [7]. However, it is usually quite difficult to collect adequate high-quality parallel corpora, since preparing such corpora is very expensive and time-consuming.

To tackle the issue where no parallel corpora are available, pivot-based NMT methods indirectly learn the alignment of the source and target languages with the help of another language [9,17,34]. Although promising results have been obtained, this kind of methods still demands large scale parallel source-pivot and pivot-target corpora. On the other hand, nowadays, we can easily find abundant monolingual text documents with rich multimedia content as the side information, e.g., text with photos or videos posted to social networking sites and

C. S. Jensen et al. (Eds.): DASFAA 2021, LNCS 12682, pp. 168–185, 2021.
https://doi.org/10.1007/978-3-030-73197-7_11

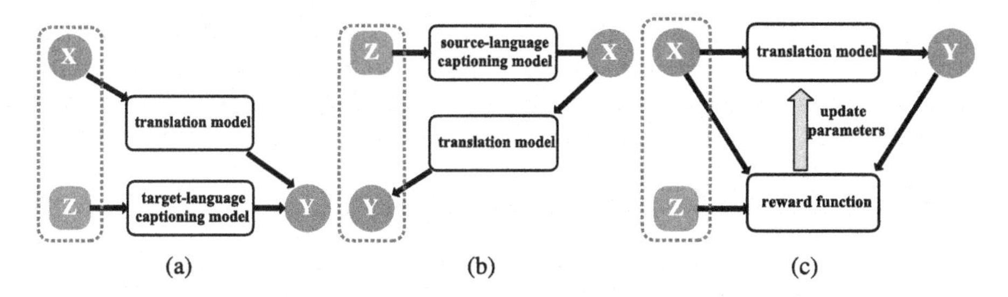

Fig. 1. (a) The teacher-student approach, (b) the 2-agent approach and (c) our reinforcement learning approach. X, Y and Z denote source, target sentences and image, respectively. We use a dashed-line box to denote that the image and sentence are paired.

blogs[1] [2]. These visual media are expected to be more or less correlated to the counterpart texts. How to utilize the multimodal content, especially image, to build NMT systems remains an open question.

To achieve modeling of source-to-target NMT using multimodal content only, the efforts in the literature can be divided into two classes. One is to learn a fixed-length modality-agnostic representation matching images and sentences in different languages in the same space [11,22]. Although this approach enables translation without parallel corpora, the use of a fixed-length vector is a bottleneck in improving translation performance [2,7]. The other class of work tries to generate pseudo parallel corpora by translating images into sentences in another language for given monolingual multimodal corpora using a pre-trained image captioning model. Then, the translation model could be learned with standard maximum likelihood estimation [6,7]. Although this kind of approach is easy to implement, the mistakes made by the image captioning model would propagate to the translation model, thus hurt the performance.

In fact, to effectively leverage the visual information for NMT, an important fact should be noticed is that we can generally understand the content of images taken in other countries regardless of which language we speak . The major challenge here is how to leverage such a fact to guide the learning of the translation model with unpaired multimodal content.

We address this challenge by casting the unpaired multimodal machine translation task as a reinforcement learning (RL) problem. Specifically, we introduce a sequence-level supervision signal by estimating the relevance between source and target sentences with the help of an image, which aims to evaluate the quality of the target sentence. Referring the translation model as *policy*, we formulate this sequence-level supervision signal as *reward* and directly optimize it. Intuitively, for a given source-language sentence and its corresponding image, a better translation in the target language should have a closer connection with the image. Based on this observation, we design two different reward functions to guide the

[1] http://blog.flickr.net/.
[2] https://mobile.twitter.com/.

learning process. Then, the policy is updated by the REINFORCE algorithm [29]. Compared with previous methods [6,7], our approach allows direct evaluation of source-target sentence pairs, without the need of translating images into sentences. Thus, this strategy avoids the problem of error propagation. Figure 1 shows the comparison between our method and previous approaches.

Our main contributions are summarized as follows:

(1) To effectively leverage the visual information for NMT, we proposed to cast the unpaired multimodal NMT task as an RL problem.
(2) We introduce sequence-level reward by estimating relevance between source and target sentences with the help of images. Specifically, we propose two kinds of reward to guide the training of the NMT model.
(3) Experiments on three translation tasks over three datasets show that the proposed rewards can provide good supervision on unlabeled multimodal corpora, and achieve better performance than existing methods.

2 Background

Neural Machine Translation. Given the source language space $\mathcal{X}$ and target language space $\mathcal{Y}$, an NMT model takes a sample from $\mathcal{X}$ as input and maps to space $\mathcal{Y}$. In common practice, the NMT model is represented by a conditional distribution $P_\theta(Y|X)$ parameterized by θ, where $X \in \mathcal{X}$ and $Y \in \mathcal{Y}$. In standard supervised learning, given a parallel corpus $\mathcal{D}_{X,Y}$, the translation model is learned by maximizing the likelihood of the training data:

$$\mathcal{L}(\theta) = \sum_{\langle X,Y\rangle \in \mathcal{D}_{X,Y}} \log P_\theta(Y|X). \tag{1}$$

Grounding Language to Visual Image. There exists monolingual multimodal content (images with text descriptions) on the Web. It is possible to ground natural language to a visual image through image captioning, which annotates a description for an input image with natural language [18,32]. Given a multimodal corpus in $\mathcal{D}_{Z,Y}$ where Z is an image and Y is a sentence describing Z in the target language, an image captioning model $P_{\phi_{Z\to Y}}(Y|Z)$ can be built, which "translates" an image to a sentence. The model parameters $\phi_{Z\to Y}$ can be learned by maximizing the log-likelihood of $\mathcal{D}_{Z,Y}$:

$$\mathcal{L}(\phi_{Z\to Y}) = \sum_{\langle Z,Y\rangle \in \mathcal{D}_{Z,Y}} \log P_{\phi_{Z\to Y}}(Y|Z) \tag{2}$$

Unpaired Multimodal NMT. In fact, parallel corpora are usually not readily available for low-resource language pairs or domains. Moreover, if these corpora are directly used for training, the linguistic dissimilarity and language mismatch between source language and target language will seriously decrease the translation performance [12]. Fortunately, it is possible to bridge the source and target languages with the multimodal information. The problem can also be called unsupervised multimodal neural machine translation [27]. Assuming that there are a source-language multimodal corpus $\mathcal{D}_{Z,X} = \{\langle Z^{(m)}, X^{(m)}\rangle\}_{m=1}^{M}$ and a target-language multimodal corpus $\mathcal{D}_{Z,Y} = \{\langle Z^{(n)}, Y^{(n)}\rangle\}_{n=1}^{N}$ where $\mathcal{D}_{Z,X}$ and $\mathcal{D}_{Z,Y}$ don't have to overlap, to achieve modeling of source-target NMT, intuitively there are two ways. First, as shown in Fig. 1(a), given $\mathcal{D}_{Z,Y}$, we can build a model $P_{\phi_{Z\to Y}}(Y|Z)$ which can translate an image to a target-language sentence. Thus, for source-language corpus $\mathcal{D}_{Z,X}$, we can translate the images to target-language sentences using $P_{\phi_{Z\to Y}}(Y|Z)$, forming pseudo source-target sentence pairs. Thus, source-target translation model can be build with Maximum Likelihood Estimation (MLE) training. This procedure is the same as the *teacher-student* approach in [6], except for replacing the pivot language with image. Similarly, for target-language corpus $\mathcal{D}_{Z,Y}$, we could form pseudo source-target corpus by translating images to source-language sentences with pre-trained image captioning model $P_{\phi_{Z\to X}}(X|Z)$, then achieve modeling of source-target translation via the *2-agent* communication game [7]. This procedure is shown in Fig. 1(b). In these approaches, the mistakes made by the image captioning model would be propagated to the translation model, thus hurt the translation performance. Different from these methods, in this paper, we propose a reinforcement learning approach to learn translation model. Below we formally define the RL training procedure, which is a general learning framework for training NMT model with unpaired multimodal documents only.

3 Methodology

3.1 Problem Definition

In this section, we define the problem of unpaired neural machine translation. On both the source and the target sides, the data comes in the paired form of $(x, z) \in \mathcal{X} \times \mathcal{Z}$ and $(y, z) \in \mathcal{Y} \times \mathcal{Z}$. Here we define two kinds of tasks. (1) **Zero-resource translation:** in this setting, the image that corresponds to x and the image that corresponds to y don't overlap, i.e., not the same image. (2) **Translation with comparable sentences:** in this settings, the source language x and target language y describe the same images. The main purpose is to learn a multi-modal translation model $X \to Y$ with the help of image Z. Note there is no explicit paired information cross two languages, making it hard to straightforwardly optimize the supervised likelihood. Our method can achieve excellent performance on both of the tasks.

3.2 Overview

In this section, we introduce a more straightforward approach to build NMT system leveraging the property that visual information can be a universal representation to ground different languages. The basic idea is that based on the property of visual information, we can estimate the relevance between source and target sentence by exploiting the relation between sentences and images, and explicitly

Source: A brown dog and a black dog are running
Generated: Ein brauner Hund und ein schwarzer Hund am Strand

Fig. 2. An example of En-De translation from the Multi30K dataset. The figure shows a picture of two dogs. The generated language of the translation model matches the image well but has a serious mismatch with the source language.

optimize the estimated relevance. Formally, we formulate the translation task as an RL problem as follows.

Specifically, the NMT model can be viewed as an *agent*, which interacts with the *environment*. The *environment* in our paper refers to the image caption model that interact with the model to produce feedback (reward). The parameters of this NMT model (agent) defines a *policy*, whose execution results in the agent picking an *action* a. In this case, an action refers to generating the next token at each time step. After taking each action, the agent updates its state. A terminal *reward* is received once the agent finished generating a complete sequence $\hat{Y}$, denoted as $R(\hat{Y})$. Note that the reward $R(\hat{Y})$ is a sentence-level reward, i.e., a scalar for each complete sentence $\hat{Y}$. Then, the goal of the training is to maximize the expected total reward. As show in Fig. 1(c), our approach is different from the teacher-student and 2-agent method since during training we don't need to use image captioning model to translate images into sentences. Instead, we leverage the alignment information between images and sentences to estimate the relevance between sentences. Thus, the problem of errors produced by the image captioning model propagating and hurting the translation model can be relieved.

3.3 Reward Computation

It is critical to set up appropriate rewards $R(\hat{Y})$ for RL training. In this section, we propose two methods to obtain the reward based on two observations respectively. To be first, the target language is a description of the image, so the generated language and the corresponding image need to be well matched.

(1) The Likelihood of the Generated Sample Given the Source Image. Our first observation is that *for an image-sentence pair* $\langle Z, X\rangle \in \mathcal{D}_{Z,X}$ *and a generated target-language sentence* $\hat{Y}$, if $\hat{Y}$ *is a good translations for* X, $\hat{Y}$ *should have close connection with* Z. Based on this observation, if $\hat{Y}$ has a closer connection with Z, it should be a better translation for X. Since an image captioning model could tell the probability of a sentence given an image, we train a target-language image captioning model $P_{\phi_{Z\to Y}}(Y|Z)$ with corpus $\mathcal{D}_{Z,Y}$. Then, the reward is set as:

$$R1(\hat{Y}) = \log P_{\phi_{Z\to Y}}(\hat{Y}|Z), \tag{3}$$

The above-proposed method can improve the consistency between the generated language and corresponding images, but it also has a serious problem. As shown in Fig. 2, the generated translation in German, which means the brown dog and the black dog are running on the beach. It matches the image well, but it has a serious mismatch between the source language. The source language focuses on the dog's movements, but the generated language focuses on the location beach, resulting in inconsistent translations. So the attention differences paid to the image areas cause the inconsistency in the translation. To tackle this problem, we propose to use the distance of attention weights of the source and generated language as the reward.

(2) The Distance of Attention Weights from Image Captioning Models. We have observed that *for an image-sentence pair* $\langle Z, X\rangle \in \mathcal{D}_{Z,X}$ *and a generated target-language sentence* $\hat{Y}$, *if* X *and* $\hat{Y}$ *are translations for each other,* $\hat{Y}$ *and* X *should have similar alignment information with* Z. Since an image captioning model with an attention mechanism can tell the alignment information between images and sentences with attention weights, we compute $R(\hat{Y})$ as the distance of the attention weights obtained from pre-trained image caption models for both languages. Specifically, we represent the source sentence and generated target sentence as $X = (x_1, x_2, \ldots, x_S)$ and $\hat{Y} = (\hat{y}_1, \hat{y}_2, \ldots, \hat{y}_T)$, where S and T denote the length of source and target sentences respectively. We use pre-trained CNNs [15] for image feature extraction and then denote the image as a matrix $Z = (\boldsymbol{z}_1, \boldsymbol{z}_2, \ldots, \boldsymbol{z}_L)$, where each of the L rows consists of a feature vector and the feature vector represents one grid in the image [4]. Then, for X and $\hat{Y}$, using soft attention computed by image captioning models $P_{\phi_{Z\to X}}(X|Z)$ and $P_{\phi_{Z\to Y}}(Y|Z)$, we could obtain the normalized alignment matrice between all the image patches and the target word to be emitted at time step, i.e., the attention weights $A_X^Z = (\boldsymbol{a}_{x,1}^Z, \boldsymbol{a}_{x,2}^Z, \ldots, \boldsymbol{a}_{x,S}^Z)$ and $A_{\hat{Y}}^Z = (\boldsymbol{a}_{\hat{y},1}^Z, \boldsymbol{a}_{\hat{y},2}^Z, \ldots, \boldsymbol{a}_{\hat{y},T}^Z)$ respectively, where each column $\boldsymbol{a}_{x,s}^Z$ or $\boldsymbol{a}_{\hat{y},t}^Z$ is a L-dimension vector representing the attention vector of the current word. Then, we compute the sum of the attention vector as:

$$\begin{aligned} \boldsymbol{\alpha}_X^Z &= \sum_{s=1}^{S} \boldsymbol{a}_{x,s}^Z, \\ \boldsymbol{\alpha}_{\hat{Y}}^Z &= \sum_{t=1}^{T} \boldsymbol{a}_{\hat{y},t}^Z. \end{aligned} \tag{4}$$

$\boldsymbol{\alpha}_X^Z$ and $\boldsymbol{\alpha}_{\hat{Y}}^Z$ are both L-dimension vectors. Since S and T have no guarantee to be equal, to make $\boldsymbol{\alpha}_X^Z$ and $\boldsymbol{\alpha}_{\hat{Y}}^Z$ comparable, we normalize them as $\hat{\boldsymbol{\alpha}}_X^Z$ and $\hat{\boldsymbol{\alpha}}_{\hat{Y}}^Z$ respectively. Then, the reward is computed as:

$$R2(\hat{Y}) = sim(\hat{\boldsymbol{\alpha}}_X^Z, \hat{\boldsymbol{\alpha}}_{\hat{Y}}^Z). \tag{5}$$

where sim is computed with cosine similarity in this work. Note that here we use a simple and effective method to calculate the attention weights similarity. There are many other potential designs. We leave these for future work.

3.4 Objective Function

Given the source-language multimodal corpus $\mathcal{D}_{Z,X} = \{\langle Z^{(m)}, X^{(m)}\rangle\}_{m=1}^{M}$, the goal of RL training is to maximize the expected reward:

$$\begin{aligned} \mathcal{O}_{RL} &= \sum_{m=1}^{M} \mathbb{E}_{\hat{Y} \sim P_\theta(\hat{Y}|X^{(m)})} R(\hat{Y}) \\ &= \sum_{m=1}^{M} \sum_{\hat{Y} \in \mathcal{Y}} P_\theta(\hat{Y}|X^{(m)}) R(\hat{Y}) \end{aligned} \tag{6}$$

where $\mathcal{Y}$ is the space of all candidate translation sentences, which is exponentially large due to the large vocabulary size, making it impossible to exactly maximize $\mathcal{O}_{RL}$. In practice, REINFORCE [29] is usually leveraged to approximate the above expectation via sampling $\hat{Y}$ from the policy P_θ, leading to the gradient of θ as:

$$\nabla_\theta \mathcal{O}_{RL} = \sum_{m=1}^{M} R(\hat{Y}) \nabla_\theta \log P_\theta(\hat{Y}|X^{(m)}) \tag{7}$$

Since the REINFORCE algorithm suffers from high variance in gradient estimation caused by using single sample $\hat{Y}$ to estimate the expectation, to reduce the variance, we subtract an average reward from the returned reward as in [24].

3.5 Training Details

Our training process consists of two steps. We first pre-train the image captioning and translation models and then train the translation model with the image captioning model fixed via RL. It is important to note that our translation model only pre-trains on the same dataset to provide an initialization that alleviates the instability of reinforcement learning [30]. We do not leverage additional data.

Specifically, since image captioning models are required when computing reward, we pre-train the captioning models with maximum likelihood estimation leveraging monolingual datasets $\mathcal{D}_{Z,X}$ and $\mathcal{D}_{Z,Y}$. For the translation model, as discussed in [24], the large action space (since the vocabulary is large) makes it extremely difficult to learn with an initial random policy. Thus, we pre-train the policy with some warm-start schemes. Since $\mathcal{D}_{Z,X}$ and $\mathcal{D}_{Z,Y}$ are not guaranteed to overlap, there may be two real-world situations: (1) $\mathcal{D}_{Z,X}$ and $\mathcal{D}_{Z,Y}$ don't overlap. In this scenario, we can pre-train the translation model with the 2-agent approach [7] and teacher-student (shorted as TS) approach [6]; (2) It is also possible in the real-world that each image is annotated with some source-language descriptions and some target-language descriptions, i.e., $\mathcal{D}_{Z,X}$ and $\mathcal{D}_{Z,Y}$ overlap with the same Z. It is worth mentioning that since each sentence is usually generated independently by different people, any source-target pair of descriptions for a given image could be considered a comparable translation pair but not translations of each other. Therefore, it is possible to use this kind of corpora either by considering the cross product of each source and target descriptions [3]. We adopt the cross product of each source and target descriptions as training corpus and use MLE to pre-train the translation model on the same datasets. The MLE objective is as follows:

$$\mathcal{O}_{MLE} = \sum_{i=1}^{M} \log P\left(Y^{(m)} \mid X^{(m)}\right) \tag{8}$$

After the pre-training of image captioning and translation models, we start the RL training process. We apply deep RL techniques [1] by adopting delayed policy with the purpose to prevent divergence. In order to further stabilize the training process, we linearly combine the MLE training objective and RL objective [21,31] as follows:

$$\mathcal{O}_{COM} = (1-\alpha) * \mathcal{O}_{MLE} + \alpha * \mathcal{O}_{RL}. \tag{9}$$

Especially, the MLE training objective is the same as that in the pre-training procedure. The entire training process is described in Algorithm 1.

Algorithm 1. REINFORCE algorithm for multimodal NMT

Require: Initial policy $P_\theta(Y|X)$, source image captioning model $P_{\phi_{Z\to X}}$ and target image captioning model $P_{\phi_{Z\to Y}}$ with random weights θ, $\phi_{Z\to X}$ and $\phi_{Z\to Y}$ respectively; a reward function $R(\hat{Y})$; monolingual multimodal corpora $\mathcal{D}_{Z,X} = \{\langle Z^{(m)}, X^{(m)}\rangle\}_{m=1}^{M}$ and $\mathcal{D}_{Z,Y} = \{\langle Z^{(n)}, Y^{(n)}\rangle\}_{n=1}^{N}$

1: Pre-train $P_{\phi_{Z\to X}}$, $P_{\phi_{Z\to Y}}$ and $P_\theta(Y|X)$
2: Initial delayed policy $P'_{\theta'}$ with the same weight: $\theta' = \theta$
3: **repeat**
4: Randomly receive an instance $\langle Z, X\rangle \in \mathcal{D}_{Z,X}$
5: Generate a sequence of actions $\hat{Y}$ from P'
6: Set the reward of the generated sequence as $r = R(\hat{Y})$
7: Update policy weight θ using the training objective in Eq. (9)
8: Update delayed policy with a constant γ: $\theta' = \gamma\theta + (1-\gamma)\theta'$
9: **until** model converged

4 Experiments

4.1 Datasets

Our method is evaluated on three publicly available multilingual multimodal datasets, i.e., Multi30K and IAPR-TC12 as in [7,22], and IKEA dataset [35]. Specifically, Multi30K [8] is a multilingual extension of Flickr30k corpus [33]. It has 29K, 1K and 1K images in the training, validation, and test splits respectively with English and German descriptions. We adopt the Multi30K task2 corpus in our experiments, which consists of 5 independently collected English and German descriptions per image, i.e., these descriptions The IAPR-TC12 dataset [13] has a total of 20K images as well as each image's multiple English descriptions and the corresponding German translations. Following [7], we use only the first description of each image and split the dataset into training, validation, and test sets with 18K, 1K, and 1K images respectively. For the IAPR-TC12 dataset, we evaluate our approach on both German-English (De-En) and English-German (En-De) tasks. To our knowledge, task2 of the Multi30K dataset and IAPR-TC12 dataset only have one language pair of English and German. To better understand the proposed method, we further evaluate our method on the English-German (En-De) and English-French (En-Fr) tasks of the IKEA dataset [35]. The data splits are the same as [35].

To fit the situation where the source and target multimodal corpora don't overlap, following [7], we randomly split the images in the training and validation datasets into two parts with equal size. Thus, the two splits have no overlapping images, and we have no direct English- German parallel corpus. The sentences in the datasets are normalized and tokenized with the Moses Toolkit [20]. For Multi30K and IKEA dataset, we adopt joint byte pair encoding [25] with 10K merge operations to reduce vocabulary size. For the IAPR-TC12 dataset, we construct the vocabulary with words appearing more than 5 times in the training set and replace the remaining words with UNK.

Table 1. BLEU scores on Multi30K German-English translation test set with testing methods Test-1 and Test-2 in zero-resource scenario.

Training strategy	Test-1	Test-2
3-way model	15.9	14.2
UMNMT	19.9	18.2
2-agent-PRE	19.5	18.0
2-agent-JOINT	20.1(+0.6)	18.2(+0.2)
TS-PRE	19.8	19.2
TS-JOINT	20.3(+0.5)	19.5(+0.3)
2-agent-PRE+RL-R1	21.2(+1.7)	19.4(+1.4)
2-agent-PRE+RL-R2	**21.5(+2.0)**	19.9(+1.9)
TS-PRE+RL-R1	20.9(+1.1)	20.1(+0.9)
TS-PRE+RL-R2	21.1(+1.3)	**20.3(+1.1)**

4.2 Baseline Methods

To demonstrate the effectiveness of our method, we compare our implementations with state-of-the-art baselines as follows.

(1) leftskip8pt *3-way model* [22]. This method adopts an end-to-end training strategy and trains the decoder with image and description.
(2) leftskip8pt *UMNMT* [27]. This method is the state-of-the-art zero-resource multimodal neural machine translation method. It adopts the Transformer model with a controllable attention mechanism that encodes both image and language and leverages cycle-consistency loss. For a fair comparison, we train the model from the used datasets and do not use the model that was not pre-trained on the tens of millions of data. We also use the same Transformer architecture with the same number of layers and feature dimensions.
(3) leftskip8pt *2-agent-PRE, 2-agent-JOINT* [7]. In this method, 2-agent-PRE keeps the captioner fixed and only trains the translator until the model converges, then 2-agent-JOINT jointly trains the captioner and translator based on 2-agent-PRE.
(4) leftskip8pt *TS-PRE, TS-JOINT* [6]. Similarly, TS-PRE keeps the captioner fixed and only trains the translator until the model converges, then TS-JOINT jointly trains the captioner and translator based on TS-PRE.

4.3 Implementation Details

To extract image features, we follow the suggestion of [3] and adopt ResNet-50 network [15] pre-trained on ImageNet without fine-tuning. We use the (14,14,1024) feature map of the res4fx (end of Block-4) layer after ReLU. Then, we vectorise this 3-tensor into a 196×1024 matrix.

We adopt the Transformer[3] model with *base* setting as defined in [28] for all the translation tasks. For pre-training, we consider two situations: (1) when the source and target multimodal corpora don't overlap, we pre-train the translation model with 2-agent-PRE and TS-PRE on both datasets in this *zero-resource* scenario. In this scenario, our proposed reward is calculated based on the source image that corresponds to the source sentence; (2) when the source and target multimodal corpora overlap with the same images, we pre-train the translation model with MLE on Multi30K dataset with these *comparable sentences*. The optimizer used for MLE is Adam [19], and we follow the same learning rate schedule in [28]. During training, roughly 4, 096 source tokens, and 4, 096 target tokens are paired in one mini-batch. Each model is trained using a single Tesla K80 GPU. For RL training, the model is initialized with parameters of the pre-trained model, and we continue training it with a learning rate of 0.0001. Hyper-parameter α is 0.5 and 0.7 for R1 and R2 respectively, and the delay constant γ is 0.1.

For evaluation, all models are quantitatively evaluated with BLEU [23]. Especially, for the Multi30K dataset, since each image is paired with 5 English and 5

[3] https://github.com/tensorflow/tensor2tensor.

Table 2. BLEU scores on IAPR-TC12 English-German and German-English translation test sets in zero-resource scenario.

Training strategy	De-En	En-De
3-way model	13.9	8.6
UMNMT	19.3	14.5
2-agent-PRE	18.7	14.4
2-agent-JOINT	19.2(+0.5)	14.6(+0.2)
TS-PRE	17.1	13.9
TS-JOINT	17.5(+0.4)	14.1(+0.2)
2-agent-PRE+RL-R1	20.1(+1.4)	15.6(+1.2)
2-agent-PRE+RL-R2	**20.5(+1.8)**	**15.7(+1.3)**
TS-PRE+RL-R1	18.9(+1.8)	15.0(+1.1)
TS-PRE+RL-R2	19.3(+2.2)	15.5(+1.6)

German descriptions in the test set, we adopt two methods to evaluate the translation models: (1) We follow the setting in [7], generating a target description for each source sentences and picking the one with the highest probability. The evaluation is performed against the corresponding 5 target descriptions. This method is denoted as *Test-1*; (2) We generate a target description for each 5 source sentences and calculate the BLEU score for each generated description against the corresponding 5 target descriptions. Then, we use the average of the calculated BLEU scores as the final result. This method is denoted as *Test-2*. During validation and testing, we set the beam search size to be 5 for the translation model.

4.4 Main Results

We first evaluate our proposed different strategies in comparison with baselines on Multi30K, IAPR-TC12, and IKEA datasets.

Zero-Resource Translation. We first show the results when the two monolingual multimodal corpora don't overlap. We first evaluate our method on the Multi30K De-En translation task with the testing methods Test-1 and Test-2. The results are shown in Table 1. In this scenario, TS-PRE and 2-agent-PRE are adopted for pre-training translation models before the RL procedure. RL-R1 and RL-R2 represent our reinforcement learning method with rewards R1 and R2 respectively. From Table 1, we can see that both the reinforcement learning method with reward R1 and R2 outperform the 2-agent and teacher-student methods across testing methods. Our best-performed methods are RL-R2 pre-trained with 2-agent-PRE for Test-1, and RL-R2 pre-trained with TS-PRE for Test-2. These two methods have an improvement of 1.4/0.8 BLEU points over the best baselines.

Table 3. BLEU scores on IKEA English-German and English-French translation test sets in zero-resource scenario.

Training strategy	En-De	En-Fr
3-way model	22.1	23.3
UMNMT	33.5	34.7
2-agent-PRE	33.2	34.4
2-agent-JOINT	33.6(+0.4)	34.6(+0.2)
TS-PRE	32.8	34.3
TS-JOINT	33.1(+0.3)	34.4(+0.1)
2-agent-PRE+RL-R1	34.4(+1.2)	35.6(+1.2)
2-agent-PRE+RL-R2	**34.7(+1.5)**	**36.0(+1.6)**
TS-PRE+RL-R1	34.2(+1.4)	35.7(+1.4)
TS-PRE+RL-R2	34.6(+1.8)	35.9(+1.6)

We also evaluate our method on IAPR-TC12 En-De and De-En translation tasks. The results are shown in Table 2. We can see that our proposed method also outperforms all the baseline approaches on both translation tasks. Specifically, our best methods have an improvement of 1.3/1.1 BLEU points on De-En/En-De translation over the best baselines.

Similarly, for the results on IKEA En-De and En-Fr translation tasks shown in Table 3, our proposed method achieves superior performance compared with baselines. Our method can obtain obvious improvement in both language pairs of En-De and En-Fr.

From the results above, we can have the following findings. First, our approach achieves the best results on all three data sets, substantially exceeding the baselines. It demonstrates the effectiveness of our method. Then we can find that our proposed R2 works better than R1 in all cases. It verifies the importance of modeling the distance of attention weights of source and generated sentences into translation models. Last, the state-of-the-art UMNMT method has achieved only a small improvement. On the one hand, it introduces too many parameters in the training process, which makes the model difficult to train. Moreover, it shows the superiority of using reinforcement learning to reinforce the consistency between images and sentences.

Translation with Comparable Sentences. We also evaluate our method when the two monolingual multimodal corpora overlap with the same images on the Multi30K dataset. The translation model is pre-trained with MLE on the cross product of each source and target sentences, which is denoted as PRE. The BLEU scores of different training strategies are shown in Table 4. We can see that our method with both rewards achieves obvious improvement over MLE pre-training, while the 2-agent and teacher-student methods can't gain any improvements in this scenario. The possible reason is that the generated sentences by

Table 4. Comparison with previous work with comparable sentences over the Multi30K test set with Test-1 and Test-2.

Training strategy	Test-1	Test-2
PRE	30.4	27.5
PRE+2-agent-JOINT	30.2	27.1
PRE+TS-JOINT	30.1	26.9
PRE+RL-R1	32.1	29.1
PRE+RL-R2	**32.8**	**29.5**

Reference image	Source (German)	Target (English)
	1. ein junges mädchen beim laufen . 2. ein mädchen rennt in einem abgesperrten bereich entlang . 3. eine läuferin läuft . 4. das mädchen rennt in einem wettspiel . 5. spielende kinder auf einer wiese neben einer sandbahn , ein mädchen lauft entlang der bahn	Ref 1: a young girl in dark shorts and a blue tank top runs on the grass near an orange cone and tape . Ref 2: a young lady wearing blue and black is running past an orange cone . Ref 3: a young woman running by an orange ribbon . Ref 4: a young girl running by herself in a park . Ref 5: a girl in a blue tank top winning a race . 2-agent-PRE: a woman running in a race . RL-R1: a woman in black is running in a race . RL-R2: a woman in a blue shirt is running in a race .
	1. ein junger mann mit schutzhelm klettert eine felswand herauf. 2. bergsteiger in weißen helm hält sich an einer lila seil . 3. ein mann beim klettern mit seil gesichert . 4. der mann mit dem weißen helm im klettergeschirr klettert an dem felsen . 5. ein bergkletterer übt an einer steinwand in geringer höhe das klettern .	Ref 1: a man in a white shirt and helmet is using climbing equipment . Ref 2: a man with a white shirt and helmet is climbing a mountain . Ref 3: a young man wearing a white helmet climbing up a rock wall Ref 4: a man in a harness climbing a rock wall Ref 5: man rock climbing looking up the rock 2-agent-PRE Hyp 1: a man is walking up a mountain . Hyp 2: a man in a red shirt is rock climbing . Hyp 3: a man is hanging from a pole . Hyp 4: a man in a white hard hat is climbing a mountain . Hyp 5: a man in a red shirt is rock climbing . RL-R2 Hyp 1: a man in a white hard hat is standing on a mountain . Hyp 2: a man in a white shirt is rock climbing . Hyp 3: a man in a white shirt is hanging from a pole . Hyp 4: a man in a white hard hat is standing on a rock . Hyp 5: a man in a white shirt is rock climbing .

Fig. 3. Examples of translations from the Multi30K test set. The first example is translated using Test-1 as [7], while the second example is translated using Test-2. For the first example, we pre-train the translation model with 2-agent-PRE, then continue training with RL-R1 and RL-R2. For the second example, we only show the translation results of the 2-agent-PRE and the better-performed RL-R2. We highlight the words that distinguish the systems' results in blue, red and green. Red words are marked for correct translations in hypotheses compared with blue words in references, and green words are marked for incorrect translations in hypotheses compared to blue words. (Color figure online)

the image captioning model in the baseline methods are low-quality compared two the cross product of each source and target sentences, thus can't help the translation model to get better performance.

4.5 Impact of Hyper-parameter

As shown in Eq. (9), the hyper-parameter α controls the trade-off between MLE and RL objectives. To show the impact of this hyper-parameter, we evaluate

Table 5. BLEU scores for different α on De-En translation Multi30K test set for RL-R1.

α	0.1	0.3	0.5	0.7	0.9
Test1	30.9	31.7	32.1	31.2	30.7
Test2	27.8	28.5	29.1	28.1	27.9

Table 6. BLEU scores for different α on De-En translation Multi30K test set for RL-R2.

α	0.1	0.3	0.5	0.7	0.9
Test1	30.4	30.7	32.1	32.8	30.9
Test2	27.7	27.8	28.7	29.5	28.0

the model performance on the De-En Multi30K test set with different α in the scenario of corpora overlapping. Specifically, for RL-R1 and RL-R2, we set α both to be [0.1, 0.3, 0.5, 0.7, 0.9] in our experiments. The results are presented in Table 5 and Table 6. We find that when α is set to be 0.5 and 0.7 for RL-R1 and RL-R2 respectively, our method achieves the best performance.

4.6 Case Study

In Fig. 3, we provide some qualitative comparisons between the translations from the pre-training method 2-agent-PRE and our RL method. In the first example, our RL-R1 properly translates the words "in black" and RL-R2 properly translates the words "in a blue shirt", while PRE didn't tell anything about the wearing. In the second example, each translation of our RL-R2 correctly translates "white shirt" or "white hard hat", while the translations of PRE tends to include the wrong words "red shirt" or include nothing about the wearing. From these examples, we can see that our RL method can guide the translation model to focus more on the image, thus correctly translate the information in the image and improve translation quality.

5 Related Work

The related research topics can be classified into the following three categories: (1) multimodal neural machine translation, (2) pivot-based neural machine translation, and (3) reinforcement learning for sequence prediction.

Multimodal Neural Machine Translation. This task aims to use images as well as parallel corpora to improve the translation performance. It has been shown that image modality can benefit NMT by relaxing ambiguity in alignment that cannot be solved by texts only [4,16,35]. This task is much easier than ours because, in its setting, multilingual descriptions for the same images are available in the training dataset, and an image is part of the query in both training and testing phases [7]. The unsupervised multimodal NMT is proposed in [27]. However, they did not consider the consistency of the image and language at the sentence level. Their introduction of too many parameters also hurts training.

Pivot-Based Neural Machine Translation. Another line of work has been to train the NMT system from non-parallel data with the help of another modality, which is called the pivot-based machine translation. Specifically, researchers have tried to build multilingual NMT systems trained by other language pairs to enable translation [9,17] with non-parallel data for the intended translation pair. In addition to the multilingual methods, several authors proposed to train the translation model in more direct ways. For example, [6] proposed a teacher-student framework under the assumption that parallel sentences have close probabilities of generating a sentence in a third language. [34] maximized the expected likelihood to train the intended source-to-target model. Nonetheless, all these methods still require that source-pivot and pivot-target parallel corpora are available. Besides languages, images are also used as the pivot to build NMT systems. Zero-resource NMT by utilizing image as a pivot was first achieved by training multimodal encoders to share common semantic representation [22]. To overcome the bottleneck of the fixed-length vector in this method, [7] proposed a 2-agent approach that jointly trains the translation and image captioning model.

Reinforcement Learning for Sequence Prediction. In the sequence prediction task, reinforcement learning is always used to learn and refine model parameters according to task-specific reward signals [5,14,30]. In [24], the authors proposed to train a neural translation model with the objective of optimizing the sentence-level BLEU score. [26] proposed to adopt minimum risk training to minimize the task-specific loss on NMT training data. Instead of the REINFORCE algorithm used in the above two works, [1] further optimizes the policy by the actor-critic algorithm.

6 Conclusion and Future Work

In this work, to tackle the challenging task of training an NMT system from just unpaired multimodal data, we successfully deploy a reinforcement learning (RL) method to build the NMT system by introducing a sequence-level supervision signal as a reward. Experiments on German-English, English-German, and English-French translation over the IAPR-TC12, Multi30K, and IKEA datasets show that our proposed reinforcement learning mechanism can significantly outperform the existing methods.

In the future, we will continue trying to ground the visual context into the translation model, such as using a shared encoder over the source, target sentences, and image, and then constraining them to be similar in order to constrain both the source and target representations to be faithful to the image. Moreover, we also would like to better understand the proposed method on larger training corpora and alternative language pairs.

Acknowledgement. This research was partially supported by grants from the National Key Research and Development Program of China (No. 2016YFB1000904) and the National Natural Science Foundation of China (Nos. 61727809, 61922073 and U20A20229).

References

1. Bahdanau, D., et al.: An actor-critic algorithm for sequence prediction (2017)
2. Bahdanau, D., Cho, K., Bengio, Y.: Neural machine translation by jointly learning to align and translate. In: 3rd International Conference on Learning Representations (2015)
3. Caglayan, O., et al.: Does multimodality help human and machine for translation and image captioning? In: 1st Conference on Machine Translation, vol. 2, pp. 627–633 (2016)
4. Calixto, I., Liu, Q., Campbell, N.: Doubly-attentive decoder for multi-modal neural machine translation. In: Proceedings of the 55th Annual Meeting of the Association for Computational Linguistics (Volume 1: Long Papers), vol. 1, pp. 1913–1924 (2017)
5. Chen, Y., Wu, L., Zaki, M.J.: Reinforcement learning based graph-to-sequence model for natural question generation. In: 8th International Conference on Learning Representations (2020)
6. Chen, Y., Liu, Y., Cheng, Y., Li, V.O.: A teacher-student framework for zero-resource neural machine translation. In: Proceedings of the 55th Annual Meeting of the Association for Computational Linguistics (Volume 1: Long Papers), vol. 1, pp. 1925–1935 (2017)
7. Chen, Y., Liu, Y., Li, V.O.: Zero-resource neural machine translation with multi-agent communication game. In: 32nd AAAI Conference on Artificial Intelligence (2018)
8. Elliott, D., Frank, S., Sima'an, K., Specia, L.: Multi30k: multilingual English-German image descriptions. In: Proceedings of the 5th Workshop on Vision and Language, pp. 70–74 (2016)
9. Firat, O., Sankaran, B., Al-Onaizan, Y., Vural, F.T.Y., Cho, K.: Zero-resource translation with multi-lingual neural machine translation. In: Proceedings of the 2016 Conference on Empirical Methods in Natural Language Processing, pp. 268–277 (2016)
10. Gehring, J., Auli, M., Grangier, D., Yarats, D., Dauphin, Y.N.: Convolutional sequence to sequence learning. In: Proceedings of the 34th International Conference on Machine Learning, vol. 70, pp. 1243–1252. JMLR.org (2017)
11. Gella, S., Sennrich, R., Keller, F., Lapata, M.: Image pivoting for learning multilingual multimodal representations. In: Proceedings of the 2017 Conference on Empirical Methods in Natural Language Processing, pp. 2839–2845 (2017)
12. Graça, Y.K.M., Ney, H.: When and why is unsupervised neural machine translation useless? In: 22nd Annual Conference of the European Association for Machine Translation, p. 35 (2020)
13. Grubinger, M., Clough, P., Müller, H., Deselaers, T.: The IAPR TC-12 benchmark: a new evaluation resource for visual information systems. In: International Workshop OntoImage, vol. 2 (2006)

14. Hashimoto, K., Tsuruoka, Y.: Accelerated reinforcement learning for sentence generation by vocabulary prediction. In: Proceedings of the 2019 Conference of the North American Chapter of the Association for Computational Linguistics: Human Language Technologies, pp. 3115–3125 (2019)
15. He, K., Zhang, X., Ren, S., Sun, J.: Deep residual learning for image recognition. In: Proceedings of the IEEE Conference on Computer Vision and Pattern Recognition, pp. 770–778 (2016)
16. Hitschler, J., Schamoni, S., Riezler, S.: Multimodal pivots for image caption translation. In: Proceedings of the 54th Annual Meeting of the Association for Computational Linguistics (Volume 1: Long Papers), vol. 1, pp. 2399–2409 (2016)
17. Johnson, M., et al.: Google's multilingual neural machine translation system: enabling zero-shot translation. Trans. Assoc. Comput. Linguist. **5**, 339–351 (2017)
18. Karpathy, A., Fei-Fei, L.: Deep visual-semantic alignments for generating image descriptions. In: Proceedings of the IEEE Conference on Computer Vision and Pattern Recognition, pp. 3128–3137 (2015)
19. Kingma, D.P., Ba, J.: Adam: a method for stochastic optimization. In: 3rd International Conference on Learning Representations (2015)
20. Koehn, P., et al.: Moses: open source toolkit for statistical machine translation. In: Proceedings of the 45th Annual Meeting of the Association for Computational Linguistics Companion Volume Proceedings of the Demo and Poster Sessions, pp. 177–180 (2007)
21. Li, J., Monroe, W., Shi, T., Jean, S., Ritter, A., Jurafsky, D.: Adversarial learning for neural dialogue generation. In: Proceedings of the 2017 Conference on Empirical Methods in Natural Language Processing, pp. 2157–2169 (2017)
22. Nakayama, H., Nishida, N.: Zero-resource machine translation by multimodal encoder-decoder network with multimedia pivot. Mach. Transl. **31**(1–2), 49–64 (2017). https://doi.org/10.1007/s10590-017-9197-z
23. Papineni, K., Roukos, S., Ward, T., Zhu, W.J.: BLEU: a method for automatic evaluation of machine translation. In: Proceedings of the 40th Annual Meeting on Association for Computational Linguistics, pp. 311–318. Association for Computational Linguistics (2002)
24. Ranzato, M., Chopra, S., Auli, M., Zaremba, W.: Sequence level training with recurrent neural networks. In: 4th International Conference on Learning Representations (2016)
25. Sennrich, R., Haddow, B., Birch, A.: Neural machine translation of rare words with subword units. In: Proceedings of the 54th Annual Meeting of the Association for Computational Linguistics (Volume 1: Long Papers), vol. 1, pp. 1715–1725 (2016)
26. Shen, S., et al.: Minimum risk training for neural machine translation. In: Proceedings of the 54th Annual Meeting of the Association for Computational Linguistics (Volume 1: Long Papers), vol. 1, pp. 1683–1692 (2016)
27. Su, Y., Fan, K., Bach, N., Kuo, C.C.J., Huang, F.: Unsupervised multi-modal neural machine translation. In: Proceedings of the IEEE Conference on Computer Vision and Pattern Recognition, pp. 10482–10491 (2019)
28. Vaswani, A., et al.: Attention is all you need. In: Advances in Neural Information Processing Systems, pp. 5998–6008 (2017)
29. Williams, R.J.: Simple statistical gradient-following algorithms for connectionist reinforcement learning. Mach. Learn. **8**(3–4), 229–256 (1992). https://doi.org/10.1007/BF00992696
30. Wu, L., Tian, F., Qin, T., Lai, J., Liu, T.Y.: A study of reinforcement learning for neural machine translation. In: Proceedings of the 2018 Conference on Empirical Methods in Natural Language Processing, pp. 3612–3621 (2018)

31. Wu, Y., et al.: Google's neural machine translation system: bridging the gap between human and machine translation. arXiv preprint arXiv:1609.08144 (2016)
32. Xu, K., et al.: Show, attend and tell: Neural image caption generation with visual attention. In: International Conference on Machine Learning, pp. 2048–2057 (2015)
33. Young, P., Lai, A., Hodosh, M., Hockenmaier, J.: From image descriptions to visual denotations: new similarity metrics for semantic inference over event descriptions. Trans. Assoc. Comput. Linguist. **2**, 67–78 (2014)
34. Zheng, H., Cheng, Y., Liu, Y.: Maximum expected likelihood estimation for zero-resource neural machine translation. In: IJCAI, pp. 4251–4257 (2017)
35. Zhou, M., Cheng, R., Lee, Y.J., Yu, Z.: A visual attention grounding neural model for multimodal machine translation. In: Proceedings of the 2018 Conference on Empirical Methods in Natural Language Processing, pp. 3643–3653 (2018)

Multimodal Named Entity Recognition with Image Attributes and Image Knowledge

Dawei Chen[1], Zhixu Li[1,2(✉)], Binbin Gu[4], and Zhigang Chen[3]

[1] School of Computer Science and Technology, Soochow University, Suzhou, China
dwchen@stu.suda.edu.cn, zhixuli@suda.edu.cn
[2] IFLYTEK Research, Suzhou, China
[3] State Key Laboratory of Cognitive Intelligence, iFLYTEK, Hefei, China
zgchen@iflytek.com
[4] University of California, Irvine, USA
binbing@uci.edu

Abstract. Multimodal named entity extraction is an emerging task which uses both textual and visual information to detect named entities and identify their entity types. The existing efforts are often flawed in two aspects. Firstly, they may easily ignore the natural prejudice of visual guidance brought by the image. Secondly, they do not further explore the knowledge contained in the image. In this paper, we novelly propose a novel neural network model which introduces both image attributes and image knowledge to help improve named entity extraction. While the image attributes are high-level abstract information of an image that could be labelled by a pre-trained model based on ImageNet, the image knowledge could be obtained from a general encyclopedia knowledge graph with multi-modal information such as DBPedia and Yago. Our emperical study conducted on real-world data collection demonstrates the effectiveness of our approach comparing with several state-of-the-art approaches.

Keywords: Named entity recognition · Multimodal learning · Social media · Knowledge graph

1 Introduction

Recent years have witnessed a dramatic growth of user-generated social media posts on various social media platforms such as Twitter, Facebook and Weibo. As an indispensable resource for many social media based tasks such as breaking news aggregation [20], the identification of cyber-attacks [21] or acquisition of user interests, there is a growing need to obtain structured information from social media. As a basic task of information extraction, Named Entity Recognition (NER) aims at discovering named entities in free text and classify them into

C. S. Jensen et al. (Eds.): DASFAA 2021, LNCS 12682, pp. 186–201, 2021.
https://doi.org/10.1007/978-3-030-73197-7_12

Tweet posts	 teachers take on top of Mount Sherman.	 Sony announced a Bad Boys in the next few years.	 Jackson is really my favorite.
Expected NER results	teachers take on top of [Mount Sherman **LOC**].	[Sony **ORG**] announced a [Bad Boys **OTHER**] in the next few years.	[Jackson **PER**] is really my favorite.
NER with text only	teachers take on top of [Mount Sherman OTHER].	[Sony ORG] announced a [Bad Boys PER] in the next few years.	[Jackson OTHER] is really my favorite.
MNER with previous methods	teachers take on top of [Mount Sherman PER].	[Sony PER] announced a [Bad Boys PER] in the next few years.	[Jackson OTHER] is really my favorite.

Fig. 1. Three example social media posts with labelled named entities

per-defined types including person (*PER*), location (*LOC*), organization (*ORG*) and other (*OTHER*).

Different from NER with plain text, NER with social media posts is defined as multimodal named entity recognition (MNER) [29], which aims to detect named entities and identify their entity types given a (post text, post image) pair. As the three example posts with images given in Fig. 1, we expect to detect "Mount Sherman" as LOC from the first post, "Sony" as an ORG and "Bad Boys" as OTHER from the second post, and "Jackson" as PER from the third post. However, the post texts are usually too short to provide enough context for named entity recognition. As a result, if we perform named entity recognition with the post text only, these mentions might be wrongly recognized as shown in the figure. Fortunately, the post images may provide necessary complementary information to help named entity recognition.

So far, plenty of efforts have been made on NER. Ealier NER systems mainly rely on feature engineering and machine learning models [9], while the state-of-the-art approaches are sequence models, which replace the handcrafted features and machine learning models with various kinds of word embeddings [6] and Deep neural network (DNN) [3]. As a variant of NER, MNER also receives much attention in recent years [1,14,17,27,29]. Some work first learns the characteristics of each modality separately, and then integrates the characteristics of different modalities with attention mechanism [14,17,29]. Some other work produces interactions between modalities with attention mechanism in the early stage of extracting different modal features [1,27].

However, the existing MNER methods are often flawed in two aspects. Firstly, they may easily ignore the natural prejudice of visual guidance brought by the image. Let's see the first tweet post with an image in Fig. 1, where the "Mount Sherman" in the post might be taken as a person given that there are several persons in the image. To allieviate the natural prejudice of visual guidance here, we need to treat the persons and the mountain in the image fairly, such that "Mount Sherman" is more likely to be associated with a mountain. Secondly, some important background knowledge about the image is yet to be obtained and furtherly explored. Let's see the second tweet post in Fig. 1, where the "Bad Boys" might be wrongly recognized as a person if we just use the shallow feature information in the image. But if we have the knowledge that the image is actually a movie poster, then "Bad Boys" in the text could be recognized as a movie instead of two persons. Similarly, we can see from the third post in Fig. 1, the director of the movie "Jackson" might be wrongly recognized as an aminal (i.e. OTHER), if we do not possess the knowledge that the image is a movie poster of the movie "King Kong".

To address the above drawbacks, we propose a novel MNER neural model integrating both *image attributes* and *image knowledge*. The image attributes are high-level abstract information of an image that are labelled by a pre-trained model based on ImageNet [22]. For instance, the labels to the image of the first post in Fig. 1 could be "person", "mountain", "sky", "cloud" and "jeans". By introducing image attributes, we could not only overcome the expression heterogeneity between text and image. More importantly, we could greatly alleviate the visual guidance bias brought by images. The knowledge about an image could be obtained from a general encyclopedia knowledge graph with multi-modal information (or MMKG for short) such as DBPedia [2] and Yago [24], which could be leveraged to better understand its meaning. However, it is nontrivial to obtain the image knowledge from MMKG, which requires us to find the entity that corresponds to the image in the MMKG firstly. It would be extreamly expensive if we search through the whole MMKG with millions of entities. Here we propose an efficient way to accomplish this task by searching the candidate entities corresponding to the entity mentions in the text, as well as their nearest neighbor entities within n-hop range.

To summarize, our main contributions are as follows:

- We introduce image attributes into MNER to alleviate the visual guidance bias brought by images and overcome the expression heterogeneity between text and image.
- We propose an efficient approach to obtain knowledge about a poster image from a large MMKG by utilizing the identified mentions in the poster text.
- We propose a novel neural model with multiple attentions to integrate both image attributes and image knowledge into our neural MNER model.

We conduct our empirical study on real-world data, which demonstrates the effctiveness of our approach comparing with several state-of-the-art approaches.

Roadmap. The rest of the paper is organized as follows: We discuss the related work in Sect. 2, and then present our approach in Sect. 3. After reporting our empirical study in Sect. 4, we finally conclude the paper in Sect. 5.

2 Related Work

In this section, we cover related work on traditional NER with text only, and MNER using image and text in recent years. Then, we present some other multi-modal tasks which inspire us deeply.

2.1 Traditional NER with Text only

The NER task has been studied for many years, and there are various mature models. Traditional approaches typically focus on designing effective features and then feed these features to different linear classifiers such as maximum entropy [5], conditional random fileds (CRF) [8] and support vector machines (SVM) [15]. Because traditional methods involve drab feature engineering, many deep learning methods for NER have emerged rapidly, such as BiLSTM-CRF [10], Bert-CRF [18], Lattice-LSTM [13]. It turns out that these neural approaches can achieve the state-of-the-art performance on formal text.

However, when using the above methods on social media tweets, the results are not satisfactory since the context of tweet texts is not rich enough. Hence, some studies propose to exploit external resources (e.g., shallow parser, Freebase dictionary and graphic characteristics) to help deal with NER task in social media text [11,12,33,34]. Indeed, the performance of these models with external resources is better than the previous work.

2.2 MNER with Image and Text

With the rapid increase of multi-modal data on social media platforms, some work starts to study using multi-modal data such as the associate images to improve the effectiveness of NER. Specifically, in order to fuse the textual and visual information, [17] proposes a multimodal NER (i.e. MNER) network with modality attention, while [29] and [14] propose an adaptive co-attention network and a gated visual attention mechanism to model the inter-modal interactions and filter out the noise in the visual context respectively. To fully capture intra-modal and cross-modal interactions, [1] extends multi-dimensional self-attention mechanism so that the proposed attention can guide visual attention module. Also, [27] proposes to leverage purely text-based entity span detection as an auxiliary module to alleviate the visual bias and designs a Unified Multimodal Transformer to guide the final predictions.

2.3 Other Multimodal Tasks

In the field of multimodal fusion, other multimodal tasks can also inspire us deeply. In VQA (Visual Question Answering) task, [25,31] introduces the attribute prediction layer as a method to incorporate high-level concepts. [4] proposes to introduce three modalities of image, text and image attributes for multi-modal irony recognition task in social media tweets. In [4], image attributes

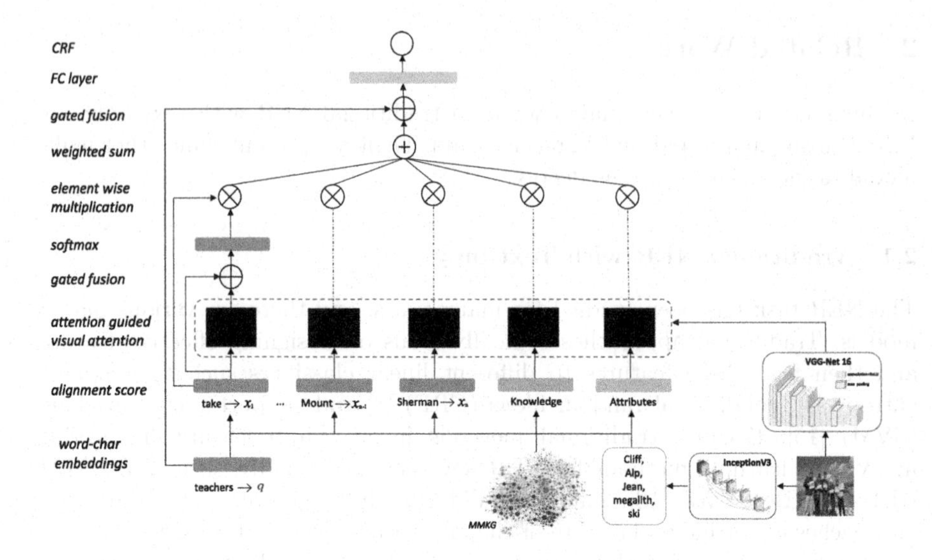

Fig. 2. The architecture of our proposed model

are used to ease the heterogeneity of image and text expression. The role of image attributes is a high-level abstract information bridging the gap between texts and images. [16,32] introduces knowledge to do some common sense reasoning and visual relation reasoning in Visual Question Answer task. Also, [23] proposes to combine external knowledge with question to solve problems where the answer is not in the image. While [7] designs modality fusion structure in order to discover the real importance of different modalities, several attention mechanisms are used to fuse text and audio [26,28]. An approach is proposed in [30] which constructs a domain-specific Multimodal Knowledge Graph (MMKG) with visual and textual information from Wikimedia Commons.

3 Our Proposed Model

In this work, we propose a novel neural network structure which includes the image attribute modality as well as image conceptual knowledge modality. This neural network uses an attention mechanism to perform the interaction among different modalities. The overall structure of our model is shown in Fig. 2. In the following, we first formulate the problem of MNER and then describe the proposed model in detail.

3.1 Problem Formulation

In this work, Multimodal Named Entity Recognition (MNER) task is formulated as a sequence labeling task. Given a text sequence $X = \{x_1, x_2, ..., x_n\}$ and associated image $Image$, MNER aims to identify entity boundaries from text first with the BIO style, and also categorize the identified entities into predefined categories including Person, Organization, Location and Other. The output of a MNER model is a sequence of tags $Y = \{y_1, y_2, ..., y_n\}$ with the input text, where $y_i \in$ {O, B-PER, I–PER, B-ORG, I-ORG, B-LOC, I-LOC, B-OTHER, I-OTHER} in this work.

3.2 Introducting Image Attributes and Knowledge

Figure 2 illustrates the framework of our model. The model introduces image knowledge using Multimodal Knowledge Graph (MMKG) and image attributes using InceptionV3.[1] We describe each part of the model respectively next.

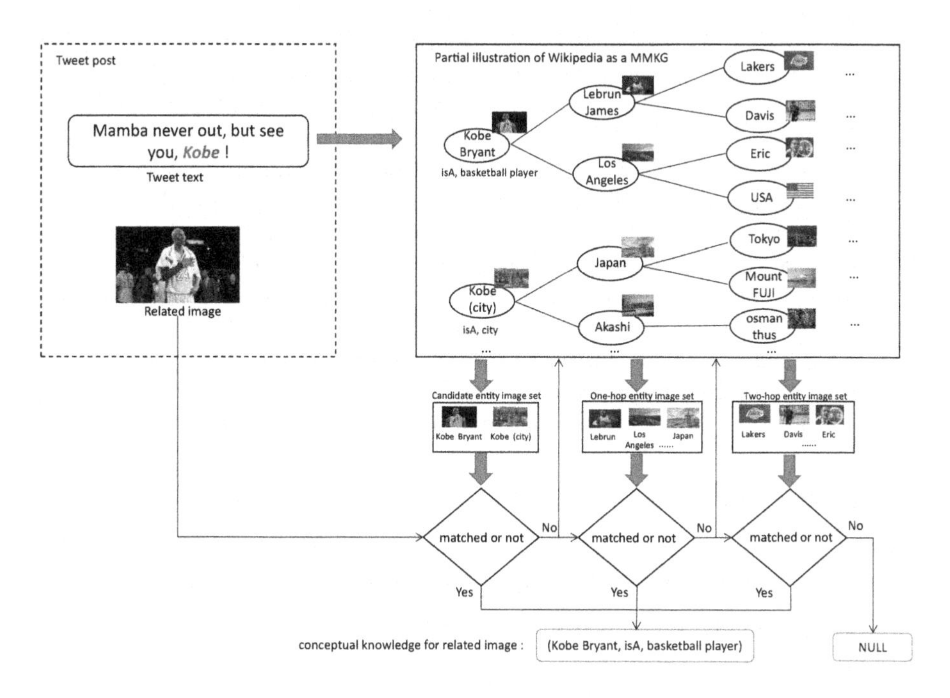

Fig. 3. The process of acquiring knowledge for an image from MMKG

Image Attributes. We use the InceptionV3 network pre-trained on ImageNet to predict the target objects in an image. Through the InceptionV3 network, we obtain the probability of a specific image corresponding to each category of 1000 categories in the ImageNet, and take the 5 category items with the highest probability value as the image attributes. Denote $IA(img)$ as the attribute set of the image img, we compute it as follows:

$$IA(img) = argsort\{p|p = InceptionV3(img)\}[1:5], p \in [0, 1] \tag{1}$$

[1] Available at: https://keras.io/api/applications/#inceptionv3.

where *argsort* sorts 1000 probability values for the output of InceptionV3, and p is the probability score returned by InceptionV3.

Image Knowledge. As for obtaining the image knowledge, we use part of WikiPedia as the MMKG, which includes entities, the triple knowledge corresponding to the entity and the images corresponding to the entity. An example MMKG is given in the upper right part of Fig. 3.

To get knowledge for an image, a straightforward but very time-consuming way is to search through the entire MMKG to find the entity who owns the image that has the highest similarity to the given image. According to our observations, most of the time, the images are often closely related to the entities mentioned in the text. Thus, in this paper we propose an efficient way to acquire image knowledge by leveraging the (mention, candidate entity) pairs between the post text and MMKG. As shown in Fig. 3, from the input text, we first recognize entity mentions, i.e., "Kobe", and its corresponding candidate entities $S_e = \{e_1, e_2 ..., e_n\}$, i.e., "Kobe Bryant" and "Kobe (city)", from the MMKG according to the fuzzywuzzy algorithm.[2] Then we first calculate the similarity between the given image in the post and all the images of these candidate entities. If some highly similar image is found, the system would output the conceptual knowledge about its corresponding entity such as (Kobe Bryant, isA, basketball player). Otherwise, we get one-hop neighbourhood entities of the candidate entities, and find if any of these entities own similar images to the input image. If yes, we return relevant conceptual triplet as the image knowledge. Otherwise, we go to the two-hop neighbourhood entities of these candidate entities. But if no matched images are found even in the two-hop neighbourhood entities, we consider that these is probably no relevant knowledge about the image in the MMKG.

3.3 Feature Extraction

In this section, we use Convolutional Neural Network to extract character features and VGG network to extract image features.

Character Feature Extraction. Social media tweets are usually informal and contain many out-of-vocabulary (OOV) words. Character-level features could alleviate informal word and OOV problems because character features can capture valid word shape information such as prefixes, suffixes and capitalization. We use 2D Convolutional Neural Network to extract character feature vectors. First, a word w is projected to a sequence of characters $c = [c_1, c_2, ..., c_n]$ where n is the word length. Next, a convolutional operation of filter size $1 \times k$ is applied to the matrix $W \in \mathbb{R}^{d_e \times n}$. At the end, the character embedding of a word w is computed by the column-wise maximum operation.

Image Feature Extraction. In order to acquire features from an image, we use a pretrained VGG16 model. Specifically, we retain features of different image regions from the last pooling layer which has a shape of $7 \times 7 \times 512$ so that we can get the spatial features of an image. Moreover, we resize it to 49×512 to

[2] Available at: https://github.com/seatgeek/fuzzywuzzy.

simplify the calculations, where 49 is the number of image regions and 512 is the dimension of the feature vector for each image region.

3.4 Modality Fusion

In this section, we use attention and gated fusion module to combine text, attributes, knowledge and image information.

Self-attention. Self-attention module is applied to compute an alignment score between elements from the same source. In NLP (natural language processing), given a sequence of word embeddings $x = [x_1, x_2, ..., x_n]$ and a query embedding q, the alignment score $h(x_i, q)$ between x_i and q can be calculated using Eq. 2.

$$h(x_i, q) = w_t \sigma(x_i W_x + q W_q) \tag{2}$$

where σ is an activation function, w_t is a vector of weights and W_q, W_x are the weight matrices. Such an alignment score $h(x_i, q)$ evaluates how important x_i is to a query q. In order to refine the impact of each feature, we compute the feature-wise score vector $h'(x_i, q)$ in the following way.

$$h'(x_i, q) = W_t \sigma(x_i W_x + q W_q) \tag{3}$$

The difference between Eq. 2 and Eq. 3 is that $W_t \in \mathbb{R}^{d_e \times d_e}$ is a matrix and $h'(x_i, q) \in \mathbb{R}^{d_e}$ is a vector with the same length as x_i so that the interaction between each dimension of x_i and each dimension of q can be studied.

The purpose of softmax applied to the output function h' is to compute the categorical distribution $p(m|x, q)$ over all tokens. To reveal the importance of each feature k in a word embedding x_i, all the dimensions of $h'(x_i, q)$ need to be normalized and the categorical distribution is calculated as:

$$p(m_k = i|x, q) = softmax([h'(x_i, q)]_k) \tag{4}$$

where $[h'(x_i, q)]_k$ represents every dimension of $[h'(x_i, q)]$. Therefore, text context C for query q can be calculated as follows:

$$C = \left[\sum_{i=1}^{n} P_{ki} x_{ki} \right]_{k=1}^{d_e} \tag{5}$$

where $P_{ki} = p(m_k = i|x, q)$.

Alignment Score. Image attributes and image conceptual knowledge can be acquired by the approach described in Sect. 3.2. We concatenate image conceptual knowledge and image attributes to the end of the tweet text. For the tweet text and knowledge, the corresponding word vector representation can be directly obtained with fasttext. We use a two-layer fully connected network to obtain the vector representation of the image attribute embeddings based on the top 5 image attributes.

Denote a_s as the alignment score between a query embedding $q \in X$ and a word embedding w_i, we compute *it* as follows:

$$a_s = h'(w_i, q) \tag{6}$$

where $w_i \in X \cup K \cup A$ and $h'(w_i, q)$ can be calculated using Eq. 3 by substituting x_i with w_i. For the three sets X, K and A, $X = \{x_1, x_2, ..., x_n\}$ represents a set of word-char embedding of tweet text, K is the word-char embedding of knowledge and A means the word-char embedding of the weighted average of image attributes.

Attention Guided Visual Attention. To obtain the visual attention matrix, we calculate a_v between a_s and image feature matrix I as follows:

$$a_v(a_s, I_j) = W_v \sigma(a_s W_s + I_j W_i) \tag{7}$$

where $a_v(a_s, I_j) \in \mathbb{R}^{d_e}$ represents a single row of the visual attention scores matrix $a_v \in \mathbb{R}^{d_e \times N}$, $a_s \in \mathbb{R}^{d_e}$, $W_i \in \mathbb{R}^{d_i \times d_e}$, W_v, $W_s \in \mathbb{R}^{d_e \times d_e}$ are the weight matrices and $I_j \in \mathbb{R}^{d_i}$ is a row vector of $I \in \mathbb{R}^{d_i \times N}$.

Gated Fusion. We normalize the score a_v by Eq. 8 to get the probability distribution of a_v, denoted by $P(a_v)$, over all regions of image.

$$P(a_v) = softmax(a_v) \tag{8}$$

The output C_v containing visual context vector for a_s is an element-wise product between $p(a_v)$ and I which is computed as follows:

$$C_v = \sum_{i=1}^{n} P_i(a_v) \odot I_i \tag{9}$$

In order to dynamically merge alignment score a_s and visual attention vectors C_v, we choose a gate function G to integrate these information to get the fused representation F_r which is calculated as:

$$G = \sigma(W_1 a_s + W_2 C_v + b) \tag{10}$$

$$F_r = G \odot C_v + (1 - G) \odot a_s \tag{11}$$

where W_1 and W_2 are the learnable parameters and b is the bias vector and $\odot$ represents element-wise product operation.

We use Eq. 4 to get a categorical distribution P for F_r over all tokens of a sequence w, where w is a sequence of tweet text, knowledge and attributes. Then, element-wise product is computed between each pair of P_i and w_i for the purpose of getting context vector $C(q)$ for query q.

$$C(q) = \sum_{i=1}^{n} P_i \odot w_i \tag{12}$$

where n is the length of w, $C(q)$ is a context vector fused text, image, image attributes and knowledge features, $C(q) \in \mathbb{R}^{d_e}$.

To deal with textual attributes component of NER, we fuse word representation x with $C(q)$ with the gated fusion in Eq. 13 which is similar to Eq. 10. Later, we compute the final output O in the following way.

$$G = \sigma(W_1 C(q) + W_2 x + b) \tag{13}$$

$$O = G \odot C(q) + (1 - G) \odot x \tag{14}$$

3.5 Conditional Random Fields

Conditional Random Fields (CRF) is the last layer in our model. It has been shown that CRF is useful to sequence labeling task in practice because CRF can detect the correlation between labels and their neighborhood.

We take $X = \{x_0, x_1, ..., x_n\}$ as an input sequence and $y = \{y_0, y_1, ..., y_n\}$ as a generic sequence of labels for X. Y represents all possible label sequences for X. Given a sequence X, all the possible label sequences y can be calculated as follows:

$$p(y|X) = \frac{\prod_{i=1}^{n} \Omega_i(y_{i-1}, y_i, X)}{\sum_{y' \in Y} \prod_{i=1}^{n} \Omega_i(y'_{i-1}, y'_i, X)} \tag{15}$$

where $\Omega_i(y_{i-1}, y_i, X)$ and $\Omega_i(y'_{i-1}, y'_i, X)$ are potential functions. Maximum conditional likelihood logarithm is used to learn parameters to maximize the log-likelihood $L(p(y|X))$. The logarithm of likelihood is given by:

$$L(p(y|X)) = \sum_{i} logp(y|X) \tag{16}$$

At the time of decoding, we predict the output sequence y_o as the one with maximal score. The formula is shown as follows.

$$y_o = argmax_{y' \in Y} P(y|X) \tag{17}$$

4 Experiments

We conduct experiments on multimodal NER dataset and compare our model with existing unimodal and multimodal approaches. Precision, Recall and F1 score are used as the evaluation metrics in this work.

4.1 Dataset

We use multimodal NER dataset Twitter2015 constructed by [29]. It contains 4 types of entities Person, Location, Organization and Other collected from 8257 tweets. Table 1 shows the number of entities for each type in the train, validate and test sets.

Table 1. Details of dataset

	Train	Validate	Test
Person	2217	552	1816
Location	2091	522	1697
Organization	928	247	839
Other	940	225	726

4.2 Implementation Details

We use 300D fasttext[3] crawl embeddings to get the word embeddings. And we get 50D character embeddings trained from scratch using a single layer 2D CNN with a kernel size of 1×3. A pre-trained 16-layer VGG network is employed to initalize the vector representation of image. We set Adam optimizer with different learning rate: 0.001, 0.01, 0.03 and 0.005. The experimental results show that we achieve the best score when the learning rate is 0.001, the batch size is 20 and the dropout is 0.5. We adopt cosine similarity to compute the similarity among images and set threshold $\theta = 0.9$ to filter out dissimilar images.

4.3 Baselines

In this part, we describe four representative text-based models and multimodal models in comparison with our method.

- *BiLSTM-CRF*: BiLSTM-CRF was proposed by [8], requiring no feature engineering or data prepocessing. Therefore, it is suitable for many sequence labeling tasks. It was reported to have achieved great result on text-based dataset.
- *T-NER*: T-NER [19] is a specific NER system on tweet post. [29] applied T-NER to train a model on Twitter2015 training set and then evaluated it using Twitter2015 testing set.
- *Adaptive Co-Attention Network*: Adaptive Co-Attention Network was proposed by [29], which defined MNER problem and constructed the dataset Twitter2015.
- *Self-attention Network*: Self-attention Network was proposed by [1], which inspired us to use self-attention to capture the relationship among tweet text, image attributes and knowledge. This model achieved state-of-the-art effect on some metrics. Thus, we take this model as an important baseline to show the effectiveness of our model.

[3] Available at: https://dl.fbaipublicfiles.com/fasttext/vectors-english/wiki-news-300d-1M.vec.zip.

Table 2. Comparison of our approach with previous state-of-the-art methods.

	PER. F1	LOC. F1	ORG. F1	OTHER F1	Overall		
					Prec.	Recall	F1
BiLSTM+CRF [8]	76.77	72.56	41.33	26.80	68.14	61.09	64.42
T-NER [19]	83.64	76.18	50.26	34.56	69.54	68.65	69.09
Adaptive co-attention network [29]	81.98	78.95	53.07	34.02	72.75	68.74	70.69
Self-attention network [1]	83.98	78.65	**59.27**	39.54	73.50	**72.33**	72.91
Our model	**84.28**	**79.43**	58.97	**41.47**	**74.78**	71.82	**73.27**

4.4 Results and Discussion

In Table 2, we report the precision(P), recall(R) and F1 score(F1) achieved by each method on Twitter2015 dataset. In Table 3, we report the F1 score achieved by our method in two different scenarios: (1) With image and (2) Without image.

First, as illustrated in Table 2, by comparing all text-based approaches with multimodal approaches, it is obvious that multimodal models outperform the other models if the dataset only contains text. This indicates that visual context is indeed quite helpful for the NER task on social media tweet posts since image can provide effective information to enrich text context.

Second, as shown in Table 2, our method outperforms the baseline by 0.78% and 1.93% in LOC and OTHER types. Both overall precision and F1 of our method are better than that of the baselines. We assume that the improvement mainly comes from the following reason: the previous methods do not learn the real meaning of some images, whereas our approach can learn deep information of image and try to understand the really effective information that image can provide to text.

Third, although we have introduced the image attributes and knowledge, we still cannot remove the image from our model. From Table 3, we can see that if we remove image but introduce image attributes and knowledge, the F1 score of Without image is lower than that of With image scenario. This is because image attributes are unable to fully represent the image features and the information we need for some images is not deep conceptual knowledge but some target objects in images.

4.5 Bad Case Analysis

In Fig. 4, we show some examples where our approach fails for sequence labeling task. Some reasons are as follows:

Table 3. Results of our method with image and without image on our dataset.

	PER. F1	LOC. F1	ORG. F1	OTHER F1	Overall F1
With image	**84.28**	**79.43**	**58.97**	**41.47**	**73.27**
Without image	83.51	77.26	58.06	37.03	72.09

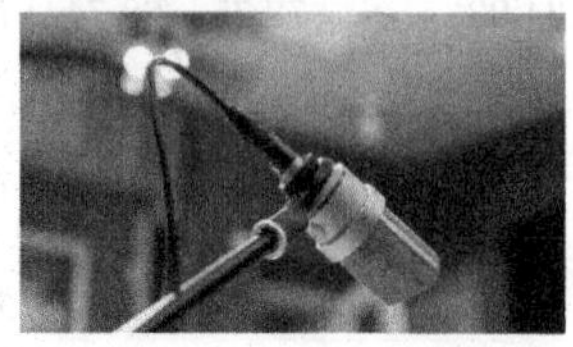

(a) [Reddit **ORG**] needs to stop pretending

(b) [Ben Davis **PER**] vs [Carmel **PER**]

Fig. 4. Two example wrong cases: (a) shows an unrelated image and a wrong prediction. (b) shows great ambiguity for text even if other information is introduced.

1) Unrelated image: Matched image do not relate with tweet text. As we can see in Fig. 4(a), "Reddit" belongs to "Other" but unrelated image could not provide valid information so that it results in wrong prediction "ORG".
2) Great ambiguity: Text is too short and has great ambiguity. As we can see in Fig. 4(b), "Ben Davis" and "Carnel" both belong to "ORG" but short tweet text has great ambiguity so that it is hard to help understand tweet even with some external information. Thus, it results in wrong prediction "PER".

5 Conclusions

In this paper, we propose a novel nerual network for multimodal NER. In our model, we use a new architecture to fuse image knowledge and image attributes. We propose an effective way to introduce image knowledge with MMKG to help us capture deep features of image to avoid error from shallow features. We introduce image attributes to help us treat the target objects in the image fairly alleviating the visual guidance bias of image naturally as well as expression heterogeneity between text and image. Experimental results show the superiority of our method compared to previous methods.

Future work includes two aspects. On the one hand, because our approach still performs not well on social media posts where text and image do not relate, we consider to identify the relevance of image and text and avoid introducing irrelevant image information to the model. On the other hand, since there are not many existing datasets and the size of the existing datasets is relatively small, we intend to build a larger and higher-quality dataset for this field.

Acknowledgment. This research is partially supported by National Key R&D Program of China (No. 2018AAA0101900), the Priority Academic Program Development of Jiangsu Higher Education Institutions, National Natural Science Foundation of China (Grant No. 62072323, 61632016), Natural Science Foundation of Jiangsu Province (No. BK20191420), and the Suda-Toycloud Data Intelligence Joint Laboratory.

References

1. Arshad, O., Gallo, I., Nawaz, S., Calefati, A.: Aiding intra-text representations with visual context for multimodal named entity recognition. In: 2019 International Conference on Document Analysis and Recognition (ICDAR), pp. 337–342. IEEE (2019)
2. Auer, S., Bizer, C., Kobilarov, G., Lehmann, J., Ives, Z.G.: DBpedia: a nucleus for a web of open data. In: Semantic Web, International Semantic Web Conference, Asian Semantic Web Conference, ISWC + ASWC, Busan, Korea, November (2007)
3. Bianco, S., Cadene, R., Celona, L., Napoletano, P.: Benchmark analysis of representative deep neural network architectures. IEEE Access **6**, 64270–64277 (2018)
4. Cai, Y., Cai, H., Wan, X.: Multi-modal sarcasm detection in twitter with hierarchical fusion model. In: Proceedings of the 57th Annual Meeting of the Association for Computational Linguistics, pp. 2506–2515 (2019)
5. Chieu, H.L., Ng, H.T.: Named entity recognition: a maximum entropy approach using global information. In: COLING 2002: The 19th International Conference on Computational Linguistics (2002)
6. Collobert, R., Weston, J., Bottou, L., Karlen, M., Kavukcuoglu, K., Kuksa, P.: Natural language processing (almost) from scratch. J. Mach. Learn. Res. **12**(ARTICLE), 2493–2537 (2011)
7. Gu, Y., Yang, K., Fu, S., Chen, S., Li, X., Marsic, I.: Multimodal affective analysis using hierarchical attention strategy with word-level alignment. In: Proceedings of the 56th Annual Meeting of the Association for Computational Linguistics (Volume 1: Long Papers) (2018)
8. Huang, Z., Xu, W., Yu, K.: Bidirectional LSTM-CRF models for sequence tagging. Computer Science. arXiv preprint arXiv:1508.01991 (2015)
9. Lafferty, J., McCallum, A., Pereira, F.C.: Conditional random fields: probabilistic models for segmenting and labeling sequence data (2001)
10. Lample, G., Ballesteros, M., Subramanian, S., Kawakami, K., Dyer, C.: Neural architectures for named entity recognition. arXiv preprint arXiv:1603.01360 (2016)
11. Limsopatham, N., Collier, N.: Bidirectional LSTM for named entity recognition in twitter messages (2016)
12. Lin, B.Y., Xu, F.F., Luo, Z., Zhu, K.: Multi-channel BiLSTM-CRF model for emerging named entity recognition in social media. In: Proceedings of the 3rd Workshop on Noisy User-generated Text, pp. 160–165 (2017)
13. Liu, C., Zhu, C., Zhu, W.: Chinese named entity recognition based on BERT with whole word masking. In: Proceedings of the 2020 6th International Conference on Computing and Artificial Intelligence, pp. 311–316 (2020)
14. Lu, D., Neves, L., Carvalho, V., Zhang, N., Ji, H.: Visual attention model for name tagging in multimodal social media. In: Proceedings of the 56th Annual Meeting of the Association for Computational Linguistics (Volume 1: Long Papers), vol. 1, pp. 1990–1999 (2018)

15. Luo, G., Huang, X., Lin, C.Y., Nie, Z.: Joint entity recognition and disambiguation. In: Proceedings of the 2015 Conference on Empirical Methods in Natural Language Processing, pp. 879–888 (2015)
16. Marino, K., Rastegari, M., Farhadi, A., Mottaghi, R.: OK-VQA: a visual question answering benchmark requiring external knowledge. In: 2019 IEEE/CVF Conference on Computer Vision and Pattern Recognition (CVPR) (2020)
17. Moon, S., Neves, L., Carvalho, V.: Multimodal named entity recognition for short social media posts. arXiv preprint arXiv:1802.07862 (2018)
18. Peng, M., Ma, R., Zhang, Q., Huang, X.: Simplify the usage of lexicon in Chinese NER. arXiv preprint arXiv:1908.05969 (2019)
19. Ritter, A., Clark, S., Etzioni, O., et al.: Named entity recognition in tweets: an experimental study. In: Proceedings of the 2011 Conference on Empirical Methods in Natural Language Processing, pp. 1524–1534 (2011)
20. Ritter, A., Etzioni, O., Clark, S.: Open domain event extraction from Twitter. In: Proceedings of the 18th ACM SIGKDD International Conference on Knowledge Discovery and Data Mining, pp. 1104–1112 (2012)
21. Ritter, A., Wright, E., Casey, W., Mitchell, T.: Weakly supervised extraction of computer security events from Twitter. In: Proceedings of the 24th International Conference on World Wide Web, pp. 896–905 (2015)
22. Russakovsky, O., et al.: ImageNet large scale visual recognition challenge. Int. J. Comput. Vis. **115**(3), 211–252 (2015)
23. Su, Z., Zhu, C., Dong, Y., Cai, D., Chen, Y., Li, J.: Learning visual knowledge memory networks for visual question answering. In: 2018 IEEE/CVF Conference on Computer Vision and Pattern Recognition (2018)
24. Suchanek, F.M., Kasneci, G., Weikum, G.: YAGO: a core of semantic knowledge. In: Proceedings of the 16th International Conference on World Wide Web, pp. 697–706 (2007)
25. Wu, Q., Shen, C., Liu, L., Dick, A., Van Den Hengel, A.: What value do explicit high level concepts have in vision to language problems? In: Proceedings of the IEEE Conference on Computer Vision and Pattern Recognition, pp. 203–212 (2016)
26. Yang, Z., Zheng, B., Li, G., Zhao, X., Zhou, X., Jensen, C.S.: Adaptive top-k overlap set similarity joins. In: ICDE, pp. 1081–1092. IEEE (2020)
27. Yu, J., Jiang, J., Yang, L., Xia, R.: Improving multimodal named entity recognition via entity span detection with unified multimodal transformer. Association for Computational Linguistics (2020)
28. Gu, Y., Yang, K., Fu, S., Chen, S., Li, X.: Hybrid attention based multimodal network for spoken language classification. In: Proceedings of the Conference. Association for Computational Linguistics. Meeting (2018)
29. Zhang, Q., Fu, J., Liu, X., Huang, X.: Adaptive co-attention network for named entity recognition in tweets. In: AAAI, pp. 5674–5681 (2018)
30. Zhang, X., Sun, X., Xie, C., Lun, B.: From vision to content: construction of domain-specific multi-modal knowledge graph. IEEE Access **7**, 108278–108294 (2019)
31. Zheng, B., et al.: Online trichromatic pickup and delivery scheduling in spatial crowdsourcing. In: ICDE, pp. 973–984. IEEE (2020)
32. Zheng, B., Su, H., Hua, W., Zheng, K., Zhou, X., Li, G.: Efficient clue-based route search on road networks. TKDE **29**(9), 1846–1859 (2017)

33. Zheng, B., Zhao, X., Weng, L., Hung, N.Q.V., Liu, H., Jensen, C.S.: PM-LSH: a fast and accurate LSH framework for high-dimensional approximate NN search. PVLDB **13**(5), 643–655 (2020)
34. Zheng, B., et al.: Answering why-not group spatial keyword queries. TKDE **32**(1), 26–39 (2020)

Multi-task Neural Shared Structure Search: A Study Based on Text Mining

Jiyi Li(✉) and Fumiyo Fukumoto

University of Yamanashi, Kofu, Japan
{jyli,fukumoto}@yamanashi.ac.jp

Abstract. Multi-task techniques are effective for handling the problem of small size of the datasets. They can leverage additional rich information from other tasks for improving the performance of the target task. One of the problems in the multi-task based methods is which resources are proper to be utilized as the auxiliary tasks and how to select the shared structures with an effective search mechanism. We propose a novel neural-based multi-task Shared Structure Encoding (SSE) to define the exploration space by which we can easily formulate the multi-task architecture search. For the search approaches, because these existing Network Architecture Search (NAS) techniques are not specially designed for the multi-task scenario, we propose two original search approaches, i.e., m-Sparse Search approach by Shared Structure encoding for neural-based Multi-Task models (m-S4MT) and Task-wise Greedy Generation Search approach by Shared Structure encoding for neural-based Multi-Task models (TGG-S3MT). The experiments based on the real text datasets with multiple text mining tasks show that SSE is effective for formulating the multi-task architecture search. Moreover, both m-S4MT and TGG-S3MT have better performance on the target aspects than the single-task method, multi-label method, naïve multi-task methods and the variant of the NAS approach from the existing works. Especially, 1-S4MT with a sparse assumption on the auxiliary tasks has good performance with very low computation cost.

Keywords: Multi-task · Shard structure search · Text mining

1 Introduction

Multi-task techniques are effective for handling the problem of small size of the datasets. They can leverage additional rich information from other tasks for improving the performance of the target task. They have been widely utilized in many natural language processing tasks, such as classification [19], summarization [7,10], parsing [8], sequence labeling [16], entity and relation [22] and natural language understanding [20]. When building the multi-task model in a main-auxiliary manner with one main task and multiple auxiliary tasks, for improving the performance of the main task, there are two important issues,

C. S. Jensen et al. (Eds.): DASFAA 2021, LNCS 12682, pp. 202–218, 2021.
https://doi.org/10.1007/978-3-030-73197-7_13

i.e., which resources (tasks) are proper to be utilized as the auxiliary resources (tasks) for sharing the useful information and how to share the information among the tasks. In these existing studies, researchers always selected specific auxiliary resources and designed hand-crafted shared structures in the models for a specific topic. However, for different datasets and main tasks, the optimal auxiliary resources and shared structures may be different.

We thus propose approaches for automatically selecting the shared structures as well as the auxiliary resources which are more beneficial for the main task. We study this topic by focusing on the models based on neural networks for text mining. There are diverse parameter sharing manners in the multi-task methods for deep neural networks [25]. How to define the exploration space of the candidate multi-task models for automatic search is a problem. We propose a Shared Structure Encoding (SSE) method in the manner of hard parameter sharing to define the exploration space. Based on SSE, we can easily formulate the automatic multi-task architecture search so that we can perform diverse search approaches to it. It is also flexible to add more auxiliary tasks to the existing multi-task models based on SSE.

For the approaches of searching multi-task architectures, on the one hand, we propose a variant of the reinforcement learning based network architecture search approaches [32] in the existing works by utilizing our SSE. On the other hand, because such existing Network Architecture Search (NAS) techniques are not specially designed for the multi-task scenario and searching models trained by the small data with high overfitting risk, we propose two original search approaches for automatically selecting well-shared network structures as well as good auxiliary resources. One is the m-Sparse Search approach by SSE for neural-based Multi-Task models (m-S4MT); the other is the Task-wise Greedy Generation Search approach (TGG-S3MT). Both of them can significantly reduce the search space and thus contribute low computation costs.

We construct the experiments based on real text datasets. The multiple tasks of text mining in the experiments are automatically predicting the review scores for academic papers on multiple review aspects such as clarity and originality based on the text content of the papers. The experimental results show that our approaches can build a multi-task model with effective structures and auxiliaries which achieve better performance compared to the single-task model, multi-label model, naïve multi-task models, and the variant of the NAS approach from existing works. Especially, 1-S4MT with the sparse assumption on the auxiliary tasks has good performance with very low computation cost. The contributions of this paper are as follows:

- We propose an approach which can search and build a neural-based multi-task model with optimal auxiliary resources and shared structures in the scenario of text mining.
- We propose a neural-based multi-task shared structure encoding method to define the exploration space by which we can easily formulate the multi-task architecture search.
- We propose two original search approaches, namely m-S4MT and TGG-S3MT, which can effectively and efficiently select well-shared network structures and good auxiliary tasks.

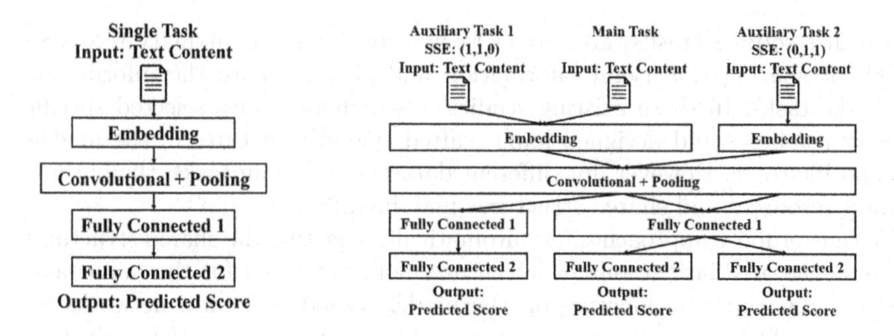

Fig. 1. Basic model CNN **Fig. 2.** Multi-task CNN with shared structure encoding

2 Our Approach

Without loss of generality, we use the basic CNN-based model for classification and regression tasks in text mining [14] as an example to facilitate the description of our multi-task approach. Figure 1 shows the architecture of the CNN model for classifying the labels or predicting the scores. It includes embedding layer, convolutional and pooling layer, and fully connected layers. The multi-task approach we propose is not limited to only utilize this model. It can be integrated with similar neural network models, e.g., XML-CNN [18] and DPCNN [11].

Let n be the number of single tasks (i.e., a label or a score) and these tasks are assumed to have the same network structures consisting of k layers. For a given task (i.e., the target label or score), we regard it as the main task and the other tasks (i.e., the other labels or scores) as the candidate auxiliary tasks, and aim to search the proper shared structures and auxiliary tasks for it.

2.1 Multi-task Shared Structure Encoding (SSE)

To automatically search the proper shared structures and auxiliary tasks, we need to define the exploration space. Because it is difficult to explore the combinations of diverse parameter sharing manners proposed in various multi-task methods [25], we utilize the typical manner of hard parameter sharing to implement our idea. Other manners of parameter sharing will be in future work.

Figure 2 illustrates an example of our Shared Structure Encoding (SSE) consisting of three tasks (one main task and two auxiliary tasks). Given a main task t_0, for each auxiliary task t_i, if the jth layer of t_i is shared with t_0, then we encode this structure as $l_{ij} = 1$; if the jth layer is not shared, then $l_{ij} = 0$. The SSE for auxiliary task t_i is $s_i = \{l_{ij}\}_j$. The SSE for the jth layer is $s^j = \{l_{ij}\}_i$. The shared structure for a given combination of auxiliary tasks $\mathcal{C} = \{t_i\}_i$ is defined as $\mathcal{S}_\mathcal{C} = \{s_i | t_i \in \mathcal{C}\}$. The last output layers (e.g., the "Fully Connected 2" layers in Fig. 1 and 2) are not shared, and thus not counted in k and not used in SSE. We do not encode the shared structures among auxiliary tasks to decrease the complexity of the multi-task model so that it is not difficult for formulating the multi-task architecture search. It is flexible to add more auxiliary tasks to a model. This advantage will be shown and utilized in the TGG-S3MT approach.

There are two special cases of this SSE. One is $l_{ij} = 1$ for all auxiliary tasks and layers. The corresponding model is equivalent to a single multi-label model for all tasks. Another is $l_{ij} = 0$ for all auxiliary tasks and layers. It is equivalent to a single-task model for the main task. In other words, in the search stage, these special models are also included. Lu et al. [21] adaptively generated the feature sharing structure by splitting the network into branches without merging. Its exploration space is a subset of our approach.

Our multi-task approach utilizes a main-auxiliary manner, rather than a manner which equally treats all tasks. The latter requires a trade-off performance among the tasks [26], which has a balancing problem and may not be able to reach optimal results for a specific task in the set of tasks. We thus use the target task as the main task and the remains as the candidates for auxiliary tasks. It decreases the size of the exploration space of possible shared structures and does not cause the balancing problem.

2.2 Shared Structure and Auxiliary Task Search

In our multi-task SSE method, the size of the exploration space for all candidates of the multi-task architectures is $2^{k(n-1)}$. It is a huge value even if the task number n and the layer number k are small. An approach that can efficiently search a good multi-task architecture is required.

Based on our SSE, the multi-task architecture search can be easily formulated as an optimization problem $\min_{\mathcal{S}\in\{0,1\}^{k(n-1)}} f(\mathcal{S})$ for which the analytic form of f is unknown. The input $\mathcal{S}$ of the objective function is the SSE. For the output value of this function, it needs to be the loss on the test set. We use the loss on both training set and validation set as the surrogate. To search a multi-task model with good test loss which is the *target loss*, we use the sum of training and validation loss as the *indicator loss* $f(\mathcal{S})$. The reason that we do not only use the loss on the validation set is the data in our scenario is too small.

2.3 Variant of Vanilla NAS Approach

Our SSE encoding approach for multi-task architecture is easy to integrate with existing search approaches in the network architecture search (NAS) area, e.g., the reinforcement learning (RL) based methods such as [17,32,33]. After this paragraph, in this paper, the term of NAS represents the vanilla reinforcement learning based NAS methods [32,33].

In the vanilla NAS, a controller for sampling a convolutional neural network is implemented by a recurrent neural network (RNN). For example, a fragment of the RNN controller predicts the hyperparameters of [filter height, filter width, stride height, stride width, number of filters] for one *isolated layer* of the convolutional neural network. The fragments of the RNN controller for each isolated layer are sequentially connected to the entire controller. A policy gradient method is used to update the parameters of the controller. To adapt the vanilla NAS approach [32] to handle the multi-task architecture search by leveraging our SSE, we can utilize a fragment of the RNN controller in the vanilla NAS

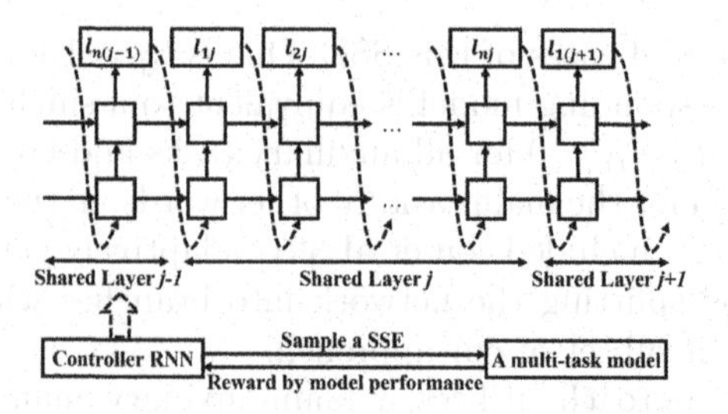

Fig. 3. NAS-MTL: shared-Layer-wise NAS for neural-based Multi-Task model search based on SSE. The RNN controller and the reinforcement learning framework.

approach to represent the SSE of each *shared layer* $s^j = \{l_{ij}\}_i$. Figure 3 shows a visual explanation.

The NAS-based approach is not exactly satisfied in our scenario for multi-task architecture search. Such a technique is not originally designed for exploring multi-task models. The multi-task architecture search does not exactly fit the underlying assumptions of these existing approaches. The SSEs in a shared layer do not hold sequential relations which are represented by the RNN controller.

2.4 *m*-*S*parse *S*earch Approach for Neural-Based *M*ulti-*t*ask Model (*m*-S4MT)

Based on the above discussions, we propose an original approach that is specially designed for searching multi-task architectures based on the SSE. The characteristics of this approach can be summarized into two-folds: (1). It has two selection components to search the shared structures and auxiliary tasks, respectively; (2). It is a heuristic method based on a sparse assumption on auxiliary tasks. In our search strategy, we denote the number of auxiliary tasks in a model as m, $m \leq n-1$. There are $\binom{n-1}{m}$ combinations of the auxiliary tasks. We named the proposed approach as m-Sparse Search approach by Shared Structure encoding for neural-based Multi-Task model (m-S4MT).

In one selection component, we select a good shared structure for each combination of the auxiliary tasks. For each combination $\mathcal{C}$ of auxiliary tasks, we search the shared structures $\mathcal{S}$ and select the one with minimum of the indicator loss, i.e., $\hat{\mathcal{S}_\mathcal{C}} = \arg\min_{\mathcal{S}_\mathcal{C}} f(\mathcal{S}_\mathcal{C})$. In the other selection component, we select a good combination of the auxiliary tasks. After selecting the shared structures for all combinations of the auxiliary tasks, we select the combination of which the *average indicator loss* $\bar{f}(\mathcal{S}_\mathcal{C})$ of all candidate shared structures is minimum, i.e., $\hat{\mathcal{C}} = \arg\min_\mathcal{C} \bar{f}(\mathcal{S}_\mathcal{C})$. Then, we can use the $\hat{\mathcal{S}_{\hat{\mathcal{C}}}}$ with $\hat{\mathcal{C}}$ as the final selection of shared structure and the combination of auxiliary tasks. For a main task, the number of candidate multi-task models is $\mathcal{N}_m = \binom{n-1}{m} \cdot 2^{km}$. When $m = n-1$, i.e., using all other tasks as the auxiliary tasks, this number is $\mathcal{N}_{n-1} = 2^{k(n-1)}$. If $m \ll n-1$, then $\mathcal{N}_m \ll \mathcal{N}_{n-1}$. Because training each candidate multi-task

Algorithm 1: Task-wise Greedy Generation Search Approach (TGG-S3MT)

```
Initialization: Ω_0 = {t_0}, Ψ_0 = Ω \ {t_0}, V_0 = ∅, r=0;
while Ψ_r ≠ ∅ do
    foreach t_i ∈ Ψ_r do
        foreach s_i do
            Ω'_r = Ω_r ∪ {t_i}, V'_r = V_r ∪ {t_i : s_i};
            Train Multi-task model with V'_r, obtain f(V'_r);
        end
    end
    t̂_i = arg min_{t_i ∈ Ψ_r} f̄(V_r ∪ {t_i : s_i});
    ŝ_i = arg min_{s_i} f(V_r ∪ {t̂_i : s_i});
    Ω_{r+1} = Ω_r ∪ {t̂_i}, Ψ_{r+1} = Ψ_r \ {t̂_i};
    V_{r+1} = V_r ∪ {t̂_i : ŝ_i}, r = r + 1;
end
```

model is the key time-consuming factor, we use the number of candidate multi-task models to measure the computation cost of the search approaches.

Another important issue is that we assume the sparsity of the auxiliary tasks, i.e., only a few auxiliary tasks mostly contribute to the main task. This assumption is rational and used in existing works for learning multiple task relations [31] which did not focus on neural-based models. For example, Zhang et al. [31] learned sparse task relation matrix for regression and classification models. They claimed that a task cannot be effective for all other tasks. Following this assumption, we set a m as a very small value, i.e., $m \in \{1, 2\}$. The corresponding versions of the approach are named as 1-S4MT and 2-S4MT.

There are several advantages of using the sparse assumption for our search approach. First, when m is small, the size of exploration space $\mathcal{N}_m$ is tractable. Second, the sparse auxiliary tasks can decrease the risk of overfitting in contrast to the dense ones. Both of these advantages can increase the possibility of selecting a good multi-task architecture with low indicator and target loss.

2.5 *T*ask-Wise *G*reedy *G*eneration *S*earch Approach for Neural-Based *M*ulti-*t*ask Model (TGG-S3MT)

Although the search space and computation time are reduced a lot, m-S4MT is still somewhat bruce-force. It reduces the search space by using the sparse assumption but still searches all candidate multi-task models in the reduced search space. When $m \geq 2$, the size of the search space is still not small. For example, for 2-S4MT, it needs to try $\mathcal{N}_2 = \binom{n-1}{2} \cdot 2^{2k}$ models. We thus propose a task-wise greedy generation method which reduces the search space from another manner that is not based on the sparse assumption. It has a smaller search space than that of m-S4MT with $m \geq 2$. Note that the search space of 1-S4MT is still smaller than TGG-S3MT.

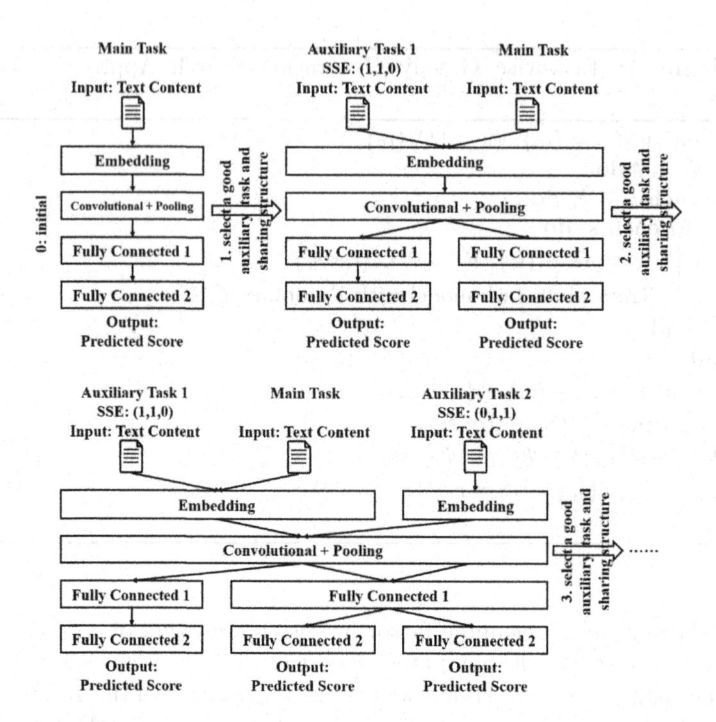

Fig. 4. A brief illustration of the generative process of TGG-S3MT. In each round, it first selects one good auxiliary task and its optimal SSE; then it attaches the selected auxiliary task to the current multi-task model.

The main idea is that we create a generation process for searching the multi-task models. The initial multi-task model is the single-task model for the main task. After that in each *round* of search, we first select one auxiliary task that has the minimum of average indicator loss if we append this task to current multi-task model; we then select its optimal shared structure; after that, we append the module of this selected auxiliary task with the shared structure to current multi-task model. After we have appended all candidate auxiliary tasks, we generate the final multi-task model. If an auxiliary task is not proper for the main task, its SSE will be all-zero. In such a way, we reach the purpose of selecting auxiliary tasks from a large number of candidate tasks. TGG-S3MT makes full use of the flexibility of SSE, i.e., it is easy to add more auxiliary tasks to a multi-task model. Figure 4 briefly illustrates the greedy generation process of this approach.

This approach can be formulated as follows. We denote $\Omega_r = \{t_i\}_i$ as the sets of tasks in the multi-task model at generation round r; the set of SSE in the multi-task model with Ω_r is $\mathcal{V}_r = \{t_i : s_i | t_i \in \Omega_r\}$. In addition, Ψ_r is the set of other tasks which are not yet used in the multi-task model; $\Omega = \Psi_r \cup \Omega_r$ is the set of all tasks including the main task. In the initial round $r = 0$, we assign $\Omega_0 = \{t_0\}$. In each round of auxiliary task selection, we select the auxiliary task $\hat{t_i} = \arg\min_{t_i \in \Psi_r} \bar{f}(\mathcal{V}_r \cup \{t_i : s_i\})$ and the shared structure $\hat{s_i} = \arg\min_{s_i} f(\mathcal{V}_r \cup \{\hat{t_i} : s_i\})$. Then we update Ω_r, Ψ_r, and $\mathcal{V}_r$ with $\hat{t_i}$ and $\hat{s_i}$. Algorithm 1 lists the computations.

The number of candidate multi-task models we need to search in TGG-S3MT is $\mathcal{N}_\mathcal{G} = (n(n-1)/2) \cdot 2^k = n(n-1) \cdot 2^{k-1}$. It is much smaller than that of m-S4MT when $n > 2$ (if $n <= 2$, we don't need to search the auxiliary tasks) and $m \geq 2$, i.e., $(n(n-1)/2) \cdot 2^k \leq (n-1) \cdot 2^k \ll \binom{n-1}{m} 2^{k(m-1)} \cdot 2^k = \binom{n-1}{m} 2^{km}$, and extremely smaller than the original number of all candidates $\mathcal{N}_{n-1} = 2^{k(n-1)} = 2^{k(n-2)} \cdot 2^k$. Especially, the first round of TGG-S3MT which searches the first auxiliary task is equivalent to 1-S4MT. Note that it is also possible to stop TGG-S3MT at a early stage with a pre-defined number of appended auxiliary tasks which is small than n to decrease the number of trained multi-task models.

3 Experiments

3.1 Datasets

We construct the experiments based on real text datasets and multiple tasks of text mining. PeerRead is a public dataset of academic peer reviews for research purposes [13]. It provides detailed peer-reviews including the aspect scores such as clarity and originality, text contents, review contents and final decisions. It raises two types of text mining tasks, i.e., paper acceptance classification and review aspect score prediction. We utilize the later type in the experiments in this paper. We use the ICLR and ACL datasets in the PeerRead because they provide the scores of the peer-review aspects. The sizes of both datasets are very small. We utilize the papers which have the review scores in some of the six aspects ($n = 6$), i.e., Clarity (*cla*), Originality (*ori*), Correctness (*cor*), Comparison (*com*), Substance (*sub*) and Impact (*imp*). The scale of these scores is from 1 to 5. The dataset splits provided by PeerRead are very unbalanced, i.e., most of the papers are in the training set and there are only several papers in the validation and test sets. We split the entire dataset with another partition, i.e., 40% training, 30% validation and 30% test data. Table 1 shows the statistics of the datasets. Because not all papers contain all six aspects in the ICLR dataset, the number of papers for each aspect are diverse.

We only used the paper text to predict the aspect scores, which is different from the works trying to predict review decisions based on the review text with sentiment analysis [6,29]. Moreover, in the PeerRead [13] article, the authors utilized the first 1,000 tokens because the paper text was extremely long; we used entire paper text in the experiments. The results obtained by our experiments and that in PeerRead are not exactly comparable.

The pre-processing is not kept exactly consistent with PeerRead. We remove the stop words and use stemming on the words in the papers. Without loss of generality, the initial word embeddings in the model are pre-trained by fastText [1,12] from each dataset. For the ground truth, we use the mean score of multiple reviews which is the general method of multiple score aggregation. Analyzing the review bias among different reviewers is out of the scope of this paper.

Table 1. Statistics of datasets

Dataset	Aspects	Train	Valid	Test
ICLR	Total	42	32	32
	Clarity	38	22	19
	Originality	34	28	26
	Correctness	30	26	18
	Comparison	12	11	12
	Substance	14	18	15
	Impact	17	26	21
ACL	Total	55	41	41

Table 2. Settings of basic models CNN and XML-CNN: Dropout rate 1 is for the embedding layer, and Dropout rate 2 is for the fully connected layers.

Settings	CNN	XMN-CNN
Input word vectors	fastText	fastText
Embedding Dimension	200	200
Stride size	1	2
Filter region size	2	2
Feature maps (m)	64	64
Pooling	Max pooling	Dynamic max pooling
Activation function	ReLU	ReLU
Hidden layers	1024	512
Batch sizes	8	8
Dropout rate 1	0.25	0.25
Dropout rate 2	0.5	0.5
Optimizer	Adam	Adam
Loss function	MSE	MSE
Epoch	40	40

3.2 Experimental Settings

We used two basic models to verify our approaches, one is the CNN model [14]; the other is one of the recent text classification methods XML-CNN [18]. Table 2 shows the hyperparameter settings of both basic models of our proposed multi-task approach. In the experiments, all cases (datasets, aspects) used the same network hyperparameter settings. We ensure the hyperparameters of the basic models are consistent for all approaches so that they can be fairly compared. Our purpose is to verify that our multi-task search approach can find good auxiliaries and share structures which attain better results compared with the baselines. The evaluation metric is the Root Mean Square Error (RMSE).

Table 3. Detailed results of shared structure selection for each combination of auxiliary tasks. Main task: clarity; basic model: CNN; dataset: ICLR; performance of single task model: 1.061. **Bold**: best performance (including single task model). *Italic*: better one between m-S4MT and AMT.

(a). $m = 1$

Auxiliary	m-S4MT (SSE)	AMT	Ain1
ori	***1.000*** (101)	1.086	1.079
cor	***1.011*** (101)	1.096	1.084
com	***0.985*** (111)	1.090	**0.985**
sub	***1.031*** (101)	1.056	1.051
imp	***0.970*** (001)	1.104	1.070

(b). $m = 2$

Auxiliaries	m-S4MT (SSE)	AMT	Ain1
ori,cor	***0.985***(101,001)	1.117	1.110
ori,com	***1.002***(101,101)	1.118	1.263
ori,sub	***1.025***(101,100)	1.099	1.198
ori,imp	***1.007***(101,100)	1.120	1.204
cor,com	***1.011***(101,000)	1.128	1.252
cor,sub	***1.011***(101,000)	1.110	1.168
cor,imp	***1.011***(101,000)	1.130	1.208
com,sub	***0.996***(101,100)	1.091	1.108
com,imp	***1.044***(101,100)	1.111	1.212
sub,imp	***1.031***(101,101)	1.107	1.164

For the NAS-based approach, we revised a public implementation.[1] We modify the rewards to fit our regression loss. The probability of random exploration was set to 0.5 so that more samples (multi-task models) were generated by the approaches, while the default value was 0.8 in the first 1,000 iterations which result in that most of the samples were generated randomly.

The target aspects of scores that are used as the main tasks are "clarity" and "comparison". The reason we select "clarity" is that it is the one that is almost only related to the plain text content. Therefore, its evaluation results have higher credibility than that of other aspects. Other aspects such as "originality" and "impact" are difficult to judge without additional knowledge. To verify our approaches with more than one aspect, we also use the aspect of "comparison" which cannot be exactly judged only from the plain text content, while plain text content may be able to somewhat support its judgment, i.e., whether the content of paper refers adequate existing work or not.

[1] https://github.com/titu1994/neural-architecture-search.

3.3 Q1: Are SSE and m-S4MT Effective?

First, we examine whether our SSE and m-S4MT method ($m \in \{1, 2\}$) can select a good shared structure for a given combination of auxiliary tasks. The baselines are as follows.

Single Task Model: It is equivalent to the case that SSEs of all auxiliary tasks are "000". It uses one network for one aspect score like the models in [3,4].

All-in-One (Ain1): It builds a single model that the main task and all auxiliary tasks share the same network as the models in [13]. It is equivalent to treating the prediction of all aspects as one task (a multi-label model) or as a multi-task that SSEs of all auxiliary tasks are "111".

Average Performance of All Explored Multi-task Models (AMT): It is equivalent to the expectation of the performance if it randomly selects a multi-task model from all candidates.

We used the detailed results in one case of settings to verify our approach. The dataset is ICLR; the basic model is CNN; the main task is "clarity". Table 3(a) shows the results in the case of $m = 1$. We can see that our method successfully builds a better model than the single task model and the model in which all tasks completely share with each other. The comparison result with AMT shows that our method can select a better shared structure from all candidate multi-task structures. Table 3(b) shows the results in the case of $m = 2$. The observation is consistent with that of $m = 1$.

Second, we examined whether our SSE and m-S4MT method can select a good combination of auxiliary tasks. After using our search strategy to select the combinations of auxiliaries, in the 1st and 2nd row of Table 4(a), m-S4MT can select the auxiliaries and shared structures with better performance. In addition, Table 4(a) also shows the performance of m-S4MT on the ACL dataset. The observation is consistent with that on the ICLR dataset.

Third, in Table 4(a), 2-S4MT performs worse than 1-S4MT on ICLR dataset and better on ACL dataset. It shows although $m = 2$ can search more models and probably reach better results, $m = 2$ has higher risk than $m = 1$ to select overfitting models. For m-S4MT, although it is still possible that there are models with $m = 5$ which is better than the best model with $m = 1$ or 2, the dense search (i.e., with large m, in contrast to our sparse search) adds a huge number of "worse" models into the exploration space. The dense search increases the risk to select an overfitting model considerably and decreases both the performance of the selected model and the speed of searching the models. Without a better search method, using a small m rather than a large m is recommended. How to adaptively decide the value of m will be in future work.

Furthermore, the performance of Ain1 model is not better than the single task model. It shows that roughly using any auxiliary information without a rational selection is not good.

3.4 Q2: Is TGG-S3MT Effective?

We then examine whether TGG-S3MT is effective. We verify it by appending all other tasks to the multi-task model, i.e., the number of iterations $\mathcal{N}_{\mathcal{G}}$ is 120. Table 4(b) illustrates the results as well as the selected auxiliary tasks and shared structures. TGG-S3MT can outperform the single and Ain1 models. It can effectively select well-shared network structures and good auxiliary resources.

Table 4. Detailed results of selecting shared structures and auxiliary tasks. Main task: Clarity; basic model: CNN; **Bold**: best performance.

(a). *m*-S4MT. *Italic*: better one between *m*-S4MT and AMT.

	m-S4MT			AMT	Single	Ain1
	m ($\mathcal{N}_m$)	Selected (SSE)	RMSE			
ICLR	1 (40)	com (111)	***0.985***	1.087	1.061	1.072
	2 (640)	com,sub (101, 100)	***0.996***	1.113		
ACL	1 (40)	cor (001)	***1.056***	1.220	1.217	1.427
	2 (640)	cor,com (001, 101)	***0.944***	1.239		

(b). TGG-S3MT. The order of the aspects for SSE is ori, cor, com, sub and imp.

	TGG-S3MT			Single	Ain1
	$\mathcal{N}_{\mathcal{G}}$	Selected (SSE)	RMSE		
ICLR	120	(000, 000, 111, 000, 000)	**0.991**	1.061	1.072
ACL	120	(101, 001, 101, 000, 000)	**0.963**	1.217	1.427

3.5 Q3: Which Search Approach Is More Efficient?

We investigate which approach is more efficient for searching the multi-task models in this experiment. We mainly focus on comparing the two original approaches m-S4MT and TGG-S3MT to a NAS-based approach which is a variant of existing NAS techniques. We evaluate the performance in diverse cases, i.e., two datasets, two target aspects, and two basic models.

For different approaches, the sizes of the search spaces (the numbers of candidate multi-task models) are different. We use this size to measure the computation cost of a search approach. $\mathcal{N}_m$ for m-S4MT is decided by m. In this paper, because the number of layers of the basic models $k = 3$, therefore, $\mathcal{N}_1 = 40$ and $\mathcal{N}_2 = 640$. $\mathcal{N}_{\mathcal{G}}$ for TTG-S3MT is decided by $n = 6$, $\mathcal{N}_{\mathcal{G}} = 120$. $\mathcal{N}_{\mathcal{L}}$ for the NAS-based approach can be any value. This is an advantage of the NAS-based approach that it is flexible to stop the search. It can stop at any number of iterations, while m-S4MT and TGG-S3MT need a fixed number of iterations. We set several values to $\mathcal{N}_{\mathcal{L}}$ which are same with $\mathcal{N}_1$, $\mathcal{N}_2$ or $\mathcal{N}_{\mathcal{G}}$ respectively so that we can fairly compare the search approaches. Table 5 lists the comparison results. The bold values show the better one of two approaches for each $\mathcal{N}$. The best performance in each row is underlined. We omit the SSE results of the selected shared structures to make it easy to observe the results in the table.

First, for columns with $\mathcal{N} = 40$ or 640, it shows that our m-S4MT approach (1-S4MT and 2-S4MT) can outperform the NAS-based approach in most of the cases. The m-S4MT based on two selection components and sparse search can effectively decrease the risk of selecting the overfitting architectures. Reinforcement learning based NAS is much more difficult to be trained than our m-S4MT approach. Because many hyperparameters of the search controller need to be carefully set so that the controller can work. Our method is easy to train as there is only one hyperparameter needs to be decided, i.e. m, and the controller does not need to be trained. Besides, our m-S4MT is easier to be executed in parallel.

Table 5. Comparison of search approaches. We mainly focus on comparing the two original approaches m-S4MT and TGG-S3MT to the NAS-based approach which is a variant of existing NAS techniques. We set several values of iterations to fairly compare the approaches. Both m-S4MT and TGG-S3MT outperforms NAS. 1-S4MT has good performance with low computation costs on search. The bold values show the better one of two approaches for each $\mathcal{N}$. The best performance in each row is underlined. We omit the SSE values of the selected shared structures because of the space limitation.

Dataset	Main	Basic model	$\mathcal{N} = 40$		$\mathcal{N} = 120$		$\mathcal{N} = 640$		Single	Ain1
			1-S4MT	NAS	TGG-S3MT	NAS	2-S4MT	NAS		
ICLR	cla	CNN	**0.985**	1.005	**0.991**	1.005	**0.996**	1.094	1.061	1.072
		XML	**1.020**	1.103	**1.060**	1.103	**1.020**	1.176	1.162	1.057
	com	CNN	**0.824**	0.904	**0.851**	0.890	0.985	**0.917**	0.756	0.842
		XML	0.967	**0.954**	1.060	**0.954**	**0.858**	0.954	1.378	1.263
ACL	cla	CNN	1.056	**0.999**	0.963	**0.923**	0.944	**0.915**	1.217	1.427
		XML	**1.229**	1.462	**1.230**	1.462	**1.229**	1.276	1.229	1.690
	com	CNN	**1.383**	1.456	**1.383**	1.407	**1.113**	1.260	1.650	1.956
		XML	**1.872**	2.006	**1.872**	1.890	1.872	**1.679**	1.872	2.000

Second, for columns with $\mathcal{N} = 120$, TGG-S3MT outperforms the NAS-based approach in most of the cases. The greedy generation process of TGG-S3MT can efficiently select the auxiliary tasks one by one with the optimal shared structures.

Third, it has shown that our original approaches m-S4MT and TGG-S3MT specially designed for the scenario of multi-task architecture search outperform the NAS-based approach which is a variant of vanilla NAS approach that is not designed for the multi-task scenario. We compare the performance and computation cost among m-S4MT ($m = 1, 2$) and TGG-S3MT. 2-S4MT outperforms TGG-S3MT in five of the eight cases, but it needs about five times of the computation cost of TGG-S3MT. 1-S4MT is not worse than TGG-S3MT on seven of the eight cases, and only use one-third of the computation cost of TGG-S3MT. 1-S4MT is not worse than 2-S4MT on five of the eight cases, and only use one-sixteenth of the computation cost of 2-S4MT. It shows that 1-S4MT has good performance with very low computation cost. The sparse assumption on the auxiliary task is effective.

In this paper, we propose approaches that can find a good multi-task model with low computation cost. However, in the entire search space, it is certain that there are better models than those found by our approaches. If there are a lot of computational resources and it is capable to try a huge number of candidate multi-task models (e.g., $\mathcal{N} > 640$). People may hope that more computation costs can pay off. The NAS-based approach which can search more candidate models cannot guarantee it. For example, the NAS-based approach with $\mathcal{N} = 640$ is worse than itself with $\mathcal{N} = 40$ on three of the eight cases. Trying more multi-task models is not exactly able to obtain better models. It is possible to increase the risk of selecting overfitting models.

4 Related Work

4.1 Multi-task Methods in Text Mining

Multi-task learning aims to learn several related tasks simultaneously for obtaining additional rich information from other tasks to reach better performance of the main task or all of the tasks. It has been widely utilized in many natural language processing tasks, such as summarization [7,10], classification [19], sequence labeling [16], question and answering [23], parsing [8], and entity and relation [22]. Recently, the natural language understanding of multiple NLP tasks for generating the pre-trained model became popular [20,28]. In these existing work, researchers always selected specific auxiliary resources and (or) designed hand-crafted shared structures in the multi-task models. However, it is not clear whether the selected auxiliary resources and shared structures are also effective for different datasets and tasks. In contrast, we propose an approach to automatically select the shared structures as well as auxiliary resources which are more beneficial for the main task.

4.2 Network Architecture Search for Multi-task Models

Recently, the research on this topic has been very popular, e.g., the Bayesian optimization based methods such as [27] and the reinforcement learning based methods such as [17,32,33]. Elsken et al. [5] surveyed on this topic. These NAS methods are not specially designed for multi-task architecture search. In this paper, we propose two original search approaches that are specially designed for the multi-task scenario.

There are a few existing works related to the multi-task model search. Lu et al. [21] adaptively generated the feature sharing structure by splitting the network into branches without merging. Its exploration space is a subset of our approach. Pasunuru et al. [24] focused on continual and life-long learning. They learned a single cell that was good at all tasks and the cell evolved when trained on new data. The objective of their work is different from that of ours.

4.3 Peer Review Prediction

Some existing work focused on the quality of the review [2,15,30]. Some existing work made an automatic prediction of the peer review outcome. For example, [9] predicted whether a paper will be accepted or rejected based solely on a paper with visual appearance.

Some recent work focused on how to make review decisions based on sentiment analysis on the review text. Wang et al. [29] proposed an abstract-based memory mechanism with multiple instance learning. Ghosal et al. [6] investigated the reviewer sentiments embedded within peer review texts to predict the peer review outcome. In contrast, our work does not use review text and only use the paper text.

5 Conclusion

In this paper, we propose approaches which can search and build a neural-based multi-task model with optimal auxiliary resources and shared structures in the scenario of text mining. We propose a multi-task shared structure encoding (SSE) method to define the exploration space by which we can easily formulate the multi-task architecture search. We propose an original m-sparse search approach (m-S4MT) and an original task-wise greedy generation search approach (TGG-S3MT) by shared structure encoding for neural-based multi-task models. The experimental results show that our original approaches can effectively and efficiently select well-shared network structures and good auxiliary resources. The multi-task model obtained by our approaches can successfully utilize the information of other aspects for improving the prediction of the target task.

There are several interesting directions for future work, e.g., how to treat the extremely long text; how to handle the aspects which cannot be directly judged by the plain text content information in the academic paper; how to improve the approaches to make more computation costs being pay off.

References

1. Bojanowski, P., Grave, E., Joulin, A., Mikolov, T.: Enriching word vectors with subword information. arXiv preprint arXiv:1607.04606 (2016)
2. De Silva, P.U.K., K. Vance, C.: Preserving the quality of scientific research: peer review of research articles. Scientific Scholarly Communication. FLS, pp. 73–99. Springer, Cham (2017). https://doi.org/10.1007/978-3-319-50627-2_6
3. Dong, F., Zhang, Y.: Automatic features for essay scoring - an empirical study. In: EMNLP, pp. 1072–1077 (2016)
4. Dong, F., Zhang, Y., Yang, J.: Attention-based recurrent convolutional neural network for automatic essay scoring. In: CoNLL, pp. 153–162 (2017)
5. Elsken, T., Metzen, J.H., Hutter, F.: Neural architecture search: a survey. J. Mach. Learn. Res. **20**, 1–21 (2019)
6. Ghosal, T., Verma, R., Ekbal, A., Bhattacharyya, P.: DeepSentiPeer: harnessing sentiment in review texts to recommend peer review decisions. In: ACL, pp. 1120–1130 (2019)

7. Guo, H., Pasunuru, R., Bansal, M.: Soft layer-specific multi-task summarization with entailment and question generation. In: ACL, pp. 687–697 (2018)
8. Hershcovich, D., Abend, O., Rappoport, A.: Multitask parsing across semantic representations. In: ACL, pp. 373–385 (2018)
9. Huang, J.B.: Deep paper gestalt. arXiv preprint arXiv:1812.08775 (2018)
10. Isonuma, M., Fujino, T., Mori, J., Matsuo, Y., Sakata, I.: Extractive summarization using multi-task learning with document classification. In: ACL, pp. 2101–2110 (2017)
11. Johnson, R., Zhang, T.: Deep pyramid convolutional neural networks for text classification. In: ACL, pp. 562–570 (2017)
12. Joulin, A., Grave, E., Bojanowski, P., Mikolov, T.: Bag of tricks for efficient text classification. arXiv preprint arXiv:1607.01759 (2016)
13. Kang, D., Ammar, W., Dalvi, B., van Zuylen, M., Kohlmeier, S., Hovy, E., Schwartz, R.: A dataset of peer reviews (PeerRead): collection, insights and NLP applications. NAACL. **1**, 1647–1661 (2018)
14. Kim, Y.: Convolutional neural networks for sentence classification. In: EMNLP. pp. 1746–1751 (2014)
15. Langford, J., Guzdial, M.: The arbitrariness of reviews, and advice for school administrators. Commun. ACM **58**(4), 12–13 (2015)
16. Lin, Y., Yang, S., Stoyanov, V., Ji, H.: A multi-lingual multi-task architecture for low-resource sequence labeling. ACL. **1**, 799–809 (2018)
17. Liu, H., Simonyan, K., Yang, Y.: Darts: Differentiable architecture search. arXiv preprint arXiv:1806.09055 (2018)
18. Liu, J., Chang, W.C., Wu, Y., Yang, Y.: Deep learning for extreme multi-label text classification. In: SIGIR, pp. 115–124 (2017)
19. Liu, P., Qiu, X., Huang, X.: Adversarial multi-task learning for text classification. In: ACL, pp. 1–10 (2017)
20. Liu, X., He, P., Chen, W., Gao, J.: Multi-task deep neural networks for natural language understanding. In: ACL, pp. 4487–4496 (2019)
21. Lu, Y., Kumar, A., Zhai, S., Cheng, Y., Javidi, T., Feris, R.S.: Fully-adaptive feature sharing in multi-task networks with applications in person attribute classification. In: CVPR, pp. 1131–1140 (2017)
22. Luan, Y., He, L., Ostendorf, M., Hajishirzi, H.: Multi-task identification of entities, relations, and coreference for scientific knowledge graph construction. In: EMNLP, pp. 3219–3232 (2018)
23. McCann, B., Keskar, N., Xiong, C., Socher, R.: The natural language decathlon: multitask learning as question answering. arXiv preprint arXiv:1806:08730 (2018)
24. Pasunuru, R., Bansal, M.: Continual and multi-task architecture search. In: ACL, pp. 1911–1922 (2019)
25. Ruder, S.: An overview of multi-task learning in deep neural networks. arXiv preprint arXiv:1706.05098 (2017)
26. Sener, O., Koltun, V.: Multi-task learning as multi-objective optimization. In: NIPS, pp. 525–536 (2018)
27. Snoek, J., Larochelle, H., Adams, R.P.: Practical Bayesian optimization of machine learning algorithms. In: NIPS, pp. 2951–2959 (2012)
28. Wang, A., Singh, A., Michael, J., Hill, F., Levy, O., Bowman, S.R.: GLUE: a multitask benchmark and analysis platform for natural language understanding. arXiv preprint arXiv:1804.07461 (2019)
29. Wang, K., Wan, X.: Sentiment analysis of peer review texts for scholarly papers. In: SIGIR, pp. 175–184 (2018)

30. Xiong, W., Litman, D.J.: Automatically predicting peer-review helpfulness. In: ACL-HLT, pp. 502–507 (2011)
31. Zhang, Y., Yang, Q.: Learning sparse task relations in multi-task learning. In: AAAI, pp. 2914–2920 (2017)
32. Zoph, B., Le, Q.V.: Neural architecture search with reinforcement learning. arXiv preprint arXiv:1611.01578 (2016)
33. Zoph, B., Vasudevan, V., Shlens, J., Le, Q.V.: Learning transferable architectures for scalable image recognition. In: CVPR, pp. 8697–8710 (2018)

A Semi-structured Data Classification Model with Integrating Tag Sequence and Ngram

Lijun Zhang[1,2(✉)], Ning Li[1,2], Wei Pan[1,2], and Zhanhuai Li[1,2]

[1] School of Computer Science, Northwestern Polytechnical University, Xi'an 710072, China
{zhanglijun,lining,panwei1002,lizhh}@nwpu.edu.cn

[2] Key Laboratory of Big Data Storage and Management, Northwestern Polytechnical University, Ministry of Industry and Information Technology, Xi'an 710072, China

Abstract. Many collaboratively building resources, such as Wikipedia, Weibo and Quora, exist in the form of semi-structured data and semi-structured data classification plays an important role in many data analysis applications. In addition to content information, semi-structured data also contain structural information. Thus, combining the structure and content features is a crucial issue in semi-structured data classification. In this paper, we propose a supervised semi-structured data classification approach that utilizes both the structural and content information. In this approach, generalized tag sequences are extracted from the structural information, and nGrams are extracted from the content information. Then the tag sequences and nGrams are combined into features called TSGram according to their link relation, and each semi-structured document is represented as a vector of TSGram features. Based on the TSGram features, a classification model is devised to improve the performance of semi-structured data classification. Because TSGram features retain the association between the structural and content information, they are helpful in improving the classification performance. Our experimental results on two real datasets show that the proposed approach is effective.

Keywords: Semi-structured data · Semi-structured data classification · XML document classification · TSGram feature · Tag sequence

1 Introduction

Over the past two decades, large amounts of information have been increasingly made available in the form of semi-structured data. EXtensible Markup

This work was supported in part by the National Natural Science Foundation of China under Grant 61972317, Grant 61672432 and Grant 61732014, and in part by the Fundamental Research Funds for the Central Universities of China under Grant 3102015JSJ0004.

C. S. Jensen et al. (Eds.): DASFAA 2021, LNCS 12682, pp. 219–234, 2021.
https://doi.org/10.1007/978-3-030-73197-7_14

Language (XML) is a widespread, flexible standard for modeling and exchanging semi-structured data as XML-formatted documents [1,2]. XML documents have been widely used in Web information management and for complex data representations. Additionally, collaboratively built semi-structured information resources, such as Wikipedia, Weibo and Quora, have become prevalent on the Web and can inherently be encoded in XML [3]. The widespread use of XML has resulted in enormous amounts of semi-structured data being generated every day. These data constitute a potentially important source of business and scientific knowledge; however they require automated processing, due to their size. Semi-structured data mining is an effective way to automatically extract knowledge from these massive volumes of data and make better use of their structure and content [4].

Classification of semi-structured data is an important and challenging task in semi-structured data mining and management. Many applications can benefit from the semi-structured data classification tasks, including online documentation, electronic commerce, data repositories, digital libraries, data exchange and information systems on the Web, and so on. In general, to implement semi-structured data classification solutions, the semi-structured documents must first be transformed into a specific representation model and then used as the input to a classification algorithm [5]. The classification algorithm and its performance largely depend on the representation model.

Feature extraction is an important step in semi-structured document representation modeling. Features extracted from semi-structured data should capture both content and structural characteristics. This feature extraction process involves several issues, such as aligning their (sub)structures, identifying similarities between such (sub)structures and their nested textual data, discovering possible mutual semantic relationships among the textual data and the (sub)structure labels [6]. In this paper, we propose TSGram as feature for representing semi-structured documents. TSGram features model the structural information as tag sequences, the content information as nGrams, and then fuse the tag sequences and nGrams into integral features that reflect the inclusion relation between structure and content. We improve the performance of semi-structured document classification by considering the similarity between tag sequences when calculating similarity between TSGram features.

Overall, the main contributions of this paper can be summarized as follows:

- We propose the concept of TSGram features to capture the relationships among components in semi-structured documents. TSGram features integrate the structural and content information extracted from semi-structured documents, and consider the relationships between structure and content, the relationships among elements in structural information, the relationships among different keywords in content information. Moreover, we provide the extraction and selection methods for TSGram features, imbuing each TSGram features with strong classification ability.

- Based on the TSGram features, we devise a distance-based semi-structured document classification model that improves the semi-structured document classification performance by using the characteristics of TSGram features.
- We illustrate the performance of our classification model against some other established competitors on two real datasets. The experimental results demonstrate the effectiveness of our proposed classification model.

The rest of this paper is organized as follows. We briefly discuss the previous related works in Sect. 2. Section 3 defines the TSGram features and provides related extraction and selection algorithms. Section 4 presents the classifier model with TSGram features and the classification process. The experimental results are analyzed in Sect. 5, and Sect. 6 concludes this paper.

2 Related Works

For semi-structured documents, document features should include both structure and content information. Therefore, the following three factors should be considered when extracting features from semi-structured data:

(1) The inclusion relation between the structure and content—reflecting that the content is organized in different structural hierarchies.
(2) The relationships among the internal elements within the structure, such as sibling, parent-child, and ancestor-descendent relationships between elements.
(3) The relationships among keywords in the content.

Most existing semi-structured data classification methods are based on the classical vector space model, which they extend it to include structural information. For example, Tran et al. extracted structure and content information from semi-structured data, represented them with a structure vector and a content vector, respectively, and then used them to calculate similarity [7]. The methods used in [8] and [9] are similar. The disadvantage of these methods is that the structure similarity and content similarity are calculated independently; thus, the relationship between the structure and the content is separated. That is, they fail to consider the first of the factors listed above.

Some methods do consider the relationship between the structure and content, such as [10], which takes the location of the keywords in the document structure into account, and [11], which considers keywords that appear in the path. While these methods consider the containment relationship between the structure and the content, the structure is modeled as a path that reflects the element hierarchy (e.g., parent-child, ancestor-descendant, etc.), but ignores the relationship between different paths, path similarity and so on. That is, these methods do not address the second factor listed above.

To capture the relationship between structure and content, Yang et al. extend vector space model to Structured Link VectorModel (SLVM) by incorporating document structures, referencing links and element similarity, and an XML document is represented as a matrix [12]. Then they used the kernel matrix to reflect the relationships between internal elements. In [13], they replace elements with structural subtrees, further strengthening these relationships. Zhao et al. further extend SLVM and apply it to the multiclass XML documents classification [14] and uncertain XML documents classification [5]. This type of method addresses the first two factors listed above, but do not clearly consider the third factor. Costa and Ortale used structure-constrained phrases to capture the relationship between structure and content [15]. Based on the structure-constrained phrases, they devised three machine-learning approaches to cluster XML documents. However, they focused on unsupervised analysis of XML documents, but not supervised analysis.

When modeling the content information of semi-structured data, most of the existing methods decompose it into individual words (keywords) that act as feature units. In the text classification field, studies have shown that in addition to single words, multiple related words, such as nGrams, itemsets and so on, can be used to improve the classification accuracy.

An nGram is a word sequence extracted from the text content, and it reflects the sequences between the words. In addition to single words, Mladenic and Grobelnik included nGrams as features of text documents and performed experiments to test the effect of nGrams with lengths no greater 5. The results showed that the text classification accuracy can be improved by using word sequences of length 2 and 3 (known as bigram and trigram, respectively) and that the accuracy is highest when using bigrams. An nGram whose length is greater than 3 provides only small effects on the classification results [16]. Furnkranz further reported that the longer nGrams can even reduce classification performance [17].

Word itemsets have also been used to improve text classification performance. A set of n words that frequently occur together in documents is called an $n-$ *itemset*. Zhang et al. used associated features (namely, frequent word itemsets) to improve the performance of Naive Bayes text classifier [18]. The authors of [19] obtained similar results using similar methods and reported that larger numbers of word itemsets improve the classification accuracy for large categories (categories that contain many documents) but reduce the classification accuracy for small categories (categories that contain fewer documents).

Tesar et al. compared these two methods (bigrams and 2-itemsets) [20]. The results showed that bigrams are more suitable for text classification than are 2-itemsets. When using these two methods, it is advantageous to choose an appropriate feature selection approach in combination with the classification model.

3 TSGram Feature

A semi-structured document can be represented as a labeled tree; then, the structure and content information can be extracted from the labeled tree. We can represent the structure information of semi-structured documents with tag sequences (for the relevant definitions of tag sequences, see [21]). To acquire the content information of the document, we can extract nGrams from it. Then, we can fuse the tag sequences and nGrams as new features, that is, TSGrams, and represent the semi-structured document as a vector composed of new features. In this section, we describe the basic definitions of the new TSGram feature and the process of building the feature space.

3.1 Basic Definitions

Definition 1. (nGram) An **nGram** is an ordered sequence of keywords. The **length of an nGram** is the number of keywords it contains; nGram with lengths of 1, 2 and 3 are called unigram, bigram and trigram, respectively.

Definition 2. (Tag Sequence support nGram) Let s be a tag sequence and g be an nGram. If a tag sequence t exists in semi-structured document d, such that $t \supseteq s$, and the text content of t contains g, we say that the tag sequence s **supports** the nGram g in d, denoted as $s \Rightarrow g$. The occurrences of g in tag sequence t are termed the **support frequency** of g related to s in d, denoted as $TSF_d(s, g)$. The definition of a tag sequence and the related operations are available in [21].

Definition 3. (TSGram Feature) A **TSGram feature** is an ordered pair $\langle s, g \rangle$, where s is a tag sequence, and g is a nGram with length of n, and there exists a semi-structured document d, such that $s \in d$ and $s \Rightarrow g$. n is also called the **length of the TSGram feature**. TSGram with lengths of 1, 2 and 3 are called TSUnigram, TSBigram and TSTrigram, respectively.

For a given semi-structured document set D, we can extract all the TSGram features to construct the feature space Ω; then any document doc_i in D can be denoted as a vector d_i:

$$d_i = \langle w_{i,1}, w_{i,2}, \cdots, w_{i,|\Omega|} \rangle$$

where $w_{i,j}$ is the weight of the TSGram feature $\langle s_j, g_j \rangle$ in document doc_i, defined as:

$$w_{i,j} = TSF_{doc_i}(s_j, g_j)$$

3.2 Constructing a TSGram Feature Space

Constructing a TSGram feature space involves two phases. First, all the candidate TSGrams are extracted from the document set, and then the TSGrams that have the strongest classification ability are selected as the features to form the feature space. To extract the TSGrams, we first extract nGrams from the content information.

Extracting nGrams. In the natural language processing field, an nGram is called a word sequence or word phrase. A word sequence is more concerned with the arrangement of words in a statistical sense, while a word phrase is more syntactically focused. Researchers have studied the compositions of these two types of nGrams [22,23]. The results of experiments show that there is no significant difference in text classification performance between the two types of nGrams [23]. An experiment reported in [24] on the Reuters-21578 dataset showed that using syntactic word phrases does not achieve better results than does using word sequences in the statistical sense. Therefore, in this paper, we adopt word sequences in the statistical sense as nGrams.

Several algorithms exist that can extract nGrams from text, such as the suffix tree used in [25]. This approach constructs a suffix tree for a text and then extracts nGrams by traversing the suffix tree. A paragraph of text may contain more than one sentence: words before and after a punctuation of a sentence are not considered a continuous sequence of words. Two data structures, parentList and hashTable, are used to assist in building a suffix tree in [25]. In this paper, we use an approach similar to that in [20] to eliminate the two data structures by repeatedly traversing the words in sentences to reduce memory utilization.

Extracting TSGrams. We can extract the tag sequences from the structural hierarchy and nGrams from the text content by traversing the semi-structured document. Then, we can combine the tag sequences and nGrams to obtain the candidate TSGram features as shown in Algorithm 1.

Algorithm 1: TSGrams Extraction Algorithm

Input: D: Semi-structured Dataset;
n: maximum length of nGrams.
Output: $TSGramSet$: the TSGram Features Set in Dataset D.

```
1  begin
2      TSGramSet ← ∅;
3      foreach Semi-structured document doc_i in D do
4          Traverse the document doc_i and represent it as a tree t_i;
5          foreach text node tn in t_i do
6              Extract the tag sequence s from root node to parent node of tn;
7              nGramSet ← extractNGrams (tn.text, n);
8              foreach nGram g in nGramSet do
9                  Add TSGram ⟨s, g⟩ to TSGramSet;
10             end
11         end
12     end
13     return TSGramSet;
14 end
```

In Algorithm 1, the semi-structured document is first represented as a tree (Line 4). Then, each text node in the tree (Line 5) is traversed to obtain the sequence of tags it contains (Line 6). Next, we extract the nGrams (Line 7) from the text content using the *extractNGrams* function, which extracts nGrams whose length is no more than n using a suffix tree similarly to [20]. Finally, each nGram extracted from the text content is combined with the tag sequence as a TSGram feature, and the TSGram features are added to the $TSGramSet$ (Lines 8–10).

Selecting TSGram Features. The feature set $TSGramSet$ obtained by the method described in Sect. 3.2 may be quite large, and not every feature has a positive effect on classification. Therefore, feature selection is necessary. There are two objectives in feature selection:

(1) Reduce the dimension of the feature space to improve the classification efficiency.
(2) Select the features that have the strongest classification ability by eliminating the features with negative effects on the classification to improve classification accuracy.

Many measurements exist for selecting features for classification, such as χ^2 (CHI), Mutual Information (MI), Information Gain (IG), Odds Ratio (OR), Document Frequency (DF) [26]. IG is a commonly used measurement, and the experiment in [26] shows that IG has a good effect on text classification; therefore, we adopt IG as a feature selection method.

Let D be a document set, and d be a document in D. The documents in D belong to k categories: $C_1, C_2, \cdots, C_k$, where C is the full set of categories, that is $C = \{C_i \mid i = 1, 2, \cdots, k\}$. Without ambiguity, we also use C_i to represent a document set consisting of documents whose category is C_i. Using an approach similar to [26], we define the IG of a TSGram feature f as:

$$\begin{aligned} IG(f) = &-\sum_{i=1}^{k} P(C_i)\log P(C_i) \\ &+ P(f)\sum_{i=1}^{k} P(C_i|f)\log P(C_i|f) \\ &+ P(\overline{f})\sum_{i=1}^{k} P(C_i|\overline{f})\log P(C_i|\overline{f}) \end{aligned} \tag{1}$$

where

$$P(C_i) = \frac{|\{\, d \mid d \in C_i \,\}|}{|D|} \tag{2}$$

$$P(f) = \frac{|\{\, d \mid f \in d \,\}|}{|D|} \tag{3}$$

$$P(\overline{f}) = 1 - P(f) \tag{4}$$

$$P(C_i|f) = \frac{|\{\, d \mid d \in C_i \wedge f \in d \,\}|}{|\{\, d \mid f \in d \,\}|} \tag{5}$$

$$P(C_i|\overline{f}) = \frac{|\{\, d \mid d \in C_i \wedge f \notin d \,\}|}{|\{\, d \mid f \notin d \,\}|} \tag{6}$$

Let $TSGramSet = TSUnigramSet \cup TSBigramSet \cup \cdots \cup TSNGramSet$ be the candidate TSGram feature set. Then the process of feature selection is as follows:

1) Calculate the IG values of all candidate TSGram features.
2) Sort the candidate TSGram features in $TSUnigramSet$ by IG values; let IG_N be the IG value of the Nth TSGram feature in sorted $TSUnigramSet$.
3) Select all the candidate TSGram features in $TSGramSet$ whose IG values are greater than IG_N.

In the above, N is a parameter that can be tuned by experiment. The TSGram features selected by the above process constitute the TSGram feature space Ω, which can be used for classification.

4 TSGram-Based Classifier

In this section, we introduce the class model based on the TSGram features and then describe the classification process using the proposed class model.

4.1 TSGrams Class Model

In the TSGram feature space Ω, different features have different classification abilities for different categories. The feature space Ω can be divided into k disjoint subsets, and all the features in each subset strongly indicate a category C_i. This subset is called the class model of category C_i, denoted as Φ_{C_i}.

The crucial issue is how to divide the feature space Ω into k subsets. MI indicates the interrelationship between features and categories; therefore, the feature space Ω can be divided by MI. Using a method similar to [26], the mutual information $MI(f, C_i)$ between the TSGram feature $f : \langle s, g \rangle$ and category C_i is defined as

$$MI(f, C_i) = \frac{P(f \wedge C_i)}{P(f)P(C_i)} \tag{7}$$

where

$$P(f \wedge C_i) = \frac{|\{\, d \mid d \in C_i \wedge f \in d \,\}|}{|D|} \tag{8}$$

and $P(f)$ and $P(C_i)$ are defined as Eq. (3) and Eq. (2), respectively. Thus,

$$MI(f, C_i) = \frac{|\{\, d \mid d \in C_i \wedge f \in d \,\}| \times |D|}{|\{\, d \mid f \in d \,\}| \times |\{\, d \mid d \in C_i \,\}|} \tag{9}$$

The TSGram feature f in Ω is partitioned into the class model Φ_{C_*} of category C_* that has the highest mutual information between them, that is,

$$C_* = Argmax_{C_i \in C} MI(f, C_i)$$

Then, the class model can be represented as a vector in the feature space Ω:

$$\phi_{C_i} = \langle w_{i,1},\ w_{i,2},\ \cdots,\ w_{i,|\Omega|} \rangle$$

where, $w_{i,j}$ is the weight of the TSGram feature f_j in the class model Φ_{C_i}. The weight is defined as the value of the mutual information between f_j and Φ_{C_i} if f_j belongs to Φ_{C_i}, otherwise it is 0. We can normalize the weight values as follows:

$$w_{i,j} = \begin{cases} \frac{MI(f_j, C_i)}{\sqrt{\sum_{k=1}^{|\Phi_{C_i}|} (MI(f_k, C_i))^2}} & f_j \in \Phi_{C_i} \\ 0 & f_j \notin \Phi_{C_i} \end{cases}$$

4.2 Classifying Documents Using the TSGrams Class Model

Any document d and class model Φ_{C_i} of category C_i can be represented as vectors in the same feature space Ω. Then we can calculate the similarity between d and Φ_{C_i}: $sim(d, \phi_{C_i})$. For a document d with an unknown category label, we can assign the category C_* that has the highest similarity as the document's category label as follows:

$$C_* = Argmax_{C_i \in C} sim(d, \phi_{C_i})$$

Several metrics can be used to calculate the similarity between d and ϕ_{C_i}, such as cosine similarity. However, simply applying the cosine similarity measure does not take the similarity of the tag sequences in the TSGram features into account. For instance, if the TSGram feature $\langle\langle article, title\rangle, \langle xml, data\rangle\rangle$ exists in document d, and the TSGram feature $\langle\langle paper, title\rangle, \langle xml, data\rangle\rangle$ exists in the class model Φ_{C_i}, then d and Φ_{C_i} are highly similar, and d is very likely to belong to the category C_i. However, if we simply apply the cosine metric, the similarity between them is 0. Therefore, the same nGrams appearing in similar tag sequences should be given special consideration. To this end, we extend cosine similarity and define the similarity between document d and the class model Φ_{C_i} of category C_i as $genCosineSim$:

$$genCosineSim(d, \phi_{C_i}) = \frac{\sum_{\langle s_j, g_l\rangle \in d} \sum_{\langle s_k, g_l\rangle \in \phi_{C_i}} w_d(\langle s_j, g_l\rangle) \times w_{\phi_{C_i}}(\langle s_k, g_l\rangle) \times sim(s_j, s_k)}{||d|| \times ||\phi_{C_i}||} \quad (10)$$

where $w_d(\langle s_j, g_l\rangle)$ is the weight of feature $\langle s_j, g_l\rangle$ in d, $w_{\phi_{C_i}}(\langle s_k, g_l\rangle)$ is the weight of feature $\langle s_k, g_l\rangle$ in class model Φ_{C_i}, $sim(s_j, s_k)$ is the similarity between the tag sequences s_j and s_k, and $||d||$ and $||\phi_{C_i}||$ are the Euclidean norms of d and ϕ_{C_i}, respectively.

If the tag sequence similarity $sim(s_j, s_k)$ is defined as follows:

$$sim(s_j, s_k) = \begin{cases} 1 & if\ s_j = s_k \\ 0 & if\ s_j \neq s_k \end{cases}$$

then Eq. (10) is a cosine metric; thus, the cosine metric is a special case of Eq. (10). However, as mentioned earlier, the cosine metric ignores the similarity between tag sequences. Consequently, we use edit distance to calculate the tag sequence similarity $sim(s_j, s_k)$.

5 Experimental Study

In this section, we compare the classification effectiveness of the TSGrams classifier with some established competitors. The effects of different parameters are also evaluated.

5.1 Experimental Setting

There are two types of semi-structured data: homogeneous (also known as document-centric) and heterogeneous (also known as data-centric) [27]. We ran experiments with our classifier and other competitors on both types using real datasets. Wikipedia is a homogeneous XML corpus proposed in the INEX contest [28] as a major benchmark for XML classification and clustering. We selected 10 categories (6,910 documents) from all 20 categories as our experimental data. Texas is a heterogeneous dataset introduced in [29]. It is generated from the XML/XSLT versions of Web pages from 20 different sites belonging to 4 different categories: automobile, movie, software and news&reference. The Texas dataset includes a total of 101 documents.

Macro-averaged effectiveness results are obtained by performing a 10-fold cross validation on each data set.

All the tests were performed on a machine running a Windows 7 operating system and equipped with an Intel Core (TM) i5 CPU @ 2.27 Ghz and 2 GB of RAM. All the algorithms were implemented in Java.

5.2 Effects of the Length and Numbers of TSGrams

The length of the TSGram features and the number of selected features affect the classification performance. When using TSGram features with the same length for classification, the feature selection parameter N is equal to the number of features.

Figure 1 and Fig. 2 show the classification precision and recall of TSGram features with lengths of 1 to 4 and different numbers of features on the Texas and Wiki datasets, respectively.

As Fig. 1 shows, on the Texas dataset, the classification performances of TSGram features with lengths of 1 (TSUniGram) and 2 (TSBigram) are better

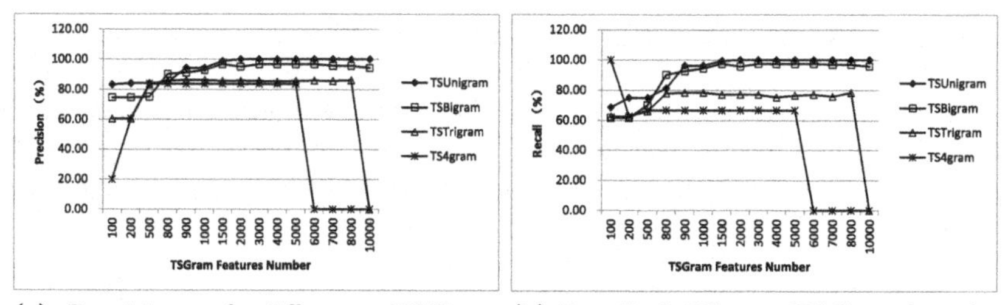

(a) Precision of different TSGram lengths (b) Recall of different TSGram lengths

Fig. 1. Effects of different lengths and numbers of TSGrams on the Texas dataset

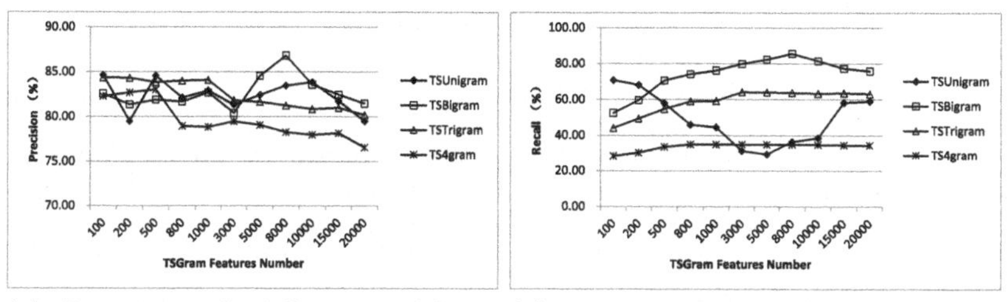

(a) Precision of different TSGram lengths (b) Recall of different TSGram lengths

Fig. 2. Effects of different lengths and numbers of TSGrams on the Wiki dataset

than those with lengths of 3 (TSTrigram) and 4 (TS4gram), and as the number of TSGram features increases, the classification performance improves. When the number of TSGram features exceeds 1,500, the precision and recall reach or approach 100%. The performance of TSBigram is slightly inferior that of TSUnigram. In the figures, when the number of TSTrigram exceeds 8,000 and the number of TS4gram exceeds 5,000, the recall and precision decrease to 0 because the number of features exceeds the maximum limit.

From Fig. 2, we can see that on the Wiki dataset, TSBigram outperforms other TSGram features, and when the number of features reaches 8,000, TSBigram achieves the best performance. TSUnigram and TSTrigram outperform TS4gram.

In summary, the classification performance of TSBigram is better, and adopting longer TSGram features is not beneficial to the classification performance. This conclusion is consistent with the nGram features in the text classification field.

5.3 Effects of TSGram Feature Selection Parameter and Feature Combination

In the experiments in Sect. 5.2, we used a fixed TSGram feature length for classification, such as TSBigram, which uses only TSGram features with a length of 2. According to [20], mixing nGram features of different lengths is helpful in improving text classification performance. Thus, we wondered whether TSGram features achieve similar results for semi-structured document classification? We conducted an experimental verification; the results are shown in Fig. 3 and 4. In the figures, TS(Uni+Bi)grams represent the classification result when using TSUnigram and TSBigram features, and so forth.

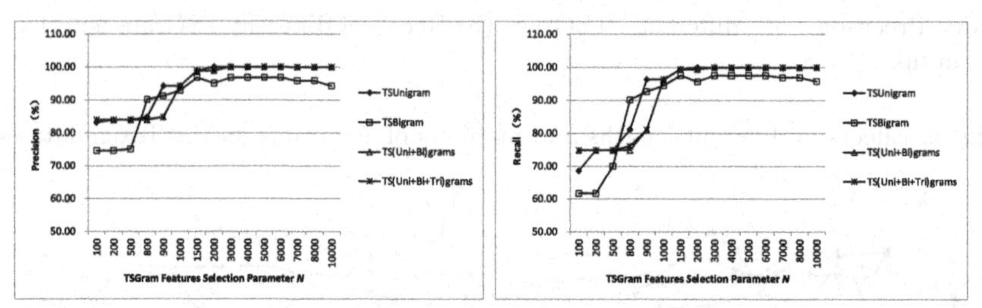

(a) Precision of various TSGram feature combinations under different feature selection parameters

(b) Recall of various TSGram feature combinations under different feature selection parameters

Fig. 3. Effects of the feature selection parameter N and feature combinations on the Texas dataset

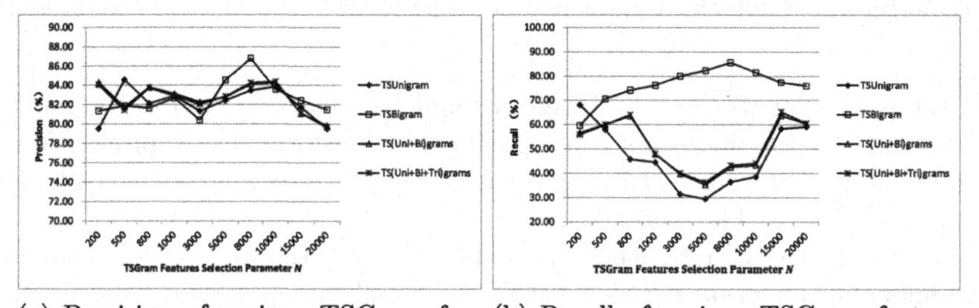

(a) Precision of various TSGram feature combinations under different feature selection parameters

(b) Recall of various TSGram feature combinations under different feature selection parameters

Fig. 4. Effects of the feature selection parameter N and feature combinations on the Wiki dataset

As the figures show, the performances of the two different TSGram feature combinations are almost identical on the Texas dataset. When the feature selection parameter N is less than 500, the feature combination performance is higher than that of the single length TSGram features. When the feature selection parameter N is within the 500–1,000 range, the feature combination performance is lower than that of the single length feature, but when N is greater than 1,500, the feature combination and TSUnigram performance are equal and slightly higher than that of TSBigram.

On the Wiki dataset, the feature combination is not helpful for improving the classification performance. The TSBigram feature still performs the best, but the feature combination performance is better than that of TSUnigram. The performances of TSGram feature combinations with different lengths vary between homogeneous and heterogeneous semi-structured documents, which contrasts with the reported results using different nGram lengths on text classification tasks.

5.4 Classification Results

We compared TSGrams with several other established competitors in terms of classification effectiveness. For the comparisons we adopted three competitive methods: VSM represents methods that use traditional vector space models to extract content information from XML documents for classification, while BottomUp and CBTS are the methods proposed in [30] and [31], respectively. TSGrams is our method. The classification results are shown in Fig. 5.

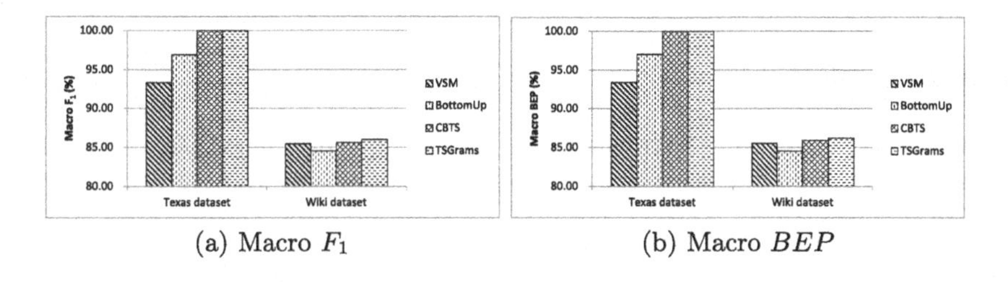

(a) Macro F_1 (b) Macro BEP

Fig. 5. Classification results

As the figure shows, on the Texas dataset, the VSM method performs the worst, while the CBTS and TSGrams methods perform the best, followed by the BottomUp method. This result may have occurred because in the data-centric Texas dataset, the document structures imply rich semantic information that provide strong hints as to the document category, but the VSM method does not take advantage of this information, resulting in the worst performance. The other three methods consider the relationship between the structure and the content and thus achieve better performances.

On the Wiki dataset, the VSM method does not behave as poorly as on the Texas dataset because the structures of all the documents in document-centric dataset Wiki are similar, and the document category is determined primarily by the content. The BottomUp method assumes that the structures of document in the same category are similar, and that the structures of different categories of documents are not similar, while in the Wiki dataset, the structures of all categories are similar; thus, the BottomUp method does not perform well on this dataset. The CBTS method mainly considers the semantics of keywords in the content information; it performs better than the VSM method. The TSGrams method considers the relationship between different keywords, so it performs the best.

6 Conclusions

Although some existing semi-structured document classification methods use both structural and content information, they either separate the relationship between the structure and the content or do not consider the relationship between the keywords in content. In this paper, we propose the TSGram concept to solve these problems. TSGram features capture the inclusion relation between the structure and content, the relationships among the internal elements within the structure, and the relationships among the keywords in the content. We present extraction and selection algorithms for TSGram features and devise a semi-structured data classification model based on TSGrams. Benefiting from the good characteristics of TSGram features, the model improves the performance of semi-structured document classification. Experiments on real datasets demonstrate the effectiveness of our approach.

References

1. Costa, G., Ortale, R.: XML clustering by structure-constrained phrases: a fully-automatic approach using contextualized N-Grams. Int. J. Artif. Intell. Tools **26**(1), 1–24 (2017)
2. Costa, G., Ortale, R.: Fully-automatic XML clustering by structure-constrained phrases. In: Proceedings IEEE 27th International Conference on Tools with Artificial Intelligence, Vietri sul Mare, Italy, pp. 146–153 (2015)
3. Tekli, J.: An overview on XML semantic disambiguation from unstructured text to semi-structured data: background, applications, and ongoing challenges. IEEE Trans. Knowl. Data Eng. **28**(6), 1383–1407 (2016)
4. Piernik, M., Brzezinski, D., Morzy, T.: Clustering XML documents by patterns. Knowl. Inf. Syst. **46**(1), 185–212 (2015). https://doi.org/10.1007/s10115-015-0820-0
5. Zhao, X., Bi, X., Wang, G., et al.: Uncertain XML documents classification using extreme learning machine. Neurocomputing **174**, 375–382 (2016)
6. Costa, G., Ortale, R.: Mining cluster patterns in XML corpora via latent topic models of content and structure. In: Proceedings 23rd Pacific-Asia Conference on Knowledge Discovery and Data Mining, Macau, China, pp. 237–248 (2019)
7. Tran, T., Nayak, R., Bruza, P.D.: Combining structure and content similarities for XML document clustering. In: Proceeedings the 7th Australasian Data Mining Conference (AusDM 2008), pp. 219–226 (2008)
8. Ghosh, S., Mitra, P.: Combining content and structure similarity for XML document classification using composite SVM Kernels. In: Proceedings 19th International Conference on Pattern Recognition (ICPR 2008), pp. 1–4 (2008)
9. Zhang, L., Li, Z., Chen, Q., Li, N.: Structure and content similarity for clustering XML documents. In: Shen, H.T., et al. (eds.) WAIM 2010. LNCS, vol. 6185, pp. 116–124. Springer, Heidelberg (2010). https://doi.org/10.1007/978-3-642-16720-1_12

10. Yuan, J., Xu, D., Bao, H.: An efficient XML documents classification method based on structure and keywords frequency. J. Comput. Res. Dev. **43**(8), 1361–1367 (2006)
11. Costa, G., Ortale, R., Ritacco, E.: Effective XML classification using content and structural information via rule learning. In: Proceedings the 23rd IEEE International Conference on Tools with Artificial Intelligence (ICTAI 2011), pp. 102–109 (2011)
12. Yang, J., Zhang, F.: XML document classification using extended VSM. In: Proceedings 6th International Workshop of the Initiative for the Evaluation of XML Retrieval, pp. 234–244 (2008)
13. Yang, J., Wang, S.: Extended VSM for XML document classification using frequent subtrees. In: Proceedings 8th International Workshop of the Initiative for the Evaluation of XML Retrieval, pp. 441–448 (2009)
14. Zhao, X., Bi, X., Qiao, B.: Probability based voting extreme learning machine for multiclass XML documents classification. World Wide Web **17**(5), 1217–1231 (2013). https://doi.org/10.1007/s11280-013-0230-8
15. Costa, G., Ortale, R.: Machine learning techniques for XML (co-)clustering by structure-constrained phrases. Inf. Retrieval J. **21**(1), 24–55 (2017). https://doi.org/10.1007/s10791-017-9314-x
16. Mladenic, D., Globelnik, M.: Word sequences as features in text learning. the 17th Electrotechnical and Computer Science Conference (ERK 1998), Slovenia, pp. 145–148 (1998)
17. Furnkranz, J.: A Study Using n-gram features for text categorization. Austrian Res. Instit. Artif. Intell. **3**, 1–10 (1998)
18. Zhang, Y., Zhang, L., Yan, J., Li, Z.: Using association features to enhance the performance of Naive Bayes text classifier. In: Proceedings the 5th International Conference on Computational Intelligence and Multimedia Applications, pp. 336–441 (2003)
19. Meretakis, D., Wuthrich, B.: Extending Naive Bayes classifiers using long itemsets. In: Proceedings the 5th ACM SIGKDD International Conference on Knowledge Discovery and Data Mining (SIGKDD 1999), pp. 165–174 (1999)
20. Tesar, R., Strnad, V., Jezek, K., Poesio, M.: Extending the single words-based document model: a comparison of bigrams and 2-itemsets. In: Proceedings the ACM Symposium on Document Engineering, pp. 138–146 (2006)
21. Zhang, L., Li, Z., Chen, Q., Li, X., Li, N., Lou, Y.: Mining frequent association tag sequences for clustering XML documents. In: Sheng, Q.Z., Wang, G., Jensen, C.S., Xu, G. (eds.) APWeb 2012. LNCS, vol. 7235, pp. 85–96. Springer, Heidelberg (2012). https://doi.org/10.1007/978-3-642-29253-8_8
22. Caropreso, M.F., Matwin, S., Sebastiani, F.: Statistical phrases in automated text categorization. Technical report IEI-B4-07-2000. Istituto di Elaborazione dell'Informazione, Pisa, Italy (2000)
23. Mitra, M., Buckley, C., Singhal, A., Cardie, C: An analysis of statistical and syntactic phrases. In: The 5th International Conference on Recherche d'Information Assistee par Ordinateur (RIAO 1997), Montreal, CA, pp. 200–214 (1997)
24. Dumais, S.T., Platt, J., Heckerman, D., Sahami, M.: Inductive learning algorithms and representations for text categorization. In: The 7th ACM International Conference on Information and Knowledge Management (CIKM 1998), New York, US, pp. 148–155. ACM Press (1998)
25. Tesar, R., Fiala, D., Rousselot, F., Jezek, K.: A comparison of two algorithms for discovering repeated word sequences. WIT transaction on information and communication technologies **35**, 121–131 (2005)

26. Yang, Y., Pedersen, J.O.: A comparative study on feature selection in text categorization. In: The 14th International Conference on Machine Learning (ICML 1997), pp. 412–420 (1997)
27. Rezk, N.G., Sarhan, A., Algergawy, A.: Clustering of XML documents based on structure and aggregated content. In: Proceedings 11th International Conference on Computer Engineering and Systems, Cairo, Egypt, pp. 93–102 (2016)
28. Denoyer, L., Gallinari, P.: Report on the XML mining track at INEX 2007 categorization and clustering of XML documents. SIGIR forum **42**, 22–28 (2008)
29. Kurt, A., Tozal, E.: Classification of XSLT-generated web documents with support vector machines. In: Nayak, R., Zaki, M.J. (eds.) KDXD 2006. LNCS, vol. 3915, pp. 33–42. Springer, Heidelberg (2006). https://doi.org/10.1007/11730262_6
30. Wu, J., Tang, J.: A bottom-up approach for XML documents classification. In: The 2008 International Symposium on Database Engineering and Applications, Coimbra, Portugal, pp. 131–137. ACM (2008)
31. Zhang, L., Li, Z., Chen, Q., et al.: Classifying XML documents based on term semantics. Jilin Daxue Xuebao/J. Jilin Univ. (Eng. Technol. Edn.) **42**(6), 1510–1514 (2012)

Inferring Deterministic Regular Expression with Unorder and Counting

Xiaofan Wang[1,2(✉)]

[1] State Key Laboratory of Computer Science, Institute of Software, Chinese Academy of Sciences, Beijing 100190, China
wangxf@ios.ac.cn
[2] University of Chinese Academy of Sciences, Beijing, China

Abstract. Schema inference has been an essential task in database management, and can be reduced to learning regular expressions from positive finite-samples. In this paper, schemata are inferred from unordered XML documents. We extend the single-occurrence regular expressions (SOREs) to single-occurrence regular expressions with unorder and counting (SOREUCs), and give an inference algorithm for SOREUCs. First, we present a *finite automaton with unorder and counting* (FAUC). Then, we construct an FAUC for recognizing a given finite sample. Next, the FAUC runs on the given finite sample to obtain counting operators. Finally we transform the FAUC to a SOREUC by introducing unordered concatenations and counting operators. Experimental results demonstrate that, SOREUCs have stronger expressive powers for modeling unordered schemata than existing works, and our algorithm can efficiently infer a concise SOREUC with better generalization ability.

1 Introduction

Unordered XML which are XML that do not restrict the order among elements have been studied in case of data-centric XML applications [1,5,6,8,18]. Unordered XML facilitates query optimization and set-oriented parallel processing [1], and the corresponding schema (unordered schema [8]) inference can be not only applied in data integration [11,14], but also in query minimization [2,10] and boosting the learning algorithms for twig queries [20]. However, in practice, many XML documents are not accompanied by a schema, or a valid schema [15,16], thus it is essential to devise algorithms for schema inference. In this paper, we focus on inferring schema from unordered XML documents.

Schema inference can be reduced to learning regular expressions from positive finite-sample. Single-occurrence regular expressions (SOREs) [3,4] are a popular model for learning XML. However, SOREs do not support unorder and counting. Regular expressions with unorder and counting, which are used in XML Schema [9,17,21], are extended from standard regular expressions with unordered concatenation % [19,21] and counting (i.e., expressions of the form $r^{[m,n]}$) [13]. % is first used in Standard Generalized Markup Language (SGML) [19], and later in a limited form in XML Schema [21]. In this paper, % is used to model unordered

C. S. Jensen et al. (Eds.): DASFAA 2021, LNCS 12682, pp. 235–252, 2021.
https://doi.org/10.1007/978-3-030-73197-7_15

schemata, we extend SOREs to single-occurrence regular expressions with unorder and counting (SOREUCs) and study the inference algorithm for SOREUCs.

For modeling schemata for unordered XML, there are two classic subclasses: disjunction multiplicity expressions (DMEs) [6,8] and disjunctive interval multiplicity expressions (DIMEs) [5]. A DME where every symbol occurs at most once does not use concatenation ($\cdot$), and forbids repetitions of symbols among the disjunctions ($|$). Furthermore, the unorder of two words u_1 and u_2 is defined as the multiset union $u_1 \uplus u_2$ [8], so does for the unorder defined by disjunctive interval multiplicity expressions (DIMEs) [5]. For instance, let $u_1 = aa$ and $u_2 = b$, $u_1 \uplus u_2 = \{aab\}$, while $u_1 \% u_2 = \{aab, baa\}$. A DIME also supports counting (called *interval* in [5]). For example, for a DME r, $\mathcal{L}(r^{[m,n]}) = \{w_1 \uplus \cdots \uplus w_i | w_1, \cdots, w_i \in \mathcal{L}(r), m \leq i \leq n\}$ [5]. DIMEs are extended from DMEs with counting, a counting operator is on a DME or a symbol $a \in \Sigma$. DMEs are a subclass of DIMEs. However, both DMEs and DIMEs have more restrictions than SOREUCs (see Definition 2). Although uSOREs which are extended from SOREs with unorder [22] can model schemata for unordered XML, uSOREs do not support counting. In addition, although algorithms $learner^{+}_{DME}$ (L^{+}_{DME} [8]) and *InfuSORE* [22] are respectively proposed to learn DMEs and uSOREs [22], the learning algorithms for DIMEs are lacking, and uSOREs are inferred by learning a kind of automata with counter (uCFA [22]) from a given sample, the unorder of any two elements only can be identified in uCFA by counting, and the number of counters in uCFA will grow exponentially if the number of unordered concatenations increases. Thus, the learning algorithms for unordered XML are still insufficient, and some new techniques are needed to learn unordered schemata.

In this paper, SOREUCs are proposed to model schemata for unordered XML. We also propose the learning algorithm for SOREUCs. Our algorithm is based on constructing an automaton and then transforming it into a SOREUC. The main contributions of this paper are as follows.

- We define a new type of automaton: finite automaton with unorder and counting (FAUC), which can recognize the language defined by a SOREUC. The membership problem[1] for FAUCs can be decidable in polynomial time.
- We devise the inference algorithm for SOREUCs. First, we construct FAUC for recognizing a given finite sample. Then, the FAUC runs on the given finite sample to obtain counting operators. Finally, we transform the FAUC to a SOREUC by introducing unordered concatenations and counting operators.
- We conduct experiments on unordered XML data. Our experiments illustrate that, SOREUCs have stronger expressive powers for modeling unordered schemata than existing works, and our algorithm can efficiently infer a concise SOREUC with better generalization ability.

The paper is structured as follows. Section 2 gives the basic definitions. Section 3 describes the FAUC and provides an example of such an automaton. Section 4 presents the inference algorithm of the SOREUC. Section 5 presents experiments. Section 6 concludes the paper.

[1] In this paper, the mentioned membership problem is the uniform version that both the string and a representation of the language are given as inputs.

2 Preliminaries

2.1 Regular Expression with Unorder and Counting

Let Σ be a finite alphabet of symbols. A standard regular expression over Σ is inductively defined as follows: ε and $a \in \Sigma$ are regular expressions, for any regular expressions r_1 and r_2, the disjunction $(r_1|r_2)$, the concatenate $(r_1 \cdot r_2)$, and the Kleene-star r_1^* are also regular expressions. Usually, we omit concatenation operators in examples. The regular expressions with unorder and counting are extended from standard regular expressions by adding the unordered concatenation $r_1\%r_2$ and counting $r_1^{[m,n]}$, where $m \in \mathbb{N}$, $n \in \mathbb{N}_{/1}$, $\mathbb{N} = \{1, 2, 3, \cdots\}$, $\mathbb{N}_{/1} = \{2, 3, \cdots\} \cup \{+\infty\}$, and $m \leq n$. Note that r^+, $r?$, and r^* are used as abbreviations of $r^{[1,+\infty]}$, $r|\varepsilon$, and $r^{[1,+\infty]}|\varepsilon$, respectively. For regular expressions $r_1, r_2, \cdots, r_k$, $\mathcal{L}(r_1\%r_2\%\cdots\%r_k) = \bigcup_{\{\tau_1,\tau_2,\cdots,\tau_k\} \in Perm(\{1,2,\cdots,k\})} \mathcal{L}(r_{\tau_1})\cdots\mathcal{L}(r_{\tau_k})$, where $1 \leq i, \tau_i \leq k$ $(k \geq 2)$ and $Perm(\{1, 2, \cdots, k\})$[2] is the set of permutations of $\{1, 2, \cdots, k\}$. For instance, $\mathcal{L}(ab\%cd\%ef) = \{abcdef, abefcd, cdabef, cdefab, efabcd, efcdab\}$. % is not associative, i.e., $\mathcal{L}((r_1\%r_2)\%r_3) \neq \mathcal{L}(r_1\%(r_2\%r_3)) \neq \mathcal{L}(r_1\%r_2\%r_3)$. $\mathcal{L}(r_1^{[m,n]}) = \{w_1 \cdots w_i | w_1, \cdots, w_i \in \mathcal{L}(r_1), m \leq i \leq n\}$. Let RE(%, #) denote the class of regular expressions with unorder and counting.

For a regular expression r, $|r|$ denotes the length of r, which is the number of symbols and operators occurring in r plus the size of the binary representations of the integers [13]. For a finite sample S, $|S|$ denotes the number of strings in S. For a set V, let $\wp(V) = 2^V$. A string s is an unordered string if $s \in u\%v$ for $u, v \in \Sigma^+$. For a directed graph (digraph) $G(V, E)$, $G.\succ(v)$ $(v \in G.V)$ denotes the set of all direct successors of v in G. $G.\prec(v)$ denotes the set of all direct predecessors of v in G. Let $P(s, a, b) \in \{0, 1\}$ for $s \in S$ and $a, b \in \Sigma$ $(a \neq b)$. $P(s, a, b) = 1$ if and only if each symbol b occurs in s and there exists a symbol a occurring before b. a can be interleaved with b for S if and only if there exists $s_1, s_2 \in S$ such that $P(s_1, a, b) = P(s_2, b, a) = 1$. Let $O(s, a, b) \in \{0, 1\}$ for $s \in S$ and $a, b \in \Sigma$ $(a \neq b)$. $O(s, a, b) = 0$ if and only if $P(s, a, b) = 1$, $P(s, b, a) = 0$ and there exists a symbol c occurring between a and b (a occurs before b) such that a or b can be interleaved with c for S. We specify that ab or ba can be an unordered word for S if a can be interleaved with b for S or there exists $s \in S$ such that $O(s, a, b) = 0$. Let $U_{\%}$ denote the set of all the tuples (a, b) where ab can be an unordered word for S. For space consideration, all omitted proofs can be found at https://github.com/GraceFun/InfSOREUC.

2.2 SORE, SOREUC, SOA and Unorder Unit

SORE is defined as follows.

[2] For instance, $Perm(\{1, 2, 3\}) = \{\{1, 2, 3\}, \{1, 3, 2\}, \{2, 1, 3\}, \{2, 3, 1\}, \{3, 1, 2\}, \{3, 2, 1\}\}$.

Definition 1 (SORE [3,4]**).** *Let Σ be a finite alphabet. A single-occurrence regular expression (SORE) is a standard regular expression over Σ in which every alphabet symbol occurs at most once.*

Since $\mathcal{L}(r^*)=\mathcal{L}((r^+)?)$, in this paper, a SORE does not use the Kleene-star operation. SOREUC which extends SORE with unorder and counting does not use the Kleene-star operation, is defined as follows.

Definition 2 (SOREUC). *Let Σ be a finite alphabet. A single-occurrence regular expression with unorder and counting (SOREUC) is a regular expression with unorder and counting over Σ in which every alphabet symbol occurs at most once.*

According to the definition of deterministic regular expressions [7], SOREUCs are deterministic by definition. In this paper, a SOREUC forbids immediately nested counters, and expressions of the forms $(r?)?$ and $(r?)^{[m,n]}$ for regular expression r.

Example 1. $a\%b$, $(c^{[1,2]}|d)^{[3,4]}$, $a?b(c^{[1,2]}\%d?)(e^{[1,+\infty]})?$, and $(a?b)\%(c|d\%e)^{[3,4]}f$ are SOREUCs, while $a(b|c)^+a$ is not a SORE, therefore not a SOREUC. However, the expressions $((a\%b)^{[3,4]})^{[1,2]}$, $((a^{[3,4]})?)^{[1,2]}$, and $((a^{[3,4]})?)?$ are forbidden.

Definition 3 (SOA [4,12]**).** *Let Σ be a finite alphabet, and let q_0, q_f be distinct symbols that do not occur in Σ. A single-occurrence automaton (SOA) over Σ is a finite directed graph $G=(V,E)$ such that (1) $q_0, q_f \in V$, and $V=\Sigma \cup \{q_0, q_f\}$. (2) q_0 has only outgoing edges, q_f has only incoming edges, and every $v \in V$ lies on a path from q_0 to q_f.*

A string $a_1 \cdots a_n$ $(n \geq 0)$ is accepted by an SOA G, if and only if there is a path $q_0 \rightarrow a_1 \rightarrow \cdots \rightarrow a_n \rightarrow q_f$ in G.

Definition 4 (unorder unit). *For a given finite sample S, an unorder unit is a list $[e_1, e_2, \cdots, e_k]$ $(k \geq 2)$, where $e_i \subset \Sigma$ and $e_i \cap e_j = \emptyset$ $(1 \leq i, j \leq k, i \neq j)$. If a symbol $u \in e_i$, there is at least one symbol $v \in e_j$ such that uv can be an unordered word for S, and there is at least one symbol $u' \in e_i$ $(u' \neq u)$ such that uu' cannot be an unordered word for S.*

Example 2. For sample $S=\{abcde, cdeab\}$, *ac*, *ad* and *ae* (resp. *bc*, *bd* and *be*) can be unordered words. But for *ab*, *cd*, *de* and *ce*, all of them cannot be unordered words. Then, $[\{a,b\},\{c,d,e\}]$ can be an unorder unit.

An unoder unit can be used to discover the substructure of an FAUC recognizing unordered strings. Such as $[\{a,b\},\{c,d,e\}]$, the corresponding substructure is in the FAUC (in Fig. 1), which can recognize the unordered string *cdeab*.

3 Finite Automaton with Unorder and Counting (FAUC)

Although finite automata with counters (FACs) defined in [17] can recognize the languages defined by RE(%, #), the membership problem for FACs is NP-hard

[17], and an FAC recognizing unordered strings does not provide any information about unorder. This implies that an FAC is hard to be learned and then transformed to a SOREUC. Therefore, for recognizing the languages defined by SOREUCs, we propose finite automata with unorder and counting (FAUCs), for which the membership problem is decidable in polynomial time. An FAUC with unorder markers running on a finite sample S not only recognizes the unordered strings from S, but also counts the minimum and maximum number of the strings or substrings (from S) that are consecutively matched by the FAUC. The counting functions of an FAUC are different from that of an FAC.

3.1 Unorder Markers, Counters and Update Instructions

For recognizing the language defined by a SOREUC r, and for the ith subexpression of the form $r_i = r_{i_1}\%r_{i_2}\%\cdots\%r_{i_k}$ $(i, k \in \mathbb{N}, k \geq 2)$ in r, there is an unorder mark $\%_i^+$ in an FAUC for recognizing the strings derived by r_i. For each subexpression r_{i_j} $(1 \leq j \leq k)$, there is a concurrent marker $||_{ij}$ in an FAUC for recognizing the symbols or strings derived by r_{i_j}. Since $\%$ is not associative, there are at most $|\Sigma|-1$ unorder markers and at most $|\Sigma|$ concurrent markers in an FAUC. Let $\mathbb{D}_\Sigma = \{1, 2, \cdots, |\Sigma|-1\}$ and $\mathbb{P}_\Sigma = \{1, 2, \cdots, |\Sigma|\}$.

An FAUC runs on a given finite sample, first, there are counters which count the numbers of the strings or substrings that are consecutively matched by the FAUC each time. Then, update instructions are used to compute the minimum and maximum of the values obtained by the counters.

Counter variables are presented as follows. Let $H(V, E)$ denote the node transition graph of an FAUC. A loop marker $+_k$ $(k \in \mathbb{N})$ which is also a node in H marks a strongly connected component (excluding singleton) in H. There are at most $2|\Sigma|-1$ loop markers. Let $\mathbb{B}_\Sigma = \{1, 2, \cdots, 2|\Sigma|-1\}$. There are corresponding counter variables for the nodes with self-loop, the markers $\%_i^+$ $(i \in \mathbb{D}_\Sigma)$ and the markers $+_k$ $(k \in \mathbb{B}_\Sigma)$. Let $V_c = \{v | v \in H.\succ(v), v \in H.V\} \cup \{\%_i^+ | i \in \mathbb{D}_\Sigma\} \cup \{+_k | k \in \mathbb{B}_\Sigma\}$. Let $\mathcal{C}$ denote the set of counter variables, and let $c(v) \in \mathcal{C}$ $(v \in V_c)$ denote a counter variable. The mapping θ: $\mathcal{C} \mapsto \mathbb{N}$ is the function assigning a value to each counter variable in $\mathcal{C}$. θ_1 denotes that $c(v) = 1$ for each $v \in V_c$.

Update instructions are introduced as follows. Let partial mapping β: $\mathcal{C} \mapsto \{\mathbf{res}, \mathbf{inc}\}$ (**res** for reset, **inc** for increment) represent an update instruction for each counter variable. β also defines mapping g_β between mappings θ. For each $v \in V_c$, if $\beta(c(v)) = \mathbf{res}$, then $g_\beta(\theta) = 1$; if $\beta(c(v)) = \mathbf{inc}$, then $g_\beta(\theta) = \theta(c(v)) + 1$. Let $[l(v), u(v)]$ denote a counting operator, where $l(v)$ and $u(v)$ are lower bound and upper bound variables, respectively. Let $T = \{(l(v), u(v)) | v \in V_c\}$. We define mapping γ: $T \mapsto \mathbb{N} \times \mathbb{N}$ as a function assigning values to lower bound and upper bound variables: $l(v)$ and $u(v)$. γ_∞ denotes all upper bound variables that are initialized to $-\infty$ and all lower bound variables that are initialized to $+\infty$. Let partial mapping α: $T \mapsto (\min(\{T.l(v) | v \in V_c\} \times \mathcal{C}), \max(\{T.u(v) | v \in V_c\} \times \mathcal{C}))$ be an update instruction for $(l(v), u(v))$. $\alpha(l(v), u(v)) = (\min(l(v), c(v)), \max(u(v), c(v)))$. α also defines the partial mapping f_α: $\gamma \times \theta \mapsto \gamma$ such that $f_\alpha(\gamma, \theta)((l(v), u(v)), c(v)) = (\min(\pi_1^2(\gamma(l(v), u(v))), \theta(c(v))), \max(\pi_2^2(\gamma(l(v), u(v))), \theta(c(v))))$. Let $\alpha = \emptyset$ (resp.

$\beta=\emptyset$) denote the empty instruction. $g_\emptyset(\theta)=\theta$ and $f_\emptyset(\gamma,\theta)=\gamma$. Let mapping λ: $H.V \mapsto \wp(\{\%_i^+\}_{i\in\mathbb{D}_\Sigma})$. $\lambda(w)$ ($w\in H.V$) is the set of the nodes $\%_i^+$ ($i\in\mathbb{D}_\Sigma$) which can reach to the node w. $\lambda(w)(k)$ ($1\leq k\leq|\lambda(w)|$) denotes the kth element in $\lambda(w)$. If $V'\subseteq H.V$, $\lambda_{V'}=\{\%_i^+\}$ (resp. $\lambda_{V'}=\{\emptyset\}$) denotes that $\lambda(v')=\{\%_i^+\}$ (resp. $\lambda(v')=\{\emptyset\}$) for each $v'\in V'$.

3.2 Finite Automata with Unorder and Counting

Definition 5 (Finite Automaton with Unorder and Counting). *A finite automaton with unorder and counting (FAUC) is a tuple $\mathcal{A}=(V,Q,\Sigma,q_0,q_f,H,\Phi)$. The members of the tuple are described as follows:*

- *Σ is a finite alphabet (non-empty).*
- *q_0 and q_f : q_0 is the unique initial state, q_f is the unique final state.*
- *$V=\Sigma\cup V_1$, where $V_1\subseteq\{+_i,\%_j^+,||_{jk}\}_{i\in\mathbb{B}_\Sigma,j\in\mathbb{D}_\Sigma,k\in\mathbb{P}_\Sigma}$.*
- *$Q=\{q_0,q_f\}\cup V_2$, where $V_2\subseteq\wp(V)$. A state $q\in Q\setminus\{q_0,q_f\}$ is a set of nodes in V.*
- *$H(V,E,R,\mathcal{C},T)$ is a node transition graph.*
  - *$H.V=\mathcal{A}.V\cup\{q_0,q_f\}$.*
  - *$H.R$: $\{+_i,\%_j^+\}_{i\in\mathbb{B}_\Sigma,j\in\mathbb{D}_\Sigma}\mapsto\wp(\Sigma)$. $H.R(+_i)$ is a set of alphabet symbols, where an alphabet symbol is the first letter of the string that can be consecutively matched by FAUC $\mathcal{A}$ from the state including the node $+_i$. $H.R(\%_j^+)$ is also a set of alphabet symbols, where an alphabet symbol is the first letter of the unordered string that can be recognized by FAUC $\mathcal{A}$ from the state including the node $\%_j^+$.*
  - *$H.\mathcal{C}$ is a set of counter variables. Let $C=\{v|v\in\{+_i,\%_j^+\}_{i\in\mathbb{B}_\Sigma,j\in\mathbb{D}_\Sigma}\vee(v\in\Sigma\wedge v\in H.\succ(v))\}$, $H.\mathcal{C}=\{c(v)|v\in C\}$. For recognizing a string by an FAUC, $c(v)$ ($v\in\Sigma$) is used to count the number of the symbol v consecutively matched by FAUC $\mathcal{A}$ each time. $c(+_i)$ (resp. $c(\%_j^+)$) is used to count the number of the strings where the first letters are in $H.R(+_i)$ (resp. the unordered strings where the first letters are in $H.R(\%_j^+)$) consecutively matched by FAUC $\mathcal{A}$ each time.*
  - *$H.T=\{(l(v),u(v))|v\in C\}$. $[l(v),u(v)]$ is a counting operator. $l(v)$ and $u(v)$ are lower bound and upper bound variables, respectively.*
- *$\Phi(H,X,z)$ where $X\subseteq C$ and $z\in H.V$ is a function returning the tuple consisting of the partial mapping of α and the partial mapping of β (α and β are update instructions) for each node in X transiting to the node z in H. $\Phi(H,X,z)=(A,B)$, where*
  - *$A=\{\emptyset\}\cup\{H.T\mapsto(\min(H.T.l(x),H.\mathcal{C}.c(x)),\max(H.T.u(x),H.\mathcal{C}.c(x)))|(x\in H.\succ(x)\wedge z\neq x)\vee(\exists x'\in H.\succ(x):x'\in\{+_i,\%_j^+\}_{i\in\mathbb{B}_\Sigma,j\in\mathbb{D}_\Sigma}\wedge z\notin H.R(x')),x\in C\wedge x\in X\}$,*
  - *$B=\{\emptyset\}\cup\{c(x)\mapsto\mathbf{res}|(x\in H.\succ(x)\wedge z\neq x)\vee(\exists x'\in H.\succ(x):x'\in\{+_i,\%_j^+\}_{i\in\mathbb{B}_\Sigma,j\in\mathbb{D}_\Sigma}\wedge z\notin H.R(x')),x\in C\wedge x\in X\}\cup\{c(x)\mapsto\mathbf{inc}|(x\in H.\succ(x)\wedge z=x)\vee(\exists x'\in H.\succ(x):x'\in\{+_i,\%_j^+\}_{i\in\mathbb{B}_\Sigma,j\in\mathbb{D}_\Sigma}\wedge z\in H.R(x')),x\in C\wedge x\in X\}\}$.*

The configuration of an FAUC is defined as follows.

Definition 6 (Configuration of an FAUC). *A configuration of an FAUC is a 4-tuple $(q,\gamma,\theta,\lambda)$, where $q\in Q$ is the current state, γ: $\mathcal{A}.H.T\mapsto\mathbb{N}\times\mathbb{N}$, θ: $\mathcal{A}.H.\mathcal{C}\mapsto\mathbb{N}$ and λ: $H.V\mapsto\wp(\{\%_i^+\}_{i\in\mathbb{D}_\Sigma})$. The initial configuration is $(q_0,\gamma_\infty,\theta_1,\lambda_{\mathcal{A}.H.V}=\{\emptyset\})$, and a configuration is final if and only if $q=q_f$.*

The transition function of an FAUC is defined as follows.

Definition 7 (Transition Function of an FAUC). *The transition function δ of an FAUC $\mathcal{A}=(V,Q,\Sigma,q_0,q_f,H,\Phi)$ is defined for any configuration $(q,\gamma,\theta,\lambda)$ and the symbol $y\in\Sigma\cup\{\dashv\}$, where $\dashv$ denotes the end symbol of a string.*

[(1)]

1. *$q=q_0$ or q is a set, where $q=\{a\}$ or $\{+_i\}$ ($a\in\Sigma,i\in\mathbb{B}_\Sigma$):*
   - *$y\in\Sigma$: $\delta((q,\gamma,\theta,\lambda),y)=\{(\{z\},f_\alpha(\gamma,\theta),g_\beta(\theta),\lambda)|z\in H.\succ(x)\wedge(z=y\vee y\in H.R(z)),x\in\{q_0,a,+_i\},z\in\{y\}\cup\{+_j\}_{j\in\mathbb{B}_\Sigma},(\alpha,\beta)=\Phi(H,\{x\},z),\lambda(z)=\emptyset\}$.*

- $y = \dashv$: $\delta((q,\gamma,\theta,\lambda),y) = \{(p, f_\alpha(\gamma,\theta), g_\beta(\theta),\lambda) | p \in H.\succ(x) \wedge p = q_f, x \in \{q_0, a\},\ (\alpha,\beta) = \Phi(H,\{x\},p), \lambda(p)=\emptyset\}$.

2. *q is a set and* $q = \{\%_i^+\}$ $(i \in \mathbb{D}_\Sigma)$: $\delta((q,\gamma,\theta,\lambda),y) = \{(p, f_\alpha(\gamma,\theta), g_\beta(\theta),\lambda) | p = H.\succ(\%_i^+), y \in H.R(\%_i^+), (\alpha,\beta)=(\emptyset,\emptyset), \lambda_p = \{\%_i^+\}\}$.
3. *q is a set and* $|q| \geq 2$:
   - $y \in \Sigma$: $\delta((q,\gamma,\theta,\lambda),y) = \bigcup_{1 \leq t \leq 3} \delta_t((q,\gamma,\theta,\lambda),y)$.
     - $\delta_1((q,\gamma,\theta,\lambda),y) = \{((q \setminus \{x\}) \cup \{z\}, f_\alpha(\gamma,\theta), g_\beta(\theta),\lambda) | z \in H.\succ(x) \wedge (z = y \vee (z \in \{\%_i^+, +_j\}_{i \in \mathbb{D}_\Sigma, j \in \mathbb{B}_\Sigma} \wedge y \in H.R(z))), x \in q \wedge (\forall x' \in q \setminus \{x\} : (x' \in \{||_{ij}\}_{i \in \mathbb{D}_\Sigma, j \in \mathbb{P}_\Sigma} \vee \exists \%_k^+ \in H.\succ(x') : \lambda(x)(1) = \%_k^+) \wedge y \notin H.\succ(x')), (\alpha,\beta) = \Phi(H,\{x\},z), \lambda(z) = \lambda(x), k \in \mathbb{D}_\Sigma\}$.
     - $\delta_2((q,\gamma,\theta,\lambda),y) = \{((q \setminus \{\%_i^+\}) \cup H.\succ(\%_i^+), f_\alpha(\gamma,\theta), g_\beta(\theta),\lambda) | \%_i^+ \in q \wedge y \in H.R(\%_i^+), (\alpha,\beta)=(\emptyset,\emptyset), \forall x \in H.\succ(\%_i^+) : \lambda(x) = \lambda(x) \cup \{\%_i^+\}, i \in \mathbb{D}_\Sigma\}$.
     - $\delta_3((q,\gamma,\theta,\lambda),y) = \{((q \setminus W) \cup \{z\}, f_\alpha(\gamma,\theta), g_\beta(\theta),\lambda) | \exists i \in \mathbb{D}_\Sigma \forall x \in W : \{z, \%_i^+\} \subseteq H.\succ(x) \wedge \%_i^+ \in \lambda(x) \wedge ((z = y \wedge y \notin R(\%_i^+)) \vee (z = \%_i^+ \wedge y \in R(\%_i^+))), W \subseteq q \wedge |W| = |\bigcup_{x' \in \mathcal{W}} H.\succ(x')| - |\mathcal{W}| + 1, \mathcal{W} = \bigcup_{w \in W} \lambda(w), (\alpha,\beta) = \Phi(H,W,y), \forall x \in W \forall x' \in H.\succ(x) : \lambda(x') = \lambda(x) \setminus H.\succ(x)\}$.
   - $y = \dashv$: $\delta((q,\gamma,\theta,\lambda),y) = \{(q_f, f_\alpha(\gamma,\theta), g_\beta(\theta),\lambda) | \exists i \in \mathbb{D}_\Sigma \forall x \in q : \%_i^+ \in \lambda(x) \wedge \%_i^+ \in H.\succ(x) \wedge |q| = |\bigcup_{x' \in \mathcal{W}} H.\succ(x')| - |\mathcal{W}| + 1, \mathcal{W} = \bigcup_{w \in q} \lambda(w), (\alpha,\beta) = \Phi(H,q,q_f), \lambda(q_f) = \emptyset\}$.

Definition 8 (Deterministic FAUC). *An FAUC* $\mathcal{A} = (V, Q, \Sigma, q_0, q_f, H, \Phi)$ *is deterministic if and only if* $|\delta((q,\gamma,\theta,\lambda),y)| \leq 1$ *for any configuration* $(q,\gamma,\theta,\lambda)$ *and the symbol* $y \in \Sigma \cup \{\dashv\}$.

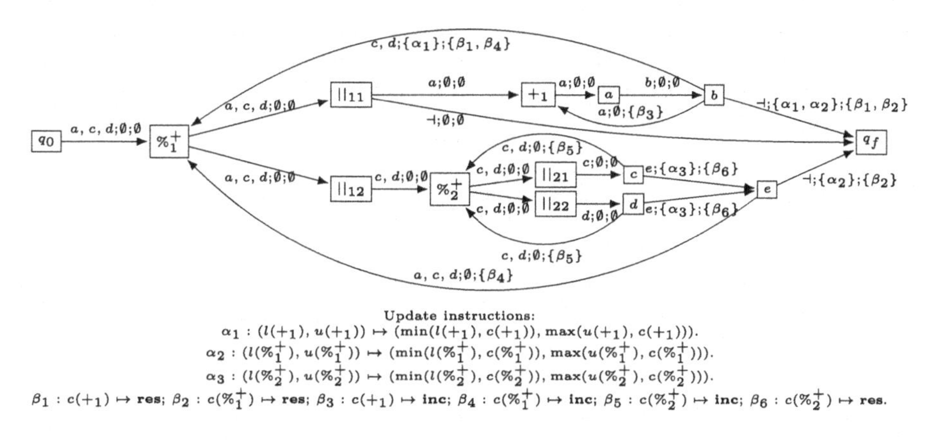

Fig. 1. The deterministic FAUC $\mathcal{A}$ for regular language $\mathcal{L}((((ab)^+)?\%(c\%d)^+e)^+)$. The label of the transition edge is $(y; A_i; B_j)$ $(i,j \in \mathbb{N})$, where $y \in \Sigma \cup \{\dashv\}$ is the current letter and A_i (resp. B_j) is the set of the update instructions from α (resp. β). $\alpha_m \in A_i$ $(m,n \in \mathbb{N})$ is an update instruction for the lower bound and upper bound variables of the counting operator, and $\beta_n \in B_j$ is an update instruction for the counter variable.

Example 3. Let $\Sigma = \{a,b,c,d,e\}$. $V = \Sigma \cup \{\%_1^+, ||_{11}, ||_{12}, \%_2^+, ||_{21}, ||_{22}, +_1\}$. $Q = \{q_0, q_f, \{\%_1^+\}, \{||_{11}, ||_{12}\}, \{||_{11}, \%_2^+\}, \{||_{11}, e\}, \{+_1, ||_{12}\}, \{a, ||_{12}\}, \{b, ||_{12}\}, \{b, \%_2^+\}, \{b, e\}, \{+_1, e\}, \{a, e\}, \{||_{11}, ||_{21}, ||_{22}\}, \{||_{11}, c, ||_{22}\}, \{||_{11}, ||_{21}, d\}, \{||_{11}, c, d\}, \{b, ||_{21}, ||_{22}\}, \{b, c, ||_{22}\}, \{b, ||_{21}, d\}, \{b, c, d\}\}$. Figure 1 illustrates a deterministic FAUC $\mathcal{A} = (V, Q, \Sigma, q_0, q_f, H, \Phi)$ recognizing the language $\mathcal{L}((((ab)^+)?\%(c\%d)^+e)^+)$. $\mathcal{A}.H.R(+_1) = \{a\}$, $\mathcal{A}.H.R(\%_1^+) = \{a,c,d\}$ and $\mathcal{A}.H.R(\%_2^+) = \{c,d\}$. Note that, the digraph H is the digraph shown in Fig. 1, and the update instructions specified by Φ are also presented in Fig. 1.

Theorem 1. *For a string s, and an FAUC $\mathcal{A}$, $\mathcal{A}$ recognizes the language defined by a SOREUC, and it can be decided in $\mathcal{O}(|s||\Sigma|^3)$ time whether $s \in \mathcal{L}(\mathcal{A})$.*

Theorem 1 implies that the membership problem for SOREUCs is also decidable in polynomial time. For a given finite sample S, we can find an FAUC $\mathcal{A}$ (resp. a SOREUC r) in polynomial time such that $\mathcal{L}(\mathcal{A}) \supseteq S$ (resp. $\mathcal{L}(r) \supseteq S$). SOREUCs and FAUCs are learnable from positive finite-samples.

4 Inference of SOREUCs

For a given finite sample S, we infer a SOREUC by learning an FAUC from S. First, we compute the set $U_{\%}$ of all the tuples (u, v) $(u \neq v)$ from S where uv can be an unordered word for S. Then, we obtain the set $P_{\%}$ of unorder units by extracting sets of the nodes (labelled by alphabet symbols) from the undirected graph (undigraph) $F(V, E)$, where $F.E = U_{\%}$. Next, we convert the SOA built for S to an FAUC by traversing the unorder units in $P_{\%}$. The constructed FAUC runs on the given finite sample S to obtain counting operators. Finally, we transform the FAUC to a SOREUC by introducing unordered concatenations and the counting operators. Our algorithm can ensure that the inferred SOREUC can recognize the given finite sample S (see Theorem 4).

Algorithm 1 is the framework of our inference algorithm. Algorithm 2T-INF [4] constructs the SOA for the finite sample S, algorithm $UnorderUnits$ is given in Sect. 4.1, algorithm $ConsFauc$ is shown in Sect. 4.2, algorithm $Running$ demonstrated in Sect. 4.3 is used to run the FAUC, algorithm $GenSoreuc$ is presented in Sect. 4.4.

Algorithm 1. *InfSoreuc*

Input: A finite sample S;
Output: A SOREUC $r : \mathcal{L}(r) \supseteq S$;
1: SOA $G = $ 2T-INF(S); $P_{\%} = \emptyset$;
2: Computing the set $U_{\%}$;
3: Constructing undirected graph $F(V, E)$: $F.E = U_{\%}$;
4: $P_{\%} = UnorderUnits(F, P_{\%})$;
5: FAUC $\mathcal{A} = ConsFauc(G, P_{\%})$;
6: **if** $Running(\mathcal{A}, S)$ **then** $r = GenSoreuc(\mathcal{A})$;
7: **return** r;

4.1 Computing Unorder Units

According to the definition of an unorder unit, an unorder unit can be used to discover the substructure of an FAUC recognizing unordered strings. Then, to learn an FAUC which can recognize all the unordered strings from a given finite sample S, for any two distinct alphabet symbols a and b that ab can be an unordered word for S (i.e., $(a, b) \in U_{\%}$), there must exist an unorder unit ut such that a and b are in different sets in ut.

Let M_F denote the set of the non-adjacent nodes selected from a given undigraph F such that the sum of all node degrees is maximum. The set $P_{\%}$ of unorder units is obtained by recursively extracting sets of nodes from the undigraph $F(V, E)$, where $F.E = U_{\%}$. Algorithm 2 shows the recursion process to extract unorder units from F.

Algorithm 2. *UnorderUnits*

Input: An undigraph $F(V, E)$, a set $P_{\%}$ of unorder units;
Output: A set $P_{\%}$ of unorder units;
1: **if** $F.V \neq \emptyset$ **then**
2:   **for each** connected component F_1 in F:
3:     Computing the set M_{F_1};
4:     Remove the nodes in M_{F_1} and their associated edges in F_1; $F_2 = F_1$;
5:     **if** $F_2.V \neq \emptyset$ and F_2 is not a connected graph **then**
6:       $P_{\%}(i) = P_{\%}(i) \cup [M_{F_1}, F_2.V]$; $i = i + 1$;
7:       Remove isolated nodes in F_2;
8:     **else** Put M_{F_1} in $P_{\%}(i)$;
9:     $P_{\%}$ =*UnorderUnits*$(F_2, P_{\%})$; $i = i + 1$;
10: **return** $P_{\%}$;

For each connected component F_1 in F, first, we compute the set M_{F_1} (line 3). Then, we obtain an undigraph F_2 by removing the nodes in M_{F_1} and their associated edges in F_1 (line 4). If F_2 ($F_2.V \neq \emptyset$) is not a connected graph, the sets M_{F_1} and $F_2.V$ can form an unorder unit (line 6), and the isolated nodes are removed from F_2 (line 7). Otherwise, the set M_{F_1} is stored in an unorder unit (line 8). We recursively extract an unorder unit or a set M_{F_2} from F_2 (line 9). For each connected component, there is a corresponding unorder unit. Note that, in Algorithm 2, i is a global variable (initially, $i = 1$).

In Algorithm 2, for a given undigraph $F(V, E)$, it takes $\mathcal{O}(|V|^2)$ time to obtain the set M_F. Since there is a corresponding unorder unit for each connected component in F, it requires $|V|$ recursions at most to obtain an unorder unit. The other processes take $\mathcal{O}(|V|)$ time. Thus, it takes $\mathcal{O}(|V|^3)$ ($|V| \leq |\Sigma|$) time to obtain the set $P_{\%}$. The time complexity of algorithm *UnorderUnits* is $\mathcal{O}(|\Sigma|^3)$.

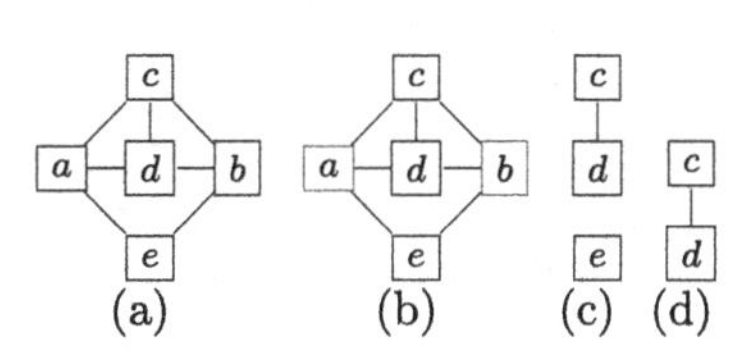

Fig. 2. The procedures computing the set $P_{\%}$ of unorder units. (a) is the undigraph F. Algorithm 2 works on (b), (b) is a connected graph, the set $M_F = \{a, b\}$. (c) is obtained by using line 4. (c) is not a connected graph, then $P_{\%}(1) = [\{a, b\}, \{c, d, e\}]$ forms an unorder unit by using line 6. (d) is obtained by using line 7. Algorithm 2 recursively works on (d) (line 9), $P_{\%}(2) = [\{c\}, \{d\}]$. The finally updated $P_{\%} = \{P_{\%}(1), P_{\%}(2)\}$.

Example 4. For sample $S = \{ababcde, dce, cdeabcdeab\}$, the computed $U_{\%} = \{(a, c), (a, d), (a, e), (b, c), (b, d), (b, e)\}$. The undigraph $F(V, E)$ ($F.E = U_{\%}$) is shown in Fig. 2(a). Figure 2 illustrates the procedures computing the set $P_{\%}$ of unorder units. The finally obtained $P_{\%} = \{[\{a, b\}, \{c, d, e\}], [\{c\}, \{d\}]\}$.

Theorem 2. *Let $P_{\%} = UnorderUnits(F(V, E), \emptyset)$ where $F.E = U_{\%}$, then for any tuple $(a, b) \in U_{\%}$, there exists an unorder unit $ut \in P_{\%}$ such that a and b are in different sets in ut.*

4.2 Constructing FAUC

For a given finite sample S, an FAUC can be constructed by building the node transition graph of an FAUC. Since an SOA built for S is a precise representation of S [12], and an unorder unit in $P_{\%}$ can be used to discover the substructure of an FAUC recognizing the unordered strings from S, first, the SOA G built for S is converted to the node transition graph of an FAUC by traversing the unorder units in $P_{\%}$. Then, we present the detailed descriptions of the constructed FAUC.

Algorithm 3 is presented to construct an FAUC. First, we remove the directed edges in G, where tails and heads are in two disjoint sets of an unorder unit in $P_{\%}$, respectively (line 1). Then, for the ith unorder unit in $P_{\%}$ ($P_{\%}(i)$, $i \in \mathbb{D}_{\Sigma}$), we identify the corresponding set of nodes from G (by using *extract*) which is the union of all the sets of nodes in $P_{\%}(i)$ to add the marker $\%_i^+$ in G (lines 3∼4). We also add edges directing to the node $\%_i^+$ to create a loop for recognizing possibly repeated and unordered strings (line 5). Additionally, for the jth set in $P_{\%}(i)$ ($P_{\%}(i)(j)$, $j \in \mathbb{P}_{\Sigma}$), we identify the corresponding set of nodes from G (by using *extract*) to add concurrent marker $||_{ij}$ in G (line 8), and by using the information provided by the obtained G_1 (line 7), the loop makers $+_k$ ($k \in \mathbb{B}_{\Sigma}$) are added into G (line 8). Note that, some edges are possibly added into G (line 10). The finally obtained G is the node transition graph of an FAUC, then the FAUC $\mathcal{A}$ is obtained (line 11). Some subroutines are described as follows.

extract on a digraph G takes a set of nodes U (of G) as input, it extracts a new digraph G_1 ($G_1.V = \{q_0, q_f\} \cup U$) from G, G_1 reserves the directed edges (in G) between any two nodes in U. All nodes in U, which have not incoming edges or have incoming edges from outside of U in G, have incoming edges from q_0 in G_1. Moreover, all nodes in U, which have not outgoing edges or have outgoing edges to outside of U in G, have outgoing edges to q_f in G_1.

Algorithm 3. *ConsFauc*

```
Input: A digraph G(V, E), a set P_% of unorder units;
Output: An FAUC A;
 1: Delete edges {(v1, v2)|v1 ∈ e1, v2 ∈ e2, e1, e2 ∈
    ut, ut ∈ P_%, e1 ≠ e2} in G;
 2: for i = 1 to |P_%| do
 3:   Let T = ∪_h P_%(i)(h); G1 = G.extract(T);
 4:   G.addnode(%_i^+, G1.≻(q0)); R(%_i^+) = G.≻(%_i^+);
 5:   Add edges {(v, %_i^+)|v ∈ G1.≺(qf)} in G;
 6:   for j = 1 to |P_%(i)| do
 7:     G1 = G.extract(P_%(i)(j));
 8:     G.addnode(||_ij, G1.≻(q0)); G.add+(G1);
 9:     if NO(S, P_%(i)(j), P_%(i)) then
10:       Add edges {(||_ij, v)|v ∈ G.≻(v'), v' ∈ G1.≺
    (qf)} in G;
11: FAUC A = (V, Q, Σ, G.q0, G.qf, H, Φ);
12: return A;
```

addnode and *add*$^+$. *addnode* on G takes a node v and a set of nodes U (of G) as inputs. It works as follows. Add a node v in G; add edges $\{(v_1, v)|v_1 \in G. \prec(v_2), v_2 \in U\}$; remove edges $\{(v_1, v_2)|v_1 \in G. \prec(v_2), v_2 \in U\}$; add edges $\{(v, v_2)|v_2 \in U\}$. *add*$^+$ on G takes a directed graph G_1 (G_1 is extracted from G) as input. It works as follows. For a strongly connected component U in G_1, let $G_2 = G_1.extract(U)$ and $G.addnode(+_k, G_2.\succ(q_0))$ (initially, $k = 1$); add edges $\{(v, +_k)|v \in G_2.\prec(q_f)\}$; let $\mathcal{R}(+_k) = G_2.\succ(q_0)$ and $k = k+1$; break the loop formed by U in G_1 by using *bend* [12] to form a new digraph; *add*$^+$ recursively works on the new digraph until an acyclic graph is obtained.

$NO(S, P_{\%}(i)(j), P_{\%}(i))$ is a bool function. Let $T = \bigcup_h P_{\%}(i)(h)$ $(1 \leq j, h \leq |P_{\%}(i)|)$. NO returns $true$ if there exists a substring t of $s \in S$ (t consists of at least one symbol from the set T.) such that t does not contain any symbols from $P_{\%}(i)(j)$ and neither lt nor tl $(l \in T)$ are substrings of s.

Then, we present the detailed descriptions of the FAUC $\mathcal{A}$.

$\mathcal{A} = (V, Q, \Sigma, G.q_0, G.q_f, H, \Phi)$, where $V = G.V \setminus \{q_0, q_f\}$, $H.V = G.V$ and $H.E = G.E$. Let $V_c' = \{v | (v \in \Sigma \wedge v \in H. \succ (v)) \vee (v \in \{+_i, \%_j^+\}_{i \in \mathbb{B}_\Sigma, j \in \mathbb{D}_\Sigma} \wedge v \in G.V)\}$, then $H.\mathcal{C} = \{c(v) | v \in V_c'\}$ and $H.T = \{(l(v), u(v)) | v \in V_c'\}$. $Q = Q' \cup \{G.q_0, G.q_f\}$ and $Q' = \bigcup_q \bigcup_y \delta((q, \gamma, \theta, \lambda), y)$ $(q \in \{G.q_0\} \cup Q'$ and $y \in \Sigma \cup \{\dashv\})$. The initial configure is $(G.q_0, \gamma_\infty, \theta_1, \lambda_{\mathcal{A}.H.V} = \{\emptyset\})$. Φ and δ can be derived from the node transition graph H, which is a parameter implied in them. Note that, $\mathcal{R}$: $\{ ||_i, \%_j^+ \}_{i \in \mathbb{B}_\Sigma, j \in \mathbb{D}_\Sigma} \mapsto \wp(\Sigma)$ (see line 4 and subroutine add^+). Thus, $H.R = \mathcal{R}$.

The SOA $G(V, E)$ and $P_{\%}$ are as inputs of Algorithm 3. It takes $\mathcal{O}(|V|^2)$ time to delete edges in G (line 1). For each unorder unit $ut \in P_{\%}$, the average time complexity of *extract* in line 3 (resp. in line 7) is $\mathcal{O}(\frac{|V||\Sigma|}{|P_{\%}|})$ (resp. $\mathcal{O}(\frac{|V||\Sigma|}{|P_{\%}||ut|})$). For *addnode* and add^+, the time complexity of them are both $\mathcal{O}(|V|)$. NO (line 10) can be obtained in $\mathcal{O}(|\Sigma|N)$ time. The other processes takes $\mathcal{O}(|V|^2)$ time in total. For an unorder unit $ut \in P_{\%}$, $|ut| \leq |\Sigma|$ and $|P_{\%}| \leq |\Sigma| - 1$. Thus, the average time complexity of constructing an FAUC is $\mathcal{O}(|V|^2 + |V||\Sigma| + \frac{|V||\Sigma|}{|P_{\%}|}|\Sigma| + \frac{|V||\Sigma|}{|P_{\%}||ut|} |\Sigma|^2 + |\Sigma|^2 V| + |\Sigma|^2 N) = \mathcal{O}(|\Sigma|^2 V| + |\Sigma|^2 N) = \mathcal{O}(|\Sigma|^2 N)$ $(|V| = |\Sigma| + 2, N > |\Sigma|)$.

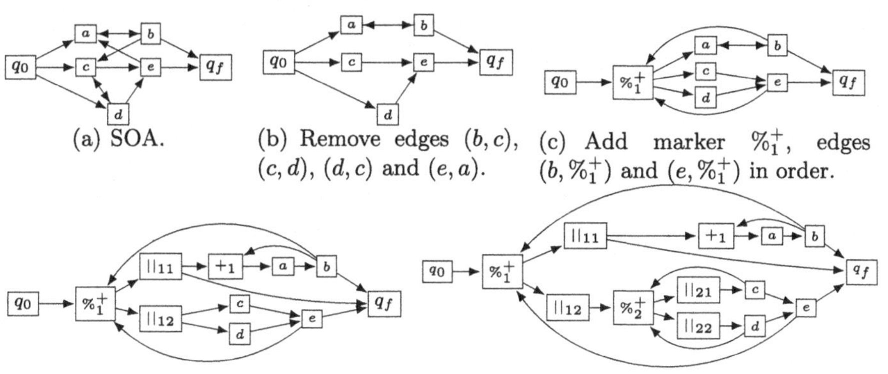

(a) SOA. (b) Remove edges (b, c), (c, d), (d, c) and (e, a). (c) Add marker $\%_1^+$, edges $(b, \%_1^+)$ and $(e, \%_1^+)$ in order.

(d) Add marks $||_{11}$, $+_1$, edge $(||_{11}, q_f)$ and marker $||_{12}$ in order. (e) Add markers $\%_2^+$, $||_{21}$ and $||_{22}$ in order.

Fig. 3. The procedures converting the SOA (in (a)) to the node transition graph (in (e)) of the FAUC $\mathcal{A}$ by traversing unorder units in $\{[\{a, b\}, \{c, d, e\}], [\{c\}, \{d\}]\}$. For $S = \{ababcde, dce, cdeabcdeab\}$, $NO(S, \{a, b\}, [\{a, b\}, \{c, d, e\}]) = true$, for neither a nor b occur in string dce. Then, in (d), the edge $(||_{11}, q_f)$ is added into G.

Example 5. For $S = \{ababcde, dce, cdeabcdeab\}$, the SOA recognizing S is shown in Fig. 3(a), $P_{\%} = \{[\{a, b\}, \{c, d, e\}], [\{c\}, \{d\}]\}$. Figure 3 illustrates the main steps to convert the SOA to the node transition graph H (shown in Fig. 3(e)) of the FAUC $\mathcal{A}$ by traversing the unorder units in $P_{\%}$. The labels on the edges of H can be seen in Fig. 1, which illustrates the finally obtained FAUC $\mathcal{A}$.

Theorem 3. *For a finite sample S, let SOA $G = 2T\text{-}INF(S)$ and $P_{\%}$ denote the result returned by UnorderUnits, let $\mathcal{A} = ConsFauc(G, P_{\%})$, then $\mathcal{L}(\mathcal{A}) \supseteq S$.*

4.3 Running FAUC

Given a finite sample S, the constructed FAUC $\mathcal{A}$ runs on S to count the minimum and maximum number of the strings or substrings (from S) that are consecutively matched by $\mathcal{A}$. Counting rules are given by the transition functions of the FAUC $\mathcal{A}$ and the update instructions returned by $\mathcal{A}.\Phi$. Let *Running*($\mathcal{A}$, S) denote the procedures of the constructed FAUC $\mathcal{A}$ running on S, if the running is terminated, *Running* returns *true*, otherwise, *false*. Then, the counting results in $\mathcal{A}.H.T$ (i.e., counting operators for a SOREUC) can be obtained when *Running* returns *true*. Note that, the initial configuration is $(q_0, \gamma_\infty, \theta_1, \lambda_{\mathcal{A}.H.V} = \{\emptyset\})$, after a string in S was recognized by $\mathcal{A}$, the configuration excluding γ is reset (i.e., the configuration becomes $(q_0, \gamma, \theta_1, \lambda_{\mathcal{A}.H.V} = \{\emptyset\})$).

For the given finite sample S, the number of strings is N and $\overline{L}$ is the average length of the strings in sample. Then, the time complexity of *Running* is $\mathcal{O}(N\overline{L})$.

Example 6. For $S = \{ababcde, dce, cdeabcdeab\}$, the FAUC $\mathcal{A}$ is obtained in Sect. 4.2, *Running* returns *true*. Table 1 lists the results in $\mathcal{A}.H.T = \{(l(v), u(v)) | v \in \{+_1, \%_1^+, \%_2^+\}\}$. For instance, $l(+_1)$ and $u(+_1)$ (resp. $l(\%_1^+)$ and $u(\%_1^+)$) are respectively the minimum and maximum number of the substring ab (resp. the unordered strings derived from $ab\%cde$ or $ab\%dce$) that can be consecutively matched by $\mathcal{A}$. Note that, neither a nor b occur in dce, but the substring ab consecutively occurs at least once in S, we initialize $l(v)$ ($v \in \{+_1, \%_1^+, \%_2^+\}$) to 1, so finally $l(+_1) = 1$, instead of $l(+_1) = 0$.

Table 1. The results in $\mathcal{A}.H.T$ after the FAUC $\mathcal{A}$ recognizing each string in S.

string	$(l(+_1), u(+_1))$	$(l(\%_1^+), u(\%_1^+))$	$(l(\%_2^+), u(\%_2^+))$
ababcde	$(2, 2)$	$(1, 1)$	$(1, 1)$
dce	$(1, 2)$	$(1, 1)$	$(1, 1)$
cdeabcdeab	$(1, 2)$	$(1, 2)$	$(1, 1)$

4.4 Generating SOREUC

In this section, we transform the FAUC $\mathcal{A}$ which ran on a given finite sample S and has been terminated (*Running*($\mathcal{A}$, S) returns *true* in Sect. 4.3) to a SOREUC by introducing unordered concatenations and counting operators.

Algorithm 4. *GenSoreuc*

Input: An FAUC $\mathcal{A}$;
Output: A SOREUC r;
1: Let $r_s = Soa2Sore(\mathcal{A}.H)$; let $\mathcal{T} = \mathcal{A}.H.T$;
2: Search all subexpressions r_b from r_s:
3: **if** $r_b = a^+$ $(a \in \Sigma)$ **then**
4: Replace r_b by $a^{[\mathcal{T}.l(a),\mathcal{T}.u(a)]}$;
5: **if** $\exists i \in \mathbb{B}_\Sigma : r_b = (+_i e)^+$ (for expression e) **then**
6: Replace r_b by $(e)^{[\mathcal{T}.l(+_i),\mathcal{T}.u(+_i)]}$;
7: **if** $\exists j \in \mathbb{D}_\Sigma \exists k \in \mathbb{P}_\Sigma : r_b = (\%_j^+(||_{j1} r_1 | \cdots | ||_{jk} r_k))^+$ (for expressions $r_1, \cdots, r_k$) **then**
8: Replace r_b by $(r_1 \% \cdots \% r_k)^{[\mathcal{T}.l(\%_j^+),\mathcal{T}.u(\%_j^+)]}$;
9: Let $r = r_s$; **return** r;

Algorithm 4 presents the procedures to transform the FAUC $\mathcal{A}$ to a SOREUC. Since an FAUC can be denoted by the corresponding node transition graph (a digraph), and algorithm *Soa2Sore* [12] can transform a digraph to an expression (the alphabet symbols are the labels of nodes), first, the node transition graph $\mathcal{A}.H$ of the FAUC $\mathcal{A}$ is transformed to an expression r_s by using algorithm *Soa2Sore* (line 1). The expression r_s contains symbols $+_i$, $\%_j^+$ and $||_{jk}$ $(i \in \mathbb{B}_\Sigma, j \in \mathbb{D}_\Sigma, k \in \mathbb{P}_\Sigma)$. Then, for all the subexpressions r_b shown in lines 3, 5 and 7, r_b is rewritten such that counting operators (resp. unordered concatenations) are introduced into r_b in lines 4, 6 and 8 (resp. line 8). Note that, $(e)^{[1,1]} = e$ for regular expression e.

Let n_h (resp. t_h) denote the number of nodes (resp. the number of transitions) in $\mathcal{A}.H$. Then, *Soa2Sore* takes $\mathcal{O}(n_h t_h)$ time to infer an expression r_s [12]. Since all the subexpressions r_b shown in lines 3, 5 and 7 can be obtained by traversing the syntax tree of r_s, it takes $\mathcal{O}(|r_s|)$ time to search and then rewrite all the subexpressions r_b. Finally, *GenSoreuc* takes $\mathcal{O}(n_h t_h + |r_s|) = \mathcal{O}(n_h t_h)$ time to generate a SOREUC. $U_\%$ can be obtained in $\mathcal{O}(|\Sigma|N)$ time, the time complexity of *UnorderUnits* is $\mathcal{O}(|\Sigma|^3)$, the average time complexity of *ConsFauc* is $\mathcal{O}(|\Sigma|^2 N)$, and the time complexity of *Running* is $\mathcal{O}(N\overline{L})$. Thus, the time complexity of *InfSoreuc* is $\mathcal{O}(|\Sigma|^2 N + N\overline{L})$ $(|\Sigma|N > n_h t_h)$.

Example 7. For $S = \{ababcde, dce, cdeabcdeab\}$, the node transition graph (H) of the FAUC $\mathcal{A}$ is presented in Fig. 3(e). The counting results in $\mathcal{A}.H.T$ are shown in Table 1. $(l(+_1), u(+_1)) = (1, 2)$, $(l(\%_1^+), u(\%_1^+)) = (1, 2)$ and $(l(\%_2^+),$ $u(\%_2^+)) = (1, 1)$. $r_s = (\%_1^+(||_{11}((+_1 ab)^+)?| \, ||_{12}(\%_2^+(||_{21}c| \, ||_{22}d))^+ e))^+$ is returned by $Soa2Sore(S)$. r_s is first rewritten to $(((+_1 ab)^+)?\%(\%_2^+(||_{21}c| \, ||_{22}d))^+ e)^{[1,2]}$, which is then rewritten to $r_s = (((+_1 ab)^+)?\%(c\%d)^{[1,1]} e)^{[1,2]}$. Finally, r_s is rewritten to $(((ab)^{[1,2]})?\%(c\%d)e)^{[1,2]}$ which is the finally obtained SOREUC.

Theorem 4. *For a finite sample S, let $r = InfSoreuc(S)$, then r is a SOREUC and $\mathcal{L}(r) \supseteq S$.*

5 Experiments

In this section, we first analyse the expressiveness of SOREUCs, then we assess the conciseness of the inferred SOREUCs, and also evaluate our algorithm on unordered XML data in terms of generalization ability and time performance. All data including XML data (unordered) used in experiments are collected from data-centric XML applications. Since the learning algorithms for DIMEs are lacking, our algorithm is mainly compared with the inference algorithms for *u*SOREs, DMEs and SOREs. Note that, the inference algorithm for SOREs is selected as *Soa2Sore*, which is the fast algorithm to infer a compact SORE [12].

5.1 Expressiveness of SOREUCs

The expressive power of a regular expression r is measured by the proportion of the other regular expressions captured by the regular expression r [5]. For regular expressions r and r', r' can be captured by r if $\mathcal{L}(r) = \mathcal{L}(r')$ [5].

For 60,578 regular expressions and 558,375 regular expressions, which are respectively extracted from DTD files and XSD files that are collected from Maven[3] and GitHub[4], Table 2 shows that the highest proportions of the regular expressions from DTDs or XSDs are captured by SOREUCs. SOREUCs have stronger expressive powers for modeling unordered schemata than existing works. Note that, since DTDs do not support counting, the proportions of the regular expressions from DTDs are captured by *u*SOREs and SOREUCs are the same.

Table 2. Proportions of regular expressions captured by each subclass.

Subclasses	DTDs (60,578 regular expressions)	XSDs (558,375 regular expressions)
DMEs	55.2%	33.6%
DIMEs	72.9%	41.3%
SOREs	83.1%	72.8%
*u*SOREs	97.5%	82.4%
SOREUCs	97.5%	93.7%

5.2 Conciseness, Generalization Ability and Time Performance

Since DTDs do not support counting, we select regular expressions from the above 558,375 regular expressions, which are collected from XSD files. Let Q_1 denote the set of 1000 selected regular expressions with alphabet size 20. We also select 1000 regular expressions for each alphabet size ranging from 10 to 100, then let Q_2 denote the set of the 10000 selected regular expressions in total. All selected regular expressions are accompanied by lots of XML data. The random sample in experiments, which is a finite set of strings, is extracted from the corresponding XML data. The size of sample is the number of the strings in sample. We evaluate our algorithm by using the datasets Q_1 and Q_2.

Conciseness. For learned regular expressions, their lengths can intuitively reflect their conciseness. We provide the statistics about lengths of the learned regular expressions in different size of samples and different size of alphabets. To

[3] https://mvnrepository.com/.
[4] https://github.com/topics/.

learn each expression in Q_1, we randomly extracted the corresponding sample, each sample size is that listed in Fig. 4(a). To learn each expression of alphabet size from 10 to 100 in Q_2, we randomly extracted the corresponding sample, of which the size is 5000 such that the language denoted by the corresponding expression can be learned. In Fig. 4(a) (resp. in Fig. 4(b)), the value for a given sample size (resp. a given alphabet size) is the average of the lengths of the 1000 learned expressions or the 1000 original expressions (original REs).

Figure 4(a) and Fig. 4(b) illustrate that, the inferred SOREUCs are not only more concise than original expressions, but also more concise than the learned DMEs. Compared with SOREUCs, *u*SOREs (resp. SOREs) just only do not support counting (resp. counting and unorder), therefore, the inferred SOREUCs are less concise than the learned *u*SOREs and SOREs.

Generalization Ability. We measure the generalization ability of our algorithm by $F = \frac{2pr}{p+r}$, where p and r are the precision and recall of the language defined by the learned regular expression, respectively. We specify that, the learning algorithm with higher F-measure has better generalization ability. The average F-measure, which is as functions of sample size, is averaged over the 1000 expressions in Q_1.

To learn each expression in Q_1, we randomly extracted the corresponding sample, each sample size is that listed in Fig. 4(c). Let r_0 denote the original expression. Then, the positive sample (S_+) is the set of all the strings accepted by r_0, and the negative sample (S_-) is the set of all the strings not accepted by r_0. p and r are computed in a finite sample. For example, the true positive sample is the set of the strings that are in S_+ and in $\mathcal{L}(r_1)^{\leq n}$, where r_1 is the learned expression, $n = 2|r_1|+1$ and $\mathcal{L}(r_1)^{\leq n} = \{s|s \in \mathcal{L}(r_1), |s| \leq n\}$ [4]. Note that, since it is trivial for deciding a string whether can be accepted by a DME [8], and we can build an equivalent receptor [17] for a SOREUC, p and r (i.e., F-measure) can be efficiently computed for learning algorithms *InfSoreuc* and L^+_{DME}.

The plots in Fig. 4(c) demonstrate that, the F-measures for *InfSoreuc* are consistently higher than that for other algorithms. In general, *InfSoreuc* has better generalization ability than other algorithms.

Time Performance. To illustrate the efficiency of algorithm *InfSoreuc*, we provide the statistics about running time in different size of samples and different size of alphabets. Figure 4(d) and Fig. 4(e) show the average running times in seconds for each learning algorithm with different inputs of sample size and with different inputs of alphabet size, respectively. To learn each expression in Q_1, we randomly extracted the corresponding sample, each sample size is that listed in Fig. 4(d). The running times listed in Fig. 4(d) are averaged over 1000 expressions of that sample size. To learn each expression in Q_2, we randomly extracted the corresponding sample, the sample size for each alphabet size listed in Fig. 4(e) is 2000. The running times listed in Fig. 4(e) are averaged over 1000 expressions of that alphabet size.

Figure 4(d) and Fig. 4(e) illustrate that, for each given sample size and alphabet size, the running times for *InfSoreuc* are less than that for *InfuSORE* and L^+_{DME}, respectively, and are closer to that for *Soa2Sore*. Thus, this implies that, *InfSoreuc* can efficiently infer SOREUCs for modeling unordered schemata.

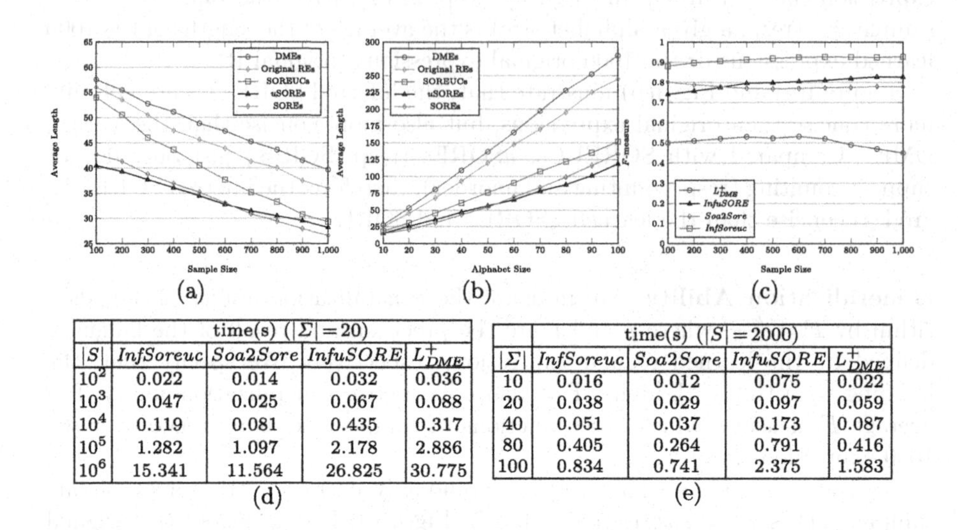

	time(s) ($\|\Sigma\|=20$)			
$\|S\|$	*InfSoreuc*	*Soa2Sore*	*InfuSORE*	L^+_{DME}
10^2	0.022	0.014	0.032	0.036
10^3	0.047	0.025	0.067	0.088
10^4	0.119	0.081	0.435	0.317
10^5	1.282	1.097	2.178	2.886
10^6	15.341	11.564	26.825	30.775

(d)

	time(s) ($\|S\|=2000$)			
$\|\Sigma\|$	*InfSoreuc*	*Soa2Sore*	*InfuSORE*	L^+_{DME}
10	0.016	0.012	0.075	0.022
20	0.038	0.029	0.097	0.059
40	0.051	0.037	0.173	0.087
80	0.405	0.264	0.791	0.416
100	0.834	0.741	2.375	1.583

(e)

Fig. 4. (a) and (b) are average lengths of the learned/original regular expressions as functions of the sample size and the alphabet size, respectively. (c) is average F-measure as function of the sample size for each algorithm. (d) and (e) are respectively average running times in seconds as the functions of sample size and alphabet size for each algorithm.

6 Conclusion

This paper proposed a subclass SOREUCs for modeling unordered schemata and the inference algorithm for SOREUCs. First, we construct an FAUC from a given finite sample. Then, the FAUC runs on the given finite sample to obtain counting operators. Finally we transform the FAUC to a SOREUC by introducing unordered concatenations and counting operators. Our experiments demonstrate that, SOREUCs have stronger expressive powers for modeling unordered schemata than existing works, and our algorithm can efficiently infer a concise SOREUC with better generalization ability. For future works, we can study more expressions with stronger expressive powers for modeling unordered schemata. We also can apply our techniques to boost the learning algorithms for twig queries.

Acknowledgements. Thanks for professor Haiming Chen, who provided much advice for preparing the final version of this paper, and this work was supported by National Nature Science Foundation of China (Nos. 61872339, 61472405).

References

1. Abiteboul, S., Bourhis, P., Vianu, V.: Highly expressive query languages for unordered data trees. Theory Comput. Syst. **57**(4), 927–966 (2015)
2. Amer-Yahia, S., Cho, S., Lakshmanan, L.V., Srivastava, D.: Tree pattern query minimization. VLDB J. **11**(4), 315–331 (2002)
3. Bex, G.J., Neven, F., Schwentick, T., Tuyls, K.: Inference of concise DTDs from XML data. In: International Conference on Very Large Data Bases, Seoul, Korea, September, pp. 115–126 (2006)
4. Bex, G.J., Neven, F., Schwentick, T., Vansummeren, S.: Inference of concise regular expressions and DTDs. ACM Trans. Database Syst. **35**(2), 1–47 (2010)
5. Boneva, I., Ciucanu, R., Staworko, S.: Schemas for unordered XML on a DIME. Theory Comput. Syst. **57**(2), 337–376 (2015)
6. Boneva, I., Ciucanu, R., Staworko, S.: Simple schemas for unordered XML. In: 16th International Workshop on the Web and Databases (WebDB) (2013)
7. Brüggemann-Klein, A., Wood, D.: One-unambiguous regular languages. Inf. Comput. **142**(2), 182–206 (1998)
8. Ciucanu, R., Staworko, S.: Learning schemas for unordered XML. arXiv preprint arXiv:1307.6348 (2013)
9. Colazzo, D., Ghelli, G., Sartiani, C.: Linear time membership in a class of regular expressions with counting, interleaving, and unordered concatenation. ACM Trans. Database Syst. (TODS) **42**(4), 1–44 (2017)
10. Czerwinski, W., Martens, W., Niewerth, M., Parys, P.: Minimization of tree pattern queries. In: Proceedings of the 35th ACM SIGMOD-SIGACT-SIGAI Symposium on Principles of Database Systems, pp. 43–54 (2016)
11. Florescu, D.: Managing semi-structured data. Queue **3**(8), 18–24 (2005)
12. Freydenberger, D.D., Kötzing, T.: Fast learning of restricted regular expressions and DTDs. Theory Comput. Syst. **57**(4), 1114–1158 (2015)
13. Gelade, W., Gyssens, M., Martens, W.: Regular expressions with counting: weak versus strong determinism. SIAM J. Comput. **41**(1), 160–190 (2012)
14. Golshan, B., Halevy, A., Mihaila, G., Tan, W.C.: Data integration: after the teenage years. In: Proceedings of the 36th ACM SIGMOD-SIGACT-SIGAI Symposium on Principles of Database Systems, pp. 101–106 (2017)
15. Grijzenhout, S., Marx, M.: The quality of the XML Web. In: Proceedings of the 20th ACM International Conference on Information and Knowledge Management, pp. 1719–1724 (2011)
16. Grijzenhout, S., Marx, M.: The quality of the XML Web. J. Web Semant **19**, 59–68 (2013)
17. Hovland, D.: The membership problem for regular expressions with unordered concatenation and numerical constraints. In: Dediu, A.-H., Martín-Vide, C. (eds.) LATA 2012. LNCS, vol. 7183, pp. 313–324. Springer, Heidelberg (2012). https://doi.org/10.1007/978-3-642-28332-1_27
18. Lohrey, M., Maneth, S., Reh, C.P.: Compression of unordered XML trees. In: 20th International Conference on Database Theory (ICDT 2017), pp. 18:1–18:17 (2017)
19. International Organization for Standardization: Information Processing: Text and Office Systems: Standard Generalized Markup Language (SGML). ISO (1986)
20. Staworko, S., Wieczorek, P.: Learning twig and path queries. In: Proceedings of the 15th International Conference on Database Theory, pp. 140–154 (2012)

21. Thompson, H., Beech, D., Maloney, M., Mendelsohn, N.: XML Schema Part 1: structures, 2nd Edn. W3C Recommendation (2004)
22. Wang, X., Chen, H.: Inferring deterministic regular expression with unorder. In: Chatzigeorgiou, A., et al. (eds.) SOFSEM 2020. LNCS, vol. 12011, pp. 325–337. Springer, Cham (2020). https://doi.org/10.1007/978-3-030-38919-2_27

MACROBERT: Maximizing Certified Region of BERT to Adversarial Word Substitutions

Fali Wang[1,2], Zheng Lin[1(✉)], Zhengxiao Liu[1,2], Mingyu Zheng[1,2], Lei Wang[1], and Daren Zha[1]

[1] Institute of Information Engineering, Chinese Academy of Sciences, Beijing, China
{wangfali,linzheng,liuzhengxiao,zhengmingyu,wanglei,zhadaren}@iie.ac.cn
[2] School of Cyber Security, University of Chinese Academy of Sciences, Beijing, China

Abstract. Deep neural networks are deemed to be powerful but vulnerable, because they will be easily fooled by carefully-crafted adversarial examples. Therefore, it is of great importance to develop models with certified robustness, which can provably guarantee that the prediction will not be easily misled by any possible attack. Recently, although a certified method based on randomized smoothing is proposed, it does not take the maximized *certified region* into account, so we develop an approach to train models with maximized certified regions via replacing the base classifier with the soft smoothed classifier which is differentiable during propagation.

Keywords: Randomized smoothing · Adversarial examples · Certified region

1 Introduction

Deep neural networks (DNNs) have gained tremendous success in various Natural Language Processing (NLP) tasks. However, previous works have shown that even state-of-the-art DNNs are vulnerable against adversarial examples that are well-crafted by adding imperceptible perturbations to inputs [5]. To enhance the robustness of textual DNNs, recent studies have focused on the adversarial texts, which are generated by methods like inserting character-level spelling errors or substituting synonyms for original words [1,7].

In this paper, we focus on *adversarial word substitution*, which is a word-level adversarial text and more difficult to defend than the other two level's attack [7]. In this setting, an attacker may replace any word in the input text with a synonym, leading to an exponentially large perturbation space. Although the perturbation space is known at training time, it is still very challenging and costly to train a model that is robust to such a large perturbation space.

Researchers have proposed a variety of defense methods to improve the robustness of neural networks. One kind of method named data augmentation

C. S. Jensen et al. (Eds.): DASFAA 2021, LNCS 12682, pp. 253–261, 2021.
https://doi.org/10.1007/978-3-030-73197-7_16

enhances the model's robustness against noise by adding augmented noisy examples during training [2]. Most of the existing data augmentation methods are based on adversarial training [9]. Nevertheless, adversarial training has two drawbacks: First, it depends on particular attack methods [12]. Second, the relative positions between word vectors of a word and its synonyms change dynamically [15], which makes a model more vulnerable against attacks because two similar words may be considered as totally different after training.

To address the problem of adversarial training, recent studies have presented several certified defense methods. The first certified robust defense method [6] propagates interval constraints layer-by-layer to compute final model output. However, this method has two limitations. For one thing, they use Interval Bound Propagation (IBP) to compute a loose outer bound and thus result in a significant performance drop on the clean data. For another, directly applying this method to pre-trained models like BERT is difficult because it needs to adjust model structures. To alleviate the dependency on model structures, [13] first introduced *randomized smoothing* technique in the NLP field and proposed a structure-free certified defense method. They leveraged the statistical properties of the randomized ensembles to compute a provable safe bound. We further define this safe bound as *certified region*, abbreviated as CR. However, they didn't consider maximizing certified region to enhance model robustness.

In this paper, we propose an attack-free and model-agnostic method to train robust textual DNNs by maximizing the certified region. By maximizing CR, we enlarge the margin between the probability of ground truth label and "runner-up" label and thus increase the percentage of samples whose $CR > 0$, which leads to better model robustness. Inspired by [4,13,14], we replace the base classifier f with a smoothed classifier g. [14] proposed a method to train a robust DNN by maximizing the certified radius in the computer vision community, which inspires us to further improve the robustness of textual DNNs by maximizing the certified region. Specifically, to compute and maximize the certified region, we directly train a smoothed classifier rather than a base classifier. [4] pointed out that to make randomized smoothing work, one must make the base classifier classify well under noise, which is another reason for us to train a smoothed classifier g instead of a base one. Moreover, our model can achieve better robustness by using higher-order neighbor information on the synonymous network [15].

We conclude our contributions as follows:

1. We propose an attack-free and theoretically-proved certified textual defense method based on maximizing certified region. The trained model possesses provable robustness that can defend arbitrary word substitution attacks within the certified region.
2. We replace the base classifier with the soft smoothed classifier to solve the problem of non-differentiability, such that our method can keep stable relative positions between the word vectors of a word and its synonyms.
3. We conduct comprehensive experiments on various datasets and NLP models. Experimental results show that our method not only maintains the equivalent performance of original models on the clean data but also achieves better model robustness than state-of-the-art defense methods.

2 Methods

Problem Setup. We consider the text classification task for giving a text input x, corresponding to a label $y \in \mathcal{Y} = \{1, 2, .., n\}$ from data distribution p_{data}, where $x = x_1, x_2, \cdots, x_L$ is a sentence consisting of L words. Let $f \in \mathcal{F}$ be the classifier that maps any $x \in \mathcal{X}$ to $\mathcal{Y}$. We call $\hat{x} = \hat{x}_1, \hat{x}_2, \cdots, \hat{x}_L$ a candidate adversarial example by perturbing at most $T \leq L$ words x_i in x to any of their synonyms $\hat{x}_i \in S(x_i), i = 1, 2, \cdots, L$. The candidate adversarial example set is denoted by $S(x) = \{\hat{x} : \|\hat{x} - x\|_0 \leq T, \hat{x} \in S(x_i)\}$. Here, 0-norm is the Hamming distance. The goal of an attacker is to find an adversarial example in $S(x)$ that makes the model assign a different label. We often use deep neural networks in text classification. Let $u(x) : \mathcal{X} \rightarrow \mathbb{R}^n$ be a neural network, whose output at an input x is a vector $(u_1(x), \cdots, u_n(x))$. The resulting network is $z(x) = (z_1(x), \cdots, z_n(x))$, which is given by softmax layer $z_c(x) = e^{u_c(x)} / \sum_{c' \in \mathcal{Y}} e^{u_{c'}(x)}$.

2.1 Certified Region

Randomized Smoothing. In this work, we use randomized smoothing technique [4,13]. It is defined as follows:

$$g(x) = \arg\max_{c \in \mathcal{Y}} P_{\hat{x} \sim \Pi(x)}(f(\hat{x}) = c), g(x, c) = P_{\hat{x} \sim \Pi(x)}(f(\hat{x}) = c) \quad (1)$$

where, $\Pi(x)$ is a perturbation probability distribution around x, see Sect. 3.2 for more details. In general, the smoothed classifier g returns the label most likely to be returned by f when its input is sampled from a perturbation distribution.

Certified Region/Certified Robustness. [13] proposed a robustness certification method that seeks to derive a sufficiently tight bound for textual neural networks. We define this lower bound as *Certified Region* and denote it by $CR(g; x, y)$, abbreviated as CR. Note that:

$$R \geq CR > 0 \Rightarrow f(\hat{x}) = y, \forall (x, y) \sim p_{data}, \hat{x} \in S(x) \quad (2)$$

Theorem 1. *Assume the perturbation set $P(x_i)$ is constructed such that $|P(x_i)| = |P(\hat{x}_i)|$ for each word x_i and its synonym $\hat{x}_i \in S(x_i)$. Define $q(x_i) = \min_{\hat{x}_i \in S(x_i)} |P(x_i) \cap P(\hat{x}_i)| / |P(x_i)|$, represents the intersection size between two perturbation set. For a given sentence $x = x_1, x_2, \cdots, , x_L$, we sort the words according to $q(x_i)$, such that $q(x_{i_1}) \leq q(x_{i_2}) \leq \cdots \leq q(x_{i_L})$. Then*

$$\min_{\hat{x} \in S(x)} g(\hat{x}, c) \geq \max(g(x, c) - q(x), 0)$$

$$\max_{\hat{x} \in S(x)} g(\hat{x}, c) \leq \min(g(x, c) + q(x), 1)$$

where, $q(x) = 1 - \prod_{j=1}^{L} q(x_{i_j})$.

$P(x_i)$ is the perturbation set extended from $S(x_i)$, i.e. a larger synonym set. We will introduce its effect in Sect. 2.2.

Proposition 1. *For a sentence x and its label y, we define the runner-up label $y_B = \arg\max_{c \in \mathcal{Y}, c \neq y} g(x, c)$. Under the condition of Theorem 1, we can certify that $g(\hat{x}) = g(x)$ for $\forall \hat{x} \in S(x)$, if*

$$CR(g; x, y) = g(x, y) - g(x, y_B) - 2q(x) > 0$$

As an illustration in Fig. 1(a), for an input sentence x, we sample k samples $\hat{x}$ through base classifier $f(\hat{x})$, where $\hat{x}$ is generated by sampling from $\Pi(x)$(the circle area). If the label y appears most in the predictions of $\hat{x}$, then $g(x)$ returns label y. Next we introduce each component of our algorithm.

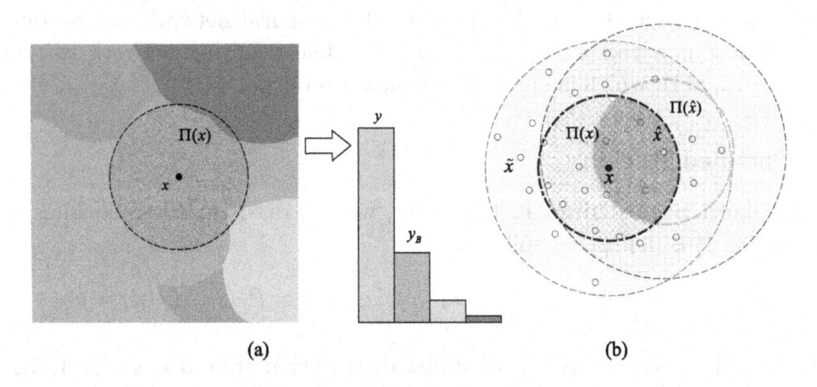

Fig. 1. Consider a sentence x and its perturbation distribution $\Pi(x)$ (projected to 2D for illustration) represented by a circle area. (Color figure online)

2.2 Perturbation Distribution Based on Multi-Hop Neighbors

Multi-Hop Neighbors. For the smoothed classifier g to classify x correctly and robustly, we firstly consider to use the candidate adversarial examples from $S(x)$ to train the smoothed classifier. In Fig. 1(b), we represent a sentence x and its perturbation distribution of $S(x)$ as the blue circle area. Assuming that the sentence x is replaced with $\hat{x}$. The sentence x and its perturbation distribution of $S(\hat{x})$ is represented by yellow circle area. If the intersection between blue and yellow circle areas is small, we cannot expect g will classify $\hat{x}$ robustly. Due to the idea of greedy filling, the difference set between the yellow circle area and the blue circle area is always predicted as a false green label. This worsens the model's robustness as their intersection set is small. To alleviate this problem, we extend the synonym set $S(x_i)$ to a larger perturbation set $P(x_i)$ using Multi-Hop neighbors. We further extend $S(x)$ to a larger candidate adversarial example set $P(x)$ using $P(x_i)$. We respectively use a larger green circle area and a larger red circle area to represent the perturbation distributions of $P(x)$ and $P(\hat{x})$, which form a bigger intersection as $\hat{x}$ is limited in the blue circle area. Specifically, an attacker only use the synonym set $S(x_i)$ for attacking. But the smoothed classifier always uses the larger perturbation set $P(x_i)$ to get the robust outputs. The construction of synonym set $S(x_i)$ and perturbation set $P(x_i)$ is as follows.

Similar to [1], we construct the synonym set $S(x_i)$ to be the set of words with cosine similarity $\geq \epsilon$ in the post-processing GloVe by Counter-fitted method [10] and all-but-the-top method [11]. We extend the synonym set $S(x_i)$ to perturbation set $P(x_i)$.

Perturbation Distribution Π(x)**.** In our work, we use the uniform distribution on a set of random word substitutions:

$$P(\Pi(x) = \tilde{x}) = \prod_{i=1}^{L} \frac{\mathbf{1}[\tilde{x}_i \in P(x_i)]}{|P(x_i)|} \tag{3}$$

where $\tilde{x}$ is the perturbation sample from $P(x)$, $\mathbf{1}[\cdot]$ is the indicator function.

2.3 Robust Training by Maximizing Certified Region

Training the Soft Smoothed Classifier $\tilde{g}$**.** To minimize the classification error of smoothed classifier g is to maximize the sum of log probability:

$$\sum_{(x,y)\sim p_{data}} \log P(g(x) = y) = \sum_{(x,y)\sim p_{data}} \log \mathbb{E}_{\tilde{x}} \mathbf{1}\left[\arg\max_{c\in\mathcal{Y}} u_c(\tilde{x}) = y\right] \tag{4}$$

Theoretically, the expectation is differentiable. However, the expectation needs to be estimated by Monte Carlo sampling $\mathbb{E}_{\tilde{x}} \mathbf{1}\left[\arg\max_{c\in\mathcal{Y}} u_c(\tilde{x}) = y\right] \approx \frac{1}{k}\sum_{j=1}^{k} \mathbf{1}\left[\arg\max_{c\in\mathcal{Y}} u_c(\tilde{x}^{(j)}) = y\right]$, where k is the number of samples. As a sum of indicator functions, this estimation is not differentiable with respect to θ, which cannot calculate gradient during optimization.

$$\tilde{g}(x) = \arg\max_{c\in\mathcal{Y}} P_{\tilde{x}\sim\Pi(x)}(z_c(\tilde{x})), \tilde{g}(x,c) = P_{\tilde{x}\sim\Pi(x)}(z_c(\tilde{x})) \tag{5}$$

The softmax function can be viewed as a continuous, differentiable approximation of argmax: $\mathbf{1}\left[\arg\max_{c\in\mathcal{Y}} u_c(\tilde{x}) = y\right] \approx z_c(\tilde{x})$. Therefore, the objective function is approximately by: $\sum_{(x,y)\sim p_{data}} \log \frac{1}{k}\sum_{j=1}^{k} z_c(\tilde{x}^{(j)})$.

Following Theorem 1, we prove Theorem 2 which shows that the calculation of the certified region for the soft smoothed classifier $\tilde{g}$ is still applicable in Appendix A.

Theorem 2. *Let the ground truth of input x be y, we can certify $\tilde{g}(\hat{x}) = \tilde{g}(x)$ for any $\hat{x} \in S(x)$ if*

$$CR(\tilde{g}; x, y) = \tilde{g}(x,y) - \max_{c\neq y, c\in\mathcal{Y}} \tilde{g}(x,c) - 2q(x) > 0$$

According to the definition of smoothed classifier g and object function, perturbation examples are sampled from a combination consisting of all synonyms. As a result, back-propagation will go through the embeddings of all synonyms, thus allowing to update all embeddings together in a coordinated way.

Maximization of Certified Region. As we can see from Proposition 1, the value of the certified region can be calculated by sampling, more importantly, it can be computed by any deep neural networks, such as BERT. It motivates us to design a training method to maximize the certified region. Therefore, we add robustness error to the classification error to form our final certified robust error:

$$L_{error} = \sum_{(x,y)\sim p_{data}} \underbrace{\mathbf{1}\left[\tilde{g}(x) \neq y\right]}_{\text{classification error}} + \underbrace{\mathbf{1}[\tilde{g}(x) = y, CR(\tilde{g}; x, y) < \gamma]}_{\text{robustness error}} \tag{6}$$

We use the hinge loss as the surrogate loss of robustness error. Our loss is:

$$L(\tilde{g}) = \sum_{(x,y)\sim p_{data}} \log \frac{1}{k} \sum_{j=1}^{k} z_y(\tilde{x}^{(j)}) + \lambda \cdot \max\{\gamma - CR(\tilde{g}; x, y), 0\} \cdot \mathbf{1}[\tilde{g}(x) \neq y] \tag{7}$$

where, γ is the parameter of hinge loss, λ is the regularization.

3 Experiment

3.1 Experimental Data and Baselines

We conduct our experiments on both IMDB [8] for sentiment analysis and SNLI [3] for natural language inference. We firstly use clean accuracy (CLN) as a metric which is the model's accuracy on clean data. Then, we evaluate the model robustness by certified accuracy (CER), which equals the percentage of samples on which $\tilde{g}$ is certified robust, which, for our method, holds when $CR > 0$. For those defense methods in which we cannot directly compute CER, we use the model's accuracy against the GA attack method [1] which can be considered as an upper bound of robust accuracy [6]. We compare our proposed method MAC-ROBERT with three defense methods: Data Augmentation (DA) [2], Interval Bound Propagation (IBP) [6], and SAFER [13].

3.2 Results and Analysis

Table 1. Results on the IMDB and SNLI dataset.

	IMDB						SNLI					
	CNN		LSTM		BERT		BOW		DecomAtt		BERT	
	CLN	CER	CLN	CER	CLN	CER	CLN	CER	CLN	CER	CLN	CER
ORIG	85.3	18.0	84.4	18.0	91.7	78.0	79.4	53.0	82.4	40.0	**90.6**	75.0
DA	85.1	20.0	84.2	25.0	91.5	76.0	79.2	65.0	**82.6**	65.0	**90.6**	75.0
IBP	81.0	65.0	76.8	60.0	–	–	75.0	58.0	77.1	58.0	–	–
SAFER	85.3	83.6	84.4	81.6	91.7	88.0	79.4	75.0	82.4	75.0	**90.6**	84.0
Ours	**85.4**	**86.0**	**85.9**	**84.4**	**92.1**	**90.0**	**79.7**	**77.0**	82.2	**86.0**	**90.6**	**89.0**

Main Experimental Results. As we can see from the experimental results of the IMDB dataset in Table 1, MACROBERT and SAFER both outperform DA and IBP due to the robustness guarantee from randomized smoothing. This fully verifies the effectiveness of randomized smoothing. MACROBERT achieves a better CER score than SAFER as we take the certified region as the optimization object to improve model robustness, which strongly demonstrates the superiority of maximizing certified region. On the CLN metric, MACROBERT consistently outperforms ORIG on all model structures except a very slight decline compared with DecomAtt. This demonstrates that our method doesn't influence the performance of the clean data as it directly combines original samples and perturbation samples to train the model. Similar results also can be found on the SNLI dataset. Based on the comprehensive experimental results of two datasets, MACROBERT has two advantages: First, the performance of MACROBERT is not easily influenced by text length, which makes MACROBERT more applicable. Second, MACROBERT is a scalable method that can be applied to arbitrary models to enhance model robustness.

Ablation Study. The results of the ablation experiment are summarized in Table 2. To reveal the impact of Multi-Hop neighbors, we generate synonyms of a given word x_i from extended $P(x_i)$ at inference time, denoted as "w/o Multi-Hop". To show the effect of the coordinated update, we train a model with a single-point update strategy in which we update the perturbation word x_j by itself during training, denoted as "w/o COORD-UPD". To verify the importance of maximization of the certified region, we train a model without considering robustness error, i.e. $\lambda = 0$, denoted as "w/o MAX-BOUND". Experimental results show MACROBERT performs better with Multi-Hop neighbors, coordinated update strategy, and maximization of the certified region.

Table 2. Clean accuracy and certified accuracy on the IMDB dataset.

Model	CLN	CER	GA
MACROBERT	85.42	86.00	89.00
w/o Multi-Hop	–	–	87.00
w/o COORD-UPD	83.19	80.40	–
w/o MAX-BOUND	85.71	81.60	–

4 Conclusion

To improve the robustness, we propose MACROBERT, a robustness certification method utilizing the maximization of the certified region of a soft smoothed classifier to defense adversarial word substitution attacks in NLP models. MACROBERT provides provably guaranteed robustness of all perturbation examples and can be extended to models with arbitrary structures including large-scale pre-trained models like BERT.

Acknowledgments. This work was supported by National Natural Science Foundation of China (No. 61976207, No. 61906187).

Appendix

A Proof of Theorem 2

Proof. Define $\mathcal{H}_{[0,1]}$ to be the set of all functions mapping from $\mathcal{X}$ to $[0,1]$. For $\forall h \in \mathcal{H}_{[0,1]}$, define $\tilde{h}(x) = \mathbf{E}_{\tilde{x}\sim\Pi(x)}[h(\tilde{x})]$. For $\forall x$ and $c \in \mathcal{Y}$, we have

$$\min_{\hat{x}\in S(x)} \tilde{g}(\hat{x},c) \geq \min_{h\in\mathcal{H}_{[0,1]}} \min_{\hat{x}\in S(x)} \{\tilde{h}(\hat{x}) \quad s.t. \quad \tilde{h}(x) = \tilde{g}(x,c)\} := \tilde{g}_{low}(x,c)$$

$$\max_{\hat{x}\in S(x)} \tilde{g}(\hat{x},c) \leq \max_{h\in\mathcal{H}_{[0,1]}} \max_{\hat{x}\in S(x)} \{\tilde{h}(\hat{x}) \quad s.t. \quad \tilde{h}(x) = \tilde{g}(x,c)\} := \tilde{g}_{up}(x,c)$$

We denote $p = \tilde{g}(x,c)$ and $\Pi_x(z) = P(\Pi(x) = z)$. Applying the Lagrange multiplier to the constraint optimization problem, we have

$$\begin{aligned}
\tilde{g}_{low}(x,c) &= \min_{\hat{x}\in S(x)} \min_{h\in\mathcal{H}_{[0,1]}} \max_{\alpha\in\mathbb{R}} \tilde{h}(\hat{x}) - \alpha\tilde{h}(x) + \alpha\tilde{g}(x,c) \\
&\geq \max_{\alpha\in\mathbb{R}} \min_{\hat{x}\in S(x)} \min_{h\in\mathcal{H}_{[0,1]}} \tilde{h}(\hat{x}) - \alpha\tilde{h}(x) + \alpha p \\
&= \max_{\alpha\in\mathbb{R}} \min_{\hat{x}\in S(x)} \min_{h\in\mathcal{H}_{[0,1]}} \int_z h(z)(d\Pi_{\hat{x}}(z) - \alpha d\Pi_x(z)) + \alpha p \\
&= -\max_{\alpha\geq 0} \max_{\hat{x}\in S(x)} \int_z (\alpha d\Pi_x(z) - d\Pi_{\hat{x}}(z))_+ + \alpha p
\end{aligned}$$

According to the lemma 2 of [13], we have

$$\begin{aligned}
&\int_z (\alpha d\Pi_x(z) - d\Pi_{\hat{x}}(z))_+ \\
&= \alpha\left[1 - \prod_{j\in[L],x_j\neq x'_j} \frac{n_{x_j,x'_j}}{n_{x_j}}\right] + \left[\prod_{j\in[L],x_j\neq x'_j} \frac{n_{x_j,x'_j}}{n_{x_j}}\right]\left[\alpha - \prod_{j\in[L],x_j\neq x'_j} \frac{n_{x'_j}}{n_{x_j}}\right]_+
\end{aligned}$$

where, $n_{x_j} = |P(x_j)|, n_{x_j,x'_j} = |P(x_j) \cap P(x'_j)|$.

According to the Lemma 3 of [13], we have x^* is the optimal solution of $\max_{\hat{x}\in S(x)} \int_z (\alpha d\Pi_{\hat{x}}(z) - d\Pi_x(z))_+$ when the adversary attack T words at most, where $\tilde{x}_i^* = \arg\min_{\tilde{x}_i\in P(x_i)} n_{x_i,\tilde{x}_i}/n_{x_i}$. Now, the lower bound becomes

$$\begin{aligned}
\tilde{g}_{low}(x,c) &= -\max_{\alpha\geq 0} \int_z (\alpha d\Pi_x(z) - d\Pi_{x^*}(z))_+ + \alpha p \\
&= \max_{\alpha\geq 0} (p - q(x))\alpha - (1 - q(x))(\alpha - 1)_+ \\
&= \max(p - q(x), 0)
\end{aligned}$$

where $q(x)$ is consistent with the definition in Theorem 1. According to [13] Proposition 1, we can easily deduce that $CR(\tilde{g}; x, y) > 0$ can reduce to be robust at x. □

References

1. Alzantot, M., Sharma, Y., Elgohary, A., Ho, B.J., Srivastava, M., Chang, K.W.: Generating natural language adversarial examples. In: Proceedings of the Conference on Empirical Methods in Natural Language Processing (2018)
2. Belinkov, Y., Bisk, Y.: Synthetic and natural noise both break neural machine translation. In: International Conference on Learning Representations (2018)
3. Bowman, S.R., Angeli, G., Potts, C., Manning, C.D.: A large annotated corpus for learning natural language inference. In: Proceedings of the 2015 Conference on Empirical Methods in Natural Language Processing, pp. 632–642, September 2015
4. Cohen, J., Rosenfeld, E., Kolter, Z.: Certified adversarial robustness via randomized smoothing. In: PMLR, vol. 97, pp. 1310–1320. PMLR (2019)
5. Goodfellow, I., Shlens, J., Szegedy, C.: Explaining and harnessing adversarial examples. In: International Conference on Learning Representations (ICLR) (2015)
6. Jia, R., Raghunathan, A., Göksel, K., Liang, P.: Certified robustness to adversarial word substitutions. In: EMNLP (2019)
7. Li, J., Ji, S., Du, T., Li, B., Wang, T.: TextBugger: generating adversarial text against real-world applications. In: NDSS (2019)
8. Maas, A.L., Daly, R.E., Pham, P.T., Huang, D., Ng, A.Y., Potts, C.: Learning word vectors for sentiment analysis. In: Proceedings of the Annual Meeting of the Association for Computational Linguistics (2011)
9. Miyato, T., Dai, A.M., Goodfellow, I.: Adversarial training methods for semi-supervised text classification. In: ICLR (2016)
10. Mrkšić, N., et al.: Counter-fitting word vectors to linguistic constraints. In: NAACL (2016)
11. Mu, J., Viswanath, P.: All-but-the-top: simple and effective postprocessing for word representations. In: ICLR (2018)
12. Raghunathan, A., Steinhardt, J., Liang, P.: Certified defenses against adversarial examples. In: International Conference on Learning Representations (2018)
13. Ye, M., Gong, C., Liu, Q.: SAFER: a structure-free approach for certified robustness to adversarial word substitutions. In: ACL (2020)
14. Zhai, R., et al.: MACER: attack-free and scalable robust training via maximizing certified radius. In: International Conference on Learning Representations (2020)
15. Zhou, Y., Zheng, X., Hsieh, C.J., Chang, K.W., Huang, X.: Defense against adversarial attacks in NLP via Dirichlet neighborhood ensemble. arXiv (2020)

A Diversity-Enhanced and Constraints-Relaxed Augmentation for Low-Resource Classification

Guang Liu, Hailong Huang, Yuzhao Mao, Weiguo Gao, Xuan Li, and Jianping Shen(✉)

Ping An Life Insurance of China, Shenzhen, China
{liuguang230,huanghailong590,maoyuzhao258,gaoweiguo801,lixuan208, shenjianping324}@pingan.com.cn

Abstract. Previous studies on Data Augmentation (DA) mostly use a fine-tuned Language Model (LM) to strengthen the constraints but ignore the fact that the potential of diversity could improve the effectiveness of generated data. To address this dilemma, we propose a **D**iversity-**E**nhanced and **C**onstraints-**R**elaxed **A**ugmentation (**DECRA**) that has two essential components on top of a transformer-based backbone model, including a **k-**β ***augmentation*** and a masked language model loss. Extensive experiments demonstrate that our DECRA outperforms state-of-the-art approaches by 3.8% in the overall score.

Keywords: Data augmentation · Regularization · Low-resource

1 Introduction

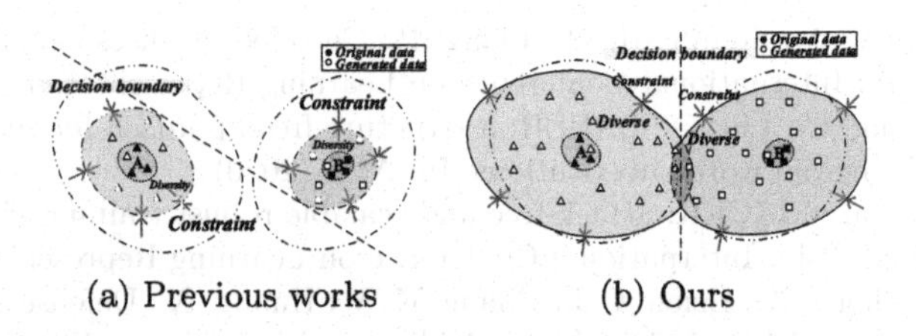

Fig. 1. The demonstration of how augmentation works in LRC.

Data Augmentation (DA) approaches [2,15] are often used to alleviate the thirst for labeled data in Low-Resource Classification (LRC). DA in text data aims to generate constrained and diversified data to improve classifier performance [5, 7,18]. Ideally, we assume data generated by DA is able to present the data distribution in every category. Thus, the generated data is supposed to extend the range of labeled data which would help the classifier make better decisions [16].

C. S. Jensen et al. (Eds.): DASFAA 2021, LNCS 12682, pp. 262–270, 2021.
https://doi.org/10.1007/978-3-030-73197-7_17

As shown in Fig. 1(a), constraints and diversity are two main concepts in DA. The constraints mainly pull the generated data towards the original one. The diversity in generating comes from partially changed labeled data. It pushes the generated data away from the original one.

However, previous studies often suffer from poor generalization ability in low-resource conditions due to strong constraints but weak diversity in augmentation. As the Language Model (LM) tends to overfit on limited data in low-resource conditions, strong constraints are formed after fine-tuning [7,17]. As a result, the generated data is pulled towards the original data. At the same time, the weakness of diversity in augmentation is often ignored. Current DA approaches mostly use the method that is identical to the masked Language Model learning in BERT [3]. In this method, diversity is influenced by the changing scope and degree of complexity in the generated data. The changing scope is proportional to the times of the DA applied. In each time of augmentation, one set of maskers is generated [7,9]. And the masked positions are the ones to be augmented. This process results in a fixed and narrow changing scope. On the other hand, the degree of complexity is related to the amount of information used in the augmenting data in masked positions. For each masked position, routinely, one sampled tokens are applied [9,17]. Therefore, it results in the low complexity of the generated data. Consequently, strong constraints but weak diversity causes the poor generalization ability in LRC.

In this paper, we propose a **D**iversity-**E**nhanced and **C**onstraints-**R**elaxed **A**ugmentation (**DECRA**). DECRA allows the generated data to be more scattered within the extended boundary. Our DECRA is based on the modified LDMAW [7], which is the state-of-the-art model in LRC. The backbone model consists of a transformer-based encoder (TBE), a language model layer (LML) and a classification layer (CL). DECRA has two essential components based on the backbone model. 1) k-β augmentation, an essential component in DECRA, will enhance the diversity in generating. It expands the changing scope by applying augmentation β times and enhances the degree of complexity by using top-k tokens to augment the masked position. 2) The regularization, masked LM loss on original data, generates more relaxed constraints compared to fine-tuning. DECRA will be trained by the combination of masked LM loss and the classification loss. Our model can learn the constraints dynamically and progressively during the training process. It will process more scattered generated data, which will reach or approach the boundary of categories, to achieve better generalization ability.

2 Model Description

Figure 2 shows the structure of **DECRA**. It has two essential components based on a backbone model. The k-β augmentation is applied to the original data to generate diversity-enhanced data. The masked Language Model (LM) loss is introduced as the regularization, which is the relaxed-constraint in generating.

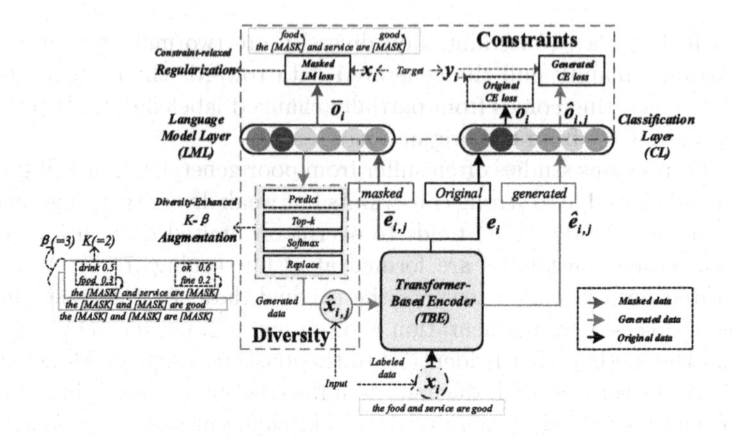

Fig. 2. The structure of diversity-enhanced and constraints-relaxed augmentation (DECRA).

2.1 Transformer-Based Encoder

Transformer-Based Encoder (TBE) stacks multiple layers of transformers [14] to encode the text data into embeddings. It is initialized by a pre-trained Language Model (LM) which is trained on large-scale multi-domain datasets. The original data $\mathbf{x}_i$ is masked into $\overline{\mathbf{x}}_i$. The original data $\mathbf{x}_i$ is encoded as follows,

$$\mathbf{e}_i = Tansformer_{\theta_t}(\mathbf{x}_i)\,. \tag{1}$$

Here, $\mathbf{e}_i \in \mathbb{R}^{T\times H}$ is the embeddings for classification, $Transformer_{\theta_t}$ represents the processing of transformers, T is the length of original data, H is the hidden size of embeddings, θ_t is the parameters of TBE. Similarly, we can get embeddings $\overline{\mathbf{e}}_i$ for the masked data $\overline{\mathbf{x}}_i$.

2.2 Language Model Layer

Language Model Layer (LML) is composed of a fully-connected layer. The fully-connected layer predicts the masked position based on its contextual embedding [3] that is fundamental for k-β augmentation. It also essential to calculate the masked Language Model loss [3] on original data as a regularization. The $\overline{\mathbf{e}}_i$ is embeddings of masked data. The prediction is calculated as,

$$\overline{p}_i = g_{\theta_a}(\overline{\mathbf{e}}_i)\,. \tag{2}$$

Here, $\overline{p}_i \in \mathbb{R}^{T\times V}$ represents the probabilities of tokens in masked positions, $g_{\theta_a}(\cdot)$ maps the embedding size vector to vocabulary size.

2.3 Classification Layer

Classification Layer (CL) takes the first position of embeddings encoded by the TBE as input, and outputs the class categories. For labeled data, we calculate the predictions as follow,

$$o_i = f_{\theta_c}(\mathbf{e}_i), \tag{3}$$

where $f_{\theta_c}(\cdot)$ represents the function of CL, θ_c is the parameters of CL, $o_i \in \mathbb{R}^C$ represents the predictions.

2.4 *K*-*β* Augmentation

k-β algorithm is designed to enhance the diversity in generating. It aims to augment the original data $\mathbf{x}_i$ β times to get the generated data $\hat{\mathbf{x}}_{i,j}, j \in [1, \beta]$.

$$\hat{\mathbf{x}}_{i,j} = \phi(\mathbf{x}_i; k, \beta, \theta_t, \theta_a), j \in [1, \beta]. \tag{4}$$

Here, ϕ is the k-β augmentation, β is the set of masks as well as the times of augmentation applied, k is the number of tokens used for replacing the masked position, θ_t and θ_a are the parameters in TBE and LML respectively. By this, the changing scope of generated data is expanded. Also, the degree of complexity of generated data is enhanced.

For each time of augmentation, the original data $\mathbf{x}_i$ is randomly masked $\overline{\mathbf{x}}_i$ for augmentation. Data augmentation consists of four steps: predict, top-k, softmax and replace.

Predict. The embedding of the randomly masked position is feed into LML to get the predictions $\overline{p}_{i,j}^t \in \mathbb{R}^V$. The predictions represent the probabilities of tokens to fit the t-th position.

Top-k. The top-k sampling, which is often used to improve the diversity in data augmentation [4], is used. The top-k probabilities tokens in $\overline{p}_{i,j}^t$ are selected as $p_{i,j}^t \in \mathbb{R}^k$.

Softmax. The top-k probabilities are feed into a softmax function to normalize the probabilities.

$$\hat{p}_{i,j}^t = softmax(p_{i,j}^t) \tag{5}$$

Here, $\hat{p}_{i,j}^t \in \mathbb{R}^k$ is the normalized top-k probabilities.

Replace. For the convenient of replacement [7], we fill the value of $\hat{p}_{i,j}^t$ into a zero vector to get $p_{i,j}^t \in \mathbb{R}^V$. Instead of only one sampled token, we use k tokens to replace the masked token $\overline{\mathbf{x}}_{i,j}^t$, and get the generated data $\hat{\mathbf{x}}_{i,j}$. Note that the number of masked tokens is not fixed. The progress is repeated β times to get $\hat{\mathbf{x}}_{i,j}, j \in [1, \beta]$. The labels $\hat{\mathbf{x}}_{i,j}$ all set to y_i as the setting in [17]. The generated data are encoded for classification $\hat{\mathbf{e}}_{i,j}, j \in [1, \beta]$ as in Eq. 1. Then, as in Eq. 3, we can get the prediction of generated data after k-β augmentation $\hat{o}_{i,j}, j \in [1, \beta]$.

2.5 Regularization

Masked Language Model (LM) loss [3] generates relaxed contextual constraints compared to fine-tuning. The labeled data is corrupted by randomly replacing some positions into maskers. Then, the model learns to predict the original token with the contextual embedding in the masked position. It takes the embeddings of the masked position $\overline{\mathbf{e}}_i$ as inputs, takes the original tokens $\mathbf{x}_i$ as labels, and calculates the loss as follow,

$$\mathcal{L}_{LM} = \frac{1}{M}\sum_{t=1}^{T} m_t \mathbf{x}_i^t \log(f_{\theta_a}(\overline{\mathbf{e}}_i^t)), \tag{6}$$

where $\mathcal{L}_{LM}$ represents the masked LM loss, $m_t = 1$ indicates the token on t position is masked, $\mathbf{x}_i^t \in \mathbb{R}^V$ is the t-th token, θ_a is the parameters of LML, M is the number of masked positions.

2.6 Training Process

The cross-entropy between the predictions and y_i is calculated as

$$\mathcal{L}_{CE} = -\frac{1}{N}\sum_{i=1}^{N} y_i \log(o_i), \tag{7}$$

where N is the total number of original data.

Similarly, we can get cross-entropy loss $\hat{\mathcal{L}}_{CE}$ for the data generated by k-β augmentation,

$$\hat{\mathcal{L}}_{CE} = -\frac{1}{N}\frac{1}{\beta}\sum_{i=1}^{N}\sum_{j=1}^{\beta} y_i log(\hat{o}_{i,j}). \tag{8}$$

Here, we average the loss calculated on β generated data which can get a more stable improvement [1].

The final loss is weighted average as follow,

$$\mathcal{L}_{final} = \mathcal{L}_{CE} + \lambda_a \hat{\mathcal{L}}_{CE} + \lambda_{lm}\mathcal{L}_{LM}\,. \tag{9}$$

Here, The λ_a and λ_{lm} are the weights for each loss term.

3 Experiments

3.1 Experimental Settings

Dataset and Baselines. To evaluate the text augmentation in low-resource classification, we use the same settings in [7]. We evaluate the UABC model based on three benchmark classification datasets, including TREC [10], SST-5 [13], and IMDB [11]. For each dataset, we randomly sample 15 small datasets. Each contains 40 samples per class for training and 5 (except SST-5 is 2) samples per class for validation. The models are evaluated on the validation set at the end of each epoch. We compare our model with six methods, including **EDA** [16], **BT**[1] [12], **Mixup** [8], **CBERT** [17], **LDMAW** [7].

[1] We implement the back translation based on MarianMT in Transformers, https://huggingface.co/transformers.

Table 1. DA extrinsic evaluation in low-resource settings. Results are reported as Mean (STD) accuracy on full test set. Experiments are repeated 15 times.† refers to the results reported in [7]

Methods	Datasets			AVG
	SST5(200)	IMDB(80)	TREC(240)	
Baseline† [3]	33.3 ± 6.2	63.6 ± 4.4	88.3 ± 2.9	61.7
EDA [16]	36.8 ± 6.1	62.8 ± 6.0	86.6 ± 4.1	62.1
BT [12]	35.8 ± 4.3	66.4 ± 4.2	86.6 ± 4.3	62.8
Mixup [8]	36.0 ± 4.0	67.3 ± 5.1	88.3 ± 3.2	63.9
CBERT† [17]	34.8 ± 6.9	63.7 ± 4.8	88.3 ± 1.1	62.3
LDMAW† [7]	37.0 ± 3.0	65.6 ± 3.7	89.2 ± 2.1	63.9
DECRA (our work)	$\mathbf{40.3 \pm 3.4}$	$\mathbf{69.0 \pm 4.0}$	$\mathbf{89.5 \pm 1.6}$	**66.3**

Table 2. The ablation results (%) of DECRA model. Results are reported as Mean (STD) accuracy on full test set. Experiments are repeated 15 times.

ID	k-β	Reg.	Dataset			AVG
			SST5(200)	IMDB(80)	TREC(240)	
1	×	×	33.3 ± 6.2	63.6 ± 4.4	88.3 ± 2.9	61.7
2	×	✓	33.8 ± 2.9	64.6 ± 4.4	86.5 ± 3.4	61.6
3	✓	×	36.5 ± 3.2	65.6 ± 5.0	89.0 ± 1.8	63.7
4	×	△	38.2 ± 3.3	68.8 ± 4.1	88.4 ± 3.0	65.1
5	✓	△	39.0 ± 5.1	68.7 ± 5.4	88.7 ± 1.9	65.5
6	✓	✓	40.3 ± 3.4	69.0 ± 4.0	89.5 ± 1.6	66.3

× indicates the component is removed form DECRA, ✓ indicates the component is added in DECRA. k-β is the k-β augmentation and Reg. is the masked LM loss. △ indicates the DECRA is pre-trained with masked LM loss and then finetuned as [6].

3.2 Main Results

Table 1 exhibits the results of all models on three datasets. Our DECRA outperforms all baselines on all three datasets. Firstly, our DECRA can improve the classification performance in LRC from 63.9% to 66.3%. When compared with LDMAW and Mixup, our model achieves a higher overall score. That benefits from the effects of our k-β augmentation which effectively enhances the diversity of generated data. Secondly, our DECRA achieves the highest mean accuracy score on every dataset. The stable improvement may benefit from the expanded changing scope in k-β augmentation. Thirdly, our DECRA has a smaller parameters-scale than LM-based approaches. When compared with CBERT and LDMAW, our model unifies the augmenter and classifier by reducing nearly half of the parameters. Noticeable that the LDMAW uses reinforcement learning to tune the augmenter(BERT) for the classifier(BERT). Our DECRA improves the overall score by a significant margin. It's 3.8% improvements against the LDMAW and 6.3% improvements against CBERT. The improvement benefits from the improvement of generalization ability.

3.3 Ablation Study

To better understand the working mechanism of the DECRA, we conduct ablation studies on all three datasets, as listed in Table 2. 1) Without augmentation, the $ID2$, which has relaxed constraints compared to $ID4$, results in lower classification accuracy. The results indicate that strong constraints are more effective in LRC without augmentation. 2) With augmentation, the $ID6$, which has relaxed-constraints compared to $ID5$, promotes the overall score from 65.5 to 66.3. The improvement of the overall score in LRC with augmentation mainly from the relaxed-constraints. 3) Besides, the $ID3$ outperforms $ID1$ due to the diversity-enhanced k-β augmentation. Also, the $ID5$ has a higher overall score (65.5) than $ID4$ (65.1). The results show the effects of k-β augmentation in LRC, which enhance the diversity of generated data.

3.4 Importance of Diversity and Constraints

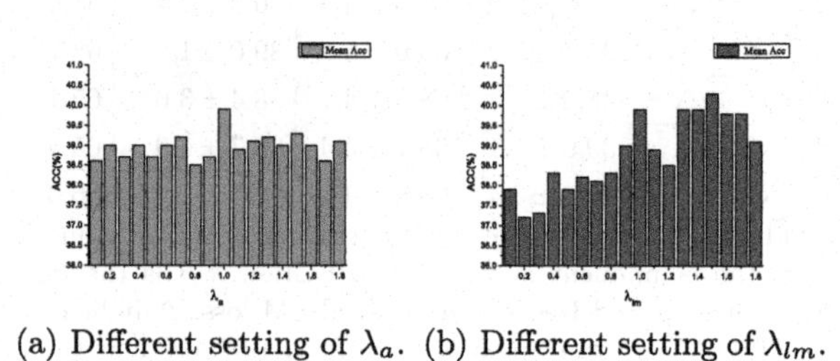

(a) Different setting of λ_a. (b) Different setting of λ_{lm}.

Fig. 3. Different loss weights on SST5.

To analyze the importance of diversity and constraints ($\tilde{\mathcal{L}}_{CE}$ and $\mathcal{L}_{LM}$), we grid search the optimal weights (λ_a and λ_{lm}) on SST5. Experiments are repeated 15 times. Firstly, the λ_{mlm} is set to 1.0 in the searching of the λ_a. Then, the λ_a is set to the optimal (1.0) in the search of λ_{lm}. Figure 3(a) describes the effects of weight λ_a for k-β augmentation $\hat{\mathcal{L}}_{CE}$. The average accuracy achieves the peak when the λ_a is 1.0. The generated data has equal importance to the original data. This setting is identical to [7]. Figure 3(b) shows the effects of weight λ_a for masked LM loss $\mathcal{L}_{LM}$. The model reaches the optimal classification performance when the λ_{lm} is 1.5. The $\mathcal{L}_{LM}$ has larger weights than $\mathcal{L}_{CE}$. It shows the importance of contextual constraints in LRC.

4 Conclusion

In this paper, we propose a Diversity-Enhanced and Constraints-Relaxed Augmentation (DECRA) that has two essential components on top of a transformer-based backbone model. We propose a k-β augmentation to enhance the diversity of generated data by expanding the changing scope and enhancing the degree of complexity in generated data. We introduce the masked Language Model loss instead of staged fine-tuning to generate relaxed-constraints. The improved diversity and relaxed constraints help to generate data scattered near or approach the category boundaries. Experimental results demonstrate that our DECRA significantly outperforms state-of-the-art in low-resource classification.

References

1. Berthelot, D., et al.: ReMixMatch: semi-supervised learning with distribution matching and augmentation anchoring. In: International Conference on Learning Representations (2019)
2. Cubuk, E.D., Zoph, B., Mane, D., Vasudevan, V., Le, Q.V.: AutoAugment: learning augmentation strategies from data. In: Proceedings of the IEEE Conference on Computer Vision and Pattern Recognition, pp. 113–123 (2019)
3. Devlin, J., Chang, M.W., Lee, K., Toutanova, K.: BERT: pre-training of deep bidirectional transformers for language understanding. In: NAACL-HLT, Volume 1 (Long and Short Papers) (2019)
4. Fadaee, M., Bisazza, A., Monz, C.: Data augmentation for low-resource neural machine translation. arXiv preprint arXiv:1705.00440 (2017)
5. Gupta, R.: Data augmentation for low resource sentiment analysis using generative adversarial networks. In: 2019 IEEE International Conference on Acoustics, Speech and Signal Processing (ICASSP), ICASSP 2019, pp. 7380–7384. IEEE (2019)
6. Gururangan, S., et al.: Don't stop pretraining: adapt language models to domains and tasks. arXiv preprint arXiv:2004.10964 (2020)
7. Hu, Z., Tan, B., Salakhutdinov, R.R., Mitchell, T.M., Xing, E.P.: Learning data manipulation for augmentation and weighting. In: Advances in Neural Information Processing Systems, pp. 15764–15775 (2019)
8. Jindal, A., Gnaneshwar, D., Sawhney, R., Shah, R.R.: Leveraging BERT with mixup for sentence classification (student abstract). In: AAAI, pp. 13829–13830 (2020)
9. Kobayashi, S.: Contextual augmentation: data augmentation by words with paradigmatic relations. arXiv preprint arXiv:1805.06201 (2018)
10. Li, X., Roth, D.: Learning question classifiers. In: The 19th International Conference on Computational Linguistics, COLING 2002 (2002)
11. Maas, A., Daly, R.E., Pham, P.T., Huang, D., Ng, A.Y., Potts, C.: Learning word vectors for sentiment analysis. In: Proceedings of the 49th Annual Meeting of the Association for Computational Linguistics: Human Language Technologies, pp. 142–150 (2011)
12. Shleifer, S.: Low resource text classification with ULMFit and backtranslation. arXiv preprint arXiv:1903.09244 (2019)
13. Socher, R., et al.: Recursive deep models for semantic compositionality over a sentiment treebank. In: Proceedings of the 2013 Conference on Empirical Methods in Natural Language Processing, pp. 1631–1642 (2013)

14. Vaswani, A., et al.: Attention is all you need. In: Advances in Neural Information Processing Systems, pp. 5998–6008 (2017)
15. Wang, Y., Yao, Q., Kwok, J.T., Ni, L.M.: Generalizing from a few examples: a survey on few-shot learning. ACM Comput. Surv. (CSUR) **53**(3), 1–34 (2020)
16. Wei, J., Zou, K.: EDA: easy data augmentation techniques for boosting performance on text classification tasks. arXiv preprint arXiv:1901.11196 (2019)
17. Wu, X., Lv, S., Zang, L., Han, J., Hu, S.: Conditional BERT contextual augmentation. In: Rodrigues, J.M.F., et al. (eds.) ICCS 2019. LNCS, vol. 11539, pp. 84–95. Springer, Cham (2019). https://doi.org/10.1007/978-3-030-22747-0_7
18. Xia, M., Kong, X., Anastasopoulos, A., Neubig, G.: Generalized data augmentation for low-resource translation. arXiv preprint arXiv:1906.03785 (2019)

Neural Demographic Prediction in Social Media with Deep Multi-view Multi-task Learning

Yantong Lai[1,2], Yijun Su[3], Cong Xue[2(✉)], and Daren Zha[2]

[1] School of Cyber Security, University of Chinese Academy of Sciences, Beijing, China
[2] Institute of Information Engineering, Chinese Academy of Sciences, Beijing, China
{laiyantong,xuecong,zhadaren}@iie.ac.cn
[3] JD.com, Beijing, China

Abstract. Utilizing the demographic information of social media users is very essential for personalized online services. However, it is difficult to collect such information in most realistic scenarios. Luckily, the reviews posted by users can provide rich clues for inferring their demographics, since users with different demographics such as gender and age usually have differences in their contents and expressing styles. In this paper, we propose a neural approach for demographic prediction based on user reviews. The core of our approach is a deep multi-view multi-task learning model. Our model first learns context representations from reviews using a context encoder, which takes semantics and syntactics into consideration. Meanwhile, we learn sentiment and topic representations from selected sentiment and topic words using a word encoder separately, which consists of a convolutional neural network to capture the local contexts of reviews in word-level. Then, we learn a unified user representation from context, sentiment and topic representations and apply multi-task learning for inferring user's gender and age simultaneously. Experimental results on three real-world datasets validate the effectiveness of our approach. To facilitate future research, we release the codes and datasets at https://github.com/icmpnorequest/DASFAA2021_DMVMT.

Keywords: Demographic prediction · Context · Sentiment and topic views · Multi-task learning

1 Introduction

User demographics have been useful for personalization and recommendation. However, collecting such information in most realistic scenarios is difficult and the collected data might not be real. Thus, how to infer effective user demographics from public available data has attracted both academia and industry.

Luckily, many researchers have studied ways to infer user demographics from social media texts. A common approach relies on lexical features [1,5,15].

C. S. Jensen et al. (Eds.): DASFAA 2021, LNCS 12682, pp. 271–279, 2021.
https://doi.org/10.1007/978-3-030-73197-7_18

For instance, Sap et al. [15] derive gender and age predictive lexica over social media for prediction. Baslie et al. [1] leverage unigrams and characters to identify author's gender and language variety. Gjurković and Šnajder [5] use Linuistic Inquiry and Word Count (LIWC) [14] (a psychological dictionary) to detect user's personality. These hand-crafted features provide good explainability and are of high quality, but they require much manual labor and may ignore rich semantics in text. Recently, deep learning and pre-trained word embeddings have been widely used for demographic prediction. For example, Bayot et al. [2] apply word2vec [12] and convolutional neural network (CNN) [9] to infer user's gender and age. More recently, Wu et al. [19] leverage hierarchical attention mechanism and Tigunova et al. [16] combine attention mechanism with CNN for demographic prediction. Despite these methods perform well, they do not consider the useful sentiment and topic information in text, which have been proved important in demographic prediction [20].

In this paper, we propose a neural approach for demographic prediction based on user reviews. Instead of merging all reviews from the same user into a long text, our approach learns user representations using a deep multi-view multi-task learning model. Our model first learns context representations with a context encoder, to represent rich semantics and syntactics in reviews. Meanwhile, we learn sentiment and topic representations from selected sentiment and topic words with a word encoder respectively. Each word encoder contains a CNN to capture the local contexts in word-level. Then, we obtain a unified user representation integrating from context, sentiment and topic representations. Since gender and age are correlated, we apply multi-task learning for capturing latent influence between them. In the end, we perform experiments on three real-world datasets and the results validate the effectiveness of our model on demographic prediction.

2 Related Work

User demographic prediction with social media text is often regarded as a classification task in natural language processing (NLP) field. Traditional demographic prediction methods mostly rely on hand-crafted linguistic features, such as lexicons [15], unigrams [1] and LIWC [5]. Despite these hand-crafted features based methods preform well, they generally require much manual labor to collect and may ignore rich semantics in text. With the development of word embeddings and deep learning, researchers [2,16] begin to infer demographics using implicit context representations. For instance, Bayot et al. [2] leverage word2vec [12] and CNN [9] to infer gender and age. Further, attention mechanism has been proposed to capture informative contents in text [16,19].

Recently, multi-task learning has been utilized in demographic prediction [17,18]. For example, [17] leverages multi-modal data from Twitter, i.e., users' profiles, following network and tweets, to infer user demographics and location. Additionally, Wang et al. [18] make use of images and user profiles for demographic prediction.

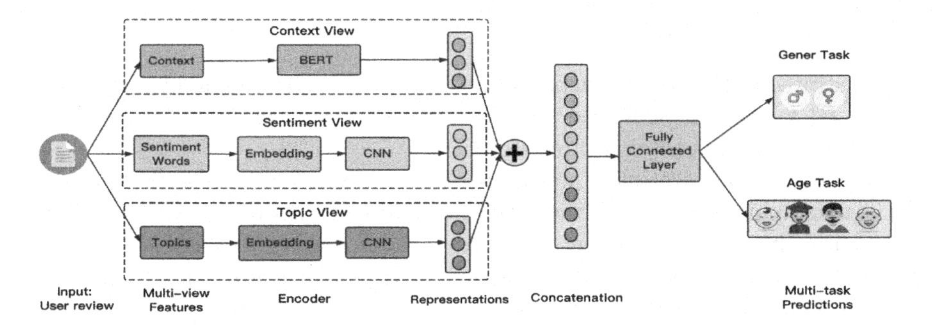

Fig. 1. Our framework for demographic prediction

In this paper, the approach we propose is different from existing methods. First, our model makes better use of context by considering both semantics and syntactics in text. Second, we capture high-order interactions from context, sentiment and topic views. Third, our approach utilizes the correlation between demographics and applies multi-task learning for better performance.

3 Methodology

In this section, we will introduce our approach in details (Fig. 1). The input of our model is a user review $s = \{w_1, w_2, ..., w_n\}$. We define a task set $U = \{u_1, u_2, ..., u_m\}$ and use ϕ_{u_i} to denote the parameters for the i-th task. The outputs of our model are probability distributions for all tasks, i.e. for task u_i is $\Pr(y_{u_i}|s, \phi_{u_i})$, where y_{u_i} represents the class in it.

3.1 Context View

As shown in Fig. 1, context view aims to capture semantics and syntactics information from a user review s, and produces a contextualized latent representation H_c. Thus, the learning process of H_c could be regarded as review-level embedding and we apply a pre-trained language model BERT [4] for encoding. The encoding procedure of BERT contains token embeddings, segment embeddings and position embeddings. Unlike static pre-trained word embeddings (e.g., word2vec [12]), token embeddings could solve polysemy and vary according to their context. Segment embeddings aim to capture inter-sentence syntactics, while position embeddings indicate the position of each words in the review s. The final context embeddings v^c of the review s are the sum of token embeddings, segment embeddings and position embeddings. In addition, the special tokens [CLS] and [SEP] are used for labeling classification tasks and separating segments respectively. Finally, we output the final hidden state of the first special token [CLS] as the context representation H_c in context view.

3.2 Sentiment View

As illustrated in Fig. 1, sentiment view is responsible to learn a hidden representation H_s for sentiment words, which are automatically extracted from a user review s. It mainly contains three steps.

The first step is sentiment words extraction. Sentiment words are indicative for demographic prediction, so we utilize a sentiment dictionary AFINN [13] to automatically extract sentiment words from the review s. If no words appear in the sentiment dictionary, we set "NA" in the sentiment words set S.

The second step is word embedding. Through this step, each sentiment word is mapped to a d-dimensional dense vector $v^s \in \mathbb{R}^{h \times d}$ using a word embedding lookup table $\mathbf{E} \in \mathbb{R}^{V \times d}$, where V is vocabulary size.

The third step is word encoding. We use a CNN as word encoder, to capture semantics from the sentiment words embeddings v^s. It learns the contextual representation through convolutional filters which slide ω-grams per step. We apply N filters to learn semantics from sentiment words embeddings v^s and obtain a feature map $\mathbf{c^s} = [\mathbf{c^s}_1, \mathbf{c^s}_2, ..., \mathbf{c^s}_N]$. The i-th feature map is $\mathbf{c^s}_i = [\mathbf{c^s}_{i_1}, \mathbf{c^s}_{i_2}, ..., \mathbf{c^s}_{i_{h-\omega+1}}]$. After generating feature map $\mathbf{c^s}$, max-over-time pooling operation is performed to capture the most important feature on each dimension of vector by taking the maximum value. For the i-th feature in $\mathbf{c^s}$, the feature after max-over-time pooling is $\hat{c_i^s} = \max\{\mathbf{c^s}_i\}$.

Finally, we concatenate all the features after max-over-time pooling operations as $H_s = [\hat{c_1^s}, \hat{c_2^s}, ..., \hat{c_N^s}]$. Additionally, H_s is the output for sentiment view.

3.3 Topic View

Topic view learns a latent representation H_t for topics (Fig. 1). The learning procedure mainly consists of three steps.

The first step is topics extraction. We make use of the traditional topic modelling method, latent Dirichlet allocation (LDA) [3], to extract topics automatically. Generally, a user review contains several topics and each topic would be formed by a set of words. Thus, a user review s could be seen as a document consisting of k topics. For each topic z in the document, a distribution φ_z on V_t is sampled from a Dirichlet function, where V_t represents a vocabulary consisting of a set of topic words. Then, LDA estimates the distribution $p(z|w)$ for $z \in s^P, w \in V_t^P$, where P denotes the set of word positions in the user review s. Finally, we get the topics set $T = \{w_1^t, w_2^t, ..., w_r^t\}$ according to the distribution $p(z|w)$, where r is the length of the sequence. Topics embedding and encoding is similar to that in sentiment view, which has been detailed described in Sect. 3.2.

Finally, we output topics latent representation $H_t = [\hat{c_1^t}, \hat{c_2^t}, ..., \hat{c_N^t}]$ for topic view, where $\hat{c_i^t}$ represents the i-th feature after max-over-time pooling.

3.4 Training Procedure

The training procedure of our framework consists of two stages: obtaining a unified user representation and multi-task learning.

Table 1. Statistics of three datasets, including number of total reviews (#Total), number of English reviews with gender and age information (#Extracted), female, male and various age categories distributions of extracted reviews.

Datasets	#Total	#Extracted	Female%	Male%	(0, 18]%	(18, 30)%	[30, 40)%	[40, 99)%
Denmark	646084	826	25.67	74.33	0.3632	8.11	31.60	59.93
US	648784	37060	39.82	60.18	0.13	11.39	29.91	58.58
UK	1424395	129175	40.38	59.62	0.18	8.06	18.59	73.17

The first stage is to obtain a unified user representation for context representation H_c, sentiment representation H_s and topic representation H_t. We apply averaging operation on context representation H_c, to get a good summary of the semantics. Then, we concatenate representations from context, sentiment and topic views as $\mathbf{H} = H_c \oplus H_s \oplus H_t$, where $\oplus$ denotes the concatenation operation.

The second stage is to predict demographics simultaneously using multi-task learning. We pass the concatenated feature $\mathbf{H}$ through a fully-connected layer and output a compressed latent feature vector $\mathbf{X}$. Then, the shared latent vector $\mathbf{X}$ is given to a task-specific layer to calculate the probability distribution $\tilde{y_{u_i}}$:

$$\tilde{y_{u_i}} = \text{softmax}(W_{u_i}^T \mathbf{X} + b_{u_i}) \tag{1}$$

where W_{u_i} and b_{u_i} represent the weights and bias for task u_i respectively.

For each task u_i, we minimize the cross-entropy of the predicted and true distributions as following:

$$L(\phi_{u_i}) = -\sum_{k=1}^{K}\sum_{c=1}^{C} y_k^c log(\tilde{y}_k^c) \tag{2}$$

where L denotes the cross-entropy loss function. y_k^c and $\tilde{y}_k^c$ are the true label and prediction probabilities for task u_i. K represents the total number of training samples and C is the total number of classes.

Finally, the total loss $\mathcal{L}$ of our approach is optimized for global objective function:

$$\mathcal{L} = \lambda_G \mathcal{L}_G + \lambda_A \mathcal{L}_A \tag{3}$$

where λ_G and λ_A are the weights for gender and age task respectively, and $\mathcal{L}_G$ and $\mathcal{L}_A$ are the losses of gender and age tasks.

4 Experiments

4.1 Experimental Setup

In this section, we conduct extensive experiments on three user reviews datasets [7] from TrustPilot (a website for online user reviews) to verify the effectiveness of our framework. The datasets collect users reviews and the reviewers' profiles (e.g., gender and birth year) from Denmark, United Kingdom and the United

States of America. We only extract English reviews with reviewer's gender and birth year information from the three datasets, using a language identifier fast-Text API[1]. Motivated by [18], we categorize user age into four categories: (0, 18] (18, 30), [30, 40) and [40,99]. The statistics of the three datasets are summarized in Table 1.

Table 2. Performance comparison of accuracy and macro-averaged F1 score on the Denmark, US and UK datasets for gender and age inference tasks. Here, gender task weight λ_G and age task weight λ_A are set to 1.

Methods	Denmark				US				UK			
	Gender		Age		Gender		Age		Gender		Age	
	Acc	Fscore	Acc	Fscore	Acc	Fscore	Acc	Fscore	Acc	Fscore	Acc	Fscore
Majority	74.33	–	59.93	–	60.18	–	58.58	–	59.62	–	73.17	–
CNN	71.95	41.84	60.98	24.74	65.52	63.62	56.93	32.92	62.92	61.37	73.06	28.43
BiLSTM	73.17	42.25	64.63	26.17	66.08	64.65	57.93	34.70	63.61	61.59	72.10	26.71
FastText	70.73	41.43	59.76	23.62	66.86	66.11	55.34	27.48	65.36	64.20	76.70	21.70
BERT	73.17	42.25	56.10	19.97	67.59	65.74	58.61	23.11	64.92	61.81	77.04	22.57
RoBERTa	67.58	35.63	56.10	23.96	60.50	37.69	57.85	18.32	60.04	37.51	77.07-	21.76
$\text{HAM}_{\text{CNN-attn}}$	68.07	45.27	52.77	24.49	63.68	60.15	57.03	24.76	52.12	49.65	67.89	24.85
HURA	70.84	44.70	51.57	25.61	66.42	57.56	54.11	22.42	58.41	41.18	75.69	22.32
Ours	**78.05**	**46.31**	**69.51**	**27.34**	**72.02**	**70.77**	**59.61**	**36.13**	**67.22**	**65.91**	**77.39**	**30.84**

We compared our proposed framework with the following methods: (1) Majority, a majority class based approach; (2) CNN [2], a CNN based model in demographics prediction; (3) BiLSTM [6], a bidirectional Long Short-Term Memory network; (4) FastText [8], a method proposed for efficient text classification; (5) BERT [4], the state-of-art method in various NLP tasks; (6) RoBERTa [11], a state-of-art robust model modified from BERT; (7) $\text{HAM}_{\text{CNN-attn}}$ [16], a state-of-art method using CNN and attention mechanism in demographic prediction; (8) HURA [19], a hierarchical demographic prediction model based on CNN and attention mechanism. As most works in user demographic prediction, we use accuracy and macro-averaged F1 score as evaluation metrics and report the average results. In all experiments, we preform 10-fold cross-validation, where 8 folds are used for training, 1 for validating and 1 for testing.

For all baselines, we adopt the optimal parameters configurations reported in their works. In our model, we use a BERT-base encoder and set max input length as 300. For sentiment and topic views, we embed sentiment words and topics into a 200-dimension vector. The filter number of word encoder is 128 and the window sizes are 2, 3, 4 and 5. We leverage Adam [10] optimizer with a dropout rate of 0.5 and set learning rate as 1e-3. Additionally, we apply an L2 weight decay 1e-3 to the loss and train our model in 20 epochs with a batch size of 64. For multi-task learning, we use equal gender task weight λ_G and age task weight λ_A as 1.

[1] https://fasttext.cc/docs/en/language-identification.html.

4.2 Experimental Results

Table 2 summarizes the comparison results between our model and baselines on three datasets. We have the following observations: (1) Our framework outperforms the statistics based baseline Majority, because our framework leverages rich information (e.g., semantics, sentiment words and topics) in text other than major class distribution of dataset; (2) Our framework also achieves better performance than neural network based methods, for we could capture context information and local semantics in text; (3) Compared with the Transformers based methods, our framework achieves better performance, especially on the macro-averaged F1 score; (4) As for the state-of-art methods $\mathrm{HAM}_{\mathrm{CNN-attn}}$ [16] and HURA [19], our model still gets better results by focusing on the effects of high-order interactions among context, sentiment and topics in text.

Table 3. Ablation study of multiple views in our framework on gender and age attributes

	Denmark				US				UK			
	Gender		Age		Gender		Age		Gender		Age	
	Acc	Fscore	Acc	Fscore	Acc	Fscore	Acc	Fscore	Acc	Fscore	Acc	Fscore
w/o Context view	69.51	41.01	58.54	18.46	65.62	61.45	57.85	29.53	60.04	58.61	76.63	21.69
w/o Sentiment view	73.17	42.25	62.20	25.56	69.13	67.81	57.69	21.34	63.61	61.37	76.70	23.11
w/o Topic view	74.39	43.45	63.54	25.25	69.24	66.55	58.61	25.45	65.48	61.98	76.84	23.49
Ours	**78.05**	**46.31**	**69.51**	**27.34**	**72.02**	**70.77**	**59.61**	**36.13**	**67.22**	**65.91**	**77.39**	**30.84**

To investigate which part contributes more to the performance, we perform ablation study. From Table 3, we could observe that: (1) Our approach performs best when leveraging context, sentiment and topic views; (2) Without context information, the performances on both gender and age tasks decrease sharply for not taking rich semantics into consideration; (3) Incorporating topics with context information performs better than using sentiment words and context. This is because topics may vary from nouns (e.g., shop and price) to adjectives (e.g., quick) and adverbs (e.g., smoothly), while sentiment words mainly consist of adjectives (e.g., helpful). Thus, we can conclude that various types of words contribute more to semantics.

5 Conclusion

In this paper, we study demographic prediction based on user reviews and propose a deep multi-view multi-task learning model. Our model first learns context representations from reviews by considering semantics and syntactics in text. At the same time, we learn sentiment and topic representations separately to capture the local contexts of reviews in word-level. Then our model integrates representations from context, sentiment and topic views and leverages the correlation between demographics to predict. Experimental results show that our model outperforms many baseline methods on gender and age predictions on three real-world datasets. Further, we perform ablation study to investigate which view contributes more to performance. In our future work, we plan to adapt our study to multilingual user reviews and explore the transferability of our model on more user attributes.

References

1. Basile, A., Dwyer, G., Medvedeva, M., Rawee, J., Haagsma, H., Nissim, M.: N-gram: new groningen author-profiling model–notebook for pan at clef 2017. In: CEUR Workshop Proceedings, vol. 1866 (2017)
2. Bayot, R.K., Gonçalves, T.: Age and gender classification of tweets using convolutional neural networks. In: Nicosia, G., Pardalos, P., Giuffrida, G., Umeton, R. (eds.) MOD 2017. LNCS, vol. 10710, pp. 337–348. Springer, Cham (2018). https://doi.org/10.1007/978-3-319-72926-8_28
3. Blei, D.M., Ng, A.Y., Jordan, M.I.: Latent dirichlet allocation. J. Mach. Learn. Res. **3**(1), 993–1022 (2003)
4. Devlin, J., Chang, M.W., Lee, K., Toutanova, K.: Bert: pre-training of deep bidirectional transformers for language understanding. In: NAACL-HLT (2019)
5. Gjurkovic, M., Šnajder, J.: Reddit: a gold mine for personality prediction. In: Second Workshop on Computational Modeling of Peoples Opinions (2018)
6. Hochreiter, S., Schmidhuber, J.: Long short-term memory. Neural Comput. **9**(8), 1735–1780 (1997)
7. Hovy, D., Johannsen, A., Søgaard, A.: User review sites as a resource for large-scale sociolinguistic studies. In: Proceedings of the 24th International Conference on World Wide Web, pp. 452–461 (2015)
8. Joulin, A., Grave, E., Bojanowski, P., Mikolov, T.: Bag of tricks for efficient text classification. arXiv preprint arXiv:1607.01759 (2016)
9. Kim, Y.: Convolutional neural networks for sentence classification. arXiv preprint arXiv:1408.5882 (2014)
10. Kingma, D.P., Ba, J.: Adam: a method for stochastic optimization. CoRR abs/1412.6980 (2014)
11. Liu, Y., et al.: Roberta: a robustly optimized bert pretraining approach. arXiv preprint arXiv:1907.11692 (2019)
12. Mikolov, T., Sutskever, I., Chen, K., Corrado, G.S., Dean, J.: Distributed representations of words and phrases and their compositionality. In: Advances in Neural Information Processing Systems, pp. 3111–3119 (2013)
13. Årup Nielsen, F.: A new anew: evaluation of a word list for sentiment analysis in microblogs (2011)
14. Pennebaker Francis, J.W.: Linguistic Inquiry and Word Count. Lawrence Erlbaum Associates Mahwah Nj (2012)
15. Sap, M., Park, G., Eichstaedt, J.C., Kern, M.L., Schwartz, A.H.: Developing age and gender predictive lexica over social media. In: Conference on Empirical Methods in Natural Language Processing (2014)
16. Tigunova, A., Yates, A., Mirza, P., Weikum, G.: Listening between the lines: learning personal attributes from conversations. In: The World Wide Web Conference, pp. 1818–1828 (2019)

17. Vijayaraghavan, P., Vosoughi, S., Roy, D.: Twitter demographic classification using deep multi-modal multi-task learning. In: Proceedings of the 55th Annual Meeting of the Association for Computational Linguistics (Volume 2: Short Papers), pp. 478–483 (2017)
18. Wang, Z., et al.: Demographic inference and representative population estimates from multilingual social media data. In: The World Wide Web Conference, pp. 2056–2067 (2019)
19. Wu, C.: Neural demographic prediction using search query. In: Proceedings of the Twelfth ACM International Conference on Web Search and Data Mining (2019)
20. Wu, C., Wu, F., Qi, T., Liu, J., Huang, Y., Xie, X.: Neural gender prediction in microblogging with emotion-aware user representation. In: Proceedings of the 28th ACM International Conference on Information and Knowledge Management, pp. 2401–2404 (2019)

An Interactive NL2SQL Approach with Reuse Strategy

Xiaxia Wang(✉), Sai Wu, Lidan Shou, and Ke Chen

ZheJiang University, Hang Zhou, China
{xiaer.wang,wusai,should,chenk}@zju.edu.cn

Abstract. This paper studies a recently proposed task that maps contextual natural language questions to SQL queries in a multi-turn interaction. Instead of synthesizing an SQL query in an end-to-end way, we propose a new model which first generates an SQL grammar tree, called Tree-SQL, as the intermediate representation, and then infers an SQL query from the Tree-SQL with domain knowledge. For semantic dependency among context-dependent questions, we propose a reuse strategy that assigns a probability for each sub-tree of historical Tree-SQLs. On the challenging contextual Text-to-SQL benchmark SParC (https://yale-lily.github.io/sparc) with the 'value selection' task which includes values in queries, our approach achieves SOTA accuracy of 48.5% in question execution accuracy and 21.6% in interaction execution accuracy. In addition, we experimentally demonstrate the significant improvements on the reuse strategy.

Keywords: Context-dependent semantic parsing · Reuse strategy · Intermediate representation

1 Introduction

The task of mapping natural language (NL) questions into queries, such as SQLs, that existing systems can process has become the hottest area. One representative Text-to-SQL task is the mapping to *cross-domain*, *nested* queries on *multi-tables*, a task relevant to the well-known benchmark dataset named Spider [6].

However, the task for Spider is not realistic, as it assumes only *context-independent* questions, namely mapping each question into an independently executable SQL query in a single-turn interaction. In a real scenario of data exploration, an analyst searches databases by asking a sequence of related questions in a multi-turn interaction. These questions are often semantically related [1] so that a question may include references/reuse of portions of previous questions.

To model semantic correlations among questions, a *context-dependent* task is introduced in [7], which comes with a new benchmark named SParC (cross-domain **S**emantic **Par**sing in **C**ontext). This task poses new challenges in handling rich contextual information and thematic relations between questions.

C. S. Jensen et al. (Eds.): DASFAA 2021, LNCS 12682, pp. 280–288, 2021.
https://doi.org/10.1007/978-3-030-73197-7_19

In addition, the task emphasizes on generalization, as each database appears for only once in either the training dataset or the dev/test dataset. It is not trivial to address *context-dependent* Text-to-SQL task, as most existing neural models fail to deliver a satisfied result.

Inspired by mismatch problems when synthesizing SQL queries directly from questions mentioned by [2] in *context-independent* task. Essentially, as NL is aimed at facilitating human communication, while SQL is designed for declarative database operations, there exists a huge gap between the implicit semantics of NL questions and the strictly formalized SQL queries. In view of this gap, we advocate a solution based on the notion of an intermediate representation called *Tree-SQL*, a tree structure complying with SQL grammar. This representation can then be used to support the value prediction, which is not considered in most work and unable to generate complete SQL for database execution.

The intuition of our solution for *context-dependent* is to transform historical queries as Tree-SQLs, and reuse sub-tree structures as context in the prediction of the next SQL query. In particular, we assign to each sub-tree of Tree-SQL a probability, which indicates the likelihood of its being reused during the prediction of the new query. For value prediction, we generate a correct prediction for specific numeric/string values in SQL queries by transforming open-domain value selection problems into closed-domain value extraction for questions.

Experiment results show that our model delivers accuracy of 48.5% in question execution and 21.6% in interaction execution. At the time of writing, we achieve state-of-the-art on complex public datasets, SParC with 'value selection' task. In addition, we experimentally demonstrate the significant improvements on the tree reuse mechanism.

2 Related Work

For NL2SQL task, *context-independent* semantic parsing has been well studied on WikiSQL [10] dataset and more complex Spider [6] dataset. Some researches have achieved promising results on both datasets. Compared to the *context-independent* parsing, the *context-dependent* semantic parsing becomes increasingly popular only in recent years. Several benchmark datasets have been proposed, such as ATIS [3], SequentialQA [4], and SParC [7]. Among those datasets, SParC is the most challenging one with complex semantic questions and database schema.

While [5] proposes a model on the ATIS single-domain dataset, which only has weak semantic and few contextual types. It maintains an interaction-level encoder and copies segments of previously predicted queries for the next prediction. Its segment-level copying strategy suffers from error propagation. To address the issue, [8] applies token-level editing strategy on the SParC, which is robust to error propagation but incurs high prediction overheads on all possible target tokens. For the reuse of historical answers, the segment-level is too coarse and the token-level is too fine. Our approach further extends a compromise reuse strategy that focuses on action-level reuse of previous Tree-SQLs to

T::=*union R R* | *intersect R R* |*except R R* |*R*
R::=*Select* | *Select Filter* |*Select Limit*|
Select Order | *Select Limit Filter* |
Select Order Filter
Select::=*A*|*A A*|*A A* ··· *A*
Limit::=*asc A V* | *desc A V*
Order::=*asc A* | *desc A*
Filter::=*and Filter Filter* | *or Filter Filter*
|> *A V* |= *A V* |≠ *A V* |< *A V* | ⩾ *A V*
| ⩽ *A V* |> *A R* |= *A R* |≠ *A R* |< *A R*
| ⩾ *A R* | ⩽ *A R* | *like A V* |*not like A V*
A::=*max S* | *min S* | *sum S* | *avg S* | *count S*
|*none S* | *max MA* | *min MA* | *sum MA*
| *avg MA* | *count MA* | *none MA*
MA::=+ *A A* | − *A A* | × *A A* | ÷ *A A*
S::=*table.column*
V::=V_t V_s V_d

Fig. 1. Grammar rules of Tree-SQL.

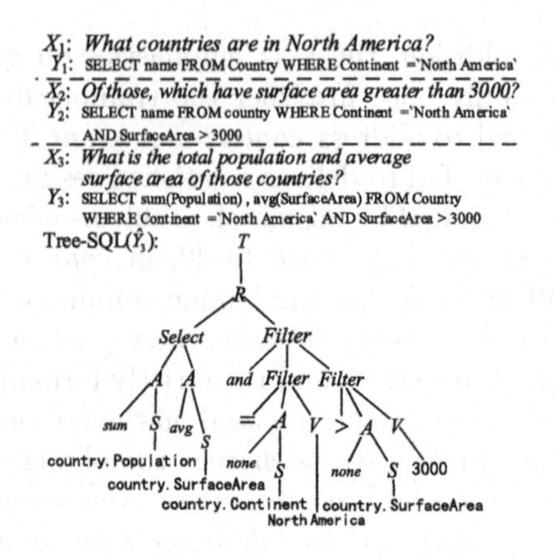

Fig. 2. An interaction from SParC and a Tree-SQL with respective SQL query at third turn.

better explore the correlations among consecutive questions, and achieves a more accurate mapping result.

3 Approach

In this section, we focus on learning a neural model for generating Tree-SQLs from multi-turn questions. We also discuss our reuse strategy for better Tree-SQL prediction.

3.1 Task Formulation

Let X and Y denote a natural language question and its corresponding SQL query respectively. We use $\hat{S} = [\hat{s}_1, \hat{s}_2, \ldots, \hat{s}_i, \ldots, \hat{s}_m]$ to represent the database schema. where $\hat{s}_i$ is a raw column, formatting as `table.column`.

In a *context-dependent* semantic parsing task, we maintain an interaction history $I_t = [(X_1, Y_1), (X_2, Y_2), \ldots, (X_{t-1}, Y_{t-1})$, consisting of $t-1$ question and SQL query pairs. So given current question X_t and database schema $\hat{S}$, our goal is to learn an optimal mapping function $f : (X_t, I_t, \hat{S}) \rightarrow Y_t$.

3.2 Tree-SQL

We propose the Tree-SQL as an intermediate representation. The grammar rules of Tree-SQL are demonstrated in Fig. 1, and Fig. 2 shows an interaction from the SParC dataset and a Tree-SQL evolves from SQL of third turn.

Currently, few work deals with "value prediction" on complex datasets, such as Spider and SParC, mainly due to the greater difficulty of value prediction, involving different sources and diversity, e.g., "North America" and "3000" in

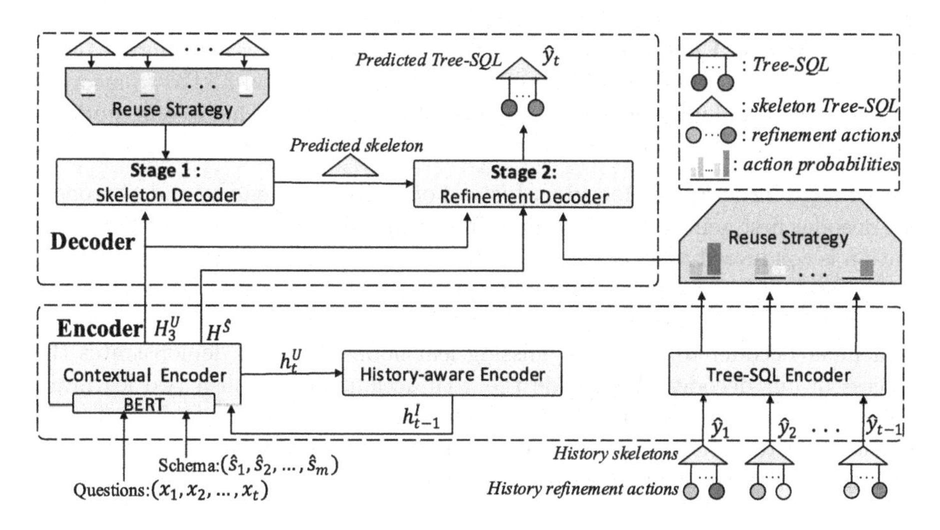

Fig. 3. Architecture of the neural model

Fig. 2. To reduce the search space, Tree-SQL transforms the open-domain value prediction problem into closed-domain value extraction in NL questions. We define three parameters to maintain the appearance of values. Suppose a value shows up in the V_t-th turn of an interaction (namely, the V_t-th question in the interaction), we use V_s and V_d to denote the starting offset and the length of the value in the question, respectively.

To infer an SQL query, we just traverse the tree structure in pre-order and map each tree node to the corresponding SQL query components based on our grammar rules defined in Fig. 1.

3.3 Basic Model

We design an SQL grammar-aware model based on the encoder-decoder structure to generate Tree-SQL. It takes questions, database schema and historical Tree-SQLs as input and outputs a Tree-SQL. An overview of the model architecture is illustrated in Fig. 3.

Contextual Encoder. We encode the question X_t and database schema $\hat{S}$ with pre-trained BERT as the first layer, which are concatenated together delimited by the `[SEP]` token: `[CLS]`,X_t, `[SEP]`,$\hat{s}_1$, `[SEP]`,$\hat{s}_2$, `[SEP]`,...,$\hat{s}_m$, `[SEP]`. Then the BERT embedding of question is fed into a Bi-LSTM layer to generate the final question embeddings $\boldsymbol{H}_t^U$. For encodings of schema items, we employ an self-attention layer to estimate the importance of different items, which can be also used to identify the primary-foreign key relationships. The final schema embeddings are the weighted combination $\boldsymbol{H}^{\hat{S}}$.

History-Aware Encoder. To reveal the correlations between different question/query pairs, we design a history-aware encoder based on LSTM to incorporate the contextual information as the interaction proceeds. The encoder updates iteratively between the current question embedding $\boldsymbol{h}_t^U$ and its hidden state $\boldsymbol{h}_t^I$.

Two-Stage Tree-SQL Decoder. Instead of applying an end-to-end approach, our decoder first generates a Tree-SQL from the embedding. The Tree-SQL generation is conducted as a *coarse-to-fine framework* prediction process, consisting of two stages. In the first stage, we apply a skeleton decoder to generate a skeleton of Tree-SQL with internal nodes only. In the second stage, we adopt a refinement decoder to fill in the missing leaf nodes. Figure 3 demonstrates the process of our decoder. To decode the k-th action, we establish two learnable embeddings, $\boldsymbol{a}_k$ for semantics of action and $\boldsymbol{b}_k$ for the type of action and learn a context vector $\boldsymbol{c}_k$. The two decoders share the same learnable parameters, the difference lies in the context vector.

$$\boldsymbol{h}_{k+1}^D = \text{LSTM}^\text{D}([\boldsymbol{a}_k; \boldsymbol{b}_k; \boldsymbol{c}_k], \boldsymbol{h}_k^D) \tag{1}$$

(a) Skeleton Decoder. The internal nodes of Tree-SQL do not involve specific raw columns and values. Therefore, the context vector $\boldsymbol{c}_k$ of skeleton decoder (denoted as LSTM^{D_1}) only comes from question information c_k^{token} via an attention between the hidden state of D_1 and all the weighted question embeddings $\boldsymbol{H}^U$, namely, $\boldsymbol{c}_k = [\boldsymbol{c}_k^{\text{token}}]$.

(b) Refinement Decoder. The refinement decoder (denoted as LSTM^{D_2}) tries to fill in the missing leaf nodes for the skeleton. We also need the schema embeddings $\boldsymbol{H}^{\hat{S}}$ to learn $\boldsymbol{c}_k^{\text{schema}}$, namely, $\boldsymbol{c}_k = [\boldsymbol{c}_k^{\text{token}}; \boldsymbol{c}_k^{\text{schema}}]$.

Based on the above definitions, we can define the loss function of the two-stage decoder as:

$$\begin{gathered} L_{D_1} = -\sum_{k=1}^{T}\sum_{j=1}^{N} y_{t,k,j}^{D_1} \log p(\hat{y}_{t,k,j}^{D_1}|X_t, I_t); L_{D_2} = -\sum_{k=1}^{T}\sum_{j=1}^{N} y_{t,k,j}^{D_2} \log p(\hat{y}_{t,k,j}^{D_2}|X_t, \hat{S}, I_t) \\ L_D = \alpha \times L_{D_1} + L_{D_2} \end{gathered} \tag{2}$$

we adopt the *Cross Entropy* as loss function of our decoders and a weight factor α to adjust the importance of two decoder, where N is the number of actions ($y_{t,k,j} = 1$ indicates that the j-th action is consistent with the ground truth one, while $y_{t,k,j} = 0$ shows a false result).

3.4 Optimization with Reuse Mechanism

To exploit the correlations of Tree-SQLs from different turns, we propose an optimization technique to improve the prediction of Tree-SQL by reusing a partial of generated answers from historical turns. The process of reuse mechanism is presented in Fig. 4.

We first generate the embedding of Tree-SQL by a encoder based on Bi-LSTM structure. Then We measure the importance of previously predicted Tree-SQLs

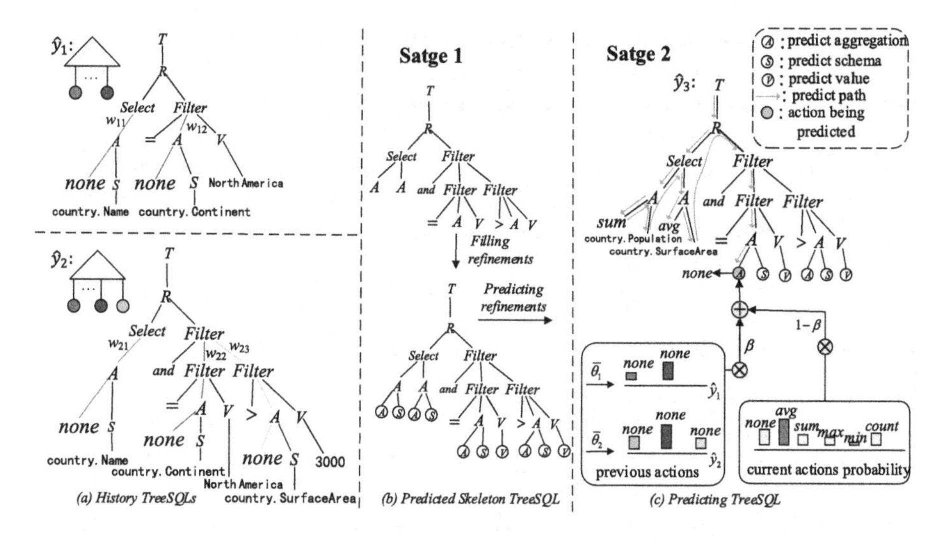

Fig. 4. Process of reuse mechanism

$\boldsymbol{c}_{l,k}^{\text{query}}$ from different turns via an attention between the hidden state of decoder and the embedding of Tree-SQL. To this end, a new context vector c_k with query embedding is defined as $\boldsymbol{c}_k = [\boldsymbol{c}_k^{\text{token}}; \boldsymbol{c}_k^{\text{schema}}; \boldsymbol{c}_k^{\text{query}}]$.

We predict a probability β_k^{reuse} of reusing previous Tree-SQLs' actions via sigmoid applied into context vector c_k. We also generate a weight $w_{l,i}$ for each sub-tree to indicate the possibility of reusing actions within the sub-tree. Then, the output distribution becomes a trade-off between the current actions and the historical Tree-SQLs of multiple turns and multiple sub-trees.

$$p(\hat{y}_{t,k}) = \beta_k^{\text{reuse}} \times \sum_{l=1}^{t-1} \bar{\theta}_l^{\text{turn}} \times p_l(\hat{y}_{l,k}) + (1 - \beta_k^{\text{reuse}}) \times p_t(\hat{y}_{t,k}) \tag{3}$$

The probability of action at current turn is denoted as p_t, while the probability of action reused from the previous Tree-SQLs is represented as p_l, where $1 <= l <= t-1$. Figure 4 shows an example of reuse strategy. Suppose we have two historical Tree-SQLs, $\hat{y}_1$ and $\hat{y}_2$. We reuse most structures and some leaf nodes from previous Tree-SQLs when generating $\hat{y}_3$.

4 Experiment

4.1 Experiment Setup

Dataset. We conduct our experiments on the SParC [7], an expert-labeled cross-domain context-dependent dataset which contains 4,298 coherent question sequences (12k+ questions paired with SQL queries) querying 200 complex databases in 138 different domains.

Metrics. For *SQL without value*, to avoid the effect of orderings, we follow the evaluation approach proposed by [6] that decomposes the predicted queries into different components, and calculates the set matching accuracy between the ground-truth and predicted queries. Similar as Zhang's work [8], we also report two metrics: the question match accuracy (QMA) and the interaction match accuracy (IMA). For *SQL with value*, we directly use the generated SQL for database execution and report question execution accuracy (QEA) and interaction execution accuracy (IEA).

4.2 Overall Results

Table 1 presents the results without value prediction. It can be observed that our model outperforms all baselines on the development and test sets. In particular, compared with state-of-the-art model EditSQL, our model outperforms it with 5.4% QMA and 4.9% IMA absolute improvement on the development set, achieving 52.6% QMA and 34.4% IMA. For test set, we achieve 48.1% QMA accuracy and 25.0% IMA accuracy. For the more complex "value prediction" tasks, our model achieves 48.5% QEA and 21.6% IEA on the test set of SParC, presented in Table 2, which is the SOTA method of value prediction so far.

Table 1. Metrics of different models on SParC without value

Approach	QMA (%)		IMA (%)	
	Dev	Test	Dev	Test
SyntaxSQL [7]	18.5	20.2	4.3	5.2
CD-Seq2Seq [7]	21.7	20.3	9.5	8.1
GuideSQL	–	34.4	–	13.1
ConcatSQL	–	46.3	–	22.4
GAZP [9]	–	45.9	–	23.5
SubTreeSQL	–	47.4	–	**25.5**
EditSQL [8]	47.2	47.9	29.5	25.3
Ours (w/o value)	**52.6**	**48.1**	**34.4**	25.0

Table 2. Metrics of models on SParC with value

Approach	QEA (%)		IEA (%)	
	Dev	Test	Dev	Test
GAZP [9]	–	44.6	–	19.7
Ours	**50.4**	**48.5**	**29.4**	**21.6**

4.3 Effectiveness of Reuse Strategy

To further understand the effectiveness of reuse strategy, we compare our approach with EditSQL on no-value task, which also supports an editing strategy. As presented in Fig. 5, applying reuse strategy benefits all turns and makes the model more stable for later turns. It can effectively reduce the error propagation by exploiting the correlations between questions.

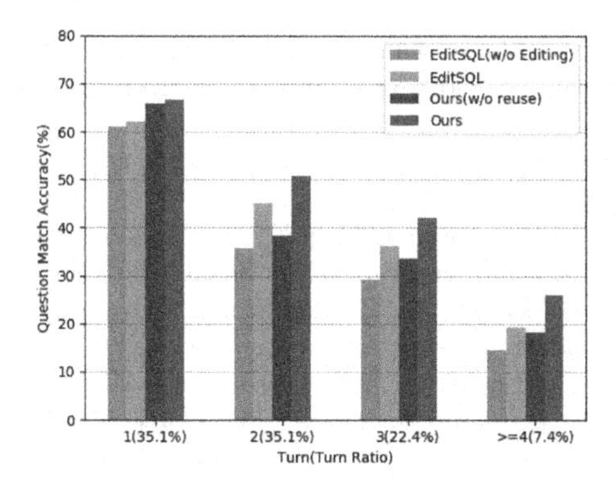

Fig. 5. The effectiveness of reuse strategy

5 Conclusions

In this paper, we propose a neural approach with encoder-decoder architecture for context-dependent cross-domain Text-to-SQL generation. We design a new tree structure, Tree-SQL, which is used as the intermediate representation for the translation. To exploit the correlations between questions of different interactions, we introduce a reuse mechanism to improve our prediction. Experimental results on the challenging SParC benchmark demonstrate the effectiveness of our model. In particular, our Tree-SQL extends the value prediction module to adapt to more real application scenarios that SQL often contains values, and our model achieves state-of-the-art on the value prediction task.

Acknowledgments. The work is supported by NSFC (grant number 61872315) and Zhejiang Provincial Natural Science Foundation (grant number LZ21F020007).

References

1. Bertomeu, N., Uszkoreit, H., Frank, A., Krieger, H.U., Jörg, B.: Contextual phenomena and thematic relations in database QA dialogues: results from a wizard-of-Oz experiment. In: Proceedings of the Interactive Question Answering Workshop at HLT-NAACL 2006, pp. 1–8 (2006)
2. Guo, J., et al.: Towards complex text-to-SQL in cross-domain database with intermediate representation. arXiv preprint arXiv:1905.08205 (2019)
3. Hemphill, C.T., Godfrey, J.J., Doddington, G.R.: The ATIS spoken language systems pilot corpus. In: Speech and Natural Language: Proceedings of a Workshop Held at Hidden Valley, Pennsylvania, 24–27 June 1990 (1990)
4. Iyyer, M., Yih, W., Chang, M.W.: Search-based neural structured learning for sequential question answering. In: Proceedings of the 55th Annual Meeting of the Association for Computational Linguistics (Volume 1: Long Papers), pp. 1821–1831 (2017)
5. Suhr, A., Iyer, S., Artzi, Y.: Learning to map context-dependent sentences to executable formal queries. arXiv preprint arXiv:1804.06868 (2018)

6. Yu, T., et al.: Spider: a large-scale human-labeled dataset for complex and cross-domain semantic parsing and text-to-SQL task. arXiv preprint arXiv:1809.08887 (2018)
7. Yu, T., et al.: SParC: cross-domain semantic parsing in context. arXiv preprint arXiv:1906.02285 (2019)
8. Zhang, R., et al.: Editing-based SQL query generation for cross-domain context-dependent questions. arXiv preprint arXiv:1909.00786 (2019)
9. Zhong, V., Lewis, M., Wang, S.I., Zettlemoyer, L.: Grounded adaptation for zero-shot executable semantic parsing. arXiv preprint arXiv:2009.07396 (2020)
10. Zhong, V., Xiong, C., Socher, R.: Seq2SQL: generating structured queries from natural language using reinforcement learning, CoRR abs/1709.00103 (2017)

Data Mining

Consistency- and Inconsistency-Aware Multi-view Subspace Clustering

Guang-Yu Zhang[1], Xiao-Wei Chen[1], Yu-Ren Zhou[1(✉)], Chang-Dong Wang[1], and Dong Huang[2]

[1] School of Computer Science and Engineering, Sun Yat-sen University, Guangzhou, China
chenxw65@mail2.sysu.edu.cn, zhouyuren@mail.sysu.edu.cn

[2] College of Mathematics and Informatics, South China Agricultural University, Guangzhou, China

Abstract. Multi-view subspace clustering has emerged as a crucial tool to solve the multi-view clustering problem. However, many of the existing methods merely focus on the consistency issue when learning the multi-view representations, failing to capture the latent inconsistency across different views (which can be caused by the view-specificity or diversity). To tackle this issue, we therefore develop a Consistency- and Inconsistency-aware Multi-view Subspace Clustering for robust clustering. In the proposed method, we decompose the multi-view representations into a view-consistent representation and a set of view-inconsistent representations, through which the multi-view consistency as well as multi-view inconsistency can be well explored. Meanwhile, our method aims to suppress the redundancy and determine the importance of different views by introducing a novel view weighting strategy. Then a unified objective function is constructed, upon which an efficient optimization algorithm based on ADMM is further performed. Additionally, we design a new way to compute the affinity matrix from both consistent and inconsistent perspectives, which makes sure that the learned affinity matrix comprehensively fit the inherent properties of multi-view data. Experimental results on multiple multi-view data sets confirm the superiority of our method.

Keywords: Multi-view subspace clustering · Multi-view representation learning · Consistency · Inconsistency · Redundancy

1 Introduction

Data in the form of multiple sources, also referred to as multi-view data, arises frequently in the fields of data mining and machine learning. Multi-view data can be obtained from heterogenous sources (or views), where each view conveys the specific physical meaning of the same object. For instance, a facial image can be influenced by different lighting conditions or input modalities; the same scene

C. S. Jensen et al. (Eds.): DASFAA 2021, LNCS 12682, pp. 291–306, 2021.
https://doi.org/10.1007/978-3-030-73197-7_20

image can be classified from multiple visual features, such as GIST, SIFT and HOG; one news report may be reported by different broadcasters or translated into multiple languages. Naturally, the increasing amount of multi-view data has led to great interest in the research of multi-view clustering. Compared with the single-view clustering methods, multi-view clustering methods have the potential to achieve more robust clustering results by capturing the rich and complementary knowledge from different views.

In the literature, researchers have proposed a variety of multi-view clustering methods [6–8,14,19,21], among which the subspace-based methods is one of the most popular research topics. In this category of methods, the primary goal relies on capturing the latent subspace structure across different views. To achieve this goal, numerous subspace-based methods have been proposed over the past few years. For example, Tang et al. [16] designed a subspace clustering method that discovers the shared multi-view subspace structure with joint graph learning. Recently, Chen et al. [3] proposed the multi-view clustering in latent space (MVCLS) method, which learns a latent embedded space in multi-view data while discovering the global cluster structure simultaneously. However, these methods mostly focus on learning the consistent subspace representation across different views, where the complementary information of multiple views hasn't been well-exploited. Motivated by this, some efforts have been made to explore the complementarity of multi-view representations. To name a few, Cao et al. [2] presented a diversity-induced method for multi-view subspace clustering, which explores the complementarity across multiple views by Hilbert-Schmidt Independence Criterion (HSIC). Further, Zhang et al. [23,25] developed a novel multi-view subspace clustering method via latent space learning. This method performs the subspace clustering in the latent space with complementarity from multiple views considered. Although these subspace-based methods have made significant progress, most of them only consider the complementarity across different views, yet neglect the multi-view consistency in subspace representation learning. More recently, Luo et al. [13] proposed a robust multi-view clustering method from both consistent and specific perspectives. However, its effectivness may be further enhanced by exploring other multi-view inconsistency (i.e., diversity) and considering the different importance among views.

In this paper, we develop a novel subspace-based method, named Consistency- and Inconsistency-aware Multi-view Subspace Clustering. In our method, the multi-view subspace representations are intuitively decomposed into a view-consistent representation and a set of view-inconsistent representations. By this means, both the consistency and inconsistency of multi-view data can be well explored. Furthermore, an adaptive view weighting strategy is incorporated into the proposed method, which helps to suppress the redundancy and determine the contributions of different views. We formulate the above components into a unified objective function, and simultaneously design an iterative algorithm to solve the resultant optimization problem. For clarity, the contributions of this paper are summarized as follows.

- This paper develops a novel subspace-based method by jointly modeling the multi-view consistency and multi-view inconsistency. The method ensures that the acquired multi-view subspace representations comprehensively fits the inherent properties of multi-view data.
- By considering the various contributions of multiple views, we design a novel strategy to generate the final affinity matrix from multi-view consistent and inconsistent perspectives.
- Extensive experiments on multiple multi-view data sets demonstrate the effectiveness of our method.

The rest of this paper is organized as follows. In Sect. 2, we review some preliminaries, and introduce the proposed objective function. In Sect. 3, we present an optimization algorithm to solve the objective function of CIMSC, followed by its complexity analysis. In Sect. 4, we evaluate the performance of our method. In Sect. 5, we conclude the whole paper.

2 Methodology

2.1 Notations

In this paper, matrices and vectors are written as uppercase letters and lowercase letters, respectively. For a matrix M, its i-th row, j-th column is denoted as m^{i} and m_{j}, respectively. The (i, j)-th element of matrix M is denoted as m_{ij}. The inverse and the transpose of matrix M are denoted as M^{-1} and M^{T}, respectively. The trace of matrix M is defined as $\mathrm{Tr}(\mathrm{M})$. Letter I represents the identity matrix and $\mathbf{1}$ represents the column vector with all entries being one. $\mathbf{0}$ represents the zero matrix.

2.2 Preliminary Knowledge

In this subsection, we briefly review some preliminary knowledge on multi-view subspace clustering before further introduction. Suppose we are given a multi-view data matrix $\mathrm{X} = [\mathrm{X}^{(1)^T}, \ldots, \mathrm{X}^{(V)^T}]^T$ with V views, and $\mathrm{X}^{(v)} = [\mathrm{x}_1^{(v)}, \cdots, \mathrm{x}_n^{(v)}] \in \mathbb{R}^{d_v \times n}$ denotes the v-th data matrix with d_v dimensionality. As a very popular research topic, multi-view subspace clustering assumes that the observed data samples obey the self-expressive property, that is, the high-dimensional data samples are usually drawn from low-dimensional subspaces. Under this assumption, we can formulate the self-expression formulation for multi-view data as follows:

$$\mathrm{X}^{(v)} = \mathrm{X}^{(v)}\mathrm{Z}^{(v)} + \mathrm{E}^{(v)}, \forall v, \tag{1}$$

where $\mathrm{Z}^{(v)}$ represents the v-th view subspace representation and $\mathrm{E}^{(v)}$ represents the v-th view error matrix. Based on the self-expression formulation in Eq. (1), the general objective function for multi-view subspace clustering can be summarized as follows:

$$\min_{Z^{(v)},E^{(v)}} \sum_{v=1}^{V} \Psi(E^{(v)}) + \lambda_1 \sum_{v=1}^{V} \Theta(Z^{(v)}) \quad (2)$$
$$\text{s.t. } X^{(v)} = X^{(v)}Z^{(v)} + E^{(v)}, \forall v,$$

where $\Psi(\cdot)$ denotes the loss function for the v-th view error matrix $E^{(v)}$ (i.e., $||E^{(v)}||_{2,1}$), and $\Theta(\cdot)$ is the regularization that imposes specific property on the v-th view subspace representation $Z^{(v)}$ (i.e., $||Z^{(v)}||_*$ or $||Z^{(v)}||_F^2$). λ_1 is a balanced parameter. Many multi-view subspace clustering methods seek to uncover the shared subspace structure of all views. Hence, if we only focus on the multi-view consistency issue, the Low-Rank Representation (LRR) [12] can be naturally extended to multi-view scenario as follows:

$$\min_{Z,E^{(v)}} \sum_{v=1}^{V} ||E^{(v)}||_{2,1} + \lambda_1 ||Z||_* \quad (3)$$
$$\text{s.t. } X^{(v)} = X^{(v)}Z + E^{(v)}, \forall v,$$

in which $||Z||_*$ is the nuclear norm on the multi-view subspace representation Z. This method is able to discover the common subspace structure across different views, however, its performance can be further improved by exploring more inherent properties of multi-view data.

2.3 The Objective Function

In this subsection, we introduce the objective function of our method CISMC in detail. Different from the previous subspace-based methods, the proposed method aims to explore more inherent properties of multi-view data (i.e., consistency and inconsistency) in a unified framework. In light of this, we consider a multi-view self-expression formulation by jointly modeling the multi-view consistency and multi-view inconsistency, i.e.,

$$X^{(v)} = X^{(v)}(Z + C^{(v)}) + E^{(v)}, \forall v. \quad (4)$$

Here Z represents the view-consistent representation across different views, and $C^{(v)}$ represents the view-inconsistent representation in the v-th view. Following the model in Eq. (3), we impose the nuclear norm on the view-consistent representation to model the structure consistence across multiple views. Based on the multi-view self-expression in Eq. (4), we consider the difference among multiple views and formulate the initial objective function by:

$$\min_{Z,E^{(v)},C^{(v)},\omega_v} \sum_{v=1}^{V} \omega_v ||E^{(v)}||_{2,1} + \lambda_1 ||Z||_* + \sum_{v=1}^{V} \Upsilon(C^{(v)}) \quad (5)$$
$$\text{s.t. } X^{(v)} = X^{(v)}(Z + C^{(v)}) + E^{(v)}, \forall v,$$
$$\omega^T \mathbf{1}_V = 1, \omega \geqslant 0,$$

where $\sum_{v=1}^{V} \omega_v ||\mathrm{E}^{(v)}||_{2,1}$ is the weighted loss function, and $\omega = [\omega_1, \omega_2, \cdots, \omega_V]^T$ is a view weighting vector which reflects the importance of different views. As can be seen, the larger weights are assigned to the error terms with smaller loss (i.e., the informative views). Here $\Upsilon(\cdot)$ denotes the regularized term on the view-inconsistent representations. In particular, this regularizer term should model the complementarity as well as view-specificity of multi-view data, which can be naturally decomposed into two parts.

The first part aims to model the complementary information across different views. Therefore, we argue that the view-inconsistent representations should be different from each other. To implement the diversity, we formulate the first part of regularized term $\Upsilon(\cdot)$ as follows:

$$\begin{aligned} &\lambda_2 \sum_{u \neq v}^{V} \mathrm{sum}\{(\omega_v \mathrm{C}^{(v)}) \circ (\omega_u \mathrm{C}^{(u)})\} \\ &= \lambda_2 \sum_{u \neq v}^{V} \omega_u \omega_v \mathrm{Tr}({\mathrm{C}^{(v)}}^T \mathrm{C}^{(u)}), \end{aligned} \tag{6}$$

in which $\circ$ denotes the Hadamard product (element-wise multiplication) of two matrices, and sum is the operator that sums all elements in a matrix. By minimizing this term, we can see that if two view-inconsistent representations $\mathrm{C}^{(v)}$ and $\mathrm{C}^{(u)}$ are diverse to each other, then their view weights ω_v and ω_u would be assigned with larger values. In this way, the correlation between $\mathrm{C}^{(v)}$ and $\mathrm{C}^{(u)}$ can be well measured, which is beneficial to promote the diversity of different views and reduce the redundancy among similar views.

Furthermore, the second part targets at modeling the view-specific property within each view. Similar to the term in Eq. (6), we formulate the second part of regularized term $\Upsilon(\cdot)$ in the following form:

$$\begin{aligned} &\lambda_3 \sum_{v=1}^{V} \mathrm{sum}\{(\omega_v \mathrm{C}^{(v)}) \circ (\omega_v \mathrm{C}^{(v)})\} \\ &= \lambda_3 \sum_{v=1}^{V} \omega_v \omega_v \mathrm{Tr}({\mathrm{C}^{(v)}}^T \mathrm{C}^{(v)}) \end{aligned} \tag{7}$$

As can be observed, the above term is equivalent to impose Frobenius norm on each view-inconsistent representations (i.e., $\lambda_3 \sum_{v=1}^{V} {\omega_v}^2 ||\mathrm{C}^{(v)}||_F^2$). To be specific, the Frobenius norm ensures the connectedness of subspace representations [12]. Hence, by minimizing this term, the v-th view weight ω_v will become larger as long as the v-th view inconsistent representation $\mathrm{C}^{(v)}$ has strong connectivity. By this means, the above term can efficiently explore the view-specificity within each view, as well as determining the different contributions of multiple views.

Consequently, integrating these two parts of regularized term $\Upsilon(\cdot)$ into the initial objective function, we have the final optimization problem:

$$\begin{aligned}\min_{\mathrm{Z},\mathrm{E}^{(v)},\mathrm{C}^{(v)},\omega_v} & \sum_{v=1}^{V} \omega_v ||\mathrm{E}^{(v)}||_{2,1} + \lambda_1 ||\mathrm{Z}||_* + \sum_{v,u=1}^{V} \mathrm{a}_{vu}\omega_v\omega_u \mathrm{Tr}(\mathrm{C}^{(v)^T}\mathrm{C}^{(u)}) \\ & \text{s.t. } \mathrm{X}^{(v)} = \mathrm{X}^{(v)}(\mathrm{Z}+\mathrm{C}^{(v)}) + \mathrm{E}^{(v)}, \forall v, \\ & \omega^T \mathbf{1}_V = 1, \omega \geqslant 0, \end{aligned} \tag{8}$$

where $\mathrm{A} = (\mathrm{a})_{ij} \in \mathbb{R}^{V\times V}$ is a parameter matrix, whose diagonal elements and nondiagonal elements being parameters λ_3 and λ_2, respectively. To represent the multi-view data more naturally, the proposed objective function jointly models the multi-view consistency and inconsistency in a unified optimization framework. Therefore, by solving the Eq. (8), a view-consistent representation and a set of view-inconsistent representations can be achieved. In Subsect. 3.1, we will describe how to compute the final affinity matrix from multi-view consistent and inconsistent perspectives.

3 Optimization

3.1 Optimization Algorithm

In this subsection, the optimization problem in Eq. (8) is solved by an ADMM-based optimization algorithm [1]. First, we introduce an auxiliary variable S (i.e., S = Z) to make the problem separable. Next, the augmented Lagrangian function can be formulated as follows:

$$\begin{aligned} & \mathcal{L}(\mathrm{S},\mathrm{Z},\{\mathrm{C}^{(v)}\}_{v=1}^{V},\{\mathrm{E}^{(v)}\}_{v=1}^{V},\omega)_{\{\omega^T\mathbf{1}_V=1,\omega\geqslant 0\}} \\ = & \sum_{v=1}^{V}\omega_v||\mathrm{E}^{(v)}||_{2,1} + \lambda_1||\mathrm{Z}||_* + \sum_{v,u=1}^{V}\mathrm{a}_{vu}\omega_v\omega_u\mathrm{Tr}(\mathrm{C}^{(v)^T}\mathrm{C}^{(u)}) + \\ & \sum_{v=1}^{V}\Phi(\mathrm{Y_1}^{(v)},\mathrm{X}^{(v)}-\mathrm{X}^{(v)}\mathrm{S}-\mathrm{X}^{(v)}\mathrm{C}^{(v)}-\mathrm{E}^{(v)}) + \Phi(\mathrm{Y}_2,\mathrm{S}-\mathrm{Z}), \end{aligned} \tag{9}$$

where $\Phi(\mathrm{A},\mathrm{B})$ is deemed as $\langle \mathrm{A},\mathrm{B}\rangle + \frac{\mu}{2}||\mathrm{B}||_F^2$, and $\langle\cdot,\cdot\rangle$ denotes the inner product of two matrices. Besides, $\{\mathrm{Y}_1^{(v)}\}_{v=1}^{V}$ and Y_2 represents the Lagrange multipliers and $\mu > 0$ is a penalty parameter. To solve the optimization problem for $\mathcal{L}$ in Eq. (9), we can divide it into following subproblems.

S-Subproblem. When fixing the other variables, we can solve the optimization problem w.r.t. variable S as follows:

$$\min_{\mathrm{S}} \sum_{v=1}^{V}\Phi(\mathrm{Y_1}^{(v)},\mathrm{X}^{(v)}-\mathrm{X}^{(v)}\mathrm{S}-\mathrm{X}^{(v)}\mathrm{C}^{(v)}-\mathrm{E}^{(v)}) + \Phi(\mathrm{Y}_2,\mathrm{S}-\mathrm{Z}). \tag{10}$$

Taking the derivative of problem (10) w.r.t. variable S and setting it to be zero, we obtain the following solution:

$$S = (\sum_{v=1}^{V}(X^{(v)^T}X^{(v)}) + I)^{-1}(\sum_{v=1}^{V}X^{(v)^T}(X^{(v)} - X^{(v)}C^{(v)} - E^{(v)} + \frac{Y_1{}^{(v)}}{\mu}) + Z - \frac{Y_2}{\mu}). \quad (11)$$

Z-Subproblem. When fixing the other variables, we can solve the optimization problem w.r.t. variable Z as follows:

$$\min_{Z} \lambda_1||Z||_* + \frac{\mu}{2}||Z - (S + \frac{Y_2}{\mu})||_F^2, \quad (12)$$

By using the Singular Value Thresholding (SVT), we have the following solution for variable Z:

$$Z = U_Z\Gamma_{\frac{\lambda_1}{\mu}}(\Sigma_Z)V_Z^T. \quad (13)$$

where $U_Z\Sigma_Z V_Z^T$ is the Singular Value Decomposition (SVD) of $S + \frac{Y_2}{\mu}$. Besides, $\Gamma_\tau(\cdot)$ denotes the SVT operator, which can be defined by:

$$\Gamma_\tau(\Sigma) = \max(0, \Sigma - \tau) + \min(0, \Sigma + \tau). \quad (14)$$

$C^{(v)}$-subproblem. When fixing the other variables, we can solve the optimization problem w.r.t. variable $C^{(v)}$ as follows:

$$\min_{C^{(v)}} \sum_{u=1}^{V} a_{vu}\omega_v\omega_u \mathrm{Tr}(C^{(v)^T}C^{(u)}) + \Phi(Y_1{}^{(v)}, X^{(v)} - X^{(v)}S - X^{(v)}C^{(v)} - E^{(v)}). \quad (15)$$

By setting the above problem w.r.t. variable $C^{(v)}$ to be zero, we can get the solution as follows:

$$C^{(v)} = (\mu X^{(v)^T}X^{(v)} + 2\lambda_3\omega_v\omega_v I)^{-1}(\mu X^{(v)^T}Q^{(v)} + X^{(v)^T}Y_1{}^{(v)} - \lambda_2\omega_v\sum_{u\neq v}\omega_u C^{(u)}), \quad (16)$$

in which $Q^{(v)} = X^{(v)} - X^{(v)}S - E^{(v)}$.

$E(v)$-subproblem. When fixing the other variables, we can solve the optimization problem w.r.t. variable $E^{(v)}$ as follows:

$$\min_{E^{(v)}} ||E^{(v)}||_{2,1} + \frac{\mu}{2}||E^{(v)} - (X^{(v)} - X^{(v)}S - X^{(v)}C^{(v)} + \frac{1}{\mu}Y_1{}^{(v)})||_F^2. \quad (17)$$

According to Lemma in [11], the above problem can be solved by:

$$e_{:j}^{(v)} = \begin{cases} \frac{||f_{:j}^{(v)}||_2 - \frac{1}{\mu}}{||f_{:j}^{(v)}||_2} f_{:j}^{(v)} & \text{if } ||f_{:j}^{(v)}||_2 > \frac{1}{\mu}, \\ 0 & \text{otherwise}, \end{cases} \quad (18)$$

where $F^{(v)} = X^{(v)} - X^{(v)}S - X^{(v)}c^{(v)} + \frac{1}{\mu}Y_1{}^{(v)}$.

ω-Subproblem. When fixing the other variables, we can solve the optimization problem w.r.t. variable ω as follows:

$$\min_{\omega} \sum_{v=1}^{V} \omega_v ||\mathrm{E}^{(v)}||_{2,1} + \sum_{v,u=1}^{V} \mathrm{a}_{vu} \omega_v \omega_u \mathrm{Tr}({\mathrm{C}^{(v)}}^T \mathrm{C}^{(u)})$$
$$\text{s.t. } \omega^T \mathbf{1}_V = 1, \omega \geqslant 0. \tag{19}$$

By introducing a matrix $\mathrm{H} \in \mathbb{R}^{V \times V}$ with its element $\mathrm{h}_{vu} = \mathrm{a}_{vu} \mathrm{Tr}({\mathrm{C}^{(v)}}^T \mathrm{C}^{(u)})$, the problem in Eq. (19) is equivalent to the following problem:

$$\omega^T \begin{bmatrix} ||\mathrm{E}^{(1)}||_{2,1} \\ \vdots \\ ||\mathrm{E}^{(V)}||_{2,1} \end{bmatrix} + \omega^T \mathrm{H} \omega, \text{ s.t. } \omega^T \mathbf{1}_V = 1, \omega \geqslant 0 \tag{20}$$

Since the matrix H is positive semidefinite, thus the problem in Eq. (20) has closed-form solution and can be solved by off-the-shelf tools.

Update the Multipliers. Finally, the multipliers $\{\mathrm{Y}_1^{(v)}\}_{v=1}^V$ and Y_2 can be updated by using the following equations:

$$\begin{cases} \mathrm{Y}_1^{(v)} = \mathrm{Y}_1^{(v)} + \mu(\mathrm{X}^{(v)} - \mathrm{X}^{(v)}\mathrm{S} - \mathrm{X}^{(v)}\mathrm{C}^{(v)} - \mathrm{E}^{(v)}) \\ \mathrm{Y}_2 = \mathrm{Y}_2 + \mu(\mathrm{S} - \mathrm{Z}) \end{cases} \tag{21}$$

In this optimization algorithm, we update the variables and multipliers iteratively until the stopping criterion is met or the number of iteration reaches the predefined threshold. For clarity, Algorithm 1 summarizes the overall algorithm of the proposed CIMSC method.

Subsequently, we construct the final affinity matrix A from the learned representation matrices and view weights as follows:

$$\mathrm{A} = \frac{|\mathrm{S}| + |\mathrm{S}|^T}{2} + \sum_{v=1}^{V} \omega_v \frac{|\mathrm{C}^{(v)}| + |\mathrm{C}^{(v)}|^T}{2} \tag{23}$$

As far as we know, this is the first attempt to compute the final affinity matrix from both multi-view consistent and inconsistent perspectives. To further suppress the redundancy, the inconsistent subspace representations are assigned with suitable weights according to their importance.

Algorithm 1. CIMSC

Input: Multi-view data set $X = \{X^{(1)}, \cdots, X^{(V)}\}$ with V views, the maximum number of iterations t_{max} (i.e., 30 here) and three balancing parameters $\lambda_1, \lambda_2, \lambda_3$.
Parameter Setup: Set parameters $\mu = 1$, $\rho = 1.5$ and $\mu_{max} = 10^6$.

1: **Initialization:** Set $t = 1$. Initialize $Z = \mathbf{0}$, $C^{(v)} = \mathbf{0}, \forall v$, $E^{(v)} = \mathbf{0}, \forall v$, ${Y_1}^{(v)} = \mathbf{0}, \forall v$ and $Y_2 = \mathbf{0}$.
2: **repeat**
3: Update S by using Eq. (11).
4: Update Z by using Eq. (13).
5: Update $C^{(v)}$ in the v-th view by using Eq. (16).
6: Update $E^{(v)}$ in the v-th view by solving problem (18).
7: Update ω by solving problem (20).
8: Update $Y_1^{(v)}$ in the t-th view and Y_2 by using Eq. (21).
9: Update μ by using $\mu = \min(\rho\mu, \mu_{\max})$.
10: $t = t + 1$.
11: **until** The following convergence conditions are meet or $t > t_{max}$.

$$\sum_{v=1}^{V} ||X^{(v)} - X^{(v)}S - X^{(v)}C^{(v)} - E^{(v)}||_\infty < \varepsilon \text{ and } ||S - Z||_\infty < \varepsilon \tag{22}$$

12: Obtain the overall affinity matrix by using Eq. (23).
13: Perform spectral clustering on the overall affinity matrix A and generate the final clustering label.
Output: The final clustering label.

3.2 Model Complexity

In this subsection, we analyze the time complexity of the proposed method. As can be observed, Algorithm 1 consists of six subproblems. The time complexities of updating multipliers and the fifth subproblem (w.r.t. variable ω) can be omitted in comparison with the other subproblems. Hence, we only focus on the left four subproblems. For the first three subproblems, the time complexities of updating variables S, Z and $C^{(v)}$ are the same, i.e., $O(n^3)$, where n is the number of data samples. For the fourth subproblem, the time complexity of updating variable $E^{(v)}$ is $O(d_v n)$, where d_v is the dimensionality of the v-th view. Therefore, the overall time complexity of Algorithm 1 is $O(TVn^3 + T\sum_{v=1}^{V} d_v n)$, where T is the desired times of Algorithm 1 and V denotes the number of views.

4 Experiments

In this section, we extensively evaluate the performance of our method from three aspects, i.e., comparison experiments, parameter analysis, and convergence analysis. All the experiments are implemented with an Intel 3.4-GHz CPU and 64-GB RAM.

4.1 Data Sets and Evaluation Measures

To validate the effectiveness of the proposed method, five benchmark data sets are selected in our experiments. These data sets are collected from real-world applications, which have been widely used for four tasks: object clustering (NUS-WIDE [9]), face clustering (VIS/NIR [17] and Yale [24]), handwritten digit recognition (UCI Digit [20]) and document clustering (Reuters [9]). For clarity, the five multi-view data sets are listed in Table 1.

Table 1. The statistics of six data sets used in the experiments.

	NUS-WIDE	UCI Digit	Reuters	VIS/NIR	Yale
View 1	CH(65)	FOU(76)	French(2000)	View 1(10000)	Intensity(2500))
View 2	CM(226)	PIX(240)	German(2000)	View 2(10000)	Gabor(6750)
View 3	CORR(145)	MOR(6)	German(2000)	–	LBP(3304)
View 4	EDH(74)	–	Spanish(2000)	–	–
View 5	WT(129)	–	Italian(2000)	–	–
# samples	2000	2000	1200	1056	165
# classes	31	10	6	40	10

Besides, the brief introduction is as follows:

- **NUS-WIDE** is a web image data set for object clustering task. Each sample in the data set can be divided into five views: 64 color histogram, 144 color correlogram, 73 edge direction histogram, 128 wavelet texture, and 225 block-wise color moment.
- **UCI Digit** is a famous image data set which records the digital numbers from 0 to 9. Same to the paper in [20], three different features are extracted from each image: 76 Fourier coefficients, 240 pixel averages and 6 morphological features.
- **Reuters** is a popular multi-view text data set that associates with six topics. This data set contains 1200 documents, where each document is reported in five different languages (views), i.e., 2000 English, 2000 French, 2000 German, 2000 Spanish and 2000 Italian.
- **VIS/NIR** is a widely-used data set that contains 1056 face images over 22 classes. In this data set, each image is available in two heterogeneous feature sets: 10000 Visible Light and 10000 Near-IR illumination.
- **Yale** is a face image data set containing 165 gray-scale images. It contains 15 classes, and each class has 11 images with different facial expressions and configurations. Following the works in [2], we extract three types of features from each image: Intensity, LBP and Gabor.

In the experiments, two well-known evaluation metrics are adopted to evaluate the quality of clustering results, namely Accuracy (Acc) [4] and Normalized Mutual Information (NMI) [4]. For these two metrics, the higher values indicate the better clustering performance.

4.2 Comparison Experiments

In order to demonstrate the superiority of our method, we empirically compare it with several other popular clustering methods. Particularly, the compared methods include two single-view clustering methods and six multi-view clustering methods. On the one hand, two classical single-view methods, i.e., Spectral Clustering (SC) and Low-Rank Recovery (LRR) [11] are employed to evaluate the performance of single-view clustering methods on multi-view data. On the other hand, six state-of-art multi-view methods, namely Self-weighted Multiview Clustering (SwMC) [15], Graph-based Multi-view Clustering (GMC) [18], Multi-view Graph Learning [22], Latent Multi-view Subspace Clustering (LMSC) [25], Consistent and Specific Multi-view Subspace Clustering (CSMSC) [13] and Dual Shared-Specific Multi-view Subspace Clustering (DSS-MSC) [26] are selected to validate the effectiveness of multi-view clustering methods. To guarantee the fairness of comparison, we tune the parameters of all methods within the candidate set $\{10^{-4}, \cdots, 10^{4}\}$ and report the scores with the best parameters. For single-view methods, we apply them on each view of the multi-view data, while reporting the best results as SC_{best} and LRR_{best}. The experiments are repeated 20 times for all clustering methods, and the average values with standard deviations are reported as the results.

Table 2 and Table 3 report the results of all clustering methods on five multi-view data sets, where the best results are highlighted in bold. Following [10], the average ranks of performance by each clustering method is also reported. According to these tables, three conclusions can be drawn as follows.

- It can be seen that multi-view clustering methods always obtain better results over the single-view clustering methods. This is probably because that multi-view clustering methods sufficiently consider the complementarity across multiple views, while single-view clustering methods only focus on the partial information within specific view.
- When it comes to image data sets (i.e., NUS-WIDE, VIS/NIR and Yale), we can observe that subspace-based methods achieve better performance than graph-based methods. This observation demonstrates the effectiveness of subspace-based methods when handing image clustering tasks.

Table 2. Comparison results in terms of Acc (%) on five benchmark data sets.

	NUS-WIDE	UCI Digit	Reuters	VIS/NIR	Yale	Avg. rank
SC_{best}	$13.9_{\pm 0.5}$	$80.9_{\pm 0.0}$	$29.4_{\pm 0.3}$	$84.5_{\pm 0.0}$	$61.6_{\pm 3.0}$	7.6
LRR_{best}	$14.3_{\pm 0.5}$	$87.5_{\pm 2.5}$	$27.4_{\pm 1.1}$	$65.8_{\pm 3.5}$	$69.7_{\pm 0.1}$	6.0
SwMC	$15.6_{\pm 0.4}$	$82.7_{\pm 0.0}$	$24.1_{\pm 0.0}$	$91.6_{\pm 0.9}$	$64.3_{\pm 0.1}$	6.0
GMC	$14.9_{\pm 0.0}$	$82.2_{\pm 0.0}$	$22.9_{\pm 0.0}$	$95.5_{\pm 0.0}$	$65.5_{\pm 0.0}$	5.8
MVGL	$14.7_{\pm 0.3}$	$84.8_{\pm 1.4}$	$25.2_{\pm 0.2}$	$94.8_{\pm 1.2}$	$64.9_{\pm 0.0}$	6.0
LMSC	$13.9_{\pm 0.7}$	$86.5_{\pm 6.2}$	$34.0_{\pm 1.2}$	$97.0_{\pm 0.6}$	$74.2_{\pm 2.4}$	4.0
CSMSC	$14.8_{\pm 0.5}$	$81.1_{\pm 0.0}$	$27.6_{\pm 0.6}$	$88.6_{\pm 0.5}$	$75.5_{\pm 1.0}$	5.6
DSS-MSC	$15.5_{\pm 0.3}$	$91.2_{\pm 0.2}$	$33.4_{\pm 0.7}$	$95.0_{\pm 1.2}$	$78.2_{\pm 1.3}$	2.8
CIMSC	$\mathbf{16.9}_{\pm 0.4}$	$\mathbf{93.6}_{\pm 0.0}$	$\mathbf{47.5}_{\pm 0.0}$	$\mathbf{98.2}_{\pm 0.1}$	$\mathbf{88.4}_{\pm 0.1}$	**1.0**

Table 3. Comparison results in terms of NMI (%) on five benchmark data sets.

	NUS-WIDE	UCI Digit	Reuters	VIS/NIR	Yale	Avg. rank
SC_{best}	$17.3_{\pm 0.3}$	$82.8_{\pm 0.0}$	$14.2_{\pm 0.2}$	$93.2_{\pm 0.0}$	$65.4_{\pm 1.2}$	7.0
LRR_{best}	$15.9_{\pm 0.3}$	$78.7_{\pm 1.2}$	$20.5_{\pm 0.9}$	$85.9_{\pm 2.7}$	$70.9_{\pm 1.1}$	6.6
SwMC	$16.3_{\pm 0.4}$	$84.3_{\pm 0.0}$	$17.4_{\pm 0.0}$	$96.1_{\pm 0.5}$	$65.6_{\pm 0.0}$	5.8
GMC	$12.9_{\pm 0.0}$	$68.5_{\pm 0.0}$	$16.5_{\pm 0.0}$	$98.5_{\pm 0.0}$	$68.9_{\pm 0.0}$	6.8
MVGL	$17.3_{\pm 0.2}$	$85.4_{\pm 1.3}$	$17.2_{\pm 1.1}$	$97.8_{\pm 0.4}$	$65.9_{\pm 0.0}$	5.0
LMSC	$17.5_{\pm 0.3}$	$78.8_{\pm 3.6}$	$16.9_{\pm 0.4}$	$\mathbf{98.8}_{\pm 3.9}$	$76.8_{\pm 1.9}$	4.6
CSMSC	$18.0_{\pm 0.2}$	$79.9_{\pm 0.0}$	$17.6_{\pm 0.2}$	$95.3_{\pm 0.2}$	$78.7_{\pm 0.4}$	4.4
DSS-MSC	$18.8_{\pm 0.2}$	$83.2_{\pm 0.2}$	$22.6_{\pm 0.8}$	$97.0_{\pm 0.6}$	$77.9_{\pm 1.2}$	3.2
CIMSC	$\mathbf{20.2}_{\pm 0.5}$	$\mathbf{87.3}_{\pm 0.0}$	$\mathbf{30.2}_{\pm 0.0}$	$98.5_{\pm 0.1}$	$\mathbf{86.7}_{\pm 0.1}$	**1.2**

- In most of cases, the proposed method outperforms its competitors including single-view methods and multi-view methods. To be specific, our method obtains very competitive results on the Reuters and Yale data sets. Take the Yale data set for an example, CIMSC makes the significant improvement over the second-best method by 10.2% and 8.8% in terms of Acc and NMI, respectively. As for the NUS-WIDE and UCI Digit data sets, although our method can not reach the same improvement as the Reuters and Yale data sets, it still ranks the best among all the testing methods. In summary, the comparison results have validated the superior of the proposed method, which also confirms the necessity of considering both multi-view consistency and inconsistency for improving multi-view subspace clustering.

4.3 Parameter Analysis and Convergence Analysis

In this subsection, we first analyze the effect of parameters λ_1, λ_2 and λ_3 on the clustering performance. In our experiments, we test these three parameters in the same range as $\{0.01, 0.1, 1, 10, 100\}$. Besides, we vary one parameter of three parameters and simultaneously fix the other two parameters in each experiment. Due to the limit of the paper length, we only report the NMI results of our method on three data sets, namely NUS-WIDE, UCI Digit and Yale.

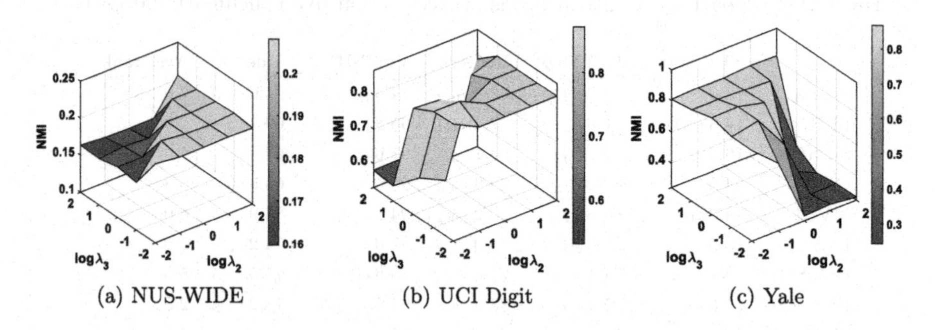

(a) NUS-WIDE (b) UCI Digit (c) Yale

Fig. 1. Parameter analysis on λ_2 and λ_3 on benchmark data sets by fixing λ_1.

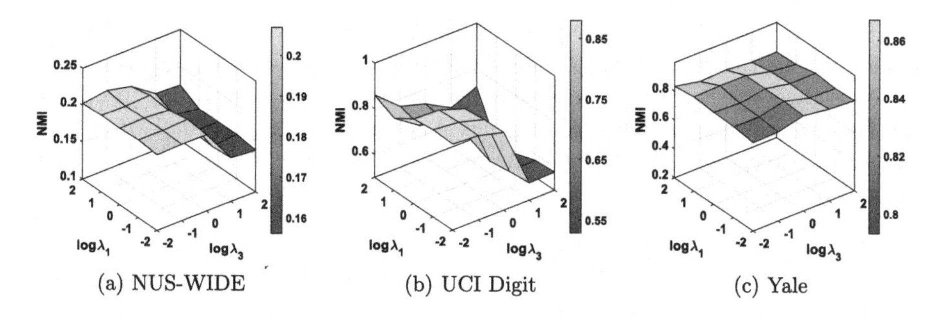

Fig. 2. Parameter analysis on λ_1 and λ_2 on benchmark data sets by fixing λ_3.

The corresponding results are shown in Fig. 1, Fig. 2 and Fig. 3, respectively. As can be seen from those figures, we have the following two findings. First, the proposed method is less sensitive to parameter λ_1, but relatively sensitive to the parameters λ_2 and λ_3. Second, when we vary the parameters λ_2 and λ_3 in the same range of (0.01,1), our method can obtain stable and relatively better

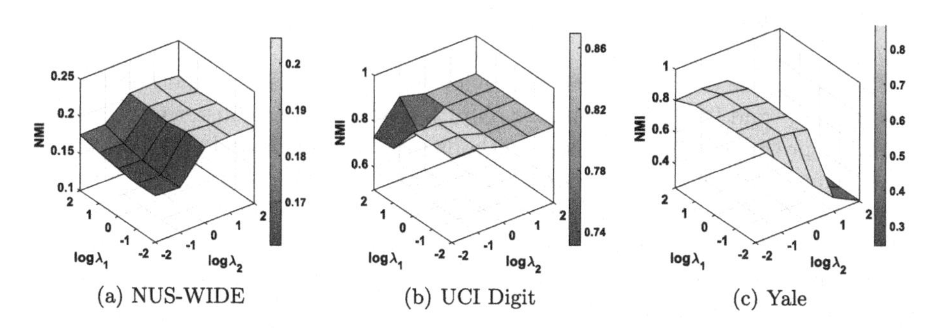

Fig. 3. Parameter analysis on λ_1 and λ_2 on benchmark data sets by fixing λ_3.

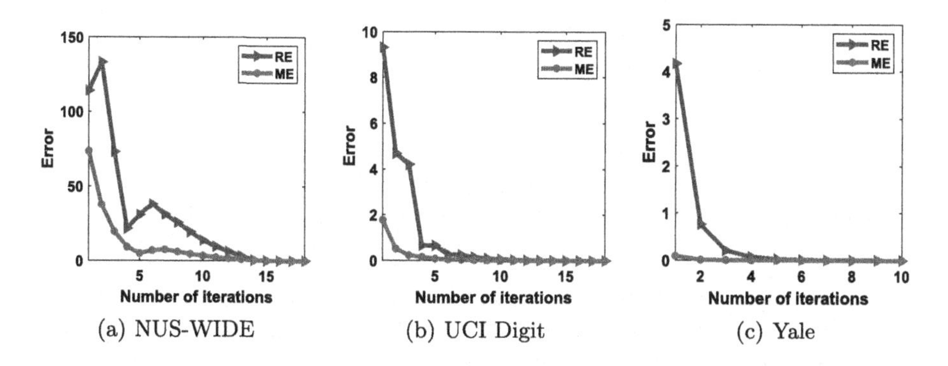

Fig. 4. Convergence analysis on three benchmark data sets.

clustering results. Hence, we suggest to set parameter λ_1 to a fixed value, and carefully tune the parameters λ_2 and λ_3 for different real-world applications.

Next, we further investigate the convergence property of the proposed method. To be specific, the convergence conditions in Algorithm 1 are determined by two error terms, namely the reconstruction error (RE) and the match error (ME). In particular, they can be defined as follows:

$$\mathrm{RE} = \sum_{v=1}^{V} ||\mathrm{X}^{(v)} - \mathrm{X}^{(v)}\mathrm{S} - \mathrm{X}^{(v)}\mathrm{C}^{(v)} - \mathrm{E}^{(v)}||_{\infty} < \varepsilon \tag{24}$$

and

$$\mathrm{ME} = ||\mathrm{S} - \mathrm{Z}||_{\infty} < \varepsilon \tag{25}$$

Figure 4 displays the convergence curves on three benchmark data sets. From this figure, we can see that our method converges relatively fast and almost within 20 iterations. Therefore, the proposed method is efficient and has strong convergence property in practice.

5 Conclusion

In this paper, we develop a new multi-view subspace clustering method termed Consistency- and Inconsistency-aware Multi-view Subspace Clustering. Different from the previous subspace-based methods, our method models the multi-view consistency and multi-view inconsistency under a unified framework. Furthermore, an adaptive view weighting strategy is incorporated into the proposed method, which aims to reduce the redundancy and determine the importance of multiple views. In particular, we design a novel strategy to construct the final affinity matrix from both multi-view consistent and inconsistent perspectives. This design ensures that the final affinity matrix better fits the real-world data sets. Finally, experimental results validate the effectiveness of the proposed method. In the future work, we plan to follow [5] and extend our method to handle a more challenging problem, namely multi-view data stream clustering problem.

References

1. Boyd, S., Parikh, N., Chu, E., Peleato, B., Eckstein, J., et al.: Distributed optimization and statistical learning via the alternating direction method of multipliers. Found. Trends® Mach. Learn. **3**(1), 1–122 (2011)
2. Cao, X., Zhang, C., Fu, H., Liu, S., Zhang, H.: Diversity-induced multi-view subspace clustering. In: Proceedings of the IEEE Conference on Computer Vision and Pattern Recognition, pp. 586–594 (2015)
3. Chen, M., Huang, L., Wang, C., Huang, D.: Multi-view clustering in latent embedding space. In: Thirty-Fourth AAAI Conference on Artificial Intelligence, pp. 3513–3520 (2020)

4. Huang, D., Wang, C.D., Wu, J., Lai, J.H., Kwoh, C.K.: Ultra-scalable spectral clustering and ensemble clustering. IEEE Trans. Knowl. Data Eng. (2019)
5. Huang, L., Wang, C.D., Chao, H.Y., Philip, S.Y.: Mvstream: multiview data stream clustering. IEEE Trans. Neural Networks Learn. Syst. (2019)
6. Huang, S., Kang, Z., Tsang, I.W., Xu, Z.: Auto-weighted multi-view clustering via kernelized graph learning. Pattern Recogn. **88**, 174–184 (2019)
7. Liang, W., et al.: Multi-view spectral clustering with high-order optimal neighborhood laplacian matrix. IEEE Trans. Knowl. Data Eng. (2020)
8. Liang, Y., Huang, D., Wang., C.D.: Consistency meets inconsistency: a unified graph learning framework for multi-view clustering. In: Proceedings of the IEEE International Conference on Data Mining (2019)
9. Liang, Y., Huang, D., Wang, C.D., Yu, P.S.: Multi-view graph learning by joint modeling of consistency and inconsistency. arXiv preprint arXiv:2008.10208 (2020)
10. Lin, K.-Y., Huang, L., Wang, C.-D., Chao, H.-Y.: Multi-view proximity learning for clustering. In: Pei, J., Manolopoulos, Y., Sadiq, S., Li, J. (eds.) DASFAA 2018. LNCS, vol. 10828, pp. 407–423. Springer, Cham (2018). https://doi.org/10.1007/978-3-319-91458-9_25
11. Liu, G., Lin, Z., Yan, S., Sun, J., Yu, Y., Ma, Y.: Robust recovery of subspace structures by low-rank representation. IEEE Trans. Pattern Anal. Mach. Intell. **35**(1), 171–184 (2012)
12. Lu, C.-Y., Min, H., Zhao, Z.-Q., Zhu, L., Huang, D.-S., Yan, S.: Robust and efficient subspace segmentation via least squares regression. In: Fitzgibbon, A., Lazebnik, S., Perona, P., Sato, Y., Schmid, C. (eds.) ECCV 2012. LNCS, vol. 7578, pp. 347–360. Springer, Heidelberg (2012). https://doi.org/10.1007/978-3-642-33786-4_26
13. Luo, S., Zhang, C., Zhang, W., Cao, X.: Consistent and specific multi-view subspace clustering. In: Thirty-Second AAAI Conference on Artificial Intelligence, pp. 3730–3737 (2018)
14. Nie, F., Cai, G., Li, X.: Multi-view clustering and semi-supervised classification with adaptive neighbours. In: Thirsty-First AAAI Conference on Artificial Intelligence, pp. 2408–2414 (2017)
15. Nie, F., Li, J., Li, X.: Self-weighted multiview clustering with multiple graphs. In: Proceedings of the Twenty-Sixth International Joint Conference on Artificial Intelligence, pp. 2564–2570 (2017)
16. Tang, C., et al.: Learning a joint affinity graph for multiview subspace clustering. IEEE Trans. Multimedia **21**(7), 1724–1736 (2019)
17. Wang, C.D., Lai, J.H., Philip, S.Y.: Multi-view clustering based on belief propagation. IEEE Trans. Knowl. Data Eng. **28**(4), 1007–1021 (2015)
18. Wang, H., Yang, Y., Liu, B.: GMC: graph-based multi-view clustering. IEEE Trans. Knowl. Data Eng. **32**, 1116–1129 (2020)
19. Wei, S., Wang, J., Yu, G., Domeniconi, C., Zhang, X.: Multi-view multiple clusterings using deep matrix factorization. In: AAAI, pp. 6348–6355 (2020)
20. Xia, R., Pan, Y., Du, L., Yin, J.: Robust multi-view spectral clustering via low-rank and sparse decomposition. In: Twenty-Eighth AAAI Conference on Artificial Intelligence, pp. 2149–2155 (2014)
21. Zhan, K., Niu, C., Chen, C., Nie, F., Zhang, C., Yang, Y.: Graph structure fusion for multiview clustering. IEEE Trans. Knowl. Data Eng. (2018)
22. Zhan, K., Zhang, C., Guan, J., Wang, J.: Graph learning for multiview clustering. IEEE Trans. Cybern. **99**, 1–9 (2017)
23. Zhang, C., et al.: Generalized latent multi-view subspace clustering. IEEE Trans. Pattern Anal. Mach. Intell. (2018)

24. Zhang, C., Fu, H., Hu, Q., Zhu, P., Cao, X.: Flexible multi-view dimensionality co-reduction. IEEE Trans. Image Process. **26**(2), 648–659 (2016)
25. Zhang, C., Hu, Q., Fu, H., Zhu, P., Cao, X.: Latent multi-view subspace clustering. In: Proceedings of the IEEE Conference on Computer Vision and Pattern Recognition, pp. 4279–4287 (2017)
26. Zhou, T., Zhang, C., Peng, X., Bhaskar, H., Yang, J.: Dual shared-specific multi-view subspace clustering. IEEE transactions on cybernetics (2019)

Discovering Collective Converging Groups of Large Scale Moving Objects in Road Networks

Jinping Jia, Ying Hu, Bin Zhao(✉), Genlin Ji, and Richen Liu

School of Computer and Electronic Information, Nanjing Normal University, Nanjing, China
{zhaobin,glji,richen.liu}@njnu.edu.cn

Abstract. Group pattern mining based on spatio-temporal trajectories have gained significant attentions due to the prevalence of location-acquisition devices and tracking technologies. Representative work includes convoy, swarm, travelling companion, gathering, and platoon. However, these works based on Euclidean space cannot handle group pattern discovery in non-planar space, such as urban road networks. In this paper, we propose a new group pattern, named converging, and its mining method in road networks. Unlike the aforementioned group patterns, a converging indicates that a group of moving objects converge from different directions for a certain time period. Motivated by this, we formalize the concept of a converging based on cluster containment relationship. Since the process of discovering convergings over large scale road network constrained trajectories is quite lengthy, we propose a density clustering algorithm based on road networks *(DCRN)* and a cluster containment join *(CCJ)* algorithm to improve the performance. Specifically, *DCRN* adopts the well-known filter-refinement-verification framework for efficiently identifying core points, which utilizes the upper bound property for ε-neighbourhood of point set on an edge to dramatically reduce the candidate core points. To process the neighbourhood queries efficiently, we further develop a vertex-neighbourhood based index, which precomputes the ε-neighbourhoods of all vertices, to facilitate neighbourhood queries of all points in road networks. In addition, to process the *CCJ* efficiently, we develop a signature tree based on road network partition index to organize the clusters in road networks hierarchically, which enable us to prune enormous unqualified candidates in an efficient way. Finally, extensive experiments with real and synthetic datasets show that our proposed methods achieve superior performance and good scalability.

Keywords: Converging pattern · Moving objects · Road networks

1 Introduction

Advances in location-acquisition devices and tracking technologies have made it possible to accumulate a large amount of spatio-temporal trajectories. Such data

C. S. Jensen et al. (Eds.): DASFAA 2021, LNCS 12682, pp. 307–324, 2021.
https://doi.org/10.1007/978-3-030-73197-7_21

provides opportunities for analysing the mobility patterns of moving objects. This problem has been extensively studied in the literature. Representative works include convoy [1], swarm [2], travelling companion [3], gathering [4], platoon [5], and so on. In spite of the significant contributions made by those work, they mainly focus on co-movement patterns [6,7]. However, we observe that there exists another common mobility pattern in real life, namely converging. A converging represents a group event that a group of moving objects gradually come to a target area from different directions and eventually form one group. Examples of convergings include traffic jams, celebrations, protests, and so on. Additionally, the existing works on mobility patterns mainly focus on group events in Euclidean space. Actually, most of them take place in urban areas, especially in road networks. For example, pedestrians walk along the streets, vehicles run on the roads, and trains run on the tracks. However, researches on mobility patterns in road networks are seldom studied in the literature. In this paper, we propose a new mobility pattern converging and its efficient mining algorithms to discover converging events in road networks.

The key property of a converging is membership variation, which is the crucial distinction compared to the existing patterns. Now we use Fig. 1 as an example to illustrate the converging pattern. There are six moving objects joining a converging event in a road network and forming one cluster in the end. Additionally, there are six clusters (i.e., c_0^1, c_0^2, c_0^3, c_1^1, c_1^2 & c_2^1) from time t_0 to t_2, which are denoted by c_j^i (ith cluster at time t_j). The clusters c_0^1 and c_0^2 move along the roads $\langle v_1, v_2 \rangle$ and $\langle v_3, v_2 \rangle$ respectively, merge into c_1^1 on the road $\langle v_2, v_5 \rangle$ at time t_1, and finally join the cluster c_2^1 with c_1^2 at t_2. Such set containment between two clusters is called *cluster containment match*, denoted by $\subseteq_c$, e.g., $c_1^1 \subseteq_c c_2^1$ & $c_1^2 \subseteq_c c_2^1$. In Fig. 1(b), we present all cluster containment matches of moving clusters in the form of a tree. The tree root c_2^1 represents a converging event of the participators $\langle o_1, o_2, o_3, o_4, o_5, o_6 \rangle$ in the road network.

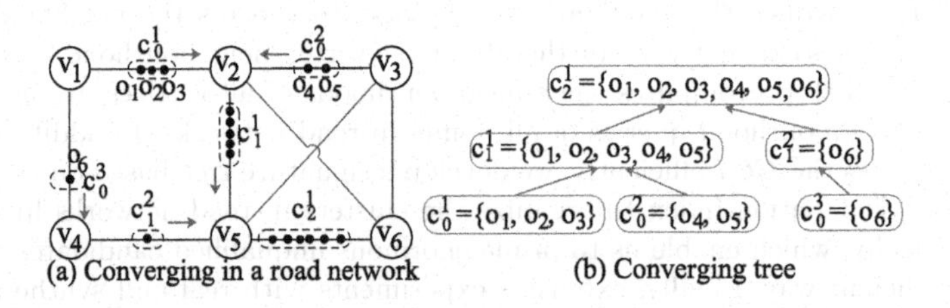

Fig. 1. Illustration of converging in road networks.

It is worth pointing out, the techniques of mobility patterns in Euclidean space cannot be applied for the same study in the settings of road networks. This is because, Euclidean space is a planar space while road network is essentially a non-planar graph. As shown in Fig. 1(a), road $\langle v_2, v_6 \rangle$ cross over road $\langle v_3, v_5 \rangle$ in the road network. Therefore, existing pattern mining methods in Euclidean space cannot distinguish the group events with different heights in urban space, e.g. pedestrian overpasses, elevated highways, and the subway network. Motivated by this, we focus on converging patterns in road networks. To the best of our knowledge, this is the first work which specifies on mobility pattern mining in road networks.

However, mining the converging patterns of large-scale moving objects in road networks is not an easy task, due to three challenges.

Modelling Method. How to define the concept of converging appropriately in order to intuitively capture the properties of the converging events. We need to deal with the increasing membership and widespread distribution in a group event, while the existing works discover moving groups with fixed membership and localized distribution. To deal with this challenge, we formalize the concept of converging using set containment and propose the converging definition based on cluster containment relationship *(CCR)*, which is able to accurately model the continuous increasing membership of converging events in urban areas.

Efficient Discovery Algorithms. How to efficiently discover convergings from large-scale trajectories in road networks. Discovering converging patterns involves two important tasks, namely *cluster discovery* and *cluster containment match*, which both have a huge search space and incur a high computation cost. As to *cluster discovery*, *DBSCAN* [8] is conventionally employed in existing group pattern mining to discover groups of moving objects. But such a popular solution involves high computational overhead since it incurs a great amount of range nearest neighbour queries. To keep the computation cost tractable, we develop a Density-based Clustering algorithm in Road Networks *(DCRN)* that adopts a Filter-Refinement-Verification *(FRV)* framework to enhance the performance of clustering objects in road networks. To improve the efficiency of *DCRN*, we further develop a vertex-neighbourhood based index, which pre-compute the neighbourhoods of all vertexes to facilitate neighbourhoods queries. As to *cluster containment match*, we propose a Cluster Containment Join *(CCJ)* algorithm to evaluate CCRs. To improve the performance of CCJ, we develop a Signature Tree based on Road Network Partition *(STRNP)* index to organize clusters in a road network hierarchically, which are able to speed up cluster containment queries. It is worth to note that, we only build the *STRNP* index once, and just update some of index nodes after dynamical updates of moving objects.

Scalability. Real life applications like traffic planning usually require mining algorithms to handle great variation of dataset size, since traffic flow varies greatly over time. However, most of the existing works in Euclidean space rely on an index structures based on spatial properties of moving objects for spatial query processing. This may result in the worse performance on scalability as the dataset size increases greatly. To tackle the scalability issue, we develop the aforementioned FRV and the spatial index structures based on edges of a road network, which enable our proposed method to scale much better even if large-scale trajectories come intensively. Extensive experiments based on both real and synthetic datasets validate the efficiency and scalability of our algorithms.

2 Overview

2.1 Basic Conception

The road network consists of intersections and roads. Taking intersections as vertices, and roads as edges, we model a road network as an undirected graph.

Definition 1 (Road Network). *A road network is an undirected graph* $G = (V, E)$*, where* $V = \{v_1, v_2, ..., v_{|V|}\}$ *is a set of vertices and* $E = \{e_1, e_2, ..., e_{|E|}\}$ *is a set of edges. A vertex in* V *is denoted by* $v = (lng, lat)$*, where* $v.lng$ *and* $v.lat$ *represent the longitude and latitude respectively. An edge* $e = (v_s, v_e)$ *in* E *has two endpoints* $e.v_s \in V, e.v_e \in V$*, and the weight* $e.len$ *is the length of* e*.*

Definition 2 (Location). *The location of an object* o *in road network* G *is defined as a point on an edge of* G*, denoted by* $o.p = (e, pos)$*, where* e *is the edge* o *located in and* pos *is the distance between* o *and* $e.v_s$*.*

Definition 3 (Road Segment). *A road segment is a section between two points on an edge, denoted by* $seg(p_1, p_2)$*, where* p_1, p_2 *are two boundary points. The length of* $seg(p_1, p_2)$ *is denoted by* $|seg(p_1, p_2)| = |p_1.pos - p_2.pos|$*.*

Location p on road segment $seg(p_1, p_2)$ is denoted by $p \in seg(p_1, p_2)$. The set of all objects on $seg(p_1, p_2)$ (edge e) is marked as $O(seg(p_1, p_2))$ $(O(e))$.

Definition 4 (Road Network Constrained Trajectory). *Given a set of moving objects* $O_{DB} = \{o_1, ..., o_m\}$*, a time domain* $T = \{t_1, ..., t_n\}$ *and a road network* G*, the trajectory of a moving object* $o \in O_{DB}$ *constrained by* G *is a finite sequence of timestamped locations* $o.traj = \langle p_{t_1}, ..., p_{t_n} \rangle$*, where* p_{t_i} *is the location of* o *sampled at* $t_i \in T$*. We mark all the trajectories constrained by* G *as* $Trj_{DB} = \{o_1.traj, ..., o_m.traj\}$*.*

Definition 5 (Network Distance). *Given a road network* G *and two locations* $p = (e, pos), p' = (e', pos')$*, the network distance between* p *and* p' *is denoted by* $d(p, p')$*, which is defined as follows: (1)* $d(p, p') \geq 0$*. (2) If* $e = e'$*, then* $d(p, p') = |pos' - pos|$*. Otherwise,* $d(p, p') = min\{d(p, v) + d(v, v') + d(v', p')\}$*, where* $v \in \{e.v_s, e.v_e\}, v' \in \{e'.v_s, e'.v_e\}$ *and* $d(v, v')$ *is the shortest distance between* v *and* v' *in* G*.*

To detect moving groups in road networks, we extend *DBSCAN* to discover clusters. We apply the idea of density-based clustering to cluster objects in road networks, and necessarily redefine some of the key notions as follows. *The* ε*-neighbourhood* of a point p is defined by $N_\varepsilon(p) = \{q | d(p, q) \leq \varepsilon\}$, where ε is a distance threshold and $d(p, q)$ is the network distance between p and q. A point p in road networks is a *core point* if $N_\varepsilon(p)$ contains at least $minPts$ points, denoted by $|N_\varepsilon(p)| \geq minPts$. A point p is *directly density-reachable* from a point q w.r.t. $\varepsilon, minPts$ if $p \in N_\varepsilon(q)$ and q is a core point. p is a border point if it is directly density-reachable from a core point but $N_\varepsilon(p) < minPts$. A point p is *density-reachable* from a point q w.r.t. $\varepsilon, minPts$ if there is a chain of points $\langle x_1, ..., x_k \rangle$, $x_1 = q, x_k = p$ such that x_{i+1} is directly density-reachable from x_i. Then a point p is said to *density-connected* to a point q w.r.t. $\varepsilon, minPts$ if there is a point x such that both p and q are density-reachable from x w.r.t. $\varepsilon, minPts$.

Definition 6 (Cluster). *Given a database of points D, a cluster c w.r.t. $\varepsilon, minPts$ is a non-empty subset L of D satisfying the following conditions: (1) Maximality. $\forall p, q$, if $p \in L$ and q is density-reachable from p w.r.t. $\varepsilon, minPts$, then $q \in L$. (2) Connectivity. $\forall p, q \in L$, p is density-connected to q w.r.t. $\varepsilon, minPts$.*

A cluster consisting of the locations of moving objects at timestamp t is said to be a snapshot cluster c_t. The set of snapshot clusters at t_i is denoted by $C_t = \{c_t^1, c_t^2, ...\}$, and the database of snapshot clusters is denoted by $C_{DB} = \{C_{t_1}, C_{t_2}, ..., C_{t_n}\}$. The snapshot cluster will be abbreviated to cluster if no ambiguity can be caused.

Definition 7 (Cluster Containment Relationship). *Given two clusters c_1 and c_2, c_2 has a cluster containment relationship r with c_1 if each moving object of c_1 is also a moving object of c_2, i.e., c_2 contains c_1, denoted by $r = c_1 \subseteq_c c_2$.*

Definition 8 (Converging Tree). *A converging tree tr is a tree that satisfies the following properties: (1) Each node of tr represents a cluster. The parent and children are clusters at consecutive timestamps, and the parent has cluster containment relationships with its children. (2) The height of tr is the length of the longest path from the root to a leaf. A tree of a single node has a height of 0.*

As mentioned in Sect. 1, a converging pattern should be able to represent the process of moving objects gradually gathering over a period of time. Therefore, we define the converging pattern from two aspects. First, we use the converging tree to represent the containment relationship of clusters at consecutive timestamps and the higher the height of the converging tree the longer the duration. Second, a converging pattern also requires a certain scale, that is, it typically involves a relatively large number of moving objects.

Definition 9 (Converging). *A converging tree tr is called a converging that satisfies the following requirements: (1) The height of tr is not less than the lifetime threshold θ_t. (2) The number of objects in $tr.root$ is not less than the scale threshold θ_m. A converging tr is called a closed converging if there isn't any cluster at next timestamp containing the root node of tr.*

2.2 Problem Definition

Given a road network G, a trajectory set of moving objects Trj_{DB} constrained by G, a time domain T, a distance threshold ε, a number threshold $minPts$, a lifetime threshold θ_t and a scale threshold θ_m, our goal is to find all the closed convergings.

2.3 Framework

In this section, we present our framework for converging pattern mining in a set of trajectories constrained by road networks. Our framework consists of three phases: cluster discovery, converging tree generation, and converging detection. Algorithm 1 outlines this process. In the first phase, we perform a density-based clustering algorithm in road networks *(DCRN)* on the trajectories at each timestamp of T to find all the clusters C_{DB} (Line 1–5). We will detail this process in Sect. 3. The second phase aims to generate the converging trees based on cluster containment relationships *(CCRs)*. We perform a cluster containment join *(CCJ)* algorithm on any two cluster sets at consecutive timestamps and return a set of *CCRs* R_{DB} (Line 6–11); then we organize the *CCRs* in chronological order to construct all the closed converging trees TR_{DB} (Line 12–17). We will detail this process in Sect. 4. In the third phase, we verify all closed converging trees in terms of the requirements in Definition 9 to get all the converging patterns P (Line 18–22). The details of this phase are omitted due to space limitation.

Algorithm 1: Converging Pattern Mining (CPM)

```
   Input  : G, Trj_DB, T, ε, minPts, θ_t, θ_m
   Output: P
 1 C_DB ← ∅ ;
 2 for i = 1 to |T| do
 3   | S_i ← SnapshotPointSet(Trj_DB, t_i) ;
 4   | C_i ← DensityClustering(G, S_i, ε, minPts);
 5   | C_DB ← C_DB ∪ {C_i};
 6 R_DB ← ∅ ;
 7 for i = 2 to |T| do
 8   | foreach q ∈ C_{i−1} do
 9   |   | r ← SearchContainmentMatch(q, C_i);
10   |   | R_i ← R_i ∪ {r} ;
11   | R_DB ← R_DB ∪ {R_i} ;
12 TR_DB ← ∅;
13 for i = 1 to |T| − 1 do
14   | foreach r ∈ R_i do
15   |   | tr ← BuildConvergingTree(R_DB, t_i) ;
16   |   | TR_i ← TR_i ∪ {tr} ;
17   | TR_DB ← TR_DB ∪ {TR_i} ;
18 P ← ∅;
19 for i = 1 to |T| − 1 do
20   | foreach tr ∈ TR_i do
21   |   | if IsValidConvergingPattern(tr, θ_t, θ_m) then
22   |   |   | P ← P ∪ {tr} ;
23 return P;
```

3 Density-Based Algorithm in Road Networks

In this section, we propose a density-based clustering algorithm in road network *(DCRN)* to discovery clusters. The basic idea is to identify core points and border points among the locations of objects, and then group them into clusters based on the connectivity in Definition 6. It is detailed in Algorithm 2. The filter-refinement-verification *(FRV)* framework is adopted to identify core points. In the filter step, an upper bound for ε-neighbourhoods of objects on every edge is calculated, which is used to excludes a large portion of invalid points (Lines 4–5). In the refinement and verification steps, the candidate points are further refined based on the properties of density-reachable and the core points are verified by ε-neighbourhood computation (Lines 6–7). After that, the core points are grouped into clusters in terms of the connectivity (Line 9), and every border point is assigned to the cluster of its corresponding core point (Lines 10–12).

3.1 Core Points Identification

In this subsection, we discuss the method of core points identification. To identify the core points, the ε-neighbourhoods of points need to be calculated, which is supported by methods of range query. However, in the existing clustering algorithms, the neighbourhoods are computed one by one, which is sensitive to the scale of objects. To tackle this problem, we design a method of core points identification based on the road network, which adopts the FRV framework to batch process the neighbourhood computation of points in unit of edge.

Algorithm 2: *DensityClustering* Function

Input : $G(V, E)$: road network, S: snapshot point set, ϵ, $minPts$
Output: C: resulting snapshot cluster set

```
1  C ← ∅ ;
2  S_core ← ∅; // S_core is a core point set
   // core points identification
3  foreach e ∈ E do
4      L ← GetPointSetInEdge(e, S) ;
5      L' ← Filter(L, S, ε, minPts) ;
6      L'' ← Refinement(L', S, ε, minPts);
7      L_core ← Verification(L'', S, ε, minPts);
8      S_core ← S_core ∪ L_core ;
9  C ← GenerateCluster(S_core, ε); // generate the clusters
10 S_border ← FindBorderPoint(C, S − S_core, ε, minPts); // find border points
11 UpdateCluster(C, S_border); // assign border points to clusters of C
12 return C;
```

Filter Step. Given an edge $e = (v_1, v_2)$, the upper bound of its ε-neighbourhood is denoted by $F(e) = |\bigcup_{p \in L_e} N_\varepsilon(p)|$, where L_e is a set of points on e. If $F(e) < minPts$, there is no core points on e. $F(e)$ is computed in two cases.

Case 1: e.len$\leq \varepsilon$. As shown in Fig. 2(a), $F(e) = |N_\varepsilon(v_1) \cup N_\varepsilon(v_2)|$, where v_1 and v_2 are two vertices of e.

Case 2: e.len $> \varepsilon$. We add some *virtual vertices* in the midpoints of road segments on e recursively until the length of each divided section doesn't exceed ε. After that, e can be handle as case 1. Note that all the added virtual vertices are considered as usual vertices in the subsequent processing. As shown in Fig. 2(b), v' is a virtual vertex on e, and $|e.seg(v_1, v')| < \varepsilon$, $|e.seg(v', v_2)| < \varepsilon$.

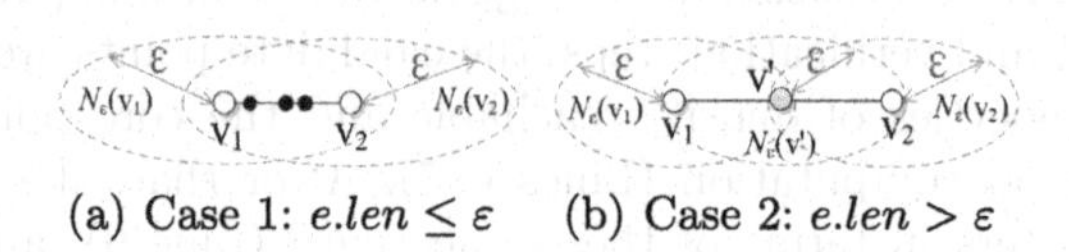

(a) Case 1: $e.len \leq \varepsilon$ (b) Case 2: $e.len > \varepsilon$

Fig. 2. Two cases of the upper bound.

Refinement and Verification Steps. Undoubtedly, the candidate points have been pruned effectively in the filter step, but for edges with upper bound more than *minPts*, the ε-neighbourhood of each point still need checking. A naive method is to compute the neighbourhoods one by one. For example, in Fig. 3 $p_1, p_2, p_3, p_4, p_5, p_6$ is verified in sequence. Obviously, this method is inefficiency. Therefore, we propose a method of checking points based on the properties of density-reachability.

Given a point sequence $\langle p_1, p_2, ..., p_n \rangle (n > 1, |seg(p_1, p_n)| \leq \varepsilon)$, two boundary points p_1, p_n are verified by computing their ε-neighbourhoods.

(1) If both p_1 and p_n are core points, other points on $seg(p_1, p_n)$ belong to the same cluster with p_1 and p_n *(Property* 1*)*. The points between p_1 and p_n are exempt from verification in clustering.
(2) If neither p_1 nor p_n is core point and $|N_\varepsilon(p_1) \cup N_\varepsilon(p_n)| < minPts$, points on $seg(p_1, p_n)$ are not core points *(Property* 2*)* and exempt from verification.
(3) Otherwise, the point sequence is divided into two sub-sequences by its median point, and they are checked respectively. This process is recursively performed until there is no unverified point.

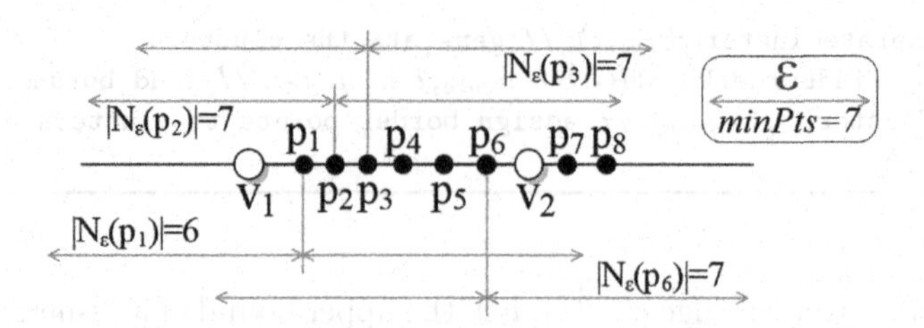

Fig. 3. Illustration of refinement and verification method of a point sequence.

As shown in Fig. 3, for segment $seg(p_1, p_6)$, point p_1 is not a core point and p_6 is, so that the median point p_3 is verified. p_3 is a core point, thus p_3, p_4, p_5, p_6 belong to one cluster and p_4, p_5 are exempt from verification. For another subsequence $\langle p_1, p_2, p_3 \rangle$, only p_2 needs to be verified as a core point.

Property 1. Given a segment shorter than ε, if its two boundary points are core points, then all points on this segment belong to the same cluster.

Proof. If two boundary points are core points and the segment is shorter than ε, then two boundary points are density-reachable from each other. They belong to the same cluster. Moreover, other points on this segment are all density-reachable from the boundary points. According to the maximality in Definition 6, all the points belong to the same cluster.

Property 2. Given a segment $seg(p_1, p_2)$ shorter than ε,if p_1, p_2 aren't core points, and $|N_\varepsilon(p_1) \cup N_\varepsilon(p_2)| < minPts$, no one point on $seg(p_1, p_2)$ is a core point.

Proof. $\forall q \in seg(p_1, p_2)$, if $|seg(p_1, p_2)| < \varepsilon$, then $N_\varepsilon(q) \subset N_\varepsilon(p_1) \cup N_\varepsilon(p_2)$. If $|N_\varepsilon(p_1) \cup N_\varepsilon(p_2)| < minPts$, then $|N_\varepsilon(q)| < minPts$, thus q is not a core point. Therefore, all points on $seg(p_1, p_2)$ are not core points.

3.2 ε-Neighbourhood Computation Method

In core points identification, two types of ε-neighbourhoods (i.e., ε-neighbourhoods of vertices and moving objects) need to be computed. However, the objects move dynamically, which leads to a great challenge to compute the ε-neighbourhoods of moving objects in the whole time domain. Observe that the ε-neighbourhoods of vertices are unchanging. Moreover, the ε-neighbourhoods of objects can be represented by that of vertices. With these observation, we propose an ε-neighbourhood computation method that represents the ε-neighbourhoods of points on an edge by that of its two vertices.

Next, we introduce our ε-neighbourhood computation method with Fig. 4. Given a road network and seven moving objects shown in Fig. 4(a), whose information is detailed in Fig. 4(d). Assume that $\varepsilon = 5, p = (e_1, 3)$, the ε-neighbourhood of object o_4 located in point p is illustrated in Fig. 4(b). It is computed in three parts, including the object set on edge e_1, the ε'-neighbourhood of $v_1(\varepsilon' = 2)$ and the ε''-neighbourhood of $v_2(\varepsilon'' = 4)$, i.e., $N_\varepsilon(p) = O(e_1) \cup N_{\varepsilon'}(v_1) \cup N_{\varepsilon''}(v_2) = \{o_3, o_4\} \cup \{o_2, o_3\} \cup \{o_3, o_4, o_5, o_6, o_7\} = \{o_2, o_3, o_4, o_5, o_6, o_7\}$.

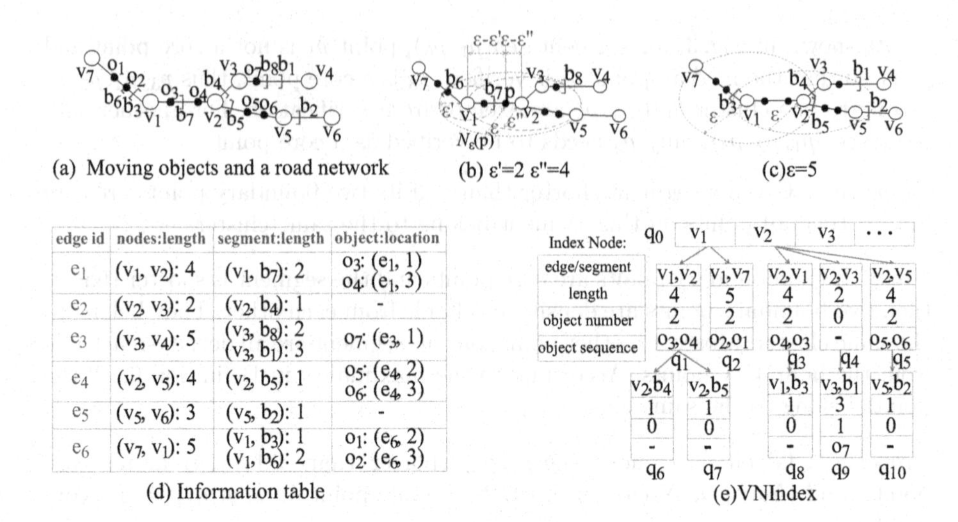

edge id	nodes:length	segment:length	object:location
e_1	(v_1, v_2): 4	(v_1, b_7): 2	o_3: $(e_1, 1)$ o_4: $(e_1, 3)$
e_2	(v_2, v_3): 2	(v_2, b_4): 1	-
e_3	(v_3, v_4): 5	(v_3, b_8): 2 (v_3, b_1): 3	o_7: $(e_3, 1)$
e_4	(v_2, v_5): 4	(v_2, b_5): 1	o_5: $(e_4, 2)$ o_6: $(e_4, 3)$
e_5	(v_5, v_6): 3	(v_5, b_2): 1	-
e_6	(v_7, v_1): 5	(v_1, b_3): 1 (v_1, b_6): 2	o_1: $(e_6, 2)$ o_2: $(e_6, 3)$

Fig. 4. An example of the network-based ε-neighbourhood computation method.

3.3 The VNIndex for ε-Neighbourhood Computation

To process the neighbourhood queries efficiently, an index is required to retrieve the neighbourhoods of points. *G-Tree* [9], *G*-Tree* [10] and *V-Tree* [11] are the latest methods of range queries in road networks. However, they are not scalable in clustering. Therefore, we design a vertex-neighbourhood based index *(VNIndex)*, which records the pre-computed ε-neighbourhoods of all vertices to facilitate neighbourhood queries of all points in road networks.

***VNIndex* Construction.** VNIndex records the ε-neighbourhood of a vertex with a structure, called *Neighbourhood Substructure (NS)*. An *NS* consists of two parts, i.e., a vertex ID and several index nodes built for the adjacent edges of the vertex. Each index node contains the endpoints, length, object number and object sequence of the edge. The index nodes in an *NS* are connected based on the topology of the road network. Additionally, VNIndex organizes the IDs of vertices by a classical B+-Tree or a Hashing index to rapidly find the corresponding *NSs*.

For example, the *VNIndex* ($\varepsilon = 5$) w.r.t. Fig. 4(a) is shown in Fig. 4(e). Easy to see that, in Fig. 4(c), the ε-neighbourhood of v_1 consists of four parts, i.e., edges $e(v_1, v_7), e(v_1, v_2)$ and segments $seg(v_2, b_4)$, $seg(v_2, b_5)$. As shown in Fig. 4(e), the index nodes built for these edges and segments are q_1, q_2, q_6, q_7. Vertex v_1 connects with index nodes q_1, q_2, and q_1 connects with q_6, q_7. Consequently, v_1, q_1, q_2, q_6 and q_7 constitute the *NS* of v_1.

***VNIndex*-Based Neighbourhood Query.** The ε-neighbourhood of a vertex is directly obtained from its *NS*. The ε-neighbourhood of a moving object is queried from the *NSs* of its two adjacent vertices.

In the example in Fig. 4(b), $N_\varepsilon(p) = O(e_1) \cup N_{\varepsilon'}(v_1) \cup N_{\varepsilon''}(v_2)$. For $O(e_1)$, the *NS* of v_1 is visited in *VNIndex* and a set of neighbours $\{o_3, o_4\}$ is obtained from index node q_1. Similarly, $O(e_1)$ can be queried from the *NS* of v_2 with the same result. For $N_{\varepsilon'}(v_1), \varepsilon' = 2$, the *NS* of v_1 is visited again and two neighbour sets $\{o_3\}, \{o_2\}$ are obtained from q_1 and q_2. As a result, $N_{\varepsilon'}(v_1) = \{o_2, o_3\}$. In the same way, $N_{\varepsilon''}(v_2) = \{o_3, o_4, o_5, o_6, o_7\}$ is obtained from the *NS* of v_2. Synthesizing the above results, $N_\varepsilon(p) = O(e_1) \cup N_{\varepsilon'}(v_1) \cup N_{\varepsilon''}(v_2) = \{o_2, o_3, o_4, o_5, o_6, o_7\}$.

***VNIndex* Update.** *VNIndex* need to be updated when the locations change. Instead of re-constructing or revising its structure, *VNIndex* only adjusts the object sequences if new locations are reported at next timestamp.

4 Cluster Containment Join

To generate the converging trees, we present a cluster containment join *(CCJ)* algorithm to find the cluster containment relationships *(CCRs)* in this section.

Definition 10 (Cluster Containment Join (CCJ)). *Given two collections of clusters C_1 and C_2, the cluster containment join $C_1 \bowtie_{\subseteq} C_2$ returns all cluster pairs $(q, s) \in C_1 \times C_2, q \in C_1, s \in C_2$ such that $q \subseteq_c s$.*

4.1 Cluster Containment Join Algorithm

The basic idea of *CCJ* is to take each cluster $q \in C_1$ as a query cluster and to find a cluster $s \in C_2$ that $q \subseteq_c s$. Intuitively, rapidly confirming the containment result of q within C_2 is the key of *CCJ*, namely the *SearchContainmentMatch* function in Algorithm 1. Two algorithms, i.e., nested-loop based clustering containment join *(NLCCJ)* and breadth-first based cluster containment join *(BFCCJ)*, can be used for *CCJ*. *NLCCJ* is designed based on the nested-loop strategy and is very costly with a time complexity of $O(|C_1| * |C_2|)$. While *BFCCJ* adopts the breadth-first search strategy to match clusters due to the spatial proximity of objects moving in a short time. However, it is inefficient to identify the mismatches. If C_2 does not have any cluster containing q, the whole network will be traversed and all the clusters in C_2 will be compared with q in a total of $O(|C_2|)$ times. To rapidly confirm the *CCR* of the clusters, we propose an index for efficiently pruning candidate clusters.

4.2 Signature Tree Based on Road Network Partition

In this subsection, we propose a signature tree based on road network partition *(STRNP)* index, which hierarchically organizes clusters based on the partition to road networks. Meanwhile, *STRNP* index also use the signature technique to rapidly confirm *CCRs*. We detail *STRNP* index in the following three aspects.

***STRNP* Index Construction.** *STRNP* index is a balanced binary tree, which organizes cluster sets into a hierarchical structure. Each leaf node of *STRNP* index contains a cluster set and a corresponding signature. For internal node, it only contains a signature obtained by OR operation of signatures in its children. *STRNP* index is constructed in a bottom-up way and used to reduce the candidate clusters for efficient queries of *CCRs*.

The cluster sets in leaf nodes come from the partitions of road networks, which are obtained by graph partitioning algorithms, e.g., multilevel k-way partitioning algorithm [12]. The signatures are generated by signature techniques.

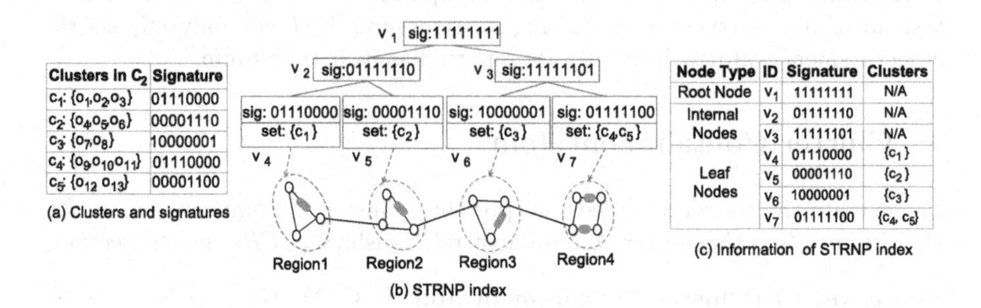

Fig. 5. An example of *STRNP* index.

For example, in Fig. 5, given a cluster set $C_2 = \{c_1, c_2, c_3, c_4, c_5\}$ with thirteen moving objects, whose information is shown in Fig. 5(a). These clusters distribute in four road network partitions (i.e., *Region*1, *Region*2, *Region*3, *Region*4). Consequently, an *STRNP* index is constructed as shown in Fig. 5(b).

Cluster Search Based on *STRNP* Index. Based on the *STRNP* index, we can prune the candidate clusters to rapidly confirm the *CCRs*. Specifically, we adopt a depth-first way to operate their signatures in *STRNP* index. When a query cluster q arriving, *STRNP* starts exploring from its root node with an *OR* operation between signatures of q and each of its index node p, denoted by $q.sig$ and $p.sig$ respectively. If $q.sig \vee p.sig = p.sig$, then *STRNP* continues to explore the child nodes of p; otherwise, it explores p's sibling nodes. Moreover, if no sibling nodes can be probed, it backtracks to the parent node of p to explore other nodes. When q is exploring a leaf node, an *OR* operation of signatures is still executed and all clusters in this leaf node are considered as candidates if it meets the requirement of the signature operation. But the correctness of *CCRs* needs further verifying because of the false positive problem in signature techniques. Above process of cluster exploration and verification is recursively performed until no index node needs probing.

Continuing with the example in Fig. 5, given a query cluster $q_1 = \{o_4, o_6\}$, $q_1.sig = 00001010$. To find a cluster in C_2 containing q_1, *STRNP* index is explored from node v_1. Since $q_1.sig \vee v_1.sig = v_1.sig$, v_2 and v_3 are explored. After signature operations, the child nodes v_4, v_5 are probed and eventually v_5 is confirmed as a valid leaf node. Next, the clusters in v_5 is further verified to get the correct *CCR* $q_1 \subseteq_c c_2$. On the contrary, given a query cluster $q_2 = \{o_1, o_7\}, q_2.sig = 01000001$, it is confirmed that a mismatch happens to q_2.

***STRNP* Index Update.** When clusters are discovered at a new timestamp, the *STRNP* index needs updating. No matter how variable the moving objects are, the structure of *STRNP* index does not need to be revised since it is constructed based on the road network partitions. Only the signatures in index nodes need revising with the moving of objects in a relatively large area. Once *STRNP* index is constructed, its structure can be reused throughout the time domain. Additionally, the storage space for *STRNP* index almost remains unchanged regardless of the number of moving objects and clusters.

5 Experiment

We conduct experiments on computers with Linux OS, Intel Xeon E5-2620 v2(2.10 GHz) CPU and 32G RAM. All the algorithms are implemented in Java.

Datasets. We evaluate our proposals on both real and synthetic datasets. The taxi dataset *(Taxi)* tracks real trajectories of 13,518 taxies sampled per minute in Shanghai. And the road network with 262,764 nodes and 286,591 edges is extracted from OpenStreetMap. Meanwhile, we also generate a synthetic dataset *(Brinkhoff)* with 100,000 trajectories via the *Brinkhoff* generator, which is widely used in the efficiency experiments of group pattern works [1,2,5,7]. These trajectories are generated on the real road network of Oldenburg with 55,994 nodes and 61,911 edges. Every location is sampled per minute when the objects move along roads at random but reasonable speeds and directions.

Evaluation Methods. We evaluate the effectiveness, efficiency and scalability of our proposals. **(1)** ***Effectiveness.*** To demonstrate the effectiveness of convergings, we present an online demo system to visualize the mined converging patterns. Additionally, we statistically analyse the spatio-temporal distribution of the mining results for a further validation, since the converging events of taxis exist in cities throughout one day with significant regularities. **(2)** ***Efficiency.*** We evaluate the efficiency of converging mining algorithms in phases. For the *Cluster Discovery* phase, we compare the running time of clustering algorithms based on different range query methods (i.e., *V-Tree* [11], *G*-Tree* [10] and *Proposal*) w.r.t. density-based clustering parameters (i.e., $\varepsilon, minPts$), where V-Tree and G*-Tree are the state-of-the-art range query methods in road networks. For the *Cluster Containment Join* phase, we compare the running time of different *CCJ* algorithms (i.e., *NLCCJ*, *BFCCJ* and *Proposal*) with same clustering results. As baselines, *NLCCJ* is a naive join algorithm based on nested-loop strategy and *BFCCJ* is a classical join algorithm based on breath-first search strategy. **(3)** ***Scalability.*** We conduct experiments on moving objects of different sizes to evaluate the scalability of the clustering and *CCJ* algorithms.

5.1 Effectiveness

As a proof-of-concept, we implement the converging mining algorithms integrated in a demo system. As shown in Fig. 6(a), several convergings are discovered from one-hour data in Taxi with parameters set as $\varepsilon = 50, minPts = 5, \theta_t = 5, \theta_m = 30$. Two representative convergings are displayed. As shown in Fig. 6(b), taxis were returning to a taxi storage yard after work. As shown in Fig. 6(c), taxis were converging at a parking lot in Shanghai Pudong International Airport for waiting guests. Besides, more convergings can be discovered using this demo system, which has been deployed on a virtual machine of Alibaba Cloud and can be accessed at http://203.195.219.39:8080/CPM/index.html.

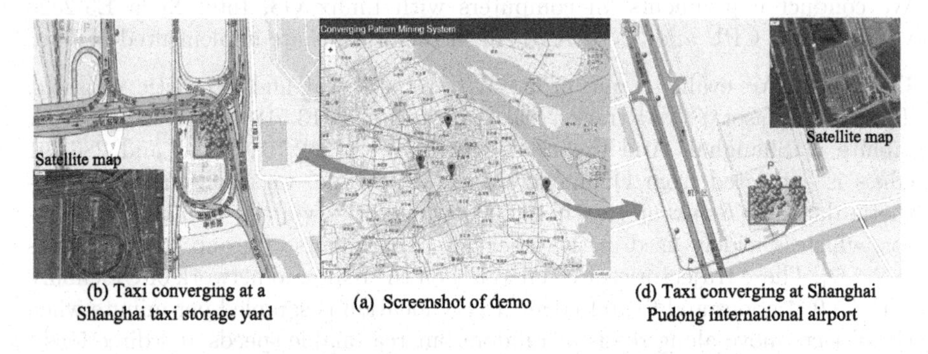

Fig. 6. Visualization of the mined convergings.

Additionally, we statistically analyse the spatio-temporal distribution of the convergings in terms of four kinds of regions of interests *(ROIs)*, i.e., transportation, leisure, office and residence. The statistical result of convergings in one day is shown in Fig. 7. The parameters are set as $\varepsilon = 200, minPts = 2, \theta_t = 10, \theta_m = 10$. Obviously, there are more convergings late at night and early in the morning. This is because, the activities range of people is relatively dense during these periods. Especially, the convergings in the residence and leisure *ROIs* increase significantly. The main reason is that, a large number of people go home or to entertainment venues by taxi, since the public transportation is out of service during these periods. In addition, we display the spatial distribution of the convergings in two typical time periods, i.e., 2am to 4am and 11am to 1pm. It is easy to see, the convergings mostly concentrate in the suburbs of Shanghai early in the morning. This is because, most taxis change shifts or stay for a long time in the suburbs during this period. On the contrary, the driving range of taxis is relatively wide due to the dispersion of passengers in the city, resulting in a scattered distribution in various regions of convergings.

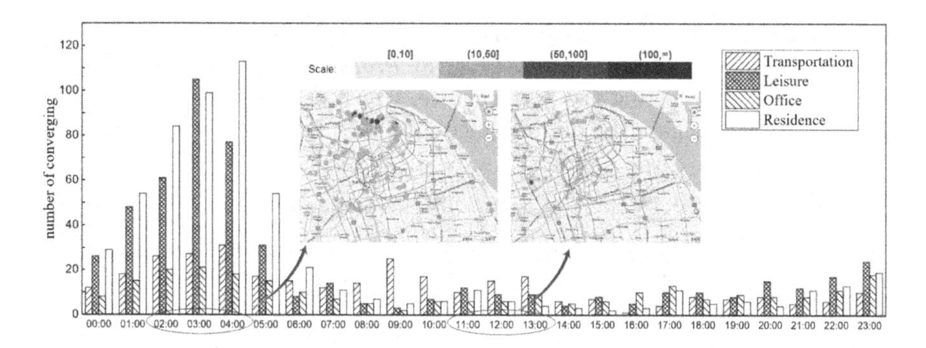

Fig. 7. The statistical result of convergings in Taxi.

5.2 Efficiency

In this subsection, we study the efficiency of our proposals. Specifically, we evaluate the performance of clustering and *CCJ* algorithms with different parameter settings (i.e., $\varepsilon = 100$, **150**, $200, 250, 300$ and $minPts =$ **2**, $3, 4, 5, 6$, where the default settings are in bold). All experiments are conducted on both real-life *Taxi* dataset and synthetic *Brinkhoff* dataset with one-hour data respectively.

Performance of Clustering Algorithms. We evaluate the clustering algorithms with different range query methods: a)*Brinkhoff*; b)*G*-Tree;* c)*Proposal*.

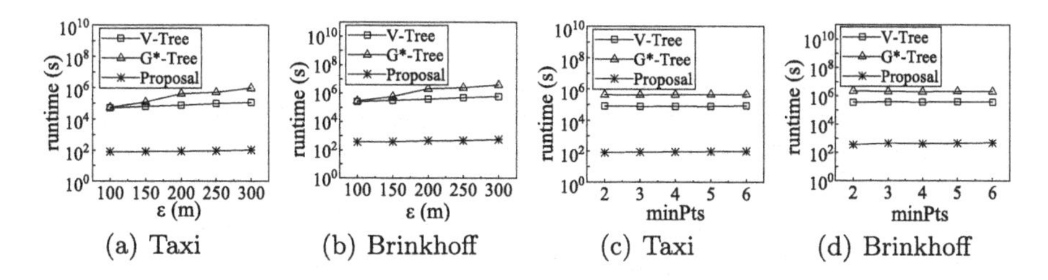

Fig. 8. Running time of clustering algorithms.

Efficiency w.r.t. ε. Figure 8(a)(b) show the running time of clustering algorithms w.r.t. ε. It is obvious that all algorithms incur higher cost as ε increases due to the enlargement of query ranges. Our proposal is several orders of magnitude faster than two baselines because invalid objects are filtered by the *FRV* framework and the neighbourhoods are rapidly queried through our scalable *VNIndex*.

Efficiency w.r.t. minPts. Figure 8(c)(d) present the performances of algorithms w.r.t. *minPts*. We can see that all algorithms are relatively insensitive to *minPts*. This is because, the computational cost relates to the number of neighbours within ε-neighbourhood rather than the number threshold.

Performance of *CCJ* Algorithms. We also evaluate the performances of *CCJ* algorithms: a) *NLCCJ*; b) *BFCCJ*; c) *Proposal*. The number of clusters is the main factor affecting the performance, which depends on ε and $minPts$. Therefore, we investigate the performances of *CCJ* algorithms w.r.t. $\varepsilon, minPts$.

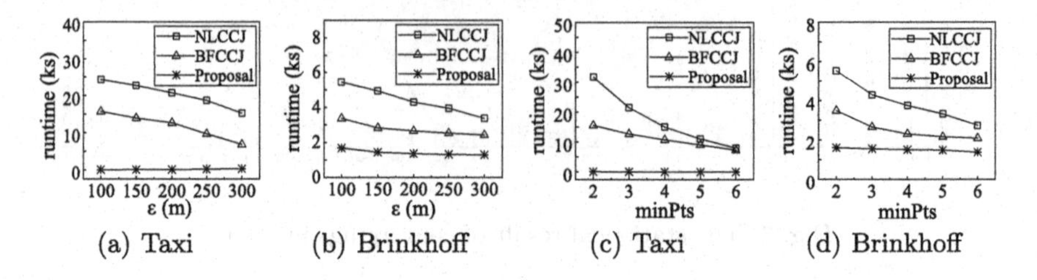

Fig. 9. Running time of *CCJ* algorithms.

Efficiency w.r.t. ε. Figure 9(a)(b) show the performances w.r.t ε. We can see our proposal is significantly more efficient than the baselines. This is because *STRNP* index effectively reduces the candidate clusters and identifies the mismatches.

Efficiency w.r.t. $minPts$. Figure 9(c)(d) present the performances w.r.t. $minPts$. As the previous experiment, our proposal has superior performance than two baselines. With the increase of $minPts$, all algorithms incur less time cost since fewer clusters are discovered.

5.3 Scalability

We evaluate the scalability of the algorithms. As shown in Fig. 10, our proposals outperform the baselines in both clustering and *CCJ*. We can also see that the performance gap between them is widened as more objects are involved. This is because, our *VNIndex* and *STRNP* index are designed based on road networks instead of objects, which can be reused over the entire time domains. Moreover, vast invalid objects and clusters are pruned to achieve better scalability.

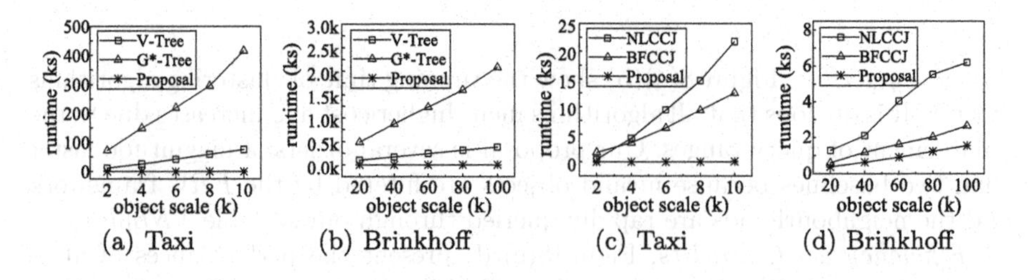

Fig. 10. Scalability of clustering and *CCJ* algorithms.

6 Related Work

Density-Based Clustering in Spatial Networks. Density-based clustering on moving objects is common in group pattern mining for discovering groups of objects. *DBSCAN* [8] is widely used for group pattern mining in Euclidean space. However, it is difficult to be directly applied in spatial networks. Yiu et al. [13] first propose an $\varepsilon-$Link algorithm to find the $\varepsilon-$neighbourhoods by expanding networks around the query points. They discover clusters correctly but inefficiently because queries are issued for all points and the graph is traversed less systematically. Chen et al. [14] propose a cluster block *(CB)* to cluster moving objects by the continuous maintenance of *CBs* and periodical construction of clusters. Nevertheless, the sampled positions of moving objects is highly variable so that it suffers from expensive overhead. Different from these works, we focus on efficient queries of neighbourhoods and *DCRN* achieves better scalability.

Query Processing in Spatial Networks. Spatial query is a basic technique of pattern mining, which has been extensively studied. Papadias et al. [15] first propose the Euclidean restriction and network expansion framework to process queries in spatial networks. A most commonly-used paradigm is to partition spatial networks into regions and to retrieve the pre-computation information in the processing of queries. *NPI* [16] partitions the road network into equal-size grids and maintains the distance data by a bound matrix. *ROAD* [17] organizes the road network as a hierarchy of *Rnets* and prunes the search space to enhance network traversal and object lookup. Recently, a height-balanced index *G-Tree* [9] is proposed based on a recursive partition to road networks. *G*-Tree* [10] builds shortcuts between some leaf nodes based on *G-Tree*. *V-tree* [11] works on efficient neighbour queries of moving objects anywhere in road networks. All aforementioned works address the issue of single query processing, which is unscalable to process a large number of queries. In contrast, our method processes range queries in batches, which achieves better efficiency and scalability.

7 Conclusion

In this paper, we propose a novel group pattern *Converging* in road networks. To efficiently discover convergings, we first propose a density-based clustering algorithm to discover clusters, which adopts an *FRV* framework and a scalable *VNIndex* to effectively identify core points. To generate converging trees, we propose an *STRNP* index to improve the efficiency of *CCJ*. The experiment results show the effectiveness, efficiency and scalability of our proposals.

Acknowledgement. This study was supported by NSFC41971343, NSFC61702271, NSF of Jiangsu Province BK20200725 and the Postgraduate Research Innovation Program of Jiangsu Province KYCX201258.

References

1. Orakzai, F., Calders, T., Pedersen, T.B.: k/2-hop: Fast mining of convoy patterns with effective pruning. PVLDB **12**(9), 948–960 (2019)
2. Li, Z., Ding, B., Han, J., Kays, R.: Swarm: mining relaxed temporal moving object clusters. PVLDB **3**(1), 723–734 (2010)
3. Tang, L., et al.: On discovery of traveling companions from streaming trajectories. In: ICDE 2012, pp. 186–197 (2012)
4. Zheng, K., Zheng, Y., Yuan, J., Shang, S.: On discovery of gathering patterns from trajectories. In: ICDE 2013, pp. 242–253 (2013)
5. Li, Y., Bailey, J., Kulik, L.: Efficient mining of platoon patterns in trajectory databases. DKE **100**, 167–187 (2015)
6. Fan, Q., Zhang, D., Wu, H., Tan, K.: A general and parallel platform for mining co-movement patterns over large-scale trajectories. PVLDB **10**(4), 313–324 (2016)
7. Chen, L., Gao, Y., Fang, Z., Miao, X., Jensen, C.S., Guo, C.: Real-time distributed co-movement pattern detection on streaming trajectories. PVLDB **12**(10), 1208–1220 (2019)
8. Ester, M., Kriegel, H., Sander, J., Xu, X.: A density-based algorithm for discovering clusters in large spatial databases with noise. In: SIGKDD 1996, pp. 226–231 (1996)
9. Zhong, R., Li, G., Tan, K., Zhou, L., Gong, Z.: G-tree: an efficient and scalable index for spatial search on road networks. TKDE **27**(8), 2175–2189 (2015)
10. Li, Z., Chen, L., Wang, Y.: G*-tree: an efficient spatial index on road networks. In: ICDE 2019, pp. 268–279 (2019)
11. Shen, B., et al.: V-Tree: efficient kNN search on moving objects with road-network constraints. In: ICDE 2017, pp. 609–620 (2017)
12. Karypis, G., Kumar, V.: Multilevel k-way partitioning scheme for irregular graphs. JPDC **48**(1), 96–129 (1998)
13. Yiu, M.L., Mamoulis, N.: Clustering objects on a spatial network. In: SIGMOD 2004, pp. 443–454 (2004)
14. Chen, J., Lai, C., Meng, X., Xu, J., Hu, H.: Clustering moving objects in spatial networks. In: Kotagiri, R., Krishna, P.R., Mohania, M., Nantajeewarawat, E. (eds.) DASFAA 2007. LNCS, vol. 4443, pp. 611–623. Springer, Heidelberg (2007). https://doi.org/10.1007/978-3-540-71703-4_52
15. Papadias, D., Zhang, J., Mamoulis, N., Tao, Y.: Query processing in spatial network database. VLDB **2003**, 802–813 (2003)
16. Sun, W., Chen, C., Zheng, B., Chen, C., Liu, P.: An air index for spatial query processing in road networks. TKDE **27**(2), 382–395 (2015)
17. Lee, K.C.K., Lee, W., Zheng, B., Tian, Y.: ROAD: a new spatial object search framework for road networks. TKDE **24**(3), 547–560 (2012)

Efficient Mining of Outlying Sequential Behavior Patterns

Yifan Xu[1], Lei Duan[1](✉), Guicai Xie[1], Min Fu[1], Longhai Li[1], and Jyrki Nummenmaa[2]

[1] School of Computer Science, Sichuan University, Chengdu, China
{xuyifan,guicaixie,fu_min,lilonghai}@stu.scu.edu.cn, leiduan@scu.edu.cn
[2] Tampere University, Tampere, Finland
jyrki.nummenmaa@tuni.fi

Abstract. Sequential patterns play an important role when observing behavior. For instance, the daily routines and practices of people can be characterized by sequences of activities. These activity sequences, in turn, can be used to find exceptional and changed behavior. Observing students' behavior changes is an effective approach to find indications of mental health problems, and changes in an elderly person's daily activities may indicate a weakening health condition. With the availability of behaviour sequential events, outlierness analysis of behavior sequences has been established as a meaningful research problem. This paper considers the mining of outlying behavior patterns (OBP) from sequential behaviors. After discussing the challenges of OBP mining, we present OBP-Miner, a heuristic method that computes OBPs by incorporating various pruning techniques. Empirical studies on two real-world datasets demonstrate that OBP-Miner is effective and efficient.

Keywords: Contrast sequence data mining · Outlying behavior pattern · Outlierness analysis

1 Introduction

Behavior can often be described as a sequence of activities. Examples include elderly people's daily activities and students' weekly routines. In both cases there are sequences from repeating time intervals, such as days or weeks. The abnormal behavior may show as different from a person's reference group and different from past typical behavior.

Such sequence data mining for abnormal behavior has important practical applications, for example, students' mental health. According to the numbers reported from NAMI[1], 72% of students experienced a mental health crisis on

This work was supported in part by the National Natural Science Foundation of China (61972268), the Sichuan Science and Technology Program (2020YFG0034), and the Academy of Finland (327352).

[1] http://www.nami.org.

C. S. Jensen et al. (Eds.): DASFAA 2021, LNCS 12682, pp. 325–341, 2021.
https://doi.org/10.1007/978-3-030-73197-7_22

campus. Moreover, among those students who experienced a mental health crisis, 64% and 90% of students stopped attending college and committed suicides respectively. However, 34% of students' college didn't know about their crisis, which means that their condition went unnoticed. In addition, the mental illness, such as depression and anxiety, can present different symptoms, depending on the person. The illness would change how those students function day-to-day, and typically for a period of time. Common symptoms include changes in sleep, changes in appetite, loss of energy, lack of interest in activities, changes in movement, etc. Based on this, observing those students' behavior changes to recognize if these students have mental issues is regarded as a breakthrough. Understanding the students' behavior changes can: (1) identify students who might have mental issues; (2) explore potential factors that cause-related behavior changes to provide some insights into solving issues.

This is a challenging task as we not only need to determine whether a student has behavior changes but also find a specific time window when there is a change in student behavior compared to that. For those students who experienced a mental health crisis, the behavior sequence of these students should be inconsistent with their groups (same dormitory or academy), and their historical behavior sequence. For instance, such a student might get depressed when he/she is unwilling to interact with others, but often sits alone or skips classes and stays in the dormitory all day. Meanwhile, students with depression may suffer from sleep disorders or loss of appetite. As a result, their sleeping and dietary habits during this period are different from their previous behavior sequences.

The above analysis indicates that discovering such behavior sequence based changes for college students is important for identifying potential students who have behavior changes and the specific period of time. We note that existing approaches are lacking in addressing the above important needs.

This leads us to a novel data mining problem. Comparing the current behavior, historical behavior, and the current behavior of the reference group we can detect outlying behavior. We say that an entity's *outlying behavior pattern* (OBP) is a behavior sequence that is different from the reference group (of similar entities), and has a change compared to the entity's historical behavior. That is to say, *outlying behavior pattern* describes the entity's behavior which may indicate the abnormality. Mining OBPs can obtain more information details to describe the outlying aspect of the entity's behavior.

To tackle the problem of mining OBPs, several technical challenges need to be addressed. First, a comprehensive and complete way is needed to represent sequential behaviors. Naturally, behavior should not only reflect the characteristics of real behavior but also conform to the actual situation of behavior change. Second, OBP is a novel representation of outlying behavior, which requires a metric of the outlierness degree for measuring the outlierness of a sequence with respect to different candidate OBPs. Third, we also need an approach to efficiently discover OBPs.

The main contributions of our work are as follows: (1) We introduce the problem of OBPs mining from behavior events, which uses OBPs to evaluate and explain the outlierness of a behavior sequence. (2) We design a heuristic method to discover OBPs of query behavior. (3) We evaluate our method by conducting an empirical study on both two real behavior datasets of students, which demonstrates that our OBPs mining algorithm is effective.

The rest of this paper is organized as follows. We review related work in Sect. 2, and formulate the problem of mining OBPs in Sect. 3. In Sect. 4, we present the framework of our OBP-Miner, and discuss the critical techniques in OBP-Miner. We report a systematic empirical study in Sect. 5, and conclude the paper in Sect. 6.

2 Related Work

Our work is related to three aspects of existing work on outlying aspect mining, distinguishing sequential pattern mining, and education data mining.

2.1 Outlying Aspect Mining

Outlying aspect mining, a topic to discover subspaces that describe how a query object stands out from the rest of objects, is the most related to this study. Among the different taxonomies which have been proposed, methods of outlying aspect mining can be categorized into three main groups, *score-and-search* based approaches, *feature-selection* based approaches, and hybrid approaches.

In *score-and-search* based approaches, *outlying aspect mining* approaches are based on measures of outlierness degree. HOS-Miner, proposed in [17], is the first work to solve the problem of outlying aspect mining, which employs a distance-based scoring measure to evaluate the outlierness degree of a given query. Nguyen *et al.* [11] presented two scoring functions that were dimensionally unbiased to compare subspaces of different dimensionalities. Duan *et al.* [4] proposed a model called OAMiner, using the rank of the probability density of an object in a subspace to measure the outlierness of the object in the subspace. The above methods are all applied to relational objects, each of which is composed of a fixed number of numerical or categorical attributes.

In *feature-selection* based approaches, the problem of outlying aspect mining is tackled as the typical feature selection problem. Liu *et al.* [9] proposed COIN, a method to explain the abnormality of outliers spotted by detectors based on outlierness score, contextual description of its neighborhoods, and attributes that contribute to the abnormality. Gupta *et al.* [5] provided explanations of outlying behaviors in multi-dimensional real-valued datasets by discovering pairwise feature plots from chosen feature subspaces. Siddiqui *et al.* [14] mined a sequence of features as an outlying subspace where the order indicates the importance with respect to causing a high outlier score for a given outlier. One issue with these approaches is that they mined sets of attributes to distinguish a given outlier detected by an existing outlier detection algorithm from other data objects.

In hybrid approaches, the problem of outlying aspect mining is addressed using a combination of *score-and-search* and *feature-selection* based approaches.

Nguyen *et al.* [10] proposed a hybrid framework called OARank, utilizing the strength of both *score-and-search* based approaches and *feature-selection* based approaches, to mine outlying aspects in very large datasets. Wang *et al.* [15] utilize the rank defined by the average probabilistic strength (*aps*) of a sequence pattern in a sequence to measure the outlierness of the sequence.

2.2 Distinguishing Sequential Pattern Mining

Distinguishing sequential pattern (DSP) mining, a task to discover patterns that best describe the significant differences between two classes of sequences, is useful in many applications. Duan *et al.* [3] investigated the problem of mining distinguishing customer focus sets from customer reviews, which can be used for online shopping decision support. Zheng *et al.* [19] developed a CSP-tree based structure to client sequential behavior analysis. Zhu *et al.* [20] introduced an approach to characterize and detect personalized and abnormal behaviors of internet users by mining user-related rare sequential topic patterns from document streams.

Our work is related to both outlying aspect mining and distinguishing sequential pattern mining, but there still exist differences between them. Most existing works discover distinguishing temporal event patterns, where each event has an associated timestamp. Our work aims to find such events with time intervals, and thus can distinguish events in different time intervals. What's more, events in each time interval is a multi-dimensional value whereby existing outlying aspect mining works cannot be applied to our work.

2.3 Education Data Mining

Education data mining has been approached from a large number of different perspectives. Hang *et al.* [6] presented a method, named EDHG, analyzing students' check-in behavior for point-of-interest prediction. Li *et al.* [8] proposed SPDN, a model with the aim of predicting students' performance in the course by analyzing students' online learning activities and internet access activities. Zhang *et al.* [18] and Cao *et al.* [2] indicated that students' academic performance is related to their behavior patterns with the analysis of smart cards. Yang *et al.* [16] proposed an algorithm called EPARS, which is devised to predict students at risk by modeling online and offline learning behaviors. Jimenez *et al.* [7] and Ameri *et al.* [1] found that the risk of dropping out is closely related to students' academic behavior. Peng *et al.* [12] discovered students' internet addiction by predicting students' daily time online. Resnik *et al.* [13] aimed to predict neuroticism and depression in college students. Zhu *et al.* [21] studied the procrastination of students based on the library borrowing records. However, to the best of our knowledge, none of the existing methods focused on the changes in students' daily behaviors, which may cause various problems, such as psychological problems. Thus, in this paper, we focus on mining *outlying behavior patterns* (OBPs) of behavior sequences, which is meaningful and necessary in real-life scenarios.

Table 1. Events of a student in a day

Student ID	Time	Event type	Value
40067	10–22 08:15:27	Breakfast	4.5
	10–22 08:30:35	Shopping	2.5
	10–22 12:16:14	Lunch	12.6
	10–22 12:26:20	Shopping	4.0
	10–22 12:35:25	Library	1.0
	10–22 20:39:27	Shopping	27.0

3 Problem Definition

Let $\mathcal{E}$ be the set of all possible *event types*. Examples of *event types* include "eating" or "reading" etc. A behavior is a triple (t, e, v), where $e \in \mathcal{E}$ is an *event type*, t is the timestamp when e occurred and v is a real number which indicates the magnitude of the behavior. Note that the value of v depends on the corresponding event type. For example, for "eating", v indicates the consumption value of "eating" at timestamp t, and for "reading", we use 1 or 0 to represent whether a student goes to the library or not.

A time interval w is an interval $w = [w.t_s, w.t_e)$, where $w.t_s$ is the start timestamp and $w.t_e$ is the end timestamp.

A student can have different event types in a time period, and the same event type can occur multiple times. Assume that we have a time interval w and a set of behaviors E which happen during w, we aggregate behaviors by creating an n-dimensional ($n = |\mathcal{E}|$) aggregation vector $A(w) = [value_1, value_2, \ldots, value_n]$, where

$$value_i = \sum_{v' \in \{v' | \forall (t', e', v') \in E, t' \in w, e' = i\}} v'.$$

Example 1. The following is an example about the aggregation vector A. Table 1 shows the four types of events $< Breakfast, Shopping, Lunch, Library >$ that occur in a student's day. We set the time interval to 2 h, from 1:00 to 23:00, thus we can get three aggregation vectors, *i.e.*, $A([7, 9)) = [4.5, 2.5, 0.0, 0.0]$, $A([11, 13)) = [0.0, 4.0, 12.6, 1.0]$, and $A([19, 21)) = [0.0, 27.0, 0.0, 0.0]$.

Daily behavior of a student can be equally divided into m behavior sequences at different time intervals. Then a student's daily behaviors can be aggregated into a *daily behavior aggregation*, which is a sequence of (*time interval, aggregation vector*) pairs: $S =< (w_1, A(w_1)), (w_2, A(w_2)), \ldots, (w_m, A(w_m)) >$. Each pair $S[i] = (w_i, A(w_i))$ is the i-th element of S $(1 \leq i \leq m)$. The *length* of S, denoted by $|S|$, is the number of elements in S. For $S[i]$, we use $S[i].w$ to denote the time interval and $S[i].A$ to denote the aggregation vector.

A method is needed to estimate the similarity of aggregation vectors. Standardized Euclidean distance is used as a distance metric for aggregation vectors, interpreting the vectors as points in n-dimensional space. Note that we only compare those aggregation vectors belonging to the identical time interval.

Given a set of aggregation vectors $\mathcal{A}$, we can calculate the mean and standard deviation of each dimension of aggregation vectors using all aggregation vectors in $\mathcal{A}$. For $A(w)$, $A'(w) \in \mathcal{A}$, the measurement of the similarity between $A(w)$ and $A'(w)$ is denoted by

$$
\begin{aligned}
&Sim(A(w), A'(w)) \\
&= \frac{1}{n}\sqrt{\sum_{k=1}^{n}\left(\frac{(value_k - m_k) - (value'_k - m_k)}{s_k}\right)^2} \\
&= \frac{1}{n}\sqrt{\sum_{k=1}^{n}\left(\frac{value_k - value'_k}{s_k}\right)^2},
\end{aligned}
\tag{1}
$$

where m_k and s_k is the mean and the standard deviation of the k-th dimension of aggregation vectors in $\mathcal{A}$, respectively. The smaller the value of $Sim(A(w), A'(w))$, $A(w)$ is more similar to $A'(w)$.

A *behavior pattern* is a sequence of pairs $P = <(w_1, A(w_1)), (w_2, A(w_2)), \ldots, (w_n, A(w_n))>$. Similarly, we denote the time interval as $P[i].w$, aggregation vector as $P[i].A$, and the length of P as $|P|$.

Given a similarity threshold α, a *daily behavior aggregation* S and a *behavior pattern* P, we say that the P *matches* S, denoted by $P \subset S$, if there exist integers $1 \leq k_1 \leq k_2 < \ldots < k_{|P|} \leq |S|$, such that

(1) $P[i].w = S[k_i].w$, and
(2) $Sim(P[i].A, S[k_i].A) < \alpha$ for $1 \leq i \leq |P|$.

Let $\mathcal{D} = \{S_1, S_2, \ldots, S_T\}$ be the set of *daily behavior aggregations* in T days. Given a support threshold γ, the support of P in $\mathcal{D}$ is the fraction of *daily behavior aggregations*, denoted by

$$Sup(P, \mathcal{D}) = \frac{|\{S \in \mathcal{D} | P \subset S\}|}{|\mathcal{D}|} \tag{2}$$

Then, a behavior pattern P is *frequent* in $\mathcal{D}$ if and only if $Sup(P, \mathcal{D}) > \gamma$.

Example 2. Here is an example of the computation of the similarity and the support. Table 2 shows a set of *daily behavior aggregations* of a student in 3 d $\mathcal{D} = \{S_1, S_2, S_3\}$. There is a sequence of interest of time intervals of a day that begins at 7:00 and ends at 21:00. Let similarity threshold $\alpha = 0.5$, support threshold $\gamma = 0.3$. For pattern $P = <([11, 13), [9, 6, 10]), ([17, 19), [8, 5, 9])>$, $P[1].w = S_1[3].w$, $Sim(P[1].A, S_1[3].A) = 0.151 < 0.5$, $P[2].w = S_1[6].w$, $Sim(P[2].A, S_1[6].A) = 0.137 < 0.5$, so $P \subset S_1$. $P[1].w = S_2[3].w$, $Sim(P[1].A, S_2[3].A) = 0.880 > 0.5$, so $P \not\subset S_2$. $P[1].w = S_3[3].w$, $Sim(P[1].A, S_3[3].A) = 0.473 < 0.5$, $P[2].w = S_3[6].w$, $Sim(P[2].A, S_3[6].A) = 1.205 > 0.5$, so $P \not\subset S_3$. Then, $Sup(P, \mathcal{D}) = \frac{1}{3} > 0.3$. Thus, P is *frequent* in $\mathcal{D}$.

Table 2. A set of *daily behavior aggregations*

Set	ID	Daily behaviors							
		Index	1	2	3	4	5	6	7
		Time interval	[7, 9)	[9, 11)	[11, 13)	[13, 15)	[15, 17)	[17, 19)	[19, 21)
$\mathcal{D}$	S_1	Aggregation vector	[5, 0, 0]	[0, 5, 0]	[10, 5, 11]	[0, 0, 10]	[0, 0, 0]	[9, 6, 9]	[0, 0, 0]
	S_2	Aggregation vector	[6, 3, 0]	[0, 5, 0]	[9, 0, 12]	[0,0, 14]	[2, 0, 0]	[7, 5, 0]	[0, 3, 0]
	S_3	Aggregation vector	[0, 0, 0]	[0, 0, 0]	[6, 6, 15]	[5, 9, 0]	[0, 0, 0]	[0, 0, 0]	[0, 0, 0]

Given a student dataset, we can divide it into two parts: a *current* dataset, denoted by $\mathcal{D}_c$, and a *history* dataset, denoted by $\mathcal{D}_h$. Both $\mathcal{D}_c$ and $\mathcal{D}_h$ denote a set of *daily behavior aggregations*. $\mathcal{D}_h$ consists of data of behaviors that took place before behaviors in $\mathcal{D}_c$, and for reasonable results, they should be comparable by nature, *e.g.*, if $\mathcal{D}_c$ contains weekdays data, then there should also be weekdays data in $\mathcal{D}_h$, and we suppose that data on weekdays and data on weekends are not comparable. Note that $\mathcal{D}_c$ and $\mathcal{D}_h$ both have a fixed maximum length (*e.g.*, one month).

We use W_h to represent the changeable time window of $\mathcal{D}_h$. That is to say, W_h is a period time of $\mathcal{D}_h$. Formally, we denote $W_h = [W_h.s, W_h.e]$, where $W_h.s$ is the start day and $W_h.e$ is the end day (e.g., [0, 14] indicates the 0–14 days of *history* dataset). Then $\mathcal{D}^u_{W_h}$ indicates the set of *daily behavior aggregations* during the time window W_h.

For a student u and his groups U (*e.g.*, students with the same dormitory or academy), the set of *daily behavior aggregations* of u in $\mathcal{D}_c$ is denoted by $\mathcal{D}^u_c$, the *set of daily behavior aggregations* of u in $\mathcal{D}_h$ is denoted by $\mathcal{D}^u_h$, the set of *daily behavior aggregations* of U in $\mathcal{D}_c$ is denoted by $\mathcal{D}^U_c$. We focus on the task of *outlying behavior pattern mining*, and formalize some definitions as follows.

Definition 1. *Given three sets of daily aggregation vectors,* $\mathcal{D}^u_c$, $\mathcal{D}^u_h$ *and* $\mathcal{D}^U_c$*, a history time window* W_h*, the outlying score of* P *targeting* $\mathcal{D}^u_c$ *against* $\mathcal{D}^u_h$ *and* $\mathcal{D}^U_c$*, denoted by* $oScore(P, W_h)$*, is*

$$oScore(P, W_h) = \frac{((Sup(P, \mathcal{D}^u_c) - Sup(P, \mathcal{D}^u_{W_h})) + (Sup(P, \mathcal{D}^u_c) - Sup(P, \mathcal{D}^U_c)))}{2} \tag{3}$$

Definition 2. (*Problem Definition*). *Given three sets of daily aggregation vectors,* $\mathcal{D}^u_c$, $\mathcal{D}^u_h$ *and* $\mathcal{D}^U_c$*, a behavior pattern* P*, and a history time window* W_h*. A tuple* (P, W_h) *is an outlying behavior pattern* (*OBP*) *of* P *if:*

(1) (*positive score*) $oScore(P, W_h) > 0$;
(2) (*score maximality*) *There does not exist another* W'_h *satisfying Condition* (1), *and*
  (*i*) $oScore(P, W'_h) \geq oScore(P, W_h)$, *or*
  (*ii*) $oScore(P, W'_h) = oScore(P, W_h)$ *and* $|W'_h| < |W_h|$.

The problem of **outlying behavior pattern mining** is to find the OBPs of candidate behavior patterns.

Table 3 lists the frequently used notations of this paper.

Table 3. Summary of notations

Notation	Description
$\mathcal{E}$	the set of all possible *event types*
w	time interval of the behavior pattern
S	a *daily behavior aggregation*
$\mathcal{D}$	a set of *daily behavior aggregations*
$Sim(A(w), A'(w))$	similarity between $A(w)$ and $A'(w)$
$Sup(P, \mathcal{D})$	support of pattern P in $\mathcal{D}$
$\mathcal{D}_c^u$, $\mathcal{D}_h^u$	student's *current*, *history* datasets resp
W_h	*history time window*
$oScore(P, W_h)$	outlying score for pattern P against W_h
α, γ	the similarity, support thresholds resp

4 Design of OBP-Miner

4.1 Framework

As defined in Definition 2, an OBP consists of a behavior pattern and its corresponding *history time window*. In brief, the OBP-Miner algorithm is divided into the following two steps in an iterative manner: (1) generating a candidate behavior pattern P, (2) for P, finding the *history time window* W_h that maximizes $oScore(P, W_h)$. In each iteration, OBP-Miner keeps the collection of OBPs discovered so far.

For the sake of efficiency, there are two critical points when designing OBP-Miner. First, how to avoid generating useless candidate behavior patterns (Sect. 4.2). Second, for each candidate behavior pattern, how to explore all possible *history time windows* efficiently (Sect. 4.3).

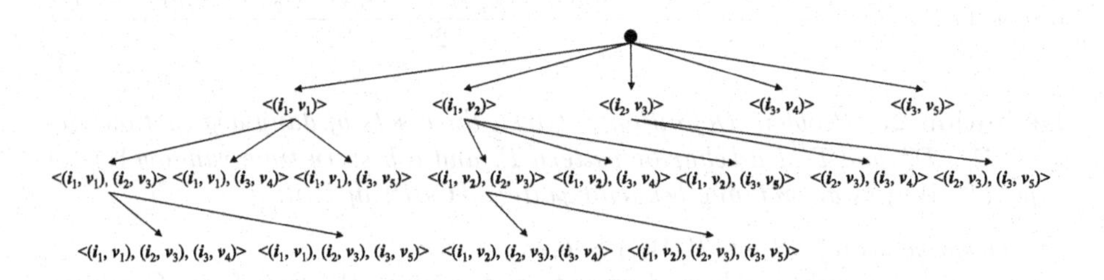

Fig. 1. An illustration of a set enumeration tree

4.2 Candidate Behavior Pattern Generation

Considering behavior patterns with fewer occurrences do not reflect well the behaviors of a student in *current* time, for each student u, we first find frequent behavior patterns in $\mathcal{D}_c^u$.

By scanning $\mathcal{D}_c^u$, we can get a set of (*time interval, aggregation vector*) pairs $\mathcal{P} = \{(\boldsymbol{i}, \boldsymbol{v}) | \exists (w, A(w)) \in S, S \in \mathcal{D}_c^u, w = \boldsymbol{i}, A(w) = \boldsymbol{v}, \forall 1 \leq j \leq |\boldsymbol{v}|\}, \boldsymbol{v}[j] \neq null\}$. To ensure that OBP-Miner can find patterns with the largest support in $\mathcal{D}_c^u$, we design a set enumeration tree to enumerate all possible behavior patterns in a systematic way. We first sort time intervals of pairs in $\mathcal{P}$ in ascending order. Figure 1 shows an example of a set enumeration tree that enumerates all patterns over $\mathcal{P} = \{(\boldsymbol{i}_1, \boldsymbol{v}_1), (\boldsymbol{i}_1, \boldsymbol{v}_2), (\boldsymbol{i}_2, \boldsymbol{v}_3), (\boldsymbol{i}_3, \boldsymbol{v}_4), (\boldsymbol{i}_3, \boldsymbol{v}_5)\}$, where each $\boldsymbol{i}_j$ priors to $\boldsymbol{i}_{j+1}$ $(1 \leq j < 3)$.

OBP-Miner generates candidate behavior patterns by traversing the enumeration tree in a depth-first manner. It is time-consuming to traverse all nodes in the pattern enumeration tree. Fortunately, Theorem 1 demonstrates the monotonicity of $Sup(P, \mathcal{D}_c^u)$ with respect to $\mathcal{D}_c^u$.

Theorem 1. *Given a set of daily behavior aggregations $\mathcal{D}$ and a similarity threshold α, we have $Sup(P, \mathcal{D}) \geq Sup(P', \mathcal{D})$ for all patterns P and P', provided that P is a subsequence of P'.*

Given a support threshold γ, Theorem 1 leads us to a useful pruning rule, which allows us to terminate the depth-first traversal of an entire branch at the current node.

Pruning Rule 1. *For pattern P, if $Sup(P, \mathcal{D}) < \gamma$, for each pattern P' such that $P \subset P'$, P' can be pruned.*

It makes sense to find patterns whose support is large early, so that the pruning methods give bigger impact. By this observation, we first compute the support of all patterns each containing exactly one single pair in $\mathcal{P}$, and sort all pairs in the descending order of support. We apply the following rule to prune pairs.

Pruning Rule 2. *A pair $(\boldsymbol{i}, \boldsymbol{v})$ in $\mathcal{P}$ can be removed from $\mathcal{P}$ without loosing any valid frequent patterns if $Sup((\boldsymbol{i}, \boldsymbol{v}), \mathcal{D}) < \gamma$.*

Given a similarity threshold α, for each $(\boldsymbol{i},\boldsymbol{v}) \in \mathcal{P}$, we construct a similarity matrix $M_{(\boldsymbol{i},\boldsymbol{v})}^{\mathcal{D}}$ for $\mathcal{D}$. $M_{(\boldsymbol{i},\boldsymbol{v})}^{\mathcal{D}}[j][k] = 1$ if $Sim(\boldsymbol{v}, S_j[k].A) > \alpha$ and $\boldsymbol{i} = S_j[k].w$, otherwise, $M_{(\boldsymbol{i},\boldsymbol{v})}^{\mathcal{D}}[j][k] = 0$.

Definition 3 *(Coverage Pair). Given a threshold α, a set of pairs $\mathcal{P}$, pairs $(\boldsymbol{i}, \boldsymbol{v}) \in \mathcal{P}$ and $(\boldsymbol{i}', \boldsymbol{v}') \in \mathcal{P}$, $(\boldsymbol{i}, \boldsymbol{v})$ is a coverage pair of $(\boldsymbol{i}', \boldsymbol{v}')$ if $M_{(\boldsymbol{i},\boldsymbol{v})}^{\mathcal{D}}[j][k] = 1$ for $\forall M_{(\boldsymbol{i}',\boldsymbol{v}')}^{\mathcal{D}}[j][k'] = 1$ where $\boldsymbol{i} = S_j[k].w$ and $\boldsymbol{i}' = S_j[k'].w$.*

Theorem 2. *Given pairs $(\boldsymbol{i}, \boldsymbol{v}) \in \mathcal{P}$ and $(\boldsymbol{i}', \boldsymbol{v}') \in \mathcal{P}$, for patterns P and P' satisfying:(1) $|P| = |P'|$; (2) if $P[j] = (\boldsymbol{i}, \boldsymbol{v})$, then $P'[j] = (\boldsymbol{i}, \boldsymbol{v})$ or $P'[j] = (\boldsymbol{i}', \boldsymbol{v}')$; (3) if $P[j] \neq (\boldsymbol{i}, \boldsymbol{v})$, then $P[j] = P'[j]$. If $(\boldsymbol{i}, \boldsymbol{v})$ is a coverage pair of $(\boldsymbol{i}', \boldsymbol{v}')$, then $Sup(P, \mathcal{D}) \geq Sup(P', \mathcal{D})$.*

Algorithm 1. OBP-Miner($\mathcal{D}_c^u$, $\mathcal{D}_h^u$, $\mathcal{D}_c^U$, α, γ)

Input: $\mathcal{D}_c^u$, $\mathcal{D}_h^u$ and $\mathcal{D}_c^U$, α: similarity threshold, γ: support threshold
Output: OBP: the set of OBPs
1: $SP \leftarrow \emptyset$;
2: generate candidate pairs $\mathcal{P}$ by scanning $\mathcal{D}_c^u$;
3: **for** each pair $p \in \mathcal{P}$ **do**
4: **if** $Sup(p, \mathcal{D}_c^u) < \gamma$ **then**
5: remove p from $\mathcal{P}$; //Pruning Rule 2
6: **else if** there exists $p' \in \mathcal{P}$ such that p' is a coverage pair of p **then**
7: remove p from $\mathcal{P}$; //Pruning Rule 3
8: **end if**
9: **end for**
10: sort all pairs in descending order of support;
11: **for** each behavior pattern P searched by traversing the behavior pattern enumeration tree in a depth-first way **do**
12: compute $Sup(P, \mathcal{D}_c^u)$ according to $\mathcal{D}_c^u$, $\mathcal{D}_h^u$ and $\mathcal{D}_c^U$;
13: **if** $Sup(P, \mathcal{D}_c^u) > \gamma$ **then**
14: generate all candidate *history time windows* $\mathcal{W}_h$ and initialize $max \leftarrow 0$;
15: **for** each *history time window* W_h in $\mathcal{W}_h$ **do**
16: **if** $oScore(P, W_h) > max$ **then**
17: $max \leftarrow oScore(P, W_h)$;
18: **end if**
19: update P regarding W_h and *oScore*; $OBP \leftarrow OBP \bigcup \{P\}$;
20: **end for**
21: else
22: perform Pruning Rule 1;
23: **end if**
24: **end for**

Proof. For given P and P' satisfying $|P| = |P'|$, and (1) if $P[j] = (\boldsymbol{i}, \boldsymbol{v})$, then $P'[j] = (\boldsymbol{i}, \boldsymbol{v})$ or $P'[j] = (\boldsymbol{i}', \boldsymbol{v}')$; (2) if $P[j] \neq (\boldsymbol{i}, \boldsymbol{v})$, then $P[j] = P'[j]$. Suppose $(\boldsymbol{i}, \boldsymbol{v})$ is a *coverage pair* of $(\boldsymbol{i}', \boldsymbol{v}')$, we have $\{S \in \mathcal{D} | P' \subset S\} \subseteq \{S \in \mathcal{D} | P \subset S\}$. Then, by Eq. 2: $\frac{|\{S \in \mathcal{D} | P \subset S\}|}{|\mathcal{D}|} \geq \frac{|\{S \in \mathcal{D} | P' \subset S\}|}{|\mathcal{D}|}$. Thus, $Sup(P, \mathcal{D}) \geq Sup(P', \mathcal{D})$. □

Using Theorem 2, once $(\boldsymbol{i}', \boldsymbol{v}')$ is removed from $\mathcal{P}$, we can reduce the generation of numerous patterns with small support, and improve the efficiency of candidate pattern generation. Thus we get the Pruning rule 3.

Pruning Rule 3. *A pair* $(\boldsymbol{i}', \boldsymbol{v}') \in \mathcal{P}$ *can be removed if there exists another* $(\boldsymbol{i}, \boldsymbol{v}) \in \mathcal{P}$ *that* $(\boldsymbol{i}, \boldsymbol{v})$ *is a coverage pair of* $(\boldsymbol{i}', \boldsymbol{v}')$.

4.3 Candidate History Time Window Generation

For each node in the behavior pattern enumeration tree traversed by OBP-Miner, a behavior pattern P is generated. The next step is to generate all candidate *history time windows* and find a W_h in $\mathcal{D}_h^u$ that maximizes $oScore(P, W_h)$ for a given fixed $|\mathcal{D}_h^u|$. A straight way is to enumerate all sub-windows of $[0, |\mathcal{D}_h^u|]$. Clearly, it's not necessary for us to generate all candidate sub-windows.

Observation 1. *There is natural periodicity in student's behavior due to organization of courses.*

Based on Observation 1, it's more reasonable for us to generate candidate *history time windows* based on the length of recent time window (*e.g.*, the recent time window is one week, then we may generate *history time windows* in one week, two weeks, etc.).

Finally, we present the pseudo-code of OBP-Miner in Algorithm 1.

5 Empirical Evaluation

5.1 Experimental Setting

Datasets. Two real datasets were used in our experiments, and they were briefly described in the following.

(1) *Subsidies* dataset. The dataset[2] was publicly available, including the following aspects of data: book borrowing data, student performance data, bursary award data, dormitory access control data, library access control data, and consumption data. Note that consumption data comes from various places such as canteens, supermarkets, and school hospitals. We chose data from the latter three aspects of 748 students for our experiment. In addition, for group comparative analysis, we treated the students from the same academy as a group according to the basic information of students extracted from student performance data.
(2) *Campus* dataset. It was a real dataset from 856 students in a university, collecting the consumption data from different places, including canteens, school buses and boiling rooms. In particular, we regarded the students in the same class as a group for comparative analysis. Table 4 lists the abbreviations of the student behaviors in the experiment about two datasets.

Table 4. Behaviors and corresponding abbreviations on both two datasets

Academic Administration (AA)	Activity Center (AC)	Boiled Water (BW)
Canteen (CT)	Dormitory (DM)	Hospital (HP)
Library Access Record (LA)	Library (LB)	Laundry Room (LR)
School Bus (SB)	Showering (SH)	Supermarket (SP)

[2] https://www.ctolib.com/datacastle_subsidy.html.

Table 5. Students in *Subsidies* dataset with the largest *oScore*

Student ID	Outlying Behavior Patterns (OBPs) with w	W_h	*oScore*
25803	(CT: 3.3, DM: 2.0) [7, 9), (CT: 7.3, DM: 2.0) [11, 13), (AC: 0.3, LB: 1.0, DM: 2.0) [13, 15), (SP: 6.6, DM: 2.0) [17, 19)	[0, 7]	0.964
	(CT: 3.3, DM: 2.0) [7, 9), (AC: 0.3, LB: 1.0, DM: 2.0) [13, 15), (SP: 10.0, DM: 2.0) [17, 19)	[0, 7]	0.964
	(CT: 3.3, DM: 2.0) [7, 9), (CT: 7.3, DM: 2.0) [11, 13), (AC: 0.3, LB: 3.0, DM: 2.0) [13, 15), (BW: 10.0, AC: 0.3, DM: 1.0) [17, 19)	[0, 7]	0.679
15701	(SB: 5.6, DM: 2.0) [17, 19)	[14, 21]	1.000
	(CT: 3.4, DM: 1.0) [7, 9), (SB: 5.6, DM: 2.0) [17, 19)	[14, 21]	0.964
	(AC: 0.2, CT: 1.0, DM: 2.0) [11, 13), (SB: 5.6, DM: 2.0) [17, 19)	[14, 21]	0.857
	(AC: 0.2, CT: 1.0, DM: 2.0) [11, 13), (SB: 2.0, DM: 2.0) [15, 17)	[0, 7]	0.857
	(AC: 0.2, CT: 1.0, DM: 2.0) [11, 13)	[0, 7]	0.857

Table 6. Students in *Campus* dataset with the largest *oScore*

Student ID	Outlying Behavior Patterns (OBPs) with w	W_h	*oScore*
11138	(AC: 5.6, BW: 1.0) [11, 13), (CT: 8.6, DM: 1.0) [17, 19)	[0, 7]	0.715
	(AC: 6.4) [17, 19)	[0, 7]	0.715
	(AC: 5.6, BW: 1.0) [11, 13), (AC: 6.4) [17, 19)	[0, 7]	0.571
	(AC: 8.6) [11, 13)	[0, 7]	0.536
11291	(AC: 5.6, BW: 1.0) [11, 13), (AC: 6.0) [17, 19)	[0, 7]	0.571
	(AC: 5.6, BW: 1.0) [11, 13), (AC: 8.6) [17, 19)	[0, 7]	0.571
	(AC: 6.0, BW: 4.0) [11, 13), (AC: 6.0) [17, 19)	[0, 7]	0.571

Parameter Setting. Recall that there are three running parameters in OBP-Miner: similarity threshold α, support threshold γ and size of *history time window* $|W_h|$ (The specific value means days). Moreover, there is one more parameter for efficiency experiment: the number of students N.

Setup. All experiments were conducted on a PC with an AMD Ryzen Threadripper 3990X 2.90 GHz CPU and 256 GB main memory, running the Windows 10 operating system. All algorithms were implemented in Java and compiled by JDK 1.8.

5.2 Effcctiveness

We first verify the usefulness and effectiveness of OBP-Miner by analyzing the OBPs of students in *Campus* dataset and *Subsidies* dataset.

Tables 5 and 6 list two students who have OBPs with the largest *oScore*, respectively. And the corresponding OBPs, time interval (w), *history time window* (W_h) and *oScore* are listed in two tables. Recall that w indicates a time interval in a day (e.g., $[11, 13)$ means time from 11:00 to 13:00), W_h indicates a *history time window* before *current* time (e.g., $[14, 21]$ means 14–21 days of the *history* dataset).

From the results in Table 5, we see that OBPs can show some interesting and personalized daily behavior patterns of students. For example, comparing the *current* time interval with the same time interval in the past 7 d (the *history time window* [0, 7] targeting student 25803 and student 15701, respectively), student 25803 goes to activity center between 13:00 and 15:00, while he/she never goes to activity center during this time within past 7 d. And for student 15701, this student takes the school bus twice between 15:00 and 17:00, and this behavior has not occurred in the same time period of the past 7 d.

In Table 6, some interesting daily behavior patterns of each student can be found. Take student 11291 as an example, he/she has a high frequency to have dinner at the activity center in recent days, and never goes to this place in *history time window*. Student 11138 prefers to have lunch in the activity center and dinner in the canteen recently.

Figures 2 and 3 present statistics on the count of OBPs and *oScore* for each student with respect to γ, α, W_h, |OBP| in *Campus* and *Subsidies* datasets, respectively. We first increase the support threshold γ. Intuitively, fewer patterns can be found when we use higher support thresholds. Then we increase the similarity threshold α. In this case, more patterns can be found with a larger similarity threshold. Considering the *oScore*, count of OBPs with respect to different *history time window* W_h, we can see that students in both two datasets have the most counts among the *history time window* in $[0, 7]$. We note that there is no clear correlation between *oScore* and *history time window*. Besides, we see that for different datasets, this count distribution varies according to different |OBP|. For *Campus* dataset, length in $\{1, 2\}$ has more OBPs while for *Subsidies* dataset, this set is $\{3, 4\}$.

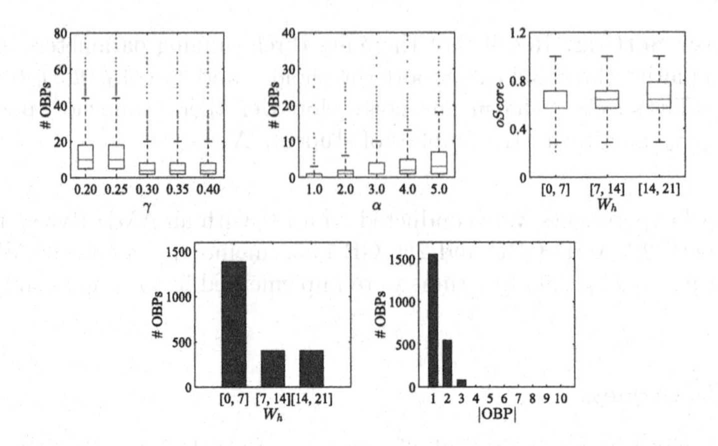

Fig. 2. Effectiveness test *w.r.t.* γ, α, W_h and |OBP| in *Campus* dataset

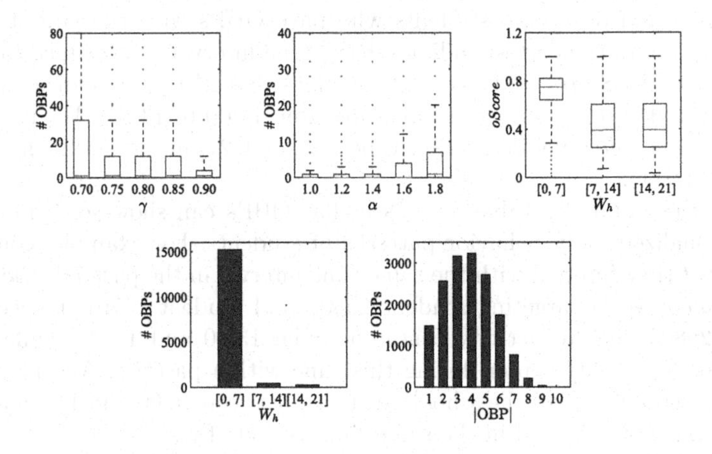

Fig. 3. Effectiveness test *w.r.t.* γ, α, W_h and |OBP| in *Subsidies* dataset

5.3 Efficiency

To the best of our knowledge, there is no previous method addressing this problem. Thus we test the efficiency of OBP-Miner compared with two of its variations, that is, Baseline and Baseline*. Baseline only adopts Pruning Rule 1 and Baseline* adopts Pruning Rules 1 and 2. In all efficiency tests, for *Campus* dataset, we set $\gamma = 0.30$, $\alpha = 3.0$ and $|W_h| = 28$ in default, and for *Subsidies* dataset we set $\gamma = 0.80$, $\alpha = 2.0$ and $|W_h| = 28$ in default. Logarithmic scale has been used for the runtime to better demonstrate the difference in the behavior between OBP-Miner and the baseline methods.

Figure 4 shows the runtime with respect to γ, α, $|W_h|$ and N. With the increase of γ and α, OBP-Miner always runs faster than Baseline and Baseline*. As γ and α increase, the runtime of both Baseline and Baseline* changes rapidly, while the runtime of OBP-Miner changes in a slow and steady way, which indi-

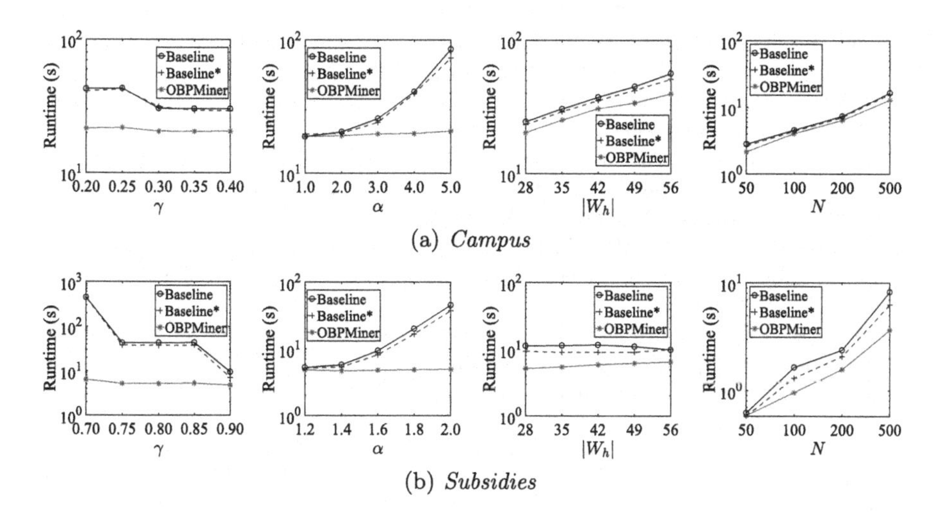

Fig. 4. Runtime *w.r.t.* α, γ, $|W_h|$ and N

cates that OBP-Miner is insensitive to those two parameters. In addition, when $|W_h|$ and N getting larger, the runtime of almost all algorithms increases, but OBP-Miner is slightly faster than Baseline and Baseline*.

Clearly, OBP-Miner runs faster than both Baseline and Baseline*, since OBP-Miner employs a *coverage pair* select strategy to reduce the generation of numerous patterns with small support. Besides, Baseline* is faster than Baseline because it uses a pruning rule to avoid generating meaningless behavior patterns. Moreover, the stability of runtime over different parameters verifies the robustness of OBP-Miner.

6 Conclusion

In this paper, we studied the novel problem of mining OBPs from student behavior sequences. We systematically developed a method with various pruning techniques. Experiments on two real datasets demonstrated that our proposed OBP-Miner is effective and efficient.

In the future, OBP-Miner can be applied to different sizes of *current* dataset and *history* dataset. Considering the huge search space candidate OBP generation, we plan to investigate various selection strategies in OBP-Miner.

References

1. Ameri, S., Fard, M.J., Chinnam, R.B., Reddy, C.K.: Survival analysis based framework for early prediction of student dropouts. In: Proceedings of the 25th ACM International Conference on Information and Knowledge Management, CIKM, pp. 903–912 (2016)

2. Cao, Y., et al.: Orderliness predicts academic performance: behavioural analysis on campus lifestyle. J. Roy. Soc. Interface **15**(146) (2018)
3. Duan, L., et al.: Mining distinguishing customer focus sets from online customer reviews. Computing **100**(4), 335–351 (2018)
4. Duan, L., Tang, G., Pei, J., Bailey, J., Campbell, A., Tang, C.: Mining outlying aspects on numeric data. Data Min. Knowl. Discov. **29**(5), 1116–1151 (2015)
5. Gupta, N., Eswaran, D., Shah, N., Akoglu, L., Faloutsos, C.: Beyond outlier detection: Lookout for pictorial explanation. Proceedings of the European Conference on Machine Learning and Knowledge Discovery in Databases, ECML PKDD. **11051**, 122–138 (2018)
6. Hang, M., Pytlarz, I., Neville, J.: Exploring student check-in behavior for improved point-of-interest prediction. In: Proceedings of the 24th ACM SIGKDD International Conference on Knowledge Discovery & Data Mining, KDD, pp. 321–330 (2018)
7. Jiménez, F., Paoletti, A., Sánchez, G., Sciavicco, G.: Predicting the risk of academic dropout with temporal multi-objective optimization. IEEE Trans. Learn. Technol. **12**(2), 225–236 (2019)
8. Li, X., Zhu, X., Zhu, X., Ji, Y., Tang, X.: Student academic performance prediction using deep multi-source behavior sequential network. In: Proceedings of the 24th Pacific-Asia Conference on Advances in Knowledge Discovery and Data Mining, PAKDD 2020, vol. 12084, pp. 567–579 (2020)
9. Liu, N., Shin, D., Hu, X.: Contextual outlier interpretation. In: Proceedings of the 27th International Joint Conference on Artificial Intelligence, IJCAI, pp. 2461–2467 (2018)
10. Nguyen, X.V., Chan, J., Bailey, J., Leckie, C., Ramamohanarao, K., Pei, J.: Scalable outlying-inlying aspects discovery via feature ranking. In: Proceedings of the 19th Pacific-Asia Conference on Advances in Knowledge Discovery and Data Mining, PAKDD, vol. 9078, pp. 422–434 (2015)
11. Nguyen, X.V., Chan, J., Romano, S., Bailey, J., Leckie, C., Ramamohanarao, K., Pei, J.: Discovering outlying aspects in large datasets. Data Min. Knowl. Discov. **30**(6), 1520–1555 (2016)
12. Peng, W., Zhang, X., Li, X.: Intelligent behavior data analysis for internet addiction. Sci. Program. 2019, 2753152:1–2753152:12 (2019)
13. Resnik, P., Garron, A., Resnik, R.: Using topic modeling to improve prediction of neuroticism and depression in college students. In: Proceedings of the 2013 Conference on Empirical Methods in Natural Language Processing, EMNLP, pp. 1348–1353 (2013)
14. Siddiqui, M.A., Fern, A., Dietterich, T.G., Wong, W.: Sequential feature explanations for anomaly detection. ACM Trans. Knowl. Discov. Data **13**(1), 1:1–1:22 (2019)
15. Wang, T., Duan, L., Dong, G., Bao, Z.: Efficient mining of outlying sequence patterns for analyzing outlierness of sequence data. ACM Trans. Knowl. Discov. Data **14**(5), 62:1–62:26 (2020)
16. Yang, Y., Wen, Z., Cao, J., Shen, J., Yin, H., Zhou, X.: EPARS: early prediction of at-risk students with online and offline learning behaviors. In: Proceedings of the 25th International Conferene on Database Systems for Advanced Applications, DASFAA 2020, vol. 12113, pp. 3–19 (2020)
17. Zhang, J., Lou, M., Ling, T.W., Wang, H.: Hos-miner: A system for detecting outlying subspaces of high-dimensional data. In: Proceedings of the 30th International Conference on Very Large Data Bases VLDB, pp. 1265–1268 (2004)

18. Zhang, X., Sun, G., Pan, Y., Sun, H., He, Y., Tan, J.: Students performance modeling based on behavior pattern. J. Ambient. Intell. Humaniz. Comput. **9**(5), 1659–1670 (2018)
19. Zheng, Z., Wei, W., Liu, C., Cao, W., Cao, L., Bhatia, M.: An effective contrast sequential pattern mining approach to taxpayer behavior analysis. World Wide Web **19**(4), 633–651 (2016)
20. Zhu, J., Wang, K., Wu, Y., Hu, Z., Wang, H.: Mining user-aware rare sequential topic patterns in document streams. IEEE Trans. Knowl. Data Eng. **28**(7), 1790–1804 (2016)
21. Zhu, Y., Zhu, H., Liu, Q., Chen, E., Li, H., Zhao, H.: Exploring the procrastination of college students: a data-driven behavioral perspective. In: Proceedings of the 21st International Conferene on Database Systems for Advanced Applications, DASFAA 2016, vol. 9642, pp. 258–273 (2016)

Clustering Mixed-Type Data with Correlation-Preserving Embedding

Luan Tran[1](✉), Liyue Fan[2], and Cyrus Shahabi[1]

[1] University of Southern California, Los Angeles, USA
{luantran,shahabi}@usc.edu

[2] University of North Carolina at Charlotte, Charlotte, USA
liyue.fan@uncc.edu

Abstract. Mixed-type data that contains both categorical and numerical features is prevalent in many real-world applications. Clustering mixed-type data is challenging, especially because of the complex relationship between categorical and numerical features. Unfortunately, widely adopted encoding methods and existing representation learning algorithms fail to capture these complex relationships. In this paper, we propose a new correlation-preserving embedding framework, COPE, to learn the representation of categorical features in mixed-type data while preserving the correlation between numerical and categorical features. Our extensive experiments with real-world datasets show that COPE generates high-quality representations and outperforms the state-of-the-art clustering algorithms by a wide margin.

Keywords: Mixed-type data · Clustering · Correlation preserving

1 Introduction

Mixed-type data, which contains both categorical and numerical features, is ubiquitous in the real world. It appears in many domains such as in network data [34] with the size of packages (numerical) and protocol type (categorical), and in personal data [26] with gender (categorical) and income information (numerical). Clustering is an important data mining task that groups data objects into clusters so that the objects in the same cluster are more similar to each other than to those in other clusters. Mixed-type data clustering has many real-world applications such as customer segmentation for differentiated targeting in marketing [17] and health data analysis [38,40]. However, most of the existing clustering algorithms have been developed for only numerical [2,12,14,18,37] or categorical data [1,4,10,22,31]. There are only a handful of algorithms [6,16,22,30] designed for mixed-type data.

A common approach to cluster mixed-type data is to generate numerical representations for categorical features, e.g., by using onehot encoding, then apply clustering algorithms designed for numerical data. The challenge of this approach is finding a good numerical representation that captures the complex

C. S. Jensen et al. (Eds.): DASFAA 2021, LNCS 12682, pp. 342–358, 2021.
https://doi.org/10.1007/978-3-030-73197-7_23

relationship between numerical and categorical features. Simple encoding methods, such as onehot, ordinal, and binary encoding, operate on individual features separately, and do not consider the relationship between features. In recent years, neural networks [27,35] have become a popular choice for representation learning because of its capacity in approximating complex functions. Autoencoder [35] is a natural choice for using a neural network to learn the data representation. It is a typical neural model with full connections between features and hidden units. However, a simple autoencoder that minimizes the reconstruction error loss may not fully capture the correlation between features.

This paper proposes a COrrelation-Preserving Embedding framework (COPE) to learn the representation for categorical features while preserving the relationship between categorical and numerical features. The COPE framework improves the representation learned by an autoencoder by incorporating two sub-networks to capture the correlation between categorical, numerical, and embedded data. By concurrently optimizing for representation learning and correlation preservation, the embedded categorical data preserves its semantics and the relationship with numerical features, thus providing more accurate clustering results.

We evaluate our proposed approach using six real-world datasets in various domains. Our extensive experimental results show that COPE outperforms other methods in clustering metrics such as Adjusted Mutual Information (AMI) [36] and Fowlkes-Mallows Index (FMI) [15]. The qualitative representation analysis using t-SNE visualization [28] depicts the effectiveness of COPE in grouping similar data into clusters. The convergence test shows that the COPE network quickly converges after a few iterations.

The remainder of this paper is organized as follows. In Sect. 2, we formally define the mixed-type data clustering problem and provide an overview of the correlation between categorical and numerical features. In Sect. 3, we discuss the current approaches for mixed-type data clustering. In Sect. 4, we introduce our proposed approach COPE. In Sect. 5, we present our experimental results in detail. We conclude the paper with discussion and future research directions in Sect. 6.

2 Background

2.1 Problem Definition

Let us denote $X = \{x^1, x^2, ..., x^N\} \in \mathcal{X}$ as a set of N objects in which each object has d_c categorical features and d_n continuous features, i.e., $\mathcal{F} = \mathcal{F}_c \cup \mathcal{F}_n$ where $\mathcal{F}_c = \{f_c^1, ..., f_c^{d_c}\}$ and $\mathcal{F}_n = \{f_n^1, ..., f_n^{d_n}\}$. Each categorical feature f_i has a discrete value domain $\mathcal{V}_i = \{v_i^1, v_i^2, ...\}$. For each data object x, its value in a continuous and categorical feature is denoted by $x_n \in \mathcal{X}_n$ and $x_c \in \mathcal{X}_c$, respectively. The problem can be defined as finding a good representation of data points, which preserves the complex relationship between the features to cluster data points accurately.

Table 1. Example mixed-type data

Area	Shape	Color
10	Triangle	Blue
12	Triangle	Blue
30	Circle	Red
50	Circle	Red
45	Diamond	Red
28	Diamond	Red
8	Square	Blue
7	Square	Blue

The challenge is that there might not be an order or apparent distances between categorical values; hence, it is impossible to compute a co-variance matrix of numerical and categorical features. We need to infer the relationship between categorical and numerical features from the data. Table 1 demonstrates one example of mixed-type data. Each row presents three features of one object. The features "Shape" and "Color" are categorical, and the feature "Area" is numerical. We assume that the order between the values in the categorical features is not known a priori. From the data, we can infer that an object with blue color and a triangle or square shape tends to have a small area (less than 20). Therefore, a triangle can be inferred to be more "similar" to a square than a circle. However, simple encoding methods, e.g., ordinal encoding: $\{$ triangle $\rightarrow 0,$ circle $\rightarrow 1,$ diamond $\rightarrow 2,$ and square $\rightarrow 3$ $\}$, do not capture that relationship. To solve this problem, we need a mechanism to measure and preserve the correlation between categorical and numerical features.

2.2 Correlation Between Numerical and Categorical Data

There are two main approaches to measure the correlation between a numerical and a categorical feature, i.e., point biserial correlation [33], and regression [21]. The point biserial correlation coefficient is a special case of Pearson's correlation coefficient [7]. It assumes the numerical variables are normally distributed and homoscedastic. The point biserial correlation is from -1 to 1. In the second approach, the intuition is that if there is a relationship between categorical and numerical features, we should be able to construct an accurate predictor of numerical features from categorical features, and vice versa. This approach does not make any assumption about the data distribution. To construct a predictor, many different models such as Linear Regression [7], SVM Regression [11], and Neural Network [32] can be used. Because of its robustness, we follow the second approach in this study.

3 Related Work

In literature, there are three main approaches for clustering mixed-type data. The first approach is finding a numerical representation of data then applying clustering algorithms designed for numerical data. Basic encoding techniques, i.e., onehot, ordinal, and binary encoding, are typically used to transform categorical features into numerical features. The basic encoding approach is fast; however, it operates on individual features, hence, does not correctly differentiate between categorical values or capture the correlation between the features. Several techniques have been introduced to address these problems. Autoencoder [3] takes the onehot encoded data as input to learn the compact representation of data. However, autoencoder alone does not fully preserve the correlations between features. DEC [41] is a variant of autoencoder, which simultaneously learns feature representations and cluster assignments. DEC first initializes its parameters with a deep autoencoder, then optimizes them by iterating between computing an auxiliary target distribution and minimizing the Kullback–Leibler (KL) divergence [19]. DEC focuses more on optimizing the discrimination between data objects. Similarly, MAI [24] learns the pair-wise relationship between features and focuses on learning the discrimination between objects. It first estimates the density between each pair of categorical and numerical feature. Then, each data object can be represented by a coupled encoding matrix. In addition, MAI also has another representation of data in the onehot encoding space. MAI takes these two representations as input and employs a neural network to learn the data representation. MAI triggers the learning process by preserving the distance orders in every set of three data objects in the onehot encoding and couple encoding space. This approach captures the pair-wise relationships between categorical and numerical features. However, it does not capture the relationship between more than two features. Besides, it preserves the order between data points in the onehot encoding space, which is generally not accurate. Moreover, the process of estimating the density of couplings is very time-consuming.

The second approach is converting numerical values into categorical values, then applying clustering techniques designed for categorical data. The numerical features are discretized into equal-size bins. K-modes [22], which is based on k-means, is a common clustering technique for categorical data. K-modes uses the Hamming distance, which is the number of features where two data objects differ. K-modes tries to minimize the sum of intra-cluster Hamming distances from the mode of clusters to their members. The mode vector consists of categorical values, each being the mode of an attribute. Several categorical clustering algorithms such as COOLCAT [4] and LIMBO [1] are based on minimizing the entropy of the whole arrangement. However, the numerical discretization process is information lossy, which can incur low clustering performance.

The third approach is applying algorithms designed for mixed-type data. ClicoT [6] and Integrate [9] cluster mixed-type data by minimizing the Huffman coding cost for coding numerical, categorical values, and model parameters. K-prototypes [22] combines k-means for numerical features and k-modes

[22] for categorical features. It minimizes the intra-cluster distances, including the Euclidean distances for numerical features and the Hamming distances for categorical features. MDBSCAN in [5] introduced distance hierarchy as a distance measure suitable for categorical and numerical attributes and then applied a modified DBSCAN [14] clustering.

4 Correlation Preserving Embedding for Categorical Features - COPE

Instead of clustering data X in the original space $\mathcal{X}$, we propose to first transform the onehot encoded values $B_c \in \mathcal{B}_c$ of categorical features $X_c \in \mathcal{X}_c$ with a non-linear mapping $f_\theta : \mathcal{B}_c \rightarrow \mathcal{Z}_c$ where θ is a list of learnable parameters and $\mathcal{Z}_c$ is the latent *embedding space*, then concatenate the numerical embedding with the normalized numerical features in $\mathcal{X}_n$. In other words, the final representation is $[f_\theta(B_c), \text{Norm}(X_n)]$, where Norm(.) is a normalization function to ensure numerical values to be in the same range with the categorical embedding. Such a representation allows to cluster the mixed-type input data with existing algorithms designed for clustering numerical data. The dimensions of $\mathcal{Z}_c$ is typically much smaller than $\mathcal{X}_c$ in order to avoid "the curse of dimensionality" [23]. To parameterize f_θ, Deep Neural Network (DNN) is a natural choice due to its theoretical function approximation property [20] and its demonstrated feature learning capability [8]. Following this approach, our proposed COPE network consists of two main components, i.e., a Deep Autoencoder [29] and two Fully Connected Neural Networks (FCNN). Figure 1 illustrates our COPE network design.

The Deep Autoencoder is a deep neural network efficient in representation learning and is used to extract latent compact features from the categorical input data. It can capture the relationship between the categorical variables. It consists of two parts, the encoder and the decoder, which can be defined as transitions $\mathcal{E}_\phi$ and $\mathcal{D}_\mu$ such that $\mathcal{E}_\phi : \mathcal{X}_c \rightarrow \mathcal{Z}_c$ and $\mathcal{D}_\mu : \mathcal{Z}_c \rightarrow \mathcal{X}_c$. The learnable parameters ϕ, μ are optimized by minimizing the reconstruction loss:

$$\mathcal{L}_{ae} = ||X_c - (\mathcal{E} o \mathcal{D})X_c||^2 \tag{1}$$

We parameterize ϕ and μ by deep fully connected networks with l layers, respectively. In the encoder, after each layer, the number of units in a fully connected layer is decreased by α. In contrast, in the decoder, after each layer, the number of units in a fully connected layer is increased by α. The embedding Z_c is computed as follows:

$$Z_c = f_l(f_{l-1}(...(X_c))) \tag{2}$$

where

$$f_i(x) = \sigma(W_i.x) \tag{3}$$

with σ is an activation function.

Our objective is to preserve the correlation between the two types of features from the plain encoding and the embedding space. Let $\mathcal{C}_\psi : \mathcal{X}_c \rightarrow \mathcal{X}_n$ and

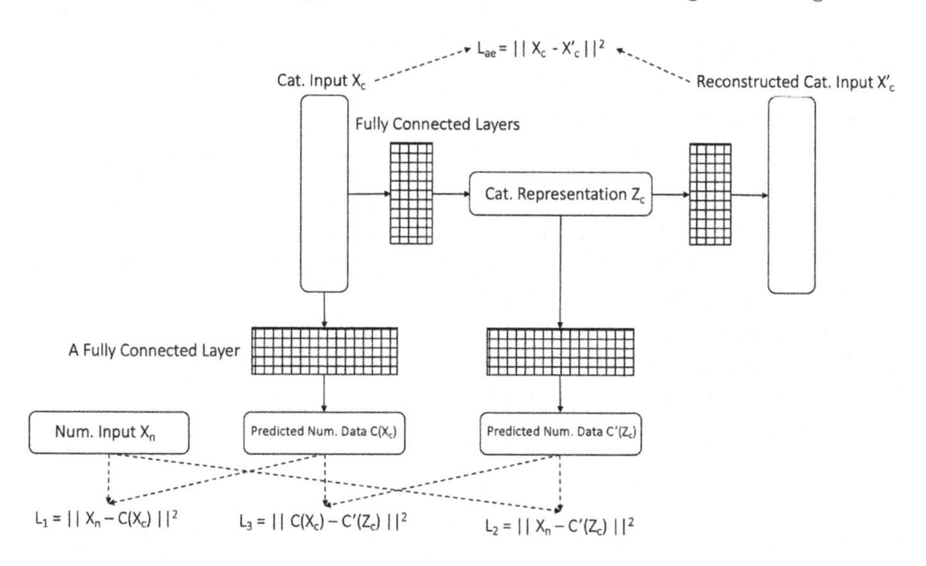

Fig. 1. The Correlation-Preserving Representation Learning Network - COPE. It contains a Deep Autoencoder and two Fully Connected Neural Networks to learn the representation of categorical features.

$\mathcal{C}'_\omega : \mathcal{Z}_c \to \mathcal{X}_n$ be the functions mapping the categorical features in the plain encoding space and embedding space to the numerical features, respectively.

The parameters ψ, ω are optimized by minimizing the loss:

$$\mathcal{L}_{cr} = \mathcal{L}_1 + \mathcal{L}_2 + \mathcal{L}_3 \tag{4}$$

where

$$\mathcal{L}_1 = ||X_n - \mathcal{C}(X_c)||^2, \mathcal{L}_2 = ||X_n - \mathcal{C}'(Z_c)||^2, \mathcal{L}_3 = ||\mathcal{C}(X_c) - \mathcal{C}'(Z_c)||^2 \tag{5}$$

The losses $\mathcal{L}_1$ and $\mathcal{L}_2$ constrain $\mathcal{C}$ and $\mathcal{C}'$ to learn the relationship between categorical features in the plain encoding and embedding space with the numerical features, respectively. The loss $\mathcal{L}_3$ constrains X_c and Z_c to have similar correlations with X_n. The intuition is that if $\mathcal{L}_3$ is small, $\mathcal{C}$ and $\mathcal{C}'$ will have similar performance in predicting X_n. Similar to the encoder and decoder, we parameterize ψ and ω by one fully connected layer, respectively. To optimize all the parameters $\theta = \{\phi, \mu, \psi, \omega\}$ concurrently, we combine the two loss $\mathcal{L}_{ae}$ and $\mathcal{L}_{cr}$ into one target loss and use Adam optimizer [25]:

$$\mathcal{L} = \mathcal{L}_{ae} + \mathcal{L}_{cr} \tag{6}$$

After the parameters are optimized, the encoder is used to compute the embedding Z_c from the categorical input data X_c. Then, the final data representation $[Z_c, \text{Norm}(X_n)]$ is obtained. Consequently, clustering algorithms for numerical data such as k-means [18] and DBSCAN [14] can be applied on the final data representation.

5 Experiment Results

5.1 Experimental Methodology

We first examined the correlation preserving and convergence capacity of COPE. Then we compared the performance of COPE with the baseline algorithms using the Adjusted Mutual Information [36] and Fowlkes-Mallows [15] scores. Finally, we compared the quality of the representations produced by algorithms using t-SNE visualization [28].

All the algorithms were implemented in Python. In COPE, the embedding dimension was set to a half of the onehot encoding dimensions for categorical features, and the parameter α was set to 1.2. In the Deep Autoencoder, in each fully connected layer, we apply a sigmoid activation function. We used the Adam optimizer with a learning rate of 0.001.

Baseline Algorithms. The baseline algorithms were selected carefully for each mixed-type data clustering approach. For the first approach of converting categorical values into numerical values, we selected Onehot, Ordinal, Binary encoding as the standard encoding methods, Autoencoder (AE) as a typical representation learning method, and MAI [24] as a state-of-the-art method. For Autoencoder, we used three variants, i.e., AE-cat, AE-all, and DEC [41]. AE-cat takes only the one-hot encoded values of categorical values as input to learn the representation for categorical features. AE-all takes both numerical features and the one-hot encoded values of categorical values as input to learn the representation for mixed-type data directly. DEC [41] is similar to AE-all but simultaneously learns feature representations and cluster assignments. We set the number of units in each layer of the encoder and decoder in these autoencoder variants similarly to COPE. For the second approach of converting numerical values into categorical values, we selected k-modes [22] because of its popularity. Each numerical feature was discretized into ten equal-size bins. For the third approach of using a clustering algorithm designed for mixed-type data, we selected k-prototypes [22] as a popular algorithm and ClicoT [6] as the state-of-the-art algorithm.

Clustering Metrics. In this study, we used the ground-truth classes to evaluate the performance of clustering algorithms. The assumption is that the members belong to the same classes are more similar than the members of different classes. Hence, the goodness of a clustering assignment can be measured as its similarity to the ground-truth classes. We used the two widely used clustering metrics, i.e., Adjusted Mutual Information (AMI) [36] and Fowlkes-Mallows Index (FMI) [15].

Adjusted Mutual Information. Let C, G denote the cluster and ground-truth class assignments, respectively, of the same N data points. The entropy of C and G is defined as follows: $H(C) = -\sum_{i=1}^{|C|} P(i)\log(P(i))$, where $P(i) = |C_i|/N$, and $H(G) = -\sum_{i=1}^{|G|} P'(i)\log(P'(i))$ where $P'(i) = |G_i|/N$.

The mutual information between C and G is calculated by: $MI(C,G) = \sum_{i=1}^{|C|}\sum_{j=1}^{|G|} P(i,j)\log\frac{P(i,j)}{P(i)P'(j)}$. The value of mutual information is adjusted for chance as follows: $AMI(C,G) = \frac{MI-E[MI]}{\text{mean}(H(C),H(G))-E[MI]}$, where $E[MI]$ is the expected value of the mutual information. The AMI value is ranged from 0 to 1. An AMI of 1 indicates two label assignments are equal.

Fowlkes-Mallows Index. The Fowlkes-Mallows index is defined as the geometric mean of pair wise precision and recall: $FMI = \frac{TP}{\sqrt{(TP+FP)(TP+FN)}}$ where TP stands for True Positive, the number of pair of data points belonging to the same classes and clusters; FP stands for False Positive, the number of pair of data points belonging to the same classes but different clusters; FN stands for False Negative, the number of pair of data points belonging to the same clusters but different classes. The FMI value is ranged from 0 to 1. A FMI of 1 indicates two label assignments are equal.

5.2 Datasets

We used six real-world UCI datasets [13] in various domains. The statistics of the datasets are reported in Table 2, including the dataset name, dataset size, the number of categorical features, numerical features, and classes. The KDD dataset was extracted from the original KDD Cup 99 dataset to obtain more balancing data. More specifically, we removed the classes with less than 1000 data points and randomly selected at most 10000 data points for each remaining class. The counts of classes are as follows: {'back.': 2203, 'ipsweep.': 1247, 'neptune.': 10000, 'normal.': 10000, 'portsweep.': 1040, 'satan.': 1589, 'smurf.': 10000, 'warezclient.': 1020} . For the Echo dataset, we removed the class with only one instance. For all datasets, we imputed missing numerical values by mean values, and categorical values by a value denoted by word "missing".

Table 2. Statistics of UCI datasets

Datasets	Description	Size	d_c	d_n	Class
KDD	Network packages of different attacks	37099	7	34	8
Income	Census income data	32561	9	6	2
ACA	Australian credit approval data	690	8	6	2
CRX	Credit card applications	690	9	6	2
Titanic	Titanic's passenger information	891	5	4	2
Echo	Patients with heart attack	131	2	8	2

5.3 COPE - Correlation Preservation and Convergence Test

We first examined the correlation preserving capacity and the convergence of our proposed method, COPE.

Correlation Preservation. Table 3 reports the training losses of COPE for all the datasets. As can be seen in this table, the losses $\mathcal{L}_1$ and $\mathcal{L}_2$ are small. It shows that the mapping functions $\mathcal{C}$ and $\mathcal{C}'$ well represents the relationship between categorical and numerical attributes in both the plain and embedding space. The loss $\mathcal{L}_3$ is very small for all the datasets, which proves the correlation preserving capacity of COPE. The autoencoder loss $\mathcal{L}_{ae}$ is small, which shows that the original data can be well reconstructed from the embedded data.

Table 3. Training losses for all datasets.

Datasets	$\mathcal{L}_1$	$\mathcal{L}_2$	$\mathcal{L}_3$	$\mathcal{L}_{ae}$
KDD	0.0058	0.0058	2.29E−07	3.35E−05
Titanic	0.0024	0.0017	1.17E−05	1.30E−04
CRX	0.012	0.0118	6.67E−06	6.50E−04
Income	0.0083	0.0083	2.42E−07	1.07E−04
ACA	0.009	0.008	3.02E−06	5.30E−05
Echo	0.0296	0.0294	1.54E−08	1.40E−04

Convergence Test. We applied k-means clustering [18] on the learned representation and reported how the loss and AMI change across epochs. We used the number of ground-truth classes to set the parameter k - the number of clusters. Figure 2 reports the AMI and the training loss with the KDD dataset in 50 epochs. Similar results can be obtained with the other datasets. As can be seen in this figure, the AMI and total training loss converged within 10 epochs.

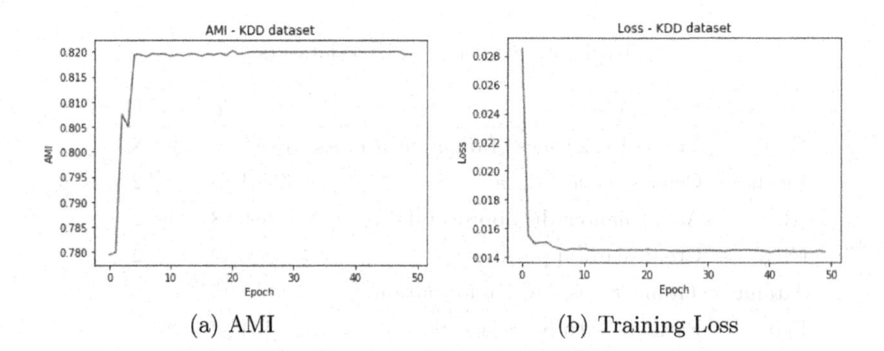

(a) AMI (b) Training Loss

Fig. 2. Convergence test on dataset KDD.

5.4 Clustering Results

We applied k-means clustering [18] on the data representation produced by the methods producing numerical representation and compared their AMIs and FMIs. We also compared them with ClicoT [6] and k-prototypes [22] designed for mixed-type data, and k-modes [22] desgined for categorical data. The number of clusters k in k-means, k-modes, and k-prototypes, was set to be the number of ground-truth classes. The results are reported in Tables 4 and 5 with all the datasets. As reported in these tables, the AMI and FMI of COPE were the highest for all the datasets. The top three performers were COPE, MAI, and DEC, which produce numerical representation. On average, COPE demonstrated approximately 37% and 30% improvement in AMI over DEC and MAI, respectively. MAI and DEC optimized the discrimination between data points and did not well preserve the relationship between categorical and numerical features. In most cases, we observed AE-all outperforms AE-cat because AE-all is able to integrate all features. However, for the KDD and Income datasets, we observed AE-cat performs better than AE-all. This is because the correlation between categorical and numerical features was not capture correctly in AE-all. ClicoT automatically determines the number of clusters, which might be different from the number of ground-truth classes. K-prototypes uses Hamming distance for categorical features, which does not capture the relationship between categorical and numerical features and offers the lowest performance. The basic encoding methods, i.e., Onehot, Ordinal, and Binary, do not consider any relationship between features, hence, they also offered low clustering performances.

Table 4. Clustering performance - AMI. The results of COPE and numerical representation methods are obtained using k-means.

Datasets	Numerical representation							Categorical representation	Designed for mixed-type data		COPE
	Onehot	Ordinal	Binary	MAI	AE-all	AE-cat	DEC	K-modes	ClicoT	K-prototypes	
KDD	0.77	0.64	0.79	0.71	0.76	0.77	0.73	0.71	0.74	0.72	**0.82**
Income	0.11	0.02	0.10	0.13	0.11	0.13	0.13	0.09	0.03	0.00	**0.17**
ACA	0.43	0.02	0.01	0.43	0.36	0.22	0.16	0.23	0.18	0.28	**0.44**
CRX	0.02	0.02	0.02	0.43	0.16	0.01	0.43	0.20	0.18	0.03	**0.44**
Titanic	**0.23**	0.02	**0.23**	0.06	**0.23**	0.08	**0.23**	0.09	0.07	0.08	**0.23**
Echo	0.11	0.21	0.01	0.32	0.44	0.01	0.44	0.11	0.26	0.37	**0.59**
Average	0.28	0.16	0.19	0.37	0.34	0.20	0.35	0.24	0.24	0.15	**0.48**

5.5 Data Representation Analysis

There are several approaches for data visualization such as PCA [39], and t-SNE [28]. They are both used for dimensionality reduction. While PCA is a linear projection, t-SNE uses the local relationship between data points to create a

Table 5. Clustering performance - FMI. The results of COPE and numerical representation methods are obtained using k-means.

Datasets	Numerical representation							Categorical representation	Designed for mixed-type data		COPE
	Onehot	Ordinal	Binary	MAI	AE-all	AE-cat	DEC	K-modes	ClicoT	K-prototypes	
KDD	0.78	0.68	0.82	0.72	0.77	0.78	0.75	0.73	0.77	0.77	**0.83**
Income	0.64	0.32	0.34	0.65	0.56	0.65	0.65	0.34	0.24	0.41	**0.67**
ACA	0.75	0.61	0.53	0.75	0.71	0.65	0.61	0.68	0.41	0.69	**0.77**
CRX	0.56	0.52	0.56	0.54	0.62	0.50	0.75	0.64	0.47	0.68	**0.77**
Titanic	**0.69**	0.52	**0.69**	0.54	**0.69**	0.63	**0.69**	0.61	0.38	0.08	**0.69**
Echo	0.56	0.52	0.53	0.79	0.80	0.53	0.80	0.64	0.71	0.37	**0.88**
Average	0.66	0.53	0.58	0.67	0.69	0.62	0.71	0.61	0.50	0.50	**0.77**

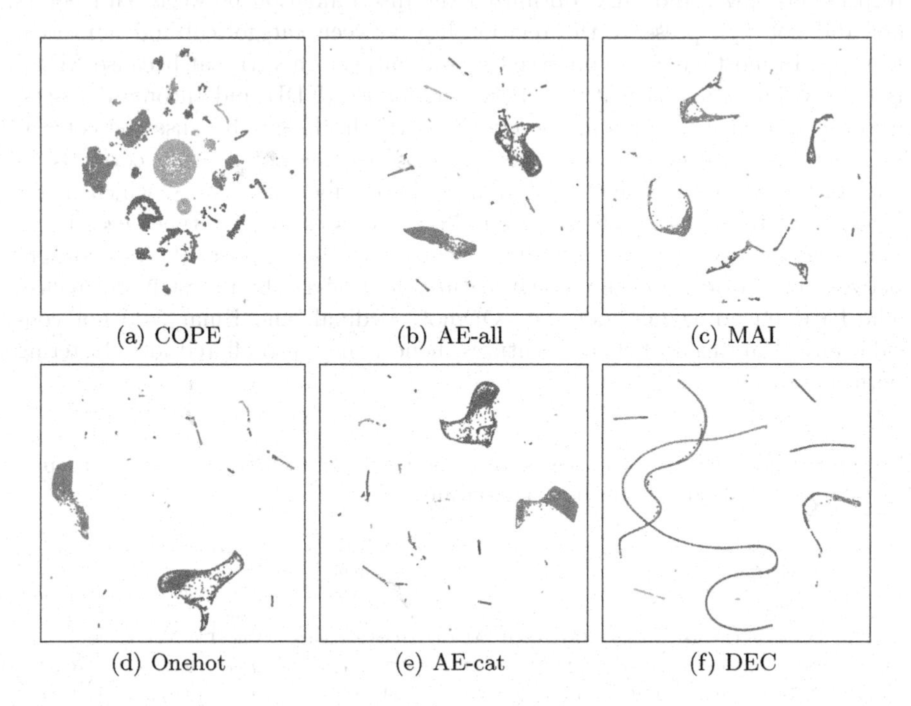

(a) COPE (b) AE-all (c) MAI

(d) Onehot (e) AE-cat (f) DEC

Fig. 3. The t-SNE visualization of data representations on the KDD dataset. (Color figure online)

low-dimensional mapping. Similar data points in the original space tend to have small distances in the t-SNE mapping. Because of the capacity to capture non-linear dependencies, we adopted t-SNE to compare the quality of the mixed-type data representations.

Here, we compared the top six methods that provide numerical representations, i.e., COPE, DEC, MAI, AE-all, Onehot, and AE-cat. In t-SNE, we set the perplexity to be 100, and the number of iterations to be 5000. Figures 3 and 4 illustrate the representations of the methods for the KDD and Echo datasets,

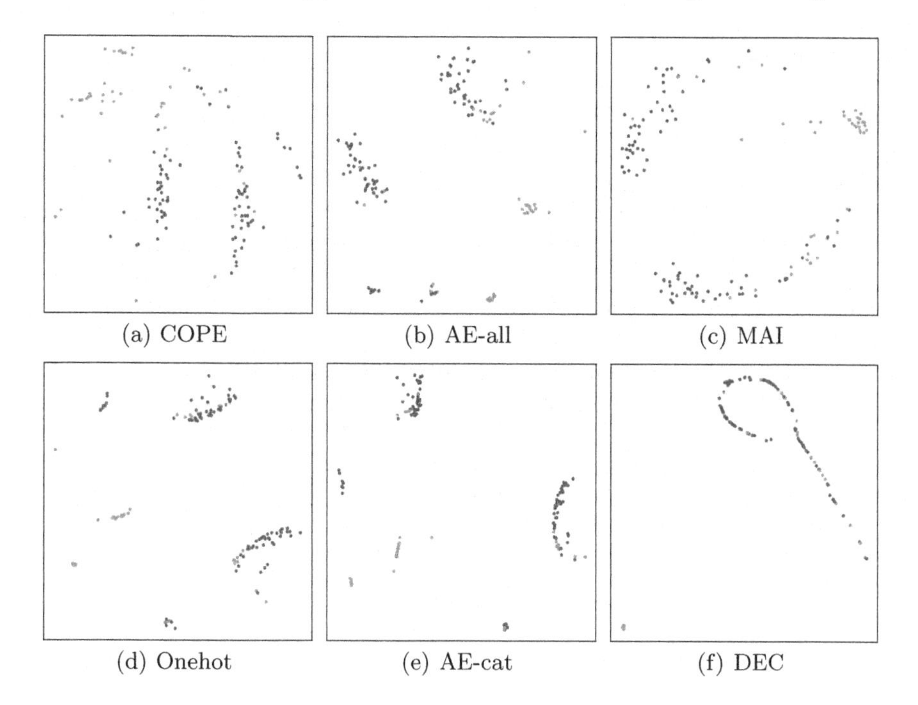

Fig. 4. The t-SNE visualization of data representations on the Echo dataset. (Color figure online)

respectively, in a two dimensional t-SNE mapping. The representations of the other datasets are in the Appendix section. Different colors represent different ground-truth classes.

The KDD dataset with eight different classes is plotted in red, blue, green, grey, yellow, pink, purple, and brown colors. As shown in Figure 3, in COPE, most data points with the same classes are grouped into clusters, and the clusters are separated quite clearly. The three largest groups are green, pink, and purple. Meanwhile, in the other methods, we only observed at most five major classes clearly with green, purple, yellow, brown, and red colors. Many data points in other colors, e.g., pink data points, were hidden among green and purple data points. Note that the group of pink data points is the largest group in the KDD dataset. It shows the much better separability of COPE compared to other methods. In MAI and DEC, we observed quite clearly groups of data points. However, each group consists of data points in different classes. In Onehot and AE-cat, more data points were scattered because they do not consider the relationship between categorical and numerical features.

For the Echo dataset, which has two different classes, as shown in Figure 4, there are two big groups of red data points in all methods. Similar to the KDD dataset, COPE has the fewest number of grey data points, which were falsely grouped with the red data points. The second best representation is AE-all. That explains why COPE and AE-all are the top two performers for the Echo dataset in clustering.

6 Conclusions

In this paper, we proposed COPE, a framework to learn representation for mixed-type data. It learns the embedding for categorical features using an autoencoder and two sub-networks to preserve the correlation between categorical and numerical features. We showed that COPE generates higher quality representation and offers better clustering results than the competing methods by more than 30% in widely used clustering metrics. As future work, COPE can be combined with techniques that refine the representation to enhance the discrimination between objects as in DEC for further improvement of the representation quality. We can also learn the embedding for numerical features using COPE by switching the roles of numerical and categorical features.

Acknowledgement. This work has been supported in part by NSF CNS-1951430, the USC Integrated Media Systems Center, and unrestricted cash gifts from Microsoft and Google. The opinions, findings, and conclusions or recommendations expressed in this material are those of the authors and do not necessarily reflect the views of the sponsors.

Appendix

See Figs. 5, 6, 7 and 8

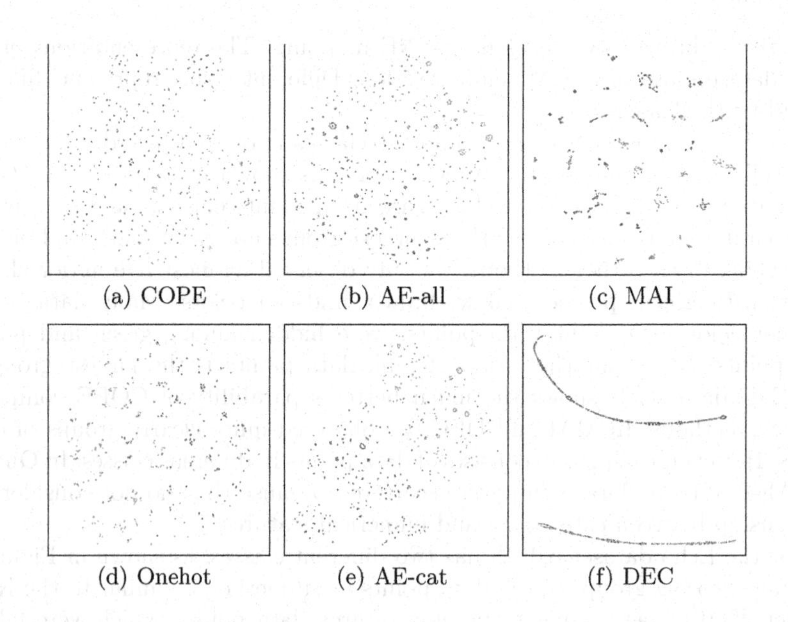

Fig. 5. The t-SNE visualization of data representations on the ACA dataset.

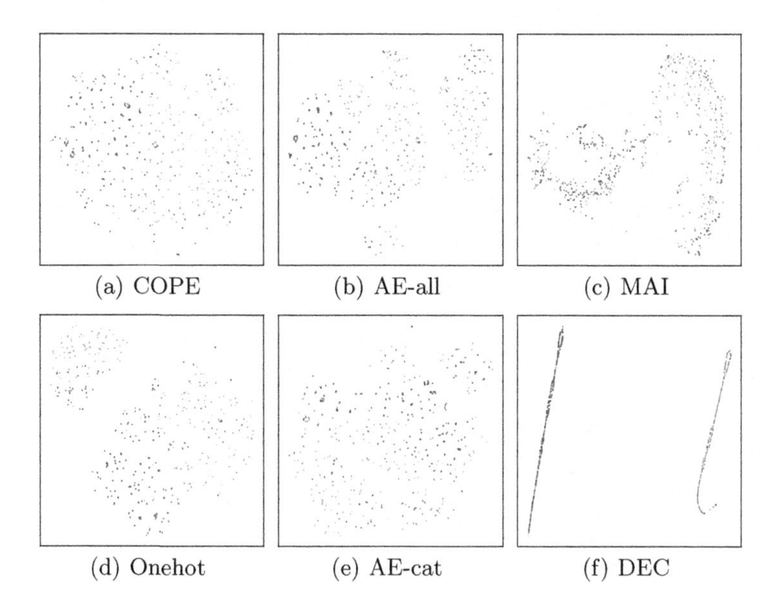

(a) COPE (b) AE-all (c) MAI

(d) Onehot (e) AE-cat (f) DEC

Fig. 6. The t-SNE visualization of data representations on the CRX dataset.

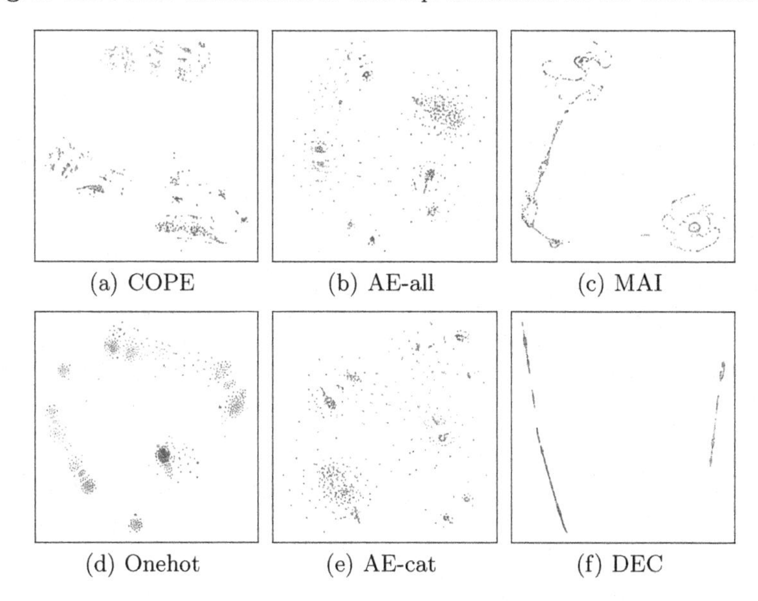

(a) COPE (b) AE-all (c) MAI

(d) Onehot (e) AE-cat (f) DEC

Fig. 7. The t-SNE visualization of data representations on the Titanic dataset.

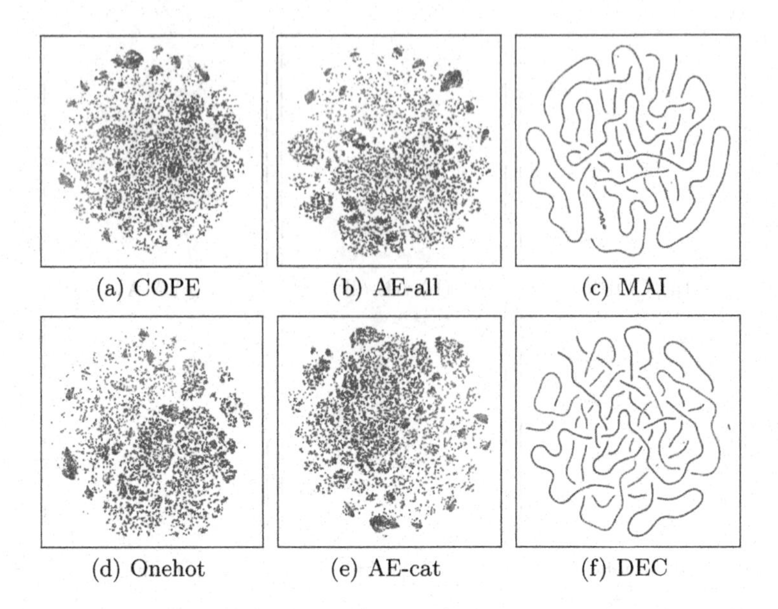

Fig. 8. The t-SNE visualization of data representations on the Income dataset.

References

1. Andritsos, P., Tsaparas, P., Miller, R.J., Sevcik, K.C.: LIMBO: scalable clustering of categorical data. In: Bertino, E., et al. (eds.) EDBT 2004. LNCS, vol. 2992, pp. 123–146. Springer, Heidelberg (2004). https://doi.org/10.1007/978-3-540-24741-8_9
2. Ankerst, M., Breunig, M.M., Kriegel, H.P., Sander, J.: Optics: ordering points to identify the clustering structure. ACM SIGMOD Rec. **28**(2), 49–60 (1999)
3. Aytekin, C., Ni, X., Cricri, F., Aksu, E.: Clustering and unsupervised anomaly detection with l 2 normalized deep auto-encoder representations. In: 2018 International Joint Conference on Neural Networks (IJCNN), pp. 1–6. IEEE (2018)
4. Barbará, D., Li, Y., Couto, J.: Coolcat: an entropy-based algorithm for categorical clustering. In: Proceedings of the Eleventh International Conference on Information and Knowledge Management, pp. 582–589 (2002)
5. Behzadi, S., Ibrahim, M.A., Plant, C.: Parameter free mixed-type density-based clustering. In: Hartmann, S., Ma, H., Hameurlain, A., Pernul, G., Wagner, R.R. (eds.) DEXA 2018. LNCS, vol. 11030, pp. 19–34. Springer, Cham (2018). https://doi.org/10.1007/978-3-319-98812-2_2
6. Behzadi, S., Müller, N.S., Plant, C., Böhm, C.: Clustering of mixed-type data considering concept hierarchies. In: Pacific-Asia Conference on Knowledge Discovery and Data Mining, pp. 555–573. Springer (2019)
7. Benesty, J., Chen, J., Huang, Y., Cohen, I.: Pearson correlation coefficient. In: Noise Reduction in Speech Processing. Springer Topics in Signal Processing, vol. 2, pp. 1–4. Springer, Heidelberg (2009). https://doi.org/10.1007/978-3-642-00296-0_5
8. Bengio, Y., Courville, A., Vincent, P.: Representation learning: a review and new perspectives. IEEE Trans. Pattern Anal. Mach. Intell. **35**(8), 1798–1828 (2013)

9. Böhm, C., Goebl, S., Oswald, A., Plant, C., Plavinski, M., Wackersreuther, B.: Integrative parameter-free clustering of data with mixed type attributes. In: Zaki, M.J., Yu, J.X., Ravindran, B., Pudi, V. (eds.) PAKDD 2010. LNCS (LNAI), vol. 6118, pp. 38–47. Springer, Heidelberg (2010). https://doi.org/10.1007/978-3-642-13657-3_7
10. Cao, F., et al.: An algorithm for clustering categorical data with set-valued features. IEEE Trans. Neural Networks Learning Syst. **29**(10), 4593–4606 (2017)
11. Cherkassky, V., Ma, Y.: Practical selection of SVM parameters and noise estimation for SVM regression. Neural Netw. **17**(1), 113–126 (2004)
12. Comaniciu, D., Meer, P.: Mean shift: a robust approach toward feature space analysis. IEEE Trans. Pattern Anal. Mach. Intell. **24**(5), 603–619 (2002)
13. Dua, D., Graff, C.: UCI machine learning repository (2017). http://archive.ics.uci.edu/ml
14. Ester, M., Kriegel, H.P., Sander, J., Xu, X., et al.: A density-based algorithm for discovering clusters in large spatial databases with noise. KDD **96**, 226–231 (1996)
15. Fowlkes, E.B., Mallows, C.L.: A method for comparing two hierarchical clusterings. J. Am. Stat. Assoc. **78**(383), 553–569 (1983)
16. Guha, S., Rastogi, R., Shim, K.: Rock: a robust clustering algorithm for categorical attributes. Inf. Syst. **25**(5), 345–366 (2000)
17. Hamka, F., Bouwman, H., De Reuver, M., Kroesen, M.: Mobile customer segmentation based on smartphone measurement. Telematics Inform. **31**(2), 220–227 (2014)
18. Hartigan, J.A., Wong, M.A.: Algorithm as 136: a k-means clustering algorithm. J. Royal Stat. Soci. Series c (applied statistics) **28**(1), 100–108 (1979)
19. Hershey, J.R., Olsen, P.A.: Approximating the kullback leibler divergence between Gaussian mixture models. In: 2007 IEEE International Conference on Acoustics, Speech and Signal Processing-ICASSP'07, vol. 4, pp. IV-317. IEEE (2007)
20. Hornik, K.: Approximation capabilities of multilayer feedforward networks. Neural Netw. **4**(2), 251–257 (1991)
21. Hosmer Jr., D.W., Lemeshow, S., Sturdivant, R.X.: Applied logistic regression, vol. 398. Wiley (2013)
22. Huang, Z.: Extensions to the k-means algorithm for clustering large data sets with categorical values. Data Min. Knowl. Disc. **2**(3), 283–304 (1998)
23. Indyk, P., Motwani, R.: Approximate nearest neighbors: towards removing the curse of dimensionality. In: Proceedings of The Thirtieth Annual ACM Symposium on Theory of Computing, pp. 604–613. ACM (1998)
24. Jian, S., Hu, L., Cao, L., Lu, K.: Metric-based auto-instructor for learning mixed data representation. In: Thirty-Second AAAI Conference on Artificial Intelligence (2018)
25. Kingma, D.P., Ba, J.: Adam: a method for stochastic optimization. arXiv preprint arXiv:1412.6980 (2014)
26. Kohavi, R.: Scaling up the accuracy of Naive-bayes classifiers: a decision-tree hybrid. KDD **96**, 202–207 (1996)
27. Krizhevsky, A., Sutskever, I., Hinton, G.E.: Imagenet classification with deep convolutional neural networks. In: Advances in Neural Information Processing Systems, pp. 1097–1105 (2012)
28. van der Maaten, L., Hinton, G.: Visualizing data using t-sne. J. Mach. Learn. Res. **9**, 2579–2605 (2008)

29. Marchi, E., Vesperini, F., Eyben, F., Squartini, S., Schuller, B.: A novel approach for automatic acoustic novelty detection using a denoising autoencoder with bidirectional LSTM neural networks. In: 2015 IEEE International Conference on Acoustics, Speech and Signal Processing (ICASSP), pp. 1996–2000. IEEE (2015)
30. Ni, X., Quadrianto, N., Wang, Y., Chen, C.: Composing tree graphical models with persistent homology features for clustering mixed-type data. In: International Conference on Machine Learning, pp. 2622–2631 (2017)
31. Salem, S.B., Naouali, S., Chtourou, Z.: A fast and effective partitional clustering algorithm for large categorical datasets using a k-means based approach. Comput. Electr. Eng. **68**, 463–483 (2018)
32. Specht, D.F., et al.: A general regression neural network. IEEE Trans. Neural Networks **2**(6), 568–576 (1991)
33. Tate, R.F.: Correlation between a discrete and a continuous variable. point-biserial correlation. Ann. Math. Stat. **25**(3), 603–607 (1954)
34. Tavallaee, M., Bagheri, E., Lu, W., Ghorbani, A.A.: A detailed analysis of the KDD cup 99 data set. In: 2009 IEEE Symposium on Computational Intelligence for Security and Defense Applications, pp. 1–6. IEEE (2009)
35. Vincent, P., Larochelle, H., Bengio, Y., Manzagol, P.A.: Extracting and composing robust features with denoising autoencoders. In: Proceedings of the 25th International Conference on Machine Learning, pp. 1096–1103 (2008)
36. Vinh, N.X., Epps, J., Bailey, J.: Information theoretic measures for clusterings comparison: variants, properties, normalization and correction for chance. J. Mach. Learn. Res. **11**, 2837–2854 (2010)
37. Von Luxburg, U.: A tutorial on spectral clustering. Stat. Comput. **17**(4), 395–416 (2007)
38. Wiwie, C., Baumbach, J., Röttger, R.: Comparing the performance of biomedical clustering methods. Nat. Methods **12**(11), 1033 (2015)
39. Wold, S., Esbensen, K., Geladi, P.: Principal component analysis. Chemom. Intell. Lab. Syst. **2**(1–3), 37–52 (1987)
40. Wu, Y., Duan, H., Du, S.: Multiple fuzzy c-means clustering algorithm in medical diagnosis. Technol. Health Care **23**(s2), S519–S527 (2015)
41. Xie, J., Girshick, R., Farhadi, A.: Unsupervised deep embedding for clustering analysis. In: International Conference on Machine Learning, pp. 478–487 (2016)

Beyond Matching: Modeling Two-Sided Multi-Behavioral Sequences for Dynamic Person-Job Fit

Bin Fu[1], Hongzhi Liu[1(✉)], Yao Zhu[2], Yang Song[5], Tao Zhang[6], and Zhonghai Wu[3,4(✉)]

[1] School of Software and Microelectronics, Peking University, Beijing, China
{binfu,liuhz}@pku.edu.cn

[2] Center for Data Science, Peking University, Beijing, China
yao.zhu@pku.edu.cn

[3] National Engineering Center of Software Engineering, Peking University, Beijing, China
wuzh@pku.edu.cn

[4] Key Lab of High Confidence Software Technologies (MOE), Peking University, Beijing, China

[5] BOSS Zhipin NLP Center, Beijing, China
songyang@kanzhun.com

[6] BOSS Zhipin, Beijing, China
kylen.zhang@kanzhun.com

Abstract. Online recruitment aims to match right talents with right jobs (Person-Job Fit, PJF) online by satisfying the preferences of both persons (job seekers) and jobs (recruiters). Recently, some research tried to solve this problem by deep semantic matching of curriculum vitaes and job postings. But those static profiles don't (fully) reflect users' personalized preferences. In addition, most existing preference learning methods are based on users' *matching* behaviors. However, *matching* behaviors are sparse due to the nature of PJF and not fine-grained enough to reflect users' dynamic preferences.

With going deep into the process of online PJF, we observed abundant auxiliary behaviors generated by both sides before achieving a *matching*, such as *click*, *invite/apply* and *chat*. To solve the above problems, we propose to collect and utilize these behaviors along the timeline to capture users' dynamic preferences. We design a Dynamic Multi-Key Value Memory Network to capture users' dynamic preferences from their multi-behavioral sequences. Furthermore, a Bilateral Cascade Multi-Task Learning framework is designed to transfer two-sided preferences learned from auxiliary behaviors to the *matching* task with consideration of their cascade relations. Offline experimental results on two real-world datasets show our method outperforms the state-of-the-art methods.

Keywords: Person-job fit · Dynamic preferences · Multi-behavioral sequence · Cascade multi-Task Learning

C. S. Jensen et al. (Eds.): DASFAA 2021, LNCS 12682, pp. 359–375, 2021.
https://doi.org/10.1007/978-3-030-73197-7_24

1 Introduction

With the rapid development of the Internet, online recruitment has become popular. It provides great convenience with low cost for both job seekers and recruiters, e.g. without regard for their locations. The core of online recruitment is Person-Job Fit (PJF), which tries to meet the preferences of both job seekers and recruiters. With the huge numbers of job postings (JPs) and curriculum vitaes (CVs) available on the Internet, the users (i.e. both job seekers and recruiters) face the problem of information overloading. For example, there are about millions of jobs and more than 675 million job seekers at LinkedIn in 2020 [1]. To solve this problem, it's vital for online recruitment to develop more effective and efficient person-job matching methods.

Due to the surge of deep learning in recent years, advanced matching approaches based on neural networks have been proposed to automatically deal with the raw data, e.g. textual profiles, in an end-to-end way. Some of them considered PJF as a profile-matching problem [2–5], i.e. deep semantic matching between curriculum vitaes and job postings based on CNN [2,4,5] or RNN [3]. Supervised by the *matching* feedbacks [2–4], they tried to learn good representations of profiles and the matching rules between CVs and JPs. However, these static profiles do not always contain nor fully reflect users' personalized preferences. For example, a person's CV usually includes education background and work experiences, but doesn't contain any preference information.

To fill this gap, some methods were proposed to profile users' preferences from their *matching* behaviors [6]. However, in real online recruitment, most job seekers and recruiters have only a few or even no historical *matching* records, because they are likely to leave the recruitment platform if they had achieved a matching. Therefore, methods only relied on the sparse *matching* feedbacks can not capture users' fine-grained and comprehensive personalized preferences.

In practice, users often generate multiple behaviors beyond *matching* on the online recruitment platform. To solve the above problems, we go deep into the interaction process of job seeker and recruiter on the largest online recruitment platform named "Boss Zhipin"[1] in China. When the job seeker/recruiter is the first time to enter the platform, he/she needs to create an account on the platform and upload his/her CV/JP. The interaction process starts with the job seeker/recruiter searching or browsing through the site or APP according to his/her expectation followed by:

(1) *The job seeker/recruiter clicks and views a JP/CV posted by the job recruiter/seeker in an impression list.* (***click***)
(2) *If the job seeker/recruiter finds the JP/CV is desirable, he/she will send an application/invitation to the job recruiter/seeker.* (***invite/apply***)
(3) *The recipient can decide to respond positively or negatively. Only if the recipient responds positively, a message exchange would begin.* (***chat***)
(4) *If they accept each other online, they achieve an online matching.* (***matching***)

[1] https://www.zhipin.com.

(5) *Finally, the job seeker/recruiter may leave the platform as having found a job or employee, or initialize new interactions with other recruiters/job seekers.*

The interaction between the job seeker and recruiter may terminate in step (1)–(3) if the job seeker/recruiter does not continue, or they finally achieve a matching (step (4)). Figure 1 shows this interaction process in the online recruitment platform except step (5). Note that the interaction can be initialized by either a job seeker or a recruiter, and a job seeker/recruiter may have multiple interactions with different job recruiters/seekers at the same time.

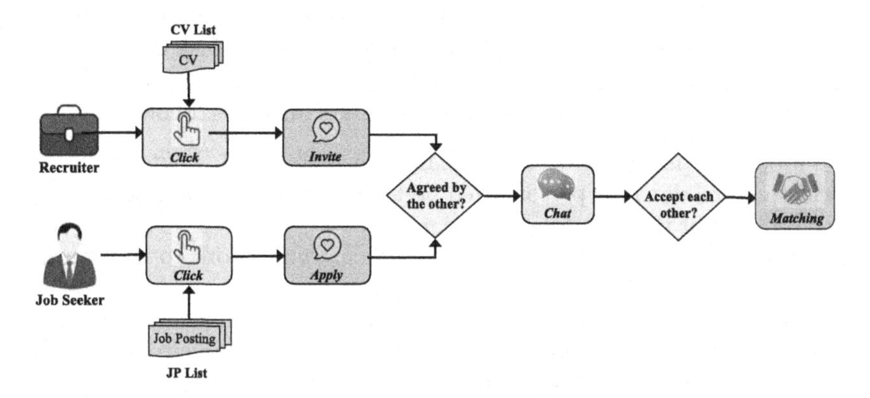

Fig. 1. The interaction process of job seekers and recruiters on the online recruitment platform "Boss Zhipin". The multi-behaviors with cascade relations are generated by job seekers and recruiters.

Figure 1 shows multiple behaviors {*click*, *invite/apply*, *chat*, *matching*} generated by users and their cascade relations. Among them, {*click*, *invite/apply*} are unilateral and {*chat*, *matching*} are bilateral. The auxiliary multi-behaviors {*click*, *invite/apply*, *chat*} are relatively abundant and reflect users' personalized preferences to some extent, even though most of them finally fail to turn into *matching*. With these auxiliary behaviors, we can capture more fine-grained and comprehensive preferences of users. Furthermore, they can alleviate the sparsity problem of the *matching* behavior. Unfortunately, these auxiliary multi-behaviors are often neglected by previous PJF methods.

Another problem is that the preferences of both job seekers and recruiters are dynamic. The online recruitment platform is a two-sided market, and the interactions between job seekers and recruiters may influence their expectations and preferences. New users often have no idea about the recruitment market at first, and try to explore and may cause matching failures. For example, if their expectations are set too high, such as the requirements of excessive salary from the job seeker side, they will fail to match any one until they change their expectations. On the contrary, if their expectations are set too low, they will adjust their expectations in the subsequent interactions to maximize their own benefits. With more knowledge of the recruitment market, their expectations become more reasonable, and matchable to their abilities/benefits. Therefore, it's necessary to consider users' dynamic preferences in PJF.

To solve these problems, we propose to learn the fine-grained and comprehensive dynamic preferences of both sides from their multi-behavioral sequences including behaviors {*click*, *invite/apply*, *chat*, *matching*}, and transfer these preferences to *matching* prediction task. A Dynamic Multi-Key Value Memory Network is designed to capture users' dynamic comprehensive preferences from their multi-behavioral sequences, which transforms the heterogeneous behaviors into a unified preference space and updates users' unified preferences dynamically according to the multi-behavioral sequences. We jointly model the prediction tasks of multi-behaviors through a Bilateral Cascade Multi-Task Learning framework with consideration of the cascade relations among behaviors.

The main contributions of this paper are summarized as follows:

- We propose to make use of the auxiliary behaviors {*click*, *invite/apply*, *chat*} to learn users' dynamic personalized preferences which can alleviate the sparsity problem of the *matching* behavior.
- We design a Dynamic Multi-Key Value Memory Network to transform the heterogeneous behaviors into a unified preference space and update users' unified preferences according to their multi-behavior sequences dynamically.
- We design a Bilateral Cascade Multi-Task Learning framework to transfer the preferences from auxiliary behaviors {*click*, *invite/apply*, *chat*} to the *matching* task with consideration of the cascade relations among behaviors.
- Experimental results on two real-world datasets show that the proposed model outperforms the state-of-the-art (SOTA) methods.

2 Related Work

Person-Job Fit is the core of recruitment which aims to match right talents (i.e. job seekers) with right jobs. The early work could date back to the1990s, and PJF was seen as a bilateral matching problem [7], which considered that, for a matched person-job pair, the skills/abilities of the person should commensurate with the requirements of the job, and the desires of the person should also be satisfied by the job's supplies. Caldwell et al. [8] tried to model and compare the competitiveness between the profiles of candidates to select the most competitive ones. The early work was based on small-sized data or simple analysis methods.

Statistical machine learning methods, such as classification, regression [9,10] and recommendation approaches [11–13], were utilized for PJF based on the handcrafted features of curriculum vitaes and job postings. However, these methods needed to design handcrafted features which were expensive and inefficient.

With the rise of deep learning, several related work based on neural networks tried to realize Person-Job Fit in an end-to-end way. Some deep semantic matching models were proposed to learn good representations of the textual profiles. Zhu et al. [2] tried to learn the representations of both CVs and JPs using CNN, and predicted the matching degrees according to cosine similarity. To capture both job seekers' experiences from CVs and job requirements from JPs, Qin et al. [3] proposed an ability-aware PJF model based on a hierarchical Recurrent Neural Network. Le et al. [5] tried to capture the intentions of both sides and learn the interdependence between CVs and JPs. However, the textual profiles CVs and JPs might not contain or fully reflect the preferences of job seekers and recruiters, respectively.

To solve this problem, Yan et al. [6] proposed to profile users' preferences from their historical *matching* behaviors. However, they only utilize the *matching* behaviors, which are sparse and not fine-grained enough to capture users' preferences. In online recruitment, users often have abundant auxiliary behaviors before achieving a *matching*, such as {*click*, *apply/invite*, *chat*} in Fig. 1, which also reflect their personalized preferences to some extent, but are ignored by previous studies. In this paper, we propose to model two-sided multi-behavioral sequences to profile the dynamic comprehensive preferences of users. To the best of our knowledge, this is the first work to utilize two-sided multi-behaviors beyond *matching* to capture the dynamic comprehensive preferences of users in PJF.

3 Problem Formulation

Let P denote the person (job seeker) set and J denote the job (recruiter) set, i.e. $P = \{p_k|k \in [1, m]\}$ and $J = \{j_k|k \in [1, n]\}$, where m and n denote the number of persons and jobs, respectively. Each person $p \in P$ has a curriculum vitae CV_p, which contains the textual description of education background, work experiences and skills. Each job $j \in J$ has a job posting JP_j, which includes the textual description of company, job requirements and benefits. Each user (i.e. person and job) has a multi-behavioral sequence, which records his/her multi-behaviors in a sequence along the timeline. The multi-behavioral sequence of person $p \in P$ is denoted as $\boldsymbol{H}_p = \{(j_t, a_t)\}$, where the tuple (j_t, a_t) denotes person p took a behavior a_t on job j_t at time t. Similarly, we have $\boldsymbol{H}_j = \{(p_t, a_t)\}$ for each job $j \in J$. $a \in A$ and $A = \{click, invite, apply, chat, matching\}$. $y_a(p, j) \in \{0, 1\}$ indicates whether there is a behavior a observed between person p and job j. Note that, for unilateral behaviors {*click*, *invite*, *apply*}, (p, j) is different from (j, p).

Given the multi-behavior sequences of all users $\boldsymbol{H} = \{\boldsymbol{H}_p|p \in P\} \cup \{\boldsymbol{H}_j|j \in J\}$ and their textual profiles $\boldsymbol{CV} = \{CV_p\,|\,p \in P\}$ and $\boldsymbol{JP} = \{JP_j\,|\,j \in J\}$, our goal is to predict the future person-job matching pairs for the target persons (jobs).

4 The Proposed Approach

We consider person-job fit as a bilateral matching problem, and model two-sided multi-behavioral sequences based on two assumptions: (1) the auxiliary behaviors {*click*, *invite*, *apply*, *chat*} also reflect users' personalized preferences, which are complementary to those of *matching* behaviors. Fusing the preferences learned from these multi-behaviors in a proper way can improve the matching prediction. (2) Users' preferences are dynamic, and modeling the multi-behaviors in a dynamic way is better than in a static way.

To capture users' dynamic comprehensive preferences, we design a Dynamic Multi-Key Value Memory Network (DMKVMN) to deal with the multi-behavior sequences. It can transform heterogeneous behaviors into a unified preference space and update users' preferences according to their multi-behavior sequences. A Bilateral Cascade Multi-Task Learning (BCMTL) framework is designed to transfer preferences from auxiliary behaviors {*click*, *invite*, *apply*, *chat*} to the *matching* task with consideration of the cascade relations between behaviors.

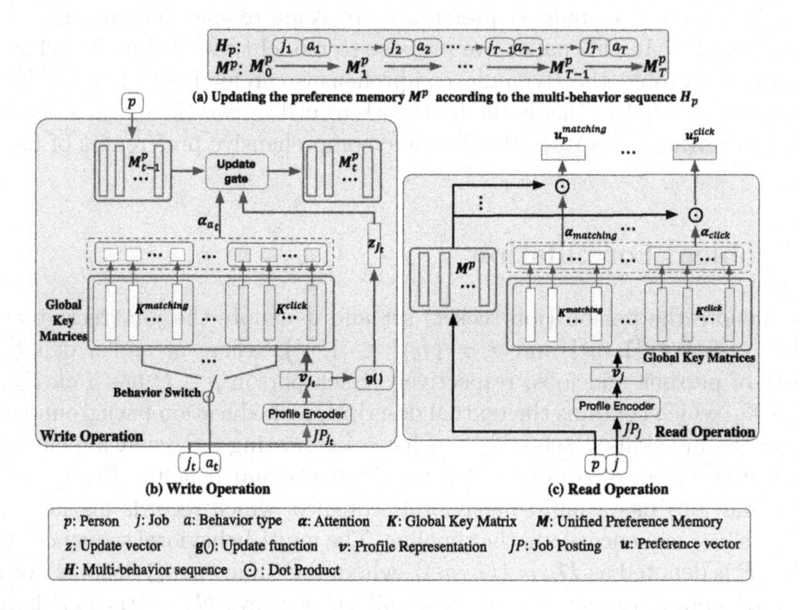

Fig. 2. Structure and operations of Dynamic Multi-Key Value Memory Network (DMKVMN). It consists of multiple global key matrices $\{\boldsymbol{K}^{matching}, \boldsymbol{K}^{chat}, \boldsymbol{K}^{apply}, \boldsymbol{K}^{invite}, \boldsymbol{K}^{click}\}$ shared by all users, and provides a private value matrix M for each user as his/her unified preference memory. It can transform the heterogeneous behaviors into the unified preference space of $\boldsymbol{M}$ based on an attention mechanism. Taking person p ($\forall p \in P$) as an example, **(a)** DMKVMN updates the unified preference memory $\boldsymbol{M}^p$ from $\boldsymbol{M}_0^p$ to $\boldsymbol{M}_T^p$ according to the multi-behavior sequence $\boldsymbol{H}_p$ via the write operation where $T = |\boldsymbol{H}_p|$, **(b)** Write Operation denotes that, for time step t, how DMKVMN updates $\boldsymbol{M}^p$ from $\boldsymbol{M}_{t-1}^p$ to $\boldsymbol{M}_t^p$ with a behavior (j_t, a_t), **(c)** Read Operation denotes how DMKVMN reads out the current preferences $\{\boldsymbol{u}_p^{matching}, \boldsymbol{u}_p^{chat}, \boldsymbol{u}_p^{apply}, \boldsymbol{u}_p^{click}\}$ from $\boldsymbol{M}^p$ *w.r.t* job j. Note that the operations on job side are similar with those on person side.

4.1 Dynamic Multi-key Value Memory Network

As shown in Fig. 1, in the online recruitment, each type of behaviors ($\forall a \in A$) has its own specific semantic meaning. For simplicity, we consider that the same type of behaviors have the same meaning for different users, and different kinds of behaviors of a user reflect his/her intrinsic preferences in different ways.

Based on these characteristics, we design a DMKVMN to transform the heterogeneous behaviors into the unified preference space and update users' preferences according to their multi-behavior sequences in a dynamic way. It consists of multiple global key matrices $\{\boldsymbol{K}^{click}, \boldsymbol{K}^{apply}, \boldsymbol{K}^{invite}, \boldsymbol{K}^{chat}, \boldsymbol{K}^{matching}\}$ shared by all users, and one private value matrix for each job seeker (e.g. $\boldsymbol{M}^p$) and recruiter (e.g. $\boldsymbol{M}^j$) as his/her unified preference memory, which is used to store the comprehensive preferences according to his/her multi-behavior sequence (e.g. $\boldsymbol{H}_p$ and $\boldsymbol{H}_j$). Each global key matrix $\boldsymbol{K}^a$ ($a \in A$) can transform the corresponding behavior (j, a) ((p, a)) of person p (job j) into the unified preference space of $\boldsymbol{M}^p$ ($\boldsymbol{M}^j$) based on an attention mechanism ($\boldsymbol{\alpha}_a$).

DMKVMN has two key operations: (1) the **write** operation is used to update the preference memory $\boldsymbol{M}^p$ ($\boldsymbol{M}^j$) according to the multi-behavior sequence $\boldsymbol{H}_p$ ($\boldsymbol{H}_j$) of person p (job j), and (2) the **read** operation is used to read out the current preferences $\{\boldsymbol{u}_p^{matching}, \boldsymbol{u}_p^{chat}, \boldsymbol{u}_p^{apply}, \boldsymbol{u}_p^{click}\}$ ($\{\boldsymbol{u}_j^{matching}, \boldsymbol{u}_j^{chat}, \boldsymbol{u}_j^{invite}, \boldsymbol{u}_j^{click}\}$) from the preference memory $\boldsymbol{M}^p$ ($\boldsymbol{M}^j$) *w.r.t* a given person-job pair (p, j) for downstream tasks. Concretely, each global key matrix $\boldsymbol{K}^a$ ($\forall a \in A$) and each private value matrix $\boldsymbol{M}$ have the same number of slots, which is set as I, i.e. $\boldsymbol{K}^a = \{\boldsymbol{k}_1^a, \boldsymbol{k}_2^a, ..., \boldsymbol{k}_I^a\}$ and $\boldsymbol{M} = \{\boldsymbol{m}_1, \boldsymbol{m}_2, ..., \boldsymbol{m}_I\}$. Each global key slot $\boldsymbol{k}_i^a \in \mathbb{R}^d$ is corresponding to the private value slot $\boldsymbol{m}_i \in \mathbb{R}^d$ where d is the dimension, which indicate the key and value on the i-th latent feature of the preference space, respectively. Note that the global key matrices are behavior-specific which are learnable parameters, and the value matrices are user-specific which are updated by rules defined in the write operation.

To show how it works, we take the person p as an example as shown in Fig. 2. DMKVMN iteratively updates his/her unified preference memory $\boldsymbol{M}^p$ from $\boldsymbol{M}_0^p$ to $\boldsymbol{M}_T^p$ according to the multi-behavior sequence $\boldsymbol{H}_p$ where $T = |\boldsymbol{H}_p|$ and the value of each element in $\boldsymbol{M}_0^p$ is initialized to 0. For each time $t \in [1, T]$, it updates the unified preference memory from $\boldsymbol{M}_{t-1}^p$ to $\boldsymbol{M}_t^p$ with the new behavior (j_t, a_t) via the write operation. In the write operation, for behavior (j_t, a_t) in $\boldsymbol{H}_p$ where the behavior type $a_t \in A$, the corresponding job posting JP_{j_t} is firstly encoded as a vector $\boldsymbol{v}_{j_t}$ via a hierarchical LSTM encoder [14]. Then, $\boldsymbol{v}_{j_t}$ is fed into the update function $\boldsymbol{g}(\cdot)$ to generate the update vector $\boldsymbol{z}_{j_t}$, i.e. $\boldsymbol{z}_{j_t} = \boldsymbol{g}(\boldsymbol{v}_{j_t})$, and $\boldsymbol{v}_{j_t}$ also multiplies with the corresponding global key matrix $\boldsymbol{K}^{a_t}$ of a_t (i.e. Behavior Switch in Write Operation of Fig. 2) to generate the attention $\boldsymbol{\alpha}_{a_t}$. Finally, DMKVMN updates the unified preference memory from $\boldsymbol{M}_{t-1}^p$ to $\boldsymbol{M}_t^p$ via an update gate. After updating $\boldsymbol{M}^p$ according to $\boldsymbol{H}_p$ via the write operations, DMKVMN can read out the current preferences $\{\boldsymbol{u}_p^{matching}, \boldsymbol{u}_p^{chat}, \boldsymbol{u}_p^{apply}, \boldsymbol{u}_p^{click}\}$ from $\boldsymbol{M}^p$ via the dot product between the attentions $\{\boldsymbol{\alpha}_{matching}, \boldsymbol{\alpha}_{chat}, \boldsymbol{\alpha}_{apply}, \boldsymbol{\alpha}_{click}\}$ and $\boldsymbol{M}^p$. The read and write operation are detailed as follows.

Write Operation. For each behavior (j_t, a_t) in $\boldsymbol{H}_p$ where $p \in P$, $j_t \in J$ and $a_t \in A$, the unified preference memory $\boldsymbol{M}^p$ is updated by the update gate of Fig. 2(b) as follows,

$$\begin{aligned} \boldsymbol{m}_{i,t}^p &= (1 - \alpha_{a_t,i}) \cdot \boldsymbol{m}_{i,t-1}^p + \alpha_{a_t,i} \cdot \boldsymbol{z}_{j_t} \\ \boldsymbol{z}_{j_t} &= \boldsymbol{g}(\boldsymbol{v}_{j_t}) = \tanh(\boldsymbol{W}\boldsymbol{v}_{j_t} + \boldsymbol{b}) \end{aligned} \tag{1}$$

where $t \in [1, |\boldsymbol{H}_p|]$ and $\boldsymbol{M}_t^p = \{\boldsymbol{m}_{1,t}^p, ..., \boldsymbol{m}_{I,t}^p\}$. $\boldsymbol{g}(\cdot)$ is the update function, and $\boldsymbol{W} \in \mathbb{R}^{d \times d}$ and $\boldsymbol{b} \in \mathbb{R}^d$ are its parameters. The attention $\boldsymbol{\alpha}_{a_t} = \{\alpha_{a_t,1}, ..., \alpha_{a_t,I}\}$ is a probability distribution over memory slots in $\boldsymbol{M}_t^p$ using softmax function, which is calculated based on $\boldsymbol{v}_{j_t}$ and the corresponding global key matrix $\boldsymbol{K}^{a_t}$ as follows,

$$\alpha_{a_t,i} = \frac{\exp\left((\boldsymbol{v}_{j_t})^T \boldsymbol{k}_i^{a_t}\right)}{\sum_{s=1}^{I} \exp\left((\boldsymbol{v}_{j_t})^T \boldsymbol{k}_s^{a_t}\right)} \quad \text{where} \quad \boldsymbol{v}_{j_t} = encode(JP_{j_t}) \tag{2}$$

where $i \in [1, I]$ and $\boldsymbol{K}^{a_t} = \{\boldsymbol{k}_1^{a_t}, ..., \boldsymbol{k}_I^{a_t}\}$. Each key slice $\boldsymbol{k}_i^{a_t}$ is a learnable vector. The job posting JP_{j_t} is encoded as a vector $\boldsymbol{v}_{j_t} \in \mathbb{R}^d$ by a hierarchical LSTM encoder [14].

Read Operation. To read out the current preferences $\{\boldsymbol{u}_p^{matching}, \boldsymbol{u}_p^{chat}, \boldsymbol{u}_p^{apply}, \boldsymbol{u}_p^{click}\}$ *w.r.t* the job posting JP_j from the unified preference memory $\boldsymbol{M}^p$, we firstly use $\boldsymbol{v}_j$, i.e. the representation of JP_j, to multiply with the corresponding global key matrices $\{\boldsymbol{K}^{matching}, \boldsymbol{K}^{chat}, \boldsymbol{K}^{apply}, \boldsymbol{K}^{click}\}$ to generate the attentions $\{\boldsymbol{\alpha}_{matching}, \boldsymbol{\alpha}_{chat}, \boldsymbol{\alpha}_{apply}, \boldsymbol{\alpha}_{click}\}$ according to Eq. 2. Then, according to the attentions, we can read out the current preference $\boldsymbol{u}_p^a$ under behavior $a \in A$ from the preference memory $\boldsymbol{M}^p$ *w.r.t* JP_j in Fig. 2(c) as follows,

$$\boldsymbol{u}_p^a = \boldsymbol{\alpha}_a \odot \boldsymbol{M}^p = \sum_{i=1}^{I} \alpha_{a,i} \boldsymbol{m}_i^p \tag{3}$$

where $\odot$ is the dot product and $\boldsymbol{M}^p = \{\boldsymbol{m}_1^p, ..., \boldsymbol{m}_I^p\}$.

With the shared global multi-key matrices $\{\boldsymbol{K}^{click}, \boldsymbol{K}^{apply}, \boldsymbol{K}^{invite}, \boldsymbol{K}^{chat}, \boldsymbol{K}^{matching}\}$ and a private value matrix for each user (e.g. $\boldsymbol{M}^p$ and $\boldsymbol{M}^j$), DMKVMN can transform the heterogeneous behaviors into the unified preference space of $\boldsymbol{M}^p$ ($\boldsymbol{M}^j$). It can not only update the unified preference memory $\boldsymbol{M}^p$ ($\boldsymbol{M}^j$) according to the multi-behavior sequence $\boldsymbol{H}_p$ ($\boldsymbol{H}_j$) via the write operation to capture the dynamic comprehensive preferences, but also can read out the current preferences *w.r.t* the input job posting (curriculum vitae) via the read operation. It relies on the learnable parameters: the parameters in the JP and CV encoders, the global key matrices $\{\boldsymbol{K}^{matching}, \boldsymbol{K}^{chat}, \boldsymbol{K}^{apply}, \boldsymbol{K}^{invite}, \boldsymbol{K}^{click}\}$, and $\{\boldsymbol{W}, \boldsymbol{b}\}$ of the update function $\boldsymbol{g}(\cdot)$. In the next section, we will introduce how to learn them via a multi-task learning framework.

4.2 Bilateral Cascade Multi-Task Learning

To transfer the preferences from auxiliary behaviors {*click*, *invite*, *apply*, *chat*} to the *matching* task, we design a Bilateral Cascade Multi-Task Learning (BCMTL) framework as shown in Fig. 3, which consists of the target *matching* task and five auxiliary tasks. The total loss function is defined as follows,

$$\begin{aligned}\mathcal{L} = \mathcal{L}_{matching} + \lambda_{chat}\mathcal{L}_{chat} + \lambda_{apply}\mathcal{L}^{apply}_{(P,J)} + \lambda_{click}\mathcal{L}^{click}_{(P,J)} \\ + \lambda_{invite}\mathcal{L}^{invite}_{(J,P)} + \lambda_{click}\mathcal{L}^{click}_{(J,P)} + \lambda||\Theta||_2^2\end{aligned} \tag{4}$$

where $_{(P,J)}$ and $_{(J,P)}$ indicate behaviors on the person and job side, respectively. The coefficients λ_{chat}, λ_{apply}, λ_{invite} and λ_{click} control the weights of auxiliary tasks *chat*, *apply*, *invite* and *click*, respectively. λ is the regularization coefficient. Θ denotes the model parameters. We use cross-entropy loss for each task.

According to Eq. 3, we can get the current preferences of person p (job j) *w.r.t* a given job j (person p), i.e. $\boldsymbol{u}_p = \{\boldsymbol{u}_p^{matching}, \boldsymbol{u}_p^{chat}, \boldsymbol{u}_p^{apply}, \boldsymbol{u}_p^{click}\}$ ($\boldsymbol{u}_j = \{\boldsymbol{u}_j^{matching}, \boldsymbol{u}_j^{chat}, \boldsymbol{u}_j^{invite}, \boldsymbol{u}_j^{click}\}$). The unilateral behaviors {*click*, *apply*, *invite*} reflect the preferences of the initial side, e.g. person p *click*s the job posting JP_j of job j since p is attracted by JP_j. The prediction functions of {*click*, *apply*, *invite*} on the person and job sides are defined as follows,

$$\begin{aligned}\hat{y}_{click}(p,j) = \phi_{click}(\boldsymbol{u}_p^{click} \otimes \boldsymbol{v}_j);\ \hat{y}_{apply}(p,j) = \phi_{apply}(\boldsymbol{u}_p^{apply} \otimes \boldsymbol{v}_j) \\ \hat{y}_{click}(j,p) = \phi_{click}(\boldsymbol{u}_j^{click} \otimes \boldsymbol{v}_p);\ \hat{y}_{invite}(j,p) = \phi_{invite}(\boldsymbol{u}_j^{invite} \otimes \boldsymbol{v}_p)\end{aligned} \tag{5}$$

where $\otimes$ is the Hadamard product. The bilateral behaviors {*chat*, *matching*} denotes that both of the person and job are attracted by each other, and their prediction functions are defined as follows,

$$\begin{aligned}\hat{y}_{chat}(p,j) = \phi_{chat}(\mathrm{CAT}([\boldsymbol{u}_p^{chat} \otimes \boldsymbol{v}_j; \boldsymbol{u}_j^{chat} \otimes \boldsymbol{v}_p; \boldsymbol{v}_p; \boldsymbol{v}_j])) \\ \hat{y}_{matching}(p,j) = \phi_{matching}(\mathrm{CAT}([\boldsymbol{u}_p^{matching} \otimes \boldsymbol{v}_j; \boldsymbol{u}_j^{matching} \otimes \boldsymbol{v}_p; \boldsymbol{v}_p; \boldsymbol{v}_j]))\end{aligned} \tag{6}$$

where we concatenate (CAT) the profile representations ($\boldsymbol{v}_p$ and $\boldsymbol{v}_j$) and the interaction features into a vector, and feed it into the prediction functions. Multi-layer Perceptron (MLP) is used as the prediction function ($\phi(\cdot)$) for each task.

Furthermore, the cascade relations among multi-behaviors in Fig. 1 may benefit the multi-task learning. For example, whether person p will *apply* the job j depends on whether p has *clicked* j before (i.e. *click*→*apply*). To utilize the cascade relations *click*→*invite/apply*→*chat*→*matching*, we design a cascade connection in the neural networks of prediction functions to transfer the information. For example, for the cascade relation *click*→*apply*, we denote the last hidden layer state of $\phi_{click}(p,j)$ as $\boldsymbol{o}_{click}(p,j) \in \mathbb{R}^{d_o}$ and let it be the additional input of $\phi_{apply}(p,j)$. Then, the prediction functions are modified as follows,

$$\hat{y}_{apply}^{cascade}(p,j) = \phi_{apply}\big(\mathrm{CAT}([\boldsymbol{u}_p^{apply} \otimes \boldsymbol{v}_j; \boldsymbol{o}_{click}(p,j)])\big)$$
$$\hat{y}_{invite}^{cascade}(j,p) = \phi_{invite}\big(\mathrm{CAT}([\boldsymbol{u}_j^{invite} \otimes \boldsymbol{v}_p; \boldsymbol{o}_{click}(j,p)])\big)$$
$$\hat{y}_{chat}^{cascade}(p,j) = \phi_{chat}\big(\mathrm{CAT}([\boldsymbol{u}_p^{chat} \otimes \boldsymbol{v}_j; \boldsymbol{u}_j^{chat} \otimes \boldsymbol{v}_p; \boldsymbol{v}_p; \boldsymbol{v}_j; \boldsymbol{o}_{apply}(p,j); \boldsymbol{o}_{invite}(j,p)])\big)$$
$$\hat{y}_{matching}^{cascade}(p,j) = \phi_{matching}\big(\mathrm{CAT}([\boldsymbol{u}_p^{matching} \otimes \boldsymbol{v}_j; \boldsymbol{u}_j^{matching} \otimes \boldsymbol{v}_p; \boldsymbol{v}_p; \boldsymbol{v}_j; \boldsymbol{o}_{chat}(p,j)])\big)$$

where $\boldsymbol{o}_{click}$, $\boldsymbol{o}_{apply}$, $\boldsymbol{o}_{invite}$ and $\boldsymbol{o}_{chat}$ denote the last hidden layer state of neural networks ϕ_{click}, ϕ_{apply}, ϕ_{invite} and ϕ_{chat}, respectively.

As shown in Fig. 3, since the cascade relations are bilateral, we call this multi-task learning as Bilateral Cascade Multi-Task Learning (BCMTL). For a person-job pair (p, j), the inputs of the model are the profiles (CV_p and JP_j) and historical multi-behavior sequences ($\boldsymbol{H}_p$ and $\boldsymbol{H}_j$). CV_p and JP_j are firstly encoded into vectors $\boldsymbol{v}_p$ and $\boldsymbol{v}_j$ by a hierarchical LSTM encoder [14]. DMKVMN iteratively updates the unified preference memories $\boldsymbol{M}^p$ and $\boldsymbol{M}^j$ according to $\boldsymbol{H}^p$ and $\boldsymbol{H}^j$, and then reads out the preferences $\boldsymbol{u}_p$={$\boldsymbol{u}_p^{matching}$, $\boldsymbol{u}_p^{chat}$, $\boldsymbol{u}_p^{apply}$, $\boldsymbol{u}_p^{click}$} and $\boldsymbol{u}_j$={$\boldsymbol{u}_j^{matching}$, $\boldsymbol{u}_j^{chat}$, $\boldsymbol{u}_j^{invite}$, $\boldsymbol{u}_j^{click}$} from $\boldsymbol{M}^p$ and $\boldsymbol{M}^j$, respectively. Finally, the interaction features between the preferences and profiles, and the profile features {$\boldsymbol{v}_p$, $\boldsymbol{v}_j$} are fed into the prediction functions with

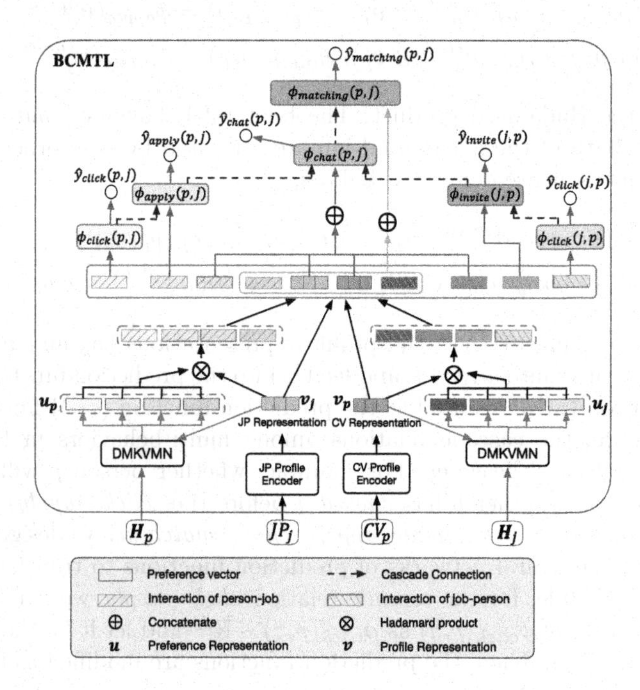

Fig. 3. Bilateral Cascade Multi-Task Learning (BCMTL). Besides modeling the two-sided multi-behaviors as multi-tasks, we also model their cascade relations by feeding the predictions of prerequisite behaviors to the current behavior tasks.

bilateral cascade connections to predict whether each type of behavior will happen between person p and job j in the very near future.

DMKVMN tries to capture the dynamic comprehensive preferences for both persons and jobs from their multi-behavior sequences, and BCMTL tries to transfer the preferences from auxiliary behaviors {*click*, *apply*, *invite*, *chat*} to the target *matching* task. We call this proposed method as ***D***ynamic ***P***erson-***J***ob ***F***it with ***M***ulti-***B***ehavioral ***S***equences (**DPJF-MBS**). The objective function in Eq. 4 is optimized by stochastic gradient descent(SGD) to learn the parameters: the JP and CV encoders, the global key matrices $\{\boldsymbol{K}^{click}, \boldsymbol{K}^{invite}, \boldsymbol{K}^{apply}, \boldsymbol{K}^{chat}, \boldsymbol{K}^{matching}\}$, the update function $\boldsymbol{g}(\cdot)$, and the neural networks $\{\phi_{click}, \phi_{invite}, \phi_{apply}, \phi_{chat}, \phi_{matching}\}$.

Table 1. Statistics of dataset *Finance* and *Technology.*

Dataset	*Finance*	*Technology*
#Multi-behaviors (A):		
#*matching*	6625	37,996
#*chat*	9495	49,142
#*apply* on person side	18,091	66,811
#*invite* on job side	13,517	56,866
#*click* on person side	25,077	98,800
#*click* on job side	21,197	72,739
#Curriculum vitae (CV)	3974	22,020
#Job posting (JP)	11,483	25,481
#Ave. sentences per curriculum vitae	12	16
#Ave. sentences per job posting	10	10
#Ave. words per sentence	22	23

5 Experiment

5.1 Datasets

Two real-world datasets *Finance* and *Technology* provided by the largest online recruitment platform "Boss Zhipin" in China are used as the experimental datasets, which contain job seekers and recruiters, and their multi-behaviors across five months in the industry of finance and technology, respectively. As shown in Fig. 1, both job seekers and recruiters have four types of behaviors with cascade relations and timestamps. Table 1 shows statistics of the two datasets. *Finance* contains 3974 persons and 11,483 jobs, and *Technology* consists of 22,020 persons and 25,481 jobs. Besides the multi-behaviors of users, they also contain users' textual profiles: curriculum vitaes (CVs) and job postings (JPs).

5.2 Experimental Settings

We split the multi-behavior data across five months into three folds: the first three months for training, the fourth month for validation and the last month for test. Each user's multi-behaviors in the train set are sorted along the timeline as a multi-behavioral sequence. For each user, we randomly select a ratio of negative samples (e.g. five times as many as the observed samples) from the non-observed samples which were not interacted by the user.

To analyze the empirical performance of the proposed model extensively, we use four widely used evaluation metrics for ranking and recommendation, including Mean Average Precision (MAP), Mean Reciprocal Rank (MRR), Normalized Discounted Cumulative Gain (nDCG) and AUC.

5.3 Comparative Methods

A neural recommender method NeuMF [15] and several SOTA PJF methods are used as baselines, including PJFNN [2], APJFNN [3], JRMPM [6] and IPJF [5].

- **Neural Collaborative Filtering(NeuMF)** [15]. It's a neural collaborative filtering method which models the non-linear interactions of users and items.
- **Person-Job Fit Neural Network(PJFNN)** [2]. It's a neural profile-matching model which tries to learn the profile representations of persons and jobs by CNN and predict the matching degrees with cosine similarity.
- **Ability-aware Person-Job Fit Neural Network(APJFNN)** [3]. It's another neural profile-matching method which models the job requirements and job seekers' experiences by a hierarchical LSTM.
- **Job-Resume Matching with Profiling Memory(JRMPM)** [6]. It's a PJF method which profiles the preferences of both persons and jobs from their interview (i.e. *matching*) behaviors.
- **Interpretable Person-Job Fit(IPJF)** [5]. It's a neural profile-matching model which tries to learn the interdependence between job postings and resumes with incorporating the intentions of persons and jobs.
- **Dynamic Person-Job Fit with Multi-behavioral Sequences(DPJF-MBS)**. It's our proposed method which models two-sided multi-behavior sequences to capture two-sided dynamic comprehensive preferences.

5.4 Implementation Details

For all the compared methods, we use mini-batch Adam [16] for optimization. Curriculum vitaes and job postings are split into sentences, and each sentence is segmented into words. The word embeddings are pre-trained by using the Skip-gram *word2vec* [17] with dimension $d_w = 128$. We keep the embedding dimension d as 64 for all the representation methods. The mini-batch size is 20. The maximum number of iterations is set to $50 * (|P| + |J|)$. The learning rate is set to 0.005.

For our DPJF-MBS[2]. in each mini-batch, we randomly sample positive and negative samples for each behavior task. We tune the value of $\{\lambda, \lambda_{click}, \lambda_{apply}, \lambda_{invite}, \lambda_{chat}\}$ in {1, 0.1, 0.01, 0.001, 1e−4, 1e−5} and I in {3, 5, 8, 10, 15}. The depth of neural network ϕ is 2 and $d_o = 32$. The profile encoders of CV and JP are hierarchical LSTM [14] including a word-level LSTM and a sentence-level LSTM.

For NeuMF[3] and JRMPM[4], we use the source code provided in the original papers. We implement PJFNN, APJFNN and IPJF according to their original papers. The kernel sizes of two layers in PJFNN are set as the same as in [2] (CV: 5-5 and JP:5-3). For IPJF, we regard the intentions as the *invite/apply* behaviors in our scenario, $\Delta = 0.05$, and tune the value of λ in {1, 0.1, 0.01, 0.001, 1e−4, 1e−5}.

5.5 Experimental Results

Overall Results. Table 2 shows the overall evaluation results of different methods according to MAP, MRR, nDCG and AUC. Our proposed method DPJF-MBS consistently outperforms NeuMF and SOTA PJF baselines. By comparing {JRMPM,IPJF,DPJF-MBS} with {PJFNN,APJFNN}, we can find that modeling the implicit preferences/intentions from users' behaviors is more effective than only matching pairs of CVs and JPs. It confirms that the profiles may not contain or fully reflect the personalized preferences. IPJF is a static profile-matching method with incorporating users' intentions, and JRMPM faces the sparsity problem of *matching* behavior. DPJF-MBS outperforms JRMPM and IPJF, which confirms that it helps to capture the dynamic comprehensive personalized preferences of users by modeling two-sided multi-behavior sequences.

Table 2. Evaluation results. * indicates $p \leq 0.01$ based on Wilcoxon signed rank test.

Dataset	Metric	Method						Improve
		NeuMF	PJFNN	APJFNN	JRMPM	IPJF	DPJF-MBS	
Finance	MAP	0.1421	0.1651	0.1662	0.1711	0.1733	**0.1803***	4.0%
	MRR	0.1889	0.2233	0.2258	0.2310	0.2344	**0.2442***	4.2%
	nDCG	0.2758	0.3094	0.3110	0.3166	0.3188	**0.3278***	2.8%
	AUC	0.5753	0.6243	0.6304	0.6410	0.6684	**0.6867***	2.7%
Technology	MAP	0.1478	0.1592	0.1532	0.1770	0.1755	**0.1978***	11.8%
	MRR	0.2086	0.2209	0.2114	0.2448	0.2428	**0.2719***	11.1%
	nDCG	0.2831	0.3049	0.2973	0.3264	0.3246	**0.3515***	7.7%
	AUC	0.5900	0.6228	0.6114	0.6603	0.6575	**0.7092***	7.4%

[2] https://github.com/BinFuPKU/DPJF-MBS.
[3] https://github.com/hexiangnan/neural_collaborative_filtering.
[4] https://github.com/leran95/JRMPM.

Do Auxiliary Behaviors and Cascade Relations Matter to PJF? To analyze the impact of different auxiliary behaviors and the cascade relations, we design several variants of DPJF-MBS with different combinations of {*matching*(A), *chat*(B), *invite/apply*(C), *click*(D), *cascade relations*} to verify their effectiveness. Since it's hard to explore all the combinations which contain 2^5 situations, we increase behavior types and cascade relations gradually in a heuristic way.

Table 3 shows the evaluation results of different combinations. Firstly, we observe that with more behavior types, all the evaluation metrics increase. It confirms that, the auxiliary behaviors {*click*, *invite/apply*, *chat*} generated by users also reflect users' personalized preferences, which are complementary to those of *matching* behaviors. Fusing the preferences learned from the multi-behaviors can improve the matching prediction (Assumption 1). Furthermore, utilizing the cascade relations (*ALL* in Table 3) among the multi-behaviors can help transfer information in multi-task learning and also improve the matching prediction.

Do Dynamic Preferences Exist and Matter in PJF? It's not easy to directly measure users' dynamic preferences (i.e. changes) based on the textual curriculum vitaes and job postings. Instead of showing several cases, we try to verify that through reverse thinking: if there is no change of user preferences, there should be no significant changes on the evaluation performance when we shuffle the order in the input multi-behavioral sequences.

To ensure the reliability of the results, we tried ten runs and calculated their average. The results *ALL+shuffle* in Table 3 show significant decreases especially on *Finance* when we shuffle the order in each user's multi-behavioral sequence.

Table 3. Effects of different combinations of behaviors and cascade relations. *A,B,C* and *D* denote *matching*, *chat*, *invite/apply* and *click*, respectively. *ALL* = *A+B+C+D+cascade relations*. *shuffle* means the time order in the multi-behavior sequence is shuffled.

Dataset	Combination	MAP(↑%)	MRR(↑%)	nDCG(↑%)	AUC(↑%)
Finance	*A*	0.1670	0.2283	0.3132	0.6397
	A+B	0.1720(3.0)	0.2333(2.2)	0.3175(1.4)	0.6582(2.9)
	A+B+C	0.1744(4.4)	0.2371(3.9)	0.3207(2.4)	0.6736(5.3)
	A+B+C+D	0.1773(6.2)	0.2405(5.3)	0.3236(3.3)	0.6805(6.4)
	ALL	**0.1803**(8.0)	**0.2442**(7.0)	**0.3278**(4.7)	**0.6867**(7.3)
	ALL+shuffle	0.1577	0.2137	0.2991	0.6561
Technology	*A*	0.1754	0.2381	0.3184	0.6548
	A+B	0.1886(7.5)	0.2575(8.1)	0.3350(5.2)	0.6704(2.4)
	A+B+C	0.1936(10.4)	0.2629(10.4)	0.3432(7.8)	0.6897(5.3)
	A+B+C+D	0.1962(11.9)	0.2680(12.5)	0.3481(9.3)	0.7073(8.0)
	ALL	**0.1978**(12.8)	**0.2719**(14.2)	**0.3515**(10.4)	**0.7092**(8.3)
	ALL+shuffle	0.1909	0.2630	0.3432	0.6957

It confirms that, in online recruitment, users' personalized preferences are dynamic and capturing the dynamic preferences is beneficial for PJF. (Assumption 2).

Parameter Sensitivity. We explore the impact of control coefficients $\{\lambda_{chat}, \lambda_{apply}(\lambda_{invite}), \lambda_{click}, \lambda\}$ in Eq. 4 and the number of memory slices I in DMKVMN. For simplicity, we let $\lambda_{apply} = \lambda_{invite}$. Figure 4 shows the impact of these control coefficients. λ_{chat} and $\lambda_{apply/invite}$ are preferred to be set to 0.1, and the best value of λ_{click} is 0.01. λ should be not too large, i.e. not more than 0.01. The number of memory slices I control the storage capacity of the memory network. Figure 5 shows that the number of memory slices impacts the model performance. The value of I should not be too small or too large. With consideration of performance and computational complexity, we suggest to set the value of I to 5.

5.6 Visualization Analysis

To show how DMKVMN transforms the heterogeneous behaviors into the unified preference space based on an attention mechanism, we try to visualize the attentions $\{\boldsymbol{\alpha}_{matching}, \boldsymbol{\alpha}_{chat}, \boldsymbol{\alpha}_{apply}, \boldsymbol{\alpha}_{invite}, \boldsymbol{\alpha}_{click}\}$ of different behavior types over the memory slices. Figure 6 shows the heatmaps of these attentions of the person and job sides on *Finance* and *Technology* where $I = 5$.

From Fig. 6, we find different behavior types have their own attention patterns. Auxiliary behaviors, such as *click*, *apply/invite* and *chat*, present relatively more dispersed distribution than that of *matching*. One reason is that auxiliary behaviors are relatively abundant and contain more diverse preferences. The attention of *matching* focuses more on one memory slice since the *matching* behaviors are sparse. Figure 6 shows that, after transformation, the preferences

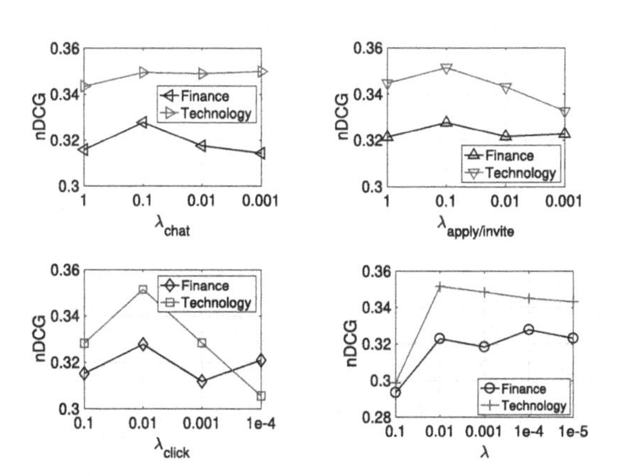

Fig. 4. The sensitivity of control coefficients λ_{chat}, $\lambda_{apply/invite}$, λ_{click} and λ.

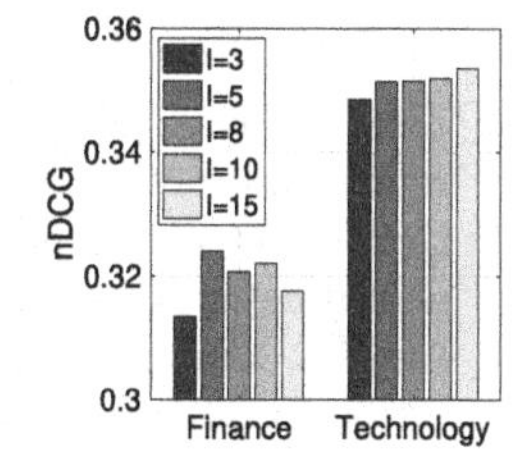

Fig. 5. Effect of I.

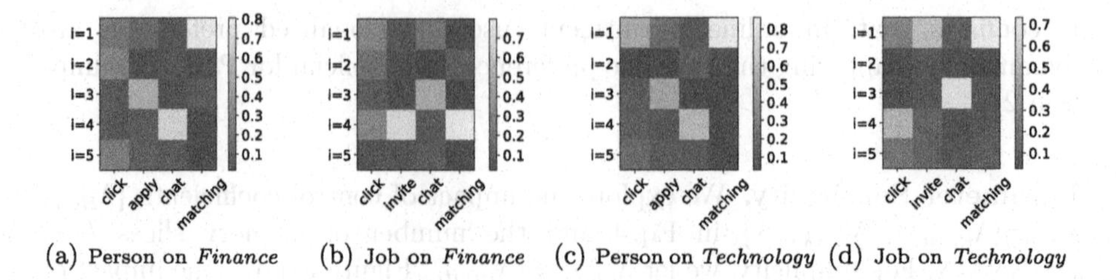

(a) Person on *Finance* (b) Job on *Finance* (c) Person on *Technology* (d) Job on *Technology*

Fig. 6. Attention distribution α of different behavior types over memory slices.

learned from heterogeneous behaviors can be fused harmoniously into the unified preference memory.

6 Conclusion

In this paper, we propose a novel dynamic PJF model for online recruitment, which tries to model two-sided multi-behavioral sequences. A Dynamic Multi-Key Value Memory Network is designed to capture the dynamic comprehensive preferences of both job seekers and recruiters from their multi-behavioral sequences. In addition, a Bilateral Cascade Multi-Task Learning framework is designed to transfer the preferences learned from multiple kinds of auxiliary behaviors to the target *matching* task with consideration of the cascade relations among behaviors. Extensive experiments confirm that: (1) both auxiliary behaviors and cascade relations are beneficial for PJF, (2) users' preferences in PJF are dynamic, (3) our proposed method is an effective dynamic PJF method. For the future work, we will try to utilize the change patterns of users' preferences.

Acknowledgement. This work was partially sponsored by National 863 Program of China (Grant No. 2015AA016009).

References

1. Li, S., et al.: Deep job understanding at linkedin. In: SIGIR, pp. 2145–2148. ACM (2020)
2. Zhu, C., et al.: Person-job fit: adapting the right talent for the right job with joint representation learning. ACM Trans. Manage. Inf. Syst. **9**(3), 12:1–12:17 (2018)
3. Qin, C., et al.: Enhancing person-job fit for talent recruitment: an ability-aware neural network approach. In: SIGIR, pp. 25–34. ACM (2018)
4. Luo, Y., Zhang, H., Wen, Y., Zhang, X.: Resumegan: an optimized deep representation learning framework for talent-job fit via adversarial learning. In: CIKM, pp. 1101–1110. ACM (2019)
5. Le, R., Hu, W., Song, Y., Zhang, T., Zhao, D., Yan, R.: Towards effective and interpretable person-job fitting. In: CIKM, pp. 1883–1892. ACM (2019)

6. Yan, R., Le, R., Song, Y., Zhang, T., Zhang, X., Zhao, D.: Interview choice reveals your preference on the market: to improve job-resume matching through profiling memories. In: KDD, pp. 914–922. ACM (2019)
7. Edwards, J.R.: Person-job fit: a conceptual integration, literature review, and methodological critique. Int. Rev. Ind. Organ. Psychol. **6**, 283–357 (1991)
8. Caldwell, D.F., Iii, C.A.O.: Measuring person-job fit with a profile-comparison process. J. Appl. Psychol. **75**(6), 648–657 (1990)
9. Paparrizos, I.K., Cambazoglu, B.B., Gionis, A.: Machine learned job recommendation. In: RecSys, pp. 325–328. ACM (2011)
10. Gupta, A., Garg, D.: Applying data mining techniques in job recommender system for considering candidate job preferences. In: ICACCI, pp. 1458–1465. IEEE (2014)
11. Hong, W., Li, L., Li, T., Pan, W.: IHR: an online recruiting system for xiamen talent service center. In: KDD, pp. 1177–1185. ACM (2013)
12. Li, L., Li, T.: MEET: a generalized framework for reciprocal recommender systems. In: CIKM, pp. 35–44. ACM (2012)
13. Yang, S., Korayem, M., AlJadda, K., Grainger, T., Natarajan, S.: Combining content-based and collaborative filtering for job recommendation system: a cost-sensitive statistical relational learning approach. Knowl. Based Syst. **136**, 37–45 (2017)
14. Li, J., Luong, M., Jurafsky, D.: A hierarchical neural autoencoder for paragraphs and documents. In: ACL (1), pp. 1106–1115 (2015)
15. He, X., Liao, L., Zhang, H., Nie, L., Hu, X., Chua, T.: Neural collaborative filtering. In: WWW, pp. 173–182. ACM (2017)
16. Kingma, D.P., Ba, J.: Adam: a method for stochastic optimization. In: ICLR (Poster) (2015)
17. Mikolov, T., Chen, K., Corrado, G., Dean, J.: Efficient estimation of word representations in vector space. In: ICLR (Workshop Poster) (2013)

A Local Similarity-Preserving Framework for Nonlinear Dimensionality Reduction with Neural Networks

Xiang Wang[1], Xiaoyong Li[1], Junxing Zhu[1(✉)], Zichen Xu[2], Kaijun Ren[1,3], Weiming Zhang[1,3], Xinwang Liu[3], and Kui Yu[4]

[1] College of Meteorology and Oceanography, National University of Defense Technology, Changsha, China
{xiangwangcn,sayingxmu,zhujunxing,renkaijun,wmzhang}@nudt.edu.cn

[2] College of Computer Science and Technology, Nanchang University, Nanchang, China
xuz@ncu.edu.cn

[3] College of Computer Science and Technology, National University of Defense Technology, Changsha, China
xinwangliu@nudt.edu.cn

[4] School of Computer Science and Information Engineering, Hefei University of Technology, Hefei, China
yukui@hfut.edu.cn

Abstract. Real-world data usually have high dimensionality and it is important to mitigate the curse of dimensionality. High-dimensional data are usually in a coherent structure and make the data in relatively small true degrees of freedom. There are global and local dimensionality reduction methods to alleviate the problem. Most of existing methods for local dimensionality reduction obtain an embedding with the eigenvalue or singular value decomposition, where the computational complexities are very high for a large amount of data. Here we propose a novel local nonlinear approach named *Vec2vec* for general purpose dimensionality reduction, which generalizes recent advancements in embedding representation learning of words to dimensionality reduction of matrices. It obtains the nonlinear embedding using a neural network with only one hidden layer to reduce the computational complexity. To train the neural network, we build the neighborhood similarity graph of a matrix and define the context of data points by exploiting the random walk properties. Experiments demenstrate that *Vec2vec* is more efficient than several state-of-the-art local dimensionality reduction methods in a large number of high-dimensional data. Extensive experiments of data classification and clustering on eight real datasets show that *Vec2vec* is better than several classical dimensionality reduction methods in the statistical hypothesis test, and it is competitive with recently developed state-of-the-art UMAP.

Keywords: Dimensionality reduction · High-dimensional data · Embedding learning · Skip-gram · Random walk

C. S. Jensen et al. (Eds.): DASFAA 2021, LNCS 12682, pp. 376–391, 2021.
https://doi.org/10.1007/978-3-030-73197-7_25

Table 1. Computational complexities of *Vec2vec* and other four state-of-the-art local dimensionality reduction (manifold learning) methods [4,12,20]. n is the number of data and k is the number of selected neighbors. D is the input dimensionality and d is the output dimensionality. $|E|$ is the number of edges in the adjacency graph.

Method	Computational	Memory
LLE	$O(n \log n \cdot D + n^2 \cdot d)$	$O(\|E\| \cdot d^2)$
LE	$O(n \log n \cdot D + n^2 \cdot d))$	$O(\|E\| \cdot d^2)$
LTSA	$O(n \log n \cdot D + n^2 \cdot d)$	$O(n^2)$
t-SNE	$O(n^2 \cdot d)$	$O(n^2)$
Vec2vec	$O(n \log n \cdot D + n \cdot d)$	$O(n^2)$

1 Introduction

Real-world data, such as natural languages, digital photographs, and speech signals, usually have high dimensionality. Moreover, coherent structure in the high-dimensional data leads to strong correlations, which makes the data in relatively small true degrees of freedom. To handle such high-dimensional real-word data effectively, it is important to reduce the dimensionality while preserving the properties of the data for data analysis, communication, visualization, and efficient storage.

Generally, there are two kinds of methods for dimensionality reduction: one is to preserve the global structure and the other is to preserve the local geometry structure [12,19]. First, for the dimensionality reduction methods of preserving the global structure of a data set, the most popular methods are Principal Components Analysis (PCA), Linear Discriminant Analysis (LDA), Multidimensional Scaling (MDS), Isometric Feature Mapping (Isomap), Autoencoders, and Sammon Mappings [4,20] and so on. We note that these global methods are either in a strong linearity assumption or can not capture local manifold intrinsic geometry structure [19]. Second, for the dimensionality reduction methods of preserving the local geometry structure of a data set, there are some typical methods like Locally Linear Embedding (LLE), Laplace Eigenmaps (LE), Local tangent space alignment (LTSA), t-SNE, LargeVis, and UMAP [8,9,12]. These methods can learn the manifold intrinsic geometry structure of a data set, which is a very useful characteristic in pattern recognition [18]. Most of them share a common construction paradigm: they first choose a neighborhood for each point and then take an eigenvalue decomposition or a singular value decomposition to find a nonlinear embedding [19]. **However, the computational complexities of obtaining an embedding with the eigenvalue decomposition or singular value decomposition ($O(n^2)$) are unbearably expensive**[1]**, especially when facing a large number of high-dimensional data.** Table 1

[1] Complexity analysis of manifold learning. https://scikit-learn.org/stable/modules/manifold.html.

shows the high computational complexities of four typical local dimensionality reduction methods comparing to *Vec2vec* [4,12,20].

In recent years, with the success of Word2vec, the embedding representation learning of words [14], documents, images, networks, knowledge graphs, biosignals, and dynamic graph are developed and successfully applied [2,3,6]. This kind of method transfers the raw data like texts and graphs to low-dimensional numerical vectors or matrices for computing. They implement dimensionality reduction of data, but they are specified in the raw data like texts or graphs [3], since they utilize the context or structure of the data points for computing. In this kind of methods, Word2vec employs neural networks with a single hidden layer to learn the low-dimensional representation of words [13]. The computational and memory complexities of obtaining an embedding in the method are linear to the number of data. **Comparing to the eigenvalue or singular value decomposition for obtaining an embedding in many manifold learning methods, it can significantly reduce the computational and memory costs.** The skip-gram model in Word2vec has been successfully applied in the embedding of graphs and networks [7,15]. Word2vec employs the contexts (co-occurrences) of words in texts to learn embedding, but there are no explicit contexts/co-occurrences for the data points in a matrix. Therefore, it cannot be applied to general purpose dimensionality reduction of matrices.

To address these problems, we propose a general-purpose dimensionality reduction approach named *Vec2vec*, which preserves the local geometry structure of high-dimensional data. It combines the advantages of local manifold dimensionality reduction methods and Word2vec. It is not specified in the raw data like texts or graphs and can be applied in the dimensionality reduction of any matrices, and it boosts the computational efficiency of obtaining an embedding simultaneously. To achieve these purposes, we generalize the skip-gram model in Word2vec to obtain the embedding and design an elaborate objective to preserve the proximity of data points. We select the neighbors of the data points to establish a neighborhood similarity graph, and define the contexts of data as the sequences of random walks in the neighborhood similarity graph to solve the objective. We conduct extensive experiments of data classification and clustering on eight typical real-world image and text datasets to evaluate the performance of our method. The experimental results demonstrate that *Vec2vec* is better than several classical well-known dimensionality reduction methods in the statistical hypothesis test. Our method is competitive with recently developed state-of-the-art UMAP but more efficient than it in high-dimensional data. Our *Vec2vec* is more efficient than LLE and LE in a dataset with both a large number of data samples and high-dimensionality.

2 Related Work

Embedding Representation Learning. With the success of Word2Vec model in word representation, embedding representation learning has been widely studied in words, documents, networks, knowledge graphs, biosignals, dynamic graph and so on [2,3,6]. Mikolov et al. [13] proposed CBOW and skip-gram models, which were widely used in many embedding methods. Pennington et al. [14] proposed the GloVe model and learned the embedding representation using matrix decomposition. Bojanowski et al. [1] proposed the FastText model to enrich word embedding representation with subword information. There are also Doc2Vec, Skip-thoughts, PTE, and Paragram-phrase models to learn the embedding of sentences and documents [11].

There are many embedding representation learning models like TADW, TriNDR, TransE, TransConv, RDF2Vec, MrMine and LINE in different kinds of networks or graphs [2,3,23]. Some methods like Node2vec, Metapath2Vec, and DeepWalk first find the neighbors of a node using random walks, and then employ the skip-gram model to learn the embedding [7,15].

However, existing methods are limited to the specified raw data like words, documents, graphs, and so on. They are designed to utilize the context of texts and the structures of graphs for representation learning of words, documents, and graphs [1,3]. Therefore, they can not be directly applied to general-purpose dimensionality reduction.

Dimensionality Reduction. Generally, there are two kinds of methods in dimensionality reduction [12,19], which preserve the global structure and local geometry structure. In recent years, there are also some methods that employ deep learning to reduce the dimensionality of data [22,24].

There are a lot of classical dimensionality reduction methods that focus on the global structure of a data set, such as PCA, LDA, MDS, Isomap, and Sammon Mapping [4,20]. The linear unsupervised PCA optimizes an object that maximizes the variance of the data representation, and there are many successful variants of PCA like Kernel PCA (KPCA) and incremental PCA algorithms [5]. The nonlinear Isomap optimizes geodesic distances between general pairs of data points in the neighborhood graph. These global methods construct dense matrices that encode global pairwise information [19].

There are also many local manifold learning methods that discover the intrinsic geometry structure of a data set like LLE, LE, LTSA, t-SNE, LargeVis, and UMAP [8,12]. These methods choose neighbors for each data point and obtain a nonlinear embedding from an eigenvalue decomposition or a singular value decomposition [19]. The nonlinear LLE preserves solely local properties of the data and considers the high-dimensional data points as a linear combination of their nearest neighbors. UMAP is an effective state-of-the-art manifold learning technology for dimension reduction based on Riemannian geometry and algebraic topology [12]. The computational complexity of UMAP is $O(n^{1.14}{\cdot}D{+}k{\cdot}n)$. Local manifold learning methods usually obtain an embedding from an eigenvalue or a singular value decomposition [19] and the computational cost is expensive.

With the success of auto-encoder, there are some deep learning methods for dimensionality reduction like Local Deep-Feature Alignment (LDFA) [24], extreme learning machine auto-encoder (ELM-AE) [10], and Deep Adaptive Exemplar AutoEncoder [16].

3 Methodology

3.1 Overview

The problem of dimensionality reduction can be defined as follows. Consider there is a dataset represented in a matrix $M \in \mathbb{R}^{n \times D}$, consisting of n data points (vectors) $x_i (i \in 1, 2, 3, \cdots, n)$ with dimensionality D. In practice, the feature dimension D is often very high. Our purpose is to transfer the matrix $M^{n \times D}$ to a low-dimensional matrix $Z^{n \times d} \in \mathbb{R}^{n \times d} (d \ll D)$ with a function f while preserving the most important information in the matrix M. Formally, we can use Eq. (1) to represent the problem.

$$Z^{n \times d} = f(M^{n \times D}) \tag{1}$$

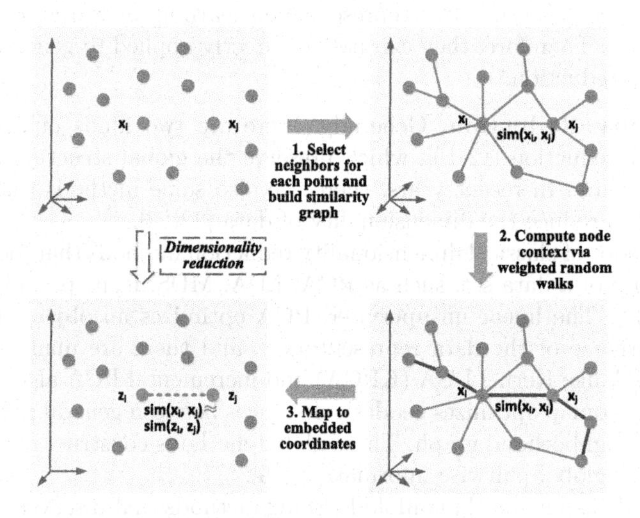

Fig. 1. The steps of *Vec2vec*: (1) Select nearest neighbors for each data point and construct an adjacency graph. Compute the similarity $sim(x_i, x_j)$ of x_i and its neighbor point x_j as the weight of the edge (x_i, x_j) in the adjacency graph. (2) Find the contexts of each data point by performing short random walks. The co-occurrences of the data points in the random sequences reflect their similarity relationships in the neighborhood graph. (3) Compute the low-dimensional embedding vector z_i while preserving the similarities of x_i and its neighbors with neural networks.

Vec2vec preserves the pairwise similarities of vectors in a matrix, which are fundamental to machine learning and data mining algorithms. It means that if $sim(x_i, x_j)$ is bigger than $sim(x_a, x_b)$, then $sim(z_i, z_j)$ is going to be bigger than $sim(z_a, z_b)$ in the low-dimensional target space.

The skip-gram model is originally developed for learning the embedding representation of words in natural languages. We generalize the representation learning model to obtain the embedding. In the skip-gram model, there is a hypothesis that words are likely to be similar if they have similar contexts in sentences. Like words in sentences, our basic hypothesis is that the data points are likely to be similar if they have similar contexts in the feature space. Given the linear nature of texts, it is natural to define the context of words as a sliding window in sentences. However, there are no contexts for data points in a matrix. To solve the problem, we build the neighborhood similarity graph of a matrix and define the context of data points in the matrix as the co-occurrences of data points in the paths of random walks of the graph.

As shown in Fig. 1, there are mainly three steps in our *Vec2vec* method for dimensionality reduction. The details of the three steps are described in the following subsections.

3.2 Building Neighborhood Similarity Graph

To define the contexts of data points in a matrix and preserve the similarity relationship between the data points, we build an adjacency graph based on their pairwise similarities. We define the similarity graph of a matrix in Definition 1.

Definition 1 (Similarity Graph (SG). A similarity graph SG of a matrix M is a weighted undirected graph, where the nodes in SG are one-to-one correspondence to the data points in M. There are edges between data points and their selected neighbors. The weights of edges are the similarities of the corresponding data points.

In Definition 1, given a matrix M with n data points, each data point x_i is represented as a node v_i in SG, thus there are n nodes in SG. Node v_i and v_j are connected by an edge if x_i (x_j) is one of the most similar vectors of x_j (x_i). There are two variations to select neighbors for a data point: (a) ϵ-neighborhoods ($\epsilon \in \mathbb{R}$). Node v_i and v_j are connected by an edge if $sim(x_i, x_j) > \epsilon$. This variation is geometrically motivated and the pairwise relationship is naturally symmetric, but it is difficult to choose ϵ. (b) *Topk* nearest neighbors ($topk \in \mathbb{N}$). Nodes v_i and v_j are connected by an edge if v_i is one of the *topk* nearest neighbors of v_j or v_j is one of the *topk* nearest neighbors of v_i. This variation is easier to implement and the relation is symmetric, but it is less geometrically intuitive comparing to the ϵ-neighborhoods variation. In our experiments, we choose the *topk* nearest neighbors variation for building the similar graph.

To compute the edge weight of two nodes, many commonly used similarity/distance functions like Euclidean distance, Minkowski distance, cosine similarity for vectors can be used. In all our experiments, we choose the similarity function "cosine measure" shown in Eq. (2) to preserve the pairwise similarity relationships of the vectors in M.

$$sim(v_i, v_j) = sim(x_i, x_j) = (x_i \cdot x_j)/(\|x_i\| \cdot \|x_j\|) \tag{2}$$

The computational complexity of building the similarity graph is $O(n^2)$. We employ the K-Nearest Neighbor method with a ball tree to build the similarity graph and the computational complexity of this step can be reduced to

$O(nlog(n) \cdot D)$. In a distributed system, this step can be further speed up using a parallel method since we only need to compute the pairwise similarities or distances of data vectors in M. In the case of $n \gg D$, we can further reduce the computational complexity of building similarity graph to $O(Dlog(D) \cdot n)$, if we build the neighborhood graph with the transpose of M ($M \in \mathbb{R}^{n \times D}$). In this case, if Z ($Z \in \mathbb{R}^{D \times d}$) is the target low-dimensional matrix of M^T, we can get the target low-dimensional matrix $\acute{Z}$ ($\acute{Z} \in \mathbb{R}^{n \times d}$) of M as $\acute{Z} = M \cdot Z$.

3.3 Node Context in Similarity Graphs

It is natural to define the context of words as a sliding window in sentences. However, the similarity graph is not linear and we need to define the notation of the contexts of data points in the similarity graph. We use random walks in the similarity graph to define the contexts of data points. Random walks have been used as a similarity measure for a variety of problems in the content recommendation, community detection, and graph representation learning [7,15]. The detection of local communities motivates us to use random walks to detect clusters of data points, and the random walk model is effortless to parallelize and several random walkers can simultaneously explore different parts of a graph. Therefore, we define the context of a data point in the similarity graph in Definition 2 based on random walks. With the definition, we define the data points around a data point in the random walk sequences as its contexts. With the linear nature of the sequences, we define the context of data points as a sliding window in the sequences.

Definition 2 (Node Context in Similarity Graphs). The node context of a data point in similarity graphs is the parts of a random walk that surround the data point.

Formally, let $(x_{w1}, x_{w2}, \cdots, x_{wl})$ denote a random walk sequence with length l. We use a small sliding window c to define the context of a data point. Then given a data point x_{wj} in the random walk sequence, we can define its node context $NC(x_{wj})$ in Eq. (3).

$$NC(x_{wj}) = \{x_{wm} | -c \leq m - j \leq c, m \in (1, 2, \cdots, l)\} \tag{3}$$

The random walk sequences in similarity graphs can be defined as follows. A random walk is a Markov chain, and the *t-th* data point only depends on the $(t-1)$-*th* data point in a random walk. The *t-th* data point x_{wt} is generated by the probability distribution defined in Eq. (4).

$$P(x_{wt} = v_a | x_{w(t-1)} = v_b) = \begin{cases} \frac{sim(v_a, v_b)}{\mathbb{Z}}, & if\ (v_a, v_b) \in E, \\ 0, & otherwise \end{cases} \tag{4}$$

where E is the edge set of the similarity graph and $sim(v_a, v_b)$ is the edge weight of v_a and v_b defined in Eq. 2). $\mathbb{Z}$ is the normalizing constant and $\mathbb{Z} = \sum_{(v_b, v_i) \in E} sim(v_b, v_i)$. For each data point in the similarity graph, we simulate a fixed number of random walks. For each random walk, we simulate it in a short fixed-length l. Our method ensures that every data point in M is sampled to the node contexts of data points.

In this paper, we assume that there are no rare vectors whose similarities with other vectors are all too small to consider. It means that there are no isolated nodes in the similarity graph. This assumption is acceptable if the dataset is not too small or too sparse, such as the image and text datasets in our experiments.

3.4 The Low-Dimensional Embedding Representation

Based on the node contexts of the data points in the similarity graph, we extend the skip-gram model to learn the embedding of the data in the matrix. The skip-gram model aims to learn continuous feature representations of words by optimizing a neighborhood preserving likelihood objective [13]. It is developed to learn the similarity of words from texts by utilizing the co-occurrence of words. The architecture of learning the low-dimensional embedding representation of metrics based on the skip-gram model is shown in Fig. 2. It is a neural network with only one hidden layer and the goal of this network is to learn the weight matrix W of the hidden layer, which is actually the target embedding matrix Z ($Z = W$) of the original high-dimensional matrix M.

We use one-hot encoding to represent the data points in the input layer. The data point x_i is represented as a vector $o_i \in \mathbb{R}^n$ in the one-hot encoding, where all elements are zero except the i-th element being one. It means that we can get the low-dimensional representation z_i of x_i using equation $z_i = o_i \cdot W$, and the output of the hidden layer of the neural network is $z_i = f(x_i)$. Given $W = [w^1, w^2, \cdots, w^n] \in \mathbb{R}^{n \times d}$, then z_i can be represented as Eq. (5).

$$z_i = f(x_i) = o_i \cdot W = w_i \tag{5}$$

The output layer is a softmax regression classifier. The input of this layer is the target embedding vector z_i of x_i. The output of this layer is the probability distribution of all data points with the input x_i. The neural network is trained by the data pairs in the node contexts of the similarity graph defined in Sect. 3.3. We formulate the dimensionality reduction of matrix M as a maximum likelihood optimization problem. The objective function we seek to optimize is shown in Eq. (6), which maximizes the log-probability of observing the node context $NC(x_i)$ of the data point x_i ($x_i \in M$) conditioned on its feature representation $z_i = f(x_i)$ with the mapping function $f : M^{n \times D} \longrightarrow \mathbb{R}^{n \times d}$ in Eq. (6). $NC(x_i)$ is defined in Eq. (3). We introduce the neighborhood similarity graph and node context to compute the $NC(x_i)$, which is different from the original skip-gram model.

$$\max_f \sum_{x_i \in M} logPr(NC(x_i)|f(x_i)) \tag{6}$$

To optimize the Eq. (6), we assume that the likelihood of observing a neighborhood data point is independent of observing any other neighborhood data

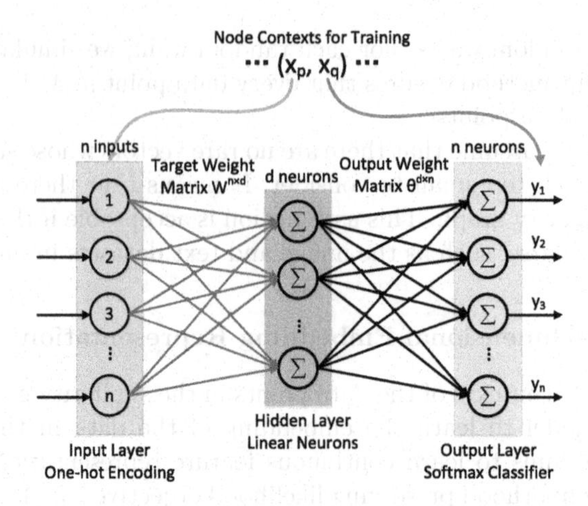

Fig. 2. The architecture of learning the low-dimensional embedding representation of metrics based on the skip-gram model. It is trained by the data point pairs in the node contexts of the similarity graph.

points given the representation of the source. Hence, the objective function of Eq. (6) can be changed to Eq. (7).

$$\max_{f} \sum_{x_i \in M} log \prod_{x_j \in NC(x_i)} Pr(x_j|f(x_i)) \tag{7}$$

In Eq. 7, the data point pair (x_i, x_j) $(x_j \in NC(x_i))$ is used to train our model. Let $\theta = [\theta_1^T, \theta_2^T, \cdots, \theta_n^T]^T$ be the output weight matrix and θ_j^T $(\theta_j \in \mathbb{R}^d)$ be the j-th columns of θ, then θ_j^T is the corresponding weight vector of the output data point x_j. As shown in Fig. 2, we employ the softmax function to compute $Pr(x_j|f(x_i))$ in Eq. 7. Then $Pr(x_j|f(x_i))$ can be calculated as Eq. (8).

$$Pr(x_j|f(x_i)) = \frac{exp(\theta_j^T \cdot f(x_i))}{\sum_{x_m \in M} exp(\theta_m^T \cdot f(x_i))} \tag{8}$$

We find that $Pr(x_j|f(x_i))$ is expensive to compute for a large dataset, since the computational cost is proportional to the number of data. Therefore, we approximate it with negative sampling for fast calculation. Let $P_n(x)$ be the noise distribution to select negative samples and k be the number of negative samples for each data sample, then $Pr(x_j|f(x_i))$ can be calculated using Eq. (9).

$$Pr(x_j|f(x_i)) = \sigma(\theta_j^T \cdot f(x_i)) \prod_{neg=1}^{k} \mathbb{E}_{x_{neg} \sim P_n(x)}[\sigma(-\theta_{neg}^T \cdot f(x_i))] \tag{9}$$

where $\sigma(w) = 1/(1 + exp(-w))$. Empirically, $P_n(x)$ can be the unigram distribution raised to the 3/4rd power [13]. In our experiments, we use the toolkit *Gensim* to implement the negative sampling and the sample threshold is set to

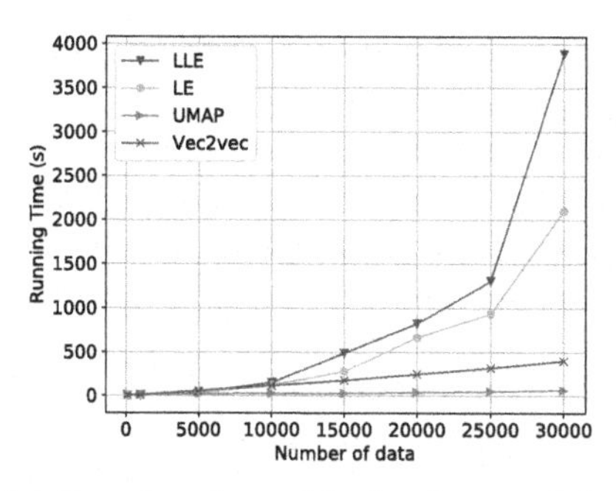

(a) Running time with the change of the number of data

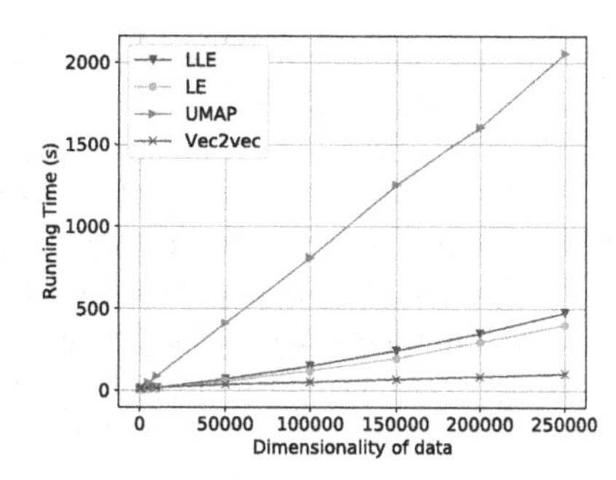

(b) Runing time with the change of the dimensionality of data

Fig. 3. The computational times of the four local dimensionality reduction methods with the change of the number and dimensionality of data. (a) The dimensionality is fixed to 3,072. (b) The number of data points is fixed to 2,000.

be 0.001. As a result, given $f(x_i) = w_i$ in Eq. (5), the objective function for learning the embedding representation can be written as Eq. (10).

$$\max_f \sum_{x_i \in M} \sum_{x_j \in NC(x_i)} [log\sigma(\theta_j^T \cdot w_i) + \sum_{neg=1}^{k} \mathbb{E}_{x_{neg} \sim P_n(x)} log\sigma(-\theta_{neg}^T \cdot w_i)] \quad (10)$$

To alleviate the problem of over fitting, we add the L2 normalization to the objective function. We finally minimize the objective function $J(W, \theta)$ in Eq. (11).

$$\begin{aligned} J(W,\theta) = & -\frac{1}{n}\{\sum_{x_i \in M} \sum_{x_j \in NC(x_i)} [log\sigma(\theta_j^T \cdot w_i) \\ & + \sum_{neg=1}^{k} \mathbb{E}_{x_{neg} \sim P_n(x)} log\sigma(-\theta_{neg}^T \cdot w_i)]\} + \frac{\lambda}{2}(||W||_2 + ||\theta||_2) \end{aligned} \quad (11)$$

We finally solve Eq. (11) with stochastic gradient descent (SGD) to train the neural network.

4 Experiments

4.1 Experimental Setup

For all the classification and clustering tasks, we compare the performance of *Vec2vec* with the six unsupervised methods: (1) PCA, (2) CMDS, (3) Isomap, (4) LLE, (5) LE, and (6) UMAP [12]. PCA, CMDS, and Isomap are typical global methods, while LLE, LE, and UMAP are typical local methods. We use the implementations of the first five methods in the "scikit-learn" toolkit for experiments, and the implementation of UMAP in Github[2]. To get the best performance of these methods, we set the number of neighbors for each point of Isomap, LLE, LE, and Umap be range from 2 to 30 with step 2 in the experiments. According to [12], UMAP is significantly more efficient than t-SNE and LargeVis when the output dimensions are larger than 3. Therefore we only compare to UMAP in this paper. We do not compare our method with embedding representation learning methods like Word2vec, Doc2vec, Node2vec, Deepwalk, and LINE [7,15,17] since they are specific in words, documents, graphs or networks. They cannot be adaptive to the general purpose dimensionality reduction.

Table 2. The details of the eight text and image datasets used in our experiments

	Dataset name	Number of data	Dimensionality
Image dataset	MNIST	5,000	784
	Coil-20	1,440	1,024
	CIFAR-10	5,000	3,072
	SVHN	5,000	3,072
Text dataset	Movie Reviews	5,000	26,197
	Google Snippets	5,000	9,561
	20 Newsgroups	2,000	374,855
	20 Newsgroups Short	2,000	13,155

In the experiments of classification and clustering, we select four typical real-world image datasets from a variety of domains as shown in Table 2. For computational reasons, we randomly select 5,000 digits of the SVHN dataset, the CIFAR-10 dataset and the MNIST dataset for our experiments like [20]. We represent each images in the datasets as a vector. To test the performance of *Vec2vec* in high-dimensional data, we select four typical text datasets as shown in Table 2. For the 20 Newsgroups short dataset, we only select the title of the articles in the 20 Newsgroups dataset. In pre-process, we represent each image or text to a vector. We perform some standard text preprocessing steps like stemming, removing stop words, lemmatization, and lowercasing on the datasets. We employ the "TFIDF" method to compute the weights of the words.

4.2 Computational Time

We compare the computational time of *Vec2vec* with the three local state-of-the-art dimensionality reduction methods with the change of the number and dimensionality of data. As shown in Fig. 3(a), the computational time of UMAP grows slowest with the growth of the number of data points, and the computational time of *Vec2vec* is the second slowest. As we know, UMAP first constructed

[2] UMAP in Github. https://github.com/lmcinnes/umap.

a weighted k-neighbor graph and then learned a low dimensional layout of the graph. The first step needs most of the computational time and UMAP optimizes it with an approximate nearest neighbor descent algorithm [12], while the implementation of our *Vec2vec* did not use a approximate algorithm (the quick approximate algorithm can also be used in our method). So *Vec2vec* is understandable to be a little slower than UMAP. The computational times of LLE and LE are smaller than UMAP when the input number of data is less than 2,000, but the running times of them increase sharply with the growth of the number of data points. The results show the computational efficiency of *Vec2vec* and UMAP in the local dimensionality reduction of large scale of data.

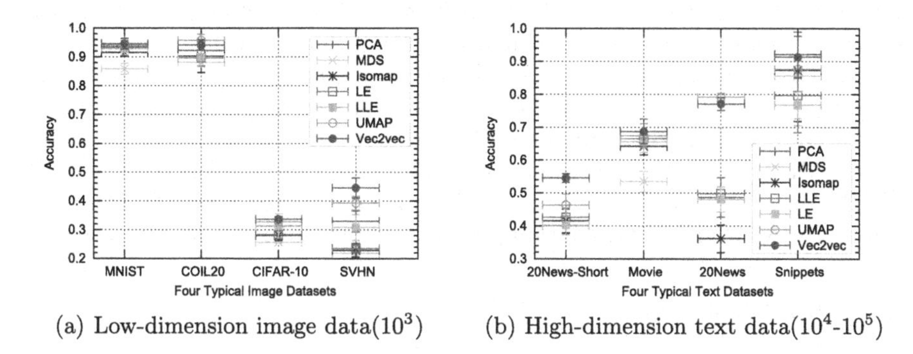

(a) Low-dimension image data(10^3) (b) High-dimension text data(10^4-10^5)

Fig. 4. The accuracy of the methods on image and text classification.

Figure 3(b) shows the computational time of the four dimensionality reduction methods with the change of the input dimensionality of data. We can find that *Vec2vec* needs the least time when dimensionality reaches nearly 20000. When the dimensionality reaches 100,000, *Vec2vec* needs less computational time than LLE, and the times of the two methods are 146.91 and 140.89. When the dimensionality reaches 150,000, *Vec2vec* needs nearly the same computational time with LLE. The experimental results show that *Vec2vec* is more suitable for dimensionality reduction of high-dimensional data than UMAP, LLE and LE.

In summary, UMAP is scalable to a large number of data, but is sensitive to the growth of data dimensionality, while *Vec2vec* is efficient in both a large number of data and high-dimensional data. LLE and LE get better computing performance in a small amount of data and low-dimensional data.

4.3 Data Classification

For all the classification experiments, we employ KNN (K-Nearest Neighbor) as our classifier like [20] and use the implementation of KNN in "Scikit-learn". For the choice of parameter $k(k = 1, 3, 5, 7, 9, 11)$ in KNN, we use "GridSearch" to find the best parameter k in all datasets. We use 4-fold cross-validation to test the performances of different methods and use the accuracy as the performance measure in all the classification experiments. We use the mean accuracy and the 95% confidence interval of the accuracy estimate (2 times the standard deviation) to be the performance measures.

UMAP is competitive with *Vec2vec* as shown in Fig. 4. We assume that the results are in Gaussian distribution and employ the Student's t-test (alpha=0.1) to test the significant difference between the two methods in the eight datasets. The H-value and p-value are 0 and 0.1543. **Therefore, UMAP is as good as *Vec2vec* in statistics and the performances of the two methods have no significant difference in the eight datasets.** The skip-gram model used in *Vec2vec* is originally developed to learn the embedding of words while preserving the similarities. The results show that it can be generalized to obtain the embedding while preserving the similarity of data points, which are important for data classification. For the local LLE and LE methods, the H-values are both 1 and the p-values are 0.0195 and 0.0122. **Therefore, *Vec2vec* is significant better than LLE and LE in the eight datasets in statistics.** For the global PCA, CMDS, and Isomap method, the p-values are 0.0819, 0.0219, and 0.0372. The H-values are all 1. **Therefore, *Vec2vec* is significantly better than the global PCA, CMDS, and Isomap in the eight datasets in statistics.**

4.4 Data Clustering

For all the clustering experiments, we employ spectral clustering with kernel RBF (Radial Basis Function) for clustering, since it is one of the best clustering methods [21]. We use the implementation of spectral clustering in the "scikit-learn" library. In our experiments, we set the number of clusters for spectral clustering to be the number of classes in the datasets and set the range of hyperparameter *gamma* to be from 10^{-6} to 10^{1}. We use the "Adjusted Rand Index (ARI)" to be the evaluation metric.

Table 3. The performance of the methods on data clustering. The evaluation metric is "Adjusted Rand Index (ARI)". The larger the value, the better the method.

Properties to preserve	Method	MNIST	COIL20	CIFAR-10	SVHN	Snippets	20News-Short	20News	Movie
Global structure	PCA	0.3682	0.6269	0.0598	0.0185	0.0070	0.0140	0.0421	0.0016
	CMDS	0.3747	0.6472	0.0579	0.0002	0.1007	0.0010	0.0040	0.0006
	ISOMAP	0.5136	0.567	0.0533	0.0063	0.0056	0.0178	0.0892	0.0216
Local structure	LLE	0.3950	0.4522	0.0443	0.0049	0.0151	0.0066	0.0780	0.0136
	LE	0.1104	0.4570	0.0113	0.0008	0.0082	0.0024	0.0065	0
	UMAP	**0.6819**	0.7376	0.0044	**0.0604**	0.3145	0.0447	0.3065	0.0020
	Vec2vec	0.5549	**0.8093**	**0.0605**	0.0200	**0.5191**	**0.1085**	**0.3066**	**0.1080**

To test the performance of *Vec2vec*, we compare the performances in the eight datasets as shown in Table 3. We can find that *Vec2vec* gets the best

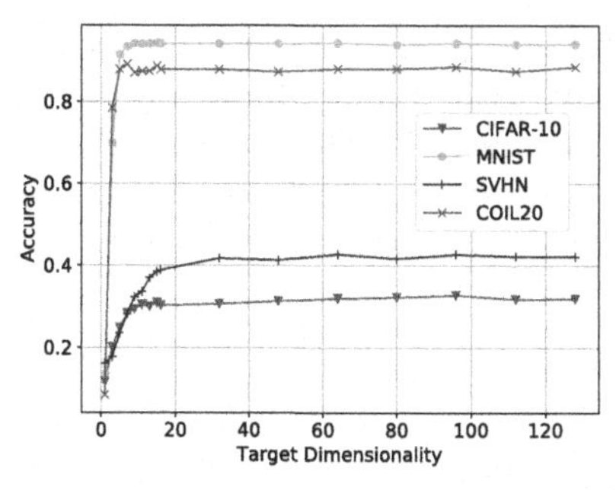

(a) Accuracies with the change of target dimensionality

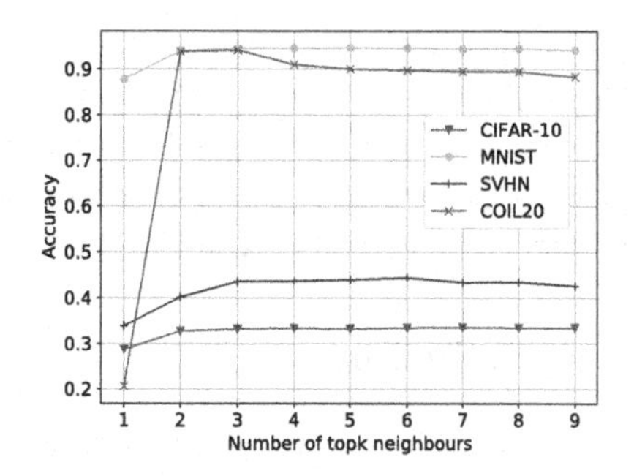

(b) Accuracies with the change of the number of neighbours

Fig. 5. Testing the parameter sensitivity of *Vec2vec* with data classification.

performance in the "COIL20" and "CIFAR-10" datasets, while UMAP gets the best performance in the other two image datasets. For the "CIFAR-10" and "SVHN" dataset, the ARI results of all seven methods are very small. We can find that *Vec2vec* gets the best ARI results in all the four text datasets.

We assume that the results are in Gaussian distribution and employ the Student's t-test to test the significant difference between the compared methods in the eight datasets. For UMAP, the H-value and p-value are 0 and 0.3118. **Therefore, UMAP is as good as *Vec2vec* in statistics and the performances of the two methods have no significant difference in the eight datasets.** For the local LLE and LE methods, the H-values are both 1 and the p-values are 0.0188 and 0.0097. **Therefore, *Vec2vec* is significantly better than LLE and LE in the eight datasets in statistics.** For the global PCA, CMDS, and Isomap method, the p-values are 0.0243, 0.0141, and 0.0411. The H-values are all 1. **Therefore, *Vec2vec* is significant better than the global PCA, CMDS, and Isomap in the eight datasets in statistics.** The experimental results show that skip-gram model is very effective to obtain the embedding while preserving the similarity of data points, which are important for clustering.

4.5 Parameter Sensitivity

The *Vec2vec* algorithm involves several parameters. We examine how the different choices of the target dimensionality d and the number of *topk* neighbors affect the performance of *Vec2vec* on the four typical image datasets. We perform 4-fold cross-validation and employ KNN as the classifier. We utilize the accuracy score as the evaluation metric. In this experiments, except for the parameter tested, all other parameters are set to be their default values.

Figure 5 shows the performance evaluation of *Vec2vec* with the change of the two parameters. For the target dimensionality d, *Vec2vec* gets its best performance when d is approximately equal to the number of classes in a dataset, which is the true dimensionality of the dataset. For the parameter of the number of *topk* neighbors in building the similarity graph, *Vec2vec* gets its best performance when *topk* is less than 5 in all the four datasets. It shows that the neighborhood similarity graph is sparse and *Vec2vec* is computationally efficient. **We can find that the performance of *Vec2vec* is stable when the two parameters reach a certain value. It is important to find that since it is easy to choose the parameters of *Vec2vec* in real applications.**

5 Conclusion

In this paper, we study the local nonlinear dimensionality reduction to relieve the curse of dimensionality problem. To reduce the computational complexity, we generalize the skip-gram model for representation learning of words to matrices. To preserve the similarities between data points in a matrix after dimensionality reduction, we select the neighbors of the data points to establish a neighborhood similarity graph. We raise a hypothesis that similar data points tend to be in similar contexts in the feature space, and define the contexts as the co-occurrences of data points in the sequences of random walks in the neighborhood graph.

We analyze the computational complexity of *Vec2vec* with the state-of-the-art local dimensionality reduction method UMAP. We find that our *Vec2vec* is efficient in datasets with both a large number of data samples and high-dimensionality, while UMAP can be scalable to datasets with a large number of data samples, but it is sensitive to high dimensionality of the data. We do extensive experiments of data classification and clustering on eight typical real-world datasets for dimensionality reduction to evaluate our method. Experimental results show that *Vec2vec* is better than several classical dimensionality reduction methods and is competitive with recently developed state-of-the-arts UMAP in the statistical hypothesis test.

Acknowledgment. We are thankful to the anonymous reviewers. This research is partially supported by the National Key R&D Program of China (Grant No. 2018YFB0203801), National Natural Science Foundation of China (Grant No. 61802424, 61702250), and ICT CAS research grant CARCHB202017.

References

1. Bojanowski, P., Grave, E., Joulin, A., Mikolov, T.: Enriching word vectors with subword information. Trans. Assn. Comput. Ling. **5**, 135–146 (2017)
2. Cai, H., Zheng, V.W., Chang, K.: A comprehensive survey of graph embedding: problems, techniques and applications. IEEE Trans. Knowl. Data Eng. **30**, 1616–1637 (2018)
3. Cui, P., Wang, X., Pei, J., Zhu, W.: A survey on network embedding. IEEE Trans. Knowl. Data Eng. 30, 1616–1637 (2018)
4. Cunningham, J.P., Ghahramani, Z.: Linear dimensionality reduction: survey, insights, and generalizations. JMLR **16**(1), 2859–2900 (2015)
5. Fujiwara, T., Chou, J.K., Shilpika, Xu, P., Ren, L., Ma, K.L.: An incremental dimensionality reduction method for visualizing streaming multidimensional data. IEEE Trans. Visualization Comput. Graph. **26**(1), 418–428 (2020)

6. Goyal, P., Chhetri, S.R., Canedo, A.: dyngraph2vec: capturing network dynamics using dynamic graph representation learning. Knowl. Based Syst. **187**, 104816 (2020)
7. Grover, A., Leskovec, J.: node2vec: calable feature learning for networks. In: Proceedings of SIGKDD, pp. 855–864. ACM (2016)
8. Huang, H., Shi, G., He, H., Duan, Y., Luo, F.: Dimensionality reduction of hyperspectral imagery based on spatial-spectral manifold learning. IEEE Trans. Syst. Man Cybern. **47**, 1–13 (2019)
9. Kang, Z., Lu, X., Lu, Y., Peng, C., Chen, W., Xu, Z.: Structure learning with similarity preserving. Neural Netw. **129**, 138–148 (2020)
10. Kasun, L.L.C., Yang, Y., Huang, G.B., Zhang, Z.: Dimension reduction with extreme learning machine. IEEE Trans. Image Process. **25**(8), 3906–3918 (2016)
11. Kiros, R., et al.: Skip-thought vectors. In: NIPS, pp. 3294–3302 (2015)
12. McInnes, L., Healy, J.: Umap: uniform manifold approximation and projection for dimension reduction. J. Open Source Softw. **3**, 861 (2018)
13. Mikolov, T., Sutskever, I., Chen, K., Corrado, G.S., Dean, J.: Distributed representations of words and phrases and their compositionality. In: Advances in Neural Information Processing Systems, pp. 3111–3119 (2013)
14. Pennington, J., Socher, R., Manning, C.: Glove: global vectors for word representation. In: Proceedings of EMNLP, pp. 1532–1543 (2014)
15. Perozzi, B., Al-Rfou, R., Skiena, S.: Deepwalk: online learning of social representations. In: Proceedings of the 20th ACM SIGKDD International Conference on Knowledge Discovery and Data Mining, pp. 701–710. ACM (2014)
16. Shao, M., Ding, Z., Zhao, H., Fu, Y.: Spectral bisection tree guided deep adaptive exemplar autoencoder for unsupervised domain adaptation. In: Proceedings of the Thirtieth AAAI Conference on Artificial Intelligence, pp. 2023–2029 (2016)
17. Tang, J., Qu, M., Wang, M., Zhang, M., Yan, J., Mei, Q.: Line: large-scale information network embedding. In: Proceedings of WWW, pp. 1067–1077 (2015)
18. Tang, J., Shao, L., Li, X., Lu, K.: A local structural descriptor for image matching via normalized graph laplacian embedding. IEEE Trans. Syst. Man Cybern. **46**(2), 410–420 (2016)
19. Ting, D., Jordan, M.I.: On nonlinear dimensionality reduction, linear smoothing and autoencoding. arXiv preprint arXiv:1803.02432 (2018)
20. Van Der Maaten, L., Postma, E., Van den Herik, J.: Dimensionality reduction: a comparative review. J. Mach. Learn. Res. **10**, 66–71 (2009)
21. Von Luxburg, U.: A tutorial on spectral clustering. Stat. Comput. **17**(4), 395–416 (2007)
22. Xu, X., Liang, T., Zhu, J., Zheng, D., Sun, T.: Review of classical dimensionality reduction and sample selection methods for large-scale data processing. Neurocomputing **328**, 5–15 (2019)
23. Zhang, D., Yin, J., Zhu, X., Zhang, C.: Network representation learning: a survey. IEEE Trans. Big Data (2018)
24. Zhang, J., Yu, J., Tao, D.: Local deep-feature alignment for unsupervised dimension reduction. IEEE Trans. Image Process. **27**(5), 2420–2432 (2018)

AE-UPCP: Seeking Potential Membership Users by Audience Expansion Combining User Preference with Consumption Pattern

Xiaokang Xu[1], Zhaohui Peng[1(✉)], Senzhang Wang[2], Shanshan Huang[3], Philip S. Yu[4], Zhenyun Hao[1], Jian Wang[1], and Xue Wang[1]

[1] School of Computer Science and Technology, Shandong University, Qingdao, China
pzh@sdu.edu.cn,
{xuxiaokang,haozhyun,jianwang9527,wangxue768}@mail.sdu.edu.cn

[2] School of Computer Science and Engineering, Central South University, Changsha, China
szwang@csu.edu.cn

[3] Juhaokan Technology Co., Ltd., Qingdao, China
huangshanshan1@hisense.com

[4] Department of Computer Science, University of Illinois at Chicago, Chicago, USA
psyu@uic.edu

Abstract. Many online video websites or platforms provide membership service, such as YouTube, Netflix and iQIYI. Identifying potential membership users and giving timely marketing activities can promote membership conversion and improve website revenue. Audience expansion is a viable way, where existing membership users are treated as seed users, and users similar to seed users are expanded as potential memberships. However, existing methods have limitations in measuring user similarity only according to user preference, and do not take into account consumption pattern which refers to aspects that users focus on when purchasing membership service. So we propose an **A**udience **E**xpansion method combining **U**ser **P**reference and **C**onsumption **P**attern (AE-UPCP) for seeking potential membership users. An autoencoder is designed to extract user personalized preference and CNN is used to learn consumption pattern. We utilize attention mechanism and propose a fusing unit to combine user preference with consumption pattern to calculate user similarity realizing audience expansion of membership users. We conduct extensive study on real datasets demonstrating the advantages of our proposed model.

Keywords: Audience expansion · Purchase prediction · User preference · Deep learning

1 Introduction

With the development of Internet technology, online video websites are growing in user scale and video resources, but not all contents are free, as some require

C. S. Jensen et al. (Eds.): DASFAA 2021, LNCS 12682, pp. 392–399, 2021.
https://doi.org/10.1007/978-3-030-73197-7_26

users to pay for membership service before they can enjoy them. For example, users who subscribe to membership service on Juhaokan of Hisense can watch the latest movies and TV plays. Identifying potential membership users can help websites to carry out more accurate marketing activities, improve the conversion rate from non-membership to membership, and thus increase revenue.

Audience expansion [1] is commonly used in advertising to find similar audiences. We take existing memberships as seed users and find potential memberships according to their similarity with seed users. However, most models [2] only consider user preference in similarity calculation, which is not comprehensive enough. We found that similar users tend to have similar consumption pattern when purchasing membership service. To calculate similarity combining user preference and consumption pattern, we mainly face the following challenges. The first challenge is to extract the personalized preferences of users. The second challenge is to model consumption pattern. We should capture aspects users pay more attention to when consuming. The last challenge is to combine user preference and consumption pattern to calculate user similarity. Which of the two factors has stronger influence needs to be taken into account in similarity calculation.

In order to effectively solve the above challenges, we propose an audience expansion model named AE-UPCP. Here we use video streaming service as an example. In AE-UPCP, we use BERT [3] to generate the text vector representation of videos and design an autoencoder to extract the user's personalized preferences from behavior sequence and user profile. We extract the consumption pattern from purchase sequence. We apply the attention mechanism [4] to learn relationship between preferences and combine user preferences and consumption pattern in the similarity calculation to achieve audience expansion. The contributions of this paper can be summarized as follows:

- We propose an audience expansion model which combines user preference and consumption pattern to seek potential membership users.
- An autoencoder is proposed to extract personalized preferences of users by combining their behavior sequences with user profiles and the extracted preferences are rich in semantic features. Consumption pattern is mined through CNN and a fusing unit is designed to combine user preference and consumption pattern so that similarity is more accurate.
- We evaluate this model on real-world datasets. The experimental comparison with several baseline methods proves that our model is more effective than others.

2 Related Work

Audience expansion methods mainly include similarity method and regression method. The similarity based methods [2,5] measure the similarity to find similar users for seed users. Regression based methods [6,7] pay more attention to prediction model, which treat seed users as positive samples and non-seed users as negative samples. The similarity between candidate users and seed users is predicted, and users with high similarity are selected as expanded audiences. In addition, LightGBM [8] models the combination of high-order features by integrating decision trees to predict. However, the disadvantage of these models is that only user portrait information is used for modeling, so the understanding of user preference is insufficient.

Click-through rate (CTR) prediction is to predict the advertisement click probability based on user preference and it can also be used for audience expansion. Wide&Deep [9] combines the memory ability of the shallow model and the generalization ability of the deep model. DeepFM [10] effectively utilizes the ability of FM [7] and DNN. [11–13] utilize the historical behavior sequence of users mining user interest to calculate the probability of users clicking on target advertisement. Although these methods can be used to calculate similarity between users and seed users, consumption pattern is not considered, which results in the deficient similarity.

3 Problem Statement

We denote user set as $U = \{u_1, u_2 \ldots u_{n_u}\}$, video set as $V = \{v_1, v_2 \ldots v_{n_v}\}$. For each user $u \in U$, we maintain a behavior sequence $S_u \subset V$ which denotes the videos user has seen, a list $P_u = \{p_1, p_2, p_3 \ldots\}$ which is the membership service purchase sequence of user u. For each video $v \in V$, its information includes text information such as ID, content introduction, classification, actor, etc.

Given membership users as seed users $U_{seed} \subset U$ and non-membership users as non-seed users $U_{non-seed} \subset U$ as well as their behavior sequences S_u and purchase sequences P_u. The task of audience expansion is to train a prediction model, in which seed users are regarded as positive samples and non-seed users as negative samples. The similarity between candidate users and seed users can be predicted and users with high similarity are potential memberships we are seeking.

4 AE-UPCP Model

The architecture of our proposed model AE-UPCP is shown in Fig. 1.

4.1 Feature Representation

User Profile Representation. User profile mainly includes the user's ID, gender, age, and demographic information, etc. Category features are encoded by one-hot, and normalized numerical features are concatenated with category features to get user profile vector x^u.

User Behavior Representation. For each video v in behavior sequence S_u, its text information includes title, introduction, actor, etc. We utilize BERT [3] to learn the text vector representation of video. According to S_u, we get the behavior sequence matrix $X_u^b \in R^{n_b \times k_v}$ of user u where n_b is the behavior sequence length and k_v is the dimension of video text vector.

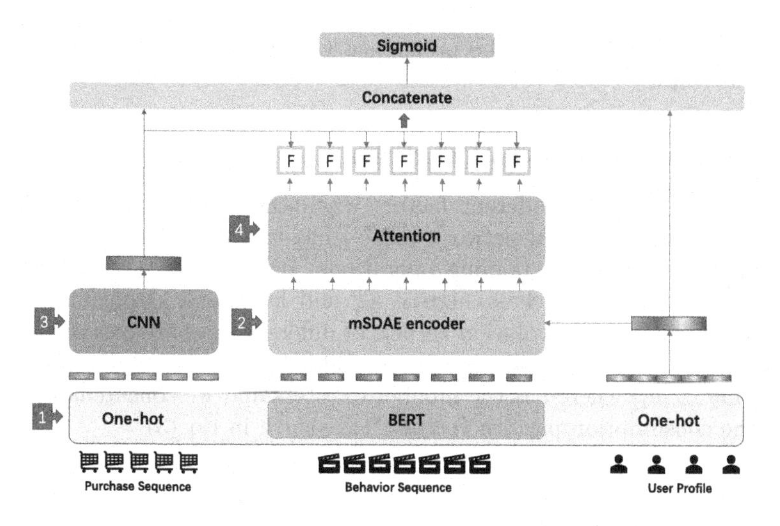

Fig. 1. The architecture of AE-UPCP model

Purchase Sequence Representation. For each purchase record $p \in P_u$, the information includes purchasing time, spending amount, consumption type and so on. We apply one-hot encoding to get the vector x^p of each record. According to the purchase sequence P_u, we get the purchase matrix $X_u^p \in R^{n_p \times k_p}$ of user u where n_p is the length of purchase sequence.

4.2 Personalized Preference Extraction

Inspired by [14,15] and the Stacked Denoising Autoencoder (SDAE) [16], we propose modified SDAE (mSDAE) to extract user preference features from the video representation vector x^v in behavior matrix X_u^b of user u and user profile representation vector x^u. After adding noise on x^v and x^u, we get x_{noise}^v, x_{noise}^u. The hidden layer h_l of the mSDAE is computed as in Eq. (1). We concate the user profile vector x_{noise}^u for each layer of the encoder and decoder to calculate the next layer where $h_0 = x_{noise}^v$ and g is the activation function which we use *ReLu* function.

$$h_l = g(W_l[h_{l-1}, x_{noise}^u] + b_l) \tag{1}$$

The output of mSDAE is h_L which is the reconstructed video representation vector and L is the total number of layers. The objective function is the mean squared error loss between the original input h_0 and their reconstruction output h_L as shown in Eq. (2):

$$loss = \sum \| h_L - h_0 \|^2 \tag{2}$$

After the autoencoder training is completed, we add the encoder to the AE-UPCP model to process vectors of behavior sequence matrix, and then fine-tune with other components as shown in Fig. 1. The user behavior sequence matrix X_u^b is encoded by the mSDAE encoder to obtain the preference matrix $X_u^r \in R^{n_b \times k_r}$

of user u where $x_j^r \in X_u^r$ is the preference vector and k_r is the dimension of preference vector.

4.3 Consumption Pattern Extraction

Different users consider different factors when they pay for the membership service. Convolution neural network (CNN) has the ability to extract features by using convolution and pooling operations. Inspired by [17], we use CNN to extract features of purchase matrix X_u^p and learn user's consumption pattern from it. Multiple convolution kernels of different sizes are convolved on the matrix X_u^p and then a Maxpooling layer is added. Then we get the pooled vector $[m_1, m_2 \ldots m_t]$ where t is the number of filters and we concatenate them to obtain the consumption pattern vector x^c as shown in Eq. (3):

$$x^c = [m_1, m_2 \ldots m_t] \tag{3}$$

4.4 Fusion Learning

There are internal relationships between user's different preferences, which should be considered when merging user's multiple preferences. Different from the methods of simply summing and averaging user preference vectors, we take attention mechanism [4] to learn the interaction between preferences.

Attention. Multi-head attention is used to learn the interrelationships between user preferences of X_u^r as in Eq. (4) and (5), and generate new preference representation vectors.

$$head_m = Attention(Q_m, K_m, V_m) = Softmax(\frac{Q_m K_m}{\sqrt{k_r}})V_m \tag{4}$$

$$MultiHead(Q, K, V) = Concat(head_1, head_2 \ldots head_m)W^O \tag{5}$$

$Q_m = K_m = V_m = W_m X_u^r$ where W_m is a mapping matrix, mapping X_u^r to different subspaces to obtain features of different subspaces and W^O is a parameter matrix. Then the results after multi-head attention are fed into a feed-forward network to capture their nonlinear relationship as in Eq. (6). After multi-head attention and feed forward network, the updated preference matrix X_u^r is obtained.

$$X_u^r = FFN(MultiHead(Q, K, V)) \tag{6}$$

Fusing Unit. When calculating similarity, which is more important between user preferences and consumption pattern should be considered. Inspired by the update gate in GRU [18], we designed fusing unit (F-Unit as shown in Fig. 1) to calculate the weight for each user's preference and consumption pattern as Eq. (7) and (8) where W^r, W^c and W^θ are weighted matrices, θ_j represents the importance of the user's consumption pattern compared with the j th preference vector.

$$\theta_j = \sigma(W^r \cdot x_j^r + W^c \cdot x^c) \tag{7}$$

$$x_j^f = (1 - \theta_j) * x_j^r + \theta_j * W^\theta \cdot x^c \tag{8}$$

The weights of user preferences that are not important to the similarity calculation can be reduced by θ. We get the fusion vector x_j^f for j th preference vector x_j^r and consumption pattern vector x^c.

$$x^s = \sum_j^{n_b} \frac{I_j}{\sum_k^{n_b} I_k} x_j^f \tag{9}$$

As shown in Eq. (9), I_j represents the playing duration time of j th behavior. All fusion vectors are weighted and summed to obtain the final similarity vector x^s which is merging user preferences and consumption pattern.

4.5 Similarity Calculation

After the above modules, we get user vector x^u, user consumption pattern vector x^c and user similarity vector x^s. After concatenating them to get x, as shown in Fig. 1, we add a full connection layer to calculate similarity $s(x)$, and the activation function is sigmoid function as shown in Eq. (10):

$$s(x) = \sigma(Wx + b) \tag{10}$$

$$L = -\frac{1}{N} \sum_i (y_i \log(s(x_i)) + (1 - y_i) \log(1 - s(x_i))) \tag{11}$$

The loss function is a binary cross entropy function as Eq. (11). $y_i \in \{0, 1\}$ is the label where 1 indicates that the user is a seed user and 0 is opposite, $s(x_i)$ is the output of our model which represents the similarity between the user and seed users, and users with high similarity exceeding the threshold are expanded as potential membership users.

5 Experiments

5.1 Experiment Setup

The statistics of two datasets are shown in Table 1, one is from Juhaokan website of Hisense and the other is from Alibaba[1]. We select the following baseline models LR [6], LightGBM [8], Wide&Deep [9], DeepFM [10], DIN [11] for comparative experiments. We use Recall and AUC to evaluate the effect of our model for seeking potential membership users.

5.2 Results and Analysis

The mSDAE carries out self-supervised training in advance, then the trained encoder is fused into AE-UPCP as shown in Fig. 1, and the parameters get fine-tuned during training continually. The experimental results are the

[1] https://tianchi.aliyun.com/dataset/dataDetail?dataId=56.

Table 1. Statistics of two datasets

Dataset	Hisense	Alibaba
Users	90253	395932
Items	68080	846811
Behaviors data	3.05 million	10.2 million
Purchase data	65681	0.6 million

Table 2. Recall and AUC of different models for seeking potential memberships on two datasets

Models	Hisense		Alibaba	
	Recall	AUC	Recall	AUC
LR	0.65	0.7496	0.11	0.6094
LightGBM	0.76	0.9012	0.24	0.6411
Wide& Deep	0.74	0.8873	0.20	0.6460
DeepFM	0.78	0.9007	0.27	0.6491
DIN	0.80	0.9225	0.29	0.6701
AE-UPCP (our model)	**0.83**	**0.9486**	**0.35**	**0.6881**

average values obtained by 5-fold cross-validation experiments and the similarity threshold is set to 0.5. As shown in Table 2, our model achieves the best results in terms of Recall and AUC. The LR method does not perform well on two datasets because it only uses the user profile information for prediction and has insufficient understanding of the user's preference. LightGBM is better than LR. DeepFM has achieved better effect compared with Wide&Deep due to modeling user's low-order and high-order features. We use DIN as an audience expansion method to achieve the best results of all baseline methods, which demonstrates that modeling user interest preference improves the performance of the model. Our model is more effective when the user's consumption pattern is added compared with DIN. Our model calculates the similarity between users and seed users, taking into account both user preference and consumption pattern, which makes the similarity calculation more sufficient.

6 Conclusion

In this paper, we propose an audience expansion model to seek potential membership users for video websites. We design an autoencoder to extract user personalized preferences combining video features with user profile features. We mine user consumption pattern by CNN and fuse user preference and consumption pattern to calculate similarity between users and seed users, which makes similarity more accurate. A large number of experiments prove that our model is more effective in seeking potential membership users compared with other methods.

Acknowledgements. This work is supported by National Natural Science Foundation of China (No. 62072282), Industrial Internet Innovation and Development Project in 2019 of China, Shandong Provincial Key Research and Development Program (Major Scientific and Technological Innovation Project) (No. 2019JZZY010105). This work is also supported in part by US NSF under Grants III-1763325, III-1909323, and SaTC-1930941.

References

1. Shen, J., Geyik, S. C., Dasdan, A.: Effective audience extension in online advertising. In: KDD, pp. 2099–2108 (2015)
2. Ma, Q., Wcn, M., Xia, Z., Chen, D.: A sub-linear, massive-scale look-alike audience extension system a massive-scale look-alike audience extension. In: BigMine, pp. 51–67 (2016)
3. Devlin, J., Chang, M.W., Lee, K., Toutanova, K.: BERT: pre-training of deep bidirectional transformers for language understanding. In: NAACL, pp. 4171–4186. NAACL (2019)
4. Vaswani, A., Shazeer, N., Parmar, N., et al.: Attention is all you need. In: NIPS, pp. 5998–6008 (2017)
5. Ma, Q., Wagh, E., Wen, J., Xia, Z., Ormandi, R., Chen, D.: Score look-alike audiences. In: ICDM, pp. 647–654 (2016)
6. McMahan, H.B., Holt, G., Sculley, D., Young, M., et al.: Ad click prediction: a view from the trenches. In: KDD, pp. 1222–1230 (2013)
7. Rendle, S.: Factorization achines. In: ICDM, pp. 995–1000 (2010)
8. Ke, G., Meng, Q., Finley, T., et al.: LightGBM: a highly efficient gradient boosting decision tree. In: NIPS, pp. 3146–3154 (2017)
9. Cheng, H. T., Koc, L., Harmsen, J., Shaked, T., et al.: Wide & deep learning for recommender systems. In: RecSys, pp. 7–10 (2016)
10. Guo, H., Tang, R., Ye, Y., Li, Z., He, X.: DeepFM: a factorization-machine based neural network for CTR prediction. In: IJCAI, pp. 1725–1731 (2017)
11. Zhou, G., Zhu, X., Song, C., Fan, Y., et al.: Deep interest network for click-through rate prediction. In: KDD, pp. 1059–1068 (2018)
12. Zhou, G., Mou, N., Fan, Y., et al.: Deep interest evolution network for click-through rate prediction. In: AAAI, pp. 5941–5948 (2019)
13. Feng, Y., Lv, F., Shen, W., Wang, M., et al.: Deep session interest network for click-through rate prediction. In: IJCAI, pp. 2301–2307 (2019)
14. Dong, X., Yu, L., Wu, Z., et al.: A hybrid collaborative filtering model with deep structure for recommender systems. In: AAAI, pp. 1309–1315 (2017)
15. Wang, S., Cao, J., Yu, P.: Deep learning for spatio-temporal data mining: a survey. IEEE Trans. Knowl. Data Eng. (2020). https://doi.org/10.1109/TKDE.2020.3025580
16. Vincent, P., Larochelle, H., Lajoie, I., Bengio, Y., et al.: Stacked denoising autoencoders: learning useful representations in a deep network with a local denoising criterion. J. Mach. Learn. Res. **11**, 3371–3408 (2010)
17. Kim, Y.: Convolutional neural networks for sentence classification. In: EMNLP, pp. 1746–1751 (2014)
18. Cho, K., Merrienboer, B. V., Gulcehre, C., et al.: Learning phrase representations using RNN encoder-decoder for statistical machine translation. In: EMNLP, pp. 1724–1734 (2014)

Self Separation and Misseparation Impact Minimization for Open-Set Domain Adaptation

Yuntao Du[1], Yikang Cao[1], Yumeng Zhou[2], Yinghao Chen[1], Ruiting Zhang[1], and Chongjun Wang[1(✉)]

[1] State Key Laboratory for Novel Software Technology at Nanjing University, Nanjing University, Nanjing 210023, China
{duyuntao,caoyikang}@smail.nju.edu.cn, chjwang@nju.edu.cn

[2] David R. Cheriton School of Computer Science, University of Waterloo, Waterloo, Canada
y443zhou@uwaterloo.ca

Abstract. Most of the existing domain adaptation algorithms assume the label space in the source domain and the target domain are exactly the same. However, such a strict assumption is difficult to satisfy. In this paper, we focus on Open Set Domain adaptation (OSDA), where the target data contains unknown classes which do not exist in the source domain. We concluded two main challenges in OSDA: (i) Separation: Accurately separating the target domain into a known domain and an unknown domain. (ii) Distribution Matching: deploying appropriate domain adaptation between the source domain and the target known domain. However, existing separation methods highly rely on the similarity of the source domain and the target domain and have ignored that the distribution information of the target domain could help up with better separation. In this paper, we propose a algorithm which explores the distribution information of the target domain to improve separation accuracy. Further, we also consider the possible misseparated samples in the distribution matching step. By maximizing the discrepancy between the target known domain and the target unknown domain, we could further reduce the impact of misseparation in distribution matching. Experiments on several benchmark datasets show our algorithm outperforms state-of-the-art methods.

1 Introduction

Domain adaptation is able to leverage a rich-labeled related domain data to facilitate learning in the unlabeled interested domain [17]. Conventional domain adaptation assumes the same label space in the source domain and the target domain. This setting is termed as Close-Set Domain Adaptation (CSDA). However, in the real-world data collection process, one can not guarantee the target domain contains the exact same classes as the source domain. A more realistic setting termed as Open Set Domain Adaptation (OSDA), which assumes the

C. S. Jensen et al. (Eds.): DASFAA 2021, LNCS 12682, pp. 400–409, 2021.
https://doi.org/10.1007/978-3-030-73197-7_27

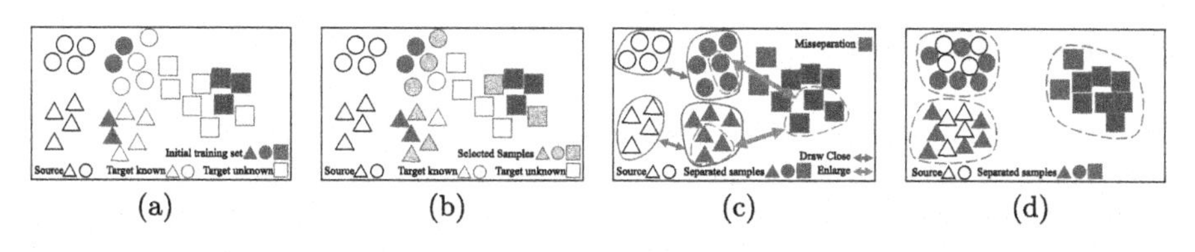

Fig. 1. (a–b) The main idea of self separation and the illustration of misseparation impact minimizing.

target domain contains extra classes that were not present in the source domain, has been studied.

For OSDA problem, the data in the source domain and the target domain follow different distributions, on the other hand, the unknown classes that only exist in the target domain, and are largely different from the known classes shared by the source domain and target domain. Previous works [8,10,18] deploy threshold strategies to separate the data in the target domain into known-class samples and unknown-class samples. They usually set a very loose threshold, so that most samples can be separated into known-class to participate in distribution matching. Meanwhile, those separation strategies are highly dependent on the distributional similarity between domains. So as the similarity decreases, the accuray of separation will drop significantly. On the other hand, previous works did not take into account the misseparated samples (unknown-class samples that are wrongly separated as known-class) in the distribution matching process.

To address the above problems, we propose Self Separation and Misseparation Impact Minimization (SSMM), a novel approach to OSDA. As illustrated in Fig. 1, the main idea of our method contains two steps: separation and distribution matching. The separation step aims to separate the target domain data into a target-known domain (target domain with only known-class samples) and a target unknown domain (target domain with only unknown-class samples), which could be regarded as a binary classification task in which we regard known-class samples as positive samples and unknown-class samples as negative samples. The target-known domain often reveals stronger similarity with the source domain and differs largely from the target-unknown domain. It can be adequately certain that part of the samples with sufficiently low or high entropy can be regarded as known-class or unknown-class confidently. Since the target domain follows the same distribution as the selected samples, we can leverage the intrinsic distribution information of the target domain to gradually separate the target domain into known and unknown parts. The distribution matching part is to minimize the distributional and geometrical discrepancy of source domain and target domain with only known-class samples and reduce the impact of misseparation simultaneously. Specificially we map both domains into a domain-invariant subspace, where the domain divergence of the source domain and the target-known domain is minimized and the distributional discrepancy of the target-known domain and the target-unknown domain is maximized.

2 Related Work

Close Set Domain Adaptation. Domain adaptation aims at reducing the domain discrepancy of the source domain and the target domain. A natural approach is to learn a transformation matrix that minimizes the statistical divergence. Transfer Component Analysis (TCA) [16] proposes to match the marginal distribution of both domains. Joint Distribution Adaptation (JDA) [13] extends TCA by matching the marginal and conditional distribution simultaneously. Deep Adaptation Networks (DAN) [11] adds adaptation layers into deep network, minimizing the marginal distribution of embedded feature representation. The researches above are all about aligning the distributional discrepancy, while the underlying geometric information cannot be revealed. The geometric information of domains, which could help with better exploring the relationship between samples, has been studied. ARTL [12] models joint distribution adaptation and manifold regularization into a unified structure risk minimization framework.

Open Set Domain Adaptation. Open Set Domain Adaptation was first proposed by [18]. This paper also proposes an approach to OSDA: Assign-and-Transform-Iteratively (ATI) separated the target domain samples in accordance with the Euclidean distance between target domain samples and source domain class centroid, then learned a linear transformation to reduce the sample-level domain distance. Seperate to Adapte (STA) [10] proposes an entropy measurement to train a binary classifier for separation, adopting the domain adversarial learning framework for domain alignment. KASE [9] realizes a Known-class Aware Recognition model based on the cross-entropy minimization principle to complete separation, then performing a continuous weights strategy in distribution matching.

3 Proposed Method

We begin with notations, given a labeled source domain $S = \{X_s, Y_s\}$ and an unlabeled target domain $T = \{X_t\}$, where $X_s \in \mathbb{R}^{d \times n_s}$ and $X_t \in \mathbb{R}^{d \times n_t}$. d is the feature dimensionality, n_s and n_t are the number of samples in the source domain and the target domain, respectively. Due to domain shift, S and T follow different distributions while their feature spaces are the same. In OSDA settings, the label spaces of domains are different. In this paper, we follow the settings in [18], which assumes the target domain contains the source domain, i.e. $C_s \subset C_t$, where C_s and C_t are the label space of the source domain and the target domain, respectively. We also use subscript k and u to denote known class (classes shared by both domains) samples and unknown class (classes only exist in target domain) samples, e.g. X_{tk} denotes known-class samples in the target domain and X_{tu} denotes unknown-class samples in the target domain.

3.1 Self Separation

As stated in Sect. 1, the separation step aims at exploring the intrinsic structure information of the target domain to separate target domain into the known domain and the unknown domain. It's worth noting that known-class samples are more related to the source domain than the unknown-class samples. Inspired by this observation, we train a basic classifier f(e.g. SVM) using X_s and deploy prediction on X_t, the result is denoted as Y_{t0}. Then we select the samples whose entropy is extremely high or low (determined by two bounds in the following), meaning that they are more pure to be known-class or unknown-class data, as the initial X_{tk} and X_{tu}. We take X_{tk} as positive samples, X_{tu} as negative samples to train a binary classifier and deploy prediction on the rest of X_t, choosing the samples with low entropy from the result and adding the chosen ones into X_{tk} and X_{tu}. Then we use the updated X_{tk} and X_{tu} as the training set to start a new epoch, repeating until all of the target domain samples are separated into either X_{tk} or X_{tu}. Specificially, we use Eq. 1 to calculate two bounds to select samples, where m is a relaxation coefficient. Samples with entropy higher than b_{upper} or lower than b_{lower} are added into training set in the initialization and samples with entropy lower than b_{lower} are added into the training set in the iterations.

$$b_{upper} = max(entropy(Y_{t0})) - m, \quad b_{lower} = min(entropy(Y_{t0})) + m \quad (1)$$

This semi-supervised-learning-like algorithm can let the target domain gradually learns the difference between known and unknown classes, and finally separate the target domain into X_{tu} and X_{tk}. Once the separation finishes, we get the well separated X_{tk} and X_{tu} to deploy distribution matching.

3.2 Distribution Matching

In this step, we propose to minimize the domain divergence between the source domain and the target-known domain by learning a feature transformation A so that the domain divergence can be reduced distributionally and geometrically. Meanwhile, we maximize the difference between the known-class and the unknown-class in the target domain, so that the misseparated samples can be as far away from the distribution of the known-class samples in the transformed space, then we can reduce the impact of misseparation in distribution matching. $\hat{X}_{tu}$ and $\hat{X}_{tk}$ denote part of the target known or unknown domain samples. We only select some very certain samples as representatives since we do not know which one is wrongly separated.

Joint Distribution Adaptation. We adpot MMD [2,4] to measure the distributional divergence between domains, which computes the mean of samples between domains in k-dimensional feature representations,

$$\arg\min_{A} \ ||\frac{1}{n_s} \sum_{x_i \in X_s} A^T x_i - \frac{1}{n_{tk}} \sum_{x_i \in X_{tk}} A^T x_j||_F^2 \quad (2)$$

n_s and n_{tk} denote the number of samples of the source domain and the target-known domain, respectively. $||\cdot||_F^2$ is the squared f-norm. However, MMD only measures the marginal distribution of domains, the conditional distribution is ignored. [13] proposes to leverage the psudo-label of the target domain, generated by source domain classifiers, to emperically estimate the conditional distribution divergence. We follow their main idea to minimize the conditional distribution divergence of domains,

$$\arg\min_{A} \sum_{c=1}^{C} ||\frac{1}{n_s^{(c)}} \sum_{x_i \in X_s^{(c)}} A^T x_i - \frac{1}{n_{tk}^{(c)}} \sum_{x_i \in X_{tk}^{(c)}} A^T x_j||_F^2 \tag{3}$$

where $c \in \{1, ..., C\}$ represents each class of known classes, $X_s^{(c)}$ denotes the samples in the source domain belonging to class c. $X_{tk}^{(c)}$ denotes the samples in the target-known domain whose pseudo-label is c. By combining Eq. 2 and Eq. 3 and applying the matrix trick, the joint distribution term $Dis(\cdot, \cdot)$ can be formulated as follows,

$$\arg\min_{A} \sum_{c=0}^{C} ||\frac{1}{n_s^{(c)}} \sum_{x_i \in X_s^{(c)}} A^T x_i - \frac{1}{n_{tk}^{(c)}} \sum_{x_i \in X_{tk}^{(c)}} A^T x_j||_F^2 = \arg\min_{A} tr(A^T X M X^T A) \tag{4}$$

where $X = [X_s, X_{tk}]$ is the samples and $M = M_0 + \sum_{c=1}^{C} M_c$ is the MMD matrix and can be found in previous work [13].

Manifold Regularization. We seek to exploit the geometric information of domains to further facilitate distribution matching. It has been proved that the intrinsic geometry of samples often reveals the underlying truth of their relationships [1], so if two samples are close in their low-dimensional manifold embeddings, they might have similar properties. The manifold regularization term $Geo(\cdot, \cdot)$ can be computed as:

$$\arg\min_{A} \sum_{i,j=0}^{n_s+n_{tk}} W_{ij}(A^T x_i - A^T x_j)^2 = \arg\min_{A} \sum_{i,j=0}^{n_s+n_{tk}} A^T x_i L_{ij} x_j A^T = \arg\min_{A} tr(A^T X L X^T A)$$

where W is the affinix matrix and L is the normalized graph Laplacian matrix, $W_{ij} = cos(x_i, x_j)$ if $x_i \in N_p(x_j)$ *or* $x_j \in N_p(x_i)$, otherwise, $W_{ij} = 0$. $N_p(x_i)$ denotes the p-nearest neighbor of sample x_i, $L = D^{-\frac{1}{2}} W D^{-\frac{1}{2}}$, where D is a diagonal matrix with $D_{ii} = \sum_{j=1}^{n} W_{ij}$.

Impact of Misseparation Minimization. Though we can get a promising separation performance in the separation step, there must be unknown-class samples which are wrongly separated as known-class (termed misseparation). As illustrated in Fig. 1(c), the misseparated samples show stronger correlations with the unknown classes (their underground truth) even though they were wrongly separated. We propose to minimize the impact of misseparated samples in domain adaptation by enlarging the discrepancy of the target-known domain and target-unknown domain, so that in the embedded feature representation those misseparated samples can move far away from the target known-domain (as shown in Fig. 1(d)). The misseparation minimizing term $Mis(\cdot,\cdot)$ can be formulated as,

$$\arg\max_{A} \ ||\frac{1}{\hat{n_s}}\sum_{x_i\in\hat{X}_{tu}} A^T x_i - \frac{1}{\hat{n_{tk}}}\sum_{x_i\in\hat{X}_{tk}} A^T x_j||_F^2 = \arg\max_{A} \ tr(A^T\hat{X}\hat{M}\hat{X}^T A) \tag{5}$$

where $\hat{X}_{tk}$ and $\hat{X}_{tu}$ represent the samples close to their own domain centroid and $\hat{X} = [\hat{X}_{tk}, \hat{X}_{tu}]$. Since enlarging the domain gap with all samples of target domain will draw the misseparated samples closer to the target-known domain, which goes against our goal, so we only choose the data that best represent the known/unknown classes.

Optimization. By incorporating Eq. 4, Eq. 3.2 and Eq. 5, we get the following objective formulation:

$$\begin{gathered}\arg\min_{A} \ tr(A^T XMX^T A) + \lambda tr(A^T XLX^T A) - \theta tr(A^T\hat{X}\hat{M}\hat{X}^T A) + \eta||A||_F^2 \\ s.t. \ A^T XHX^T A = I\end{gathered} \tag{6}$$

We add the regularization term $||A||_F^2$ to control the complexity of A. λ, η and θ are the trade-off parameters. The constraint term $A^T XHX^T A = I$ is defined to avoid trivial solutions, e.g. $A = 0$. $H = I - \frac{1}{n}\mathbf{1}$ is the centering matrix, $\mathbf{1} \in \mathbb{R}^{n\times n}$ is the matrix with all 1s and n denotes the number of all samples. This optimization can be transformed as,

$$(X(M+\lambda L)X^T + \theta\hat{X}\hat{M}\hat{X}^T + \eta I)A = XHX^T A\Phi \tag{7}$$

where $\Phi = [\phi_1, ..., \phi_k] \in \mathbb{R}^{k\times k}$ is the lagrange multipier, The above function could be solved through eigenvalues decomposition. Then the k eigenvectors corresponding to the k smallest eigenvalues of 7 are the solution of A.

4 Experiments

4.1 Data Preparation

Office-31 [19] is a standard benchmark dataset in domain adaptation, widely-used in previous works [14]. It consists of 3 real-world object domains: Webcam (W), DSLR (D) and Amazon (A), each domain contains 31 classes and 4,652 images in total. We follow the previous work, taking 0–10 classes as shared classes, 11–20 classes as unknown classes in the source domain and 21–31 classes as unknown classes in the target domain, all classes are sorted by alphabetical order. The features are extracted by Resnet-50 [7].

Office-Home [21] is a more challenging dataset which consists of 15,550 images from 4 domains: Artistic (Ar), Clipart (Cl), Product (Pr) and Real-World (Rw), with each domain containing 65 classes. Following the previous work, we

Table 1. Classification Accuracy(%) of OSDA tasks on Office-31 (Resnet50)

Method	A→W		A→D		W→A		W→D		D→A		D→W		Avg	
	OS	OS*	OS	OS*	OS	OS*	OS	OS*	OS	OS*	OS	OS*	OS	OS*
ResNet50	82.5	82.7	85.2	85.5	75.5	75.2	96.6	97.0	71.6	71.5	94.1	94.3	84.2	84.4
RTN	85.6	88.1	89.5	90.1	73.5	73.9	97.1	98.7	72.3	72.8	94.8	96.2	85.4	86.8
DANN	85.3	87.7	86.5	87.7	74.9	75.6	**99.5**	**100.0**	75.7	76.2	97.5	98.3	86.6	87.6
OpenMax	87.4	87.5	87.1	88.4	82.8	82.8	98.4	98.5	83.4	82.1	96.1	96.2	89.0	89.3
ATI-λ	87.4	88.9	84.3	86.6	80.4	81.4	96.5	98.7	78.0	79.6	93.6	95.3	86.7	88.4
OSBP	86.5	87.6	88.6	89.2	85.8	84.9	97.9	98.7	88.9	90.6	97.0	96.5	90.8	91.3
STA	89.5	92.1	93.7	96.1	87.9	87.4	**99.5**	99.6	89.1	93.5	97.5	96.5	92.9	94.1
TI	91.3	93.2	94.2	**97.1**	**88.7**	**88.1**	**99.5**	99.4	**90.1**	**91.5**	96.5	97.4	93.4	**94.5**
SSMM	**94.2**	**94.7**	**96.0**	96.6	87.0	87.7	98.6	**100.0**	87.7	88.1	**98.5**	**100.0**	**93.7**	**94.5**

Table 2. Classification Accuracy OS(%) of OSDA tasks on Office-Home (Resnet50)

Method	Ar→Cl	Ar→Pr	Ar→Rw	Cl→Ar	Cl→Pr	Cl→Rw	Pr→Ar	Pr→Cl	Pr→Rw	Rw→Ar	Rw→Cl	Rw→Pr	Avg
ResNet50	53.4	69.3	78.7	61.4	61.8	71.0	64.0	52.7	74.9	70.0	51.9	74.1	65.3
DANN	54.6	69.5	80.2	61.9	63.5	71.7	63.3	49.7	74.2	71.3	51.9	72.9	65.4
OpenMax	56.5	69.1	80.3	**64.1**	64.8	73.0	64.0	52.9	76.9	71.2	53.7	74.5	66.7
ATI-λ	55.2	69.1	79.2	61.7	63.5	72.9	64.5	52.6	75.8	70.7	53.5	74.1	66.1
OSBP	56.7	67.5	80.6	62.5	65.5	74.7	64.8	51.5	71.5	69.3	49.2	74.0	65.7
STA	58.1	71.6	**85.0**	63.4	69.3	**75.8**	65.2	53.1	80.8	**74.9**	54.4	81.9	69.5
TI	60.1	70.9	83.2	64.0	70.0	75.7	**66.1**	54.2	81.3	**74.9**	56.2	78.6	69.6
SSMM	**64.8**	**82.3**	84.3	58.3	**73.2**	75.1	64.8	**57.5**	**81.9**	73.4	**64.2**	**85.1**	**72.1**

take the first 25 classes as the shared classes, 26–65 classes as the unknown classes in the target domain, all classes are sorted by alphabetical order. The features are extracted by Resnet-50 [7].

4.2 Setup

Baselines: We compare our approach with several state-of-the-art OSDA methods. **OpenMax** [3], **ATI-λ** [18], **OSBP** [20], **STA** [10], **TI** [8], and **MTS** [5], We also compare our approach with three close set domain adaptation methods: **DAN** [11], **RTN** [15] and **DANN** [6]. For close set based methods, we follow the previous work, regarding the unknown classes as one class for evaluation and adopting confidence thresholding for outlier rejection.

Parameter Setting: We set the number of iterations $T = 10$ for both separation step and distribution matching step. To get all samples separated, we gradually enlarge the relaxation coefficient m from 0 to the average of predicted entropy, i.e. $(max(Ent) + min(Ent))/2$, each epoch we enlarge $(max(Ent) + min(Ent))/(2 * T)$. The threshold for selecting domain representatives in 5 is set to 50%, i.e. the closest half to the domain centroid. The trade-off parameters λ, θ and η are set as $\lambda = 0.1, \theta = 0.1$ *and* $\eta = 0.1$ in Office-31 dataset, $\lambda = 0.5, \theta = 0.5$ *and* $\eta = 1.0$ in Office-Home dataset. The manifold feature dimension d is set to 10. The subspace bases k is set to 100.

Evaluation Metric: In line with [18], we adpot two metrics: **OS**: normalized accuracy for all classes including the accuracy on the unknown class(considering the unknown classes on the target domain as one single class). **OS***: normalized accuracy only on the known classes.

Table 3. Separation accuracy of unknown samples on Office31 dataset (Resnet50)

Method	A→W	A→D	W→A	W→D	D→A	D→W	Avg
ATI-λ	72.4	61.3	70.4	74.5	62	76.6	69.5
OSBP	75.5	82.6	**94.8**	89.9	71.9	**100**	86.1
STA	63.5	69.7	92.9	98.5	45.1	**100**	79.5
TI	72.3	73.2	94.7	**100**	76.1	87.5	84.1
MTS	48.4	59.7	67.7	87.9	65.6	81.9	67.8
SSMM-N	85.7	87.7	81.5	84.5	**84.4**	84.6	84.7
SSMM	**89.2**	**90**	80	84.6	83.7	97.5	**87.5**

4.3 Results

The experiment results on Office-31 dataset are shown in Table 1. Our method shows comparable performance against state-of-the-art methods and achieves the best performance on some tasks. Noting that close-set based methods performs worse than open-set based methods generally, which further proved the unknown-class samples will cause negative transfer. Previous works recognize the unknown-class samples mainly depend on the similarity of domains, hence the performance drop in some large domain gap tasks (e.g. A→W).

Table 2 shows the results on Office-Home dataset. Our method outperforms existing methods on most tasks under large domain gap situation. Due to space constraints we only report **OS** metric. Some of the algorithms done by separating and distribution matching iteratively perform worse (e.g. ATI-λ) on some tasks, we explain this observation as the misseparated target samples will lead the distribution matching to a wrong direction, which will aggravate the negative transfer iteration after iteration. Our method optimize the two tasks independently, avoiding such a problem.

4.4 Abation Analysis

Self Separation: To verify the superiority of our separation part, we compare the separation accuracy of the unknown samples against existing methods on Office31 dataset. To further verify the effectiveness of gradually separation strategy, we propose a variant of separation stratgy. **SSMM-N**, separate all target samples in one iteration. The result is shown in Table 3, the results of previous works are calculated by the relationship between **OS** and **OS*** ($OS \times (|C| + 1) = OS^* \times |C| + Acc_{unk}$, where $|C|$ denotes the known class number and Acc_{unk} denotes the accuracy of unknown class). Table 3 shows our self separation stratgy outperforms previous works. Accurate separation will facilitate the distribution matching since less misseparated samples participant in the distribution matching step. So some tasks which we improve the separation accuracy a lot (e.g. A–>W and A–>D) also shows great improvement on the classification tasks (corresponding results in Table 1). **SSMM-N** directly separates the target domain in one iteration, which means it does not learn the structure information from target domain, merely applying the knownledge from the source domain. So **SSMM-N** performs worse than **SSMM**, which proves the effectiveness of our gradually separation stratgy.

5 Conclusion

In this papar, we propose Self Separation in OSDA settings. By exploiting the intrinsic structure information of the target domain, we reduce the dependence of the source domain in the separation step. We also take into account the impact of misseparated samples in distribution matching, which could further reduce negative transfer caused by the unknown samples. Extensive experiments show that our method performs well and shows strong robustness.

Acknowledgements. This paper is supported by the National Key Research and Development Program of China (Grant No. 2018YFB1403400), the National Natural Science Foundation of China (Grant No. 61876080), the Collaborative Innovation Center of Novel Software Technology and Industrialization at Nanjing University.

References

1. Belkin, M., Niyogi, P., Sindhwani, V.: Manifold regularization: a geometric framework for learning from labeled and unlabeled examples. JMLR **7**(Nov), 2399–2434 (2006)
2. Ben-David, S., Blitzer, J., Crammer, K., Pereira, F.: Analysis of representations for domain adaptation. In: NeurIPS (2007)
3. Bendale, A., Boult, T.E.: Towards open set deep networks. In: CVPR (2016)
4. Borgwardt, K.M., Gretton, A., Rasch, M.J., Kriegel, H.P., Schölkopf, B., Smola, A.J.: Integrating structured biological data by kernel maximum mean discrepancy. Bioinformatics **22**(14), e49–e57 (2006)
5. Chang, D., Sain, A., Ma, Z., Song, Y.Z., Guo, J.: Mind the gap: Enlarging the domain gap in open set domain adaptation. arXiv preprint arXiv:2003.03787 (2020)
6. Ganin, Y., et al.: Domain-adversarial training of neural networks. JMLR **17**(1), 2030–2096 (2016)
7. He, K., Zhang, X., Ren, S., Sun, J.: Deep residual learning for image recognition. In: CVPR (2016)
8. Kundu, J.N., Venkat, N., Revanur, A., Babu, R.V., et al.: Towards inheritable models for open-set domain adaptation. In: CVPR (2020)
9. Lian, Q., Li, W., Chen, L., Duan, L.: Known-class aware self-ensemble for open set domain adaptation. arXiv preprint arXiv:1905.01068 (2019)

10. Liu, H., Cao, Z., Long, M., Wang, J., Yang, Q.: Separate to adapt: open set domain adaptation via progressive separation. In: CVPR (2019)
11. Long, M., Cao, Y., Wang, J., Jordan, M.: Learning transferable features with deep adaptation networks. In: ICML (2015)
12. Long, M., Wang, J., Ding, G., Pan, S.J., Philip, S.Y.: Adaptation regularization: a general framework for transfer learning. TKDE **26**(5), 1076–1089 (2013)
13. Long, M., Wang, J., Ding, G., Sun, J., Yu, P.S.: Transfer feature learning with joint distribution adaptation. In: ICCV (2013)
14. Long, M., Wang, J., Ding, G., Sun, J., Yu, P.S.: Transfer joint matching for unsupervised domain adaptation. In: CVPR (2014)
15. Long, M., Zhu, H., Wang, J., Jordan, M.I.: Unsupervised domain adaptation with residual transfer networks. In: NeurIPS (2016)
16. Pan, S.J., Tsang, I.W., Kwok, J.T., Yang, Q.: Domain adaptation via transfer component analysis. TNNLS **22**(2), 199–210 (2010)
17. Pan, S.J., Yang, Q.: A survey on transfer learning. TKDE **22**(10), 1345–1359 (2009)
18. Panareda Busto, P., Gall, J.: Open set domain adaptation. In: ICCV, pp. 754–763 (2017)
19. Saenko, K., Kulis, B., Fritz, M., Darrell, T.: Adapting visual category models to new domains. In: Daniilidis, K., Maragos, P., Paragios, N. (eds.) ECCV 2010. LNCS, vol. 6314, pp. 213–226. Springer, Heidelberg (2010). https://doi.org/10.1007/978-3-642-15561-1_16
20. Saito, K., Yamamoto, S., Ushiku, Y., Harada, T.: Open set domain adaptation by backpropagation. In: ECCV (2018)
21. Venkateswara, H., Eusebio, J., Chakraborty, S., Panchanathan, S.: Deep hashing network for unsupervised domain adaptation. In: CVPR (2017)

Machine Learning

Partial Modal Conditioned GANs for Multi-modal Multi-label Learning with Arbitrary Modal-Missing

Yi Zhang, Jundong Shen, Zhecheng Zhang, and Chongjun Wang(✉)

National Key Laboratory for Novel Software Technology at Nanjing University, Department of Computer Science and Technology, Nanjing University, Nanjing, China
{njuzhangy,jdshen,zzc}@smail.nju.edu.cn,
chjwang@nju.edu.cn

Abstract. Multi-modal multi-label (MMML) learning serves an important framework to learn from objects with multiple representations and annotations. Previous MMML approaches assume that all instances are with complete modalities, which usually does not hold for real-world MMML data. Meanwhile, most existing works focus on data generation using GAN, while few of them explore the downstream tasks, such as multi-modal multi-label learning. The major challenge is how to jointly model complex modal correlation and label correlation in a mutually beneficial way, especially under the arbitrary modal-missing pattern. Aim at addressing the aforementioned research challenges, we propose a novel framework named Partial Modal Conditioned Generative Adversarial Networks (PMC-GANs) for MMML learning with arbitrary modal-missing. The proposed model contains a modal completion part and a multi-modal multi-label learning part. Firstly, in order to strike a balance between consistency and complementary across different modalities, PMC-GANs incorporates all available modalities during training and generates high-quality missing modality in an efficient way. After that, PMC-GANs exploits label correlation by leveraging shared information from all modalities and specific information of each individual modality. Empirical studies on 3 MMML datasets clearly show the superior performance of PMC-GANs against other state-of-the-art approaches.

Keywords: Multi-modal multi-label · GAN · Arbitrary modal-missing · Label correlation

1 Introduction

As a learning framework that handles objects characterized with multiple modal features and annotated with multiple labels simultaneously, Multi-Modal Multi-Label (MMML) learning [28] has been widely applied in many real-world applications. For example, a piece of web news can be represented with different

The original version of this chapter was revised: the acknowledgement section has been corrected. The correction to this chapter is available at
https://doi.org/10.1007/978-3-030-73197-7_54

C. S. Jensen et al. (Eds.): DASFAA 2021, LNCS 12682, pp. 413–428, 2021.
https://doi.org/10.1007/978-3-030-73197-7_28

modalities which consist of *text*, *image*, and *video*, while annotated with multiple labels including *science*, *sports* and *economic* at the same time.

In conventional MMML studies, it is commonly assumed that all modalities have been collected completely for each instance. Nonetheless, there are usually incomplete modalities for MMML data in many practical scenarios [12,15]. Taking webpage news analysis as an example, some webpages may contain *text*, *image*, and *video*, but others may only have one or two types, result in MMML data with arbitrary missing modalities. Furthermore, we formalize the corresponding learning task as the problem of Partial Multi-Modal Multi-Label (PMMML) learning. Intuitively, the problem PMMML can be solved by manually grouping samples according to the availability of data sources [24] and subsequently learning multiple models on these groups for late fusion. However, the grouping strategy is not flexible especially for the data with a large number of modalities.

The key challenge of PMMML problem is how to integrate available modalities in an efficient and mutually benefit way. In this paper, a novel Generative Adversarial Networks (GAN) [5] based approach is proposed to capture the relationship between the available modalities and the missing one, which complete missing modalities conditioned on available modalities. PMC-GANs combines modal completion process and multi-modal multi-label learning to resolve the arbitrary modal-missing pattern. As for each modality, the missing modality is completed by mapping the available through a generator, and a discriminator is used to distinguish the true modality and the generated one. In this way, the relationship among modalities is captured by the generative network and the missing modalities can be imputed. Afterward, we connect the imputation part with multi-modal multi-label learning by exploiting modal correlation and label correlation. Considering each modality has its own contribution to the multi-label learning, we enforce orthogonal constraints w.r.t. the shared subspace to exploit specific information of each individual modality. And then, we exploit label correlation with the shared and specific modal information.

The main contributions of this paper are three-fold:

- A novel framework named Partial Modal Conditioned Generative Adversarial Networks (PMC-GANs) for multi-modal multi-label (MMML) learning with arbitrary missing modalities is proposed.
- In order to achieve consistency and complementary across different modalities, PMC-GANs generates missing modalities with GAN conditioned on other available modalities. In addition to the available modalities, multiple labels are used for conditional variants of GAN. What's more, for label correlation exploitation, PMC-GANs extracts shared information among all modalities and specific information of each individual modality.
- Comprehensive experiments on 3 benchmark multi-modal multi-label datasets shows that PMC-GANs achieves highly competitive performance against other state-of-the-art MMML approaches.

The rest of this paper is organized as follows. Section 2 discusses existing works related to generative adversarial network and multi-modal multi-label learning. Section 3 presents the technical details of the proposed PMC-GANs. Section 4 analyses comparative experimental results to show the superiority of PMC-GANs. Finally, Sect. 5 presents the conclusion and future work.

2 Related Work

When it comes to integrating information from multiple sources with missing data, [10] decomposes the multi-source block-wise missing data into the multiple completed sub-matrix; SLIM [22] utilizes the intrinsic modal consistencies, which can also learn the most discriminative classifiers for each modality separately. To generate high quality modality, researches based on Generative Adversarial Networks (GAN) [5] are developed to learn the mapping among modalities. The basic GAN framework consists of a generator network and a discriminator network and the generator and the discriminator play a minmax game. Recently, [2,16] are proposed to impute missing data with the generative ability of GAN models. Recently, there has been increasing interest in conditional variants of GAN because they can improve the generation quality and stability. As a representative of class-conditional GANs, AC-GAN [13] incorporates class labels by introducing an auxiliary classifier in the original GAN framework. However, few of them explore the downstream tasks, such as multi-modal multi-label learning. And then, we briefly review some state-of-the-art approaches with regard to both multi-modal learning and multi-label learning.

For multi-label learning, the most straightforward way is to decompose the problem into a set of independent binary classification tasks [1], while it neglects label correlation. To compensate for this deficiency, the exploitation of label correlation has been widely studied [4], for example, [29] exploits global and local label correlation simultaneously, [7] learns both shared features and label-specific features. These approaches take label correlation as prior knowledge, which may not correctly characterize the real relationships among labels. And then [3] first propose a novel method to learn the label correlation via sparse reconstruction in the label space.

For multi-modal learning, the representative methods are Canonical Correlation Analysis (CCA) based, i.e., deep neural networks based CCA [17]. For partial multi-modal learning, a natural way is to complete missing modalities and then the on-shelf multi-modal learning approaches could be adopted. The imputation methods [15] complete the missing modalities by deep neural networks. In addition, the grouping strategy [24] divides all instances by availability of data sources, then multiple classifiers are learned for late fusion.

For multi-modal multi-label learning, [29] exploits the consensus among different modalities, where multi-view latent spaces are correlated by Hilbert–Schmidt Independence Criterion (HSIC). To further exploit communication among various modalities, SMISFL [18] jointly learns multiple modal-individual transformations and one sharable transformation. SIMM [20] leverages shared subspace exploitation and modal-specific information extraction. Nevertheless, previous approaches rarely consider the label correlation. CS3G approach [23] handles types of interactions between multiple labels, while no interaction between features from different modalities. To make each modality interacts and further reduce modal extraction cost, MCC [28] extends Classifier Chains to exploit label correlation with partial modalities.

3 Methodology

3.1 Problem Formulation

In Multi-modal Multi-label (MMML) learning with missing modalities, an instance can be characterized by partial multiple modal features and associated with multiple labels. The task of this paper is to learn a function $h : \mathcal{X} \rightarrow 2^{\mathcal{Y}}$, given N data samples $\mathcal{D} = \{(\boldsymbol{X}_i, \boldsymbol{Y}_i)\}_{i=1}^{N}$ with P modalities and L labels. $\boldsymbol{X}_i = (\boldsymbol{X}_i^1, \cdots, \boldsymbol{X}_i^m, \cdots, \boldsymbol{X}_i^P)$ is the feature vector, where $\boldsymbol{X}_i^m \in \mathcal{R}^{d_m}$ denotes the m-th modality of the i-th instance with dimension d_m. It is notable that each instance may have incomplete modalities, $\boldsymbol{X}_i^m = \emptyset$ if the m-th modality of the i-th instance is missing. $\boldsymbol{Y}_i = [y_i^1, y_i^2, \cdots, y_i^L] \in \{-1, 1\}^L$ denotes the label vector of the i-th instance $\boldsymbol{X}_i$.

Afterward, we introduce Partial Modal Conditioned Generative Adversarial Networks (PMC-GANs) in detail. The whole framework of PMC-GANs is shown in Fig. 1, and there are 3 major steps of PMC-GANs: modal completion, modal information extraction and label correlation exploitation.

3.2 Modal Completion

PMC-GANs consists of P generators and P discriminators, which are all conditioned on other available modalities. Each of the generator is associated with one of the modalities. As illustrated in Fig. 2, for the m-th modality, we maintain the following two adversarial models, which are both conditioned on $\boldsymbol{A}^m = (\boldsymbol{X}^1, \cdots, \boldsymbol{X}^{m-1}, \boldsymbol{X}^{m+1}, \cdots, \boldsymbol{X}^P)$:

A Generative Model. G^m that captures the data distribution $\mathbb{P}_{g^m}$ to infer the m-th missing modality $\hat{\boldsymbol{X}}^m$ conditioned on corresponding available modalities $\boldsymbol{A}^m$. When training G^m, a content loss function is employed to encourage the generated $\hat{\boldsymbol{X}}^m = G^m(\boldsymbol{z}^m|\boldsymbol{A}^m)$ to be similar to the true modality $\boldsymbol{X}^m$. To minimize Euclidean distance between the generated and the true data, the mean square error loss $\mathcal{L}_{MSE}^m$ is calculated as:

$$\mathcal{L}_{MSE}^m = \|\hat{\boldsymbol{X}}^m - \boldsymbol{X}^m\|_2^2 \tag{1}$$

Moreover, G^m tries to fool the discriminator D^m. Based on WGAN-GP [19], the objective function for G^m is defined as:

$$\mathcal{L}_G^m = - \mathop{\mathbb{E}}_{\hat{\boldsymbol{X}}^m \sim \mathbb{P}_{g^m}} [D^m(\hat{\boldsymbol{X}}^m|\boldsymbol{A}^m)] \tag{2}$$

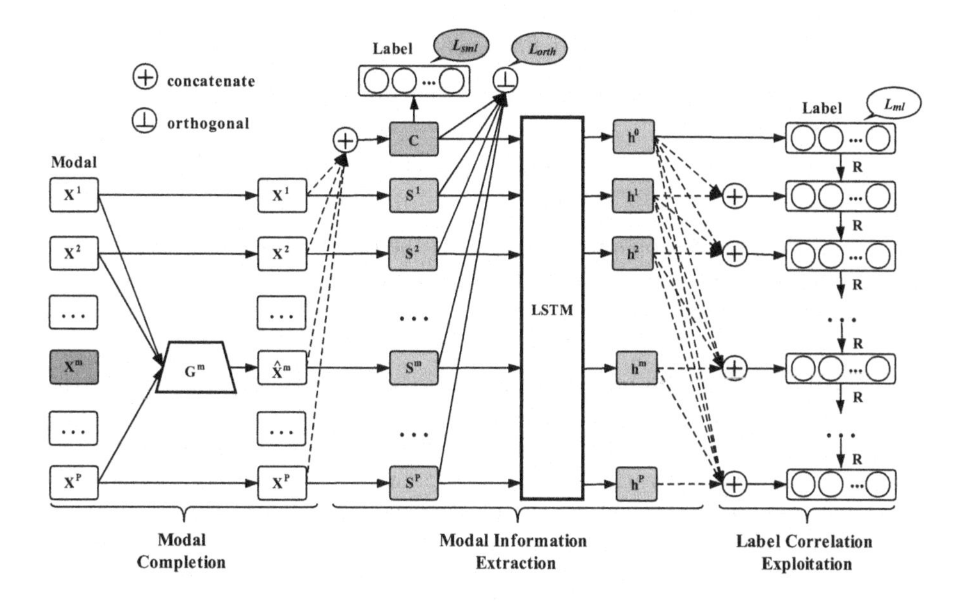

Fig. 1. Diagrammatic illustration of PMC-GANs. Firstly, PMC-GANs complete missing modality with GAN conditioned on available modalities. Secondly, PMC-GANs enforces orthogonal constraint to extract shared and specific information. Finally, shared and specific information are synergized in the LSTM network for label correlation exploitation.

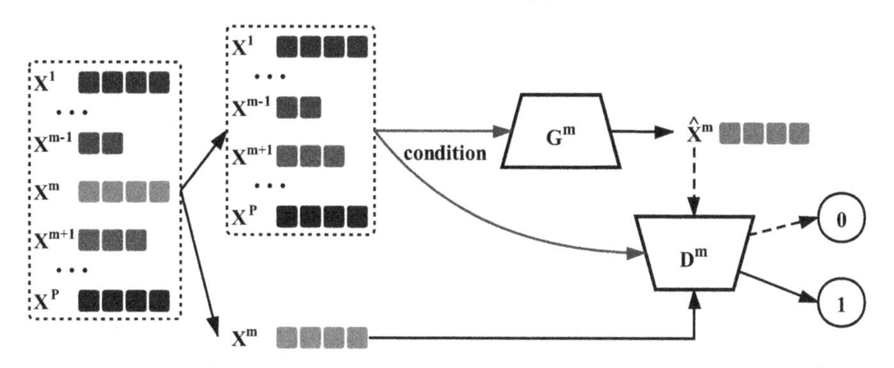

Fig. 2. Structure of conditional GANs for the m-th modality. Both the generator G^m and the discriminator D^m are conditioned on $\boldsymbol{A}^m = (\boldsymbol{X}^1, \cdots, \boldsymbol{X}^{m-1}, \boldsymbol{X}^{m+1}, \cdots, \boldsymbol{X}^P)$. And $\hat{\boldsymbol{X}}^m = G^m(\boldsymbol{z}^m|\boldsymbol{A}^m)$.

A Discriminative Model. D^m that tries to identify whether the m-th modality is available from the training data or completed by the generator G^m. Each discriminator D^m is treated as a binary classification network. The m-th true modality $\boldsymbol{X}^m$ is set with label 1, and the generated modality $\hat{\boldsymbol{X}}^m$ is set with label 0. Based on Wasserstein GANs with gradient penalty term (WGAN-GP), the objective function for D^m is defined as:

$$\begin{aligned}\mathcal{L}_D^m = & \mathop{\mathbb{E}}_{\boldsymbol{X}^m \sim \mathbb{P}_{r^m}} [D^m(\boldsymbol{X}^m | \boldsymbol{A}^m)] - \mathop{\mathbb{E}}_{\hat{\boldsymbol{X}}^m \sim \mathbb{P}_{g^m}} [D^m(\hat{\boldsymbol{X}}^m | \boldsymbol{A}^m)] \\ & + \lambda \mathop{\mathbb{E}}_{\widetilde{\boldsymbol{X}}^m \sim \mathbb{P}_{y^m}} [(\|\nabla D^m(\widetilde{\boldsymbol{X}}^m | \boldsymbol{A}^m)\|_2 - 1)^2]\end{aligned} \tag{3}$$

where ∇ is the gradient operator and $\mathbb{P}_{y^m}$ is the distribution obtained by sampling uniformly along straight lines between points from the real and fake data distributions $\mathbb{P}_{r^m}$ and $\mathbb{P}_{g^m}$.

An Auxiliary Classifier. $\boldsymbol{f}^m$ that prompts the generator G^m to generate the m-th modality for given labels. For instances from different labels, the relationship between the missing modalities and the available modalities can be different. In order to improve generation quality and stability, it's necessary to take multiple labels into consideration for the missing modal completion task. Inspired by AC-GAN [13], we employ an auxiliary classification loss in the discriminator D^m to distinguish the different labels of inputs. The discriminator D^m not only produces the probability distribution of real/generated, but also the probability distribution over multiple labels given multiple modalities $\boldsymbol{X}^m \oplus \boldsymbol{A}^m$ or $\hat{\boldsymbol{X}}^m \oplus \boldsymbol{A}^m$. Consequently, $\boldsymbol{f}^m$ learns to assign proper labels to both the real modality and the generated modality. The auxiliary classification loss is composed of two parts: the cross entropy loss for the real modality and the cross entropy loss for the generated modality, which can be defined as follows:

$$\mathcal{L}_{AC}^m = \sum_{l=1}^{L} \mathcal{L}_{ce}\big(D^m(\boldsymbol{X}^m \oplus \boldsymbol{A}^m, y^l)\big) + \mathcal{L}_{ce}\big(D^m(\hat{\boldsymbol{X}}^m \oplus \boldsymbol{A}^m, y^l)\big) \tag{4}$$

where $\oplus$ is the concatenation operator and y^l is the l-th label of $\boldsymbol{X}$. Meanwhile, $\mathcal{L}_{ce}(c, \hat{c}) = -\big(clog(\hat{c}) + (1-c)log(1-\hat{c})\big)$.

Objective. Above all, the discriminator D^m is trained by maximizing $\mathcal{L}_D^m - \lambda_{AC}\mathcal{L}_{AC}^m$, while the generator G^m is trained by minimizing $\mathcal{L}_G^m + \lambda_{MSE}\mathcal{L}_{MSE}^m + \lambda_{AC}\mathcal{L}_{AC}^m$. λ_{MSE}, λ_{AC} are the tradoff parameters for each loss. Parameters of D^m and G^m are updated using Adam optimization algorithm [9].

3.3 Modal Information Extraction

After completing the missing modality through the generators $\{G^m\}_{m=1}^P$ in Sect. 3.2, we perform the multi-modal multi-label learning on the whole training dataset. The goal of multi-modal multi-label learning is to fully integrate various representations of a complex object and assign a set of proper labels to it. Intrinsically, multiple modalities share consistency with each other and can provide complementary information together. Different modalities often contains shared information and specific information. Therefore, two key steps of modal information extraction are designed as follows:

Shared Information Extraction: aiming at exploiting shared subspace representation of all modalities, we concatenate all modalities in $\boldsymbol{X}_i$ to formalize a new single modal $\boldsymbol{X}_i^0 = [\boldsymbol{X}_i^1, \boldsymbol{X}_i^2, \cdots, \boldsymbol{X}_i^P] \in \mathbb{R}^{d_{all}}$, where $d_{all} = d_1 + d_2 + \cdots + d_P$. And then we add a dense layer to transform $\boldsymbol{X}_i^0$ to $\boldsymbol{C}_i \in \mathbb{R}^d$ according to $\boldsymbol{C}_i = ReLU(\boldsymbol{X}_i^0 \boldsymbol{U}_0 + \boldsymbol{b}_0)$, where $\boldsymbol{U}_0 \in \mathbb{R}^{d_{all} \times d}$ is the weight vector, $\boldsymbol{b}_0 \in \mathbb{R}^{1 \times d}$ is the bias vector.

Furthermore, the corresponding prediction model between common representation and multiple labels can be represented by $\boldsymbol{f}(\boldsymbol{C}_i) = \sigma(\boldsymbol{C}_i \boldsymbol{W}_C + \boldsymbol{b}_C)$, where $\boldsymbol{W}_C \in \mathbb{R}^{d \times L}$ is the weight vector, $\boldsymbol{b}_C \in \mathbb{R}^{1 \times L}$ is the bias vector. And then, we use shared information multi-label loss $\mathcal{L}_{sml}$ to guarantee that $\boldsymbol{C}_i$ contains certain semantics, $\mathcal{L}_{sml}$ can be formed as:

$$\mathcal{L}_{sml} = -\sum_{i=1}^{N_b} \sum_{k=1}^{L} \left(y_i^k log z_i^k + (1 - y_i^k) log(1 - z_i^k) \right) \tag{5}$$

where y_i^k is the ground-truth of $\boldsymbol{X}_i$ on the k-th label. $y_i^k = 1$ if k-th label is the relevant label, 0 otherwise. $z_i^k = \boldsymbol{f}(\boldsymbol{C}_i)$ is the prediction with the shared modal vector $\boldsymbol{C}_i$, which is extracted from $\boldsymbol{X}_i$. And N_b is the batch size.

Specific Information Extraction: each modality contains its own specific contribution to the multi-label prediction. Firstly, we add a dense layer to transform the m-th specific modality $\boldsymbol{X}_i^m \in \mathbb{R}^{d_m}$ to $\boldsymbol{S}_i^m \in \mathbb{R}^d$. And $\boldsymbol{S}_i^m = ReLU(\boldsymbol{X}_i^m \boldsymbol{U}_m + \boldsymbol{b}_m), m = 1, \cdots, P$, where $\boldsymbol{U}_m \in \mathbb{R}^{d_m \times d}$ is weight vector, $\boldsymbol{b}_m \in \mathbb{R}^{1 \times d}$ is bias vector.

It is difficult to define which is the specific information of a particular modality, but we can extract it from original $\{\boldsymbol{X}_i^m\}_{m=1}^P$ by eliminating shared information. We penalize the independence between shared information $\boldsymbol{C}_i$ and each specific information $\boldsymbol{S}_i^m$ with orthogonal loss function:

$$\mathcal{L}_{orth} = \sum_{i=1}^{N_b} \sum_{m=1}^{P} \|\boldsymbol{C}_i^T \boldsymbol{S}_i^m\|_2^2 \tag{6}$$

where $\|\cdot\|_2$ is the L_2-norm. $\mathcal{L}_{orth}$ encourages $\boldsymbol{S}_i^m$ extracted from the original m-th modal vector $\boldsymbol{X}_i^m$ to be as discriminative from $\boldsymbol{C}_i$ as possible.

After preparing share information $\boldsymbol{C}_i$ and specific information $\{\boldsymbol{S}_i^m\}_{m=1}^P$, which are all fixed in the same d dimension, then we input $\boldsymbol{Q}_i = \{\boldsymbol{C}_i, \boldsymbol{S}_i^1, \cdots, \boldsymbol{S}_i^P\}$ to the LSTM network [6] in order. At t-th step, the hidden features of $\boldsymbol{X}_i$ in LSTM structure can be represented as $\boldsymbol{h}_i^t \in \mathbb{R}^{d_h}$. In traditional multi-modal learning, the shared subspace exploitation is often implemented in an independent way, while the communication among all modalities is neglected. To better exploit relationship among different modalities, we stack all the previous hidden outputs as $\boldsymbol{H}_i^t = [\boldsymbol{h}_i^0, \boldsymbol{h}_i^1, \cdots, \boldsymbol{h}_i^t] \in \mathbb{R}^{(t+1)d_h}$, where d_h is the dimension of the hidden layer. All the parameters in LSTM structure are denoted as $\boldsymbol{\Psi}$.

3.4 Label Correlation Exploitation

It is well-known that exploiting label correlation is crucially important in multi-label learning and each modality contains its own specific contribution to the multi-label prediction. Meanwhile, complementary information among different modalities is of great importance. Thus, PMC-GANs models label correlation with extracted modal information stored in the memory of LSTM. At the t-th step, we add a fully connected structure between hidden layer and label prediction layer, which makes label prediction with stacked hidden outputs $\boldsymbol{H}_i^t$. The final label prediction is composed of the prediction of the current modality and the prediction of modality used in the last step. And then we predict multiple labels at the t-th step by a nonlinear softmax function:

$$\boldsymbol{F}^t(\boldsymbol{H}_i^t) = \begin{cases} \sigma(\boldsymbol{H}_i^t \boldsymbol{W}_L^t + \boldsymbol{b}_L^t) \; t = 0 \\ \sigma(\boldsymbol{H}_i^t \boldsymbol{W}_L^t + \boldsymbol{F}^{t-1}(\boldsymbol{H}_i^{t-1})\boldsymbol{R} + \boldsymbol{b}_L^t) \; t > 0 \end{cases} \tag{7}$$

where $\boldsymbol{W}_L^t \in \mathbb{R}^{((t+1)d_h)\times L}$ denotes the fully connected weights between $\boldsymbol{H}_i^t$ and label prediction layer, $\boldsymbol{b}_L^t \in \mathbb{R}^{1\times L}$ is the bias vector. $\boldsymbol{H}_i^t \boldsymbol{W}_L^t$ is similar to BR, which predicts each label independently. $\boldsymbol{F}^{t-1}(\boldsymbol{H}_i^{t-1})\boldsymbol{R}^T$ is the prediction of other labels, in which $\boldsymbol{F}^{t-1}(\boldsymbol{H}_i^{t-1}) \in \mathbb{R}^{1\times L}$ denotes label prediction at the $(t-1)$-th step. Meanwhile, we learn label correlation matrix $\boldsymbol{R} \in \mathbb{R}^{L\times L}$. The k-th row, j-th column of $\boldsymbol{R}$ represents the contribution of the k-th label prediction in $(t-1)$-th step to j-th label, which is denoted as R_{kj}.

Furthermore, we design binary cross-entropy loss function for final label prediction at t-th step by Eq. 8.

$$\mathcal{L}_{ml}^{i,t} = -\sum_{k=1}^{L} \left(y_i^k log\hat{y}_i^{k,t} + (1 - y_i^k) log(1 - \hat{y}_i^{k,t})\right) \tag{8}$$

where $\hat{y}_i^{k,t}$ the prediction of $\boldsymbol{X}_i$ on the k-th label at t-th step, predicted by $\boldsymbol{F}^t$ in Eq. 7.

Above all, we combine shared multi-label loss, orthogonal loss and final prediction loss function together to compute the overall loss function $\mathcal{L}$:

$$\mathcal{L} = \left(\sum_{i=1}^{N_b}\sum_{t=0}^{P} \mathcal{L}_{ml}^{i,t}\right) + \alpha\mathcal{L}_{orth} + \beta\mathcal{L}_{sml} \tag{9}$$

where α and β control the trade-off.

$\boldsymbol{\Theta} = [\boldsymbol{U}_0, \boldsymbol{b}_0, \boldsymbol{U}_m, \boldsymbol{b}_m, \boldsymbol{W}_C, \boldsymbol{b}_C, \boldsymbol{\Psi}, \boldsymbol{R}, \boldsymbol{W}_L^t, \boldsymbol{b}_L^t]$ denotes all the parameters need to be updated, where $m = 1, \cdots, P$, $t = 0, \cdots, P$. Then we adopt popular optimization algorithm Adam [9] to update parameters in $\boldsymbol{\Theta}$. The pseudo code of PMC-GANs in the training phase is presented in Algorithm 1.

Algorithm 1. Training algorithm for PMC-GANs

Input:
$\mathcal{D} = \{(\boldsymbol{X}_i, \boldsymbol{Y}_i)\}_{i=1}^N$: Training dataset; N_{critic}: number of iterations of the critic per generator iteration; M_b: minibatch size; N_b: batch size

Output:
$\boldsymbol{F}^P$: classifier trained with extracted modal sequence

1: **for** $m = 1 : P$ **do**
2:   **repeat**
3:     **for** $k = 1 : N_{critic}$ **do**
4:       Sample minibatch of M_b samples from $\mathcal{D}$
5:       Update the discriminator D^m by maximizing $\mathcal{L}_D^m - \lambda_{AC}\mathcal{L}_{AC}^m$
6:     **end for**
7:     Sample minibatch of M_b samples from $\mathcal{D}$
8:     Update the generator G^m by minimizing $\mathcal{L}_G^m + \lambda_{MSE}\mathcal{L}_{MSE}^m + \lambda_{AC}\mathcal{L}_{AC}^m$
9:   **until** Converge
10: **end for**
11: Complete missing modalities with $\{G^m\}_{m=1}^P$
12: **repeat**
13:   Sample N_b instances from $\mathcal{D}$ without replacement
14:   **for** $i = 1 : N_b$ **do**
15:     Extract shared $\boldsymbol{C}_i$ and specific $\{\boldsymbol{S}_i^m\}_{m=1}^P$
16:     $\boldsymbol{Q}_i = \{\boldsymbol{C}_i, \boldsymbol{S}_i^1, \cdots, \boldsymbol{S}_i^P\}$
17:     **for** $t = 0 : P$ **do**
18:       Input $\boldsymbol{Q}_i^t$ to LSTM cell
19:       Stack hidden output $\boldsymbol{H}_i^t = [\boldsymbol{h}_i^0, \boldsymbol{h}_i^1, \cdots, \boldsymbol{h}_i^t]$
20:       Compute label prediction $\boldsymbol{F}^t(\boldsymbol{H}_i^t)$ with Eq. 7
21:       Compute label loss function $\mathcal{L}_{ml}^{i,t}$ with Eq. 8
22:     **end for**
23:   **end for**
24:   Compute $\mathcal{L}_{sml}$, $\mathcal{L}_{orth}$, $\mathcal{L}$ with Eq. 5, 6, 9 respectively
25:   Update parameters in $\boldsymbol{\Theta}$
26: **until** converge
27: **return** $\boldsymbol{F}^P$

4 Experiment

4.1 Experimental Setup

Dataset Description. Table 1 summarizes the description of 3 multi-modal multi-label datasets, collected or generated as follows. *ML2000*: is an image dataset with 2000 images from 5 categories [26]. And 3 types of features: BoW, FV, and HOG are extracted for each image. *FCVID*: is a subset of Fudan-Columbia Video Dataset [8], composed of 4388 videos with most frequent category names. 5 modalities [23] are extracted for each video. *MSRA*: is a subset of a salient object recognition database [11], which contains 15000 images from 50 categories. And 7 modalities are extracted for each image, including RGB color histogram features, dimension block-wise color moments, HSV color histogram, color correlogram, distribution histogram, wavelet features, and face features.

Table 1. Characteristic of the real-world multi-modal multi-label datasets. N, L and P denote the number of instances, labels and modalities in each dataset, respectively. D shows the dimensionality of each modality.

Dataset	N	L	P	D
ML2000	2000	5	3	[500, 1040, 576]
FCVID	4388	28	5	[400, 400, 400, 400, 400]
MSRA	15000	50	7	[256, 225, 64, 144, 75, 128, 7]

Evaluation Metrics. To have a fair comparison, we employ 4 widely-used multi-label evaluation metrics, including Hamming Loss, Micro F1, Example F1, Subset Accuracy [27,28].

Compared Approaches. The performance of PMC-GANs is compared with 4 state-of-the-art approaches, including:

- DMP: is a serialized multi-modal algorithm [21], which can automatically extract an instance-specifically discriminative modal sequence within a limited modal acquisition budget. For each instance, we set the extraction cost of missing modality higher than that of available modality. Meanwhile, we treat each label independently.
- CAMEL(C) & CAMEL(B): a multi-label algorithm CAMEL [3] with two types of feature inputs. (C) denotes the concatenation of all modalities. (B) stands for the best performance obtained from the best single modality.
- MCC: is a novel multi-modal multi-label approach [28], which can make convince prediction with partial modalities. As for missing modality, the extraction cost is set higher.

4.2 Experimental Results

For all approaches, we tune the parameter with 10-fold cross validation and report the mean values as well as standard deviations. We set trade-off parameter $\lambda_{MSE} = 0.5$, $\lambda_{AC} = 0.1$ in the modal completion part, $\alpha = 0.1, \beta = 100$, minibatch size $M_b = 32$, batch size $N_b = 64$ in the multi-modal multi-label learning part. Furthermore, we drop out 40% modal features at each step to avoid over-fitting [14].

For all benchmark datasets, the missing modalities are randomly selected by guaranteeing at least one modality is available for each instance. And missing rate $\eta = \frac{\sum_m N_m}{P \times N}$, where N_m denotes the number of samples without the m-th modality [25]. As a result, partial multi-modal multi-label dataset are obtained with diverse missing patterns. For all the compared approaches, the missing modalities are filled with vector $\mathbf{0}$. What's more, for DMP and MCC approaches, we set the extraction cost higher with missing modalities.

Table 2. Predictive performance of each compared approach (mean ± std. deviation) on the multi-modal multi-label datasets (with missing rate $\eta = 0.2$). ↑ / ↓ indicates that the larger/smaller the better. The best performance is bolded.

ML2000				
Compared	Evaluation metrics			
Approaches	Hamming loss↓	Micro F1↑	Example F1↑	Subset accuracy↑
DMP	0.132 ± 0.009	0.670 ± 0.021	0.580 ± 0.022	0.489 ± 0.025
CAMEL(C)	0.119 ± 0.010	0.737 ± 0.024	0.690 ± 0.026	0.563 ± 0.026
CAMEL(B)	0.119 ± 0.010	0.718 ± 0.021	0.629 ± 0.026	0.525 ± 0.037
MCC	0.146 ± 0.014	0.686 ± 0.032	0.685 ± 0.034	0.563 ± 0.033
PMC-GANs	**0.113 ± 0.011**	**0.765 ± 0.022**	**0.766 ± 0.020**	**0.685 ± 0.035**
FCVID				
Compared	Evaluation metrics			
Approaches	Hamming loss↓	Micro F1↑	Example F1↑	Subset accuracy↑
DMP	0.033 ± 0.002	0.583 ± 0.026	0.553 ± 0.033	0.424 ± 0.038
CAMEL(C)	0.024 ± 0.001	0.553 ± 0.015	0.389 ± 0.015	0.375 ± 0.017
CAMEL(B)	0.022 ± 0.001	0.595 ± 0.021	0.434 ± 0.022	0.421 ± 0.024
MCC	0.032 ± 0.002	0.593 ± 0.026	0.567 ± 0.021	0.444 ± 0.019
PMC-GANs	**0.016 ± 0.001**	**0.780 ± 0.020**	**0.760 ± 0.021**	**0.741 ± 0.021**
MSRA				
Compared	Evaluation metrics			
Approaches	Hamming loss↓	Micro F1↑	Example F1↑	Subset accuracy↑
DMP	0.047 ± 0.001	0.273 ± 0.010	0.191 ± 0.009	0.046 ± 0.003
CAMEL(C)	0.048 ± 0.002	0.297 ± 0.010	0.055 ± 0.003	0.049 ± 0.007
CAMEL(B)	0.048 ± 0.002	0.264 ± 0.015	0.051 ± 0.003	0.044 ± 0.008
MCC	0.049 ± 0.001	0.323 ± 0.009	0.238 ± 0.009	0.065 ± 0.005
PMC-GANs	**0.048 ± 0.002**	**0.386 ± 0.019**	**0.295 ± 0.018**	**0.122 ± 0.013**

Table 2 reports detailed experimental results with 0.2 missing modalities. The experimental results across different datasets and different evaluation metrics demonstrate that PMC-GANs outperforms all the compared approaches. In addition, we evaluate PMC-GANs by investigating the performance with respect to varying missing rate. And the comparison results on the *FCVID* dataset are shown in Fig. 3.

Based on the results in Table 2 and Fig. 3, we have the following observations: 1) Without missing modalities ($\eta = 0$), PMC-GANs achieves very competitive performance compared with other approaches, which validates the effectiveness of modal information extraction and label correlation exploitation. 2) With the increase of missing rate, PMC-GANs achieves best and degrades more slowly, which demonstrates excellent stability of PMC-GANs when dealing with large missing rate. 3) When the missing rate substantially increases, PMC-GANs usually performs relatively promising, which shows the robustness.

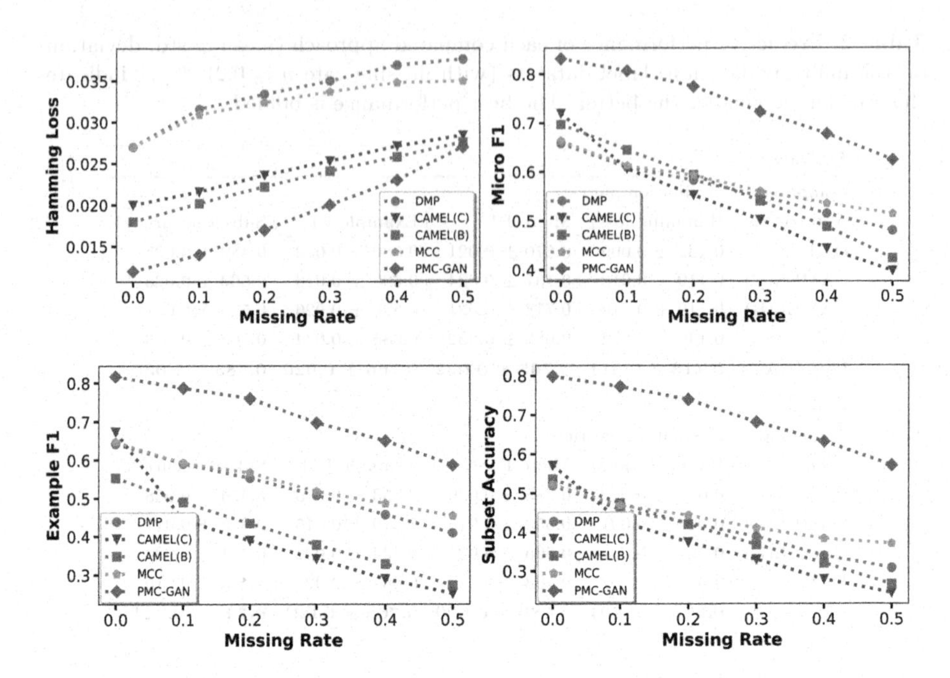

Fig. 3. Performance comparison on the *FCVID* dataset under different missing rate $\eta = \{0, 0.1, 0.2, 0.3, 0.4, 0.5\}$.

Table 3. Predictive performance of PMConcat and PMC-GANs (mean ± std. deviation) on the multi-modal multi-label datasets (with missing rate $\eta = 0.2$). ↑ / ↓ indicates that the larger/smaller the better. The best performance is bolded.

ML2000				
Compared	Evaluation metrics			
Approaches	Hamming loss↓	Micro F1↑	Example F1↑	Subset accuracy↑
PMConcat	0.119 ± 0.015	0.754 ± 0.030	0.764 ± 0.029	0.667 ± 0.051
PMC-GANs	**0.113 ± 0.011**	**0.765 ± 0.022**	**0.766 ± 0.020**	**0.685 ± 0.035**
FCVID				
Compared	Evaluation metrics			
Approaches	Hamming loss↓	Micro F1↑	Example F1↑	Subset accuracy↑
PMConcat	0.019 ± 0.002	0.730 ± 0.024	0.700 ± 0.022	0.688 ± 0.024
PMC-GANs	**0.016 ± 0.001**	**0.780 ± 0.020**	**0.760 ± 0.021**	**0.741 ± 0.021**
MSRA				
Compared	Evaluation metrics			
Approaches	Hamming loss↓	Micro F1↑	Example F1↑	Subset accuracy↑
PMConcat	0.049 ± 0.001	0.346 ± 0.013	0.254 ± 0.013	0.108 ± 0.010
PMC-GANs	**0.048 ± 0.002**	**0.386 ± 0.019**	**0.295 ± 0.018**	**0.122 ± 0.013**

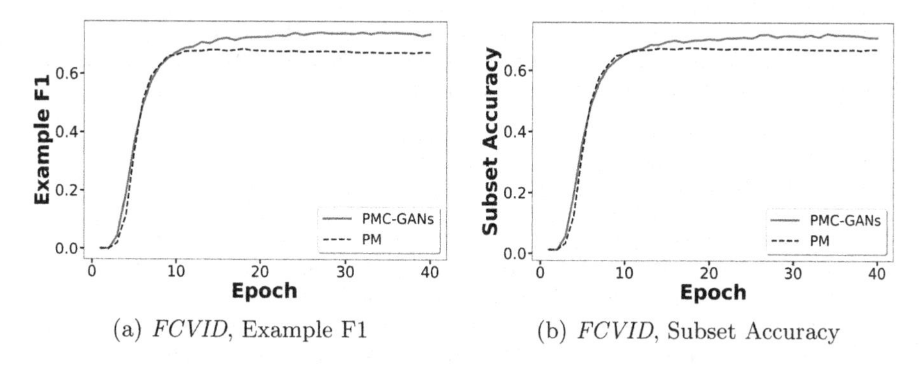

(a) *FCVID*, Example F1 (b) *FCVID*, Subset Accuracy

Fig. 4. Comparison of PMC-GANs against modal completion without conditional GANs (PM) on the *FCVID* dataset.

4.3 Performance Analysis

In this section, we examine the performance of modal completion part and multi-modal multi-label part respectively.

Performance of Modal Completion Part. In general, if the completion modalities are in high quality, the downstream multi-modal multi-label task will be better. By remaining the modal completion part of PMC-GANs, we add a fully connected layer for multi-label prediction with all modalities concatenated, which is denoted as PMConcat. In this case, we set the missing rate η as 0.2. Combining Table 2 and Table 3, PMConcat performs better than other state-of-the-art approaches. What's more, PMC-GANs performs better than PMConcat, which indicates it's necessary to carry on multi-modal multi-label learning part.

In order to examine the effectiveness of modal completion based on conditional GANs, we remain the basic structure of PMC-GANs while complete missing modalities with vector $\mathbf{0}$, which is denoted as PM. It is clearly shown in Fig. 4 that (1) both PMC-GANs and PM can converge fast within a small number of epochs, (2) the model performs better with modal completion based on conditional GANs.

Performance of Multi-modal Multi-label Part. PMC-GANs makes full use of modal information to better characterize label correlation, which polishes the performance of multi-modal multi-label learning. We keep the basic structure of PMC-GANs, and only input $\boldsymbol{X}_i$ instead of $\boldsymbol{Q}_i$ to the LSTM network, denoted as PMSeq. Here, we set the missing rate η as 0. As shown in Table 4, PMC-GANs performs better than PMSeq, which validates PMC-GANs of great effectiveness to extract shared and specific information for better label correlation exploitation. In other words, it's not enough to merely fuse cross-modal interaction.

Table 4. Predictive performance of each compared approach (mean ± std. deviation) on the multi-modal multi-label datasets (with missing rate $\eta = 0$). ↑ / ↓ indicates that the larger/smaller the better. The best performance is bolded.

ML2000				
Compared	Evaluation metrics			
Approaches	Hamming loss↓	Micro F1↑	Example F1↑	Subset accuracy↑
DMP	0.103 ± 0.010	0.783 ± 0.021	0.753 ± 0.026	0.617 ± 0.031
CAMEL(C)	0.098 ± 0.011	0.783 ± 0.024	0.739 ± 0.029	0.633 ± 0.036
CAMEL(B)	0.089 ± 0.011	0.806 ± 0.024	0.769 ± 0.031	0.655 ± 0.040
MCC	0.105 ± 0.012	0.784 ± 0.024	0.787 ± 0.022	0.662 ± 0.032
PMSeq	0.091 ± 0.012	0.813 ± 0.022	0.815 ± 0.024	0.731 ± 0.032
PMC-GANs	**0.086 ± 0.011**	**0.826 ± 0.021**	**0.827 ± 0.022**	**0.745 ± 0.034**
FCVID				
Compared	Evaluation metrics			
Approaches	Hamming loss↓	Micro F1↑	Example F1↑	Subset accuracy↑
DMP	0.027 ± 0.002	0.658 ± 0.023	0.642 ± 0.026	0.520 ± 0.033
CAMEL(C)	0.020 ± 0.001	0.658 ± 0.016	0.503 ± 0.017	0.485 ± 0.019
CAMEL(B)	0.018 ± 0.001	0.695 ± 0.024	0.551 ± 0.031	0.534 ± 0.030
MCC	0.027 ± 0.001	0.663 ± 0.014	0.647 ± 0.019	0.522 ± 0.024
PMSeq	0.015 ± 0.001	0.777 ± 0.015	0.718 ± 0.021	0.708 ± 0.024
PMC-GANs	**0.012 ± 0.001**	**0.828 ± 0.021**	**0.796 ± 0.026**	**0.783 ± 0.026**
MSRA				
Compared	Evaluation metrics			
Approaches	Hamming loss↓	Micro F1↑	Example F1↑	Subset accuracy↑
DMP	0.046 ± 0.001	0.311 ± 0.009	0.216 ± 0.007	0.053 ± 0.005
CAMEL(C)	0.045 ± 0.001	0.349 ± 0.010	0.248 ± 0.006	0.066 ± 0.010
CAMEL(B)	0.046 ± 0.001	0.329 ± 0.014	0.232 ± 0.010	0.057 ± 0.010
MCC	0.048 ± 0.001	0.359 ± 0.006	0.266 ± 0.005	0.076 ± 0.003
PMSeq	0.046 ± 0.001	0.353 ± 0.015	0.259 ± 0.016	0.115 ± 0.007
PMC-GANs	**0.046 ± 0.001**	**0.430 ± 0.012**	**0.341 ± 0.013**	**0.148 ± 0.010**

5 Conclusion

In this paper, a novel Partial Modal Conditioned Generative Adversarial Networks (PMC-GANs) is proposed to solve the multi-modal multi-label problem with arbitrary missing modalities, which can jointly exploit modal correlation and label correlation with available modalities. Specifically, PMC-GANs mainly consists of two parts: one is conditional Generative Adversarial Networks for generating missing modalities, the other is a modal-oriented information extraction network for exploiting label correlation. In addition, we verify the effectiveness and robustness of PMC-GANs via comprehensive experiments on benchmark multi-modal multi-label datasets with various missing rates. In the future, it is interesting to place the missing modalities generation problem into the reinforcement learning environment.

Acknowledgment. This paper is supported by the National Key Research and Development Program of China (Grant No. 2018YFB1403400), the National Natural Science Foundation of China (Grant No. 61876080), the Key Research and Development Program of Jiangsu (Grant No. BE2019105), the Collaborative Innovation Center of Novel Software Technology and Industrialization at Nanjing University.

References

1. Boutell, M.R., Luo, J., Shen, X., Brown, C.M.: Learning multi-label scene classification. Pattern Recogn. **37**(9), 1757–1771 (2004)
2. Cai, L., Wang, Z., Gao, H., Shen, D., Ji, S.: Deep adversarial learning for multi-modality missing data completion. In: Proceedings of the 24th ACM SIGKDD International Conference on Knowledge Discovery & Data Mining, pp. 1158–1166 (2018)
3. Feng, L., An, B., He, S.: Collaboration based multi-label learning. In: Thirty-Third AAAI Conference on Artificial Intelligence, pp. 3550–3557 (2019)
4. Gibaja, E., Ventura, S.: A tutorial on multilabel learning. ACM Comput. Surv. (CSUR) **47**(3), 52 (2015)
5. Goodfellow, I., et al.: Generative adversarial nets. In: Advances in Neural Information Processing Systems, pp. 2672–2680 (2014)
6. Hochreiter, S., Schmidhuber, J.: Long short-term memory. Neural Comput. **9**(8), 1735–1780 (1997)
7. Huang, J., Li, G., Huang, Q., Wu, X.: Joint feature selection and classification for multilabel learning. IEEE Trans. Cybern. **48**(3), 876–889 (2018)
8. Jiang, Y.G., Wu, Z., Wang, J., Xue, X., Chang, S.F.: Exploiting feature and class relationships in video categorization with regularized deep neural networks. IEEE Trans. Pattern Anal. Mach. Intell. **2**, 352–364 (2018)
9. Kingma, D.P., Ba, J.: Adam: a method for stochastic optimization. In: Proceedings of the 3rd International Conference on Learning Representations (2015)
10. Li, Y., Yang, T., Zhou, J., Ye, J.: Multi-task learning based survival analysis for predicting Alzheimer's disease progression with multi-source block-wise missing data. In: Proceedings of the 2018 SIAM International Conference on Data Mining, pp. 288–296. SIAM (2018)
11. Liu, T., et al.: Learning to detect a salient object. IEEE Trans. Pattern Anal. Mach. Intell. **33**(2), 353 (2011)
12. Liu, X., et al.: Late fusion incomplete multi-view clustering. IEEE Trans. Pattern Anal. Mach. Intell. **41**(10), 2410–2423 (2018)
13. Odena, A., Olah, C., Shlens, J.: Conditional image synthesis with auxiliary classifier gans. In: Proceedings of the 34th International Conference on Machine Learning-vol. 70, pp. 2642–2651. JMLR. org (2017)
14. Srivastava, N., Hinton, G., Krizhevsky, A., Sutskever, I., Salakhutdinov, R.: Dropout: a simple way to prevent neural networks from overfitting. J. Mach. Learn. Res. **15**(1), 1929–1958 (2014)
15. Tran, L., Liu, X., Zhou, J., Jin, R.: Missing modalities imputation via cascaded residual autoencoder. In: Proceedings of the IEEE Conference on Computer Vision and Pattern Recognition, pp. 1405–1414 (2017)
16. Wang, Q., Ding, Z., Tao, Z., Gao, Q., Fu, Y.: Partial multi-view clustering via consistent GAN. In: 2018 IEEE International Conference on Data Mining (ICDM), pp. 1290–1295. IEEE (2018)

17. Wang, W., Arora, R., Livescu, K., Bilmes, J.: On deep multi-view representation learning. In: International Conference on Machine Learning, pp. 1083–1092 (2015)
18. Wu, F., et al.: Semi-supervised multi-view individual and sharable feature learning for webpage classification. In: The World Wide Web Conference, pp. 3349–3355. ACM (2019)
19. Wu, J., Huang, Z., Thoma, J., Acharya, D., Van Gool, L.: Wasserstein divergence for GANS. In: Proceedings of the European Conference on Computer Vision (ECCV), pp. 653–668 (2018)
20. Wu, X., et al.: Multi-view multi-label learning with view-specific information extraction. In: Proceedings of the 28th International Joint Conference on Artificial Intelligence, pp. 3884–3890. AAAI Press, California (2019)
21. Yang, Y., Zhan, D.C., Fan, Y., Jiang, Y.: Instance specific discriminative modal pursuit: a serialized approach. In: Asian Conference on Machine Learning, pp. 65–80 (2017)
22. Yang, Y., Zhan, D.C., Sheng, X.R., Jiang, Y.: Semi-supervised multi-modal learning with incomplete modalities. In: IJCAI, pp. 2998–3004 (2018)
23. Ye, H.J., Zhan, D.C., Li, X., Huang, Z.C., Jiang, Y.: College student scholarships and subsidies granting: A multi-modal multi-label approach. In: 2016 IEEE 16th International Conference on Data Mining (ICDM), pp. 559–568. IEEE (2016)
24. Yuan, L., Wang, Y., Thompson, P.M., Narayan, V.A., Ye, J.: Multi-source learning for joint analysis of incomplete multi-modality neuroimaging data. In: Proceedings of the 18th ACM SIGKDD International Conference on Knowledge Discovery and Data Mining, pp. 1149–1157. ACM (2012)
25. Zhang, C., Han, Z., Fu, H., Zhou, J.T., Hu, Q., et al.: CPM-Nets: Cross partial multi-view networks. In: Advances in Neural Information Processing Systems, pp. 557–567 (2019)
26. Zhang, M.L., Zhou, Z.H.: ML-KNN: a lazy learning approach to multi-label learning. Pattern Recogn. **40**(7), 2038–2048 (2007)
27. Zhang, M.L., Zhou, Z.H.: A review on multi-label learning algorithms. IEEE Trans. Knowl. Data Eng. **26**(8), 1819–1837 (2014)
28. Zhang, Y., Zeng, C., Cheng, H., Wang, C., Zhang, L.: Many could be better than all: a novel instance-oriented algorithm for multi-modal multi-label problem. In: 2019 IEEE International Conference on Multimedia and Expo (ICME), pp. 838–843. IEEE (2019)
29. Zhu, Y., Kwok, J.T., Zhou, Z.H.: Multi-label learning with global and local label correlation. IEEE Trans. Knowl. Data Eng. **6**, 1081–1094 (2018)

Cross-Domain Error Minimization for Unsupervised Domain Adaptation

Yuntao Du, Yinghao Chen, Fengli Cui, Xiaowen Zhang, and Chongjun Wang(✉)

State Key Laboratory for Novel Software Technology at Nanjing University, Nanjing University, Nanjing 210023, China
{duyuntao,zhangxw}@smail.nju.edu.cn, chjwang@nju.edu.cn

Abstract. Unsupervised domain adaptation aims to transfer knowledge from a labeled source domain to an unlabeled target domain. Previous methods focus on learning domain-invariant features to decrease the discrepancy between the feature distributions as well as minimizing the source error and have made remarkable progress. However, a recently proposed theory reveals that such a strategy is not sufficient for a successful domain adaptation. It shows that besides a small source error, both the discrepancy between the feature distributions and the discrepancy between the labeling functions should be small across domains. The discrepancy between the labeling functions is essentially the **cross-domain errors** which are ignored by existing methods. To overcome this issue, in this paper, a novel method is proposed to integrate all the objectives into a unified optimization framework. Moreover, the incorrect pseudo labels widely used in previous methods can lead to error accumulation during learning. To alleviate this problem, the pseudo labels are obtained by utilizing structural information of the target domain besides source classifier and we propose a curriculum learning based strategy to select the target samples with more accurate pseudo-labels during training. Comprehensive experiments are conducted, and the results validate that our approach outperforms state-of-the-art methods.

Keywords: Transfer learning · Domain adaptation · Cross-domain errors

1 Introduction

Traditional machine learning methods have achieved significant progress in various application scenarios [14,33]. Training a model usually requires a large amount of labeled data. However, it is difficult to collect annotated data in some scenarios, such as medical image recognition [30] and automatic driving [42]. Such a case may lead to performance degradation for traditional machine learning methods. *Unsupervised domain adaptation* aims to overcome such challenge by transferring knowledge from a different but related domain (source

Y. Chen and F. Cui—Equal contribution.

C. S. Jensen et al. (Eds.): DASFAA 2021, LNCS 12682, pp. 429–448, 2021.
https://doi.org/10.1007/978-3-030-73197-7_29

domain) with labeled samples to a target domain with unlabeled samples [28]. And unsupervised domain adaptation based methods have achieved remarkable progress in many fields, such as image classification [45], automatic driving [42] and medical image precessing [30].

According to a classical theory of domain adaptation [1], the error of a hypothesis h in the target domain $\varepsilon_t(h)$ is bounded by three terms: the empirical error in the source domain $\hat{\varepsilon}_s(h)$, the distribution discrepancy across domains $d(\mathcal{D}_s, \mathcal{D}_t)$ and the ideal joint error λ^*:

$$\varepsilon_t(h) \leq \hat{\varepsilon}_s(h) + d(\mathcal{D}_s, \mathcal{D}_t) + \lambda^* \tag{1}$$

Note that $\mathcal{D}_s, \mathcal{D}_t$ denotes the source domain and the target domain, respectively. $\lambda^* = \varepsilon_s(h^*) + \varepsilon_t(h^*)$ is the ideal joint error and $h^* := \arg\min_{h \in \mathcal{H}} \varepsilon_s(h) + \varepsilon_t(h)$ is the ideal joint hypothesis. It is usually assumed that there is an ideal joint hypothesis h^* which can achieve good performance in both domains, making λ^* becoming a small and constant term. Therefore, besides minimizing the source empirical error, many methods focus on learning domain-invariant representations, i.e., intermediate features whose distributions are similar in the source and the target domain to achieve a small target error [6,20,29,34,36,39,44]. In shallow domain adaptation, *distribution alignment* is a widely used strategy for domain adaptation [21,22,27,35,38]. These methods assume that there exists a common space where the distributions of two domains are similar and they concentrate on finding a feature transformation matrix that projects the features of two domains into a common subspace with less distribution discrepancy.

Although having achieved remarkable progress, recent researches show that transforming the feature representations to be domain-invariant may inevitably distort the original feature distributions and enlarge the error of the ideal joint hypothesis [5,18]. It reminds us that the error of the ideal joint error λ^* can not be ignored. However, it is usually intractable to compute the ideal joint error λ^*, because there are no labeled data in the target domain. Recently, a general and interpretable generalization upper bound without the pessimistic term λ^* for domain adaptation has been proposed in [47]:

$$\varepsilon_t(h) \leq \hat{\varepsilon}_s(h) + d(\mathcal{D}_s, \mathcal{D}_t) + \min\{E_{\mathcal{D}_s}[|f_s - f_t|], E_{\mathcal{D}_t}[|f_s - f_t|]\} \tag{2}$$

where f_s and f_t are the labeling functions (i.e., the classifiers to be learned) in both domains. The first two terms in Eq. (2) are similar compared with Eq. (1), while the third term is different. The third term measures the discrepancy between the labeling functions from the source and the target domain. Obviously, $E_{\mathcal{D}_s}[|f_s - f_t|] = \varepsilon_s(f_t)$ and $E_{\mathcal{D}_t}[|f_s - f_t|] = \varepsilon_t(f_s)$. As a result, the discrepancy between the labeling functions is essentially the **cross-domain errors**. Specifically, the cross-domain errors are the classification error of the source classifier in the target domain and the classification error of the target classifier in the source domain. Altogether, the newly proposed theory provides a sufficient condition for the success of domain adaptation: besides a small source error, not only the discrepancy between the feature distributions but also the cross-domain errors need to be small across domains, while the cross-domain errors are ignored by existing methods.

Besides, estimating the classifier errors is important for domain adaptation. Various classifiers such as $k-$NN, linear classifier and SVMs have been used in shallow domain adaptation [3,21,22,27]. Recently, some methods adopt the *prototype classifier* [12] for classification in domain adaptation. The prototype classifier is a non-parametric classifier, where one class can be represented by one or more prototypes. And a sample can be classified according to the distances between the sample and the class prototypes.

In this paper, we propose a general framework named *Cross-Domain Error Minimization* (CDEM) based on the prototype classifier. CDEM aims to simultaneously learn domain-invariant features and minimize the cross-domain errors, besides minimizing the source classification error. To minimize the cross-domain errors, we maintain a classifier for each domain separately, instead of assuming that there is an ideal joint classifier that can perform well in both domains. Moreover, we conduct discriminative feature learning for better classification. To sum up, as shown in Fig. 1, there are four objectives in the proposed method. (*i*) Minimizing the classification errors in both domains to optimize the empirical errors. (*ii*) Performing distribution alignment to decrease the discrepancy between feature distributions. (*iii*) Minimizing the cross-domain errors to decrease the discrepancy between the labeling functions across domains. (*iv*) Performing discriminative learning to learn discriminative features. Note that the objectives (*i*), (*ii*) and (*iv*) have been explored in previous methods [27,37,38], while the objective (*iii*) is ignored by existing methods. We integrate the four objectives into a unified optimization problem to learn a feature transformation matrix via a closed-form solution. After transformation, the discrepancy between the feature distributions and the cross-domain errors will be small, and the source classifier can generalize well in the target domain.

Since the labels are unavailable in the target domain, we use *pseudo labels* instead in the learning process. Inevitably, there are some incorrect pseudo labels, which will cause error accumulation during learning [39]. To alleviate this problem, the pseudo labels of the target samples are obtained based on the structural information in the target domain and the source classifier, in this way, the pseudo labels are likely to be more accurate. Moreover, we propose to use *curriculum learning* [2] based strategy to select target samples with high prediction confidence during training. We regard the samples with high prediction confidence as "easy" samples and the samples with low prediction confidence as "hard" samples. The strategy is to learn the transformation matrix with "easy" samples at the early stage and with "hard" samples at the later stage. With the iterations going on, we gradually add more and more target samples to the training process.

Note that CDEM is composed of two processes: learning transformation matrix and selecting target samples. We perform these two processes in an alternative manner for better adaptation. Comprehensive experiments are conducted on three real-world object datasets. The results show that CDEM outperforms the state-of-the-art adaptation methods on most of the tasks (16 out of 24), which validates the substantial effects of simultaneously learning domain-invariant features and minimizing cross-domain errors for domain adaptation.

2 Related Work

Domain Adaptation Theory. The theory in [1] is one of the pioneering theoretical works in this field. A new statistics named $\mathcal{H}\Delta\mathcal{H}$-divergence is proposed as a substitution of traditional distribution discrepancies (e.g. L_1 distance, KL-divergence) and a generalization error bound is presented. The theory shows that the target error is bounded by the source error and the distribution discrepancy across domains, so most domain adaptation methods aim to minimize the source error and reduce the distribution discrepancy across domains. A general class of loss functions satisfying symmetry and subadditivity are considered in [25] and a new generalization theory with respect to the newly proposed discrepancy distance is developed. A margin-aware generalization bound based on asymmetric margin loss is proposed in [25] and reveals the trade-off between generalization error and the choice of margin. Recently, a theory considering labeling functions is proposed in [46], which shows that the error of the target domain is bounded by three terms: the source error, the discrepancy in feature distributions and the discrepancy between the labeling functions across domains. The discrepancy between the labeling functions are essentially the cross-domain errors which are ignored by existing methods. CDEM is able to optimize all the objectives simultaneously.

Domain Adaptation Algorithm. The mostly used shallow domain adaptation approaches include instance reweighting [3,7,16] and distribution alignment [22,27,34,37,38].

The instance reweighting methods assume that a certain portion of the data in the source domain can be reused for learning in the target domain and the samples in the source domain can be reweighted according to the relevance with the target domain. Tradaboost [7] is the most representative method which is inspired by Adaboost [41]. The source samples classified correctly by the target classifier have larger weight while the samples classified wrongly have less weight. LDML [16] also evaluates each sample and makes full use of the pivotal samples to filter out outliers. DMM [3] learns a transfer support vector machine via extracting domain-invariant feature representations and estimating unbiased instance weights to jointly minimize the distribution discrepancy. In fact, the strategy for selecting target samples based on *curriculum learning* can be regarded as a special case of instance reweighting, where the weight of selected samples is 1, while the weight of unselected samples is 0.

The distribution alignment methods assume that there exists a common space where the distributions of two domains are similar and focus on finding a feature transformation that projects features of two domains into another latent shared subspace with less distribution discrepancy. TCA [27] tries to align marginal distribution across domains, which learns a domain-invariant representation during feature mapping. Based on TCA, JDA [22] tries to align marginal distribution and conditional distribution simultaneously. Considering the balance between the marginal distribution and conditional distribution discrepancy, both BDA [37] and MEDA [38] adopt a balance factor to leverage the importance of different distributions. However, these methods all focus on learning domain-invariant features across domains and ignore the cross-domain errors. While our proposed method takes the cross-domain errors into consideration.

3 Motivation

3.1 Problem Definition

In this paper, we focus on unsupervised domain adaptation. There are a source domain $\mathcal{D}_s = \{(x_s^i, y_s^i)\}_{i=1}^{n_s}$ with n_s labeled source examples and a target domain $\mathcal{D}_t = \{x_t^j\}_{j=1}^{n_t}$ with n_t unlabeled target examples. It is assumed that the feature space and the label space are the same across domains, i.e., $\mathcal{X}_s = \mathcal{X}_t \in \mathbb{R}^d$, $\mathcal{Y}_s = \mathcal{Y}_t = \{1, 2, ..., C\}$, while the source examples and target examples are drawn from different joint distributions $P(\mathcal{X}_s, \mathcal{Y}_s)$ and $Q(\mathcal{X}_t, \mathcal{Y}_t)$, respectively. The goal of CDEM is to learn a feature transformation matrix $P \in R^{d \times k}$, which projects the features of both domains into a common space to reduce the shift in the joint distribution across domains, such that the target error $\varepsilon_t(h) = E_{(x,y)\sim Q}[h(x) \neq y]$ can be minimized, where h is the classifier to be learned.

3.2 Main Idea

As shown in Fig. 1(a), there is a large discrepancy across domains before adaptation. Previous methods only focus on minimizing the source error and performing distribution alignment to reduce the domain discrepancy (Fig. 1(b–c)). As the new theory revealed [47], in addition to minimizing the source error and learning domain-invariant features, it is also important to minimize the cross-domain errors. As shown in Fig. 1(d), although performing distribution alignment can reduce the domain discrepancy, the samples near the decision boundary are easy to be misclassified. Because performing distribution alignment only considers the discrepancy between the feature distributions, while the cross-domain errors are ignored. In the proposed method, minimizing the cross-domain errors can pull the decision boundaries across domains close, so that we can obtain a further reduced domain discrepancy. Moreover, we also perform discriminative learning to learn discriminative features (Fig. 1(e)). Eventually, the domain discrepancy can be reduced and the classifier in the source domain can generalize well in the target domain (Fig. 1(f)).

To sum up, we propose a general framework named *cross-domain error minimization* (CDEM), which is composed of four objectives:

$$h = \arg\min_{h \in \mathcal{H}} \sum_{i=1}^{n_s+n_t} l(h(x_i), y_i) + l_d(\mathcal{D}_s, \mathcal{D}_t) + l_f(\mathcal{D}_s, \mathcal{D}_t) + l_m(\mathcal{D}_s, \mathcal{D}_t) \quad (3)$$

where $l(h(x_i), y_i)$ is the classification errors in both domains. $l_d(\mathcal{D}_s, \mathcal{D}_t)$ and $l_f(\mathcal{D}_s, \mathcal{D}_t)$ represent the discrepancy between the feature distributions and the discrepancy between the labeling functions across domains, respectively. $l_m(\mathcal{D}_s, \mathcal{D}_t)$ is the discriminative objective to learn discriminative features. Note that CDEM is a shallow domain adaptation method and use the prototype classifier as the classifier, where no extra parameters are learned except the transformation matrix P. The framework is general and can generalize to other methods such as deep models.

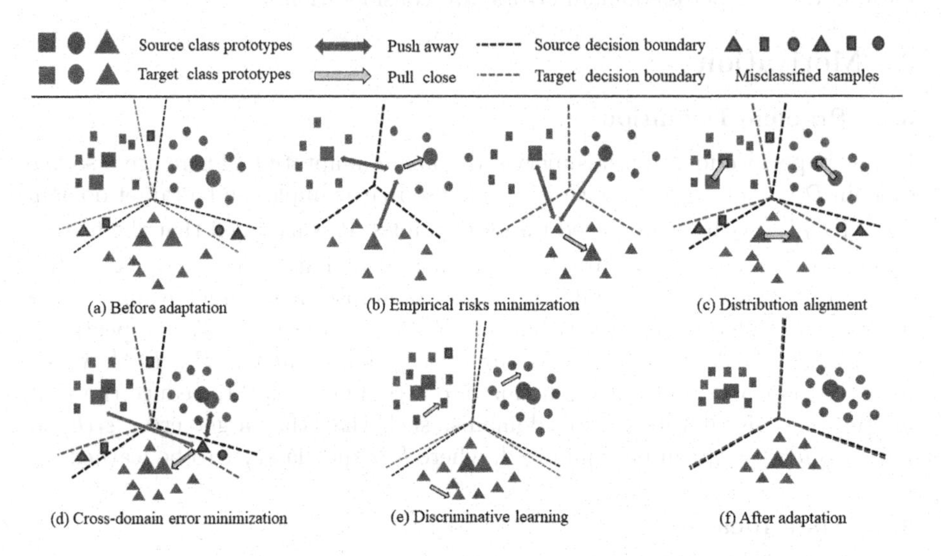

Fig. 1. An overview of the proposed method. In this paper, we use the prototype classifier as the basic classifier. There is one prototype in each class, and we choose the class center as the prototype. We use the distances between samples and prototypes to calculate the classification error. (a) Before adaptation, the classifier trained in the source domain can not generalize well in the target domain. (b–e) We aim to minimize empirical errors in both domains, perform distribution alignment to learn domain-invariant features, minimize the cross-domain errors to pull the decision boundaries across domains close, and perform discriminative learning to learn discriminative features. (f) After adaptation, the discrepancy across domains is reduced, so that the target samples can be classified correctly by the source classifier. Best viewed in color. (Color figure online)

As the labels in the target domain are unavailable, the *pseudo labels* for the target data are used for training instead. However, they are always some incorrect pseudo labels and may lead to catastrophic error accumulation during learning. To alleviate this problem, we use the *curriculum learning* based strategy to select the target samples with more accurate pseudo labels which are obtained by taking advantage of source classifier and structural information of the target domain. With the iterations going on, we gradually add more and more target samples to the training process.

3.3 Classification Error

In this paper, we choose the prototype classifier [12] as the classifiers in both domains since the prototype classifier is a non-parametric classifier and is widely used in many tasks. As shown in Fig. 1, we maintain one prototype for each class and adopt prototype matching for classification. The class centers $\{\mu_c\}_{c=1}^{C}$ are used as the prototype of each class in this paper. And we denote the classifier in the source domain as f_s and the classifier in the target domain as f_t. Given a training set $\mathcal{D} = \{x_i, y_i\}_{i=1}^{|\mathcal{D}|}$ with $|\mathcal{D}|$ samples, a sample $x \in \mathcal{D}$, the class center (prototype) for each class is defined as $\mu_c = \frac{1}{n_c}\sum_{x_i \in \mathcal{D}_c} x_i$, where $\mathcal{D}_c = \{x_i : x_i \in \mathcal{D}, y(x_i) = c\}$ and $n_c = |\mathcal{D}_c|$. We can derive the conditional probability of a given sample x belonging to class y as:

$$p(y|x) = \frac{exp(-||x - \mu_y||)}{\sum_{c=1}^{C} exp(-||x - \mu_c||)} \tag{4}$$

Assume the sample x belongs to class c, it is expected that the conditional probability $p(y|x)$ is close to $[0, 0, ..., 1, ..., 0]$, which is a C-dimensional one hot vector with the c-th dimension to be 1. Our goal is to pull the sample close to the center of c-th class while push the sample away from other $C-1$ class centers. Note that instead of pushing samples directly away from $C-1$ centers, we view the data of other $C-1$ classes as a whole, and use the center of the $C-1$ classes $\widehat{\mu_c}$ to calculate the distance. As a result, the algorithm complexity can be reduced and the proposed algorithm can be accelerated. The objective of minimizing classification error can be represented as,

$$\min \sum_{(x,c)\sim\mathcal{D}} ||x - \mu_c||_2^2 - \beta||x - \widehat{\mu_c}||_2^2 \tag{5}$$

where $\widehat{\mu_c} = \frac{1}{n_c^\star}\sum_{x_i \in \mathcal{D}/\mathcal{D}_c} x_i$ and $\widehat{\mu_c}$ is the center of all classes except class c in the training set, $n_c^\star = |\mathcal{D}/\mathcal{D}_c|$, β is the regularization parameter.

4 Method

In this section, we will describe all the objectives and the method to select target samples separately.

4.1 Empirical Error Minimization

For classifying the samples correctly, the first objective of CDEM is to minimize the empirical errors in both domains. Since there are no labeled data in the target domain, we use the pseudo labels [22] instead. The empirical errors in both domains are represented as,

$$\sum_{i=1}^{n_s+n_t} l(h(x_i), y_i) = \varepsilon_s(f_s) + \varepsilon_t(f_t)$$
$$= \sum_{c=1}^{C} \sum_{x_i \in \mathcal{D}_{s,c}} \left(||P^T(x_i - \mu_{s,c})||_2^2 - \beta ||P^T(x_i - \widehat{\mu_{s,c}})||_2^2 \right) \quad (6)$$
$$+ \sum_{c=1}^{C} \sum_{x_j \in \mathcal{D}_{t,c}} \left(||P^T(x_j - \mu_{t,c})||_2^2 - \beta ||P^T(x_j - \widehat{\mu_{t,c}})||_2^2 \right)$$

where $\mathcal{D}_{s,c} = \{x_i : x_i \in \mathcal{D}_s, y(x_i) = c\}$ is the set of examples belonging to class c in the source domain and $y(x_i)$ is the true label of x_i. Correspondingly, $\mathcal{D}_{t,c} = \{x_j : x_j \in \mathcal{D}_t, \hat{y}(x_j) = c\}$ is the set of examples belonging to class c in the target domain, where $\hat{y}(x_j)$ is the pseudo label of x_j. $\mu_{s,c} = \frac{1}{n_{s,c}} \sum_{x_i \in \mathcal{D}_{s,c}} x_i$ and $\mu_{t,c} = \frac{1}{n_{t,c}} \sum_{x_j \in \mathcal{D}_{t,c}} x_j$ are the centers of c-th class in the source domain and the target domain respectively, where $n_{s,c} = |\mathcal{D}_{s,c}|$ and $n_{t,c} = |\mathcal{D}_{t,c}|$. Similarly, $\widehat{\mu_{s,c}} = \frac{1}{n^{\star}_{s,c}} \sum_{x_i \in \mathcal{D}_s/\mathcal{D}_{s,c}} x_i$ and $\widehat{\mu_{t,c}} = \frac{1}{n^{\star}_{t,c}} \sum_{x_j \in \mathcal{D}_t/\mathcal{D}_{t,c}} x_j$ are the centers of all classes except class c in the source domain and the target domain respectively, where $n^{\star}_{s,c} = |\mathcal{D}_s/\mathcal{D}_{s,c}|, n^{\star}_{t,c} = |\mathcal{D}_t/\mathcal{D}_{t,c}|$.

We further rewrite the first term of the objective function in Eq. (6) as follows,

$$\sum_{c=1}^{C} \sum_{x_i \in \mathcal{D}_{s,c}} \left(||P^T(x_i - \mu_{s,c})||_2^2 - \beta ||P^T(x_i - \widehat{\mu_{s,c}})||_2^2 \right)$$
$$= \sum_{c=1}^{C} \left((1-\beta) \sum_{x_i \in \mathcal{D}_{s,c}} ||P^T(x_i - \mu_{s,c})||_2^2 - \beta n_{s,c} ||P^T(\mu_{s,c} - \widehat{\mu_{s,c}})||_2^2 \right) \quad (7)$$
$$= (1-\beta) \sum_{c=1}^{C} \sum_{x_i \in \mathcal{D}_{s,c}} ||P^T(x_i - \mu_{s,c})||_2^2 - \beta \sum_{c=1}^{C} n_{s,c} ||P^T(\mu_{s,c} - \widehat{\mu_{s,c}})||_2^2$$

Inspired by Linear Discriminant Analysis (LDA) [26] and follow previous method [17], we further transform the two terms, which can be considered as intra-class variance in Eq. (7), into similar expressions as Eq. (8).

$$(1-\beta) \sum_{c=1}^{C} \sum_{x_i \in \mathcal{D}_{s,c}} ||P^T(x_i - \mu_{s,c})||_2^2 - \beta \sum_{c=1}^{C} n_{s,c} ||P^T(\mu_{s,c} - \widehat{\mu_{s,c}})||_2^2$$
$$= \mathrm{tr}(P^T X_s (I - Y_s (Y_s^T Y_s)^{-1} Y_s^T) X_s^T P) - \beta \sum_{c=1}^{C} n_{s,c} \mathrm{tr}(P^T X_s \widehat{Q_{s,c}} X_s^T P) \quad (8)$$

Where $X_s \in \mathbb{R}^{d \times n_s}$ and $Y_s \in \mathbb{R}^{n_s \times C}$ are the samples and labels in the source domain. $tr(\cdot)$ is the trace of a matrix. By using target samples $X_t \in \mathbb{R}^{d \times n_s}$ and pseudo labels $\hat{Y}_t \in \mathbb{R}^{n_t \times C}$, the same strategy is also used to transform the second term in Eq. (6). Denote $X = X_s \cup X_t \in \mathbb{R}^{d\times(n_s+n_t)}$, the objective of minimizing empirical errors can be written as,

$$\varepsilon_s(f_s) + \varepsilon_t(f_t) = (1-\beta)(\mathrm{tr}(P^T X Q^Y X^T P) - \beta \sum_{c=1}^{C} \mathrm{tr}(P^T X \widehat{Q^c} X^T P)) \quad (9)$$

where

$$Q^Y = \begin{bmatrix} I - Y_s(Y_s^T Y_s)^{-1} Y_s^T & \mathbf{0} \\ \mathbf{0} & I - \hat{Y}_t(\hat{Y}_t^T \hat{Y}_t)^{-1} \hat{Y}_t^T \end{bmatrix}, \widehat{Q^c} = \begin{bmatrix} n_{s,c}\widehat{Q_{s,c}} & \mathbf{0} \\ \mathbf{0} & n_{t,c}\widehat{Q_{t,c}} \end{bmatrix} \quad (10)$$

$$(\widehat{Q_{s,c}})_{ij} = \begin{cases} \frac{1}{n_{s,c}n_{s,c}}, & x_i, x_j \in \mathcal{D}_{s,c} \\ \frac{1}{n^\star_{s,c}n^\star_{s,c}}, & x_i, x_j \in \mathcal{D}_s/\mathcal{D}_{s,c} \\ -\frac{1}{n_{s,c}n^\star_{s,c}}, & otherwise \end{cases}, (\widehat{Q_{t,c}})_{ij} = \begin{cases} \frac{1}{n_{t,c}n_{t,c}}, & x_i, x_j \in \mathcal{D}_{t,c} \\ \frac{1}{n^\star_{t,c}n^\star_{t,c}}, & x_i, x_j \in \mathcal{D}_t/\mathcal{D}_{t,c} \\ -\frac{1}{n_{t,c}n^\star_{t,c}}, & otherwise \end{cases} \quad (11)$$

4.2 Distribution Alignment

As there are feature distribution discrepancy across domains, the second objective of CDEM is to learn domain-invariant features for decreasing the discrepancy between feature distributions across domains. Distribution alignment is a popular method in domain adaptation [22,27,38]. To reduce the shift between feature distributions across domains, we follow [19] and adopt *Maximum Mean Discrepancy* (MMD) as the distance measure to compute marginal distribution discrepancy $d_m(\mathcal{D}_s, \mathcal{D}_t)$ across domains based on the distance between the sample means of two domains in the feature embeddings:

$$d_m(\mathcal{D}_s, \mathcal{D}_t) = ||\frac{1}{n_s} \sum_{x_i \in \mathcal{D}_s} P^T x_i - \frac{1}{n_t} \sum_{x_j \in \mathcal{D}_t} P^T x_j||^2 = \mathrm{tr}(P^T X M_0 X^T P) \quad (12)$$

Based on the pseudo labels of the target data, we minimize the conditional distribution discrepancy $d_c(\mathcal{D}_s, \mathcal{D}_t)$ between domains:

$$d_c(\mathcal{D}_s, \mathcal{D}_t) = \sum_{c=1}^{C} ||\frac{1}{n_{s,c}} \sum_{x_i \in \mathcal{D}_{s,c}} P^T x_i - \frac{1}{n_{t,c}} \sum_{x_j \in \mathcal{D}_{t,c}} \cdot P^T x_j||^2 = \sum_{c=1}^{C} \mathrm{tr}(P^T X M_c X^T P) \quad (13)$$

where,

$$(\boldsymbol{M}_0)_{ij} = \begin{cases} \frac{1}{n_s^2}, & x_i, x_j \in \mathcal{D}_s \\ \frac{1}{n_t^2}, & x_i, x_j \in \mathcal{D}_t \\ -\frac{1}{n_s n_t}, & otherwise \end{cases}, (\boldsymbol{M}_c)_{ij} = \begin{cases} \frac{1}{n_{s,c}^2}, & x_i, x_j \in \mathcal{D}_{s,c} \\ \frac{1}{n_{t,c}^2}, & x_i, x_j \in \mathcal{D}_{t,c} \\ -\frac{1}{n_{s,c}n_{t,c}}, & \begin{cases} x_i \in \mathcal{D}_{s,c}, x_j \in \mathcal{D}_{t,c} \\ x_j \in \mathcal{D}_{s,c}, x_i \in \mathcal{D}_{t,c} \end{cases} \\ 0, & otherwise \end{cases} \quad (14)$$

Denote $M = M_0 + \sum_{c=1}^{C} M_c$, then the objective of distribution alignment is equal to:

$$l_d(\mathcal{D}_s, \mathcal{D}_t) = d_m(\mathcal{D}_s, \mathcal{D}_t) + d_c(\mathcal{D}_s, \mathcal{D}_t) = \mathrm{tr}(P^T X M X^T P) \tag{15}$$

4.3 Cross-domain Error Minimization

Although performing distribution alignment can pull the two domains close, it is not enough for a good adaptation across domains. The discrepancy between the labeling functions, which is essentially the cross-domain errors, is another factor leading to the domain discrepancy [47] while is ignored by existing methods. Thus, the third objective of CDEM is to minimize cross-domain errors, by which the decision boundaries across domains can be close and the samples near the decision boundaries can be classified correctly, achieving a further reduced domain discrepancy and better adaptation.

It is noticed that the cross-domain errors are the performance of the source classifier in the target domain and the performance of the target classifier in the source domain. As we use the prototype classifier, the cross-domain error in each domain is represented by the distances between the source samples (target samples) and the corresponding class centers in the target domain (source domain). For example, the cross-domain error in the source domain $\varepsilon_s(f_t)$ is the empirical error of applying the target classifier f_t to the source domain $\mathcal{D}_s$. Technically, the cross-domain errors in both domains are represented as,

$$\begin{aligned} l_f(\mathcal{D}_s, \mathcal{D}_t) &= \varepsilon_s(f_t) + \varepsilon_t(f_s) \\ &= \sum_{c=1}^{C} \sum_{x_i \in \mathcal{D}_{s,c}} \left(\|P^T(x_i - \mu_{t,c})\|_2^2 - \beta \|P^T(x_i - \widehat{\mu_{t,c}})\|^2 \right) \\ &+ \sum_{c=1}^{C} \sum_{x_j \in \mathcal{D}_{t,c}} \left(\|P^T(x_j - \mu_{s,c})\|_2^2 - \beta \|P^T(x_j - \widehat{\mu_{s,c}})\|^2 \right) \end{aligned} \tag{16}$$

Similar to the first objective, we transform the formula in Eq. (16) as the following,

$$\begin{aligned} \varepsilon_s(f_t) + \varepsilon_t(f_s) =& (1-\beta)\mathrm{tr}(P^T X Q^Y X^T P) + \sum_{c=1}^{C} n^c \mathrm{tr}(P^T X M_c X^T P) \\ &- \beta \sum_{c=1}^{C} \mathrm{tr}(P^T X (n_{s,c} \widehat{Q_{s,t}^c} + n_{t,c} \widehat{Q_{t,s}^c}) X^T P) \end{aligned} \tag{17}$$

where

$$(\widehat{Q_{s,t}^c})_{ij} = \begin{cases} \frac{1}{n_{s,c} n_{s,c}}, & x_i, x_j \in \mathcal{D}_{s,c} \\ \frac{1}{n_{t,c}^\star n_{t,c}^\star}, & x_i, x_j \in \mathcal{D}_{t,c} \\ -\frac{1}{n_{s,c} n_{t,c}^\star}, & \begin{cases} x_i \in \mathcal{D}_{s,c}, x_j \in \mathcal{D}_{t,c} \\ x_j \in \mathcal{D}_{s,c}, x_i \in \mathcal{D}_{t,c} \end{cases} \\ 0 & otherwise \end{cases} \quad (\widehat{Q_{t,s}^c})_{ij} = \begin{cases} \frac{1}{n_{t,c} n_{t,c}}, & x_i, x_j \in \mathcal{D}_{t,c} \\ \frac{1}{n_{s,c}^\star n_{s,c}^\star}, & x_i, x_j \in \mathcal{D}_{s,c} \\ -\frac{1}{n_{t,c} n_{s,c}^\star}, & \begin{cases} x_i \in \mathcal{D}_{s,c}, x_j \in \mathcal{D}_{t,c} \\ x_j \in \mathcal{D}_{s,c}, x_i \in \mathcal{D}_{t,c} \end{cases} \\ 0 & otherwise \end{cases} \tag{18}$$

4.4 Discriminative Feature Learning

Learning domain-invariant features to reduce the domain discrepancy may harm the discriminability of the features [43]. So the fourth objective of CDEM is to perform discriminative learning to enhance the discriminability of the features [4]. To be specific, we resort to explore the structural information of all the samples to make the samples belonging to the same class close, which is useful for classification. Thus, the discriminative objective is,

$$l_m(\mathcal{D}_s, \mathcal{D}_t) = \sum_{x_i, x_j \in X} ||P^T x_i - P^T x_j||_2^2 W_{ij} \tag{19}$$

where $W \in \mathbb{R}^{(n_s+n_t)\times(n_s+n_t)}$ is the similarity matrix, which is defined as follows,

$$W_{ij} = \begin{cases} 1, & y_i(\hat{y}_i) = y_j(\hat{y}_j) \\ 0, & y_i(\hat{y}_i) \neq y_j(\hat{y}_j) \end{cases} \tag{20}$$

This objective can be transformed as follows,

$$\begin{aligned} \sum_{x_i, x_j \in X} ||P^T x_i - P^T x_j||_2^2 W_{ij} &= \mathrm{tr}(\sum_{x_i \in X} P^T x_i B_{ii} x_i^T P - \sum_{x_i, x_j \in X} P^T x_i W_{ij} x_j^T P) \\ &= \mathrm{tr}(P^T X B X^T P - P^T X W X^T P) = \mathrm{tr}(P^T X L X^T P) \end{aligned} \tag{21}$$

Where, $L = B - W \in \mathbb{R}^{(n_s+n_t)\times(n_s+n_t)}$ is the laplacian matrix, and $B \in \mathbb{R}^{(n_s+n_t)\times(n_s+n_t)}$ is a diagonal matrix with $(B)_{ii} = \sum_j (W)_{ij}$.

4.5 Optimization

Combining the four objectives together, we get the following optimization problem,

$$\begin{aligned} L(p) &= \mathrm{tr}(P^T X Q^Y X^T P) - \beta \sum_{c=1}^{C} \mathrm{tr}(P^T X \widehat{Q^c} X^T P) + \lambda \mathrm{tr}(P^T X M X^T P) \\ &- \gamma \sum_{c=1}^{C} \mathrm{tr}(P^T X (\widehat{Q^c_{s,t}} + \widehat{Q^c_{t,s}}) X^T P)) + \eta \mathrm{tr}(P^T X L X^T P) + \delta ||P||_F^2 \\ &= \mathrm{tr}(P^T X \Omega X^T P) + \delta ||P||_F^2 \\ s.t.\quad & P^T X H X^T P = I \end{aligned} \tag{22}$$

where $\Omega = Q^Y + \lambda M + \eta L - \sum_{c=1}^{C} (\beta \widehat{Q^c} + \gamma \widehat{Q^c_{s,t}} + \gamma \widehat{Q^c_{t,s}})$ and $H = \mathbf{I} - \frac{1}{n_s+n_t}\mathbf{1}$ is the centering matrix. According to the constrained theory, we denote $\Theta = diag(\theta_1, ..., \theta_k) \in R^{k\times k}$ as the Langrange multiplier, and derive the Langrange function for problem (22) as,

$$L = \mathrm{tr}(P^T X \Omega X^T P) + \delta ||P||_F^2 + \mathrm{tr}((I - P^T X H X^T P)\Theta) \tag{23}$$

Setting $\frac{\partial L}{\partial P} = \mathbf{0}$, we get generalized eigendecomposition,

$$(X\Omega X^T + \delta I)P = XHX^T P\Theta \tag{24}$$

Finally, finding the optimal feature transformation matrix P is reduced to solving Eq. (24) for the k smallest eigenvectors.

4.6 Selective Target Samples

To avoid the catastrophic error accumulation caused by the incorrect pseudo labels, we predict the pseudo labels for the target samples via exploring the structural information of the target domain and source classifier. Moreover, based on curriculum learning, we propose a strategy to select a part of target samples, whose pseudo labels are more likely to be correct, to participate in the next iteration for learning the transformation matrix. One simple way to predict pseudo labels for target samples is to use the source class centers $\{\mu_c\}_{c=1}^C$ (the prototypes for each class) to classify the target samples. Therefore the conditional probability of a given target sample x_t belonging to class y is defined as:

$$p_s(y|x_t) = \frac{exp(-P^T||x_t - \mu_{s,y}||)}{\sum_{c=1}^C exp(-P^T||x_t - \mu_{s,c}||)} \tag{25}$$

Because there exists distribution discrepancy across domains, only using source prototypes is not enough for pseudo-labeling, which will lead to some incorrect pseudo labels. We further consider the structural information in the target domain, which can be exploited by unsupervised clustering. In this paper, K-Means clustering is used in the target domain. The cluster center $\mu_{t,c}$ is initialized with corresponding class center $\mu_{s,c}$ in the source domain, which ensures one-to-one mapping for each class. Thus, based on target clustering, the conditional probability of a given target sample x_t belonging to class y is defined by:

$$p_t(y|x_t) = \frac{exp(-P^T||x_t - \mu_{t,y}||)}{\sum_{c=1}^C exp(-P^T||x_t - \mu_{t,c}||)} \tag{26}$$

After getting $p_s(y|x_t)$ and $p_t(y|x_t)$, we can obtain two different kinds of pseudo labels $\hat{y}_s^t$ and $\hat{y}_t^t$ for target samples x_t:

$$\hat{y}_s^t = \arg\max_{y \in Y_t} p_s(y|x_t) \quad \hat{y}_t^t = \arg\max_{y \in Y_t} p_t(y|x_t) \tag{27}$$

Based on these two kinds of pseudo labels, a curriculum learning based strategy is proposed to select a part of target samples for training. We firstly select the target samples whose pseudo labels predicted by $p_s(y|x_t)$ and $p_t(y|x_t)$ are the same (i.e., $\hat{y}_s^t = \hat{y}_t^t$). And these samples are considered to satisfy the *label consistency* and are likely to be correct. Then, we progressively select a subset containing top tn_t/T samples with highest prediction probabilities from the samples satisfying the label consistency, where T is the number of total iterations and t is the number of current iteration. Finally, we combine $p_s(y|x_t)$ and

$p_t(y|x_t)$ in an iterative weighting method. Formally, the final class conditional probability and the pseudo label for x_t are as follows:

$$\begin{aligned} p(y|x_t) &= (1-t/T) \times p_s(y|x_t) + t/T \times p_t(y|x_t) \\ \hat{y}_t &= \arg\max_{y \in Y_t} p(y|x_t) \end{aligned} \tag{28}$$

To avoid the class imbalance problem when selecting samples, we take the class-wise selection into consideration to ensure that each class will have a certain proportion of samples to be selected, namely,

$$N_{t,c} = \min(n_{t,c} \times t/T, n_{t,c}^{con}) \tag{29}$$

where $N_{t,c}$ is the number of target samples being selected of class c, $n_{t,c}^{con}$ denotes the number of target samples satisfying the label consistency in the class c and t is the current epoch.

Remark: CDEM is composed of two processes: learning transformation matrix P and selecting target samples. We firstly learn the transformation matrix P via solving the optimization problem (24). Then, we select the target samples in the transformed feature space. We perform the two processes in an alternative manner as previous method [39].

5 Experiment

In this section, we evaluate the performance of CDEM by extensive experiments on three widely-used common datasets. The source code of CDEM is available at https://github.com/yuntaodu/CDEM.

5.1 Data Preparation

The **Office-Caltech** dataset [11] consists of images from 10 overlapping object classes between Office31 and Caltech-256 [13]. Specifically, we have four domains, **C** (*Caltech-256*), **A** (*Amazon*), **W** (*Webcam*), and **D** (*DSLR*). By randomly selecting two different domains as the source domain and target domain respectively, we construct $3 \times 4 = 12$ cross-domain object tasks, e.g. **C** → **A**, **C** → **W**,..., **D** → **W**.

The **Office-31** dataset [31] is a popular benchmark for visual domain adaptation. The dataset contains three real-world object domains, *Amazon* (**A**, images downloaded from online merchants), *Webcom* (**W**, low-resolution images by a web camera), and *DSLR* (**D**, high-resolution images by a digital camera). It has 4652 images of 31 classes. We evaluate all methods on six transfer tasks: **A** → **W**, **A** → **D**, **W** → **A**, **W** → **D**, **D** → **A**, and **D** → **W**.

ImageCLEF-DA[1] is a dataset organized by selecting 12 common classes shared by three public datasets, each is considered as a domain: *Caltech-256*

[1] http://imageclef.org/2014/adaptation.

(**C**), *ImageNet ILSVRC 2012* (**I**), and *Pascal VOC 2012* (**P**). We evaluate all methods on six transfer tasks: **I** → **P**, **P** → **I**, **I** → **C**, **C** → **I**, **C** → **P**, and **P** → **C**.

5.2 Baseline Methods

We compare the performance of CDEM with several state-of-the-art traditional and deep domain adaptation methods:

- Traditional domain adaptation methods: **1NN** [8],**SVM** [10] and **PCA** [15], Transfer Component Analysis (**TCA**) [27], Joint Distribution Alignment (**JDA**) [22], CORrelation Alignment (**CORAL**) [34], Joint Geometrical and Statistical Alignment (**JGSA**) [43], Manifold Embedded Distribution Alignment (**MEDA**) [38], Confidence-Aware Pseudo Label Selection (**CAPLS**) [40] and Selective Pseudo-Labeling (**SPL**) [39].
- Deep domain adaptation methods: Deep Domain Confusion (**DDC**) [36], Deep Adaptation Network (**DAN**) [19], Deep CORAL (**DCORAL**) [35], Residual Transfer Network (**RTN**) [23], Multi Adversarial Domain Adaptation(**MADA**) [29], Conditional Domain Adversarial Network (**CDAN**) [20], Incremental CAN (**iCAN**) [44], Domain Symmetric Networks (**SymNets**) [45], Generate To Adapt (**GTA**) [32] and Joint Domain alignment and Discriminative feature learning (**JDDA**) [4].

Table 1. Classification accuracy (%) on Office-Caltech dataset using Decaf6 features.

Method	C→A	C→W	C→D	A→C	A→W	A→D	W→C	W→A	W→D	D→C	D→A	D→W	Average
DDC [36]	91.9	85.4	88.8	85.0	86.1	89.0	78.0	84.9	**100.0**	81.1	89.5	98.2	88.2
DAN [19]	92.0	90.6	89.3	84.1	91.8	91.7	81.2	92.1	**100.0**	80.3	90.0	98.5	90.1
DCORAL [35]	92.4	91.1	91.4	84.7	–	–	79.3	–	–	82.8	–	–	–
1NN [8]	87.3	72.5	79.6	71.7	68.1	74.5	55.3	62.6	98.1	42.1	50.0	91.5	71.1
SVM [10]	91.6	80.7	86.0	82.2	71.9	80.9	67.9	73.4	**100.0**	72.8	78.7	98.3	82.0
PCA [15]	88.1	83.4	84.1	79.3	70.9	82.2	70.3	73.5	99.4	71.7	79.2	98.0	81.7
TCA [27]	89.8	78.3	85.4	82.6	74.2	81.5	80.4	84.1	**100.0**	82.3	89.1	99.7	85.6
JDA [22]	89.6	85.1	89.8	83.6	78.3	80.3	84.8	90.3	100.0	85.5	91.7	99.7	88.2
CORAL[34]	92.0	80.0	84.7	83.2	74.6	84.1	75.5	81.2	**100.0**	76.8	85.5	99.3	84.7
JGSA[43]	91.4	86.8	93.6	84.9	81.0	88.5	85.0	90.7	**100.0**	86.2	92.0	99.7	90.0
MEDA[38]	93.4	95.6	91.1	87.4	88.1	88.1	**93.2**	**99.4**	99.4	87.5	93.2	97.6	92.8
CAPLS [40]	90.8	85.4	95.5	86.1	87.1	94.9	88.2	92.3	**100.0**	88.8	93.0	**100.0**	91.8
SPL[39]	92.7	93.2	**98.7**	87.4	95.3	89.2	87.0	92.0	**100.0**	88.6	92.9	98.6	93.0
CDEM (Ours)	**93.5**	**97.0**	96.2	**88.7**	**98.0**	**95.5**	89.1	93.5	**100.0**	**90.1**	**93.4**	99.7	**94.6**

5.3 Experimental Setup

To fairly compare our method with the state-of-the-art methods, we adopt the deep features commonly used in existing unsupervised domain adaption methods. Specifically, DeCaf6 [9] features (activations of the 6th fully connected layer of a convolutional neural network trained on ImageNet, $d = 4096$) are used for Office-Caltech dataset, ResNet50 [14] features ($d = 2048$) are used for Office-31 dataset and ImageCLEF-DA dataset. In this way, we can compare our proposed method with these deep models.

In our experiments, we adopt the PCA algorithm to decrease the dimension of the data before learning to accelerate the proposed method. We set the dimensionality of PCA space $m = 128$ for Office-Caltech dataset and $m = 256$ for Office-31 and ImageCLEF-DA datasets. For the dimensionality of the transformation matrix P, we set $k = 32, 128$ and 64 for Office-Caltech, Office-31 and ImageCLEF-DA respectively. The number of iterations for CDEM to converge is $T = 11$ for all datasets. For regularization parameter δ, we set $\delta = 1$ for Office-Caltech and ImageCLEF-DA datasets and $\delta = 0.1$ for Office-31 dataset. As for the other hyper-parameters, we set β, λ, γ and η by searching through the grid with a range of $\{0.0001, 0.001, 0.01, 0.1, 1, 10\}$. In addition, the coming experiment on parameter sensitivity shows that our method can keep robustness with a wide range of parameter values.

5.4 Results and Analysis

The results on the *Office-Caltech* dataset are reported in Table 1, where the highest accuracy of each cross-domain task is boldfaced. The results of baselines are directly reported from original papers if the protocol is the same. The CDEM method significantly outperforms all the baseline methods on most transfer tasks (7 out of 12) in this dataset. It is desirable that CDEM promotes the classification accuracies significantly on hard transfer tasks, e.g., **A**→**D** and **A** →**W**, where the source and target domains are substantially different [31]. Note that CDEM performs better than SPL in most tasks, which only learns domain-invariant features across domains.

The results on *Office-31* dataset are reported in Table 2. The CDEM method outperforms the comparison methods on most transfer tasks. Compared with the best shallow baseline method (CAPLS), the accuracy is improved by 1.7%. Note that the CDEM method outperforms some deep domain adaptation methods, which implies the performance of CDEM in domain adaptation is better than several deep methods.

The results on *ImageCLEF-DA* dataset are reported in Table 3. The CDEM method substantially outperforms the comparison methods on most transfer tasks, and with more rooms for improvement. An interpretation is that the three domains in *ImageCLEF-DA* are visually dissimilar with each other, and are difficult in each domain with much lower in-domain classification accuracy [22]. MEDA and SPL are the representative shallow domain adaptation methods, which both focus on learning domain-invariant features. Moreover, SPL also uses selective target samples for adaptation. Consequently, the better performance of CDEM implies that minimizing cross-domain errors can further reduce the discrepancy across domains and achieve better adaptation.

Table 2. Accuracy (%) on Office-31 dataset using either ResNet50 features or ResNet50 based deep models.

Method	A→W	D→W	W→D	A→D	D→A	W→A	Avg
RTN [23]	84.5	96.8	99.4	77.5	66.2	64.8	81.6
MADA [29]	90.0	97.4	99.6	87.8	70.3	66.4	85.2
GTA [32]	89.5	97.9	<u>99.8</u>	87.7	72.8	71.4	86.5
iCAN [44]	92.5	**98.8**	**100.0**	90.1	72.1	69.9	87.2
CDAN-E [20]	<u>94.1</u>	98.6	**100.0**	92.9	71.0	69.3	87.7
JDDA [4]	82.6	95.2	99.7	79.8	57.4	66.7	80.2
SymNets [45]	90.8	**98.8**	**100.0**	<u>93.9</u>	74.6	72.5	<u>88.4</u>
TADA [18]	**94.3**	<u>98.7</u>	<u>99.8</u>	91.6	72.9	73.0	<u>88.4</u>
MEDA [38]	86.2	97.2	99.4	85.3	72.4	74.0	85.7
CAPLS [40]	90.6	98.6	99.6	88.6	<u>75.4</u>	<u>76.3</u>	88.2
CDEM (Ours)	91.1	98.4	99.2	**94.0**	**77.1**	**79.4**	**89.9**

Table 3. Accuracy (%) on ImageCLEF-DA dataset using either ResNet50 features or ResNet50 based deep models.

Method	I→P	P→I	I→C	C→I	C→P	P→C	Avg
RTN [23]	75.6	86.8	95.3	86.9	72.7	92.2	84.9
MADA [29]	75.0	87.9	96.0	88.8	75.2	92.2	85.8
iCAN [44]	79.5	89.7	94.7	89.9	78.5	92.0	87.4
CDAN-E [20]	77.7	90.7	**97.7**	91.3	74.2	94.3	87.7
SymNets [45]	<u>80.2</u>	93.6	97.0	93.4	78.7	<u>96.4</u>	89.9
MEDA [38]	79.7	92.5	95.7	92.2	78.5	95.5	89.0
SPL [39]	78.3	<u>94.5</u>	96.7	<u>95.7</u>	<u>80.5</u>	96.3	<u>90.3</u>
CDEM (ours)	**80.5**	**96.0**	<u>97.2</u>	**96.3**	**82.1**	**96.8**	**91.5**

5.5 Effectiveness Analysis

Ablation Study. We conduct an ablation study to analyse how different components of our method contribute to the final performance. When learning the final classifier, CDEM involves four components: the empirical error minimization (ERM), the distribution alignment (DA), the cross-domain error minimization (CDE) and discriminative feature learning (DFL). We empirically evaluate the importance of each component. To this end, we investigate different combinations of four components and report average classification accuracy on three datasets in Table 4. Note that the result of the first setting (only ERM used) is like the result of the source-only method, where no adaptation is performed across domains. It can be observed that methods with distribution alignment or cross-domain error minimization outperform those without distribution alignment or cross-domain error minimization. Moreover, discriminative learning can further improve performance and CDE achieves the biggest improvement compared with other components. Summarily, using all the terms together achieves the best performance in all tasks.

Table 4. Results of ablation study.

Method				Office-Caltech	Office31	ImageCLEF-DA
ERM	DA	CDE	DFL			
✓	✗	✗	✗	90.2	86.6	87.5
✓	✓	✗	✗	91.5	87.2	88.6
✓	✓	✓	✗	94.0	89.2	90.8
✓	✓	✓	✓	94.6	89.9	91.5

Evaluation of Selective Target Samples. We further perform experiments to show the effectiveness of selective target samples. We compare several variants of the proposed method: **a**) No selection: We use all the target samples for training without any samples removed. **b**) Only label consistency: We only select the samples where the predicted label by $p_s(x_t)$ is the same with $p_t(x_t)$. **c**) Only high probabilities: We only select the target samples with high prediction confidence. **d**) The proposed method. As shown in Fig. 2(a), "No selection" leads to a model with the worst performance due to the catastrophic error accumulation. The "Only label consistency" and "Only high probabilities" achieve significantly better results than "No selection", but are still worse than the proposed method, which verifies that our method of explicitly selecting easier samples can make the model more adaptive and less likely to be affected by the incorrect pseudo labels.

Feature Visualization. In Fig. 2(b–d), we visualize the feature representations of task **A**→**D** (10 classes) by t-SNE [24] as previous methods [39] using JDA and CDEM. Before adaptation, we can see that there is a large discrepancy across domains. After adaptation, JDA learns domain-invariant features which can reduce distribution discrepancy, the source domain and the target domain can become closer. While CDEM further considers the cross-domain errors, achieving a better performance.

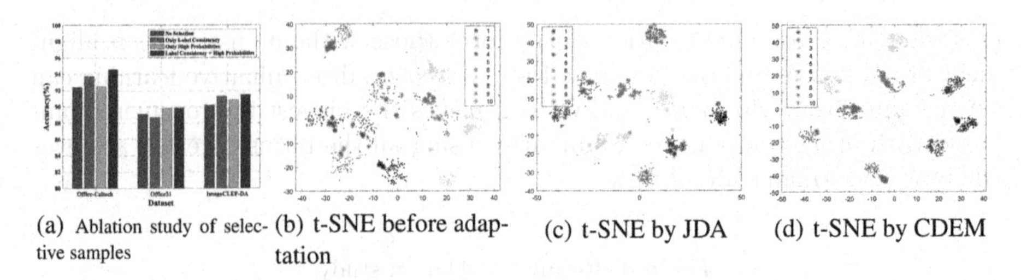

(a) Ablation study of selective samples (b) t-SNE before adaptation (c) t-SNE by JDA (d) t-SNE by CDEM

Fig. 2. Ablation study of selective samples, t-SNE visualization and parameter sensitivity

6 Conclusion

In this paper, we propose the *Cross-Domain Error Minimization* (CDEM), which not only learns domain-invariant features across domains but also performs cross-domain error minimization. These two goals complement each other and contribute to better domain adaptation. Apart from these two goals, we also integrate the empirical error minimization and discriminative learning into a unified learning process. Moreover, we propose a method to select the target samples to alleviate error accumulation problem caused by incorrect pseudo labels. Through a large number of experiments, it is proved that our method is superior to other strong baseline methods.

Acknowledgements. This paper is supported by the National Key Research and Development Program of China (Grant No. 2018YFB1403400), the National Natural Science Foundation of China (Grant No. 61876080), the Collaborative Innovation Center of Novel Software Technology and Industrialization at Nanjing University.

References

1. Ben-David, S., Blitzer, J., Crammer, K., Kulesza, A., Pereira, F., Vaughan, J.W.: A theory of learning from different domains. Mach. Learn. 151–175 (2009). https://doi.org/10.1007/s10994-009-5152-4
2. Bengio, Y., Louradour, J., Collobert, R., Weston, J.: Curriculum learning. In: ICML 2009 (2009)
3. Cao, Y., Long, M., Wang, J.: Unsupervised domain adaptation with distribution matching machines. AAAI (2018)
4. Chen, C., Chen, Z., Jiang, B., Jin, X.: Joint domain alignment and discriminative feature learning for unsupervised deep domain adaptation. AAAI (2019)
5. Chen, C., et al.: Progressive feature alignment for unsupervised domain adaptation. In: CVPR, pp. 627–636 (2018)
6. Chen, Q., Du, Y., Tan, Z., Zhang, Y., Wang, C.: Unsupervised domain adaptation with joint domain-adversarial reconstruction networks. In: ECML/PKDD (2020)
7. Dai, W., Yang, Q., Xue, G.R., Yu, Y.: Boosting for transfer learning. In: ICML 2007 (2007)
8. Delany, S.J.: k-nearest neighbour classifiers (2007)

9. Donahue, J., et al.: DeCAF: a deep convolutional activation feature for generic visual recognition. In: ICML (2014)
10. Evgeniou, T., Pontil, M.: Support vector machines: theory and applications. In: Paliouras, G., Karkaletsis, V., Spyropoulos, C.D. (eds.) ACAI 1999. LNCS (LNAI), vol. 2049, pp. 249–257. Springer, Heidelberg (2001). https://doi.org/10.1007/3-540-44673-7_12
11. Gong, B., Shi, Y., Sha, F., Grauman, K.: Geodesic flow kernel for unsupervised domain adaptation. In: CVPR, pp. 2066–2073 (2012)
12. Graf, A.B.A., Bousquet, O., Rätsch, G., Schölkopf, B.: Prototype classification: insights from machine learning. Neural Comput. **21**(1), 272–300 (2009)
13. Griffin, G., Holub, A., Perona, P.: Caltech-256 object category dataset (2007)
14. He, K., Zhang, X., Ren, S., Sun, J.: Deep residual learning for image recognition. In: CVPR, pp. 770–778 (2016)
15. Hotelling, H.: Analysis of a complex of statistical variables into principal components. J. Educ. Psychol. **24**, 498–520 (1933)
16. Jing, M., Li, J., Zhao, J., Lu, K.: Learning distribution-matched landmarks for unsupervised domain adaptation. In: DASFAA (2018)
17. Liang, J., He, R., Sun, Z., Tan, T.: Aggregating randomized clustering-promoting invariant projections for domain adaptation. IEEE Trans. Pattern Anal. Mach. Intell. **41**, 1027–1042 (2019)
18. Liu, H., Long, M., Wang, J., Jordan, M.I.: Transferable adversarial training: a general approach to adapting deep classifiers. In: ICML (2019)
19. Long, M., Cao, Y., Wang, J., Jordan, M.I.: Learning transferable features with deep adaptation networks. In: ICML (2015)
20. Long, M., Cao, Z., Wang, J., Jordan, M.I.: Conditional adversarial domain adaptation. In: NeurIPS (2018)
21. Long, M., Wang, J., Ding, G., Pan, S.J., Yu, P.S.: Adaptation regularization: a general framework for transfer learning. IEEE TKDE **26**, 1076–1089 (2014)
22. Long, M., Wang, J., Ding, G., Sun, J.G., Yu, P.S.: Transfer feature learning with joint distribution adaptation. In: ICCV, pp. 2200–2207 (2013)
23. Long, M., Zhu, H., Wang, J., Jordan, M.I.: Unsupervised domain adaptation with residual transfer networks. In: NIPS (2016)
24. Maaten, L.V.D., Hinton, G.E.: Visualizing data using T-SNE. J. Mach. Learn. Res. **9**, 2579–2605 (2008)
25. Mansour, Y., Mohri, M., Rostamizadeh, A.: Domain adaptation: learning bounds and algorithms. In: NeurIPS (2009)
26. Mika, S., Rätsch, G., Weston, J., Scholkopf, B., Mullers, K.R.: Fisher discriminant analysis with kernels. In: Neural Networks for Signal Processing IX, pp. 41–48 (1999)
27. Pan, S.J., Tsang, I.W.H., Kwok, J.T., Yang, Q.: Domain adaptation via transfer component analysis. IEEE Trans. Neural Netw. **22**, 199–210 (2011)
28. Pan, S.J., Yang, Q.: A survey on transfer learning. IEEE TKDE **22**(10), 1345–1359 (2010)
29. Pei, Z., Cao, Z., Long, M., Wang, J.: Multi-adversarial domain adaptation. AAAI (2018)
30. Raghu, M., Zhang, C., Kleinberg, J.M., Bengio, S.: Transfusion: understanding transfer learning for medical imaging. In: NeurIPS (2019)
31. Saenko, K., Kulis, B., Fritz, M., Darrell, T.: Adapting visual category models to new domains. In: Daniilidis, K., Maragos, P., Paragios, N. (eds.) ECCV 2010. LNCS, vol. 6314, pp. 213–226. Springer, Heidelberg (2010). https://doi.org/10.1007/978-3-642-15561-1_16

32. Sankaranarayanan, S., Balaji, Y., Castillo, C.D., Chellappa, R.: Generate to adapt: aligning domains using generative adversarial networks. In: CVPR, pp. 8503–8512 (2018)
33. Smith, N., Gales, M.J.F.: Speech recognition using SVMs. In: NIPS (2001)
34. Sun, B., Feng, J., Saenko, K.: Return of frustratingly easy domain adaptation. AAAI (2015)
35. Sun, B., Saenko, K.: Deep CORAL: correlation alignment for deep domain adaptation. In: Hua, G., Jégou, H. (eds.) ECCV 2016. LNCS, vol. 9915, pp. 443–450. Springer, Cham (2016). https://doi.org/10.1007/978-3-319-49409-8_35
36. Tzeng, E., Hoffman, J., Zhang, N., Saenko, K., Darrell, T.: Deep domain confusion: maximizing for domain invariance. arXiv:1412.3474 (2014)
37. Wang, J., Chen, Y., Hao, S., Feng, W., Shen, Z.: Balanced distribution adaptation for transfer learning. In: ICDM, pp. 1129–1134 (2017)
38. Wang, J., Feng, W., Chen, Y., Yu, H., Huang, M., Yu, P.S.: Visual domain adaptation with manifold embedded distribution alignment. In: MM 2018 (2018)
39. Wang, Q., Breckon, T.P.: Unsupervised domain adaptation via structured prediction based selective pseudo-labeling. AAAI (2020)
40. Wang, Q., Bu, P., Breckon, T.P.: Unifying unsupervised domain adaptation and zero-shot visual recognition. In: IJCNN, pp. 1–8 (2019)
41. Wyner, A.J., Olson, M., Bleich, J., Mease, D.: Explaining the success of AdaBoost and random forests as interpolating classifiers. J. Mach. Learn. Res. **18**, 48:1–48:33 (2017)
42. Yang, L., Liang, X., Wang, T., Xing, E.: Real-to-virtual domain unification for end-to-end autonomous driving. In: Ferrari, V., Hebert, M., Sminchisescu, C., Weiss, Y. (eds.) ECCV 2018. LNCS, vol. 11208, pp. 553–570. Springer, Cham (2018). https://doi.org/10.1007/978-3-030-01225-0_33
43. Zhang, J., Li, W., Ogunbona, P.: Joint geometrical and statistical alignment for visual domain adaptation. In: CVPR, pp. 5150–5158 (2017)
44. Zhang, W., Ouyang, W., Li, W., Xu, D.: Collaborative and adversarial network for unsupervised domain adaptation. In: CVPR, pp. 3801–3809 (2018)
45. Zhang, Y., Tang, H., Jia, K., Tan, M.: Domain-symmetric networks for adversarial domain adaptation. In: CVPR, pp. 5026–5035 (2019)
46. Zhang, Y., Liu, T., Long, M., Jordan, M.I.: Bridging theory and algorithm for domain adaptation. In: ICML (2019)
47. Zhao, H., des Combes, R.T., Zhang, K., Gordon, G.J.: On learning invariant representation for domain adaptation. In: ICML (2019)

Unsupervised Domain Adaptation with Unified Joint Distribution Alignment

Yuntao Du, Zhiwen Tan, Xiaowen Zhang, Yirong Yao, Hualei Yu, and Chongjun Wang(✉)

State Key Laboratory for Novel Software Technology at Nanjing University, Nanjing University, Nanjing 210023, China
{duyuntao,zhangxw,yaotx,hlyu}@smail.nju.edu.cn
chjwang@nju.edu.cn

Abstract. Unsupervised domain adaptation aims at transferring knowledge from a labeled source domain to an unlabeled target domain. Recently, domain-adversarial learning has become an increasingly popular method to tackle this task, which bridges the source domain and target domain by adversarially learning domain-invariant representations. Despite the great success in domain-adversarial learning, these methods fail to achieve the invariance of representations at a class level, which may lead to incorrect distribution alignment. To address this problem, in this paper, we propose a method called *domain adaptation with Unified Joint Distribution Alignment* (UJDA) to perform both domain-level and class-level alignments simultaneously in a unified learning process. Instead of adopting the classical domain discriminator, two novel components named joint classifiers, which are provided with both domain information and label information in both domains, are adopted in UJDA. Single joint classifier plays the min-max game with the feature extractor by the joint adversarial loss to align the class-level alignment. Besides, two joint classifiers as a whole also play the min-max game with the feature extractor by the disagreement loss to achieve the domain-level alignment. Comprehensive experiments on two real-world datasets verify that our method outperforms several state-of-the-art domain adaptation methods.

Keywords: Transfer learning · Domain adaptation · Distribution alignment

1 Introduction

Deep neural networks have achieved remarkable success in many applications [11,32]. However, it requires a large amount of labeled data to train the model for a good generalization. Collecting and annotating sufficient data are very expensive and timeconsuming. It is a natural idea to utilize annotated data from a similar domain to help improve the performance in the target domain, which is the goal of transfer learning. Generally, transfer learning aims at leveraging knowledge from a labeled source domain to an unlabeled target domain [19]. Unsupervised domain adaptation is a sub-filed of transfer learning, in which the feature space and label space in the source domain and target domain are the same, but the data distribution is different [19].

C. S. Jensen et al. (Eds.): DASFAA 2021, LNCS 12682, pp. 449–464, 2021.
https://doi.org/10.1007/978-3-030-73197-7_30

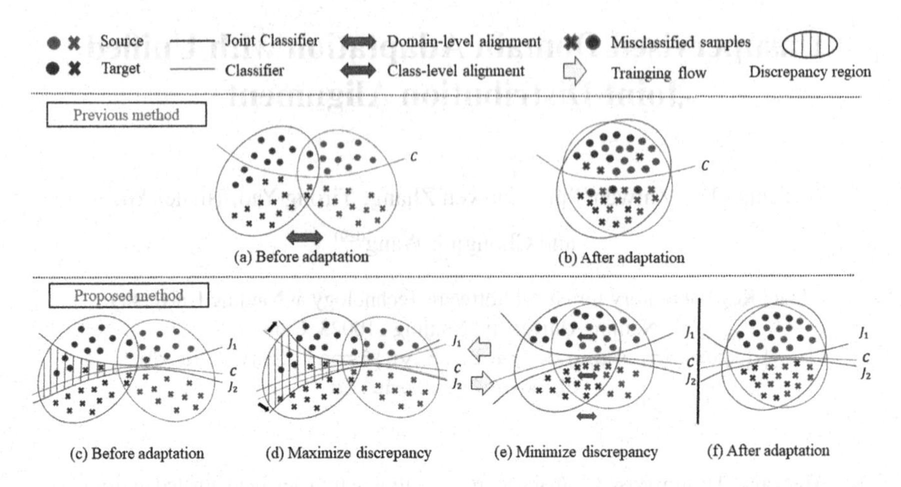

Fig. 1. An overview of the proposed method. Instead of using binary classifier as the discriminator, we use two joint classifiers with $2K$-dimensional output as the discriminator, which can not only distinguish the domain label but also can classify the training data into a certain class. The discrepancy between classifiers refers to the disagreement between the prediction of two classifiers. **Top**: Previous methods can achieve domain-level alignment. However, due to the ignored class-level alignment, there exist some mismatchings between different classes. **Bottom**: Single joint classifier plays a min-max game with the feature extractor by a joint adversarial loss to perform the class-level alignment. Besides, two joint classifiers (J_1, J_2) also play a min-max game with the feature extractor by the disagreement loss between the joint classifiers, achieving domain-level alignment. We design a unified adversarial learning process, thus making the modules in this method can be trained alternately. Best viewed in color. (Color figure online)

It is crucial for domain adaptation to reduce the distribution discrepancy across domains [1,35]. Early methods focus on shallow methods such as sample reweighting which reweights the source samples according to the relation with the target samples [2,5] or statistics matching, e.g., Maximum Mean Discrepancy (MMD) [10], Correlation Alignment (CORAL) [27]. Some experiments have indicated that deep neural networks can learn transferable features [33]. Thus, many deep domain adaptation methods have been proposed to learn domain-invariant features by embedding domain adaptation modules in the pipeline of deep feature learning [13,16]. Recently, many adversarial domain adaptation methods inspired by the generative adversarial networks [8] have been proposed. Generally, there exists a min-max game between the domain discriminator and the feature extractor [7]. The domain discriminator is trained to distinguish the source domain from the target domain, while the feature extractor is trained to learn domain-invariant representations to confuse the discriminator. Theoretically, domain alignment is achieved when the min-max optimization reach an equilibrium.

Although achieving remarkable progress, most previous methods focus on domain-invariant representation, in other words, they only focus on aligning the marginal distributions of two domains, which is referred as domain-level distribution alignment, but the alignment of class conditional distributions across domains is ignored, which is referred as class-level distribution alignment. A perfect domain-level distribution alignment does not imply a fine-grained class-to-class overlap. As is shown in the top of Fig. 1, although the learned domain-invariant features can reduce the distribution discrepancy, there may exist some mismatchings between different classes in both domains which will lead to incorrect distribution matching. The lack of class-level distribution alignment is a major cause of performance reduction [14]. Thus, it is necessary to pursue the class-level and domain-level distribution alignments simultaneously under the absence of true target labels.

To tackle the aforementioned problems, in this paper, we propose a method called *domain adaptation with Unified Joint Distribution Alignment (UJDA)*. As is shown in Fig. 1(c–f), UJDA aims to perform domain-level and class-level alignments in a unified learning progress. Instead of using the classical binary classifier as the discriminator, UJDA adopts the joint classifiers [4] with $2K$-dimensional output as the discriminator, which is provided with domain and class information in both domains. The first K-dimensional outputs of joint classifier represent the known source classes, and the last K-dimensional outputs represent the target classes. The joint adversarial loss is proposed to train the model (shown in Fig. 2). Each single joint classifier aims to minimize the joint adversarial loss to classify source and target samples correctly, while the feature extractor aims to maximize the joint adversarial loss to classify the data in one domain to another domain while keeping the label unchanged at the same time (e.g., classifying dogs in the source domain to dogs in the target domain and vice versa). Thus, this learning process can achieve explicit class-level adaptation and implicit domain-level adaptation. To further perform explicit domain-level adaptation, inspired by MCD [24], two joint classifiers are adopted in this paper. As is shown in Fig. 1(c–f), two joint classifiers as a whole also play a min-max game with the feature extractor by the disagreement loss between these two joint classifiers. On the one hand, two joint classifiers are trained to maximize the prediction discrepancy (Fig. 1(d)), so that the target samples outside the support of the source domain can be detected. On the other hand, the feature extractor is trained to confuse the joint classifiers, which encourages the target samples to be generated inside the support of the source. By this min-max game, we can achieve explicit domain-level adaptation. We design a unified adversarial learning progress and the modules in UJDA can be trained alternately.

As the labels in the target domain are unavailable, in this work, we use pesudo labels instead. The class predictor trained in the source domain is adopted to predict the pseudo labels for the target data. Furthermore, we use semi-supervised learning (SSL) regularization to extract more discriminative features. Comprehensive experiments on two real-world image datasets are conducted and the results verify the effectiveness of our proposed method. Briefly, our contributions lie in three folds:

- Domain-level alignment and class-level alignment are simultaneously explored in a unified adversarial learning progress. Moreover, SSL regularization is used to make the extracted features more discriminative.
- There exist two complementary mix-max games between the feature extractor and two joint classifiers to achieve implicit domain-level and class-level adaptation, respectively.
- We conduct extensive experiments on two real-world datasets and the results validate the effectiveness of our proposed method.

2 Related Work

Unsupervised domain adaptation is a sub-field of transfer learning, where there are abundant labeled data in the source domain and some unlabeled data in the target domain. Early studies focus on shallow (traditional) domain adaptation. Recently, more and more works pay attention to deep domain adaptation and adversarial domain adaptation.

Shallow Domain Adaptation. The most common strategy in shallow learning is *distribution alignment*. The distribution discrepancy across domains includes marginal distribution discrepancy and conditional distribution discrepancy [6,15,20,30,31]. TCA [20] tries to align marginal distribution across domains, which learns domain-invariant representations during feature mapping. Based on TCA, JDA [15] tries to algin marginal distribution and conditional distribution simultaneously. Considering the balance between marginal distribution and conditional distribution discrepancy, BDA [30] proposes a balance factor to leverage the importance of different distributions. MEDA [31] is able to dynamically evaluate the balance factor and has achieved promising performance.

Deep Domain Adaptation. Most deep domain adaptation methods are based on discrepancy measure [13,27,29,34]. DDC [29] embeds a domain adaptation layer into the network and minimizes Maximum Mean Discrepancy(MMD) between features of this layer. DAN [13] minimizes the feature discrepancy between the last three layers and the mutil-kernel MMD is used to better approximate the discrepancy. Other measures are also adopted such as Kullback-Leibler (KL) divergence, Correlation Alignment (CORAL) [27] and Central Moment Discrepancy (CMD) [34]. These methods can utilize the deep neural network to extract more transferable features and also have achieved promising performance.

Adversarial Domain Adaptation. Recently, adversarial learning is widely used in domain adaptation [3,7,24,28,35]. DANN [7] uses a discriminator to distinguish the source data from the target data, while the feature extractor learns domain-invariant features to confuse the discriminator. Based on the theory in [1], when maximizing the error of discriminator, it is actually approximating the H-distance, and minimizing the error of discriminator is actually minimizing the discrepancy across domains. ADDA [28] designs a symmetrical structure where two feature extractors are adopted. Different from DANN, MCD [24] proposes a method to minimize the $H\Delta H$-distance across domains in an adversarial way. A new theory using margin loss is proposed in [35] for mutli-class doamin adaptation problem, based on this theory, MDD is designed to minimize the *disparity discrepancy* across domains.

3 Method

3.1 Problem Setting

In unsupervised domain adaptation, we are given a source domain $D_s = \{(x_i, y_i)\}_{i=1}^{n_s}$ of n_s labeled source examples and a target domain $D_t = \{(x_i)\}_{i=1}^{n_t}$ of n_t unlabeled target examples. The source data are drawn from the distribution $P(x_s, y_s)$ and the

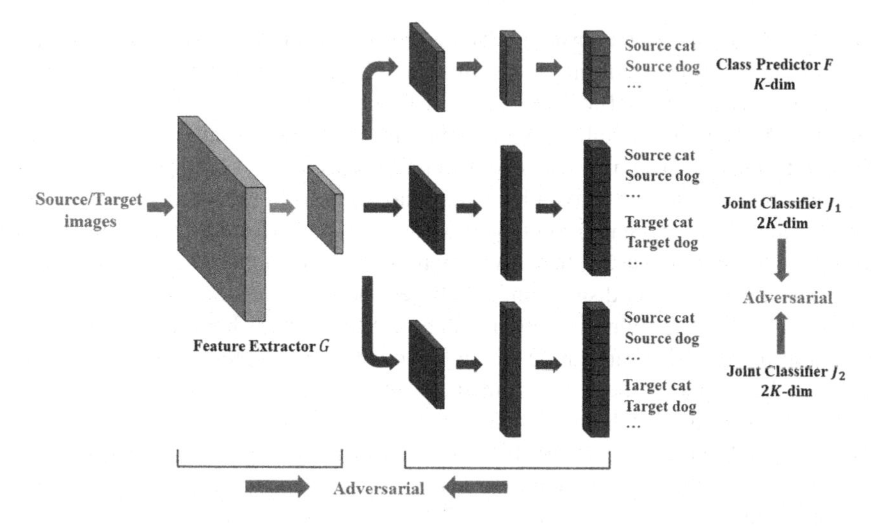

Fig. 2. The structure of UJDA algorithm. The output of class predictor is K-dimensional while the output of joint classifier is $2K$-dimensional. Each joint classifier distinguishes the domain as well as the class of the training data and plays a min-max game with the features extractor G with the joint adversarial loss to perform class-level alignments. Two joint classifiers (J_1,J_2) are trained to maximize the disagreement of two joint classifiers, while the feature are trained to minimize the disagreement to learn domain-invariant features and achieve domain-level adaptation. Domain-level alignment and class-level alignment is performed simultaneously, which can avoid mode collapse [4,21]. The class predictor is used to classify source examples as well as predict pseudo labels for the target data.

target data are drawn from the distribution $Q(x_t, y_t)$. Note that the i.i.d. assumption is violated, where $P(x_s, y_s) \neq Q(x_t, y_t)$. Both distributions are defined on feature space and label space $\mathcal{X} \times \mathcal{Y}$, where $\mathcal{Y} = \{1, 2, ..., K\}$. Our goal is to design a deep network $f : \mathcal{X} \mapsto \mathcal{Y}$ to reduce the distribution discrepancy across domains in order that the generalization error $\epsilon_t(f)$ in the target domain can be bounded by source risk $\epsilon_s(f)$ plus the distribution discrepancy across domains [1], where

$$\epsilon_s(f) = E_{(x_s,y_s)\sim P}[f(x_s) \neq y_s], \epsilon_t(f) = E_{(x_t,y_t)\sim Q}[f(x_t) \neq y_t] \quad (1)$$

3.2 Overall

Previous adversarial-based methods adopt the binary classifier as the discriminator. While recent experiments have shown that the informative discriminator which accesses the domain information and label information in both domains is able to preserve the complex multimodal information and high semantic information in both domains [4]. In this paper, we adopt the joint classifiers [4] which are provided with both domain and label information to adapt the features across domains. As is shown in Fig. 2, UJDA consists of a feature extractor G, a class predictor F and two joint classifiers J_1, J_2. Note that the output of the class predictor is K-dimensional while the outputs of the joint classifiers are $2K$-dimensional. There are two complementary adversarial process in UJDA. For the first one, each joint classifier pits against the feature extractor with a $2K$-way adversarial loss to learn a distribution over domains and classes variables, which can perform both implicit domain-level adaptation and explicit class-level alignment simultaneously in a single joint classifier (Sect. 3.4). For the second one, two joint classifiers J_1, J_2 as a whole play a min-max game with the feature extractor G to learn the domain-invariant representations, so that we can perform explicit domain-level adaptation across domains. In this adversarial process, two joint classifiers are trained to increase the prediction disaggrement between the joint classifiers while the feature extractor is trained to minimizie the prediction disagreement between the joint classifiers. During the adversarial process, ambiguous target samples can be detected and pushed in the support of the source domain and the domain-level adaptation is achieved (Sect. 3.5). Since the labels in the target doamin are unavailable, we use the pseudo labels instead. We use the class predictor F trained in the source domain to predict the pseudo labels for the target domain. To make the representations more discriminative and the pseudo labels more accurate, we introduce semi-supervised learning regularization, where the entropy minimization and *Virtual Adversarial Training* (VAT) are adopted (Sect. 3.6).

3.3 Class Predictor Loss

The class predictor F is trained to classify the source samples correctly. During the training process, it is also used to predict *pseudo labels* for the target domain. The output of the class predictor can be written as,

$$f(x) = F(G(x)) \in R^K \tag{2}$$

We train the network to minimize the cross entropy loss. The source classification loss of the class predictor is as follows:

$$\ell_{sc}(F) = E_{(x_s,y_s)\sim P} l_{CE}(f(x_s), y_s) \tag{3}$$

where, the cross-entropy loss is calculated with one-hot ground-truth labels y_s and label estimates $f(x)$. For a target example x_t, its predicted label according to the class predictor is [1]

$$\hat{y} = \arg\max_k f(x_t)[k] \tag{4}$$

3.4 Class-Level Alignment Loss

The conditional distribution mismatching means that the source classes are wrongly matched to target classes and vice versa. For example, the source dog data samples matched to the target cat samples, which may lead a poor performance in the target domain. Previous methods adopt the domain discriminator for adopting the features

[1] We use the notation x[k] for indexing the value at the kth index of the vector x.

across domains, but they only can perform domain-level (global level) adaptation while they can not promise the class-level adaptation.

Recent experiments have shown that the informative discriminator that accesses the domain information and label information in both domains is able to preserve the complex multimodal information and high semantic information in both domains [4,21]. In this paper, instead of adopting the binary classifier as the discriminator, we use the joint classifier [4] with $2K$-dimensional output for adaptation. The joint classifier is trained by a $2K$-way adversarial loss. The first K-dimensional outputs are for the source classes, and the last K-dimensional outputs are for the unknown classes. Such a component can learn a distribution over domain and class variables and can perform both explicit class-level alignment and implicit domain-level alignment in a single component. The output of the joint classifier, taking J_1 as an example, can be written as,

$$f_{J_1}(x) = J_1(G(x)) \in R^{2K} \tag{5}$$

For the labeled source examples, we train the joint classifiers with the same classification loss to classify the source samples correctly. The source classification loss of the joint classifier is defined as,

$$\ell_{jsc}(J_1) = E_{(x_s,y_s)\sim P} l_{CE}(J_1(G(x_s)), [\boldsymbol{y}_s, \mathbf{0}]) \tag{6}$$

where $\mathbf{0}$ is the zero vector of size K, chosen to make the last K joint probabilities zero for the source samples. $\boldsymbol{y}_s \in \{0,1\}^K$ is the one-hot label.

Similarly, to capture the label information in the target domain, we also train the joint classifier using the target examples. Since the labels for the target data are not known, we use pseudo labels $\hat{y}_t$ instead. The target classification loss of the joint classifier is,

$$\ell_{dtc}(J_1) = E_{x_t\sim Q} l_{CE}(J_1(x_t), [\mathbf{0}, \hat{\boldsymbol{y}}_t]) \tag{7}$$

Here, it is assumed that the source-only model can achieve reasonable performance in the target domain. In experiments, where the source-only model has poor performance initially, we use this loss after training the class predictor for a period of time.

The feature extractor G is designed to confuse the joint classifiers as in DANN [7]. The basis idea is that the feature extractor can confuse the joint classifier with the domain information, but keep label information unchanged. For example, given a source example x_s with label y_s, the correct output of the joint classifier should be $[\boldsymbol{y}_s, \mathbf{0}]$, while the feature extractor is trained to fool the joint classifier, making the sample to be classified from the target domain but with the same label y_s, which is $[\mathbf{0}, \boldsymbol{y}_s]$ formally.

The source alignment loss of the joint classifier is defined by changing the label from $[\boldsymbol{y}_s, \mathbf{0}]$ to $[\mathbf{0}, \boldsymbol{y}_s]$,

$$\ell_{dsa1}(G) = E_{(x_s,y_s)\sim P} l_{CE}(J_1(G(x_s)), [\mathbf{0}, \boldsymbol{y}_s]) \tag{8}$$

Similarly, the target alignment loss of joint classifier is defined by changing the pseudo-label from $[\mathbf{0}, \hat{\boldsymbol{y}}_t]$ to $[\hat{\boldsymbol{y}}_t, \mathbf{0}]$,

$$\ell_{dta1}(G) = E_{x_t\sim Q} l_{CE}(J_1(G(x_t)), [\hat{\boldsymbol{y}}_t, \mathbf{0}]) \tag{9}$$

The above two losses are minimized only by the feature extractor G. The same adversarial process is also applied in joint classifier J_2.

3.5 Domain-Level Alignment Loss

By the adversarial process in Sect. 3.4, we can achieve explicit class-level adaptation and implicit domain-level adaptation. To further perform explicit domain-level adaptation, the second adversarial process between the joint classifiers and the feature extractor is proposed. Moreover, as is shown in Fig. 1, the relationship between the samples and the decision boundary is considered to detect the target samples outside the support of the source domain. Maximizing the disagreement of two joint classifiers can help better detect the target samples outside the support of the source domain while minimizing the disagreement can push these target samples inside the support of the source domain [7]. The above adversarial process consider all the samples as a whole and can perform the domain-level (global level) alignment [7].

In order to detect the target samples outside the support of the source domain, we propose to utilize the disagreement of two joint classifiers on the prediction for both domains. The disagreement of the two joint classifiers is defined by utilizing the absolute values of the difference between the probabilistic outputs as discrepancy loss:

$$d(f_{J_1}(x), f_{J_2}(x)) = \frac{1}{K}\sum_{k=1}^{K}|f_{J_1}(x)[k] - f_{J_2}(x)[k]| \tag{10}$$

We firstly train the joint classifiers to increase their discrepancy. It can not only help different joint classifiers to capture different information, but also detect the target samples excluded by the support of the source [24]. The objective is as follows,

$$\max_{J_1, J_2} \ell_d \tag{11}$$

$$\ell_d = E_{x_s \sim P}[d(f_{J_1}(x_s), f_{J_2}(x_s))] + E_{x_t \sim Q}[d(f_{J_1}(x_t), f_{J_2}(x_t))] \tag{12}$$

Moreover, the feature extractor is trained to minimize the discrepancy for fixed joint classifiers. On the one hand, minimizing the discrepancy can make these two joint classifiers not too far away from each other, thus making them similar. On the other hand, minimizing the discrepancy can avoid generating target features outside the support of the source domain [24]. The objective is as follows,

$$\min_{G} \ell_d \tag{13}$$

Note that although the domain-level adversarial process is similar with MCD [7], where two classifiers are also adopted, there are difference between MCD and the proposed method. On the one hand, MCD aims only perform domain-level adaptation, while the proposed method focus on domain-level adaptation and class-level adaptation simultaneously. On the other hand, the classifier in MCD is trained with only source samples, while the classifier in the proposed method is trained with both source samples and target samples, thus proposed method can achieve better domain-level adaptation.

3.6 SSL Regularization Loss

After the distribution alignment, the discrepancy across domains can be smaller. In this case, we can approximate the unsupervised domain adaptation as a semi-supervised learning problem. On this bisis, many previous works have explored semi-supervised learning (SSL) regularization in domain adaptation [4, 12] and made sufficient improvements. Although lacking of labels, a large amount of unlabeled data can be used to bias the classifier boundaries to pass through the regions containing low density data. Thus, the learned representation can become more discriminative. Entropy minimization [9] is a widely used regularization method to achieve this goal. In our method, the class predictor is also trained to minimize the target entropy loss, which is defined as follows,

$$\ell_{te}(F) = E_{(x_t, y_t)\sim Q} H(f(x_t)) \tag{14}$$

where $H(f(x_t)) = -\sum_k f(x_t)[k] \cdot \log f(x_t)[k]$. However, minimizing entropy is only applicable to locally-Lipschitz classifiers [9]. So we propose to explicitly incorporate the locally-Lipschitz condition via *virtual adversarial training*(VAT) [18] and add the following losses to the objective,

$$\ell_{svat}(F) = E_{x_s\sim P}[\max_{||r||\leq\epsilon} \ell_{CE}(f(x_s)||f(x_s + r))] \tag{15}$$

$$\ell_{tvat}(F) = E_{x_t\sim Q}[\max_{||r||\leq\epsilon} \ell_{CE}(f(x_t)||f(x_t + r))] \tag{16}$$

3.7 Overall Objective

We combine the objective functions discussed in section3.4-3.6 and divide our training procedure into three steps.

Step 1. We only use the source data to train the feature extractor G, the class predictor F as well as the joint classifiers J_1, J_2. We minimize the source classification loss of the class predictor and joint classifiers. After the predictor and joint classifiers are trained to classify the source samples correctly, we will go on the next step. The objective of this step is shown as follows,

$$\min_{G,F,J_1,J_2} \ell_{sc}(F) + \lambda_{jsc1}\ell_{jsc}(J_1) + \lambda_{jsc2}\ell_{jsc}(J_2) \tag{17}$$

Step 2. We fix the feature extractor, and update the class predictor as well as the joint classifiers. We use both the source and target data to train the model. This process corresponds to Step 2 in Fig. 3. We have three sub-objectives. The first one is to minimize the source and target classification loss of the joint classifiers. The second one is to minimize the source classification loss of the class predictor as well as the SSL regularization loss. Without this loss, we experimentally found that the performance dropped. The last one is to increase the discrepancy between the joint classifiers. The objective of this step is shown as follows,

$$\min_{F,J_1,J_2} \ell_F + \ell_{J_1} + \ell_{J_2} - \lambda_d \ell_d \tag{18}$$

$$\ell_{J_1} = \lambda_{jsc1}\ell_{jsc}(J_1) + \lambda_{jtc1}\ell_{jtc}(J_1) \tag{19}$$

$$\ell_{J_2} = \lambda_{jsc2}\ell_{jsc}(J_2) + \lambda_{jtc2}\ell_{jtc}(J_2) \tag{20}$$

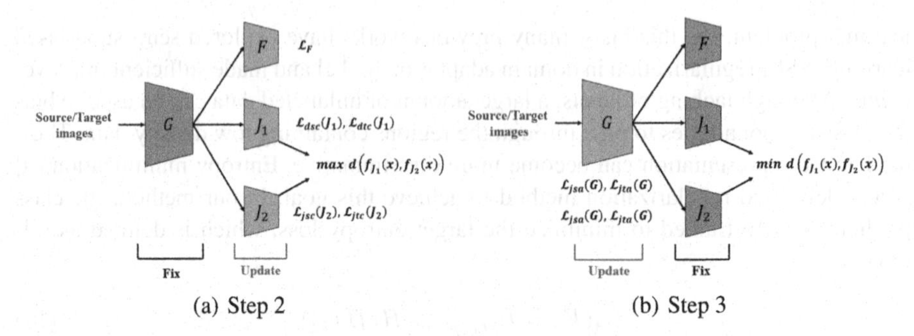

(a) Step 2 (b) Step 3

Fig. 3. Adversarial training steps of our method. There are three steps in total, step 2 and step 3 are shown in this figure. In step 2, the class predictor and two discriminators minimize the classification loss. Besides, the two discriminators pit against each other to increase the discrepancy between discriminators. In step 3, the feature extractor learns to minimize the discrepancy between discriminators as well as to confuse the discriminator in both domain and class level.

$$\ell_F = \ell_{sc}(F) + \lambda_{svat}\ell_{svat}(F) + \lambda_{te}\ell_{te}(F) + \lambda_{tvat}\ell_{tvat}(F) \tag{21}$$

Step 3. We fix the class predictor as well as the joint classifiers, and update the feature extractor. We train the model by minimizing the joint adversarial loss of joint classifiers as well as the discrepancy between joint classifiers. The objective of this step is shown as follows,

$$\min_G \lambda_{jsa1}\ell_{jsa1}(G) + \lambda_{jta1}\ell_{jta1}(G) + \lambda_{jsa2}\ell_{jsa2}(G) + \lambda_{jta2}\ell_{jta2}(G) + \lambda_d\ell_d \tag{22}$$

Note that Step2 and Step3 are repeated alternately in our method. We are concerned on that the feature extractor, class predictor and joint classifiers are trained in an adversarial manner so that they can classify the source samples correctly as well as promote the cross-domain discrepancy decreasing.

4 Experiments

We evaluate the proposed method with many state-of-the-art domain adaptation methods on two image datasets. Codes will be available at https://github.com/yaoyueduzhen/UJDA.

4.1 Setup

Office-31 [23] is the most widely used dataset for domain adaptation, with 4,652 images and 31 categories collected from three distinct domains: Amazon (**A**), Webcam (**W**) and DSLR (**D**). From this dataset, we build six transfer tasks: **A** → **W**, **D** → **W**, **W** → **D**, **A** → **D**, **D** → **A**, and **W** → **A**.

ImageCLEF-DA [16] is a dataset organized by selecting the 12 common classes shared by three public datasets (domains): Caltech-256 (**C**), ImageNet ILSVRC 2012 (**I**), and Pascal VOC 2012 (**P**). We evaluate all methods on six transfer tasks: **I** → **P**, **P** → **I**, **I** → **C**, **C** → **I**, **C** → **P**, and **P** → **C**.

We compare Unified Joint Distribution Alignment (**UJDA**) with several state-of-the-art domain adaptation methods:

- Deep Adaptation Network (**DAN**) [13]. DAN aims to learn domain-invariant features by mininizing the Multi-Kernel *Mean Maximum Mean* distance across domains.
- Domain Adversarial Neural Network (**DANN**) [7]. DANN is the first adversarial based domain adaptation method, which only performs domain-level adaptation.
- Adversarial Discriminative Domain Adaptation (**ADDA**) [28]. ADDA adopts two different feature extractors for different domains and only performs domain-level adaptation.
- Multi-Adversarial Domain Adaptation (**MADA**) [21]. MADA adopts multiple domain discriminators for adaptation, each for one class. This method only foucues on class-level adaptation.
- Virtual Adversarial Domain Adaptation (**VADA**) [26]. VADA aims to optimize two different classifiers for both domains and only focus on domain-level adaptation.
- Generate to Adapt (**GTA**) [25]. GTA aims to generate target-like samples with labels by generative adversarial network.
- Maximum Classifier Discrepancy (**MCD**) [24]. MCD adopts two classfiers for minimizing the $H\delta H$ distance across domains, which also performs domain-level adaptation.
- Conditional Domain Adversarial Network (**CDAN**) [14]. CDAN is an extension of DANN, where the input of domain discriminator is both the feature and the prediction, so it can perform class-level adaptation.
- Transferable Adversarial Training (**TAT**) [12]. TAT aims to generate transferable adversarial samples for a safe training for adversarial based method.
- Regularized Conditional Alignment (**RCA**) [4]. RCA adopts single joint classifier to perform class-level adaptation across domains.

4.2 Implementation Details

Following the standard protocols for unsupervised domain adaptation [7, 16], all labeled source samples and unlabeled target samples participate in the training stage. We compare the average classification accuracy based on five random experiments. The results of other methods are reported in the corresponding papers except RCA which is reimplemented by ourselves.

We use PyTorch to implement our method and use **ResNet-50** [11] pretrained on ImageNet [22] as the feature extractor. The class predictor and two joint classifiers are both two-layer fully connected networks with a width of 1024. We train these new layers and feature extractor using back-propagation, where the learning rates of these new layers are 10 times that of the feature extractor. We adopt mini-batch SGD with the momentum of 0.9 and use the same learning rate annealing strategy as [7]: the learning rate is adjusted by $\eta_p = \eta_0(1+\alpha p)^{-\beta}$, where p is the training progress changing from 0 to 1, and $\eta_0 = 0.04, \alpha = 10, \beta = 0.75$.

Table 1. Classification accuracy (%) on Office-31 for unsupervised domain adaptation with ResNet-50.

Method	A → W	D → W	W → D	A → D	D → A	W → A	Avg
ResNet-50 [11]	68.4±0.2	96.7±0.1	99.3±0.1	68.9±0.2	62.5±0.3	60.7±0.3	76.1
DAN [13]	80.5±0.4	97.1±0.2	99.6±0.1	78.6±0.2	63.6±0.3	62.8±0.2	80.4
DANN [7]	82.6±0.4	96.9±0.2	99.3±0.2	81.5±0.4	68.4±0.5	67.5±0.5	82.7
ADDA [28]	86.2±0.5	96.2±0.3	98.4±0.3	77.8±0.3	69.5±0.4	68.9±0.5	82.9
MADA [21]	90.0±0.1	97.4±0.1	99.6±0.1	87.8±0.2	70.3±0.3	66.4±0.3	85.2
VADA [26]	86.5±0.5	98.2±0.4	99.7±0.2	86.7±0.4	70.1±0.4	70.5±0.4	85.4
GTA [25]	89.5±0.5	97.9±0.3	99.7±0.2	87.7±0.5	**72.8**±0.3	71.4±0.4	86.5
MCD [24]	88.6±0.2	98.5±0.1	100.0±.0	92.2±0.2	69.5±0.1	69.7±0.3	86.5
RCA [4]	93.8±0.2	98.4±0.1	100.0±.0	91.6±0.2	68.0±0.2	70.2±0.2	87.0
CDAN [14]	93.1±0.1	98.6±0.1	100.0±.0	92.9±0.2	71.0±0.3	69.3±0.3	87.5
UJDA (one joint classifier)	93.2±0.2	97.9±0.2	100.0±.0	92.2±0.2	68.0±0.1	69.5±0.2	86.8
UJDA (without SSL)	94.0±0.2	98.4±0.2	100.0±.0	91.6±0.2	68.3±0.1	69.7±0.2	87.0
UJDA	**95.0**±0.2	**98.7**±0.2	**100.0**±.0	**94.0**±0.3	70.5±0.3	**71.5**±0.2	**88.4**

Table 2. Classification accuracy (%) on ImageCLEF-DA for unsupervised domain adaptation with ResNet-50.

Method	I→P	P→I	I→C	C→I	C→P	P→C	Avg
ResNet-50 [11]	74.8±0.3	83.9±0.1	91.5±0.3	78.0±0.2	65.5±0.3	91.2±0.3	80.7
DAN [13]	74.5±0.4	82.2±0.2	92.8±0.2	86.3±0.4	69.2±0.4	89.8±0.4	82.5
DANN [7]	75.0±0.3	86.0±0.3	96.2±0.4	87.0±0.5	74.3±0.5	91.5±0.6	85.0
MADA [21]	75.0±0.3	87.9±0.2	96.0±0.3	88.8±0.3	75.2±0.2	92.2±0.3	85.8
CDAN [14]	76.7±0.3	90.6±0.3	97.0±0.4	90.5±0.4	74.5±0.3	93.5±0.4	87.1
TAT [12]	78.8±0.2	92.0±0.2	97.5±0.3	92.0±0.3	**78.2**±0.4	94.7±0.4	88.9
RCA [4]	78.7±0.2	92.8±0.2	97.7±0.3	92.0±0.2	77.0±0.3	**95.0**±0.3	88.9
UJDA	**79.8**±0.3	**94.6**±0.2	**98.6**±0.2	**92.8**±0.2	77.8±0.3	**95.0**±0.3	**89.9**

We fix $\lambda_d = 1.0$ and search the rest hyperparameters over $\lambda_{jsc1}, \lambda_{jsc2} \in \{0.1, 0.5, 1.0\}$, $\lambda_{jtc1}, \lambda_{jtc2} \in \{0.1, 1.0, 10.0\}$, $\lambda_{svat}, \lambda_{tvat} \in \{0.0, 0.1, 1.0\}$, $\lambda_{te} \in \{0.1, 1.0\}$, $\lambda_{jsa1}, \lambda_{jta1}, \lambda_{jsa2}, \lambda_{jta2} \in \{0.1, 1.0\}$. We also search for the upper bound of the adversarial perturbation in VAT, where $\epsilon \in \{0.5, 1.0, 2.0\}$.

4.3 Results

The results on Office-31 dataset are shown in Table 1. As we can see, our method outperforms baseline methods in most tasks and achieves the best result in average accuracy. Compared with DANN and ADDA, which only perform domain-level alignment using a binary class discriminator, our method performs not only domain-level but also class-level alignments and outperforms them. MADA considers domain-level and class-level alignment, but it constrains each discriminator to be responsible for only one class. Our method avoids this limitation by adopting $2K$-dimensional discriminators where the classes can share information. The $2K$-dimensional joint classifier is also used in RCA, but we train two discriminators in an adversarial manner so that they can provide complementary information for each other. Moreover, we clearly observe that our method can also perform well on **A** $\rightarrow$ **D** and **W** $\rightarrow$ **A** with relatively large domain shift and imbalanced domain scales.

The results on ImageCLEF-DA are shown in Table 2. We have several findings based on the results. Firstly, all methods are better than ResNet-50, which is a source-only model trained without exploiting the target data in a standard supervised learning setting. Our method increases the accuracy from 80.7% to 89.9%. Secondly, the above comparisons with baseline methods on Office-31 are also the same on ImageCLEF-DA, which verifies the effectiveness of our method. Thirdly, UJDA outperforms the baseline methods on most transfer tasks, but with less room for improvement. This is reasonable since the three domains in ImageCLEF-DA are of equal size and balanced in each category, which make domain adaptation easy.

4.4 Analysis

Ablation Study. We study the effect of SSL regularization by removing entropy minimization and VAT losses from our method ($\lambda_{te} = \lambda_{svat} = \lambda_{tvat} = 0$), which is denoted by UJDA (without SSL). Moreover, we study the effectiveness of using two joint classifiers by using only one joint classifier, which is denoted by UJDA (one joint classifier). The results on Office-31 dataset are reported in Table 1. Results show that without SSL regularization, our method can perform better in two tasks (**A**$\rightarrow$**W**, **W**$\rightarrow$**D**) than baseline methods, but the average accuracy of all tasks is decreased by 1.4% compared to the proposed UJDA. The results validate the effectiveness of SSL regularization. Besides, using one joint classifier can only perform class-level alignment while the domain-level adaptation is ignored, the results show that the proposed UJDA achieve better performance by further performing domain-level alignment.

Feature Visualization. In Fig. 4, we visualize the feature representations of task **A**$\rightarrow$**W**(31 classes) by t-SNE [17] using the source-only method and UJDA. The source-only method is trained without exploiting the target training data in a standard supervised learning setting using the same learning procedure. As we can see, source and target samples are better aligned for UJDA than the source-only method. This shows the advance of our method in discriminative prediction.

Distribution Discrepancy. The A-distance is a measure of distribution discrepancy, defined as $dist_A = 2(1-2\epsilon)$, where ϵ is the test error of a classifier trained to distinguish the source from the target. We use A-distance as a measure of the transferability of feature representations. Table 3 shows the cross-domain A-distance for tasks **A**$\rightarrow$**W**, **W**$\rightarrow$**D**. We compute the A-distance of our method based on the output of the feature extractor G, which turns out to be the smallest of all methods.

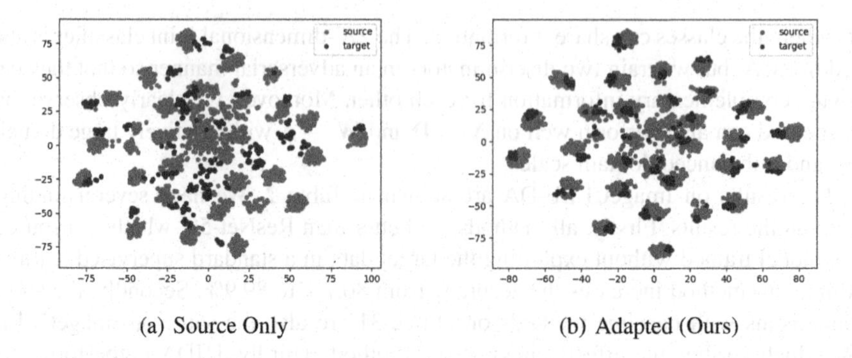

(a) Source Only (b) Adapted (Ours)

Fig. 4. Visualization of features obtained from the feature extractor of task **A→W** using t-SNE [17]. Red and blue points indicate the source and target samples respectively. We can see that applying our method makes the target samples more discriminative. (Color figure online)

Table 3. Cross-domain A-distance of different approaches.

Method	D→W	W→A
ResNet-50 [11]	1.27	1.86
DANN [7]	1.23	1.44
MCD [24]	1.22	1.60
UJDA	**1.14**	**1.18**

5 Conclusion

In this paper, we propose a method called *unsupervised domain adaptation with Unified Joint Distribution Alignment* (**UJDA**), which is able to perform both domain-level and class-level alignments simultaneously in a unified adversarial learning process. Single joint classifier provided with both domain and label information in both domains plays a min-max game with the feature extractor by a joint adversarial loss, which can perform the class-level alignment. Two joint classifiers as a whole also play a minimax game with the feature extractor by the prediction disagreement between two joint classifiers, which can perform the domain-level alignment. These modules are trained in the unified adversarial learning process, and they can provide complementary information for each other to avoid mode collapse. Moreover, SSL regularization is used to make the representations more discriminative so that the predicted pseudo labels can be more accurate. We conduct comprehensive experiments and the results verify the effectiveness of our proposed method.

Acknowledgements. This paper is supported by the National Key Research and Development Program of China (Grant No. 2018YFB1403400), the National Natural Science Foundation of China (Grant No. 61876080), the Collaborative Innovation Center of Novel Software Technology and Industrialization at Nanjing University.

References

1. Ben-David, S., Blitzer, J., Crammer, K., Kulesza, A., Pereira, F.C., Vaughan, J.W.: A theory of learning from different domains. Mach. Learn. **79**, 151–175 (2009)
2. Cao, Y., Long, M., Wang, J.: Unsupervised domain adaptation with distribution matching machines. In: AAAI (2018)
3. Chen, Q., Du, Y., Tan, Z., Zhang, Y., Wang, C.: Unsupervised domain adaptation with joint domain-adversarial reconstruction networks. In: ECML/PKDD (2020)
4. Cicek, S., Soatto, S.: Unsupervised domain adaptation via regularized conditional alignment. In: 2019 IEEE/CVF International Conference on Computer Vision (ICCV), pp. 1416–1425 (2019)
5. Dai, W., Yang, Q., Xue, G.R., Yu, Y.: Boosting for transfer learning. In: ICML 2007 (2007)
6. Du, Y., Tan, Z., Chen, Q., Zhang, Y., Wang, C.J.: Homogeneous online transfer learning with online distribution discrepancy minimization. In: ECAI (2020)
7. Ganin, Y., et al.: Domain-adversarial training of neural networks. J. Mach. Learn. Res. **17**, 59:1–59:35 (2016)
8. Goodfellow, I.J., et al.: Generative adversarial nets. In: NIPS (2014)
9. Grandvalet, Y., Bengio, Y.: Semi-supervised learning by entropy minimization. In: NIPS (2005)
10. Gretton, A., Borgwardt, K.M., Rasch, M.J., Schölkopf, B., Smola, A.J.: A kernel method for the two-sample-problem. In: NIPS (2006)
11. He, K., Zhang, X., Ren, S., Sun, J.: Deep residual learning for image recognition. In: 2016 IEEE Conference on Computer Vision and Pattern Recognition (CVPR), pp. 770–778 (2016)
12. Liu, H., Long, M., Wang, J., Jordan, M.I.: Transferable adversarial training: a general approach to adapting deep classifiers. In: ICML (2019)
13. Long, M., Cao, Y., Wang, J., Jordan, M.I.: Learning transferable features with deep adaptation networks. In: ICML (2015)
14. Long, M., Cao, Z., Wang, J., Jordan, M.I.: Conditional adversarial domain adaptation. In: NeurIPS (2018)
15. Long, M., Wang, J., Ding, G., Sun, J.G., Yu, P.S.: Transfer feature learning with joint distribution adaptation. In: ICCV, pp. 2200–2207 (2013)
16. Long, M., Zhu, H., Wang, J., Jordan, M.I.: Deep transfer learning with joint adaptation networks. In: ICML (2017)
17. Maaten, L.V.D., Hinton, G.E.: Visualizing data using T-SNE. J. Mach. Learn. Res. **9**, 2579–2605 (2008)
18. Miyato, T., Maeda, S., Koyama, M., Ishii, S.: Virtual adversarial training: a regularization method for supervised and semi-supervised learning. IEEE Trans. Pattern Anal. Mach. Intell. **41**, 1979–1993 (2019)
19. Pan, S., Yang, Q.: A survey on transfer learning. IEEE Trans. Knowl. Data Eng. **22**, 1345–1359 (2010)
20. Pan, S.J., Tsang, I.W.H., Kwok, J.T., Yang, Q.: Domain adaptation via transfer component analysis. IEEE Trans. Neural Networks **22**, 199–210 (2011)
21. Pei, Z., Cao, Z., Long, M., Wang, J.: Multi-adversarial domain adaptation. In: AAAI (2018)
22. Russakovsky, O., Deng, J., Su, H., Krause, J., Satheesh, S., Ma, S., Huang, Z., Karpathy, A., Khosla, A., Bernstein, M.S., Berg, A., Fei-Fei, L.: Imagenet large scale visual recognition challenge. Int. J. Comput. Vision **115**, 211–252 (2015)
23. Saenko, K., Kulis, B., Fritz, M., Darrell, T.: Adapting visual category models to new domains. In: Daniilidis, K., Maragos, P., Paragios, N. (eds.) ECCV 2010. LNCS, vol. 6314, pp. 213–226. Springer, Heidelberg (2010). https://doi.org/10.1007/978-3-642-15561-1_16

24. Saito, K., Watanabe, K., Ushiku, Y., Harada, T.: Maximum classifier discrepancy for unsupervised domain adaptation. In: 2018 IEEE/CVF Conference on Computer Vision and Pattern Recognition, pp. 3723–3732 (2018)
25. Sankaranarayanan, S., Balaji, Y., Castillo, C.D., Chellappa, R.: Generate to adapt: Aligning domains using generative adversarial networks. In: 2018 IEEE/CVF Conference on Computer Vision and Pattern Recognition, pp. 8503–8512 (2018)
26. Shu, R., Bui, H.H., Narui, H., Ermon, S.: A dirt-t approach to unsupervised domain adaptation. In: ICLR (2018)
27. Sun, B., Saenko, K.: Deep CORAL: correlation alignment for deep domain adaptation. In: Hua, G., Jégou, H. (eds.) ECCV 2016. LNCS, vol. 9915, pp. 443–450. Springer, Cham (2016). https://doi.org/10.1007/978-3-319-49409-8_35
28. Tzeng, E., Hoffman, J., Saenko, K., Darrell, T.: Adversarial discriminative domain adaptation. In: 2017 IEEE Conference on Computer Vision and Pattern Recognition (CVPR), pp. 2962–2971 (2017)
29. Tzeng, E., Hoffman, J., Zhang, N., Saenko, K., Darrell, T.: Deep domain confusion: Maximizing for domain invariance. ArXiv abs/1412.3474 (2014)
30. Wang, J., Chen, Y., Hao, S., Feng, W., Shen, Z.: Balanced distribution adaptation for transfer learning. In: ICDM, pp. 1129–1134 (2017)
31. Wang, J., Feng, W., Chen, Y., Yu, H., Huang, M., Yu, P.S.: Visual domain adaptation with manifold embedded distribution alignment. In: MM 18 (2018)
32. Wu, X., et al.: Top 10 algorithms in data mining. Knowl. Inf. Syst. **14**, 1–37 (2007)
33. Yosinski, J., Clune, J., Bengio, Y., Lipson, H.: How transferable are features in deep neural networks? In: NIPS (2014)
34. Zellinger, W., Grubinger, T., Lughofer, E., Natschläger, T., Saminger-Platz, S.: Central moment discrepancy (CMD) for domain-invariant representation learning. In: ICLR (2017)
35. Zhang, Y., Liu, T., Long, M., Jordan, M.I.: Bridging theory and algorithm for domain adaptation. In: ICML (2019)

Relation-Aware Alignment Attention Network for Multi-view Multi-label Learning

Yi Zhang, Jundong Shen, Cheng Yu, and Chongjun Wang(✉)

National Key Laboratory for Novel Software Technology at Nanjing University,
Department of Computer Science and Technology,
Nanjing University, Nanjing, China
{njuzhangy,jdshen,mf1833089}@smail.nju.edu.cn, chjwang@nju.edu.cn

Abstract. Multi-View Multi-Label (MVML) learning refers to complex objects represented by multi-view features and associated with multiple labels simultaneously. Modeling flexible view consistency is recently demanded, yet existing approaches cannot fully exploit the complementary information across multiple views and meanwhile preserve view-specific properties. Additionally, each label has heterogeneous features from multiple views and probably correlates with other labels via common views. Traditional strategy tends to select features that are distinguishable for all labels. However, globally shared features cannot handle the label heterogeneity. Furthermore, previous studies model view consistency and label correlations independently, where interactions between views and labels are not fully exploited. In this paper, we propose a novel MVML learning approach named **R**elation-aware **A**lignment attent**I**on **N**etwork (RAIN), where three types of relationships are considered. Specifically, 1) view interactions: capture diverse and complementary information for deep correlated subspace learning; 2) label correlations: adopt multi-head attention to learn semantic label embedding; 3) label-view dependence: dynamically extracts label-specific representation with the guidance of learned label embedding. Experiments on various MVML datasets demonstrate the effectiveness of RAIN compared with state-of-the-arts. We also experiment on one real-world *Herbs* dataset, which shows promising results for clinical decision support.

Keywords: Multi-view multi-label · View interactions · Label correlations · Label-view dependence · Alignment attention

1 Introduction

Multi-label learning deals with the problem where an instance may be associated with multiple semantic meanings. It has been widely applied in many real-world

---

The original version of this chapter was revised: the acknowledgement section has been corrected. The correction to this chapter is available at
https://doi.org/10.1007/978-3-030-73197-7_54

C. S. Jensen et al. (Eds.): DASFAA 2021, LNCS 12682, pp. 465–482, 2021.
https://doi.org/10.1007/978-3-030-73197-7_31

applications, such as image annotation [22], document categorization [12], information retrieval [7], and bioinformatics [27]. The most straightforward approach is to treat each label independently, then decompose the problem into a set of independent binary classification tasks, one for each label. Although easy to implement, it is limited by neglecting the relationships among labels. To compensate for this deficiency, the exploitation of label correlations has been widely accepted as a key component for effective multi-label learning [29].

Most existing multi-label learning approaches only consider single view data, however, the information obtained from an individual view cannot comprehensively describe all labels. Therefore, it has become popular to leverage the information collected from various feature extractors or diverse domains, which has resulted in Multi-view (or Multi-modal) Learning [2,20]. For example, a natural scene image can be characterized by heterogeneous visual descriptors, e.g., HSV color histogram, globe feature (GIST) and scale invariant feature transform (SIFT). A straightforward way is to concatenate all the views and then treat them as a single view task. It seems unreasonable since the concatenation of different views either causes high dimensional feature vectors or neglect the information derived from the connections and differences among multiple views. Different views have shared and independent information, omitting such shared and private nature of multi-view data would limit the performance of classification. So far, many approaches have been developed to integrate multiple features from diverse views and exploit the underlying interactions among views, which is an important clue to make the view complement each other and improve the discriminating power of multi-view representation.

To further take multi-label information into account, multi-view multi-label learning provides a fundamental framework to solve the above problems, i.e., data are often represented by heterogeneous features from multiple views and associated with multiple labels simultaneously. Recent researches have gradually shifted their emphasis from the problems with single-heterogeneity to the ones with dual-heterogeneity. Multi-view features comprehensively describe objects' characteristics from distinct perspectives, as a result, combining the heterogeneous properties from different views can better characterize objects and improve the overall learning performance. Each label might be determined by some specific features of its own and these features are the most pertinent and discriminative features to the corresponding label. For example, appearance characteristics have strong discriminability in judging whether a person is a basketball player. These features could be considered as label-specific features to the corresponding label. To exploit information from both related views and related labels, a common strategy is to model label relatedness and view consistency independently, which have not fully exploited the associated relationships during the learning process.

To address the above issues, we propose a novel multi-view multi-label learning framework to jointly make use of the specific and complementary information from multiple views, and extract label-specific representation guided by label embedding. Specifically, we first learn the proper subspace representation that is suitable for multi-label classification, which embeds various view-specific information and model deep interactive information among different views. Considering labels in multi-label datasets are not independent but inherently correlated, e.g., an image is likely to be annotated as *cloud* if it has the label *blue sky*, while it is not likely to be tagged with *smog*. Generally, the relationships among different labels could be positively related, negatively related, or unrelated. Thus, it is necessary to capture label correlations to guide the label-view dependence learning process. Inspired by multi-head attention prototype in Transformer [16], we propose a novel multi-head attention module to learn label semantic embeddings, which reflects the high-order collaborative relationships among labels. Multi-view multi-label datasets usually emerge in high-dimensionality, where only a subset of features are useful, and the noisy features might reduce the prediction performance. Furthermore, we design an alignment attention mechanism to automatically extract customized representation for each label severally.

The main contributions of this paper can be summarized as follows:

- A novel **R**elation-aware **A**lignment attent**I**on **N**etwork (RAIN) is proposed. RAIN exploits view interactions, label correlations, and label-view dependence collaboratively, which can be quantized to improve the model interpretability.
- RAIN learns the proper subspace representation by exploiting view-specific information as well as capturing the interactions across multiple views. It can strengthen the diversity of various views and further discover the complementary information.
- Label correlations are automatically exploited with multi-head attention, which are further aligned with the enhanced multi-view subspace representations to capture label-specific representations.
- Experiments on one real-world and five public multi-view multi-label datasets validate the superiority of the proposed model over the state-of-the-arts.

The rest of this paper is organized as follows: Sect. 2 briefly reviews some related works of multi-view multi-label learning. Section 3 introduces the problem formulation and presents our proposed approach. Section 4 reports experimental results and analysis on several multi-view multi-label datasets. Finally, Sect. 5 concludes this paper.

2 Related Work

Our work is related to two branches: multi-label learning and multi-view learning. In this section, we briefly review some state-of-the-art approaches in the two fields.

Multi-label classification [29] has received intensive attention in recent years. Generally, based on the order of label correlations considered by the system, multi-label learning algorithms can be categorized into following three strategies. First-order strategy copes with multi-label learning problem in a label-by-label manner. Binary Relevance (BR) [1] takes each label independently and decomposes it into multiple binary classification tasks. However, BR neglects the relationship among labels. Second-order strategy introduces pairwise relations among multiple labels, such as the ranking between the relevant and irrelevant labels [6]. Calibrated Label Ranking (CLR) [5] firstly transforms the multi-label learning problem into label ranking problem by introducing the pairwise comparison. High-order strategy builds more complex relations among labels. Classifier Chain (CC) [13] transforms into a chain of binary classification problems, where the quality is dependent on the label order in the chain. Ensemble Classifier Chains (ECC) [13] constructs multiple CCs by using different random label orders. A boosting approach Multi-label Hypothesis Reuse (MLHR) [10] is proposed to exploit label correlations with a hypothesis reuse mechanism. Considering the potential association between paired labels, Dual-Set Multi-Label Learning (DSML) [11] exploits pairwise inter-set label relationships for assisting multi-label learning. And Collaboration based Multi-Label Learning (CAMEL) [4] is proposed to learn the label correlations via sparse reconstruction in the label space.

In real-word applications, data is usually represented with different views, including multiple modalities or multiple types of features. Multi-view learning can be embedded into multi-label learning naturally to further improve the classification performance by exploit a multi-view latent space [32]. MvNNcor [21] is a novel multi-view learning framework, which seamlessly embeds various view-specific information and deep interaction information, and introduces a new multi-view loss fusion strategy to jointly make decisions and infer categories. There have been some researches for multi-view multi-label learning. [3] proposes a new classification framework using the multi-label correlation information to address the problem of simultaneously combining multiple views and maximum margin classification. LLSF [9] performs joint label specific feature selection and take the label correlation matrix as prior knowledge for model training. Considering that label heterogeneity and feature heterogeneity often co-exist, [23] proposes a novel graph-based model for Learning with both Label and Feature heterogeneity (L^2F), which imposes the view consistency by requiring that view-based classifiers generate similar predictions on the same examples. Multi-Label Co-Training (MLCT) [19] introduces a predictive reliability measure to select samples, and applies label-wise filtering to confidently communicate labels of selected samples among co-training classifiers. CS3G approach [24] handles types of interactions between multiple labels, while no interaction between features from different views in the model training phase. To make each view interacts and further reduce the extraction cost, Multi-modal Classifier Chains (MCC) [31] extends Classifier Chains to exploit label correlations with partial views. TMV-LE [26] use multiple views to more comprehensively mine the topological structure in the feature space and migrate it to the label space to obtain the label distribution. In addition, there are also some high-order approaches that exploit label correlations on the hypothesis space, Latent Semantic Aware Multi-view Multi-label Learning (LSA-MML) [25] implicitly encodes label correlations by the common representation based on the uncovering latent semantic bases and the relations among them. CoDiSP [30] learns low-dimensional common representation with all modalities, and extracts discriminative information of each modality by enforcing orthogonal constraint. GLOCAL [33] aims to remain consensus on multi-view latent spaces by Hilbert-Schmidt independence criterion during the mapping procedure. However, there is no communication among various views. Hence, SIMM [18] is proposed to leverage shared subspace exploitation and view-specific information extraction. CoDiSP [30] exploits relationship among different views and label correlations with the help of extracted common and specific view features.

3 Methodology

To begin, we present a formal definition of Multi-View Multi-Label (MVML) problems. Let $\mathcal{X} = \mathbb{R}^{d_1} \times \cdots \times \mathbb{R}^{d_v} \times \cdots \times \mathbb{R}^{d_M}$ be the feature space of M views, where $d_v (1 \leq v \leq M)$ is the dimensionality of the v-th view. Let $\mathcal{Y} = \{y_k\}_{k=1}^{L}$ ($y_k \in \{-1, 1\}$) be the label space with L labels. Given the training dataset with N data samples $\mathcal{D} = \{(\boldsymbol{X}_i, \boldsymbol{Y}_i)\}_{i=1}^{N}$, where $\boldsymbol{X}_i = [\boldsymbol{X}_i^1, \cdots, \boldsymbol{X}_i^v, \cdots, \boldsymbol{X}_i^M] \in \mathcal{X}$ ($\boldsymbol{X}_i^v \in \mathbb{R}^{d_v}$) is the feature vector and $\boldsymbol{Y}_i \in \mathcal{Y}$ is the label vector of the i-th

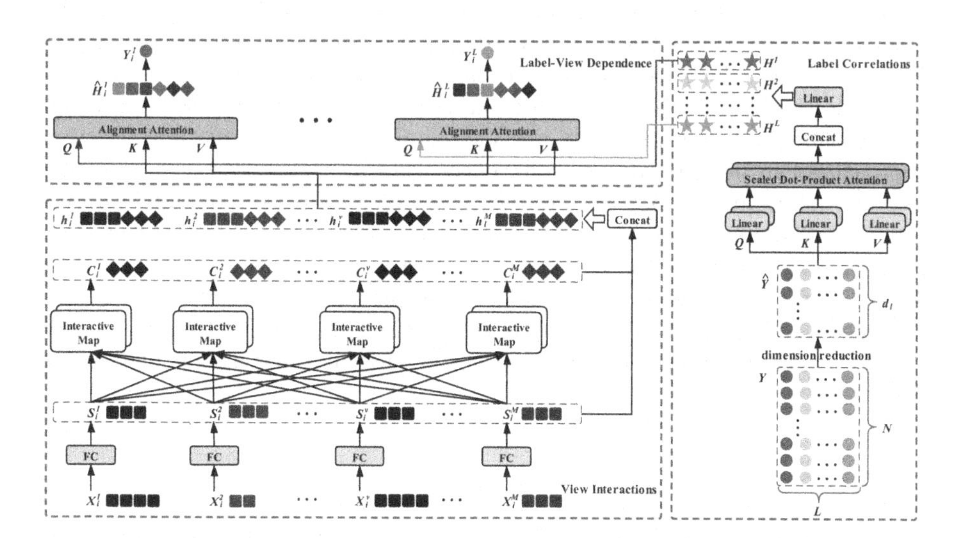

Fig. 1. The overall architecture of our proposed RAIN, where view interactions, label correlations, label-view dependence are exploited in a unified framework. For the i-th instance, we firstly excavate diversity and complementarity from the original views $\boldsymbol{X}_i^1, \boldsymbol{X}_i^2, \cdots, \boldsymbol{X}_i^M$, where view-specific information $\boldsymbol{S}_i^1, \boldsymbol{S}_i^2, \cdots, \boldsymbol{S}_i^M$ and complementary information $\boldsymbol{C}_i^1, \boldsymbol{C}_i^2, \cdots, \boldsymbol{C}_i^M$ are explicitly captured. By concatenating $\boldsymbol{S}_i^v$ and $\boldsymbol{C}_i^v$, we obtain the integrated multi-view representation $\boldsymbol{h}_i = [\boldsymbol{h}_i^1, \boldsymbol{h}_i^2, \cdots, \boldsymbol{h}_i^M]$. Apart from the view interactions, label correlations, and label-view dependence could also be employed to guide the learning process. Secondly, we adapt multi-head attention mechanism to exploit label correlations and achieve new label embedding $\boldsymbol{H} = [\boldsymbol{H}^1; \boldsymbol{H}^2; \cdots; \boldsymbol{H}^L]$. Eventually, the new label embedding are identified to dynamically extract label-specific information for each label by applying alignment attention, which amalgamates the relevant information filtered from the deep correlated multi-view representation $\boldsymbol{h}_i$.

instance $\boldsymbol{X}_i$. The goal of multi-view multi-label learning is to learn a mapping function $\boldsymbol{f} : \mathcal{X} \rightarrow \mathcal{Y}$ from $\mathcal{D}$, which can assign a set of proper labels for the unseen instance.

Next we will introduce our proposed **R**elation-aware **A**lignment attent**I**on **N**etwork (RAIN) in details and Fig. 1 displays the overall architecture. It models complex interactions among heterogeneous variables from the following three aspects: deep correlated subspace learning, label correlations learning and label-view dependence learning. Deep correlated subspace learning layer captures the shared multi-view representation for all the labels. Meanwhile, the new label embedding obtained by label correlations learning module can be used to guide the label-view dependence learning.

3.1 Deep Correlated Subspace Learning with View Interactions

In order to obtain the proper multi-view data representation for multi-label classification, it is necessary to consider the following two factors. On the one hand, for each instance, various feature vectors from M views may show tremendous diversity and complementarity in heterogeneous feature space. On the other hand, view interactions contain abundant descriptions of relations among multiple views, which can be used to enhance the effectiveness of multi-view subspace learning. Furthermore, different views contribute distinct importance to each label and multiple views provide supplementary information. Existing approaches can be improved by sufficiently utilizing view-specific and complementary information. As a result, two types of information should be interactively considered: view-specific information and complementary information. The deep correlated subspace learning module enables multi-view information to be shared across different labels, where multiple labels are correlated via common views. Within the latent space, RAIN employs specific and complementary information to capture both the consensus and particular information of different views.

View-Specific Information
Each view in the i-th original vector $\boldsymbol{X}_i$ can be used to extract discriminative information. For the v-th ($v = 1, \cdots, M$) original view $\boldsymbol{X}_i^v$ with d_v dimension, we add a fully connected layer and transform $\boldsymbol{X}_i^v$ to d_s dimension view-specific representation $\boldsymbol{S}_i^v$,

$$\boldsymbol{S}_i^v = \mathrm{ReLU}(\boldsymbol{X}_i^v \boldsymbol{U}_v + \boldsymbol{b}_v) \tag{1}$$

where $\boldsymbol{U}_v \in \mathbb{R}^{d_v \times d_s}$ is weight vector, $\boldsymbol{b}_v \in \mathbb{R}^{d_s}$ is bias vector. As a result, we can significantly improve the discriminating power of the subspace representation.

Complementary Information
Multi-view learning aims to integrate multiple features and discover consistent information among different views, which makes full use of the shared information from all views to enhance view consistency. Recently, it is demanded to

promote flexible view consistency for that different views provide complementary information rather than strictly consistent with each other. The key is to explore the interactive information shared among different view-specific representations, which can explicitly model the relationship among multiple views and explore the potential consistent information. In this way, irrelevant features are removed by relying on the complementary information from multiple views.

Interactive Map. Aiming at enhancing the interactions among different views, for the v-th view, we first calculate the interactive matrix between $\boldsymbol{S}_i^v$ and $\boldsymbol{S}_i^c$. As a result, we obtain a two-dimensional interactive map $\mathcal{CM}_i^{v,c} \in \mathbb{R}^{d_s \times d_s}$ by Eq. 2,

$$\mathcal{CM}_i^{v,c} = \mathbb{E}[(\boldsymbol{S}_i^v)^T \boldsymbol{S}_i^c] \tag{2}$$

where $c = 1, \cdots, M$ and $c \neq v$.

After calculating interactive map on every view pairs, we further design an interactive network to project each $\mathcal{CM}_i^{v,c}$ into an embedded space $\mathbb{R}^{d_c}$. It learns the deep interactive information and makes it incorporate with $\boldsymbol{S}_i^v$:

$$\boldsymbol{C}_i^{v,c} = \sigma\big(vec(\mathcal{CM}_i^{v,c})^T \boldsymbol{W}_{v,c} + \boldsymbol{b}_{v,c}\big) \tag{3}$$

where $\boldsymbol{W}_{v,c} \in \mathbb{R}^{d_s^2 \times d_c}$, $\boldsymbol{b}_{v,c} \in \mathbb{R}^{d_c}$, and $vec(\cdot)$ denotes the vectorization of a matrix.

Then, we concatenate $\boldsymbol{C}_i^{v,c}$ as the complementary representation for the v-th view,

$$\boldsymbol{C}_i^v = [\boldsymbol{C}_i^{v,1}, \cdots, \boldsymbol{C}_i^{v,v-1}, \boldsymbol{C}_i^{v,v+1}, \cdots, \boldsymbol{C}_i^{v,M}] \tag{4}$$

where $\boldsymbol{C}_i^v \in \mathbb{R}^{(M-1)d_c}$.

Above all, we combine both view-specific representation and complementary representation together to formalize enhanced multi-view subspace representation,

$$\boldsymbol{h}_i^v = [\boldsymbol{S}_i^v, \boldsymbol{C}_i^v] \tag{5}$$

where $\boldsymbol{h}_i^v \in \mathbb{R}^{d_h}$ and $d_h = d_s + (M-1)d_c$.

3.2 Label Correlations Learning with Multi-head Attention

It is well-known that exploiting label correlations is crucially important for multi-label learning and each label contains its own specific contribution to the multi-label prediction. The key factor of multi-label learning is the sharing and specific scheme among different labels, for that different labels lie in different distributions. Previous approaches take label correlations as prior knowledge and exploit label correlations by manipulating the hypothesis space, which may not correctly characterize the real relationships among labels. To address the above limitation, we employ multi-head attention mechanism to learn new label embedding, which characterizes label correlations efficiently and tailors label-specific information. Compared with the standard additive attention mechanism which is implemented by using a one-layer feed-forward neural network, multi-head attention [16] allows the model to jointly attend to information from different representation subspaces at different positions.

Firstly, we conduct dimension reduction to transform the sparse label vector $\boldsymbol{Y} \in \mathbb{R}^{N \times L}$ into a low-dimension vector $\hat{\boldsymbol{Y}} \in \mathbb{R}^{d_l \times L}$, where $d_l = Md_h$.

And then h parallel heads are employed to focus on different parts of the label vector, which can exploit label correlations. We linearly project $\hat{\boldsymbol{Y}}$ to queries ($\boldsymbol{Q}$), keys ($\boldsymbol{K}$) and values ($\boldsymbol{V}$) h times, respectively, where $\boldsymbol{Q} = \boldsymbol{K} = \boldsymbol{V} = \hat{\boldsymbol{Y}}^T$. On each of these projected versions of queries, keys and values, we then perform the attention function in parallel, yielding d_v-dimensional output values. For the t-th head ($t = 1, 2, \cdots, h$), we capture label correlations via the similarity among L labels:

$$\boldsymbol{R} = \frac{(\boldsymbol{Q}_t \boldsymbol{W}_t^Q)(\boldsymbol{K}_t \boldsymbol{W}_t^K)^T}{\sqrt{d_l/h}} \tag{6}$$

where $\boldsymbol{Q}_t, \boldsymbol{K}_t \in \mathbb{R}^{L \times d_l/h}$, $\boldsymbol{W}_t^Q, \boldsymbol{W}_t^K \in \mathbb{R}^{d_l/h \times d_l/h}$, $\boldsymbol{R} \in \mathbb{R}^{L \times L}$.

The new label representation is computed based on the following equation:

$$\boldsymbol{head}_t = \text{softmax}(\boldsymbol{R})(\boldsymbol{V}_t \boldsymbol{W}_t^V) \tag{7}$$

where $\boldsymbol{V}_t \in \mathbb{R}^{L \times d_l/h}$, $\boldsymbol{W}_t^V \in \mathbb{R}^{d_l/h \times d_l/h}$, $\boldsymbol{head}_t \in \mathbb{R}^{L \times d_l/h}$.

Then, we concatenate the resulting h heads and multiply with weight matrix $\boldsymbol{W}_O \in \mathbb{R}^{d_l \times d_l}$ to produce the new label embedding:

$$\boldsymbol{Head} = \text{Concat}(\boldsymbol{head}_1, \cdots, \boldsymbol{head}_h)\boldsymbol{W}_O. \tag{8}$$

where $\boldsymbol{Head} \in \mathbb{R}^{L \times d_l}$.

Finally, we employ residual connection [8] and layer normalization. The new label representation is computed as:

$$\boldsymbol{H} = \text{LayerNorm}(\hat{\boldsymbol{Y}}^T + \boldsymbol{Head}). \tag{9}$$

where $\boldsymbol{H} = [\boldsymbol{H}^1; \boldsymbol{H}^2; \cdots; \boldsymbol{H}^L] \in \mathbb{R}^{L \times d_l}$.

3.3 Label-View Dependence Learning with Alignment Attention

Multiple labels are correlated through multi-view representation and different views are of various importance under specific circumstance. However, it is possible that a few views become useless for certain labels due to noise pollution. After modeling view interactions and label correlations, we utilize alignment attention to capture the dependence relationship between label embedding and deep correlated multi-view representation. The new label embedding obtained by label correlations learning module is utilized as queries that supervise the model to assign appropriate attention weights to different view representations.

Training Phase. Given the new label embedding $\boldsymbol{H} = [\boldsymbol{H}^1; \cdots; \boldsymbol{H}^k; \cdots; \boldsymbol{H}^L] \in \mathbb{R}^{L \times d_l}$ and multi-view representation $\boldsymbol{h}_i = [\boldsymbol{h}_i^1, \boldsymbol{h}_i^2, \cdots, \boldsymbol{h}_i^M] \in \mathbb{R}^{Md_h}$, where $d_l = Md_h$, the alignment attention weight vector for the k-th label ($k = 1, 2, \cdots, L$) is:

$$\boldsymbol{\alpha}_i^k = \text{softmax}(\frac{\boldsymbol{H}^k(\boldsymbol{h}_i)^T}{\sqrt{d_l}}) \tag{10}$$

which control the contribution of different views information.

And then, the new label-specific representation can be computed by adding up the value over the corresponding dimension of the multi-view representation vector:

$$\hat{\boldsymbol{H}}_i^k = \text{LayerNorm}(\boldsymbol{H}^k + \boldsymbol{\alpha}_i^k \boldsymbol{h}_i). \tag{11}$$

RAIN extracts customized representation $[\hat{\boldsymbol{H}}_i^1; \cdots; \hat{\boldsymbol{H}}_i^k; \cdots; \hat{\boldsymbol{H}}_i^L]$ for different labels, and eventually masters the complete or a subset of expertise from all views.

Testing Phase. For the i-th instance in the testing phase, we firstly compute multi-view representation $\boldsymbol{h}_i$ by Eq. 5 and then directly capture label-specific representation by the learned alignment weight vector for each label by Eq. 10. After obtaining α_i^k and $\boldsymbol{h}_i$, the k-th label representation is computed by Eq. 11.

By encoding the label correlations into the multi-view representation, each attention mask automatically determines the importance of the shared multi-view subspace representation for the respective label, allowing learning of both label-shared and label-specific features in an end-to-end manner.

3.4 Label Prediction Layer

Based on the above three subsections, we predict the k-th label:

$$\boldsymbol{f}_k(\hat{\boldsymbol{H}}_i^k) = \text{softmax}(\hat{\boldsymbol{H}}_i^k \boldsymbol{W}_L^k + \boldsymbol{b}_L^k) \tag{12}$$

where $\boldsymbol{W}_L^k \in \mathbb{R}^{d_l \times 1}$ denotes the fully connected weight between label-specific representation vector $\hat{\boldsymbol{H}}_i^k$ and label prediction layer, $\boldsymbol{b}_L^k \in \mathbb{R}^{1 \times 1}$ is the bias vector.

Furthermore, we design binary cross-entropy loss function for final label prediction:

$$\mathcal{L} = -\sum_{i=1}^{N_b} \sum_{k=1}^{L} (y_i^k log \hat{y}_i^k + (1 - y_i^k) log(1 - \hat{y}_i^k)) \tag{13}$$

where N_b is the batch size. $\hat{y}_i^k$ is the prediction of $\boldsymbol{X}_i$ on the k-th label, predicted by $\boldsymbol{f}_k(\cdot)$ in Eq. 12.

4 Experiments

In this section, the performance of our proposed RAIN is evaluated. We begin by introducing details on datasets, evaluation metrics, and baselines.

Table 1. Characteristic of multi-view multi-label datasets, where $\#N$, $\#M$ and $\#L$ denote the number of instances, views, and labels, respectively. $\#d$ shows the dimensionality of each view.

Dataset	$\#N$	$\#M$	$\#L$	$\#d$
Yeast	2417	2	14	[24, 79]
Emotions	593	2	6	[32, 32, 8]
MSRC	591	3	24	[500,1040,576]
Taobao	2079	4	30	[500,48,81,24]
Scene	2407	6	6	[49, 49, 49, 49, 49, 49]
Herbs	11104	5	29	[13, 653, 433, 768, 36]

4.1 Experimental Setting

Dataset Description. We employ six multi-view multi-label datasets for performance evaluation including five public datasets and one real-world *Herbs* dataset. Table 1 summarizes the statistics of these datasets.

- *Yeast* [18,28] has two views including the genetic expression (79 attributes) and the phylogenetic profile of gene attributes (24 attributes).
- *Emotions* [31] is a publicly available multi-label dataset with 8 rhythmic attributes and 64 timbre attributes.
- *MSRC* [14] is used for object class recognition. As for each image, there are 3 types of views including: BoW, FV and HOG.
- *Taobao* [24] is used for shopping items classification. Description images of items are crawled from a shopping website, and four types of features, i.e., BoW, Gabor, HOG, HSVHist, are extracted to construct 4 views of data. Corresponding categories path of an item provides the label sets.
- *Scene* [1,31] is a public multi-label dataset with 6 views.
- *Herbs* is a real-world multi-view multi-label dataset. Each herb consists of multiple view features, i.e., *indications*, *function*, *dosage*, *channel tropism*, *property and flavor*. All of these features are appeared in Chinese and a descriptive way. Meanwhile, each herb is linked with a number of efficacies and we aim to predict efficacy categories of each herb.

Evaluation Metrics. For performance evaluation, we employ six widely-used multi-label evaluation metrics, including Hamming Loss (HamLoss), Ranking Loss (RankLoss), Subset Accuracy (SubsetAcc), Example-F1, Micro-F1 and Macro-F1. All their values vary within the interval $[0, 1]$. In addition, for the first two metrics, the smaller values indicate better performance. While for the other four metrics, the larger values indicate better performance. More concrete metric definitions can be found in [17,29].

Table 2. Experimental results (mean ± std) of RAIN compared with other state-of-the-art approaches on six multi-view multi-label datasets. The best performance for each criterion is bolded. ↑ / ↓ indicates the larger/smaller the better of a criterion.

Dataset	Approaches	Evaluation metrics					
		HamLoss↓	RankLoss↓	SubsetAcc↑	Example-F1↑	Micro-F1↑	Macro-F1↑
MSRC	CAMEL(B)	0.059±0.009	0.033±0.009	0.322±0.058	0.805±0.032	0.814±0.029	0.680±0.043
	CAMEL(C)	0.106±0.020	0.153±0.074	0.070±0.073	0.618±0.074	0.629±0.071	0.208±0.041
	LLSF	0.066±0.007	0.043±0.009	0.276±0.063	0.782±0.023	0.790±0.025	0.607±0.040
	CS3G	0.077±0.008	0.043±0.010	0.205±0.043	0.729±0.024	0.744±0.026	0.506±0.040
	MCC	0.068±0.008	0.054±0.011	0.345±0.088	0.798±0.024	0.799±0.023	0.652±0.024
	SIMM	0.058±0.008	**0.033±0.005**	0.382±0.034	0.827±0.024	0.829±0.024	0.730±0.035
	MvNNcor	0.062±0.007	0.042±0.007	0.355±0.046	0.815±0.024	0.816±0.023	0.710±0.032
	RAIN	**0.050±0.005**	0.035±0.010	**0.455±0.048**	**0.848±0.014**	**0.852±0.013**	**0.754±0.043**
Yeast	CAMEL(B)	0.193±0.007	0.162±0.011	0.173±0.022	0.618±0.014	0.642±0.012	0.392±0.025
	CAMEL(C)	0.189±0.007	0.163±0.012	0.201±0.018	0.628±0.013	0.658±0.012	**0.458±0.020**
	LLSF	0.295±0.006	0.354±0.015	0.003±0.004	0.049±0.006	0.056±0.005	0.046±0.003
	CS3G	0.255±0.008	0.211±0.012	0.048±0.009	0.549±0.011	0.566±0.011	0.217±0.017
	MCC	0.213±0.009	0.224±0.016	0.190±0.034	0.585±0.025	0.615±0.021	0.340±0.018
	SIMM	0.191±0.009	0.160±0.008	0.213±0.026	0.634±0.015	0.660±0.015	0.446±0.025
	MvNNcor	0.196±0.007	0.167±0.009	0.206±0.025	0.632±0.017	0.652±0.015	0.380±0.019
	RAIN	**0.185±0.008**	**0.157±0.011**	**0.225±0.028**	**0.640±0.021**	**0.666±0.019**	0.456±0.035
Taobao	CAMEL(B)	0.032±0.001	0.151±0.014	0.150±0.020	0.054±0.012	0.101±0.023	0.034±0.008
	CAMEL(C)	0.033±0.002	0.159±0.012	0.238±0.046	0.266±0.044	0.373±0.053	0.182±0.028
	LLSF	0.036±0.001	0.161±0.015	0.052±0.014	0.058±0.013	0.106±0.023	0.047±0.013
	CS3G	0.064±0.003	0.171±0.009	0.104±0.017	0.323±0.025	0.332±0.026	0.125±0.015
	MCC	0.054±0.002	0.235±0.016	0.213±0.023	0.333±0.026	0.354±0.021	0.222±0.026
	SIMM	0.028±0.001	0.099±0.008	0.414±0.026	0.417±0.036	0.523±0.028	0.298±0.047
	MvNNcor	0.029±0.001	0.120±0.019	0.382±0.059	0.372±0.070	0.477±0.062	0.228±0.053
	RAIN	**0.024±0.002**	**0.095±0.012**	**0.520±0.027**	**0.529±0.027**	**0.616±0.027**	**0.400±0.040**
Emotions	CAMEL(B)	0.218±0.014	0.176±0.023	0.248±0.034	0.534±0.039	0.607±0.038	0.590±0.039
	CAMEL(C)	0.207±0.025	0.179±0.038	0.272±0.048	0.581±0.049	0.637±0.048	0.615±0.058
	LLSF	0.207±0.014	0.174±0.021	0.254±0.049	0.594±0.039	0.641±0.033	0.616±0.033
	CS3G	0.290±0.013	0.225±0.018	0.165±0.041	0.538±0.028	0.574±0.031	0.461±0.023
	MCC	0.214±0.023	0.178±0.029	0.299±0.037	0.624±0.043	0.653±0.041	0.621±0.039
	SIMM	0.180±0.006	**0.128±0.015**	0.356±0.025	0.658±0.016	0.701±0.018	0.692±0.020
	MvNNcor	0.187±0.022	0.141±0.024	0.349±0.080	0.636±0.067	0.684±0.048	0.656±0.079
	RAIN	**0.165±0.012**	0.136±0.023	**0.415±0.040**	**0.687±0.029**	**0.725±0.027**	**0.718±0.027**
Scene	CAMEL(B)	0.144±0.007	0.158±0.018	0.300±0.033	0.341±0.035	0.457±0.037	0.458±0.033
	CAMEL(C)	0.076±0.006	0.057±0.011	0.646±0.024	0.695±0.027	0.763±0.019	0.772±0.021
	LLSF	0.106±0.006	0.095±0.010	0.487±0.028	0.536±0.034	0.643±0.027	0.644±0.027
	CS3G	0.217±0.013	0.289±0.030	0.327±0.037	0.366±0.036	0.372±0.036	0.282±0.034
	MCC	0.102±0.010	0.101±0.011	0.662±0.038	0.713±0.029	0.709±0.027	0.718±0.026
	SIMM	0.071±0.005	0.054±0.008	0.727±0.014	0.769±0.022	0.795±0.017	0.805±0.021
	MvNNcor	0.074±0.003	0.061±0.006	0.733±0.013	0.762±0.013	0.784±0.010	0.792±0.008
	RAIN	**0.063±0.009**	**0.051±0.009**	**0.767±0.027**	**0.805±0.025**	**0.820±0.023**	**0.828±0.019**
Herbs	CAMEL(B)	0.015±0.002	0.019±0.008	0.250±0.056	0.334±0.096	0.464±0.077	0.159±0.029
	CAMEL(C)	0.013±0.003	**0.017±0.006**	0.306±0.109	0.330±0.102	0.503±0.050	0.129±0.017
	LLSF	0.012±0.000	0.018±0.002	0.358±0.020	0.355±0.012	0.540±0.014	0.138±0.006
	CS3G	0.013±0.000	0.019±0.001	0.340±0.012	0.342±0.011	0.526±0.013	0.133±0.007
	MCC	0.013±0.000	0.059±0.004	0.400±0.018	0.378±0.012	0.579±0.010	0.147±0.017
	SIMM	0.012±0.000	0.023±0.004	0.411±0.018	0.375±0.017	0.576±0.014	0.142±0.011
	MvNNcor	0.011±0.000	0.031±0.004	0.417±0.020	0.371±0.028	0.578±0.021	0.141±0.033
	RAIN	**0.011±0.000**	0.020±0.001	**0.423±0.015**	**0.382±0.010**	**0.582±0.011**	**0.166±0.018**

Compared Approaches. Considering RAIN is related to multi-view multi-label learning, the performance of RAIN is compared against 7 approaches, including a state-of-the-art multi-label learning approach with two types of feature inputs, one label-specific multi-label approach, and 5 multi-view multi-label approaches.

- CAMEL(B) & CAMEL(C): CAMEL [4] is a novel multi-label learning approach that exploits label correlations via sparse reconstruction in the label space and integrates the learned label correlations into model training. CAMEL(B) stands for the best performance obtained from the best single view, while CAMEL(C) simply concatenating the multi-view inputs as a new single-view input.
- LLSF [9]: A second-order multi-label algorithm which performs joint label-specific feature selection and take the label correlation matrix as prior knowledge.
- MvNNcor [21]: A novel multi-view learning framework that utilize different deep neural networks to learn multiple view-specific representation, and model deep interactive information through a shared interactive network. Finally, we adapt the fusion strategy to make a joint decision for multi-label classification.
- CS3G [24]: A multi-view multi-label approach utilizes multi-view information in a privacy-preserving style, which treats each view unequally and has the ability to extract the most useful modal features for the final prediction.
- MCC [31]: A novel multi-view multi-label approach that makes great use of views, and can make a convince prediction with many instead of all views.
- SIMM [18]: A novel multi-view multi-label learning approach, which leverages shared subspace exploitation and view-specific information extraction. For shared subspace exploitation, SIMM jointly minimizes confusion adversarial loss and multi-label loss to utilize shared information from all views.

4.2 Experimental Results

For all these approaches, we report the best results of the optimal parameters in terms of classification performance. 10-fold cross-validation is conducted on these datasets, where the mean metric (with standard deviations) are recorded for all approaches. We set the batch size $N_b = 64$. For the label embedding module, the head number of *Emotions* and *MSRC* are set to 4, while the other datasets are set to 6. Furthermore, we drop out at a rate of 40% at each step to avoid over-fitting [15].

Table 2 presents the comparative results of all approaches within the same environmental settings, where the best results have been highlighted. Based on the experimental results, we obtain the following observations: 1) From the results of CAMLE(C) approach, it is obviously shown that roughly concatenating all views as a single view may not always be a wise choice. Because the concatenation of all the views may confuse the view-specific information and miss the interactive information. 2) RAIN achieves the best performance compared with several state-of-the-art approaches on all datasets, which reveals the priority of RAIN in dealing with MVML learning problem.

4.3 Experimental Analysis

Ablation Study. To reveal the contribution of each component, we test the performance of RAIN by removing different parts. As for the effectiveness of complementary information exploration in deep correlated subspace learning, we keep the basic structure of RAIN and remove the complementary representation $\boldsymbol{C}^v$, denoted as RAIN-NC. As for the efficacy of label correlations, we remove label embedding, denoted as RAIN-NL. Furthermore, we remove the alignment attention structure in label-view dependence learning, which is denoted as RAIN-NA.

As shown in Table 3, RAIN performs better than RAIN-NC, RAIN-NL, and RAIN-NA, which shows that: 1) the enhanced complementary information by exploiting view interactions is helpful to multi-view subspace learning; 2) with the guidance of label semantic embeddings, the discovered label-specific representation can better match the annotated semantic labels; 3) the benefits of customizing label-specific representation compared to the globally shared ones.

Attention Visualization. As shown in Table 4, different views encode different properties of data. To understand the role of the proposed attention modules, we visualize the learned label correlations, and label-view dependence matrices to illustrate the ability of capturing semantic dependence. As illustrated in Fig. 2, we can see a clear difference in label-view attention between two labels, which can provide richer semantic from different perspectives. In Fig. 2(b), all labels puts less emphasis on View 2, which accords with the worst performance of View 2 in Table 4.

Convergence Analysis. We conduct convergence experiments to validate the convergence of RAIN approach. Due to the page limit, we only give the convergence results on the *Emotions*, *Scene*, and *Herbs* datasets. As shown in Fig. 3, it is clear that RAIN can converge fast within a small number of epochs.

Table 3. Comparison results (mean ± std) of RAIN from three aspects: view interactions, label correlations, label-view dependence. The best performance for each criterion is bolded.

Dataset	Approaches	Evaluation metrics					
		HamLoss↓	RankLoss↓	SubsetAcc↑	Example-F1↑	Micro-F1↑	Macro-F1↑
MSRC	RAIN-NC	0.056±0.008	0.046±0.009	0.372±0.056	0.825±0.029	0.831±0.026	0.722±0.032
	RAIN-NL	0.054±0.007	0.042±0.011	0.404±0.051	0.837±0.024	0.840±0.022	0.737±0.057
	RAIN-NA	0.055±0.008	0.036±0.007	0.415±0.055	0.837±0.031	0.838±0.027	0.729±0.050
	RAIN	**0.050±0.005**	**0.035±0.010**	**0.455±0.048**	**0.848±0.014**	**0.852±0.013**	**0.754±0.043**
Yeast	RAIN-NC	0.191±0.008	0.164±0.013	0.197±0.022	0.639±0.019	0.661±0.016	0.412±0.044
	RAIN-NL	0.190±0.007	0.165±0.013	0.208±0.021	0.635±0.014	0.662±0.015	0.425±0.048
	RAIN-NA	0.192±0.010	0.165±0.013	0.198±0.029	0.635±0.019	0.658±0.019	0.401±0.030
	RAIN	**0.185±0.008**	**0.157±0.011**	**0.225±0.028**	**0.640±0.021**	**0.666±0.019**	**0.456±0.035**
Taobao	RAIN-NC	0.025±0.002	0.107±0.014	0.485±0.030	0.486±0.025	0.587±0.028	0.361±0.046
	RAIN-NL	0.025±0.001	0.114±0.010	0.507±0.018	0.509±0.021	0.594±0.021	0.371±0.038
	RAIN-NA	0.027±0.001	0.102±0.010	0.478±0.036	0.487±0.041	0.567±0.029	0.361±0.043
	RAIN	**0.024±0.002**	**0.095±0.012**	**0.520±0.027**	**0.529±0.027**	**0.616±0.027**	**0.400±0.040**
Emotions	RAIN-NC	0.182±0.017	0.161±0.023	0.371±0.048	0.648±0.041	0.690±0.030	0.675±0.045
	RAIN-NL	0.168±0.013	0.146±0.029	0.390±0.053	0.681±0.030	0.720±0.029	0.708±0.034
	RAIN-NA	0.170±0.020	0.143±0.033	0.390±0.065	0.676±0.047	0.718±0.030	0.703±0.042
	RAIN	**0.165±0.012**	**0.136±0.023**	**0.415±0.040**	**0.687±0.029**	**0.725±0.027**	**0.718±0.027**
Scene	RAIN-NC	0.073±0.007	0.067±0.009	0.717±0.035	0.758±0.018	0.789±0.018	0.798±0.013
	RAIN-NL	0.065±0.008	0.058±0.013	0.753±0.035	0.790±0.028	0.813±0.024	0.822±0.021
	RAIN-NA	0.066±0.008	0.054±0.008	0.748±0.033	0.785±0.033	0.807±0.025	0.816±0.027
	RAIN	**0.063±0.009**	**0.051±0.009**	**0.767±0.027**	**0.805±0.025**	**0.820±0.023**	**0.828±0.019**
Herbs	RAIN-NC	0.012±0.000	0.023±0.002	0.398±0.017	0.367±0.017	0.568±0.017	0.141±0.018
	RAIN-NL	0.012±0.000	0.038±0.007	0.401±0.017	0.361±0.033	0.571±0.024	0.138±0.034
	RAIN-NA	0.012±0.000	0.032±0.002	0.389±0.016	0.367±0.025	0.565±0.023	0.146±0.020
	RAIN	**0.011±0.000**	**0.020±0.001**	**0.423±0.015**	**0.382±0.010**	**0.582±0.011**	**0.166±0.018**

Table 4. Experimental results (mean ± std) of CAMEL with each individual view severally on the *Emotions* and *Scene* datasets. The best performance for each criterion is bolded.

Dataset	Approaches	Evaluation metrics					
		HamLoss↓	RankLoss↓	SubsetAcc↑	Example-F1↑	Micro-F1↑	Macro-F1↑
Emotions	View 0	**0.218±0.014**	**0.176±0.023**	**0.248±0.034**	**0.534±0.039**	**0.607±0.038**	**0.590±0.039**
	View 1	0.225±0.021	0.202±0.032	0.238±0.035	0.533±0.051	0.588±0.043	0.566±0.048
	View 2	0.298±0.029	0.366±0.048	0.061±0.042	0.188±0.054	0.237±0.064	0.183±0.060
Scene	View 0	0.159±0.004	0.220±0.015	0.217±0.013	0.236±0.014	0.342±0.020	0.344±0.019
	View 1	0.171±0.005	0.250±0.016	0.148±0.028	0.157±0.027	0.240±0.033	0.241±0.038
	View 2	0.154±0.007	0.177±0.020	0.238±0.023	0.263±0.023	0.374±0.028	0.392±0.027
	View 3	0.173±0.006	0.290±0.017	0.089±0.022	0.093±0.023	0.154±0.037	0.148±0.036
	View 4	**0.144±0.007**	**0.158±0.018**	**0.300±0.033**	**0.341±0.035**	**0.457±0.037**	**0.458±0.033**
	View 5	0.167±0.007	0.275±0.019	0.128±0.021	0.138±0.023	0.222±0.036	0.218±0.032

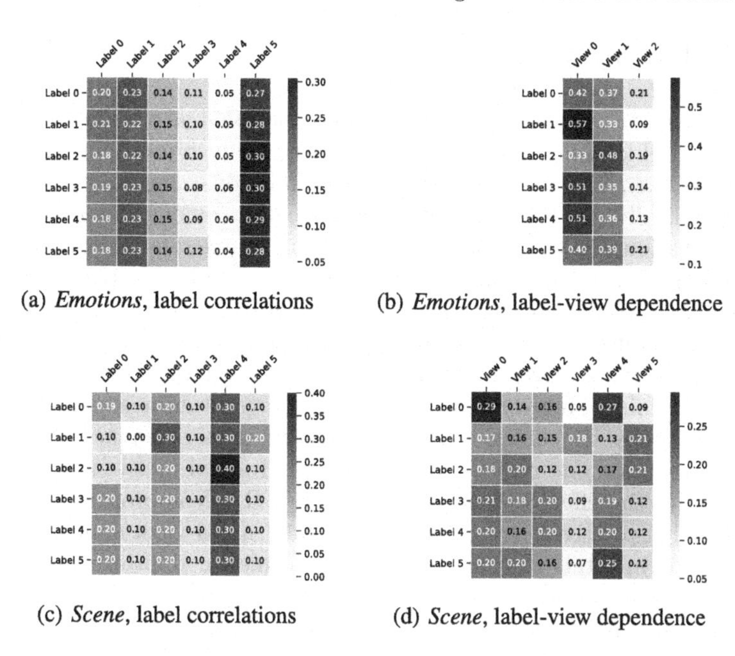

Fig. 2. Attention visualization. A higher blue intensity value indicates a stronger correlation. (Color figure online)

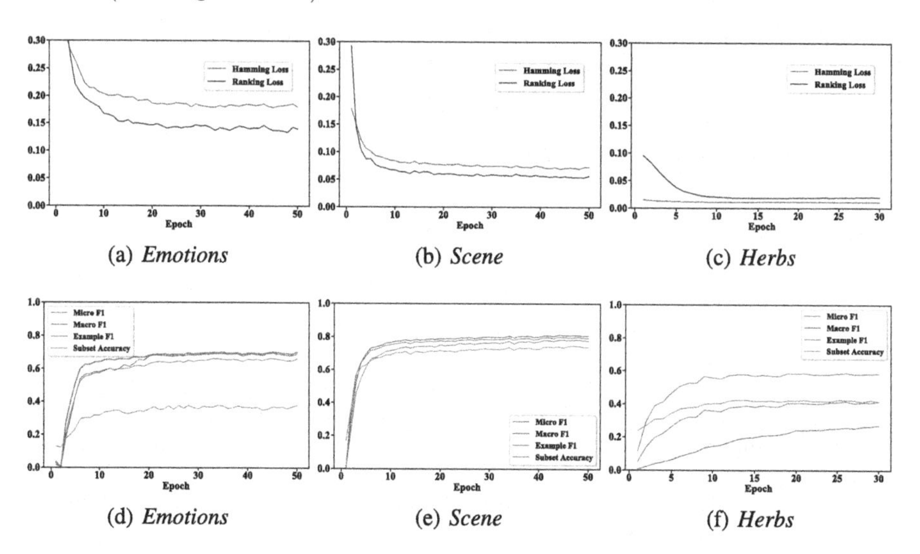

Fig. 3. Convergence analysis of RAIN on the *Emotions*, *Scene*, and *Herbs* datasets.

5 Conclusion

In this paper, we propose a novel mutual relational attention-based multi-view multi-label learning framework named RAIN, which jointly leverages three relationships, i.e., view interactions, label correlations, and label-view dependence. Specifically, we first enhance communication among different views by embedding various view-specific information and interactive information. Then, we adaptively learn label embedding with multi-head attention mechanism, which encodes the label correlations. Last, in order to extract discriminative view information for different labels, we seamlessly integrate the learned label embedding into multi-view representation by label-view alignment attention mechanism. The learning procedure of multi-view representation and label embedding can mutually benefit each other for maximum performance gain. Extensive experiments on one real-world and five widely-used multi-view multi-label datasets verify the effectiveness of our proposed RAIN approach. Three types of relationships positively contribute to the final multi-view multi-label learning. The experimental analysis indicates performance improvement not only by sufficiently utilizing complementary view-specific information, but also label embedding with multi-head attention mechanism. Furthermore, it also reveals the effectiveness of label-view alignment attention, due to introducing more guiding information.

Acknowledgment. This paper is supported by the National Key Research and Development Program of China (Grant No. 2018YFB1403400), the National Natural Science Foundation of China (Grant No. 61876080), the Key Research and Development Program of Jiangsu (Grant No. BE2019105), the Collaborative Innovation Center of Novel Software Technology and Industrialization at Nanjing University.

References

1. Boutell, M.R., Luo, J., Shen, X., Brown, C.M.: Learning multi-label scene classification. Pattern Recogn. **37**(9), 1757–1771 (2004)
2. Cao, X., Zhang, C., Fu, H., Liu, S., Zhang, H.: Diversity-Induced Multi-View Subspace Clustering, In: CVPR. pp. 586–594. IEEE Computer Society (2015)
3. Fang, Z., Zhang, Z.M.: Simultaneously combining multi-view multi-label learning with maximum margin classification. In: ICDM, pp. 864–869. IEEE Computer Society (2012)
4. Feng, L., An, B., He, S.: Collaboration based multi-label learning. In: AAAI, pp. 3550–3557. AAAI Press (2019)
5. Fürnkranz, J., Hüllermeier, E., Mencía, E.L., Brinker, K.: Multilabel classification via calibrated label ranking. Mach. Learn. **73**(2), 133–153 (2008)
6. Ghamrawi, N., McCallum, A.: Collective multi-label classification. In: CIKM, pp. 195–200. ACM (2005)
7. Gopal, S., Yang, Y.: Multilabel classification with meta-level features. In: SIGIR, pp. 315–322. ACM (2010)
8. He, K., Zhang, X., Ren, S., Sun, J.: Deep residual learning for image recognition. In: CVPR, pp. 770–778. IEEE Computer Society (2016)

9. Huang, J., Li, G., Huang, Q., Wu, X.: Learning label-specific features and class-dependent labels for multi-label classification. IEEE Trans. Knowl. Data Eng. **28**(12), 3309–3323 (2016)
10. Huang, S., Yu, Y., Zhou, Z.: Multi-label hypothesis reuse. In: KDD, pp. 525–533. ACM (2012)
11. Liu, C., Zhao, P., Huang, S., Jiang, Y., Zhou, Z.: Dual set multi-label learning. In: AAAI, pp. 3635–3642. AAAI Press (2018)
12. Peng, H., et al.: Hierarchical taxonomy-aware and attentional graph capsule RCNNs for large-scale multi-label text classification. In: TKDE (2019)
13. Read, J., Pfahringer, B., Holmes, G., Frank, E.: Classifier chains for multi-label classification. Mach. Learn. **85**(3), 333–359 (2011)
14. Schroff, F., Criminisi, A., Zisserman, A.: Harvesting image databases from the web. TPAMI **33**(4), 754–766 (2010)
15. Srivastava, N., Hinton, G.E., Krizhevsky, A., Sutskever, I., Salakhutdinov, R.: Dropout: a simple way to prevent neural networks from overfitting. J. Mach. Learn. Res. **15**(1), 1929–1958 (2014)
16. Vaswani, A., et al.: Attention is all you need. In: NIPS, pp. 5998–6008 (2017)
17. Wu, X., Zhou, Z.: A unified view of multi-label performance measures. In: ICML. Proceedings of Machine Learning Research, vol. 70, pp. 3780–3788. PMLR (2017)
18. Wu, X., et al.: Multi-view multi-label learning with view-specific information extraction. In: IJCAI, pp. 3884–3890. ijcai.org (2019)
19. Xing, Y., Yu, G., Domeniconi, C., Wang, J., Zhang, Z.: Multi-label co-training. In: IJCAI, pp. 2882–2888. ijcai.org (2018)
20. Xu, C., Tao, D., Xu, C.: Multi-view intact space learning. TPAMI **37**(12), 2531–2544 (2015)
21. Xu, J., Li, W., Liu, X., Zhang, D., Liu, J., Han, J.: Deep embedded complementary and interactive information for multi-view classification. In: AAAI, pp. 6494–6501. AAAI Press (2020)
22. Yang, H., Zhou, J.T., Zhang, Y., Gao, B., Wu, J., Cai, J.: Exploit bounding box annotations for multi-label object recognition. In: CVPR, pp. 280–288. IEEE Computer Society (2016)
23. Yang, P., et al.: Jointly modeling label and feature heterogeneity in medical informatics. ACM Trans. Knowl. Discov. Data **10**(4), 39:1–39:25 (2016)
24. Ye, H., Zhan, D., Li, X., Huang, Z., Jiang, Y.: College student scholarships and subsidies granting: A multi-modal multi-label approach. In: ICDM, pp. 559–568. IEEE Computer Society (2016)
25. Zhang, C., Yu, Z., Hu, Q., Zhu, P., Liu, X., Wang, X.: Latent semantic aware multi-view multi-label classification. In: AAAI, pp. 4414–4421. AAAI Press (2018)
26. Zhang, F., Jia, X., Li, W.: Tensor-based multi-view label enhancement for multi-label learning. In: IJCAI, pp. 2369–2375. ijcai.org (2020)
27. Zhang, M.L., Zhou, Z.H.: Multilabel neural networks with applications to functional genomics and text categorization. TKDE **18**(10), 1338–1351 (2006)
28. Zhang, M., Zhou, Z.: ML-KNN: a lazy learning approach to multi-label learning. Pattern Recogn. **40**(7), 2038–2048 (2007)
29. Zhang, M., Zhou, Z.: A review on multi-label learning algorithms. IEEE Trans. Knowl. Data Eng. **26**(8), 1819–1837 (2014)
30. Zhang, Y., Shen, J., Zhang, Z., Wang, C.: Common and discriminative semantic pursuit for multi-modal multi-label learning. In: ECAI. Frontiers in Artificial Intelligence and Applications, vol. 325, pp. 1666–1673. IOS Press (2020)

31. Zhang, Y., Zeng, C., Cheng, H., Wang, C., Zhang, L.: Many could be better than all: a novel instance-oriented algorithm for multi-modal multi-label problem. In: ICME, pp. 838–843. IEEE (2019)
32. Zhou, T., Zhang, C., Gong, C., Bhaskar, H., Yang, J.: Multiview latent space learning with feature redundancy minimization. IEEE Trans. Cybern. **50**(4), 1655–1668 (2020)
33. Zhu, Y., Kwok, J.T., Zhou, Z.H.: Multi-label learning with global and local label correlation. TKDE **30**(6), 1081–1094 (2017)

BIRL: Bidirectional-Interaction Reinforcement Learning Framework for Joint Relation and Entity Extraction

Yashen Wang(✉) and Huanhuan Zhang

National Engineering Laboratory for Risk Perception and Prevention (RPP), China Academy of Electronics and Information Technology of CETC, Beijing 100041, China
yswang@bit.edu.cn, huanhuanz_bit@139.com

Abstract. Joint relation and entity extraction is a crucial technology to construct a knowledge graph. However, most existing methods (i) can *not* fully capture the beneficial connections between relation extraction and entity extraction tasks, and (ii) can *not* combat the noisy data in the training dataset. To overcome these problems, this paper proposes a novel **B**idirectional-**I**nteraction **R**einforcement **L**earning (**BIRL**) framework, to extract entities and relations from plain text. Especially, we apply a relation calibration RL policy to (i) measure relation consistency and enhance the bidirectional interaction between entity mentions and relation types; and (ii) guide a dynamic selection strategy to remove noise from training dataset. Moreover, we also introduce a data augmentation module for bridging the gap of data-efficiency and generalization. Empirical studies on two real-world datasets confirm the superiority of the proposed model.

Keywords: Joint relation and entity extraction · Reinforcement learning · Bidirectional-interaction · Representation learning

1 Introduction

Relation extraction (RE) and Entity Extraction (EE) are two fundamental tasks in natural language processing (NLP) applications. In practice, these two tasks are often to be solved simultaneously. Recently, relation and entity joint extraction has become a fundamental task in NLP, which can facilitate many other tasks, including knowledge graph construction, question answering, and automatic text summarization. The goal of this task is to extract triples (e_h, r, e_t) from the unstructured texts, wherein e_h indicates the head entity, e_t represents the tail entity, and r is the semantic relation between e_h and e_t.

Recently, with the advancement of reinforcement learning (RL) in the field of NLP, RL has been adopted by many previous studies for relation extraction and entity extraction. [4] used a RL agent to remove noisy data from the training dataset, and they used the likelihood of the sentence as a reward to guide the

C. S. Jensen et al. (Eds.): DASFAA 2021, LNCS 12682, pp. 483–499, 2021.
https://doi.org/10.1007/978-3-030-73197-7_32

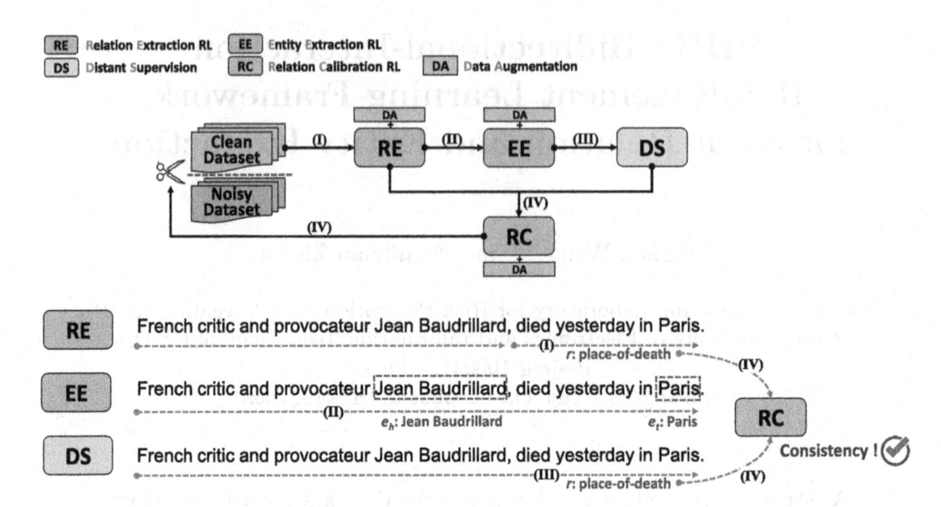

Fig. 1. The overall architecture of the proposed bidirectional interaction reinforcement learning framework.

agent's training. Similarly, [21] used RL strategy to generate the false-positive indicator. The reward in these studies was derived from the performance change of the joint network for different epochs. Unlike their research aiming to generate a false-positive indicator, this paper focuses on enhancing the anti-noise robustness of the model by leveraging a dynamic selection strategy to remove noise from training dataset. [5] modeled the joint extraction as a two-step decision process and employed the Q-Learning algorithm to get the control policy network in the two-step decision process. Also, [28] modeled relation and entity extraction as two levels of the RL process and use a hierarchical RL framework to extract relations and entities. However, this work replied heavily on the relation indicator, whose identifying procedure is unexplainable. Besides, the interaction between relation extraction and entity extraction in this work is unidirectional, as well as [37], hence the beneficial connections between them is not unfortunately fully modeled. Different from the two studies mentioned above, this study investigates the bidirectional interaction between relation extraction and entity extraction processes to fully capture the beneficial connections between them, and also models the removal of noisy data as a reinforcement learning process.

In this paper, we propose a novel joint extraction framework based on bidirectional interaction RL paradigm, wherein we first detect a relation, extract the corresponding entities as the arguments of this relation, then a relation is inversely generated based on these entities, and finally relation consistency is calibrated for refining the training data. Hence, the proposed framework is composed of five parts:

(I) **R**elation **E**xtraction (**RE**) RL: identifies relation from the given text;
(II) **E**ntity **E**xtraction (**EE**) RL: identifies the participating entities for this relation;

(III) **D**istant **S**upervision (**DS**): inversely identifies a relation respect to these entities;
(IV) **R**elation **C**alibration (**RC**) RL: refines the training dataset based on the consistency between the relations obtained in aforementioned procedures (I) and (III).

As shown in Fig. 1, the relation extraction procedure makes sequential scans from the beginning of a sentence, and detect relation (I). If a certain relation is identified, an entity extraction process is triggered to identify the corresponding entities for that relation (II). Then, a distant supervision model is leveraged here for generating relation for these entities (III). Finally, a relation calibration process measures the consistency between the relation from relation extraction process and the relation from distant supervision process (IV), to finally guide the refine of the training dataset for filtering noise. Then the relation extraction RL process continues its scan to search for the next relation in the sentence. Note that, relations are generated twice and then calibrated for consistency in aforementioned procedure, hence the proposed framework could fully model and capture the bidirectional interaction between relation extraction and entity extraction processes.

In conclusion, our work has strengths in dealing with two issues existing in prior studies:

(i) Most traditional models [8,9,19] determine a relation only after all the entities have been recognized, whereas the interaction between the two tasks is *not* fully captured. In some sense, these methods are aligning a relation to entity pairs, and therefore, they may introduce additional noise since a sentence containing an entity pair may not truly mention the relation [34], or may describe multiple relations [27]. In our work, the dependency between entity mentions and relation types is formulated through designing the state representations and rewards in the different RL processes (as sketched in Fig. 1). The bidirectional interaction is well captured since the relation extraction RL's reward is passed to the entity extraction RL (in (II)), then distant supervision model is launched to generate relation (in (III)), and finally the consistency of these relations acts as the reward for relation calibration RL (in (IV)). In this manner, the interaction between relation types and entity mentions can be better modeled bidirectionally.

(ii) Many existing studies on information extraction adopted both supervised [2,36] and distant supervised [18] methods. Although distant supervision method is effective and cheap, it inevitably introduces a large amount of *noisy data* (false-positive). That is because its assumption is too strict, and the same entity pair may *not* express the desired relation-types. Therefore, we argue that a model's performance will be affected severely if it is trained on a noisy dataset. This paper focuses on combatting the noisy data in the training dataset, and exploring the possibility of using dynamic selection strategies to remove them. Especially, the performance-driven relation calibration RL (in (IV)) is designed to recognize the noisy sentences and take appropriate actions to remove them. During the training, the relation calibration RL agent interacts with the other RLs continuously and divides the training dataset into two parts: clean dataset and noisy dataset. In each interaction, the relation calibration RL agent updates its policy based on the relation consistency performance received from the relation extraction policy (in (I)) and distant supervision model (in (III)). The interaction process will come to an end when the other RLs trained on the clean dataset achieves the best performance.

Moreover, bridging the gap of *data-efficiency* and *generalization* has been demonstrated pivotal to the real-world applicability of RL [14]. Hence, this paper investigates the utility of data augmentations in model-free RL by processing text observations with stochastic augmentations before passing them to the RL agent (in (I) (II) (IV)) for training. Therefore, we propose a simple plug-and-play module that can enhance RL algorithm with augmented data, *without* any changes to the underlying RL algorithm. We show that data augmentations such as Synonym, Edit-Distance, Insertion and Deletion, can enable RL algorithms, in terms of data-efficiency and generalization.

2 Methodology

2.1 Relation Extraction with RL

The **R**elation **E**xtraction (**RE**) RL policy π_{RE} aims to detect the relations in the given sentence $s = \{w_1, w_2, \cdots, w_{|s|}\}$, which can be regarded as a conventional RL policy over actions. A following entity extraction RL process (Sect. 2.2) will be launched once an action is executed by the this RL agent.

Action: The action a_t is selected from $\mathcal{A}_{\mathsf{RE}} = \mathrm{R} \bigcup \mathtt{None}$, wherein notation None indicates no relation, and R is the relation-type set. When a relation calibration RL process (Sect. 2.3) enters a terminal state, the control of the agent will be taken over to the relation extraction RL process to execute the next actions.

State: The state $s_t^{\mathsf{RE}} \in \mathcal{S}_{\mathsf{RE}}$ of the relation extraction RL process at time step t, is represented by: (i) the current hidden state $\mathbf{h}_t$, (ii) the relation-type vector $\mathbf{a}_t^{\mathsf{RE}}$ (the embedding of the latest action a_{t*}^{RE} that $a_{t*}^{\mathsf{RE}} \neq \mathtt{NONE}$, a learnable parameter), and (iii) the state from the last time step $\mathbf{s}_{t-1}^{\mathsf{RE}}$, formally represented as follows:

$$\mathbf{s}_t^{\mathsf{RE}} = f_{\mathsf{RE}}(\mathbf{W}_{\mathcal{S}_{\mathsf{RE}}}[\mathbf{h}_t; \mathbf{a}_t^{\mathsf{RE}}; \mathbf{s}_{t-1}^{\mathsf{RE}}]) \tag{1}$$

where $f_{\mathsf{RE}}(\cdot)$ is a non-linear function implemented by MLP. To obtain the hidden state $\mathbf{h}_t$, this paper introduces a sequence Bi-LSTM over the current input word embedding $\mathbf{w}_t$, character embedding $\mathbf{c}_t$, and position embedding $\mathbf{p}_t$:

$$\begin{aligned} \overrightarrow{\mathbf{h}_t} &= \overrightarrow{\mathtt{LSTM}}(\overrightarrow{\mathbf{h}_{t-1}}, \mathbf{w}_t, \mathbf{c}_t, \mathbf{p}_t) \\ \overleftarrow{\mathbf{h}_t} &= \overleftarrow{\mathtt{LSTM}}(\overleftarrow{\mathbf{h}_{t+1}}, \mathbf{w}_t, \mathbf{c}_t, \mathbf{p}_t) \\ \mathbf{h}_t &= [\overrightarrow{\mathbf{h}_t}; \overleftarrow{\mathbf{h}_t}] \end{aligned} \tag{2}$$

Policy: The stochastic policy for relation detection $\pi_{\mathsf{RE}} : \mathcal{S}_{\mathsf{RE}} \rightarrow \mathcal{A}_{\mathsf{RE}}$, which specifies a probability distribution over actions:

$$a_t^{\mathsf{RE}} \sim \pi_{\mathsf{RE}}(a_t^{\mathsf{RE}} | s_t^{\mathsf{RE}}) = SoftMax(\mathbf{W}_{\pi_{\mathsf{RE}}} \mathbf{s}_t^{\mathsf{RE}}) \tag{3}$$

Reward: The environment provides intermediate reward r_t^{RE} to estimate the future return when executing action a_t^{RE}. The reward is computed as follows:

$$r_t^{\mathsf{RE}} = \begin{cases} 1, & a_t^{\mathsf{RE}} \text{ in } s, \\ 0, & a_t^{\mathsf{RE}} = \texttt{None}, \\ \text{-1}, & a_t^{\mathsf{RE}} \text{ not in } s. \end{cases} \tag{4}$$

If a_t^{RE} equals to None at certain time step t, the agent transfers to a new relation extraction state at the next time step $t+1$. Otherwise, the entity extraction policy (Sect. 2.2) will execute the entity extraction process, and then distant supervision model and relation calibration RL will be triggered continuously (Sect. 2.3). The state will *not* transfer until the relation calibration task over current option a_t^{RE} is done. Such a semi-Markov process continues until the *last* action about the *last* word $w_{|s|}$ of current sentence s is sampled. Finally, a final reward r_*^{RE} is obtained to measure the sentence-level relation extraction performance that the relation extraction RL policy π_{RE} detects, which is obtained by the weighted harmonic mean of precision and recall in terms of the relations in given sentence s. The final reward r_*^{RE} is defined as weighted harmonic mean, as follows:

$$r_*^{\mathsf{RE}} = \frac{(1+\beta^2) \cdot Prec_s \cdot Rec_s}{\beta^2 \cdot Prec_s + Rec_s} \tag{5}$$

Wherein, notation $Prec_s$ and Rec_s indicate the precision value and recall value respectively, computed over the current sentence s.

2.2 Entity Extraction (EE) with RL

Once the relation extraction policy has predicted a non-NONE relation-type (Sect. 2.1), the **E**ntity **E**xtraction (**EE**) RL policy π_{EE} will extract the participating entities for the corresponding relation. Note that, to make the predicted relation-type accessible in the entity extraction process, the *latest* action a_{t*}^{RE} from the relation extraction RL is taken as additional input throughout the entity extraction process in this section.

Action: The action at each time step t is to assign an entity-label to the current word. The action space, i.e., entity-label space $\mathcal{A}_{\mathsf{EE}} = (\{\mathtt{H}, \mathtt{T}, \mathtt{O}\}) \times (\{\mathtt{B}, \mathtt{I}\}) \bigcup \{\mathtt{N}\}$, where in label H represents the participating **h**ead entity, label T for the **t**ail one, label O for the entities that are not associated with the predicted relation-type a_{t*}^{RE} from the aforementioned relation extraction RL (details in Sect. 2.1), and N for the words which are **n**ot the entity-mentions. Note that, the same entity mention may be assigned with different H/T/O labels according to different relation-types concerned at the moment. In this way, the proposed model could deal with the problem of the *overlapping* relation-types. Moreover, label B and label I are used to indicate the beginning word and the inside word(s) of an entity, respectively.

State: The entity extraction RL state $s_t^{\mathsf{EE}} \in \mathcal{S}_{\mathsf{EE}}$ is represented by four parts as follows: (i) the hidden state $\mathbf{h}_t$ of current word embedding $\mathbf{w}_t$, (ii) the entity label vector $\mathbf{a}_t^{\mathsf{EE}}$ which is a learnable embedding of a_t^{EE}, (iii) the previous state $\mathbf{s}_{t-1}^{\mathsf{EE}}$ from previous time step $t-1$, and finally (iv) the relational state representation assigned to the *latest* state $\mathbf{s}_{t*}^{\mathsf{RE}}$ in Eq. (1), as follows:

$$\mathbf{s}_t^{\mathsf{EE}} = f_{\mathsf{EE}}(\mathbf{W}_{\mathcal{S}_{\mathsf{EE}}}[\mathbf{h}_t; \mathbf{a}_t^{\mathsf{EE}}; \mathbf{s}_{t-1}^{\mathsf{EE}}; \mathbf{s}_{t*}^{\mathsf{RE}}]) \tag{6}$$

Wherein $\mathbf{h}_t$ is the hidden state obtained from the Bi-LSTM module similar to Eq. (2), and $f_{\mathsf{EE}}(\cdot)$ is the non-linear function implemented by MLP.

Policy: The stochastic policy for entity extraction RL $\pi_{\mathsf{EE}} : \mathcal{S}_{\mathsf{EE}} \rightarrow \mathcal{A}_{\mathsf{EE}}$ outputs an action distribution given state $\mathbf{s}_t^{\mathsf{EE}}$ and the previous relation extraction RL action a_{t*}^{RE} that launches the current entity extraction process, as follows:

$$a_t^{\mathsf{EE}} \sim \pi_{\mathsf{RE}}(a_t^{\mathsf{EE}}|\mathbf{s}_t^{\mathsf{EE}}; a_{t*}^{\mathsf{RE}}) = \mathrm{SoftMax}(\mathbf{W}_{\pi_{\mathsf{EE}}}[a_{t*}^{\mathsf{RE}}; \cdot]\mathbf{s}_t^{\mathsf{EE}}) \tag{7}$$

Obviously, the aforementioned equation reveals the interactive between relation extraction process (Sect. 2.1) and entity extraction process (described in this section). Wherein, $\mathbf{W}_{\pi_{\mathsf{EE}}}$ represents an array of $|\mathrm{R}|$ matrices.

Reward: Given the relation-type a_{t*}^{RE}, the entity label for each word can be easily obtained by sampling actions from the policy. Therefore, an immediate reward r_t^{EE} is provided when the action a_t^{EE} is sampled by simply measuring the prediction error over gold-standard annotation, as follows:

$$r_t^{\mathsf{EE}} = \psi(l_t(a_{t*}^{\mathsf{RE}})) \cdot \mathbf{sgn}(a_t^{\mathsf{EE}} = l_t(a_{t*}^{\mathsf{RE}})) \tag{8}$$

Where in $\mathbf{sgn}(\cdot)$ indicates the sign function, and $l_t(a_{t*}^{\mathsf{RE}})$ is the ground-truth entity label conditioned on the predicted relation-type a_{t*}^{RE} from the relation extraction process described in Sect. 2.1. Besides, $\psi(\cdot)$ is a bias weight for down-weighing non-entity label (respect to label N), defined as follows:

$$l_t(a_{t*}^{\mathsf{RE}}) = \begin{cases} 1, & a_{t*}^{\mathsf{RE}} \neq \mathtt{N}, \\ \eta, & a_{t*}^{\mathsf{RE}} = \mathtt{N}. \end{cases} \tag{9}$$

Intuitively, the smaller threshold η leads to less reward on words that are not entities. In this manner, the model avoids to learn a *trivial* policy that predicts all words as N (non-entity words). When all the actions are sampled, an additional *final* reward r_*^{EE} is computed. If *all* the entity labels are predicted correctly, then the agent receives +1 reward, otherwise -1.

2.3 Relation Calibration (RC) with Distant Supervision

With efforts above, entities corresponding to the relation-type a_{t*}^{RE} are obtained by entity extraction RL policy in last Sect. 2.2. To release the bidirectionally interaction between relation extraction process and entity extraction process, this section aims at inversely re-detecting the relation based on these entities, via **D**istant **S**upervision (**DS**) mechanisms, and finally calibrating the relation by comparing it with the relation generated in former relation extraction RL (described in Sect. 2.1).

Note that, any distantly-supervised models for relation extraction [18,21,27], could be adopted here, on the condition that the head entity and tail entity in the current sentence s have been provided in last Sect. 2.2. For simplicity, [18] is utilized here, with input of $\{a_1^{\mathsf{EE}}, a_2^{\mathsf{EE}}, \cdots, a_{|s|}^{\mathsf{EE}}\}$, and it generate a relation $a_{t*}^{\mathsf{DS}} \in \{\mathrm{R} \bigcup \mathtt{None}\}$. Intuitively, we would like to compare the a_{t*}^{RE} released in relation extraction RL (Sect. 2.1) and the a_{t*}^{DS} provided by distant supervision model in this section, to verify the relation's *consistency*. Because, this paper have decoded the relation a_{t*}^{RE} into the corresponding entities in Sect. 2.2, and this section transfers these entities back to a relation a_{t*}^{DS} again. If the proposed model is robust enough, a_{t*}^{DS} should be *equal to* its ancestry a_{t*}^{RE}. Hence, we could add another constraint or reward along this mentality. Therefore, we define a score function, as follows:

$$r_t^{\mathsf{RC}} = \begin{cases} +1, & a_{t*}^{\mathsf{DS}} = a_{t*}^{\mathsf{RE}}, \\ \text{-1}, & a_{t*}^{\mathsf{DS}} \neq a_{t*}^{\mathsf{RE}}. \end{cases} \tag{10}$$

If inconsistency occurs, we argue that it is necessary to take right actions to remove false-positive samples according to the value of r_t^{RC}.

Following [21], this study casts the noisy data removing as a RL problem and aims at obtaining an agent that can take right actions to remove false positive sentences. The agent obtained will interact with the external environment and updates its parameters based on the reward r_t^{RC} mentioned above. The definitions of this **R**elation **C**alibration (**RC**) RL model are elaborated as follows.

Action: There are two actions in relation calibration RL: *retaining* and *removing*, i.e., $\mathcal{A}_{\mathsf{RC}} = \{\mathtt{Retain}, \mathtt{Remove}\}$. For each sentence of the training dataset, the agent takes an action a_t^{RC} to determine whether it should be *removed or not.* The value of the action a_t^{RC} is obtained from the policy network π_{RC}.

State: The relation calibration RL state $s_t^{\mathsf{RC}} \in \mathcal{S}_{\mathsf{RC}}$ includes the information of the current sentence and the sentences that have been removed in early states. To represent the state s^{RC} as a continuous vector, this paper utilizes: (i) word embedding, (ii) character embedding, and (iii) position embedding, to convert the sentences into vectors. With the sentence vectors, the current state $\mathbf{s}_t^{\mathsf{RC}}$ is concatenated by the vector of the current input sentence and the averaged vector of the removed sentences.

Policy: For the relation calibration RL agent, it is formulated as a policy network $\pi_{\mathsf{RC}} : \mathcal{S}_{\mathsf{RC}} \rightarrow \mathcal{A}_{\mathsf{RC}}$. The policy network is used to indicate whether a sentence should be *removed or not* according to the state. Thus, it is a binary classifier, and this study adopts a simple CNN model as the policy network which was used as a text classifier in previous researches [12]: After the feature map was obtained, a max-overtime pooling operation [37] is applied to capture the most import feature; Finally, these features are passed to a fully connected layer to generate the probability distribution over labels.

Reward: The goal of the relation calibration RL model is to filter out the noisy data from the training dataset, resulting in better performance consistency between a_{t*}^{DS} and a_{t*}^{RC} trained on the clean dataset. Therefore, the reward is calculated through the performance consistency measurement, and we intuitively select the r_t^{RC} (E.g. (10)) as the evaluation criterion to reflect the comprehensive performance consistency of relation extraction RL (details in Sect. 2.1) and distantly-supervised model (in this section). In this paper, we use a simple set {+1,-1} as a reward because the reward is used to label the removing dataset. If the reward is +1, then the removed dataset is labeled as removing; otherwise, the removed dataset is labeled as retraining.

2.4 RL Policy Learning

To optimize the relation extraction (RE) RL policy (Sect. 2.1), we aim to maximize the expected cumulative rewards at each time step t as the agent samples trajectories following the relation extraction RL π_{RE}, which can be computed discount factor $\gamma \in [0, 1)$, as follows:

$$J(\pi_{\mathsf{RE}}, t) = \mathbb{E}_{s^{\mathsf{RE}}, a^{\mathsf{RE}}, r^{\mathsf{RE}} \sim \pi_{\mathsf{RE}}(a^{\mathsf{RE}}|s^{\mathsf{RE}})}[\sum_{i=t}^{|s|} \gamma^{i-t} \cdot r_i^{\mathsf{RE}}] \tag{11}$$

Where in the whole sampling process π_{RE} takes T time steps before it terminates. Similarly, we learn the entity extraction (EE) RL policy (Sect. 2.2) by maximizing the expected cumulative rewards from the entity extraction task over action a_{t*}^{RE} when the agent samples along entity extraction policy $\pi_{\mathsf{EE}}(\cdot; a_{t*}^{\mathsf{RE}})$ at time step t:

$$\begin{aligned} J(\pi_{\mathsf{EE}}, t; a_{t*}^{\mathsf{EE}}) =& \mathbb{E}_{s^{\mathsf{EE}}, a^{\mathsf{EE}}, r^{\mathsf{EE}} \sim \pi_{\mathsf{EE}}(a^{\mathsf{EE}}|s^{\mathsf{EE}}; a_{t*}^{\mathsf{RE}})} \\ & [\sum_{i=t}^{|s|} \gamma_{i-t} \cdot r_i^{\mathsf{EE}}] \end{aligned} \tag{12}$$

The optimization procedure of relation calibration RL (Sect. 2.3) is a continuous interaction with relation extraction RL and distantly-supervised model, which receives reward from their performance consistency to guide the all agents' training. Practically, the detail of the training process is depicted as follows. The experience rollout dataset Ω is composed of two parts: *positive* (clean) samples and *negative* (noisy) samples, i.e., $\Omega = \Omega_{\mathsf{pos}} \cup \Omega_{\mathsf{neg}}$. Before the training procedure, we assume that *all* the samples in the original training dataset are positive. Therefore, the positive dataset Ω_{pos} is equal to the original dataset Ω, and the negative dataset Ω_{neg} is initialized as zero, i.e., $\varnothing$.

The relation calibration RL agent's job is to filter out the negative samples in the original dataset Ω. During the training procedure, in each epoch, the agent removes a noisy set Ω' from the original training dataset according to the policy π_{RC}. Because the set Ω' is regarded as a negative dataset, then we obtain a new positive set and a negative sample dataset, according to $\Omega_{\mathsf{pos}} = \Omega_{\mathsf{pos}} - \Omega'$ and $\Omega_{\mathsf{neg}} = \Omega_{\mathsf{neg}} + \Omega'$, respectively. Then we utilize positive data Ω_{pos} to train: (i) the relation extraction RL policy π_{RE} (in Sect. 2.1); (ii) entity extraction RL policy π_{EE} (in Sect. 2.2), and (iii) the distantly-supervised model (in Sect. 2.3). When these models are trained to converge, we use the validation set to measure their performance. The relation consistency between a_{t*}^{RE} and a_{t*}^{DS} (Eq. 10) is calculated and used to calculate the reward for relation calibration RL, which will be used to train a more robust agent (details in Sect. 2.3).

As discussed above, during training policy π_{RC}, a fixed number of sentences with the lowest scores will be filtered out from the training dataset Ω_{pos} in every epoch based on the scores predicted by the RL agent. The remained clean parts Ω_{pos} in different epochs are the determinant of relation consistency: (i) If the consistency occurs (i.e., $a_{t*}^{\mathsf{DS}} = a_{t*}^{\mathsf{RE}}$) in epoch i, it means that the agent takes reasonable actions in epoch i to remove the negative samples. In other words, the removed dataset should be labeled as removing which means to be removed. (ii) Conversely, if the inconsistency occurs (i.e., $a_{t*}^{\mathsf{DS}} \neq a_{t*}^{\mathsf{RE}}$) in epoch i, it reflects the agent takes unreasonable actions to removed data and the removed data should be labeled as retraining. Therefore, we use the removed dataset Ω' to retrain the policy in epoch i where the label of the dataset Ω' comes from the reward r_t^{RC} received by the agent: If the agent receives a reward with a value of $+1$, the dataset is labeled as removing, and if the agent received a reward with a value of -1, the dataset is labeled as retaining. The optimization objective for training relation calibration RL policy, can be formulated as follows:

$$J(\pi_{\mathsf{RC}}) = \mathbb{E}_{s^{\mathsf{RC}}, a^{\mathsf{RC}}, r^{\mathsf{RC}} \sim \pi_{\mathsf{RC}}(a^{\mathsf{RC}} | s^{\mathsf{RC}}; a_{t*}^{\mathsf{RE}}, a_{t*}^{\mathsf{DS}})} \left[\sum^{\Omega'} \gamma \cdot r^{\mathsf{RC}}\right] \tag{13}$$

2.5 Data Augmentation for RL

While recent achievements are truly impressive, RL is notoriously plagued with poor *data-efficiency* and *generalization* capabilities. [14] proposed a plug-and-play module that enhanced RL algorithm with augmented data for image-based task.

Inspired by the impact of data augmentation in computer vision field, we present a technique to incorporate data augmentations on input observations for RL pipelines in aforementioned sections. We investigate the utility of data augmentations in model-free RL by processing text observations with stochastic augmentations before passing them to the agent for training. Through this module, we ensure that the agent is learning on multiple views (or augmentations) of the same input. This allows the agent to improve on two key capabilities mentioned above: (i) data-efficiency: learning to quickly master the task at hand with drastically fewer experience rollouts; (ii) generalization: improving transfer to unseen tasks or levels simply by training on more diversely augmented samples. More importantly, through this mechanism, we present the extensive study of the use of data augmentation techniques for RL with *no* changes to the underlying RL algorithm and *no* additional assumptions about the domain other than the knowledge that the RL agent operates from text-based observations.

When applying data augmentation module to aforementioned RLs (i.e., relation extraction RL in Sect. 2.1, entity extraction RL in Sect. 2.2 and relation calibration RL in Sect. 2.3), our data augmentations are applied to the observation passed to the corresponding policies, i.e., π_{RE}, π_{EE} and π_{RC}.

During training, we sample observations from either a replay buffer or a recent trajectory, and augment the texts within the mini-batch. Crucially, augmentations are applied randomly across the batch but *consistently* across time. This enables the augmentation to retain *temporal* information present across time. Across our experiments, we investigate and ablate the following types of data augmentations:

(i) **Synonym:** replaces a random word from the input with its synonym [1,11,13,22].
(ii) **Edit-Distance:** extracts a random word and deliberately misspelling this word [3,6,15].
(iii) **Insertion:** determines the most important words in the input and then used heuristics to generate perturbed inputs by adding important words [10].
(iv) **Deletion:** Another variant of "Insertion" where instead of adding important words, the determined important words deleted [20,25].

Note that, these types of data augmentations chosen in this paper, are also inspired by the research about adversarial examples and evaluations towards NLP model's robustness [11,20]. Obviously, other kinds of data augmentations, could be utilized here according to the flexibility of the proposed model.

3 Experiments

3.1 Datasets and Metrics

In this section, we conduct experiments to evaluate our model on two public datasets NYT [24] and WebNLG [7]. NYT dataset was originally produced by a distant supervision method. It consists of 1.18M sentences with 24 predefined relation-types. WebNLG dataset was created by Natural Language Generation (NLG) tasks and adapted by [33] for relational triple extraction task. It contains 246 predefined relation classes. For a fair comparison, we directly use the preprocessed datasets provided by [33]. For both datasets, we follow the evaluation setting used in previous works. A triple (e_h, r, e_t) is regarded as correct if the relation type and the two corresponding entities are all correct. We report Precision, Recall and F1-score for all the compared models. The statistics of the datasets are summarized in Table 1. For each dataset, we randomly chose 0.5% data from the training set for validation.

Table 1. Distribution of splits on dataset NYT and dataset WebNLG.

Dataset	#Train	#Dev	#Test
NYT	56,195	5,000	5,000
WebNLG	5,019	500	703

3.2 Baselines

The baselines in this research included four categories: pipeline models, joint learning models, tagging scheme models, and RL-based models.

(i) The chosen pipeline models are **FCM** [12] and **LINE** [8]. **FCM** is a compositional embedding model by combining hand-crafted features with learned word embedding for relation extraction. **LINE** is a network embedding method which can embed very large information networks into low-dimensional vectors. Both of them obtain the NER results by CoType [23], and then the results are fed into the two models to predict the relation-type.

(ii) The joint learning models used in this research include feature-based methods (**DS-Joint** [31], **MultiR** [9] and **CoType** [23]), and neural-based methods (**SPTree** [16] and **CopyR** [33]). **DS-Joint** is an incremental joint framework which extracts entities and relations based on structured perceptron and beam-search. **MultiR** is a joint extracting approach for multi-instance learning with overlapping relations. **CoType** extracts entities and relations by jointly embedding entity mentions, relation mentions, text features, and type labels into two meaningful representations. **SPTree** is a joint learning model that represents both word sequence and dependency tree structures using bidirectional sequential and tree-structured LSTM-RNNs. **CopyR** is a Seq2Seq learning framework with a copy mechanism for joint extraction.

(iii) When comparing with the tagging mechanism, we choose **Tagging-BiLSTM** [35] and **Tagging-Graph** [29] as baselines. **Tagging-BiLSTM** gets the context representation of the input sentences through a BiLSTM network and uses an LSTM network to decode the tag sequences. The **Tagging-Graph** converts the joint task into a directed graph by designing a novel graph scheme.

(iv) When comparing with the RL based models, we choose **HRL** [28], **JRL** [37] and **Seq2SeqRL** [32]. **HRL** is a hierarchical reinforcement learning framework which decomposes the whole extraction process into a hierarchy of two-level RL policies for relation extraction and entity extraction, respectively. **JRL** consists of two components: a joint network (extracts entities and relations simultaneously) and a reinforcement learning agent (refines the training dataset for anti-noise). **Seq2SeqRL** applies RL into a sequence-to-sequence model to take the extraction order into consideration.

3.3 Experimental Settings

All hyper-parameters are tuned on the validation set. The dimension of all vectors in Eq. (1), Eq. (2) and Eq. (6) is 300. The word vectors are initialized using Word2Vec vectors [17] and are updated during training. Both relation-type vectors and entity label vectors (in Eq. (1) and Eq. (6)) are initialized randomly. The learning rate is $2e^{-4}$ for both datasets, the mini-batch size is 16, $\eta = 0.1$ in Eq. (9), $\beta = 0.9$ in Eq. (5), and the discount factor γ in Sect. 2.4 is set as 0.95. Policy gradient methods [26] with REINFORCE algorithm [30] are used here to optimize both all the policies (discussed in Sect. 2.4).

Table 2. Performance comparison of different models on the benchmark datasets. Average results over 5 runs are reported. The best performance is bold-typed.

Model	NYT			WebNLG		
	Prec	Recall	F1	Prec	Recall	F1
FCM [12]	–	–	–	0.472	0.072	0.124
LINE [8]	–	–	–	0.286	0.153	0.193
MultiR [9]	-	-	-	0.289	0.152	0.193
DS-Joint [31]	–	–	–	0.490	0.119	0.189
CoType [23]	–	–	–	0.423	0.175	0.241
SPTree [16]	0.492	0.557	0.496	0.414	0.339	0.357
CopyR [33]	0.569	0.452	0.483	0.479	0.275	0.338
Tagging-BiLSTM [35]	0.624	0.317	0.408	0.525	0.193	0.276
Tagging-Graph [29]	–	–	–	0.528	0.194	0.277
HRL [28]	0.714	0.586	0.616	0.601	0.357	0.432
JRL [37]	0.691	0.549	0.612	0.581	0.334	0.410
Seq2SeqRL [32]	**0.779**	0.672	0.690	0.633	0.599	0.587
BIRL (Ours)	0.756	**0.706**	**0.697**	**0.660**	**0.636**	**0.617**

3.4 Performance Comparison

We now show the results on NYT and WebNLG datasets in Table 2. It can be seen that the proposed model consistently outperforms all previous models in most cases. Especially even the strong baseline models **HRL** and **JRL** significantly surpass **Tagging-BiLSTM** and **CopyR**, revealing the superiority of RL-based methods over encoder-decoder based methods. Compared with **HRL** and **JRL**, the proposed **BIRL** improves the F1 score by 13.04% and 13.85% on NYT dataset, respectively. Compared with **Seq2SeqRL**, our **BIRL** improves the F1 score by 5.12% on WebNLG dataset. To notice that, **SPTree** shows a low F1 performance of 0.496 and 0.357 on the two datasets. As discussed above, previous models *cannot* address the cases where more than one relations exist between two entities. It may be the main reason for the low performance. We also find that **BIRL** has significantly outperformed **Tagging-BiLSTM** and **Tagging-Graph** on the two datasets. This improvement proves the effectiveness of the beneficial interaction between relation extraction process and entity extraction process used in **BIRL**, which is also adopted in **HRL**. However, **BIRL** leverages bidirectional interaction rather than one-way interaction used in **HRL**, which capture more connections between relation extraction process and entity extraction process.

3.5 Ablation Study

Table 3. F1 performance on different ablation models.

Models	NYT	WebNLG
BIRL	0.697	0.617
-RC	0.647	0.454
-DA	0.659	0.462
-Synonym	0.665	0.467
-Edit-Distance	0.664	0.466
-Insertion	0.681	0.598
-Deletion	0.695	0.610
-ALL	0.617	0.430

In this section, we perform the ablation study on the proposed **BIRL**. The ablated models are formed by (i) removing different types of data augmentations (described in Sect. 2.5); and (ii) removing relation calibration RL mechanism (described in Sect. 2.3). The results are shown in Table 3. Wherein, "-DA" indicates the ablated model which removes **d**ata **a**ugmentation module holistically, and "-RC" indicates the ablated model which removes **r**elation **c**alibration mechanism, and "-ALL" denotes the condition that removes both of them.

We find that the performance of **BIRL** deteriorates as we remove different kinds of data augmentations. By considering data augmentation and relation calibration (respect to bidirectional interaction), **BIRL** achieves the best F1 performance in WebNLG and NYT datasets. The F1 performances have been significantly improved compared to strong baselines such as **HRL** and **Seq2SeqRL**. However, the ablated models are defeated by these baselines, and especially "-ALL" release the most significant performance drops. The performance improvement achieved by leveraging data augmentation, proves the effectiveness of the multiple modeling views in the proposed model. Specifically, "-RC" underperforms relative to **BIRL** on both datasets, which making the proposed

Table 4. Case study for the proposed model and the comparative models.

Case	Model	Result
#1: A cult of victimology arose and was happily exploited by clever radicals among Europes Muslims, especially certain religious leaders like Imam Ahmad Abu Laban in Denmark and Mullah Krekar in Norway.	**Golden**	(Europe, contains, Denmark) (Europe, contains, Norway)
	HRL	(Europe, contains, Denmark)√ (Europe, contains, Norway)√
	Seq2SeqRL	(Europe, contains, Denmark)√ (Europe, contains, Norway)√
	BIRL (Ours)	(Europe, contains, Denmark)√ (Europe, contains, Norway)√
#2: Scott (No rating, 75 min) Engulfed by nightmares, blackouts and the anxieties of the age, a Texas woman flees homeland insecurity for a New York vision quest in this acute, resourceful and bracingly ambitious debut film.	**Golden**	(York, contains, Scott)
	HRL	(York, contains, Scott)√
	Seq2SeqRL	(York, contains, Texas)×
	BIRL (Ours)	(York, contains, Scott)√
#3: For as long as Stephen Harper, is prime minister of Canada, I vow to send him every two weeks, mailed on a Monday, a book that has been known to expand stillness.	**Golden**	(Stephen_Harper, nationality, Canada)
	HRL	(Stephen_Harper, place_lived, Canada) ×
	Seq2SeqRL	(Stephen_Harper, nationality, Canada)√
	BIRL (Ours)	(Stephen_Harper, nationality, Canada) √

model deteriorates to **HRL** [28], suggesting the importance of modeling the relation consistency for performance improvement and robust, as well as interaction of relation extraction procedure and entity extraction procedure.

We observe that, the performance of "-ALL" underperforms relative to both "-RC" and "-DA", indicating the fact that both relation calibration (Sect. 2.3) and data augmentation (Sect. 2.5) contribute to refining the performances, which demonstrates the motivation and hypotheses held in this research. Besides, the performance deterioration is larger by removing relation calibration, comparing to removing data augmentation. This observation sufficiently proves that it is the modeling with relation calibration that plays more important role in refining the performances. We also find that "-Insertion" and "-Deletion" outperform relative to "-Synonym" and "-Edit-Distance", indicating the fact that synonym and edit-distance play a more important role in data augmentation.

3.6 Case Study

In this subsection, some representative examples are given to illustrate the effectiveness of the RL model. The cases are shown in Table 4. Each case contains four rows, the gold results, the results produced by the proposed model **BIRL**, and the results generated from the comparative baselines (i.e., **HRL** and **Seq2SeqRL**).

From the case #2, we observe that both ours and **HRL** correctly extract the relational triple "(York, /location/location/contains, Scott)". However, **Seq2SeqRL** identifies "Texas" as an entity by error while our **BIRL** correctly extracts the entity "Scott" that involves in the relation "/location/location/contains". This fact suggests that the proposed **BIRL** is capable of leveraging the prediction state of relation extraction to refine its entity extraction, and vice versa, and hence is prone to extract the word which involves the relational triple as an entity. In the case #3, the correct relationship between the two entities is "/people/person/nationality". The **HRL** model, which doesn't refine the training data, trains on the original dataset can identify the two entities correctly, but the relationship between them is mispredicted while the our model with the assistance of relation calibration RL generate the triple correctly.

4 Conclusion

In this paper, we aimed at handing (i) relation consistency, (ii) noisy training data, and (iii) data-efficiency of RL-based model for jointly extracting entities and relations, and propose a novel bidirectional interaction RL model. Especially, the proposed model is capable of leveraging the prediction state of relation extraction to refine its entity extraction, and vice versa, by fully modeling and capturing their interaction.

Acknowledgement. We thank anonymous reviewers for valuable comments. This work is funded by: (i) the National Natural Science Foundation of China (No. U19B2026); (ii) the New Generation of Artificial Intelligence Special Action Project (No. AI20191125008); and (iii) the National Integrated Big Data Center Pilot Project (No. 20500908, 17111001,17111002).

References

1. Alzantot, M., Sharma, Y., Elgohary, A., Ho, B.J., Srivastava, M.B., Chang, K.W.: Generating natural language adversarial examples. ArXiv abs/1804.07998 (2018)

2. Bunescu, R.C., Mooney, R.J.: Subsequence kernels for relation extraction. In: NIPS (2005)
3. Ebrahimi, J., Rao, A., Lowd, D., Dou, D.: Hotflip: white-box adversarial examples for text classification. In: ACL (2018)
4. Feng, J., Huang, M., Zhao, L., Yang, Y., Zhu, X.: Reinforcement learning for relation classification from noisy data. ArXiv abs/1808.08013 (2018)
5. Feng, Y., Zhang, H., Hao, W., Chen, G.: Joint extraction of entities and relations using reinforcement learning and deep learning. Comput. Intell. Neurosci. **2017**, 1–11 (2017). Article ID 7643065
6. Gao, J., Lanchantin, J., Soffa, M.L., Qi, Y.: Black-box generation of adversarial text sequences to evade deep learning classifiers. In: 2018 IEEE Security and Privacy Workshops (SPW), pp. 50–56 (2018)
7. Gardent, C., Shimorina, A., Narayan, S., Perez-Beltrachini, L.: Creating training corpora for nlg micro-planners. In: ACL (2017)
8. Gormley, M.R., Yu, M., Dredze, M.: Improved relation extraction with feature-rich compositional embedding models. In: EMNLP (2015)
9. Hoffmann, R., Zhang, C., Ling, X., Zettlemoyer, L., Weld, D.S.: Knowledge-based weak supervision for information extraction of overlapping relations. In: ACL (2011)
10. Jia, R., Liang, P.: Adversarial examples for evaluating reading comprehension systems. In: EMNLP (2017)
11. Jin, D., Jin, Z., Zhou, J.T., Szolovits, P.: Is bert really robust? a strong baseline for natural language attack on text classification and entailment. arXiv: Computation and Language (2019)
12. Kim, Y.: Convolutional neural networks for sentence classification. In: EMNLP (2014)
13. Kuleshov, V., Thakoor, S., Lau, T., Ermon, S.: Adversarial examples for natural language classification problems (2018)
14. Laskin, M., Lee, K., Stooke, A., Pinto, L., Abbeel, P., Srinivas, A.: Reinforcement learning with augmented data. ArXiv abs/2004.14990 (2020)
15. Li, J., Ji, S., Du, T., Li, B., Wang, T.: Textbugger: generating adversarial text against real-world applications. ArXiv abs/1812.05271 (2018)
16. Li, Q., Ji, H.: Incremental joint extraction of entity mentions and relations. In: ACL (2014)
17. Mikolov, T., Sutskever, I., Chen, K., Corrado, G., Dean, J.: Distributed representations of words and phrases and their compositionality. Adv. Neural Inf. Process. Syst. **26**, 3111–3119 (2013)
18. Mintz, M., Bills, S., Snow, R., Jurafsky, D.: Distant supervision for relation extraction without labeled data. In: Joint Conference of the Meeting of the Acl and the International Joint Conference on Natural Language Processing of the AFNLP: volume (2009)
19. Miwa, M., Bansal, M.: End-to-end relation extraction using lstms on sequences and tree structures. ArXiv abs/1601.00770 (2016)
20. Morris, J., Lifland, E., Lanchantin, J., Ji, Y., Qi, Y.: Reevaluating adversarial examples in natural language. ArXiv abs/2004.14174 (2020)
21. Qin, P., Xu, W., Wang, W.Y.: Robust distant supervision relation extraction via deep reinforcement learning. In: ACL (2018)
22. Ren, S., Deng, Y., He, K., Che, W.: Generating natural language adversarial examples through probability weighted word saliency. In: ACL (2019)

23. Ren, X., et al.: Cotype: joint extraction of typed entities and relations with knowledge bases. In: Proceedings of the 26th International Conference on World Wide Web (2017)
24. Riedel, S., Yao, L., McCallum, A.: Modeling relations and their mentions without labeled text. In: ECML/PKDD (2010)
25. Samanta, S., Mehta, S.: Towards crafting text adversarial samples. ArXiv abs/1707.02812 (2017)
26. Sutton, R.S., McAllester, D.A., Singh, S.P., Mansour, Y.: Policy gradient methods for reinforcement learning with function approximation. In: NIPS (1999)
27. Takamatsu, S., Sato, I., Nakagawa, H.: Reducing wrong labels in distant supervision for relation extraction. In: ACL (2012)
28. Takanobu, R., Zhang, T., Liu, J., Huang, M.: A hierarchical framework for relation extraction with reinforcement learning. In: AAAI (2018)
29. Wang, S., Zhang, Y., Che, W., Liu, T.: Joint extraction of entities and relations based on a novel graph scheme. In: IJCAI (2018)
30. Williams, R.J.: Simple statistical gradient-following algorithms for connectionist reinforcement learning. Mach. Learn. **8**, 229–256 (1992)
31. Yu, X., Lam, W.: Jointly identifying entities and extracting relations in encyclopedia text via a graphical model approach. In: COLING (2010)
32. Zeng, X., He, S., Zeng, D., Liu, K., Liu, S., Zhao, J.: Learning the extraction order of multiple relational facts in a sentence with reinforcement learning. In: EMNLP/IJCNLP (2019)
33. Zeng, X., Zeng, D., He, S., Liu, K., Zhao, J.: Extracting relational facts by an end-to-end neural model with copy mechanism. In: ACL (2018)
34. Zhang, X., Zhang, J., Zeng, J., Yan, J., Chen, Z., Sui, Z.: Towards accurate distant supervision for relational facts extraction. In: ACL (2013)
35. Zheng, S., Wang, F., Bao, H., Hao, Y., Zhou, P., Xu, B.: Joint extraction of entities and relations based on a novel tagging scheme. ArXiv abs/1706.05075 (2017)
36. Zhou, G., Su, J., Zhang, J., Zhang, M.: Exploring various knowledge in relation extraction. In: ACL (2005)
37. Zhou, X., Liu, L., Luo, X., Chen, H., Qing, L., He, X.: Joint entity and relation extraction based on reinforcement learning. IEEE Access **7**, 125688–125699 (2019)

DFILAN: Domain-Based Feature Interactions Learning via Attention Networks for CTR Prediction

Yongliang Han[1], Yingyuan Xiao[1(✉)], Hongya Wang[2], Wenguang Zheng[1(✉)], and Ke Zhu[1]

[1] Engineering Research Center of Learning-Based Intelligent System, Ministry of Education, Tianjin University of Technology, Tianjin 300384, China
{yyxiao,wenguangz}@tjut.edu.cn

[2] College of Computer Science and Technology, Donghua University, Shanghai 201620, China

Abstract. Click-Through Rate (CTR) prediction has become an important part of many enterprise applications, such as recommendation systems and online advertising. In recent years, some models based on deep learning have been applied to the CTR prediction systems. Although the accuracy is improving, the complexity of the model is constantly increasing. In this paper, we propose a novel model called Domain-based Feature Interactions Learning via Attention Networks (DFILAN), which can effectively reduce model complexity and automatically learn the importance of feature interactions. On the one hand, the DFILAN divides the input features into several domains to reduce the time complexity of the model in the interaction process. On the other hand, the DFILAN interacts at the embedding vector dimension level to improve the feature interactions effect and leverages the attention network to automatically learn the importance of feature interactions. Extensive experiments conducted on the two public datasets show that DFILAN is effective and outperforms the state-of-the-art models.

Keywords: Click-through rate · Domain · Feature interactions · Attention networks

1 Introduction

Advertising revenue is the main source of income for Internet companies. The core technique of advertising business is the CTR prediction. CTR prediction leverages the interactive information between users and ads to predict whether the user clicks on the advertisement. Due to the CTR prediction can increase advertising revenue, the research of CTR prediction has the great significance. Many models have been applied to the CTR prediction such as Matrix factorization (MF) [1], Factorization Machine (FM) [2], Gradient Boosting Decision Tree (GBDT) [3], and FM-based models [4,5,17]. Due to the high fitting ability

C. S. Jensen et al. (Eds.): DASFAA 2021, LNCS 12682, pp. 500–515, 2021.
https://doi.org/10.1007/978-3-030-73197-7_33

of neural networks, many DNN-based models [6–18] are proposed to improve the prediction accuracy, and some models [19–23] use attention networks to automatically learn the importance of feature interactions. The combination of neural networks and attention networks has been a research trend in the CTR field. Although the current work has achieved some results, the CTR prediction model based on deep learning still has the following challenges: 1) The time complexity of models become larger and larger, and the number of parameters generated during the model training process increase sharply. 2) The feature interactions method is too simple to fully learn the relationship between features. Therefore, this article mainly studies how to improve the CTR prediction model based on deep learning.

In this paper, we propose a domain-based feature interactions model, which is called DFILAN. The correlation between features and label is different, and the interaction of features with a large correlation gap will reduce the impact of the strong correlation feature on the label. For example, the interaction between the feature "occupation" and the feature "hobby" will reduce the influence of the "occupation" on the label "income" and increase the influence of the "hobby" on the label "income". Taking this into consideration, we use the concept of domain and divide the input features into several domains according to the correlation between features and the label. In the interaction process, the independent interaction of inter-domain and intra-domain can effectively reduce the complexity of the model. Besides, feature interaction is a core part of the field of CTR prediction and the commonly used interaction method is the inner product or Hadamard product. We propose a new interaction method, called vector-dims product, which performs feature interactions at the embedding vector dimensions level, and the interaction process is more subtle. At the same time, to improve the effectiveness of feature interactions, the attention network [21] is added to the interaction layer and the importance of feature interactions is automatically learned through the attention network.

Our main contributions are listed as follows:

- We leverage the concept of domain and divide the input features into several domains according to the correlation between features and the label, which can greatly reduce the time complexity of the model.
- We propose a new interactive method called vector-dims product at the feature interactions layer, which can get better interactive effects compared to the commonly used interactive method Hadamard product.
- We apply the attention network to automatically learn the importance of feature interactions, which can greatly improve the effectiveness of feature interactions, and combine vector-dims product to achieve better interaction results.
- We conduct extensive experiments on the two public datasets Avazu and MovieLens, and the results show our model DFILAN outperforms the state-of-the-art models.

The rest of this paper is organized as follows. In Sect. 2, we review related works that are relevant to our proposed model. Section 3 describes the details of our DFILAN. The experimental evaluation is discussed in Sect. 4. Section 5 concludes this paper.

2 Related Work

In recent years, many research efforts have been devoted to the CTR prediction problem. In this section, we briefly divide related research work in the CTR prediction into two categories: FM-based CTR models and deep learning-based CTR models.

2.1 FM-based CTR Models

FM [2] is the most commonly used model for CTR prediction tasks. It has lower time complexity and is suitable for large sparse data, but the prediction accuracy is lower. Field-aware Factorization Machine (FFM) [4] introduces the concept of field to improve the degree of feature interactions. Operation-aware Neural Networks (ONN) [17] combines FFM [4] and Product-based Neural Network (PNN) [8] to improve the accuracy of the model prediction. It considers that each feature should learn different embedding vectors for different operations to improve the expression effect of the feature. However, ONN requires very large memory requirements in the process of implementation and will generate a huge number of interactions during the feature interactions process. The complexity of this model is too high to be easily used in Internet companies.

2.2 Deep Learning-Based CTR Models

With the successful application of deep learning in other fields, such as CV [24], NLP [25], many models based on deep learning are applied in CTR prediction systems. Neural Factorization Machines (NFM) [7] uses the second-order interaction of the original embedding as the input of the DNN and applies the Hadamard product as the feature interaction method. Wide & Deep model (WDL) [11] has both the advantages of memory and generalization. The linear structure provides the model with the ability to remember, and at the same time uses the generalization ability of Multilayer Perceptron (MLP) to learn the high-order feature interactions. While preserving the advantages of WDL, DeepFM [12] designs the feature interactions module and applies the vector inner product to obtain the second-order interaction results of the features. Deep & Cross Network (DCN) [13] designs a cross-network to carry out the explicit interaction process between features and the cross-network is simple and effective. xDeepFM [14] combines explicit and implicit feature interactions and proposes Compressed Interaction Network (CIN) to get a better interaction result, but the time complexity is so heavy that it is difficult to apply the model to the actual industry. FLEN [18] uses Field-Wise Bi-Interaction (FwBI) to reduce the time complexity of feature interactions. But it ignores that the input features do not have specific meanings, and the interaction method is too simple to get good results. Deep Interest

Network (DIN) [20] designs the activation unit to learn the interaction weight between a target item and user behavior. Attentional Factorization Machines (AFM) [21] uses the attention network to learn the weight of feature interactions. Relying on neural networks and attention networks have become an important development trend of CTR prediction models. However, the high complexity and multi-parameter problems of deep learning itself have not been effectively solved. How to reduce the complexity of the model while obtaining effective feature interactions to ensure that the accuracy of predictions is improved has become the primary issue for enterprises.

3 Our Proposed Model

In this section, we describe the architecture of the DFILAN as depicted in Fig. 1. The DFILAN model consists of the following parts: input layer, embedding layer, linear unit, interaction layer, MLP component, and prediction layer. The input features are divided into several domains in the input layer, and the embedding layer adopts a sparse representation for input features and embeds the raw feature input into a dense vector. The linear unit, interaction layer and MLP component are designed to obtain feature linear interaction, second-order interaction and advanced interaction respectively. The prediction layer combines all the interactive results to output the prediction score.

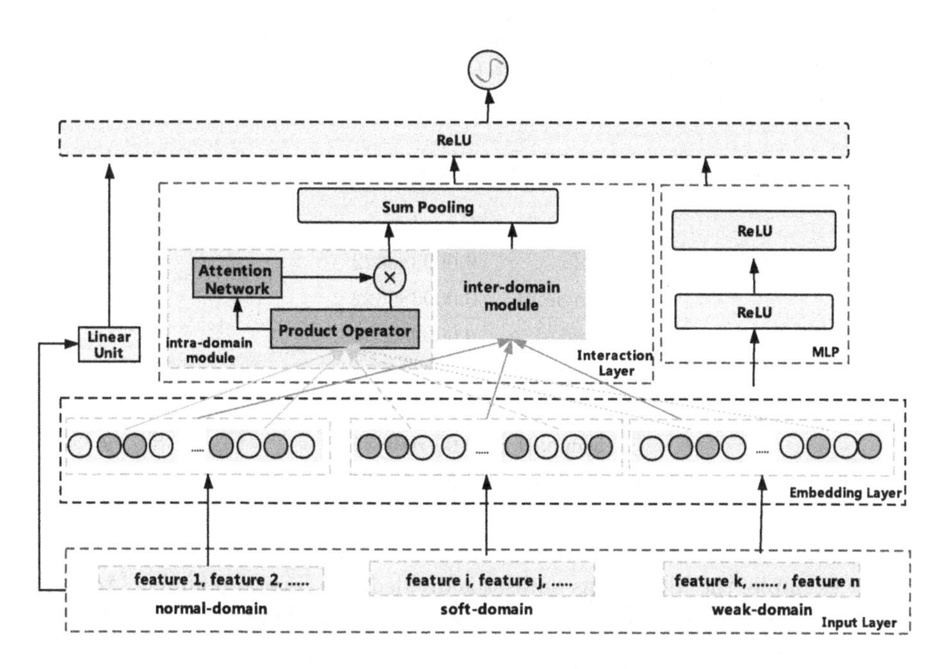

Fig. 1. The architecture of our proposed DFILAN.

3.1 Input Layer

We introduce the concept of domain and divide input features into several domains to reduce the time spent on feature interactions. Domain is the grouping of input features, each domain contains several different features. Assume that the number of input features is n. If the domain is not divided, the number of interactions is $A = n(n-1)/2$. If the domain is divided, the number of domains is m, each domain contains p features, $m * p = n$, and the number of interactions is $B = (np^2(p-1) + n(np))/2p^2$. Only when $p = 1$ and $p = n$, $A = B$, otherwise $A > B$. Therefore, dividing the input features into several domains, and the independent interactions between inter-domain and intra-domain can reduce the interaction time.

We regard that the interaction between features with large differences in correlation will weaken the influence of the features on the predicted value of the label. Therefore, according to the correlation, the input features are divided into several domains. In the CTR field, the input features are mainly discrete data of different types. For discrete data, the information gain rate is applied to measure the correlation between each feature and label. The formula for information gain rate is defined as:

$$Gain - ratio(label, C) = \frac{Gain(label, C)}{IV(C)} \tag{1}$$

where $Gain(label, C)$ represents the information gain of feature "C" relative to the label, and $IV(C)$ denotes the number of different values of feature "C". The greater the value of $Gain - ratio(label, C)$, the stronger the correlation between feature "C" and the label.

Table 1. The correlation of features and the label on the MovieLens dataset.

Feature	Correlation
title	0.009386822
movie_id	0.009386822
user_id	0.006265287
genres	0.005102952
age	0.001148495
occupation	0.000480635
gender	0.000213156

For example, the MovieLens dataset is divided into several domains. According to Formula 1, the correlation between each feature and label in the MovieLens dataset is obtained, and the features are sorted in descending order according to the correlation. The results are shown in Table 1. The division result of MovieLens dataset is "normal-domain: title, movie_id, user_id; weak-domain: genres, age, occupation, gender".

3.2 Embedding Layer

One-hot encoding is used for the expression of input features. For example, one input instance "[user_id = 3, gender = male, hobby = comedy&rock]" is transformed into a high-dimensional sparse vector via one-hot encoding: "[[0,0,1,...,0], [1,0], [0,1,0,1,...,0]]". After using one-hot encoding to express the input data, $x = [x_1, x_2, ..., x_n]$, n is the number of input features, it will become very sparse. We apply embedding technology to convert high-dimensional sparse input data into low-dimensional dense embedding vectors. The embedding layer is shown in Fig. 2. The result of the embedding layer is a wide concatenated vector:

$$e = [e_1, e_2, \ldots\ldots, e_n] \tag{2}$$

where e_i represents the embedding vector of i-th feature, and each feature embedding vector is D-dimensional. Although the length of the input features may be different, they share an embedding matrix with a size of $n * D$, where D is the embedding dimension.

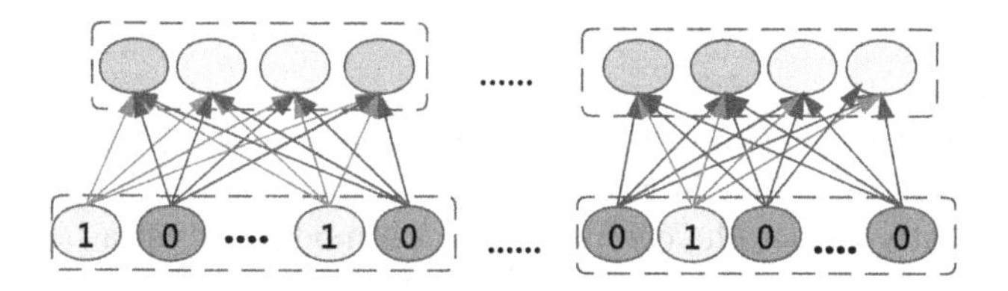

Fig. 2. Illustration of embedding layer with D = 4.

After the embedding vector of the input feature is obtained, the expression result of m domains is defined as:

$$E = [E_1, E_2, \ldots\ldots, E_m] \tag{3}$$

$$E_j = \sum_{i=1}^{p} F(e_i|j) \tag{4}$$

where $j \in [1, m]$, p represents the number of features contained in each domain, $F(e_i|j)$ which means that the i-th feature embedding vector belongs to the j-th domain.

3.3 Interaction Layer

We can clearly find that dividing the input features into different domains can improve the efficiency of feature interactions. After dividing the domains, the feature interactions are divided into two parts: inter-domain interactions and intra-domain interactions.

Inter-domain Interactions Module. The Hadamard product is applied for the inter-domain interactions, and the interactions process is shown in Formula 5, where $O_{inter-domain}$ represents the result of the inter-domain interactions and $\odot$ represents the Hadamard product. The interaction process of the Hadamard product is shown in Formula 6, where $\cdot$ represents the product of two numbers.

$$O_{inter-damain} = \sum_{i=1}^{m}\sum_{j=1}^{m} E_i \odot E_j \tag{5}$$

$$E_i \odot E_j = \sum_{d=1}^{D} E_{i,d} \cdot E_{j,d} \tag{6}$$

In order to get a better interaction effect, the value of m is often set relatively small, generally 3 or 4. As shown in Fig. 1, we divide the n features into 3 domains, namely normal-domain, soft-domain, and weak-domain. Dividing features into several domains can not only reduce the complexity of the model, but also make the model framework more extensible, and other factors such as time or position can be well added. The value of the domain is generally small, so in the interaction layer, we mainly consider the intra-domain interaction process.

Intra-domain Interactions Module. After the division of domains, the number of features in each domain will be much less than that in the input features, which can greatly reduce the complexity of the model. To get a better interaction result within the domain, we design a product operator module to use two different interaction rules to get the interaction result. One is the Hadamard product, another is the vector-dims product.

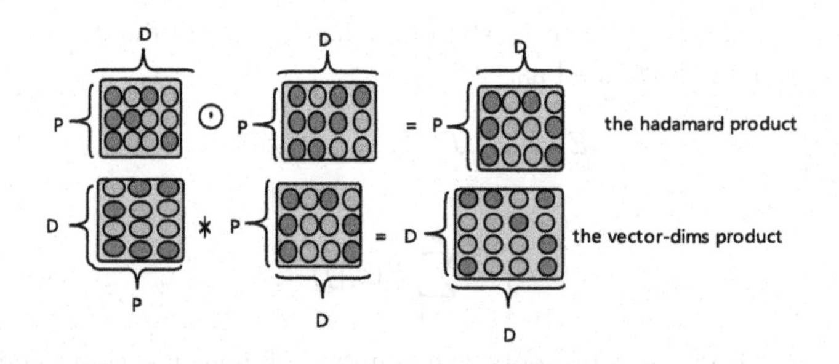

Fig. 3. The different methods to calculate the feature interactions.

a. Hadamard product. The interactions process is shown in Formula 7, where $i \in [1,p], j \in [1,p]$, e_i and e_j represents the embedding vector of the i-th and j-th feature respectively.

$$I_{ij} = e_i \odot e_j \tag{7}$$

As shown in Formula 6, the Hadamard product is the multiplication of the corresponding dimensions of the embedding vector, which takes into account the feature embedding vector as a result of partial interaction, the interaction process is subtle. But it is only interaction in the same dimension of the embedding vector and does not consider the interaction process between the dimensions of the embedding vector. In response to this problem, we combine vector inner product operations to perform feature interactions from the perspective of embedding vector dimensions, called vector-dims product.

b. Vector-dims product. As shown in Fig. 3, there are P features, each feature is D dimensions. We get a $P * D$ two-dimensional matrix and use the vector inner product operation to get a new $D * D$ matrix. In the new matrix, the i-th row and j-th column represents an interactive result of the i-th dimension and the j-th dimension of the embedding vector of P features. Formally, the vector-dims product is defined as:

$$I_{ij} = \sum_{k=1}^{P} e_{ik} e_{kj} \tag{8}$$

Where $i \in [1, D], j \in [1, D]$, e_{jk} represents the i-th dimension of the k-th feature emebedding vector.

Even after the division of domains, there may still be useless feature interactions in a domain. For example, there is such a domain in which there is occupation feature, gender feature, and age feature. As far as we know, the impact of occupation and age is much greater than that of gender. The importance of feature interactions is learned by an attention network [21]. Formally, the attention network is defined as:

$$\alpha'_{ij} = q^T ReLU(W(I_{ij}) + b) \tag{9}$$

$$\alpha_{ij} = softmax(\alpha'_{ij}) = \frac{exp(\alpha'_{ij})}{\sum exp(\alpha'_{ij})} \tag{10}$$

where $W \in \mathbb{R}^{t \times D}, b \in \mathbb{R}^t, q \in \mathbb{R}^t$ are model parameters, and t denotes the hidden layer size of the attention network, which we call "attention unit". The attention scores are normalized through the softmax function. ReLU as the activation function.

The attention network learns different weights for different interaction methods. $O_{intra-domain}$ represents the result of intra-domain interactions. The weight of the interaction vetcor is learned for Hadamard product, $O_{intra-domain} = \sum_{i=1}^{P} \sum_{j=1}^{P} I_{ij} \alpha_{ij}$ and the weight of each dimension in the interaction vector is learned for vector-dims product, $O_{intra-domain} = \sum_{i=1}^{D} \sum_{j=1}^{D} I_{ij} \alpha_{ij}$. Starting from the embedding dimension, the feature interaction process is more subtle. We combine the interaction results of inter-domain and intra-domain as the result of the interaction layer:

$$O_{In} = O_{inter-domain} + O_{intra-domain} \tag{11}$$

Complexity Analysis. It is worth pointing out that our DFILAN is much more memory-efficient. The number of feature interactions is $m(m-1)/2+mp(p-1)/2$, where $m*p = n$, $m << n$, and the number of parameters is $Dt+t$ in the attention networks. Therefore, the DFILAN model complexity of feature interactions can be efficiently computed in $O(m^2 + p^2 + Dt)$. Because the values of D and t are often relatively small, the model complexity is approximately considered as $O(m^2 + p^2)$, which is much smaller than $O(n^2)$.

3.4 Linear Unit and MLP Component

We design a separate linear unit to obtain the first-order interactive results of the input features, $O_{Linear} = w_0 + \sum_{i=1}^{N} w_i x_i$. An MLP is employed to capture non-linear, high-order feature interactions. The input is simply a concatenation of input features embedding vector. $h_0 = concat(e_1, e_2, ..., e_n)$. A stack of fully connected layers is constructed on the input h_0. Formally, the definition of fully connected layers are as follows:

$$h_{l+1} = \sigma(W_l h_l + b_l) \tag{12}$$

Where l denotes the number of hidden layers, W_l, b_l and σ denotes the weight matrix, bias vector and activation function for the l-th layer, respectively. ReLU is applied as the activation function for each layer.

3.5 Prediction Layer

The output vector of the last hidden MLP layer h_l is concatenated with O_{In} and O_{Linear} to form Q, $Q = concat(h_l, O_{In}, O_{Linear})$. Sigmoid activation function is used to make predictions, and the prediction layer is defined as follow:

$$\hat{y} = \frac{1}{1+e^{-(W^T Q)}} \tag{13}$$

Our loss function is cross entropy, which is defined as follows:

$$loss = -\frac{1}{S}\sum_{i=1}^{S}(y_i log(\hat{y}_i) + (1 - y_i) * log(1 - \hat{y}_i)) \tag{14}$$

Where $\hat{y}_i$ is the predicted value, $\hat{y}_i \in (0, 1)$ via the sigmoid funcation, y_i is the ground truth of i-th instance, and S is the total size of samples.

4 Experiments

We call the model using the Hadamard product DFILAN_I, and the model using the vector-dims product DFILAN_II. In this section, extensive experiments are conducted to answer the following questions:

- (RQ1) How does our model DFILAN_I perform as compared to the state-of-the-art methods for CTR prediction?
- (RQ2) Is the DFILAN_II model better than DFILAN_I?
- (RQ3) How do some model hyperparameters affect DFILAN_II?

4.1 Experimental Setup

Dataset. 1) Avazu. The Avazu dataset was originally used for Kaggle competitions. It has 24 fields, and the label of each sample is 0 or 1. We randomly select 1 million pieces of data for the experiment. 2) MovieLens. The MovieLens dataset was originally designed for user rating prediction. It has 7 fields, and the label of each sample is 1 to 5. The label is 3–5 points is set to 1, and the label is 1–2 points is set to 0. We select MovieLens-1M dataset for the experiment.

Evaluation Metric. 1) AUC: Area under ROC evaluation indicators are widely used in classification problems. Its actual meaning is the probability that the positive example score is ranked before the negative example when the model is scored. The upper limit of AUC is 1, the larger the better. 2) Logloss: Logloss is a widely used metric in binary classification to measure the distance between two distributions. The lower limit of Logloss is 0, the smaller the value, the better.

Baseline Methods. We compare DFILAN with DeepFM [12], DCN [13], ONN [17], xDeepFM [14], FLEN [18]. As introduced and discussed in Sect. 2, these models are highly related to our DFILAN. Note that an improvement of 1‰ in AUC is usually regarded as significant for the CTR prediction because it will bring a large increase in a company's revenue if the company has a very large user base.

Implementation Details. We implement all the models with Tensorflow in our experiments. For the division of domains, the number of domains is set to 3 for the Avazu dataset and 2 for the MovieLens dataset. For the embedding layer, the dimension of the embedding vector is set to 8 for the Avazu dataset and 4 for the MovieLens dataset. For all models, Adam [26] is applied as the optimization method with a mini-batch size of 256 for Avazu and MovieLens datasets, and the learning rate is set to 0.0001. For all models, in the MLP module, the depth of the network layer is set to 2, the number of neurons in each layer is 128, and the activation function is ReLU. For the attention network part, the number of attention unit is set to 16 for Avazu and MovieLens datasets.

4.2 Performance Comparison (RQ1)

In this subsection, we compare the proposed DFILAN_I model with the model in baseline methods, and the results are shown in Table 2, Table 3. For simplicity, we assume that the number of network layers and the number of units in each layer of all models are equal.

Where N represents the size of the input features, M represents the number of divided domain, P represents the number of feature in each domain, $M * P = N$ and $M << N$, D represents the size of the embedding dimension, L represents the number of layers in the MLP component, T represents the number of attention units, and H represents the number of neurons in each layer. By observing the experimental results, some conclusions are as follows.

Table 2. The overall performance of models on Avazu and MovieLens datasets.

Model	Avazu		MovieLens	
	Logloss	AUC	Logloss	AUC
DeepFM	0.3989	0.7519	0.3389	0.8381
DCN	0.3966	0.7526	0.3381	0.8386
XDeepFM	0.3953	0.7545	0.3379	0.8388
ONN	0.4129	0.7449	0.353	0.8512
FLEN	0.3945	0.756	0.3341	0.8394
DFILAN_I	**0.3913**	**0.7574**	**0.3332**	**0.8481**

Table 3. The complexity of all models on Avazu and MovieLens datasets.

Model	Model complexity	Parameter size
DeepFM	$O(ND + N^2 + NLH)$	6,918,072
DCN	$O(ND + LN^2 + NLH)$	6,952,408
XDeepFM	$O(ND + L^2N^2 + NLH)$	7,824,584
FLEN	$O(ND + M^2 + P^2 + NLH)$	6,909,888
ONN	$O(N^2D + N^3 + NLH)$	48,878,688
DFILAN_I	$O(ND + M^2 + P^2 + DT + NLH)$	**6,910,208**

- Table 2 shows the AUC and Logloss of all models on the Avazu and MovieLens datasets. Obviously, the DFILAN_I model we proposed has achieved the best performance on both AUC and Logloss. Table 3 shows the model complexity of all models and the number of parameters generated during their training in the Avazu datasets. Obviously, the DFILAN_I model we proposed is dominant in model complexity.
- From Table 2, we can find that compared with the latest FLEN [18] model, the DFILAN_I model we proposed on the Avazu dataset has a 0.32% increase in Logloss and a 0.14% increase in AUC. On the MovieLens dataset, Logloss increased by 0.09% and AUC increased by 0.16%.
- From Table 3 we can find that the DFILAN_I model we proposed has advantages in reducing model complexity. Compared with the xDeepFM [14], the DFILAN_I model can effectively reduce the size of model parameters, thereby reducing the cost of enterprises. Although our DFILAN_I model has larger model parameters than the FLEN [18] model, combined with the performance of the model in Table 2 on AUC, we believe that the slightly larger model parameters are negligible.

4.3 Different Product Operator Comparision (RQ2)

In this subsection, we will compare the DFILAN_I model with the Hadamard product as the interaction rule and the DFILAN_II model with the vector-dims product as the interaction rule. The comparison results are shown in Table 4.

Table 4. The performance of DFILAN_I and DFILAN_II on the Avazu and MovieLens datasets.

Model	Avazu		MovieLens		Parameter size
	Logloss	AUC	Logloss	AUC	
DFILAN_I	0.3913	0.7574	0.3332	0.841	6,910,208
DFILAN_II	**0.3889**	**0.7591**	**0.3316**	**0.8426**	**6,926,336**

From Table 4, we can find that the DFILAN_II model with vector-dims as the interaction rule has better performance on the Avazu and MovieLens datasets. Comparing DFILAN_I, on the Avazu dataset, Logloss increased by 0.24% and AUC increased by 0.17%, while on the MovieLens dataset, Logloss increased by 0.16% and AUC increased by 0.16%. Comparing the parameter sizes of the models on the Avazu dataset, it can be found that the DFILAN_II does not increase too many model parameters, which are within the acceptable range. So we can get the result that the vector-dims product interaction rule we proposed is better than the Hadamard product interaction rule.

4.4 Hyper-parameter Tuning (RQ3)

In this subsection, some hyper-parameter investigations will be conducted in our model. Because the number of attention units is very small, the impact of attention units is not considered. We change the following hyper-parameters: 1) the number of domains; 2) the dimension of embeddings; 3) the number of units per layer in DNN, and the depth of DNN. Unless specially mentioned in our paper, the default parameter of our network is set as Sect. 4.1.

Table 5. The performance of different number of domains on MovieLens dataset.

Number of domains	Logloss	AUC
1	0.3322	0.8425
2	**0.3281**	**0.8447**
3	0.3322	0.8425

Part of the Domains. On the Avazu dataset, we change the domain size from 2 to 7 and on the MovieLens dataset, the domain size increase from 1 to 3. From Table 6, it can be obviouly found that in the process of increasing the domain size, the performance of the DFILAN_II model on the Avazu dataset is always better than the latest model. When the number of domains is 3, the DFILAN_II model has the best performance. From Table 5, it can be clearly found that in the process of increasing the domain, the performance of the DFILAN_II model on the MovieLens dataset is always better than the latest model, when the number of domains is 2, the DFILAN_II model has the best performance. We consider that the MovieLens dataset has too few features, which does not reflect the performance improvement brought by the division of the domain.

Table 6. The performance of different number of domains on Avazu dataset.

Number of domains	Logloss	AUC
2	0.3794	0.7574
3	**0.3784**	**0.7588**
4	0.3806	0.7578
5	0.3787	0.7578
6	0.379	0.7582
7	0.3807	0.7577

Embedding Part. The dimension of the embedding vector is not fixed, and we change the embedding dimension from 4 to 64 on the Avazu and MovieLens datasets. From Table 7, some observations are as follows:

- As the dimension is expanded from 4 to 64, our DFILAN_II obtains a substantial improvement on the MovieLens dataset. When the embedding dimension is 4, the model performance reaches the best.
- When we increase the embedding dimension on the Avazu dataset, the performance will fluctuate, and when the embedding size is 16, the performance reaches the best. We hold that the embedding size will affect whether the information of the feature itself is adequately represented.

Table 7. The performance of different embedding dimensions on Avazu and MovieLens datasets.

Embedding dimensions	Avazu		MovieLens	
	Logloss	AUC	Logloss	AUC
4	0.3814	0.7531	**0.3316**	**0.8427**
8	**0.3811**	**0.7533**	0.3379	0.8382
16	0.3831	0.7527	0.3529	0.8343
32	0.3838	0.7496	0.3588	0.831
64	0.3845	0.7477	0.344	0.8362

MLP Part. We study the impact of different neural units per layer and different depths in the MLP part. We can observe from Fig. 4 that an increasing number of layers improves model performance at the beginning. However, the performance is degraded if the number of layers keeps increasing. This is because an over-complicated model is easy to overfit. It's a good choice that the number of hidden layers is set to 2 for the Avazu dataset and MovieLens dataset. Likewise, increasing the number of neurons per layer introduces complexity. In Fig. 5, we find that it is better to set 128 neurons per layer for the Avazu dataset and MovieLens dataset.

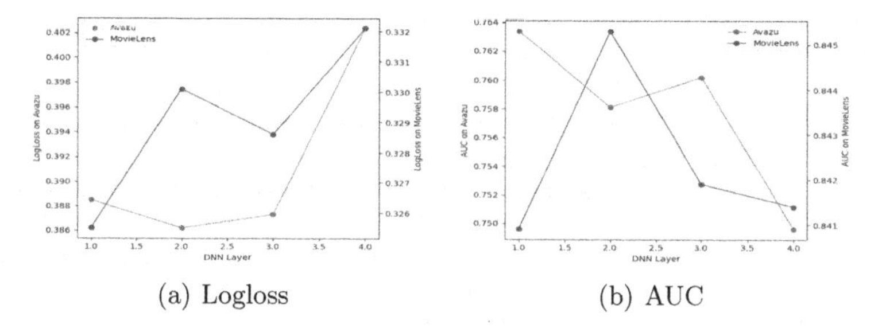

(a) Logloss (b) AUC

Fig. 4. The performance of different numbers of layer in DNN.

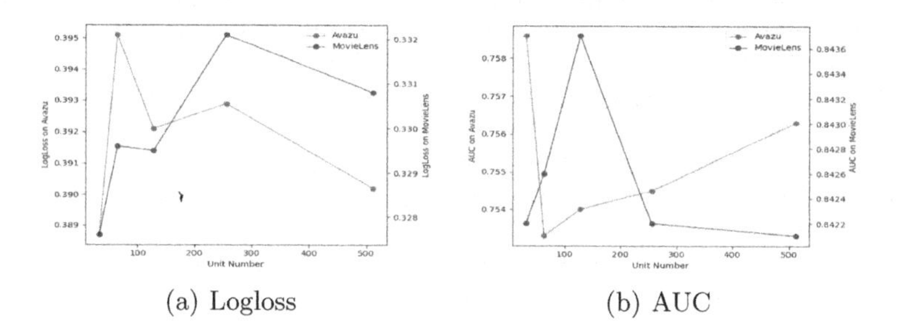

(a) Logloss (b) AUC

Fig. 5. The performance of different unit numbers of each layer in DNN.

5 Conclusion

Motivated by the drawbacks of the state-of-the-art models, we propose a new model named DFILAN and aim to reduce model time complexity and improve feature interaction effects. The input features are divided into several domains according to the correlation between features and the label. During the interaction, the independent interaction of inter-domain and intra-domain can effectively reduce feature interactions time. In the feature interactions layer, we propose a new interaction method, named vector-dims. Performing inner product operation at the level of embedding vector dimensions can obtain the interaction relationship between dimensions, making the interaction process more subtle and obtaining better interaction results. Besides, we apply the attention network to learn the importance of feature interactions automatically. Experiments on two public datasets, we can clearly find that our proposed DFILAN model has the best performance. In the future, we will consider combining the relevant content of the knowledge graph to further improve the accuracy of the model prediction.

Acknowledgements. This work is supported by Tianjin "Project + Team" Key Training Project (XC202022), the National Nature Science Foundation of China (61702368), and the Natural Science Foundation of Tianjin (18JCQNJC00700).

References

1. Koren, Y., Bell, R., Volinsky, C.: Matrix factorization techniques for recommender systems. Computer **42**(8), 30–37 (2009)
2. Rendle, S.: Factorization machines. In: ICDM 2010, the 10th IEEE International Conference on Data Mining, pp. 995–1000. IEEE Computer Society (2010)
3. He, X., et al.: Practical lessons from predicting clicks on ads at Facebook. In: Proceedings of the Eighth International Workshop on Data Mining for Online Advertising, pp. 1–9. ADKDD 2014 (2014)
4. Juan, Y., Lefortier, D., Chapelle, O.: Field-aware factorization machines in a real-world online advertising system. In: Proceedings of the 26th International Conference on World Wide Web Companion, pp. 680–688. ACM (2017)
5. Pan, J., et al.: Field-weighted factorization machines for click-through rate prediction in display advertising. In: Proceedings of the 2018 World Wide Web Conference on World Wide Web, pp. 1349–1357. ACM (2018)
6. Zhang, W., Du, T., Wang, J.: Deep learning over multi-field categorical data. In: Ferro, N., et al. (eds.) ECIR 2016. LNCS, vol. 9626, pp. 45–57. Springer, Cham (2016). https://doi.org/10.1007/978-3-319-30671-1_4
7. He, X., Chua, T.: Neural factorization machines for sparse predictive analytics. In: Proceedings of the 40th International ACM SIGIR Conference on Research and Development in Information Retrieval, pp. 355–364. ACM (2017)
8. Qu, Y., et al.: Product-based neural networks for user response prediction. In: IEEE 16th International Conference on Data Mining, pp. 1149–1154. IEEE Computer Society (2016)
9. Liu, B., Tang, R., Chen, Y., Yu, J., Guo, H., Zhang, Y.: Feature generation by convolutional neural network for click-through rate prediction. In: The World Wide Web Conference, pp. 1119–1129. ACM (2019)
10. Xin, X., Chen, B., He, X., Wang, D., Ding, Y., Jose, J.: CFM: convolutional factorization machines for context-aware recommendation. In: Proceedings of the Twenty-Eighth International Joint Conference on Artificial Intelligence, pp. 3926–3932 (2019). https://www.ijcai.org/
11. Cheng, H., et al.: Wide & deep learning for recommender systems. In: Proceedings of the 1st Workshop on Deep Learning for Recommender Systems, pp. 7–10. ACM (2016)
12. Guo, H., Tang, R., Ye, Y., Li, Z., He, X.: DeepFM: a factorization-machine based neural network for CTR prediction. In: Proceedings of the Twenty-Sixth International Joint Conference on Artificial Intelligence, pp. 1725–1731 (2017). https://www.ijcai.org/

13. Wang, R., Fu, B., Fu, G., Wang, M.: Deep & cross network for ad click predictions. In: Proceedings of the ADKDD 2017, pp. 12:1–12:7. ACM (2017)
14. Lian, J., Zhou, X., Zhang, F., Chen, Z., Xie, X., Sun, G.: xDeepFM: combining explicit and implicit feature interactions for recommender systems. In: Proceedings of the 24th ACM SIGKDD International Conference on Knowledge Discovery & Data Mining, pp. 1754–1763. ACM (2018)
15. Ouyang, W., Zhang, X., Ren, S., Qi, C., Liu, Z., Du, Y.: Representation learning-assisted click-through rate prediction. In: Proceedings of the Twenty-Eighth International Joint Conference on Artificial Intelligence, pp. 4561–4567 (2019). https://www.ijcai.org/
16. Hong, F., Huang, D., Chen, G.: Interaction-aware factorization machines for recommender systems. In: The Thirty-Third AAAI Conference on Artificial Intelligence, AAAI, pp. 3804–3811. AAAI Press (2019)
17. Yang, Y., Xu, B., Shen, S., Shen, F., Zhao, J.: Operation-aware neural networks for user response prediction. Neural Netw. **121**, 161–168 (2020)
18. Chen, W., Zhan, L., Ci, Y., Lin, C.: FLEN: leveraging field for scalable CTR prediction. arXiv preprint arXiv:1911.04690 (2019)
19. Zhou, G., et al.: Deep interest network for click-through rate prediction. In: Proceedings of the 24th ACM SIGKDD International Conference on Knowledge Discovery & Data Mining, pp. 1059–1068. ACM (2018)
20. Zhou, G., et al.: Deep interest evolution network for click-through rate prediction. In: The Thirty-Third AAAI Conference on Artificial Intelligence, pp. 5941–5948. AAAI Press (2019)
21. Xiao, J., Ye, H., He, X., Zhang, H., Wu, F., Chua, T.: Attentional factorization machines: learning the weight of feature interactions via attention networks. In: Proceedings of the Twenty-Sixth International Joint Conference on Artificial Intelligence, pp. 3119–3125 (2017). https://www.ijcai.org/
22. Huang, T., Zhang, Z., Zhang, J.: FiBiNET: combining feature importance and bilinear feature interaction for click-through rate prediction. In: Proceedings of the 13th ACM Conference on Recommender Systems, pp. 169–177. ACM (2019)
23. Song, W., et al.: AutoInt: automatic feature interaction learning via self-attentive neural networks. In: Proceedings of the 28th ACM International Conference on Information and Knowledge Management, pp. 1161–1170. ACM (2019)
24. He, K., Zhang, X., Ren, S., Sun, J.: Deep residual learning for image recognition. In: 2016 IEEE Conference on Computer Vision and Pattern Recognition, pp. 770–778. IEEE Computer Society (2016)
25. Vaswani, A., et al.: Attention is all you need. In: Advances in Neural Information Processing Systems 30: Annual Conference on Neural Information Processing Systems, pp. 5998–6008 (2017)
26. Diederik P., Ba, J.: Adam: A method for stochastic optimization. arXiv preprint arXiv:1412.6980 (2014)

Double Ensemble Soft Transfer Network for Unsupervised Domain Adaptation

Manliang Cao[1], Xiangdong Zhou[1(✉)], Lan Lin[2], and Bo Yao[1]

[1] School of Computer Science, Fudan University, Shanghai, China
{mlcao17,xdzhou,byao16}@fudan.edu.cn

[2] Department of Electronic Science and Technology, Tongji University, Shanghai, China
linlan@tongji.edu.cn

Abstract. Domain adaptation aims to transfer the enriched label knowledge from large amounts of source data to unlabeled target data. Recent methods start to solve the class-wise domain adaptation problem by incorporating the soft labels to each target data. Although the soft label strategy could alleviate the negative influence caused by the hard label strategy to some extent, the improper propagation sequence ignoring the labeling difficulties of different target examples will lead to confusing probabilities problem. Moreover, the instability of a single propagation model in dealing with various data may also hinder the performance of target label inference. To address these limitations, we propose a Double Ensemble Soft Transfer Network (DESTN) to jointly optimize the class-wise adaptation and learn the discriminative domain-invariant features with clear soft target labels. Our motivation is to construct a Label Propagation Ensemble (LPE) model by various feature subspaces so as to get robust and clear soft target labels for class-wise domain adaptation. Meanwhile, the other Classifiers Ensemble Framework (CEF) is trained on the labeled source samples and the reliable pseudo-labeled target samples for learning the discriminative features during the iteration. Extensive experiments show that DESTN significantly outperforms several state-of-the-art methods.

Keywords: Unsupervised domain adaptation · Deep ensemble network · Soft labels · Discriminative feature learning

1 Introduction

Deep neural networks can be trained very well with sufficient labeled data, which have shown great success in multimedia applications [24,25,29]. However, collecting the well-annotated datasets is exceedingly expensive and time-consuming. Domain adaptation (DA) models [33,35] can leverage the off-the-shelf data from a different but related source domain. This will boost the task in the new target domain and reduce the labeling consumption as well. Depending on the availability of labeled data in the target domain, domain adaptation can be divided

C. S. Jensen et al. (Eds.): DASFAA 2021, LNCS 12682, pp. 516–532, 2021.
https://doi.org/10.1007/978-3-030-73197-7_34

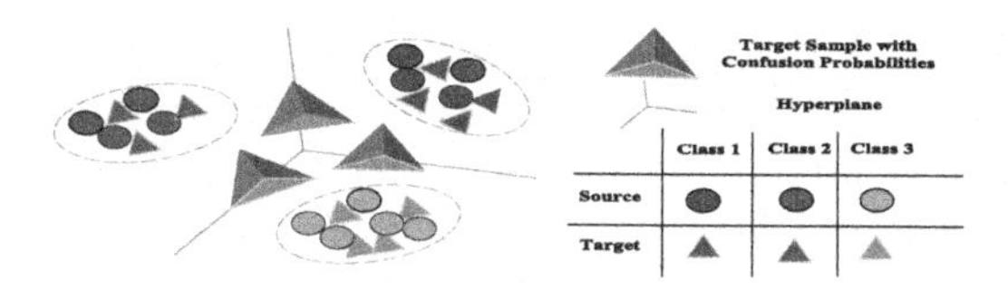

Fig. 1. It is difficult to decide which class the target samples with confusion probabilities near the hyperplane belong to.

into semi-supervised and unsupervised domain adaptation (UDA). In this paper, we focus on UDA problem. Most domain adaptation techniques [18,38] focus on the class-wise adaptation problem, which iteratively matches the marginal and conditional distributions to extract the domain invariant features and obtains the pseudo target labels via constructing a classifier on source data. For the hard pseudo label guided methods (assigning only one category to each target sample) [4,6,23,36], the class-wise methods try to tackle this issue by incorporating the hard pseudo target labels to better reduce the cross-domain distribution shifts. Since the accuracy of hard assignment of each unlabeled target sample to only one category cannot be guaranteed explicitly, these approaches are vulnerable to the error accumulation and hence unable to preserve cross-domain category consistency. As a result, the following adaptation performance could be degraded. Recent graph-based or label propagation (LP) guided methods [8,20,39] start to solve this problem by introducing the soft labels (a probability distribution over all the categories) [37] to each target data.

Although the soft label strategy can alleviate the negative influence caused by hard label strategy to some extent, it still faces two limitations. (1) the propagation sequence adopted by existing methods is completely governed by the connectivity among examples in the graph, namely the label information will be transferred from one labeled example to another as long as there is an edge between them. This propagation sequence is sometimes problematic because it does not explicitly consider the propagation difficulty or reliability of different unlabeled examples, especially the ones that are near the hyperplane. If the label information is compulsory transferred to the unlabeled samples, as shown in Fig. 1, it will lead to confusing probabilities (unclear soft target labels) problem. (2) There is not a propagation model that can perfectly handle all the practical situations for the strength of one propagation approach is very limited. For example, some tough outliers can misled the propagation process, so utilizing a single method is not enough reliable for achieving accurate propagation, which is inevitable to hinder the performance of target label inference.

Motivated by these issues, as shown in Fig. 2, we propose a Double Ensemble Soft Transfer Network (DESTN) to align the discriminative features across domains progressively and effectively via exploring the clear soft target labels. Specifically, to satisfy the class-wise domain adaptation, we need labels from both domains, while target labels are unknown in UDA situation. For label propagation guided methods, a natural way is to estimate the target labels by network outputs during training. However, the instability of a single propagation method in dealing with various data may include some misclassified ones with high confidence. Therefore, we propose to construct a label propagation ensemble (LPE) framework to get reliable labels. In LPE, random subspace method is introduced to partition the feature space into multiple subspaces, then several label propagation models are constructed on corresponding subspaces. Finally, the results of different label propagation models are fused at decision level, and only the unlabeled target samples whose label propagation results are the same will then be assigned with pseudo labels for the following class-wise domain adaptation.

The LPE structure is expected to get more clear soft target labels as much as possible so as to obtain reliable pseudo target labels for the next iteration training, which is determined by the propagation easiness of different unlabeled target examples. The easier the propagation of target samples, the more clear the soft target labels can be. However, the target samples distributed near the hyperplane are likely to be difficult for label propagation, which will lead to confusing probabilities problem. Therefore, the learned domain-invariant features need to be discriminative enough such that these tough target samples can be pushed away from the hyperplane and the propagation difficulty of these unlabeled examples can be reduced. To this end, we introduce the other Classifiers Ensemble Framework (CEF) to be trained on the labeled source samples and the reliable pseudo-labeled target samples during the iteration. It can keep better intra-class compactness and inter-class separability.

LPE and CEF can complement each other iteratively and alternatively. LPE will facilitate the final prediction which boosts the robustness of CEF by providing reliable pseudo target labels, and the discriminative domain-invariant features learned by CEF can effectively regularize LPE to alleviate the propagation difficulty of those unlabeled target samples. As training goes on, an increasing number of reliable labeled target samples will be chosen to learn the model in turn. Such progressive learning can promote DESTN to capture more accurate statistics of data distributions. We summarize our contributions as follows.

- We propose a model DESTN, where Label Propagation Ensemble (LPE) structure is to introduce reliable pseudo target labels for robust class-wise adaptation and Classifiers Ensemble Framework (CEF) is applied to alleviate these negative influence caused by confusing probabilities so as to boost the end-to-end training with LPE.
- DESTN aims to learn both reliable domain invariant and class discriminative features by simultaneously exploring the class-wise adaptation and exploiting the class-level relations in the same domain with robust soft label strategy.
- The experimental results show that compared with the state-of-the-art methods, our method achieves the best performance on the Office-31 dataset and competitive performance on Digital dataset and the challenging Office-Home dataset.

2 Related Work

A common practice for unsupervised domain adaptation is to reduce the domain shifts between the source and target domain distributions [2,27,28,31] so as to obtain domain-invariant features. In this section, we briefly review on domain adaptation methods, which are closely related to our method.

Encouraged by the deep frameworks developed in recent years, an increasing interest in ConvNets [15] methods comes up to alleviate the discrepancy between domains. These approaches are trained to simultaneously minimize a classification loss and maximize domain confusion. The model JAN [19] employs the Maximum Mean Discrepancy (MMD) [10] as the measure of domain discrepancy to achieve domain confusion. In [5], the authors apply the center loss [34] to guarantee the class relations in the source domain with better intra-class compactness and inter-class separability for domain adaptation. Inspired generative adversarial nets (GANs) [9], the method ADDA [32] converts a domain confusion task into a min-max optimization, with the aim of classifying the source samples and getting domain-invariant features through adversarial training. All of these approaches only consider the global domain distribution alignment, without exploring the class-level relations between source and target samples as the target samples are unlabeled.

Recently, several works [6,13,21,36] start to utilize the hard pseudo labels to compensate the lack of categorical information in the target domain. Since no label information in target domain for class-wise adaptation, these methods iteratively select pseudo labeled target samples based on the source-domain classifier from the previous training epoch and update the model by using the enlarged training set. Nevertheless, the direct use of such hard pseudo labels might not be preferable due to possible domain mismatch. To solve this issue, some works [8,37] resort to soft label strategy. In GAKT [8], the authors utilize a label propagation (LP) structure to assign each class with probability for the target sample when matching the domain distributions. Similarly, Yuan *et al.* propose STN [37] for heterogeneous domain adaptation, which exploits the training classifier to learn soft target labels for class-wise adaptation. A^2LP [39] improves LP via generation of unlabeled virtual instances, which uses weights computed by the entropy of the propagated soft cluster assignments to get high-confidence label predictions.

3 The Proposed Method

We focus on the problem of Unsupervised Domain Adaptation (UDA). A domain $\mathcal{D}$ is composed of a feature space χ and a marginal probability distribution $P(x)$, *i.e.*, $x \in \chi$. For a specific domain, a task $\mathcal{T}$ consists of a C-cardinality label set $\mathcal{Y}$ and a classifier $f(x)$, *i.e.*, $\mathcal{T} = \{\mathcal{Y}, f(x)\}$, where $y \in \mathcal{Y}$, and $f(x) = \mathcal{Q}(y|x)$ is the conditional probability distribution. Given a labeled source domain $\mathcal{D}_s = \{x_i^s, y_i^s\}_{i=1}^{n_s} = \{X_s, Y_s\}$ with n_s labeled samples, where $x_i^s \in \mathbb{R}^{d_s}$ is the feature vector. Define an unlabeled target domain $\mathcal{D}_t = \left\{x_j^t\right\}_{j=1}^{n_t} = X_T$

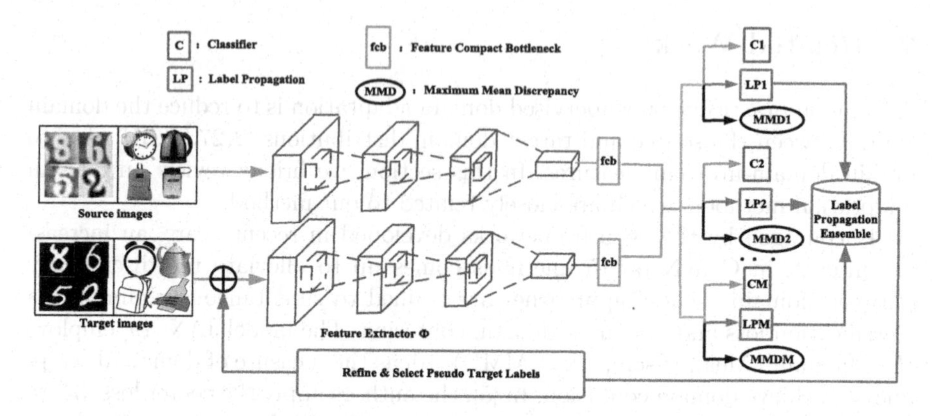

Fig. 2. Architectures of DESTN. The ensemble structure LPE is constructed by various feature subspaces and its goal is to get reliable target labels for class-wise domain adaptation; to boost the robustness of LPE, CEF will alleviate the propagation difficulty of those unlabeled target samples near the hyperplane so as to reduce the negative influence of confusion probabilities. A bottleneck layer *fcb* generated by different feature subspaces is added behind the last fully-connected layer for safer transfer representation learning.

with n_t unlabeled target samples, where $x_i^t \in \mathbb{R}^{d_t}$. Note that $\mathcal{Y}_s = \mathcal{Y}_t$, $\chi_S \neq \chi_T$, $\mathcal{P}(\chi_s) \neq \mathcal{P}(\chi_t)$, $\mathcal{Q}(\mathcal{Y}_s|\chi_s) \neq \mathcal{Q}(\mathcal{Y}_t|\chi_t)$. The goal of DESTN is to learn an ensemble model C that can maximize the classification accuracy on X_t with the following properties: 1) preserving the data manifolds and learning the reliable target labels by label propagation ensemble structure; 2) making the domain classes more discriminative with classifiers ensemble framework; 3) matching feature distributions.

3.1 Label Propagation Ensemble

To preserve the cross-domain data manifold structure and alleviate the bias caused by those underlying false pseudo labels when exploring the class-wise adaptation, we propose a graph-based label propagation optimization strategy for refining the target labels with soft ones. Formally,

$$\begin{aligned}\mathcal{L}(Z) = \frac{\nu}{2} \sum_{i,j=1}^{n_s+n_t} W_{ij} \left\| \frac{1}{\sqrt{D_{ii}}} z_i - \frac{1}{\sqrt{D_{jj}}} z_j \right\|^2 \\ + (1-\nu) \sum_{j=1}^{C} \sum_{i=1}^{n_s+n_t} \left\| Z_{ij} - Y_{ij} \right\|^2,\end{aligned} \tag{1}$$

where $Y \in \mathbb{R}^{(n_s+n_t)\times C}$ is the initial label matrix, the first n_s rows of Y are corresponding to one-hot encoded source labels and the rest rows for the target data are zero. $Z \in \mathbb{R}^{(n_s+n_t)\times C}$ is the learning soft label matrix, $z_i \in \mathbb{R}^c$ is the i-th row of matrix Z, in which every element $z_{i,c}(z_{i,c} \geqslant 0 \ and \ \sum_{c=1}^{C} z_{i,c} = 1)$ means the probability for the i-th data point belonging to the c-th category. $\|\bullet\|_F$ represents the Frobenius norm and $\nu \in [0,1)$ is a parameter. $W \in \mathbb{R}^{(n_s+n_t)\times(n_s+n_t)}$ is the symmetric nonnegative affinity matrix constructed by source and target transformed samples, which are extracted by the feature extraction network $\Phi_\theta : X_{s/t} \rightarrow R^d$ mapping the input into a feature vector or descriptor. This matrix can preserve the geometric structure information among domain samples. D denotes a diagonal matrix with $D_{ii} = \sum_{j=1}^{n_s+n_t} W_{ij}$. Note that in Eq. (1), the first term encourages smoothness such that nearby examples get the same predictions, while the second term attempts to maintain predictions for the labeled examples. We choose the Gaussian similarity function to calculate W_{ij},

$$W_{ij} = \begin{cases} exp(\frac{-(\|h_i - h_j\|^2)}{2\sigma^2}), if \ i \neq j \wedge h_i \in NN_k(h_j) \\ 0, \qquad\qquad\qquad otherwise \end{cases} \tag{2}$$

where h_i or h_j is the transformed sample, NN_k denotes the set of k nearest neighbors of h_j, σ is the length scale parameter and is set as 1 in this paper.

A single label propagation (LP) model is not always reliable, so we propose to construct a Label Propagation Ensemble (LPE) structure. It is easy to find the feature space vector of the transformed sample directly affects the result of Eq. (2). Hence, we apply different feature subspaces to measure similarities so as to perform different LP models. Formally,

$$\mathcal{L}_{LPE}(X_s, X_t, \Phi) = \sum_{m=1}^{M} \mathcal{L}_m(Z), \tag{3}$$

where $L_m(Z)$ denotes the m-th label propagation model. Then, the training unlabeled target samples with consistent label propagation results will be assigned with pseudo labels.

3.2 Classifiers Ensemble Framework

Classifiers Learning. To further regularize LPE and learn more robust and clear soft target labels, we introduce the other Classifiers Ensemble Framework (CEF) to be trained on the labeled and the reliable pseudo-labeled target samples during the iteration. Given labeled training instances $X_{train} = \{(x^i, y^i)\}_{i=1}^{N}$, the deep ensemble neural network with M classification functions, $\{C_m \circ \Phi(x)\}_{m=1}^{M}$, can be trained by minimizing the following negative *log*-likelihood loss function:

$$\mathcal{L}_{CL}(\omega_1, \omega_2, \cdots, \omega_M) = \frac{1}{N}\sum_{i=1}^{N}\sum_{m=1}^{M} l_m\left(u_m^i, y^i\right), \tag{4}$$

where $(\omega_1, \omega_2, \cdots, \omega_M)$ are the model parameters, $u_m^i = C_m \circ \Phi(x)$ is the m-th classifiers prediction vector of instance x^i in its feature subspace, and $l_m(.,.)$ denotes a negative *log*-likelihood loss function. Although initially the labeled training data only contain the labeled source instances, such that $X_{train} = X_s$ and $N = n_s$, the classifiers will improve its transfer capability and more reliable pseudo target samples can be selected by LPE as training goes on.

Class Discriminative Constraint. We propose Class Discriminative Constraint (CDC) to further regularize the CEF so as to promote those tough target samples to be away from the hyperplane. The MMD [10] has been proven to be a powerful tool for exploring the statistics of samples. Inspired by this, we propose the following formula to describe the constraint of target samples:

$$\mathcal{L}^{k1k2}(Z_t, \Phi) = f1 + f2 - 2f3, \tag{5}$$

where

$$f1 = \sum_{i=1,j=1}^{n_t} \frac{V_{k1k1}(z^t_{i,c1}, z^t_{j,c2}) K(\Phi(x^i_t), \Phi(x^j_t))}{\sum_{i=1,j=1}^{n_t} V_{k1k1}(z^t_{i,c1}, z^t_{j,c2})}$$

$$f2 = \sum_{i=1,j=1}^{n_t} \frac{V_{k2k2}(z^t_{i,c1}, z^t_{j,c2}) K(\Phi(x^i_t), \Phi(x^j_t))}{\sum_{i=1,j=1}^{n_t} V_{k2k2}(z^t_{i,c1}, z^t_{j,c2})} \tag{6}$$

$$f3 = \sum_{i=1,j=1}^{n_t} \frac{V_{k1k2}(z^t_{i,c1}, z^t_{j,c2}) K(\Phi(x^i_t), \Phi(x^j_t))}{\sum_{i=1,j=1}^{n_t} V_{k1k2}(z^t_{i,c1}, z^t_{j,c2})}$$

Note that $V_{ab}(y_1, y_2) = y_1 \times y_2$, $Z_t \in \mathbb{R}^{n_t \times C}$ denotes the soft target label matrix constructed by the target n_t rows of the learning soft label matrix Z, and the element like $z^t_{i,c1}$ is the corresponding class probability of i-th target sample. Gaussian kernel K is adopted here. Equation (6) defines two kinds of class-level relation in target domain. When $k1 = k2$, it compacts the intra-class variations; When $k1 \neq k2$, it enlarges inter-class difference. Based on this analysis, using CDC to measure the relations of all the soft target labels can be formulated as:

$$\mathcal{L}^c_t(X_t, Z_t) = \sum_{m=1}^{M} \sum_{k1=1}^{C} \frac{\mathcal{L}^{k1k1}_m(Z_t, \Phi)}{C} - \sum_{\substack{k1=1,k2=1,\\ k1\neq k2}}^{C} \frac{\mathcal{L}^{k1k2}_m(Z_t, \Phi)}{C(C-1)}, \tag{7}$$

where $\mathcal{L}^{k1k1}_m$ is the m-th CDC. With Eq. (7), the intr-class and the inter-class relations will be optimized in the opposite direction. We can get the similar formula $\mathcal{L}^c_s$ to show the relations for source samples.

Finally we can get the classifiers ensemble loss: $\mathcal{L}_{CEF} = \mathcal{L}^c_s + \mathcal{L}^c_t + \mathcal{L}_{CL}$.

Final Target Labels Learning. With the multiple classification functions learned in CEF, we can integrate the M classification functions to perform class prediction on each unlabeled target instance x_t with majority voting strategy,

$$y^i_t = argmax \sum_{m=1}^{M} C_m(x^i_t). \tag{8}$$

3.3 Class-Wise Adaptation

As aforementioned, the primary goal is to learn a domain-invariant feature space by matching domain distributions. MMD is also a widely used approach to alleviating marginal distribution disparity [19], which computes the distance between expectations of source and target data in the projected feature space. Formally,

$$\mathcal{L}_{Mar}(X_s, X_t, \Phi) = \left\| \frac{1}{n_s} \sum_{i=1}^{n_s} \Phi(x_i^s) - \frac{1}{n_t} \sum_{j=1}^{n_t} \Phi(x_j^t) \right\|^2 . \tag{9}$$

The MMD strategy in Eq. (9) can reduce the difference of the marginal distributions, but it fails to guarantee that the divergency of conditional distributions is minimized. Since no label information in target domain for matching conditional distributions, instead of using hard target labels, we propose to predict clear soft labels for unlabeled target data. This will develop a probabilistic class-wise adaptation formula to effectively guide the intrinsic knowledge transfer. Formally,

$$\mathcal{L}_{Con}(X_s, X_t, \Phi) = \sum_{c=1}^{C} \left\| \frac{1}{n_s^c} \sum_{x_i \in X_s^c} \Phi(x_i) - \frac{1}{n_t^c} \sum_{x_j \in X_t} z_{j,c}^t \Phi(x_j) \right\|^2 , \tag{10}$$

where $X_s^c = \{x_i : x_i \in X_s \wedge argmax(y(x_i)) = c\}$ denotes the set of source domain samples belonging to the c-th class , $y(x_i)$ represents the true one-hot label of x_i in the learning soft label matrix Z, and $n_s^c = |X_s^c|$. $z_{j,c}^t$ is the probability for the j-th unlabeled target sample belonging to the c-th class in Z. Thus, the target sample size n_t^c can be approximately computed by $n_t^c = \sum_{j=1}^{n_t} z_{j,c}^t$. Finally, the class-wise adaptation term for all the feature subspace can be formulated as:

$$\mathcal{L}_{CA}(X_s, X_t, \Phi) = \sum_{m=1}^{M} \mathcal{L}_{Mar}^m(X_s, X_t, \Phi) + \mathcal{L}_{Con}^m(X_s, X_t, \Phi), \tag{11}$$

where $\mathcal{L}_{Mar}^m$ and $\mathcal{L}_{Con}^m$ are the marginal and conditional distributions of the corresponding feature subspace, respectively.

3.4 Overall Object Function

Considering all the above discussions, we have the overall objective function:

$$\begin{aligned} \mathcal{L}(X_s, Y_s, X_t, Z_t, \Phi, \omega_1, \omega_2, \cdots, \omega_M) &= \mathcal{L}_{CA}(X_s, X_t, \Phi) \\ + \beta \mathcal{L}_{LPE}(X_s, X_t, \Phi) &+ \gamma \mathcal{L}_{CEF}(X_s, X_t, \Phi, \omega_1, \omega_2, \cdots, \omega_M), \end{aligned} \tag{12}$$

where (β, γ) are weights that control the interaction of losses to achieve a better trade-off between LPE and CEF.

3.5 Theoretical Analysis

We provide a theoretical analysis to show the relations between our method and the existing theory of domain adaptation [1]. The theory presents the target expected error $R_{\mathcal{T}}(h)$ which is bounded by three terms as follows,

$$\forall\, h \in \mathcal{H}, \mathcal{R}_{\mathcal{T}}(h) \leq \mathcal{R}_{\mathcal{S}}(h) + \frac{1}{2} d_{\mathcal{H}\Delta\mathcal{H}}(\mathcal{S}, \mathcal{T}) + E, \tag{13}$$

where $\mathcal{R}_{\mathcal{S}}(h)$ is the expected error on the labeled source samples, $d_{\mathcal{H}\Delta\mathcal{H}}(\mathcal{S}, \mathcal{T})$ denotes the domain divergence measured by a discrepancy distance between two domain distributions $\mathcal{S}$ and $\mathcal{T}$ *w.r.t.* a hypothesis set $\mathcal{H}$. E is the shared error of the ideal joint hypothesis and is often overlooked by previous approaches [17,32] because it is regarded as to be negligible small. However, when the cross-domain category alignment dose not explicitly enforce, E will become large. This leads to a problem that a small $\mathcal{R}_{\mathcal{S}}(h)$ and a small $d_{\mathcal{H}\Delta\mathcal{H}}(\mathcal{S}, \mathcal{T})$ cannot guarantee small $\mathcal{R}_{\mathcal{T}}(h)$. Hence, E needs to be explored as well. We resort to soft labels to realize this:

$$E = \mathcal{R}_{\mathcal{S}}(h^*, f_{\mathcal{S}}) + \mathcal{R}_{\mathcal{T}}(h^*, f_{\mathcal{T}}), \tag{14}$$

where $h^* = \underset{h \in \mathcal{H}}{argmin}\, \mathcal{R}_{\mathcal{S}}(h, f_{\mathcal{S}}) + \mathcal{R}_{\mathcal{T}}(h, f_{\mathcal{T}})$, $f_{\mathcal{S}}$ and $f_{\mathcal{T}}$ are the labeling functions for source and target domains, respectively. With the triangle inequality for classification, then

$$\begin{aligned} E &\leq \min_{h \in \mathcal{H}} \mathcal{R}_{\mathcal{S}}(h, f_{\mathcal{S}}) + \mathcal{R}_{\mathcal{T}}(h, f_{\mathcal{S}}) + \mathcal{R}_{\mathcal{T}}(f_{\mathcal{S}}, f_{\mathcal{T}}) \\ &\leq \min_{h \in \mathcal{H}} \mathcal{R}_{\mathcal{S}}(h, f_{\mathcal{S}}) + \mathcal{R}_{\mathcal{T}}(h, f_{\mathcal{S}}) + \mathcal{R}_{\mathcal{T}}(f_{\mathcal{S}}, f_{\tilde{\mathcal{T}}}) + \mathcal{R}_{\mathcal{T}}(f_{\mathcal{T}}, f_{\tilde{\mathcal{T}}}). \end{aligned} \tag{15}$$

Since we have source labels, we can easily get a proper h to approximate the labeling function $f_{\mathcal{S}}$, which satisfies the first and second term. The last term is the confusion probabilities of soft labels rate, obviously, the proposed LPE aims to get more reliable soft labels in target domain so that $\mathcal{R}_{\mathcal{T}}(f_{\mathcal{T}}, f_{\tilde{\mathcal{T}}})$ can be minimized. Now we focus on the third term. Formally,

$$\begin{aligned} \mathcal{R}_{\mathcal{T}}(f_{\mathcal{S}}, f_{\tilde{\mathcal{T}}}) &= \mathbb{E}_{m \in [1,M]} \mathbb{E}_{x \sim X_t} \left[\psi(C_m^{\mathcal{S}}(\Phi(x)) - C_m^{\tilde{\mathcal{T}}}(\Phi(x))) \right] \\ &= \mathbb{E}_{m \in [1,M]} \mathbb{E}_{x \sim X_t} \left[\left| \psi(C_m^{\mathcal{S}}(\Phi(x), z_1) - \psi(C_m^{\tilde{\mathcal{T}}}(\Phi(x), z_2) \right| \right], \end{aligned} \tag{16}$$

where

$$\left| \psi(C_m^{\mathcal{S}}(\Phi(x), z_1) - \psi(C_m^{\tilde{\mathcal{T}}}(\Phi(x), z_2) \right| = \begin{cases} 1, if \quad z_1 \neq z_2. \\ 0, \quad otherwise. \end{cases} \tag{17}$$

When $z_1 = z_2$, it means the categories between the source ones and the soft target ones are aligned. Thus, the $\mathcal{R}_{\mathcal{T}}(f_{\mathcal{S}}, f_{\tilde{\mathcal{T}}})$ can be minimized.

4 Experiments

4.1 Experimental Settings

Datasets. We use three popular DA datasets: 1) **Office31** [26] contains 4652 images and 31 categories collected from 3 domains: Amazon(A), Webcam(W)

Table 1. Accuracy (%) on Office-31 for UDA (ResNet).

Method	A → W	D → W	W → D	A → D	D → A	W → A	Avg
ResNet-50	68.4	93.2	97.3	68.9	62.5	60.7	75.1
JDDA [5]	82.6	95.2	99.7	79.8	57.4	66.7	80.2
MADA [23]	90.0	97.4	99.6	87.8	70.3	66.4	85.2
TAT [16]	92.5	99.3	**100.0**	93.2	73.1	72.1	88.4
DSR [3]	93.1	98.7	99.8	92.4	73.5	73.9	88.6
CAN [13]	94.5	99.1	99.8	95.0	78.0	77.0	90.6
STN [37]	94.3	98.8	99.5	95.2	77.8	77.4	90.6
ALDA [7]	95.6	97.7	**100.0**	94.0	72.2	72.5	88.7
A^{2}LP [39]	93.4	98.8	**100.0**	96.1	78.1	77.6	90.7
DESTN	**95.5**	**99.8**	**100.0**	**96.3**	**78.8**	**77.9**	**91.4**

Table 2. Accuracy (%) for cross-domain experiments on Office-Home (ResNet).

Method	Ar ↓ Cl	Ar ↓ Pr	Ar ↓ Rw	Cl ↓ Ar	Cl ↓ Pr	Cl ↓ Rw	Pr ↓ Ar	Pr ↓ Cl	Pr ↓ Rw	Rw ↓ Ar	Rw ↓ Cl	Rw ↓ Pr	Avg
GAKT [8]	34.5	43.6	55.3	36.1	52.7	53.2	31.6	40.6	61.4	45.6	44.6	64.9	47.0
TAT [16]	51.6	69.5	75.4	59.4	69.5	68.6	59.5	50.5	76.8	70.9	56.6	81.6	65.8
DSR [3]	53.4	71.6	77.4	57.1	66.8	69.3	56.7	49.2	75.7	68.0	54.0	79.5	64.9
STN [37]	53.8	71.9	77.8	58.4	67.8	70.2	57.8	50.1	75.5	68.4	54.9	80.2	65.5
ALDA [7]	53.7	70.1	76.4	**60.2**	72.6	71.5	56.8	51.9	77.1	70.2	56.3	82.1	66.6
DESTN	**54.6**	**73.9**	**78.5**	59.3	**72.8**	**72.3**	**59.9**	**52.5**	**78.6**	**72.5**	**57.1**	**84.3**	**68.0**

and DSLR(D). The low resolution images in Webcam are captured with a web camera. The medium resolution images in Amazon are downloaded from amazon.com. DSLR consists of high resolution images collected by a SLR camera. 2) **Office-Home**[1] is a more challenge dataset for domain adaptation, which consists of 15,500 images in total from 65 categories of common objects in office and home settings. These images come from 4 significantly different domains: Artistic images (Ar), Clip Art (Cl), Product images (Pr) and Real-World images (Rw). 3) We apply MNIST [15], SVHN [22] and USPS [12] as **Digital** recognition dataset. These datasets include 10 classes.

Baseline Methods. We compare DESTN with the state-of-the-art transfer learning methods. The global distribution matching guided methods: ADDA [32], JDDA [5], TAT [16] and DSR [3] match the marginal distributions for domain adaptation. Hard pseudo-target label guided methods: MSTN [36], MADA [23], PFAN [6], CAN [13], and ALDA [7] resort to pseudo target labels to capture multimode structures to enable fine-grained alignment. We further compare with several soft target label guided approaches: A^{2}LP [39], GAKT [8] and STN [37] use a single label propagation strategy to perform the class-wise adaptation. We implement STN using the released code and cite the performance of other methods in their corresponding papers.

[1] https://hemanthdv.github.io/officehome-dataset/.

4.2 Implementation Details

We follow standard evaluation protocols for UDA [14,30]. For all the baseline methods, we follow their original model selection procedures. We adopt transfer cross-validation to select parameters for the DESTN models. We examine the influence of deep representations for domain adaptation by exploring ResNet-50 [11] which has been pre-trained on ImageNet for Office-31 and Office-Home, and employ the modified LeNet by [32] for the digital datasets. We fine-tune all convolutional and pooling layers and a bottleneck layer fcb is added behind the last fully-connected layer. The classifiers can be constructed with any kind of deep layers followed by a softmax output.

Table 3. Accuracy (%) on digital dataset for UDA (LeNet).

Method	SVHN → MNIST	USPS → MNIST
ADDA [32]	76.0 ± 1.8	89.4± 0.2
MSTN [36]	91.7 ± 1.5	–
PFAN [6]	93.9 ± 0.8	–
STN [37]	94.8 ± 0.3	95.6± 0.1
ALDA [7]	**98.7** ± 0.4	98.6± 0.1
DESTN	98.5± 0.3	**98.8**± 0.2

The model is implemented with TensorFlow. For the fcb layer we set different neuron numbers to make sure we have different feature subspaces, starting at 128 and incrementing by 128. We resize all images to 224×224 for Office-31 and Office-home (32×32 for digits). The batch size is set to 128 for each domain. We adopt the stochastic gradient descent (SGD) with momentum of 0.95. We employ the similar learning rate strategy implemented in MSTN [36], which computes the learning rate by formula: $\eta_p = \frac{\eta_0}{(1+\nu p)^{\omega}}$, where $\eta_0 = 0.01$, $\nu = 10$, $\omega = 0.75$ and p is the training progress linearly changing from 0 to 1. The weight balance parameters are set as β =0.5, γ =0.8 and the feature subspace number M is 30.

4.3 Results and Discussion

The classification results on **Office-31** are shown in Table 1. The DESTN outperforms all comparison methods on all transfer tasks. It is noteworthy that DESTN promotes accuracies substantially on four hard transfer tasks: A→W, A→D, D→A, and W→A, where the source domain is remarkably different from the target.

The classification accuracies on the **Office-Home** dataset are reported in Table 2. Obviously, our DESTN model outperforms the shallow comparison method GAKT on all the transfer tasks and boosts the accuracy by an absolute (21.0%) on average. It improves the state-of-the-art result from 66.6% to

68.0% on average. Compared with those recent deep methods, *i.e.*, TAT, DSR and ALDA, our model gets quite competitive results and performs best in 11 out of 12 tasks.

The performance of **Digital** dataset is shown in Table 3. Compared with other methods, DESTN achieves competitive results on all tasks, especially on the challenging one SVHN→MNIST.

All of the above results reveal several observations. (1) Taking class information of the target samples into account is beneficial to the adaptation. It can be seen that the hard and soft guided methods achieve better performance than those class-agnostic approaches, *i.e.*, ADDA and JDDA. (2) The hard label guided methods like MADA, PFAN and CAN assign a hard label for every target sample, which may deteriorate the following training when the target samples are assigned with wrong labels. On the contrary, ADAL introduces confusion matrix to alleviate this issue and the soft label guided approaches *i.e.*, our model DESTN, A^{2}LP and STN apply the soft target labels to explore the intrinsic structure to benefit the final performance. (3) DESTN achieves competitive results to some recent best performance deep models like TAT and STN. The reason is that only DSETN adopts the strategy of learning robust soft target labels with ensemble strategy and conducting discriminative feature learning to benefit each other for effective knowledge transfer.

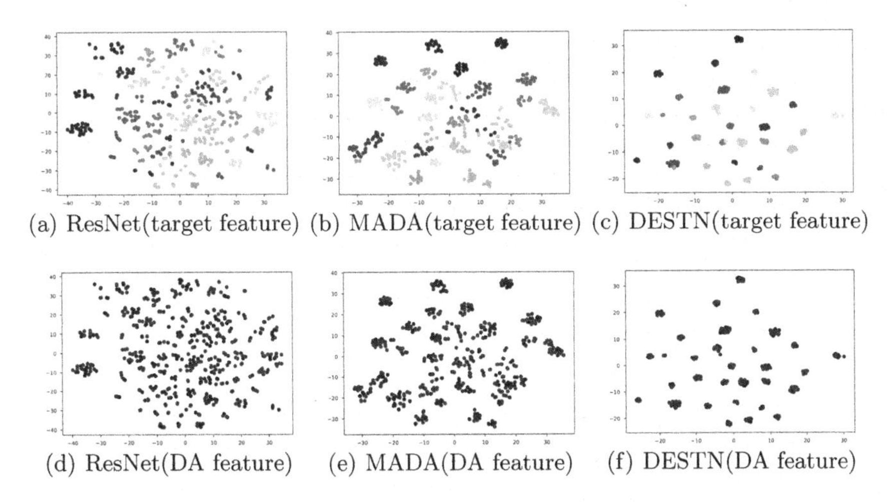

(a) ResNet(target feature) (b) MADA(target feature) (c) DESTN(target feature)

(d) ResNet(DA feature) (e) MADA(DA feature) (f) DESTN(DA feature)

Fig. 3. The t-SNE visualization of features extracted by different models for D → W task. (a)-(c) represent the distribution from category perspective and each color denotes a category. (d)-(f) represent the distribution from domain alignment(DA) perspective, where red and blue points represent samples of source and target domains, respectively. (Best Viewed in Color). (Color figure online)

4.4 Empirical Analysis

Ablation Study. Table 4 examines two key components of DESTN, *i.e.*, label propagation ensemble (LPE) structure and classifiers ensemble framework (CEF). We conduct ablation study on several tasks with different components ablation. The method "DESTN-LPE" means the model learns the discriminative domain-invariant features with hard pseudo target labels using CEF. "DESTN-CEF" learns the features with soft target labels but without adopting the regularizer CEF. Interestingly, "DESTN-CEF" achieves better than "DESTN-LPE", which verifies target samples assigned with robust soft labels can alleviate some wrong information caused by false hard pseudo-target labels. The results of DESTN confirm all the components can complement each other.

Table 4. The effect of alternative optimization (LPE) and (CEF).

Method	A → W	D → W	Ar → Pr	Ar → Rw
DESTN-LPE	92.5	97.0	72.1	75.9
DESTN-CEF	94.4	98.7	73.0	77.3
DESTN	95.5	99.8	73.9	78.5

Feature Visualization. We apply t-SNE to visualize the features on task D → W learned by Source Only model (ResNet), MADA and DESTN, respectively. In Fig. 3(a)–(c), we plot the feature distribution from category perspective and each color denotes a category. Figure 3(d)–(f) show the domain alignment information, where red and blue points represent samples of source and target domain, respectively. The visualization results reveal the following observations. (1) Compared with Fig. 3(b), the distributions of Fig. 3(a) have more scattered points distributed in the inter-class gap, which verifies that features learned by the hard target label guided model MADA are discriminated much better than that learned by the no domain adaptation metric model (ResNet). Besides,

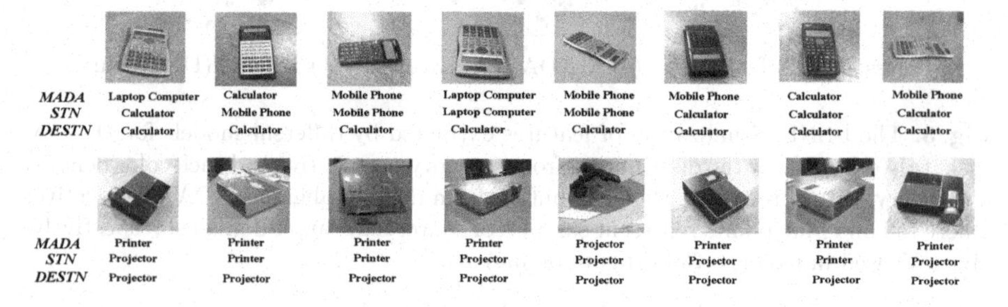

Fig. 4. Misclassified samples analysis of MADA and STN for task A→W of Office-31 with respect to classes "Calculator" and "Projector". Red and black are the misclassified and correct samples, respectively. (Color figure online)

Fig. 3(c) shows the representations learned by our method that features in the same class are much more compact and features with different classes are well separated. (2) From domain alignment perspective, in Fig. 3(d) and 3(e), the source and target domain distributions are made indistinguishable, but different categories are not aligned very well across domains. However, with features learned by DESTN, the categories are aligned much better.

Misclassified Samples Analysis. Figure 4 shows randomly selected misclassified samples of MADA and STN for the task A $\rightarrow$ W with respect to the classes "Calculator" and "Projector". MADA will misclassify most target samples that are much similar to other classes in source domain, *i.e.*, "Printor" and "Mobile Phone", which means learning hard pseudo target labels can enhance the error accumulation. Instead, STN with soft label strategy can alleviate this issue to some extend. But without considering the discriminative structures and alleviating the negative information caused by those confusion probabilities of soft labels, STN will still mix up some source and target samples. By contrast, DESTN can distinguish those similar samples.

Confusion Matrices. We draw the confusion matrices in Fig. 5(a)–(c) to intuitively illustrate the efficacy of our approach. For the method MSTN, there are many wrong digit predictions. For instance, most samples of class "6" are mistakenly predicted into "0" which reveals the tremendously large difference among domains. STN performs better, but in some cases it is quite possible to

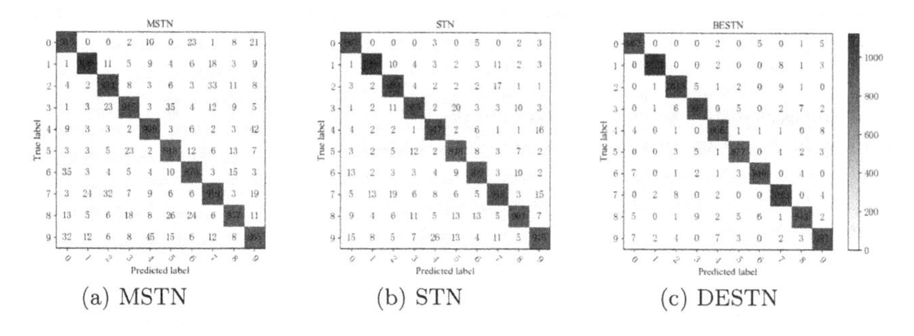

(a) MSTN (b) STN (c) DESTN

Fig. 5. Confusion matrices for visualization of the performance of MSTN, STN and DESTN for task SVHN $\rightarrow$ MNIST. (Best Viewed in Color). (Color figure online)

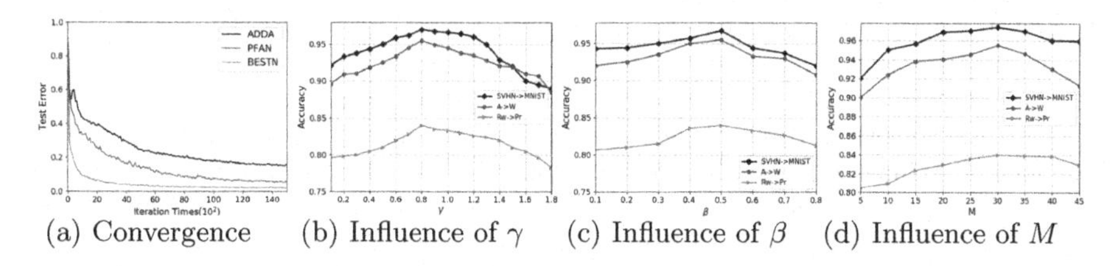

(a) Convergence (b) Influence of γ (c) Influence of β (d) Influence of M

Fig. 6. Empirical analysis: (a) Convergence performance, (b)$\sim$(d) The influence of parameter settings on the classification accuracy.

be misclassified, particularly when testing similar digits like "2" and "7", "4" and "9". By contrast, much more right predictions appear in the diagonal using DESTN, which proves the domain discrepancies could be effectively mitigated by the model.

Parameter Sensitivity. Figure 6(a) shows the convergence curves of the test error of different methods on USPS→MNIST, which reveals that the DESTN converges fastest and reaches to the lowest error. We further run DESTN with different values of γ and β on several random tasks. From Fig. 6(b)–(c), we find that the accuracy increases first and then decreases as the parameters increase. This confirms the validity of jointly learning the domain-invariant features and exploring the soft class information, since a good trade-off between them can enhance feature transferability. Figure 6(d) investigates the sensitivity of the number M of classifiers derived from different feature subspaces and the best number is 30.

5 Conclusion

We propose a simple, yet effective model for domain adaptation. The key idea is to jointly optimize the soft target labels and learn the discriminative domain-invariant features. The effective results on several cross-domain datasets benefit from the interest of applying label propagation ensemble strategy and classifiers ensemble structure together to guarantee the geometric structure of the underlying data manifold and keep the discriminative properties of the learned features.

Acknowledgement. This work was supported by the National Key Research and Development Program of China, No.2018YFB1402600.

References

1. Ben-David, S., Blitzer, J., Crammer, K., Kulesza, A., Pereira, F. and Vaughan, J.W.: A theory of learning from different domains. Mach. Learn. **79**(1), 151-175 (2009). https://doi.org/10.1007/s10994-009-5152-4
2. Bermúdez Chacón, R., Salzmann, M., Fua, P.: Domain-adaptive multibranch networks. In: ICLR, No. CONF (2020)
3. Cai, R., Li, Z., Wei, P., Qiao, J., Zhang, K.: Learning disentangled semantic representation for domain adaptation. In: IJCAI (2019)
4. Cao, M., Zhou, X., Xu, Y., Pang, Y., Yao, B.: Adversarial domain adaptation with semantic consistency for cross-domain image classification. In: CIKM (2019)
5. Chen, C., Chen, Z., Jiang, B., Jin, X.: Joint domain alignment and discriminative feature learning for unsupervised deep domain adaptation. In: AAAI (2019)
6. Chen, C., et al.: Progressive feature alignment for unsupervised domain adaptation. In: CVPR (2019)
7. Chen, M., Zhao, S., Liu, H., Cai, D.: Adversarial-learned loss for domain adaptation. In: AAAI (2020)

8. Ding, Z., Li, S., Shao, M., Fu, Y.: Graph adaptive knowledge transfer for unsupervised domain adaptation. In: ECCV, pp. 37–52 (2018)
9. Goodfellow, I., et al.: Generative adversarial nets. In: NeurIPS (2014)
10. Gretton, A., Borgwardt, K., Rasch, M., Schölkopf, B., Smola, A.J.: A kernel method for the two-sample-problem. In: NeurIPS, pp. 513–520 (2007)
11. He, K., Zhang, X., Ren, S., Sun, J.: Deep residual learning for image recognition. In: CVPR, pp. 770–778 (2016)
12. Hull, J.J.: A database for handwritten text recognition research. IEEE Trans. Pattern Anal. Mach. Intell. **16**(5), 550–554 (2002)
13. Kang, G., Jiang, L., Yang, Y., Hauptmann, A.G.: Contrastive adaptation network for unsupervised domain adaptation. In: CVPR, pp. 4893–4902 (2019)
14. Laradji, I.H., Babanezhad, R.: M-adda: Unsupervised domain adaptation with deep metric learning. In: ICML (2018)
15. LeCun, Y., Bottou, L., Bengio, Y., Haffner, P.: Gradient-based learning applied to document recognition. Proc. IEEE **86**(11), 2278–2324 (1998)
16. Liu, H., Long, M., Wang, J., Jordan, M.: Transferable adversarial training: a general approach to adapting deep classifiers. In: ICML, pp. 4013–4022 (2019)
17. Long, M., Cao, Y., Wang, J., Jordan, M.I.: Learning transferable features with deep adaptation networks. In: ICML (2015)
18. Long, M., Cao, Z., Wang, J., Jordan, M.I.: Conditional adversarial domain adaptation. In: NeurIPS, pp. 1640–1650 (2018)
19. Long, M., Zhu, H., Wang, J., Jordan, M.I.: Deep transfer learning with joint adaptation networks. In: ICML, pp. 2208–2217 (2017)
20. Luo, L., Wang, X., Hu, S., Wang, C., Tang, Y., Chen, L.: Close yet distinctive domain adaptation. In: ICCV (2017)
21. Luo, Y., Zheng, L., Guan, T., Yu, J., Yang, Y.: Taking a closer look at domain shift: category-level adversaries for semantics consistent domain adaptation. In: CVPR, pp. 2507–2516 (2019)
22. Netzer, Y., Wang, T., Coates, A., Bissacco, A., Wu, B., Ng, A.Y.: Reading digits in natural images with unsupervised feature learning. In: NeurIPS (2011)
23. Pei, Z., Cao, Z., Long, M., Wang, J.: Multi-adversarial domain adaptation. In: AAAI (2018)
24. Philip, J., Gharbi, M., Zhou, T., Efros, A.A., Drettakis, G.: Multi-view relighting using a geometry-aware network. TOG **38**(4), 1–14 (2019)
25. Ro, H., Park, Y.J., Byun, J.H., Han, T.D.: Display methods of projection augmented reality based on deep learning pose estimation. In: SIGGRAPH (2019)
26. Saenko, K., Kulis, B., Fritz, M., Darrell, T.: Adapting visual category models to new domains. In: Daniilidis, K., Maragos, P., Paragios, N. (eds.) ECCV 2010. LNCS, vol. 6314, pp. 213–226. Springer, Heidelberg (2010). https://doi.org/10.1007/978-3-642-15561-1_16
27. Soh, J.W., Cho, S., Cho, N.I.: Meta-transfer learning for zero-shot super-resolution. In: CVPR (2020)
28. Sohn, K., Shang, W., Yu, X., Chandraker, M.: Unsupervised domain adaptation for distance metric learning. In: ICLR (2019)
29. Sun, X., Nasrabadi, N.M., Tran, T.D.: Supervised deep sparse coding networks for image classification. TIP **29**, 405–418 (2019)
30. Tang, H., Jia, K.: Discriminative adversarial domain adaptation. In: AAAI (2020)
31. Tseng, H.Y., Lee, H.Y., Huang, J.B., Yang, M.H.: Cross-domain few-shot classification via learned feature-wise transformation. In: ICLR (2020)
32. Tzeng, E., Hoffman, J., Saenko, K., Darrell, T.: Adversarial discriminative domain adaptation. In: CVPR, pp. 7167–7176 (2017)

33. Wang, H., Shen, T., Zhang, W., Duan, L., Mei, T.: Classes matter: a fine-grained adversarial approach to cross-domain semantic segmentation. In: ECCV (2020)
34. Wen, Y., Zhang, K., Li, Z., Qiao, Yu.: A discriminative feature learning approach for deep face recognition. In: Leibe, B., Matas, J., Sebe, N., Welling, M. (eds.) ECCV 2016. LNCS, vol. 9911, pp. 499–515. Springer, Cham (2016). https://doi.org/10.1007/978-3-319-46478-7_31
35. Wu, X., et al.: A unified adversarial learning framework for semi-supervised multi-target domain adaptation. In: DASFAA (2020)
36. Xie, S., Zheng, Z., Chen, L., Chen, C.: Learning semantic representations for unsupervised domain adaptation. In: ICML, pp. 5419–5428 (2018)
37. Yao, Y., Zhang, Y., Li, X., Ye, Y.: Heterogeneous domain adaptation via soft transfer network. In: ACM MM, pp. 1578–1586 (2019)
38. Zhang, W., Ouyang, W., Li, W., Xu, D.: Collaborative and adversarial network for unsupervised domain adaptation. In: CVPR, pp. 3801–3809 (2018)
39. Zhang, Y., Deng, B., Jia, K., Zhang, L.: Label propagation with augmented anchors: A simple semi-supervised learning baseline for unsupervised domain adaptation. In: ECCV (2020)

Attention-Based Multimodal Entity Linking with High-Quality Images

Li Zhang[1], Zhixu Li[1,2](✉), and Qiang Yang[3]

[1] School of Computer Science and Technology, Soochow University, Suzhou, China
lzhang1997@stu.suda.edu.cn, zhixuli@suda.edu.cn
[2] IFLYTEK Research, Suzhou, China
[3] King Abdullah University of Science and Technology, Jeddah, Saudi Arabia
qiang.yang@kaust.edu.sa

Abstract. Multimodal entity linking (MEL) is an emerging research field which uses both textual and visual information to map an ambiguous mention to an entity in a knowledge base (KB). However, images do not always help, which may also backfire if they are irrelevant to the textual content at all. Besides, the existing efforts mainly focus on learning a representation of both mentions and entities from their textual and visual contexts, without considering the negative impact brought by noisy irrelevant images, which happens frequently with social media posts. In this paper, we propose a novel MEL model, which not only removes the negative impact of noisy images, but also uses multiple attention mechanism to better capture the connection between mention representation and its corresponding entity representation. Our empirical study on a large real data collection demonstrates the effectiveness of our approach.

1 Introduction

Entity linking is a crucial task in information extraction, mapping a disambiguating named mention into a target knowledge base [19]. The problem has been studied extensively on using either "local" information (contextual information of the mention in the text) [4] or "global" information (relations among candidate entities) [17] for entity linking.

In recent years, with the rapid development of social medias such as Facebook, Twitter, and Sina Weibo, a large volume of user-generated posts are emerged, which present new challenges as well as big opportunities to entity linking. On the one hand, most social media posts have relatively short text. On the other hand, many posts have images attached. For deep semantic parsing to these posts, an emerging task called multimodal entity linking (MEL) is proposed, which uses both textual and visual information to map an ambiguous mention to an entity in a knowledge base (KB) [1,15]. The usage of image information could better capture the relationship among mentions, context and candidate entities [15]. There has been work leveraging both semantic textual and visual information extracted from mention and entity contexts [1], which greatly improves the accuracy of entity linking work on social media datasets [1,15].

C. S. Jensen et al. (Eds.): DASFAA 2021, LNCS 12682, pp. 533–548, 2021.
https://doi.org/10.1007/978-3-030-73197-7_35

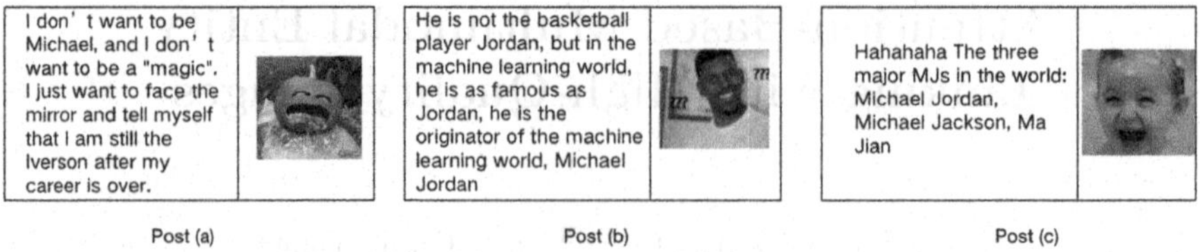

I don' t want to be Michael, and I don' t want to be a "magic". I just want to face the mirror and tell myself that I am still the Iverson after my career is over.

Post (a)

He is not the basketball player Jordan, but in the machine learning world, he is as famous as Jordan, he is the originator of the machine learning world, Michael Jordan

Post (b)

Hahahaha The three major MJs in the world: Michael Jordan, Michael Jackson, Ma Jian

Post (c)

Fig. 1. Example irrelevant images attached with social network posts

However, there are at least two drawbacks with the existing efforts. Firstly, images are not always helpful, which may also backfire if they are irrelevant to the textual content at all. Case One: Sometimes the attached image has nothing to do with the textual content of the posts. As the examples given in Fig. 1, these images are not quite relevant to the content of the post texts, but are some popularly-used images which may only express the user's mood. Case Two: Although the attached image is relevant to the text, it may also bring more confusion to entity disambiguation. Let's see the example post (a) in Fig. 2, the "Michael" here actually refers to "Michael Jackson". But since the person in the image "Kobe" is usually more relevant to "Michael Jordan" in the knowledge graph, the MEL results may more likely take the "Michael" here as "Michael Jordon". Secondly, the existing MEL methods never consider to use the attention mechanism to capture the interaction between mention representation and its corresponding entity representation. Thus, there is a great space left for the improvement in the accuracy of MEL results based on social media data.

Given the above, we propose a novel MEL model, which not only removes the negative impact of noisy images, but also uses multiple attention mechanisms to get richer information from the text and images with mentions and their corresponding candidate entities. Specifically, to reduce the the negative impact of noisy images, we design a two-stage image and text correlation mechanism to filter out the irrelevant images based on the predefined threshold. Also, we use multiple attention mechanisms to capture important information in the mention representation and entity representation by quering multi-hop entities around the mention's candidate entities. Let's see the example post (b) in Fig. 2, we first look for entities around the candidate entity "Michael Jordan" which contain the mention text's information. Then we could find the entity "Fruitville Station", such that the multi-modal representation of the entity could be used to strengthen both the mention and the entity's representation. The main contribut-ions of this paper are listed as follows:

- We first propose to remove the bad effect brought by noisy images to MEL task, by identifying noisy images with a two-stage image and text correlation mechanism.
- We design multiple attention mechanism to better capture the connection between mention representation and its corresponding entity through multi-hop query.

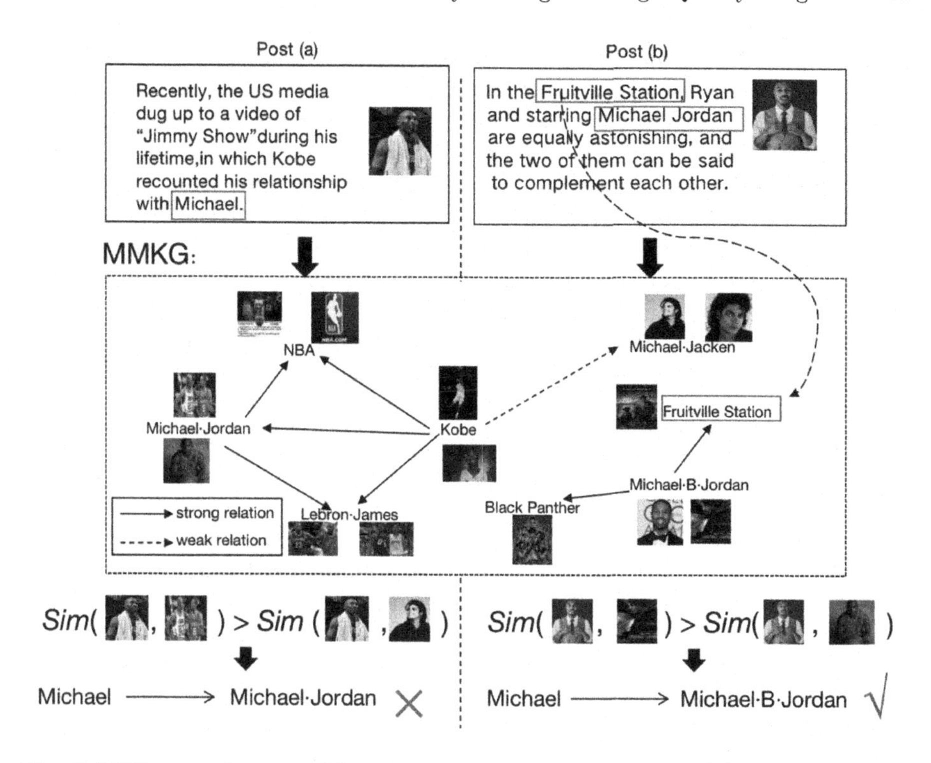

Fig. 2. MEL examples: post (a) is a negative example, while post (b) is a positive one

- We construct the first Chinese data set for MEL task based on Weibo data, which would be published later for public use.

We conduct our empirical study on a large real data collection from WeiBo, which demonstrates that our approach could outperform several state-of-art MEL models.

Roadmap. We first introduce the related work in Sect. 2. The proposed model is introduced in Sect. 3, followed by the experiment section in Sect. 4. We conclude our work in Sect. 5.

2 Related Work

2.1 Muiltimodal Representation Learning

Since BERT [6] was proposed, it has greatly improved the benchmark performance of various NLP tasks by virtue of the powerful feature learning capabilities of Transformer and the two-way encoding realized through the masking language model. Recently, many researchers have turned their attention to Bert-based multimodal representation learning. Lu proposes ViLBERT [28], which ext-ends the popular BERT architecture to a multi-modal model that supports two stream inputs. It preprocesses visual and textual inputs in the two streams separately, and makes them interacted in the joint attention transformer layer. Tan proposes the LXMERT framework to learn the connection between language and vision, through masking cross-modal language modeling, masking target prediction, cross-modal matching, etc. to learn the connection between multiple modaliti-es [29]. Li proposes VisualBERT, which includes a set of stacked Transformer layers and implicitly aligns the elements in a piece of input text with the regions in a related input image with the help of self-attention [27]. Su proposes VL-BERT which takes the simple and effective Transformer model as the backbone and expands it by using visual and language embedded features at the same time [24].

2.2 Entity Linking on Social Data

Recently, several research efforts propose to solve the challenges posed by the EL task on social data. Collective approaches are preferred where the information about social data in relation with the target mention is leveraged. [5,14] extract discriminative features of a mention from its context and an entity from its description, then links a mention to the entity which is most similar. Shen determines the user's topics of interest from all its posted tweets to collectively link all its named entity mentions [20]. Huang adopts a cascade approach to identity links between mentions in microblog and entities in the knowledge base [10]. Liu resolves a set of mentions by aggregating all their related posts to compute mention-mention and mention-entity similarities [11]. Ma adds effective topic semantics on Siamese network to learn representations of context, mention and entity, and ranks the mention-entity similarity [13]. Additionally, with the help of Bert model, Yin improves the entity linking task with the powerful pre-trained general language model by deliberately tackling its potential shortcoming of learning literally [26]. Hua considers social (user's interest + popularity) and temporal contexts [9]. Other collective approaches include the EL model with the non-textual features. For example, Fang and Chong use global information of posts that are close in space and time to the post of the target mention [3,8]. Dreze proposes a joint cross-document co-reference resolution and disambiguation approach including temporal information associated with their corpus to improve EL performance [7]. While these works yield the interesting results using non-textual features, they often depend on the availability of social data and do not exploit visual information. Recently, some work has considered the use of visual information, Omar leverages both semantic textual and visual information extracted from mention and entity contexts [1]. Seungwhan proposes a zero-shot multimodal entity linking solution [15]. Their work has greatly impro-ved the accuracy of entity linking work on social media datasets.

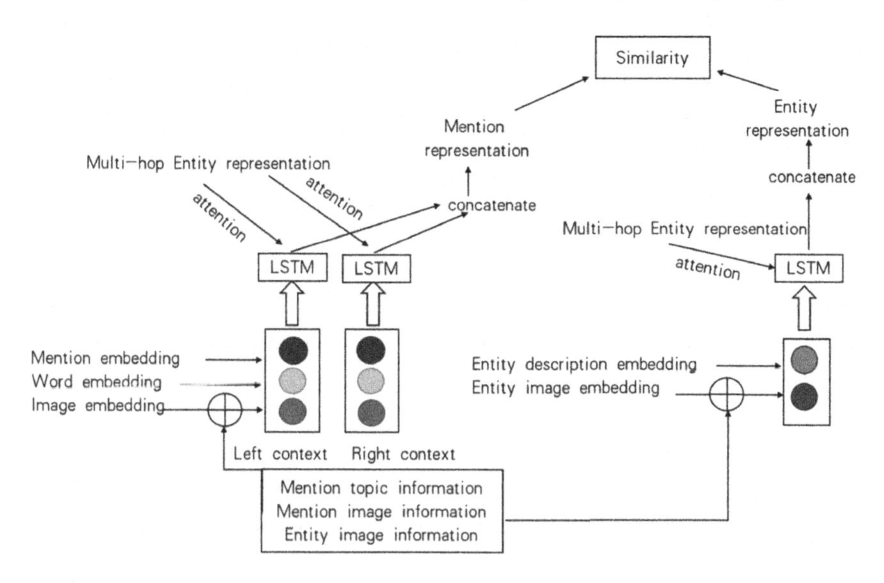

Fig. 3. The architecture of our model

3 Proposed Approach

In this section, we first define our MLE problem, and then introduce how to learn the mention representation followed by the MMKG embedding. Next, we propose to use a two-stage image and text correlation mechanism to filter out the irrelevant images. Then, we novelly propose an improved attention mechanism to capture important information shared by the mention through querying multi-hop entities around candidate entities. Finally, we rank the candidate entities based on a score function. The architecture of our model is illustrated in Fig. 3. We first make respective representations of the mention (mention embedding + word embedding) and the entity (description embedding), based on the plain text entity linking method. Then we use a two-stage image and text correlation mechanism to determine whether to introduce images attached to the text and images attached to the entity. If images are introduced, we will add image embeddings to the representation of mention and entity. The details on how to get these embeddings will be introduced in Sect. 3.2 and Sect. 3.3. Finally, we obtain the important information (multi-hop entity representation) between mention and entities through the multi-hop query, then use the attention mechanism to strengthen the interaction of the common information between the text's mention and the candidate entity.

3.1 Problem Definition

Given the input text-image pair (t, v) where the ambiguous name m_i are extracted from t by searching for names with common English surnames, our goal is to find the most similar entity $e^*(m_i)$ for m_i using the constructed multimodal knowledge graph $MMKG$ which consists of the entity-relation-entity tuples, $\{(e_h, r, e_t)|e_h, e_t \in \mathcal{E}, r \in \mathcal{R}\}$ where $\mathcal{E}$ and $\mathcal{R}$ are the set of entities and relations respectively. Formally, we select the entity $e(m_i)$ which maximizes the similarity between the disambiguated name m_i and the candidate entities $Cand(m_i)$ to get the unambiguous mention as follows:

$$e^*(m_i) = \underset{e_j \in Cand(m_i)}{\arg\max} \quad score(m_i, e_j) \tag{1}$$

where $Cand(m_i)$ is constructed by finding the entities in $\mathcal{E}$ which have the same surname with m_i and $sim(\dot{;})$ is the function to calculate the score function, which will be introduced in Sect. 3.6.

3.2 Mention Representation

In this section, we introduce how to learn the mention representation. For the text-image pair (t, v), we use textual features and visual features from t and v respectively to represent the extracted disambiguated name m_i.

For the textual representation of the mention, we decompose the mention m_i and its context t as a whole into the pre-training Bert model [2] due to its power to understand the semantics of the textual data. Particularly, we obtain the textual representation of the mention, which is denoted as $C_m(t)$, by concatenating the mention embedding and word embedding as follows:

$$C_m(t) = concat(Bert(m_i); Bert(t)) \tag{2}$$

where $Bert(\cdot)$ is used to get the sentence or phrase embedding from the pre-trained embedding, and $concat(\cdot)$ is used to concatenate two emebddings. Similarly, we get the visual representation of the mention, denoted as $C_m(v)$, with the pre-trained model, i.e., VGG16 [21] which is trained with huge amount of image data, to extract the visual features as follows:

$$C_m(v) = VGG(v) \tag{3}$$

where $VGG(\cdot)$ is the function to extract image features and v is the relevant image to the mention.

To capture the semantics of mention completely, we design two strategies to get the mention representation. If the relevant image is not filtered out by our designed two-stage of the calculation of image and text correlation, we will use the concatenation of $C_m(v)$ and $C_m(t)$ as the final representation of the mention C_m. Otherwise, we only use the textual representation of the mention as the final representation of the mention C_m. Specially, the representation of mention is calculated as follows:

$$C_m = \begin{cases} Concat(C_m(v); C_m(t)), & v! = NULL \\ C_m(t), & otherwise \end{cases} \tag{4}$$

The details of the two-stage of the calculation of image and text correlation will be introduced later.

3.3 MMKG and Candidate Entity Representation

MMKG is mainly used to combine different entities and images in different knowledge graphs to perform relational reasoning [16,18,23,25,30], which contains multi-relational link prediction and entity matching. Existing work [1] has shown the improvement of linking effects which combines textual representation and visual representation of the entity by using MMKG. while in this paper we use the multi-hop relationship between entities in MMKG to enhance the performance of entity linking task.

Similar to previous work [12], we construct our MMKG on the basis of the knowledge graph containing all entity attributes, triple knowledge and (linked to) images, as well as knowledge graphs alignment between entities. However, there exist the entities in the MMKG which share the same surnames, i.e., ambiguous entities. We denote the ambiguous entities as the candidate entities $Cand(m_i)$ by matching the entity in $\mathcal{E}$.

For the representation of candidate entities, we adopt a method similar to that of the mention. Specifically, for the textual representation of candidate entities, since each entity in the MMKG has more than one corresponding attributes-attribute values and a long text description, we put their embedding together to form the entity context (similar to the context of mention), and then input the candidate entity and its "context" into the pre-trained Bert model, which is denoted as $C_e(t)$. For the visual information representation of candidate entities, we also use the VGG16 to extract features of the entity image. We unify the entity representation as $C_e(v)$. We adopt the similar way to compute the representation of candidate entities to the mention representation, which is denoted as C_e.

3.4 A Two-Stage Image and Text Correlation Mechanism

In this section, we introduce how to calculate the similarity between the topical information of the mention's text and the category information of the mention's image, the similarity between the category information of the mention's image and the category information of the entity's image. Considering that the topic of the mention's text represents the semantic information of the mention's text while the image category information (image topic) can also represent what the image focus on, therefore, we try to get the topics of mention's text and the category of images to calculate their similarity. Specifically, we use LDA model to get the topic information of the mention text, which is denoted as T_m, and the pre-trained Inception-v3 model [22] to get the category information of the mention image and the entity image, denoted as I_m and I_e respectively, with the help of image classification task. Note that there are 1000 categories in the ImageNet, but we only take the 5 category items with the highest probability value as the image topics.

After that, to determine whether the text's image is highly related to the mention, we adopt a two-stage image and text correlation mechanism. In the first stage, if the correlation is larger than the predefined similarity threshold τ_1, the text's image will be retained. Otherwise, it will be eliminated. The similarity between the topical information of mention's text and the category information of the mention's image is computed as follows:

$$sim(T_m, I_m) = \frac{Bert(T_m) \cdot VGG(I_m)}{\sqrt{Bert(T_m)^2 + VGG(I_m)^2}} \tag{5}$$

Otherwise, we go to the second stage. Based on our observation on the data, we find that given the randomness of the content posted by social media users, some of the data is like this: the text content (containing topical information) has a low degree of relevance to the mention, but the text's image has a high degree of relevance to the mention. The images in this type of data were eliminated in the previous stage of the matching calculation process, but it is clear that these images should be retained. As shown in the example post (b) in Fig. 2, the similarity between the text's topical information and the category information of the text's image is not up to the threshold, but it is obvious that the image containing the "correct" Michael Jordan (Michael·B·Jordan) is helpful for us to link the mention to the correct entity. So if this situation is satisfied, we calculate the similarity between the category information of the mention's image and the category information of entity's image as follow:

$$sim(I_m, I_e) = \frac{VGG(I_m) \cdot VGG(I_e)}{\sqrt{VGG(I_m)^2 + VGG(I_e)^2}} \tag{6}$$

If the relevance is larger than the threshold τ_2, the mention's image and entity's image will be retained.

3.5 Attention Mechanism Based on Finding Useful Multi-hop Entities

Although we have learned the mention representation and the entity representation through the combination of Bert and VGG networks, we input the mention representation into the LSTM network based on this consideration: given that the text often contains a lot of irrelevant information, we only need to pay attention to the surrounding text information of the mention that we need to disambiguate. Using LSTM network can help us better capture the key information of the mention context. Specifically, we divide the C_m into the left and right parts according to the position of the mention in the text: $C_{m_{left}}$ and $C_{m_{right}}$. We take $C_{m_{left}}$ as an example, $C_{m_{left}} = C^1_{m_{left}}, C^2_{m_{left}}, ..., C^n_{m_{left}}$, where $C^t_{m_{left}}$ is a concatenation of the mention embedding ,word embedding and image embedding to be passed as input at time t, we take $C^t_{m_{left}}$ as x_t, the LSTM unit will output h_t for each time step t. The hidden vector h_t is computed as follows:

$$i_t = \sigma(W_{xi}x_t + W_{hi}h_{t-1} + W_{ci}c_{t-1} + b_i) \tag{7}$$

$$f_t = \sigma(W_{fi}x_t + W_{hf}h_{t-1} + W_{cf}c_{t-1} + b_f) \tag{8}$$

$$c_t = f_t c_{t-1} + i_t \tanh(W_{xc}x_t + W_{hc}h_{t-1} + b_c) \tag{9}$$

$$o_t = \sigma(W_{xo}x_t + W_{ho}h_{t-1} + W_{co}c_{t-1} + b_o) \tag{10}$$

$$h_t = o_t \tanh(c_t) \tag{11}$$

where i_t,f_t,o_t,c_t are the input gate, forget gate, output gate, and cell memory at position t respectively. σ denotes the sigmoid function. W_c,W_h,W_x are weighted matrices, and b_i,b_f,b_c,b_o represent the biases of the LSTM network. All the above parameters need to be learned during training.

Though we have got the representation of the mention through LSTM network, given the randomness of the content of the Weibo data (even if noisy images have been eliminated), we may still be unable to capture important information to help our linking work. To solve this problem, we propose to design an attention mechanism that can capture the connection between the mention's representation and its corresponding entity. In additon, we also use the multi-hop query during using the attention mechanism. Before we go to the details, we first define one-hop entity and two-hop entity in MMKG.

Definition 1. *One-hop entities. The candidate entities in the MMKG contain a large number of triples where these associated entities, i.e., the head entity or tail entity corresponding to candidate entities, are denoted as the one-hop. In these one-hop entities, we look for the name of entities that have appeared in the mention's text.*

Definition 2. *Multi-hop entities. In the process of searching for one-hop entities, if no one-hop entities that have appeared in the mention's text are found, we will look for the related entities of these one-hop entities, that is, the two-hop entities of candidate entities. By that analogy, we denote these kinds of entities which are connected with candidate entities with multiple hops as the multi-hop entities.*

Specifically, we inject the representation of these one-hop or multi-hop entities as attention vectors into the mention representation and the entity representation. It is worth noting that in the process of searching for a multi-hop entities, we find more than one-hop entities or two-hop entities have appeared in the mention's text. In this paper, we take these multi-hop entities as $Q_1, Q_2, ..., Q_N$, which are used as different queries to pay attention to mention's representation and candidate entity's representation, equalling to repeating multiple single-layer attention. The way that one-hop entity is used to contract important information in the mention's representation is shown in Fig. 4. In detail, we calculate the attention scores as follows:

$$\begin{gathered} Attention(\boldsymbol{Q}, \boldsymbol{K}, \boldsymbol{V}) = softmax(\frac{\boldsymbol{Q}\boldsymbol{K}^T}{\sqrt{d}})\boldsymbol{V} \\ Q_i = QW_i^Q, K_i = KW_i^K, V_i = VW_i^V, i = 1, 2, ..., N \end{gathered} \tag{12}$$

$$\begin{gathered} MultiHead(\boldsymbol{Q}, \boldsymbol{K}, \boldsymbol{V}) = Concat(head_1, ..., head_N)W^o \\ where\, head_i = Attention(QW_i^Q, KW_i^K, VW_i^V) \end{gathered} \tag{13}$$

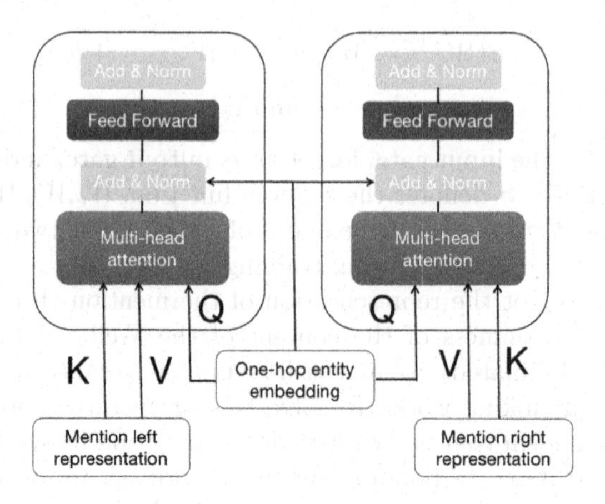

Fig. 4. Modified attention layer

In the example post (b) shown in Fig. 2, through multi-hop quering around Michael·B·Jordan, we find that the entity Fruitville Station appeared in the mention's text (only Fruitville Station appeared in the surrounding one-hop entities), so we use the multi-modal representation of the entity Fruitville Station (the representation method is the same as that of the candidate entity) as an attention vector to extract important information in the mention representation and the entity representation.

3.6 Candidate Entities Ranking

The mention and entity representations are concatenated and then forwarded to a fully connected layer. The output of the fully connected layer is a single node, denoting the similarity score after processed by a sigmoid function. Suppose s is the final similarity score and g represents whether the entity is the true entity (ground truth). The training objective is to minimize the following loss value:

$$L(s, g) = g \log(s) + (1 - g) \log(1 - s) \tag{14}$$

where g is the ground truth label with the value 1 or 0.

The candidate entities are not ranked solely based on multi-modal representation's similarity. Instead, the final score of each candidate entity is the combination of similarity score and prior probability $p(e|m_i)$ of an entity e, which denotes the possibility that the entity e is the true one given a specific mention m_i. The specific values of prior probabilities are derived from a frequency dictionary which we will introduce in the experiment, while entities not in the frequency dictionary are assigned with frequency value of 0. Formally, the ranking score of mention-entity pair (m, e) is:

$$Score(m, e) = \theta sim(m, e) + \eta p(e, m) \tag{15}$$

where θ and η are coefficients balancing the weights of similarity and prior probability.

Table 1. Neural network parameter settings

Parameters	Values
Window size of mention context	10
Window size of entity description	150
LSTM hidden state size for mention context	288
LSTM hidden state size for entity description	96
Output size for hidden layer	200
Activation function for hidden layer	tanh
Number of epochs	50
Batch size	128
Optimizer	Adam

4 Experiments

4.1 Experimantal Settings

MMKG: The MMKG is derived from Baidu Encyclopedia. Specifically, we utilize Mongodb to store the knowledge due to its popularity and simplicity, as well as limited demand for relation information in our work. We conduct basic SQL operations to keep the useful information needed for MEL task, which consists of entity ID, entity name, entity description and entity image. In all, there are about 1,500,000 entities in our MMKG.

Frequency Dictionary: We construct a name dictionary for formalizing irregul-ar forms of mentions. Specifically, the elements in the dictionary are obtained from Baidu Encyclopedia pages. Aside from normal words, Baidu Encyclopedia pages also contain anchor texts, which are attached with links directing at other pages. Since each Baidu Encyclopedia page represents a unique entity, the anchor text accordingly could be regarded as surface form of the entity it points to. As thus, we can attain a dictionary consisting of surface forms and their possible referent entities.

Settings for the Neural Network: The specific hyperparameters in the deep neural network are shown in Table 1. For each mention, we extract the context with a window size of 10, and the entity description text include 150 words in its Baidu Encyclopedia page. The word embeddings in the neural network are stable during the model training until using the attention mechanism, and only the parameters in the neural network are learned. Due to limitation of computational resources, we use negative samples to accelerate the training process. Specifically, for each true entity, 5 negative entities are created by replacing the correct entity from the mention's candidate entities.

4.2 Dataset and Baselines

We first construct a dictionary of common English surnames (in the form of Chinese), which containes about 200 common English surnames. We index these 200 common English surnames and crawl 50,000 text-image data on Weibo. For the crawled data, we only keep the data containing the last name instead of the full name, whose size is about 20,000, and use manual annotations to link the names in these data to the corresponding names in the MMKG. We split the dataset randomly into train (70%), validation (15%), and test sets (15%).

Since our approach is not directly comparable with previous works, we compa-re the results with different configurations and baselines. We first conduct a few baseline experiments, which are based on popularity-based candidate entity ranking, text-based similarity-based candidate entity ranking, and image similar-ity-based entity ranking. For popularity-based candidate entity ranking, we use the frequency dictionary we have set up, the input is the text context, mention, some attributes of a certain entity, and the output is the confidence that the mention points to the entity, and use the rank to select the most credible entity. For text-based similarity-based candidate entity ranking, we input the mention texutual representation and the entity textual representation to the nerual network we have designed (without the two-stage image and text correlat-ion mechanism and attention mechanism) to calculate their similarity. For image-based similarity-based candidate entity ranking, we calculate the similarity of the visual representations of the mention and the entity which are obtained by VGG16 through the method similar to the second stage which is introduced in Sect. 3.4. In order to test the effect of the model designed in this article, we carry multi-modal entity linking, and in the multi-modal entity linking work, we test whether to introduce the calculation of image and text correlation, whether to introduce the attention mechanism. Which features and models we have adopted are summarized in Table 2.

4.3 Results

Table 3 reports the accuracy on the validation and test sets for the candidate entity ranking task. We find that our proposed model achieves the best performance compared with all the baselines. Specifically, the conventional multi-modal entity linking model performs better than the single modal data models, i.e., text-based and image-based. And our model outperforms the conventional multi-modal entity linking model with 11% improvement. This illustrates our two-stage image and text correlation mechanism and the attention mechanism contribute a lot. In details, we find that attention mechanism improves the accuracy by 4%, which proves that multi-hop entities can capture important information in the mention representation and the entity representation. Two-stage correlation strategy improves the accuracy by 6% compared with the only first-stage strategy, which proves that some images that can be helpful to MEL task are eliminated in the first stage.

Table 2. Features and models used in our experiments.

Features	Description
Popularity	Baseline feature where the most popular entity is selected
Text	Similarity measured between the mention text's embedding and the entity description's embedding
Image	Similarity measured between the mention image's embedding and the entity image's embedding
Text+image	Combine the text embedding and the image embedding
One-stage (text-image similarity)	Calculate the similarity between the topical information of the mention's text and the category information of the mention's image
Two-stage (text-image similarity+image-text similarity)	Calculate the similarity between the topical information of the mention's text, the category information of the mention's image and the category information of the entity's image
One-stage (text-image similarity)+Att	Add the attention mechanism based on the one-stage
Two-stage (text-image similarity+image-text similarity)+Att	Add the attention mechanism based on the two-stage

Table 3. Multimedia entity linking results (accuracy)

Features and models	Valid	Test
Popularity	0.352	0.470
Text	0.463	0.535
Image	0.182	0.196
Text+image	0.635	0.664
One-stage (text-image similarity)	0.632	0.668
Two-stage (text-image similarity+image-image similarity)	0.693	0.737
One-stage (text-image similarity)+Att	0.676	0.702
Two-stage (text-image similarity+image-image similarity)+Att	0.743	0.796

Fig. 5. An example of failed entity linking

4.4 Error Analysis

By observing the examples of linking errors, we find that linking errors are only in the following two cases. The first case is that the text of the mention is very short and does not contain important semantic information, and the text's image is not related to the mention. In this case, whether or not leaving the image has no effect on the representation of the mention. As shown in Fig. 5, we can't obtain useful information from the text or the image to link Michael Jordan to the correct entity. The second case is that we do not know what the mention refers to based on the text and the image, that is, when we manually label the mention, we cannot determine which entity it belongs to in MMKG.

5 Conclusions and Future Work

Based on the conventional multi-modal entity linking method, this paper designs a two-stage image and text correlation mechanism to eliminate images that are not helpful for the linking task, and introduces an attention mechanism to capture important information in the mention representation and entity representation. Our experiments on the Weibo dataset prove that this method of only retaining high-quality images and using the attention mechanism can significantly improve the accuracy of entity linking.

In future work, we will improve the model design. For the text of the removed image, we will generate an image for the text that conforms to the text theme, so as to complete higher-quality multi-modal entity linking.

Acknowledgments. This research is supported by National Key R&D Program of China (No. 2018-AAA0101900), the Priority Academic Program Development of Jiangsu Higher Education Institutions, National Natural Science Foundation of China (Grant No. 62072323, 61632016), Natural Science Foundation of Jiangsu Province (No. BK20191420), and the Suda-Toycloud Data Intelligence Joint Laboratory.

References

1. Adjali, O., Besançon, R., Ferret, O., Le Borgne, H., Grau, B.: Multimodal entity linking for Tweets. In: Jose, J.M., et al. (eds.) ECIR 2020, Part I. LNCS, vol. 12035, pp. 463–478. Springer, Cham (2020). https://doi.org/10.1007/978-3-030-45439-5_31
2. Cheng, J., et al.: Entity linking for Chinese short texts based on BERT and entity name embeddings. In: China Conference on Knowledge Graph and Semantic Computing (CCKS) (2019). https://conference.bj.bcebos.com/ccks2019/eval/webpage/pdfs/eval_paper_2_1.pdf
3. Chong, W.-H., Lim, E.-P., Cohen, W.: Collective entity linking in Tweets over space and time. In: Jose, J.M., Hauff, C., Altıngovde, I.S., Song, D., Albakour, D., Watt, S., Tait, J. (eds.) ECIR 2017. LNCS, vol. 10193, pp. 82–94. Springer, Cham (2017). https://doi.org/10.1007/978-3-319-56608-5_7
4. Csomai, A., Mihalcea, R.: Linking documents to encyclopedic knowledge. IEEE Intell. Syst. **23**(5), 34–41 (2008)
5. Cucerzan, S.: Large-scale named entity disambiguation based on Wikipedia data. In: Proceedings of the 2007 Joint Conference on Empirical Methods in Natural Language Processing and Computational Natural Language Learning (EMNLP-CoNLL), pp. 708–716 (2007)
6. Devlin, J., Chang, M.W., Lee, K., Toutanova, K.: BERT: Pre-training of deep bidirectional transformers for language understanding. arXiv preprint arXiv:1810.04805 (2018)
7. Dredze, M., Andrews, N., Deyoung, J.: Twitter at the Grammys: a social media corpus for entity linking and disambiguation. In: International Workshop on Natural Language Processing for Social Media (2016)
8. Fang, Y., Chang, M.W.: Entity linking on microblogs with spatial and temporal signals. Trans. Assoc. Comput. Linguist. **2**, 259–272 (2014)
9. Hua, W., Zheng, K., Zhou, X.: Microblog entity linking with social temporal context, pp. 1761–1775 (2015)
10. Huang, D., Wang, J.: An approach on Chinese microblog entity linking combining Baidu Encyclopaedia and word2vec. Procedia Comput. Sci. **111**, 37–45 (2017)
11. Liu, X., Li, Y., Wu, H., Ming, Z., Yi, L.: Entity linking for Tweets. In: Meeting of the Association for Computational Linguistics (2017)
12. Liu, Y., Li, H., Garcia-Duran, A., Niepert, M., Onoro-Rubio, D., Rosenblum, D.S.: MMKG: multi-modal knowledge graphs. In: Hitzler, P., Hitzler, P., et al. (eds.) ESWC 2019. LNCS, vol. 11503, pp. 459–474. Springer, Cham (2019). https://doi.org/10.1007/978-3-030-21348-0_30
13. Ma, C., Sha, Y., Tan, J., Guo, L., Peng, H.: Chinese social media entity linking based on effective context with topic semantics. In: 2019 IEEE 43rd Annual Computer Software and Applications Conference (COMPSAC), vol. 1, pp. 386–395. IEEE (2019)
14. Mihalcea, R., Csomai, A.: Wikify! linking documents to encyclopedic knowledge. In: Proceedings of the Sixteenth ACM Conference on Conference on Information and Knowledge Management, pp. 233–242 (2007)
15. Moon, S., Neves, L., Carvalho, V.: Multimodal named entity disambiguation for noisy social media posts. In: Proceedings of the 56th Annual Meeting of the Association for Computational Linguistics (Volume 1: Long Papers), pp. 2000–2008 (2018)

16. Mousselly-Sergieh, H., Botschen, T., Gurevych, I., Roth, S.: A multimodal translation-based approach for knowledge graph representation learning. In: Proceedings of the Seventh Joint Conference on Lexical and Computational Semantics, pp. 225–234 (2018)
17. Nguyen, T.H., Fauceglia, N.R., Muro, M.R., Hassanzadeh, O., Gliozzo, A., Sadoghi, M.: Joint learning of local and global features for entity linking via neural networks. In: Proceedings of COLING 2016, the 26th International Conference on Computational Linguistics: Technical Papers, pp. 2310–2320 (2016)
18. Pezeshkpour, P., Chen, L., Singh, S.: Embedding multimodal relational data for knowledge base completion. arXiv preprint arXiv:1809.01341 (2018)
19. Shen, W., Wang, J., Han, J.: Entity linking with a knowledge base: issues, techniques, and solutions. IEEE Trans. Knowl. Data Eng. **27**(2), 443–460 (2014)
20. Shen, W., Wang, J., Luo, P., Wang, M.: Linking named entities in tweets with knowledge base via user interest modeling. In: ACM SIGKDD International Conference on Knowledge Discovery & Data Mining (2013)
21. Simonyan, K., Zisserman, A.: Very deep convolutional networks for large-scale image recognition. arXiv preprint arXiv:1409.1556 (2014)
22. Szegedy, C., Vanhoucke, V., Ioffe, S., Shlens, J., Wojna, Z.: Rethinking the inception architecture for computer vision. In: Proceedings of the IEEE Conference on Computer Vision and Pattern Recognition, pp. 2818–2826 (2016)
23. Tao, Z., Wei, Y., Wang, X., He, X., Huang, X., Chua, T.S.: MGAT: multimodal graph attention network for recommendation. Inf. Process. Manage. **57**(5), 102277 (2020)
24. Yang, Z., Zheng, B., Li, G., Zhao, X., Zhou, X., Jensen, C.S.: Adaptive top-k overlap set similarity joins. In: 2020 IEEE 36th International Conference on Data Engineering (ICDE), pp. 1081–1092. IEEE (2020)
25. Yen, A.Z., Huang, H.H., Chen, H.H.: Multimodal joint learning for personal knowledge base construction from Twitter-based lifelogs. Inf. Process. Manage. **57**(6), 102148 (2019)
26. Yin, X., Huang, Y., Zhou, B., Li, A., Lan, L., Jia, Y.: Deep entity linking via eliminating semantic ambiguity with BERT. IEEE Access **7**, 169434–169445 (2019)
27. Zheng, B., et al.: Online trichromatic pickup and delivery scheduling in spatial crowdsourcing. In: 2020 IEEE 36th International Conference on Data Engineering (ICDE), pp. 973–984. IEEE (2020)
28. Zheng, B., Zhao, X., Weng, L., Hung, N.Q.V., Liu, H., Jensen, C.S.: PM-LSH: a fast and accurate lSH framework for high-dimensional approximate NN search. Proceedings of the VLDB Endow. **13**(5), 643–655 (2020)
29. Zheng, B., et al.: Answering why-not group spatial keyword queries. IEEE Trans. Knowl. Data Eng. **32**(1), 26–39 (2018)
30. Zhu, Y., Zhang, C., Ré, C., Fei-Fei, L.: Building a large-scale multimodal knowledge base system for answering visual queries. arXiv preprint arXiv:1507.05670 (2015)

Learning to Label with Active Learning and Reinforcement Learning

Xiu Tang, Sai Wu(✉), Gang Chen, Ke Chen, and Lidan Shou

College of Computer Science and Technology, Zhejiang University, Hangzhou, Zhejiang, China
{tangxiu,wusai,cg,chenk,should}@zju.edu.cn

Abstract. Training data labelling is financially expensive in domain-specific learning applications, which heavily relies on the intelligence from domain experts. Thus, with budget constraint, it is important to judiciously select high-quality training data for labelling in order to prevent over-fitting. In this paper, we propose a learning-to-label (L2L) framework leveraging active learning and reinforcement learning to iteratively select data to label for Name Entity Recognition (NER) task. Experimental results show that our approach is more effective than strong previous methods using heuristics and reinforcement learning. With the same number of labeled data, our approach improves the accuracy of NER by 11.91%. Our approach is superior to state-of-the-art learning-to-label method, with an improvement of accuracy by 6.49%.

Keywords: Active learning · Reinforcement learning · Name Entity Recognition

1 Introduction

Most neural models are built on top of a few open datasets with well-defined labels, such as ImageNet [1], Coco [7] and Wikipedia dataset. Those models cannot be directly applied to a new domain and hence, we need to train the neural model using domain-specific labels. However, it is very expensive or even impossible to create a large training dataset for domain-specific applications.

To address the problem of lacking labeled training data, two different approaches have been proposed, the co-training technique [11] and the transfer learning technique [3]. However, the performance of co-training relies on the assumption that different models will generate correct labels for different portions of data. This may not be true in the real case. On the other hand, transfer learning establishes a model on the source domain and transfers the knowledge to a target domain. If the target domain and source domain do not share many common features, the results are not very promising.

In this paper, we extend the idea of transfer learning by exploiting the active learning and reinforcement learning techniques. We propose learning to label approach, L2L, to rank the data for labeling for the Name Entity Recognition

C. S. Jensen et al. (Eds.): DASFAA 2021, LNCS 12682, pp. 549–557, 2021.
https://doi.org/10.1007/978-3-030-73197-7_36

(NER) application. L2L consists of two models, a transfer learning model and an active learning model designed by a reinforcement learning process, named as T-model and A-model, respectively. The contributions of the paper are as follows:

- We propose a new L2L architecture to reduce the number of required labels for training a domain-specific neural model.
- We design T-model by using an adversarial neural network to domain adaptation for the NER task. A multi-granularity attention model is proposed, which can catch both the local features and global features.
- We propose a NAS-powered learning to rank model to estimate the importance of sentences to the T-model, so that we can improve the T-model effectively with fewer labeled data.

Experimental results show that our approach is more effective than previous methods using heuristics and reinforcement learning. We obtain an improvement over the previous state-of-the-art active learning approach [3] by 6.49%.

2 Related Work

Our work is related to two lines of research: transfer learning and active learning.

Transfer Learning. For the lack of labeled data, domain adaptation of transfer learning is a solution [4]. Adversarial learning is a widely used model to solve the problem of image generation [2] and domain adaptation [5]. However, due to the poor commonality of features between different domains, the results of these methods are not satisfactory. Therefore, we propose to extend the idea of transfer learning by using active learning technology.

Active Learning. Active learning is a technique that chooses fewer datasets for annotation to obtain a better classification effect [8]. An active learning algorithm based on a deep Q-network is designed, where actions correspond to binary annotation decisions applied to the data stream [3]. However, due to the complex data distribution, these methods cannot guarantee the validity of the selected data. Therefore, we propose to use reinforcement learning to design the best active learning model.

3 L2L Approach

The architecture of the L2L is illustrated in Fig. 1. The system consists of three main components: NER model, multi-granularity attention, and learning to rank. In this section, we will explicitly explain the three parts of the proposed approach and training approaches.

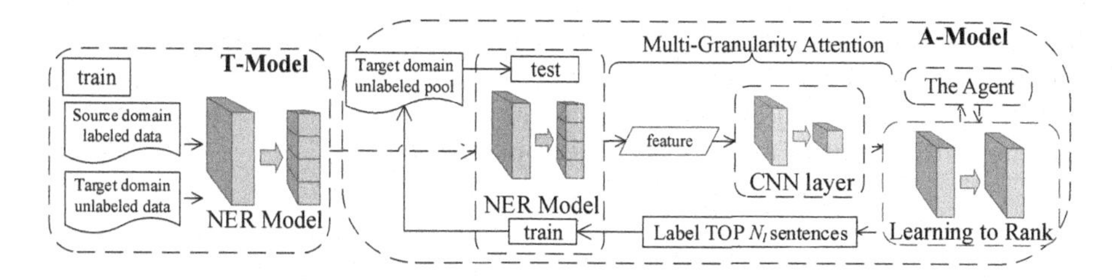

Fig. 1. The basic flow of the L2L model

3.1 Domain Adaptation with Multi-granularity Attention

We design T-model by using an adversarial neural network to domain adaption for the NER task. Our T-model has four modules: feature extractor, entity classifier, domain discriminator and target domain autoencoder. The T-model is based on the target preserved adversarial neural network (TPANN) [6], improves the TPANN for the NER task by using the character encoding layer and the CRF layer for classification. The character encoding layer is used to handle out-of-vocabulary words. The CRF layer allows us to capture the contextual connection for sequential data and the Viterbi algorithm is used to represent the probability of path planning.

The feature extractor F module uses CNN (Convolutional Neural Network) to extract the features of character embedding, which effectively solves the problem of out-of-vocabulary words. Then, we connect the character embedding and the word embedding as the input of the subsequent biLSTM layer, which is used to model sentences. And the hidden states h of the biLSTM become the feature that will be transferred to the next three models, which are Entity Classifier P, Domain Discriminator Q and Target Domain Autoencoder R.

The entity classifier P and domain discriminator Q are both feed-forward neural networks. P is enhanced with a CRF layer to predict labels for entities. Q is trained to discriminate domain labels to make the prediction domain-invariant. The parameter θ_f is optimized to make the domain discriminator Q can not discriminate the label of the domain. Namely, the feature extractor F constructs the common feature between the source domain and target domain. The establishment of an autoencoder module to maintain domain-specific functions is important for entity classification. According to the above process, our model learns the common features between the source domain and the target domain, while preserving specific feature of target domain.

The sentence feature representation extracted by T-model is enhanced with multi-granularity attentions. We propose the manufactured hierarchies based on batch length, the larger batch length extracts global features for entity recognition, whereas the smaller batch length maintains local features. The overall architecture of the multi-granularity attention network is shown in Fig. 2.

We can simultaneously train N_m NER models. Suppose each batch has K_s sentences. The ith model is trained to represent data characteristics, prediction label and prediction probability based on $\frac{K_s}{2^i}$ sentences. In other words, K_s

sentences will be split into 2^i mini-batches and the model is trained to represent a mini-batch. In our current settings, $K_s = 20$ and $N_m = 3$, i.e., in Fig. 2, the big batch, middle batch and small batch models are trained for representing 20 sentences, 10 sentences and 5 sentences respectively.

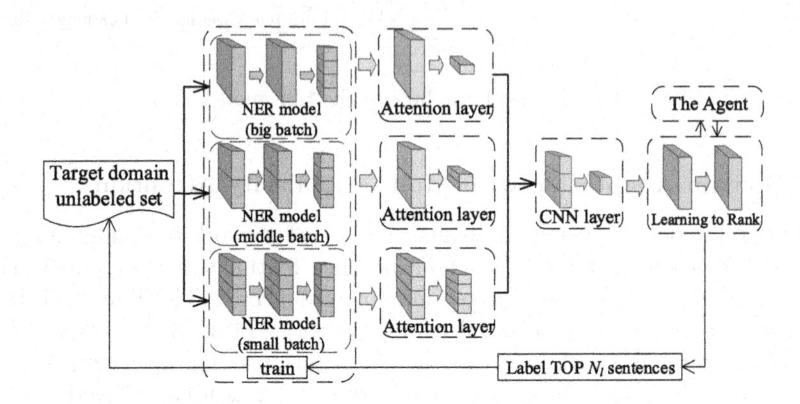

Fig. 2. The general architecture of the multi-granularity attention model

To make the local and global features more significant, we propose an attention module, which is shared among all active learning models, to obtain the features that should be paid more attention to in the local and global features respectively. After the informative feature is acquired through the attention mechanism, the local feature and the global feature are integrated through CNN to obtain the final sentence feature representation.

3.2 Learning to Ranking

In this section, we describe our active learning model in our A-model. A NAS-powered learning to rank model is proposed to estimate the importance of sentences to the T-model.

Model Description. In our task, we are given an unlabeled set of N sentences $\{s_i\}_{i=1}^N$, each represented by a D-dimensional feature vector $s_i \in \mathbb{R}^D$. The goal of the learning to rank model is to learn nonlinear function $Lr : (s_i, s_j) \mapsto z \in \{0, 1\}$ from input space $\mathbb{R}^D$ to label space $\{0, 1\}$ using deep neural networks, which encodes sentences s into compact feature $z = Lr(s_i, s_j)$ such that the information of pairwise importance between the given pairs can be preserved in the compact features. The neural model is defined as a label classifier: $Lr(s_i, s_j) = [z_{ij}, p_i]^+$, s_i and s_j is the vector of the sentence pair. z_{ij} indicates the representation of the importance rank and p_i is the probability that the importance rank belongs to the class.

To construct feature vectors, the learning to rank model is composed of several CNN networks. The learning to rank model aims to judge pairwise relative importance following loss function:

$$L_{rank} = -\sum_{i=1}^{N}\sum_{j=1}^{N}\{z_{ij}\log\hat{z_{ij}} + (1-z_{ij})\log(1-\hat{z_{ij}})\}, \quad (1)$$

where z_{ij} is the ground truth of the pairwise importance label for sample pair (s_i, s_j) and $\hat{z_{ij}}$ is the predicted label probability. N sample pairs are retrieved from the target domain.

The architecture accepts pairwise input sentences $\{(s_i, s_j, z_{ij})\}$ and processes them by an end-to-end deep learning model: (i) a convolution network (CNN) is used to learn deep representation of each sentence s_i, (ii) an activation function is used to map feature representation to relative importance $z_{ij} \in \{0, 1\}$.

Designing Convolutional Cells. We propose a selection model based on reinforcement learning to automatically generate the CNN architecture with high performance for our learning to rank model. The model is an agent which can sequentially select CNN layers. The agent is trained by using Q-learning model, while using experience replay and ε-greedy exploration strategy. The agent explores possible architectures based on experience, which constitute a large but finite space, and iteratively discovers architectures with better performance.

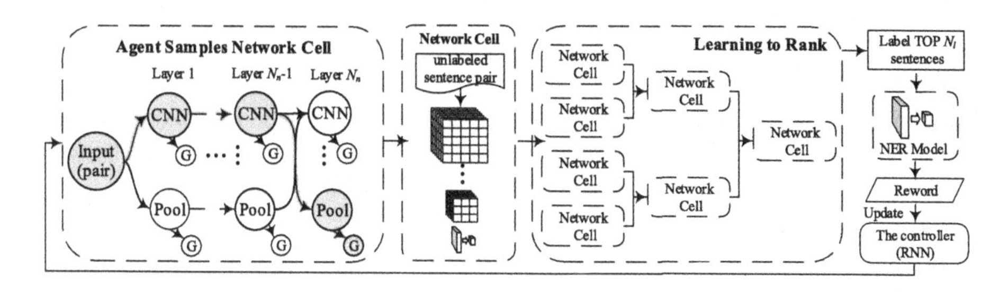

Fig. 3. The training procedure of the learning to rank model

The basic flow of the learning to rank model with the controller is shown in Fig. 3. G represents a termination state (softmax). For each layer of the network cell, the controller selects the layer type and the corresponding parameters. In Fig. 3, feasible states (each layer) and action spaces (lines) and a possible selection trajectory of the agent and the corresponding CNN architecture of the trajectory are shown. The figure shows a trajectory which highlighted in green to indicate the path that the agent may select and the CNN architecture defined by the path.

Training Procedure: In the model training process, the intermediate reward at each step is defined as the change in validation accuracy, and the reward is $\mathcal{R}(e_{i-1}, a) = Acc(\phi_i) - Acc(\phi_{i-1})$, where Acc is the F1 score representing the accuracy of the prediction, and ϕ_i is the NER model trained according to the learning to rank model after performing action a.

We use experience replay memory M to maintain every transition (e, a, r, e') during the training process. Then, a mini-batch of transitions is sampled from the memory to optimize the parameters of the model. And the loss function is as follows:

$$L_{agent} = E_{e,a,r,e'}[(y_i(r, e') - Q(e, a; \theta_q))^2], \tag{2}$$

where $y_i(r, e') = r + \gamma max_{a'} Q(e', a'; \theta_{q_{i-1}})$ is the expected Q-value of the current neural parameters $\theta_{q_{i-1}}$, and the target is over the minibatch.

4 Experiments

The experiments are conducted in two steps. First, we use the T-model trained from the source domain and transfer it into the target domain as our NER model. Then, we train an A-model to rank data for labeling. For comparison, we use other baseline methods to select the same number of samples for labeling and compare the F1 score with our method.

4.1 Evaluation Plan

The datasets required for the training of the proposed model include resource-rich labeled data from out-of-domain, a large amount of unlabeled data and a small amount of labeled data from in-domain. Therefore, three kinds of datasets are used in our work:

Labeled out-of-domain data: As source corpora, we use CoNLL2003 shared tasks[1] for the English dataset. For the Chinese dataset, we use a standard benchmark dataset for adversarial NER, the tagged corpus of People's Daily[2].

Unlabeled in-domain data: As target corpora, we use NER corpora from OntoNotes-5.0[3] for the English dataset. For the Chinese dataset, we use NER corpora from the MSRA corpus[4] and the Financial News (Real dataset that we build).

Labeled in-domain data: We adopt labeled in-domain data for training and evaluating the active learning model. 1,000 sentences from the datasets are used to train the model, and another 2,000 sentences are for testing.

We compare with the following active learning approaches: Random sampling, Diversity sampling [10], Uncertainty-based sampling [9], Policy based Active Learning (PAL) [3].

[1] http://www.cnts.ua.ac.be/conll2003/ner/.

[2] https://www.lancaster.ac.uk/fass/projects/corpus/pdcorpus/pdcorpus.htm.

[3] https://catalog.ldc.upenn.edu/LDC2013T19.

[4] http://sighan.cs.uchicago.edu/bakeoff2006/download.html.

4.2 Results

In this section, we report the results of the experiment and give detailed analyses of the results.

Figure 4 shows the F1 score of the NER model using labeled data selected by different active learning approaches on the corpus of the OntoNotes-5.0, Financial news and MSRA. "L2R" refers to the learning to rank model. "L2L" refers to the learning to rank model with the multi-granularity attention method.

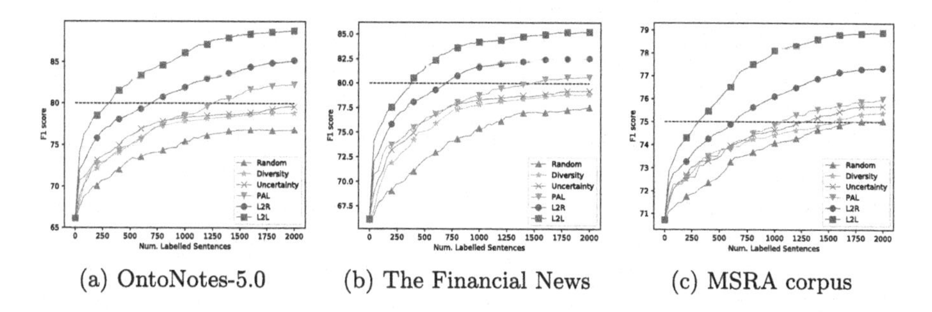

(a) OntoNotes-5.0 (b) The Financial News (c) MSRA corpus

Fig. 4. The F1 score on the test dataset

Using the same number of labeled data, the L2L model leads to a much better prediction result, indicating that it is capable of selecting the most important sentences for the NER model. And if we want to achieve intermediate accuracy, our model can highly reduce the number of labeled data. For the OntoNotes-5.0 corpus, we obtain an improvement over the closest competitor by 6.49%. As Fig. 4(a), our L2L model labels 48% less data than other active learning models for achieving 80% accuracy. For the Financial corpus, we obtain an improvement over the closest competitor by 4.62%. As Fig. 4(b), our L2L model labels 54% less data than other active learning models for achieving 80% accuracy. For the MSAR corpus, we obtain an improvement over the closest competitor by 2.9%. As Fig. 4(c), our L2L model labels 36% less data than other active learning models for achieving 75% accuracy.

Table 1. Results from transfer learning using the different methods

Methods	OntoNotes-5.0	Financial	MSRA
TPANN	62.35	62.21	66.85
T-Model	66.12	65.82	70.72

We report the detailed results for all approaches on the four corpora in Table 1 and Table 2. Table 1 shows the results with the cold-starting setting. Namely, we directly transfer the NER models trained using source domain to the target

Table 2. Results from active learning using the different methods

Methods	OntoNotes-5.0		Financial		MSRA	
	F1	C/R	F1	C/R	F1	C/R
Random	76.83	100	77.53	100	75.04	100
Diversity [10]	78.82	36	78.82	46	75.39	78
Uncertainty [9]	79.61	30	79.21	36	75.68	66
PAL [3]	82.25	36	80.60	36	75.99	44
L2R	85.19	14	82.53	18	77.33	30
L2L model	**88.74**	**8**	**85.22**	**10**	**78.89**	**18**

domain. The F1 score of the T-Model in our task is better than the one of the TPANN. In Table 2, the datasets of four target domains are displayed as columns, reporting the F1 score (%) of validation accuracy and the relative cost reduction (%) compared to the performance of Random method.

In all datasets, L2L outperforms the heuristic methods, including Uncertainty Sampling. We speculate this may be due to these two heuristics not being able to capture the polarity information during the data selection process. And L2L also outperforms PAL. We attribute this to the ineffectiveness of the RL-based approach for learning a reasonable AL query strategy. Compared with the previous state-of-the-art approach, our method achieves an improvement of 6.49%. By applying a better representation approach, we extract more semantic features and hence, our approach can provide better performance than the baseline one with the same number of labeled data. Table 2 also reports the cost reduction versus random sampling, showing that the PAL methods can reduce the annotation burden to as low as 8%.

5 Conclusion

In this work, we propose a novel approach combining active learning and reinforcement learning techniques for selective data labeling. The idea is to first transfer a learning model from a source domain to a target domain, and then apply the active learning to gradually improve the performance of the model using as few labeled data in the target domain as possible. Experimental results show that our approach is more effective than strong previous methods using heuristics and reinforcement learning.

Acknowledgments. The work is supported by NSFC (grant number 61872315) and Zhejiang Provincial Natural Science Foundation (grant number LZ21F020007).

References

1. Deng, J., Dong, W., Socher, R., Li, L.J., Li, K., Fei-Fei, L.: ImageNet: a large-scale hierarchical image database. In: CVPR, pp. 248–255. IEEE (2009)

2. Denton, E.L., Chintala, S., Fergus, R., et al.: Deep generative image models using a Laplacian pyramid of adversarial networks. In: NIPS, pp. 1486–1494 (2015)
3. Fang, M., Li, Y., Cohn, T.: Learning how to active learn: a deep reinforcement learning approach. In: EMNLP (2017)
4. Ganin, Y., Lempitsky, V.: Unsupervised domain adaptation by backpropagation. In: ICML (2015)
5. Ganin, Y., et al.: Domain-adversarial training of neural networks. J. Mach. Learn. Res. **17**, 59:1–59:35 (2016)
6. Gui, T., Zhang, Q., Huang, H., Peng, M., Huang, X.: Part-of-speech tagging for Twitter with adversarial neural networks. In: EMNLP, pp. 2411–2420 (2017)
7. Lin, T.-Y., et al.: Microsoft COCO: common objects in context. In: Fleet, D., Pajdla, T., Schiele, B., Tuytelaars, T. (eds.) ECCV 2014, Part V. LNCS, vol. 8693, pp. 740–755. Springer, Cham (2014). https://doi.org/10.1007/978-3-319-10602-1_48
8. Sener, O., Savarese, S.: Active learning for convolutional neural networks: a core-set approach. In: ICLR 2018 (2018)
9. Settles, B., Craven, M.: An analysis of active learning strategies for sequence labeling tasks. In: EMNLP, pp. 1070–1079 (2008)
10. Shen, D., Zhang, J., Su, J., Zhou, G., Tan, C.L.: Multi-criteria-based active learning for named entity recognition. In: ACL, pp. 589–596 (2004)
11. Xia, R., Wang, C., Dai, X.Y., Li, T.: Co-training for semi-supervised sentiment classification based on dual-view bags-of-words representation. In: ACL, pp. 1054–1063 (2015)

Entity Resolution with Hybrid Attention-Based Networks

Chenchen Sun[1(✉)] and Derong Shen[2]

[1] Key Laboratory of Computer Vision and System (Ministry of Education), Tianjin University of Technology, Tianjin, China
[2] School of Computer Science and Engineering, Northeastern University, Shenyang, China
shendr@mail.neu.edu.cn

Abstract. Entity resolution (ER) is an important step of data preprocessing. Deep learning based entity resolution is a growing topic in research communities. Considering that record structure is hierarchical: token, attribute, record, we propose a hybrid attention-based network framework for entity resolution. It synthesizes information from different abstract levels of record hierarchy. Systematic attention mechanisms are exploited in several aspects of ER: self-attention for internal dependency capture, inter-attention for alignments, and multi-dimensional weight attention for importance discrimination. Also attribute order is taken into account in ER learning for better similarity representations. Moreover, we tackle ER over low-quality data by hybrid soft token alignments. Extensive experiments on 4 datasets are conducted, and the resultsshow that our approach surpasses existing ER approaches.

Keywords: Entity resolution · Attention mechanism · Deep learning · Hybrid neural network

1 Introduction

Entity resolution (ER) is a fundamental problem of data preprocessing and data integration, and is also known as entity match, record linkage, and duplicate detection [1]. ER distinguishes records referring to the same real-world entity in datasets, and is widely applied in healthcare, e-commerce, criminal investigation *et al.* There are two kinds of classical ER approaches: rule based ER and machine learning based ER [1]. Recently deep learning (DL) introduces a new chance to ER, and DL based ER (deep ER for short) shows remarkable competitive advantages over classical ER [2, 3]. DL is able to provide an end-to-end solution to ER [4], substantially reducing manual involvements. DL also is able to deeply capture semantic similarity for textual data.

Although DL based approaches substantially accelerated ER research recently [2, 3, 5–7], there still exists much improvement space for deep ER. A record is typically hierarchical in structure: token to attribute to record, as shown in example 1. Schema, consisted of attributes, is a key factor in organization of record hierarchy. We have following observations. First, a basic problem is what DL architecture can properly

C. S. Jensen et al. (Eds.): DASFAA 2021, LNCS 12682, pp. 558–565, 2021.
https://doi.org/10.1007/978-3-030-73197-7_37

handle data with attributes in ER setting. Different levels (i.e. token and attribute) contain different abstractions of knowledge in records. Existing works usually focus on one level for similarity computation: [3, 5–7] at token level or [2] at record level. The complex structural information of record is not fully taken into account. Second, there are hidden relations between attributes in schema, which matter in representations and comparisons of deep ER, but such relations are ignored in previous works. Taking r_6 in example 1–(3) as an example, Title and Type are closely related and should be put together when comparison. Third, how to sufficiently improve representations and comparisons with attention mechanisms. Previous works utilize attentions in the way of text matching [3, 5, 7], but neglect ER adaptation and systematicness. Fourth, how to well resolve low-quality data: 1) data with misplaced values and injected noise, as illustrated in example 1–(2); 2) data with not well-matched heterogeneous schema, as illustrated in example 1–(3). Previous works [5, 7] solve such problems by regarding a record as a long sequence and diluting or even abandoning schema information, which harms usability and accuracy.

Example 1. There are three examples of data as follows.

(1) $r_1 = \{$Name = William Joe, City = LA, Age = 21$\}$, $r_2 = \{$Name = Will Joe, City = Los Angeles, Age = 21$\}$. Pair $< r_1, r_2 >$ is of *standard structured data.*
(2) $r_3 = \{$Name = Apple IPhone 12, Brand =, Price = 999$\}$, $r_4 = \{$Name = IPhon 12, Brand = Apple Inc, Price = 999.00$\}$. Pair $< r_3, r_4 >$ is of *low-quality data with misplaced values.*
(3) $r_5 = \{$Name = Surface Pro 7, QWU-00001, Category = Microsoft Computers, Price = 869.00, Size = 12.3$\}$, $r_6 = \{$Title = Micrsoft Surface Pro 7, 12.3″, Price = 869.00, Type = Computers QWU-00001, Location = Redmond, WA$\}$. Pair $< r_5, r_6 >$ is of *low-quality data with not well-matched schemas*: Name-Title, Category-Type, Price-Price, Size-, -Location. In fact, Name is related to Title & Type, Category is related to Title & Type, and Size is related to Title. So there are N-N relations, which is very difficult for perfect schema matching.

In this work, we create an end-to-end solution for ER: **h**ybrid **a**ttention-based **n**etworks (HAN). HAN exploits different levels of record hierarchy for similarity computation by setting up two sub-networks. In this way, two level similarity representations are generated: one is for concrete low-level (token) knowledge and the other is for abstract high-level (attribute) knowledge. We sort attributes according to relations between them for better deep similarities. A family of attention mechanisms is introduced to systematically enhance hierarchical representations and comparisons in HAN. Self-attention captures internal dependencies in a sequence of tokens (or attributes), inter-attention aligns two sequences of tokens, and multi-dimensional weight attention discriminates importance of tokens (or attributes) in a multi-perspective way. To effectively resolve low-quality data, we design hybrid soft alignments between tokens, i.e., aligning tokens both intra and inter aligned attributes. In this setting, tokens can be compared across attributes if necessary. Experimental evaluations on 4 datasets show that HAN apparently outperforms existing works.

Organization of the rest work is as follows. Section 2 reviews related work. Section 3 elaborates the hybrid attention-based network framework. Section 4 introduces experimental evaluations. Section 5 concludes the whole work.

2 Related Work

Entity resolution is a long-stand topic in data mining, database, and artificial intelligence [1]. Classical ER belong to one of two categories: rule based approaches and machine learning (ML) based approaches. Inspired by the revolutionary DL based success in NLP [4], there comes a growing trend of DL based ER recently. As a pioneer work, DeepER learns a distributed representation for each record, and then compute record similarities with record representation vectors [2]. DeepMatcher summarizes and compares attributes with bidirectional RNN and inter-attention, and then aggregate attribute similarities to make ER decisions [3]. Seq2SeqMatcher is a deep sequence-to-sequence ER method, which regards a record as a long sequence [6]. HierMatcher builds a hierarchical matching network for ER, and is an improvement to Seq2SeqMatcher by limitedly considering attribute information [7]. MCA is an integrated multi-context attention framework for ER [5]. MCA tries to fully exploit the semantic context and capture the highly discriminative terms for a pair of attribute values with 3 types of attentions.

3 Hybrid Attention-Based Networks

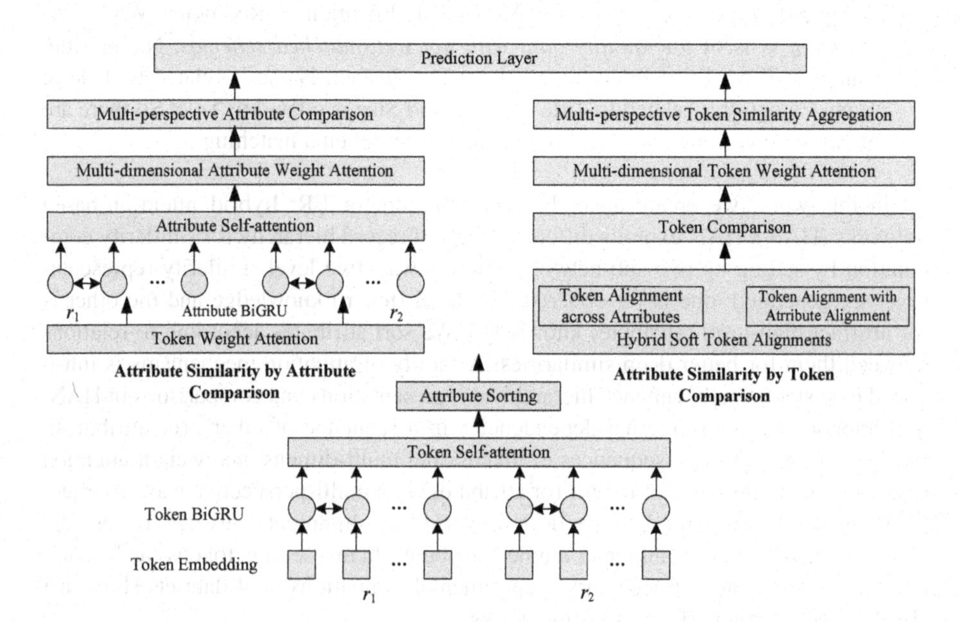

Fig. 1. Overview of our hybrid attention-based network framework for ER

Given two records $r_1 = \{a_{11}, \ldots, a_{1m}\}$ from data source R_1 and $r_2 = \{a_{21}, \ldots, a_{2n}\}$ from data source R_2, *entity resolution* (*ER*) decides whether r_1 and r_2 refer to the same real-world entity. An attribute value a_{1i} is a token sequence.

The hierarchical structure of a record is token to attribute to record. Inspired by hierarchical DL solutions to document classification [8] and text matching [9], we design a hybrid attention-based network (HAN) framework for ER over data with various qualities. HAN synthesizes information from different record structure levels: token and attribute. Three kinds of attention mechanisms are utilized in different steps of deep ER. The overview of our bottom-up solution is illustrated in Fig. 1. The neural computation between r_1 and r_2 is bi-directional. We just present $r_1 \rightarrow r_2$, and it is easy to get $r_1 \leftarrow r_2$.

Attribute Similarity by Token Comparison. The sketch workflow is token representation to token similarity to attribute similarity, as illustrated in Fig. 1. Attributes are converted into sequential token representations, and are contextualized with token self-attention. To address low-quality data problem, hybrid soft token alignment is introduced. With multi-dimensional weight attention, token level similarities are aggregated into token based attribute similarities.

Token Representation and Contextualization. Given an attribute value a_{1i} of record r_1, we embed it with fastText [10], and then utilize bi-directional GRU (BiGRU) to incorporate the sequential contextual information from both directions of tokens [11], arriving at $\boldsymbol{h}_{1i}$. In a sequence, self-attention establishes the connection between any two tokens regardless of their distance [12], which is important contextual information. Self-attended representation $\boldsymbol{h}^s_{1i}$ is generated by putting the self-attention matrix $\boldsymbol{\alpha}_{1i}{}^{\mathrm{T}}$ upon $\boldsymbol{h}_{1i}$. $\boldsymbol{\alpha}_{1i}$ is computed by softmax over dot product of $\boldsymbol{h}_{1i}$ and its transpose.

Hybrid Soft Token Alignment. Inter-attention softly aligns tokens between two sequences, and builds inter-sequence token connections [12]. To tackle low-quality data, we conduct two kinds of token alignments, and combine them with a flexible gating mechanism (GM). Token alignment with attribute alignment (TAw) focuses on standard structured data, while token alignment across attributes (TAac) focuses on corrupted data. (1) *TAw*. Given a pair of well-matched attribute values a_{1i} and a_{2j}, their self-attended representations $\boldsymbol{h}^s_{1i}$ and $\boldsymbol{h}^s_{2j}$ are used to compute inter-attention between them. $\hat{\boldsymbol{h}}_{1i}$ is the inter-attended representation of a_{1i}, and is the dot product of $\boldsymbol{h}^s_{2j}$ and $\boldsymbol{\beta}_{1i\rightarrow 2j}$. $\boldsymbol{\beta}_{1i\rightarrow 2j}$ is the alignment matrix between token pairs from a_{1i} and a_{2j}, and each entry $\boldsymbol{\beta}_{1i\rightarrow 2j}(k, l)$ is a connection weight between tokens a_{1ik} and a_{2jl}. (2) *TAac*. For corrupted data, attributes are not well matched or values are misplaced, so token alignment should break attribute boundaries. a_{1i} is aligned against the whole record r_2 (i.e., $[a_{21}, \ldots, a_{2n}]$). (3) *GM*. GM is invented to flexibly control information flow from different representations [13]. We use GM to combine the two different soft alignments $\hat{\boldsymbol{h}}_{1i}$ and $\hat{\boldsymbol{h}}'_{1i}$, arriving at $\hat{\boldsymbol{h}}''_{1i}$.

Token Comparison. Given a token a_{1ik} of a_{1i}, we put $\boldsymbol{h}^s_{1i}(k)$, $\hat{\boldsymbol{h}}''_{1i}(k)$, their element-wise absolute difference $\left|\boldsymbol{h}^s_{1i}(k) - \hat{\boldsymbol{h}}''_{1i}(k)\right|$, and their element-wise multiplication $\boldsymbol{h}^s_{1i}(k) \odot$

$\hat{\boldsymbol{h}}''_{1i}(k)$ together into a highway network, and get the unidirectional token similarity from a_{1ik} to r_2, $\boldsymbol{c}_{1i}(k)$.

Multi-perspective Token Similarity Aggregation. For an attribute a_{1i}, we aggregate token similarities into token based attribute similarities from different perspectives. We introduce multi-dimensional weight attention Γ_{1i} to represent token importance in a sequence [12]. Γ_{1i} is of q_{1i} dimensions. With q_{1i} weight attentions, we arrive at q_{1i} token based attribute similarities by weighted aggregations, where each $\boldsymbol{c}_{1i} \cdot (\Gamma_{1i}(l))^{\mathrm{T}}, l \in [1, q_{1i}]$ captures a unique semantic aspect of attribute sequence a_{1i}. All token based attribute similarities are concatenated into a multi-perspective token based attribute similarity vector ϵ_{1i}^{mp}.

Attribute Similarity by Attribute Comparison. The sketch workflow is token representation to attribute representation to attribute similarity, as shown in Fig. 1. We propose an attribute sorting algorithm, which puts related attributes close to each other. Attribute representation is generated by aggregating token representations with BiGRU, and is contextualized with attribute self-attention. With multi-dimensional attribute weight attention, attribute based attribute similarities are aggregated into record similarities.

Attribute Sorting. A record consists of several attributes with no intrinsic order, and each attribute value is a token sequence. We observe that good attribute orders help create good contexts for attributes. We propose an attribute self-attention based attribute sorting method. In ER, the key of attribute sorting is to put related attributes close to each other. Thus, we utilize attribute self-attention to measure relatednesses between attributes, and then sort attributes with relatednesses. The most related attributes are iteratively merged until all attributes are sorted. Note that attributes should be sorted separately in each of the two data sources.

Attribute Representation and Contextualization. After the attribute order is fixed, we use an attribute level BiGRU to capture sequential contextual information. For an attribute a_{1i}, its bidirectional sequential contextual state is $\boldsymbol{u}_1(i)$. Then self-attended attribute representation $\boldsymbol{u}_1^{\mathrm{s}}(i)$ is dot production of $\boldsymbol{u}_1(i)$ and $(\boldsymbol{\alpha}_1(i))^{\mathrm{T}}$, where $(\boldsymbol{\alpha}_1(i))^{\mathrm{T}}$ is self-attention matrix over $\boldsymbol{u}_1(i)$.

Multi-perspective Attribute Comparison. Given a pair of matched attribute values a_{1i} and a_{2j}, their attribute representations are $\boldsymbol{u}_1(i)$ and $\boldsymbol{u}_2(j)$ respectively. The concatenation of $\boldsymbol{u}_1(i)$, $\boldsymbol{u}_2(j)$, their element-wise absolute difference $|\boldsymbol{u}_1(i) - \boldsymbol{u}_2(j)|$, and their element-wise multiplication $\boldsymbol{u}_1(i) \odot \boldsymbol{u}_2(j)$ is put into a highway network for attribute level comparison ($a_{1i} \rightarrow a_{2j}$), arriving at $\boldsymbol{c}_1(i)$, the attribute based attribute similarity of $a_{1i} \rightarrow a_{2j}$. We introduce multi-dimensional weight attention Γ_1 for attributes, where each dimension is targeted at a specific semantic aspect of a record. Then we apply multiple different weights (in Γ_1) to $\boldsymbol{c}_1$, which contains all original attribute based attribute similarities of $a_1 \rightarrow a_2$, and arrive at a multi-perspective attribute based attribute similarity vector ϵ_1^{mp}.

Prediction Layer and Model Learning. We regard ER as a binary classification problem. Overall record similarities are generated from token based attribute similarities &

attribute based attribute similarities, and are put into a deep classifier for ER decisions. Finally our deep ER model is trained by minimizing cross entropy.

Token based similarity of $r_1 \rightarrow r_2$ is derived from ϵ_{1i}^{mp}, $i \in [1, m]$, denoted as $s_{1\rightarrow 2}^{tk}$. Each token based attribute similarity ϵ_{1i}^{mp} is weighted, and $(\frac{1}{q_1}\sum_{i=1}^{q_1}\Gamma_1(i))^{\mathrm{T}}$ is an average attribute weight vector. Similarly, token based similarity of $r_1 \leftarrow r_2$ is $s_{1\leftarrow 2}^{tk}$. Attribute based similarity of $r_1 \rightarrow r_2$ is $s_{1\rightarrow 2}^{att} = \epsilon_1^{mp}$. Similarly, attribute based similarity of $r_1 \leftarrow r_2$ is $s_{1\leftarrow 2}^{att}$. Overall record similarity of $r_1 \leftrightarrow r_2$ is concatenation of above 4 similarities, denoted as $s_{1\leftrightarrow 2}$. To get the ER decision, the overall record similarity $s_{1\leftrightarrow 2}$ is fed into a two-layer fully-connected highway network succeeded by a softmax layer. The highway network aggregates $s_{1\leftrightarrow 2}$ into a compact representation $s_{1\leftrightarrow 2}^{'}$. Finally a softmax layer takes $s_{1\leftrightarrow 2}^{'}$ as input, and generates the matching distribution $P(z|r_1, r_2)$.

4 Experimental Evaluation

Datasets. We conduct experiments over 4 datasets from various domains. There are 2 standard structured datasets and 2 low-quality datasets, as illustrated in Table 1. The table columns contain the type of a dataset, dataset, application domain, the number of records (Size), the number of matches (# Pos.), and the number of attributes (# Att.). All datasets are from open datasets published by the DeepMatcher research group [3], but some datasets are edited for special features.

The two datasets AG and BR are the same as in [3]. In low-quality data, either values are misplaced, or schemas are not well matched. Based on Walmart-Amazon$_1$ in [3], WA1 is generated by moving a value from one attribute to another attribute in the same record with 50% probability. Based on Walmart-Amazon$_1$ in [3], WA2 is generated by merging attributes brand and model from Walmart to get a new attribute brand-model, and merging attributes category and brand from Amazon to get a new attribute category-brand. For both datasets, a selected value is injected into a spelling error with 20% probability, and the spelling error is changing a token into another token with edit distance 2.

Table 1. Datasets for experimental evaluation

Type	Dataset	Domain	Size	# Pos.	# Att.
Structured	Amazon-Google (AG)	Software	11,460	1,167	3
	BeerAdvo-RateBeer (BR)	Beer	450	68	4
Low-quality data	Walmart-Amazon^1 (WA1)	Electronics	10,242	962	5
	Walmart-Amazon^2 (WA2)	electronics	10,242	962	4:4

Settings, Parameters and Metric. The approaches are implemented with PyTorch. The experiments are run on a server with 8 CPU cores (Intel(R) E5–2667, 3.2 GHz), 64G memory, and NVIDIA GeForce GTX 980 Ti.

The dimension of fastText is 300. Adam is used as optimization algorithms for all deep learning models. The number of hidden units in BiGRU is 150. The number of epochs, mini-batch size, and dropout rate are 15, 16, and 0.1, respectively. Given a dataset, it is partitioned into 3: 1: 1 for training, validation, and testing, respectively.

F_1 is used to measure ER accuracy, and is defined as $2PR/(P + R)$ [3], where P is precision and R is recall.

Comparison with Existing Works. We compare our approach with 4 existing works: DeepMatcher [3], DeepER [2], hierarchical matching network (HierMatcher) base approach [7], and multi-context attention (MCA) based approach [5].

Test with Standard Structured Data. Figure 2 presents the accuracy of 5 ER approaches over 2 standard structured datasets. Generally, our HAN outperforms the other 4 ER approaches. On AG, the F_1 gap (ΔF_1) between HAN and DeepMatcher, DeepER, HierMatcher, & MCA are 10.5%, 17.7%, 5.2%, and 8.4%; on BR, ΔF_1 between HAN and DeepMatcher, DeepER, HierMatcher, & MCA are 13.6%, 14.2%, 8.1%, and 6.3%. Compared to existing works, we think HAN benefits from 3 aspects: (1) HAN captures information from different levels of record hierarchy, and appropriately combines them together; (2) HAN properly weights attributes and tokens with multi-dimensional attentions; (3) HAN provides a meaningful attribute order for representations and comparisons in deep ER.

Test with Low-Quality Data. Figure 3 illustrates the accuracy of 5 ER approaches over 2 low-quality datasets. Obviously, HAN has more superiority over the 4 existing works on low-quality data than on standard structured data. On WA1, ΔF_1 between HAN and DeepMatcher, DeepER, HierMatcher, & MCA are 24.3%, 49.8%, 10.7%, and 14.5%; on WA2, ΔF_1 between HAN and DeepMatcher, DeepER, HierMatcher, & MCA are 28.3%, 41.9%, 12.3%, and 18.1%. WA1 is low-quality data with misplaced values and error injections; WA2 is low-quality data with not well matched schemas and error injections. Compared to existing works, we deem, HAN well handles low-quality data, because HAN synthesizes token comparisons with aligned attributes and across attributes, in addition to the 3 aspects mentioned in the preceding paragraph. In this way, HAN captures different kinds of complementary semantic similarities.

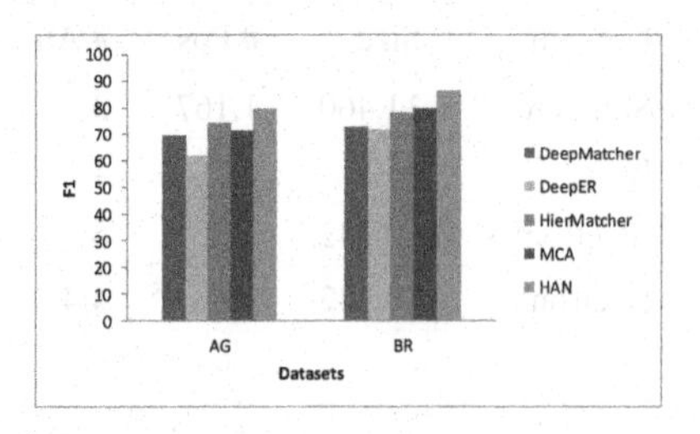

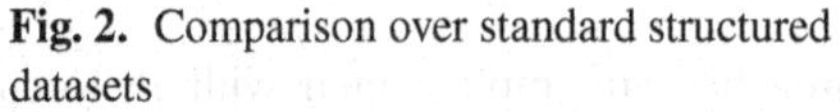

Fig. 2. Comparison over standard structured datasets

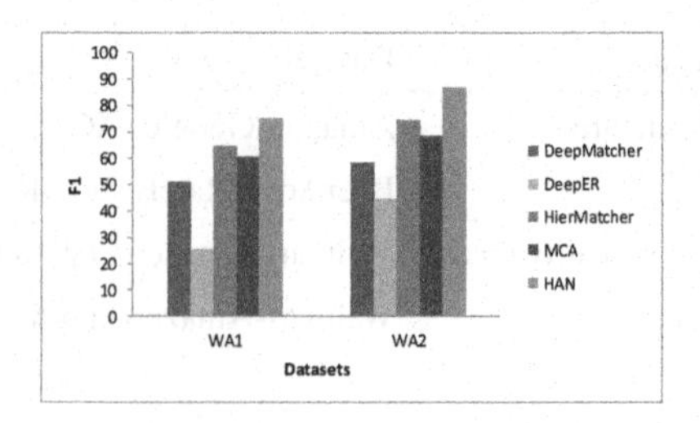

Fig. 3. Comparison over low-quality datasets

5 Conclusion

This work creates a hybrid attention-based network framework for deep ER. HAN exploits a deep understanding of record hierarchy, and captures sematic similarities from both concrete low-level information (token) and abstract high-level information (attribute). Extensive experimental evaluations on 4 datasets show that HAN has obvious advantages over previous ER approaches.

Acknowledgements. This work is supported by the National Natural Science Foundation of China under Grants (62002262, 61672142, 61602103, 62072086, 62072084), and the National Key Research & Development Project under Grant (2018YFB1003404).

References

1. Elmagarmid, A.K., Ipeirotis, P.G., Verykios, V.S.: Duplicate record detection: a survey. IEEE Trans. Knowl. Data Eng. **19**(1), 1–16 (2007)
2. Ebraheem, M., Thirumuruganathan, S., Joty, S., Ouzzani, M., Tang, N.: Distributed representations of tuples for entity resolution. Proc. VLDB Endowment **11**(11), 1454–1467 (2018)
3. Mudgal, S., et al.: Deep learning for entity matching: a design space exploration. In: Proceedings of the 2018 International Conference on Management of Data, pp. 19–34 (2018)
4. LeCun, Y., Bengio, Y., Hinton, G.: Deep learning. Nature **521**(7553), 436–444 (2015)
5. Zhang, D., Nie, Y., Wu, S., Shen, Y., Tan, K.L.: Multi-context attention for entity matching. Proc. Web Conf. **2020**, 2634–2640 (2020)
6. Nie, H, et al.: Deep sequence-to-sequence entity matching for heterogeneous entity resolution. In: Proceedings of the 28th ACM International Conference on Information and Knowledge Management, pp. 629–638 (2019)
7. Fu, C., Han, X., He, J., Sun, L.: Hierarchical matching network for heterogeneous entity resolution. In: Proceedings of the 29th International Joint Conference on Artificial Intelligence, pp. 3665–3671 (2020)
8. Yang, Z., Yang, D., Dyer, C., He, X., Smola, A., Hovy, E.: Hierarchical attention networks for document classification. In: Proceedings of the 2016 conference of the North American chapter of the association for computational linguistics, pp. 1480–1489 (2016)
9. Jiang, J.Y., Zhang, M., Li, C., Bendersky, M., Golbandi, N., Najork, M.: Semantic text matching for long-form documents. In: The World Wide Web Conference 2019, pp. 795–806 (2019)
10. Bojanowski, P., Grave, E., Joulin, A., Mikolov, T.: Enriching word vectors with subword information. Trans. Assoc. Comput. Linguistics **5**, 135–146 (2017)
11. Cho, K, et al.: Learning phrase representations using RNN encoder-decoder for statistical machine translation. In: Proceedings of the 2014 Conference on Empirical Methods in Natural Language Processing (EMNLP), pp. 1724–1734 (2014)
12. Hu, D.: An introductory survey on attention mechanisms in NLP problems. In: Proceedings of SAI Intelligent Systems Conference, pp. 432–448 (2019)
13. Tang, M., Cai, J., Zhuo, H.: Multi-matching network for multiple choice reading comprehension. In: Proceedings of the AAAI Conference on Artificial Intelligence 2019, pp. 7088–7095 (2019)

Information Retrieval and Search

MLSH: Mixed Hash Function Family for Approximate Nearest Neighbor Search in Multiple Fractional Metrics

Kejing Lu(✉) and Mineichi Kudo

Graduate School of Information Science and Technology Hokkaido University, Sapporo, Japan
{lkejing,mine}@ist.hokudai.ac.jp

Abstract. In recent years, the approximate nearest neighbor search in multiple ℓ_p metrics (MMS) has gained much attention owing to its wide applications. Currently, LazyLSH, the state-of-the-art method only supports limited values of p and cannot always achieve high accuracy. In this paper, we design a mixed hash function family consisting of two types of functions generated in different metric spaces to solve MMS problem in the external memory. In order to make the mixed hash function family work properly, we also design a novel searching strategy to ensure the theoretical guarantee on the query accuracy. Based on the given scenario, MLSH constructs the corresponding mixed hash function family automatically by determining the proportion of two types of hash functions. Experimental results show that MLSH can meet the different user-specified recall rates and outperforms other state-of-the-art methods on various datasets.

Keywords: Locality sensitive hashing · Fractional metrics

1 Introduction

The nearest neighbor search (NNS) in Euclidean spaces is of great importance in areas such as database, information retrieval, data mining, pattern recognition and machine learning [2,8,13]. In high-dimensional spaces, this problem is quite challenging due to *the curse of dimensionality* [12]. In order to circumvent this difficulty, many researchers turned to solve *the approximate nearest neighbor search* (ANNS) problem in high-dimensional spaces and proposed various ANNS methods. Among these methods, *the locality sensitive hashing technique* (LSH) [5] has gained particular attention since, compared with other ANNS methods, LSH owns attractive query performance and success probability guarantee in theory, and finds broad applications in practice [3,7,11,16,17]. It is notable that, for most LSH variants, ANNS in ℓ_2 metric is of great interest due to the practical importance. In contrast, the research on ANNS in *fractional distances*, i.e., ℓ_p metrics with $0 < p \leq 1$, is limited, although it also plays

C. S. Jensen et al. (Eds.): DASFAA 2021, LNCS 12682, pp. 569–584, 2021.
https://doi.org/10.1007/978-3-030-73197-7_38

an important role in many applications. As pointed out in [1,4,9], ℓ_p metrics ($0 < p \leq 1$) can provide more insightful results from both theoretical and empirical perspectives for data mining and content-based image retrievals than ℓ_2 metric. In general, the optimal value of p (of ℓ_p metric) is application-dependent and requires to be tuned or adjusted [1,4,6,9]. Motivated by these practical requirements, the following multi-metric search (MMS) problem in factional distances has been raised in [18]: given query q and a group of metrics $\{\ell_{p_j}\}_{j=1}^{b}$, where $0 < p_j \leq 1$, retrieve efficiently the approximate nearest neighbors of q's in all ℓ_{p_j} metrics at one time. Note that this problem is non-trivial even for a single ℓ_p metric. Although, LSH technique can support ANNS in arbitrary fractional distance in theory due to the existence of p-stable distributions for $0 < p \leq 1$ [12], it is hard to be implemented in practice. Besides, due to the lack of the closed form of the probability density function (PDF) of the general p-table distribution, the collision probability (in the same hash bucket) of two arbitrary objects cannot be computed analytically.

To the best of authors' knowledge, LazyLSH [18] is the first method which could solve MMS problems with the probability guarantee. An important fact on which LazyLSH depends is that, for arbitrary two close objects o_1 and o_2 in ℓ_p metric, it is highly possible that they are also close in $\ell_{p'}$ metric for $0 < p \neq p' \leq 1$. Based on this fact, LazyLSH only generates hash functions under the 1-stable distribution, i.e. the Cauchy distribution, in the indexing phase. This corresponds to the case of $p_0 = 1$. In the query phase, LazyLSH searches for the nearest neighbors of queries in ℓ_{p_j} ($1 \leq j \leq b$) metrics simultaneously by means of ℓ_{p_0} metric ($p_0 = 1$) based hash functions. By carefully setting the filtering criteria and the termination condition, the probability that LazyLSH returns the true results in all ℓ_{p_j} metrics ($1 \leq j \leq b$) is given in theory.

Although LazyLSH offers a practicable solution to MMS problems, it has following two limitations. (1) The range of p's in which LazyLSH works is limited. Specifically, when $p_0 = 1$, that is, ℓ_1 is chosen as the base metric, LazyLSH can do ANNS only in ℓ_p metric of $0.44 < p \leq 1$. For the other values of p, LazyLSH loses the effectiveness due to the lack of the locality sensitive property. In reality, this limitation is independent of the choice of p_0, and only related to the searching scheme of LazyLSH. (2) LazyLSH lacks of flexibility. In practice, the range of p of interest may be different in various applications. Obviously, hash functions generated in ℓ_1 metric are not appropriate if small values of p_j's are required by the applications.

In order to overcome the limitations mentioned above, in this paper, we propose a novel method called *Multi-metric LSH* (MLSH) to solve the MMS problem. Different from existing LSH-based methods, MLSH works on a mixed hash function family consisting of two types of hash functions which are generated in $\ell_{0.5}$ metric and $\ell_{1.0}$ metric, respectively. Such a mixed hash function family is designed exclusively for solving MMS problems since it can support different user-specified ranges of $\{p_j\}_{j=1}^{b}$ efficiently by adjusting the proportion of two types of hash functions. In order to make the mixed hash function family work properly, we also design a corresponding searching strategy to have the probability guarantee on query results.

Although the idea of mixed hash function family is straightforward, its realization is challenging due to the following two difficulties. (1) Since the PDFs of 0.5 and 1.0 stable distributions are different, MLSH should utilize these two types of hash functions coordinately to ensure the probability guarantee. (2) Given the user-specified range of p_j's, the proportion of two types of hash functions should be determined suitably to make the searching efficient. In order to clarify how to overcome these two difficulties, we will introduce the working mechanism of MLSH in two steps. First, we propose a novel search method called *Single-metric LSH* (SLSH), which works on a single type of hash functions, to solve MMS problem. Then, we discuss the construction way of mixed hash function family and propose MLSH based on SLSH.

Our contributions are summarized as follows.

(1) Compared with LazyLSH, SLSH can support ANNS in arbitrary ℓ_p metric ($0 < p \leq 1$), which significantly expands the application scope.
(2) Based on the framework of SLSH, we propose its improved version called MLSH which works on a mixed hash function family consisting of hash functions generated both in $\ell_{0.5}$ metric and $\ell_{1.0}$ metric. In addition to inheriting the merits of SLSH, MLSH could automatically determine the proportion of two types of hash functions, and construct the efficient mixed hash function family for the user-specified fractional distances.
(3) Extensive experiments on real datasets show that MLSH owns stable performance and outperforms the other state-of-the-art methods in both I/O cost and search time on real datasets.

2 Related Work

In this paper, we focus on the solution of ANNS in multiple fractional distances with probability guarantee. Thus, we only review those methods highly related to this topic. A summary of ANNS methods in ℓ_2 metric can be found in [14].

LSH is the most widely used technique for solving ANNS problems owing to its remarkable theoretical guarantees and empirical performance. E2LSH [5], the classical LSH implementations, cannot solve the c-ANNS problem efficiently because it requires a prohibitively large space in the indexing. To cope with this problem, the authors in [7] proposed C2LSH to reduce the storage cost. Different from E2LSH, C2LSH uses *virtual rehashing* and *dynamic counting* techniques to update the count frequency of each object by extending hash buckets step by step. The correctness of C2LSH is based on the following fact: the higher count frequency is, the more likely to be NN the corresponding object is. Later, QALSH [11] was proposed to further improve the search accuracy of C2LSH. Unlike C2LSH, QALSH uses query-centric hash buckets such that the near neighbors of queries on each vector can be tested more accurately. It is notable that, although QALSH and C2LSH were proposed originally for the search in ℓ_2 space, they can be easily modified to support the search in other ℓ_p metrics, as pointed out in [10].

Independently of QALSH, LazyLSH [18] was proposed to solve MMS problem with the probability guarantee. Similar to QALSH, LazyLSH also uses query-centric hash buckets and thus achieves better performance than C2LSH in ℓ_1 space. The core idea of LazyLSH is to build a single index structure and conduct the search in multiple metric spaces simultaneously. On such an index, by carefully setting the number of hash functions and the count threshold, the results in other fractional distances can also be returned with the probability guarantee. In addition, similar to other LSH variants, LazyLSH can achieve different accuracy-efficiency tradeoffs by adjusting the approximation ratio c.

3 Working Mechanism of SLSH

For ease of presentation, in this section, we mainly focus on the case of a single base metric and only one nearest neighbor to be found, and postpone the general case to Sect. 3.3: given p_0, p_1 $(0 < p_0, p_1 \leq 1)$, query q, dataset $\mathcal{D}$, error rate δ, and the hash function family $\{h_i^{p_0}\}_{i=1}^m$ in ℓ_{p_0} metric, find the nearest neighbor o_* of q in ℓ_{p_1} metric with probability guarantee. Although this problem has been solved by LazyLSH under some constraints, here, we present a different algorithm called SLSH which does not require the constraint that $p > 0.44$ as LazyLSH has.

3.1 Preparations

A Special Case: $p_0 = p_1$ First, we give the following fundamental result of LSH technique [5].

Lemma 1. *For two objects o and q with ℓ_p distance $\|o - q\|_p = s$ $(o, q \in \mathbb{R}^d)$, $\Pr_{a \sim X_p}\{|h_a(o) - h_a(q)| \leq w\} = \int_{-w/s}^{w/s} f(x)dx$, where w is a positive number, h is a hash function generated randomly and $f(x)$ is the PDF of p-stable distribution X_p.*

We assume $p_0 = p_1 = p$ and try to find a candidate set S such that o_* is included in S with probability at least $1-\delta$, where o_* is the true nearest neighbor of q in ℓ_p metric. Given m hash functions generated in ℓ_p metric, we first define width w_*:

$$w_* = \|o_*\| \cdot f_p^{-1}(\frac{\tau}{m} + \sqrt{\frac{1}{2m} \ln \frac{1}{\delta}}), \tag{1}$$

where f_p is the probability density function of the p-stable distribution. Then, by the Hoeffding's inequality, we have the following lemma, which is derived in a similar way in Theorem 1 in [15].

Lemma 2. *For any $\delta > 0$ and positive integers m, τ $(m > \tau)$, two points o_1 and o_2 in $\mathbb{R}^d$ of distance $s = \|o_1 - o_2\|_p$ collide in the same hash bucket τ times or more in the m query centric hash buckets, with probability at least $1-\delta$ if the half-bucket size w is set to w^*.*

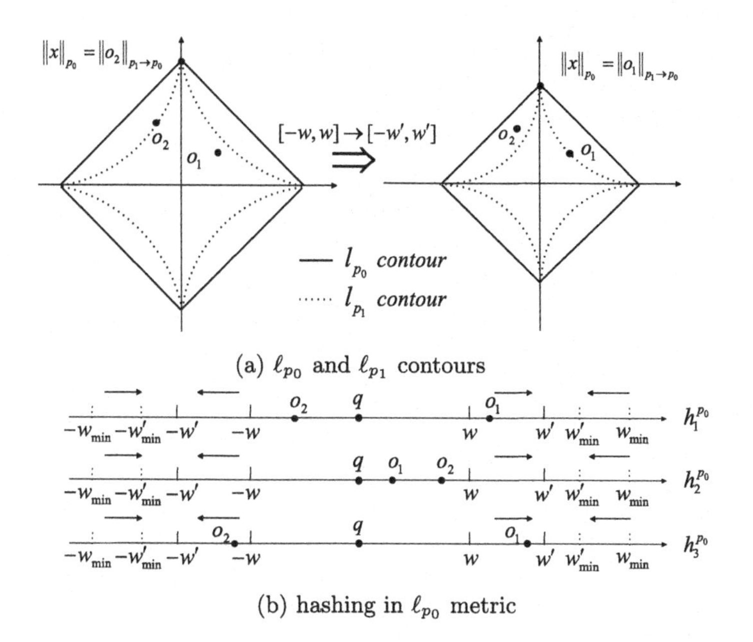

Fig. 1. A running example of SLSH for finding approximate nearest neighbors of q in ℓ_{p_1} metric using ℓ_{p_0} hashing ($p_1 < p_0$).

Clearly, $\tau \in (0, m)$ is the count threshold. $S = \{o \in \mathcal{D} | \#Col(o; w_*) \geq \tau\}$ is the candidate set that we desire.

Some Definitions and Notations. Next, we introduce some notations for the later use. For two different metrics ℓ_{p_0} and ℓ_{p_1}, we define a norm $\|x\|_{p_1 \to p_0}$ as follows:

$$\|x\|_{p_1 \to p_0} = \max_{y \in \mathbb{R}^d} \{\|y\|_{p_0} \mid \|y\|_{p_1} = \|x\|_{p_1}\}. \tag{2}$$

Then, we can derive following two lemmas by elementary computation.

Lemma 3. *For* $0 < p_0, p_1 \leq 1$, $\|x\|_{p_1 \to p_0}$ *can be expressed as follows:*

$$\|x\|_{p_1 \to p_0} = \begin{cases} \|x\|_{p_1} & 0 < p_1 < p_0 \leq 1 \\ d^{1/p_0 - 1/p_1} \|x\|_{p_1} & 0 < p_0 < p_1 \leq 1 \end{cases}$$

Lemma 4. *For* $0 < p_0, p_1 \leq 1$, $\|y\|_{p_0} \leq \|x\|_{p_0} \to \|y\|_{p_1} \leq \|x\|_{p_0 \to p_1}$.

3.2 The Algorithm

After preparations above, we are ready to show the details of SLSH (Algorithm 1). The basic idea behind Algorithm 1 is illustrated in Fig. 1. Here, for simplicity, we only show the case $p_1 < p_0 = 1$. From Fig. 1, we can see that the width of search windows is gradually increased in the searching process. If some object o

satisfies the candidate condition, that is, $\#Col(o;w) \geq \tau$, we view it as a candidate and compute its exact distance with q in ℓ_{p_1} metric. During this process, we maintain the nearest one we have checked, denoted by $o_{\min}$. By Lemma 3 and Lemma 4, it is easy to see that the set $V = \{o \in \mathcal{D} | \, \|o\|_{p_0} \leq \|o_{\min}\|_{p_1 \to p_0}\}$ contains $o_{\min}$ and o_* as well. Thus, it suffices to find a candidate condition such that the probability that an arbitrary object in V satisfies this count condition is at least $1-\delta$. To this end, we introduce $w_{\min}$, which is analogous to w_* in (1), as follows.

$$w_{\min} = \|o_{\min}\|_{p_1 \to p_0} \cdot f_{p_0}^{-1}(\frac{\tau}{m} + \sqrt{\frac{1}{2m} \ln \frac{1}{\delta}}). \tag{3}$$

Let $U(w_{\min}, \tau) = \{o \in \mathcal{D} | \#Col(o; w_{\min}) \geq \tau\}$. According to Eq. (1) and the discussion in Sect. 3.1, we have

$$\Pr\{o_* \in U(w_{\min}, \tau)\} \geq 1 - \delta. \tag{4}$$

Thus, $U(w_{\min}, \tau)$ is viewed as the candidate set, where $w_{\min}$ is always known during the searching process due to the existence of $o_{\min}$. Clearly, for any $w \geq w_{\min}$, $U(w, \tau) \geq U(w_{\min}, \tau)$. Therefore, we can write the candidate condition and the termination condition as follows, where w is the current half-width of search windows.

Candidate condition: $\#Col(o; w) \geq \tau$, where $o \in \mathcal{D}$.
Termination condition: $w \geq w_{\min}$.

According to the discussion above, the candidate condition and the termination condition work together to ensure that all objects in the candidate set can be checked. When $w_{\min}$ is updated during the extension of search windows, we also need to check whether the termination condition is satisfied or not. Note that, the termination condition becomes tighter as the searching proceeds because the value of $w_{\min}$ is non-decreasing. Figure 1 illustrates this situation.

Up to now, all results are obtained only in the case $c = 1$. For the general case $c \geq 1$, it suffices to rewrite $w_{\min}$ as follows:

$$w_{\min} = (\|o_{\min}\|_{p_1 \to p_0} / c) \cdot f_{p_0}^{-1}(\frac{\tau}{m} + \sqrt{\frac{1}{2m} \ln \frac{1}{\delta}}) \tag{5}$$

The validity of this setting of $w_{\min}$ is shown in the following theorem.

Theorem 1. *For any query $q \in \Re^d$, the probability that algorithm 1 returns a c-approximate nearest neighbor of q in l_{p_1} metric with success probability at least $1 - \delta$.*

Proof. **Case 1:** ($c = 1$). According to the discussion, the probability that o_* is checked is at least $1 - \delta$. Thus, we conclude.

Case 2: ($c > 1$). According to the expression of $w_{\min}$, if the algorithm terminates before w reaches w_*, at least one c-approximate nearest neighbor has been found. On the other hand, if $w \geq w_*$, the probability that o_* is found is at least $1 - \delta$ (Case 1). Combining these two situations, we conclude (Note that o_* is of course a c-approximate nearest neighbor).

Algorithm 1: SLSH

Input: $\mathcal{D}$ is the dataset; d is the dimension of $\mathcal{D}$; q is the query; c is the approximation ratio; δ is the error rate; ℓ_{p_0} is the base metric; m is the number of hash functions in ℓ_{p_0} metric; λ is the parameter controlling count threshold τ;

Output: $o_{\min}$: the approximate nearest neighbor of q in ℓ_{p_1} metric

1. generate m hash functions $\{h_i\}_{i=1}^{m}$ in ℓ_{p_0};
2. $w \leftarrow 0$, $o_{\min} \leftarrow NULL$;
3. $\tau \leftarrow \left\lfloor m\lambda - \sqrt{m \ln(1/\delta)/2} \right\rfloor$;
4. $w_{\min} = (\|o_{\min}\|_{p_1 \to p_0}/c) \cdot f_{p_0}^{-1}(\lambda)$;
5. **while** $w \le w_{\min}$ **do**
6. $\quad w \leftarrow w + \Delta w$ $(\Delta w > 0)$;
7. $\quad \forall o \in \mathcal{D}$ update $\#Col(o; w)$ if necessary;
8. $\quad$ **if** *o is not visited and* $\#Col(o; w) \ge \tau$ **then**
9. $\quad\quad$ **if** $\|q - o\|_{p_1} < \|o_{\min}\|_{p_1}$ **then**
10. $\quad\quad\quad$ Update $o_{\min}$;
11. $\quad\quad\quad$ $w_{\min} = (\|o_{\min}\|_{p_1 \to p_0}/c) \cdot f_{p_0}^{-1}(\lambda)$;
12. Return $o_{\min}$

3.3 More Analysis

In the discussion above, we only consider a single objective metric l_{p_1}. Here, we consider the case of multiple objective metrics, that is $\ell_{p_1}, \ell_{p_2}, \ldots, \ell_{p_b}$. Suppose that we have generated m hash functions $\{h_i\}_{i=1}^{m}$ in the base metric ℓ_{p_0}. During the searching process, we need to keep and update the current nearest neighbors in different metrics, denoted by $o_{\min}^{p_1}, \cdots, o_{\min}^{p_b}$, and compute $w_{\min}^{p_1}, \cdots, w_{\min}^{p_b}$, respectively in a similar way of $w_{\min}$ in (5). Let $w_{\min} = \max_{1 \le j \le b}\{w_{\min}^{p_j}\}$. Then, we only need to replace the termination condition in Algorithm 1 by $w \ge w_{\min}$.

4 Working Mechanism of MLSH

From now on, we consider the following generalized MMS problem: given dataset $\mathcal{D}$, query q and error rate δ, find c-approximate nearest neighbors of q in b different metrics $\{\ell_{p_j}\}_{j=1}^{b}$ at once with probability at least $1-\delta$, where $0 < p_1 < p_2 < \cdots < p_b \le 1$.

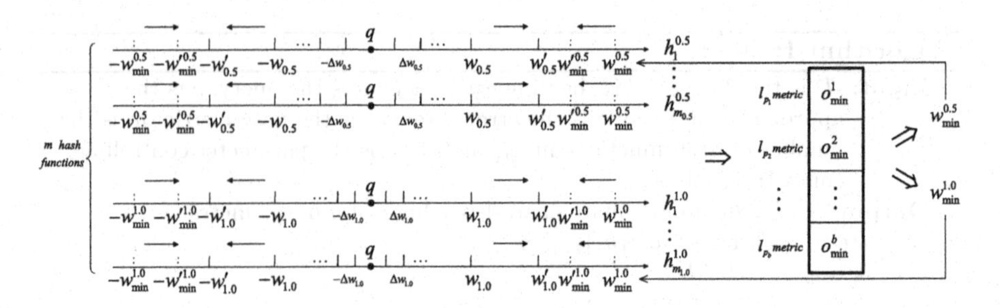

Fig. 2. A running example of MLSH.

4.1 The Mixed Hash Function Family

We present the details of the construction of the mixed hash function family from the following four aspects.

(1) **Selection of hash functions**. We choose two sets of hash functions; one in $\ell_{0.5}$ metric and another in $\ell_{1.0}$ metric, because (1) probability density functions of 0.5-stable and 1-stable distributions have their closed-forms, and (2) random variables following these two distributions can be easily generated. Thus, we generate $m_{0.5}$ hash functions randomly under the 0.5-stable (Levy) distribution and generate $m_{1.0}$ hash functions randomly under the 1-stable (Cauchy) distribution, where $m = m_{0.5} + m_{1.0}$ is the total number of hash functions.

(2) **The upper bound of window size**. Since MLSH generates two sets of hash functions, $w_{\min}$ cannot be directly used. Thus, we need the following modifications. Let o_*^j and $o_{\min}^j$ be the true nearest neighbor and the nearest neighbor among probed objects in ℓ_{p_j} metric, respectively. Given ℓ_{p_j} metric $(1 \leq j \leq b)$, we introduce notation $w_{\min}^{i,j}$ analogous to $w_{\min}$ in (5) ($i =$ 0.5, 1.0), which is defined as follows:

$$w_{\min}^{i,j} = \|o_{\min}\|_{p_j \to p_i} \cdot f_{p_i}^{-1}\left(\frac{\tau}{m} + \sqrt{\frac{1}{2m}\ln\frac{1}{\delta}}\right) \tag{6}$$

Let $w_{0.5}$ and $w_{1.0}$ be the half-widths of search windows on $\{h_j^{0.5}\}_{j=1}^{m_{0.5}}$ and $\{h_j^{1.0}\}_{j=1}^{m_{1.0}}$, respectively. According to the analysis in the preceding section, if $w_{1.0} \geq w_{\min}^{1.0,j}$ holds for functions $\{h_j^{1.0}\}_{j=1}^{m}$, the probability that the count frequency of o_*^j is not less than τ is at least λ. Since the value of j varies from 1 to b, we need to keep the value $w_{\min}^{1.0}$ which is defined as follows:

$$w_{\min}^{1.0} = \max_j\{w_{\min}^{1.0,j}\} \tag{7}$$

When $w_{1.0} \geq w_{\min}^{1.0}$, we terminate the extension of search windows $[-w_{1.0}, w_{1.0}]$. Similarly, we can define $w_{\min}^{0.5}$ and treat $w_{0.5} \geq w_{\min}^{0.5}$ as the termination condition on search windows $[-w_{0.5}, w_{0.5}]$. By such settings of termination condition, we can easily derive the probability guarantee on the mixed hash function family, as will be shown in Theorem 2 later.

(3) **The increment of window size.** For SLSH, we use a single uniform increment Δw in the extension of the windows. For MLSH, since there are two types of hash functions, we use $\Delta w_{0.5}$ and $\Delta w_{1.0}$ for $w_{0.5}$ and $w_{1.0}$, respectively. It is easy to see that, for an arbitrary object o, the ideal value of $\Delta w_{0.5}/\Delta w_{1.0}$ is $(f_{0.5}^{-1}(\lambda)\,\|o\|_{0.5})/(f_{1.0}^{-1}(\lambda)\,\|o\|_{1.0})$ since such setting ensures that the probabilities that o lies in the search windows of two types of functions are always the same $(=\lambda)$ during the extension of search windows. In other words, under such a setting, we can avoid too large or too small collision probabilities between o and q. Although the value of $\|o\|_{0.5} / \|o\|_{1.0}$ varies for different objects o, it can be bounded by $1 \leq \|o\|_{0.5} / \|o\|_{1.0} \leq d$. Thus we set it to a moderate value $\sqrt{d}$ and accordingly set $\Delta w_{0.5}/\Delta w_{1.0}$ to $\sqrt{d} f_{0.5}^{-1}(\lambda)/f_{1.0}^{-1}(\lambda)$. Note that under such setting, $w_{0.5}$ may reach $w_{\min}^{0.5}$ after $w_{1.0}$ reaches $w_{\min}^{1.0}$. If this happens, we only increase $w_{0.5}$ by $\Delta w_{0.5}$ stepwise afterwards (the operation in the other case is similar).

(4) **The setting of $m_{0.5}$ and $m_{1.0}$.** Given m, p_1 and p_b, we discuss how to determine $m_{0.5}$ and $m_{1.0}$ in the following two steps.

Step 1: First, according to the values of p_1 and p_b, we need to choose either SLSH or MLSH. For example, if p_b (the largest) is less than 0.5, it is clear that only hash functions generated in $\ell_{0.5}$ are enough. Thus, we directly choose SLSH with base metric $\ell_{0.5}$. Otherwise, we examine the following criteria.

Let $t_j = w_{\min}^{0.5,j}/w_{\min}^{1.0,j}$. If $t_j < \Delta w_{0.5}/\Delta w_{1.0}$, we say that hash functions in $\ell_{0.5}$ metric are more efficient than hash functions in $\ell_{1.0}$ metric for the ANN search in ℓ_{p_j} metric. Otherwise, we say that hash functions in $\ell_{1.0}$ metric are more efficient. This is because if $t_j < \Delta w_{0.5}/\Delta w_{1.0}$, $w_{0.5}$ can reach $w_{\min}^{0.5,j}$ before $w_{1.0}$ reaches $w_{\min}^{1.0,j}$. In other words, the following extension of $[-w_{1.0}, w_{1.0}]$ is redundant in practice. Thus, it is better to choose hash functions in $\ell_{0.5}$ metric in this case. The analysis under $t_j > \Delta w_{0.5}/\Delta w_{1.0}$ is similar.

For multiple metrics $\ell_{p_1}, \ell_{p_2}, \cdots, \ell_{p_b}$, we only need to consider metrics ℓ_{p_1} and ℓ_{p_b} because t_j is increasing as j increases from 1 to b. According to this fact and the discussion above, we use SLSH with base metric $\ell_{0.5}$ if $t_b \leq \Delta w_{0.5}/\Delta w_{1.0}$, and use SLSH with base metric $\ell_{1.0}$ if $t_1 \geq \Delta w_{0.5}/\Delta w_{1.0}$. If $t_1 < \Delta w_{0.5}/\Delta w_{1.0} < t_b$, we use MLSH.

Step 2: Once we determine to use MLSH, the next step is to determine the values of $m_{0.5}$ and $m_{1.0}$. Similar to the discussion in step 1, we only consider ℓ_{p_1} metric and ℓ_{p_b} metric. By means of some elementary computation, it is easy to see that, when $w_{1.0}$ reaches $w_{\min}^{1.0,j}$, and $w_{0.5}$ reaches $w_{\min}^{0.5,j}$, the probability P_1 that $o_{\min}^{1.0}$ lies in the search window of a randomly selected hash function (from m hash functions) can be expressed as follows:

$$P_1 = (\lambda m_{1.0} + f_{1.0}(\frac{\Delta w_{0.5}}{\Delta w_{1.0} t_1} f_{1.0}^{-1}(\lambda) m_{0.5}))/(m_{0.5} + m_{1.0}). \tag{8}$$

Algorithm 2: MLSH

Input: $\mathcal{D}$ is the dataset; d is the dimension of $\mathcal{D}$; q is the query; c is the approximation ratio; δ is the error rate; m is the number of hash functions; λ is the parameter related to the counting threshold; $\{\ell_{p_j}\}_{j=1}^{b}$ is the fractional metrics of interest;

Output: $\{o_{\min}^{j}\}_{j=1}^{b}$: return nearest neighbors of q in $\{\ell_{p_j}\}_{j=1}^{b}$ metrics

1 Compute $m_{0.5}$ and $m_{1.0}$ based on (10);
2 $w_{0.5} \leftarrow 0$, $w_{1.0} \leftarrow 0$, $s_{\min} \leftarrow \infty$, $o_{\min} \leftarrow NULL$, $w_{\min}^{0.5} \leftarrow \infty$, $w_{\min}^{1.0} \leftarrow \infty$;
3 $\tau \leftarrow \left\lfloor m\lambda - \sqrt{m\ln(1/\delta)/2} \right\rfloor$;
4 **while** $w_{0.5} \leq w_{\min}^{0.5}$ or $w_{1.0} \leq w_{\min}^{1.0}$ **do**
5   **if** $w_i \leq w_{\min}^{i}$ **then**
6         $w_i \leftarrow w_i + \Delta w_i$ ($\Delta w_i > 0, i = 0.5, 1$);
7     $\forall o \in \mathcal{D}$ update $\#Col(o)$ if necessary;
8   **if** *o is not visited and* $\#Col(o) \geq \tau$ **then**
9   **if** $\|q - o\|_{p_j} < s_{\min}^{j}$ **then**
10   Update $o_{\min}^{j}$ and $s_{\min}^{j}$;
11   Compute $w_{\min}^{0.5}$ and $w_{\min}^{1.0}$ by (7);
12 Return $\{o_{\min}^{j}\}_{j=1}^{b}$

Similarly, we can define P_b as follows:

$$P_b = (\lambda m_{0.5} + f_{0.5}(\frac{\Delta w_{1.0} t_b}{\Delta w_{0.5}} f_{0.5}^{-1}(\lambda) m_{1.0}))/(m_{0.5} + m_{1.0}). \tag{9}$$

In order to make the collision probability between q and $o_{\min}^{1}$ or $o_{\min}^{b}$ as large as possible, we maximize the lower bound of these two probabilities, that is,

$$(m_{0.5}, m_{1.0})^{*} = \arg\max_{m_{0.5}, m_{1.0}} \{\min\{P_1, P_b\}\}, \tag{10}$$

where $m_{0.5}, m_{1.0} > 0$ and $m_{0.5} + m_{1.0} = m$. In practice, the equation above can be solved fast by brute force because $m_{0.5}$ and $m_{1.0}$ are positive integers.

According to the discussion above, an example of MLSH is shown in Fig. 2 and its work flow is shown in Algorithm 2. In addition, we have the following theorem on MLSH, which is a straightforward consequence of Theorem 1.

Theorem 2. *Given $q \in \Re^d$, the probability that algorithm 2 returns a c-approximate nearest neighbor of q in l_{p_j} metric is at least $1 - \delta$ for every j $(1 \leq j \leq b)$.*

5 Parameter Setting and Complexity Analysis

As for the setting of λ, we solve the following function:

$$\lambda = \arg\min_{\lambda}\{\lambda - (m_{0.5} \times f_{0.5}(f_{0.5}^{-1}(\lambda)/c) + m_{1.0} \times f_{1.0}(f_{1.0}^{-1}(\lambda)/c))/m. \tag{11}$$

Here, λ can be interpreted as the expected collision probability between q and the true nearest neighbor in an arbitrary search window and the subtracted term denotes the expected collision probability between q and a c-approximate nearest neighbor in an arbitrary search window. Thus, we make the difference as large as possible in order to distinguish true and false objects better.

For the choice of m, if we choose a larger value of m, we can reduce the number of probed objects at the expense of more sequential I/O in index search. Thus, the optimal value of m depends on the specification of PC on which SLSH/MLSH works. In this paper, we fix m to 400 in all experiments for SLSH/MLSH, which is found suitable on our PC.

Next, let us analyze the complexity of MLSH. In the indexing phase, since the time complexities of building B^+ trees with regard to hash functions in different metrics are the same, the time complexity in the index construction is $O(mn \log n)$. In the searching phase, MLSH needs to scan those objects close to the query on each projection vector and calculate the exact distances of a small number of candidates to the query. Thus, the time complexity of MLSH is $O(mn + \alpha nd)$, where αn denotes the number of candidates. Although we do not limit the value of α for ensuring the probability guarantee, the value of α is expected to be very small on real datasets. In experiments described later, the value of αn was less than 1K in general. In addition, since α is decreasing as c goes higher, users are free to control the number of candidates by adjusting the value of c.

6 Experiments

SLSH/MLSH was implemented in C++. All the experiments were performed on a PC with Xeon E5-165, 3.60 GHz six-core processor with 128 GB RAM, in Ubuntu 16.04. The page size was fixed to 4 KB.

6.1 Experimental Setup

Benchmark Methods Selection

- **LazyLSH.** We used the original implementation of LazyLSH offered by its authors. In the experiments, all internal parameters of LazyLSH were set properly according to the suggestions of its author. In addition, we adopted the multi-query optimization to achieve its best performance.
- **MLSH.** The error rate δ and the number of hash functions m were fixed to 0.1 and 400, respectively. Parameter λ was computed by (11).
- **SLSH(0.5)** denotes SLSH with base metric $\ell_{0.5}$. Here, where m was fixed to 400 and λ was adjusted to be experimentally optimal.

It is notable that (1) SLSH(1.0) was excluded due to its difficultly in tuning c. In fact, for SLSH(1.0), the width of search windows on hash functions generated in ℓ_1 space should be very large for the searching in those ℓ_p spaces with small p, which makes SLSH(1.0) spends 100X-400X running time of MLSH to achieve the same practical target recall on each dataset. (2) QALSH was excluded in the experiments because it could not solve MMS problems directly.

Table 1. Real datasets

Dataset	Dimension	Size	Type
Msong	420	992,272	Text
Deep	256	1,000,000	Image
ImageNet	150	2,340,373	Image
Sift10M	128	11,164,666	Image
Deep1B	96	1,000,000,000	Image

Table 2. Comparison on index sizes (GB)

Method	Msong	Deep	ImageNet	Sift10M	Deep1B
MLSH	1.6	1.6	3.2	17.7	1653
SLSH	1.6	1.6	3.2	17.7	1653
LazyLSH	11.3	9.9	19.1	63.1	\

Real Datasets. We chose five real datasets (Table 1). For each dataset except for Deep1B, we randomly sampled 50 data objects to form the query set. In the following experiments, we take the average value over all queries for each performance metric. Since LazyLSH could not work on billion-scale datasets due to overlarge memory consumption, we postpone the comparison on the results of SLSH/MLSH on Deep1B to Sect. 6.4.

In the following experiments, we set two objective ranges of p. The first range is $[0.2, 0.7]$ ($b = 6$) with interval 0.1 and the second range is $[0.5, 1]$ ($b = 6$) with interval 0.1. Here, [0.5, 1] is also the range of p adopted in the paper of LazyLSH. In addition, the default value of K (top-K) was set to 100.

6.2 Experimental Results on MLSH

The Verification of Stability. Figure 3 (1st row) shows the recalls of MLSH under different approximation ratios given $p \in [0.5, 1.0]$. According to the results, we have following observations. (1) For the same approximation ratio, the performance of MLSH is stable against the change of p. (2) MLSH can achieve different accuracies by adjusting the value of approximation ratio c. To be specific, as c decreases, the recall of MLSH increases.

We also show the recalls of MLSH for another range $p \in [0.2, 0.7]$ in Fig. 3 (2nd row). Except for Sift10M, the performance of MLSH is again stable against p. With regard to Sift10M, the recall in $\ell_{0.2}$ metric is 25% less than that in $\ell_{0.5}$ metric. This demonstrates that the stability of MLSH is affected by the actual data distribution to some extent. Thus, we recommend users to adopt a smaller c for hard datasets.

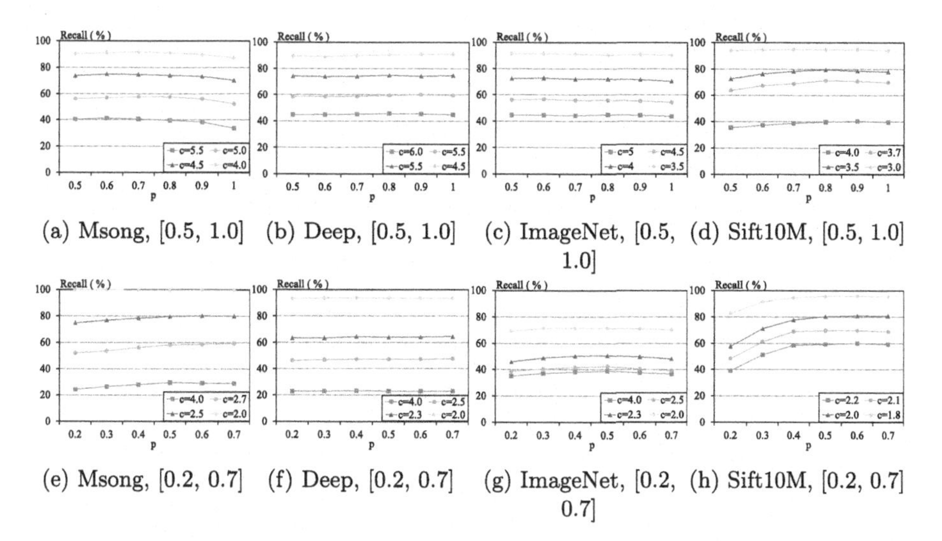

(a) Msong, [0.5, 1.0] (b) Deep, [0.5, 1.0] (c) ImageNet, [0.5, 1.0] (d) Sift10M, [0.5, 1.0]

(e) Msong, [0.2, 0.7] (f) Deep, [0.2, 0.7] (g) ImageNet, [0.2, 0.7] (h) Sift10M, [0.2, 0.7]

Fig. 3. Recalls of MLSH for different ℓ_p metrics

6.3 The Comparison Study

In this section, we compare MLSH with the other methods on various performance metrics. LazyLSH was excluded in the comparison when $p \in [0.2, 0.7]$ since the value of p in LazyLSH could not be less than 0.44.

Comparison on Index Sizes. In Table 2, we compared the index sizes of three methods. Because the index size of LazyLSH depends on the value of c, we only show its results under the recommended value of c, that is, $c = 3$. From the results, we can see that the index sizes of MLSH and SLSH are equal because both of them generate 400 hash functions. On the other hand, LazyLSH requires more space to store the index than the other two methods because LazyLSH generates around or over 1000 hash functions on each dataset to ensure its probability guarantee.

Comparison on Probed Objects. In Fig. 4 (1st row), we compared the number of probed objects of SLSH(0.5) and MLSH given $p \in [0.5, 1]$. LazyLSH was excluded in this experiment because different from the other two methods, LazyLSH fixes the number of probe objects beforehand and varies the number of hash functions to achieve different accuracies. From the results, we have following observations. (1) On three real datasets, MLSH probed less objects than SLSH to achieve the same precision level, which shows a less sensitivity of the mixed hash function family. (2) The advantage of MLSH over SLSH becomes more obvious as the target recall increases, which shows the superiority of the mixed hash function family for high-accuracy oriented applications.

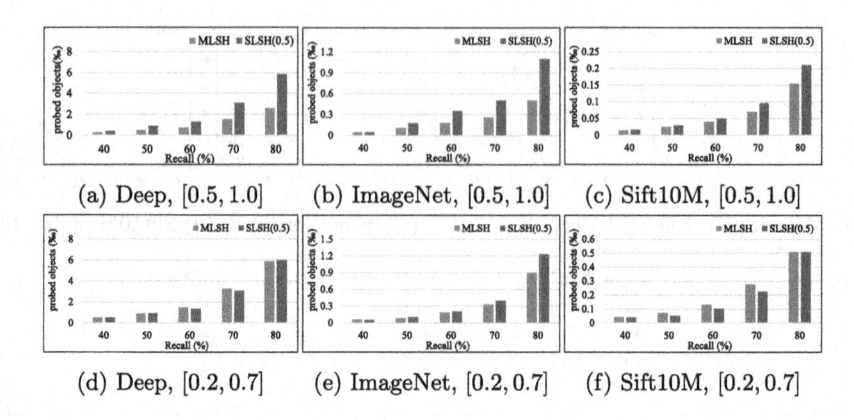

(a) Deep, [0.5, 1.0] (b) ImageNet, [0.5, 1.0] (c) Sift10M, [0.5, 1.0]

(d) Deep, [0.2, 0.7] (e) ImageNet, [0.2, 0.7] (f) Sift10M, [0.2, 0.7]

Fig. 4. Comparison on probed objects between SLSH(0.5) and MLSH

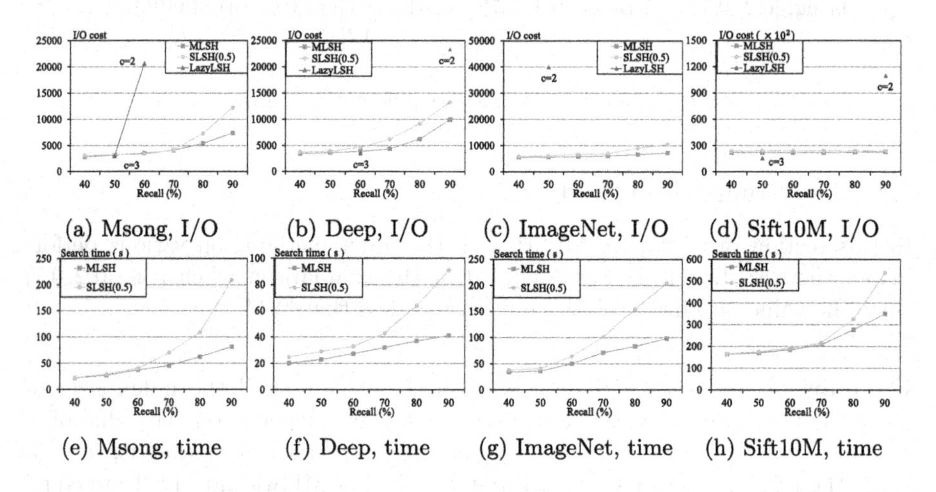

(a) Msong, I/O (b) Deep, I/O (c) ImageNet, I/O (d) Sift10M, I/O

(e) Msong, time (f) Deep, time (g) ImageNet, time (h) Sift10M, time

Fig. 5. Comparison on time/I/O cost - recall curves ($p \in [0.5, 1]$)

We also show the results under $p \in [0.2, 0.7]$ in Fig. 4 (2nd row). From the results, we can see that the number of probed objects of SLSH(0.5) and MLSH are very close for each target recall. This is because MLSH automatically generates more hash functions in $\ell_{0.5}$ metric and avoids the poor performance of hash functions in ℓ_1 metric for those objective ℓ_p metrics with small p's.

Comparison on I/O Cost. The results on I/O-recall tradeoff are shown in Fig. 5 (1st row). Since c of LazyLSH should be chosen as an integer greater than 1, we only plot two points for each curve of LazyLSH: one is for $c = 3$ (recommended setting) and the other one is for $c = 2$ (the highest accuracy). According to the results, we have following observations. (1) Overall, the curve of MLSH is lowest on each dataset, which shows the high efficiency of MLSH in saving I/O cost. (2) Although LazyLSH achieves the comparable performance

Table 3. SLSH (0.5) vs. MLSH on Deep1B

Recall	Probed objects (‰)		I/O cost ($\times 10^5$)	
60%	MLSH	0.10	MLSH	23
	SLSH(0.5)	0.12	SLSH(0.5)	26
80%	MLSH	0.33	MLSH	26
	SLSH(0.5)	0.52	SLSH(0.5)	28

of MLSH under $c = 3$, its performance under $c = 2$ is much worse than the other two methods. This is because, for LazyLSH, the required number of hash functions under $c = 2$ is around 7 times more than that under $c = 3$, but its search accuracy is not improved significantly as c varies from 3 to 2. (3) The advantage of MLSH over SLSH(0.5) becomes more obvious as the target recall increases, which shows again the superiority of MLSH in achieving the high accuracy.

Comparison on Search Time. In Fig. 5 (2nd row), we compared the search times of MLSH and SLSH. With regard to the search time of LazyLSH, we report its results under $c = 3$ as follows. LazyLSH incurs 2.8X, 9.4X,14.4X and 16.6X search time of MLSH(target recall 90%) on Msong, Deep, ImageNet and Sift10M, respectively. According to the results, we have following observations: (1) SLSH and MLSH run much faster than LazyLSH, which shows our searching strategy is efficient in time consumption. (2) Compared with SLSH, MLSH spends less search time on each dataset, which is consistent with the results in I/O cost and probed objects.

6.4 Results on Deep1B

In Table 3, we show the results on Deep1B, where the range of p is $[0.2, 0.7]$ and the target recall is fixed to 60% or 80%. Since the searching on billion-scale datasets is quite time-consuming, we sampled 10 queries from Deep1B to ensure that the search can finish within the controllable time. From the results, we can see that while the total I/O costs of two methods are very close, MLSH requires less probed objects to achieve the target recall, which confirms the advantage of MLSH over SLSH on billion-scale datasets.

7 Conclusion

In this paper, we proposed a mixed hash function family called MLSH for solving multi-metric search (MMS) problem. Compared with traditional algorithms, the new hash function family can be used for the search in multiple metrics (fractional distances) more efficiently by choosing a suitable proportion of two types hash functions with the user-specified probability guarantee. Experiments show that MLSH performs well in various situations of different objective fractional metrics.

Acknowledgment. This work was partially supported by JSPS KAKENHI Grant Number 19H04128.

References

1. Aggarwal, C.C., Hinneburg, A., Keim, D.A.: On the surprising behavior of distance metrics in high dimensional spaces. In: ICDT, pp. 420–434 (2001)
2. Aref, W.G., et al.: A video database management system for advancing video database research. In: Multimedia Information Systems, pp. 8–17 (2002)
3. Zheng, B., Zhao, X., Weng, L., Hung, N.Q.V., Liu, H., Jensen, C.S.: PM-LSH: A fast and accurate LSH framework for high-dimensional approximate NN search. PVLDB **13**(5), 643–655 (2020)
4. Cormode, G., Indyk, P., Koudas, N., Muthukrishnan, S.: Fast mining of massive tabular data via approximate distance computations. In: ICDE, pp. 605–614 (2002)
5. Datar, M., Immorlica, N., Indyk, P., Mirrokni, V.S.: Locality-sensitive hashing scheme based on p-stable distributions. In: SoCG, pp. 253–262 (2004)
6. Francois, D., Wertz, V., Verleysen, M.: The concentration of fractional distances. IEEE Trans. Knowl. Data Eng. **19**(7), 873–886 (2007)
7. Gan, J., Feng, J., Fang, Q., Ng, W.: Locality-sensitive hashing scheme based on dynamic collision counting. In: SIGMOD, pp. 541–552 (2012)
8. He, J., Kumar, S., Chang, S.F.: On the difficulty of nearest neighbor search. In: ICML, pp. 1127–1134 (2012)
9. Howarth, P., Ruger, S.: Fractional distance measures for content-based image retrieval. In: ECIR, pp. 447–456 (2005)
10. Huang, Q., Feng, J., Fang, Q., Ng, W., Wang, W.: Query-aware locality-sensitive hashing scheme for l_p norm. VLDB J. **26**(5), 683–708 (2017)
11. Huang, Q., Feng, J., Zhang, Y., Fang, Q., Ng, W.: Query-aware locality-sensitive hashing for approximate nearest neighbor search. PVLDB **9**(1), 1–12 (2015)
12. Indyk, P., Motwani, R.: Approximate nearest neighbors: Towards removing the curse of dimensionality. In: STOC, pp. 604–613 (1998)
13. Ke, Y., Sukthankar, R., Huston, L.: An efficient parts-based near-duplicate and sub-image retrieval system. In: ACM Multimedia, pp. 869–876 (2004)
14. Li, W., Zhang, Y., Sun, Y., Wang, W., Zhang, W., Lin, X.: Approximate nearest neighbor search on high dimensional data - experiments, analyses, and improvement. IEEE Trans. Knowl. Data Eng. **32**(8), 1475–1488 (2020)
15. Lu, K., Kudo, M.: R2LSH: A nearest neighbor search scheme based on two-dimensional projected spaces. In: ICDE, pp. 1045–1056 (2020)
16. Lu, K., Wang, H., Wang, W., Kudo, M.: VHP: Approximate nearest neighbor search via virtual hypersphere partitioning. PVLDB **13**(9), 1443–1455 (2020)
17. Sun, Y., Wang, W., Qin, J., Zhang, Y., Lin, X.: SRS: solving c-approximate nearest neighbor queries in high dimensional Euclidean space with a tiny index. PVLDB **8**(1), 1–12 (2014)
18. Zheng, Y., Guo, Q., Tung, A.K.H., Wu, S.: LazyLSH: Approximate nearest neighbor search for multiple distance functions with a single index. In: SIGMOD, pp. 2023–2037 (2016)

Quantum-Inspired Keyword Search on Multi-model Databases

Gongsheng Yuan[1,2], Jiaheng Lu[1](✉), and Peifeng Su[3]

[1] Department of Computer Science, University of Helsinki, 00014 Helsinki, Finland
{gongsheng.yuan,jiaheng.lu}@helsinki.fi
[2] School of Information, Renmin University of China, Beijing 100872, China
[3] Department of Geosciences and Geography, University of Helsinki, 00014 Helsinki, Finland
peifeng.su@helsinki.fi

Abstract. With the rising applications implemented in different domains, it is inevitable to require databases to adopt corresponding appropriate data models to store and exchange data derived from various sources. To handle these data models in a single platform, the community of databases introduces a multi-model database. And many vendors are improving their products from supporting a single data model to being multi-model databases. Although this brings benefits, spending lots of enthusiasm to master one of the multi-model query languages for exploring a database is unfriendly to most users. Therefore, we study using keyword searches as an alternative way to explore and query multi-model databases. In this paper, we attempt to utilize quantum physics's probabilistic formalism to bring the problem into vector spaces and represent events (e.g., words) as subspaces. Then we employ a density matrix to encapsulate all the information over these subspaces and use density matrices to measure the divergence between query and candidate answers for finding top-k the most relevant results. In this process, we propose using pattern mining to identify compounds for improving accuracy and using dimensionality reduction for reducing complexity. Finally, empirical experiments demonstrate the performance superiority of our approaches over the state-of-the-art approaches.

1 Introduction

In the past decades, due to an explosion of applications with the goal of helping users address various transactions in different domains, there are increasing needs to store and query data produced by these applications efficiently. As a result, researchers have proposed diverse data models for handling these data, including structured, semi-structured, and graph models. Recently, to manage these data models better, the community of databases introduces an emerging concept, multi-model databases [16], which not only embraces a single and unified platform to manage both well-structured and NoSQL data, but also satisfies the system demands for performance, scalability, and fault tolerance.

C. S. Jensen et al. (Eds.): DASFAA 2021, LNCS 12682, pp. 585–602, 2021.
https://doi.org/10.1007/978-3-030-73197-7_39

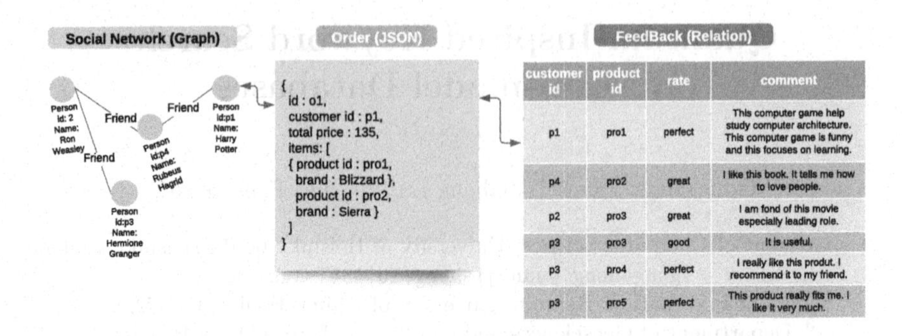

Fig. 1. An Example of multi-model data.

Although multi-model database systems provide a way to handle various data models in a unified platform, users have to learn corresponding specific multi-model query languages to access different databases (e.g., using AQL to access ArangoDB [2], SQL++ for AsterixDB [1], and OrientDB SQL for OrientDB [19]). Moreover, users also need to understand complex and possibly evolving multi-model data schemas as background knowledge for using these query language. This is unfriendly to most users because it usually with a steeper learning curve. For example, Fig. 1 depicts multi-model data in social commerce. *Suppose we want to find Rubeus Hagrid's friends who have bought Blizzard and given a perfect rating.* It won't be an easy job to write a multi-model query involving social network (Graph), order (JSON), and feedback (Relation) for novices to achieve this goal. Therefore, in this paper, we study using keyword searches as an alternative way to explore and query multi-model databases, which does not require users to have strong background knowledge.

After reviewing the literature, we find that most existing works [27] only restrict keyword searches over a specific database supporting a single data model (e.g., relational, XML, and graph databases). Unfortunately, there is a lack of relevant research literature for the issue of performing keyword searches on multi-model databases. However, we think it is a promising research topic that remains a big challenge. This is because a trivial solution, which firstly performs a keyword search on individual data models by conventional methods and then combines results by assembling the previous results, cannot work well. The reason is that it may miss results that consist of multiple models simultaneously.

Facing this challenge, previous researchers used graph methods [14] to solve it. However, there are several matters needing attention in this method. Firstly, when we use a graph to represent these heterogeneous data, this graph may be vast and complex. So we need to divide this graph into many subgraphs, which is a graph partition problem. And if we perform keyword searches on these graphs to find answers, this means we need to find some subgraphs relating to partial keywords or all as the answer. Therefore, this involves subgraph matching and subgraph relevance ranking problems. And lastly, we need to consider how to take the dependencies among keywords and schema information into consideration when doing keyword searches. We could see all these problems have a significant influence on the final returned results. This means we should be careful to choose the corresponding solutions for these graph problems.

To avoid these graph issues, we start by introducing the "*quantum-inspired*" framework into the database community [28], in which we utilize the probabilistic formalism of quantum physics to do keyword searches. In quantum probability, the probabilistic space is naturally encapsulated in a vector space. Based on the notion of information need vector space, we could regard the data in multi-model databases as the information (*statements*) collection, define events (e.g., words) as subspaces, use a density matrix (probability distribution) to encapsulate all the information over these subspaces for measuring the relevance of the candidate answers to the user's information need.

For this framework, the idea behind the quantum-inspired approach to disciplines other than physics is that, although macroscopic objects cannot exhibit the quantum properties manifested by particles such as photons, some phenomena can be described by the language or have some features of the phenomena (e.g., superposition) represented by the quantum mechanical framework in physics [24]. Therefore, except for the above theoretical method introduction, there are two other reasons underlying this attempt. One is that the similarity between the quantum mechanical framework to predict the values which can only be observed in conditions of uncertainty [24] and the decision about the relevance of the content of a text to an information need is subject to uncertainty [18]. Another one is that increasing works support the notion that quantum-like phenomena exist in human natural language and text, cognition and decision making [22], all related to the critical features of keyword searches.

Besides, the pioneering work [24] formalized quantum theory as a formal language to describe the objects and processes in information retrieval. Based on this idea, we use this mathematical language to describe relational, JSON, and graph data in the database community as information collection. Next, we transform keyword searches from a querying-matching work into a calculating-ranking task over this information collection and return the most relevant top-k results. And we take the possible relevance among input query keywords and database schema information into consideration, which helps the framework understand the user's goal better. Now, we summarize our contributions as follows:

1. Based on quantum theory, we attempt to use a quantum-inspired framework to do keyword searches on multi-model databases, utilizing quantum physics' probabilistic formalism to bring the problem into information need vector space. We want to take advantage of the ability of quantum-inspired framework in capturing potential semantics of candidate answers and query keywords for improving query accuracy.
2. In this process, we introduce the co-location concept to identify compounds for enhancing the topic of statements. We want to use it to improve query performance (precision, recall, and F-measure).
3. By analyzing the existing quantum-inspired method, we propose constructing a query density vector instead of a density matrix to reduce the framework's

complexity. And we present an algorithm to perform keyword searches on multi-model databases.
4. Finally, we perform extensive empirical experiments that demonstrate our approaches' overall performance is better than the state-of-the-art approaches. The F-measure is at least nearly doubled.

2 Preliminaries

For the sake of simplicity, we assume the vector space is $\mathbb{R}^n$ in this paper. A unit vector $\vec{u} \in \mathbb{R}^n$ is defined as $|u\rangle$ and termed as *ket* on the basis of Dirac notation. Its conjugate transpose $\vec{u}^H = \vec{u}^T$ is written as $\langle u|$ and called *bra*. The inner product between two vectors is denoted as $\langle u|v\rangle = \sum_{i=1}^{n} u_i v_i$. The outer product between two vectors is denoted as $|u\rangle\langle v|$ and called *dyad*. When the vector has a unitary length (i.e., $\|\vec{u}\|_2 = 1$), a special operator can be defined as the outer product between two vectors. We call this operator *projector*. To explain this, suppose $|u\rangle$ is a vector; the projector corresponding to this vector is a dyad written as $|u\rangle\langle u|$. For example, if there is $|u_1\rangle = (1,0)^T$, the projector corresponding to this ket is transforming it into $|u_1\rangle\langle u_1| = \begin{pmatrix} 1 & 0 \\ 0 & 0 \end{pmatrix}$. Due to the projector, $|u\rangle$ could be mapped to the generalized probability space. In this space, each rank-one dyad $|u\rangle\langle u|$ can represent a quantum *elementary event* and each dyad $|\kappa\rangle\langle\kappa|$ represent a *superposition event*, where $|\kappa\rangle = \sum_{i=1}^{p} \sigma_i |u_i\rangle$, the coefficients $\sigma_i \in \mathbb{R}$ satisfy $\sum_i \sigma_i^2 = 1$. And density matrices ρ are interpreted as generalized probability distributions over the set of dyads when its dimension greater than 2 according to Gleason's Theorem [7]. A real density matrix ρ is a positive semidefinite ($\rho \geq 0$) Hermitian matrix ($\rho = \rho^H = \rho^T$) and has trace 1 ($Tr(\rho) = 1$). It assigns a generalized probability to each dyad $|u\rangle\langle u|$, whose formula is:

$$\mu_\rho(|u\rangle\langle u|) = Tr(\rho\,|u\rangle\langle u|). \tag{1}$$

For example, $\rho_1 = \begin{pmatrix} 0.75 & 0 \\ 0 & 0.25 \end{pmatrix}$, $\rho_2 = \begin{pmatrix} 0.5 & 0.5 \\ 0.5 & 0.5 \end{pmatrix}$, density matrix ρ_2 assigns a probability value $Tr(\rho_2 |u_1\rangle\langle u_1|) = 0.5$ to the event $|u_1\rangle\langle u_1|$. If ρ is unknown,

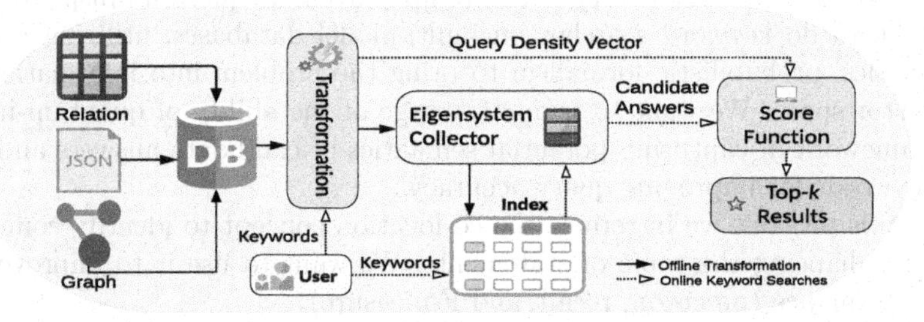

Fig. 2. The framework of quantum-inspired keyword search

we could utilize the *Maximum Likelihood* (MaxLik) estimation to get it. Finally, through the value of negative Von-Neumann Divergence (VND) between ρ_q (query) and ρ_c (candidate answer), we could get their difference. Its formalization is:

$$-\Delta_{VN}(\rho_q||\rho_c) \overset{rank}{=} \sum_i \lambda_{q_i} \sum_j (\log \lambda_{c_j} \langle q_i|c_j\rangle^2). \quad (2)$$

3 The Framework of the Quantum-Inspired Keyword Search

An overview of this entire framework is shown in Fig. 2, which has two functional modules. One is offline transformation. We will use quantum language to represent heterogeneous data in a uniform format. Another one is online keyword searches with generalized probabilities for finding the most relevant results to answer keyword searches.

3.1 Transformation

Here we consider three kinds of data. They are relational, JSON, and graph data. For the **relational** data model, it is natural to think of a tuple with schema in tables as a *statement* (a piece of information). To avoid information explosion (caused by information fragmentation), we observe the **JSON** file in a coarse granularity and treat each object in JSON as an integral *statement*. For each node in the **graph** data, we put information about all of its neighbor nodes in the same *statement*, including itself information. In this way, it can keep complete neighbor relationships and node information.

Only considering these discrete data is not enough. We need to take join into consideration to get complete information in our framework. For this problem, there are many methods. For example, we could mine the relationship among different data models with [5], then join them. Or we could choose meaningfully related domains to do equi-join operations based on expert knowledge, which is a quite straightforward and easy way for finding meaningful joins and cutting down the search space by avoiding generating a lot of useless intermediate results. Since this part is not the focus of this paper, we will use the later one to continue our work.

Now we have gotten a statement collection. Next, we use the mathematical language of quantum theory to represent these statements. To achieve this goal, we first use an elementary event to represent a single word and use a superposition event to represent a compound. In this process, to construct proper superposition events to enhance the topic of statement for improving query accuracy (i.e., choose appropriate compounds κ from a given statement and determine the value of each coefficient σ_i in κ), we introduce a special spatial pattern concept called *co-location pattern* [12]. After this process, we could get an event set $\mathscr{P}_{st}$ for each statement. Then, we learn a density matrix from each event set $\mathscr{P}_{st}$ with MaxLik estimation and use this density matrix to represent the corresponding statement. Here, we use a $R\rho R$ algorithm, which is an iterative schema of MaxLik outlined in the [17]. And this iterative schema has been used in the [23] for getting density matrices. Finally, we could use VND to measure the difference between different statements by their density matrices and use VND' values to rank them.

To get an event set $\mathscr{P}_{st}$ for each statement, we introduce a special spatial pattern concept called *co-location pattern* [12] to help construct compounds κ. Next, we will start by redefining some concepts in the field of co-location to adapt our problem for identifying compounds. When a word set $c = \{w_1, ..., w_p\}$, $\|c\| = p$, appears in any order in a fixed-window of given length L, we say there is a relationship R among these words. For any word set c, if c satisfies R in a statement, we call c co-location compound.

Definition 1. ***Participation ratio (PR)*** *$PR(c, w_i)$ represents the participation ratio of word w_i in a co-location compound $c = \{w_1, ..., w_p\}$, i.e.*

$$PR(c, w_i) = \frac{T(c)}{T(w_i)} \quad w_i \in c, \tag{3}$$

where $T(c)$ is how many times the c is observed together in the statement , $T(w_i)$ is the number of word w_i in the whole statement.

Table 1. Example about identifying compound κ

$\mathscr{W}_{st}$ = {*This computer game help study computer architecture this computer game is funny and this focuses on learning.* }				
c_1 = { *computer, game* }	PR(c_1, computer) PR(c_1, game)	$\frac{2}{3}$ $\frac{2}{2}$	PI(c_1)	$\frac{2}{3}$
c_2 = { *game, architecture* }	PR(c_2, game) PR(c_2, architecture)	$\frac{1}{2}$ $\frac{1}{1}$	PI(c_2)	$\frac{1}{2}$
c_3 = { *computer, architecture* }	PR(c_3, computer) PR(c_3, architecture)	$\frac{1}{3}$ $\frac{1}{1}$	PI(c_3)	$\frac{1}{3}$
c_4 = { *computer, game, architecture* }	PR(c_4, computer) PR(c_4, game) PR(c_4, architecture)	$\frac{1}{3}$ $\frac{1}{2}$ $\frac{1}{1}$	PI(c_4)	$\frac{1}{3}$
PI(c_1) $\geq$ *min_threshold* = 0.6, c_1 is a compound κ				

Definition 2. ***Participation index (PI)*** *$PI(c)$ of a co-location compound c is defined as:*

$$PI(c) = \min_{i=1}^{p}\{PR(c, w_i)\}. \tag{4}$$

Given a minimum threshold *min_threshold*, a co-location compound c is a compound κ if and only if $PI(c) \geq min\_threshold$. It is obvious that when $min\_threshold = 0$, all the co-location compounds are compounds. Based on the method of estimating whether c is a compound κ, with each appearance of word $w_i \in c$ in the statement, there has been a great effect on calculating value of PR which determines whether $PI(c)$ is greater than or equal to $min\_threshold$. This is to say each appearance of word w_i plays an important role in the statement for expressing information. Therefore, when we set a value for the coefficient σ_i of $|w_i\rangle$ in κ , we need to take the above situation into consideration. A natural way is to set $\sigma_i^2 = T(w_i)/\sum_{j=1}^{p} T(w_j)$ for each w_i in a compound $\kappa = \{w_1, ..., w_p\}, 1 \leq i \leq p$, where $T(w_i)$ is the number of times that word w_i occurs in this statement. We call it *co-location weight.*

For example, we assume "This computer game help study ..." is a statement $\mathscr{W}_{st}$, which is in the table FeedBack of Fig. 1. Given c_1, c_2, c_3, and c_4 (see Table 1), if $min\_threshold = 0.6$, thus c_1 is a compound. This helps our framework to have a better understanding of this statement whose theme is about the computer game, not computer architecture. Meanwhile, with co-location weight we could set $\sigma_{computer}^2 = 3/5$, $\sigma_{game}^2 = 2/5$ for the $\kappa = \{computer,\ game\}$. According to the related state-of-the-art work [4], the fixed-window of Length L could be set to $l||c||$. And it provides a robust pool for l. Here, we select $l = 1$ from the robust pool. Because if one decides to increase l, more inaccurate relations will be detected, and performance will deteriorate.

Next, we could use an iterative schema of $MaxLik_{matrix}$ [23] to learn a density matrix from each event set $\mathscr{P}_{st}$. However, unfortunately, both dyads and density matrices are $n \times n$ matrices, i.e., their dimensions depend on n, the size of word space $\mathscr{V}$. When n is bigger, the cost of calculating is higher, especially in the field of databases. Therefore, we need to find a way to reduce complexity.

According to the Eq. (1), the probability of quantum event Π is $Tr(\rho\Pi)$. Through $Tr(\rho\Pi)$, we could get the dot product of two vectors, i.e.,

$$Tr(\rho\Pi) = \vec{\rho}.\vec{\Pi}^2, \tag{5}$$

where $\vec{\rho} = (\beta_1,\ \beta_2,\ \ldots,\ \beta_n)$, we call it *density vector.* If we assume $\vec{\Pi} = ((\sum_{i=1}^{n} u_{i1}v_{i1}),\ (\sum_{i=1}^{n} u_{i1}v_{i2}),\ \ldots,\ (\sum_{i=1}^{n} u_{i1}v_{in}))$, then $\vec{\Pi}^2 = ((\sum_{i=1}^{n} u_{i1}v_{i1})^2$, $(\sum_{i=1}^{n} u_{i1}v_{i2})^2,\ \ldots,\ (\sum_{i=1}^{n} u_{i1}v_{in})^2)$.

Definition 3. *Density vector* *A density vector is a real vector $\vec{\rho} = (\beta_1, \beta_2, \ldots, \beta_n)$, noted by $\langle\rho|$, where $\beta_i \geq 0$, and $\beta_1 + \beta_2 + \ldots + \beta_n = 1$.*

For example, density vector $\langle\rho_3| = (0.2, 0.1, 0.7)$. Next, we will use a density vector instead of a density matrix in the rest of this paper. This is because it has three advantages.

- Firstly, density vectors have a more concise way of calculating quantum probabilities than density matrices' (comparing Eq. 1 with Eq. 6);
- Secondly, when one calculates VND between two density vectors, it is faster to get the value of VND than density matrices do (comparing Eq. 2 with Eq. 7);

- Thirdly, comparing with density matrices, learning a query density vector from the input keywords is easier.

Based on the generated density matrices, we could get *val* (the list of all eigenvalues) and *vec* (the list of eigenvectors, each eigenvector corresponding to its own eigenvalue) through eigendecomposition of density matrices. To simplify our framework further, we use the Principal Component Analysis (PCA) method to take the first h largest eigenvalues and require that the sum of these values is greater than or equal to a *threshold*. Based on the research concerning PCA [10], we could set $threshold = 85\%$, which is enough for our work. This is because if one increases *threshold*, a higher *threshold* may have a higher chance to cause the increase of the number of val_i instead of precision of framework and deteriorate the performance of the framework. Then, we store h eigenvalues and corresponding eigenvectors vec_i (termed as $\{val, vec\}$) into the set of density systems, S_{denSys} (the component, "Eigensystem Collector" in Fig. 2).

We have now transformed all kinds of data into eigenvalues (density vectors) and eigenvectors (mapping directions) and converted querying-matching problem into calculating-ranking. In this way, we eliminate heterogeneous data and convert the querying-matching problem into calculating-ranking. Next, we will consider how to do keyword searches on these density vectors online.

3.2 Online Keyword Search Query

In this module, when users submit their queries, there are two data flows about queries.

- The first one is from users to "Transformation", which will construct query density vectors from input keywords. This step will consider the relationship between input keywords instead of treating keywords as mutually independent entities.
- Another one is from users to index structure for getting candidate answers, which will reduce the query scope through our index structure (inverted list).

In the above data flows, it involves constructing query density vectors. Before doing it, we propose the new representation for single words and compounds in query keywords. Through analyzing Eq. (5), we could represent each word $w_i \in \mathscr{V}$ by $|e_i\rangle$, where $|e_i\rangle$, the standard basis vector, is an one-hot vector. Based on this new representation, we present compound by $|\kappa\rangle = \sum_{i=1}^{p} \sigma_i |e_{w_i}\rangle$, where $\kappa = \{w_1, ..., w_p\}$, the coefficients $\sigma_i \in \mathbb{R}$ satisfy $\sum_{i=1}^{p} \sigma_i^2 = 1$ to guarantee the proper normalization of $|\kappa\rangle$.

For example, Considering $n = 3$ and $\mathscr{V} = \{computer, science, department\}$, if $\kappa_{cs} = \{computer, science\}$ and $|\kappa_{cs}\rangle = \sqrt{2/3}\,|e_c\rangle + \sqrt{1/3}\,|e_s\rangle$, then we could get:

$$|e_{computer}\rangle = \begin{pmatrix}1\\0\\0\end{pmatrix}, \quad |e_{science}\rangle = \begin{pmatrix}0\\1\\0\end{pmatrix}, \quad |\kappa_{cs}\rangle = \sqrt{\frac{2}{3}}\begin{pmatrix}1\\0\\0\end{pmatrix} + \sqrt{\frac{1}{3}}\begin{pmatrix}0\\1\\0\end{pmatrix} = \begin{pmatrix}\sqrt{\frac{2}{3}}\\\sqrt{\frac{1}{3}}\\0\end{pmatrix}.$$

Next, we still use the iterative schema of *Maximum Likelihood* (MaxLik) estimation for getting the query density vector. We would also use the following formula to get the quantum probability of event $|\Pi\rangle$ at $MaxLik$ estimation.

$$Probability_{\Pi} = \langle \rho | \Pi^2 \rangle . \tag{6}$$

Finally, we propose Algorithm 1 to do online keyword searches, in which we use line 1–9 to get candidate density system set C_{denSys} through the index. Then we use them to help get corresponding query density vectors and regard them as the score function (VND) inputs to answer queries (line 10–26).

Firstly, we could classify the input keywords into two groups according to whether or not w_i is relevant to the database schema. For the group K_{schema}, we use the union to get a candidate density system set C_{schema} in line 4, where C_{w_i} can be gotten by the inverted list, which is relevant to the input word w_i. For the K_{non_schema}, we use intersection to get C_{non_schema}. Then we could get the candidate density system set C_{denSys} through difference (line 9). We use these operations to take advantage of the schema for helping users explore more potential possibilities about answers.

Line 10–26 are going to get top-k results. Firstly, we use every mapping direction vec_{st} of $\{val, vec\}_{st}$ in the candidate density systems C_{denSys} to transform original events into new representations so that the algorithm could learn a query density vector from these new events. The above transforming could guarantee that each constructed query density vector and corresponding density vector val_{st} in one coordinate system. Then we could calculate the divergence between $\langle \rho |_q$ and $\langle \rho |_{st}$ (val_{st}) by Eq. 7 in line 23.

$$-\Delta_{VN}(\langle \rho|_q \,||\, \langle \rho|_{st}) \overset{rank}{=} \sum_i \beta_{q_i} \log \beta_{st_i}, \; where \; \beta_{q_i} \in \langle \rho|_q , \beta_{st_i} \in \langle \rho|_{st} . \tag{7}$$

Considering the previous problem "*Suppose we want to find Rubeus Hagrid's friends who have bought Blizzard and given a perfect rating.*", we take {Rubeus Hagrid friends Blizzard perfect} as Algorithm 1 inputs and perform keyword searches on the Fig. 1. And we could get the result in which a person named Harry Potter bought Blizzard and gave a perfect rating, and he is Rubeus Hagrid's friend. The original result is "*social network person id p1 name harry potter friend person id p4 name rubeus hagrid order id o1 custom id p1 total price 135 item product id pro1 brand blizzard feedback rate perfect comment this computer game help study computer architecture this computer game is funny and this focuses on learning*". And its score is –2.665.

Algorithm 1. Answer Keyword Searches with Density Vectors

Input: Input keywords $K = \{w_1, \ldots, w_t\}$
Output: the top-k most relevant statements about queries
1: $\{K_{schema}, K_{non\_schema}\} = Classification(K)$
2: $C_{schema} \leftarrow \Phi$
3: **for** each $w_i \in K_{schema}$ **do**
4: $\quad C_{schema} \leftarrow C_{schema} \cup (C_{w_i} \subset S_{denSys})$
5: **end for**
6: **for** all $w_j \in K_{non\_schema}$ **do**
7: $\quad C_{non\_schema} \leftarrow C_{w_1} \cap, \ldots, \cap C_{w_j}$
8: **end for**
9: $C_{denSys} \leftarrow C_{non\_schema}$ - C_{schema}
10: $S_{Result} \leftarrow \Phi$
11: **for** each $\{val, vec\}_{st} \in C_{denSys}$ **do**
12: $\quad \mathscr{P}_q \leftarrow \Phi$
13: $\quad$ **for** each $w_i \in K$ **do**
14: $\quad\quad$ Get event $\vec{\Pi}_{w_i}$ through rotating the $|e_i\rangle$ (w_i) into a new coordinate by vec_{st}
15: $\quad\quad \mathscr{P}_q \leftarrow \mathscr{P}_q \cup \vec{\Pi}_{w_i}$
16: $\quad$ **end for**
17: $\quad$ **for** each c *in* K **do**
18: $\quad\quad$ **if** $PI(c) \geq min\_threshold$ **then**
19: $\quad\quad\quad \mathscr{P}_q \leftarrow \mathscr{P}_q \cup |\kappa_c\rangle$
20: $\quad\quad$ **end if**
21: $\quad$ **end for**
22: $\quad$ Learn a query density vector $\langle\rho|_q$ from $\mathscr{P}_q$ by $MaxLik$
23: $\quad score_{st} = ScoreFunction(\langle\rho|_q, val_{st})$
24: $\quad S_{Result} \leftarrow S_{Result} \cup score_{st}$
25: **end for**
26: Sort S_{Result} and return top-k results

4 Experiment

4.1 Data Sets

We use synthetic data (UniBench) and real data (IMDB and DBLP) to evaluate our approaches. The statistics about them are listed in Table 2.

UniBench [29] is a multi-model benchmark, including data of relational, JSON, and graph models. It simulates a social commerce scenario that combines the social network with the E-commerce context. The relational model

Table 2. The number of records/objects in different data models

	Relational	JSON	Graph-entity	Graph-relation
UniBench	150 000	142 257	9 949	375 620
IMDB	494 295	84 309	113 858	833 178
DBLP	1 182 391	435 469	512 768	3 492 502

Table 3. Queries employed in the experiments

ID	Queries
(a) Queries on DBLP	
Q_1	Soni Darmawan friends
Q_2	Gerd Hoff friends' rank 1 paper
Q_3	Slawomir Zadrozny rank 1 paper
Q_4	Phebe Vayanos phdthesis paper
Q_5	Neural sequence model
Q_6	Brian Peach 2019 papers
Q_7	Performance of D2D underlay and overlay for multi-class elastic traffic. authors
Q_8	Carmen Heine rank Modell zur Produktion von Online-Hilfen
Q_9	Exploring DSCP modification pathologies in the Internet
Q_{10}	The papers of Frank Niessink
(b) Queries on UniBench	
Q_{11}	Abdul Rahman Budjana friends BURRDA feedback perfect
Q_{12}	Shmaryahu Alhouthi order Li-Ning Powertec Fitness Roman Chair
Q_{13}	Kamel Abderrahmane Topeak Dual Touch Bike Storage Stand
Q_{14}	Alexandru Bittman whether has friend Ivan Popov
Q_{15}	Mohammad Ali Forouhar Oakley Radar Path Sunglasses
Q_{16}	Soft Air Thompson 1928 AEG Airsoft Gun and Genuine Italian Officer's Wool Blanket
Q_{17}	Roberto Castillo Total Gym XLS Trainer and Reebok
Q_{18}	Who Kettler, Volkl and Zero Tolerance Combat Folding Knife
Q_{19}	Francois Nath Nemo Cosmo Air with Pillowtop Sleeping Pad
Q_{20}	Hernaldo Zuniga Advanced Elements AdvancedFrame Expedition Kayak and TRYMAX
(c) Queries on IMDB	
Q_{21}	Lock, Stock and Two Smoking Barrels actors
Q_{22}	Forrest Gump
Q_{23}	The title and imdbVote of films of Bruce Willis
Q_{24}	The films of director Robert Zemeckis
Q_{25}	Films in 1997 Genre Action, Adventure, Family
Q_{26}	The Legend of 1900 awards
Q_{27}	Scent of a Woman imdbRating
Q_{28}	The film of Dustin Hoffman with Tom Cruise
Q_{29}	Morgan Freeman friends
Q_{30}	Aamir Khan films

includes the structured feedback information; The JSON model contains the semi-structured orders; The social network is modeled as a graph, which contains one entity and one relation, i.e., customer, and person knows person. These also have correlations across the data models. For instance, the customer makes transactions (Graph correlates with JSON).

The IMDB dataset is crawled from website[1] by OMDB API. Through extracting inner relationships and potential information, we generate several data models to represent the original data. The relational data includes performing information and rating information, which are stored in different tables; The JSON model is made up of film information (e.g., imdbID, title, and year); The graph is about cooperative information, where two actors would be linked together if they have ever worked for the same movie.

The DBLP[2] data consists of bibliography records in computer science. Each record in DBLP is associated with several attributes such as authors, year, and title. The raw data is in XML format. Here we describe it in three data models. The publication records are presented in relational data, including author id, paper id, and the author's rank in the author list. A subset of the papers' attributes (e.g., paper id, key, and title) are represented in JSON. We also construct a co-authorship (friend) graph where two authors are connected if they publish at least one paper together.

Table 4. Presion, Recall, F-measure on DBLP

AKSDV								EASE
min_threshold	0	0.2	0.4	0.6	0.8	1.0	Non	
Precision	0.830	0.847	0.847	0.847	0.847	0.847	0.797	0.090
Recall	0.867	0.917	0.917	0.917	0.917	0.917	0.817	0.500
F-measure	0.834	0.861	0.861	0.861	0.861	0.861	0.794	0.141

4.2 Queries and Answers

In the experiments, three groups of keyword queries, as shown in Table 3, are proposed by a few people randomly to evaluate our methods on the DBLP, UniBench, and IMDB datasets, respectively. Each keyword query involves one or more than one data model to test the ability of our methods in capturing potential semantics of keywords and search accuracy. Besides, the corresponding AQL query for each Q_i is issued in ArangoDB, and the output answers are used to evaluate the results produced by the algorithm, Answer Keyword Searches with Density Vectors (AKSDV), and EASE [14]. EASE models heterogeneous data as graphs and aims at finding r-radius Steiner graphs as query results. In this method, each returned Steiner graph includes at least two keywords.

The experiments are implemented in Java except for eigendecomposition by Matlab (offline work). The experiments are run on a desktop PC with an Intel(R) Core(TM) i5-6500 CPU of 3.19 GHz and 16 GB RAM. Note that all operations

[1] https://www.omdbapi.com/.
[2] https://dblp.uni-trier.de/xml/.

Table 5. Presion, Recall, F-measure on UniBench

AKSDV								EASE
min_threshold	0	0.2	0.4	0.6	0.8	1.0	Non	
Precision	0.902	0.902	0.917	0.922	0.922	0.922	0.897	0.220
Recall	0.883	0.883	0.885	0.886	0.886	0.886	0.882	0.136
F-measure	0.844	0.844	0.848	0.850	0.850	0.850	0.842	0.061

Table 6. Presion, Recall, F-measure on IMDB

AKSDV								EASE
min_threshold	0	0.2	0.4	0.6	0.8	1.0	Non	
Precision	0.753	0.753	0.753	0.758	0.758	0.758	0.758	0.548
Recall	0.782	0.782	0.782	0.784	0.784	0.784	0.784	0.466
F-measure	0.657	0.657	0.657	0.661	0.661	0.661	0.661	0.269

are done in memory, and the standard NLP pre-processing such as dropping the stop words and stemming are conducted in advance. In the experiments, we measure the precision, recall, and F-measure for the top-20 returned results.

4.3 Results Analysis

Table 4 shows the comparison of the average precision, recall, and F-measure of AKSDV with EASE's. This comparison result demonstrates that our proposed methods outperform EASE on the DBLP dataset. Table 5 and Table 6 show that the performance of AKSDV also outperforms EASE's on the UniBench and IDMB dataset. And the F-measure of AKSDV is at least nearly twice EASE's on these datasets. These high accuracy values show that our framework could understand the potential semantics underlying the statements and get the most relevant statements about queries.

For example, Q_9 wants to find all information about the paper "Exploring DSCP modification pathologies in the Internet". EASE returns a Steiner graph consisting of a single node that includes the paper name itself. AKSDV could find all of the information about this paper; Q_{11} wants to look for Abdul Rahman's friends who have bought BURRDA and given a perfect rating. For EASE, it returns answers mainly about "Abdul Rahman". But AKSDV could return the relevant information which users want to find.

In these three tables, the "*min_threshold*" decides which co-location compounds will be regarded as compounds $\kappa = \{w_1, ..., w_p\}$. Each column corresponds to the performance of keyword searches when assigned a value to *min_threshold*. For example, in Table 4, the second column illustrates that when we set *min_threshold* $= 0$, the values of average precision, recall, and F-measure of AKSDV on the DBLP data set are 0.830, 0.867, and 0.834, respectively.

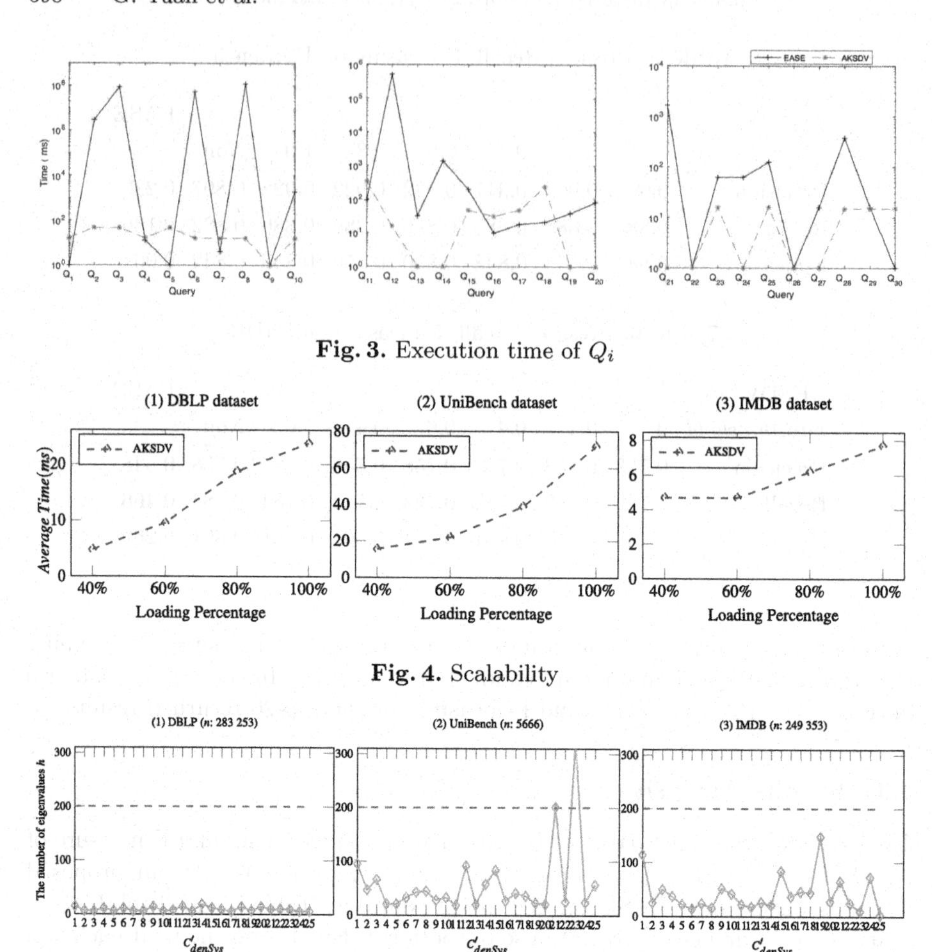

Fig. 3. Execution time of Q_i

Fig. 4. Scalability

Fig. 5. The change in value of h on different datasets

In general, our method of identifying compounds works well on these datasets for improving query performance.

Column "non" means that we assign an average weight ($\sigma_i^2 = 1/p$, $||\kappa|| = p$) to each w_i in the compound κ instead of considering which word will have more contributions to the compound (without using co-location weight). Column "0" and "non" regard all the co-location compounds as compounds. The difference between them is whether using co-location weight when constructing compounds. Table 4 and Table 5 demonstrate our co-location weight is better than the average weight method on the DBLP and UniBench dataset. In Table 6, the performance of column "0" is little less than the column "non"s.

4.4 Time and Scalability Analysis

In this part of the experiments, firstly, we analyze the time cost of AKSDV and EASE. The reported time cost values are collected by executing each query several times to take the median value. Figure 3 gives executing time of AKSDV ($min\_threshold = 0.6$) and EASE. In Fig. 3, there is a different fluctuation in the scales of time about different queries, which is caused by the queries of complexity involving different data models. But in general, comparing with EASE, AKSDV has a good time performance.

Now, we focus on the scalability of AKSDV. To test the scalability, we load different proportions of metadata into our framework, respectively. And we let the correct results increase as loading more data for each query. Then, we perform all queries in Table 3 on these datasets having different sizes to get a median time after executing several times. Finally, we calculate the average time on these datasets, respectively. Figure 4 (1)–(2) show the results of experiments on the DBLP, Unibench, and IMDB dataset. In general, AKSDV has a good scalability performance on these datasets. The UniBench dataset, due to existing mass joins as the loading percentage increases, shows the query time increases faster than others. But generally, it is within an acceptable range.

4.5 The Dimension of Density Vectors Analysis

Unlike the word embedding in the machine learning community, which needs a toolkit using lots of time and samples to train vector space models, we only need to extend the events' dimension. And elementary events still keep disjoint feature. Although the dimensions of events have increased, the query's cost still depends on the h (the number of eigenvalues of density matrices). The Fig. 5 shows the change in values of h on the candidate statement set C'_{eigen}, where $C'_{eigen} \subset C_{eigen}$ and C_{eigen} is made of candidate statements about all the queries on different datasets in Table 3, and C'_{eigen} is made of selected 25 candidate answers from the corresponding C_{eigen} randomly. In Fig. 5, we can see the values of h are much less than the word space n, even less than 200 in most situations. It guarantees the complexity of our framework at the appropriate level.

5 Related Works

Keyword search has been well known as a user-friendly way of satisfying users' information needs with few keywords in diverse fields such as Information Retrieval (IR) and database. Unfortunately, finding a fundamental axiomatic formulation of the IR field has proven tricky, maybe because of the intrinsic role humans play in the process [3]. However, due to information having proven considerably easier to capture mathematically than "meaning", researchers are investigating how to apply quantum theory to attack the various IR challenges, which can be traced back to [21]. Piwowarski et al. [20] used the probabilistic formalism of quantum theory to build a principled interactive IR framework.

Frommholz et al. [6] utilized poly representation and quantum measurement [25] to do keyword searches in a quantum-inspired interactive framework.

Considering the power of quantum-inspired framework and the similarity of keyword searches between databases and IR, we attempt to use the quantum-inspired method to support keyword searches on multi-model databases. Hence our research work also conforms to the current trend of seamlessly integrating database and information retrieval [26]. The keyword search on the database is particularly appealing. From relation, XML to graph databases, there are already many exciting proposals in the scientific literature [8,9,11,13]. However, most of the existing keyword search works are designed for specific data models. Through review literature, the only complete relevant current work is EASE [14]. They returned r-radius Steiner graphs as results to users for answering keyword searches on heterogeneous data. In addition, there is another work [15], which presented an architecture to support keyword searches over diverse data sources. Due to lacking complete performance analysis, we temporarily do not compare it in this paper.

6 Conclusion

This paper has proposed using the quantum probability mechanism to solve the keyword search problem on multi-model databases. To reduce complexity and improve performance, we introduced new representations of events and the density vector concept. We also used the spatial pattern to help improve query accuracy and used PCA to support the framework to work well further. Finally, extensive experiments have demonstrated the superiority of our methods over state-of-the-art works.

Acknowledgements. The work is partially supported by the China Scholarship Council and the Academy of Finland project (No. 310321). We would also like to thank all the reviewers for their valuable comments and helpful suggestions.

References

1. Altwaijry, S., et al.: Asterixdb: a scalable, open source BDMS. In: Proceedings of the VLDB Endowment, vol. 7, No.14 (2014)
2. ArangoDB: three major nosql data models in one open-source database (2016), http://www.arangodb.com/
3. Ashoori, E., Rudolph, T.: Commentary on Quantum-Inspired Information Retrieval. arXiv e-prints arXiv:1809.05685 (September 2018)
4. Bendersky, M., Croft, W.B.: Modeling higher-order term dependencies in information retrieval using query hypergraphs. In: SIGIR (2012)
5. Bergamaschi, S., Domnori, E., Guerra, F., Lado, R.T., Velegrakis, Y.: Keyword search over relational databases: a metadata approach. In: SIGMOD 2011 (2011)
6. Frommholz, I., Larsen, B., Piwowarski, B., Lalmas, M., Ingwersen, P., Van Rijsbergen, K.: Supporting polyrepresentation in a quantum-inspired geometrical retrieval framework. In: Proceedings of the Third Symposium on Information Interaction in Context, pp. 115–124 (2010)

7. Gleason, A.M.: Measures on the closed subspaces of a Hilbert space. J. Math. Mech. **6**, 885–893 (1957)
8. Guo, L., Shao, F., Botev, C., Shanmugasundaram, J.: Xrank: ranked keyword search over xml documents. In: Proceedings of the 2003 ACM SIGMOD International Conference on Management of Data, pp. 16–27 (2003)
9. He, H., Wang, H., Yang, J., Yu, P.S.: Blinks: ranked keyword searches on graphs. In: Proceedings of the 2007 ACM SIGMOD International Conference on Management of Data, pp. 305–316 (2007)
10. Holland, S.M.: Principal components analysis (PCA), pp. 30602–32501. Department of Geology, University of Georgia, Athens, GA (2008)
11. Hristidis, V., Papakonstantinou, Y., Balmin, A.: Keyword proximity search on xml graphs. In: Proceedings 19th International Conference on Data Engineering (Cat. No. 03CH37405), pp. 367–378. IEEE (2003)
12. Huang, Y., Shekhar, S., Xiong, H.: Discovering colocation patterns from spatial data sets: a general approach. IEEE Trans. Knowl. Data Eng. **16**(12), 1472–1485 (2004)
13. Kargar, M., An, A., Cercone, N., Godfrey, P., Szlichta, J., Yu, X.: Meaningful keyword search in relational databases with large and complex schema. In: 2015 IEEE 31st International Conference on Data Engineering, pp. 411–422. IEEE (2015)
14. Li, G., Feng, J., Ooi, B.C., Wang, J., Zhou, L.: An effective 3-in-1 keyword search method over heterogeneous data sources. Inf. Syst. **36**(2), 248–266 (2011)
15. Lin, C., Wang, J., Rong, C.: Towards heterogeneous keyword search. In: Proceedings of the ACM Turing 50th Celebration Conference-China, pp. 1–6 (2017)
16. Lu, J., Holubová, I.: Multi-model databases: a new journey to handle the variety of data. ACM Comput. Surv. (CSUR) **52**(3), 1–38 (2019)
17. Lvovsky, A.: Iterative maximum-likelihood reconstruction in quantum homodyne tomography. J. Optics B Quantum Semiclassical Optics **6**(6), S556 (2004)
18. Melucci, M.: Deriving a quantum information retrieval basis. Comput. J. **56**(11), 1279–1291 (2013)
19. OrientDB, D.: Orientdb. hybrid document-store and graph noSQL database (2017)
20. Piwowarski, B., Frommholz, I., Lalmas, M., Van Rijsbergen, K.: What can quantum theory bring to information retrieval. In: Proceedings of the 19th ACM International Conference on Information and Knowledge Management, pp. 59–68 (2010)
21. van Rijsbergen, C.J.: Towards an information logic. In: Proceedings of the 12th Annual International ACM SIGIR Conference on Research and Development in Information Retrieval, pp. 77–86 (1989)
22. Song, D., et al.: How quantum theory is developing the field of information retrieval. In: 2010 AAAI Fall Symposium Series (2010)
23. Sordoni, A., Nie, J.Y., Bengio, Y.: Modeling term dependencies with quantum language models for IR. In: Proceedings of the 36th International ACM SIGIR Conference on Research and Development in Information Retrieval, pp. 653–662 (2013)
24. Van Rijsbergen, C.J.: The Geometry of Information Retrieval. Cambridge University Press, New York (2004)
25. Wang, J., Song, D., Kaliciak, L.: Tensor product of correlated text and visual features. In: QI 2010 (2010)
26. Weikum, G.: Db & JR: both sides now (keynote). In: Proceedings of the 2007 ACM SIGMOD International Conference on Management of Data (2007)
27. Yu, J.X., Qin, L., Chang, L.: Keyword search in databases. Synth. Lect. Data Manage. **1**(1), 1–155 (2009)

28. Yuan, G.: How the quantum-inspired framework supports keyword searches on multi-model databases. In: Proceedings of the 29th ACM International Conference on Information & Knowledge Management, pp. 3257–3260 (2020)
29. Zhang, C., Lu, J., Xu, P., Chen, Y.: UniBench: a benchmark for multi-model database management systems. In: Nambiar, R., Poess, M. (eds.) TPCTC 2018. LNCS, vol. 11135, pp. 7–23. Springer, Cham (2019). https://doi.org/10.1007/978-3-030-11404-6_2

ZH-NER: Chinese Named Entity Recognition with Adversarial Multi-task Learning and Self-Attentions

Peng Zhu[1,2], Dawei Cheng[3], Fangzhou Yang[4], Yifeng Luo[1,2(✉)], Weining Qian[1,2], and Aoying Zhou[1,2]

[1] School of Data Science and Engineering, East China Normal University, Shanghai, China
yfluo@dase.ecnu.edu.cn
[2] Shanghai Engineering Research Center of Big Data Management, Shanghai, China
[3] Department of Computer Science and Technology, Tongji University, Shanghai, China
[4] SeekData Inc., Shanghai, China

Abstract. NER is challenging because of the semantic ambiguities in academic literature, especially for non-Latin languages. Besides, recognizing Chinese named entities needs to consider word boundary information, as words contained in Chinese texts are not separated with spaces. Leveraging word boundary information could help to determine entity boundaries and thus improve entity recognition performance. In this paper, we propose to combine word boundary information and semantic information for named entity recognition based on multi-task adversarial learning. We learn common shared boundary information of entities from multiple kinds of tasks, including Chinese word segmentation (CWS), part-of-speech (POS) tagging and entity recognition, with adversarial learning. We learn task-specific semantic information of words from these tasks, and combine the learned boundary information with the semantic information to improve entity recognition, with multi-task learning. We conduct extensive experiments to demonstrate that our model achieves considerable performance improvements.

Keywords: Named entity recognition · Chinese word segmentation · Part-of-speech tagging · Adversarial learning · Multi-task learning

1 Introduction

Named entity recognition (NER) is a preliminary task in natural language processing [1,2,4], aiming to identify multiple types of entities from unstructured texts, such as person names, places, organizations and dates etc. NER is usually deemed as a sequence-to-sequence tagging task, and many downstream application tasks rely on NER to implement their task objectives [3,5,8]. It is well-known that it is challenging to identify Chinese named entities, because there

C. S. Jensen et al. (Eds.): DASFAA 2021, LNCS 12682, pp. 603–611, 2021.
https://doi.org/10.1007/978-3-030-73197-7_40

exist no explicit word boundaries in Chinese. People usually should perform Chinese word segmentation (CWS) to determine word boundaries before executing other Chinese text processing tasks. What's more, a Chinese entity may consist of multiple segmented words, while determining the segmented words belonging to an entity is non-trivial, as it is hard to determine the relationships between segmented words. Employing CWS information could help to identify word boundaries, and employing information concerned with segmented words' relationships could help to properly pin closely related words together for entity identification. The part-of-speech (POS) tagging information is easy to obtain and could be employed to infer the semantic relationships of contiguous words. So we combine CWS and POS tagging information to improve NER performance in this paper.

In this paper, we propose an NER model, namely ZH-NER, to combine word boundary information and semantic information for named entity recognition. Specifically, we learn common shared boundary information of entities from multiple kinds of tasks, including Chinese word segmentation (CWS), part-of-speech (POS) tagging and entity recognition, with adversarial learning. We learn task-specific semantic information of words from these tasks, and combine the learned boundary information with the semantic information to improve entity recognition, with multi-task learning. The contributions of this paper are summarized as follows: 1) To the best of our knowledge, ZH-NER is the first to improve Chinese NER with CWS and POS tagging information, based on adversarial multi-task learning. 2) Multiple self-attentions are employed to learn and integrate the key features concerned about entity boundaries and semantic information, learned from different kinds of tasks with various kinds of labels. 3) We conducormance of ZH-NER on four public and the results demonstrate the effectiveness of the proposed model.

2 Related Work

The extensive studies on NER mainly include rule-based methods, statistical machine learning-based methods, deep learning-based methods, and methods based on attention mechanisms, transfer learning, semi-supervised learning etc. The rule-based methods often use linguistic expertise to manually create rule templates, considering various kinds of features, and match the created rules for NER predictions. These systems rely on the creation of knowledge bases and dictionaries. The methods based on statistical machine learning mainly include: hidden Markov model (HMM), maxmium entropy model (MEM), support vector machine (SVM) and conditional random fields (CRF). Deep neural networks have recently been applied to improve NER performance by learning representative word embeddings [9] and sequential semantic information [7]. These methods are based on the classic LSTM-CRF architecture, where LSTM is used to learn hidden representations of characters, and CRF is used for joint label decoding. Researchers recently endeavor to improve Chinese NER by learning intrinsic semantic meanings, with semi-supervised learning and transfer learning etc. [10]

proposes a hybrid semi-Markov conditional random field (SCRF) architecture for neural sequence labeling, where word-level labels are utilized to derive segment scores in SCRFs. [11] investigates a lattice-structured LSTM model for Chinese NER, which encodes a sequence of input characters and all potential words that match a lexicon.

3 ZH-NER Model

The architecture of our proposed model is illustrated in Fig. 1, and we will describe each part of the model in detail in the rest of this section.

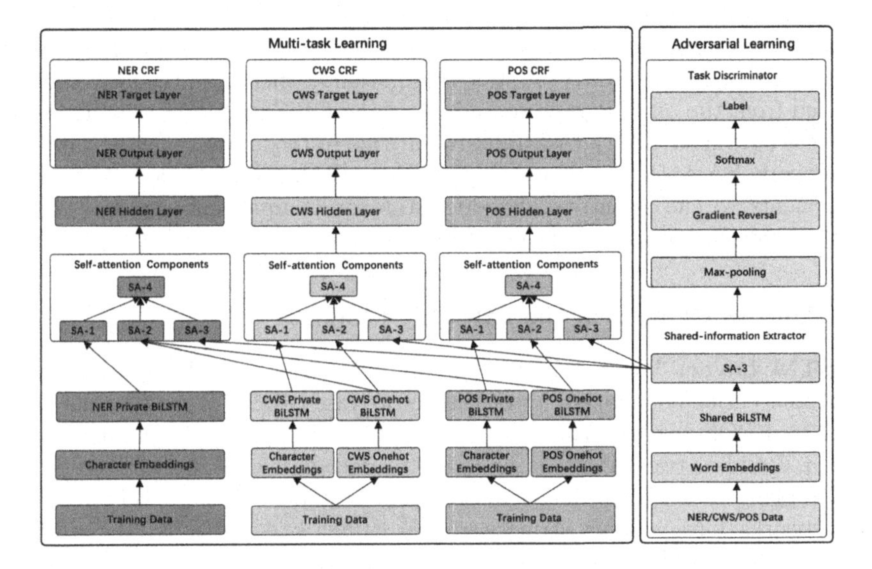

Fig. 1. The architecture of the ZH-NER model.

3.1 Character and Label Encoding

The model training corpora consist of two kinds of datasets, datasets used for adversarial learning, and datasets used for multi-task learning. The datasets used for adversarial learning include multiple NER datasets, a CWS dataset and a POS tagging dataset, and datasets used for multi-task learning include multiple NER datasets. The NER datasets used for adversarial learning are the same datasets used for multi-task learning. We denote $t = (c_1, c_2, \cdots, c_N)$ as a sentence contained within a dataset, and denote a dataset C with q sentences as $C = (t_1, t_2, \cdots, t_q)$. We map a character c_i to its distributed representation $x_i \in \mathbb{R}^{d_e}$, via looking up the pre-trained embedding matrix, where d_e is the dimensionality of the pre-trained character embeddings, and word embeddings are mapped likewise.

3.2 Adversarial Multi-task Learning

We first align sentences contained in a dataset for adversarial learning with those for multi-task learning. Each aligned sentence pair is fed into our model for task training, where sentences from adversarial learning datasets are fed for adversarial learning, and sentences from multi-task learning dataset are fed for multi-task learning. In adversarial learning, a sentence is first fed to the Shared-information Extractor for hidden state encoding, via a shared BiLSTM neural network and the SA-3 self-attention neural network. The sentence's encoded hidden state is then fed to the Task Discriminator, which endeavors to discriminate which dataset the sentence comes from. In multi-task learning, sentences are first fed to the task's private BiLSTM neural network for hidden state encoding, and the encoded hidden states are fed to the SA-1 self-attention neural network, to extract the contained key features. Each training task also contains a second self-attention neural network, called SA-2, to integrate the information learned from the hidden states encoded from the one-hot representations of the CWS and POS tagging labels. Then the outputs of SA-1, SA-2 and SA-3 are fed to the fourth self-attention neural network, SA-4, for information integration. With the multiple self-attention components, we can get an overall hidden state for an input sentence, and we then use the hidden state for task-specific label prediction.

3.3 Different Layers of the ZH-NER Model

BiLSTM Layer. We use the Shared BiLSTM to learn task-shared information about entity boundaries, and use three private BiLSTMs to train the NER, CWS, and POS tagging tasks, for learning task-specific word boundary information. Besides, two BiLSTMs are used to encode each sentence's CWS and POS tagging labels, represented as one-hot vectors. All these BiLSTM neural networks have similar structures. We define $k \in \{NER, CWS, POS\}$, $k^{'} \in \{onehot\ CWS, onehot\ POS\}$, and then we can compute hidden states for a sentence as: $\overrightarrow{\mathrm{h}_i} = \overrightarrow{\mathrm{LSTM}}(\overrightarrow{\mathrm{h}_{i-1}}, \mathrm{x}_i)$, $\overleftarrow{\mathrm{h}_i} = \overleftarrow{\mathrm{LSTM}}(\overleftarrow{\mathrm{h}_{i+1}}, \mathrm{x}_i)$, $\mathrm{h}_i = \overrightarrow{\mathrm{h}_i} \oplus \overleftarrow{\mathrm{h}_i}$, $\mathrm{s}_i^k = \mathrm{BiLSTM}(\mathrm{x}_i^k, \mathrm{s}_{i-1}^k; \theta_s)$, $\mathrm{p}_i^k = \mathrm{BiLSTM}(\mathrm{x}_i^k, \mathrm{p}_{i-1}^k; \theta_k)$ and $\mathrm{o}_i^{k^{'}} = \mathrm{BiLSTM}(x_i^{k^{'}}, o_{i-1}^{k^{'}}; \theta_{k'})$, where $\overrightarrow{\mathrm{h}_i} \in \mathbb{R}^{d_h}$ and $\overleftarrow{\mathrm{h}_i} \in \mathbb{R}^{d_h}$ are the contextualized states of the forward and backward LSTM at position i, $\oplus$ denotes the concatenation operation, d_h is the dimensionality of the hidden units, θ_s, θ_k and $\theta_{k'}$ respectively denote the parameters of the Shared, private and one-hot BiLSTMs.

Multiple Self-attentions. SA-1, SA-2, SA-3 and SA-4 are employed for information extraction and integration. All these self-attentions have similar structures, and we take SA-1 as an example to illustrate how these self-attentions work, without loss of generality. We denote the output of the Shared BiLSTM as $\mathrm{S} = (s_1, s_2, \cdots, s_N)$, the output of a private BiLSTM as $\mathrm{P} = (p_1, p_2, \cdots, p_N)$, and the output of a one-hot BiLSTM as $\mathrm{O} = (o_1, o_2, \cdots, o_N)$. The scaled dot-product attention can be described as follows: $\mathrm{Attention}(\mathrm{Q}, \mathrm{K}, \mathrm{V}) = \mathrm{softmax}(\frac{\mathrm{QK}^{\mathrm{T}}}{\sqrt{\mathrm{d}}})\mathrm{V}$, where $\mathrm{Q} \in \mathbb{R}^{N\times 2d_h}$, $\mathrm{K} \in \mathbb{R}^{N\times 2d_h}$ and $\mathrm{V} \in \mathbb{R}^{N\times 2d_h}$ are the query matrix, key matrix and value matrix respectively. In our setting, $\mathrm{Q} = \mathrm{K} = \mathrm{V} = \mathrm{P}$, and the dimensionality of the hidden units contained in a BiLSTM neural network is equal to $2d_h$. Formally, a multi-head attention can be expressed as follows: $head_i = \mathrm{Attention}(\mathrm{QW}_i^{\mathrm{Q}}, \mathrm{KW}_i^{\mathrm{K}}, \mathrm{VW}_i^{\mathrm{V}})$ and $\mathrm{P}'_{(\mathrm{SA}-1)} = (head_i \oplus \cdots \oplus head_h)\mathrm{W}_o$, where $\mathrm{W}_i^{\mathrm{Q}} \in \mathbb{R}^{2d_h\times d_k}$, $\mathrm{W}_i^{\mathrm{K}} \in \mathbb{R}^{2d_h\times d_k}$, $\mathrm{W}_i^{\mathrm{V}} \in \mathbb{R}^{2d_h\times d_k}$ and $\mathrm{W}_o \in \mathbb{R}^{2d_h\times 2d_h}$ are trainable projection parameters and $d_k = 2d_h/h$. For a sentence in task k's training dataset, we compute its final representation via SA-4 as: $head_i = \mathrm{Attention}(\mathrm{QW}_i^{\mathrm{Q}}, \mathrm{KW}_i^{\mathrm{K}}, \mathrm{VW}_i^{\mathrm{V}})$ and $\mathrm{P}'_{(\mathrm{SA}-4)} = (head_i \oplus \cdots \oplus head_h)\mathrm{W}_o$, where $\mathrm{W}_i^{\mathrm{Q}} \in \mathbb{R}^{6d_h\times d_k}$, $\mathrm{W}_i^{\mathrm{K}} \in \mathbb{R}^{6d_h\times d_k}$, $\mathrm{W}_i^{\mathrm{V}} \in \mathbb{R}^{6d_h\times d_k}$ and $\mathrm{W}_o \in \mathbb{R}^{6d_h\times 6d_h}$ are trainable projection parameters and $d_k = 6d_h/h$.

CRF Layer. Considering that NER, CWS and POS tagging are sequence labeling problems, we leverage CRF (conditional random field) to generate the label sequence for the input sentence. Given a sentence $t = (c_1, c_2, \cdots, c_N)$ with a predicted label sequence $\mathrm{y} = (y_1, y_2, \cdots, y_N)$, the CRF labeling process can be formalized as: $\mathrm{score}_i = \mathrm{W}_p \mathrm{p}'_{i(\mathrm{SA}-4)} + \mathrm{b}_p$, $result(x, y) = \sum_{i=1}^{N}(score_{i,y_i} + \mathrm{T}_{y_{i-1},y_i})$ and $\bar{\mathrm{y}} = \arg\max_{\mathrm{y}\in \mathrm{Y_x}} result(x, y)$, where $\mathrm{W}_p \in \mathbb{R}^{d_l\times 6d_h}$ and $\mathrm{b}_p \in \mathbb{R}^{d_l}$ are trainable parameters, d_l denotes the number of output labels, and $score_{i,y_i}$ represents the score of the y_i−th label of character c_i. Here T is a transition matrix which defines the scores of three successive labels, and Y_x represents all candidate label sequences for a given sentence. We use the Viterbi algorithm to generate the predicted label sequence $\bar{\mathrm{y}}$. For training, we exploit negative log-likelihood objective as the loss function. The probability of the ground-truth label sequence is computed by: $p(\hat{\mathrm{y}}|\mathrm{x}) = \frac{e^{result(x,\hat{y})}}{\sum_{\tilde{y}\in Y_x} e^{result(x,\tilde{y})}}$, where $\hat{\mathrm{y}}$ denotes the ground-truth label sequence. Given X training samples $(x^{(i)}; (\hat{\mathrm{y}}^{(\mathrm{i})})$, we use gradient back-propagation to minimize the loss function, and the loss function L_{Task} can be defined as follows: $L_{Task} = -\sum_{i=1}^{X} \log\, p(\hat{\mathrm{y}}^{(i)}|x^{(i)})$.

Task Discriminator. In order to guarantee that task-specific information is not extracted as common information, we endeavor to have Task Discriminator not be able to clearly discriminate which task dateset a sentence comes from, and the Task Discriminator can be defined as: $\mathrm{s}'^k = \mathrm{Maxpooling}(\mathrm{S}'^k)$ and $D(\mathrm{s}'^k; \theta_d) = \mathrm{Softmax}(\mathrm{W}_d \mathrm{s}'^k + \mathrm{b}_d)$, where θ_d denotes the parameters of the task discriminator, S'^k is the shared self-attention of task k, $\mathrm{W}_d \in \mathbb{R}^{K\times 3d_h}$ and $\mathrm{b}_d \in \mathbb{R}^K$ are trainable parameters, and K is the number of trainable tasks. Besides the task loss L_{Task}, we introduce the adversarial loss L_{Adv} to prevent task-specific information of the CWS and the POS tagging tasks from being extracted as task-shared information, and the adversarial loss can be computed as: $L_{Adv} = \min_{\theta_s}(\max_{\theta_d} \sum_{k=1}^{K} \sum_{i=1}^{X_k} logD(E_s(x_k^{(i)})))$, where θ_s denotes the trainable parameters of the Shared BiLSTM, E_s denotes the Shared-information Extractor, X_k denotes the number of training samples of task k, $x_k^{(i)}$ denotes the i-th sample of task k. The minimax optimization objective could mislead the task discriminator. We add a gradient reversal layer [6] before the softmax layer to address the minimax optimization problem. Gradients throughout the gradient reversal layer will become opposed signs of encouraging the Shared-information Extractor to learn task-shared information.

3.4 Model Training

The final objective function of ZH-NER can be formulated as: $L = L_{NER} \cdot I(x) \cdot I'(x) + L_{CWS} \cdot (1 - I(x)) \cdot I'(x) + L_{POS} \cdot (1 - I(x)) \cdot (1 - I'(x)) + \lambda L_{Adv}$, where λ is the loss weight coefficient, L_{NER}, L_{CWS} and L_{POS} can be computed via Eq. 7. Here $I(x)$ and $I'(x)$ are switching functions for identifying which task an input sentence comes from, and they are defined as follows:

$$I(x) = \begin{cases} 1, if \in C_{NER} \\ 0, if \in C_{CWS} \\ 0, if \in C_{POS} \end{cases}, I'(x) = \begin{cases} 1, if \in C_{NER} \\ 1, if \in C_{CWS} \\ 0, if \in C_{POS} \end{cases} \quad (1)$$

where C_{NER}, C_{CWS} and C_{POS} respectively denote the training corpora of Chinese NER, CWS and POS tagging tasks. We use the Adam algorithm to optimize the final objective function. Since NER, CWS and POS tagging tasks have different convergence rates, we repeat the model training iterations until early stopping.

Table 1. Results achieved on the Weibo NER dataset.

Models	Named entity			Named mention			Overall
	P(%)	R(%)	F1(%)	P(%)	R(%)	F1(%)	F1(%)
Peng and Dredze (2015)	74.78	39.81	51.96	71.92	53.03	61.05	56.05
Peng and Dredze (2016)	66.67	47.22	55.28	74.48	54.55	62.97	58.99
He and Sun (2017a)	66.93	40.67	50.60	66.46	53.57	59.32	54.82
He and Sun (2017b)	61.68	48.82	54.50	74.13	53.54	62.17	58.23
Cao et al. (2018)	59.51	50.00	54.34	71.43	47.90	57.35	58.70
Lattice	–	–	53.04	–	–	62.25	58.79
CAN-NER	–	–	55.38	–	–	62.98	59.31
WC-LSTM	–	–	52.55	–	–	67.41	59.84
LGN	–	–	55.34	–	–	64.98	60.21
Lexicon	67.31	48.61	56.45	75.15	62.63	68.32	63.09
Ding et al. (2019)	–	–	63.10	–	–	56.30	59.50
TENER	–	–	–	–	–	–	58.17
LR-CNN	–	–	57.14	–	–	66.67	59.92
PLTE	–	–	62.21	–	–	49.54	55.15
FLAT	–	–	–	–	–	–	60.32
Peng et al. (2020)	–	–	56.99	–	–	61.41	61.24
ZH-NER	69.81	56.63	62.54	75.78	56.07	64.45	**63.41**

Table 2. Model performance on the MSRA dataset.

MSRA dataset							
Models	P(%)	R(%)	F1(%)	Models	P(%)	R(%)	F1(%)
Chen et al. (2006)	91.22	81.71	86.20	Zhang et al. (2006)	92.20	90.18	91.18
Zhou et al. (2013)	91.86	88.75	90.28	Lu et al. (2016)	–	–	87.94
Dong et al. (2016)	91.28	90.62	90.95	Cao et al. (2018)	91.30	89.58	90.64
Yang et al. (2018)	92.04	91.31	91.67	Lattice	93.57	92.79	93.18
CAN-NER	93.53	92.42	92.97	WC-LSTM	94.58	92.91	93.74
LGN	94.19	92.73	93.46	Lexicon	94.01	92.93	93.47
Ding et al. (2019)	94.60	94.20	94.40	TENER	–	–	92.74
LR-CNN	94.50	92.93	93.71	PLTE	94.25	92.30	93.26
FLAT	–	–	94.12	Peng et al. (2020)	95.36	93.44	93.50
ZH-NER	94.50	93.05	**93.77**				

4 Experimental Evaluations

4.1 Evaluation Datasets and Experimental Settings

We evaluate our model on the Weibo NER, MSRA, OntoNotes4, Chinese Resume datasets. We use the MSRA dataset to obtain CWS information and UD1 dataset to obtain POS tagging information. We adjust hyper parameters according to NER performance achieved on the development set of the Chinese NER task. The initial learning rate is set to 0.001, and we use Adam to optimize all trainable parameters. The dimensionality of BiLSTM hidden states d_h, the number of self-attention units and self-attention heads are set to 120, 240 and 10. To avoid overfitting, we set the dropout rate to 0.3, and the training batch size to 80. The loss weight coefficient λ is set to 0.06. We use the Jieba toolkit to generate CWS and POS tagging labels. The character embeddings used in our experiments are pre-trained. We use precision(P), recall(R) and F1 score as the performance evaluation metrics.

4.2 Baseline Models and Experimental Results

We compare our model with 24 NER models on different datasets, depending on the availability of source codes and publicity of datasets. These models include two traditional models Che et al. and Wang et al. (2013); three multi-task learning models Peng and Dredze (2015), Peng and Dredze (2016), and Cao et al.; two semi-supervised learning models He and Sun (2017a) and He and Sun (2017b); a model based on neural feature combination Yang et al. (2016); two CNN-based models CAN-NER and LR-CNN, two models based on word-character information Lattice and WC-LSTM; seven models based on lexicon and graph network LGN, Lexicon, Ding et al. (2019), TENER, PLTE, FLAT and Peng et al. (2020);

Table 3. Model performance on the OntoNotes dataset and Chinese resume dataset.

OntoNotes dataset							
Models	P(%)	R(%)	F1(%)	Models	P(%)	R(%)	F1(%)
Che et al. (2013)	77.71	72.51	75.02	Wang et al. (2013)	76.43	72.32	74.32
Yang et al. (2016)	72.98	80.15	76.40	Lattice	76.35	71.56	73.88
CAN-NER	75.05	72.29	73.64	WC-LSTM	76.09	72.85	74.43
LGN	76.13	73.68	74.89	Lexicon	75.06	74.52	74.79
Ding et al. (2019)	75.40	76.60	76.00	TENER	–	–	72.43
LR-CNN	76.40	72.60	74.45	PLTE	76.78	72.54	74.60
FLAT	–	–	76.45	Peng et al. (2020)	77.31	73.85	75.54
ZH-NER	77.01	76.39	**76.70**				
Chinese resume dataset							
Models	P(%)	R(%)	F1(%)	Models	P(%)	R(%)	F1(%)
Lattice	94.81	94.1	94.46	CAN-NER	95.05	94.82	94.94
WC-LSTM	95.27	95.15	95.21	LGN	95.28	95.46	95.37
TENER	–	–	95.00	LR-CNN	95.37	94.84	95.11
PLTE	95.34	95.46	95.40	FLAT	–	–	95.45
Peng et al. (2020)	95.53	95.64	95.59	ZH-NER	95.18	96.23	**95.70**

other five models Chen et al. (2006), Zhang et al. (2006), Zhou et al. (2013), Lu et al. (2016) and Dong et al. (2016).

The NER performance evaluation results on the Weibo NER dataset are presented in Table 1, where "NE", "NM" and "Overall" respectively indicate F1-scores for identifying named entities, nominal entities (excluding named entities) and both. We can see that the Lexicon model and the Peng et al. model achieve the highest and second highest F1-scores in all baseline models, which are 63.09% and 61.24% respectively, while ZH-NER improves the overall F1-score to 63.41%. The NER performance evaluation results on the MSRA dataset are presented in Table 2. We can see that the two most recent models, Ding et al. model and the FLAT model, achieve the highest and second highest F1-scores in all baseline models, which are 94.40% and 94.12% respectively. ZH-NER achieves the third F1-score of 93.77%. The NER performance evaluation results on the OntoNotes4 dataset are presented in Table 3. Che et al. (2013) and Wang et al. (2013) achieve F1-scores of 74.32% and 75.02%, and FLAT achieves an F1-score of 76.45%, which is the highest F1-score in the baseline models, and ZH-NER achieves the highest F1-score of 76.70%, outperforming the three recent models by substantial margins. The NER performance evaluation results on the Chinese Resume dataset are presented in Table 3. [11] achieves an F1-score of 94.46%. The F1-scores of FLAT and Peng et al. (2020) are 95.45% and 95.59%, and ZH-NER achieves the highest F1-score.

5 Conclusion

In this paper, we propose to extract shared information concerned with entity boundaries across NER, CWS and POS tagging tasks, with adversarial multi-task learning. Each of the NER, CWS and POS tagging tasks is trained with multiple self-attentions, to extract task-specific information, and properly combine it with the learned boundary information. Experimental results demonstrate that ZH-NER achieves better performance than other state-of-the-art methods.

Acknowledgments. This work was supported by National Key R&D Program of China (2018YFC0831904), the National Natural Science Foundation of China (U1711262, 62072185), and the Joint Research Program of SeekData Inc. and ECNU.

References

1. Allen, J.F.: Natural language processing. In: Encyclopedia of Computer Science, pp. 1218–1222 (2003)
2. Bunescu, R.C., Mooney, R.J.: A shortest path dependency kernel for relation extraction. In: HLT-EMNLP, pp. 724–731 (2005)
3. Cheng, D., Niu, Z., Zhang, Y.: Contagious chain risk rating for networked-guarantee loans. In: KDD, pp. 2715–2723 (2020)
4. Cheng, D., Wang, X., Zhang, Y., Zhang, L.: Graph neural network for fraud detection via spatial-temporal attention. TKDE (2020)
5. Fan, M., et al.: Fusing global domain information and local semantic information to classify financial documents. In: CIKM, pp. 2413–2420 (2020)
6. Ganin, Y., Lempitsky, V.: Unsupervised domain adaptation by backpropagation. In: ICML, pp. 1180–1189 (2014)
7. Lample, G., Ballesteros, M., Subramanian, S., Kawakami, K., Dyer, C.: Neural architectures for named entity recognition. In: NAACL, pp. 260–270 (2016)
8. Liang, X., Cheng, D., Yang, F., Luo, Y., Qian, W., Zhou, A.: F-HMTC: detecting financial events for investment decisions based on neural hierarchical multi-label text classification. In: IJCAI-20, pp. 4490–4496 (2020)
9. Peng, N., Dredze, M.: Improving named entity recognition for Chinese social media with word segmentation representation learning. In: ACL, pp. 149–155 (2016)
10. Ye, Z., Ling, Z.H.: Hybrid semi-Markov CRF for neural sequence labeling. In: ACL, pp. 235–240 (2018)
11. Zhang, Y., Yang, J.: Chinese NER using lattice LSTM. In: ACL, pp. 1554–1564 (2018)

Drug-Drug Interaction Extraction via Attentive Capsule Network with an Improved Sliding-Margin Loss

Dongsheng Wang[1], Hongjie Fan[2](✉), and Junfei Liu[3]

[1] School of Software and Microelectronics, Peking University, Beijing, China
wangdsh@pku.edu.cn
[2] The Department of Science and Technology, China University of Political Science and Law, Beijing, China
hjfan@cupl.edu.cn
[3] School of Electronics Engineering and Computer Science, Peking University, Beijing, China
liujunfei@pku.edu.cn

Abstract. Relation extraction (RE) is an important task in information extraction. Drug-drug interaction (DDI) extraction is a subtask of RE in the biomedical field. Existing DDI extraction methods are usually based on recurrent neural network (RNN) or convolution neural network (CNN) which have finite feature extraction capability. Therefore, we propose a new approach for addressing the task of DDI extraction with consideration of sequence features and dependency characteristics. A sequence feature extractor is used to collect features between words, and a dependency feature extractor is designed to mine knowledge from the dependency graph of sentence. Moreover, we use an attention-based capsule network for DDI relation classification, and an improved sliding-margin loss is proposed to well learn relations. Experiments demonstrate that incorporating capsule network and improved sliding-margin loss can effectively improve the performance of DDI extraction.

Keywords: RE · DDI · Capsule network

1 Introduction

A drug-drug interaction (DDI) arises when two or more drugs are taken at the same time that leads to the disturbance of their function. Adverse drug reactions (ADRs) may bring some serious unexpected consequences such as iron-deficiency anemia and toxic reaction. The more DDIs doctors know, the fewer ADRs occur. Therefore, increasing our knowledge of DDIs is conducive to reduce medical accidents. At present, several drug knowledge databases like DrugBank [12] have been constructed to summarize DDIs and guide doctors for avoiding ADRs. In parallel with the development of drug-related databases, the automatic

C. S. Jensen et al. (Eds.): DASFAA 2021, LNCS 12682, pp. 612–619, 2021.
https://doi.org/10.1007/978-3-030-73197-7_41

DDI extraction from unstructured biomedical literature has become a trend in biomedical text mining and health informatics.

DDI extraction is a classic RE task in biomedical domain. After the success of DDIExtraction 2013 task[1], more and more machine learning-based methods have been proposed on the public corpora. These methods can be divided into non-deep-learning [4,9,16] and deep learning-based methods [1,15,17]. However, despite the success of these methods, some challenges remain: (A) Existing methods heavily rely on the representation of sentence. Only using one aspect of information such as word embedding is limited as language is complex and diversiform. (B) Both CNN and RNN fail to transmit spatial relationship to high-level parts. Just handling sentence with fixed structures or directions makes it difficult to cluster relation features of different positions.

In this paper, we propose a novel method to utilize both textual and dependency information for DDI extraction from texts. On one hand, we employ an RNN module to learn the textual features. On the other hand, we utilize a graph convolutional network (GCN) to obtain the representations of dependency graphs of sentences. Then, the textual features and dependency representations are merged together and sent to a capsule network. To the best of our knowledge, it is the first research to use capsule network with integrated features in DDI extraction. The contributions of this paper can be summarized as follows:

- We propose a novel neural method to extract DDIs with the textual features and the dependency representations at the same time.
- We apply capsule network to precisely represent and deliver semantic meanings of DDIs. We show that the improved sliding-margin loss is better than the original margin loss in model training.
- Experimental results demonstrate that our proposed method achieves a new state-of-the-art result for DDI extraction on the DDIExtraction 2013 dataset.

2 Background

Graph Neural Network. Lots of applications employ graph structure for representing various types of data. In order to mining these data, graph neural network (GNN) is proposed to capture the dependences of graph through message passing between graph nodes, and numerous GNN models are developed based on various theories [3,6]. GCN is a typical model in GNN which generalizes convolutions to the graph domain.

Capsule Network. Capsule network (CapsNet) proposed by Sabour [10] is a novel structure for deep neural network. CapsNet abandons pooling operations and applies capsules to represent various properties. The most popular version of CapsNet uses an algorithm called "routing by agreement" which replaces pooling in CNN and a vector output replacing scalar outputs in CNN. Since CapsNet has recently proposed to use dynamic routing and achieved better performance than CNN, CapsNet and its improvement are widely used in tasks such as knowledge graph completion [7] and relation extraction[14].

[1] https://www.cs.york.ac.uk/semeval-2013/task9/.

3 Proposed Method

Our model architecture is shown in Fig. 1. The BiLSTM layer learns the textual feature representations from a combined embedding. The GCN part obtains the representations of dependency relationships between words by using the adjacent matrix and the output of BiLSTM layer. These two types of representations are merged and sent to the capsule network for DDI extraction.

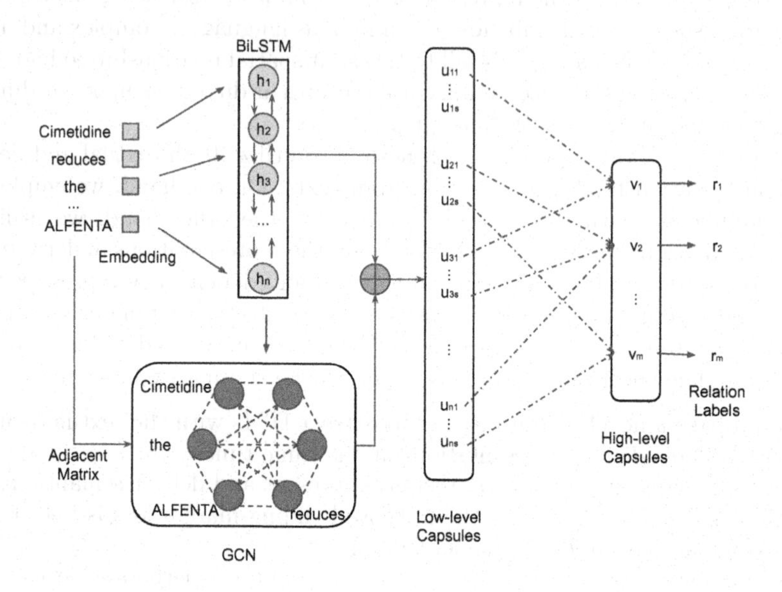

Fig. 1. The overall architecture of our method.

3.1 Extract Textual Features

The input features of our model include word embeddings, part of speech (POS) embeddings and position embeddings. Word embeddings are initialized with pre-trained word embeddings[2]. The POS tags of each sentence and the relative distances from tokens to entities are initialized by xavier initialization [2]. These three input features are concatenated into one feature vector as the input. Then, we employ a BiLSTM layer to exploit sentence features. The final output of each token w_i is the sum of the forward hidden state $\overrightarrow{h_i}$ and the backward hidden state $\overleftarrow{h_i}$.

3.2 Obtain Dependency Representations

For each sentence $s_i = (w_1, w_2, \cdots, w_n)$, it has a dependency graph. A tree is constructed based on this dependency graph, then the tree is converted to an

[2] http://bio.nlplab.org/.

adjacent matrix M. The element M_{ij} is set to be one if node i and node j have a connection in the dependency graph. After that, we utilize a GCN layer to iteratively transfer dependency information between node and its neighbors. The symbol g_i^l represents the output vector of node i in the l-th layer. The first layer of GCN initialized with BiLSTM output is denoted as $g_1^0, g_2^0, \cdots, g_n^0$, and the last layer of GCN is denoted as $g_1^T, g_2^T, \cdots, g_n^T$. The output of all vectors in the l-th layer is demonstrated with the following equation.

$$g^l = \sigma(\sum_{j=1}^{n} \overline{M}_{ij} W^l g^{l-1}/d_i + b^l) \tag{1}$$

$$d_i = \sum_{j=1}^{n} \overline{M}_{ij} \tag{2}$$

where σ is a nonlinear function (e.g., Leaky ReLu), W^l and b^l are the learning parameters of linear transformation. $\overline{M}$ is calculated by adding M and I, where I is an identity matrix and d_i is the degree of node i. Each node absorbs information from its neighbors and updates its node representation l times.

3.3 Attentive Capsule Network

Primary Capsule Layer. We combine all the low-level sequence features extracted by the BiLSTM layer and the dependency information generated by the GCN layer. The integrated low-level features m_t is the sum of the t-th hidden vector h_t in the output of BiLSTM and the t-th dependency representation g_t in the output of GCN. We split the integrated features into numerous low-level capsules denoted as u. Hence, each word is represented by s low-level capsules.

$$m_t = (\hat{u}_{t1}; \cdots; \hat{u}_{ts}) \tag{3}$$

In order to make the length of the output vector of a capsule represent the probability that the feature is present in current input, we apply each low-level capsule with a non-linear "squashing" function f to shrunk its length between zero and one. All capsules in m_t are computed with the following equation:

$$u_{ts} = f(\hat{u}_{ts}) = \frac{\|\hat{u}_{ts}\|^2}{1 + \|\hat{u}_{ts}\|^2} \frac{\hat{u}_{ts}}{\|\hat{u}_{ts}\|} \tag{4}$$

Several low-level capsules are combined together to create a high-level capsule. The input to a high-level capsule v_j is a weighted sum of all output vectors from the previous level capsules that each is calculated by multiplying the output u_i of a capsule in the layer below by a weight matrix W_{ij}.

$$v_j = f(\sum_{i} c_{ij} W_{ij} u_i) \tag{5}$$

where c_{ij} are coupling coefficients that are determined by an iterative dynamic routing process.

Attention-Based Routing Algorithm. However, the original routing algorithm proposed by Sabour [10] does not pay attention to the head and tail entities. Hence, we adopt an attention mechanism firstly used in [14] to focus on entity tokens. The coupling coefficients c between the i-th capsule and all the previous level capsules are calculated as follows:

$$c_{ij} = \frac{exp(b_{ij})}{\sum_{\hat{j}} exp(b_{i\hat{j}})} \tag{6}$$

where the initial logits b_{ij} are log prior probabilities that capsule u_i should be coupled to capsule v_j. The attention weights β are utilized to maximize the weights of low-level capsules with important words and minimize that of trivial capsules. The weight of capsule u_i is calculated by the integrated entity features h_e and hidden states h_t^i from which u_i comes from.

$$\beta_i = \sigma(h_e^T h_t^i) \tag{7}$$

where h_e is the sum of hidden states of head entity and tail entity, σ is the sigmoid function. The whole routing iteration is shown in Algorithm 1.

Algorithm 1. Attention-based Routing Algorithm

Input: low-level capsule u, iterative number r, entity states h_e and hidden states h_t
Output: high-level capsule v
1: **for** all capsules u_i and capsules v_i **do**
2: init the logits of coupling coefficients $b_{ij} = 0$;
3: **for** r iterations **do**
4: **for** all capsule i in layer l and capsule j in layer $l+1$ **do**
5: $c_{ij} = softmax(b_{ij})$
6: **for** all capsule j in layer $l+1$ **do**
7: $\beta_i = \sigma(h_e^T h_t^i)$; $v_j = f(\sum_i c_{ij}\beta_i W_j u_i)$
8: **for** all capsule i in layer l and capsule j in layer $l+1$ **do**
9: $b_{ij} = b_{ij} + W_j u_i v_j$
10: **return** v

3.4 Weighted Exponential Sliding-Margin Loss

Sabour et al. [10] and Zhang et al. [14] equally treat all predictions that are beyond the upper boundary and lower boundary, which stop optimizing the entire model. Therefore, we propose a non-zero loss. The loss function of the k-th relation can be calculated by:

$$L_k = \alpha Y_k (e^{m^+ - \|v_k\|} - e^{-m^-})^2 + \lambda(1 - Y_k)(e^{\|v_k\| - m^-} - e^{-m^-})^2 \tag{8}$$

where $Y_k = 1$ if a relation of class k is present, and $Y_k = 0$ if not. m^+ is the upper boundary which is the sum of the threshold of the margin and the width of the margin, denoted as B and δ. B is a learnable variable and is initialized by

0.5, and δ is a hyperparameter set by 0.4. m^{-} is the lower boundary which is the difference between B and δ. λ is used to down-weight the loss for absent relation classes and it is set by 0.5. We apply α to deal with the problem of imbalance of the training data and manually set it to be $[3, 5, 2, 10, 1]$, which corresponds to five relations (mechanism, advice, effect, int, false) respectively.

4 Experiments

4.1 Dataset and Experimental Settings

We evaluate our method on the DDIExtraction 2013 corpus[3]. In order to relieve the problem of class imbalance, we filter out some invalid negative instances under several rules proposed by previous works [5,11]. The statistics of the corpus after processing are shown in Table 1. We use Adam optimizer with learning rate 0.004, batch size 64, BiLSTM hidden size 128, word embedding size 200, POS embedding size 30, position embedding size 30. In GCN, we use 2 layers with output size 128 and dropout 0.5. The dimension of capsule is 16, the routing iteration is 3. We use micro-averaged F_1 score to evaluate our model.

Table 1. Statistics of the DDI 2013 extraction corpus

Corpus	Training set		Test set	
	Original	Processed	Original	Processed
Effect	1687	1676	360	358
Mechanism	1319	1309	302	301
Advice	826	824	221	221
Int	188	187	96	96
Negative DDIs	23772	19342	4737	3896

4.2 Overall Performance

Table 2 shows the comparative results on the DDIExtraction 2013 dataset. We use † to represent non-deep-learning methods and ‡ to represent deep-learning methods. Neural networks can learn useful features automatically, so most deep-learning methods have better performance.

As indicated by the table, we also conduct the F-score of each DDI type to assess the difficulty of detecting different interactions. Our model performs the best on *Advice* type and worst on *Int* type. These types have obvious different quantity in training data, thus making an apparent difference in F-score. By comparing with other models, our model achieves better performance. On one hand, we use both sequence features and dependency characteristics, which have great significance to DDI extraction. On the other hand, we utilize low-level capsules to merge basic features and cluster these capsules to form high-level capsules, which are helpful to the classification of DDI types.

[3] https://www.cs.york.ac.uk/semeval-2013/task9/.

Table 2. Comparative results of our model and baseline methods

	Method	F-score of four DDI types				Overall performance		
		Advice	Effect	Mechanism	Int	Precision	Recall	F-score
†	Kim [4]	72.50	66.20	69.30	48.30	–	–	67.00
	Zheng [16]	71.40	71.30	66.90	51.60	–	–	68.40
	Raihani [9]	77.40	69.60	73.60	52.40	73.70	68.70	71.10
‡	Liu [5]	77.72	69.32	70.23	46.37	75.70	64.66	69.75
	Asada [1]	81.62	71.03	73.83	45.83	73.31	71.81	72.55
	Zhang [15]	80.30	71.80	74.00	54.30	74.10	71.80	72.90
	PM-BLSTM [17]	81.60	71.28	74.42	48.57	75.80	70.38	72.99
	RHCNN [11]	80.54	73.49	78.25	**58.90**	77.30	73.75	75.48
	AGCN [8]	**86.22**	74.18	78.74	52.55	78.17	**75.59**	76.86
	Xiong [13]	83.50	75.80	79.40	51.40	80.10	74.00	77.00
	Our method	83.37	**79.23**	**79.65**	55.94	**82.62**	74.77	**78.50**

4.3 Effect of Various Modules

The ablation studies are shown in Table 3. First, without dependency characteristics, the performance of our model sharply drops by 2.77%. This drop shows that the dependency characteristics extracted by GCN plays a vital role in finding DDI interactions. Next, we replace the improved sliding-margin loss with the original margin loss [10]. Our improved loss's removal results in poor performance across all metrics. Moreover, the results of taking away filter rules prove that the problem of data imbalance does exist in the original dataset.

Table 3. Ablation study of GCN, improved sliding-margin loss and filter rules

Methods	Precision	Recall	F-score
Our method	82.62	74.77	78.50
- GCN	82.53	69.97	75.73
- Improved sliding-margin loss	82.16	71.50	76.46
- Filter rules	79.31	74.77	76.97

5 Conclusion

In this paper, we introduce a new method using capsule network for the task of DDI extraction. We combine textual features and dependency representations in our model. These representations form a complete semantic representation. The low-level capsules in the capsule network are able to extract low-level semantic meanings, and the high-level capsules clustered by a routing algorithm represent relation features. Moreover, we propose weighted exponential sliding-margin loss to enhance the model performance. Experimental results show our method outperform previous methods on the overall performance. The study of various modules also confirms the effectiveness of those modules in our model.

References

1. Asada, M., Miwa, M., Sasaki, Y.: Enhancing drug-drug interaction extraction from texts by molecular structure information. In: ACL (2), pp. 680–685 (2018)
2. Glorot, X., Bengio, Y.: Understanding the difficulty of training deep feedforward neural networks. In: AISTATS, pp. 249–256 (2010)
3. Jia, Z., et al.: Graphsleepnet: adaptive spatial-temporal graph convolutional networks for sleep stage classification. In: IJCAI, pp. 1324–1330 (2020)
4. Kim, S., et al.: Extracting drug-drug interactions from literature using a rich feature-based linear kernel approach. J. Biomed. Inform. **55**, 23–30 (2015)
5. Liu, S., et al.: Drug-drug interaction extraction via convolutional neural networks. Comput. Math. Meth. Med. 6918381:1–6918381:8 (2016)
6. Liu, X., You, X., Zhang, X., Wu, J., Lv, P.: Tensor graph convolutional networks for text classification. In: AAAI, pp. 8409–8416 (2020)
7. Nguyen, D.Q., Vu, T., Nguyen, T.D., Nguyen, D.Q., Phung, D.Q.: A capsule network-based embedding model for knowledge graph completion and search personalization. In: NAACL-HLT (1), pp. 2180–2189 (2019)
8. Park, C., Park, J., Park, S.: AGCN: attention-based graph convolutional networks for drug-drug interaction extraction. Expert Syst. Appl. **159**, 113538 (2020)
9. Raihani, A., Laachfoubi, N.: Extracting drug-drug interactions from biomedical text using a feature-based kernel approach. J. Theor. Appl. Inf. Technol. **92**(1), 109 (2016)
10. Sabour, S., Frosst, N., Hinton, G.E.: Dynamic routing between capsules. In: NIPS, pp. 3856–3866 (2017)
11. Sun, X., et al.: Drug-drug interaction extraction via recurrent hybrid convolutional neural networks with an improved focal loss. Entropy **21**(1), 37 (2019)
12. Wishart, D.S., et al.: Drugbank 5.0: a major update to the drugbank database for 2018. Nucleic Acids Res. 46(Database-Issue), D1074–D1082 (2018)
13. Xiong, W., et al.: Extracting drug-drug interactions with a dependency-based graph convolution neural network. In: BIBM, pp. 755–759 (2019)
14. Zhang, X., Li, P., Jia, W., Zhao, H.: Multi-labeled relation extraction with attentive capsule network. In: AAAI, pp. 7484–7491 (2019)
15. Zhang, Y., et al.: Drug-drug interaction extraction via hierarchical RNNS on sequence and shortest dependency paths. Bioinform. **34**(5), 828–835 (2018)
16. Zheng, W., et al.: A graph kernel based on context vectors for extracting drug-drug interactions. J. Biomed. Inform. **61**, 34–43 (2016)
17. Zhou, D., Miao, L., He, Y.: Position-aware deep multi-task learning for drug-drug interaction extraction. Artif. Intell. Med. **87**, 1–8 (2018)

Span-Based Nested Named Entity Recognition with Pretrained Language Model

Chenxu Liu[1], Hongjie Fan[2(✉)], and Junfei Liu[1]

[1] Peking University, Beijing, China
{chenxuliu,liujunfei}@pku.edu.cn
[2] China University of Political Science and Law, Beijing, China
hjfan@cupl.edu.cn

Abstract. Named Entity Recognition (NER) is generally regarded as a sequence labeling task, which faces a serious problem when the named entities are nested. In this paper, we propose a span-based model for nested NER, which enumerates all possible spans as potential entity mentions in a sentence and classifies them with pretrained BERT model. In view of the phenomenon that there are too many negative samples in all spans, we propose a multi-task learning method, which divides NER task into entity identification and entity classification task. In addition, we propose the entity IoU loss function to focus our model on the hard negative samples. We evaluate our model on three standard nested NER datasets: GENIA, ACE2004 and ACE2005, and the results show that our model outperforms other state-of-the-art models with the same pretrained language model, achieving 79.46%, 87.30% and 85.24% respectively in terms of F1 score.

Keywords: Nested NER · Span-based model · Multi-task learning

1 Introduction

Named entity recognition (NER) is the task of identifying named entities like person, organization, biological protein, drug, etc. NER is generally treated as a sequence labeling problem [7], where each token is tagged with a label that is composed of entity boundary label and categorical label. However, when named entities contain nested entities, as illustrated in Fig. 1, traditional sequence labeling models are hard to handle it.

Various approaches to extract nested entities have been proposed in recent years. There is a straightforward model which enumerates all possible regions or spans as potential entity mentions and classifies them into the correct category, and we call it span-based model. Span-based model is not limited by the number of nested layers and different nested entity categories. However, the span-based model faces many challenges, which affect the performance of this model. In this

C. S. Jensen et al. (Eds.): DASFAA 2021, LNCS 12682, pp. 620–628, 2021.
https://doi.org/10.1007/978-3-030-73197-7_42

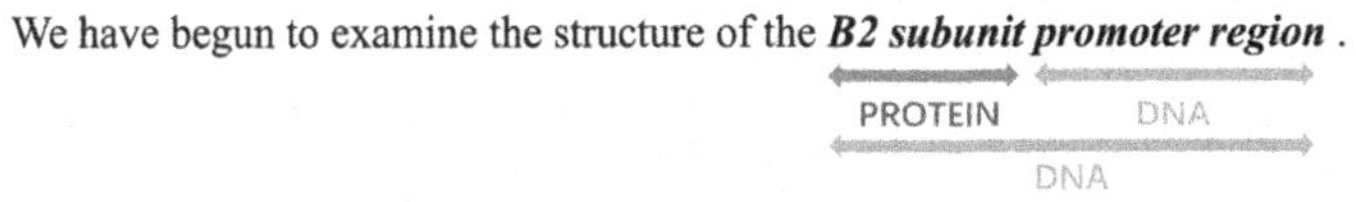

We have begun to examine the structure of the ***B2 subunit promoter region*** .

PROTEIN DNA

DNA

Fig. 1. An example of nested entities from GENIA corpus.

paper, we aim to solve these challenges and build an effective span-based model for nested NER.

Original span-based models [11,15] encode sentences via LSTM or CNN, which cannot learn good semantic information to represent spans. Pretrained language models such as BERT [1] have recently attracted strong attention in NLP research area, and many methods [13,17] use the pretrained BERT model to improve span representation for nested NER. Instead of fine-tuning BERT, these approaches just adopt BERT to enrich token embeddings, which limits the performance of the models. To get a better span representation, we employ the pretrained BERT model [1] to encode the context information and add extra features to enhance the representation.

Besides, the span-based model comes with an extremely imbalanced dataset where the number of non-entity spans far exceeds the entity spans. For example, there are 88 non-entity spans and only 3 entity spans in the sentence in Fig. 1, so the positive/negative sample ratio is about 1:30. None of the previous span-based models have considered the negative effect of having too many negative samples, which may affect the performance of the model. To mitigate the class imbalance problem, we propose a multi-task learning approach, which first divides the entity recognition task into a binary classification task for entity identification and a multiple classification task for entity classification, and then trains a model to do these two tasks simultaneously.

Another challenge for the span-based method is the hard negative samples problem. For example, both the span "we have" and "the B2 subunit" are negative samples, but the first does not overlap with any entities while the second overlap with the entity "B2 subunit". Within our knowledge, it's hard to classify the second span as a negative sample while the first is easy for the model. We call the negative spans which overlap with entity as hard negative samples.

We introduce Intersection over Union (IoU), which is a common concept in computer version, to quantitatively analyze hard negative samples. We define the entity IoU for a span as the intersection length of this span and the entity in a sentence divided by the length of the union. Intuitively, the harder negative samples have higher entity IoU. To focus on the hard negative samples, we propose the entity IoU loss, which could assign higher weights to the hard negative samples according to the IoU between the negative samples with entities. This loss function is a dynamically scaled cross entropy loss, where the scaling factor increases as the IoU between negative samples with entities increases.

In summary, we make the following contributions:

- We propose a span-based nested NER model, which utilizes the pretrained BERT to encode the context information and represent spans.
- We propose a multi-task learning approach to mitigate the data imbalance problem, and we design the entity IoU loss to focus the model on hard negative samples in the training step.
- Empirically, experiments show our model outperforms previous state-of-the-art models with the same pretrained language model on three datasets.

2 Releated Work

Span-based approaches detect entities by identifying over subsequences of a sentence respectively, and nested mentions can be detected because they correspond to different subsequences. For this, Xu et al. [15] utilize a local detection approach to classify every possible span. Sohrab and Miwa [11] propose a simple deep neural model for nested NER, which enumerates all possible spans and then classifies them with LSTM. Besides, Wang et al. [14] propose a transition-based method to construct nested mentions via a sequence of specially designed actions. Tan et al. [13] incorporate an additional boundary detection task to predict entity boundary in addition to classify the span. Generally, these approaches do not consider the problem of class imbalance, which may affect the model's performance.

Pretrained language models are applied for nested NER to improve performance in the last two years. Straková et al. [12] encode the nested labels using a linearized scheme, so the nested NER task can be treated as a sequence labeling task. Fisher and Vlachos [3] introduce a novel architecture that merges

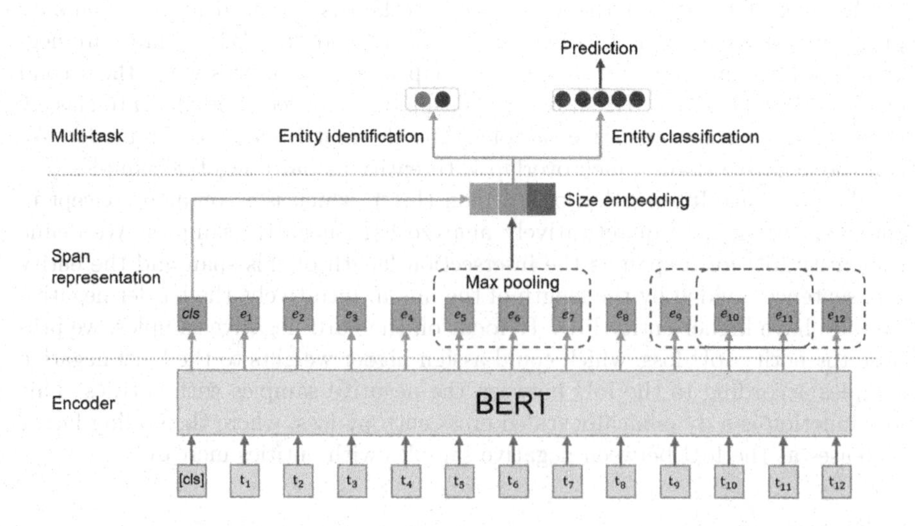

Fig. 2. The overall framework of our proposed model. (Color figure online)

tokens and/or entities into entities forming nested structures and then labels them. Shibuya and Hovy [10] treat the tag sequence for nested entities as the second-best path within the span of their parent entity based on BERT. Yu et al. [17] use a biaffine model to assign scores to all possible spans in a sentence based on document-level BERT feature.

Our work is inspired by SpERT [2], a span-based joint entity and relation extraction model and Ye et al. [16], who divide the relation extraction task into relation identification task and relation classification task.

3 Proposed Method

The overall framework of our proposed model is shown in Fig. 2. The model consists of three layers: Encoder layer, Span representation layer and Multi-task layer. Given an input sentence, a pretrained BERT encoder is first adopted to encode the context information and represent the tokens. Then the span representation layer enumerates all possible spans and represents these spans with the output of the encoder layer. Finally, the multi-task layer trains the model to do entity identification task with the entity IoU loss and entity classification task with the pairwise ranking loss simultaneously.

3.1 Encoder Layer

Given an input sentence consisting of n tokens $(t_1, t_2, \ldots, t_n)$, the encoder layer extracts context information from the sentence, which will represent spans. We employ a pretrained BERT model to encode the context information.

For limiting the vocabulary and mapping OOV words and rare words, BERT uses a byte-pair encoder (BPE) tokenizer [9]. We get a sequence of m BPE tokens after the input sentence is tokenized, where $m \geq n$. Then the BPE tokens sequence is passed through BERT, we obtain an embedding sequence $e = (cls, e_1, e_2, \ldots, e_m)$ of length $m + 1$, where cls represents a special classifier token encoding the overall sentence context.

3.2 Span Representation Layer

The span representation layer represents spans with BERT's output e. We generate all possible spans with sizes less than or equal to the maximum span size L, which is a pre-defined parameter for limiting the model's complexity. We use the $span(i, j)$ to represent the span from i to j, where $1 \leq i \leq j \leq m$ and $j - i < L$.

The span representation consists of three parts:

- **Span' s Context Encoding** (Fig. 2, blue): For a $span(i, j)$ and its BPE token context encoding $(e_i, e_{i+1}, \ldots, e_j)$, we utilize a fusion operation to convert the token context encoding to a fix-sized vector $v_{i,j}$. We find max-over-time pooling to get the best performance.

- **Sentence Context Encoding** (Fig. 2, purple): The categories of entities in a sentence are likely to be related to the overall context information of a sentence. For this reason, we add *cls* encoding.
- **Width Embedding** (Fig. 2, yellow): Within our knowledge, spans that are too long are unlikely to represent entities. We adopt width embedding to put this prior knowledge into our model. Given a span with width k, we look-up a width embedding from an embedding matrix. The embedding matrix is initialized randomly and learned in the training step.

In summary, we obtain the representation $s(i,j)$ of the $span(i,j)$ as follows:

$$s(i,j) = [cls; v_{i,j}; w_{j-i+1}] \tag{1}$$

where w_{j-i+1} denotes the width embedding and [;] denotes concatenation.

3.3 Multi-task Layer

Entity Identification with Entity IoU Loss. For the binary classification task of entity identification, we propose an entity IoU loss function. In a sentence, Intersection over Union (IoU) between two spans means the length of intersection divided by the length of union. As illustrated in Fig. 1, The union of the span "of the B2 subunit promoter" and the entity span "B2 subunit promoter region" is 6 and the intersection is 3, so we can calculate that the IoU is 0.5. Formally, we can compute the IoU as follows:

$$\mathrm{IoU}(span(i_1,j_1), span(i_2,j_2)) = \begin{cases} \frac{\min(j_1,j_2)-\max(i_1,i_2)+1}{\max(j_1,j_2)-\min(i_1,i_2)+1}, & \min(j_1,j_2) \geq \max(i_1,i_2) \\ 0, & otherwise \end{cases} \tag{2}$$

We define ESS as the set consisting of all entity spans in a sentence and $\mathrm{ENIoU}(s)$ as the max IoU between the span s and all entity spans in the sentence.

$$\mathrm{ENIoU}(s) = \max(\mathrm{IoU}(s, en), en \in ESS) \tag{3}$$

Then we can define the entity IoU loss as follows:

$$\mathrm{EI}(p,y) = \begin{cases} -(1-\alpha)\log(p), & if\ y = 1 \\ -\alpha(1 + ENIoU)\log(1-p), & otherwise \end{cases} \tag{4}$$

In the above $y \in 0, 1$ specifies the span's ground-truth class (0 for non-entity and 1 for entity), $\alpha \in [0,1]$ is a balance factor for addressing class imbalance, $p \in [0,1]$ is the model's estimated probability for the class with label $y = 1$ and $ENIoU$ is the span's entity IoU.

This loss function is a dynamically scaled cross entropy loss, where α and $ENIoU$ determine the scaling factor. We set $\alpha < 0.5$ to pay more attention to the positive samples which are less in all samples. For the negative samples, the loss is bigger when the $ENIoU$ is bigger, which could focus the model on the hard negative samples in the training step.

Then we can compute the entity identification loss as follows:

$$Loss_1 = \sum \mathrm{EI}(p,y) \tag{5}$$

Entity Classification with Ranking Loss. To increase the loss diversity in multi-task learning, we adopt the pairwise ranking loss proposed by dos et al. [8] for the entity classification task. We define $Loss_2$ as the entity classification ranking loss.

Finally, the loss function of our multi-task framework is formulated as:

$$Loss = \beta Loss_1 + Loss_2 \tag{6}$$

We learn the parameters of entity identification and entity classification as well as the width embeddings and fine-tune BERT in the training step by minimizing $Loss$ in Eq. 6.

3.4 Prediction

In the prediction stage, to avoid error propagation, we only use the class score in the multiple entity classification task, while the binary entity identification task is only used for optimizing the network parameters.

4 Experiment

4.1 Experimental Setup

We mainly evaluate our model on three standard nested NER datasets: GENIA [6], ACE2004 and ACE2005. The dataset split is following previous work [5] and we report micro precision, recall and F1 score to evaluate the performance.

We compare our model with several previous state-of-the-art models:

Seq2seq [12]**:** Straková et al. encode the nested labels using a linearized scheme to treat the nested NER as a sequence to sequence problem.

Second-path [10]**:** Shibuya and Hovy treat the tag sequence for nested entities as the second-best path within the span of their parent entity based on BERT.

Boundary[1] [13]**:** A model that incorporates an additional boundary detection task to predict entity boundary in addition to classify the span.

Biaffine-NER[2] [17]**:** A biaffine span-based model.

Pyramid [5]**:** A pyramid architecture to stack flat NER layers for nested NER.

[1] This model is based on BERT-base and they use different dataset splits. Except for this model, other models are based on BERT-large.

[2] This model also uses the fasttext embedding, for the sake of fairness, we re-train this model without the fasttext embedding.

4.2 Overall Evaluation and Ablation Study

Table 1 shows the comparison of our model with several previous state-of-the-art nested NER models on the test datasets. Our model outperforms the state-of-the-art models with 79.46%, 87.30% and 85.24% in terms of F1 score, achieving the new state-of-the-art performance in the nested NER tasks. Our model gains the highest improvement of F1 score with 1.02% on ACE2004 while 0.58% and 0.27% on GENIA and ACE2005. A possible reason accounts for it, it is that ACE2004 has more nested entities (46%) compared with ACE2005 (38%) and GENIA (18%) and our span-based model is good at handling nested entities.

We also conduct the ablation study to verify the effectiveness of our multi-task learning method and entity IoU loss. We first replace the entity IoU loss with cross-entropy loss. Then, we remove the entity IoU loss and multi-task learning method and just use ranking loss for entity classification.

From the results shown in Table 1, we can see: our multi-task learning method improves performance by 0.65%, 0.58% and 0.25% and the entity IoU improves the performance by −0.12%, 0.80% and 0.83% in GENIA, ACE2004 and ACE2005. We notice entity IoU loss does not affect GENIA. One possible reason is that the words in biomedical entities are usually terminologies, the model could distinguish the hard negative samples easily without entity IoU loss.

Table 1. Performance comparison of the state-of-the-art nested NER models on the test dataset and results when ablating away the entity IoU loss (EIL) and multi-task learning method (MTL).

	GENIA			ACE2004			ACE2005		
Model	P(%)	R(%)	F1(%)	P(%)	R(%)	F1(%)	P(%)	R(%)	F1(%)
Seq2seq	–	–	78.31	–	–	84.40	–	–	84.33
Second-path	78.70	75.74	77.19	85.23	84.72	84.97	83.30	84.69	83.99
Boundary	79.2	77.4	78.3	85.8	84.8	85.3	83.8	83.9	83.9
Biaffine-NER	79.30	78.99	79.14	86.00	84.92	85.46	83.93	84.79	84.36
Pyramid	79.45	78.94	79.19	86.08	86.48	86.28	83.95	85.39	84.66
Our model	**79.53**	79.37	79.46	**87.51**	**87.08**	**87.30**	84.75	**85.73**	**85.24**
w/o EIL	78.71	**80.47**	**79.58**	87.33	85.69	86.50	**85.91**	82.96	84.41
w/o EIL& MTL	79.17	78.70	78.93	87.44	84.44	85.92	83.51	84.81	84.16

4.3 Running Time

Although our model needs to classify all possible spans with a max length in a sentence, the spans can be processed in a batch with GPU to reduce the training and inference time and classifying spans is a lightweight operation compared with encoding. Figure 3 shows the inference speed of our model, BERT-Tagger, Boundary-aware [18], Neural-layered [4] and Biaffine-NER [17]. These experiments are performed on the same machine (one NVIDIA 1080ti GPU with Intel i7-6700K CPU). BERT-Tagger's speed is the BERT-based models' upper bound speed. Our model is about 1/4 slower than BERT-Tagger because our model needs to classify spans while BERT-Tagger just needs to classify tokens. Our model is about 3 times faster than Boundary-aware, 9 times faster than Neural-layered and about 700 times faster than Biaffine-NER.

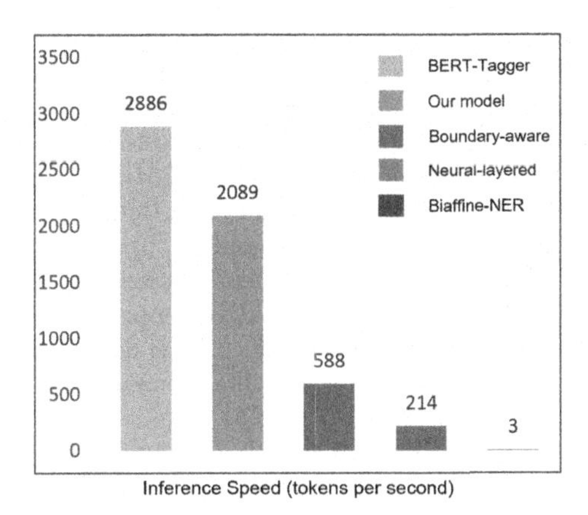

Fig. 3. The inference speed of our model and compared models.

5 Conclusion

In this paper, we build an effective span-based nested NER model based on pretrained BERT. We propose a multi-task learning approach to deal with the class imbalance problem and entity IoU loss to focus on hard negative samples. Experiments show that our model outperforms previous state-of-the-art models on three standard datasets with a competitive running speed.

References

1. Devlin, J., Chang, M.W., Lee, K., Toutanova, K.: BERT: pre-training of deep bidirectional transformers for language understanding. In: Proceedings of NAACL, pp. 4171–4186 (2019)
2. Eberts, M., Ulges, A.: Span-based joint entity and relation extraction with transformer pre-training. arXiv preprint arXiv:1909.07755 (2019)
3. Fisher, J., Vlachos, A.: Merge and label: a novel neural network architecture for nested ner. In: Proceedings of ACL, pp. 5840–5850 (2019)
4. Ju, M., Miwa, M., Ananiadou, S.: A neural layered model for nested named entity recognition. In: Proceedings of NAACL, pp. 1446–1459 (2018)
5. Jue, W., Shou, L., Chen, K., Chen, G.: Pyramid: a layered model for nested named entity recognition. In: Proceedings of ACL, pp. 5918–5928 (2020)
6. Kim, J.D., Ohta, T., Tateisi, Y., Tsujii, J.: Genia corpus–a semantically annotated corpus for bio-text mining. Bioinformatics **19**(suppl_1), i180–i182 (2003)

7. Lafferty, J., McCallum, A., Pereira, F.C.: Conditional random fields: Probabilistic models for segmenting and labeling sequence data (2001)
8. dos Santos, C., Xiang, B., Zhou, B.: Classifying relations by ranking with convolutional neural networks. In: Proceedings of ACL, pp. 626–634 (2015)
9. Sennrich, R., Haddow, B., Birch, A.: Neural machine translation of rare words with subword units. arXiv preprint arXiv:1508.07909 (2015)
10. Shibuya, T., Hovy, E.: Nested named entity recognition via second-best sequence learning and decoding. arXiv preprint arXiv:1909.02250 (2019)
11. Sohrab, M.G., Miwa, M.: Deep exhaustive model for nested named entity recognition. In: Proceedings of EMNLP, pp. 2843–2849 (2018)
12. Straková, J., Straka, M., Hajic, J.: Neural architectures for nested NER through linearization. In: Proceedings of ACL pp. 5326–5331 (2019)
13. Tan, C., Qiu, W., Chen, M., Wang, R., Huang, F.: Boundary enhanced neural span classification for nested named entity recognition, pp. 9016–9023. AAAI (2020)
14. Wang, B., Lu, W., Wang, Y., Jin, H.: A neural transition-based model for nested mention recognition. In: Proceedings of EMNLP, pp. 1011–1017 (2018)
15. Xu, M., Jiang, H., Watcharawittayakul, S.: A local detection approach for named entity recognition and mention detection. In: Proceedings of ACL, pp. 1237–1247 (2017)
16. Ye, W., Li, B., Xie, R., Sheng, Z., Chen, L., Zhang, S.: Exploiting entity bio tag embeddings and multi-task learning for relation extraction with imbalanced data. In: Proceedings of ACL, pp. 1351–1360 (2019)
17. Yu, J., Bohnet, B., Poesio, M.: Named entity recognition as dependency parsing. arXiv preprint arXiv:2005.07150 (2020)
18. Zheng, C., Cai, Y., Xu, J., Leung, H.f., Xu, G.: A boundary-aware neural model for nested named entity recognition. In: Proceedings of EMNLP, pp. 357–366 (2019)

Poetic Expression Through Scenery: Sentimental Chinese Classical Poetry Generation from Images

Haotian Li[1], Jiatao Zhu[1], Sichen Cao[1], Xiangyu Li[1], Jiajun Zeng[1], and Peng Wang[1,2,3](✉)

[1] School of Artificial Intelligence, Southeast University, Nanjing, China
pwang@seu.edu.cn
[2] School of Computer Science and Engineering, Southeast University, Nanjing, China
[3] School of Cyber Science and Engineering, Southeast University, Nanjing, China

Abstract. Most Chinese poetry generation methods only accept texts or user-specified words as input, which contradicts with the fact that ancient Chinese wrote poems inspired by visions, hearings and feelings. This paper proposes a method to generate sentimental Chinese classical poetry automatically from images based on convolutional neural networks and the language model. First, our method extracts visual information from the image and maps it to initial keywords by two parallel image classification models, then filters and extends these keywords to form a keywords set which is finally input into the poetry generation model to generate poems of different genres. A bi-directional generation algorithm and two fluency checkers are proposed to ensure the diversity and quality of generated poems, respectively. Besides, we constrain the range of optional keywords and define three sentiment-related keywords dictionary to avoid modern words that lead to incoherent content as well as ensure the emotional consistency with given images. Both human and automatic evaluation results demonstrate that our method can reach a better performance on quality and diversity of generated poems.

Keywords: Information extraction and summarization · Poetry generation · Vision-driven modelling · Natural language processing

1 Introduction

Traditional Chinese classical poetry is fascinating and important in Chinese literature. Automatic poetry generation is an interesting challenge and has caught

The work is supported by National Key R&D Program of China (2018YFD1100302), National Natural Science Foundation of China (No. 61972082, No. 72001213), and All-Army Common Information System Equipment Pre-Research Project (No. 31511110310, No. 31514020501, No. 31514020503).

C. S. Jensen et al. (Eds.): DASFAA 2021, LNCS 12682, pp. 629–637, 2021.
https://doi.org/10.1007/978-3-030-73197-7_43

increasing attention in recent years because of its significant research value in automatic language analysis and computer creativity. Various attempts have been made on this task. Early work is based on rules and templates [4]. Then statistical machine learning methods are applied to poetry generation [8]. With the development of neural network, poetry generation models based on recurrent neural network [14] and its variants [13] have been widely used. In recent years, models based on language models like Generative Pre-training (GPT) have realized generating complete poetry under given keywords without any additional modifications [5]. However, works on generating sentimental Chinese classical poetry from images are deficient. Only few works address the problem of image-related poetry generation [6,15], and in some cases these methods may generate poems with bad tone patterns, vague emotional tendencies, as well as incoherent contents when modern words are involved.

Poetry can be generated from diverse inspirations, where vision (images) is of great significance. Poetic feelings may occur when one contemplates an image or a painting which represents natural scenery. In our work we mimic this process of poetry writing from images. Besides, Chinese classical poetry has different genres. All genres have strict regulations on tonal patterns and rhyme schemes. In our work we mainly focus on two types among them: *Jueju* (poems of four lines, five or seven characters per line, i.e., Wujue and Qijue) and *Lvshi* (poems of eight lines, five or seven characters per line, i.e., Wulv and Qilv).

Using images as inspirations for poetry generation has many advantages. On the one hand, an image contains richer information than words which enables varieties of generated poetry. On the other hand, different people may have different interpretations and feelings towards the same images, thus using images as inspiration may bring surprising and impressive results. In this paper, we propose a method for sentimental Chinese classical poetry generation from images, including information extraction, keyword extension and poetry generation.

In conclusion, our contributions are as follows:

- We design an innovative pipeline for poetry generation from images and propose a bi-directional generation algorithm and two fluency checkers which are effective in generating diverse poems consistent with the given image.
- We map visual information to keywords and build three sentiment-related keyword dictionaries to extend keywords with specific emotional tendencies, which ensures the generated poems are emotionally related to given images.
- From human and automatic evaluation results, our method can reach a better performance on quality and diversity of generated poems.

2 Methodology

2.1 Problem Formulation and Overview

Let P be a poem with l lines $x_1, x_2, ..., x_l$, and each line has n characters $x_i = (x_{i,1}, x_{i,2}, ..., x_{i,n})$. For image input I , We first extract visual information and map it into a keywords set $K = (E_1, E_2, ..., E_l)$, where E_1 to E_{l-1} represent $l-1$

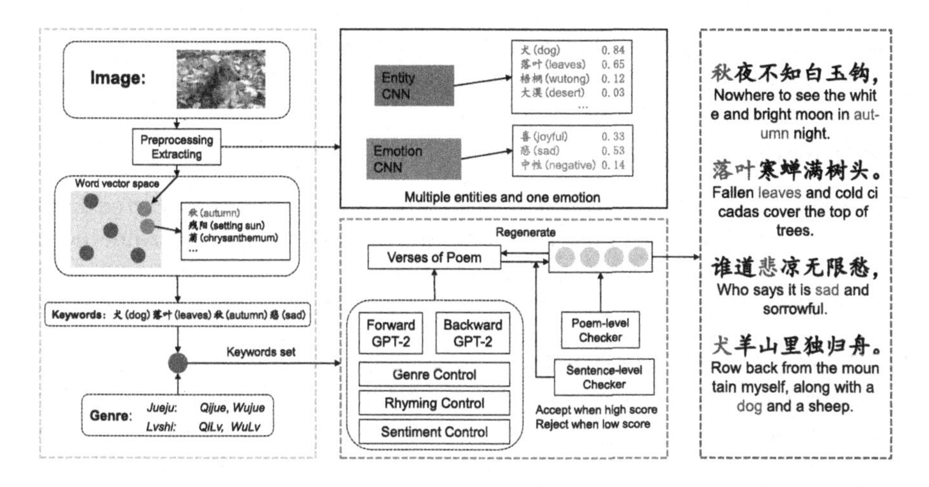

Fig. 1. Overview of the proposed method for poetry generation from images.

entities and E_l is the emotion. Then these keywords are utilized as the initial seed for every sentence to generate a poem $P = (x_1, x_2, ..., x_l)$. Overview of our method is shown in Fig. 1. In our work, we utilize extracted information from the given image to generate a four-line poem ($l = 4$, the most common length in Chinese classical poetry). Firstly the CNNs extract initial keywords (entities and the emotion, e.g., *dog*, *leaves*, *sad*) from a given image. Then in keyword extension module, we get word vectors of these extracted keywords and extend them to form the final keywords set (e.g., *dog*, *leaves*, *autumn*, *sad*). Finally, each keyword in the set works as an initial seed to generate each sentence in the final poem by the GPT-2 based generation model. In the following discussion, the role of the emotion is both an sentiment label (guide the generation of sentiment-related keywords and verses) and an emotional keyword (occur in generated poems and express feelings explicitly).

2.2 Information Extraction

We define the extraction of visual information as an image classification problem and design two parallel CNN models with different parameters to extract entities and the emotion, respectively. Different from modern poetry generation [2], modern words (i.e., words those do not exist in ancient times like *neural network*) in classical poetry generation bring serious problems. If a modern word is input to generate classical poetry, the generated results are possible to have incoherent content and confusing logic. For poetry generation from images, this problem can be solved if we ensure that the range of extracted keywords exclude modern concepts. So we choose all candidate keywords from common imageries in Chinese classical poetry (e.g., *cloud*, *pavilion* and *the setting sun*) based on their frequencies. Though excluding modern words may cause the omission of some information, we retrieve the loss to some extent in keyword extension module.

The emotional words in our method have three types: *joyful*, *sad* and *neutral*, which are basic keynotes for most Chinese classical poems.

2.3 Keyword Extension

The goal of keyword extension is to regularize the number of extracted initial keywords and output the final keywords set. Concretely, for a four-line poem, if the number of initial keywords is more than 4, the emotion is retained and other 3 entities are randomly selected to increase the diversity. Otherwise extra keywords are added according to word similarity [9] based on word vectors [11]. We build three sentiment-related keyword dictionaries by respective statistics of word frequency in sentimental poetry corpus. Extra keywords are selected from corresponding sentimental dictionary to ensure the consistency with the emotion. Notice that if the extracted emotion is *neutral*, it will be removed from keywords set and replaced with another extended entity but still work as sentiment label in generation model.

2.4 Poetry Generation

Generation Model. We choose Generative Pre-Training 2 (GPT-2) as the basic forward generation model. For input $K = (E_1, E_2, E_3, E_4)$, firstly E_1 is input to generate the first line, then the second line is generated based on E_2 and previously generated contents, as well as the third and last line. With this forward generation method, however, all keywords will occur only at the beginning of each line. To solve this problem, we train an extra backward GPT-2 model with inverse corpus and propose a bi-directional generation algorithm. Concretely, for generation of i-th line, we first put E_i in random position $x_{i,j}$ and use backward GPT-2 to generate $(x_{i,j-1}, x_{i,j-2}, ..., x_{i,1})$, then we input all generated characters $(x_1, x_2, ..., x_{i,1}, ..., x_{i,j})$ into forward GPT-2 to generate the rest contents $(x_{i,j+1}, x_{i,j+2}, ..., x_{i,n})$. Besides, we adopt truncated top-k sampling strategy to select candidate characters to increase the diversity. We also simplify the rhyming rules by controlling the tone of the last character of every sentence.

Sentiment Control. The generation model is pretrained with a huge number of poetry corpus and fine-tuned with a small sentiment-labelled poetry corpus. The format of poetry in fine-tuning process is *sentiment-label#poetry-contents*. The extracted emotional words will work as the sentiment label during generating process and the format of input when generating the i-th character will be *emotion#generated-contents*. With the guidance of sentiment label, the model tends to generate poetry with contents emotionally related to given images.

Fluency Checkers. Using truncated top-k sampling strategy to select randomly from candidate characters may break the rules of grammar and destroy the fluency of poems. To avoid this problem, we design a sentence-level grammar checker and a poem-level content checker to evaluate grammar and fluency scores of generated sentences and poems, respectively. We utilize a 5-gram model with POS tagged corpus and 3-gram model with masterpieces corpus to calculate these scores and reject contents with low scores. These checkers help exclude poor sentences and enable our model to generate poems of high quality.

3 Experiment

3.1 Dataset

In visual information extraction module, we collect images of 44 candidate entities to train the entity CNN. The dataset we choose to train the emotion CNN is *GAPED*, which is a new database of 730 images with negative, neural or positive contents [3]. In keyword extension and poetry generation modules, we use *CCPC* (Chinese Classical Poetry Corpus) [15] to pre-train models and *FSPC* (Fine-grained Sentimental Poetry Corpus) [1] to build sentiment-related keywords dictionary and fine-tune the poetry generation model. Details of datasets are shown in Table 1. Notice that all these models are trained separately.

Table 1. Statistics of datasets.

Dataset	Types	Average	Total
Entity images	44 entities	404	17800
GAPED	3 emotion	243	730
CCPC	Mostly Jueju	–	170787
FSPC	3 emotion	1700	5000

3.2 Evaluation Metrics

In this experiment, we evaluate the generated results from five aspects: **Poeticness** (If generated poems follow the rhyme and tone regulations and meet requirements of different genres), **Diversity** (If extracted keywords and generated poems are different from the same image), **Relevance** (If generated poems are relevant to the given image), **Meaning** (If contents are coherent and logical) and **Emotional expression** (If generated poems express some emotions).

3.3 Model Variants and Baselines

In addition to the proposed method of GPT-2 based poetry generation from images (**GPGI**), we also evaluate three variants of the model to study the effects of bi-directional generation algorithm and two fluency checkers, including **GPGI (origin)** (original GPT-2 without any additional components), **GPGI (w/o backward)** (the proposed method without bi-directional generation), **GPGI (w/o checkers)** (the proposed method without two fluency checkers).

To further investigate the effectiveness of the proposed methods, we choose two state-of-the-art methods which are accessible on the Internet as baselines for comparisons. **Jiuge** is a human-machine collaborative Chinese classical poetry generation system [15]. It uses the Aliyun image recognition tool and knowledge graph to extract keywords from images and generate poems by a GRU based encoder-decoder model. **Yuefu** is a simple yet effective system for generating high quality classical Chinese poetry from images with image recognition methods and generative pre-trained language model (GPT) [5].

Table 2. Results of human and automatic evaluation on different methods. Scores of poeticness, diversity, relevance, meaning and expression are recalculated by mapping original scores to the range [0, 1] in order to figure out the average performance.

Method	Poeticness	Diversity	Relevance	Meaning	Emotion	BLEU	Average	Recall
GPGI (full)	**0.94**	**0.85**	**0.83**	0.75	**0.76**	0.68	0.80	**46.3%**
Jiuge	**0.94**	0.80	0.81	**0.83**	0.72	**0.75**	**0.81**	33.2%
Yuefu	0.92	**0.85**	0.77	0.73	0.70	0.63	0.77	23.3%
GPGI (origin)	0.82	0.80	–	0.68	0.68	0.55	0.71	–
GPGI (w/o checkers)	0.82	0.81	–	0.71	0.72	0.57	0.73	–
GPGI (w/o backward)	0.92	0.82	–	0.72	0.72	0.65	0.76	–

3.4 Human and Automatic Evaluation

Human and automatic evaluation metrics are combined to conduct the experiment. An human assessment tool is designed for human evaluation. To obtain rich feedback without user bias, we randomly choose 20 assessors from various career fields with a bachelor degree and requisite knowledge of poetry to assess generated poems. 30 images are randomly sampled as testing set, each of which will be utilized to generate several poems by different models. Assessors can rate these poems on a scale of one to five. Besides, We select some masterpieces from dataset as references and calculate BLEU scores [7,10] of generated poems. Table 2 summarizes the results.

Overall Performance. The average scores in Table 2 indicate that the proposed model GPGI outperforms *Yuefu* and has similar performance to *Jiuge*. Concretely, *Jiuge* succeeds in coherence and automatic evaluation because of its complex contents control mechanism, whereas our method works better than two state-of-art baselines in terms of poeticness, diversity, relevance and emotional expression. *Yuefu* get the lowest score of poeticness and meaning due to lack of relevant control components. Though *Jiuge* can generate poems of the best meaning, the diversity decrease accordingly as diverse contents mean the risk of incoherence because of random sampling strategy, whereas our methods maintain a good balance between these aspects. Besides, our method achieve the highest emotional expression score, which may attribute to emotion extraction CNN, sentiment-related keywords dictionary and GPT-2 model fine-tuned on sentimental poetry corpus.

Automatic Evaluation on Relevance. Unlike texts and words based poetry generation methods, relevance of the generated poem to the given image is an important metric for image inspired poetry generation. Therefore, in addition to subjective evaluation, we also conduct an automatic evaluation in terms of relevance referring to [12] by computing the recall rate of the key concepts in an image that are described in the generated poem. The results are also shown in the last column of Table 2. Both the subjective and quantitative evaluation on relevance indicate that manually collected dataset of common imageries as well as keyword extension are efficient and effective methods to extract key information from an image.

Case	1	2	3	4	
Image					
Keywords Set	犬 (dog) 落叶 (leaves) 秋 (autumn) 悲 (sad)	大漠 (desert) 白云 (cloud) 玉门关 (Yumen pass) 喜 (joyful)	桃花 (peach blossom) 青山 (green hills) 梧桐 (wutong) 喜 (joyful)	白云 (cloud) 乡村 (countryside) 行人 (pedestrian) 喜 (joyful)	白云 (cloud) 乡村 (countryside) 松韵 (tree rhyme) 喜 (joyful)
Results	秋夜不知白玉钩， Nowhere to see the white and bright moon in autumn night. 落叶寒蝉满树头。 Fallen leaves and cold cicadas cover the top of trees. 谁道悲凉无限愁， Who says it is sad and sorrowful. 犬羊山里独归舟。 Row back from the mountain myself, along with a dog and a sheep.	大漠不知人事多， There are many people and anecdotes in the desert, which are not known by people. 唯将衰老喜如何。 Only to get old without happiness. 白云归去无心过， When the white clouds go back, they are unwilling to get through. 落日相随玉门关。 The setting sun follows me through Yumen pass.	桃花飘去不知是， Peach blossom floating away, unknown where to go. 遥望青山色尽啼。 From far away, I can hear songs around the green hills. 若陪梧桐落花夜， If I accompany the Chinese wutong whose flowers fall down in the night. 不见春风喜见君。 I will meet you happily even if I can not feel the spring breeze.	白云上山月不知， Moon does not know it that white clouds float up to the mountain. 钟白日前乡村归。 I return to the countryside under the sun. 长叹行人无处休， Sighing that pedestrians have nowhere to rest. 何然喜江充元思。 The river flows happily just like my thoughts.	白云台上见明天， See dawn on cloud platform. 却入乡村作客船。 Then go into the countryside as a guest ship. 松韵不知歌此卷， Surrounded by tree rhyme ,I do not know how to sing this volume. 山光喜合醉中眠。 Mountain light brings me happiness and I fall asleep drunk.
Evaluation (BLEU)	0.68	0.62	0.71	0.64	0.60

Fig. 2. Cases and automatic evaluation of the generated poems. We take 4 cases for example to show that the generated poems meet requirements of content and diversity.

Analysis of Model Variants. Results in the bottom rows of Table 2 correspond to the variants including GPGI (origin), GPGI (w/o checkers) and GPGI (w/o backward). Notice that relevance is not evaluated as all variants share the same visual information extraction module. From studies on the effect of proposed generation algorithm and two fluency checkers, we find that both two parts help improve the performance, which means appropriate human crafted rules or features and additional neural components are essential for high quality of generated poems. Moreover, the fluency checkers seem to have bigger impact than the bi-directional generation algorithm. The main reason is that contents generated by backward GPT-2 only take a small proportion of whole poem.

3.5 Case Study

To further illustrate the quality and diversity of generated poems, we study several cases of the poems generated with corresponding images. From cases in

Fig. 2, we find that these poems have relatively high relation to the given images because their contents include the main visual information in given images like *desert*, *cloud*, *flowers* as well as semantic concepts such as *joyful*, *happiness*. As shown in case 4, the keywords set and the generated poems are different for a certain image, which demonstrates that the generated poems are diverse and flexible. Evaluation results of these poems are also provided.

4 Conclusion and Future Work

This paper proposes an innovative hierarchical model to generate fluent and high image-related poems with CNN based information extraction, vector based keyword extension and GPT-2 based poetry generation, as well as a bi-directional generation algorithm and two fluency checkers. However, there are still some aspects we can improve. To avoid constraining the emotion of a given image based on low-level visual feature, which would ignore the fact that different people may have different feelings towards the same image, new methods will be adopted to get more meaningful and coherent content in the future work.

References

1. Chen, H., Yi, X., Sun, M., Yang, C., Li, W., Guo, Z.: Sentiment-controllable Chinese poetry generation. In: Proceedings of the Twenty-Eighth International Joint Conference on Artificial Intelligence, Macao, China (2019)
2. Cheng, W., Wu, C., Song, R., Fu, J., Xie, X., Nie, J.: Image inspired poetry generation in Xiaoice. arXiv Artificial Intelligence (2018)
3. Danglauser, E.S., Scherer, K.R.: The Geneva affective picture database (GAPED): a new 730-picture database focusing on valence and normative significance. Behav. Res. Methods **43**(2), 468–477 (2011). https://doi.org/10.3758/s13428-011-0064-1
4. Gervas, P.: An expert system for the composition of formal Spanish poetry. Knowl. Based Syst. **14**(3/4), 181–188 (2001)
5. Liao, Y., Wang, Y., Liu, Q., Jiang, X.: GPT-based generation for classical Chinese poetry. arXiv Computation and Language (2019)
6. Liu, B., Fu, J., Kato, M.P., Yoshikawa, M.: Beyond narrative description: generating poetry from images by multi-adversarial training. In: ACM Multimedia (2018)
7. Madnani, N.: iBLEU: interactively debugging and scoring statistical machine translation systems. In: IEEE International Conference Semantic Computing (2011)
8. Manurung, H.M.: An evolutionary algorithm approach to poetry generation. Ph.D. thesis, University of Edinburgh (2003)
9. Mikolov, T., Chen, K., Corrado, G.S., Dean, J.: Efficient estimation of word representations in vector space. In: International Conference on Learning Representations (2013)
10. Papineni, K., Roukos, S., Ward, T., Zhu, W.: BLEU: a method for automatic evaluation of machine translation. In: Meeting of the Association for Computational Linguistics (2002)
11. Rong, X.: Word2vec parameter learning explained. arXiv Computation and Language (2014)

12. Xu, L., Jiang, L., Qin, C., Wang, Z., Du, D.: How images inspire poems: generating classical Chinese poetry from images with memory networks. In: National Conference on Artificial Intelligence (2018)
13. Yi, X., Sun, M., Li, R., Yang, Z.: Chinese poetry generation with a working memory model. In: International Joint Conference on Artificial Intelligence (2018)
14. Zhang, X., Lapata, M.: Chinese poetry generation with recurrent neural networks. In: Empirical Methods in Natural Language Processing (2014)
15. Zhipeng, G., Yi, X., Sun, M., Li, W., Li, R.: Jiuge: a human-machine collaborative Chinese classical poetry generation system. In: Proceedings of the 57th Annual Meeting of the Association for Computational Linguistics: System Demonstrations (2019)

Social Network

SCHC: Incorporating Social Contagion and Hashtag Consistency for Topic-Oriented Social Summarization

Ruifang He[1,2,3(✉)], Huanyu Liu[1,2], and Liangliang Zhao[1,2]

[1] College of Intelligence and Computing, Tianjin University, Tianjin 300350, China
{rfhe,huanyuliu,liangliangzhao}@tju.edu.cn

[2] Tianjin Key Laboratory of Cognitive Computing and Application, Tianjin 300350, China

[3] State Key Laboratory of Communication Content Cognition, People's Daily Online, Beijing 100733, China

Abstract. The boom of social media platforms like Twitter brings the large scale short, noisy and redundant messages, making it difficult for people to obtain essential information. We study extractive topic-oriented social summarization to help people grasp the core information on social media quickly. Previous methods mainly extract salient content based on textual information and shallow social signals. They ignore that user generated messages propagate along the social network and affect users on their dissemination path, leading to user-level redundancy. Besides, hashtags on social media are a special kind of social signals, which can be regarded as keywords of a post and contain abundant semantics. In this paper, we propose to leverage social theories and social signals (i.e. multi-order social relations and hashtags) to address the redundancy problem and extract diverse summaries. Specifically, we propose a novel unsupervised social summarization framework which considers **S**ocial **C**ontagion and **H**ashtag **C**onsistency (**SCHC**) theories. To model relations among tweets, two relation graphs are constructed based on user-level and hashtag-level interaction among tweets. These social relations are further integrated into a sparse reconstruction framework to alleviate the user-level and hashtag-level redundancy respectively. Experimental results on the CTS dataset prove that our approach is effective.

Keywords: Social summarization · Social network analysis · Natural language processing

1 Introduction

Social media platforms like Twitter have become a popular way for users to freely produce content (called tweets) on their interested topics. However, the rapid growth of tweets makes it difficult for people to quickly grasp useful information. Social summarization is an non-trivial and challenging task, which aims

C. S. Jensen et al. (Eds.): DASFAA 2021, LNCS 12682, pp. 641–657, 2021.
https://doi.org/10.1007/978-3-030-73197-7_44

to generate a concise summary to reveal the essential content of a tremendous amount of tweets in a given topic. Different from traditional document (like news articles), it is more challenging to deal with tweets on social media due to their short and informal nature. Besides, since tweets are generated by independent users, a large number of tweets may contain similar or even the same information, leading to extremely serious information redundancy on social media. Therefore, it becomes particularly important to eliminate redundancy and produce more diverse summaries in social summarization.

The serious redundant information on social media is mostly caused by the frequent interaction and information propagation of active users. These information redundancy on social media can be explained from the following three aspects: **(1) Intra-user redundancy**: People are more likely to keep the same sentiment and opinions on a specific topic in a short period, leading to self-redundant tweets; **(2) Inter-user redundancy**: Information spreads along connected users on social networks, which could bring about mutual influence and make their opinions more and more similar. Therefore, users with relations are more likely to generate redundant information. **(3) Hashtag-level redundancy**: Tweets sharing the same hashtags tend to contain more similar content. Essentially, these phenomena are consistent with the social theories proposed by [12], which reveal the reciprocal influence of networked information.

There have been many researches trying to generate good summaries for social media text, which can be roughly divided into three categories: (a) **Content-based methods** [13] regard each post as a sentence and directly adopt the traditional summarization methods on tweets. However, the special characteristics of tweets make these traditional methods unsuitable and perform poorly when migrated to social media data. (b) **Static social signals** [17,18] are further considered such as number of replies, number of retweets, author popularity and so on. Nonetheless, these methods treat each post as an independent sentence and extract sentence one by one from the original corpus, while ignoring the relationship among tweets, thus failing to filter redundant content. (c) **Social relations based approaches** [2,3] are proposed to incorporate network information from user-level, assuming that high authority users are more likely to post salient tweets. Others [7] model the tweet-level relationship through pairwise friends (i.e., one-hop neighbors) to alleviate the redundancy caused by directly connected users, yet ignoring the deeper user relationships.

Actually, salient tweets have a deeper spread on social networks and thus affecting more nodes in the network. Only considering one-hop neighbors is obviously insufficient since redundant information not only exists between adjacent users, but may also exist between users who are not directly connected. Hence the user interactions with the deeper neighborhood scope also bring the valuable clues for summarization.

In this paper, we try to explore the redundancy problem by introducing **social contagion** and **hashtag consistency** theories. Specifically, we model the relationship between tweets from two different perspectives: social interaction and hashtag co-occurrence. We construct two social relation graphs according to these two kinds of relations, based on which we verify the existence of these two social theories. We further propose an unsupervised social summarization model which leverages **S**ocial **C**ontagion and **H**ashtag **C**onsistency (**SCHC**) to address user-level and hashtag-level redundancy simultaneously. In summary, the main contributions include:

- Propose to model tweet relationships from two different perspectives (i.e. social interaction and hashtag co-occurrence) respectively, and construct two tweet relation graphs accordingly.
- Verify the existence of social contagion and hashtag consistency phenomenon in real world social media data based on the constructed social relation graphs;
- Based on social contagion and hashtag consistency, we incorporate the two kinds of tweet relationships into a group sparse reconstruction framework to address the user-level and hashtag-level redundancy problem of social summaries simultaneously.
- Our model outperforms the SOTA model on CTS dataset in both automatic and human evaluation, which proves the effectiveness of the proposed SCHC model.

2 Related Work

2.1 Social Summarization

Most content-based social summarization approaches are based on traditional document summarization techniques [13] or extended versions. [20] proposes a PageRank based model to extract event summaries using status updates in the sports domain. [32] extends the TextRank algorithm through extracting bigrams from the relevant tweets. [9] and [16] propose to conduct multi-document summarization from the data reconstruction perspective. However, these methods only consider text information. Some static social features such as user influence [18], the retweet count [3], temporal signal [8] and user behaviors in conveying relevant content [19] have been proved to be useful for social summarization. Nonetheless, they ignore that social data is networked through social users' connections. Recently, [7] regards social summarization as an unsupervised sparse reconstruction optimization problem through integrating pair-wise social relations (i.e., one-hop neighbors). Our method tries to explore the deeper user behaviors which may contain more valuable summarization clues.

2.2 Social Network Influence

Social network influence also known as social propagation or network influence has been researched in several domains, such as topic detection [1], topic identification [30], and network inference. Existing studies are based on social users or tweets. [10] studies the social user influence based on two-way reciprocal relationship prediction. Socialized languages models are developed to search problems

Table 1. The CTS dataset statistics.

Topics	Osama	Mavs	Casey	Obama	Oslo	SDCC
# Of Tweets	4780	3859	6241	4888	4571	5817
# Of Users	1309	1780	1318	2009	1026	442
# Of Hashtags	7634	8393	10791	8264	6469	19664
Max degree of users	69	76	74	142	77	81
Min degree of users	1	1	1	1	1	2
P-value (Second-order contagion)	2.83E-160	2.54E-92	3.98E-165	8.33E-05	3.38E-08	1.23E-07
P-value (Hashtag consistency)	9.36E-28	4.14E-30	8.20E-22	2.65E-65	3.82E-37	6.33E-36

[23,28]. [33] proposes to deal with phenomena of social language use and demonstrates that social media language understanding requires the help of social networks. In this article, we will explore how the spread of social media content in social networks influences social summarization.

2.3 Hashtags in Social Media

Hashtags are keyword-based tags provided by social media services, through which users can insert topic information into posts conveniently. They describe the semantic of tweets and assist in understanding short social texts from the topic perspective [31]. Hashtag mining researches include hashtag popularity prediction [27], sentiment analysis [29], hashtag diffusion [22] and so on. [26] proposes a learning-to-rank approach for modeling hashtag relevance to address the real-time recommendation of hashtags to streaming news articles. [24] creates a dataset named HSpam14 for hashtag-oriented spam filtering in tweets. In this paper, we model the tweet-hashtag relations to capture the hashtag-level redundant information by introducing hashtag consistency into social summarization.

3 Task Description and Observations

3.1 Task and Dataset

Given a collection of n tweets about a topic denoted as $S = [s_1, s_2, ..., s_n]$ where s_i represents the i-th tweet, it is represented as a weighted term frequency inverse tweet frequency matrix, denoted as $X = [x_1, x_2, \ldots, x_n] \in R^{m \times n}$, where m is the size of vocabulary, n is the total number of tweets. Each column $x_i \in \mathbb{R}^{m \times 1}$ stands for a tweet vector. The goal of social summarization is to extract l summary tweets where $l \ll n$.

We use the CTS corpus provided in [7] as our dataset to validate the proposed method. It contains 12 popular topics happened in May, June and July 2012, including various topics such as politics, science and technology, sports, natural disasters, terrorist attacks and entertainment gossips, whose raw data comes from the public Twitter data collected by University of Illinois. Each topic has 25 tweets individually selected by four volunteers as a golden summary respectively, altogether 48 expert summaries. Due to the limited space, we show the statistics of 6 topics in Table 1.

3.2 Verification of Sociological Theories

Social theories indicate users often exhibit correlated behavior due to network diffusion, which have been proved to be beneficial for social media mining. Contagion means that friends can influence each other [12,25]. Consistency demonstrates that social behaviors conducted by the same person tend to keep consistent in a short period of time [21]. Expression consistency and contagion are observed in the CTS corpus [7]. However, they only verify the contagion theory between pair-wise users. In our work, we further investigate higher-order contagion phenomenon and hashtag consistency theory in CTS corpus, which are defined as follows:

- **Higher-order contagion**: Whether the tweets posted by users with the common first-order neighbors are more similar than two randomly selected tweets?
- **Hashtag consistency**: Whether the semantics of two tweets sharing the same hashtags are more consistent than two randomly selected tweets?

To verify the above two assumptions, we use cosine similarity to measure the distance between two tweets: $D_{ij} = cos(x_i, x_j)$, where x_i and x_j denote the vector representation of the i-th and the j-th tweet respectively. The more similar the two tweets are, the more D_{ij} tends to be 1. For the first observation, we define the vector $cont_c$ as the distance of two tweets posted by users with the common first-order neighbor, and the vector $cont_r$ as the distance of two randomly selected tweets. Then we conduct the two sample t-test on the two vectors $cont_c$ and $cont_r$. The null hypothesis $H_0 : cont_c = cont_r$, shows that there is no difference between two tweets posted by users with the common first-order neighbor and those randomly selected tweets. The alternative hypothesis $H_1 : cont_c < cont_r$, shows that the distance between two tweets posted by users with the common first-order neighbor is shorter than those randomly selected tweets.

Similarly, to ask the second observation, we conduct two sample t-test where null hypothesis $H_0 : conh_c = conh_r$ means there is no difference between tweets with the same hashtags and those randomly selected tweets. The alternative hypothesis $H_1 : conh_c < conh_r$ means the distance between tweets with the same hashtags is shorter than that of randomly selected ones. For all the topics, the second-order contagion null hypothesis and the hashtag consistency null hypothesis are rejected at significance level $\alpha = 0.01$. Due to the space constraint, we only display the p-values of 6 topics in the last two rows of Table 1.

4 Our Method

4.1 Coverage and Sparsity

Essentially, reconstruction and summarization have the similar objective. We regard social summarization as an issue of sparse reconstruction problem, whose intuition is that the selected summary tweets should be able to reconstruct the original corpus. Meanwhile, since not all tweets have an contribution on reconstructing one certain tweet, given the original corpus X, we regard each tweet as a group and formulate this task as a reconstruction process with group sparse constraint:

$$\min_{W} \frac{1}{2} \|X - XW\|_F^2 + \lambda \|W\|_{2,1} \quad s.t. \quad W \geq 0,\ diag(W) = 0 \tag{1}$$

$$\|W\|_{2,1} = \sum_{i=1}^{n} \|W(i,:)\|_2 \tag{2}$$

where $W = [W_{*1}, W_{*2}, \ldots, W_{*n}] \in R^{n \times n}$ is the reconstruction coefficient matrix. The element w_{ij} in W denotes the importance of the i-*th* tweet to the j-*th* tweet. The second term is $l_{2,1}$ -norm regularization on weight matrix W, which causes some of coefficients to be exactly zero. The λ denotes the weight of sparse regularization term. Here an additional inequality constraint is introduced to avoid the numerically trivial solution ($W \approx I$) in practice, by forcing the diagonal elements to be zeros. Hence, our goal is to learn an optimal reconstruction coefficient matrix W and use it to produce a summary accordingly.

4.2 Social Contagion

The frequent interaction and information propagation of active users on social media bring strong redundant information. To alleviate user-level redundancy, we propose to model multi-order social relationship from the perspective of user interactions by transforming the user-tweet relations and user-user relations into tweet-tweet correlation graph. Formally, let $U \in \mathbb{R}^{n \times d}$ denotes the adjacency matrix of tweet-user graph, where $U_{i,j} = 1$ denotes that the i-th tweet is posted by the j-th user and d is the number of users. $F \in \mathbb{R}^{d \times d}$ represents the adjacency matrix of user-user graph, where $F_{i,j} = 1$ indicates that user u_i is connected with user u_j. Then, the self-contagion, first-order and higher-order contagion are defined as follows respectively:

Self-contagion means that social behaviors conducted by the same user keep consistent in a short period, and can be defined as Eq. (3):

$$T^{self} = U \times U^T \tag{3}$$

where $\cdot^T$ indicates the transpose operation. $T^{self} \in \mathbb{R}^{n \times n}$ represents tweet-tweet matrix for self-contagion and $T_{i,j}^{self} > 0$ indicates that tweet i and tweet j are generated by the same user. This relation graph is used to alleviate intra-user redundancy.

First-order contagion discloses that tweets posted by friends have a higher probability to be similar than those randomly selected [25], and can be defined as Eq. (4):

$$T^{first} = U \times F \times U^T \tag{4}$$

$T^{first} \in \mathbb{R}^{n \times n}$ represents tweet-tweet matrix for first-order contagion, where $T^{first}_{i,j} > 0$ denotes that tweet i and tweet j are published by directly connected friends. This is used to avoid inter-user redundancy.

Higher-order Contagion demonstrates that users who are not directly connected in network but share common friends have a higher probability to share similar content, which can also cause inter-user level redundancy. Hence, we further model the deeper user behaviors to alleviate inter-user level redundancy. The higher-order social relation graph is constructed as Eq. (5) and Eq. (6):

$$S = F \times F^T \tag{5}$$

$$T^{second} = U \times S \times U^T \tag{6}$$

where $S \in \mathbb{R}^{d \times d}$ represents the indirect relationship between users, where $S_{ij} = k$ represents that there are k paths from u_i to u_j whose length equals to 2. $T^{second}_{i,j} \in \mathbb{R}^{n \times n}$ represents tweet-tweet matrix for second-order contagion, where $T^{second}_{i,j} > 0$ denotes that tweet i and tweet j are generated by those potentially connected users.

Finally, we integrate the above different levels of social contagion into a unified relation graph as Eq. (7):

$$T = T^{self} + d_1 T^{first} + d_2 T^{second} \tag{7}$$

where d_1 and d_2 are parameters of T^{first} and T^{second} respectively. Hence, the social regularization can be mathematically formulated as Eq. (8):

$$\begin{aligned} R_{social} &= \frac{1}{2} \sum_{i=1}^{n} \sum_{j=1}^{n} T_{ij} \left\| \hat{X}_{*i} - \hat{X}_{*j} \right\|_F \\ &= \sum_{i=1}^{m} \hat{X}_{i*} \| D_{social} - T \|_F \hat{X}_{i*}^T \\ &= tr \left(X W \mathcal{L}_{social} W^T X^T \right) \end{aligned} \tag{8}$$

where $\hat{X}$ denotes the reconstruction matrix of X ($\hat{X} = XW$), $tr(\cdot)$ denotes the trace of matrix, $\mathcal{L}_{social} = D_{social} - T$ is the laplacian matrix. The $D_{social} \in \mathbb{R}^{n \times n}$ is the degree matrix of T with $D_{ii} = \sum_{j=1}^{n} T_{ij}$ indicating the degree of the i-th tweet in the combined relation graph.

4.3 Hashtag Consistency

To address the hashtag-level redundancy, we capture relations between tweets from the view of hashtags. Intuitively, tweets sharing more same hashtags tend to share more similar content. Therefore, we construct a tweet relation graph according to hashtag co-occurrence relationship. Formally, given a tweet-hashtag relation matrix denoted as $P \in \mathbb{R}^{n \times t}$, where $P_{ij} = 1$ indicates that i-th tweet contains the j-th hashtag, and t is the total number of hashtags. We transform the tweet-hashtag relations into a tweet-tweet hashtag co-occurrence relation graph as follows:

$$M = P \times P^T \tag{9}$$

where $M \in \mathbb{R}^{n \times n}$ represents the adjacency matrix of hashtag co-occurrence relation graph among tweets. Each element M_{ij} denotes the number of common hashtags shared by tweet x_i and x_j. The hashtag consistency regularization denoted as $R_{hashtag}$ is further modeled as Eq. (10):

$$\begin{aligned} R_{hashtag} &= \frac{1}{2}\sum_{i=1}^{n}\sum_{j=1}^{n} M_{ij} \left\| \hat{X}_{*i} - \hat{X}_{*j} \right\|_F \\ &= \sum_{i=1}^{m} \hat{X}_{i*} \| D_{hashtag} - M \|_F \hat{X}_{i*}^T \\ &= tr\left(XW\mathcal{L}_{hashtag}W^T X^T\right) \end{aligned} \tag{10}$$

where $D_{hashtag}$ and $\mathcal{L}_{hashtag}$ are degree matrix and laplacian matrix of M respectively. Intuitively, the more hashtags two tweets share, the more similar their content tend to be. Hashtag consistency could make the model select tweets with more various hashtags and further help to avoid redundancy.

To further avoid redundancy, we utilize a diversity regularization with a relatively simple but effective cosine similarity matrix ∇ to prevent tweets from being reconstructed by those tweets pretty similar to them. $\nabla_{ij} \in [0, 1]$ denotes the cosine similarity between tweet x_i and tweet x_j.

$$\nabla_{ij} = \begin{cases} 1 & \text{if} \quad \nabla_{ij} \geq \theta \\ 0 & \text{otherwise} \end{cases} \tag{11}$$

$$tr(\nabla^T W) = \sum_{i=1}^{n}\sum_{j=1}^{n} \nabla_{ij} W_{ij} \tag{12}$$

where θ is the similarity threshold. Through integrating social, hashtag and diversity regularizations, the tweet reconstruction objective function Eq. (1) can be transformed as:

$$\begin{aligned} \min_W f(W) = &\ \frac{1}{2}\|X - XW\|_F^2 + \lambda\|W\|_{2,1} \\ &+ \alpha R_{social} + \beta R_{hashtag} + \gamma tr(\nabla^T W) \\ s.t. \quad & W \geq 0, diag(W) = 0 \end{aligned} \tag{13}$$

where α, β and γ are the corresponding parameters to control the contributions of three regularization terms. Finally, we select tweets according to the ranking score of the tweet x_i that is calculated as Eq. (14) by solving Eq. (13) to form the social summary.

$$score(x_i) = \|W(i,:)\|_2 \tag{14}$$

4.4 Sparse Optimization for Social Summarization

Inspired by [6], we present an efficient algorithm to solve this optimization problem. The objective function can be reformulated as:

$$\begin{aligned} \min_{W} f(W) = \frac{1}{2}\|X - XW\|_F^2 + \gamma tr(\nabla^T W) \\ + \alpha R_{social} + \beta R_{hashtag} \\ s.t. \quad W \geq 0, diag(W) = 0, \|W\|_{2,1} \leq z \end{aligned} \tag{15}$$

where $\mathcal{Z} = \{\|W\|_{2,1} \leq z\}$, $z \geq 0$ is the radius of the $\ell_{2,1}$−ball, and there is a one-to-one correspondence between λ and z.

Due to the limited space, we omit the details of our mathematical derivations. Interested readers may reference [11] and SLEP package[1] (Sparse Learning with Efficient Projects). The entire algorithm is described in Algorithm 1, where

$$S_{\epsilon/lr_t}(U_t[j,*]) = max(1 - \frac{\epsilon}{lr\|U_t[j,*]\|_2}, 0)U_t[j,*] \tag{16}$$

$$G_{lr_t,V_t}(W) = f(V_t) + \langle \nabla f(W_t), W - V_t \rangle + \frac{lr_t}{2}\|W - V_t\|_F^2 \tag{17}$$

and $U_t[j,*]$ is the j-th row of U_t.

The convergence rate of Algorithm 1 is elaborated in the following theorem.

Theorem 1. Assuming that W_t is the reconstruction coefficient matrix generated by Algorithm 1. Then for any $t \geq 1$ we have:

$$f(W_t) - f(W^*) \leq \frac{2\hat{L}_f\|W^* - W_1\|_F^2}{(t+1)^2} \tag{18}$$

where $W^* = \arg\min_W f(W)$, $\hat{L}_f = \max(2L_f, L_0)$, L_0 is an initial guess of the lipschitz continuous gradient L_f of $f(W)$ and W^* is the solution of $f(W)$. This theorem shows that the convergence rate of Algorithm 1 is $O\left(\frac{1}{\sqrt{\varepsilon}}\right)$, where ε is the desired accuracy. Finally, we discuss time complexity of the proposed method SCHC. First, for the reconstruction with group sparse Eq. (1) , it costs $O(n^2m + n^2)$ floating point operations for calculating the function value and gradient of the objective function. Second, for the regularization R_{social} and $R_{hashtag}$, the time complexity is $O(2n^2m)$. Third, for the diversity regularization part, the time complexity is $O(n^2)$ based on the Euclidean projection algorithm. Therefore, we can solve the objective function in Eq. (13) with a time complexity of $O\left(\frac{1}{\sqrt{\varepsilon}}\left(n^2m + 2n^2m + 2n^2\right)\right) \approx O\left(\frac{1}{\sqrt{\varepsilon}}\left(n^2m\right)\right)$.

[1] http://www.yelab.net/software/SLEP/.

Algorithm 1. A Sparse Optimization Algorithm for SCHC

Input: $\{X, T, M, U, \nabla, W_0, \alpha, \beta, \gamma, \varepsilon, lr_1\}$
Output: W

1: Initialize $\mu_0 = 0, \mu_1 = 1, W_1 = W_0, t = 1$
2: Construct Laplacian matrix $\mathcal{L}$ from T and M
3: **while** Not convergent **do**
4: Set $V_t = W_t + \frac{\mu_{t-1}-1}{\mu_t}(W_t - W_{t-1})$
5: $\frac{\partial f(W_t)}{\partial W_t} = X^T X W_t - X^T X + \gamma \nabla + \alpha X^T X W_t L_{social} + \beta X^T X W_t L_{hashtag}$
6: **loop**
7: Set $U_t = V_t - \frac{1}{lr_t}\frac{\partial f(W_t)}{\partial W_t}$
8: **for** each row $U_t[j,*]$ in U_t
9: Set $W_{t+1}[j,*] = S_{\epsilon/lr_t}(U_t[j,*])$
10: **end for**
11: **if** $f(W_{t+1}) \le G_{lr_t, V_t}(W_{t+1})$ **then**
12: $lr_{t+1} = lr_t$, break
13: **end if**
14: $lr_t = 2 \times lr_t$
15: **end loop**
16: $W = W_{t+1}$
17: **if** stopping criteria satisfied **then**
18: break
19: **end if**
20: Set $\mu_{t+1} = \frac{1+\sqrt{1+4\mu_t}}{2}$
21: Set $t = t + 1$
22: **end while**
23: $W = W_{t+1}$

5 Experiments

5.1 Research Questions

To verify the effectiveness of the proposed SCHC model, we list the following research questions to guide our experiments: **RQ1**: What is the overall performance of SCHC? Does it outperform other baselines? **RQ2**: Is our model effective in human evaluation experiments? **RQ3**: What is the generalization ability of our method on different topics? **RQ4**: What's the influence of each core component and how does it affect the performance? **RQ5**: How to determine the optimal parameters setting?

5.2 Evaluation Metrics

As shown in Sect. 3, we use the CTS dataset [7] and set the number of system summary tweets for each topic as 25, which is consistent with that of expert summary. We adopt (1) automatic metrics [15] and (2) human evaluations to validate the proposed method. For the automatic evaluation, we use ROUGE-N to evaluate the generated summaries. In our experiments, we report the F-measure of

ROUGE-1, ROUGE-2, ROUGE-L, which respectively measures word-overlap, bigram-overlap, and the longest common subsequence between system output and reference summaries. We also report the ROUGE-SU* which measures the overlapping of skip-bigrams. For the human evaluation, we ask three highly-educated volunteers to score system summary based on coverage, informativeness and diversity in range [1,5] individually under the condition of reading expert summary. The averaged scores are reported in our experiments. In order to obtain the consistency of human evaluations, we also calculate the Fleiss' kappa [14] among volunteers.

5.3 Compared Methods

Considering the SCHC is an unsupervised extractive summarization model, we only compare with the relevant systems, including the following traditional text and sparse reconstruction based methods: (1) **Expert** averages mutual assessment of expert summaries. This is the upper bound of the human summary. (2) **Random** selects tweets randomly to construct summaries. (3) **LexRank** ranks tweets by the PageRank like algorithm [4]. (4) **LSA** exploits SVD to decompose the TF-IDF matrix and then selects the highest-ranked tweets from each singular vector [5]. (5) **MDS-Sparse** presents a two-level sparse reconstruction model for multi-document summarization [16]. (6) **DSDR** represents each sentence as a non-negative linear combination of the summary sentences under a sparse reconstruction framework [9]. (7) **SNSR** proposes an approach for social summarization by integrating social network and sparse reconstruction [7].

6 Experimental Results

6.1 The Overall Performance

Automatic Evaluation. To start, we address the research question **RQ1**. The performance of the model on each topic is shown in Fig. 1, where x axis is the keyword of each topic. The average performance on the dataset is shown in Table 2, from which we have the following observations:

1) Our method SCHC consistently outperforms the other baseline methods and is below the upper bound. The performance gaps between the SCHC and the expert summary are 2.07%, 1.93%, 1.97%, 1.39% in terms of ROUGE-1, ROUGE-2, ROUGE-L and ROUGE-SU*. And we also notice the reconstruction-based methods are better than traditional approaches. One of the reasons may be that reconstruction-based methods can better integrate the deeper user behaviors and network structure information, and hence achieve the higher content coverage.

2) The SCHC achieves 1.00%, 1.07%, 0.80% and 0.86% increment over the state-of-the-art method SNSR in terms of ROUGE-1, ROUGE-2, ROUGE-L and ROUGE-SU*. The reasons mainly come from two aspects: **(i)** Social contagion captures both direct and latent influence among users to address the intra-user and inter-user level redundancy caused by information propagation. **(ii)** Hashtag

Table 2. ROUGE scores comparison between baselines

Methods	ROUGE-1	ROUGE-2	ROUGE-L	ROUGE-SU*
Expert	0.47388	0.15973	0.45114	0.21555
Random	0.41574	0.09440	0.39227	0.16596
LexRank	0.42132	0.13302	0.39965	0.18192
LSA	0.43524	0.13077	0.41347	0.18197
MDS-Sparse	0.42119	0.10059	0.40101	0.16686
DSDR	0.43335	0.13106	0.41055	0.17264
SNSR	0.44886	0.13891	0.42800	0.19990
SCHC	**0.45318**	**0.14040**	**0.43143**	**0.20162**

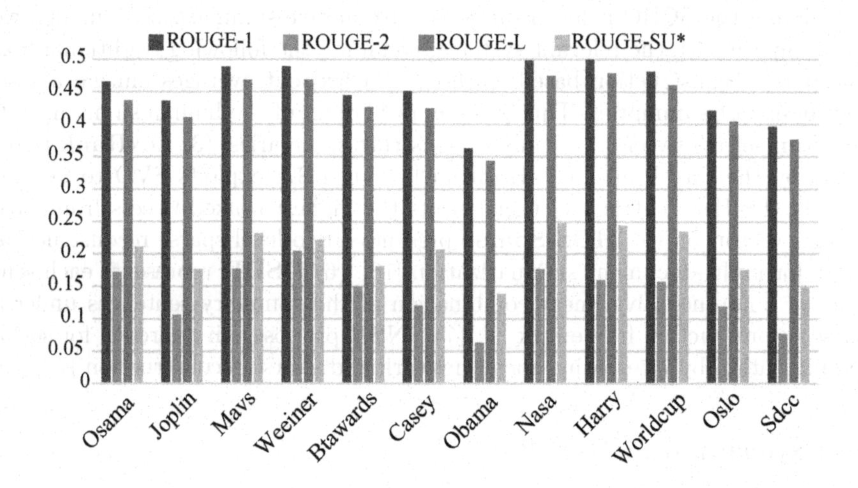

Fig. 1. The ROUGE performance of SCHS on twelve topics.

consistency can capture hashtag level redundancy among tweets. Actually, it acts like a cluster, which forces tweets that share the same hashtags to be closer so as to more accurately reconstruct the original content.

3) It is necessary to point out that all baselines achieve a good performance, especially for ROUGE-1. One of the reasons is that the corpus is topic-oriented, and most of the tweets is about a specific topic, which leads to a relatively large overlap of words. It also may be due to that we conduct an effective preprocessing.

Human Evaluation. To address the research question **RQ2**, we choose LSA and SNSR as baselines since their performance is relatively higher than other methods. The result is shown in Table 3. We make the following observations:

1) The SCHC and SNSR methods perform better than LSA, especially in terms of coverage. This maybe because the goal of reconstruction is to select essential features from the data to reduce reconstruction error and thus the

Table 3. Human evaluation comparison with baselines.

	Coverage	Informativeness	Diversity	Kappa
LSA	2.33	2.41	2.25	0.48
SNSR	3.29	2.86	2.56	0.56
SCHC	**3.86**	**3.32**	**3.01**	0.58

selected features could cover the original data as much as possible from multiple aspects.

2) The SCHC achieves the higher informativeness and diversity scores than baselines. This maybe because social contagion is able to address both intra-user and inter-user level redundancy, while hashtag consistency can force the model to filter redundant hashtags and select more diverse tweets.

3) To prove the consistency of human evaluation, we also conduct the Fleiss' kappa statistics, which demonstrates that the volunteers have a moderate agreement [14].

Next, we turn to the research question **RQ3**. Figure 2 displays the ROUGE scores of our SCHC method under twelve different topics. The horizontal axis indicates the abbreviation of the topic name. It is worth noting that the ROUGE indicators under these twelve topics are comparable. Based on the above statistics and analysis, our SCHC method has strong robustness and generalization ability for different topics.

6.2 Ablation Study

Then, we turn to **RQ4**. The results of ablation experiments are shown in Fig. 2, through which we have the following observations:

(1) Removing any component will degrade system performance, which demonstrates the necessity of each component in SCHC. The total decrease caused by removing hashtag consistency (-T) or social contagion (-S) respectively, is less than the decrease caused by removing these two components at the same time(-T-S). This shows that hashtag consistency and social contagion are not independent, and can promote each other.

(2) The performance drops sharply when removing the entire social contagion item. This indicates that different level of social relations can influence the content reconstruction and help to address both intra-user and inter-user redundancy. This makes sense since information travels along the social network and affects the users it passes through. Meanwhile, the drop in scores from SCHC to SCHC-T shows that hashtag consistency can alleviate hashtag level redundancy to promote the performance.

(3) When removing second-order contagion from SCHC-T, the scores of ROUGE-1, ROUGE-2, ROUGE-L, ROUGE-SU* drop by 0.904%, 1.590%, 0.985%, 0.972% respectively, which demonstrates the importance of second-order

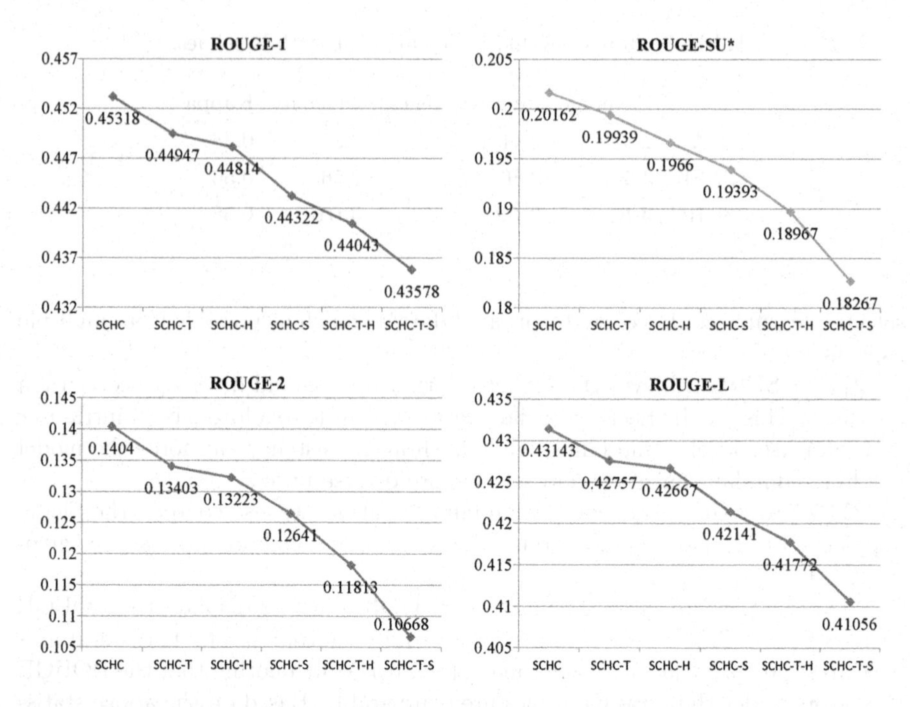

Fig. 2. Performance of ROUGE for different ablation models. -T, -H and -S denote deleting hashtag consistency, second-order social contagion and social contagion respectively.

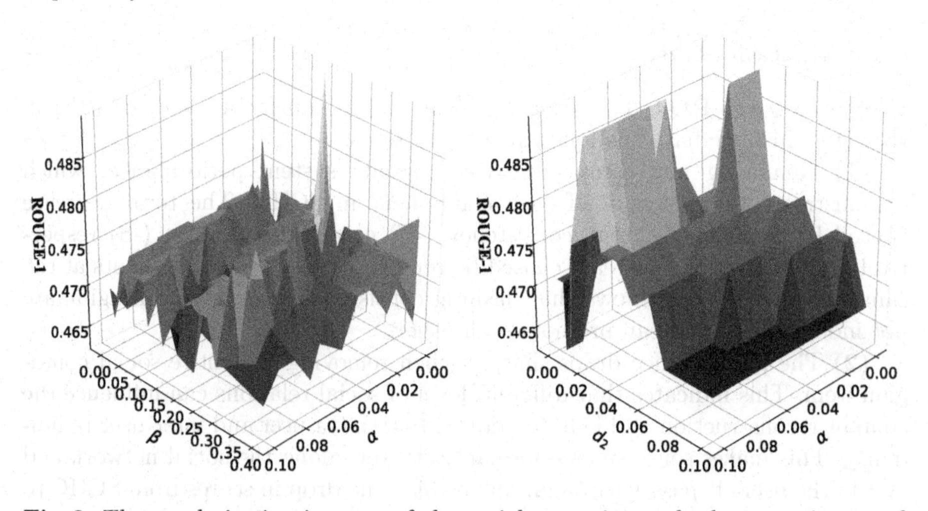

Fig. 3. The regularization impacts of the social contagion α, hashtag consistency β and second-order social contagion d_2.

contagion. Besides, the difference between SCHC-H and SCHC-S demonstrates the emphasis of self-contagion and first-order relation among users.

6.3 Parameters Settings and Analysis

Finally, we turn to the research question **RQ5**. In our experiment, we mainly focus on second-order contagion and hashtag consistency to explore how these signals influence the content of tweets. Therefore, we empirically fix $\gamma = 1, \lambda = 1$ in Eq. 13, $\theta = 0.1$ in Eq. 11 and $d_1 = 1$ in Eq. 7, and search for the optimal values of α, β in Eq. 13 and d_2 in Eq. 7 using grid search. To investigate the sensitivity of the model to the important parameters α, β, d_2, we further visualize the relationships between performance and these factors on one topic (weiner), as shown in Fig. 3. It can be seen that the model performs best when d_2 is around 0.003. With the increase of d_2, the performance drops sharply. Finally, we set $\alpha = 0.01, \beta = 0.2$ and $d_2 = 0.003$.

7 Conclusion

The propagation of tweets and interaction between users cause serious redundant information on social media, making it challenging to generate succinct summaries for social text. In this paper, we propose to leverage social theories including multi-order social contagion and hashtag consistency to alleviate the serious redundancy problem on social media data. The two social assumptions are verified on real world corpus, based on which we construct two kinds of social relation graphs to capture the relationship between tweets from the perspective of user interaction and hashtag co-occurrence. These relations are further integrated into the sparse reconstruction framework to address the intra-user, inter-user and hashtag-level redundancy at the same time. Experiments are conducted on CTS dataset and the results demonstrate the effectiveness of our model.

Acknowledgement. We thank the anonymous reviewers for their valuable feedback. Our work is supported by the National Natural Science Foundation of China (61976154), the National Key R&D Program of China (2019YFC1521200), the Tianjin Natural Science Foundation (18JCYBJC15500), and the State Key Laboratory of Communication Content Cognition, People's Daily Online (No.A32003).

References

1. Bi, B., Tian, Y., Sismanis, Y., Balmin, A., Cho, J.: Scalable topic-specific influence analysis on microblogs. In: Proceedings of the 7th ACM International Conference on Web Search and Data Mining, pp. 513–522 (2014)
2. Chang, Y., Tang, J., Yin, D., Yamada, M., Liu, Y.: Timeline summarization from social media with life cycle models. In: Proceedings of the 25th International Joint Conference on Artificial Intelligence, pp. 3698–3704 (2016)
3. Duan, Y., Chen, Z., Wei, F., Zhou, M., Shum, H.Y.: Twitter topic summarization by ranking tweets using social influence and content quality. In: Proceedings of the 24th International Conference on Computational Linguistics, pp. 763–780 (2012)

4. Erkan, G., Radev, D.R.: Lexrank: graph-based lexical centrality as salience in text summarization. J. Artif. Intell. Res. **22**(1), 457–479 (2004)
5. Gong, Y., Liu, X.: Generic text summarization using relevance measure and latent semantic analysis. In: Proceedings of the 24th Annual International ACM SIGIR Conference on Research and Development in Information Retrieval, pp. 19–25 (2001)
6. Gu, Q., Han, J.: Towards feature selection in network. In: Proceedings of the 20th ACM International Conference on Information and Knowledge Management, pp. 1175–1184 (2011)
7. He, R., Duan, X.: Twitter summarization based on social network and sparse reconstruction. In: Proceedings of the 32th AAAI Conference on Artificial Intelligence, pp. 5787–5794 (2018)
8. He, R., Liu, Y., Yu, G., Tang, J., Hu, Q., Dang, J.: Twitter summarization with social-temporal context. World Wide Web **20**(2), 267–290 (2016). https://doi.org/10.1007/s11280-016-0386-0
9. He, Z., et al.: Document summarization based on data reconstruction. In: Proceedings of the 26th AAAI Conference on Artificial Intelligence, pp. 620–626 (2012)
10. Hopcroft, J.E., Lou, T., Tang, J.: Who will follow you back? reciprocal relationship prediction. In: Proceedings of the 20th ACM International Conference on Information and Knowledge Management, pp. 1137–1146 (2011)
11. Hu, X., Tang, L., Tang, J., Liu, H.: Exploiting social relations for sentiment analysis in microblogging. In: Proceedings of the 6th ACM International Conference on Web Search and Data Mining, pp. 537–546 (2013)
12. Iacopini, I., Petri, G., Barrat, A., Latora, V.: Simplicial models of social contagion. Nature Commun. **10**(1), 2485 (2019)
13. Inouye, D., Kalita, J.K.: Comparing twitter summarization algorithms for multiple post summaries. In: 2011 IEEE Third international conference on privacy, Security, Risk and Trust and 2011 IEEE Third International Conference on Social Computing, pp. 298–306 (2011)
14. Landis, Koch, G.G.: The measurement of observer agreement for categorical data. Biometrics **33**(1), 159–174 (1977)
15. Lin, C.Y.: ROUGE: A package for automatic evaluation of summaries. In: Workshop on Text Summarization Branches Out, Post-Conference Workshop of ACL, vol. 2004, pp. 74–81 (2004)
16. Liu, H., Yu, H., Deng, Z.H.: Multi-document summarization based on two-level sparse representation model. In: Proceedings of the 29th AAAI Conference on Artificial Intelligence, pp. 196–202 (2015)
17. Liu, X., Li, Y., Wei, F., Zhou, M.: Graph-based multi-tweet summarization using social signals. In: Proceedings of the 24th International Conference on Computational Linguistics, pp. 1699–1714 (2012)
18. Nguyen, M., Tran, D., Nguyen, L., Phan, X.: Exploiting user posts for web document summarization. ACM Trans. Knowl. Discovery Data **12**(4), 49 (2018)
19. Nguyen, M., Tran, V.C., Nguyen, X.H., Nguyen, L.: Web document summarization by exploiting social context with matrix co-factorization. Inf. Process. Manage. **56**(3), 495–515 (2019)
20. Nichols, J., Mahmud, J., Drews, C.: Summarizing sporting events using twitter. In: Proceedings of the 2012 ACM International Conference on Intelligent User Interfaces, pp. 189–198 (2012)
21. Abelson, R.P.: Whatever Became of Consistency Theory?. Pers. Soc. Psychol. Bull. **9**(1), 37–54 (1983)

22. Romero, D.M., Meeder, B., Kleinberg, J.: Differences in the mechanics of information diffusion across topics: idioms, political hashtags, and complex contagion on twitter. In: Proceedings of the 20th International Conference on World Wide Web, pp. 695–704 (2011)
23. Rui, Y., Li, C.T., Hsieh, H.P., Hu, P., Hu, X., He, T.: Socialized language model smoothing via bi-directional influence propagation on social networks. In: Proceedings of the 25th International Conference on World Wide Web, pp. 1395–1406 (2016)
24. Sedhai, S., Sun, A.: Hspam14: A collection of 14 million tweets for hashtag-oriented spam research. In: Proceedings of the 38th International ACM SIGIR Conference on Research and Development in Information Retrieval, pp. 223–232 (2015)
25. Shalizi, C.R., Thomas, A.C.: Homophily and contagion are generically confounded in observational social network. Sociol. Methods Res. **40**(2), 211–239 (2011)
26. Shi, B., Ifrim, G., Hurley, N.: Learning-to-rank for real-time high-precision hashtag recommendation for streaming news. In: Proceedings of the 25th International Conference on World Wide Web, pp. 1191–1202 (2016)
27. Tsur, O., Rappoport, A.: What's in a hashtag? content based prediction of the spread of ideas in microblogging communities. In: Proceedings of the Fifth ACM International Conference on Web Search and Data Mining, pp. 643–652 (2012)
28. Vosecky, J., Leung, K.W.T., Ng, W.: Collaborative personalized twitter search with topic-language models. In: Proceedings of the 37th International ACM SIGIR Conference on Research & Development in Information Retrieval, pp. 53–62 (2014)
29. Wang, X., Wei, F., Liu, X., Zhou, M., Zhang, M.: Topic sentiment analysis in twitter: a graph-based hashtag sentiment classification approach. In: Proceedings of the 20th ACM International Conference on Information and Knowledge Management, pp. 1031–1040 (2011)
30. Wang, X., Wang, Y., Zuo, W., Cai, G.: Exploring social context for topic identification in short and noisy texts. In: Proceedings of the 29th AAAI Conference on Artificial Intelligence, pp. 1868–1874 (2015)
31. Wang, Y., Li, J., King, I., Lyu, M.R., Shi, S.: Microblog hashtag generation via encoding conversation contexts. In: Proceedings of the 2019 Conference of the North American Chapter of the Association for Computational Linguistics: Human Language Technologies, pp. 1624–1633 (2019)
32. Wu, Y., Zhang, H., Xu, B., Hao, H., Liu, C.: TR-LDA: a cascaded key-bigram extractor for microblog summarization. Int. J. Mach. Learn. Comput. **5**(3), 172–178 (2015)
33. Zeng, Z., Yin, Y., Song, Y., Zhang, M.: Socialized word embeddings. In: Proceedings of the Twenty-Sixth International Joint Conference on Artificial Intelligence, pp. 3915–3921 (2017)

Image-Enhanced Multi-Modal Representation for Local Topic Detection from Social Media

Junsha Chen[1,2,3], Neng Gao[1,3], Yifei Zhang[1,2,3], and Chenyang Tu[1,3](✉)

[1] State Key Laboratory of Information Security, Chinese Academy of Sciences, Beijing, China
[2] School of Cyber Security, University of Chinese Academy of Sciences, Beijing, China
[3] Institute of Information Engineering, Chinese Academy of Sciences, Beijing, China
{chenjunsha,gaoneng,zhangyifei,tuchenyang}@iie.ac.cn

Abstract. Detecting local topics from social media is an important task for many applications, ranging from event tracking to emergency warning. Recent years have witnessed growing interest in leveraging multi-modal social media information for local topic detection. However, existing methods suffer great limitation in capturing comprehensive semantics from social media and fall short in bridging semantic gaps among multi-modal contents, i.e., some of them overlook visual information which contains rich semantics, others neglect indirect semantic correlation among multi-modal information. To deal with above problems, we propose an effective local topic detection method with two major modules, called IEMM-LTD. The first module is an image-enhanced multi-modal embedding learner to generate embeddings for words and images, which can capture comprehensive semantics and preserve both direct and indirect semantic correlations. The second module is an embedding based topic model to detect local topics represented by both words and images, which adopts different prior distributions to model multi-modal information separately and can find the number of topics automatically. We evaluate the effectiveness of IEMM-LTD on two real-world tweet datasets, the experimental results show that IEMM-LTD has achieved the best performance compared to the existing state-of-the-art methods.

1 Introduction

With the rapid development of social media (e.g., Twitter), more and more people tend to generate and share rich multi-modal contents online. As shown in Fig. 1, both tweets contain multi-modal information, which can reflect semantics from different aspects. In real-world scenarios, most of these multi-modal contents uploaded by users are related to some specific topics, therefore, detecting topics from social media has attracted extensive research interests. Compared to global topics, local topics happen in a local area and during a certain period

C. S. Jensen et al. (Eds.): DASFAA 2021, LNCS 12682, pp. 658–674, 2021.
https://doi.org/10.1007/978-3-030-73197-7_45

Text: By finalists do you mean runners up to DJ LeMahieu who is the AL MVP by holding together an entire Yankees team all year ?
Time: 2018.03.29-15:26
Location: (40.829, -73.925)

Image:

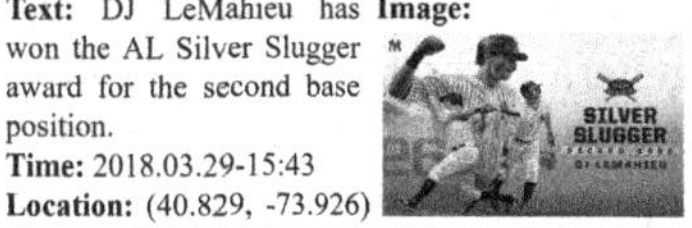

Text: DJ LeMahieu has won the AL Silver Slugger award for the second base position.
Time: 2018.03.29-15:43
Location: (40.829, -73.926)
Image:
SILVER SLUGGER

Fig. 1. Two examples of tweets, both containing multi-modal information.

of time. Detecting local topics is important for many applications [9,23], such as local event discovery, activity recommendation and emergency warning.

Many efforts have already been done for topic detection from social media, however, there exist some challenges that largely limit the performance of existing methods: **1) Capturing comprehensive multi-modal semantics from social media.** Social media contains various modal contents, different modal contents convey information in different aspects. Only by combining as much multi-modal information as possible, can we obtain the comprehensive semantics. Existing methods, however, fall short in capturing comprehensive semantics. They either overlook spatio-temporal information [19,26], resulting in local topic detection impossible, or neglect visual information [5,25], which also contains rich semantics. **2) Bridging semantic gaps among multi-modal information.** Multi-modal contents talking about the same topic are semantically correlated, but there exist semantic gaps among them due to discrepant expressions. In order to achieve the semantic complement effect among multi-modal information, it's essential to maintain both direct semantic correlation (within single-modal content) and indirect semantic correlation (among multi-modal contents). Nevertheless, most existing methods are able to maintain direct semantic correlation, but lack the ability to well preserve indirect semantic correlation [11,15]. **3) Determining the number of topics automatically.** The number of topics happening in social media is uncertain, in order to ensure high quality of detected topics, it's vital to generate the number of topics automatically. However, due to the lack of prior knowledge, many existing methods need to manually specify the number of topics in advance [24,26], which causes the detected topics to be less accurate and pragmatic.

To address above issues, we propose Image-Enhanced Multi-Modal Representation for Local Topic Detection, namely **IEMM-LTD**. Our method has two major modules: the first module is an image-enhanced multi-modal embedding learner to generate highly semantically related embeddings for both words and images. This module can jointly model multi-modal information to capture comprehensive semantics, and map texts and images into the latent space with both direct and indirect semantic correlations preserved. The motivation is that, text and image in the same tweet reflect similar semantics by different expressions, thus, we use the semantic similarity among textual and visual information as bridge, and present a unified structure using both self- and cross- attention to capture semantic correlations. Meanwhile, spatio-temporal information is also injected into textual and visual information to maintain spatio-temporal similarity, which is important for local topic detection. Based on the word and image embeddings generated by the first module, the second module is topic generation, which aims at determining the number of topics automatically and finding high-quality local topics. To achieve this goal, we develop a novel topic model (probabilistic graph model) that models word embeddings by von Mises-Fisher (vMF) distribution and models image embeddings by multivariate Gaussian distribution. By modeling textual and visual information jointly in the same topic model, both direct and indirect semantic correlations can be further maintained. Moreover, the generated local topics are represented by both words and images, which can provide high coherence and interpretability for understanding.

Our main contributions are summarized as follows:

(1) We design an embedding module which takes texts, images, time and location into consideration, to capture comprehensive semantics and preserve both direct and indirect semantic correlations from multi-modal information. The generated embeddings of words and images are highly semantically related, and can be directly used in subsequent topic model.
(2) We propose a topic model which can determine the number of topics automatically and model both word and image embeddings concurrently, to preserve both direct and indirect semantic correlations. These generated local topics represented by both words and images have high quality.
(3) We collect two real-world tweet datasets, including millions of tweets with spatio-temporal information and images. Experiments show that IEMM-LTD outperforms the existing state-of-the-art methods by a large margin.

2 Related Work

Semantic Extraction. Semantic extraction refers to the processing techniques that extract abstract semantic representation from unstructured contents, such as texts and images. Extracting abstract semantics from texts is an important task in NLP. Most prior studies [2,12] try to extract textual representation based on the framework of the combination of RNN and LSTM, which has achieved certain effects but lack of parallelism. Thus, Ashish Vaswani et al. [22] propose an effective framework, called Transformer, which can generate highly abstract semantic representations in parallel and achieve SOTA results in most NLP tasks. Based on Transformer, Jacob Devlin et al. [8] present Bert, an efficient pre-training language model, which can be fine-tuned for amounts of downstream applications. Except texts, Images also contain rich semantics, more and more researchers focus on visual semantic extraction area, trying to analyze and extract the semantics from images. Some work [4,7] concentrates on extracting visual semantics from the whole images. Among them, VggNet [20] has been proved to be an effective way to extract semantic features from images, and many studies [18,21] are proposed to process images using VggNet and prepare for downstream applications. Other work [13,16] pays attention to extracting visual semantics from image segmentations. Guang Li et al. [14] select the salient regions as the region proposals by following the settings in Up-Down [1], and encode the whole image to a sequence of visual representation.

Topic Detection. As the rapid development of social media, lots of researches have been done to utilize social media information for topic detection. Among them, probabilistic graph model based topic detection methods are proved to perform very well. Twitter-LDA [27] and BTM [24] are specially designed for topic detection from short texts, like tweets, however, these early methods only take texts into account. Recently, lots of studies have emerged to utilize multi-modal information for topic detection. mmETM [19] is a multi-modal topic model, which takes both texts and images into consideration. Huan Liu et al. [15] present a probabilistic topic model LTM to extract local topics by using tweets within the corresponding location. IncrAdapTL [11] is a spatio-temporal topic model proposed by Giannakopoulos et al., which also use probability graph model for local topic detection. HTDF [26] is a novel topic detection framework with integration of image and short text information from twitter. However, above traditional probabilistic topic models regard information as independent item without considering semantic correlations among contents. Therefore, some studies introduce representation learning into probabilistic graph model in order to capture the correlation among information. Chen et al. [5] propose multi-layer network embedding to map location, time, and words into a latent space, then present a probabilistic topic model based on embeddings to detect local topics. Above multi-modal topic detection methods can detect relatively good topics, nevertheless, they lack the ability to maintain comprehensive semantics and preserve both direct and indirect semantic correlations.

3 Preliminaries

3.1 Problem Description

Defination 1 (Local topic). *A local topic is defined as a set of words and related images* $\{w_1, w_2, w_3, ..., w_n\} \cup \{v_1, v_2, ..., v_m\}$*, which can reflect a hot local event happening in a local area and during a certain period of time.*

A local topic is something unusual that occurs at specific place and time, which often results in relevant tweets posted around its happening time and occurring location. Given a tweet corpus $TW = \{d_1, d_2, d_3, ..., d_n\}$, each tweet d is represented by a quadruple (x_d, t_d, l_d, v_d), where x_d denotes the text of tweet message, t_d is its post time, l_d is its geo-location, and v_d is the attached image. Consider a query window $Q = [t_s, t_e]$, where t_s and t_e are the start and end timestamps satisfying $t_{d_1} \leq t_s < t_e \leq t_{d_n}$. Our local topic detection task aims at finding high-quality local topics that occur during query time window Q.

3.2 The Framework of IEMM-LTD

As mentioned above, a local topic often leading to posting relevant tweets around its occurring place within a certain period of time, and attached with relevant images. For instance, suppose a baseball game is held at Yankee Stadium in New York City, many participants will post tweets on the spot during the game time to share information (like the examples shown in Fig. 1), using words such as 'yankee', 'team' and enclosing with related images. These words and images can form a word-image cluster as a local topic describing this baseball game.

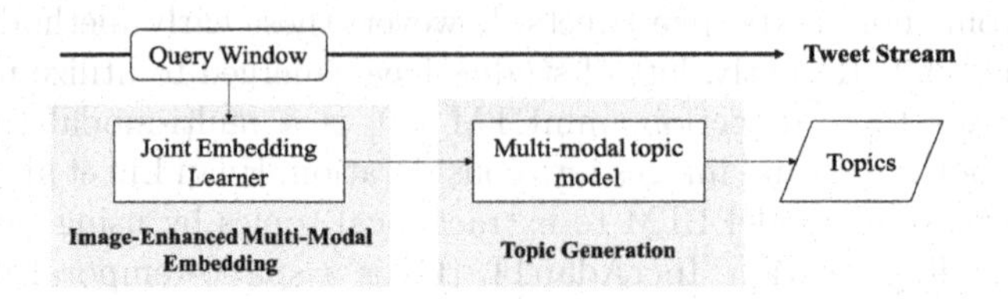

Fig. 2. The framework of IEMM-LTD.

In this paper, we propose an unsupervised method called IEMM-LTD to detect local topics from Twitter. As shown in Fig. 2, IEMM-LTD consists of two major modules: 1) Image-Enhanced Multi-Modal Embedding, and 2) Topic Generation. The first module can jointly model multi-modal information to capture comprehensive semantics, and generate semantically related embeddings for words and images. By introducing self-attention from Bert encoder [8], the direct semantic correlation can be maintained in single-modal information; and by performing cross-attention and maximizing multi-modal semantic similarity, the indirect semantic correlation can also be well captured. Based on the generated embeddings, the second module is a topic model, which can jointly model word and image embeddings to produce semantically coherent word-image clusters as local topics. The vMF distribution and multivariate Guassian distribution are adopted to model textual and visual information separately.

4 Image-Enhanced Multi-modal Embedding

The image-enhanced multi-modal embedding module (shown in Fig. 3) jointly models text, image, time and location, which can well preserve both direct and indirect semantic correlations by the utilization of self- and cross- attention. The objective of this module is to generate highly semantically related word embeddings and image embeddings, that can be directly used in the following topic generation module to produce high-quality local topics. Specifically, for a given tweet containing text, image, time and location, our embedding module processes these multi-modal information in two parts: *textual and visual encoding* and *multi-modal embedding.*

4.1 Textual and Visual Encoding

We present textual and visual encoder to map the original textual and visual inputs into highly abstract semantic representations separately. For textual encoding, we adopt Bert encoder, which is originally designed for sequence modeling and has been proved to be effective in sequence semantic extraction. For

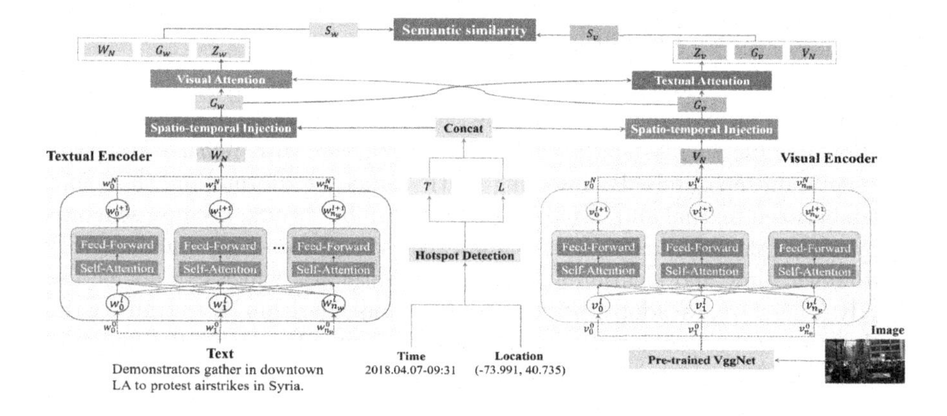

Fig. 3. The structure of image-enhanced multi-modal embedding.

visual encoding, we also use Bert encoder. In most cases, CNN-related models are considered for encoding visual information, however, the Bert encoder with sophisticated design has been shown to achieve better results in exploring both inter- and intra- relationships between the visual entities [14]. Both textual and visual encoder can preserve direct semantic correlation in single-modal information. Each encoder is self-attentive and stacked of N identical blocks.

Take the outputs of l-th $(0 \leq l < N)$ block O^l as an example, they are first fed into the multi-head self-attention layer in the $(l+1)$-th block:

$$M^{l+1} = MultiHead(O^l, O^l, O^l), \tag{1}$$

where M^{l+1} is the hidden state calculated by multi-head attention. The query, key and value matrices have the same shape. Notice that the O^0 is the output of the initial embedding layer.

The subsequent sub-layer is feed-forward network(FFN) which consists of two linear transformations with a ReLU activation in between, $FFN(x) = \boldsymbol{A_2} \cdot ReLU(\boldsymbol{A_1} \cdot x + \boldsymbol{b_1}) + \boldsymbol{b_2}$.

$$O^{l+1} = [FFN(M^{l+1}_{\cdot,1}); ...; FFN(M^{l+1}_{\cdot,n})], \tag{2}$$

where O^{l+1} is the outputs of the $(l+1)$-th block, and $M^{l+1}_{\cdot,i}$ represents the column i of matrix M, i.e., the i-th feature vector. We omit residual connection and layer normalization for a concise explanation.

The structure described above can be used for encoding both textual and visual information. Textual information is composed of a sequence of words, that can be directly used as inputs of textual encoder. Specifically, n_w one-hot word representations are projected into W_0 by an embedding layer. As for visual information, whose original representation can not be used as the inputs of visual encoder. Therefore, we adopt [1] to extract a sequence of visual regions from the original image, and map these visual regions into V_0 by pre-trained VggNet [20].

4.2 Multi-modal Embedding

Based on textual and visual encoding, the abstract semantic representations (W_N and V_N) of text and image can be obtained. We devise a spatio-temporal injection module to fuse spatio-temporal information with textual and visual representation separately. By injecting spatio-temporal information, the semantic gap between textual and visual information can be shrunk, because semantics conveyed by text and image in the same tweet have spatio-temporal similarity.

However, time and location are continuous variables, which can not be used directly. Therefore, we adopt the hotspot detection algorithm [6] to transfer time and location to temporal and spatial representations (T and L). By combining T and L together, we can get spatio-temporal representation G_{tl}.

$$G_{tl} = concat([\alpha T, \beta L]) \tag{3}$$

Where α and β are trade-off parameters, i.e., if we focus on temporal locality of local topics, α should be big and β should be small; if we focus on spatial locality, a small α and a big β are suitable.

Then, the spatio-temporal information is injected into textual and visual information separately by attention mechanism.

$$G_w = Attention(G_{tl}, W_N, W_N), \quad G_v = Attention(G_{tl}, V_N, V_N) \tag{4}$$

where $Attention(Q, K, V) = Softmax(\frac{QK^T}{\sqrt{d}})V$, and d is the width of the input feature vectors.

Most previous studies only perform attention within single modality, which fail to leverage the semantic complementary nature among multi-modal information. Therefore, we perform cross-attention to further preserve indirect semantic correlation.

$$\begin{aligned} Z_w = CrossAttention(G_v, G_w, G_w), \quad Z_v = CrossAttention(G_w, G_v, G_v), \\ CrossAttention(Q, K, V) = concat([H_1, H_2, ..., H_h])\boldsymbol{A^O}, \end{aligned} \tag{5}$$

here $H_i = Attention(Q\boldsymbol{A_i^Q}, K\boldsymbol{A_i^K}, V\boldsymbol{A_i^V})$, $\boldsymbol{A^O}$ denotes the linear transformation, and $\boldsymbol{A_i^Q}$, $\boldsymbol{A_i^K}$, $\boldsymbol{A_i^V}$ are the independent head projection matrices.

Then, the obtained W_N, G_w and Z_w are combined to get final textual representation, the final visual representation is obtained in the same way, where $\sigma(\cdot)$ denotes the sigmoid function.

$$S_w = \sigma(\boldsymbol{A_w} \cdot [W_N, G_w, Z_w]), \quad S_v = \sigma(\boldsymbol{A_v} \cdot [V_N, G_v, Z_v]), \tag{6}$$

After obtaining final textual and visual representations (S_w and S_v), we fine-tune the parameters in two encoders and train other parameters in this module by maximizing the semantic similarity between S_w and S_v. The reason why maximizing semantic similarity is that, 1) the multi-modal information in the same tweet expresses similar semantics in different expressions; 2) the textual information, talking about the same topic but expressing using different words, usually share identical or similar images, time and location; 3) the visual information, discussing the same topic but presenting with diverse images, usually attached with identical or similar texts, time and location. Thus, by maximizing semantic similarity, this module can fuse multi-modal information to capture both direct and indirect semantic correlations. Specifically, we adopt $exp(\cdot)$ function to measure the similarity between textual and visual representation. Because the experiments show that, $exp(\cdot)$ is more suitable for measuring sentence-level semantic similarity, while $cosine(\cdot)$ is more suitable for measuring lexical-level semantic similarity.

At the end of training process, the trained parameters from the embedding layer in textual encoder are taken as the word embeddings; and the image embedding is obtained by averaging the outputs of the last block in visual encoder.

5 Topic Generation

Compared to other clustering methods, which mainly focus on textual domain and detect topics by simply measuring vector similarity, topic model based methods can process multi-modal contents by using different prior distributions to capture different latent patterns, which are more flexible. Thus, we develop a novel embedding-based topic model, which can preserve both direct and indirect semantic correlations between texts and images, and can group semantically relevant words and images in the query window Q into a number of word-image clusters as the final local topics. Consider each tweet d as a tuple (W_d, v_d). Here, $W_d = \{w_1, w_2, w_3, \ldots\}$ is set of p-dimensional embeddings of words in tweet d, and v_d is the q-dimensional embedding of the corresponding image.

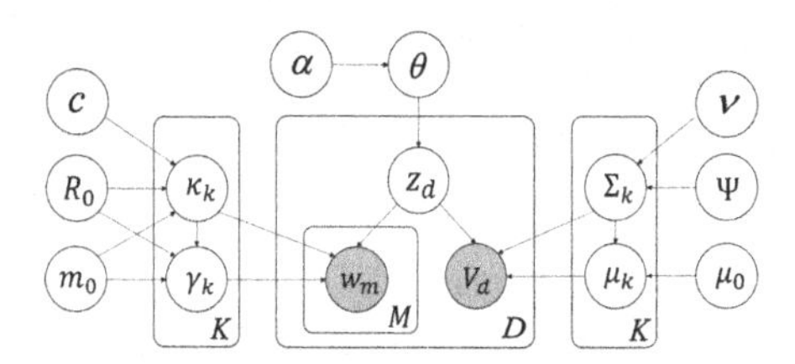

Fig. 4. The structure of multi-modal topic model.

5.1 Embedding-Based Topic Model

The generative process for all the words and images of tweets in the query window Q is shown in Fig. 4. We assume that there are at most K clusters, note that assuming the maximum number of clusters is a weak assumption that can be readily met in practice. At the end of clustering process, some of these K clusters may become empty, and the appropriate number of topics can be automatically discovered. Besides, due to the limited contents in each tweet, we assume that each tweet is represented by only one topic rather than a mixture of topics. As shown, we first draw a multinomial distribution θ from a Dirichlet prior $Dirichlet(.|\alpha)$. Meanwhile, for modeling image embeddings, we draw K multivariate Gaussian distributions; and for modeling word embedding, we draw K von Mises-Fisher (vMF) distributions. For each tweet $d \in Q$, we first draw its cluster membership z_d from θ. Once the cluster membership is determined, we draw its image embedding v_d from the respective multivariate Gaussian distribution, and draw its word embeddings $\{w_m\}$ from the respective vMF distribution. Above generative process is summarized as follows, where $\Lambda = \{\alpha, \Psi, \nu, \mu_0, m_0, R_0, c\}$ are the hyper-parameters.

$$\theta \propto Dirichlet(.|\alpha)$$
$$\{\mu_k\ \Sigma_k\} \propto N(.|\mu_0, \nu, \Psi), \quad \{\gamma_k, \kappa_k\} \propto \Phi(.|m_0, R_0, c), \quad k = 1, 2, ..., K$$
$$z_d \propto Categorical(.|\theta), \quad d \in Q$$
$$v_d \propto Gaussian(.|\mu_{z_d}, \Sigma_{z_d}), \quad w_m \propto vMF(.|\gamma_{z_d}, \kappa_{z_d}), \quad d \in Q,\ w_m \in d$$

The superiority of vMF distribution over other distributions for modeling textual embeddings has been demonstrated in recent study [3]. As for modeling visual embeddings, we adopt multivariate Gaussian distribution and some other distributions for experiments, the results show that the multivariate Gaussian distribution can achieve the best effects.

5.2 Parameter Estimation

We introduce Gibbs sampling to estimate the parameters. The notations used in this subsection are summarized in Fig. 5. Since we have chosen conjugate priors for all the hyper-parameters Λ, these parameters can be integrated out during Gibbs sampling, resulting in a collapsed Gibbs sampling procedure. The key of Gibbs sampling is to obtain the posterior distribution for z_d. Due to space limitation, we directly give the conditional probabilities for z_d:

$$p(z_d = k|X, \Gamma, Z^{\neg d, \Lambda}) \propto p(z_d = k|Z^{\neg d}, \alpha) \cdot p(v_d|\Gamma^{\neg d}, Z^{\neg d}, z_d = k, \Lambda) \cdot \prod_{w_m \in d} p(w_m|X^{\neg m}, Z^{\neg d}, z_d = k, \Lambda) \tag{7}$$

Notation	Description
X	the set of word embeddings in Q
Z	the set of cluster memberships for tweets in Q
Γ	the set of image embeddings in Q
$A^{\neg e}$	the subset of any set A excluding element e
w_m	the embedding vector of word m
V_d	the embedding vector of image of tweet d
n^k	the number of tweets in cluster k
$n^{k,\neg d}$	the number of tweets in cluster k excluding tweet d
$w^{k,\neg m}$	the sum of word embeddings in cluster k excluding word m

Fig. 5. The structure of multi-modal topic model.

The three quantities in Eq. 7 are given by:

$$p(z_d = k|.) \propto (n^{k,\neg d} + \alpha) \; (10.1), \quad p(v_d|.) \propto Tr(v_d|\mu_k, \frac{\tau_k + 1}{\tau}\Sigma_k) \; (10.2),$$
$$p(w_m|.) \propto \frac{C_D(\kappa_k)C_D(\|\kappa_k(R_0 m_0 + \boldsymbol{w}^{k,\neg m})\|_2)}{C_D(\|\kappa_k(R_0 m_0 + \boldsymbol{w}^{k,\neg m} + w_m)\|_2)} \; (10.3) \tag{8}$$

where $Tr(\cdot)$ is the multivariate Student's t-distribution for Gaussian sampling, and $C_D(\kappa) = \frac{\kappa^{D/2-1}}{I_{D/2-1}(\kappa)}$, where $I_{D/2-1}(\kappa)$ is the modified Bessel function.

From Eq. 8, we observe that the proposed topic model enjoys several nice properties: 1) With Eq. 10.1, tweet d tends to join a cluster membership that has more members, resulting in a rich-get-richer effect; 2) With Eq. 10.2, tweet d tends to join a cluster that is more visually close to its image embedding v_d, leading to visually compact clusters; 3) With Eq. 10.3, tweet d tends to join a cluster that is more semantically similar to its word embeddings $\{w_m\}$, resulting in textual coherent clusters.

6 Experiments

6.1 Experimental Setups

Datasets. Since there is a lack of publicly available multi-modal social media dataset for local topic detection, we collect two real-world tweet datasets using Twitter API[1] from 2018.03.21 to 2018.04.27. We will publish the datasets in the near future. The two datasets we collected are called NY and LA: the first data set (NY), consists of 1.7 million spatio-temporal tweets with corresponding images in New York; the second data set (LA), consists of 1.9 million spatio-temporal tweets with corresponding images in Los Angeles.

Evaluation Metrics. We use two typical metrics (*topic coherence* and *topic interpretability*) to evaluate the performance of IEMM-LTD, which are often considered in topic detection task [10,17] to assess the quality of detected topics.

Topic coherence is an effective way used to evaluate topic models by measuring the coherence degree of topics. A good topic model will generate coherent topics, which means high-quality topics have high topic coherence values. We average topic coherence of all the topics detected from each method as the coherence value of this method, where $PMI = \log \frac{P(w_i,w_j)+\epsilon}{P(w_i)\cdot P(w_j)}$.

$$\boldsymbol{coherence} = (\sum_{t=1}^{N_{topics}} c_t)/N_{topics}, \quad c_t = \sum_{i=1}^{N_{topic_t}-1} \sum_{j=i+1}^{N_{topic_t}} PMI(w_i, w_j) \tag{9}$$

Topic interpretability is another important indicator to evaluate topic quality, however, there is no suitable metric to measure quantitatively. Therefore, we use manual scoring to quantitatively measure the topic interpretability. Specifically,

[1] http://docs.tweepy.org/en/latest/.

we invite five subjects who are graduates and familiar with topic detection. We select the top 100 topics from each method by sorting coherence values in descending order, and manually score each topic with value between 0 and 5. The larger the value, the higher the interpretability of the topic. Then average the scores of all topics by all subjects under a certain method, and use the average value as the interpretable value of this method, where $score_{ij} = s, s \in [0, 5]$.

$$\boldsymbol{interpretability} = (\sum_{j=1}^{N_{subjects}} \sum_{i=1}^{N_{topics}} score_{ij})/N_{subjects} * N_{topics} \tag{10}$$

Comparison Methods. To demonstrate the performance of our proposed method IEMM-LTD, we compare it with some existing state-of-the-art topic detection methods. We detail the comparison methods as follows.

- Twitter-LDA [27] is a topic model proposed specifically for Twitter. It addresses the noisy nature of tweets by capturing background words in tweets.
- BTM [24] is a topic model, specifically for modeling short texts. It learns topics by directly modeling the generation of word co-occurrence patterns.
- mmETM [19] is multi-modal topic model, which can effectively model social media documents, including texts with related images.
- IncrAdapTL [11] is a spatio-temporal topic model, which captures both spatial and temporal information in the same probabilistic graph model.
- LTM [15] is a spatio-temporal topic model proposed for local topic detection, which captures both spatial and temporal aspects.
- HNE-BMC [5] is a topic model, which maps words, location and time into a latent space, and develop a Bayesian mixture model to find local topics.
- HTDF [26] is a four-stage framework, proposed to improve the performance of topic detection with integration of both image and short text information.

Method	Text	Time	Location	Image
Twitter-LDA (2011)	√			
BTM (2013)	√			
mmETM (2016)	√			√
LTM (2018)	√	√	√	
IncrAdapTL (2018)	√	√	√	
HNE-BMC (2019)	√	√	√	
HTDF (2019)	√			√
IEMM(-TL)-LTD	√			√
IEMM(-T)-LTD	√		√	√
IEMM(-L)-LTD	√	√		√
IEMM-LTD(-V)	√	√	√	√
IEMM-LTD	√	√	√	√

Fig. 6. The input-modal details of comparison methods.

Besides, we also conduct some ablation experiments to further show how each part of our method affects the final results.

- IEMM(-TL)-LTD removes time and location inputs, and deletes spatio-temporal injection from the first module.
- IEMM(-T)-LTD removes time input, and simplifies spatio-temporal injection to spatial injection in the first module.
- IEMM(-L)-LTD removes location input, and simplifies spatio-temporal injection to temporal injection in the first module.
- IEMM-LTD(-V) removes visual information modeling from the second module, and the first module remains the same.

Figure 6 shows the input-modal details of all baselines that we use to compare with our proposed method IEMM-LTD.

6.2 Experimental Results

During experiments, we set the query window size to 1, 3, 5 and 7 days separately, and slide the query window on both datasets to detect local topics. Experimental results show that, compared to baselines, IEMM-LTD achieves the best results in different window size on both datasets. Due to the space limitation, we only present the results in detail when the query window is set to 3 days.

Illustrative Cases. Fig. 7 shows two exemplifying local topics detected by IEMM-LTD on LA and NY, respectively. Due to space limitation, for each detected local topic, we show top four images and list top seven words. Figure 7(a) refers to a baseball game between Angels and Dodgers; and Fig. 7(b) corresponds to the pillow fight held in Washington Park. As we can see, the detected local topics are of high quality - the images and words in each cluster are highly meaningful and semantically related, which have better interpretability than the topics represented by only words.

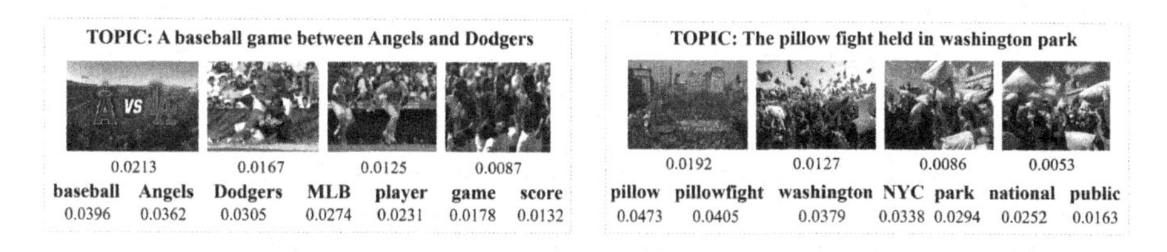

Fig. 7. Two detect local topics by IEMM-LTD on both real-world tweet data sets.

Fig. 8. Given the query word/image, list the most similar words and images by computing the cosine similarity.

To further understand why IEMM-LTD is capable of grouping relevant words and images into the same cluster, we perform similarity queries based on the generated word and image embeddings, shown in Fig. 8. As we can see, the retrieved words and images are meaningful. For instance, given the query word 'protest', the retrieved words are all highly related to the protest, such as 'violence', 'against', etc.; and the retrieved images are also highly connected to the protest, which are all about 'people march on the street holding a sign'. Given the query image about a concert, the retrieved words all reflect the information of a concert, such as 'concert', 'stage', etc.; and the retrieved images are all visually similar to the query image, which are all reflect 'concert-related activities'.

Quantitative Evaluation. We measure the performance of topic detection task with comparison of both topic coherence and topic interpretability. Figure 9 reports the comparative results of all the methods on both datasets (LA and NY) when the query window size is set to 3 days.

As we can see, BTM and Twitter-LDA show inferior performance. This is due to both of them are text-only topic detection methods without considering other multi-modal information. And mmETM improves the topic coherence and interpretability of detected topics significantly, because mmETM takes visual information into account, which can enhance semantics for short texts like tweets. However, mmETM ignores spatio-temporal information, which falls short in local topic detection. Furthermore, LTM, IncrAdapTL and HNE-BMC take spatio-temporal information into consideration, and make local topic detection possible. And these three methods can obtain more coherent topics by capturing semantic correlation between textual and spatio-temporal information. Nevertheless, LTM and IncrAdapTL have lower interpretability than mmETM because they overlook the visual information. And the interpretability of HNE-BMC is a little higher than mmETM but much lower than HTDF, because HTDF adopts visual semantic translation to better enrich the semantics than mmETM. Although HTDF takes visual information into account, it neglects spatio-temporal information, so its topic coherence is lower than HNE-BMC, and it can only detect global topics rather than local topics.

Compared to above methods, IEMM-LTD achieves the best performance, because it takes all texts, images, time and location into modeling to capture comprehensive semantics and well preserve both direct and indirect semantic correlations. Specifically, in LA, compared with the strongest baselines, IEMM-LTD yields around 11% improvement in topic coherence (compared to HNE-BMC), and 18% improvement in topic interpretability (compared to HTDF).

After performing above six multi-modal methods to detect topics in NY from 2018.04.16 to 2018.04.18, Fig. 10 lists the results of the top topic with highest coherence value detected by each method. Among all baselines, only mmETM and HTDF consider both texts and images. Between them, mmETM can produce topics with both words and images, while HTDF can only produce topics with words. However, these two methods neglect spatio-temporal information and are tend to find global topics. Moreover, LTM, IncrAdapTL and HNE-BMC take spatio-temporal information into account, and aim at detecting local topics.

Method	LA		NY	
	coherence	interpretability	coherence	interpretability
Twitter-LDA (2011)	0.912	1.86	0.885	1.73
BTM (2013)	0.942	1.35	0.905	1.57
mmETM (2016)	1.874	2.98	1.886	2.81
LTM (2018)	2.162	2.57	2.274	2.64
IncrAdapTL (2018)	2.315	2.68	2.276	2.79
HNE-BMC (2019)	2.713	3.06	2.676	2.96
HTDF (2019)	2.683	3.37	2.608	3.55
IEMM(-TL)-LTD	2.183	2.46	2.084	2.32
IEMM(-T)-LTD	2.328	2.63	2.295	2.74
IEMM(-L)-LTD	2.403	2.87	2.376	2.92
IEMM-LTD(-V)	2.845	3.11	2.773	3.06
IEMM-LTD	**3.015**	**3.97**	**2.928**	**3.89**

Fig. 9. Set the query window size to 3, performance of comparison methods on both tweet data sets.

Method	Detected Topic		Coherence / Interpretability	Global or Local
	Keyword	Image		
mmETM	shop, restaurant, dinner, date, show		2.286 / 3.32	global
LTM	bacon, party, beach, beer, barbecue	——	2.539 / 3.23	local
IncrAdapTL	mets, game, brewers, baseball, win	——	2.663 / 3.75	local
HNE-BMC	rain, fall, hard, subway, drench	——	3.051 / 3.97	local
HTDF	traffic, street, jam, car, clogged	——	2.816 / 4.28	global
IEMM-LTD	**subway, waterfall, drown, rain, heavy**		**3.357 / 4.62**	**local**

Fig. 10. Comparison of the top topic detected by different multi-modal methods in NY from April 16 to April 18.

As shown, the detected topics are relatively good to understand, nevertheless, they fall short in well preserving the indirect semantic correlation, leading to the generated topics less coherent. Besides, they overlook visual information, which also contains rich semantics. Among all methods, the topic detected by IEMM-LTD has highest coherence and interpretability values, because IEMM-LTD considers all texts, images, time and location to capture comprehensive semantics, and can well preserve both direct and indirect semantic correlations.

Ablation Experiments. We use four simplified versions to compare with original IEMM-LTD, shown in Fig. 9. We can get some valuable conclusions:

(1) The more multi-modal information the method incorporates, the better the local topic detection perform, which agrees with the intuition. As shown, the coherence and interpretability values of IEMM(-T)-LTD, IEMM(-L)-LTD are higher than IEMM(-TL)-LTD, while lower than original IEMM-LTD. Because different modal information contains different aspect of semantics, thus, incorporating more modal information can better enrich semantics.
(2) The spatial locality is more coherent than temporal locality for local topic detection. As shown, the performance of IEMM(-T)-LTD is better than IEMM(-L)-LTD. Because when a local topic emerges, its occurring location

is often confined to a limited area while its happening time sometimes lasts for a long periods. Thus, spatial similarity contributes more than temporal similarity.

(3) Incorporating visual information in topic model can detect local topics with higher quality. As shown, the performance of IEMM-LTD(-V) is much lower than original IEMM-LTD. Because by considering both textual and visual information, the topic model can preserve both direct and indirect semantic correlations and generate local topics represented by both words and images.

Since omitting parts of multi-modal information causes the performance loss, thus, taking more multi-modal information as inputs is essential for local topic detection task. Therefore, IEMM-LTD can achieve satisfactory results by capturing comprehensive semantics from multi-modal information and preserving both direct and indirect semantic correlations in both modules.

Efficiency Comparison. We show log-likelihood changing as the number of Gibbs sampling iterations increases in Fig. 11(a), the log-likelihood quickly converges after a few iterations, so it's sufficient to set the number of iterations to a relatively small value for better efficiency. Then, we perform efficiency comparison. Figure 11(b) shows the average running time variation as the query window size increases. We observe that IEMM-LTD is slower than mmETM, HTDF and HNE-BMC, because mmETM and HTDF only consider texts and images but overlook spatio-temporal information, while HNE-BMC takes textual and spatio-temporal information into account but neglects images. As for IEMM-LTD, it incorporates texts, images, time and location into modeling, which takes more time to process extra modal information. Compared to LTM and IncrAdapTL, IEMM-LTD is more efficient. Although IEMM-LTD detects topics by two modules, the first module can generate embeddings efficiently by fine-tuning, and the second module is a topic model with two-plate structure, which can perform sampling very quickly. On the contrary, the topic models in LTM and IncrAdapTL are very complicated, which are all multi-plate (more than four plates) structure, so it's very time-consuming when performing sampling.

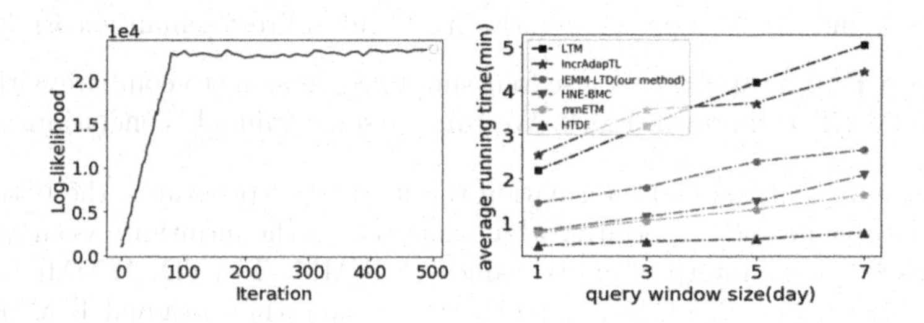

Fig. 11. (a) Gibbs sampling convergence. (b) Comparison of running time.

7 Conclusion

In this paper, we propose IEMM-LTD to detect local topics represented by both words and images from Twitter. With the image-enhanced multi-modal embedding learner, our method can capture comprehensive semantics from all multi-model information and preserve both direct and indirect semantic correlations, which can ensure that the generated embeddings are semantically related. Then, the embeddings of words and images can be directly used in our proposed multi-modal topic model, which can further preserve direct and indirect semantic correlations and produce high-quality local topics. The extensive experiments show that IEMM-LTD improves the performance of local topic detection task significantly.

References

1. Anderson, P., et al.: Bottom-up and top-down attention for image captioning and visual question answering. In: CVPR, pp. 6077–6086. IEEE Computer Society (2018)
2. Bao, J., Duan, N., Zhou, M.: Knowledge-based question answering as machine translation. In: ACL (1). The Association for Computer Linguistics (2014)
3. Batmanghelich, K.N., Saeedi, A., Narasimhan, K., Gershman, S.: Nonparametric spherical topic modeling with word embeddings. In: ACL (2). The Association for Computer Linguistics (2016)
4. Chen, F., Xie, S., Li, X., Li, S.: What topics do images say: A neural image captioning model with topic representation. In: ICME Workshops, IEEE (2019)
5. Chen, J., Gao, N., Xue, C., Tu, C., Zha, D.: Perceiving topic bubbles: local topic detection in spatio-temporal tweet stream. In: Li, G., Yang, J., Gama, J., Natwichai, J., Tong, Y. (eds.) DASFAA 2019. LNCS, vol. 11447, pp. 730–747. Springer, Cham (2019). https://doi.org/10.1007/978-3-030-18579-4_43
6. Chen, J., Gao, N., Xue, C., Zhang, Y.: The application of network based embedding in local topic detection from social media. In: ICTAI, pp. 1311–1319. IEEE (2019)
7. Cheng, Z., Bai, F., Xu, Y., Zheng, G., Pu, S., Zhou, S.: Focusing attention: Towards accurate text recognition in natural images. In: ICCV, pp. 5086–5094. IEEE Computer Society (2017)
8. Devlin, J., Chang, M., Lee, K., Toutanova, K.: BERT: pre-training of deep bidirectional transformers for language understanding. In: NAACL-HLT (2019)
9. Duan, J., Ai, Y., Li, X.: LDA topic model for microblog recommendation. In: IALP, pp. 185–188. IEEE (2015)
10. Fang, A., Macdonald, C., Ounis, I., Habel, P.: Topics in tweets: a user study of topic coherence metrics for twitter data. In: Ferro, N. (ed.) ECIR 2016. LNCS, vol. 9626, pp. 492–504. Springer, Cham (2016). https://doi.org/10.1007/978-3-319-30671-1_36
11. Giannakopoulos, K., Chen, L.: Incremental and adaptive topic detection over social media. In: Pei, J., Manolopoulos, Y., Sadiq, S., Li, J. (eds.) DASFAA 2018. LNCS, vol. 10827, pp. 460–473. Springer, Cham (2018). https://doi.org/10.1007/978-3-319-91452-7_30
12. Huang, S., Chen, H., Dai, X., Chen, J.: Non-linear learning for statistical machine translation. In: ACL (1). The Association for Computer Linguistics (2015)

13. Jégou, S., Drozdzal, M., Vázquez, D., Romero, A., Bengio, Y.: The one hundred layers tiramisu: Fully convolutional densenets for semantic segmentation. In: CVPR Workshops, pp. 1175–1183 (2017)
14. Li, G., Zhu, L., Liu, P., Yang, Y.: Entangled transformer for image captioning. In: ICCV, pp. 8927–8936. IEEE (2019)
15. Liu, H., Ge, Y., Zheng, Q., Lin, R., Li, H.: Detecting global and local topics via mining twitter data. Neurocomputing **273** 120-132 (2018)
16. Long, J., Shelhamer, E., Darrell, T.: Fully convolutional networks for semantic segmentation. In: CVPR, pp. 3431–3440 (2015)
17. Morstatter, F., Liu, H.: In search of coherence and consensus: measuring the interpretability of statistical topics. J. Mach. Learn. Res. **18**(169), 1–32 (2017)
18. Muhammad, U., Wang, W., Chattha, S.P., Ali, S.: Pre-trained vggnet architecture for remote-sensing image scene classification. In: ICPR, pp. 1622–1627. IEEE Computer Society (2018)
19. Qian, S., Zhang, T., Xu, C., Shao, J.: Multi-modal event topic model for social event analysis. IEEE Trans. Multimedia **18**(2), 233–246 (2016)
20. Simonyan, K., Zisserman, A.: Very deep convolutional networks for large-scale image recognition. In: ICLR (2015)
21. Singh, V.K., et al.: Classification of breast cancer molecular subtypes from their micro-texture in mammograms using a vggnet-based convolutional neural network. In: CCIA, vol. 300, pp. 76–85. IOS Press (2017)
22. Vaswani, A., et al.: Attention is all you need. In: NIPS, pp. 5998–6008 (2017)
23. Wold, H.M., Vikre, L., Gulla, J.A., Özgöbek, Ö., Su, X.: Twitter topic modeling for breaking news detection. In: WEBIST (2), pp. 211–218. SciTePress (2016)
24. Yan, X., Guo, J., Lan, Y., Cheng, X.: A biterm topic model for short texts. In: WWW, pp. 1445–1456. International World Wide Web Conferences Steering Committee/ACM (2013)
25. Zhang, C., et al.: Regions, periods, activities: Uncovering urban dynamics via cross-modal representation learning. In: WWW, pp. 361–370. ACM (2017)
26. Zhang, C., Lu, S., Zhang, C., Xiao, X., Wang, Q., Chen, G.: A novel hot topic detection framework with integration of image and short text information from twitter. IEEE Access **7**, 9225–9231 (2019)
27. Zhao, W.X., et al.: Comparing Twitter and traditional media using topic models. In: Clough, P. (ed.) ECIR 2011. LNCS, vol. 6611, pp. 338–349. Springer, Heidelberg (2011). https://doi.org/10.1007/978-3-642-20161-5_34

A Semi-supervised Framework with Efficient Feature Extraction and Network Alignment for User Identity Linkage

Zehua Hu, Jiahai Wang(✉), Siyuan Chen, and Xin Du

School of Computer Science and Engineering, Sun Yat-sen University, Guangzhou 510006, China
wangjiah@mail.sysu.edu.cn

Abstract. Nowadays, people tend to join multiple social networks to enjoy different kinds of services. User identity linkage across social networks is of great importance to cross-domain recommendation, network fusion, criminal behaviour detection, etc. Because of the high cost of manually labeled identity linkages, the semi-supervised methods attract more attention from researchers. Different from previous methods linking identities at the pair-wise sample level, some semi-supervised methods view all identities in a social network as a whole, and align different networks at the distribution level. Sufficient experiments show that these distribution-level methods significantly outperform the sample-level methods. However, they still face challenges in extracting features and processing sample-level information. This paper proposes a novel semi-supervised framework with efficient feature extraction and network alignment for user identity linkage. The feature extraction model learns node embeddings from the topology space and feature space simultaneously with the help of dynamic hypergraph neural network. Then, these node embeddings are fed to the network alignment model, a Wasserstein generative adversarial network with a new sampling strategy, to produce candidate identity pairs. The proposed framework is evaluated on real social network data, and the results demonstrate its superiority over the state-of-the-art methods.

Keywords: User identity linkage · Semi-supervised framework · Feature extraction · Network alignment

1 Introduction

Online social networks are becoming increasingly important. Most people participate in different social networks for different purposes. For example, a user gets news on Twitter, shares photos on Instagram, and seeks job opportunities on LinkedIn. Different platforms have diverse functionalities, and a platform usually can only reflect one aspect of users. Combining the information from multiple

C. S. Jensen et al. (Eds.): DASFAA 2021, LNCS 12682, pp. 675–691, 2021.
https://doi.org/10.1007/978-3-030-73197-7_46

social networks provides a more comprehensive view, and helps to improve many downstream applications (e.g., cross-domain recommendation, network fusion, criminal behaviour detection, etc.). The fundamental task of cross-network data mining is to link the identities belonging to the same individual across multiple social networks, which is called user identity linkage (UIL) [17,20,32], across social networks user identification [25] or social anchor link prediction [4,16,28].

For UIL, most existing methods are supervised and need a large number of annotations to guide the learning process, but the cost of manually labeling data is extremely high. Therefore, semi-supervised methods attract more attention from researchers. Early semi-supervised studies [13,27] usually take advantage of the discriminative features in user profiles (e.g., username, location, affiliation, etc.). However, some important profiles are often missing due to privacy considerations, and thus these methods inevitably suffer from data insufficiency in practice. Researchers [9,29,30] further suggest that the information of neighbors is helpful to UIL. But in semi-supervised learning, most of the neighbor pairs are unlabeled, which prevents us from directly calculating the metrics like Jaccard's coefficient based on the shared neighbors.

With the advances in the network embedding, some embedding-based methods [17,33,34] handle semi-supervised UIL at the pair-wise sample level. They first utilize a feature extraction model to encapsulate network structures and user features into low-dimensional node embeddings, and then learn the mapping among different networks through these embeddings. However, researchers [11,12] point out that performing UIL at the sample level cannot maintain the isomorphism among networks, and thus they introduce generative adversarial networks (GAN) for semi-supervised UIL. Specifically, after learning the node embeddings, they consider all the identities in a social network as a whole, and align different networks through Wasserstein GAN (WGAN), a variant of GAN. These WGAN-based frameworks still face two challenges. Firstly, there is a lack of the semi-supervised feature extraction model suitable for UIL. Previous WGAN-based methods [10,11] use unsupervised feature extraction models to learn node embeddings. Secondly, WGAN is performed at the distribution level to align networks, but UIL is essentially a sample-level task, and annotations also preserve the guidance information in the sample level [10]. Considering the different granularity, it is necessary to place more focus on the sample-level information when aligning networks through WGAN.

Therefore, a new semi-supervised framework, named SSF, is proposed for UIL in this paper, and it can solve the two problems mentioned above. SSF contains two models, a feature extraction model, and a network alignment model. To design the feature extraction model, we first analyze the requirements for well-trained node embeddings. According to these requirements, SSF propagates node features over both the topology space and feature space of networks [24]. Technically, the network structure is taken as the structural graph of topology space, and the k-nearest neighbor (kNN) graph generated by node features is used as the structural graph of feature space. Moreover, SSF utilizes dynamic hypergraph neural network (DHGNN) [3,5,7] to capture the global dependency in graphs. In the network alignment model, a new sampling strategy is proposed for WGAN. This sampling strategy is based on uniform sampling, and further uses the output of the discriminator to dynamically adjust the sampling weight of nodes. Our sampling strategy enables WGAN to pay more attention to the sample-level information during training.

The main contributions of this work are summarized as follows:

- We design a new semi-supervised framework for UIL, named SSF. The proposed SSF consists of the feature extraction model and network alignment model. Extensive experiments are conducted on the real-world dataset to validate the effectiveness of SSF, and the results demonstrate its superiority over the state-of-the-art methods.
- The feature extraction model extracts features from both the topology space and feature space to obtain the underlying structure information about these two spaces. Moreover, SSF integrates the hypergraph information to capture the global dependency in graphs.
- The network alignment model utilizes WGAN to perform network alignment at the distribution level. SSF proposes a new sampling strategy for WGAN, which enables WGAN to focus more on the sample-level information during the training process.

2 Related Work

Most existing methods for UIL consist of two major phases: feature extraction and network alignment [20]. According to what information is used in feature extraction, existing methods are classified into profile-based [13,27], content-based [15,18] and network-based [9,30] methods. According to the use of labeled and unlabeled data, existing methods are divided into supervised [16,17], semi-supervised [11,22] and unsupervised methods [12,13]. For a comprehensive review, please see [20,25]. In this paper, we focus on semi-supervised methods and divide them into propagation methods and embedding methods as follows [20].

Propagation Methods. Propagation methods iteratively generate candidate identity pairs from seed identity linkages. This kind of methods [15,19,31,35] compute the matching degree of user identities by the neighbor information or discriminative profiles. Zhou et al. [35] propose a friend relationship-based user identification algorithm, which can capture friend relationships and network structures. Shen et al. [19] develop an information control mechanism against UIL, and they combine distance-based profile features and neighborhood-based network structures to identify unknown identity pairs iteratively. Liu et al. [15] propose a multi-objective framework, which models the structure consistency and heterogeneous behaviors simultaneously. Essentially, propagation methods are based on inference, and they need to infer the labels of neighbor pairs during the iterative process, and thus these methods usually face error propagation.

Embedding Methods. Inspired by the advances in network embedding, researchers [14,22,32–34] pay more attention to the application of network embedding in UIL. Tan et al. [22] map the nodes from different networks into a common latent space with the help of the hypergraph, and then they measure the similarity of node embeddings to generate candidate identity pairs. Zhou et al. [33] use the skip-gram model to obtain the social representation of nodes, and employ a dual-learning mechanism to boost the training of UIL. Liu et al. [14] incorporate UIL into the probabilistic model based on the follower-followee relationships among identities. Different from previous studies that train the mapping among networks at the sample level, some studies [10–12] propose to align networks at the distribution level through WGAN. Li et al. [12] propose an unsupervised framework, which first learns node embeddings by an unsupervised network embedding model [26], and then aligns the node embeddings from different networks by WGAN. Li et al. [11] further incorporate a few annotations to improve the alignment performance of WGAN.

The proposed SSF follows the semi-supervised embedding methods based on WGAN, and it designs a new feature extraction model suitable for semi-supervised UIL, and improves the original network alignment model from the perspective of sampling strategy.

3 The Proposed Framework

Let $N = \{V, A, X\}$ denotes a social network, where V is the set of user identities, A is the adjacency matrix that records social connections, and X is the user feature matrix extracted from user profiles through data preprocessing.

Given a pair of networks $N_1 = \{V_1, A_1, X_1\}$ and $N_2 = \{V_2, A_2, X_2\}$, UIL aims to predict the identity linkages between these two networks with the help of annotated identity linkages S. S is formally defined as $S = \{(u, v) \subseteq V_1 \times V_2 | u$ and v belong to the same individual$\}$. Practically, for each node in N_1, SSF will output several nodes of N_2 to form candidate identity pairs.

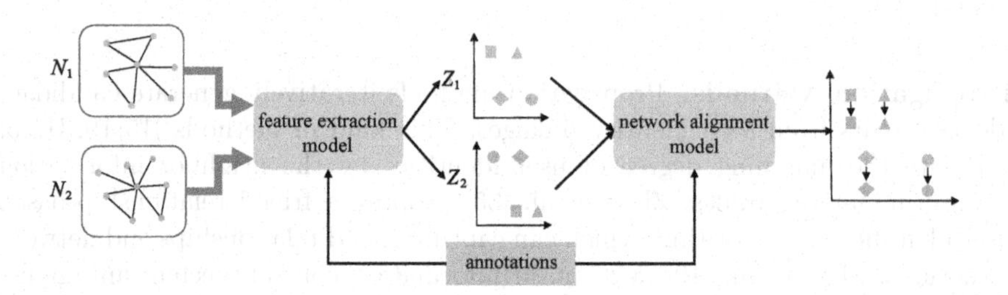

Fig. 1. The overall framework of the proposed SSF. It contains two main models: feature extraction model (FEM) and network alignment model (NAM). For networks N_1 and N_2, the purpose of FEM is to generate desired node embeddings Z_1 and Z_2, respectively, which can be represented in the latent space as shown in the figure. Then, NAM aligns the two networks by learning the projection of node embeddings.

The overall framework is shown in Fig. 1. In the first step, from the perspective of structure and feature proximity, the feature extraction model (FEM) utilizes user feature matrices and adjacency matrices to learn the desired node embeddings Z_1 and Z_2. In the second step, the network alignment model (NAM) considers all the identities in a social network as a whole, and align the networks N_1 and N_2 at the distribution level. The identity pairs with high similarity are taken as our prediction. These two models will be introduced in the following subsections.

3.1 Feature Extraction Model

FEM aims to learn the well-trained node embeddings for UIL, and thus we first figure out three requirements (R1, R2, R3) that these embeddings should meet:

R1. They can reflect the structural proximity. If a pair of nodes are adjacent to each other in the network structure, and then their embeddings in the latent space should be close.
R2. They can reflect the proximity of user features. Similar to R1, if a pair of nodes have similar features, and then their embeddings in the latent space should also be close.
R3. They can be used to effectively distinguish whether two identities from different networks belong to the same individual.

To satisfy requirements R1 and R2, it is necessary to extract information from the topology space and feature space simultaneously. In this way, the underlying structure information about the topology space and feature space can be captured at the same time. Inspired by graph auto-encoders [8], FEM reconstructs the proximity of network structures and user features by the inner product of node embeddings. As shown in Fig. 2, FEM adopts an encoder-decoder architecture. Since FEM treats networks N_1 and N_2 in the same way, it suffices to describe the feature extraction procedure on network N_1.

In the encoders, the kNN graph constructed from user features is used to capture the underlying structure information of feature space. Specifically, FEM first calculates the cosine similarity between any two nodes. Then for each node, the top k similar nodes are taken as its neighbors, and the adjacency matrix of the induced feature graph is denoted as A_1'. FEM takes the user feature matrix X_1 and the adjacency matrix A_1' as the feature graph. Similarly, FEM takes the user feature matrix X_1 and the adjacency matrix A_1 as the topology graph.

FEM adopts graph attention networks (GAT) as the main building block to aggregate user features. For topology space, with the input embeddings T^{l-1}, the corresponding embedding of node i in the l-th GAT layer's output can be represented as follows:

$$T_i^l = \sigma(\sum_{j \in N_i^T} \alpha_{ij} W_T^l T_j^{l-1}), \tag{1}$$

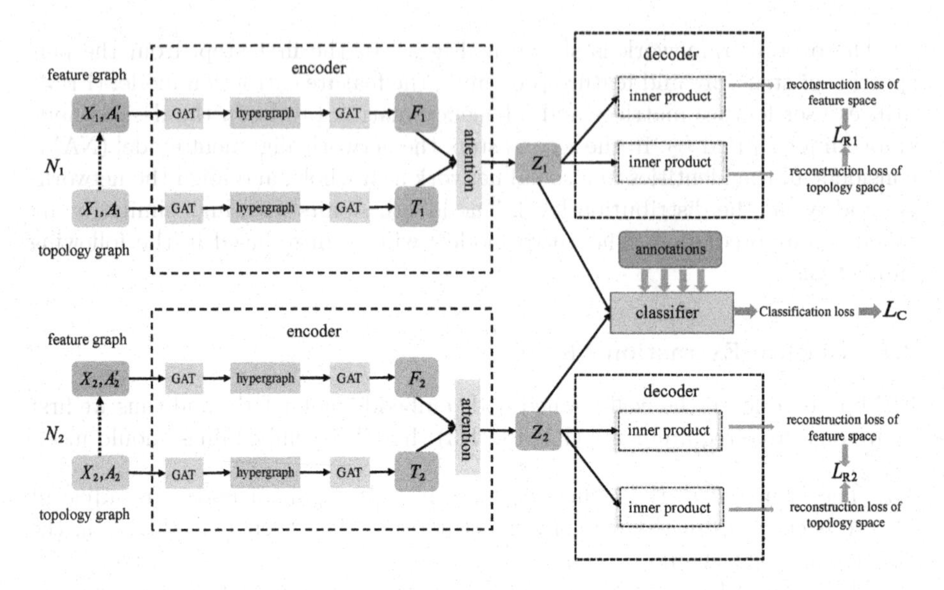

Fig. 2. Feature extraction model. A_1' and A_2' are the adjacency matrices of feature graphs constructed from X_1 and X_2, respectively. For each network, FEM first utilizes both topology graph and feature graph to extract features. F_1 and F_2 are the output embeddings of the feature graphs, and T_1 and T_2 are the output embeddings of the topology graphs. Then, FEM integrates the output embeddings of the topology graph and feature graph to obtain the desired node embeddings Z_1 and Z_2.

where σ is the sigmoid function; W_T^l represents the weight matrix of the l-th layer; T_j^{l-1} is the corresponding embedding of node j in the input embeddings T^{l-1}; α_{ij} is the attention coefficient between the node i and its neighbor $j \in N_i^T$, and it is defined as follows:

$$\alpha_{ij} = \frac{\exp(\text{LeakyReLU}(a^T[W_T^l T_i^l || W_T^l T_j^l]))}{\sum_{k \in N_i^T} \exp(\text{LeakyReLU}(a^T[W_T^l T_i^l || W_T^l T_k^l]))}, \tag{2}$$

where $||$ is the concatenation operation, and a is the weight vector of edges. For feature space, FEM calculates the output embeddings F^l of the GAT layer in the same way.

To capture the global dependency in graphs, FEM utilizes dynamic hypergraph neural network (DHGNN) between GAT layers. Different from the traditional graphs based on pair-wise relations, hypergraph allows multiple nodes on an edge, and this kind of edge is named hyperedge. Technically, for node i, its k nearest neighbors, along with this node, are selected to form the hyperedge $e = \{i, i_1, i_2, \ldots, i_k\}$, and the node features corresponding to this hyperedge is $E = [T_i^l; T_{i_1}^l; T_{i_2}^l; \ldots; T_{i_k}^l]$. For topology space, the hypergraph convolution on this hyperedge is defined as follows:

$$w = g(E), \tag{3}$$

$$T_i^{l+1} = \sum_{j=1}^{k+1} w_j E_j, \tag{4}$$

where $g(\cdot)$ is a multi-layer perceptron (MLP). FEM first obtains the node weight $w \in \mathbb{R}^{k+1}$ for the node embeddings on the hyperedge e, and then it applies a one-dimensional convolution to incorporate these weighted node embeddings into the output embedding T_i^{l+1}. For feature space, the output embeddings F^{l+1} of the DHGNN layer can be calculated in the same way. Moreover, DHGNN can dynamically adjust hyperedges according to the output embeddings of the previous GAT layer.

After extracting features through the combination of GAT and DHGNN, FEM integrates the output embeddings of the topology space and feature space through an attention mechanism to obtain the desired node embeddings Z_1 and Z_2 for networks N_1 and N_2, respectively. Taking Z_1 as an example, the attention mechanism is formulated as follows:

$$Z_1 = \frac{h(T_1)}{h(T_1) + h(F_1)} T_1 + \frac{h(F_1)}{h(T_1) + h(F_1)} F_1, \tag{5}$$

where $h(\cdot)$ is a scoring function for the attention mechanism.

In the decoders, the requirements R1 and R2 are implemented by reducing the difference between the original graph and the graph reconstructed from node embeddings. The adjacency matrices of two networks' reconstructed graphs $\widehat{A_1}$ and $\widehat{A_2}$ can be obtained as follows:

$$\widehat{A_1} = \sigma(Z_1 \cdot Z_1^T), \tag{6}$$

$$\widehat{A_2} = \sigma(Z_2 \cdot Z_2^T). \tag{7}$$

For each network, FEM utilizes the graph-based loss function [6] to make the reconstructed graph as similar as possible to the original topology graph and feature graph simultaneously. Practically, the graph-based loss function encourages that the representations of nearby nodes are similar, and requires that the representations of disparate nodes are distinct. Taking the network N_1 for example, the corresponding objective function is defined as follows:

$$\begin{aligned} L_{\mathrm{R1}} = & \frac{\lambda_{\mathrm{R1}}}{N_{A_1}} \sum_{A_1(u,v)=1} - \log(\widehat{A_1}(u,v)) - Q \cdot \mathbb{E}_{v_n \sim P_n(v)} \log(1 - \widehat{A_1}(u, v_n)) \\ & + \frac{\lambda'_{\mathrm{R1}}}{N_{A'_1}} \sum_{A'_1(u,v)=1} - \log(\widehat{A_1}(u,v)) - Q \cdot \mathbb{E}_{v_n \sim P_n(v)} \log(1 - \widehat{A_1}(u, v_n)), \end{aligned} \tag{8}$$

where λ_{R1} and λ'_{R1} are the hyperparameters weighting the reconstruction loss of the topology graph and feature graph, respectively; N_{A_1} and $N_{A'_1}$ are the number of non-zero elements in matrices A_1 and A'_1, respectively; Q defines the number of negative sample, and P_n is the negative sampling distribution.

Similarly, for network N_2, the objective function of the reconstruction loss is defined as follows:

$$\begin{aligned} L_{\text{R2}} =& \frac{\lambda_{\text{R2}}}{N_{A_2}} \sum_{A_2(u,v)=1} -\log(\widehat{A_2}(u,v)) - Q \cdot \mathbb{E}_{v_n \sim P_n(v)} \log(1 - \widehat{A_2}(u,v_n)) \\ &+ \frac{\lambda'_{\text{R2}}}{N_{A'_2}} \sum_{A'_2(u,v)=1} -\log(\widehat{A_2}(u,v)) - Q \cdot \mathbb{E}_{v_n \sim P_n(v)} \log(1 - \widehat{A_2}(u,v_n)). \end{aligned} \tag{9}$$

As described in requirement R3, the node embeddings Z_1 and Z_2 should have superior classification performance. This requirement is considered as a binary classification task, and FEM processes annotated identity linkages as follows:

$$L_{\text{C}} = \frac{1}{|S|} \sum_{(i,j)\in S} -\log(\sigma(p(z_1^i \| z_2^j))) - Q \cdot \mathbb{E}_{k \sim P_n(j)} \log(1 - \sigma(p(z_1^i \| z_2^k))), \tag{10}$$

where $z_1^i \in Z_1$ and $z_2^j \in Z_2$; $z_2^k \in Z_2$ is the node sampled from the negative sampling distribution; $p(\cdot)$ is the scoring function measuring the likelihood of two identities belonging to the same individual. Combining the classification loss and reconstruction losses, the overall objective function of FEM is given as follows:

$$L = L_{\text{R1}} + L_{\text{R2}} + L_{\text{C}}. \tag{11}$$

To sum up, we first figure out three requirements for well-trained node embeddings. Based on the requirements R1 and R2, SSF adopts an encoder-decoder architecture for FEM. The encoders extract information from the feature space and topology space simultaneously with the help of the hypergraph. The decoders guide the learning process by making the reconstruction graph of node embeddings close to the feature graph and topology graph simultaneously. Moreover, FEM utilizes a classifier to meet the task-specific requirement R3 for UIL.

3.2 Network Alignment Model

Since it is difficult for sample-level network alignment methods to maintain the isomorphism among networks, WGAN-based methods are proposed to perform network alignment at the distribution level. WGAN-based methods can take advantage of the isomorphic information in networks to effectively align unlabeled nodes, and thus the need for annotations is further reduced.

Figure 3 shows the framework of WGAN-based network alignment model. The generator G acts as a projection function $\mathbf{\Phi}$. Given a source node i with its node embedding $z_1^i \in Z_1$, its projected point is defined as: $\mathbf{\Phi}(z_1^i) = Gz_1^i$. Following previous works [16,17], the linear transformation is chosen as the projection function. The discriminator D aims to measure the distance between the projected source distribution $\mathbb{P}^{G(N_1)}$ and the target distribution $\mathbb{P}^{N_2}$. Here, Wasserstein Distance (WD) is taken as the distance metric. Compared with other

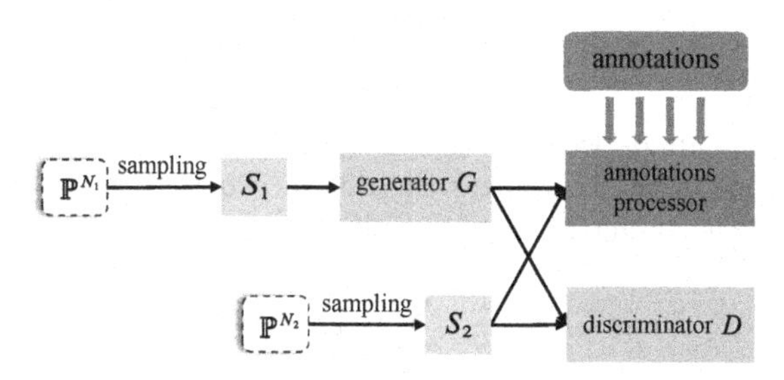

Fig. 3. WGAN-based network alignment model. The users of network N_1 are viewed as the source distribution $\mathbb{P}^{N_1} = \sum_i P_{S_1^i} \delta_{S_1^i}$, in which S_1^i is a sample in the source distribution, and $P_{S_1^i}$ is its corresponding probability and $\delta_{S_1^i}$ is the Dirac delta function. Similarly, the users of network N_2 are defined as the target distribution $\mathbb{P}^{N_2}$. $S_1 \subseteq V_1$ and $S_2 \subseteq V_2$ are the batches sampled from these two distributions during the training process. The generator G projects the source distribution $\mathbb{P}^{N_1}$ into the projected source distribution $\mathbb{P}^{G(N_1)}$, and the discriminator D estimates the Wasserstein distance between $\mathbb{P}^{G(N_1)}$ and the target distribution $\mathbb{P}^{N_2}$.

metrics such as Kullback-Leibler divergence or Jensen–Shannon divergence, WD is able to measure the distance between two distributions even if they do not overlap. The objective function of WGAN can be formally defined as follows:

$$\min_G \mathrm{WD}(\mathbb{P}^{G(N_1)}, \mathbb{P}^{N_2})) = \inf_{\gamma \in \Gamma(\mathbb{P}^{G(N_1)}, \mathbb{P}^{N_2})} \mathbb{E}_{(Gz_1^i, z_2^j) \sim \gamma}[d(Gz_1^i, z_2^j)], \tag{12}$$

where i and j are the nodes sampled from the source distribution $\mathbb{P}^{N_1}$ and the target distribution $\mathbb{P}^{N_2}$, respectively; $\inf_{\gamma \in \Gamma(\mathbb{P}^{G(N_1)}, \mathbb{P}^{N_2})}$ is the expectation infimum of the joint probability distribution $\Gamma(\mathbb{P}^{G(N_1)}, \mathbb{P}^{N_2})$. As traversing all the possible joint distributions is difficult, a simple version of the WD minimization [23] is proposed when $d(\cdot, \cdot)$ is the Euclidean distance, which is defined as follows:

$$\mathrm{WD} = \frac{1}{K} \sup_{\|f\| \leq K} \mathbb{E}_{z_2^j \sim \mathbb{P}^{N_2}} f(z_2^j) - \mathbb{E}_{Gz_1^i \sim \mathbb{P}^{G(N_1)}} f(Gz_1^i), \tag{13}$$

where f is the K-Lipschitz continuous function. This objective function aims to find the supremum over all the possible K-Lipschitz functions. MLP owns powerful approximation capabilities, and NAM employs it as the discriminator to approximate the function f. The objective function of the discriminator is given as follows:

$$\max_{\theta: \|f_\theta\| \leq K} L_D = \mathbb{E}_{z_2^j \sim \mathbb{P}^{N_2}} f_\theta(z_2^j) - \mathbb{E}_{Gz_1^i \sim \mathbb{P}^{G(N_1)}} f_\theta(Gz_1^i), \tag{14}$$

where θ is the parameter set of the discriminator.

The generator achieves network alignment by minimizing the estimated distance between the projected source distribution $\mathbb{P}^{G(N_1)}$ and the target distribution $\mathbb{P}^{N_2}$. As shown in Eq. (14), it only exists in the second term of the right-hand side. Therefore, the objective function of the generator is minimized as follows:

$$\min_{G} L_G = -\mathbb{E}_{Gz_1^i \sim \mathbb{P}^{G(N_1)}} f_\theta(Gz_1^i). \tag{15}$$

Meanwhile, the annotations S are used to guide the learning of the generator. For any $(i, j) \in S$, NAM minimizes the distance between the projected source point $G(z_1^i)$ and the target point z_2^j as follows:

$$\min_{G} L_S = \frac{1}{|S|} \sum_{(i,j) \in S} (1 - c(Gz_1^i, z_2^j)), \tag{16}$$

where $c(\cdot, \cdot)$ means the cosine similarity.

Imposing orthogonal constraint to the generator contributes to network alignment [17]. Specifically, FEM reconstructs original source distribution $\mathbb{P}^{N_1}$ from the projected distribution $\mathbb{P}^{G(N_1)}$ with the transpose matrix G^T, and the difference between the original source distribution and the reconstructed one is minimized as follows:

$$\min_{G} L_O = \mathbb{E}_{z_1^i \sim \mathbb{P}^{N_1}} [d(z_1^i, G^T G z_1^i)]. \tag{17}$$

In the training process of WGAN, the entire objective of the generator is the weighted sum of L_G, L_S and L_O as follows:

$$\min_{G} \lambda_O L_O + \lambda_S L_S + L_G, \tag{18}$$

where λ_O and λ_S are hyperparameters that control the weights of L_O and L_S, respectively.

UIL is essentially a sample-level task, and the annotations also preserve the guidance information in the sample level. Therefore, more attention should be paid to sample-level matching than distribution-level alignment when using WGAN for network alignment. SSF proposes a new sampling strategy, which enables WGAN to focus more on the sample-level information during training.

Previous studies [10–12] sample batches according to the influence of nodes in the network, and this sampling method is called degree sampling. In degree sampling, the nodes with higher degrees are more likely to be sampled, and thus degree sampling is more beneficial to distribution-level tasks. Our sampling strategy is based on uniform sampling. In uniform sampling, all nodes are given the same weight, and thus they are equally important in the training process. Moreover, SSF combines the reweighting strategy [1] to guide the learning of WGAN. The reweighting strategy can dynamically adjust the sampling weights of nodes according to the output of the discriminator. For a misclassified node, the discriminator's output for it is larger than those of correctly classified nodes, and thus it should be assigned a higher weight in the subsequent training process. Taking network N_2 for example, let $f_\theta(z_2^i)$ be the discriminator output of node i, its corresponding weight is defined as follows:

$$\frac{\exp(\beta \times f_\theta(z_2^i))}{\sum_{j=0}^{|V_1|} \exp(\beta \times f_\theta(z_2^j))}, \tag{19}$$

where β is the hyperparameter controls the weight for nodes with low prediction scores. The reweighting strategy assigns higher weights to misclassified nodes than correctly classified ones with $\beta > 1$.

To sum up, We first introduce how NAM aligns networks through WGAN, and then propose a new sampling strategy that enables WGAN to focus more on the sample-level information.

4 Experimental Results

4.1 Dataset and Experimental Settings

Dataset. This paper uses a dataset collected from two social networks, Douban (https://www.douban.com) and Weibo (https://www.weibo.com). Douban is a popular Chinese social website that allows users to comment on movies or share ideas, and Weibo is the most influential microblogging platform of China on which users post or forward blogs. In these networks, the follower-followee relationships are regarded as directed edges. Besides, some Douban users address their Weibo links on home pages, and thus the annotated identity linkages can be obtained according to them. The statistics are listed in Table 1.

Table 1. Statistics of Douban-Weibo Dataset.

Network	Users	Edges	Annotations
Douban	4399	204591	2309
Weibo	4414	202048	

Parameter Settings. In data preprocessing, the word segmentation toolkit LTP [2] is used to process Chinese contents. All the posts of a user are combined into a document, and then processed by the bag-of-words model. In FEM, as feature and structure are considered equally important, all hyperparameters (λ_{R1}, λ'_{R1}, λ_{R2}, λ'_{R2}) in Eq. (8) and Eq. (9) are set to 0.5. For all methods, the dimension of node embeddings is set to 100 for fair comparison. In NAM, the projection matrix G is initialized as an orthogonal matrix. As in previous work [10,11], the frequency of the alternating training of generator and discriminator is $1:5$. The annotation component's weight λ_S is set to 0.5, and the orthogonal component's weight λ_O is set to 0.5. For each source identity, the top k candidate target identities are selected to evaluate the effectiveness of SSF, where k ranges from 10 to 50. The ratio of the training set δ is set to 5%.

Metric. Following previous works [11,12,17], the hit-precision is selected as the evaluation metric for SSF. For each source identity x, its hit-precision is defined as follows:

$$h(x) = \frac{k - (\text{hit}(x) - 1)}{k}, \tag{20}$$

where $\text{hit}(x)$ is the position of correct target identity in the returned top-k candidate identity pairs. After that, the hit-precision is calculated by $\frac{1}{N}\sum_{i=1}^{N} h(x_i)$, where N is the number of source identities.

4.2 Baselines

To evaluate the performance of the proposed SSF, we compare our framework with the following state-of-the-art methods:

1. SNNA [11]: SNNA is a WGAN-based method. It learns node embeddings via a text-associated network embedding method [26], and then performs network alignment with the help of the WGAN.
2. CoLink [32]: CoLink is a weakly-supervised method. It proposes an attribute-based model with sequence-to-sequence [21], and employs the co-training of the attribute model and relationship model to generate high-quality linkages.
3. DeepLink [33]: Deeplink is an embedding based method. It takes advantage of deep neural networks to learn the latent semantics of user features and network structures, and then uses a dual-learning process to improve the identity linkage performance.
4. ULink [17]: ULink is a supervised method. It maps the identities from different networks into a latent space, and generates matching identity pairs by minimizing the distance between the user identities of the same individual.

The codes of ULink[1] and DeepLink[2] are public and thus directly used in our experiments. Other baselines are implemented by ourselves according to the original papers. Code and data of the proposed framework SSF will be made publicly available online.

4.3 Comparisons with Baselines

We compare the proposed framework SFF with the baselines in terms of the hit-precision, and the results are summarized in Table 2. SSF achieves the best performance among all methods, and outperforms the best baseline CoLink by 8.15% on average. CoLink performs the best among all baselines, maybe because it carefully designs an objective function to incorporate the attributes. But CoLink still suffers from error propagation in the later training stage. SNNA achieves undesirable performance, maybe because it does not take full advantage of the sample-level information when using WGAN to align networks at the distribution level. DeepLink attempts to learn a bijective mapping between two networks,

[1] http://www.lamda.nju.edu.cn/code_ULink.ashx.

[2] https://github.com/KDD-HIEPT/DeepLink.

Table 2. Comparisons with baselines (hit-precision).

Number of candidates k	10	20	30	40	50
SNNA	26.16%	32.30%	36.42%	39.70%	42.28%
CoLink	28.81%	33.71%	37.08%	39.64%	41.81%
DeepLink	16.94%	21.88%	25.22%	27.83%	30.01%
Ulink	14.27%	18.65%	21.8%	24.33%	26.45%
Ours (SSF)	**33.13%**	**41.10%**	**44.66%**	**50.01%**	**52.93%**

which is limited by the overlapping degree of different networks. ULink performs the worst maybe because it needs a large portion of annotations to achieve satisfactory performance. Interestingly, the hit-precisions of all methods improve by 5.84% on average when k changes from 10 to 20, while the improvement reduces to 2.39% when k changes from 40 to 50. This indicates a moderate increase of k will greatly improve the hit-precision, while the improvement is less significant for sufficiently large k.

4.4 Effect of Feature Extraction

To verify the effectiveness of the components in FEM, some variants of SSF are introduced as follows:

1. SSF-FEM$_\mathrm{d}$: it trains FEM without the reconstruction losses. This variant is used to verify the effectiveness of requirements R1 and R2.
2. SSF-FEM$_\mathrm{c}$: it trains FEM without the classification loss. This variant is used to study the effectiveness of the requirement R3.
3. SSF-FEM$_\mathrm{f}$: it only extracts features from the topology space in the encoders. This variant is introduced to explore the benefits of feature space.
4. SSF-FEM$_\mathrm{h}$: it removes the DHGNN layers from encoders. This variant is used to verify the effectiveness of DHGNN layers.

Figure 4 shows the hit-precisions of SSF and its variants mentioned above. Compared with SSF, the hit-precisions of SSF-FEM$_\mathrm{d}$, SSF-FEM$_\mathrm{c}$, SSF-FEM$_\mathrm{f}$ and SSF-FEM$_\mathrm{h}$ decrease by 15.80%, 11.88%, 2.99% and 3.77% on average, respectively, which indicates that each component of SSF is indispensable. Among all variants, the hit-precision decreases the most by ignoring the reconstruction losses during the training procedure, demonstrating the effectiveness of requirements R1 and R2. Compared with SSF-FEM$_\mathrm{c}$, SSF achieves higher performance with the help of requirement R3, which indicates that SSF can effectively utilize annotations in feature extraction. The performance of SSF-FEM$_\mathrm{f}$ is also unsatisfactory, which suggests that the information from the feature space dominates the performance of feature extraction.

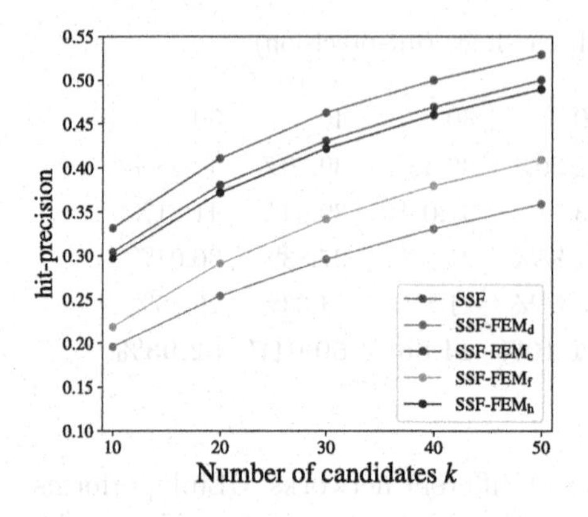

Fig. 4. Hit-precision on different k.

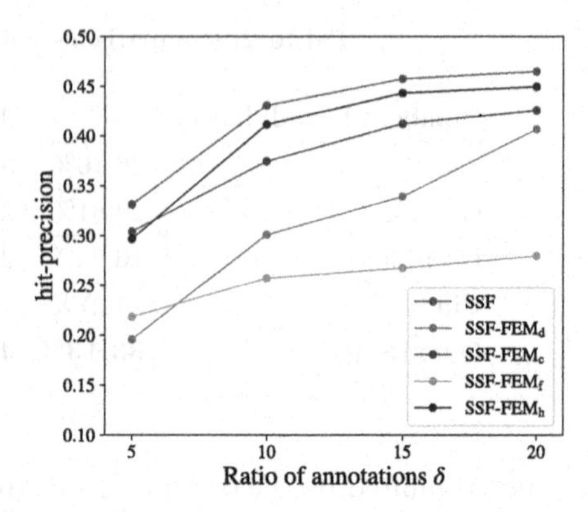

Fig. 5. Hit-precision on different δ.

To understand how annotations affect the feature extraction, we fix $k = 10$ and evaluate the performance of SSF and its variants with the ratio of annotations ranging from 5% to 20%. As shown in Fig. 5, SSF achieves higher hit-precisions than SSF-FEM$_\text{d}$, SSF-FEM$_\text{c}$, SSF-FEM$_\text{f}$ and SSF-FEM$_\text{h}$, which further validates the effectiveness of different components of FEM. Note that for SSF, SSF-FEM$_\text{c}$, SSF-FEM$_\text{f}$ and SSF-FEM$_\text{h}$, the marginal gains are diminishing with the increase of δ. However, for SSF-FEM$_\text{d}$, the marginal gains become more significant. That means the requirements R1 and R2 are more effective with fewer annotations.

4.5 Effect of Network Alignment

To verify the effectiveness of the proposed sampling strategy for WGAN-based UIL, variants with two different sampling strategies are introduced as follows:

1. SSF-NAM$_\text{d}$: it adopts the degree sampling strategy, which is used by previous sampling methods. It is used to verify the effectiveness of the proposed sampling strategy compared with previous methods.
2. SSF-NAM$_\text{u}$: it adopts a uniform sampling strategy without reweighting. It is used to study the influence of the reweighting strategy.

Table 3. Comparisons with variants (hit-precision).

Number of candidates k	10	20	30	40	50
SSF-NAM$_\text{d}$	32.04%	39.80%	44.84%	48.55%	51.41%
SSF-NAM$_\text{u}$	32.16%	40.37%	45.53%	49.51%	52.58%
Ours (SSF)	**33.13%**	**41.09%**	**46.26%**	**50.00%**	**52.92%**

We compare the hit-precisions of SSF-NAM$_d$, SSF-NAM$_u$ and SSF that adopt different sampling strategies, and the results are shown in Table 3. SSF outperforms SSF-NAM$_d$ by 1.35% on average in terms of the hit-precision, which indicates that our sampling strategy is better than previous sampling strategies. Besides, SSF-NAM$_u$ outperforms SSF-NAM$_d$ by 0.70% on average, verifying that the uniform sampling is more effective than the degree sampling for WGAN-based UIL. Without reweighting, the hit-precision decreases by 0.65% on average, indicating that the uniform sampling combined with the reweighting strategy can better guide the learning of WGAN.

5 Conclusion

This paper proposes a new semi-supervised framework for user identity linkage with the feature extraction model and network alignment model. The feature extraction model learns node embeddings from topology space and feature space with the help of dynamic hypergraph neural network. The network alignment model uses a new sampling strategy in the training process of WGAN, which focuses more on the sample-level information. The results of extensive experiments on the real-world dataset validate the effectiveness of the proposed framework.

There are several directions to be investigated in the future. First, we only consider the users' textual content in SSF. Other heterogeneous contents (e.g., images, videos, etc.) can also be incorporated into our framework to extract information from multiple content spaces. Second, the performance of semi-supervised UIL is mainly limited by the lack of annotations. Therefore, it is also in our interest to explore the self-supervised learning for UIL, which can extract information from the large-scale unsupervised data by using pretext tasks.

Acknowledgement. This work is supported by the National Key R&D Program of China (2018AAA0101203), and the National Natural Science Foundation of China (62072483).

References

1. Burnel, J.C., Fatras, K., Courty, N.: Generating natural adversarial hyperspectral examples with a modified Wasserstein GAN. arXiv preprint arXiv:2001.09993 (2020)
2. Che, W., Li, Z., Liu, T.: Ltp: a Chinese language technology platform. In: COLING, pp. 13–16 (2010)
3. Chen, H., Yin, H., Sun, X., Chen, T., Gabrys, B., Musial, K.: Multi-level graph convolutional networks for cross-platform anchor link prediction. In: KDD, pp. 1503–1511 (2020)
4. Cheng, A., et al.: Deep active learning for anchor user prediction. In: IJCAI, pp. 2151–2157 (2019)
5. Feng, Y., You, H., Zhang, Z., Ji, R., Gao, Y.: Hypergraph neural networks. In: AAAI, pp. 3558–3565 (2019)

6. Hamilton, W.L., Ying, R., Leskovec, J.: Inductive representation learning on large graphs. In: NIPS, pp. 1025–1035 (2017)
7. Jiang, J., Wei, Y., Feng, Y., Cao, J., Gao, Y.: Dynamic hypergraph neural networks. In: IJCAI, pp. 2635–2641 (2019)
8. Kipf, T., Welling, M.: Variational graph auto-encoders. In: NIPS Workshop on Bayesian Deep Learning (2016)
9. Kong, X., Zhang, J., Yu, P.: Inferring anchor links across multiple heterogeneous social networks. In: CIKM, pp. 179–188 (2013)
10. Li, C., et al.: Partially shared adversarial learning for semi-supervised multi-platform user identity linkage. In: CIKM, pp. 249–258 (2019)
11. Li, C., et al.: Adversarial learning for weakly-supervised social network alignment. In: AAAI, pp. 996–1003 (2019)
12. Li, C., et al.: Distribution distance minimization for unsupervised user identity linkage. In: CIKM, pp. 447–456 (2018)
13. Liu, J., Zhang, F., Song, X., Song, Y.I., Lin, C.Y., Hon, H.: What's in a name?: an unsupervised approach to link users across communities. In: WSDM, pp. 495–504 (2013)
14. Liu, L., Cheung, W., Li, X., Liao, L.: Aligning users across social networks using network embedding. In: IJCAI, pp. 1774–1780 (2016)
15. Liu, S., Wang, S., Zhu, F., Zhang, J., Krishnan, R.: Hydra: Large-scale social identity linkage via heterogeneous behavior modeling. In: SIGMOD, pp. 51–62 (2014)
16. Man, T., Shen, H., Liu, S., Jin, X., Cheng, X.: Predict anchor links across social networks via an embedding approach. In: IJCAI, pp. 1823–1829 (2016)
17. Mu, X., Zhu, F., Lim, E., Xiao, J., Wang, J., Zhou, Z.: User identity linkage by latent user space modelling. In: KDD, pp. 1775–1784 (2016)
18. Nie, Y., Jia, Y., Li, S., Zhu, X., Li, A., Zhou, B.: Identifying users across social networks based on dynamic core interests. Neurocomputing **210**, 107–115 (2016)
19. Shen, Y., Jin, H.: Controllable information sharing for user accounts linkage across multiple online social networks. In: CIKM, pp. 381–390 (2014)
20. Shu, K., Wang, S., Tang, J., Zafarani, R., Liu, H.: User identity linkage across online social networks: a review. ACM SIGKDD Explor. Newslett. **18**(2), 5–17 (2017)
21. Sutskever, I., Vinyals, O., Le, Q.V.: Sequence to sequence learning with neural networks. In: NIPS, pp. 3104–3112 (2014)
22. Tan, S., Guan, Z., Cai, D., Qin, X., Bu, J., Chen, C.: Mapping users across networks by manifold alignment on hypergraph. In: AAAI, pp. 159–165 (2014)
23. Villani, C.: Optimal Transport: Old and New, vol. 338. Springer Science & Business Media, Heidelberg (2008)
24. Wang, X., Zhu, M., Bo, D., Cui, P., Shi, C., Pei, J.: AM-GCN: adaptive multi-channel graph convolutional networks. In: SIGKDD, pp. 1243–1253 (2020)
25. Xing, L., Deng, K., Wu, H., Xie, P., Zhao, H.V., Gao, F.: A survey of across social networks user identification. IEEE Access **7**, 137472–137488 (2019)
26. Yang, C., Liu, Z., Zhao, D., Sun, M., Chang, E.: Network representation learning with rich text information. In: IJCAI, pp. 2111–2117 (2015)
27. Zafarani, R., Liu, H.: Connecting users across social media sites: a behavioral-modeling approach. In: KDD, pp. 41–49 (2013)
28. Zhang, J., Philip, S.Y.: Integrated anchor and social link predictions across social networks. In: IJCAI, pp. 2125–2131 (2015)
29. Zhang, S., Tong, H.: Final: Fast attributed network alignment. In: KDD, pp. 1345–1354 (2016)

30. Zhang, Y., Tang, J., Yang, Z., Pei, J., Yu, P.: COSNET: connecting heterogeneous social networks with local and global consistency. In: KDD, pp. 1485–1494 (2015)
31. Zhang, Y., Wang, L., Li, X., Xiao, C.: Social identity link across incomplete social information sources using anchor link expansion. In: Bailey, J., Khan, L., Washio, T., Dobbie, G., Huang, J.Z., Wang, R. (eds.) PAKDD 2016. LNCS (LNAI), vol. 9651, pp. 395–408. Springer, Cham (2016). https://doi.org/10.1007/978-3-319-31753-3_32
32. Zhong, Z., Cao, Y., Guo, M., Nie, Z.: CoLink: an unsupervised framework for user identity linkage. In: AAAI, pp. 5714–5721 (2018)
33. Zhou, F., Liu, L., Zhang, K., Trajcevski, G., Wu, J., Zhong, T.: DeepLink: a deep learning approach for user identity linkage. In: INFOCOM, pp. 1313–1321 (2018)
34. Zhou, J., Fan, J.: TransLink: user identity linkage across heterogeneous social networks via translating embeddings. In: INFOCOM, pp. 2116–2124 (2019)
35. Zhou, X., Liang, X., Zhang, H., Ma, Y.: Cross-platform identification of anonymous identical users in multiple social media networks. IEEE Trans. Knowl. Data Eng. **28**(2), 411–424 (2015)

Personality Traits Prediction Based on Sparse Digital Footprints via Discriminative Matrix Factorization

Shipeng Wang[1], Daokun Zhang[2], Lizhen Cui[1,3], Xudong Lu[1,3], Lei Liu[1,3], and Qingzhong Li[1,3](✉)

[1] School of Software, Shandong University, Jinan, China
wangshipeng95@mail.sdu.edu.cn, {clz,dongxul,l.liu,lqz}@sdu.edu.cn
[2] Discipline of Business Analytics, The University of Sydney, Sydney, Australia
daokun.zhang@sydney.edu.au
[3] Joint SDU-NTU Centre for Artificial Intelligence Research (C-FAIR), Shandong University, Jinan, China

Abstract. Identifying individuals' personality traits from their digital footprints has been proved able to improve the service of online platforms. However, due to the privacy concerns and legal restrictions, only some sparse, incomplete and anonymous digital footprints can be accessed, which seriously challenges the existing personality traits identification methods. To make the best of the available sparse digital footprints, we propose a novel personality traits prediction algorithm through jointly learning discriminative latent features for individuals and a personality traits predictor performed on the learned features. By formulating a discriminative matrix factorization problem, we seamlessly integrate the discriminative individual feature learning and personality traits predictor learning together. To solve the discriminative matrix factorization problem, we develop an alternative optimization based solution, which is efficient and easy to be parallelized for large-scale data. Experiments are conducted on the real-world Facebook like digital footprints. The results show that the proposed algorithm outperforms the state-of-the-art personality traits prediction methods significantly.

Keywords: Personality traits prediction · Digital footprints · Discriminative matrix factorization · Alternative optimization

1 Introduction

Human personality is a lasting feature that can reflect the nature of human behavior and influence peoples behavioral decisions. Just like sentiment analysis[2,3], analyzing people's personalities is quite indispensable to identifying mental disorders, developing the e-commerce platforms and search engines, and improving the collaboration efficiency of the network of crowd intelligence [8].

C. S. Jensen et al. (Eds.): DASFAA 2021, LNCS 12682, pp. 692–700, 2021.
https://doi.org/10.1007/978-3-030-73197-7_47

Personality identification and measurement in traditional psychology mainly relies on questionnaire method and interview method according to psychological measures such as Five Factor Model (FFM) [7]. Nevertheless, these two methods require a lot of manpower, material resources and financial costs and inevitably involve some subjective bias. Fortunately, the widely existing people's digital footprints on the Internet provide the possibility to analyze people's personality traits in a much more efficient way, which have been proved able to help estimate personality traits accurately [12]. However, most of the existing digital footprints based methods [4,6,12] rely heavily on the availability of rich digital footprints. In reality, due to the privacy protection consideration and legal restrictions, it is hard to access rich, complete and high-quality human digital footprint records. Comparatively, only some incomplete digital footprints are available, which seriously challenges the digital footprints based personality traits prediction methods, with properties of high dimensionality, sparsity and anonymousness.

To make the best of the available incomplete digital footprints, we propose a novel Discriminative Matrix Factorization based Personality Traits Prediction (DMF-PTP) algorithm. The proposed DMF-PTP algorithm jointly learns discriminative latent features from people's high-dimensional, sparse and anonymous digital footprints with matrix factorization, and a personality traits predictor performed on the learned latent digital footprint features simultaneously. To solve the discriminative matrix factorization problem, we develop an alternative optimization based solution, which is efficient and easy to be parallelized for large-scale training data. The contributions of this paper can be summarized as follows:

- We propose a novel personality traits prediction algorithm from incomplete digital footprints by formulating it as a discrimiative matrix factorization problem and develop an efficient alternate optimization based solution.
- We conduct personality traits prediction experiments on the anonymized Facebook like digital footprints dataset. Compared with the state-of-the-art baseline methods, our proposed DMF-PTP algorithm achieves significantly better personality trait prediction performance.

2 Related Work

Existing research work on personality traits prediction can be categorized into two main groups: **1) feature engineering based methods** [4,10] that first extract digital footprint features indicative to personality traits and then predict personality traits with the traditional machine learning algorithms, and **2) personality traits modeling based methods** [1,11] that predict personality traits by directly modeling the relation between raw digital footprint features and personality traits. However, all of those methods rely heavily on the availability of rich digital footprints and fail to perform well for predicting personality traits based on sparse digital footprints. To make the best of the available sparse digital footprints for personality trait prediction, we leverage the power of matrix factorization, which provides an elegant way for learning informative representations from sparse raw features [9,13].

3 Problem Definition

Suppose we are given a set of anonymized sparse digital footprint records formed by some social platform users, which are denoted by $\boldsymbol{M} \in \mathbb{R}^{m \times n}$, with m being the number of social platform users and n being the number of recorded digital footprints. The element at the i-th row and j-th column of $\boldsymbol{M}$, $\boldsymbol{M}_{ij} \in \{0,1\}$, indicates whether user i has the j-th digital footprint, with 1 for true and 0 for false. In addition, users are attached with personality trait scores $Y \in \mathbb{R}^m$, with Y_i being the personality trait score of the i-th user. According to the task, the personality trait score can be measured on five dimensions [4].

Given a set of labeled personality trait scores $\boldsymbol{Y}_{\mathcal{L}}$, with $\mathcal{L}$ being the set of indices of labeled users, and the sparse user digital footprint records $\boldsymbol{M}$, we aim to learn discriminative low-dimensional latent user digital footprint representations $\boldsymbol{u}_i \in \mathbb{R}^k$ for each user i, and simultaneously a mapping function $f(\cdot)$ from the learned latent users' representations to their personality trait scores. With the learned latent representations and the mapping function, for each unlabeled user i, his/her personality trait score can be predicted by $f(\boldsymbol{u}_i)$, *i.e.*, applying the learned mapping function $f(\cdot)$ to the learned latent digital footprint representation $\boldsymbol{u}_i$ of unlabeled user i.

4 Personality Traits Prediction with Discriminative Matrix Factorization

4.1 Objective Formulation

Our formulation combines the learning of latent user digital footprint representations with matrix factorization, and the learning of personality traits predictor into a unified objective. Mathematically, our learning objective is

$$\min_{\boldsymbol{U},\boldsymbol{V},\boldsymbol{w}} \mathcal{J}(\boldsymbol{U},\boldsymbol{V},\boldsymbol{w}), \tag{1}$$

where

$$\begin{aligned} \mathcal{J}(\boldsymbol{U},\boldsymbol{V},\boldsymbol{w}) = \frac{1}{2}\left\|\boldsymbol{M} - \boldsymbol{U}\boldsymbol{V}^\top\right\|_F^2 + \frac{\gamma}{2}\sum_{i\in\mathcal{L}}(Y_i - \boldsymbol{u}_i^\top \boldsymbol{w})^2 \\ + \frac{\lambda}{2}\left(\|\boldsymbol{U}\|_F^2 + \|\boldsymbol{V}\|_F^2 + \|\boldsymbol{w}\|_2^2\right). \end{aligned} \tag{2}$$

In Eq. (2), the sparse user digital footprints matrix $\boldsymbol{M}$ is factorized into two low-rank matrices—$\boldsymbol{U} \in \mathbb{R}^{m\times k}$ and $\boldsymbol{V} \in \mathbb{R}^{n\times k}$—that carry the latent representations of users and digital footprints, with $\boldsymbol{u}_i \in \mathbb{R}^k$, the transpose of the i-th row of $\boldsymbol{U}$, being the latent representation of the i-th user, and $\boldsymbol{v}_j \in \mathbb{R}^k$, the transpose of the j-th row of $\boldsymbol{V}$, being the latent representation of the j-th digital footprint.

Together with the matrix factorization, the personality traits of labeled users, $Y_{\mathcal{L}}$, are predicted via the linear regression model performed on the learned user representations, $\boldsymbol{u}_i$, by minimizing the squared loss $\sum_{i \in \mathcal{L}}(Y_i - \boldsymbol{u}_i^\top \boldsymbol{w})^2$, with $\boldsymbol{w}$ being the regression coefficients. To ensure the low rank property of $\boldsymbol{U}$ and $\boldsymbol{V}$, and avoid overfitting the regression parameter $\boldsymbol{w}$, we add the regularizer term $\|\boldsymbol{U}\|_F^2 + \|\boldsymbol{V}\|_F^2 + \|\boldsymbol{w}\|_2^2$ in Eq. (2), with γ and λ used to weight different components.

4.2 Solving the Optimization Problem

Our objective is to minimize $\mathcal{J}(\boldsymbol{U}, \boldsymbol{V}, \boldsymbol{w})$ with regards to $\boldsymbol{U}$, $\boldsymbol{V}$ and $\boldsymbol{w}$. As $\mathcal{J}(\boldsymbol{U}, \boldsymbol{V}, \boldsymbol{w})$ is not convex with regards to $\boldsymbol{U}$, $\boldsymbol{V}$ and $\boldsymbol{w}$, it is hard to find its global optima analytically. However, it can be easily proved that, with any two of $\boldsymbol{U}$, $\boldsymbol{V}$ and $\boldsymbol{w}$ fixed, $\mathcal{J}(\boldsymbol{U}, \boldsymbol{V}, \boldsymbol{w})$ is convex with the remainder variable. Thus, after randomly initializing $\boldsymbol{U}, \boldsymbol{V}$ and $\boldsymbol{w}$ as $\boldsymbol{U}^{(0)}, \boldsymbol{V}^{(0)}$, and $\boldsymbol{w}^{(0)}$ respectively, we can update $\boldsymbol{U}, \boldsymbol{V}, \boldsymbol{w}$ alternately towards a local minimum of $\mathcal{J}(\boldsymbol{U}, \boldsymbol{V}, \boldsymbol{w})$:

$$\boldsymbol{U}^{(t)} \leftarrow \arg\min_{\boldsymbol{U}} \mathcal{J}(\boldsymbol{U}, \boldsymbol{V}^{(t-1)}, \boldsymbol{w}^{(t-1)}); \tag{3}$$

$$\boldsymbol{V}^{(t)} \leftarrow \arg\min_{\boldsymbol{V}} \mathcal{J}(\boldsymbol{U}^{(t)}, \boldsymbol{V}, \boldsymbol{w}^{(t-1)}); \tag{4}$$

$$\boldsymbol{w}^{(t)} \leftarrow \arg\min_{\boldsymbol{w}} \mathcal{J}(\boldsymbol{U}^{(t)}, \boldsymbol{V}^{(t)}, \boldsymbol{w}). \tag{5}$$

Here, $\boldsymbol{U}^{(t)}$, $\boldsymbol{V}^{(t)}$ and $\boldsymbol{w}^{(t)}$ respectively denote the values of $\boldsymbol{U}$, $\boldsymbol{V}$ and $\boldsymbol{w}$ after the t-th update. $\boldsymbol{U}$, $\boldsymbol{V}$ and $\boldsymbol{w}$ are updated with Eq. (3), (4) and (5) respectively.

Updating $\boldsymbol{U}$. With $\boldsymbol{V}$ and $\boldsymbol{w}$ fixed, updating $\boldsymbol{U}$ can be performed by separately updating each $\boldsymbol{u}_i$:

$$\boldsymbol{u}_i \leftarrow \arg\min_{\boldsymbol{u}} f_i(\boldsymbol{u}), \tag{6}$$

where

$$f_i(\boldsymbol{u}) = \begin{cases} \frac{1}{2}\sum_{j=1}^{n}(\boldsymbol{M}_{ij} - \boldsymbol{u}^\top \boldsymbol{v}_j)^2 + \frac{\gamma}{2}(Y_i - \boldsymbol{u}^\top \boldsymbol{w})^2 + \frac{\lambda}{2}\boldsymbol{u}^\top \boldsymbol{u}, & \text{if } i \in \mathcal{L}; \\ \frac{1}{2}\sum_{j=1}^{n}(\boldsymbol{M}_{ij} - \boldsymbol{u}^\top \boldsymbol{v}_j)^2 + \frac{\lambda}{2}\boldsymbol{u}^\top \boldsymbol{u}, & \text{if } i \notin \mathcal{L}. \end{cases} \tag{7}$$

As is shown in Eq. (7), $f_i(\boldsymbol{u})$ is s a convex function of $\boldsymbol{u}$. We can solve the optimization problem in Eq. (6) with the Conjugate Gradient method, after calculating its gradient and Hessian Matrix as follows:

$$\nabla f_i(\boldsymbol{u}) = \begin{cases} -\sum_{j=1}^{n}(\boldsymbol{M}_{ij} - \boldsymbol{u}^\top \boldsymbol{v}_j)\boldsymbol{v}_j - \gamma(\boldsymbol{Y}_i - \boldsymbol{u}^\top \boldsymbol{w})\boldsymbol{w} + \lambda \boldsymbol{u}, & \text{if } i \in \mathcal{L}; \\ -\sum_{j=1}^{n}(\boldsymbol{M}_{ij} - \boldsymbol{u}^\top \boldsymbol{v}_j)\boldsymbol{v}_j + \lambda \boldsymbol{u}, & \text{if } i \notin \mathcal{L}. \end{cases} \tag{8}$$

$$\nabla^2 f_i(\boldsymbol{u}) = \begin{cases} \sum_{j=1}^{n} \boldsymbol{v}_j \boldsymbol{v}_j^\top + \gamma \boldsymbol{w}\boldsymbol{w}^\top + \lambda \boldsymbol{I}, & \text{if } i \in \mathcal{L}; \\ \sum_{j=1}^{n} \boldsymbol{v}_j \boldsymbol{v}_j^\top + \lambda \boldsymbol{I}, & \text{if } i \notin \mathcal{L}. \end{cases} \tag{9}$$

In Eq. (9), $\boldsymbol{I}$ is the $k \times k$-dimensional identity matrix.

Updating $\boldsymbol{V}$. When $\boldsymbol{U}$ and $\boldsymbol{w}$ are fixed, we can update $\boldsymbol{V}$ by separately updating each $\boldsymbol{v}_j$:

$$\boldsymbol{v}_j \leftarrow \arg\min_{\boldsymbol{v}} g_j(\boldsymbol{v}), \tag{10}$$

where

$$g_j(\boldsymbol{v}) = \frac{1}{2}\sum_{i=1}^{m}(\boldsymbol{M}_{ij} - \boldsymbol{u}_i^\top \boldsymbol{v})^2 + \frac{\lambda}{2}\boldsymbol{v}^\top \boldsymbol{v}. \tag{11}$$

Similarly, with the convex function $g_j(\boldsymbol{v})$, we use the Conjugate Gradient method to solve the optimization problem in Eq. (10), after calculating $g_j(\boldsymbol{v})$'s gradient and Hessian Matrix as:

$$\nabla g_j(\boldsymbol{v}) = -\sum_{i=1}^{m}(M_{ij} - \boldsymbol{u}_i^\top \boldsymbol{v})\boldsymbol{u}_i + \lambda \boldsymbol{v},\ \nabla^2 g_j(\boldsymbol{v}) = \sum_{i=1}^{m} \boldsymbol{u}_i \boldsymbol{u}_i^\top + \lambda \boldsymbol{I}. \tag{12}$$

Updating $\boldsymbol{w}$. We can update the regression coefficients $\boldsymbol{w}$ with $\boldsymbol{U}$ and $\boldsymbol{V}$ fixed as following:

$$\boldsymbol{w} \leftarrow \arg\min_{\boldsymbol{w}} h(\boldsymbol{w}), \tag{13}$$

where

$$h(\boldsymbol{w}) = \frac{\gamma}{2}\sum_{i \in \mathcal{L}}(\boldsymbol{Y}_i - \boldsymbol{u}_i^\top \boldsymbol{w})^2 + \frac{\lambda}{2}\boldsymbol{w}^\top \boldsymbol{w}. \tag{14}$$

Similarly, we use the Conjugate Gradient method to solve the optimization problem in Eq. (13), after calculating the gradient and Hessian matrix of the convex function $h(\boldsymbol{w})$ as

$$\nabla h(\boldsymbol{w}) = -\gamma \sum_{i \in \mathcal{L}}(\boldsymbol{Y}_i - \boldsymbol{u}_i^\top \boldsymbol{w})\boldsymbol{w} + \lambda \boldsymbol{w},\ \nabla^2 h(\boldsymbol{w}) = \gamma \sum_{i \in \mathcal{L}} \boldsymbol{u}_i \boldsymbol{u}_i^\top + \lambda \boldsymbol{I}. \tag{15}$$

Algorithm 1: Solving the optimization problem in Eq. (1)

Input:
Sparse digital footprints matrix $\boldsymbol{M}$ and labeled personality traits scores $\boldsymbol{y}_{\mathcal{L}}$;
Output:
Matrix $\boldsymbol{U}$, $\boldsymbol{V}$ and vector $\boldsymbol{w}$;
1: $(\boldsymbol{U}, \boldsymbol{V}, \boldsymbol{w}) \leftarrow$ Random initialization;
2: **repeat**
3: $\quad \boldsymbol{U} \leftarrow$ Updating by solving Eq. (6);
4: $\quad \boldsymbol{V} \leftarrow$ Updating by solving Eq. (10);
5: $\quad \boldsymbol{w} \leftarrow$ Updating by solving Eq. (13);
6: **until** convergence or a fixed number of iterations expire;
7: **return** $\boldsymbol{U}$, $\boldsymbol{V}$ and $\boldsymbol{w}$;

Algorithm 1 gives the procedure for solving the optimization problem in Eq. (1) with alternative optimization. In Algorithm 1, the time complexity of updating $\boldsymbol{U}$ is $\mathcal{O}(k^2mn)$, by taking the number of iterations of Conjugate gradient descent as a constant. Similarly, the time complexity of updating $\boldsymbol{V}$ is $\mathcal{O}(k^2mn)$, and updating $\boldsymbol{w}$ requires $\mathcal{O}(k^2|\mathcal{L}|)$ time. By taking the number of iterations of alternative parameter update as a constant, the overall time complexity of our algorithm is $\mathcal{O}(k^2mn)$.

4.3 Inferring Personality Traits for Unlabeled Users

With the learned latent user representations $\boldsymbol{u}_i$ and the linear regression coefficients $\boldsymbol{w}$, for each unlabeled user $i \notin \mathcal{L}$, his or her personality trait score can be predicted by

$$\hat{Y}_i = \boldsymbol{u}_i^\top \boldsymbol{w}, \;\; i \notin \mathcal{L}. \tag{16}$$

5 Experiments

5.1 Dataset and Experimental Setting

To verify the effectiveness of our method, we conduct personality traits prediction experiments on Facebook likes digital footprints data[1], formed by 110,728 Facebook users with 1,580,284 likes and their personality trait scores measured by five dimensions: extraversion, agreeableness, conscientiousness, neuroticism and openness. Following [5], we preprocess the dataset by selecting users with at least 100 likes and selecting likes visited by more than 200 users.

[1] http://mypersonality.org.

We compare DMF-PTP with two competitive baseline methods: Raw+LR and SVD+LR [4], which respectively predict personality traits using linear regression with raw user digital footprint features $\boldsymbol{M}$ and the latent user features extracted from $\boldsymbol{M}$ by SVD. We used two criteria to evaluate personality traits prediction performance: Pearson Correlation Coefficient (PCC) and Root Mean Square Error (RMSE).

We randomly select 80% of users as training set and the rest users as test set. We repeat the random training/test set split for 10 times and report the averaged PCC and RMSE. We set the number of iterations for alternate optimization to 10. The parameters k, λ, and γ are respectively set to 10, 10, and 1000.

5.2 Experimental Results

Table 1. Personality traits prediction results of three methods

Measure	Openness		Conscientiousness		Extraversion		Agreeableness		Neuroticism	
	PCC ↑	RMSE ↓	PCC ↑	RMSE ↓	PCC ↑	RMSE ↓	PCC ↑	RMSE ↓	PCC ↑	RMSE ↓
Raw+LR	0.279	0.941	0.117	1.191	0.148	1.106	0.075	1.401	0.182	1.114
SVD+LR	0.317	**0.925**	0.120	1.278	0.160	1.126	0.156	**0.952**	0.197	1.124
DMF-PTP	**0.450**	0.970	**0.192**	**0.981**	**0.247**	**0.955**	**0.225**	0.962	**0.266**	**0.978**

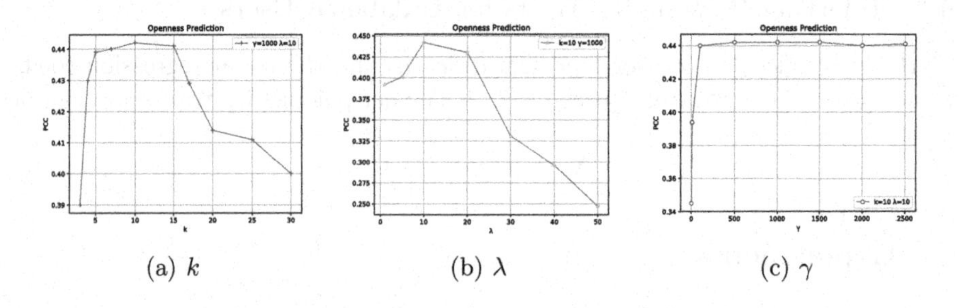

(a) k (b) λ (c) γ

Fig. 1. Parameter sensitivity on predicting openness

Table 1 compares the personality traits prediction performance of three compared methods on five measures. For each measure, the best PCC and RMSE scores are highlighted by **bold**. From Table 1, we can see that DMF-PTP outperforms Raw+LR and SVD+LR model in most cases, except for the RMSE scores on openness and agreeableness, where DMF-PTP is slightly worse than the best performer. This attributes to DMF-PTP's special capacity of learning personality traits indicative user latent digital footprint representations, through coupling it with personality traits predictor learning task.

5.3 Parameter Sensitivity

We also study the sensitivity of our method with regards to the three parameters: k, γ and λ, by evaluating the openness prediction PCC. Figure 1 plots the PCC scores with the changes of the three parameters. DMF-PTP achieves the best PCC scores when k takes values from 5 to 15. With the increase of λ, the PCC score of DMF-PTP first gradually rises to the peak at $\lambda = 10$ and then starts to decrease. In comparison, the performance of DMF-PTP is stable when γ changes in a proper range (larger than 100).

6 Conclusion

In this paper, we propose a novel algorithm DMF-PTP to predict personality traits from users' sparse digital footprints, which jointly learns discriminative latent features for individuals and a personality traits predictor on the learned features. By formulating a discrimative matrix factorization problem, we seamlessly integrated the discriminative individual feature learning and the personality traits predictor learning together. We developed an alternate optimization based algorithm to solve the formulated discriminative matrix factorization problem, which is efficient and easy to be parallelized for large-scale data. Experiments on real-world Facebook likes digital footprints data were conducted. The results confirm the effectiveness of DMF-PTP.

Acknowledgements. This work is partially supported by the National Key R&D Program No.2017YFB1400100; the NSFC No.91846205; the Innovation Method Fund of China No.2018IM020200; the Shandong Key R&D Program No.2018YFJH0506, 2019JZZY011007.

References

1. Basu, A., Dasgupta, A., Thyagharajan, A., Routray, A., Guha, R., Mitra, P.: A portable personality recognizer based on affective state classification using spectral fusion of features. IEEE Trans. Affect. Comput. **9**(3), 330–342 (2018)
2. He, J., Liu, H., Zheng, Y., Tang, S., He, W., Du, X.: Bi-labeled LDA: Inferring interest tags for non-famous users in social network. Data Sci. Eng. **5**(1), 27–47 (2020)
3. Ito, T., Tsubouchi, K., Sakaji, H., Yamashita, T., Izumi, K.: Contextual sentiment neural network for document sentiment analysis. Data Sci. Eng. **5**(2), 180–192 (2020)
4. Kosinski, M., Stillwell, D., Graepel, T.: Private traits and attributes are predictable from digital records of human behavior. Proc. Nat. Acad. Sci. **110**(15), 5802–5805 (2013)
5. Kosinski, M., Wang, Y., Lakkaraju, H., Leskovec, J.: Mining big data to extract patterns and predict real-life outcomes. Psychol. Methods **21**(4), 493 (2016)
6. Liu, X., Zhu, T.: Deep learning for constructing microblog behavior representation to identify social media user's personality. PeerJ Comput. Sci. **2**, e81 (2016)

7. McCrae, R.R., John, O.P.: An introduction to the five-factor model and its applications. J. Pers. **60**(2), 175–215 (1992)
8. Wang, S., Cui, L., Liu, L., Lu, X., Li, Q.: Projecting real world into crowdintell network: a methodology. Int. J. Crowd Sci. **3**(2), 138–154 (2019)
9. Wang, S., Pedrycz, W., Zhu, Q., Zhu, W.: Subspace learning for unsupervised feature selection via matrix factorization. Pattern Recogn. **48**(1), 10–19 (2015)
10. Wei, H., et al.: Beyond the words: Predicting user personality from heterogeneous information. In: Proceedings of the 10th ACM International Conference on Web Search and Data Mining, pp. 305–314 (2017)
11. Ye, Z., Du, Y., Zhao, L.: Predicting personality traits of users in social networks. In: Yin, H. (ed.) IDEAL 2017. LNCS, vol. 10585, pp. 181–191. Springer, Cham (2017). https://doi.org/10.1007/978-3-319-68935-7_21
12. Youyou, W., Kosinski, M., Stillwell, D.: Computer-based personality judgments are more accurate than those made by humans. Proc. Nat. Acad. Sci. **112**(4), 1036–1040 (2015)
13. Zhang, D., Yin, J., Zhu, X., Zhang, C.: Collective classification via discriminative matrix factorization on sparsely labeled networks. In: Proceedings of the 25th ACM International on Conference on Information and Knowledge Management, pp. 1563–1572 (2016)

A Reinforcement Learning Model for Influence Maximization in Social Networks

Chao Wang, Yiming Liu, Xiaofeng Gao(✉), and Guihai Chen

Shanghai Key Laboratory of Scalable Computing and Systems,
Department of Computer Science and Engineering, Shanghai Jiao Tong University,
Shanghai, China
{wangchao.2014,lucien}@sjtu.edu.cn, {gao-xf,gchen}@cs.sjtu.edu.cn

Abstract. Social influence maximization problem has been widely studied by the industrial and theoretical researchers over the years. However, with the skyrocketing scale of networks and growing complexity of application scenarios, traditional approximation approaches suffer from weak approximation guarantees and bad empirical performances. What's more, they can't be applied to new users in dynamic network. To tackle those problems, we introduce a social influence maximization algorithm via graph embedding and reinforcement learning. Nodes are represented in the graph with their embedding, and then we formulate a reinforcement learning model where both the states and the actions can be represented with vectors in low dimensional space. Now we can deal with graphs under various scenarios and sizes, just by learning parameters for the deep neural network. Hence, our model can be applied to both large-scale and dynamic social networks. Extensive real-world experiments show that our model significantly outperforms baselines across various data sets, and the algorithm learned on small-scale graphs can be generalized to large-scale ones.

Keywords: Social influence maximization · Reinforcement learning · Graph neural network

1 Introduction

Nowadays, online social network is everywhere around us. We have never communicated with each other so frequently before. Today, social network is not only a platform for chatting, it also has great economic values, e.g., viral marketing [4].

This work was supported by the National Key R&D Program of China [2020YFB1707903]; the National Natural Science Foundation of China [61872238, 61972254], the Huawei Cloud [TC20201127009], the CCF-Tencent Open Fund [RAGR20200105], and the Tencent Marketing Solution Rhino-Bird Focused Research Program [FR202001]. The authors also would like to thank Ge Yan for his contribution on the early version of this paper.

C. S. Jensen et al. (Eds.): DASFAA 2021, LNCS 12682, pp. 701–709, 2021.
https://doi.org/10.1007/978-3-030-73197-7_48

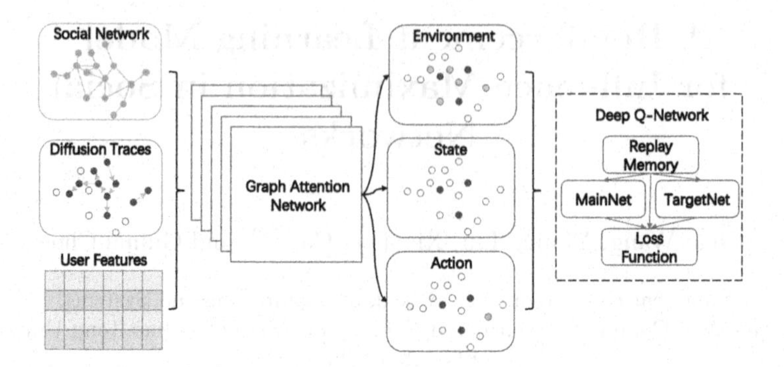

Fig. 1. Illustration of our proposed framework.

Motivated by various applications, social influence maximization problem has been widely studied by a lot of researchers over the last decade.

Influence Maximization problem was first proposed by Kempe et al. [2]. They worked out the greedy algorithm framework with the theoretic guarantee of approximation ratio of $1 - 1/e$. Generally, with the gigantic scale of network, traditional approaches based on enumeration or branch-and-bound are ineffective. On the contrary, heuristics are fast and effective algorithms, but they may require substantial problem-specific researches and trial-and-error for algorithm designers. Although approximation algorithms are mostly used for solving those problems, they may suffer from weak approximation ratio and bad empirical performance.

In addition, previous methods in IM problems didn't pay attention to how edge weights are generated. Generally, the probability influence between users is randomly generated in approximation algorithms. Although some studies pay attention to influence estimation rather than influence maximization, few existing works jointly captures the diffusion traces and user features. In particular, existing models cannot effectively learn probabilities for the edges without observed influence propagation. Similar to our work, [6] solve the combinatorial optimization with Graph Convolutional Network (GCN) and guided tree search, but they lack strategies to maximize user influence.

To tackle the above problems, we purpose a social influence maximization algorithm via graph embedding and reinforcement learning, illustrated in Fig. 1. Unlike much previous work that analyzes node attributes and graph structures according to specific problems, we use a method based on graph embedding network to capture the local network structure information for each node. By this method, the model is applicable to networks of different structures and sizes. If a new user is added to the social network or a novel friend relationship is formed, our model can easily handle it by ignoring the differences of neighbor nodes compared with GAT. Therefore, our model can be applied to dynamic social networks. Based on user representation and social influence prediction, we can use reinforcement learning to learn a greedy policy and select the seed set to maximize social influence. With those methods, the policy is generalized so that it can be used to solve problems in different social networks. We test our model by using it to select seeds on different real-world data sets: Facebook, Twitter, YouTube, Weibo, and Flickr. By analyzing the results, we find that our model outperforms baselines across various graph types. Extensive experiments show that the learned algorithm using small-scale graphs can generalize to test large-scale ones.

2 Problem Formulation

In this section, we will show the formulation of influence maximization and the influence function. We also give a clearly description of our framework.

2.1 Influence Maximization in Social Networks

Assume there is a kind of selection in social networks with a seed budget k. We construct a weighted digraph $G = (V, E)$, where V represents the set of users in the network, E represents the set of relationships in the network. Denote $\phi(S)$ as the overall influence spread. Our goal is to find a seed set S which contains no more than k users to maximize the influence spread. Now we formulate the IM problem.

Definition 1 (Influence Maximization (IM)). *Given a social graph $G = (V, E)$ and seed budget k, the goal is to find a small seed set of nodes S^* such that*

$$S^* = \arg\max_S \{\phi(S)\}$$

where $S \subseteq V$, $k \leq |V|$, $|S| \leq k$.

The IM problem can be proved to be NP-hard by taking it as a special case of the hitting set problem, and it is proved that $\phi(\cdot)$ is submodular function [4].

2.2 Graph Embedding for Representing Users

We wish to quantify to what extent an user u influences the probability that user v favorites (or buys) a item after user u disseminates the information of this item. Rather than the weight matrix W of the network G, we prefer node-based information representation to edge-based information representation for the reason that the number of edges in social networks is much larger than that of nodes. Hence, we need to represent users as useful and predictive embeddings.

Considering both the feature of the user and the local structure of a node v can be very complex and may depend on complicated statistics such as global/local degree distribution, distance to tagged nodes in these problems, we will use a deep neural network architecture to represent users. Graph Attention Network (GAT) [11] is a recent proposed technique that introduces the attention mechanism into Graph Convolutional Networks. As opposed to GCNs,

GAT allows us to assign different importance to nodes connecting to a same neighbor.

In order to transform the input features into higher-level features, we perform self-attention $atte : \mathbb{R}^{F'} \times \mathbb{R}^{F'} \rightarrow \mathbb{R}$ on every node to compute the attention coefficients to measure the importance of j's features to node i on graph G.

$$e_{ij} = atte\left(\mathbf{W}\mathrm{h}_i, \mathbf{W}\mathrm{h}_j\right) = \text{ LeakyReLU } \left(\mathbf{a}^T \left[\mathbf{W}\mathrm{h}_i \| \mathbf{W}\mathrm{h}_j\right]\right) \tag{1}$$

where $atte(\cdot, \cdot)$ is the attention mechanism function, which can be implemented by a neural network with weight vector $\mathbf{a}$ and a LeakyReLU nonlinearity like Velickovic et al. $\cdot \parallel \cdot$ is the concatenation operation.

Then, we normalize the coefficients using the softmax function to make them comparable across different nodes

$$\alpha_{ij} = \text{softmax}_j(e_{ij}) = \frac{\exp\left(\text{ LeakyReLU } \left(\mathbf{a}^T \left[\mathbf{W}\mathrm{h}_i \| \mathbf{W}\mathrm{h}_j\right]\right)\right)}{\sum_{l \in \mathcal{N}_i} \exp\left(\text{ LeakyReLU } \left(\mathbf{a}^T \left[\mathbf{W}\mathrm{h}_i \| \mathbf{W}\mathrm{h}_l\right]\right)\right)} \tag{2}$$

where $\mathcal{N}_i$ is the neighbors of node i.

After the coefficients are obtained, we can compute a linear combination of the features corresponding to them, and get the output embedding z_j. In practice, to stabilize the learning process of self-attention, [10] has found extending the mechanism to employ multi-head attention to be beneficial. Specifically, K independent attention mechanisms execute the transformation, and then their features are concatenated.

3 Reinforcement Learning Model for Selecting Seed Nodes

In this section we present the training part of our model. We use deep Q learning to train the model and select seeds.

3.1 Reinforcement Learning Model Formulation

In detail, we define the states, actions, rewards, and policy in the reinforcement learning framework as follows:

1. A state s is represented by the embeddings of seed nodes in $S = (v_1, v_2, \cdots, v_i)$.
2. An action a is to add a node v to current state. We will represent actions as their corresponding p-dimensional node embedding $\mathbf{z}_v$.
3. A reward function $r(s, a)$ is to evaluate the transition reward of an action. In social influence maximization, we regard the nodes influenced by current state $\phi(S)$ as $c(S, G)$, and the difference caused by an action a as $r(s, a)$.

4. Training deep Q-network and selecting seeds. A greedy algorithm selects a node v to add next such that v maximizes an evaluation function $Q(s, a)$. Then, the solution S will be extended as

$$S = (S, v^*), \text{where } v^* = \arg\max_{v \in V - S} Q(s, v) \quad (3)$$

Thus, we can define node selection function according to ϵ−greedy policy. Then, we use DDQN to learn parameters of $Q(s, a)$ and further select nodes to solve the real-world problem.

3.2 Parameters Learning

For parameters learning, we refer to a neural network function approximator with weights θ as a Q-network, as illustrated in Algorithm 1. The Q-network can be trained by minimizing a sequence of loss functions

$$(\sum_{i=0}^{n-1} r\left(S_{t+i}, a_{t+i}\right) + \gamma \max_{a'} Q'\left(S_{t+n}, a', \theta^-\right) - Q(S, a_t, \theta))^2 \quad (4)$$

where γ is reward discount factor, $Q(S, a, \theta)$ represent the action-value function of MainNet and $Q'(S, a, \theta^-)$ represent the action-value function of TargetNet.

Algorithm 1. n-step DQN for node selection

1: Initialize experience replay memory D to capacity N
2: Initialize main action-value function Q with random weights θ
3: Initialize target action-value function Q' with weights $\theta^- = \theta$
4: **repeat**
5: $t \leftarrow 1$
6: Initialize the state to empty $S_t = (\ )$
7: **repeat**
8: Select node by node selection function over Eq.(??)
9: Add v_t to current solution $S_{t+1} \leftarrow (S_t, v_t)$
10: $t \leftarrow t + 1$
11: **if** $t \geq n$ **then**
12: Store transition $(S_{t-n}, v_{t-n}, r_{t-n,t}, S_t)$ in D
13: Sample random minibatch from D
14: Update θ by SGD over Eq.(4)
15: Every C steps reset $Q' = Q$
16: **end if**
17: **until** terminal S_t **or** $t = t_{max}$
18: **until** the number of episodes reach e_{max}
19: **return** θ

4 Experiments

In the following section, we will describe our datasets and discuss empirical results in two aspects: social spread prediction and seed set selection.

4.1 Datasets

We use five social media datasets to carry out experiments—SNAP Social (Facebook, Twitter, and YouTube) dataset [7], Weibo [12], and Flickr [1]. In these datasets, retweet, like and other similar social behaviors can be defined as being influenced. A user who was influenced at some time, we generate a positive instance. Next, for each neighbor of the influenced user, if it was never observed to be active in our observation window, we create a negative instance. Our target is to distinguish positive instances from negative ones. Thus, the social influence prediction problem can be considered as a classification problem.

The exact size of our datasets is shown in Table 1.

Table 1. Summary of datasets

Name	# of nodes	# of edges	Type
Facebook	4k	88k	Undirected, dense
Twitter	10k	203k	Directed, dense
YouTube	10k	29k	Undirected, sparse
Weibo	1.8M	308M	Directed, dense
Flickr	2.5M	33M	Directed, dense

4.2 Social Spread Prediction

We compare the prediction performance of our model and three methods on the four datasets and the results is shown in Fig. 2. Our model—**I**nfluence **M**aximization via **G**raph **E**mbedding and **R**einforcement learning (IMGER) achieves the best performance over baselines in terms of AUC, Recall, and F1-score, which indicates the effectiveness of our model.we compare our embedding model with several baselines in social influence prediction. Logistic Regression (LR) is a simple machine learning classification model with the initial features. PATCHY-SAN (PSCN) [9] is a state-of-the-art graph classification model. Inf2vec [3] is a state-of-the-art latent representation model for social influence embedding.

Inf2vec performs equally well in terms of Precision or even slightly better on Facebook dataset, and almost outperforms both LR and PSCN in all respects, which demonstrates that graph embedding methods are better than previous model mining features in social network analysis. Understandably, the embedding approaches that consider the history propagation track can help to effectively identify the hidden influence relationships among users. For instance, given that user u_a can influence user u_b, and user u_b can affect both user u_c and user u_d, then user u_a probably is also able to influence u_d. However, such relationships cannot be explicitly captured by previous prediction models, such as LR. Compared with Inf2vec and PSCN, our model can jointly capture the diffusion traces and user features, and dig out the implicit relationship between propagation and attributes.

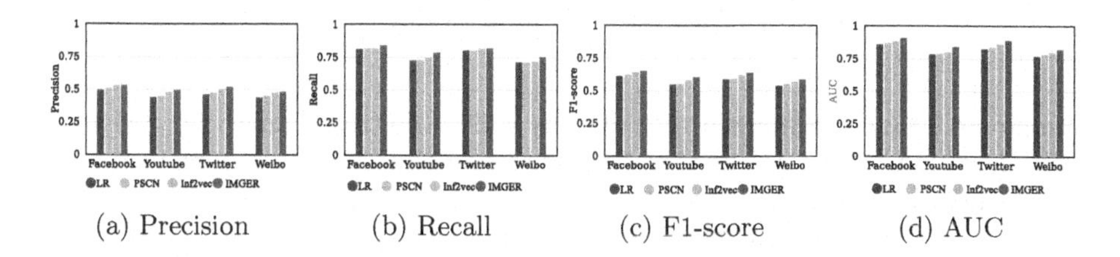

(a) Precision (b) Recall (c) F1-score (d) AUC

Fig. 2. Prediction performance of different methods

In conclusion, our model get significantly gain over both traditional model and the-state-of-the-art model.

4.3 Seed Set Selection

To evaluate the solution quality, we define the efficiency ratio as the ratio of the number of influenced nodes to the size of seeds set. Because almost all approximation algorithms are close to the optimal solution when the number of seeds is close to the number of nodes, we just pay more attention to the situation where the number of seeds is not very large. We compare our reinforcement learning model with other models to select seed nodes that can maximize social influence. Naive Greedy (NG) is a simple model that sort our nodes by their influences, and select the largest k nodes as seed nodes. Stop-and-Stare Algorithm (SSA) [8] is one of the best current influence maximization approximation algorithms, provides a $(1 - 1/e - \epsilon)$-approximation guarantee. Structure2Vec Deep Q-learning (SVDN) [5] is a state-of-the-art learning method, which can learn combinatorial optimization algorithms over graphs. As shown in Fig. 3, our results successfully demonstrate state-of-the-art performance being achieved across all three datasets. The results also directly give us a viewpoint on the basic network property that a small fraction of nodes can influence a very large portion of the networks.

The results of a simple greedy algorithm is the worst on all datasets. When the number of seeds is very small, the efficiency ratios of SSA, SVDN, and IMGER are almost equal. The experimental results of SSA and SVDN are

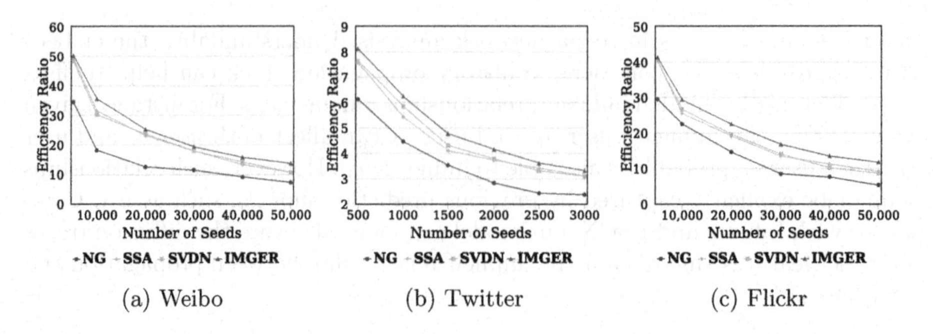

(a) Weibo (b) Twitter (c) Flickr

Fig. 3. Influence efficiency ratio on three datasets

similar and the performance of SVDN is a little better than the approximation algorithm, which means that heuristic algorithm is not worse than that of approximation algorithm, although it has no theoretical guarantee.

In conclusion, the results show that our reinforcement learning method is better than the previous algorithm in practice.

5 Conclusion

In this paper, we propose IMGER, an influence maximization model via graph embedding and reinforcement learning, to select nodes and maximize social influence. Since we represent nodes in the graph with their embeddings, both the state and the action are vectors in low dimensional space. Thus, our model is applicable to solve IM problem in large-scale social networks. Together with the deep Q-learning algorithm, algorithm learned on small graphs can be applied to graphs of larger sizes. We conduct numerical experiments to test how our model works in different real-life networks.

References

1. Cha, M., Mislove, A., Gummadi, K.P.: A measurement-driven analysis of information propagation in the Flickr social network. In: International World Wide Web Conference (WWW) (2009)
2. Domingos, P., Richardson, M.: Mining the network value of customers. In: Conference on Knowledge Discovery and Data Mining (KDD), pp. 57–66 (2001)
3. Feng, S., Cong, G., Khan, A., Li, X., Liu, Y., Chee, Y.M.: Inf2vec: latent representation model for social influence embedding. In: International Conference on Data Engineering (ICDE), pp. 941–952 (2018)
4. Kempe, D., Kleinberg, J., Tardos, É.: Maximizing the spread of influence through a social network. In: ACM International Conference on Knowledge Discovery and Data Dining (KDD), pp. 137–146 (2003)
5. Khalil, E., Dai, H., Zhang, Y., Dilkina, B., Song, L.: Learning combinatorial optimization algorithms over graphs. In: Advances in Neural Information Processing Systems (NeurIPS), pp. 6348–6358 (2017)

6. Li, Z., Chen, Q., Koltun, V.: Combinatorial optimization with graph convolutional networks and guided tree search, pp. 537–546 (2018)
7. McAuley, J.J., Leskovec, J.: Learning to discover social circles in ego networks. In: Annual Conference on Neural Information Processing Systems (NeurIPS), pp. 548–556 (2012)
8. Nguyen, H.T., Thai, M.T., Dinh, T.N.: Stop-and-stare: optimal sampling algorithms for viral marketing in billion-scale networks. In: ACM International Conference on Management of Data (SIGMOD), pp. 695–710. ACM (2016)
9. Niepert, M., Ahmed, M., Kutzkov, K.: Learning convolutional neural networks for graphs. In: International Conference on Machine Learning, (ICML), pp. 2014–2023 (2016)
10. Vaswani, A., et al.: Attention is all you need. In: Annual Conference on Neural Information Processing Systems (NeurIPS), pp. 5998–6008 (2017)
11. Velickovic, P., Cucurull, G., Casanova, A., Romero, A., Liò, P., Bengio, Y.: Graph attention networks. In: International Conference on Learning Representations (ICLR) (2018)
12. Zhang, J., Liu, B., Tang, J., Chen, T., Li, J.: Social influence locality for modeling retweeting behaviors. In: International Joint Conference on Artificial Intelligence (IJCAI), pp. 2761–2767 (2013)

A Multilevel Inference Mechanism for User Attributes over Social Networks

Hang Zhang[1,2], Yajun Yang[1,2(✉)], Xin Wang[1], Hong Gao[3], Qinghua Hu[1,2], and Dan Yin[4]

[1] College of Intelligence and Computing, Tianjin University, Tianjin, China
{aronzhang,yjyang,wangx,huqinghua}@tju.edu.cn
[2] State Key Laboratory of Communication Content Cognition, Beijing, China
[3] Harbin Institute of Technology, Harbin, China
honggao@hit.edu.cn
[4] Harbin Engineering University, Harbin, China
yindan@hrbeu.edu.cn

Abstract. In a real social network, each user has attributes for self-description called user attributes which are semantically hierarchical. With these attributes, we can implement personalized services such as user classification and targeted recommendations. Most traditional approaches mainly focus on the flat inference problem without considering the semantic hierarchy of user attributes which will cause serious inconsistency in multilevel tasks. To address these issues, in this paper, we propose a cross-level model called IWM. It is based on the theory of maximum entropy which collects attribute information by mining the global graph structure. Meanwhile, we propose a correction method based on the predefined hierarchy to realize the mutual correction between different layers of attributes. Finally, we conduct extensive verification experiments on the DBLP data set and it has been proved that compared with other algorithms, our method has a superior effect.

Keywords: Attribute inference · Multilevel inference · Social network

1 Introduction

In a social network, each user has a series of labels used to describe their characteristics called user attributes. However, for a certain type of attributes, they are not flat but hierarchical. The most existing methods [4,5] mainly focus on the single-level attribute inference and it will bring some problems for hierarchical structures as shown in Fig. 1. Even though utilizing the same method for every single-level, the attributes of different level may be conflicted for the same user, attributes at the same level may be indeterminate, and the results of a certain layer may be missing.

In this paper, we propose a multi-level inference model named IWM to solve the problems mentioned above. This model can infer hierarchical attributes for unknown users by collecting attributes from nearby users under maximum

C. S. Jensen et al. (Eds.): DASFAA 2021, LNCS 12682, pp. 710–718, 2021.
https://doi.org/10.1007/978-3-030-73197-7_49

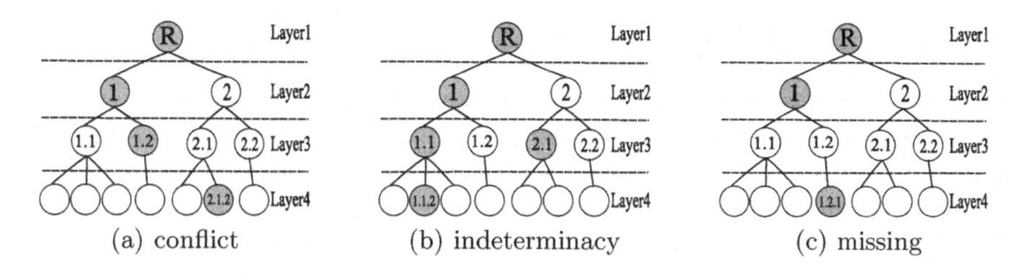

Fig. 1. Problems of labeling in real social networks

entropy random walk. Meanwhile, we propose a correction method based on the predefined hierarchy of attributes to revise the results. Finally, we conduct the experiments on real datasets to validate the effectiveness of our method.

The rest of the paper is organized as follows. Section 2 defines the problem. Section 3 proposes the multilevel inference model. Algorithm is given in Sect. 4. The experimental results and analysis are presented in Sect. 5. The related works are introduced in Sect. 6. Finally, we conclude this paper in Sect. 7.

2 Problem Definition

2.1 Semantic Tree

The semantic tree T is a predefined structure which is semantically exists used to describe the hierarchical relationship between different user attributes. We use T_g to represent the user attributes at T's gth layer.

2.2 Labeled Graph

Labeled graph is a simple undirected graph, denoted as $G = (V, E, T, L)$, where V is the set of vertices and E is the set of edges. T is the semantic tree of attributes in G. L is a function mapping V to a cartesian product of the attributes in T defined as $L : V \to T_1 \times T_2 \times \cdots \times T_m$, where m is the depth of T.

Problem Statement: Given a labeled graph $G(V, E, T, L)$ and labeled vertices set $V_s \subset V$, where V_s is the set of vertices with complete attributes. So for every vertex $v_s \in V_s, L(v_s) = \{l_1, l_2, \cdots, l_m\}$, where $l_1 \in T_1, l_2 \in T_2, \cdots, l_m \in T_m$. The input of the problem is $L(v_s)$ for every vertex $v_s \in V_s$ and the output is $L(v_u)$ for every vertex $v_u \in V_u$, where $V_u = V - V_s$.

3 Attribute Inference Model

Our attribute inference model can be divided into two parts. The first part is called the information propagation model. Based on the maximum entropy theory and one step random walk, vertices in V_s spread their own attributes to other vertices layer by layer. The second part is a correction model based on the semantic tree. This model realizes the mutual correction between different layers of attributes. These two models are described in detail below.

3.1 Information Propagation Model

The information propagation model is an extension of the model proposed in [7]. The main idea is that the higher the entropy value of the vertex, the stronger the uncertainty of its own user attributes, so more information should be collected. The attributes of v_j's each layer can be represented by $L_g(v_j) = \{l_x, w_x(v_j), l_x \in T_g\}$. Then the entropy value of v_j's gth layer $H_g(v_j)$ can be calculated as blow.

$$H_g(v_j) = -\sum_{l_x \in T_g} w_x(v_j) \times \ln w_x(v_j) \tag{1}$$

If v_i is a neighbor of v_j, then the transition probability $P_g(v_i, v_j)$ from v_i to v_j at gth layer is computed as follows.

$$P_g(v_i, v_j) = \frac{H_g(v_j)}{\sum_{v_j \in N(v_i)} H_g(v_j)} \tag{2}$$

Where $N(v_i)$ is the set of neighbors of v_j.

Next, we use the following equation to normalize the attribute probability obtained by different vertices.

$$w_x(v_j) = \frac{\sum_{v_i \in N(v_j)} P_g(v_i, v_j) \times w_x(v_i)}{\sum_{l_y \in T_g} \sum_{v_i \in N(v_j)} P_g(v_i, v_j) \times w_y(v_i)} \tag{3}$$

$L_g(v_j)$ will be updated through $w_x(v_j)$. In this way, the attribute information is spread hierarchically in the graph.

3.2 Attribute Correction Model

The formal definitions of the concepts involved in this section are given below.

Definition 1. *Define the following relationships in the semantic tree:*

(1) If x_2 is a child node of x_1, then x_1, x_2 have a relationship called $Child(x_1, x_2)$.
(2) Say that x_1, x_2 have a descendant relationship called $Descendant(x_1, x_2)$,if $Child(x_1, x_2) \cup \exists x_3(Child(x_1, x_3) \cap Descendant(x_3, x_2))$.
(3) If x_2 is a brother node of x_1, then x_1, x_2 have a relationship called $Brother(x_1, x_2)$.

Definition 2 ***(Descendant vertex set).*** *For a node x_1, its descendant node set is defined as $DesSet(x_1) = \{x | Descendant(x_1, x)\}$.*

Definition 3 ***(Brother vertex set).*** *For a node x_1, its brother node set is defined as $BroSet(x_1) = \{x | Brother(x_1, x)\}$.*

For the attribute l_x in the middle layer of the semantic tree, its existence depends on both $Parent(x)$ and $DesSet(x)$, so $w_x(v_j)$ can be corrected by Eq. (4).

$$w_x(v_j) = w_{Parent(x)}(v_j) \times \frac{(1-\alpha) \times w_x(v_j) + \alpha \times \sum_{y \in DesSet(x)} w_y(v_j)}{\sum_z (1-\alpha) \times w_z(v_j) + \alpha \times \sum_{y \in DesSet(z)} w_y(v_j)} \quad (4)$$

where $z \in BroSet(x)$ and α represents a correction strength. When the value of α is large, the result is inclined to the hierarchy of the semantic tree, otherwise, it is more inclined to the information collected by propagation.

There is another case that the highest layer attributes don't have any child node, so they can be corrected as follows.

$$w_x(v_j) = w_{Parent(x)}(v_j) \times \frac{w_x(v_j)}{\sum_{z \in BroSet(x)} w_z(v_j)} \quad (5)$$

4 Attribute Inference Algorithm

4.1 Algorithm Description

The detailed steps of the algorithm are shown in Algorithm 1. Firstly, we use Eq. (1) to calculate entropy $H_g(v_u)$ for all $v_u \in V_u$ layer by layer (line 1 to 3). Line 4 to 9 start inferring hierarchically. After all layers' information are collected, correction can be performed by Eq. (4) or Eq. (5) (line 10 to 11).

Algorithm 1. Cross-level Attribute Inference(G, V_s)

Input: $G(V, E, T, L)$ and V_s.
Output: $L(v_u)$ for every vertex $v_u \in V_u$.

1: **for** every layer g in T **do**
2:   **for** every vertex $v_u \in V_u$ **do**
3:     compute $H_g(v_u)$
4: **for** every vertex $v_u \in V_u$ **do**
5:   **for** every layer g in T **do**
6:     **for** every vertex $v_i \in N(v_u)$ **do**
7:       compute $P_g(v_i, v_u)$
8:     **for** every attribute $l_x \in T_g$ **do**
9:       compute $w_x(v_u)$
10:   **for** every attribute $l_x \in T$ **do**
11:     correct $w_x(v_u)$
12: **if** $\sum_{v_u \in V_u} \sum_{l_x \in T} |diffw_x(v_j)| \leq |V_u| \times |T| \times \sigma$ **then**
13:   **return** $L(v_u)$ for every vertex $v_u \in V_u$
14: **else**
15:   **return** step 1

The algorithm terminates when the convergence is satisfied. The condition of convergence is given by the following equation.

$$\sum_{v_u \in V_u} \sum_{l_x \in T} |diffw_x(v_u)| \leq |V_u| \times |T| \times \sigma \tag{6}$$

where $diff(w_x(v_u))$ is the difference on $w_x(v_u)$ after the inference algorithm is executed, and σ is a threshold to control the number of iterations.

4.2 Time Complexity

We assume that the labeled graph G has n vertices and p attributes, the semantic tree has m layers. So the time complexity of information propagation is $O(m|V_u| + mnd + pnd) = O(mnd + pnd)$, where d is the average degree of all the vertices in G. After that, we need to modify every attribute for each user by the complexity of $O(pn)$. To sum up, the total time complexity of our algorithm for one iteration is $O(mnd + pn)$.

5 Experiment

The experiments are performed on a Windows 10 PC with Intel Core i5 CPU and 8 GB memory. Our algorithms are implemented in Python 3.7. The default parameter values in the experiment are $\alpha = 0.5$, $\sigma = 0.0001$.

5.1 Experimental Settings

Dataset. We will study the performance on DBLP dataset. DBLP is a computer literature database system. Each author is a vertex and their research field is used as the attributes to be inferred. We extract 63 representative attributes and define a 4-layer semantic tree in advance.

Baselines and Evaluation Metrics. We compare our method IWM with three classic attribute inference baselines which are SVM, Community Detection (CD) [6] and Traditional Random Walk (TRW) [7].

We use five commonly metrics to make a comprehensive evaluation of the inference results. The calculation method of these metrics are shown below.

$$Precison = \frac{\sum_{l \in T} |\{v_u | v_u \in V_u \wedge l \in Predict(v_u) \cap Real(v_u)\}|}{\sum_{l \in T} |\{v_u | v_u \in V_u \wedge l \in Predict(v_u)\}|} \tag{7}$$

$$Recall = \frac{\sum_{l \in T} |\{v_u | v_u \in V_u \wedge l \in Predict(v_u) \cap Real(v_u)\}|}{\sum_{l \in T} |\{v_u | v_u \in V_u \wedge l \in Real(v_u)\}|} \tag{8}$$

$$F_1 = \frac{2 \times Precision \times Recall}{Precision + Recall} \tag{9}$$

$$Accuracy = \frac{1}{|V_u|} \times |\{v_u | v_u \in V_u \wedge Predict(v_u) = Real(v_u)\}| \tag{10}$$

$$Jaccard = \frac{1}{|V_u|} \times \sum_{v_u \in V_u} \frac{|Predict(v_u) \cap Real(v_u)|}{|Predict(v_u) \cup Real(v_u)|} \tag{11}$$

where $Predict(v_u)$ and $Real(v_u)$ respectively represent the inference result set and real original attribute set of v_u. For all metrics, the larger value means the better performance.

5.2 Results and Analysis

Exp1-Impact of Vertex Size. We conduct the first experiment in coauthor relationship networks with 5,000, 10,000, 20,000, and 40,000 vertices. The proportion of unknown vertices is 30% (Table 1).

Table 1. Inference performance on different vertex size.

Vertex size	Method	Precision				Recall				F1				Mean-Acc	Jaccard
		Layer2	Layer3	Layer4	Mean-Pre	Layer2	Layer3	Layer4	Mean-Rec	Layer2	Layer3	Layer4	Mean-F1		
5000	SVM	0.6410	0.5460	0.4700	0.5640	0.5630	0.5070	0.4450	0.5260	0.5810	0.5100	0.4300	0.5200	0.5021	0.6611
	CD	0.8428	0.6347	0.2117	0.4384	0.8307	0.6839	0.5718	0.6949	0.8364	0.6581	0.3078	0.5368	0.5180	0.5888
	TRW	0.8721	0.6423	0.6423	0.4099	**0.8754**	**0.7554**	**0.7317**	**0.7867**	0.8735	0.6931	0.3153	0.5377	0.6171	0.6446
	IWM	**0.9552**	**0.8310**	**0.8310**	**0.7773**	0.8629	0.7518	0.6867	0.7666	**0.9067**	**0.7892**	**0.6364**	**0.7718**	**0.7604**	**0.7187**
10000	SVM	0.8070	0.5800	0.4870	0.5180	0.6090	0.4860	0.4400	0.4480	0.6640	0.4990	0.4340	0.4490	0.4650	0.6314
	CD	0.7852	0.6427	0.2074	0.4488	0.7591	0.6106	0.4596	0.6103	0.7720	0.6259	0.2848	0.5164	0.3871	0.5109
	TRW	0.8309	0.6388	0.6388	0.3505	**0.8632**	0.7269	**0.7099**	**0.7653**	0.8466	0.6798	0.2583	0.4803	0.5815	0.6181
	IWM	**0.9492**	**0.8373**	**0.8373**	**0.7655**	0.8465	**0.7354**	0.6769	0.7526	**0.8949**	**0.7830**	**0.6170**	**0.7591**	**0.7288**	**0.7003**
20000	SVM	0.7400	0.5440	0.4460	0.5220	0.5320	0.4620	0.3920	0.4290	0.5820	0.4730	0.3980	0.4440	0.4260	0.6058
	CD	0.7602	0.6099	0.1888	0.4176	0.7332	0.6020	0.4423	0.5935	0.7463	0.6053	0.2634	0.4895	0.3579	0.4848
	TRW	0.8294	0.6063	0.6063	0.3143	**0.8392**	**0.7418**	**0.6817**	**0.7446**	0.8342	0.6561	0.2243	0.4418	0.5296	0.5810
	IWM	**0.9396**	**0.8170**	**0.8170**	**0.7436**	0.8372	0.7218	0.6526	0.7372	**0.8854**	**0.7664**	**0.5895**	**0.7403**	**0.6924**	**0.6688**
40000	SVM	0.7489	0.6167	0.4311	0.4850	0.4811	0.4522	0.3589	0.3950	0.5444	0.4911	0.3622	0.4050	0.3378	0.5473
	CD	0.7458	0.5333	0.1928	0.4547	0.6855	0.4669	0.2568	0.4710	0.7143	0.4979	0.2200	0.4626	0.1579	0.3797
	TRW	0.8093	0.5888	0.5888	0.2870	**0.8347**	0.7059	**0.6629**	**0.7340**	0.8214	0.6419	0.2006	0.4125	0.4652	0.5572
	IWM	**0.9360**	**0.8061**	**0.8061**	**0.7270**	0.8344	**0.7169**	0.6349	0.7284	**0.8817**	**0.7587**	**0.5667**	**0.7276**	**0.6561**	**0.6642**

It is obvious that our method shows the best performance on different evaluation indicators. For examplewhen it comes to a 20,000 vertices network, our model improves over the strongest baseline 22.2%, 35.1%, 16.3% and 6.3% on Precision, F1, Accuracy, and Jaccard index, separately. In terms of recall, our method does not have obvious advantages over TRW.

Exp2-Impact of the Proportion of Unknown Vertices. In Exp2 the vertex scale of the network is 20,000 and we set the unlabeled scale 10%, 20%, 30%, and 50% respectively.

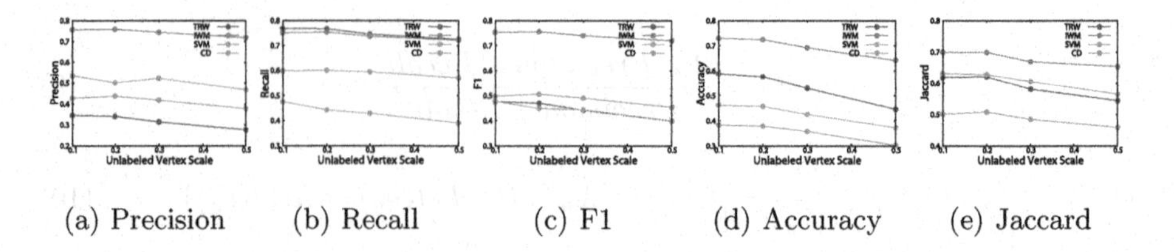

(a) Precision (b) Recall (c) F1 (d) Accuracy (e) Jaccard

Fig. 2. Inference performance on different proportion of unknown vertices

We can analyze the results to get that as the proportion of unknown vertices increases, the decline tendency of our method is much slower than other methods. It is interesting to see that the five evaluate indicators of our method are 71.77%, 72.17%, 71.96%, 64.21% and 65.43% at the condition of 50% vertices lack of attributes which can show that it has a great value in practical applications.

Exp3-Real Case Study. In Table 2 we present partial results of the experiment which gives a clear comparison between our method and TRW. We use these examples to demonstrate the effectiveness of our method.

Table 2. Comparison of inference results by TRW and IWM.

Author	True label			TRW result			IWM result		
	Layer2	Layer3	Layer4	Layer2	Layer3	Layer4	Layer2	Layer3	Layer4
Chris Stolte	**Data**	Database	Query	**Unknown**	Database	Query	**Data**	Database	Query
Marcel Kyas	Network	**Wireless**	Localization	Network	**Databse**	Localization	Network	**Wireless**	Localization
William Deitrick	**Data**	Mining	Clusters	**Network, Data**	Unknown	Clusters	**Data**	Mining	Clusters
V. Dhanalakshmi	Learning	Language	Extraction	Unknown	Classification	Speech	Learning	Language	Speech

For Chris Stolte, IWM can complement the missing information which can't be inferred by TRW. For Marcel Kyas, our method modify the error information on Layer 3 and obtain the correct result. TRW causes indeterminacy problem on Layer2 of William Deitrick, while IWM can select more relevant attributes. However, for V. Dhanalakshmi, due to its special structure, when most of the collected information is interference, IWM can't make correct inference either.

6 Related Work

There has been an increasing interest in the inference of single-layer user attributes over the last several years.

Firstly, based on resource content there are [1,11] which utilize the user's text content for inference. [3] constructs a social-behavior-attribute network and design a vote distribution algorithm to perform inference. There are also methods based on the analysis of graph structure such as Local Community Detection [6] and Label Propagation [12]. [10] discovers the correlation between item recommendation and attribute reasoning, so they use an Adaptive Graph Convolutional Network to joint these two tasks. However, these methods don't explore the relationship existing in the attribute hierarchy, which will greatly reduce the effectiveness in our multilevel problem.

Another method is to build a classifier to treat the inference problem as a multilevel classification problem. [2] trains a binary classifier for each attribute. [8] trains a multi-classifier for each parent node in the hierarchy. [9] trains a classifier for each layer in the hierarchical structure, and use it in combination with [8] to solve the inconsistency. However, classifier-based approaches have a high requirement for data quality. It will make the construction of the classifier complicated and the amount of calculation for training is huge.

7 Conclusion

In this paper, we study the multilevel user attribute inference problem. We first define the problem and propose the concept of semantic tree and labeled graph. We present a new method to solve this problem. The information propagation model is proposed to collect attributes for preliminary inference. The attribute correction model is proposed to conduct a cross-level correction. Experimental results on real-world data sets have demonstrated the superior performance of our new method. In future work, we will improve our method for multi-category attributes and do more works on optimizing the algorithm to save more time.

Acknowledgement. This work is supported by the National Key Research and Development Program of China No. 2019YFB2101903, the State Key Laboratory of Communication Content Cognition Funded Project No. A32003, the National Natural Science Foundation of China No. 61702132 and U1736103.

References

1. Choi, D., Lee, Y., Kim, S., Kang, P.: Private attribute inference from facebook's public text metadata: a case study of korean users. Ind. Manage. Data Syst. **117**(8), 1687–1706 (2017)
2. Fagni, T., Sebastiani, F.: Selecting negative examples for hierarchical text classification: an experimental comparison. J. Am. Soc. Inf. Sci. Technol. **61**(11), 2256–2265 (2010)
3. Gong, N.Z., Liu, B.: Attribute inference attacks in online social networks. ACM Trans. Priv. Secur. **21**(1), 3:1–3:30 (2018)
4. Krestel, R., Fankhauser, P., Nejdl, W.: Latent dirichlet allocation for tag recommendation. In: Proceedings of the 2009 ACM Conference on Recommender Systems, pp. 61–68. ACM (2009)
5. Lu, Y., Yu, S., Chang, T., Hsu, J.Y.: A content-based method to enhance tag recommendation. In: Proceedings of the 21st International Joint Conference on Artificial Intelligence, pp. 2064–2069 (2009)
6. Mislove, A., Viswanath, B., Gummadi, P.K., Druschel, P.: You are who you know: inferring user profiles in online social networks. In: Proceedings of the Third International Conference on Web Search and Web Data Mining, pp. 251–260. ACM (2010)

7. Pan, J., Yang, Y., Hu, Q., Shi, H.: A label inference method based on maximal entropy random walk over graphs. In: Li, F., Shim, K., Zheng, K., Liu, G. (eds.) APWeb 2016. LNCS, vol. 9931, pp. 506–518. Springer, Cham (2016). https://doi.org/10.1007/978-3-319-45814-4_41
8. Secker, A.D., Davies, M.N., Freitas, A.A., Timmis, J., Mendao, M., Flower, D.R.: An experimental comparison of classification algorithms for hierarchical prediction of protein function. Expert Update (Mag. Br. Comput. Soc. Spec. Group AI) **9**(3), 17–22 (2007)
9. Taksa, I.: David Taniar: research and trends in data mining technologies and applications. Inf. Retr. **11**(2), 165–167 (2008)
10. Wu, L., Yang, Y., Zhang, K., Hong, R., Fu, Y., Wang, M.: Joint item recommendation and attribute inference: an adaptive graph convolutional network approach. In: Proceedings of the 43rd International ACM SIGIR conference on research and development in Information Retrieval, pp. 679–688. ACM (2020)
11. Yo, T., Sasahara, K.: Inference of personal attributes from tweets using machine learning. In: 2017 IEEE International Conference on Big Data, pp. 3168–3174. IEEE Computer Society (2017)
12. Zhu, X., Ghahramani, Z.: Learning from labeled and unlabeled data with label propagation (2003)

Query Processing

Accurate Cardinality Estimation of Co-occurring Words Using Suffix Trees

Jens Willkomm(✉), Martin Schäler, and Klemens Böhm

Karlsruhe Institute of Technology (KIT), Karlsruhe, Germany
{jens.willkomm,martin.schaeler,klemens.boehm}@kit.edu

Abstract. Estimating the cost of a query plan is one of the hardest problems in query optimization. This includes cardinality estimates of string search patterns, of multi-word strings like phrases or text snippets in particular. At first sight, suffix trees address this problem. To curb the memory usage of a suffix tree, one often prunes the tree to a certain depth. But this pruning method "takes away" more information from long strings than from short ones. This problem is particularly severe with sets of long strings, the setting studied here. In this article, we propose respective pruning techniques. Our approaches remove characters with low information value. The various variants determine a character's information value in different ways, e.g., by using conditional entropy with respect to previous characters in the string. Our experiments show that, in contrast to the well-known pruned suffix tree, our technique provides significantly better estimations when the tree size is reduced by 60% or less. Due to the redundancy of natural language, our pruning techniques yield hardly any error for tree-size reductions of up to 50%.

Keywords: Query optimization · Cardinality estimation · Suffix tree

1 Introduction

Query optimization and accurate cost estimation in particular continue to be fundamentally important features of modern database technology [27,43]. While cardinality estimation for numerical attributes is relatively well understood [34], estimating the cardinality of textual attributes remains challenging [9,23,40]. This is particularly true when the query (1) contains regular expressions, e.g., aims to find the singular of a word and the plural, (2) searches for word chains, e.g., `high noon`, or (3) a combination of both, i.e., a regular expression involving several words. One application where this is necessary is text mining. Various text mining tasks query co-occurring words (co-occurrences), i.e., words alongside each other in a certain order [29,30]. Such queries are dubbed *co-occurrence queries*. Co-occurrence queries are important because, in linguistic contexts, co-occurrences indicate semantic proximity or idiomatic expressions [24]. Optimizing such queries requires estimates of the cardinality of co-occurrences.

C. S. Jensen et al. (Eds.): DASFAA 2021, LNCS 12682, pp. 721–737, 2021.
https://doi.org/10.1007/978-3-030-73197-7_50

Problem Statement. To query co-occurrences, one uses search patterns like `the␣emancipation␣of␣*` or `emancipation␣*␣Catholics`. A particularity of co-occurrences is that they only exist in chains of several words (word chain), like phrases or text snippets. This calls for cardinality estimates on a set of word chains. Compared to individual words, word chains form significantly longer strings, with a higher variance in their lengths. Our focus in this article is on the accuracy of such estimates with little memory consumption at the same time.

State-of-the-Art. One approach to index string attributes is to split the strings into trigrams, i.e., break them up into sequences of three characters [1,9,23]. This seems to work well to index individual words. However, the trigram approach will not reflect the connection between words that are part of a word chain. Another method to index string attributes is the suffix tree [23,28]. Since suffix trees tend to be large, respective pruning techniques have been proposed. A common technique is to prune the tree to a maximal depth [23]. Since the size of a suffix tree depends on the length of the strings [37], other pruning methods work similarly. We refer to approaches that limit the depth of the tree as *horizontal pruning*. With horizontal pruning however, all branches are shortened to the same length. So horizontal pruning "takes away" more information from long strings than from short ones. This leads to poor estimation accuracy for long strings and more uncertainty compared to short ones.

Challenges. Designing a pruning approach for long strings faces the following challenges: First, the reduction of the tree size should be independent of the length of the strings. Second, one needs to prune, i.e., remove information, from both short and long strings to the same extent rather than only trimming long strings. Third, the pruning approach should provide a way to quantify the information loss or, even better, provide a method to correct estimation errors.

Contribution. In this work, we propose what we call *vertical pruning*. In contrast to horizontal pruning that reduces any tree branch to the same maximal height, vertical pruning aims at reducing the number of branches, rather than their length. The idea is to map several strings to the same branch of the tree, to reduce the number of branches and nodes. This reduction of tree branches makes the tree *thinner*. So we dub the result of pruning with our approach *thin suffix tree* (TST). A TST removes characters from words based on the information content of the characters. We propose different ways to determine the information content of a character based on empirical entropy and conditional entropy. Our evaluation shows that our pruning approach reduces the size of the suffix tree depending on the character distribution in natural language (rather than depending on the length of the strings). TST prunes both short and long strings to the same extent. Our evaluation also shows that TST provides significantly better cardinality estimations than a pruned suffix tree when the tree size is reduced by 60% or less. Due to the redundancy of natural language, TST yields hardly any error for tree-size reductions of up to 50%.

Paper Outline. Section 2 features related work. We introduce the thin suffix tree in Sect. 3. We say how to correct estimation errors in Sect. 4. Our evaluation is in Sect. 5.[1]

2 Related Work

This section is split into three parts. First, we summarize lossless methods to compress strings and suffix trees. These methods reduce the memory consumption without loss of quality. Such methods allow for a perfect reconstruction of the data and provide exact results. Second, we turn to pruning methods for suffix trees. Pruning is a lossy compression method that approximates the original data. Finally, we review empirical entropy in string applications.

String Compression Methods. There exist various methods to compress strings. One is statistical compression, like Huffman coding [20] and Hu-Tucker coding [19]. Second, there are compressed text self-indexes, like FM-index [13]. Third, there is dictionary-based compression, like the Lempel and Ziv (LZ) compression family [3,42,44]. Fourth, there are grammar-based compression methods, like Re-Pair [26]. However, these compression methods are either incompatible with pattern matching or orthogonal to our work.

Suffix Tree Compression Methods. A suffix tree (or trie) is a data structure to index text. It efficiently implements many important string operations, e.g., matching regular expressions. To reduce the memory requirements of the suffix tree, there exist approaches to compress the tree based on its structure [35]. Earlier approaches are path compression [21,22] and level compression [2]. More recent compression methods and trie transformations are path decompositions [17], top trees [4], and hash tables [36]. Our approach in turn works on the input strings rather than on the tree structure.

In addition to structure-based compression, there exist alphabet-based compression techniques. Examples are the compressed suffix tree [18], the sparse suffix tree [25] and the idea of alphabet sampling [10,16]. These methods reduce the tree size at the expense of the query time. All these methods provide exact results, i.e., are lossless compression methods. Lossless tree compression is applicable in addition to tree pruning, i.e., such methods work orthogonal to our approach.

Suffix Tree Pruning. Suffix trees allow to estimate the cardinality of string predicates, i.e., the number of occurrences of strings of arbitrary length [41]. With large string databases in particular, a drawback of suffix trees is their memory requirement [12,41]. To reduce the memory requirements of the suffix tree, variants of it save space by removing some information from the tree [23].

[1] For further information, like the procedures *insert* and *query*, see the extended version of this article; available at https://doi.org/10.5445/IR/1000128678.

We are aware of three approaches to select the information to be removed: A first category is data-insensitive, application-independent approaches. This includes shortening suffixes to a maximum length [23]. Second, there are data-sensitive, application-independent pruning approaches that exploit statistics and features of the data, like removing infrequent suffixes [15]. Third, there are data-sensitive, application-dependent approaches. They make assumptions on the suffixes which are important or of interest for a specific application. Based on the application, less useful suffixes are removed [38]. For example, suffixes with typos or optical character recognition errors are less useful for most linguistic applications. It is also possible to combine different pruning approaches. In this work, we focus on data-sensitive and application-independent pruning.

Horizontal Pruning Approaches. Existing pruning techniques usually reduce the height of the tree, by pruning nodes that are deeper than a threshold depth [23]. Another perspective is that all nodes deeper than the threshold are merged into one node. We in turn propose a pruning technique that reduces the width of the tree rather than the depth.

Empirical Entropy in String Applications. The usage frequency of characters in natural language is unevenly distributed [39]. For this reason, the empirical entropy is an essential tool in text and string compression. It is used in optimal statistical coding [20] and many data compression methods [3,14].

3 The Thin Suffix Tree

To store word chains as long strings in a suffix tree efficiently, we propose the thin suffix tree (TST). In contrast to horizontal pruning approaches, it aims at reducing the number of branches in the tree, rather than their length. We refer to this as *vertical pruning*. The idea is to conflate branches of the tree to reduce its memory consumption. This means that one branch stands for more than one suffix. As usual, the degree of conflation is a trade-off between memory usage and accuracy. In this section, we (1) present the specifics of TST and (2) define interesting map functions that specify which branches to conflate.

3.1 Our Vertical Pruning Approach

To realize the tree pruning, we propose a map function that discerns the input words from the strings inserted into the tree. For every input word (preimage) that one adds to the tree, the map function returns the string (image) that is actually inserted. TST stores the image of every suffix. This map function is the same for the entire tree, i.e., we apply the same function to any suffix to be inserted (or to queries). Thinning occurs when the function maps several words to the same string. The map function is surjective.

Fixing a map function affects the search conditions of the suffix tree. A suffix tree and suffix tree approximation techniques usually search for exactly the given

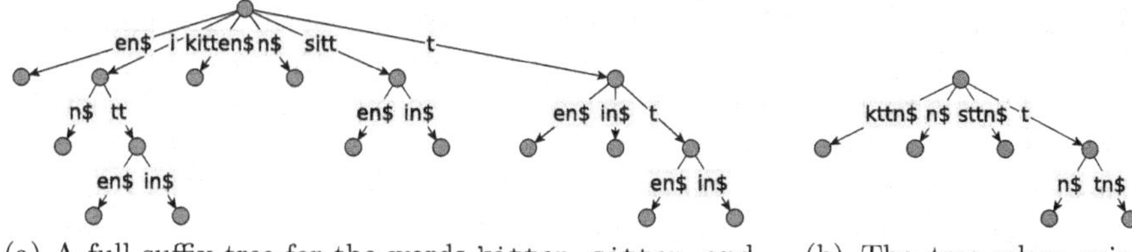

(a) A full suffix tree for the words `kitten`, `sitten`, and `sittin`.

(b) The tree when using a map function that removes `i` and `e`.

Fig. 1. The impact of character-removing map functions on a suffix tree.

suffix, i.e., all characters in the given order. Our approximation approach relaxes this condition to words that contain the given characters in the given order, but may additionally contain characters that the map function has removed. For example, instead of querying for the number of words that contain the exact suffix `mnt`, one queries the number of words that contain the characters `m`, `n`, and `t` in this order, i.e., a more general pattern. We implement this by using a map function that removes characters from the input strings. We see the following two advantages of such a map functions: (1) Branches of similar suffixes conflate. This reduces the number of nodes. (2) The suffix string gets shorter. This reduces the number of characters. Both features save memory usage.

3.2 Character-Removing Map Functions

A character-removing map function removes characters from a given string.

Definition 1. *A* character-removing map function *is a function that maps a preimage string to an image string by removing a selection of specific characters.*

To remove characters systematically, we consider the information value of the characters. According to Shannon's information theory, common characters carry less information than rare ones [8,33]. This is known as Shannon entropy of the alphabet. The occurrence probability $P(c)$ of a character c is its occurrence frequency relative to the one of all characters of the alphabet Σ: $P(c) = \frac{\text{frequency}(c)}{\sum_{\sigma \in \Sigma} \text{frequency}(\sigma)}$. The information content of a character c is inversely proportional to its occurrence probability $P(c)$: $\mathrm{I}(c) = \frac{1}{P(c)}$. According to Zipf's law, the occurrence probability of each c is inversely proportional to its rank in the frequency table [39]: $P(c) \sim \frac{1}{\text{rank}(c)}$. Thus, the information content of character c is proportional to its rank: $\mathrm{I}(c) \sim \text{rank}(c)$. To create an tree of approximation level n, a map function removes the n most frequent characters in descending order of their frequency.

Example 1. The characters `e`, `t`, and `a` (in this order) are the three most frequent characters in English text. At approximation degree 3, a map function maps the input word `requirements` to the string `rquirmns`.

Example 2. Figure 1a shows the full suffix tree for the words `kitten`, `sitten`, and `sittin`. Its size is 288 bytes. The exact result of the regular expression `*itten` is 2 counts. For illustrative purposes, we keep character `t` and show the impact of a map function that removes characters `e` and `i`. Figure 1b shows the corresponding tree. It has a size of 192 bytes and, thus, saves 33% of memory usage. When we query the regular expression `*itten`, the result is 3 counts. So it now overestimates the cardinality.

3.3 More Complex Map Functions

When removing characters of the input words, one can also think of more complex map functions. We see two directions to develop such more complex functions: (1) Consider character chains instead of single characters or (2) respect previous characters. We will discuss both directions in the following.

Removing Character Chains. The map function considers the information content of combinations of several consecutive characters, i.e., character chains. Take a string $s = c_1c_2c_3$ that consists of characters c_1, c_2, and c_3. We consider character chains of length o and create a frequency table of character chains of this length. The information content of a character chain $c_1 \dots c_o$ is proportional to its rank: $\mathrm{I}(c_1 \dots c_o) \sim \mathrm{rank}(c_1 \dots c_o)$. Our character-chain-removing map function is a character-removing map function that removes every occurrence of the n most frequent character chains of the English language of length o.

Example 3. A character-chain-removing map function removing the chain `re` maps the input `requirements` to `quiments`.

Using Conditional Entropy. We define a character-removing map function that respects one or more previous characters to determine the information content of a character c_1. Given a string $s = c_0c_1c_2$, instead of using the character's occurrence probability $P(c_1)$, a conditional-character removing map function considers the conditional probability $P(c_1|c_0)$. Using Bayes' theorem, we can express the conditional probability as $P(c_1|c_0) = \frac{\mathrm{frequency}(c_0c_1)}{\mathrm{frequency}(c_0)}$. Since the frequencies for single characters and character chains are roughly known (for the English language for instance), we compute all possible conditional probabilities beforehand. So we can identify the n most probable characters with respect to previous characters to arrive at a tree approximation level n.

Example 4. A map function removes `e` if it follows `r`. Thus, the function maps the input word `requirements` to `rquirments`.

3.4 A General Character-Removing Map Function

In this section, we develop a general representation of the map functions proposed so far, in Sects. 3.2 and 3.3. All map functions have in common that they remove characters from a string based on a condition. Our generalized map function has two parameters:

Observe. The length of the character chain to observe, i.e., the characters we use to determine the entropy. We refer to this as o. Each map function requires a character chain of at least one character, i.e., $o > 0$.

Remove. The length of the character chain to remove. We refer to this as r with $0 < r \leq o$. For $r < o$, we always remove the characters from the right of the chain observed. In more detail, when observing a character chain c_0c_1, we determine the conditional occurrence probability of character c_1 using $P(c_1|c_0)$. Therefore, we remove character c_1, i.e., the rightmost character in the chain observed.

We refer to a map function that observes and removes single characters as `o1r1`, to ones that additionally observe one previous character and remove one character as `o2r1`, and so on.

3.5 Cases of Approximation Errors

Character-removing map functions may lead to two sources of approximation error, by (1) conflating tree branches and (2) by the character reduction itself.

Case 1: Branch Conflation. In most cases, when removing characters, words still map to different strings and, thus, are represented differently in the tree. But in some cases, different words map to the same string and these words correspond to the same node. For example, when removing the characters `e` and `t`, the map function maps `water` and `war` to the same string `war`. The occurrences of the two words are counted in the same node, the one of `war`. Thus, the tree node of `war` stores the sum of the numbers of occurrences of `war` and `water`. So TST estimates the same cardinality for both words.

Case 2: Character Reduction. A character-removing map function shortens most words. But since it removes the most frequent characters/character chains first, it tends to keep characters with high information value. However, with a very high approximation degree, a word may be mapped to the empty string. Our estimation in this case is the count of the root node, i.e., the total number of strings in the tree.

4 Our Approach for Error Correction

Since we investigate the causes of estimation errors, we now develop an approach to correct them. Our approach is to count the number of different input strings that conflate to a tree node. Put differently, we count the number of different preimages of a node. To estimate the string cardinality, we use the multiplicative inverse of the number of different input strings as a correction factor. For example, think of a map function that maps two words to one node., i.e., the node counts the cardinality of both strings. TST estimates half of the node's count for both words.

4.1 Counting the Branch Conflations

To compute the correction factor, we need the number of different input words that map to a node. To prevent the tree from double counting input words, each node has to record words already counted. A first idea to do this may be to use a hash table. However, this would store the full strings in the tree nodes and increase the memory usage by much. The directed acyclic word graph (DAWG) [6] seems to be an alternative. It is a deterministic acyclic finite state automaton that represents a set of strings. However, even a DAWG becomes unreasonably large, to be stored in every node [7]. There also are approximate methods to store a set of strings. In the end, we choose the Bloom filter [5] for the following reasons: Firstly, it needs significantly less memory than exact alternatives. Its memory usage is independent of the number as well as of the length of the strings. Secondly, the only errors are false positives. In our case, this means that we may miss to count a preimage. This can lead to a slightly higher correcting factor, e.g., $\frac{1}{3}$ instead of $\frac{1}{4}$ or $\frac{1}{38}$ instead of $\frac{1}{40}$. As a result, the approximate correction factor is always larger than or equal to the true factor. This means that our correction factor only affects the estimation in one direction, i.e., it only corrects an overestimated count.

To sum up, our approach to bring down estimation errors has the following features: For ambiguous suffixes, i.e., ones that collide, it yields a correction factor in the range $(0, 1)$. This improves the estimation compared to no correction. For unambiguous suffixes, i.e., no collision, the correction factor is exactly 1. This means that error correction does not falsify the estimate.

4.2 Counting Fewer Input Strings

TST stores image strings, while our error correction relies on preimage strings. Since the preimage and the image often have different numbers of suffixes (they differ by the number of removed characters), our error correction may count too many different suffixes mapped to a node. For example, take the preimage `water`. A map function that removes characters `a` and `t` returns the image `war`. The preimage `water` consists of 5 suffixes, while the image `war` consists of 3. This renders the correction factor too small and may result in an underestimation.

We count too many preimages iff a preimage suffix maps to the same image as its next shorter suffix. This is the case for every preimage suffix that starts with a character that is removed by the map function. We call the set of characters a map function removes *trim characters*. This lets us discern between two cases: First, the map function removes the first character of the suffix (and maybe others). Second, the map function keeps the first character of the suffix and removes none, exactly one or several characters within it. To distinguish between the two cases, we check whether the first character of the preimage suffix is a trim character. This differentiation also applies to complex map functions that reduce multiple characters from the beginning of the suffix. To solve the issue of counting too many different preimages, our error correction only counts preimage strings which do not start with a trim character.

5 Experimental Evaluation

In this section, we evaluate our thin suffix tree approach. We (1) define the objectives of our evaluation, (2) describe the experimental setup, and (3) present and discuss the results.

5.1 Objectives

The important points of our experiments are as follows.

Memory Usage. We examine the impact of our map functions on the size of the suffix tree.
Map Function. We study the effects of our map functions on the estimation accuracy and analyze the source of estimation errors.
Accuracy. We investigate the estimation accuracy as function of the tree size.
Query Run Time. We evaluate the query run times, i.e., the average and the distribution.

We rely on two performance indicators: *memory usage*, i.e., the total tree size, and the *accuracy* of the estimations. To quantify the accuracy, we use the q-error, a maximum multiplicative error commonly used to properly measure the cardinality estimation accuracy [11,31]. The q-error is the preferred error metric for cardinality estimation, since it is directly connected to costs and optimality of query plans [32]. Given the true cardinality $\hat{f}$ and the estimated cardinality f, the q-error is defined as: $\max\left(\frac{f}{\hat{f}}, \frac{\hat{f}}{f}\right)$.

5.2 Setup

Our intent is to benchmark the approaches in a real-world scenario. In addition, we inspect and evaluate the impact of different character-removing map functions on the tree. We now describe the database and the queries used in the experiments.

Database. For pattern search on word chains, we use the 5-grams from the English part of the Google Books Ngram corpus.[2] We filter all words that contain a special character, like a digit.[3] At the end, we randomly sample the data set to contain 1 million 5-grams.

[2] The Google Books Ngram corpus is available at http://storage.googleapis.com/books/ngrams/books/datasetsv2.html.
[3] We use Java's definition of special characters. See function `isLetter()` of class `Java.lang.Character` for the full definition.

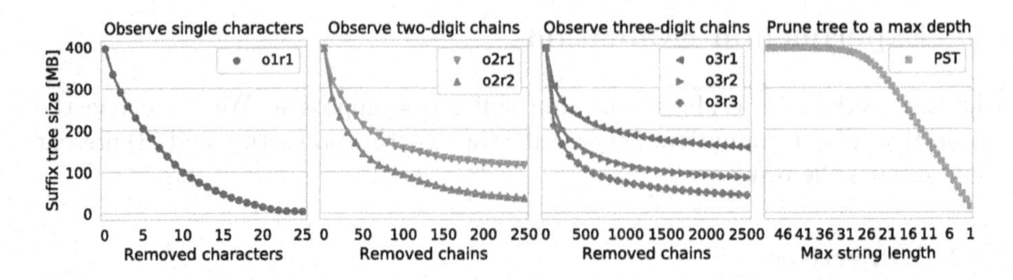

Fig. 2. TST's memory usage for various map functions and approximation levels.

Queries. Users may be interested in querying the number of 5-grams that start with a specific word or a specific word chain. Others may want to query for the number of different words that are used together with a specific word or word chain. The answer to both types of question is the cardinality of a search pattern. For our evaluation, we create 1000 queries requesting the cardinality of the 1000 most common nouns in the English language. For example, we query the number of 5-grams containing words like `way`, `people`, or `information`.

Parametrization. Each TST uses a single character-removing map function. The indicators o and r specify the character-removing map function, i.e., how a map function works. Each function has a character frequency list that stores all characters in descending order of their frequency. Hence, we use the list provided by Peter Norvig in *English Letter Frequency Counts: Mayzner Revisited or ETAOIN SRHLDCU*[4]. To specify the approximation level of the suffix tree, we parametrize the chosen character-removing map function with the number of characters x that are removed by it. This means that the function takes the first x characters from the character frequency list to specify the set of characters removed by the map function. The competitor has the level of approximation as parameter. The parameter specifies the maximal length of the suffixes to store in the tree. Depending on the string lengths of the database, reasonable parameter values lie between 50 and 1 [39].

Technical Details. Our implementation makes use of SeqAn[5], a fast and robust C++ library. It includes an efficient implementation of the suffix tree and of the suffix array together with state-of-the-art optimizations. We run our experiments on a AMD EPYC 7551 32-Core Processor with 125 GB of RAM. The machine's operating system is an Ubuntu 18.04.4 LTS on a Linux 4.15.0-99-generic kernel. We use the C++ compiler and linker from the GNU Compiler Collection in version 5.4.0.

[4] The article and the list are available at https://norvig.com/mayzner.html.
[5] The SeqAn library is available at https://www.seqan.de.

5.3 Experiments

We now present and discuss the results of our experiments.

Experiment 1: Memory Usage. In Experiment 1, we investigate how a character-removing map functions affects the memory consumption of a TST. We also look at the memory needs of a pruned suffix tree (PST) and compare the two. In our evaluation, we inspect chains of lengths up to 3 ($o = \{1, 2, 3\}$) and in each case remove 1 to o characters. Figure 2 shows the memory usage of the TST for various map functions and approximation levels. The figure contains four plots. The first three (from the left) show the memory usage for map functions that observe character chains of length 1, 2, or 3. The right plot shows the memory usage of the pruned suffix tree contingent on the maximal length of the suffixes. Note that there are different scales on the x-axis. The database we use in this evaluation includes 179 different characters, 3,889 different character chains of length two, and 44,198 different chains of length three.

The Effect of Character-Removing Map Functions. For a deeper insight into our pruning approach, we now study the effect of character-removing map functions on the memory consumption of a suffix tree in more detail. As discussed in Sect. 3.5, there are two effects that reduce memory consumption: *branch conflation* and *character reduction*. See Fig. 2. All map functions yield a similar curve: They decrease exponentially. Hence, removing the five most frequent characters halves the memory usage of a TST with map function `o1r1`.

The frequency of characters and character chains in natural language follows Zipf's law, i.e., the Zeta distribution [39]. Zipf's law describes a relation between the frequency of a character and its rank. According to the distribution, the first most frequent character nearly occurs twice as often as the second one and so on. All character-removing map functions in Fig. 2 show a similar behavior: Each approximation level saves nearly half of the memory as the approximation level before. For example, map function `o1r1` saves nearly 65 MB from approximation level 0 to 1 and nearly 40 MB from approximation level 1 to 2. This shows the expected behavior, i.e., character reduction has more impact on the memory usage of a TST than branch conflation.

The Effect of Horizontal Pruning. The right plot of Fig. 2 shows the memory usage of a pruned suffix tree. The lengths of words in natural language are Poisson distributed [39]. In our database, the strings have an average length of 25.9 characters with a standard deviation of 4.7 characters. Since the pruning only affects strings longer than the maximal length, the memory usage of the pruned suffix tree follows the cumulative distribution function of a Poisson distribution.

Summary. Experiment 1 reveals two points. First, the tree size of TST and of the pruned suffix are markedly different for the various approximation levels. Second, the memory reduction of a TST is independent of the length of the

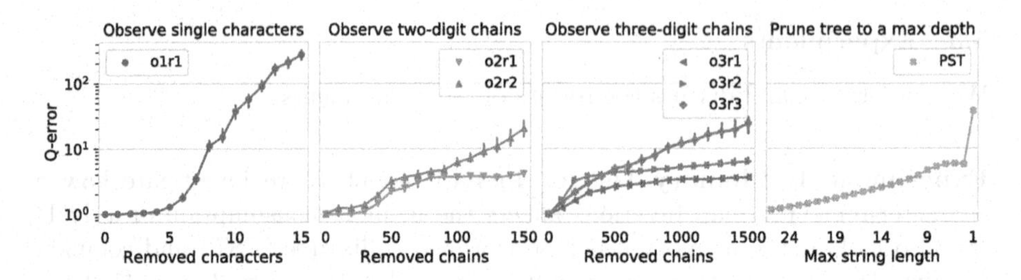

Fig. 3. TST's q-error for various map functions and approximation levels.

strings in the database. Its memory reduction depends on the usage frequency of characters in natural language. Third, the tree sizes for the different character-removing map functions tend to be similar, except for one detail: The shorter the chain of observed characters, i.e., the smaller o, the more linear the reduction of the tree size over the approximation levels.

Experiment 2: Map Functions. In Experiment 2, we investigate the impact of map functions on estimation accuracy. We take the map functions from Experiment 1 and measure the q-error at several approximation levels. See Fig. 3. The right plot is the q-error with the pruned suffix tree. The points are the median q-error of 1000 queries. The error bars show the 95% confidence interval of the estimation. Note that Fig. 3 shows the estimation performance as a function of the approximation level of the respective map function. So one cannot compare the absolute performance of different map functions, but study their behavior. The plots show the following. First, the map functions behave differently. At low approximation levels, the map function `o1r1` has a very low q-error. The q-error for this function begins to increase slower than for map functions considering three-digit character chains. At high approximation levels, the q-error of map function `o1r1` is significantly higher than for all the other map functions. The other map functions, the ones that consider three-digit character chains in particular, only show a small increase of the q-error for higher approximation levels. Second, the sizes of the confidence interval differ. With map functions that remove characters independently from previous characters, i.e., `o1r1`, `o2r2`, and `o3r3`, the confidence interval becomes larger with a larger approximation level. For map functions that remove characters depending of previous characters, i.e., `o2r1`, `o3r1`, or `o3r2`, the confidence interval is smaller. This means that, for lower approximation levels and map functions that remove characters depending on previous characters in particular, our experiments yield reliable results. We expect results to be the same on other data. Third, the accuracy of the pruned suffix tree increases nearly linear with increasing approximation levels until it sharply increases for very short maximal strings lengths. This means that every character that is removed from the back of the suffix contributes a similar extent of error to an estimation.

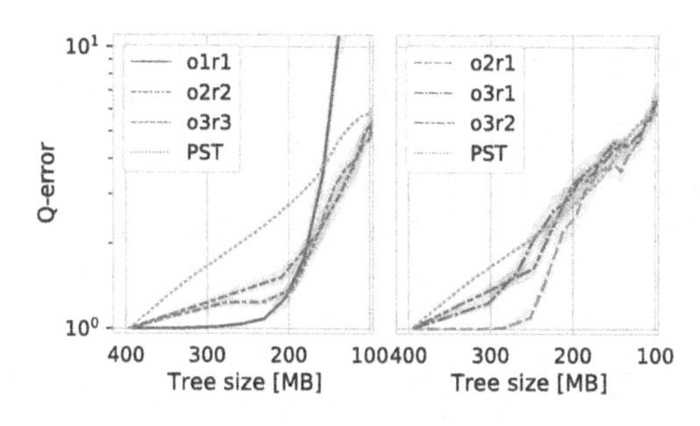

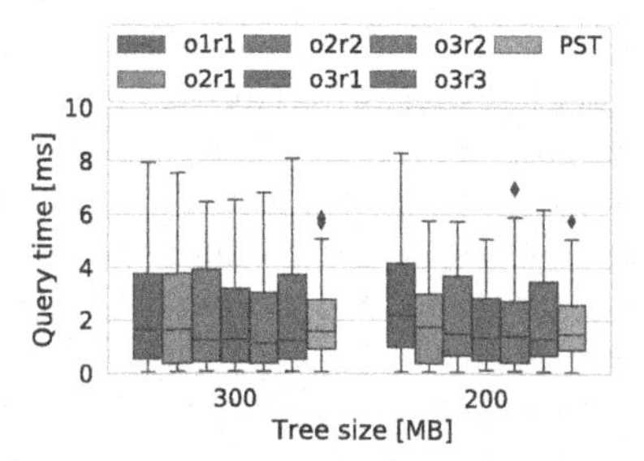

Fig. 4. The estimation accuracy as function of the tree size.

Fig. 5. The query run time of the TST.

Summary. None of our map functions dominates all the other ones. It seems to work best to remove single characters dependent on either 0, 1, or 2 previous characters, i.e., map functions `o1r1`, `o2r1`, or `o3r1`. For low approximation levels, say up to a reduction of 50% of the tree size, map function `o1r1` performs well. For high approximation levels, say starting from a reduction of 50% of the tree size, one should use a map function that removes characters dependent on previous ones, i.e., map function `o2r1` or `o3r1`.

The first five approximation levels of map function `o1r1` are of particular interest, as they show good performance in Experiments 1 and 2. In the first approximation levels, this map function yields a very low q-error, see Fig. 3, while the tree size goes down very rapidly, see Fig. 2. In Experiments 1 and 2, we inspect (1) the tree size depending on the approximation and (2) the q-error depending on the approximation level. In many applications, one is interested in the q-error depending on the tree size rather than on the approximation level. In our next experiment, we compare the q-error of our map functions for the same tree sizes.

Experiment 3: Accuracy. In Experiment 3, we compare the map functions from Experiment 1 against each other and against existing horizontal pruning. Figure 4 shows the q-error as function of the tree size. For the sake of clarity, there are two plots for this experiment: The plot on the left side shows the map functions that remove characters independently from previous characters. The plot on the right side is for the remaining map functions.

Vertical Pruning vs. Horizontal Pruning. Figure 4 shows the following: TST produces a significantly lower q-error than the pruned suffix tree for all map functions used and for tree sizes larger than 60% of the one of the full tree. For smaller sizes, the accuracy of most map functions does not become much worse than the one of the pruned suffix tree. The only exception is `o1r1`. At a tree size of 50%, the q-error of `o1r1` starts to increase exponentially with decreasing tree size.

Summary. As Experiments 1 and 2 already indicate, map function `o1r1` achieves a very low q-error to tree size ratio, for tree size reductions of up to 60%. For this map function, TST yields a significantly lower q-error than the pruned suffix tree for comparable tree sizes. The intuition behind this result is that our vertical pruning respects the redundancy of natural language. Due to this redundancy, map function `o1r1` keeps most input words unique. This results in almost no errors for reductions of the tree size that are less than 50%. TST also shows a higher degree of confidence, i.e., a smaller 95% confidence interval, than the pruned suffix tree for reductions that are less than 40%.

Experiment 4: Query Run Time. In Experiment 4, we compare the query run time of the TST using different map functions with the one of a pruned suffix tree. We consider sample tree sizes of 300 and 200 MB. Figure 5 shows the average and distribution of the run time for all queries. There is no significant difference in the run time. TST potentially needs a slightly higher run time than a pruned suffix tree. This is because TST is potentially deeper than a pruned suffix tree and additionally executes a map function. To conclude, the additional work of TST is of little importance for the query run time compared to a pruned suffix tree.

6 Conclusions

Cardinality estimation for string attributes is challenging, for long strings in particular. Suffix trees allow fast implementations of many important string operations, including this estimation. But since they tend to use much memory, they usually are pruned down to a certain size. In this work, we propose a novel pruning technique for suffix trees for long strings. Existing pruning methods mostly are horizontal, i.e., prune the tree to a maximum depth. Here we propose what we call *vertical pruning.* It reduces the number of branches by merging them. We define map functions that remove characters from the strings based on the entropy or conditional entropy of characters in natural language. Our experiments show that our thin suffix tree approach does result in almost no error for tree size reductions of up to 50% and a lower error than horizontal pruning for reductions of up to 60%.

References

1. Adams, E., Meltzer, A.: Trigrams as index element in full text retrieval: observations and experimental results. In: CSC, pp. 433–439. ACM (1993). https://doi.org/10.1145/170791.170891
2. Andersson, A., Nilsson, S.: Improved behaviour of tries by adaptive branching. Inf. Proc. Lett. **46**, 295–300 (1993). https://doi.org/10.1016/0020-0190(93)90068-k
3. Arz, J., Fischer, J.: LZ-compressed string dictionaries. In: DCC, IEEE (2014). https://doi.org/10.1109/dcc.2014.36

4. Bille, P., Fernstrøm, F., Gørtz, I.L.: Tight bounds for top tree compression. In: Fici, G., Sciortino, M., Venturini, R. (eds.) SPIRE 2017. LNCS, vol. 10508, pp. 97–102. Springer, Cham (2017). https://doi.org/10.1007/978-3-319-67428-5_9
5. Bloom, B.: Space/time trade-offs in hash coding with allowable errors. Commun. ACM 422–426 (1970). https://doi.org/10.1145/362686.362692
6. Blumer, A., Blumer, J., Haussler, D., Ehrenfeucht, A., Chen, M., Seiferas, J.: The smallest automation recognizing the subwords of a text. Theor. Comput. Sci. 31–55 (1985). https://doi.org/10.1016/0304-3975(85)90157-4
7. Blumer, A., Ehrenfeucht, A., Haussler, D.: Average sizes of suffix trees and DAWGs. Discrete Appl. Math. 37–45 (1989). https://doi.org/10.1016/0166-218x(92)90270-k
8. Brown, P., Della, V., Mercer, R., Pietra, S., Lai, J.: An estimate of an upper bound for the entropy of English. Comput. Linguist. **18**(1), 31–40 (1992)
9. Chaudhuri, S., Ganti, V., Gravano, L.: Selectivity estimation for string predicates: Overcoming the underestimation problem. ICDE. IEEE (2004). https://doi.org/10.1109/icde.2004.1319999
10. Claude, F., Navarro, G., Peltola, H., Salmela, L., Tarhio, J.: String matching with alphabet sampling. J. Discrete Algorithms 37–50 (2012). https://doi.org/10.1016/j.jda.2010.09.004
11. Cormode, G., Garofalakis, M., Haas, P., Jermaine, C.: Synopses for massive data: Samples, histograms, wavelets, sketches. Found. Trends Databases **4**, 1–294 (2011). https://doi.org/10.1561/1900000004
12. Dorohonceanu, B., Nevill-Manning, C.: Accelerating protein classification using suffix trees. In: ISMB, pp. 128–133 (2000)
13. Ferragina, P., Manzini, G., Mäkinen, V., Navarro, G.: Compressed representations of sequences and full-text indexes. ACM Trans. Algorithms 20 (2007). https://doi.org/10.1145/1240233.1240243
14. Ferragina, P., Venturini, R.: The compressed permuterm index. ACM Trans. Algorithms 1–21 (2010). https://doi.org/10.1145/1868237.1868248
15. Gog, S., Moffat, A., Culpepper, S., Turpin, A., Wirth, A.: Large-scale pattern search using reduced-space on-disk suffix arrays. TKDE 1918–1931 (2014). https://doi.org/10.1109/tkde.2013.129
16. Grabowski, S., Raniszewski, M.: Sampling the suffix array with minimizers. In: Iliopoulos, C., Puglisi, S., Yilmaz, E. (eds.) SPIRE 2015. LNCS, vol. 9309, pp. 287–298. Springer, Cham (2015). https://doi.org/10.1007/978-3-319-23826-5_28
17. Grossi, R., Ottaviano, G.: Fast compressed tries through path decompositions. J. Exp. Algorithmics 11–120 (2015). https://doi.org/10.1145/2656332
18. Grossi, R., Vitter, J.: Compressed suffix arrays and suffix trees with applications to text indexing and string matching. SIAM J. Comput. 378–407 (2005). https://doi.org/10.1137/s0097539702402354
19. Hu, T., Tucker, A.: Optimal computer search trees and variable-length alphabetical codes. SIAM J. Appl. Math. 514–532 (1971). https://doi.org/10.1137/0121057
20. Huffman, D.: A method for the construction of minimum-redundancy codes. IRE 1098–1101 (1952). https://doi.org/10.1109/jrproc.1952.273898
21. Kanda, S., Morita, K., Fuketa, M.: Practical implementation of space-efficient dynamic keyword dictionaries. In: Fici, G., Sciortino, M., Venturini, R. (eds.) SPIRE 2017. LNCS, vol. 10508, pp. 221–233. Springer, Cham (2017). https://doi.org/10.1007/978-3-319-67428-5_19
22. Kirschenhofer, P., Prodinger, H.: Some further results on digital search trees. In: Kott, L. (ed.) ICALP 1986. LNCS, vol. 226, pp. 177–185. Springer, Heidelberg (1986). https://doi.org/10.1007/3-540-16761-7_67

23. Krishnan, P., Vitter, J., Iyer, B.: Estimating alphanumeric selectivity in the presence of wildcards. In: ACM SIGMOD Record, pp. 282–293 (1996). https://doi.org/10.1145/235968.233341
24. Kroeger, P.: Analyzing Grammar: An Introduction. Cambridge University Press, Cambridge (2015)
25. Kärkkäinen, J., Ukkonen, E.: Sparse suffix trees. In: Cai, J.-Y., Wong, C.K. (eds.) COCOON 1996. LNCS, vol. 1090, pp. 219–230. Springer, Heidelberg (1996). https://doi.org/10.1007/3-540-61332-3_155
26. Larsson, N., Moffat, A.: Off-line dictionary-based compression. IEEE 1722–1732 (2000). https://doi.org/10.1109/5.892708
27. Leis, V., Gubichev, A., Mirchev, A., Boncz, P., Kemper, A., Neumann, T.: How good are query optimizers, really? VLDB Endowment 204–215 (2015). https://doi.org/10.14778/2850583.2850594
28. Li, D., Zhang, Q., Liang, X., Guan, J., Xu, Y.: Selectivity estimation for string predicates based on modified pruned count-suffix tree. CJE 76–82 (2015). https://doi.org/10.1049/cje.2015.01.013
29. Manning, C., Schütze, H.: Foundations of Statistical Natural Language Processing. MIT Press, Cambridge (1999)
30. Miner, G., Elder, J., Fast, A., Hill, T., Nisbet, R., Delen, D.: Practical Text Mining and Statistical Analysis for Non-structured Text Data Applications. Academic Press, Waltham (2012)
31. Moerkotte, G., DeHaan, D., May, N., Nica, A., Boehm, A.: Exploiting ordered dictionaries to efficiently construct histograms with q-error guarantees in SAP HANA. In: ACM SIGMOD (2014). https://doi.org/10.1145/2588555.2595629
32. Moerkotte, G., Neumann, T., Steidl, G.: Preventing bad plans by bounding the impact of cardinality estimation errors. VLDB Endowment 982–993 (2009). https://doi.org/10.14778/1687627.1687738
33. Moradi, H., Grzymala-Busse, J., Roberts, J.: Entropy of english text: Experiments with humans and a machine learning system based on rough sets. Inf. Sci. 31–47 (1998). https://doi.org/10.1016/s0020-0255(97)00074-1
34. Müller, M., Moerkotte, G., Kolb, O.: Improved selectivity estimation by combining knowledge from sampling and synopses. VLDB Endowment 1016–1028 (2018). https://doi.org/10.14778/3213880.3213882
35. Nilsson, S., Tikkanen, M.: An experimental study of compression methods for dynamic tries. Algorithmica **33**, 19–33 (2002). https://doi.org/10.1007/s00453-001-0102-y
36. Poyias, A., Raman, R.: Improved practical compact dynamic tries. In: Iliopoulos, C., Puglisi, S., Yilmaz, E. (eds.) SPIRE 2015. LNCS, vol. 9309, pp. 324–336. Springer, Cham (2015). https://doi.org/10.1007/978-3-319-23826-5_31
37. Sadakane, K.: Compressed suffix trees with full functionality. Theor. Comput. Syst. **41**, 589–607 (2007). https://doi.org/10.1007/s00224-006-1198-x
38. Sautter, G., Abba, C., Böhm, K.: Improved count suffix trees for natural language data. IDEAS. ACM (2008). https://doi.org/10.1145/1451940.1451972
39. Sigurd, B., Eeg-Olofsson, M., van Weijer, J.: Word length, sentence length and frequency - zipf revisited. Studia Linguistica **58**, 37–52 (2004). https://doi.org/10.1111/j.0039-3193.2004.00109.x
40. Sun, J., Li, G.: An end-to-end learning-based cost estimator (2019)
41. Vitale, L., Martín, Á., Seroussi, G.: Space-efficient representation of truncated suffix trees, with applications to markov order estimation. Theor. Comput. Sci. **595**, 34–45 (2015). https://doi.org/10.1016/j.tcs.2015.06.013

42. Welch, T.: A technique for high-performance data compression. Computer 8–19 (1984). https://doi.org/10.1109/mc.1984.1659158
43. Wu, W., Chi, Y., Zhu, S., Tatemura, J., Hacigümüs, H., Naughton, J.: Predicting query execution time: Are optimizer cost models really unusable? In: ICDE. IEEE (2013). https://doi.org/10.1109/icde.2013.6544899
44. Ziv, J., Lempel, A.: A universal algorithm for sequential data compression. IEEE Trans. Inf. Theor. **23**, 337–343 (1977). https://doi.org/10.1109/tit.1977.1055714

Shadow: Answering Why-Not Questions on Top-K Spatial Keyword Queries over Moving Objects

Wang Zhang[1], Yanhong Li[1](✉), Lihchyun Shu[2], Changyin Luo[3], and Jianjun Li[4]

[1] College of Computer Science, South-Central University for Nationalities, Wuhan, China
wang.zhang@scuec.edu.cn, liyanhong@mail.scuec.edu.cn
[2] College of Management, National Cheng Kung University, Taiwan, Republic of China
shulc@mail.ncku.edu.tw
[3] School of Computer, Central China Normal University, Wuhan, China
changyinluo@mail.ccnu.edu.cn
[4] School of Computer Science and Technology, Huazhong University of Science and Technology, Wuhan, China
jianjunli@hust.edu.cn

Abstract. The popularity of mobile terminals has generated massive moving objects with spatio-textual characteristics. A top-k spatial keyword query over moving objects (Top-k SKM query) returns the top-k objects, moving or static, based on a ranking function that considers spatial distance and textual similarity between the query and objects. To the best of our knowledge, there hasn't been any research into the why-not questions on Top-k SKM queries. Aiming at this kind of why-not questions, a two-level index called Shadow and a three-phase query refinement approach based on Shadow are proposed. The first phase is to generate some promising refined queries with different query requirements and filter those unpromising refined queries before executing any promising refined queries. The second phase is to reduce the irrelevant search space in the level 1 of Shadow as much as possible based on the spatial filtering technique, so as to obtain the promising static objects, and to capture promising moving objects in the level 2 of Shadow as fast as possible based on the probability filtering technique. The third phase is to determine which promising refined query will be returned to the user. Finally, a series of experiments are conducted on three datasets to verify the feasibility of our method.

Keywords: Top-k spatial keyword queries · Why-not questions · Moving objects · Shadow

C. S. Jensen et al. (Eds.): DASFAA 2021, LNCS 12682, pp. 738–760, 2021.
https://doi.org/10.1007/978-3-030-73197-7_51

1 Introduction

With the widespread use of mobile devices and the popularity of location based services (LBS), a large number of spatio-textual objects have been generated [15]. The top-k spatial keyword (SK) query, as one of the most important query forms in LBS, has been widely studied in academia and industry [17,19]. A top-k SK query takes a spatial coordinate, a set of keywords as the query requirements, and returns k objects that best match the query requirements.

In some cases, after a user initiates a top-k SK query, some of user-desired objects (missing objects) may not appear in the query result set. The user may then wonder why these objects disappeared, whether any other unknown relevant objects disappeared, or even question the whole query result. Therefore, it is necessary to explain the reasons for the loss of these expected objects and to provide a refined query that returns all the missing objects and the original query result objects.

Example 1. John is thirsty in the office and wants a cup of milk tea. Then he launches a query to find the top-3 milk tea shops nearby. However, he unexpectedly finds that both a nice mobile milk tea stall and the milk tea shop he used to visit are not in the query result. John wants to know why his desired objects are missing and how to obtain a refined query so that all missing shops and other possible better options appear in the refined query result set.

John's questions in the above example are called why-not questions. *Query Refinement* can retrieve all missing objects for the user by adjusting the original query requirements, compared to other methods on answering why-not questions, such as *Manipulation Identification*, *Ontology* and *Database Modification*. Chen *et al.* [4] answered why-not questions on spatial keyword top-k queries by adjusting the weight between spatial relevance and textual similarity. Later, Zhao *et al.* [25], Wang *et al.* [18] and Zheng *et al.* [26] dealt with why-not questions on geo-social spatial keyword queries, SPARQL queries and group queries, respectively.

However, existing researches mainly focus on why-not questions in top-k SK queries over static objects [4,25], and as far as we know, there is no research on the why-not questions in SK queries over moving objects. Four factors make it more challenging to answer why-not questions on top-k spatial keyword queries over moving objects (Top-k WSKM queries). First of all, different moving objects have different motion patterns, such as moving direction and moving speed. How to reasonably define the motion pattern of moving objects is a challenge. Secondly, the probability of a moving object appearing in a region at a certain time is a continuous variable. How to set up the probability density function to calculate the probability that a moving objects will appear in a region over a period of time is also a problem worth considering. Thirdly, during refined query processing, some of original query result objects and missing objects may move away from the query point. How to prune the search space as much as possible while ensuring that these moving objects are not pruned is also worth consideration.

Finally, a large number of objects move around the system, which means that object inserts and deletes frequently occur in the index before the refined queries are executed. Frequent inserts and deletes can be a waste of time, but users need to obtain query results as quickly as possible, which is a contradiction.

To address these challenges, we first define the movement patterns of moving objects. Assuming that a moving object has a probability density of appearing at a point in an area at a given time, we can then calculate the probabilities of moving objects appearing in that area at a time point and over a period of time, respectively. We formulate the why-not questions on spatial keyword top-k queries over moving objects (Top-k WSKM queries), and propose a cost model to measure the modification degree of refined queries compared with the original query, so as to obtain the refined query with the minimum modification cost. To deal with Top-k WSKM queries efficiently, a two-level index called Shadow and a three-phase query processing method based on Shadow are proposed. The first phase is to generate some promising refined queries with different query requirements and filter those unpromising refined queries before executing any promising refined queries. The second phase is to reduce the search space as much as possible by analyzing the spatial relationship among *OA* (Original Query Result Area), *AA* (Actual Refined Query Result Area) and *IA* (Irrelevant Area), and by using the principle of 3σ in normal distribution. The third phase is to determine which promising refined query will be returned to the user.

The key contributions are summarized as follows:

- We formulate why-not questions on top-k spatial keyword queries over moving objects (Top-k WSKM queries). To our knowledge, this is the first work on this issue.
- A novel index *Shadow* is proposed to organize the textual, spatial and motion pattern information of the objects. *Shadow* calculates the probability of moving objects appearing in a certain area within a certain period of time by analyzing the motion pattern of moving objects, helping users to capture the refined queries with the minimum cost as fast as possible.
- Extensive experiments are conducted on three real datasets to show the feasibility of *Shadow*.

2 Related Work

2.1 Spatial Keyword Query

To deal with top-k SKQ queries, Li *et al.* [12] proposed a hybrid index named IR-tree, which introduces inverted files into each leaf node of R-tree. Zhang *et al.* [24] proposed a novel index called I^3 to capture k best objects for the user. In addition, Zhang *et al.* [23] also proposed an index called IL-Quadtree to organize objects and handle top-k SKQ queries effectively. In recent years, Yang *et al.* [21] proposed HGR-tree to process spatial keyword search with fuzzy token matching, which considers approximate keyword matching rather than exact keyword matching. Cui *et al.* [7] focused on processing Boolean-Boolean spatial keyword queries while preserving the confidential information of users. In addition, Chen *et al.* [6] studied the Cluster-based continuous spatial keyword queries based on the publish/subscribe problem and proposed a novel solution to efficiently cluster, feed, and summarize a stream of geo-textual messages. Yao *et al.* [22] studied the computation of trajectory similarity, and proposed a method called NeuTraj to further improve the efficiency of neural-network based trajectory similarity computation.

2.2 Moving Objects Query

Dittrich *et al.* [8] proposed a mobile object indexing technique called MOVIES, which has no complex index structure, but constructs short-term discarded indexes with simple concepts to achieve high query processing speed and high index update rate. Heendaliya *et al.* [11] summarized the early work of mobile object processing in Euclidean space and road network, and provided solutions for the management and processing of mobile objects in road network environment. Shen *et al.* [16] proposed a balanced search tree named V-tree for dealing with k nearest neighbor (kNN) queries of moving objects with road-network constraints. Cao *et al.* [1] proposed a scalable and in-memory kNN query processing technique, which uses an R-tree to store the topology of the road network and a hierarchical grid model to process the moving objects in non-uniform distribution. In addition, Dong *et al.* [9] answered direction-aware KNN queries for moving objects in the road network. Xu *et al.* [20] summarized the research on moving objects with transportation modes in the last decade.

2.3 Why-Not Question

The why-not question was first proposed by Chapman *et al.* [2]. To deal with why-not questions, *Manipulation Identification Ontology Database Modification* [27] and *Query Refinement* [10] are presented. *Query Refinement* is widely used to answer why-not questions, since the missing objects cannot become the query result objects returned to the user by adopting *Manipulation Identification* or *Ontology* and the user usually does not have the right to modify the data in the database when using *Database Modification* [13]. *Query Refinement* was adopted by Chen *et al.* [4] to first answer why-not questions on top-k spatial keyword queries. In [5], they tried to find more accurate query keywords to fit all the missing objects into the refined query result set, while keeping other original query requirements unchanged. In addition, they answered why-not questions on direction-aware spatial keyword top-k queries by modifying the original query direction [3]. In recent years, Zhao *et al.*, Zheng *et al.* and Miao *et al.* answered why-not questions on top-k geo-social keyword queries in road networks [25], group spatial keyword queries [26] and range-based skyline queries in road networks [14], respectively.

3 Preliminaries and Problem Formulation

3.1 Top-*k* Spatial Keyword Query over Moving Objects

In existing related work, a spatio-textual object (object for short) $o \in O$ is usually defined as $o = (o.loc, o.doc)$, where $o.loc$ is a two-dimensional spatial point and $o.doc$ is a set of keywords. In reality, however, objects do not always stand still; they may keep moving. The moving object o^m usually has certain motion characteristics, such as the maximum speed $v_{max}(o^m)$, the minimum speed $v_{min}(o^m)$, and the actual speed $v^t(o^m)$ at time t. These characteristics are called the moving ability of object o^m at time t, $MA^t(o^m)$, which is defined as $(v_{max}(o^m), v_{min}(o^m), v^t(o^m))$. Note that in the real world, moving objects usually have their destinations, so we assume that each moving object moves towards its destination without changing direction or returning.

Thus given a spatial point $o.loc$, a set of keywords $o.doc$ and the moving ability $MA^t(o)$), an object $o \in O$ can be expressed as $o = (o.loc, o.doc, MA^t(o))$. Note that when $MA^t(o_i)$ is empty, the object o_i is static.

Given a query point $q.loc$, a keyword set $q.doc$, and two values α and k, then a Top-k spatial keyword query over moving objects (Top-k SKM query) $q = \langle q.loc, q.doc, \alpha, k\rangle$ retrieves k best objects from O, based on a ranking function that considers both the spatial proximity and textual similarity between the query q and objects, where α is the smooth parameter that satisfies $0 \leq \alpha \leq 1$.

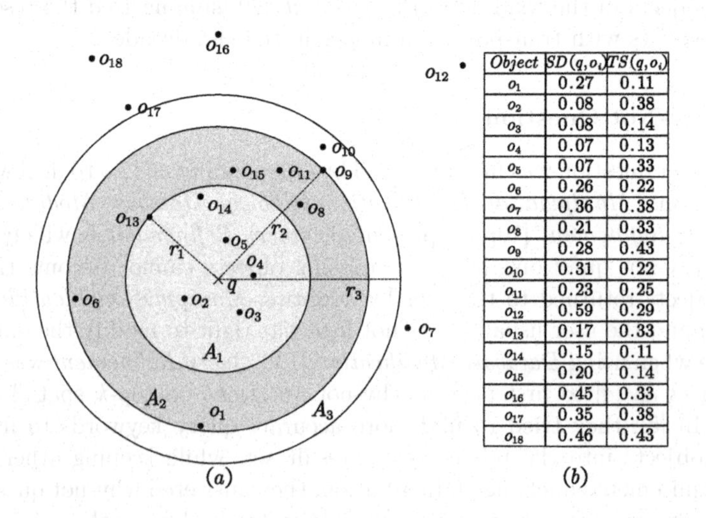

Object	$SD(q,o_i)$	$TS(q,o_i)$
o_1	0.27	0.11
o_2	0.08	0.38
o_3	0.08	0.14
o_4	0.07	0.13
o_5	0.07	0.33
o_6	0.26	0.22
o_7	0.36	0.38
o_8	0.21	0.33
o_9	0.28	0.43
o_{10}	0.31	0.22
o_{11}	0.23	0.25
o_{12}	0.59	0.29
o_{13}	0.17	0.33
o_{14}	0.15	0.11
o_{15}	0.20	0.14
o_{16}	0.45	0.33
o_{17}	0.35	0.38
o_{18}	0.46	0.43

Fig. 1. An example of the top-k spatial keyword query over moving objects

We use a widely used ranking function to measure the similarity between query q and object o as follows:

$$Rank(q, o) = \alpha \cdot (1 - SD(o, q)) + (1 - \alpha)\, ST(o, q) \tag{1}$$

where $SD(o, q)$ and $ST(o, q)$ are normalized spatial distance and textual similarity, respectively. $ST(o, q) = \frac{|q.doc \bigcap o.doc|}{|q.doc \bigcup o.doc|}$, and $SD(o, q) = \frac{D_E(q,o)}{MaxD_E}$, where $D_E(q, o)$ is the Euclidean distance between q and o, and $MaxD_E$ represents the maximal distance between any two objects in O.

Example 2. As shown in Fig. 1(a), a query q and 18 objects are distributed in the search space, and the normalized spatial distances and textual similarity information between the query and objects are shown in Fig. 1(b). When the user initiates a top-3 SKM query with $\alpha = 0.5$ as the input, objects o_2, o_5 and o_{13} are returned as the query results, since these objects have higher ranking scores than other objects.

3.2 Why-Not Questions on Top-k Spatial Keyword Query over Moving Objects

When the user initiates a top-k spatial keyword query over moving objects (Top-k WSK query) $q = (loc, doc, k, \alpha)$, k objects are returned to form the original query result set ^{o}R. If some query parameters are improperly set, one or more user-desired objects may unexpectedly disappear, which are called missing objects. We denote the missing object set as ^{m}O. In existing research on why-not questions, all objects in O are treated as static objects. Our model can deal directly with this simple case by leaving the moving ability $MA^t(o)$ empty, and the corresponding refined query result set is called the static refined query result set ^{s}R.

Example 3. Continuing with Example 2, the user obtains an original result set $^o\text{R} = \{o_2, o_5, o_{13}\}$, and we assume $^m\text{O} = \{o_4, o_9\}$. If all objects in Fig. 1 are static, then existing methods (such as [4]) can be used to return the user a refined query $q = (loc, doc, 7, 0.6)$ to get the static refined query result set $^s\text{R} = \{o_2, o_5, o_{13}, o_4, o_3, o_8, o_9\}$.

However, some objects are often constantly moving, and existing methods cannot directly handle the why-not questions on Top-k SKM queries. To this end, we focus on why-not questions on top-k spatial keyword query over moving objects (Top-k WSKM query for short). Since query users often do not know how to choose an initial value to balance the relative importance between spatial proximity and textual similarity, the initial α value often does not accurately express users' preference. Hence, we adjust α to capture a refined query result set ^{r}R containing ^{o}R and ^{m}O, some of which are moving objects. Since the refined query result set must include all original query results and missing objects, its size k' is greater than k. Note that we can adjust other query parameters to achieve our goal, but this paper only explores how to adjust α to achieve that goal. For ease of expression, we first give the following definitions.

Definition 1. *(Original Query Result Area, OA for short). Given a Top-k SKM query* $q = (loc, doc, k, \alpha)$ *with a query result set* ^oR*, there is an object* $o_i \in {^oR}$*,* $\forall o \in {^oR} - \{o_i\}$*,* $d(q, o) \leq d(q, o_i)$*. Then the original query result area can be* *defined as a circular area with the query point q.loc as the center and* $d(q, o_i)$ *as the radius.*

Definition 2. *(Actual Refined Query Result Area, AA for short). Given a Top-k SKM query q with a query result set* ^oR *and a missing object set* ^mO*, there are a moving object* $o_k^m \in {}^oR \cup {}^mO$ *and a static object* $o_j \in {}^oR \cup {}^mO$*,* $\forall$ *moving objects* $o^m \in {}^oR \cup {}^mO - \{o_k^m\}$*, we have* $MAX\{v_{min}(o_k^m) \cdot (t_t - t_s) + d(q, o_k^m), d(q, o_j)\} \geq v_{max}(o^m) \cdot (t_t - t_s) + d(q, o^m)$*, where* t_s *is the starting time of q and* t_t *is the starting time of the refined query of q.* $MAX\{a, b\}$ *returns the maximum value between a and b. Then the actual refined query result area can be defined as a circular area with the query point q.loc as the center and* $MAX\{v_{min}(o_k^m) \cdot (t_t - t_s) + d(q, o_k^m), d(q, o_j)\}$ *as the radius.*

Definition 3. *(Irrelevant Area, IA for short). Given AA, the irrelevant area is defined as the region outside of AA.*

To measure the modification degree of a refined query $q' = (loc, doc, k', \alpha')$ relative to the original query $q = (loc, doc, k_0, \alpha)$, a modification cost model is defined as follows.

$$cost(q, q') = \beta \cdot \frac{|\alpha' - \alpha|}{\Delta\alpha_{max}} + (1 - \beta) \cdot \frac{|k' - k_0|}{\Delta k} \tag{2}$$

where $\beta \in [0, 1]$ is a weight to represent the user's preference for adjusting α or k, and $\Delta\alpha_{max}$ and Δk are the maximum possible modifications of α and k, respectively. We then formulate the why-not questions on top-k spatial keyword query over moving objects as follows.

Definition 4. *(Why-Not questions on Top-k Spatial Keyword Query over Moving Objects, Top-k WSKM query for short). Given an original Top-k SKM query* $q = (loc, doc, k_0, \alpha)$*, an original query result set* ^oR *and a missing object set* ^mO*, the Top-k WSKM query returns a refined query* $q' = (loc, doc, k', \alpha')$ *with the minimum modified cost according to Eq. (2) and its result set* $^rO \supseteq {}^oR \cup {}^mO$*.*

4 Moving Objects and Probability Distribution Function

4.1 Moving Ability

As discussed in the previous section, each moving object has a destination and its moving ability. Since a moving object o_i^m has the maximum speed $v_{max}(o_i^m)$ and the minimum speed $v_{min}(o_i^m)$, the moving distance of o_i^m in the time interval Δt is within the range of $[v_{min}(o_i^m) \cdot \Delta t, v_{max}(o_i^m) \cdot \Delta t]$. For the sake of discussion, we further assume that a moving object, no matter how fast it moves, moves towards its destination and arrives at its destination at a specified time point.

As shown in Fig. 2(a), a moving object o_i^m moves from point A to point B. If it keeps moving at its maximum speed, $v_{max}(o_i^m)$, and it needs to reach point B at time t_t, then its trajectory is arc AB, denoted as $\widehat{AB}$. Similarly, the trajectory of this object moving at the minimum speed $v_{min}(o_i^m)$ is the line AB. When moving at speed $v^t(o_i^m)$, the movement of object o_i^m at time t will be limited to the region bounded by the two $\widehat{AB}$ arcs, which is called the moving region of o_i^m. Note that $t \in [t_s, t_t]$ and $v^t(o_i^m) \in [v_{min}(o_i^m), v_{max}(o_i^m)]$. Therefore, the length of line AB is $v_{min}(o_i^m) \cdot (t_t - t_s)$ and the length of $\widehat{AB}$ is approximately equal to $v_{max}(o_i^m) \cdot (t_t - t_s)$.

Now, we discuss how to calculate the area of the moving region mentioned above. As shown in Fig. 2(b), γ represents half of the circumference angle corresponding to $\widehat{AB}$. Then we have $2\pi \cdot \frac{v_{min}(o_i^m)\cdot(t_t-t_s)}{2sin\gamma} \cdot \frac{2\gamma}{2\pi} = v_{max}(o_i^m) \cdot (t_t - t_s)$. By simplifying it, we have $\frac{sin\gamma}{\gamma} = \frac{v_{min}(o_i^m)}{v_{max}(o_i^m)}$. Since $\gamma \in (0, \frac{\pi}{2})$, then we can find a value $\gamma_1 \in (0, \frac{\pi}{2})$ to make sure that $\frac{sin\gamma_1}{\gamma_1} = \frac{v_{min}(o_i^m)}{v_{max}(o_i^m)}$ as shown in Fig. 2(c). Then the whole area of o_i^m's moving region can be calculated as follows: $2 \cdot \pi \cdot (\frac{v_{min}(o_i^m)\cdot(t_t-t_s)}{2sin\gamma_1})^2 \cdot \frac{2\gamma_1}{2\pi}$ - $2 \cdot \frac{1}{2} \cdot v_{min}(o_i^m) \cdot (t_t - t_s) \cdot \frac{v_{min}(o_i^m)\cdot(t_t-t_s)}{2tan\gamma_1} = \frac{(v_{min}(o_i^m)\cdot(t_t-t_s))^2}{2sin\gamma_1} \cdot (\frac{v_{max}(o_i^m)}{v_{min}(o_i^m)} - cos\gamma_1)$.

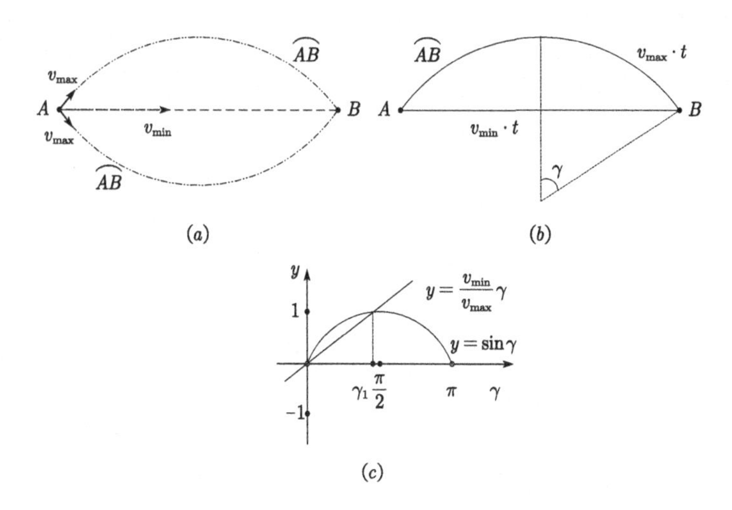

Fig. 2. The moving ability of a moving object

By the way, we can also calculate the function expressions of the two arcs $\widehat{AB}$. Due to space limitation, we do not give the detailed calculation process, but denote them as $f_1^t(o_i^m)$ and $f_2^t(o_i^m)$. Then the whole area of o_i^m's moving region can be calculated in another way as follows: $\int_{t_s}^{t_t} |f_1^t(o_i^m) - f_2^t(o_i^m)|\, dt$.

4.2 Probability Density and Probability

Since the speed of a moving object o_i^m is not fixed, the position that o_i^m moves to at time t should be on a curve l. As we all known, spatial indexing is based on dividing the whole search space into basic units, which we call basic cells.

Therefore, for any given spatial index, the curve l may appear in one or more basic cells.

Suppose the curve l appears in n basic cells at time t, called $C_1, C_2, ..., C_n$, and the parts of the curve that appear in these basic cells are called $l_1, l_2, ..., l_n$, respectively. Then $\sum_{i=1}^{n} l_i = 1$ and $l_i > 0$. If the probability density of o_i^m appearing at a point on curve l at time t is $p_l^t(o_i^m)$, then the probability of o_i^m appearing on part l_i is $P_{l_i}^t(o_i^m) = \int_{l_i} p_l^t(o_i^m)dl$. Note that $\sum_{i=1}^{n} P_{l_i}^t(o_i^m) = \sum_{i=1}^{n} \int_{l_i} p_l^t(o_i^m)dl = 1$.

Since the possible positions of the moving object o_i^m at different time points consist of different curves, the intersecting parts of different curves with a basic cell can be obtained. Hence, we can calculate the probability of o_i^m appearing in a basic cell during time t_1 to t_2 as follows: $P_{l_i}^{t_2-t_1} = \frac{\int_{t_1}^{t_2} P_{l_i}^t(o_i^m)dt}{\int_{t_1}^{t_2} l_i dt}$.

Figure 3 shows the movement of the moving object o_i^m from t_s to t_t. It appears in the basic cell C_2 and C_1 at time t_1 and t_2, respectively. At time t_3, the part of the position curve of o_i^m appearing in C_1 and C_2 is l_1 and l_2, respectively. Then if the probability density of l_1 and l_2 is $p_l^t(o_i^m)$, the probability of o_i^m appearing on l_1 and l_2 is $P_{l_1}^t(o_i^m) = \int_{l_1} p_l^t(o_i^m)dl$ and $P_{l_2}^t(o_i^m) = \int_{l_2} p_l^t(o_i^m)dl$, respectively. Since the probability of o_i^m appearing in C_1 is different at different time point $t \in [t_2, t_3]$, the probability of o_i^m appearing in C_1 from t_2 to t_3 is $P_{l_1}^{t_3-t_2} = \frac{\int_{t_2}^{t_3} P_{l_1}^t(o_i^m)dt}{\int_{t_2}^{t_3} l_1 dt}$.

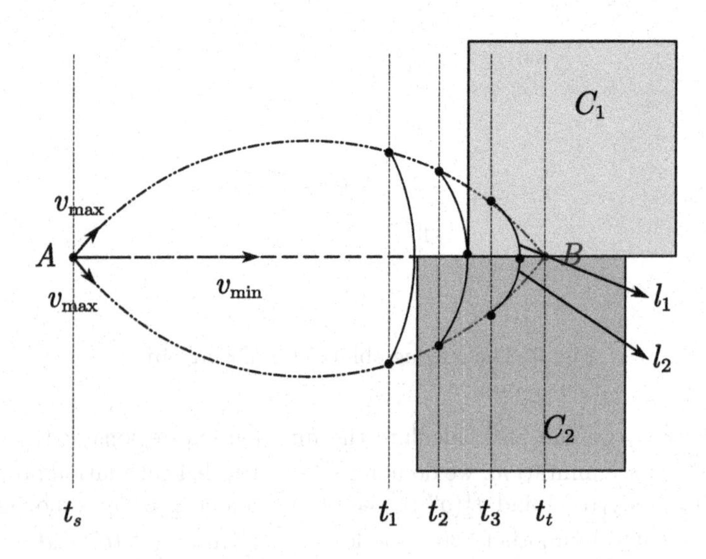

Fig. 3. An example of a moving object and two basic cells

4.3 Probability Distribution Function and Normal Distribution

After discussing how to calculate the probability of a moving object appearing in a basic cell over a period of time, we give a probability distribution function as follows.

$$P_{l_i}^{t-t_s} = \frac{\int_{t_s}^{t} \int_{l_i} p_l^t(o_i^m) dl dt}{\int_{t_s}^{t} l_i dt} \tag{3}$$

This probability distribution function can be used in two aspects: 1) Given two time points t_i and t_j, where $t_s \leq t_i < t_j \leq t_t$, the probability of a moving object appearing in a basic cell during t_i to t_j can be calculated; 2) Given a starting time t_s, if we want the probability of a moving object o_i^m appearing in a basic cell c_i to be greater than a certain value, a critical time t_k can be calculated so that the probability of o_i^m appearing in c_i is greater than that value during t_s to t, where $t \in (t_k, t_t)$.

As mentioned in Sect. 4.2, at time t, the curve l is divided into n parts: $l_1, l_2, ..., l_n$, each of which has a certain probability that object o_i^m appears on it. In the same way, we can calculate the probabilities of o_i^m appearing in $C_1, C_2, ..., C_n$ during t_i to t_j using Eq. (3), where $t_s \leq t_i < t_j \leq t_t$. It means that during the period t_i to t_j, each moving object has different probabilities of appearing in different basic cells. Then if a missing object is a moving object, a refined query must return it along with its probabilities in different locations. Note that the probability of a moving object appearing at a certain point during a period of time is a probability density and has no practical meaning. In the experiment, we calculate the probability of a moving object appearing in a basic cell and use it as the probability of the object appearing at any point in the cell.

However, a contradiction arises when the probability of a moving object appearing in a basic cell is very small and the basic cell is very far from the query point over a period of time. On the one hand, if we want to ensure that each refined query captures the user-desired moving object with a 100% probability, the basic cells far away from the query point and other unnecessary search space need to be accessed, which takes time. On the other hand, if a refine query cannot access all the basic cells that the moving object appears with a certain probability, there is a probability that the refined query cannot retrieve the user-expected object. To make sure that the probability is as small as possible and to solve this contradiction, we use the principle of 3σ in normal distribution.

For a normal distribution $\frac{1}{\sqrt{2\pi}\sigma} e^{-[\frac{(x-\mu)^2}{2\sigma^2}]}$, where μ is the mean value and σ is the standard deviation value, the principle of 3σ means the probability of a variable $x \notin (\mu - 3\sigma, \mu + 3\sigma]$ is a small probability event that can't happen under certain circumstances. Hence, if the sum of the probabilities that a moving object o_i^m appears in some basic cells is greater than $P(\mu - 3\sigma < x \leq \mu + 3\sigma)$, there is no need to access other basic cells where o_i^m appears with a certain probability. This can be used to filter out unnecessary search space to help us improve query processing efficiency, which is described further in the next section.

5 Shadow-Based Method on Answering Top-k WSKM Queries

When we know how to calculate the probability of a moving object appearing in a basic cell over a period of time, and learn the fact that a refined query does not need to find user-desired moving objects with a 100% probability, it is important to design a novel index to help us store and process objects efficiently. If the existing methods are used to keep the objects, static or moving, to answer Top-k WSKM queries, we need to first delete or insert moving objects from the index, then update the index accordingly, and finally execute a refined query on the modified index. There are two disadvantages: 1) time consuming, especially when a large number of moving objects are far from the query point; 2) If we delete, insert moving objects along with their information and update the index structure, some moving objects will move to other locations during the execution of the refined query, which will affect the accuracy of refined query results. To overcome these shortcomings, we propose an index named Shadow, which will be discussed in more detail.

5.1 The Index Structure of Shadow

The Shadow has two levels. In the first level, the whole search space is divided into several basic cells as described earlier, and all static objects in the search space are indexed using a quadtree. A leaf node in level 1 of Shadow stores the static objects in a basic cell corresponding to the leaf node. Therefore, no matter how the moving object moves, it does not affect the level 1 of Shadow.

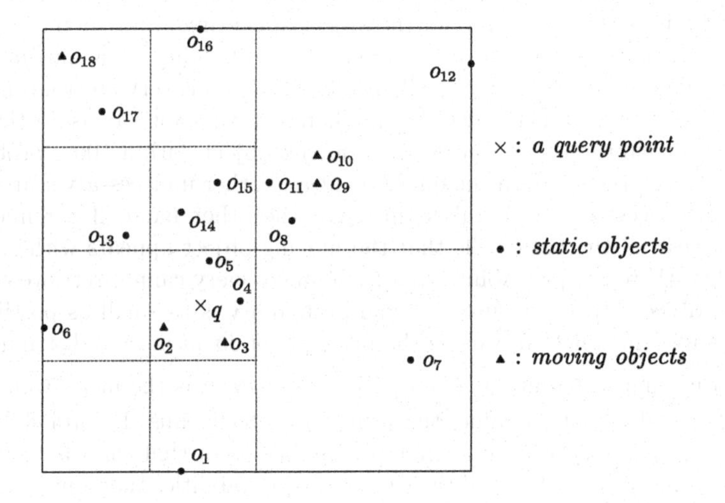

Fig. 4. The space partition result of Fig. 1 using Shadow

In the second level, all moving objects are organized. To insert the first moving object into the Shadow, we first need to find the leaf node to be inserted in the level 1, assuming that the object is static. Then we create a new node in the level 2 with a different node id and the same node location information as the corresponding leaf node in layer 1. Finally, the moving object is inserted into the new node. When there are other moving objects to be inserted into the shadow, we need to determine if the node in level 2 to be inserted exists. If so, we simply insert the object into this node. Otherwise, a new node needs to be built in level 2, as mentioned earlier. Note that if the number of objects in a node of level 2 exceeds the node maximum capacity, the node needs to be split into four children, the same as in a quadtree.

Each node in level 2 stores the information about the moving objects it contains, including the object identify $o^m.id$, the object location $o^m.loc$, object keywords $o^m.doc$ and the probability $P_{l_k}^{t_j-t_i}(o^m)$ of object o^m appearing in the basic cell corresponding to that node during a time period $[t_i, t_j]$. After inserting all moving objects into the level 2 of Shadow, all the nodes in level 2 are assigned to multiple groups according to the distances between the nodes and the groups. For each group, we store the following information: the group identify $G_a.id$, the group location $G_a.loc$ and the group pointer to its adjacent group $G_a.p$. The length of the group can be calculated according to the location information $G_a.loc$, and the distance of any two objects in the group will not exceed the length. Note that each group has the same upper limit of group length. When a group reaches the length limit, no matter how close a node in level 2 is to the group, it will not be assigned to the group. If and only if the level 1 of Shadow has no nodes, the moving object cannot found the corresponding leaf node. This means that all objects are moving objects. If so, a node, pointing to all the groups at level 2 of Shadow, is built at level 1.

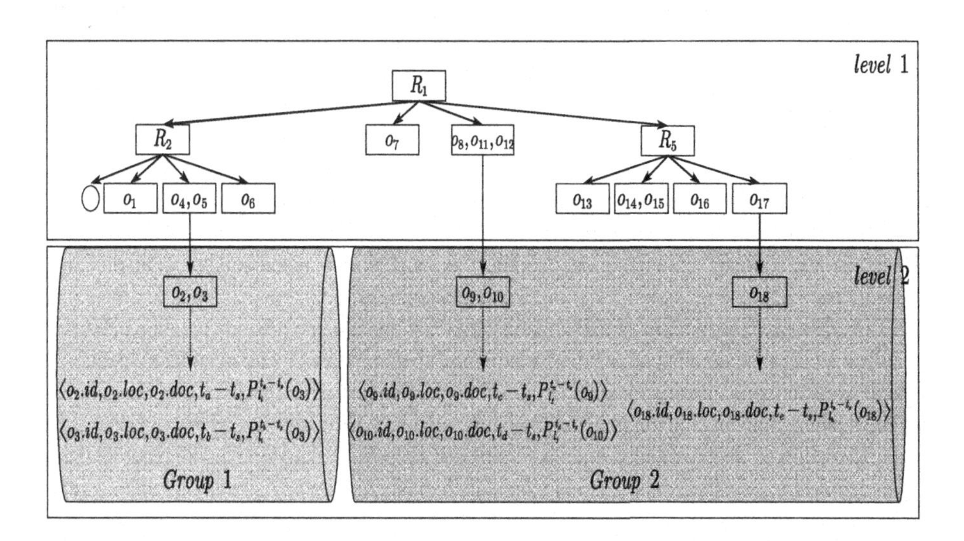

Fig. 5. The Shadow index structure

Algorithm 1: Creating Shadow Algorithm

1 Input: an object set O, a queue Q storing the moving objects, the upper limit Gl_{max} of group length;
2 Output: Shadow;
3 **begin**
4 Build a quadtree for static objects in O to form the level 1 of Shadow;
5 **while** *Q is not empty* **do**
6 o_i^m=Out_Queue(Q);
7 Find the left node R_i whose corresponding basic cell includes o_i^m assuming that o_i^m is a static object;
8 **if** *R_i has a pointer linked to the node $R_i^{'}$ in level 2* **then**
9 Insert o_i^m into $R_i^{'}$;
10 **else**
11 Create a node $R_i^{'}$ linked by R_i in level 2, and insert o_i^m into $R_i^{'}$;
12 Group all nodes in level 2 to ensure that the length of each group does not exceed Gl_{max};
13 Return Shadow;

Figure 4 shows the partition results of all objects in Fig. 1, where o_2, o_3, o_9, o_{10} and o_{18} are moving objects, and others are static objects. As shown in Fig. 4, each static object is allocated to its basic cell and each moving object also can find its corresponding basic cell. Note that in this example, we assume that each basic cell can hold information about up to four objects.

Figure 5 shows the Shadow index structure for all objects in Fig. 1, with level 1 and level 2 storing information for static and moving objects, respectively. Each leaf node in level 1 and level 2 corresponds to a basic cell in Fig. 4. Since o_9 and o_{10} are in the same basic cell as o_8, o_{11} and o_{12}, we need to first find the leaf node that stores o_8, o_{11} and o_{12} in level 1, and then create a node in level 2 to store o_9 and o_{10}. A pointer needs to be built from the node storing o_8, o_{11} and o_{12} to the node storing o_9 and o_{10} as one of the entrances to the level 2 of Shadow. Since the storage node (A) for o_9 and o_{10} is closer to o_{18}'s storage node (B) than the one (C) for o_2 and o_3, then A and C are assigned to group 2 and B to group 1.

The detailed steps for creating the Shadow index are shown in Algorithm 1. It takes as inputs an object set O containing static objects and moving objects, a queue Q storing all moving objects, and the group length upper limit Gl_{max} used to limit the space size of a group, and outputs the Shadow index structure. The level 1 and level 2 of the Shadow are built on Line 4 and Lines 5–12, respectively.

5.2 Pruning Techniques

We then introduce several lemmas to efficiently prune the unnecessary search space for each refined query q' to be examined. Firstly, we use AA (Actual

Refined Query Result Area) defined in Sect. 3.2 to prune unpromising nodes in level 1 of Shadow.

Lemma 1. *Given a node R_i in level 1 of Shadow and an AA of a refined query q', if R_i and AA do not intersect with each other, then R_i and its child nodes will be pruned.*

Proof. Assume that R_i contains the result object o_i, then $d(q', o_i)$ is not greater than the radius of AA. Since R_i does not intersect with AA, all objects in R_i are outside the scope of AA, and so is o_i. Thus, $d(q', o_i)$ is greater than the radius of AA, which contradicts the hypothesis. As a result, R_i and its children can be pruned safely.

Note that this pruning technique can also be used in level 2 of Shadow to filter unpromising nodes and groups. Secondly, we can use the principle of 3σ in normal distribution to filter unnecessary nodes and groups in level 2 of Shadow.

Lemma 2. *Given a node set $\{R_1, R_2, ..., R_i\}$ in level 2 of Shadow, where a user-desired moving object o_i^m appears with different probabilities in the basic cells corresponding to these nodes during a time period, if the sum of the probability values of some nodes in the node set is greater than $P(\mu - 3\sigma < x \leq \mu + 3\sigma)$ of a normal distribution, no other nodes and groups need to be accessed for the processing of o_i^m .*

Proof. The proof can be obtained directly from the 3σ principle of normal distribution and hence is omitted.

5.3 Answering Top-k WSKM Queries

As mentioned earlier, some user-desired objects may disappear from the query result set after executing an original query. Our main goal is to obtain a refined query with the minimum modification cost, whose result set contains all original query result objects and all missing objects required by the user.

Algorithm 2 gives the processing steps of using Shadow to answer Top-k WSKM queries. Since some existing methods (e.g. [4]) have studied how to adjust α and k to answer why-not questions on spatial keyword top-k queries when all objects are static, we mainly discuss the different part of our algorithm from the existing static methods.

It takes as inputs a Shadow index, an original query $q = (loc, doc, k, \alpha)$, a promising refined query set, an actual refined query result area AA, an original query result set ^{o}R, a missing object set ^{m}O, a time t of executing the current refined query, a starting time t_s, an ending time t_t, and outputs a refined query q' and its query results. Note that by using an existing method [4], a promising refined query set can be generated by setting different query requirements.

Firstly, the refined query result set ^{r}R and a node set RS are set to be empty to keep the objects which satisfy the refined query requirements and the leaf node of level 1 of Shadow as the entrance to level 2 of Shadow, respectively

Algorithm 2: Shadow Based Algorithm

```
1  Input: Shadow, q, a promising refined query set, AA, ᵒR, ᵐO, t, t_s, t_t;
2  Output: Best refined query q' = (loc, doc, k', α'), ʳR;
3  begin
4      set ʳR=∅;RS = ∅;
5      Execute a refined query in the promising refined query set, whose cost is
       less than any other refined queries previously executed;
6      Prune the irrelevant nodes in level 1 of Shadow corresponding to the search
       space that has no intersection with AA;
7      Find promising static objects and the node set RS of the leaf nodes
       containing these objects;
8      while RS is not empty do
9          Take out a leaf node R_i from RS and access the level 2 of Shadow;
10         Find promising moving objects that are not in ʳR and satisfy the
           refined query requirements, and insert them into RS according to the
           requirement of 3σ principle ;
11     Computer the ranking scores of all objects in ʳR and retain top-k' objects;
12     Return q' = (loc, doc, k', α'), ʳR;
```

(line 4). Next, a refined query is taken out from the promising refined query set and the cost of this refined query will be calculated. The refined query will be executed if its cost is less than any other refined queries which have been executed before. Otherwise, the execution of this refined query is terminated and the next promising query is selected to execute (line 5).

Using lemma 1, we can filter the irrelevant branches and nodes in level 1 of Shadow corresponding to the search space that has no intersection with AA (line 6). And then the promising static objects and the node set RS of the leaf nodes containing these objects can be found (line 7). Now all the promising static objects are captured, and desired moving objects will be obtained according to the remaining steps of Algorithm 2.

If RS is not empty, a leaf node R_i in RS is taken out as an entrance (using the pointer stored at this leaf node) to access nodes and groups in level 2 of Shadow. Then the moving objects that satisfy the refined query requirements are found in the node linked by R_i. If the probabilities of these objects appearing in the search space corresponding to the node are greater than $P(\mu - 3\sigma < x \leq \mu + 3\sigma)$ of a normal distribution, these moving objects are added to ^{r}R. If one of the objects is not satisfied, the other nodes in the same group that the node is in will be accessed first. Then, the nodes in other groups close to that group will be accessed until the probability requirement of the object is met. (lines 8–10).

When RS is empty, the ranking scores of all objects in ^{r}R are calculated, and k best objects are retained. Finally, the refined query and its ^{r}R are returned.

6 Experiments

6.1 Experimental Setup

System Setup and Metrics. The experiments are conducted on a PC with Inter Core i7 1.80 GHz CPU and 8 GB RAM, running Windows 10 OS. All the algorithms are implemented in C++. For each group of experiments, 1000 queries were randomly selected and the average processing time was reported. Since the probability of a moving object appearing in a basic cell at time t is a probability density, which approaches 0 and does not have practical significance, we calculate the probability of a moving object appearing in a basic cell over a period of time. When building the level 2 of shadow, we store the probability calculated along with object's keywords and the object's location which is randomly generated in the basic cell. Hence, in our experiments, a moving object retrieved by a refined query has a spatial location, a series of keywords and a probability.

Datasets. We use three datasets[1], US, CN and AF, which contain the geographical object ids, names as keywords, latitudes and longitudes as the location information of objects, respectively. We keep objects' ids, location information and keywords, and clean up the other tuples, such as "feature class" and "feature code". The detailed statistics for the datasets are shown in Table 1.

For each dataset, we select 1% of the objects as moving objects, which are evenly distributed among all objects. The speed range of 90%, 9% and 1% of moving objects is 1–5 km/h, 5–15 km/h and 15–50 km/h, respectively. We also generate a destination for each moving object. For each moving object with a speed range of 1–5 km/h, we draw a virtual circle with the object's location as the center and 1 km as the radius. The destination of a moving object is then randomly generated at any point in the circle. Similarly, the virtual circle radius of the moving object with a speed range of 5–15 km/h and 15–50 km/h is 5 km and 15 km, respectively.

Table 1. Dataset information

Dataset	*US*	*CN*	*AF*
# of objects	2,237,870	777,759	75,649
# of moving objects (1–5 km/h)	20,141	7,000	681
# of moving objects (3–15 km/h)	2,014	700	68
# of moving objects (15–50 km/h)	224	78	8
# of distinct words	266,327	46,280	6,752
avg.# of keywords per object	4	2	3

[1] http://www.geonames.org/.

Parameters Setting. We evaluate the performance of our method by varying k_O, the number of keywords in the original query, the original rank of the missing object, the number of missing objects, the α value and the dataset size. The parameters and their values are summarized in Table 2.

Table 2. Parameter setting

Parameter	Setting	Default
k_O	5,10,15,20	10
# of keywords	3,6,9,12	6
Original rank of the missing object	51, 101, 201, 301	51
α	0.1,0.3,0.5,0.7,0.9	0.5
# of missing objects	1,2,3,4	1
Data Size(M)	0.2,0.6,1.0,1.4,1.8	0.6

6.2 Experimental Result

For one original query among 1,000 randomly selected queries, several promising refined queries are generated by using different query requirements. In these promising refined queries, if some queries succeed in retrieving all the original query results and missing objects, and the others do not, the lowest-cost query that successfully retrieves all the required objects is returned to the user. Conversely, if all of these promising refined queries fail to retrieve all of the original query results and missing objects, the user fails to receive the desired refined query and expected objects, which is called *Failed Retrieval.* Theoretically, this is a very small probability event, and we can simply calculate it as follows. Assume that n promising refined queries are generated for each original query, and the $P(\mu - 3\sigma < x \leq \mu + 3\sigma)$ of a normal distribution is approximately equal to 99.7%, then the probability of *Failed Retrieval* for an original query is 0.003^n. The larger the value of n, the smaller the probability.

However, some promising refined queries can obtain all of the original query result objects and missing objects, but they take a lot of time to retrieve these objects. Therefore, in the experiment, we give an upper limit of query time. If the promising refined query takes longer than the upper limit to retrieve the desired objects, the query will be terminated, also known as *Failed Retrieval.* Therefore, we also examine the relationship between the number of *Failed Retrievals* and different query requirements. The first, second and last two-sets of experiments are conducted on CN, AF and US dataset, respectively. For these three datasets, we set the upper limit of query time to 10 s, 2 s and 1 min, respectively.

In addition, if previous related works are used to deal with our proposed problems, their approaches will spend a lot of time on adjusting the index structures.

This is because moving objects may appear in different places at different times, causing their index structures to be updated frequently. However, in our method, the Shadow does not need to be updated. Therefore, in the following experiments, we only measure the performance of Shadow under different parameters and do not compare the performance with other methods.

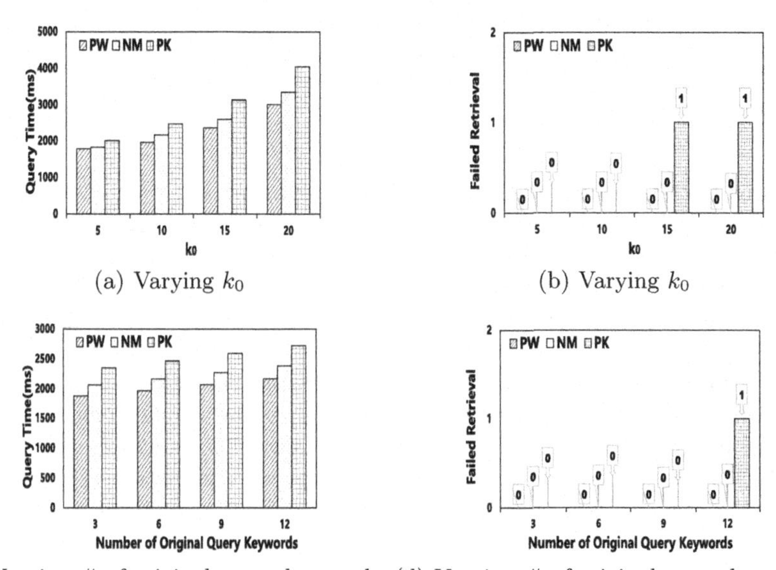

(a) Varying k_0 (b) Varying k_0

(c) Varying # of original query keywords (d) Varying # of original query keywords

Fig. 6. The experiment results on the CN dataset

Varying. $\boldsymbol{k_0}$ We first evaluate the performance of our scheme by varying the value of k_0. Figure 6(a) shows the query time of the algorithm under different cost options, where PW represents "prefer modifying the weight between spatial distance and textual similarity ($\beta = 0.9$)"; PK stands for "prefer modifying k ($\beta = 0.1$)"; and NM represents "never mind ($\beta = 0.5$)". The query time of these three cases increases with the enlargement of k_0. On the one hand, the size of the refined query result set usually increases as k_0 rises, resulting in a longer processing time. On the other hand, an increase in k_0 means that the missing objects rank worse in the original query. Note that in the experiments, the missing objects we chose ranked $5 \cdot k_0 + 1$ in the original query. The lower the ranking score of the missing object in the original query, the more difficult it is to find a refined query to capture it into the refined query result set. Compared with PW and NM, PK performs worst because it takes more time when the algorithm relies more on increasing k to obtain the missing objects. For example, if k_0 is 10, the missing object ranks 51 among all the objects evaluated by the original query, and 40 unknown objects need to be processed, which takes time. PW performs best, and when $k_0 = 10$, its performance is 1.10 and 1.2 times that of NM and PK, respectively. This is because PW can prune unpromising refined queries and reduces unnecessary calculations by adjusting α. In addition, Fig. 6(b) shows the relationship between *Failed Retrieval* and different k_0. A large k_0 value means that the refined queries relies more on adjusting k, thus : 1) a large number of objects will be accessed, which will affect the pruning effect in level 1 of Shadow; 2) the missing objects may be far away from the query point, so more groups and nodes need to be accessed. Both of these scenarios are time consuming, and may cause the refined query time to exceed the query time limit, leading to *Failed Retrieval.*

Varying the Number of Original Query Keywords. Next, we study the impact of varying the number of original query keywords on algorithm performance in the three cases. As shown in Fig. 6(c), the increase in the number of original keywords do not have a obvious effect on the query time of PW, NM and PK. The reason lies in two aspects. First, as the number of original query keywords increases, the time to calculate the textual similarity between the refined query keywords and object keywords may increase. Secondly, the textual similarity calculation of the method is relatively simple and it relies more on spatial filtering and probability filtering. Figure 6(d) shows that using PK will result in 1 *Failed Retrieval* when the number of original query keywords ia 12. It could be accidental, or it could be that the large amount of textual similarity calculations between the refine query keywords and the object keywords waste some time, causing the refining query time to exceed the given time limit.

Varying the Original Rank of the Missing Object. We investigate the performance of our method in answering why-not questions for missing objects with different ranks in the original query. In other sets of experiments, the missing object selected ranks 51st by default among the objects evaluated by the original query. In this set of experiments, we expand the rank range of the missing objects in the original query to 101, 201 and 301, respectively. Figure 7(a) shows that the Top-k WSKM queries take more time to obtain the moving object that ranks poorly in the original query. Moreover, poor ranking can mean that the object is far away from the query point. Especially if the missing object is a moving object, it can be stored in a group on level 2 of Shadow away from the query point, which will take more time to retrieve it. Figure 7(b) shows that the number of *Failed Retrievals* increases with the enlargement of the original rank of the missing object. The reasons why NM and PK have 1 and 2 *Failed Retrievals*, respectively, are similar to those discussed in the experiment of Varying k_0, which are omitted here.

Varying the Number of Missing Objects. We then study the effect of the number of missing objects on algorithm performance, and Fig. 7(c) shows that as the number of missing objects increases, the processing cost of our method

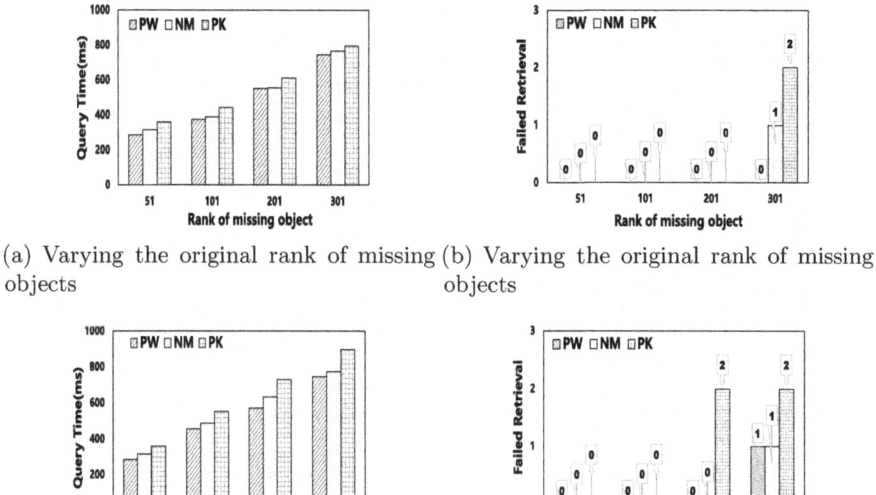

(a) Varying the original rank of missing objects (b) Varying the original rank of missing objects

(c) Varying # of missing objects (d) Varying # of missing objects

Fig. 7. The experiment results on the AF dataset

increases, which is basically in line with our expectations. Since our default original query is a top-10 query, all missing objects are randomly selected from the 6th to 51st objects of the original query. On the one hand, the more the missing objects, the more search space needs to be accessed. On the other hand, as the number of missing objects increases, the probability of moving objects in the missing object set increases. Retrieving moving objects takes more time than searching static objects, because more groups and nodes will be accessed and some probability calculations will be performed. In addition, in Fig. 7(d), the more the number of moving objects, the higher the probability that a moving object has a high speed (15–50 km/h). If one of the missing objects is moving at a high speed, such as 50 km/h, the object may move away from the query point in a short time. It would take a lot of time to capture it, exceeding the upper limit of query time.

Varying the Value α of the Original Query. We then study the performance of the method on the US dataset by varying the value of α in the original Top-k SKM query. As shown in Fig. 8(a), the query time of the three cases keeps basically unchanged. Since different α means spatial proximity or textual similarity has different filtering capability for unnecessary search space and unpromising objects, our method generates refined queries with different α to answer Top-k WSKM queries. However, the experimental results show that no matter what the value of α in the original query is, it will have no significant impact on the refined query performance. The *Failed Retrieval* data are shown in Fig. 8(b) as the increase of α. It can be seen from this figure, only when $\alpha = 0.3$, PK has 1 *Failed Retrieval.* For all other α values, the *Failed Retrieval* of PM, NM and *PK* is 0. Therefore, we consider this *Failed Retrieval* to be an accident.

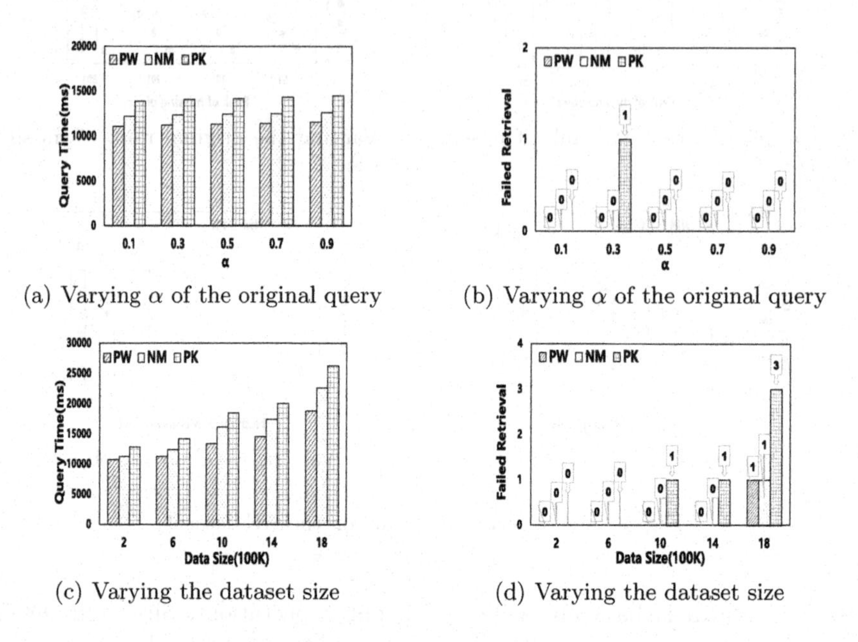

(a) Varying α of the original query (b) Varying α of the original query

(c) Varying the dataset size (d) Varying the dataset size

Fig. 8. The experiment results on the US dataset

Varying the Dataset Size. Finally, we select different numbers of spatial objects from the US dataset, ranging from 0.2M to 1.8M, to study the stability of the method. As shown in Fig. 8(c), PW and NM are superior to PK in query time. As the size of the dataset increases, there is no sharp increase in query time, which means that all methods scale well. As can be seen from Fig. 8(d), the number of *Failed Retrievals* increases obviously with the increase in data size. Since the number of *Failed Retrievals* divided by 1000 is less than or close to $P(x \leq \mu - 3\sigma \text{ or } x > \mu + 3\sigma)$, we can treat it as a small probability event. On the whole, our method is still effective.

7 Conclusion and Future Work

This paper explores how to answer the why-not questions on top-k spatial keyword queries over moving objects (Top-k WSKM queries). A novel index called Shadow is proposed to efficiently organize the textual, spatial and motion pattern information of objects. By analyzing the spatial location and velocity of original query result objects and missing objects, the maximum query search range is determined. By using this search range, the unnecessary branches and nodes in level 1 of Shadow are pruned, and static objects that satisfy the refined query requirements are obtained. In addition, the probability filtering technique is used to capture the user-desired moving objects with a certain probability. Finally, extensive experiments on three datasets demonstrate the feasibility of our method. There are several interesting directions to be studied in the future. First, the spatial environment can be extended to the road network. Second, the private protection issue should be concerned when answering the why-not questions.

Acknowledgments. This research is supported by Natural Science Foundation of China (Grant No.61572215), Natural Science Foundation of Hubei Province of China (Grant No.2017CFB135), the Ministry of education of Humanities and Social Science project of China (Grant No.20YJCZH111), and the Fundamental Research Funds for the Central Universities (CCNU: Grant No. CCNU20ZT013).

References

1. Cao, B., et al.: SIMkNN: a scalable method for in-memoryknn search over moving objects in road networks. IEEE Trans. Knowl. Data Eng. **30**(10), 1957–1970 (2018)
2. Chapman, A., Jagadish, H.V.: Why not. In: Proceedings Acm Sigmod, pp. 523–534 (2009)
3. Chen, L., Li, Y., Xu, J., Jensen, C.S.: Direction-aware why-not spatial keyword top-k queries. In: Proceedings of the IEEE 33rd International Conference on Data Engineering, pp. 107–110 (2017)
4. Chen, L., Lin, X., Hu, H., Jensen, C.S., Xu, J.: Answering why-not questions on spatial keyword top-k queries. In: Proceedings of the IEEE 31st International Conference on Data Engineering, pp. 279–290 (2015)
5. Chen, L., Xu, J., Lin, X., Jensen, C.S., Hu, H.: Answering why-not spatial keyword top-k queries via keyword adaption. In: Proceedings of the IEEE 32nd International Conference on Data Engineering, pp. 697–708 (2016)
6. Chen, L., Shang, S., Zheng, K., Kalnis, P.: Cluster-based subscription matching for geo-textual data streams. In: Proceedings of the IEEE 35th International Conference on Data Engineering, pp. 890–901 (2019)
7. Cui, N., Li, J., Yang, X., Wang, B., Reynolds, M., Xiang, Y.: When geo-text meets security: privacy-preserving boolean spatial keyword queries. In: Proceedings of the IEEE 35th International Conference on Data Engineering, pp. 1046–1057 (2019)
8. Dittrich, J., Blunschi, L., Vaz Salles, M.A.: Indexing moving objects using short-lived throwaway indexes. In: Mamoulis, N., Seidl, T., Pedersen, T.B., Torp, K., Assent, I. (eds.) SSTD 2009. LNCS, vol. 5644, pp. 189–207. Springer, Heidelberg (2009). https://doi.org/10.1007/978-3-642-02982-0_14
9. Tianyang, D., Lulu, Y., Qiang, C., Bin, C., Jing, F.: Direction-aware KNN queries for moving objects in a road network. World Wide Web **22**(4), 1765–1797 (2019). https://doi.org/10.1007/s11280-019-00657-1
10. He, Z., Lo, E.: Answering why-not questions on top-k queries. In: Proceedings of the IEEE 28th International Conference on Data Engineering, pp. 750–761 (2012)
11. Heendaliya, L., Wisely, M., Lin, D., Sarvestani, S.S., Hurson, A.R.: Indexing and querying techniques for moving objects in both euclidean space and road network. Adv. Comput. **102**, 111–170 (2016)

12. Li, Z., Lee, K.C.K., Zheng, B., Lee, W., Lee, D.L., Wang, X.: Ir-tree: an efficient index for geographic document search. IEEE Trans. Knowl. Data Eng. **23**(4), 585–599 (2011)
13. Liu, Q., Gao, Y.: Survey of database usability for query results. J. Comput. Res. Dev. **54**(6), 1198–1212 (2017)
14. Miao, X., Gao, Y., Guo, S., Chen, G.: On efficiently answering why-not range-based skyline queries in road networks (extended abstract). In: Proceedings of the IEEE 35th International Conference on Data Engineering, pp. 2131–2132 (2019)
15. Salgado, C., Cheema, M.A., Ali, M.E.: Continuous monitoring of range spatial keyword query over moving objects. World Wide Web **21**(3), 687–712 (2017). https://doi.org/10.1007/s11280-017-0488-3
16. Shen, B., et al.: V-tree: efficient knn search on moving objects with road-network constraints. In: Proceedings of the IEEE 33rd International Conference on Data Engineering, pp. 609–620 (2017)
17. Su, S., Teng, Y., Cheng, X., Xiao, K., Li, G., Chen, J.: Privacy-preserving top-k spatial keyword queries in untrusted cloud environments. IEEE Trans. Serv. Comput. **11**(5), 796–809 (2018)
18. Wang, M., Liu, J., Wei, B., Yao, S., Zeng, H., Shi, L.: Answering why-not questions on SPARQL queries. Knowl. Inform. Syst. **58**(1), 169–208 (2018). https://doi.org/10.1007/s10115-018-1155-4
19. Wu, D., Choi, B., Xu, J., Jensen, C.S.: Authentication of moving top-k spatial keyword queries. IEEE Trans. Knowl. Data Eng. **27**(4), 922–935 (2015)
20. Xu, J., Güting, R.H., Zheng, Y., Wolfson, O.: Moving objects with transportation modes: a survey. J. Comput. Sci. Technol. **34**(4), 709–726 (2019)
21. Yang, J., Zhang, Y., Zhou, X., Wang, J., Hu, H., Xing, C.: A hierarchical framework for top-k location-aware error-tolerant keyword search. In: Proceedings of the IEEE 35th International Conference on Data Engineering, pp. 986–997 (2019)
22. Yao, D., Cong, G., Zhang, C., Bi, J.: Computing trajectory similarity in linear time: a generic seed-guided neural metric learning approach. In: Proceedings of the IEEE 35th International Conference on Data Engineering, pp. 1358–1369 (2019)
23. Zhang, C., Zhang, Y., Zhang, W., Lin, X.: Inverted linear quadtree: efficient top k spatial keyword search. In: Proceedings of the IEEE 29th International Conference on Data Engineering, pp. 901–912 (2013)
24. Zhang, D., Tan, K.L., Tung, A.K.H.: Scalable top-k spatial keyword search. In: Proceedings of the ACM 16th International Conference EDBT, pp. 359–370 (2013)
25. Zhao, J., Gao, Y., Chen, G., Chen, R.: Why-not questions on top-k geo-social keyword queries in road networks. In: Proceedings of the IEEE 34th International Conference on Data Engineering, pp. 965–976 (2018)
26. Zheng, B., et al.: Answering why-not group spatial keyword queries (extended abstract). In: Proceedings of the IEEE 35th International Conference on Data Engineering, pp. 2155–2156 (2019)
27. Zong, C., Wang, B., Sun, J., Yang, X.: Minimizing explanations of why-not questions. In: Han, W.-S., Lee, M.L., Muliantara, A., Sanjaya, N.A., Thalheim, B., Zhou, S. (eds.) DASFAA 2014. LNCS, vol. 8505, pp. 230–242. Springer, Heidelberg (2014). https://doi.org/10.1007/978-3-662-43984-5_17

DBL: Efficient Reachability Queries on Dynamic Graphs

Qiuyi Lyu[1(✉)], Yuchen Li[2], Bingsheng He[3], and Bin Gong[1]

[1] Shandong University, Jinan, China
gb@sdu.edu.cn
[2] Singapore Management University, Singapore, Singapore
yuchenli@smu.edu.sg
[3] National University of Singapore, Singapore, Singapore
hebs@comp.nus.edu.sg

Abstract. Reachability query is a fundamental problem on graphs, which has been extensively studied in academia and industry. Since graphs are subject to frequent updates in many applications, it is essential to support efficient graph updates while offering good performance in reachability queries. Existing solutions compress the original graph with the Directed Acyclic Graph (DAG) and propose efficient query processing and index update techniques. However, they focus on optimizing the scenarios where the Strong Connected Components (SCCs) remain unchanged and have overlooked the prohibitively high cost of the DAG maintenance when SCCs are updated. In this paper, we propose DBL, an efficient DAG-free index to support the reachability query on dynamic graphs with insertion-only updates. DBL builds on two complementary indexes: Dynamic Landmark (DL) label and Bidirectional Leaf (BL) label. The former leverages landmark nodes to quickly determine reachable pairs whereas the latter prunes unreachable pairs by indexing the leaf nodes in the graph. We evaluate DBL against the state-of-the-art approaches on dynamic reachability index with extensive experiments on real-world datasets. The results have demonstrated that DBL achieves orders of magnitude speedup in terms of index update, while still producing competitive query efficiency.

1 Introduction

Given a graph G and a pair of vertices u and v, reachability query (denoted as $q(u, v)$) is a fundamental graph operation that answers whether there exists a path from u to v on G. This operation is a core component in supporting numerous applications in practice, such as those in social networks, biological complexes, knowledge graphs, and transportation networks. A plethora of index-based approaches have been developed over a decade [5,16,17,19–21,23–25] and demonstrated great success in handling reachability query on *static* graphs with millions of vertices and edges. However, in many cases, graphs are highly

The original version of this chapter was revised: affiliation of second author has been corrected and on page 775 the citation of Figure 6 was corrected. The correction to this chapter is available at https://doi.org/10.1007/978-3-030-73197-7_54

C. S. Jensen et al. (Eds.): DASFAA 2021, LNCS 12682, pp. 761–777, 2021.
https://doi.org/10.1007/978-3-030-73197-7_52

dynamic [22]: New friendships continuously form on social networks like Facebook and Twitter; knowledge graphs are constantly updated with new entities and relations; and transportation networks are subject to changes when road constructions and temporary traffic controls occur. In those applications, it is essential to support efficient graph updates while offering good performance in reachability queries.

There have been some efforts in developing reachability index to support graph updates [4,7,8,14–16]. However, there is a major assumption made in those works: the Strongly Connected Components (SCCs) in the underlying graph remain unchanged after the graph gets updated. The Directed Acyclic Graph (DAG) collapses the SCCs into vertices and the reachability query is then processed on a significantly smaller graph than the original. The state-of-the-art solutions [23,26] thus rely on the DAG to design an index for efficient query processing, yet their index maintenance mechanisms only support the update which does not trigger SCC merge/split in the DAG. However, such an assumption can be invalid in practice, as edge insertions could lead to updates of the SCCs in the DAG. In other words, the overhead of the DAG maintenance has been mostly overlooked in the previous studies.

One potential solution is to adopt existing DAG maintenance algorithms such as [25]. Unfortunately, this DAG maintenance is a prohibitively time-consuming process, as also demonstrated in the experiments. For instance, in our experiments, the time taken to update the DAG on one edge insertion in the LiveJournal dataset is two-fold more than the time taken to process *1 million queries* for the state-of-the-art methods. Therefore, we need a new index scheme with a low maintenance cost while efficiently answering reachability queries.

In this paper, we propose a DAG-free dynamic reachability index framework (DBL) that enables efficient index update and supports fast query processing at the same time on large scale graphs. We focus on insert-only dynamic graphs with new edges and vertices continuously added. This is because the number of deletions are often significantly smaller than the number of insertions, and deletions are handled with lazy updates in many graph applications [2,3]. Instead of maintaining the DAG, we index the reachability information around two sets of vertices: the "landmark" nodes with high centrality and the "leaf" nodes with low centrality (e.g., nodes with zero in-degree or out-degree). As the reachability information of the landmark nodes and the leaf nodes remain relatively stable against graph updates, it enables efficient index update opportunities compared with approaches using the DAG. Hence, DBL is built on the top of two simple and effective index components: (1) a Dynamic Landmark (DL) label, and (2) a Bidirectional Leaf (BL) label. Combining DL and BL in the DBL ensures efficient index maintenance while achieves competitive query processing performance.

Efficient Query Processing: DL is inspired by the landmark index approach [5]. The proposed DL label maintains a small set of the landmark nodes as the label for each vertex in the graph. Given a query $q(u, v)$, if both the DL labels of u and v contain a common landmark node, we can immediately determine that u reaches v. Otherwise, we need to invoke Breadth-First Search(BFS) to process $q(u, v)$. We devise BL label to quickly prune vertex pairs that are not reachable to limit the number of costly BFS. BL complements DL and it focuses on building labels around the leaf nodes in the graph. The leaf nodes form an exclusive set apart from the landmark node set. BL label of a vertex u is defined to be the leaf nodes which can either reach u or u can reach them. Hence, u does not reach v if there exists one leaf node in u's BL label which does not appear in the BL label of v. In summary, DL can quickly determine reachable pairs while BL, which complements DL, prunes disconnected pairs to remedy the ones that cannot be immediately determined by DL.

Efficient Index Maintenance: Both DL and BL labels are lightweight indexes where each vertex only stores a constant size label. When new edges are inserted, efficient pruned BFS is employed and only the vertices where their labels need update will be visited. In particular, once the label of a vertex is unaffected by the edge updates, we safely prune the vertex as well as its descendants from the BFS, which enables efficient index update.

To better utilize the computation power of modern architectures, we implement DL and BL with simple and compact bitwise operations. Our implementations are based on OpenMP and CUDA in order to exploit parallel architectures multi-core CPUs and GPUs (Graphics Processing Units), respectively.

Hereby, we summarize the contributions as the following:

- We introduce the DBL framework which combines two complementary DL and BL labels to enable efficient reachability query processing on large graphs.
- We propose novel index update algorithms for DL and BL. To the best of our knowledge, this is the first solution for dynamic reachability index without maintaining the DAG. In addition, the algorithms can be easily implemented with parallel interfaces.
- We conduct extensive experiments to validate the performance of DBL in comparison with the state-of-the-art dynamic methods [23,26]. DBL achieves competitive query performance and orders of magnitude speedup for index update. We also implement DBL on multi-cores and GPU-enabled system and demonstrate significant performance boost compared with our sequential implementation.

The remaining part of this paper is organized as follows. Section 2 presents the preliminaries and background. Section 3 presents the related work. Section 4 presents the index definition as well as query processing. Sections 5 demonstrate the update mechanism of DL and BL labels. Section 6 reports the experimental results. Finally, we conclude the paper in Sect. 7.

2 Preliminaries

A directed graph is defined as $G = (V, E)$, where V is the vertex set and E is the edge set with $n = |V|$ and $m = |E|$. We denote an edge from vertex u to vertex v as (u, v). A path from u to v in G is denoted as $Path(u, v) = (u, w_1, w_2, w_3, \dots, v)$ where $w_i \in V$ and the adjacent vertices on the path are connected by an edge in G. We say that v is reachable by u when there exists a $Path(u, v)$ in G.

Table 1. Common notations in this paper

Notation	Description
$G(V, E)$	The vertex set V and the edge set E of a directed graph G
G'	The reverse graph of G
n	The number of vertex in G
m	The number of edges in G
$Suc(u)$	The set of u's out-neighbors
$Pre(u)$	The set of u's in-neighbors
$Des(u)$	The set of u's descendants including u
$Anc(u)$	The set of u's ancestors including u
$Path(u, v)$	A path from vertex u to vertex v
$q(u, v)$	The reachability query from u to v
k	The size of DL label for one vertex
k'	The size of BL label for one vertex
$\mathtt{DL}_{in}(u)$	The label that keeps all the landmark nodes that could reach u
$\mathtt{DL}_{out}(u)$	The label that keeps all the landmark nodes that could be reached by u
$\mathtt{BL}_{in}(u)$	The label that keeps the hash value of the leaf nodes that could reach u
$\mathtt{BL}_{out}(u)$	The label that keeps the hash value of the leaf nodes that could be reached by u
$h(u)$	The hash function that hash node u to a value

In addition, we use $Suc(u)$ to denote the direct successors of u and the direct predecessors of u are denoted as $Pre(u)$. Similarly, we denote all the ancestors of u (including u) as $Anc(u)$ and all the descendants of u (including u) as $Des(u)$. We denote the reversed graph of G as $G' = (V, E')$ where all the edges of G are in the opposite direction of G'. In this paper, the forward direction refers to traversing on the edges in G. Symmetrically, the backward direction refers to traversing on the edges in G'. We denote $q(u, v)$ as a reachability query from u to v. In this paper, we study the dynamic scenario where edges can be inserted into the graph. Common notations are summarized in Table 1.

3 Related Work

There have been some studies on dynamic graph [4,7,8,14–16]. Yildirim et al. propose DAGGER [25] which maintains the graph as a DAG after insertions and deletions. The index is constructed on the DAG to facilitate reachability query processing. The main operation for the DAG maintenance is the merge and split of the *Strongly Connected Component* (SCC). Unfortunately, it has been shown that DAGGER exhibits unsatisfactory query processing performance on handling large graphs (even with just millions of vertices [26]).

The state-of-the-art approaches: TOL [26] and IP [23] follow the maintenance method for the DAG from DAGGER and propose novel dynamic index on the DAG to improve the query processing performance. We note that TOL and IP are only applicable to the scenarios where the SCC/s in the DAG remains unchanged against updates. In the case of SCC merges/collapses, DAGGER is still required to recover the SCC/s. For instance, TOL and IP can handle edge insertions (v_1, v_5) in

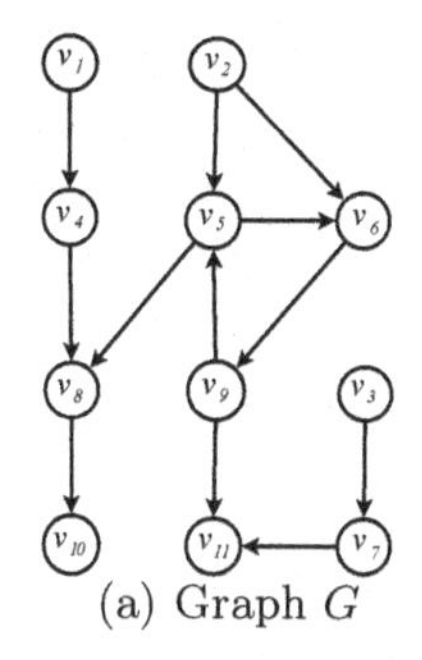

(a) Graph G

v	DL_{in}	DL_{out}
v_1	$\emptyset$	$\{v_8\}$
v_2	$\emptyset$	$\{v_5, v_8\}$
v_3	$\emptyset$	$\emptyset$
v_4	$\emptyset$	$\{v_8\}$
v_5	$\{v_5\}$	$\{v_5, v_8\}$
v_6	$\{v_5\}$	$\{v_5, v_8\}$
v_7	$\emptyset$	$\emptyset$
v_8	$\{v_5, v_8\}$	$\{v_8\}$
v_9	$\{v_5\}$	$\{v_5, v_8\}$
v_{10}	$\{v_5, v_8\}$	$\emptyset$
v_{11}	$\{v_5\}$	$\emptyset$

(b) DL label for G

v	BL_{in}	$h(\mathrm{BL}_{in})$	BL_{out}	$h(\mathrm{BL}_{out})$
v_1	$\{v_1\}$	$\{0\}$	$\{v_{10}\}$	$\{0\}$
v_2	$\{v_2\}$	$\{1\}$	$\{v_{10}, v_{11}\}$	$\{0,1\}$
v_3	$\{v_3\}$	$\{1\}$	$\{v_{11}\}$	$\{1\}$
v_4	$\{v_1\}$	$\{0\}$	$\{v_{10}\}$	$\{0\}$
v_5	$\{v_2\}$	$\{1\}$	$\{v_{10}, v_{11}\}$	$\{0,1\}$
v_6	$\{v_2\}$	$\{1\}$	$\{v_{10}, v_{11}\}$	$\{0,1\}$
v_7	$\{v_3\}$	$\{1\}$	$\{v_{11}\}$	$\{1\}$
v_8	$\{v_1, v_2\}$	$\{0,1\}$	$\{v_{10}\}$	$\{0\}$
v_9	$\{v_2\}$	$\{1\}$	$\{v_{10}, v_{11}\}$	$\{0,1\}$
v_{10}	$\{v_1, v_2\}$	$\{0,1\}$	$\{v_{10}\}$	$\{0\}$
v_{11}	$\{v_2, v_3\}$	$\{1\}$	$\{v_{11}\}$	$\{1\}$

(c) BL label for G

Fig. 1. A running example of graph G

Fig. 1(a), without invoking DAGGER. However, when inserting (v_9, v_2), two SCC/s $\{v_2\}$ and $\{v_5, v_6, v_9\}$ will be merged into one larger SCC $\{v_2, v_5, v_6, v_9\}$. For such cases, TOL and IP rely on DAGGER for maintaining the DAG first and then perform their respective methods for index maintenance and query processing. However, the overheads of the SCC maintenance are excluded in their experiments [23,26] and such overheads is in fact non-negligible [13,25].

In this paper, we propose the DBL framework which only maintains the labels for all vertices in the graph without constructing the DAG. That means, DBL can effectively avoid the costly DAG maintenance upon graph updates. DBL achieves competitive query processing performance with the state-of-the-art solutions (i.e., TOL and IP) while offering orders of magnitude speedup in terms of index updates.

4 DBL Framework

The DBL framework is consist of DL and BL label which have their independent query and update components. In this section, we introduce the DL and BL label. Then, we devise the query processing algorithm that builds upon DBL index.

4.1 Definitions and Construction

We propose the DBL framework that consists of two index components: DL and BL.

Definition 1. (DL **label**). *Given a landmark vertex set $L \subset V$ and $|L| = k$, we define two labels for each vertex $v \in V$: $\mathrm{DL}_{in}(v)$ and $\mathrm{DL}_{out}(v)$. $\mathrm{DL}_{in}(v)$ is a subset of nodes in L that could reach v and $\mathrm{DL}_{out}(v)$ is a subset of nodes in L that v could reach.*

It is noted that DL label is a subset of the 2-hop label [5]. In fact, 2-Hop label is a special case for DL label when the landmark set $L = V$. Nevertheless, we find that maintaining 2-Hop label in the dynamic graph scenario leads to index explosion. Thus, we propose to only choose a subset of vertices as the landmark set L to index DL label. In this way, DL label has up to $O(n|L|)$ space complexity and the index size can be easily controlled by tunning the selection of L. The following lemma shows an important property of DL label for reachability query processing.

Lemma 1. *Given two vertices u,v and their corresponding* DL *label,* $\text{DL}_{out}(u) \cap \text{DL}_{in}(v) \neq \emptyset$ *deduces u reaches v but not vice versa.*

Example 1. We show an running example in Fig. 1(a). Assuming the landmark set is chosen as $\{v_5, v_8\}$, the corresponding DL label is shown in Fig. 1(b). $q(v_1, v_{10})$ returns true since $\text{DL}_{out}(v_1) \cap \text{DL}_{in}(v_{10}) = \{v_8\}$. However, the labels cannot give negative answer to $q(v_3, v_{11})$ despite $\text{DL}_{out}(v_3) \cap \text{DL}_{in}(v_{11}) = \emptyset$. This is because the intermediate vertex v_7 on the path from v_3 to v_{11} is not included in the landmark set.

To achieve good query processing performance, we need to select a set of vertices as the landmarks such that they cover most of the reachable vertex pairs in the graph, i.e., $\text{DL}_{out}(u) \cap \text{DL}_{in}(v)$ contains at least one landmark node for any reachable vertex pair u and v. The optimal landmark selection has been proved to be NP-hard [12]. In this paper, we adopt a heuristic method for selecting DL label nodes following existing works [1,12]. In particular, we rank vertices with $M(u) = |Pre(u)| \cdot |Suc(u)|$ to approximate their centrality and select top-k vertices. Other landmark selection methods are also evaluated in our extended version [11]. We also discuss how to choose the number of landmarks (k) in the experimental evaluation.

Definition 2. (BL label). BL *introduces two labels for each vertex* $v \in V$: $\text{BL}_{in}(v)$ *and* $\text{BL}_{out}(v)$. $\text{BL}_{in}(v)$ *contains all the zero in-degrees vertices that can reach v, and* $\text{BL}_{out}(v)$ *contains all the zero out-degrees vertices that could be reached by v. For convenience, we refer to vertices with either zero in-degree or out-degree as the leaf nodes.*

Lemma 2. *Given two vertices u,v and their corresponding* BL *label, u does not reach v in G if* $\text{BL}_{out}(v) \nsubseteq \text{BL}_{out}(u)$ *or* $\text{BL}_{in}(u) \nsubseteq \text{BL}_{in}(v)$.

BL label can give negative answer to $q(u, v)$. This is because if u could reach v, then u could reach all the leaf nodes that v could reach, and all the leaf nodes that reach u should also reach v. DL label is efficient for giving positive answer to a reachability query whereas BL label plays a complementary role by pruning unreachable pairs. In this paper, we take vertices with zero in-degree/out-degree as the leaf nodes. Other leaf nodes selection methods are also evaluated in our extended version [11].

Example 2. Figure 1(c) shows BL label for the running example. BL label gives negative answer to $q(v_4, v_6)$ since $\text{BL}_{in}(v_4)$ is not contained by $\text{BL}_{in}(v_6)$. Intuitively, vertex v_1 reaches vertex v_4 but cannot reach v_6 which indicates v_4 should not reach v_6. BL label cannot give positive answer. Take $q(v_5, v_2)$ for an example, the labels satisfy the containment condition but positive answer cannot be given.

Algorithm 1. DL label Batch Construction

Input: Graph $G(V, E)$, Landmark Set D
Output: DL label for G

```
1: for i = 0; i < k; i++ do
2:     //Forward BFS
3:     S ← D[i]
4:     enqueue S to an empty queue Q
5:     while Q not empty do
6:         p ← pop Q
7:         for x ∈ Suc(p) do
8:             DL_in(x) ← DL_in(x) ∪ {S};
9:             enqueue x to Q
10:    //Symmetrical Backward BFS is performed.
```

The number of BL label nodes could be huge. To develop efficient index operations, we build a hash set of size k' for BL as follows. Both $\mathtt{BL}_{in}$ and $\mathtt{BL}_{out}$ are a subset of $\{1, 2, \ldots, k'\}$ where k' is a user-defined label size, and they are stored in bit vectors. A hash function is used to map the leaf nodes to a corresponding bit. For our example, the leaves are $\{v_1, v_2, v_3, v_{10}, v_{11}\}$. When $k' = 2$, all leaves are hashed to two unique values. Assume $h(v_1) = h(v_{10}) = 0$, $h(v_2) = h(v_3) = h(v_{11}) = 1$. We show the hashed BL label set in Fig. 1(c) which are denoted as $h(\mathtt{BL}_{in})$ and $h(\mathtt{BL}_{out})$. In the rest of the paper, we directly use $\mathtt{BL}_{in}$ and $\mathtt{BL}_{out}$ to denote the hash sets of the corresponding labels. It is noted that one can still use Lemma 2 to prune unreachable pairs with the hashed BL label.

We briefly discuss the batch index construction of DBL as the focus of this work is on the dynamic scenario. The construction of DL label is presented in Algorithm 1, which follows existing works on 2-hop label [5]. For each landmark node $D[i]$, we start a BFS from S (Line 4) and include S in $\mathtt{DL}_{in}$ label of every vertices that S can reach (Lines 5–9). For constructing $\mathtt{DL}_{out}$, we execute a BFS on the reversed graph G' symmetrically (Line 10). To construct BL label, we simply replace the landmark set D as the leaf set D' and replace S with all leaf nodes that are *hashed* to bucket i (Line 3) in Algorithm 1. The complexity of building DBL is that $O((k + k')(m + n))$.

Note that although we use [5] for offline index construction, the contribution of our work is that we construct DL and BL as complementary indices for efficient query processing. Furthermore, we are the first work to support efficient dynamic reachability index maintenance without assuming SCC/s remain unchanged.

Space Complexity. The space complexities of DL and BL labels are $O(kn)$ and $O(k'n)$, respectively.

4.2 Query Processing

With the two indexes, Algorithm 2 illustrates the query processing framework of DBL. Given a reachability query $q(u, v)$, we return the answer immediately if the labels are sufficient to determine the reachability (Lines 6–9). By the definitions

Algorithm 2. Query Processing Framework for DBL

```
Input: Graph G(V, E), DL label, BL label, q(u, v)
Output: Answer of the query.
1: function DL_Intersec(x,y)
2:     return (DL_out(x) ∩ DL_in(y));
3: function BL_Contain(x,y)
4:     return (BL_in(x) ⊆ BL_in(y) and BL_out(y) ⊆ BL_out(x));
5: procedure QUERY(u,v)
6:     if DL_Intersec(u,v) then
7:         return true;
8:     if not BL_Contain(u,v) then
9:         return false;
10:    if DL_Intersec(v,u) then
11:        return false;
12:    if DL_Intersec(u,u) or DL_Intersec(v,v) then
13:        return false;
14:    Enqueue u for BFS;
15:    while queue not empty do
16:        w ← pop queue;
17:        for vertex x ∈ Suc(w) do
18:            if x = v then
19:                return true;
20:            if DL_Intersec(u,x) then
21:                continue;
22:            if not BL_Contain(x,v) then
23:                continue;
24:            Enqueue x;
25:    return false;
```

of DL and BL labels, u reaches v if their DL label overlaps (Line 6) where u does not reach v if their BL label does not overlap (Line 9). Furthermore, there are two early termination rules implemented in Lines 10 and 12, respectively. Line 10 makes use of the properties that all vertices in a SCC contain at least one common landmark node. Line 12 takes advantage of the scenario when either u or v share the same SCC with a landmark node l then u reaches v if and only if l appeared in the DL label of u and v. We prove their correctness in Theorem 1 and Theorem 2 respectively. Otherwise, we turn to BFS search with efficient pruning. The pruning within BFS is performed as follows. Upon visiting a vertex q, the procedure will determine whether the vertex q should be enqueued in Lines 20 and 22. BL and DL labels will judge whether the destination vertex v will be in the $Des(w)$. If not, q will be pruned from BFS to quickly answer the query before traversing the graph with BFS.

Theorem 1. *In Algorithm 2, when* DL_*Intersec(x,y) returns false and* DL_*Intersec(y,x) returns true, then x cannot reach y.*

Algorithm 3. DL_{in} label update for edge insertion

Input: Graph $G(V, E)$, DL label, Inserted edge (u, v)
Output: Updated DL label
1: **if** $\text{DL}_{out}(u) \cap \text{DL}_{in}(v) == \emptyset$ **then**
2: Initialize an empty queue and enqueue v
3: **while** queue is not empty **do**
4: $p \leftarrow$ pop queue
5: **for** vertex $x \in Suc(p)$ **do**
6: **if** $\text{DL}_{in}(u) \nsubseteq \text{DL}_{in}(x)$ **then**
7: $\text{DL}_{in}(x) \leftarrow \text{DL}_{in}(x) \cup \text{DL}_{in}(u)$
8: enqueue x

Proof. DL_$Intersec(y, x)$ returns true indicates that vertex y reaches x. If vertex x reaches vertex y, then y and x must be in the same SCC (according to the definition of the SCC). As all the vertices in the SCC are reachable to each other, the landmark nodes in $\text{DL}_{out}(y) \cap \text{DL}_{in}(x)$ should also be included in DL_{out} and DL_{in} label for all vertices in the same SCC. This means DL_$Intersec(x, y)$ should return true. Therefore x cannot reach y otherwise it contradicts with the fact that DL_$Intersec(x, y)$ returns false.

Theorem 2. *In Algorithm 2, if* DL_$Intersec(x, y)$ *returns false and* DL_$Intersec(x,x)$ *or* DL_$Intersec(y,y)$ *returns true then vertex x cannot reach y.*

Proof. If DL_$Intersec(x,x)$ returns true, it means that vertex x is a landmark or x is in the same SCC with a landmark. If x is in the same SCC with landmark l, vertex x and vertex l should have the same reachability information. As landmark l will push its label element l to DL_{out} label for all the vertices in $Anc(l)$ and to DL_{in} label for all the vertices in $Des(l)$. The reachability information for landmark l will be fully covered. It means that x's reachability information is also fully covered. Thus DL label is enough to answer the query without BFS. Hence y is not reachable by x if DL_$Intersec(x, y)$ returns false. The proving process is similar for the case when DL_$Intersec(y, y)$ returns true.

Query Complexity. Given a query $q(u, v)$, the time complexity is $O(k + k')$ when the query can be directly answered by DL and BL labels. Otherwise, we turn to the pruned BFS search, which has a worst case time complexity of $O((k+k')(m + n))$. Let ρ denote the ratio of vertex pairs whose reachability could be directly answered by the label. The amortized time complexity is $O(\rho(k + k') + (1 - \rho)(k + k')(m + n))$. Empirically, ρ is over 95% according to our experiments (Table 3 in Sect. 6), which implies efficient query processing.

5 DL and BL Update for Edge Insertions

When inserting a new edge (u, v), all vertices in $Anc(u)$ can reach all vertices in $Des(v)$. On a high level, all landmark nodes that could reach u should also reach vertices in $Des(v)$. In other words, all the landmark nodes that could be reached

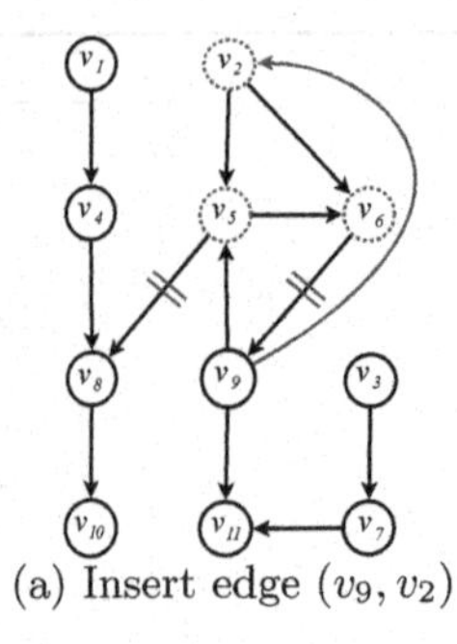

(a) Insert edge (v_9, v_2)

v	$\mathtt{DL}_{in}$
v_1	$\emptyset \rightarrow \emptyset$
v_2	$\emptyset \rightarrow \{v_5\}$
v_3	$\emptyset \rightarrow \emptyset$
v_4	$\emptyset \rightarrow \emptyset$
v_5	$\{v_5\} \rightarrow \{v_5\}$
v_6	$\{v_5\} \rightarrow \{v_5\}$
v_7	$\emptyset \rightarrow \emptyset$
v_8	$\{v_5, v_8\} \rightarrow \{v_5, v_8\}$
v_9	$\{v_5\} \rightarrow \{v_5\}$
v_{10}	$\{v_5, v_8\} \rightarrow \{v_5, v_8\}$
v_{11}	$\{v_5\} \rightarrow \{v_5\}$

(b) $\mathtt{DL}_{in}$ label update

v	$h(\mathtt{BL}_{in})$
v_1	$\{0\} \rightarrow \{0\}$
v_2	$\{1\} \rightarrow \{1\}$
v_3	$\{1\} \rightarrow \{1\}$
v_4	$\{0\} \rightarrow \{0\}$
v_5	$\{1\} \rightarrow \{1\}$
v_6	$\{1\} \rightarrow \{1\}$
v_7	$\{1\} \rightarrow \{1\}$
v_8	$\{0, 1\} \rightarrow \{0, 1\}$
v_9	$\{1\} \rightarrow \{1\}$
v_{10}	$\{0, 1\} \rightarrow \{0, 1\}$
v_{11}	$\{1\} \rightarrow \{1\}$

(c) $\mathtt{BL}_{in}$ label update

Fig. 2. Label update for inserting edge (v_9, v_2)

Table 2. Dataset statistics

Dataset	$\|V\|$	$\|E\|$	d_{avg}	Diameter	Connectivity (%)	DAG-$\|V\|$	DAG-$\|E\|$	DAGCONSTRUCT (ms)
LJ	4,847,571	68,993,773	14.23	16	78.9	971,232	1,024,140	2368
Web	875,713	5,105,039	5.83	21	44.0	371,764	517,805	191
Email	265,214	420,045	1.58	14	13.8	231,000	223,004	17
Wiki	2,394,385	5,021,410	2.09	9	26.9	2,281,879	2,311,570	360
BerkStan	685,231	7,600,595	11.09	514	48.8	109,406	583,771	1134
Pokec	1,632,803	30,622,564	18.75	11	80.0	325,892	379,628	86
Twitter	2,881,151	6,439,178	2.23	24	1.9	2,357,437	3,472,200	481
Reddit	2,628,904	57,493,332	21.86	15	69.2	800,001	857,716	1844

by v should also be reached by vertices in $Anc(u)$. Thus, we update the label by 1) adding $\mathtt{DL}_{in}(u)$ into $\mathtt{DL}_{in}(x)$ for all $x \in Des(v)$; and 2) adding $\mathtt{DL}_{out}(v)$ into $\mathtt{DL}_{out}(x)$ for all $x \in Anc(u)$.

Algorithm 3 depicts the edge insertion scenario for $\mathtt{DL}_{in}$. We omit the update for $\mathtt{DL}_{out}$, which is symmetrical to $\mathtt{DL}_{in}$. If DL label can determine that vertex v is reachable by vertex u in the original graph before the edge insertion, the insertion will not trigger any label update (Line 1). Lines 2–8 describe a BFS process with pruning. For a visited vertex x, we prune x without traversing $Des(x)$ iff $\mathtt{DL}_{in}(u) \subseteq \mathtt{DL}_{in}(x)$, because all the vertices in $Des(x)$ are deemed to be unaffected as their $\mathtt{DL}_{in}$ labels are supersets of $\mathtt{DL}_{in}(x)$.

Example 3. Figure 2(a) shows an example of edge insertion. Figure 2(b) shows the corresponding $\mathtt{DL}_{in}$ label update process. $\mathtt{DL}_{in}$ label is presented with brackets. Give an edge (v_9, v_2) inserted, $\mathtt{DL}_{in}(v_9)$ is copied to $\mathtt{DL}_{in}(v_2)$. Then an inspection will be processed on $\mathtt{DL}_{in}(v_5)$ and $\mathtt{DL}_{in}(v_6)$. Since $\mathtt{DL}_{in}(v_9)$ is a subset of $\mathtt{DL}_{in}(v_5)$ and $\mathtt{DL}_{in}(v_6)$, vertex v_5 and vertex v_6 are pruned from the BFS. The update progress is then terminated.

DL label only gives positive answer to a reachability query. In poorly connected graphs, DL will degrade to expensive BFS search. Thus, we employ the Bidirectional Leaf (BL) label to complement DL and quickly identify vertex pairs which are not reachable. We omit the update algorithm of BL, as they are very similar to those of DL, except the updates are applied to $\mathtt{BL}_{in}$ and $\mathtt{BL}_{out}$ labels.

Figure 2(c) shows the update of $\mathtt{BL}_{in}$ label. Similar to the DL label, the update process will be early terminated as the $\mathtt{BL}_{in}(v_2)$ is totally unaffected after edge insertion. Thus, no $\mathtt{BL}_{in}$ label will be updated in this case.

Update Complexity of DBL. In the worst case, all the vertices that reach or are reachable to the updating edges will be visited. Thus, the time complexity of DL and BL is $O((k+k')(m+n))$ where $(m+n)$ is the cost on the BFS. Empirically, as the BFS procedure will prune a large number of vertices, the actual update process is much more efficient than a plain BFS.

6 Experimental Evaluation

In this section, we conduct experiments by comparing the proposed DBL framework with the state-of-the-art approaches on reachability query for dynamic graphs.

6.1 Experimental Setup

Environment: Our experiments are conducted on a server with an Intel Xeon CPU E5-2640 v4 2.4 GHz, 256 GB RAM and a Tesla P100 PCIe version GPU.

Datasets: We conduct experiments on 8 real-world datasets (see Table 2). We have collected the following datasets from SNAP [9]. LJ and Pokec are two social networks, which are power-law graphs in nature. BerkStan and Web are web graphs in which nodes represent web pages and directed edges represent hyperlinks between them. Wiki and Email are communication networks. Reddit and Twitter are two social network datasets obtained from [22].

6.2 Effectiveness of DL+BL

Table 3 shows the percentages of queries answered by DL label, BL label (*when the other label is disabled*) and DBL label. All the queries are randomly generated. The results show that DL is effective for dense and highly connected graphs (LJ, Pokec and Reddit) whereas BL is effective for sparse and poorly connected graphs (Email, Wiki and Twitter). However, we still incur the expensive BFS if the label is disabled. By combining the merits of both indexes, our proposal leads to a significantly better performance. DBL could answer much more queries than DL and BL label. The results have validated our claim that DL and BL are complementary to each other. We note that the query processing for DBL is able to handle one million queries with sub-second latency for most datasets, which shows outstanding performance.

Impact of Label Size: On the query processing of DBL. There are two labels in DBL: both DL and BL store labels in bit vectors. The size of DL label depends on the number of selected landmark nodes whereas the size of BL label is determined by how many hash values are chosen to index the leaf nodes. We evaluate all the datasets to show the performance trend of varying DL and BL label sizes per vertex (by processing 1 million queries) in Table 4.

When varying DL label size k, the performance of most datasets remain stable before a certain size (e.g., 64) and deteriorates thereafter. This means that extra landmark nodes will cover little extra reachability information. Thus, selecting more landmark nodes does not necessarily lead to better overall performance since the cost of processing the additional bits incur additional cache misses. BerkStan gets benefit from increasing the DL label size to 128 since 64 landmarks are not enough to cover enough reachable pairs.

Table 3. Percentages of queries answered by DL label, BL label (when the other label is disabled) and DBL label respectively. We also include the time for DBL to process 1 million queries

Dataset	DL Label	BL Label	DBL Label	DBL time
LJ	97.5%	20.8%	99.8%	108 ms
Web	79.5%	54.3%	98.3%	139 ms
Email	31.9%	85.4%	99.2%	36 ms
Wiki	10.6%	94.3%	99.6%	157 ms
Pokec	97.6%	19.9%	99.9%	35 ms
BerkStan	87.5%	43.3%	95.0%	1590 ms
Twitter	6.6%	94.8%	96.7%	709 ms
Reddit	93.7%	30.6%	99.9%	61 ms

Compared with DL label, some of the datasets get a sweet spot when varying the size of BL label. This is because there are two conflicting factors which affect the overall performance. With increasing BL label size and more hash values incorporated, we can quickly prune more unreachable vertex pairs by examining BL label without traversing the graph with BFS. Besides, larger BL size also provides better pruning power of the BFS even if it fails to directly answer the query (Algorithm 2). Nevertheless, the cost of label processing increases with increased BL label size. According to our parameter study, we set Wiki's DL and BL label size as 64 and 256, BerkStan's DL and BL label size as 128 and 64. For the remaining datasets, both DL and BL label sizes are set as 64.

6.3 General Graph Updates

In this section, we evaluate DBL's performance on general graph update. As DAGGER is the only method that could handle general update, we compare DBL against DAGGER in Fig. 3. Ten thousand edge insertion and 1 million queries are randomly generated and performed, respectively. Different from DAGGER, DBL don't need to maintain the DAG, thus, in all the datasets, DBL could achieve great performance lift compared with DAGGER. For both edge insertion and query, DBL is orders of magnitude faster than DAGGER. The minimum performance gap lies in BerkStan. This is because BerkStan has a large diameter. As DBL rely on

Table 4. Query performance (ms) with varying DL and BL label sizes

(a) Varying BL label sizes

Dataset	16	32	64	128	256
LJ	136.1	131.9	108.1	107.4	110.3
Web	177.2	128.5	152.9	156.6	174.3
Email	77.4	53.9	38.3	41.1	44.4
Wiki	911.6	481.4	273.7	181.3	157.4
Pokec	54.8	43.7	38.6	40.6	53.6
BerkStan	4876.1	4958.9	4862.9	5099.1	5544.3
Twitter	1085.3	845.7	708.2	652.7	673.2
Reddit	117.1	80.4	67.3	63.5	67.9

(b) Varying DL label sizes

Dataset	16	32	64	128	256
LJ	108.2	110.3	106.9	120.2	125.5
Web	154.0	152.5	151.1	158.8	167.8
Email	37.9	39.5	35.8	39.8	43.7
Wiki	274.5	282.6	272.4	274.8	281.1
Pokec	38.1	40.6	36.3	49.7	55.6
BerkStan	6369.8	5853.1	4756.3	1628.3	1735.2
Twitter	716.1	724.4	695.3	707.1	716.9
Reddit	64.6	65.9	62.9	75.4	81.4

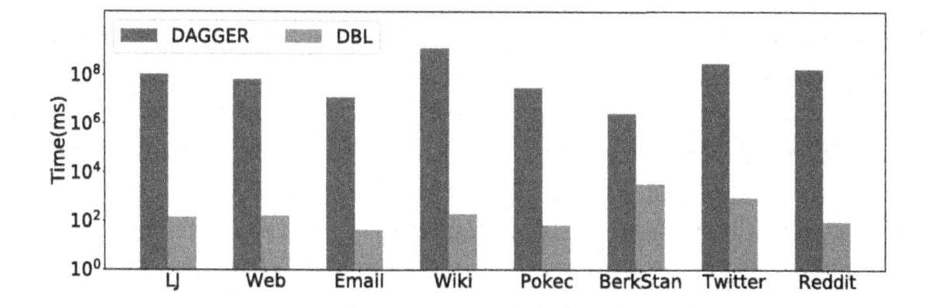

Fig. 3. The execution time for insert 10000 edges as well as 1 million queries

BFS traversal to update the index. The traversal overheads is crucial for it's performance. BerkStan's diameter is large, it means, during index update, DBL need to traversal extra hops to update the index which will greatly degrade the performance.

6.4 Synthetic Graph Updates

In this section, we compare our method with IP and TOL. Different from DBL, which could handle real world update, IP and TOL could only handle synthetic edge update that will not trigger DAG maintaining. Thus, for IP and TOL, we follow their experimental setups depict in their paper [23,26]. Specifically, we randomly select 10,000 edges from the DAG and delete them. Then, we will insert the same edges back. In this way, we could get the edge insertion performance without trigger DAG maintenance. For DBL, we stick to general graph updates. The edge insertion will be randomly generated and performed. One million queries will be executed after that. It needs to be noted that, although both IP and TOL claim they can handle dynamic graph, due to their special pre-condition, their methods are in fact of limited use in real world scenario.

The results are shown in Fig. 4. DBL outperforms other baselines in most cases except on three data sets (Wiki, BerkStan and Twitter) where IP could achieve a better performance. Nevertheless, DBL outperforms IP and TOL by 4.4x and 21.2x, respectively with respect to geometric mean performance. We analyze the reason that DBL can be slower than IP on Wiki, BerkStan and Twitter.

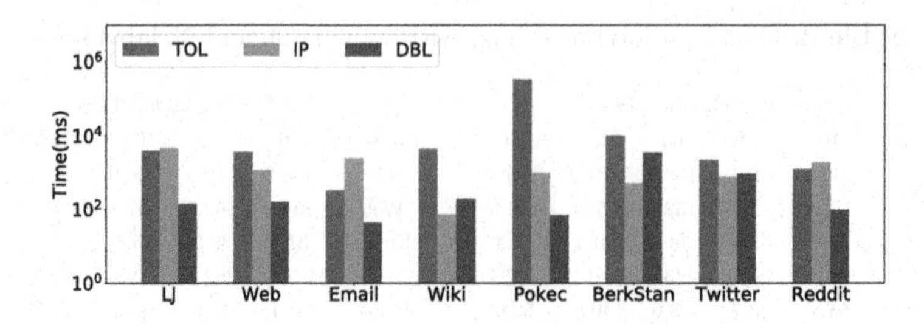

Fig. 4. The execution time for insert 10000 edges as well as 1 million queries, for TOL and IP, the updates are synthetic that will not trigger SCC update

As we aforementioned, DBL relies on the pruned BFS to update the index, the BFS traversal speed will determine the worst-case update performance. Berkstan has the largest diameter as 514 and Twitter has the second largest diameter as 24, which dramatically degrade the update procedure in DBL. For Wiki, DBL could still achieve a better update performance than IP. However, IP is much more efficiency in query processing which lead to better overall performance.

Although this experimental scenario has been used in previous studies, the comparison is unfair for DBL. As both IP and TOL rely on the DAG to process queries and updates, **their synthetic update exclude the DAG maintaining procedure/overheads from the experiments.** However, DAG maintenance is essential for their method to handle real world edge updates, as we have shown in Fig. 3, the overheads is nonnegligible.

6.5 Parallel Performance

We implement DBL with OpenMP and CUDA (DBL-P and DBL-G respectively) to demonstrate the deployment on multi-core CPUs and GPUs achieves encouraging speedup for query processing. We follow existing GPU-based graph processing pipeline by batching the queries and updates [6,10,18]. Note that the transfer time can be overlapped with GPU processing to minimize data communication costs. Both CPU and GPU implementations are based on the vertex centric framework.

To validate the scalability of the parallel approach, we vary the number of threads used in DBL-P and show its performance trend in Fig. 5. DBL-P achieves almost linear scalability against increasing number of threads. The linear trend of scalability tends to disappear when the number of threads is beyond 14. We attribute this observation as the memory bandwidth bound nature of the processing tasks. DBL invokes the BFS traversal once the labels are unable to answer the query and the efficiency of the BFS is largely bounded by CPU memory bandwidth. This memory bandwidth bound issue of CPUs can be resolved by using GPUs which provide memory bandwidth boost.

The compared query processing performance is shown in Fig. 6. Bidirectional BFS (B-BFS) query is listed as a baseline. We also compare our parallel solutions with a home-grown OpenMP implementation of IP (denoted as IP-P). Twenty

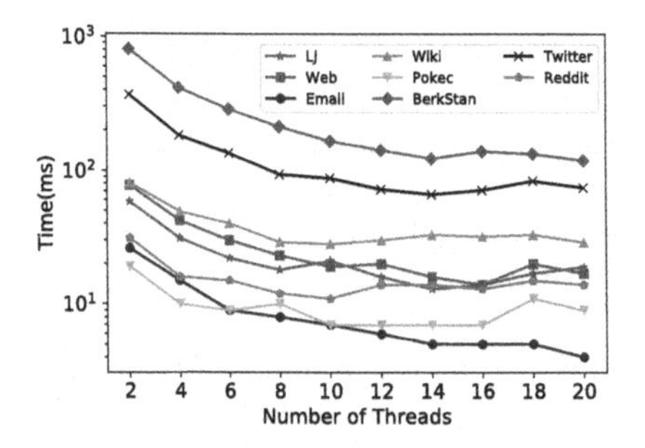

Fig. 5. Scalability of DBL on CPU

Dataset	TOL	IP	IP-P	DBL	DBL-P	DBL-G	B-BFS
LJ	46.6	50.7	24.9	108.1	16.4	**6.1**	555561
Web	40.6	39.7	22.6	139.2	**12.4**	14.2	236892
Email	26.8	21.6	9.4	36.4	4.1	**2.8**	10168
Wiki	74.9	12.7	**4.1**	157.2	28.4	14.8	61113
Pokec	27.2	37.6	23.2	34.8	9.0	**3.1**	253936
BerkStan	37.2	31.6	**16.4**	1590.0	131.0	835.1	598127
Twitter	64.6	30.2	**7.1**	709.1	79.4	202.1	78496
Reddit	56.7	44.7	19.56	61.2	14.6	**3.1**	273935

Fig. 6. The query performance (ms) on CPU and GPU architectures. B-BFS means the bidirectional BFS

threads are used in the OpenMp implementation. We note that IP has to invoke a pruned DFS if its labels fail to determine the query result. DFS is a sequential process in nature and cannot be efficiently parallelized. For our parallel implementation IP-P, we assign a thread to handle one query. We have the following observations.

First, DBL is built on the pruned BFS which can be efficiently parallelized with the vertex-centric paradigm. We have observed significant performance improvement by parallelized executions. DBL-P (CPUs) gets 4x to 10x speedup across all datasets. DBL-G (GPUs) shows an even better performance. In contrast, as DFS incurs frequent random accesses in IP-P, the performance is bounded by memory bandwidth. Thus, parallelization does not bring much performance gain to IP-P compared with its sequential counterpart.

Second, DBL provides competitive efficiency against IP-P but DBL can be slower than TOL and IP when comparing the single thread performance. However, this is achieved by assuming the DAG structure but the DAG-based approaches incur prohibitively high cost of index update. In contrast, DBL achieves sub-second query processing performance for handling 1 million queries while still support efficient updates without using the DAG (Fig. 6).

Third, there are cases where DBL-P outperforms DBL-G, i.e., Web, Berkstan and Twitter. This is because these datasets have a higher diameter than the rest of the datasets and the pruned BFS needs to traverse extra hops to determine the reachability. Thus, we incur more random accesses, which do not suit the GPU architecture.

7 Conclusion

In this work, we propose DBL, an indexing framework to support dynamic reachability query processing on incremental graphs. To our best knowledge, DBL is the first solution which avoids maintaining DAG structure to construct and build reachability index. DBL leverages two complementary index components to answer the reachability query. The experimental evaluation has demonstrated that the sequential version of DBL outperforms the state-of-the-art solutions with orders of magnitude speedups in terms of index update while exhibits competitive query processing performance. The parallel implementation of DBL on multi-cores and GPUs further boost the performance over our sequential implementation. As future work, we are interested in extending DBL to support deletions, which will be lazily supported in many applications.

Acknowledgement. Yuchen Li's work was supported by the Ministry of Education, Singapore, under its Academic Research Fund Tier 2 (Award No.: MOE2019-T2-2-065).

References

1. Akiba, T., Iwata, Y., Yoshida, Y.: Fast exact shortest-path distance queries on large networks by pruned landmark labeling. In: Proceedings of the 2013 ACM SIGMOD International Conference on Management of Data, pp. 349–360. ACM (2013)
2. Akiba, T., Iwata, Y., Yoshida, Y.: Dynamic and historical shortest-path distance queries on large evolving networks by pruned landmark labeling. In: Proceedings of the 23rd International Conference on World Wide Web, pp. 237–248 (2014)
3. Boccaletti, S., Latora, V., Moreno, Y., Chavez, M., Hwang, D.U.: Complex networks: structure and dynamics. Phys. Rep. **424**(4–5), 175–308 (2006)
4. Bramandia, R., Choi, B., Ng, W.K.: Incremental maintenance of 2-hop labeling of large graphs. TKDE **22**(5), 682–698 (2010)
5. Cohen, E., Halperin, E., Kaplan, H., Zwick, U.: Reachability and distance queries via 2-hop labels. SICOMP **32**(5), 1338–1355 (2003)
6. Guo, W., Li, Y., Sha, M., Tan, K.L.: Parallel personalized pagerank on dynamic graphs. PVLDB **11**(1), 93–106 (2017)
7. Henzinger, M.R., King, V.: Fully dynamic biconnectivity and transitive closure. In: FOCS, pp. 664–672. IEEE (1995)
8. Jin, R., Ruan, N., Xiang, Y., Wang, H.: Path-tree: an efficient reachability indexing scheme for large directed graphs. TODS **36**(1), 7 (2011)
9. Leskovec, J., Krevl, A.: SNAP Datasets: Stanford large network dataset collection (June 2014). http://snap.stanford.edu/data
10. Li, Y., Zhu, Q., Lyu, Z., Huang, Z., Sun, J.: Dycuckoo: dynamic hash tables on gpus. In: ICDE (2021)
11. Lyu, Q., Li, Y., He, B., Gong, B.: DBL: Efficient reachability queries on dynamic graphs (complete version) (2021)
12. Potamias, M., Bonchi, F., Castillo, C., Gionis, A.: Fast shortest path distance estimation in large networks. In: CIKM, pp. 867–876. ACM (2009)
13. Qiu, X., et al.: Real-time constrained cycle detection in large dynamic graphs. In: Proceedings of the VLDB Endowment, vol. 11, pp. 1876–1888 (08 2018). https://doi.org/10.14778/3229863.3229874
14. Roditty, L.: Decremental maintenance of strongly connected components. In: SODA, pp. 1143–1150. SIAM (2013)
15. Roditty, L., Zwick, U.: A fully dynamic reachability algorithm for directed graphs with an almost linear update time. SICOMP **45**(3), 712–733 (2016)
16. Schenkel, R., Theobald, A., Weikum, G.: Efficient creation and incremental maintenance of the HOPI index for complex xml document collections. In: ICDE, pp. 360–371. IEEE (2005)

17. Seufert, S., Anand, A., Bedathur, S., Weikum, G.: Ferrari: Flexible and efficient reachability range assignment for graph indexing. In: ICDE, pp. 1009–1020. IEEE (2013)
18. Sha, M., Li, Y., He, B., Tan, K.L.: Accelerating dynamic graph analytics on GPUs. PVLDB **11**(1), (2017)
19. Su, J., Zhu, Q., Wei, H., Yu, J.X.: Reachability querying: can it be even faster? TKDE **1**, 1 (2017)
20. Valstar, L.D., Fletcher, G.H., Yoshida, Y.: Landmark indexing for evaluation of label-constrained reachability queries. In: Proceedings of the 2017 ACM International Conference on Management of Data, pp. 345–358 (2017)
21. Wang, H., He, H., Yang, J., Yu, P.S., Yu, J.X.: Dual labeling: answering graph reachability queries in constant time. In: ICDE, p. 75. IEEE (2006)
22. Wang, Y., Fan, Q., Li, Y., Tan, K.L.: Real-time influence maximization on dynamic social streams. Proc. VLDB Endowment **10**(7), 805–816 (2017)
23. Wei, H., Yu, J.X., Lu, C., Jin, R.: Reachability querying: an independent permutation labeling approach. VLDB J. **27**(1), 1–26 (2017). https://doi.org/10.1007/s00778-017-0468-3
24. Yildirim, H., Chaoji, V., Zaki, M.J.: Grail: scalable reachability index for large graphs. PVLDB **3**(1–2), 276–284 (2010)
25. Yildirim, H., Chaoji, V., Zaki, M.J.: Dagger: a scalable index for reachability queries in large dynamic graphs. arXiv preprint arXiv:1301.0977 (2013)
26. Zhu, A.D., Lin, W., Wang, S., Xiao, X.: Reachability queries on large dynamic graphs: a total order approach. In: SIGMOD, pp. 1323–1334. ACM (2014)

Towards Expectation-Maximization by SQL in RDBMS

Kangfei Zhao[1], Jeffrey Xu Yu[1(✉)], Yu Rong[2], Ming Liao[1], and Junzhou Huang[3]

[1] The Chinese University of Hong Kong, Hong Kong S.A.R., China
{kfzhao,yu,mliao}@se.cuhk.edu.hk
[2] Tencent AI Lab, Bellevue, China
[3] University of Texas at Arlington, Arlington, USA
jzhuang@uta.edu

Abstract. Integrating machine learning techniques into *RDBMS*s is an important task since many real applications require modeling (e.g., business intelligence, strategic analysis) as well as querying data in *RDBMS*s. Without integration, it needs to export the data from *RDBMS*s to build a model using specialized *ML* toolkits and frameworks, and import the model trained back to *RDBMS*s for further querying. Such a process is not desirable since it is time-consuming and needs to repeat when data is changed. In this paper, we provide an *SQL* solution that has the potential to support different *ML* models in *RDBMS*s. We study how to support unsupervised probabilistic modeling, that has a wide range of applications in clustering, density estimation, and data summarization, and focus on Expectation-Maximization (EM) algorithms, which is a general technique for finding maximum likelihood estimators. To train a model by EM, it needs to update the model parameters by an E-step and an M-step in a while-loop iteratively until it converges to a level controlled by some thresholds or repeats a certain number of iterations. To support EM in *RDBMS*s, we show our solutions to the matrix/vectors representations in *RDBMS*s, the relational algebra operations to support the linear algebra operations required by EM, parameters update by relational algebra, and the support of a while-loop by *SQL* recursion. It is important to note that the *SQL*'99 recursion cannot be used to handle such a while-loop since the M-step is non-monotonic. In addition, with a model trained by an EM algorithm, we further design an automatic in-database model maintenance mechanism to maintain the model when the underlying training data changes. We have conducted experimental studies and will report our findings in this paper.

1 Introduction

Machine learning (*ML*) plays a leading role in predictive and estimation tasks. It needs to fully explore how to build, utilize and manage *ML* models in *RDBMS*s for the following reasons. First, data are stored in a database system. It is time-consuming of exporting the data from the database system and then feeding it

C. S. Jensen et al. (Eds.): DASFAA 2021, LNCS 12682, pp. 778–794, 2021.
https://doi.org/10.1007/978-3-030-73197-7_53

into models, as well as importing the prediction and estimation results back to the database system. Second, users need to build a model as to query data in *RDBMSs*, and query their data by exploiting the analysis result of the models trained as a part of a query seamlessly in *RDBMSs*. A flexible way is needed to train/query an *ML* model together with data querying by a high-level query language (e.g., *SQL*). Third, the data maintained in *RDBMSs* are supposed to change dynamically. The analysis result of the *ML* models trained may be outdated, which requires repeating the process of exporting data from *RDBMSs* followed by importing the model trained into *RDBMSs*. It needs to support *ML* model update automatically in *RDBMSs*.

There are efforts to support *ML* in *RDBMSs* [12,20]. Model-based views [8, 13] are proposed to support classification and regression analysis in database systems. In brief, [8,13] use an ad-hoc create view statement to declare a classification view. In this create view statement, [8] specifies the model by an as...fit...bases clause, and the training data is fed by an *SQL* query, while [13] specifies a model explicitly with using svm clause, where the features and labels are fed by feature function and labels, respectively. Here, feature function takes database attributes as the input features, and labels are database attributes. Although these approaches provide an optimized implementation for classification models, their create view statement is lack of generality and deviates from the regular *SQL* syntax. In addition, the models supported are limited and implemented in a low-level form in a database system, which makes it difficult for ordinary database end-users to develop new models swiftly. In this work, we demonstrate that our SQL recursive query can define a model-based view in an explicit fashion to support many *ML* models. Different from [8,13], we focus on unsupervised models in the application of in-database clustering, density estimation, and data summarization.

The main contributions of this work are summarized below. First, we study how to support Expectation-Maximization (EM) algorithms [15] in *RDBMSs*, which is a general technique for finding maximum likelihood estimators. We discuss how to represent data in *RDBMSs*, how to compute E-step and M-step of EM using relational algebra operations, how to update parameters using relational algebra operations, and how to support its while-loop using *SQL* recursive queries. Note that mutual recursion is needed for both E/M-step in EM, where the E-step is to compute the conditional posterior probability by Bayesian inference, which is a monotonic operation, whereas the M-step is to compute and update the parameters of the model given a closed-form updating formula, which can be non-monotonic. This fact suggests that *SQL*'99 recursion cannot be used to support EM, since *SQL*'99 recursion (e.g., recursive with) only supports stratified negation, and therefore cannot support non-monotonic operations. We adopt the enhanced *SQL* recursion (e.g., with+) given in [27] based on *XY*-stratified [4,25,26], to handle non-monotonic operations. We have implemented our approach on top of *PostgreSQL*. We show how to train a batch of classical statistical models [5], including Gaussian Mixture model, Bernoulli Mixture model, the mixture of linear regression, Hidden Markov model,

Algorithm 1: EM Algorithm for Mixture Gaussian Model

1: Initialize the means $\boldsymbol{\mu}$, covariances $\boldsymbol{\sigma}$ and mixing coefficients $\boldsymbol{\pi}$;
2: Compute the initial log-likelihood L; $i \leftarrow 0$;
3: **while** $\Delta L > \epsilon$ **or** $i <$ maxrecursion **do**
4: E-step: compute the responsibilities $p(z_{ik})$ based on current $\boldsymbol{\mu}$, $\boldsymbol{\sigma}$ and $\boldsymbol{\pi}$ by Eq. (3);
5: M-step: re-estimate $\boldsymbol{\mu}$, $\boldsymbol{\sigma}$ and $\boldsymbol{\pi}$ by Eq. (4)-(6);
6: re-compute the log-likelihood L; $i \leftarrow i+1$;
7: **end while**
8: **return** $\boldsymbol{\mu}$, $\boldsymbol{\sigma}$, $\boldsymbol{\pi}$;

Mixtures of Experts. Third, given a model trained by EM, we further design an automatic in-database model maintenance mechanism to maintain the model when the underlying training data changes. Inspired by the online and incremental EM algorithms [14,17], we show how to obtain the sufficient statistics of the models to achieve the incremental even decremental model updating, rather than re-training the model by all data. Our setting is different from the incremental EM algorithms which are designed to accelerate the convergence of EM. We maintain the model dynamically by the sufficient statistics of partial original data in addition to the delta part. Fourth, we have conducted experimental studies and will report our findings in this paper.

Organization. In Sect. 2, we introduce the preliminaries including the EM algorithm and the requirements to support it in database systems. We sketch our solution in Sect. 3 and discuss EM training in Sect. 4. In Sect. 5, we design a view update mechanism, which is facilitated by triggers. We conduct extensive experimental studies in Sect. 6 and conclude the paper in Sect. 7.

2 Preliminaries

In this paper, we focus on unsupervised probabilistic modeling, which has broad applications in clustering, density estimation and data summarization in database and data mining area. Specifically, the unsupervised models aim to reveal the relationship between the observed data and some latent variables by maximizing the data likelihood. The expectation-maximization (EM) algorithm, first introduced in [7], is a general technique for finding maximum likelihood estimators. It has a solid statistical basis, robust to noisy data and its complexity is linear in data size. Here, we use the Gaussian mixture model [5], a widely used model in data mining, pattern recognition, and machine learning, as an example to illustrate the EM algorithm and our approach throughout this paper.

Suppose we have an observed dataset $X = \{x_1, x_2, \cdots, x_n\}$ of n data points where $x_i \in \mathbb{R}^d$. Given $\mathcal{N}(x|\boldsymbol{\mu}, \boldsymbol{\sigma})$ is the probability density function of a Gaussian distribution with mean $\boldsymbol{\mu} \in \mathbb{R}^d$ and covariance $\boldsymbol{\sigma} \in \mathbb{R}^{d \times d}$, the density of

Table 1. The relation representations

K	π	μ	σ
1	π_1	μ_1	σ_1
2	π_2	μ_2	σ_2

(a) relation GMM

ID	x
1	x_1
2	x_2

(b) relation X

ID	K	p
1	1	$p(z_{11})$
1	2	$p(z_{12})$
2	1	$p(z_{21})$
2	2	$p(z_{22})$

(c) relation R

ID	x
1	[1.0, 2.0]
2	[3.0, 4.0]

(d) row-major representation X_c

Gaussian mixture model is a simple linear super-position of K different Gaussian components in the form of Eq. (1).

$$p(x_i) = \sum_{k=1}^{K} \pi_k \mathcal{N}(x_i|\boldsymbol{\mu}_k, \boldsymbol{\sigma}_k) \tag{1}$$

Here, $\pi_k \in \mathbb{R}$ is the mixing coefficient, i.e., the prior of a data point belonging to component k and satisfies $\sum_{i=1}^{K} \pi_k = 1$. To model this dataset X using a mixture of Gaussians, the objective is to maximize the log-likelihood function in Eq. (2).

$$lnp(\boldsymbol{X}|\boldsymbol{\pi}, \boldsymbol{\mu}, \boldsymbol{\sigma}) = \sum_{i=1}^{n} ln[\sum_{k=1}^{K} \pi_k \mathcal{N}(x_i|\boldsymbol{\mu}_k, \boldsymbol{\sigma}_k)] \tag{2}$$

Algorithm 1 sketches the EM algorithm for training the Gaussian Mixture Model. First, in line 1–2, the means $\boldsymbol{\mu}_k$, covariances $\boldsymbol{\sigma}_k$ and the mixing coefficients $\boldsymbol{\pi}_k$ of K Gaussian distributions are initialized, and the initial value of the log-likelihood (Eq. (2)) is computed. In the while loop of line 3–7, the Expectation-step (E-step) and Maximization-step (M-step) are executed alternatively. In the E-step, we compute the responsibilities, i.e., the conditional probability that x_i belongs to component k, denoted as $p(z_{ik})$ by fixing the parameters based on the Bayes rule in Eq. (3).

$$p(z_{ik}) = \frac{\pi_k \mathcal{N}(x_i|\boldsymbol{\mu}_k, \boldsymbol{\sigma}_k)}{\sum_{j=1}^{K} \pi_j \mathcal{N}(x_i|\boldsymbol{\mu}_j, \boldsymbol{\sigma}_j)} \tag{3}$$

In the M-step, we re-estimate a new set of parameters using the current responsibilities by maximizing the log-likelihood (Eq. (2)) as follows.

$$\boldsymbol{\mu}_k^{new} = \frac{1}{n_k} \sum_{i=1}^{n} p(z_{ik}) x_i \tag{4}$$

$$\boldsymbol{\sigma}_k^{new} = \frac{1}{n_k} \sum_{i=1}^{n} p(z_{ik})(x_i - \boldsymbol{\mu}_k^{new})(x_i - \boldsymbol{\mu}_k^{new})^T \tag{5}$$

$$\boldsymbol{\pi}_k^{new} = \frac{n_k}{n} \tag{6}$$

where $n_k = \sum_{i=1}^{n} p(z_{ik})$. At the end of each iteration, the new value of log-likelihood is evaluated and used for checking convergence. The algorithm ends when the log-likelihood converges or a given iteration time is reached. In *RDBMS*, the learned model, namely the parameters of K components, can be persisted in a relation of K rows as shown in Table 1(a). Suppose 1-dimensional dataset X as Table 1(b) is given, the posterior probability of x_i belongs to component k can be computed as Table 1(c) and clustering can be conducted by assigning x_i to component with the maximum $p(z_{ik})$.

To fulfill the EM algorithm in database systems, there are several important issues that need to be concerned, including (1) the representation and storage of high dimensional data in the database, (2) the relation algebra operation used to perform linear algebra computation in EM, (3) the approach for iterative parameter updating, (4) the way to express and control the iteration of EM algorithm, and (5) the mechanism to maintain the existing model when underlying data evolves.

3 Our Solution

In this section, we give a complete solution to deal with above issues in applying the EM algorithm and building model-based views inside *RDBMS*.

High Dimensional Representation. Regarding the issue of high dimensional data, different from [19], we adopt the row-major representation, as shown in Table 1(d), which is endorsed by allowing array/vector data type in the database. With it, we can use the vector/matrix operations to support complicated linear algebra computation in a concise *SQL* query.

Consider computing the means $\boldsymbol{\mu}$ in the M-step (Eq. (4)) with the representation X_c, in Table 1(d). Suppose the responsibilities are in relation $R(ID, K, p)$, where ID, K and p are the identifier of data point, component, and the value of $p(z_{ik})$. The relational algebra expression to compute Eq. (4) is shown in Eq. (7).

$$M_c \leftarrow \rho_{(K,\text{mean})}({}_K\mathcal{G}_{\mathsf{sum}(p\cdot x)}(R \underset{R.ID=X_c.ID}{\bowtie} X_c)) \tag{7}$$

It joins X and R on the ID attribute to compute $p(z_{ik}) \cdot x_i$ using the operator $\cdot$ which denotes a scalar-vector multiplication. With the row-major representation which nesting separate dimension attributes into one vector-type attribute, the $\cdot$ operator for vector computation, Eq. (4) is expressed efficiently (Eq. (7)).

Relational Algebra to Linear Algebra. On the basis of array/vector data type and the derived statistical function and linear algebra operations, the complicated linear algebra computation can be expressed by basic relational algebra operations (selection (σ), projection (Π), union ($\cup$), Cartesian product ($\times$), and rename (ρ)), together with group-by & aggregation. Let V and E (E') be the relation representation of vector and matrix, such that $V(ID, v)$ and $E(F, T, e)$.

Here *ID* is the tuple identifier in V. F and T, standing for the two indices of a matrix. [27] introduces two new operations to support the multiplication between a matrix and a vector (Eq. (8)) and between two matrices (Eq. (9)) in their relation representation.

$$E \overset{\oplus(\odot)}{\underset{T=ID}{\bowtie}} V = {}_{F}\mathcal{G}_{\oplus(\odot)}(E \underset{T=ID}{\bowtie} V) \tag{8}$$

$$E \overset{\oplus(\odot)}{\underset{E.T=E'.F}{\bowtie}} E' = {}_{E.F,E'.T}\mathcal{G}_{\oplus(\odot)}(E \underset{E.T=E'.F}{\bowtie} E') \tag{9}$$

The matrix-vector multiplication (Eq. (8)) consists of two steps. The first step is computing $v \odot e$ between a tuple in E and a tuple in V under the join condition $E.T = V.ID$. The second step is aggregating all the $\odot$ results by the operation of $\oplus$ for every group by grouping-by the attribute $E.F$. Similarly, the matrix-matrix multiplication (Eq. (9)) is done in two steps. The first step computes $\odot$ between a tuple in E and a tuple in E' under the join condition $E.T = E'.F$. The second step aggregates all the $\odot$ results by the operation of $\oplus$ for every group-by grouping by the attributes $E.F$ and $E'.T$. The formula of re-estimating the means $\boldsymbol{\mu}$ (Eq. (4)) is a matrix-vector multiplication if data is 1-dimensional or a matrix-matrix multiplication otherwise. When high dimensional data is nested as the row-major representation (Table 1(d)), the matrix-matrix multiplication is reduced to matrix-vector multiplication, as shown in Eq. (7).

Re-estimating the covariance/standard deviation $\boldsymbol{\sigma}$ (Eq. (5)) involves the element-wise matrix multiplication if data is 1-dimensional or a tensor-matrix multiplication otherwise. The element-wise matrix multiplication can be expressed by joining two matrices on their two indices to compute $E.e \odot E'.e$. An extra aggregation is required to aggregate on each component k as shown in Eq. (10).

$$E \overset{\oplus(\odot)}{\underset{\substack{E.F=E'.F\\ E.T=E'.T}}{\bowtie}} E' = {}_{E.F}\mathcal{G}_{\oplus(\odot)}(E \underset{\substack{E.F=E'.F\\ E.T=E'.T}}{\bowtie} E') \tag{10}$$

Similarly, when $\odot$ and $\oplus$ are vector operation and vector aggregation, Eq. (10) is reduced to high dimensional tensor-matrix multiplication.

Value Updating. We need to deal with parameter update when training the model in multiple iterations. We use union by update, denoted as $\uplus$, defined in [27] (Eq. (11)) to address value update.

$$R \uplus_A S = (R - (R \underset{R.A=S.A}{\ltimes} S)) \cup S \tag{11}$$

Suppose t_r is a tuple in R and t_s is a tuple in S. The union by update updates t_r by t_s if t_r and t_s are identical by some attributes A. If t_s does not match any t_r, t_s is merged into the resulting relation. We use $\uplus$ to update the relation of parameters in Table 1(a).

Iterative Evaluation. *RDBMS*s have provided the functionality to support recursive queries, based on *SQL*'99 [11,16], using the with clause in *SQL*. This with clause restricts the recursion to be a stratified program, where non-monotonic operation (e.g., $\uplus$) is not allowed. To support iterative model update, we follow [27], which proves $\uplus$ (union by update) leads to a fixpoint in the enhanced recursive *SQL* queries by *XY*-stratification. We prove that the vector/-matrix data type, introduced in this paper, can be used in *XY*-stratification. We omit the details due to the limited space.

```
with R as
    select ··· from R_{1,j}, ··· computed by ···  (Q_1)
    union by update
    select ··· from R_{2,j}, ··· computed by ···  (Q_2)
```

Fig. 1. The general form of the enhanced recursive with

The general syntax of the enhanced recursive with is sketched in Fig. 1. It allows union by update to union the result of initial query Q_1 and recursive query Q_2. Here, the computed by statement in the enhanced with allows users to specify how a relation $R_{i,j}$ is computed by a sequence of queries. The queries wrapped in computed by must be non-recursive.

4 The EM Training

We show the details of supporting the model-based view using *SQL*. First, we present the relational algebra expressions needed followed by the enhanced recursive query. Second, we introduce the queries for model inference.

Parameter Estimation: For simplicity, here we consider the training data point x_i is 1-dimensional scalar. It is natural to extend the query to high dimensional input data when matrix/vector data type and functions are supported by the database system. We represent the input data by a relation $X(ID, x)$, where *ID* is the tuple identifier for data point x_i and x is a numeric value. The model-based view, which is persisted in the relation GMM(K, pie, mean, cov), where K is the identifier of the k-th component, and 'pie', 'mean', and 'cov' denote the corresponding parameters, i.e., mixing coefficients, means and covariances (standard deviations), respectively. The relation representations are shown in Table 1. The following relational algebra expressions describe the E-step (Eq. (12)), M-step (Eq. (13)–(16)), and parameter updating (Eq. (17)) in one iteration.

$$R \leftarrow \rho_{(ID,K,p)} \Pi_{(ID,K,f)}(GMM \times X) \tag{12}$$

$$N \leftarrow \rho_{(K,\text{pie})}(R \underset{R.ID=X.ID}{\overset{\mathsf{sum}(p)}{\bowtie}} X) \tag{13}$$

$$M \leftarrow \rho_{(K,\text{mean})}(R \underset{R.ID=X.ID}{\overset{\mathsf{sum}(p*x)/\mathsf{sum}(p)}{\bowtie}} X) \tag{14}$$

$$T \leftarrow \Pi_{ID,K,\mathsf{pow}(x-\text{mean})}(X \times N) \tag{15}$$

$$C \leftarrow \rho_{(K,\text{cov})} {}_K\mathcal{G}_{\mathsf{sum}(p*t)}(T \underset{\substack{R.ID=T.ID\\ R.K=T.K}}{\bowtie} R) \tag{16}$$

$$GMM \leftarrow \rho_{(K,\text{pie},\text{mean},\text{cov})}(N \underset{N.K=M.K}{\bowtie} M \underset{M.K=C.K}{\bowtie} C) \tag{17}$$

In Eq. (12), by performing a Cartesian product of GMM and X, each data point is associated with the parameters of each component. The responsibilities are evaluated by applying an analytical function f to compute the normalized probability density (Eq. (3)) for each tuple, which is the E-step. The resulting relation $R(ID, K, p)$ is shown in Fig. 1(c). For the M-step, the mixing coefficients 'pie' (Eq. (13)), the means 'mean' (Eq. (14)) and the covariances 'cov' (Eq. (15)–(16)) are re-estimated based on their update formulas in Eq. (4)–(6), respectively. In the end, in Eq. (17), the temporary relations N, M and C are joined on attribute K to merge the parameters. The result is assigned to the recursive relation GMM.

```
1.  with
2.  GMM(K, pie, mean, cov) as (
3.        (select K, pie, mean, cov from INIT_PARA)
4.        union by update K
5.        (select N.K, pie/n, mean, sqrt (cov/pie)
6.        from N, C where N.K = C.K
7.        computed by
8.        R(ID, K, p) as select ID, k, norm(x, mean, cov) * pie /
9.                     (sum(norm(x, mean, cov) * pie) over (partition by ID))
10.                    from GMM, X
11.       N(K, pie, mean) as select K, sum(p), sum(p * x) / sum(p)
12.                      from R, X where R.ID = X.ID
13.                      group by K
14.       C(K, cov) as select R.K, sum(p * T.val) from
15.                    (select ID, K, pow(x-mean) as val from X, N) as T, R
16.                    where T.ID = R.ID and T.K = R.K
17.                    group by R.K)
18.       maxrecursion 10)
19. select * from GMM
```

Fig. 2. The enhanced recursive *SQL* for Gaussian Mixtures

Figure 2 shows the enhanced with query to support Gaussian Mixture Model by EM algorithm. The recursive relation GMM specifies the parameters of k Gaussian distributions. In line 3, the initial query loads the initial parameters from relation INI_PARA. The new parameters are selected by the recursive query (line 5–6) evaluated by the computed by statement and update the recursive relation by union by update w.r.t. the component index K. It wraps the queries to compute E-step and M-step of one iteration EM.

We elaborate on the queries in the computed by statement (line 8–17). Specifically, the query in line 8–10 performs the E-step, as the relational algebra in Eq. (12). Here, norm is the Gaussian (Normal) probability density function of data point x given the mean and covariance as input. We can use the window function, introduced in *SQL'*03 to compute the responsibility by Bayes rule in Eq. (3). In line 9, sum() over (partition by()) is the window function performing calculation across a set of rows that are related to the current row. As it does not group rows, where each row retains its separate identity, many *RDBMSs* allow to use it in the recursive query, e.g., *PostgreSQL* and *Oracle*. The window function partitions rows of the Cartesian product results in partitions of the same *ID* and computes the denominator of Eq. (3). In line 11–13, the query computes the means (Eq. (4)) and the mixing coefficients together by a matrix-matrix multiplication due to their common join of R and X. Then, line 14–17 computes the covariances of Eq. (5). First, we compute the square of $x_i - \boldsymbol{\mu}_k$ for each x_i and k, which requires a Cartesian product of N and R (Eq. (15)). Second, the value is weighted by the responsibility and aggregated as specified in Eq. (16). The new parameters in the temporary relation N and C will be merged by joining on the component index K in line 6. The depth of the recursion can be controlled by maxrecursion clause, adapted from *SQL Server* [2]. The maxrecursion clause can effectively prevent infinite recursion caused by infinite fix point, e.g., '**with** $R(n)$ **as** ((**select values**(0)) **union all** (**select** $n+1$ **from** R))' , a legal *SQL'*99 recursion.

Model Inference: Once the model is trained by the recursive query in Fig. 2, it can be materialized in a view for online inference. In the phase of inference, users can query the view by *SQL* to perform clustering, classification and density estimation. Given a batch of data in relation X and a view GMM computed by Fig. 2, the query below computes the posterior probability that the component K generated the data with index *ID*. The query is similar to computing the E-step (Eq. (3)) in line 5–7 of Fig. 2.

create table R **as select** *ID*, K,
norm(x, mean, cov) * pie / (sum(norm(x, mean, cov) * pie)
over (**partition by** *ID*)) **as** p **from** GMM, X

Based on relation $R(ID, K, p)$ above, we can further assign the data into K clusters, where x_i is assigned to cluster k if the posterior probability $p(z_{ik})$ is the maximum among the $\{p(z_{i1}), \cdots, p(z_{iK})\}$. The query below creates a relation CLU(*ID*, K) to persist the clustering result where *ID* and K are the index of the data point and its assigned cluster, respectively. It first finds the maximum $p(z_{ik})$ for each data point by a subquery on relation R. The result is renamed as T and is joined with R on the condition of $R.ID = T.ID$ and $R.p = T.p$ to find the corresponding k.

create table CLU **as select** *ID*, *K* **from** *R*,
(**select** *ID*, max (*p*) **as** *p* **from** *R* **group by** *ID*) **as** *T*,
where *R.ID* = *T.ID* **and** *R.p* = *T.p*

It is worth noting that both of the queries above only access the data exactly once. Thereby, it is possible to perform the inference on-the-fly and only for interested data. Besides density estimation and clustering, result evaluation, e.g., computing the purity, normalized mutual information (NMI) and Rand Index can be conducted in the database by *SQL* queries.

5 Model Maintenance

In this section, we investigate the automatic model/view updating. When the underlying data X changes, a straightforward way is to re-estimate the model over the updated data. However, when only a small portion of the training data are updated, the changes of the corresponding model are slight, it is inefficient to re-estimate the model on-the-fly. Hence, a natural idea arises that whether we can update the existing model by exploring the "incremental variant" of the EM algorithm. And this variant can be maintained by the newly arriving data and a small portion of data extracted from the original dataset. As the statistical model trained by an *SQL* query can be represented by its sufficient statistics, the model is updated by maintaining the model and sufficient statistics.

Maintaining Sufficient Statistics: The sufficient statistic is a function of data X that contains all of the information relevant to estimate the model parameters. As the model is updated, the statistics of data is also updated followed by the changing of the posterior probability $p(z_{ik})$. This process repeats until the statistics converge. We elaborate on the sufficient statistics updating rules below.

Suppose the training dataset of model $\boldsymbol{\theta}$ is $\{x_1, x_2, \cdots, x_n\}$. Let $\boldsymbol{s}$ be the sufficient statistics of $\boldsymbol{\theta}$, based on the Factorization Theorem [10], we can obtain

$$\boldsymbol{s} = \sum_{i=1}^{n} \sum_{\boldsymbol{z}} p(\boldsymbol{z}|x_i, \boldsymbol{\theta}) \phi(x_i, \boldsymbol{z}) \tag{18}$$

where $\boldsymbol{z}$ is the unobserved variable, ϕ denotes the mapping function from an instance $(x_i, \boldsymbol{z})$ to the sufficient statistics contributed by x_i. The inserted data is $\{x_{n+1}, x_{n+2}, \cdots, x_m\}$. Let the model for overall data $\{x_1, \cdots, x_n, x_{n+1}, \cdots, x_m\}$ be $\widetilde{\boldsymbol{\theta}}$ and the corresponding sufficient statistics be $\widetilde{\boldsymbol{s}}$. The difference of $\widetilde{\boldsymbol{s}} - \boldsymbol{s}$, denoted as $\Delta \boldsymbol{s}$ is

$$\begin{aligned} \Delta \boldsymbol{s} &= \sum_{i=1}^{n+m} \sum_{\boldsymbol{z}} p(\boldsymbol{z}|x_i, \widetilde{\boldsymbol{\theta}}) \phi(x_i, \boldsymbol{z}) - \sum_{i=1}^{n} \sum_{\boldsymbol{z}} p(\boldsymbol{z}|x_i, \boldsymbol{\theta}) \phi(x_i, \boldsymbol{z}) \\ &= \sum_{i=1}^{n+m} \sum_{\boldsymbol{z}} [p(\boldsymbol{z}|x_i, \widetilde{\boldsymbol{\theta}}) - p(\boldsymbol{z}|x_i, \boldsymbol{\theta})] \phi(x_i, \boldsymbol{z}) && (19) \\ &+ \sum_{i=n+1}^{m} \sum_{\boldsymbol{z}} p(\boldsymbol{z}|x_i, \boldsymbol{\theta}) \phi(x_i, \boldsymbol{z}) && (20) \end{aligned}$$

According to above equations, we observe that the delta part of the sufficient statistics $\Delta \boldsymbol{s}$ consists of two parts: (1) changes of the sufficient statistics for the overall data points $\{x_1, x_2 \cdots x_m\}$ in Eq. (19), and (2) the additional sufficient statistics for the newly inserted data points $\{x_{n+1}, \cdots x_m\}$ in Eq. (20). Consider to retrain a new model $\widetilde{\boldsymbol{\theta}}$ over $\{x_1, x_2, \cdots, x_m\}$ in T iterations by taking $\boldsymbol{\theta}$ as the initial parameter, i.e., $\boldsymbol{\theta}^{(0)} = \boldsymbol{\theta}$ and $\boldsymbol{\theta}^{(T)} = \widetilde{\boldsymbol{\theta}}$. We have

$$\Delta \boldsymbol{s} = \sum_{i=1}^{n+m} \sum_{\boldsymbol{z}} [p(\boldsymbol{z}|x_i, \boldsymbol{\theta}^{(T)}) - p(\boldsymbol{z}|x_i, \boldsymbol{\theta}^{(0)})] \phi(x_i, \boldsymbol{z}) \tag{21}$$

$$+ \sum_{i=n+1}^{m} \sum_{\boldsymbol{z}} p(\boldsymbol{z}|x_i, \boldsymbol{\theta}^{(0)}) \phi(x_i, \boldsymbol{z}) \tag{22}$$

$$= \sum_{t=1}^{T} \sum_{i=1}^{n+m} \sum_{\boldsymbol{z}} [p(\boldsymbol{z}|x_i, \boldsymbol{\theta}^{(t)}) - p(\boldsymbol{z}|x_i, \boldsymbol{\theta}^{(t-1)})] \phi(x_i, \boldsymbol{z}) \tag{23}$$

$$+ \sum_{i=n+1}^{m} \sum_{\boldsymbol{z}} p(\boldsymbol{z}|x_i, \boldsymbol{\theta}^{(0)}) \phi(x_i, \boldsymbol{z}) \tag{24}$$

Above equations indict how to compute $\Delta \boldsymbol{s}$. For the inserted data $\{x_{n+1}, \cdots x_m\}$, the delta can be directly computed by evaluating the original model $\boldsymbol{\theta}^{(0)}$ as Eq. (24), while for original data, the delta can be computed by updating the model $\boldsymbol{\theta}^{(t)}$ iteratively using all the data $\{x_1, x_2 \cdots x_m\}$ as Eq. (23). Since most of the computational cost is on the iteration of Eq. (23), we approximate the computation. First, we use the stochastic approximation algorithm, where the parameters are updated after the sufficient statistics of each new data point x_i is computed, instead of the full batch dataset [14,17,21]. The second is discarding the data points which are not likely to change their cluster in the future, as the scaling clustering algorithms adopt for speedup [6]. We discuss our strategy of selecting partial original data in $\{x_1, x_2, \cdots, x_n\}$ for model update. There is a tradeoff between the accuracy of the model and the updating cost. We have two strategies: a distance-based and a density-based strategy. For the distance-based strategy, we use Mahalanobis distance [9] to measure the distance between a data point and a distribution. For each data x_i, we compute the Mahalanobis distance, $D_k(x_i)$, to the k-th component with mean $\boldsymbol{\mu}_k$ and covariance $\boldsymbol{\sigma}_k$.

$$D_k(x_i) = \sqrt{(x_i - \boldsymbol{\mu}_k)^T \boldsymbol{\sigma}_k^{-1} (x_i - \boldsymbol{\mu}_k)} \tag{25}$$

We can filter the data within a given thresholding radius with any component. Another density-based measurement is the entropy of the posterior probability for data x_i as in Eq. (26), where $p(z_{ik})$ is evaluated by parameter $\boldsymbol{\theta}^{(0)}$. The larger the entropy, the lower the possibility of assigning x_i to any one of the components.

$$E(x_i) = -\sum_{k=1}^{K} p(z_{ik}) ln\ p(z_{ik}) \tag{26}$$

Similarly, considering deleting m data points $\{x_{n-m+1}, \cdots x_n\}$ from $\{x_1, x_2 \cdots x_n\}$, the difference of the sufficient statistics, $\Delta \boldsymbol{s}$ is

$$\Delta \boldsymbol{s} = \sum_{t=1}^{T} \sum_{i=1}^{n-m} \sum_{\boldsymbol{z}} [p(\boldsymbol{z}|x_i, \boldsymbol{\theta}^{(t)}) - p(\boldsymbol{z}|x_i, \boldsymbol{\theta}^{(t-1)})] \phi(x_i, \boldsymbol{z}) - \sum_{i=n-m+1}^{n} \sum_{\boldsymbol{z}} p(\boldsymbol{z}|x_i, \boldsymbol{\theta}^{(0)}) \phi(x_i, \boldsymbol{z}) \quad (27)$$

```
1. create trigger T1 before insert on X
2. for each statement
3. execute procedure DATA_SELECTION

4. create trigger T2 before insert on X
5. for each row
6. execute procedure DATA_INSERTION

7. create trigger T3 after insert on X
8. for each statement
9. execute procedure MODEL_UPDATE
```

Fig. 3. The triggers for incremental update

A Trigger-based Implementation: In *RDBMS*s, the automatic model updating mechanism is enabled by triggers built on the relation of the input data. There are three triggers built on the relation of training data X, whose definitions are shown in Fig. 3. Before executing the insertion operation, two triggers T1 (line 1–3 in Fig. 3) and T2 (line 4–6 in Fig. 3) prepare the data for model updating in a temporary relation X'. Here, T1 performs on each row to select a subset from original data in $\{x_1, x_2, \cdots, x_n\}$ based on a selection criterion. Additionally, T2 inserts all the newly arrived data $\{x_{1+n}, x_2, \cdots, x_m\}$ to relation X'. After the data preparation finished, another trigger T3 (line 7–9 in Fig. 3) will call a *PSM* to compute the $\Delta \boldsymbol{s}$ by X'. In the *PSM*, first, the delta of the newly inserted data (Eq. (23)) is computed to reinitialize the parameters of the model. Then, T iterations of scanning relation X' is performed. Where in each iteration. X' is randomly shuffled and each data point is used to update the sufficient statistics it contributes as well as the model instantly. It is worth mentioning that the data selection in trigger T1 can be performed offline, i.e., persisting a subset of training data with a fixed budget size for model updating in the future. In addition, the sufficient statistics for original model $\boldsymbol{\theta}^0$ can be precomputed. Those will improve the efficiency of online model maintenance significantly. The actions of these triggers are transparent to the database users.

6 Experimental Studies

In this section, we present our experimental studies of supporting model-based view training, inference, and maintenance in *RDBMS*. We conduct extensive experiments to investigate the following: (a) to compare the performance of our enhanced with and looping control by a host language, (b) to test the scalability of the recursive queries for different models, and (c) to validate the efficiency of our model maintenance mechanism.

Experimental Setup: We report our performance studies on a PC with Intel(R) Xeon(R) CPU E5-2697 v3 (2.60 GHz) with 96 GB RAM running Linux CentOS 7.5 64 bit. We tested the enhanced recursive query on *PostgreSQL* 10.10 [3]. The statistical function and matrix/vector computation function are supported by Apache MADlib 1.16 [12]. All the queries we tested are evaluated in a single thread *PostgreSQL* instance.

with+ **vs.** ***Psycopg2***: We compare the enhanced with, which translates the recursive *SQL* query to *SQL/PSM* with the implementation of using a host language to control the looping, which is adopted in previous EM implementation [19]. We implement the latter by *Psycopg2* [1], a popular *PostgreSQL* adapter for the python language. Regarding the EM algorithm, the E-step, M-step, and parameter updating are wrapped in a python for-loop, and executed by a cursor alternatively. We compare the running time of these two implementations, i.e., enhanced with and *Psycopg2* for training Gaussian Mixture Model by varying the dimension d of data point (Fig. 4(b)), the scale of the training data n (Fig. 4(c)), the number of components k (Fig. 4(a)) and the number of iterations (Fig. 4(d)). The training data is evenly generated from 10 Gaussian distributions.

The evaluated time is the pure query execution time where the costs of database connection, data loading and parameter initialization are excluded. The experiments show that enhanced with outperforms *Psycopg2* significantly, not only for multiple iterators in Fig. 4(d) but also for per iteration in Fig. 4(b)–4(a). For one thing, the implementation of *Psycopg2* calls the database multiple times per iteration, incurring much client-server communication and context switch costs. For the other, the issued queries from client to server will be parsed, optimized and planned on-the-fly. These are the general problems of calling *SQL* queries by any host language. Meanwhile, we implement the hybrid strategy of *SQLEM* [18] in *PostgreSQL*. For Gaussian Mixture model, one iteration for 10,000 data points with 10 dimensions fails to terminate within 1 h. In their implementation, $2k$ separate *SQL* queries evaluate the means and variances of k components respectively, which is a performance bottleneck.

Experiments on Synthetic Data: We train Gaussian Mixture model (GMM) [5], mixture of linear regression (MLR) [22] and a neural network model, mixture of experts (MOE) [24] by evaluating *SQL* recursive queries in *PostgreSQL*. Given the observed dataset as $\{(x_1, y_1), (x_2, y_2), \cdots, (x_n, y_n)\}$, where

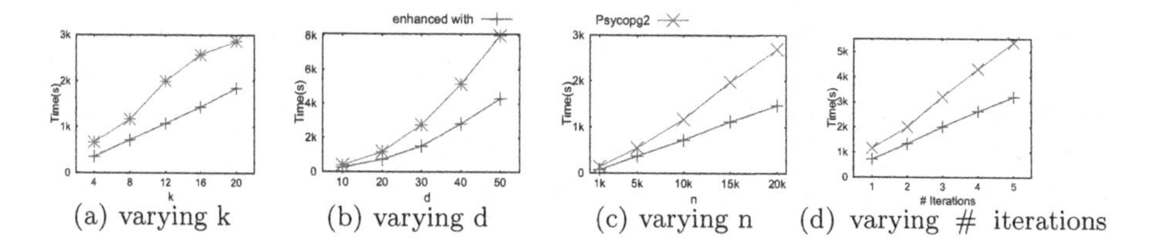

(a) varying k (b) varying d (c) varying n (d) varying # iterations

Fig. 4. with+ vs. *Psycopg2*

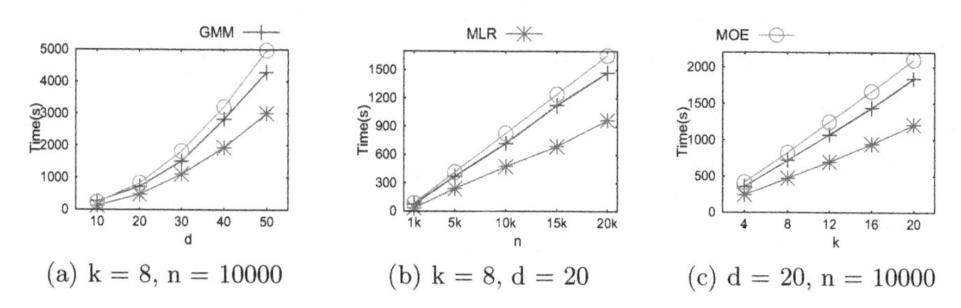

(a) k = 8, n = 10000 (b) k = 8, d = 20 (c) d = 20, n = 10000

Fig. 5. Scalability test

$x_i \in \mathbb{R}^d$ and $y_i \in \mathbb{R}$, the MLR models the density of y given x as

$$p(y_i|x_i) = \sum_{k=1}^{K} \pi_k \mathcal{N}(y_i|x_i^T \boldsymbol{\beta}_k, \boldsymbol{\sigma}_k) \tag{28}$$

And the MOE models the density of y given x as

$$p(y_i|x_i) = \sum_{k=1}^{K} g_k(x_i) \mathcal{N}(y_i|x_i^T \boldsymbol{\beta}_k, \boldsymbol{\sigma}_k) \tag{29}$$

where $\boldsymbol{\beta}_k \in \mathbb{R}^d$ is the parameters of a linear transformer, $\mathcal{N}$ is the probability density function of a Gaussian given mean $x_i^T \boldsymbol{\beta}_k \in \mathbb{R}$ and standard deviation $\boldsymbol{\sigma}_k \in \mathbb{R}$. In Eq. (29), $g_k(x)$ is called the gating function, given by computing the softmax in Eq. (30) where $\boldsymbol{\theta} \in \mathbb{R}^d$ is a set of linear weights on x_i.

$$g_k(x_i) = \frac{e^{x_i \boldsymbol{\theta}_k}}{\sum_{j=1}^{K} e^{x_i \boldsymbol{\theta}_j}} \tag{30}$$

The intuition behind the gating functions is a set of 'soft' learnable weights which determine the mixture of K local models. We adopt the single loop EM algorithm [23] to estimate the parameters of MOE, which uses least-square regression to compute the gating network directly. For GMM, the training data is evenly drawn from 10 Gaussian distributions. For MLR and MOE, the training data is generated from 10 linear functions with Gaussian noise. The parameters of the Gaussians and the linear functions are drawn from the uniform distribution $[0, 10]$. And the initial parameters are also randomly drawn from $[0, 10]$.

Figure 5 displays the training time per iteration of the 3 models by varying the data dimension d (Fig. 5(a)), the scale of the training data n (Fig. 5(b)) and the number of clusters k (Fig. 5(c)). In general, for the 3 models, the training time grows linearly as n and k increase, while the increment of data dimension d has a more remarkable impact on the training time. When increasing n and k, the size of intermediate relations, e.g., relation R for computing the responsibilities in Eq. (12) grow linearly. Therefore the training cost grows linearly with regards to n and k. However, in the 3 models, we need to deal with $d \times d$ dimensional matrices in the M-step. For GMM, it needs to compute the probability density of the multivariable Gaussians and reestimate the covariance matrices. For MLR and MOE, they need to compute the matrix inversion and least square regression. The training cost grows regarding the size of the matrix. The comparison shows it is still hard to scale high-dimensional analysis in a traditional database system. However, efficiency can be improved on a parallel/distributed platform and new hardware.

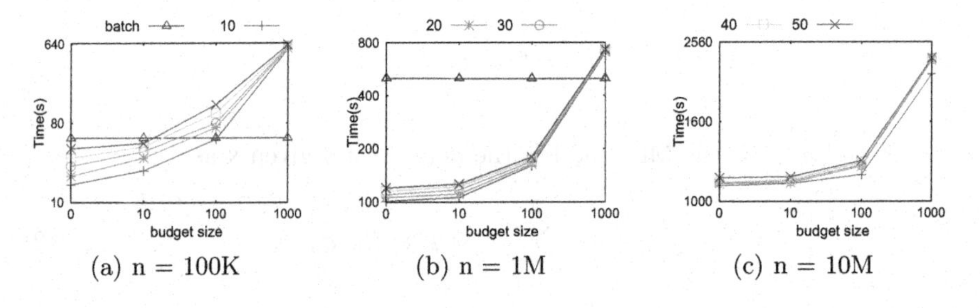

(a) n = 100K (b) n = 1M (c) n = 10M

Fig. 6. Insert maintenance

Incremental Maintenance: Finally, we test the performance of our trigger-based model updating mechanism. First, we train GMM for 1-dimensional data generated from 2 Gaussian distributions. The original models are trained over 100k, 1M and 10M data points, respectively with 15 iterations. The overall training time is recorded as the 'batch' mode training time, which is 54 s, 501 s and 4,841 s respectively. After the model is trained and persisted. We insert 10, 20, 30, 40, 50 data points to the underlying data by varying the budget size of selected data from 0 to 1,000.

Figure 6 shows the insertion time w.r.t. the budget size of the selected data for the 3 models. The insertion time is the collapsed time from the insert command issuing to the transaction commit, including the cost of data selection with the density-based strategy and computing initial sufficient statistics. As the number of processed tuples increases, the insertion time grows linearly. Compare to the retraining cost, i.e., the batch training time, it is not always efficient to update the existing model. The choice depends on two factors, the size of overall data points, and the budget size plus insertion size, i.e., the numbers of data points to be processed in the updating. The updating mechanism may not be efficient and effective when the overall data size is small or there is a large volume of insertion. That is because, for the batch training mode, computation of parameter evaluation dominates the cost. While for the model updating, since the sufficient statistics and the model are updated when processing each data point, the updating overhead becomes a main overhead. Meanwhile, we notice that the collapsed time of data selection and computing initial sufficient statistics take about 10 s, 100 s and 1,000 s for data size of 100k, 1M and 10M, respectively. Precomputing and persisting these results will benefit for a larger dataset.

7 Conclusion

Integrating machine learning techniques into database systems facilitates a wide range of applications in industrial and academic fields. In this paper, we focus on supporting EM algorithm in *RDBMS*. Different from the previous approach, our approach wraps the E-step and M-step in an enhanced *SQL* recursive query to reach an iterative fix point. We materialize the learned model as a database view to query. Furthermore, to handle model updates, we propose an automatic view updating mechanism by exploiting the incremental variant of the EM algorithm. The extensive experiments we conducted show that our approach outperforms the previous approach significantly, and can support multiple mixture models by EM algorithm, as well as the efficiency of the incremental model update. The *SQL* implementation can be migrated to parallel and distributed platforms, e.g., *Hadoop* and *Spark*, to deploy large scale machine learning applications. These directions deserve future explorations.

Acknowledgement. This work is supported by the Research Grants Council of Hong Kong, China under No. 14203618, No. 14202919 and No. 14205520.

References

1. http://initd.org/psycopg/docs/index.html
2. Microsoft SQL documentation. https://docs.microsoft.com/en-us/sql/
3. Postgresql. https://www.postgresql.org
4. Arni, F., Ong, K., Tsur, S., Wang, H., Zaniolo, C.: The deductive database system LDL++. TPLP **3**(1) (2003)
5. Bishop, C.M.: Pattern Recognition and Machine Learning. Information Science and Statistics, 5th edn. Springer, Heidelberg (2007)
6. Bradley, P.S., Fayyad, U.M., Reina, C.: Scaling clustering algorithms to large databases. In Proceedings of KDD 1998, pp. 9–15 (1998)
7. Dempster, A.P.: Maximum likelihood estimation from incomplete data via the em algorithm. J. Roy. Stat. Soc. Ser. B (Stat. Methodol) **39**, 1–38 (1977)
8. Deshpande, A., Madden, S.: Mauvedb: supporting model-based user views in database systems. In Proceedings of SIGMOD 2006, pp. 73–84 (2006)
9. Duda, R.O., Hart, P.E.: Pattern Classification and Scene Analysis. A Wiley-Interscience publication, Hoboken (1973)

10. Duda, R.O., Hart, P.E., Stork, D.G.: Pattern Classification, 2nd edn. Wiley, New York (2001)
11. Finkelstein, S.J., Mattos, N., Mumick, I., Pirahesh, H.: Expressing recursive queries in SQL. ISO-IEC JTC1/SC21 WG3 DBL MCI, (X3H2-96-075) (1996)
12. Hellerstein, J.M., et al.: The madlib analytics library or MAD skills, the SQL. PVLDB **5**(12), 1700–1711 (2012)
13. Koc, M.L., Ré, C.: Incrementally maintaining classification using an RDBMS. PVLDB **4**(5), 302–313 (2011)
14. Liang, P., Klein, D.: Online EM for unsupervised models. In: Proceedings of NAACL 2009, pp. 611–619 (2009)
15. McLachlan, G., Krishnan, T.: The EM Algorithm and Extensions, vol. 382. John Wiley & Sons, Hoboken (2007)
16. Melton, J., Simon, A.R.: SQL: 1999: Understanding Relational Language Components. Morgan Kaufmann, Burlington (2001)
17. Neal, R.M., Hinton, G.E.: A view of the EM algorithm that justifies incremental, sparse, and other variants. In: Jordan, M.I. (ed.) Learning in Graphical Models. NATO ASI Series (Series D: Behavioural and Social Sciences), vol. 89, pp. 355–368. Springer, Heidelberg (1998). https://doi.org/10.1007/978-94-011-5014-9_12
18. Ordonez, C.: Optimization of linear recursive queries in SQL. IEEE Trans. Knowl. Data Eng. **22**(2), 264–277 (2010)
19. Ordonez, C., Cereghini, P.:SQLEM: fast clustering in SQL using the EM algorithm. In: Proceedings of SIGMOD, pp. 559–570 (2000)
20. Tamayo, P., et al.: Oracle data mining - data mining in the database environment. In: The Data Mining and Knowledge Discovery Handbook, pp. 1315–1329 (2005)
21. Titterington, D.M.: Recursive parameter estimation using incomplete data. J. Roy. Stat. Soc. Ser. B (Methodol.) **46**(2), 257–267 (1984)
22. Viele, K., Tong, B.: Modeling with mixtures of linear regressions. Stat. Comput. **12**(4), 315–330 (2002)
23. Yang, Y., Ma, J.: A single loop EM algorithm for the mixture of experts architecture. In Advances in Neural Networks - ISNN 2009, 6th International Symposium on Neural Networks, Proceedings, Part II, pp. 959–968 (2009)
24. Yuksel, S.E., Wilson, J.N., Gader, P.D.: Twenty years of mixture of experts. IEEE Trans. Neural Netw. Learn. Syst. **23**(8), 1177–1193 (2012)
25. Zaniolo, C., Arni, N., Ong, K.:Negation and aggregates in recursive rules: the LDL++ approach. In: Proceedings of DOOD (1993)
26. Zaniolo, C., et al.: Advanced Database Systems. Morgan Kaufmann, Burlington (1997)
27. Zhao, K., Yu, J.X.: All-in-one: graph processing in RDBMSS revisited. In: Proceedings of SIGMOD 2017 (2017)

Correction to: Database Systems for Advanced Applications

Christian S. Jensen, Ee-Peng Lim, De-Nian Yang,
Wang-Chien Lee, Vincent S. Tseng, Vana Kalogeraki,
Jen-Wei Huang, and Chih-Ya Shen

Correction to:
C. S. Jensen et al. (Eds.): *Database Systems for Advanced Applications*, LNCS 12682, https://doi.org/10.1007/978-3-030-73197-7

The original version of the book was inadvertently published with incorrect acknowledgements in chapters 28 and 31.
The acknowledgements have been corrected and read as follows:
Acknowledgement: "This paper is supported by the National Key Research and Development Program of China (Grant No. 2018YFB1403400), the National Natural Science Foundation of China (Grant No. 61876080), the Key Research and Development Program of Jiangsu (Grant No. BE2019105), the Collaborative Innovation Center of Novel Software Technology and Industrialization at Nanjing University."

In chapter 52, the original version of this chapter affiliation of second author has been corrected and on page 775 the citation of Figure 6 was inadvertently published as "Table 6". This has been corrected.

The updated version of these chapters can be found at
https://doi.org/10.1007/978-3-030-73197-7_28
https://doi.org/10.1007/978-3-030-73197-7_31
https://doi.org/10.1007/978-3-030-73197-7_52

C. S. Jensen et al. (Eds.): DASFAA 2021, LNCS 12682, p. C1, 2021.
https://doi.org/10.1007/978-3-030-73197-7_54

Author Index

Akintande, Olalekan J. III-633
Ao, Xiang II-120

Bai, Jiyang I-123
Bao, Siqi III-516
Bao, Xuguang III-608
Bao, Zhifeng I-439
Böhm, Klemens II-721
Bonfitto, Sara III-643

Cai, Shunting I-323
Cai, Tianchi III-499
Cai, Xiangrui I-667, III-413, III-445
Cao, Manliang II-516
Cao, Meng I-375
Cao, Sichen II-629
Cao, Yang I-474
Cao, Yikang II-400
Chai, Yunpeng I-53
Chang, Chao III-165
Chang, Liang III-608
Chao, Pingfu I-658
Chen, Cen II-152
Chen, Chen I-158
Chen, Dawei II-186
Chen, Enhong I-3, II-168, III-461
Chen, Gang II-549
Chen, Guihai I-609, II-701, III-20, III-330
Chen, Hong II-20
Chen, Huan III-511
Chen, Jessica II-70
Chen, Jiayin II-3
Chen, Jiaze III-529
Chen, Junsha II-658
Chen, Ke II-280, II-549
Chen, Lei I-356, I-591, II-104, III-375, III-516
Chen, Qinhui III-600
Chen, Siyuan II-675
Chen, Xi II-3
Chen, Xiaojun III-541
Chen, Xiao-Wei II-291
Chen, Xu I-422
Chen, Xuanhao I-641
Chen, Yinghao II-400, II-429
Chen, Zhenyang III-165
Chen, Zhigang II-186
Chen, Zhihao III-341
Chen, Zihao I-307, III-627
Cheng, Dawei II-603
Cheng, Daxi III-499
Cheng, Hao I-375
Choudhury, Farhana M. I-439
Čontoš, Pavel III-647
Cui, Fengli II-429
Cui, Jie I-71
Cui, Lizhen II-692
Cui, Qing III-579
Cui, Yue III-148

Dai, Feifei III-321
Dai, Shaojie I-558
Deng, Geng I-307
Deng, Liwei I-641, III-148
Deng, Sijia III-622
Deng, Zhiyi I-207
Diao, Yupeng I-422
Ding, Ling III-541
Dong, Junyu I-558
Dong, Qian II-88
Dong, Xinzhou III-315
Dong, Yu III-553
Dong, Yucheng II-70
Dou, Jiaheng I-240
Du, Xiaofan III-341
Du, Xin II-675
Du, Yang I-405
Du, Yichao III-461
Du, Yuntao I-375, II-400, II-429, II-449
Du, Zhijuan I-290
Duan, Lei I-307, II-325
Duan, Liang I-224

Fan, Hao I-558
Fan, Hongjie II-612, II-620
Fan, Liyue II-342
Fan, Wei III-622
Fang, Junhua I-650, III-36

Fei, Xingjian I-174
Feng, Hao I-96, II-120
Feng, Shi I-141
Fu, Bin II-359
Fu, Min II-325
Fu, Yanan I-37
Fukumoto, Fumiyo II-202

Gao, Guoju I-405
Gao, Hong II-710
Gao, Neng II-658
Gao, Qianfeng III-244
Gao, Weiguo II-262
Gao, Xiaofeng I-609, II-701, III-3, III-20, III-330
Gao, Yichen III-622
Gao, Yuanning I-609
Gao, Yucen I-609
Ge, Ningchao III-595
Geng, Haoyu III-3
Gong, Bin II-761
Gong, Chen I-526
Gong, Xiaolong II-3
Gong, Zengyang I-491
Gong, Zhiguo I-394
Gu, Binbin II-186
Gu, Chengjie I-71
Gu, Hansu III-69
Gu, Jinjie III-499
Gu, Lihong III-499
Gu, Ning III-69, III-358
Gu, Tianlong III-608
Gu, Xiaoyan III-321
Guo, Gaoyang III-664
Guo, Qingsong III-659
Guo, Zhiqiang III-279

Han, Siyuan III-375
Han, Yongliang II-500
Hao, Shufeng III-100
Hao, Yongjing III-115, III-211
Hao, Zhenyun II-392
Hao, Zhuolin III-529
Haritsa, Jayant R. I-105
He, Bingsheng II-761
He, Dan I-658
He, Huang III-516
He, Liangliang I-340
He, Ming III-297, III-306
He, Ruifang II-641
He, Zhenying III-244
Hong, Xudong III-461
Hsieh, Hsun-Ping III-667
Hu, Binbin III-553
Hu, Changjun I-394
Hu, Guoyong III-165
Hu, Huiqi III-478
Hu, Qinghua II-710
Hu, Tianxiang II-37
Hu, Tianyu III-195
Hu, Yikun I-385
Hu, Ying II-307
Hu, Zehua II-675
Hua, Liping III-600
Hua, Yifan III-393
Huang, Bingyang I-96
Huang, Chenchen III-478
Huang, Dong II-291
Huang, Hailong II-262
Huang, He I-405
Huang, Jen-Wei III-652
Huang, Jun II-152
Huang, Junheng III-85
Huang, Junzhou II-778
Huang, Kaixin III-393
Huang, Linpeng III-393
Huang, Shanshan II-392, III-228
Huang, Xin III-617
Huang, Xinjing II-152
Huang, Zhangmin I-37
Huo, Chengfu III-553

Iosifidis, Vasileios III-617

Ji, Daomin III-262
Ji, Feng II-152
Ji, Genlin II-307
Ji, Hangxu I-20
Jia, Jinping II-307
Jia, Shengbin III-541
Jiang, Di III-516
Jiang, Nan III-622
Jiang, Peng III-165
Jiang, Shimin II-70
Jiao, Pengfei II-53
Jin, Beihong III-315
Jin, Cheqing III-341, III-622
Jing, Yinan III-244

Kang, U. III-662
Kou, Yue I-575
Kudo, Mineichi II-569

Lai, Kunfeng I-174
Lai, Yantong II-271
Li, Beibei III-315
Li, Bo I-526, III-321
Li, Changheng III-211
Li, Cuiping II-20
Li, Dongsheng III-69
Li, Guohui III-279
Li, Haotian II-629
Li, Haoyang I-356
Li, Huiyuan III-53
Li, Jianjun II-738, III-279
Li, Jianyu I-224
Li, Jie I-224
Li, Jiyi II-202
Li, Kenli I-385
Li, Lei I-658
Li, Longfei III-579
Li, Longhai II-325
Li, Meng III-579
Li, Ning II-219
Li, Qinghua II-20
Li, Qingzhong II-692
Li, Renhao I-307
Li, Shuang I-509
Li, Wenbin I-457
Li, Xiang I-609, III-330
Li, Xiangyu II-629
Li, Xiao I-340
Li, Xiaoyong II-376
Li, Xinyi I-272
Li, Xuan II-262
Li, Yanhong II-738
Li, Yuanxiang III-262
Li, Yuchen II-761
Li, Yuhao I-609
Li, Yunyi III-179
Li, Zhanhuai II-219
Li, Zhi I-3
Li, Zhixu II-186, II-533
Lian, Rongzhong III-516
Lian, Xiang I-591
Liang, Chen III-499
Liang, Jingxi II-37
Liang, Shuang I-207
Liao, Chung-Shou III-604
Liao, Ming II-778
Lin, Fandel III-667
Lin, Lan II-516
Lin, Li I-509
Lin, Xueling I-356
Lin, Zheng II-253
Link, Sebastian I-113
Litvinenko, Ilya I-113
Liu, An I-405, I-650
Liu, Baozhu I-323
Liu, Binghao III-132
Liu, Chaofan III-358
Liu, Chengfei III-36
Liu, Chenxu II-620
Liu, Guang II-262
Liu, Hao III-566
Liu, Haobing I-526
Liu, Hongtao II-53
Liu, Hongzhi II-359
Liu, Huanyu II-641
Liu, Junfei II-612, II-620
Liu, Lei II-692
Liu, Lu I-71
Liu, Ning III-429
Liu, Pengkai I-323
Liu, Qi I-3, II-168
Liu, Qiyu I-591
Liu, Richen II-307
Liu, Shuncheng I-422
Liu, Shushu I-650
Liu, Wei I-457, III-604
Liu, Xinwang II-376
Liu, Xuefeng I-37
Liu, Yanchi III-115, III-132, III-179, III-211
Liu, Yiming II-701
Liu, Yuling I-385
Liu, Yun II-104
Liu, Zhen III-617
Liu, Zhengxiao II-253
Liu, Zhidan I-491
Liu, Ziqi III-499
Lou, Yunkai III-664
Lu, Guangben III-20
Lu, Jiaheng II-585, III-659
Lu, Jinjie III-608
Lu, Kejing II-569
Lu, Tun III-69, III-358
Lu, Xudong II-692
Luo, Changyin II-738
Luo, Chong I-375

Luo, Fangyuan III-85
Luo, Minnan I-272
Luo, Pengfei III-461
Luo, Yifeng II-603
Lv, Yao III-36
Lyu, Qiuyi II-761

Ma, Hao-Shang III-652
Ma, Zhiyi II-3
Ma, Ziyi I-385
Mao, Yuzhao II-262
Meng, Lingkang III-100
Mondal, Anirban III-487

Ni, Shiguang III-529
Nie, Tiezheng I-575
Niu, Jianwei I-37
Niu, Shuzi II-88, III-53
Nummenmaa, Jyrki I-307, II-325

Ouyang, Jiawei III-413

Pan, Peng III-279
Pan, Wei II-219
Pan, Xuan I-667
Pandey, Sarvesh III-638
Peng, Hao I-174
Peng, Jinhua III-516
Peng, Peng III-595
Peng, Zhaohui II-392, III-228
Peng, Zhiyong I-474
Pu, Xiao III-511

Qi, Guilin I-256
Qi, Xiaodong III-341
Qi, Zhixin I-88
Qian, Chen I-509
Qian, Mingda III-321
Qian, Weining II-603
Qiao, Lianpeng I-158
Qin, Gang III-622
Qin, Zheng III-595
Qiu, Minghui II-152
Qiu, Ye II-3

Reddy, P. Krishna III-487
Ren, Chao III-461
Ren, Kaijun II-376
Ren, Weijun III-553
Ren, Yuxiang I-123
Richly, Keven I-542
Rong, Yu II-778

Sanghi, Anupam I-105
Santhanam, Rajkumar I-105
Schäler, Martin II-721
Shahabi, Cyrus II-342
Shanker, Udai III-638
Shao, Jie I-207
Shao, Ruihang I-394
Shen, Derong I-575, II-558
Shen, Jianping I-174, II-262
Shen, Jundong II-413, II-465
Shen, Yuan I-474
Sheng, Victor S. III-115, III-132, III-179, III-211
Shi, Chongyang III-100
Shi, Gengyuan I-96
Shi, Jieming I-625
Shi, Meihui I-575
Shi, Qitao III-579
Shi, Tianyao III-330
Shijia, E. III-541
Shou, Lidan II-280, II-549
Shu, Lihchyun II-738
Siddiqie, Shadaab III-487
Song, Jinbo III-165
Song, Kaisong I-141
Song, Kehui III-445
Song, Shuangyong III-511
Song, Sujing III-612
Song, Wei I-474
Song, Yan II-120
Song, Yang II-359
Song, Yuanfeng III-516
Song, Zhenqiao III-529
Su, Han I-422
Su, Peifeng II-585
Su, Xiangrui III-100
Su, Yijun II-271
Sun, Chenchen II-558
Sun, Fei III-165
Sun, Hao III-148
Sun, Jie III-612
Sun, Xuehan III-330
Sun, Yu-E I-405
Sung, Chang-Wei III-604
Surbiryala, Jayachander III-617

Tan, Zhiwen II-449
Tang, Haibo III-622
Tang, Jintao I-340
Tang, Jiuyang I-272
Tang, Mengfei I-474
Tang, Shaojie I-37
Tang, Xiu II-549
Tang, Xuejiao III-617
Tian, Bing I-240
Tian, Xintao III-115
Tong, Xing III-622
Tran, Luan II-342
Tu, Chenyang II-658

Wang, Binbin I-96, III-664
Wang, Chang-Dong II-291
Wang, Changping I-394, III-664
Wang, Chao II-701, III-511
Wang, Chaokun I-96, I-394, III-664
Wang, Chaoyang III-279
Wang, Chengyu II-152
Wang, Chongjun I-375, II-400, II-413, II-429, II-449, II-465
Wang, Daling I-141
Wang, Deqing III-115, III-179
Wang, Dongsheng II-612
Wang, Fali II-253
Wang, Guoren I-20, I-158
Wang, Hongya II-500
Wang, Hongzhi I-88
Wang, Jiahai II-675
Wang, Jian II-392
Wang, Jianmin I-509
Wang, Jianyong III-429
Wang, Jiaqi II-136
Wang, Jingbing III-529
Wang, Jinghao III-529
Wang, Jun I-3
Wang, Lei II-253
Wang, Lizhen III-608
Wang, Peng II-629
Wang, Qian I-474
Wang, Senzhang II-392, III-228
Wang, Shaohua II-70
Wang, Shipeng II-692
Wang, Ting I-340
Wang, Weiping III-321
Wang, Wenjun II-53
Wang, X. Sean III-244
Wang, Xiang II-376
Wang, Xiaofan II-235
Wang, Xiaxia II-280
Wang, Xin I-323, II-710
Wang, Xiting II-120
Wang, Xiuchong III-553
Wang, Xue II-392
Wang, Yan II-70
Wang, Yangyang I-53
Wang, Yashen II-483
Wang, Yijun II-168
Wang, Zehui III-511
Wang, Zheng I-394
Wang, Zhuo III-321, III-664
Wei, Junjie III-600
Wei, Tianxin II-168
Wei, Ziheng I-113
Wen, Han III-297, III-306
Wen, Lijie I-509
Wen, Xinyu III-228
Wen, Yanlong I-667
Willkomm, Jens II-721
Wu, Fangwei I-625
Wu, Gang I-20
Wu, Hua III-516
Wu, Jun III-85
Wu, Kaishun I-491
Wu, Likang I-3
Wu, Longcan I-141
Wu, Sai II-280, II-549
Wu, Shaozhi I-422
Wu, Tianxing I-256
Wu, Zhonghai II-359
Wu, Zijian III-148

Xian, Xuefeng III-132, III-179
Xiang, Yang III-541
Xiang, Zhenglong III-262
Xiao, Shan I-307
Xiao, Yingyuan II-500
Xie, Guicai I-307, II-325
Xie, Huizhi III-499
Xie, Xike I-625
Xing, Chunxiao I-240
Xiong, Hui III-566
Xu, Chen III-627
Xu, Chi II-120
Xu, Chuanyu III-553
Xu, Hongyan II-53
Xu, Jiajie III-36
Xu, Jianliang I-457

Xu, Jianqiu III-612
Xu, Liang I-191
Xu, Tong III-461, III-566
Xu, Xiaokang II-392
Xu, Yifan II-325
Xu, Zheng III-358
Xu, Zhi I-422
Xu, Zhizhen III-627
Xu, Zichen II-376
Xu, Zihuan III-375
Xue, Cong II-271
Xue, Taofeng III-315

Yadamjav, Munkh-Erdene I-439
Yan, Qiben I-474
Yang, Fangzhou II-603
Yang, Haiqin I-174
Yang, Jiandong III-529
Yang, Jianye I-385
Yang, Min II-120
Yang, Qiang II-533
Yang, Wei III-195
Yang, Xinghao III-604
Yang, Xinxing III-579
Yang, Yajun II-710
Yang, Yingjie III-622
Yang, Yiying I-174
Yao, Bo II-516
Yao, Junjie I-191
Yao, LingLing III-541
Yao, Yirong II-449
Ye, Wei II-37
Yin, Dan II-710
Yin, Jian I-457
Yin, Xi I-174
Yu, Cheng II-465
Yu, Ge I-141, I-575
Yu, Guoxin II-120
Yu, Hualei I-375, II-449
Yu, Jeffrey Xu II-778
Yu, Kui II-376
Yu, Philip S. II-392, III-228
Yu, Shenglong III-445
Yu, Shuodian III-3
Yu, Yanwei I-558
Yuan, Gongsheng II-585
Yuan, Tao III-53
Yuan, Xiaojie I-667, III-413, III-445
Yuan, Ye I-20, I-158
Yue, Kun I-224

Zang, Tianzi I-526
Zeng, Jiajun II-629
Zeng, Weixin I-272
Zeng, Xiaodong III-499
Zha, Daren II-253, II-271
Zha, Rui III-566
Zhang, Bo II-104
Zhang, Boyu I-405
Zhang, Chao III-659
Zhang, Chen III-516
Zhang, Daokun II-692
Zhang, Guang-Yu II-291
Zhang, Haiwei III-445
Zhang, Hang II-710
Zhang, Hanyu III-297, III-306
Zhang, Haodi I-491
Zhang, Huanhuan II-483
Zhang, Ji III-617
Zhang, Jiangwei I-667
Zhang, Jiasheng I-207
Zhang, Jiatao I-256
Zhang, Jiawei I-123
Zhang, Jing I-71
Zhang, Kai III-244
Zhang, Le III-566
Zhang, Li II-136, II-533
Zhang, Lijun II-219
Zhang, Mengdi I-3
Zhang, Mengxuan I-658
Zhang, Mingli III-617
Zhang, Peng III-69, III-358
Zhang, Rui III-429
Zhang, Ruiting II-400
Zhang, Shikun II-37
Zhang, Shuxun III-659
Zhu, Junchao III-664
Zhang, Tao II-359
Zhang, Wang II-738
Zhang, Wei III-429
Zhang, Weiming II-376
Zhang, Wenbin III-617
Zhang, Xiaoming II-104
Zhang, Xiaowen II-429, II-449
Zhang, Xue II-20
Zhang, Ya-Lin III-579
Zhang, Yi II-413, II-465
Zhang, Yifei I-141, II-658
Zhang, Yin II-152
Zhang, Ying I-667, III-413, III-445
Zhang, Yong I-240

Zhang, Yuhao III-413
Zhang, Zhao III-341, III-622
Zhang, Zhe III-461
Zhang, Zhecheng II-413
Zhang, Zhenghao III-69
Zhang, Zhiqiang III-499
Zhang, Zhiwei I-158
Zhao, Bin II-307
Zhao, Gang III-600
Zhao, Hongke I-3
Zhao, Hui III-600
Zhao, Kangfei II-778
Zhao, Liangliang II-641
Zhao, Pengpeng III-115, III-132, III-179, III-211
Zhao, Xiang I-272
Zhao, Yan I-641, III-148
Zhao, Yuhai I-20
Zhao, Ziheng III-20
Zheng, Baihua I-439
Zheng, Junhong I-491
Zheng, Kai I-422, I-641, III-148
Zheng, Libin I-591
Zheng, Mingyu II-253
Zheng, Qinghua I-272
Zheng, Shengan III-393
Zheng, Wenguang II-500
Zheng, Yanan II-70
Zheng, Yi III-461
Zhong, Hong I-71
Zhou, Aoying II-603, III-478
Zhou, Da III-553
Zhou, Ding III-566
Zhou, Hao III-529
Zhou, Jun III-579
Zhou, Kaijie I-174
Zhou, Minping III-529
Zhou, Rui III-36
Zhou, Xiangdong II-516
Zhou, Xiaofang I-658
Zhou, Xu I-385
Zhou, Yumeng II-400
Zhou, Yu-Ren II-291
Zhu, Fei I-650
Zhu, Huaijie I-457
Zhu, Jiatao II-629
Zhu, Junchao III-664
Zhu, Junxing II-376
Zhu, Ke II-500
Zhu, Peng II-603
Zhu, Yanmin I-526
Zhu, Yao II-359
Zhuang, Fuzhen III-132, III-211
Zhuo, Wei III-315
Zong, Zan I-509
Zou, Lei III-595

Printed in the United States
by Baker & Taylor Publisher Services